GEOLOGY OF THE
U.S.S.R

GEOLOGY OF THE
U.S.S.R

D. V. NALIVKIN

Member of the Academy of Sciences of the U.S.S.R.
and formerly Professor of Historical Geology
Institute of Mining, Leningrad

Translated from the Russian by
N. RAST, B.Sc., Ph.D.
Professor of Geology
University of New Brunswick, Canada

and edited by N. RAST and
T. S. WESTOLL, F.R.S.
Professor of Geology
The University of Newcastle upon Tyne

1973

OLIVER & BOYD
EDINBURGH

OLIVER & BOYD

Croythorn House, 23 Ravelston Terrace, Edinburgh, EH4 3TJ

(A Division of Longman Group Ltd.)

This is a translation of ГЕОЛОГИЯ СССР by D. V. Nalivkin,
published in 1962 by The Academy of Sciences U.S.S.R.
Moscow – Leningrad.

ISBN 0 05 002197 4

Printed in Great Britain by
Western Printing Services Ltd., Bristol

CONTENTS

Part III
PALAEOZOIC GEOSYNCLINES

Part IV

MESO-CAINOZOIC GEOSYNCLINES

LIST OF TABLES

AUTHOR'S PREFACE

The basic aim of this book *Geology of the U.S.S.R.* is to prepare students for geological surveying and the exploration of useful deposits. Such a preparation has two aspects: the first being the study of the known geological data for a particular region, while the second is the understanding of the methods of regional geological investigation.

The known data are conventionally arranged in the following order. After a short general outline of the region follows a similarly short description of its orography, which indicates the basic geomorphological information. Stratigraphy, which is the most important section, is given more fully and in detail. Then follows the tectonic history, which is also important, and leads to discussion of magmatism, since the time, the position, and the character of magmatic phenomena depend largely on tectonics.

In regional geological memoirs (igneous) petrography is sometimes placed after stratigraphy while the description of tectonics is given the third position. Such an arrangement is technically erroneous, since it limits the quality of stratigraphical description, the content of which is considerably wider than the description of the sedimentary deposits. In addition, such an arrangement substitutes petrography— the description of rocks—for magmatism—a complex of geological phenomena, associated with these magmatic rocks. This substitution lowers the quality of the work. The description of magmatism cannot be given at an appropriate level without connecting it to the tectonics. Thus tectonics should precede magmatism.

Useful deposits are described somewhat briefly, and the principal aim is to clarify their interrelations with other geological phenomena such as stratigraphy tectonics and magmatism. The economically useful deposits are dealt with in another textbook.

It should never be forgotten that the basic problem of all geological research is the generation, growth and expansion of the mineral basis of our socialist society. Amongst all the objects of geological investigations the useful deposits should stand in the first place.

Methodological matters are not given separate headings in the regional descriptions, but are distributed throughout the text. They are especially abundant where controversial problems are discussed. However, such methodological discussions are not restricted to the controversial problems. They are introduced when the history of investigation is being clarified and to show why one point of view has been replaced by another. They are also introduced in other subdivisions of the textbook. The knowledge of methodology is no less significant than the knowledge of the factual data.

The second aim of the book is to serve as a reference for geologists working in industry. For this reason there is concentrated in it—in excess of what is necessary for elementary students—a considerable amount of factual data.

In the Soviet Union there are more than 25,000 geologists. They are trained in numerous cities of our homeland by teachers representing various schools of geology. The territory of the U.S.S.R. equals one sixth of the dry surface of the earth and includes nearly all the known types of geological provinces. With such a diversity of views and factual data it is natural for lively controversies to arise about most diverse problems.

When the controversial problems are being investigated three cases may arise. The first is when enough factual material has been collected and a majority of geologists have reached a definite conclusion. Only a minority are objecting and even then often on the basis of inertia. In such cases adherence to the majority is correct, and especially so when the problem has been solved collectively at an authoritative conference.

The second case arises when new data, which differ sharply from the previously known ones, come to light. The new data are supported only by a few innovators, whereas the majority either expresses opposition or is undecided. The support of the innovators is of course correct, but it is not always easy to express such support openly.

The third case arises when the existing data are inadequate to achieve a final solution and the supporters of varying points of view are divided into two approximately equal groups. In this instance the adherence to one or another groups often depends on subjective causes. It is sometimes most correct to refuse accepting a final solution and to demand the collection of new additional data which would prove on which side lies the truth.

Now, a few words about the completeness and up to date character of the data collected in this book. The author has taken all the measures to make these data as complete as possible, but the quantity of the new data obtained by Soviet geologists grows every year and does so very rapidly. Consequently omissions, mistakes and inadequacies are inevitable. The author will be very grateful for any remarks and would invite them to be sent to the following address: Chair of Historical Geology, Institute of Mining, Leningrad.

In conclusion I would like to thank the comrades who have helped during the formulation of the book. Some of them have reported new data and have presented their data for my use, others made very important critical remarks, the third group have expressed significant wishes. To all of them my deepest thanks. In particular I am grateful to my colleagues in the department, whose help was very valuable.

EDITORIAL PREFACE

The area of the Soviet Union is about one sixth of the total land surface of the Earth. Much of this vast ground until recently has been effectively unknown. Furthermore because of its central position in Eurasia the geology of the U.S.S.R. is of a fundamental importance in interpreting the geology of the largest continental mass in existence. In Siberia, in particular, substantial geological research only began in the twenties and did not really yield great results until the thirties. Then the World War intervened and in the immediately post-war era the state of international politics was such that most of the advances in the elucidation of Siberian geology remained unavailable to western readers. Because of their comparative isolation Soviet geologists both in approach and terminology developed in different ways from those working in the west. However, even in the Soviet Union because of rapid expansion of geological research few publications appeared dealing on an adequate basis with the country as a whole. NALIVKIN'S *Geology of the U.S.S.R.* from our point of view fills the two gaps. Firstly it summarizes the geology of the Soviet Union and secondly, since it is primarily a text-book, it covers fairly comprehensively the usage of Soviet geologists both as regards the terminology and the conceptual framework. Thus although the book now is some ten years old it is nevertheless an almost unique document and therefore had to be translated.

In the process of translation the non-parallelism of the Soviet and Western methods of description had to be constantly borne in mind and therefore I attempted to translate somewhat freely avoiding such terms as 'regularities' which normally disfigure translations from Russian. Also some references to purely internal politically-oriented patriotic notions were left aside, although it should be stated that in this respect Nalivkin's original is remarkably restrained. Some specifically Russian terms such as 'tectonic storey' and 'deep faults' have been retained since they convey useful notions. In transliterating Russian stratigraphic terms, as far as possible, the adjectival terminations were made appropriate to the Soviet type localities. Only when the original ended in the not untypical Russian 'sk' or the derivation was obscure did I employ the adjectival termination 'skian'.

The original book is peppered with numerous references to geographical localities and geographical formations etc.; to avoid too many capital letters in general I used the lower case letters to denote geographical features as for example: lake Baikal, Stanovoi range, etc. I used, however, capital 'p' for peninsula, as in Kola Peninsula; and since peninsulas as geographical subdivision of the U.S.S.R. are generally unknown in the west they are worth singling out. All the place specific geological terms have been capitalized.

In transliterating Russian specific words the scheme recommended by the Royal

Society of London has been employed throughout the volume. Occasionally this practice leads to difficulties. For instance the name of the famous geologist V. V. Belousov gets transliterated as V. V. Byelousov, but in general, inconsistencies are few. A more considerable difficulty arises with the transliteration of originally non-Russian Soviet names – is it Shmidt or Schmidt? I felt that in general Soviet names should be transliterated from Russian and not from the original German, Baltic Tadzhik or other names. Even greater difficulty arose when the author after using Kara-Kum halfway through switched off to Karakum. I decided to use the latter.

Regarding the indices, after an initial attempt, it became clear that the organization of completely paginated stratigraphic and structural indices is not practicable. The Russian Platform for instance gets used on over 200 pages and therefore instead of the traditional indices, lists of stratigraphic terms and of structural terms have been prepared referring to the page where a comprehensive definition can be found. The age of Russian stratigraphic units has also been included in the list, since the complexity and multiplicity of various Russian stratigraphic units is such that this aid to a western reader is essential.

Lastly in view of the fact that the book is already some ten years old I have been persuaded to include a brief summary of up-to-date ideas in Soviet geology. This section under the heading of Translator's Editorial Introduction has been included at the beginning of the volume as a position paper with few, if any, ideas of my own. This section exists merely as a guide to someone who having read the main text, decides to delve further into the mysteries of Soviet geology. The editor-translator does not claim either any great originality or a comprehensive character for his review, but hopes that it may serve some useful purpose in singling out main recent advances in Soviet geological sciences.

I would like to express my gratitude to my wife, Mrs D. E. Rast, who prepared the indices and helped with the editing, to Professor T. S. Westoll who was the principal editor of the first few chapters and to Oliver and Boyd our publishers who patiently followed and encouraged the translation and editing of the book over several years.

Nicholas Rast
Fredericton, New Brunswick
April 1973

TRANSLATOR'S EDITORIAL INTRODUCTION

The Geology of the U.S.S.R. by Nalivkin was published in Russian some ten years ago. It nevertheless reflects the modern trends of Soviet geology and, in a fairly comprehensive manner, describes the geology of a country spread over one sixth of the Earth. Even in the U.S.S.R. the book is one of the few one-volume publications summarizing not only the historical geology of the Soviet Union, but also general information on its petrology, tectonics and mineral deposits. In the west the existing notions about Soviet geology are fragmentary. In one respect the book is remiss: it does not, even partially, describe the results of modern geophysical investigations in the U.S.S.R. Admittedly some of the investigations, such as those concerned with gravity, are classified material as indeed is all the geodetic work; yet much seismic and magnetic information is available. In this editorial introduction the salient advances of Soviet geology in the past ten years are reviewed and some of the geophysical data introduced, although in the context of geotectonics. The plan of this review follows the plan of the original volume, but again the tectonics are especially emphasized.

Organization of Geological Research and Publications

The basis of the organization of geological sciences in the U.S.S.R. has hardly changed since the data of the original publication of Nalivkin's book. Perhaps the most important feature of the last ten years has been the expansion of the Siberian Academy of Sciences, centred on Novosibirsk and of the affiliated centre of Khabarovsk. These institutions have become centres of geological research in Soviet Central Asia and Siberia. As in the western countries, research is pursued both in government-run and academic institutions. Because of the complex nature of work, the government-sponsored research is divided between the institutes of the Ministry of Geology as well as institutes of other ministries. The academic research is carried out in the institutes of Academies of Sciences or by universities. It should, however, be pointed out that most of the scientific personnel in universities is normally attached to specialized scientific institutes or field-projects organized by the Ministry of Geology. The entirely independent research of the western world in practice does not exist.

There are approximately 100,000 practising geologists in the Soviet Union. To maintain, replenish and expand geological work a threefold comprehensive system of geological education has been devised. Firstly, there are special secondary technikums which provide elementary education in geology, geological prospecting, mining, and geophysics. The aim of the technikums is to produce geological technicians.

According to a recent paper of Gorshkov *et al.* (1972) there are some 50 technikums offering specialization in geology.

The institutions of higher education again offer two different options—an early applied orientation in mining, petroleum and other institutes or a more general (pure) scientific education in universities. Some of the mining institutes such as that at Leningrad, mentioned by Nalivkin, are of especial importance in this historical contribution to the development of Soviet geology. The total number of Soviet universities specializing in geology is relatively small, partly owing to the fact that Soviet universities on average tend to be larger than in the west. Nevertheless, in 1970 Soviet universities produced 5,000 graduates in geology and associated sciences. Consequently, it is not surprising that in the U.S.S.R. geological literature is extensive and about 35 per cent of total world geoscience literature is published in Russian. To classify, record and retrieve this volume of literature, as well as that published outside the Soviet Union, it is necessary to organize a proper cataloguing and retrieval system, which is now being set up by V.S.E.G.E.I. (The All Union Geological Research Institute). The preliminary description of the system (D.R.S.) is given by Moshkin *et al.* (1972).

Much of the Soviet geoscience literature is available in the west through the auspices of Collet's book-shop in London, U.K. Some of the recently published Soviet 'synoptic' maps are also available and can be purchased through 'Mezhdunarodnaya Kniga, Moskva, G–200, U.S.S.R. The most useful recent maps are as follows:

> Geological Map of the U.S.S.R. 1:5000000 – 1967 (V.S.E.G.E.I.)
> Map of the Quaternary of the U.S.S.R. 1:5000000 – 1969 (V.S.E.G.E.I.)
> Metallogenic Map of the U.S.S.R. 1:2500000 – 1971 (V.S.E.G.E.I.)
> Magmatic Formations Map of the U.S.S.R. 1:2500000 – 1971 (V.S.E.G.E.I.)
> Tectonic Map of Oil and Gas Regions of the U.S.S.R. 1:2500000 – 1970
>   (V.N.I.G.R.I.)
> Magnetic Anomalies Map of the U.S.S.R. 1:10000000 – 1971 (V.S.E.G.E.I.)
> Structural Formation Map of the S.W. part of the Pacific Ocean Mobile Belt
>   1:1500000 – 1972 (V.S.E.G.E.I.)
> Geological Map of the Russian Platform and its Framework 1:1500000 – 1970
>   (V.S.E.G.E.I.)
> Metamorphic Facies of the Eastern Central Asia Map 1:1500000 – 1971
>   (Kirghiz Akad. Nauk)

In addition, smaller-scale maps for various parts of the world are produced and can be obtained from the same source.

The Soviet geological literature is so extensive that it is impossible to summarize it briefly. Perhaps the two most important periodicals are the *Doklady* and *Izvestia* of the Academy of Sciences, of which the former is totally and the latter is partially translated into English in the U.S.A. Some journals such as *Geological Journal* (U.K.) and *Journal of Geological Society of London* (U.K.), *Journal of Geology* (U.S.A.), *Earth Science Reviews* (Holland) from time to time carry substantial reviews of Soviet work. Yet there are a large number of other periodicals, reports of symposia and colloquia. The most important among the latter are announced and often reviewed in *Izvestia*, which functions as a source of general information as well as publishing learned papers. Many Soviet papers of special interest are published in the *International Geological Reviews* selected and assembled by the American Geological Institute.

The perusal of the Soviet literature over the last decade indicates the gradual emergence of new trends in Soviet geological research. These trends and new conclusions are now only all too briefly summarized.

Stratigraphic Research in the U.S.S.R.

Much of the stratigraphic research in the U.S.S.R. is concerned with detailed delineation and mapping of geological formations. In this respect Soviet geologists are no different from their western colleagues. In two matters related to the Precambrian geology Soviet geologists, however, have made advances which hitherto have not been entirely accepted in the west. Firstly, in the Upper Precambrian and Lower Palaeozoic formations Russians have used stromatolites as indices of correlation. Raaben (1969), who is one of the foremost Soviet palaeontologists working on stromatolites, has summarized in English the recent advances, and Semikhatov & Komar (1965) have produced an earlier statement as to the stratigraphic use of these algal remains. Certain investigators claim that stromatolites continuously grade from the Rhiphean, through the Vendian to the Cambrian (Keller, 1971; Raaben, 1971), although this is denied by Sokolov (1972) on general stratigraphic grounds. On the other hand, Semikhatov (1972), another important Soviet investigator, claims that the Upper Proterozoic (Rhiphean) and the Lower Proterozoic (Aphebian) deposits can be distinguished on the basis of their stromatolites. In the Rhiphean he proposes a fourfold zonation of world-wide extent. Semikhatov points out that while the pre-Rhiphean stromatolites did not show appreciable evolutionary changes, the Rhiphean stromatolites started evolving rapidly, thus allowing widespread correlation of Rhiphean rocks.

Partly as a result of their interest in stromatolites, Soviet geologists have also made numerous attempts to establish a unified Precambrian time-scale. In this respect, however, even in the U.S.S.R. there are considerable divergences of views. Table I presents some of the modern correlations proposed. The differences amongst the proposed schemes reflect partly the uncertainty in isotopic ages, partly the preoccupation of the authors with regions of greatest familiarity and partly the confusion that exists about the boundary between Proterozoic and Archean rocks. Clearly, if the very old (pre-3000 myr.) rocks now recognized in the Kola Peninsula (Katarchean) and the Aldan Shield (Aldanian) are to be separated as Archean, then the boundary between Archean and Proterozoic must be at a level of 3000 myr +. Nevertheless, it seems clear that in general the main problem of Precambrian correlation has been solved by means of a very large number of isotopic age determinations. The general facilities for isotopic age dating in U.S.S.R. are excellent, and although sometimes to the self-important sophisticates in the west they do not appear as progressive as our own, in so far as they are essentially geared to the practical business of dating the Precambrian and Phanerozoic rocks, Soviet laboratories yield important results. I have reviewed aspects of this work elsewhere (Rast, 1971).

Much of the Soviet stratigraphical and palaeolontological research of Phanerozoic formations, however, tends to be somewhat old-fashioned. Papers suggestive of palaeogeographic inferences involving continental drift started appearing only very recently and, with few exceptions, little palaeosedimentology is done.

The general modern tendency in the U.S.S.R. is to produce extensive stratigraphic schemes of regional character depending on geophysical as well as geological data. Much of the territory has been deeply drilled, resulting in an extensive stratigraphic coverage.

TABLE I. *Stratigraphic and Strato-tectonic Divisions of the Precambrian*

Isotopic Time	Yanshin et al. (1966) Tectonic Complexes	Salop 1964 1967	Semikhatov 1972	Sokolov 1972	Palei 1963 Baltic Shield	Kosygin et al. 1972 Siberia	Semenenko et al. 1972 Ukraine
500	Cambrian	Cambrian	Cambrian				
570		Eocambrian		Vendian		Vendian	
	Baikalids						Precambrian V
1000	Satpurids Sveco-Norwegian Regeneration	Neoproterozoic	Upper Proterozoic Rhiphean		Rhiphean	Rhiphean	
1500	Kaielids	Neoproterozoic			Jotnian		Precambrian IV
					Karelids	Ulcanian	Precambrian III
2000	Byelomorids	Mesoproterozoic	Lower Proterozoic Aphebean		Byelomorids	Udocanian	
2500	Saamids				Saamids		Precambrian II
		Paleoproterozoic	Archean				Precambrian I
3000	Katarchean				Ancient Blocks		
3500		Pre-3500 myr. Archean				Pre-3500 myr. Aldarian	3500

The significance of the work on Precambrian shields can be assessed from research on individual units, but, even better than that, from regionally correlative papers such as that by Semenenko (1972), or Parfenov (1970), in which they discuss the chronology and the distribution of Precambrian formations respectively. Semenenko in particular extends the S cycle Precambrian geology over the world. At the same time, much geophysical work has been done and, in this respect, deep geophysics are particularly important. Owing to its economic significance the Ukrainian shield has been widely studied. Its gravitational anomalies have been investigated by Golizdra & Akhmetshina (1972), the relation between the gravity and magnetic fields by Krutikhovskaya & Pashkevich (1970), while the whole subject of the magnetic anomalies in Precambrian shields and the depths of their sources has been very recently reviewed by Krutikhovskaya, Pashkevich & Simonenko (1973). The tendency to produce broad generalizations unifying the Precambrian is certainly a feature of Soviet approach to the stratigraphy and structure of these rocks.

Tectonics in the U.S.S.R.

Tectonic notions in the U.S.S.R. until recently have been guided by the ideas of Professor N. S. Shatskii (1946) who differentiated two types of platforms, the ancient and the young as the principal constituents of continental structure. The ancient platforms, themselves polygenetic, were taken to be margined by Palaeozoic geosynclinal fold belts separated from the platforms by the so-called 'seams'. In this book Nalivkin's classification of the platforms closely follows that of Shatskii. The young platforms, part-Precambrian and part-Palaeozoic are similarly margined by Meso-Cainozoic fold belts.

As pointed out by Nalivkin, within the platforms Shatskii recognized a system of anticlises and syneclises as well as tectonic ridges and depressions. In addition, in a posthumously-published paper, Shatskii (1964) introduced the notion of aulocogens as narrow elongated graben-like mobile depressions with great thicknesses of sediments. As examples of these structures he suggested the Donyets Basin, the Pachelma Trough and also the Timan. These depressions were associated with large-scale faults breaking up the basements of the older platforms into separate massifs. V. D. Nalivkin (1963) had also recognized such structures. It is suggested by Yanshin *et al.* (1966, p. 294) that with regard to orogenic cycles the Russian Platform had suffered successive depressions and elevations indicated by transgressions and regressions in association with the Baskakid, Caledonian, Hercynian and Alpine orogenies. Yanshin *et al.* (1966) also point out that the sedimentary covers of the ancient and younger platforms have different characters. The existence of the larger syneclises on the platforms has convinced many Soviet geologists of the reality of substantial vertical movements causing major deformations in the crust. Consequently, over the last ten to fifteen years Soviet geology has witnessed arguments for and against major horizontal displacements in the crust. The main proponent of vertical movements, V. V. Byelousov, is well-known in the west as a geological activist who has consistently and strongly argued against the lateral (horizontal) displacements of large magnitude. His point of view is in the U.S.S.R. known as fixist and his publications are too numerous and too well-known to be recounted here. His views, however, can be gleaned from his papers published in English (1970, 1971) and his translated book (1962), although a more recent version of it is available in Russian. Some of the foremost Russian geotectonicians, such as V. E. Khain (1970) and M. V. Muratov (1972), in their generalizations appear to favour Byelousov's standpoint, although the former has recently (1972) conceded the possibility of large horizontal displacements, which he claims are governed by deep faults, which existed from Early Precambrian times.

The other point of view which admits large horizontal tectonic displacements has been promoted and supported by A. V. Peive, who first as far back as 1945 introduced the notion of deep faults. Recently Peive (1969, 1971, 1972) has moved to advocate the ensimatic origin of all eugeosynclines of the continental U.S.S.R. including the Urals. He particularly recognized the significance of the ophiolite suite as the remnant of oceanic crust within the continents. Similar ideas have been expressed by Zonenshain (1972). He has been concerned with the interpretations of the Mongolian part of the so-called Ural-Mongolian orogenic belt, which is now adopted as a single unit combining both the Urals and Angara Geosyncline of Nalivkin.

The idea of deep faults has been widely popularized in the U.S.S.R. Deep faults are major surfaces or zones of movement between large mobile blocks; such faults often developed over long periods of time involving numerous phases of movement. The idea became so much of a band waggon that any discontinuity whether geological or even geophysical became a deep fault. Nevertheless, vulcanicity and mineralization, according to Russian authors, were often associated with such deep faults. Suvorov (1968) has written a comprehensive monograph on this subject. In his comprehensive category of deep faults Suvorov includes such structures as subduction zones as well as major continental rifts. Thus his list includes: 1) platform lineaments; 2) periocenic faults; 3) oceanic transform faults; 4) pericratonic faults; 5) block faults within ancient platforms; 6) geosynclinal margin faults; 7) block faults within geosynclines and geoanticlines.

The deep faults are not uncommonly identified on the basis of combined geological and geophysical work. Godin (1959), for instance, has pointed out that, gravity evidence apart, such structures can be identified seismically since seismic waves are either refracted by them or are completely absorbed at the interface. Similar results have been noted by Grachev et al. (1960). By now both refraction and reflection methods have been used with, occasionally, such spectacular results as the evidence for the displacement of the M-discontinuity (Vashchilov, 1963).

Bulina (1964) has used aeromagnetic maps to demonstrate major deep faults in Siberia. In a rather similar way Hamilton (1970) has used the aeromagnetic map to show the separation of the Siberian and Russian plates.

In the context of these paragraphs on the modern tectonic achievements in the U.S.S.R., it ought perhaps to be mentioned that the volume of geophysical research in the U.S.S.R. is enormous, and much of it is oriented towards practical problems in searching for oil, gas and mineral resources (see Rast, 1962). In the process large territories of the Soviet Union have been surveyed but, unlike geological maps, geophysical data have a much more restricted circulation. However, some of the most accurate work has been carried out at the Pacific Ocean margin where the deep trenches, island arcs and the inland seas from Kamchatka to Japan have been extensively studied (Gnibidenko, 1970; Kochergin et al., 1970; Sychev, 1969; Tuyezov, 1969; Vassilkovskii et al., 1971; Andreyev, 1973) with some authors still arguing for substantial vertical movements. The most recent information is that a 'Geological Map of the Pacific Mobile Belt and Pacific Ocean' to the scale of 1:10000000 is being prepared (Krashyi et al., 1972).

Magmatism

The work on magmatism in the Soviet Union is perhaps less advanced than on other branches of geology. This is despite the fact that in specific branches, such as vulcanology, Soviet scientists have a high reputation. The cause is the paucity of academic field-research unoriented towards specific economic aims. On the other hand, where there are economic returns, excellent work is being done (viz., Sobolev, 1972 on Kimberlite pipes). Geochemistry, on the other hand, is widely pursued in the Soviet Union, as can be gathered from the perusal of their principal journal, *Geokhimiya*. In one respect, however, a new petrological magmatic concept has been developed in the U.S.S.R. and it is known as the concept of Volcano-Plutonic Formation. The concept was introduced by Ustiyev (1963, 1970) and has since been widely pursued. In essence it advocates the study of consanguineous extrusive and intrusive rocks petrologically and structurally in specific settings. One of the basic premises of the concept has been that, in different tectonic and time settings, entirely different coexisting intrusions and extrusions are generated. In a sense, therefore, the idea is much more akin to plate tectonic notions. Thus it is not surprising there are major reappraisals from a plate tectonic point of view. The recent paper by Mossakovskii (1972), reprinted in translation by the *International Geological Reviews*, deals extensively with orogenic vulcanism in Eurasia.

Of course, the significant aspect of such studies has been the practical line of the Russian approach to magmatism: namely the association of useful mineral deposits. It is the finding of hydrothermal minerals in the Verkhoyansk-Okhotsk fold belt that gave impetus to the volcano-plutonic associations. Because of their partial separation from the mainstream of western science, Soviet scientists have generated a specialized terminology, a summary of which by Ustiyev (1971) is available in English.

Useful Mineral Deposits

The Soviet Union is in practice self-sufficient regarding useful mineral deposits. Even diamonds, which until recently were not found in the U.S.S.R., are now obtained in sufficient quantities for industrial needs. The shortage of bauxite is supplemented by the extraction of aluminium from nepheline obtained from the per-alkaline intrusives of the Kola Peninsula. The most intensive search in the last few years has been for new deposits of petroleum since energy requirements have been rising rapidly.

In general, much recent work is published in the Russian journal, *Geologia Rudnykh Mestorozhdenii*, some articles from which get translated in the *International Geological Reviews*. Applied geophysics is also catered for in *Prikladnaya Geofizika*. Of course the more general papers appear in other Soviet periodic publications. Specialized publications on petroleum deposits abound and include *Nefty-anoye Khozyaistvo* and *Geologiya Nefti i Gaza* as well as a large number of monographs, proceedings of symposia and books. The regional approach used in the U.S.S.R. can be gathered from for instance the paper by Dolenko (1972) who discusses the oil-bearing regions of Ukraine. V. D. Nalivkin (1973) on the basis of the assumption that certain structures are likely to contain petroleum has recently proposed that the deposits in the U.S.S.R. three times exceed those in the U.S.A. In this respect the recent Soviet exploration of the Western Siberian Lowlands is especially important. This region is considered as a megabasin (Vassoyevich *et al.*, 1972) and is expected to yield as much as 300 × 10 tons by 1980.

Oil and gas are not the only deposits considered from a geotectonic and regional point of view. Khain (1964), for instance, in his well-known book on geotectonics discusses the relationships between geotectonics and metalliferous deposits. This approach clearly has much to recommend it although the Russian terminology sometimes seems unusual to a western scientist.

Nicholas Rast
June 11, 1973

REFERENCES

ANDREYEV, A. A., 1973. Certain general problems in Terrestrial crust structure of Sakhalin in the light of DSS and gravimetric data. *Internat. Geology Rev.*, **15**, 328–333.

BULINA, L. V., 1964. On the identification of faults in the Siberian Platform according to the air-survey data. *Geol. i Geofiz. No. 2*.

BELOUSSOV, V. V., 1970. Against the hypothesis of ocean-floor spreading. *Tectonophysics*, **9**, 499–511.

——, 1971. On possible forms of relationship between magmatism and tectonogenesis. *J. Geol. Soc. Lond.* **127**, 57–68.

DOLENKO, G. N., 1972. Regularities of oil and gas accumulation on the territory of the Ukrainian S.S.R., *24th Internat. Geol. Congr.*, **5**, 75–81.

GNIBIDENKO, N. S., 1970. On the basement of the northwest sector of the Pacific belt. *Tectonophysics*, **9**, 513–523.

GODIN, Yu. N., 1958. *Ref. sketch (Fig. 1) in Suvorov*.

GOLIZDRA, G. Ya. & AKHMETSHINA, A. K., 1972. On the nature of the extensive gravitational anomalies of the Ukrainian Shield. *Izv. Earth Phys.* (*transl.*), No. 8, 91–99.

GORSHKOV, G. P., SLAVIN, V. I. & BAZHENOV, B. P., 1972. Training in geology in the U.S.S.R., *24th Internat. Geol. Congr.*, **17**, 73–79.

GRACHEV, V. N., *et al.*, 1960. Deep geophysical investigations over the Baltic Shield. *21st Internat. Geol. Congr. Dokl. Sov. Geol. 2.* Moscow Gosgeoltekhizdat.

GROSSGEIM, V. A., 1972. Structure and conditions of formation of flysch. *Geotectonics* (*transl.*), 22–25.

HAMILTON, W., 1970. The Uralides and the motion of the Russian and Siberian platforms. *Geol. Soc. Amer. Bull.*, **81**, 2553–2576.

KELLER, B. M., 1971. Vendian and Udomian. *Bull. Moscow Nat. Hist. Soc.*, **46**, 19–27.

KHAIN, V. E., 1964. *General Geotectonics*, Nedra. Moscow. 477pp.

——, 1970. About the relationship between ancient platforms, young platforms and the so-called regions of completed folding. *Bull. Moscow Nat. Hist. Soc.* **45**, 18–30.

——, 1972. Main trends in the development of the Earth's crust (lithosphere). *24th Internat. Geol. Congr.*, **3**, 58–63.

KOCHERGIN, E. V., *et al.*, 1970. The anomalous geomagnetic fields in the northwest Pacific mobile belt and its relation to the crustal structure. *Geol. i Geofiz.* No. 12, 77–79.

KOSYGIN, Yu. A., 1969. *Tectonics*, Nedra. Moscow. 616pp.

——, *et al.*, 1972. General features of Precambrian tectonics of continents. *24th Internat. Geol. Congr.*, **1**, 342–347.

KRASNYI, L. I., *et al.*, 1972. Biostratigraphy and palaeogeography of the Pacific mobile belt and Pacific Ocean (abstract). *24th Internat. Geol. Congr.*, **8**, 124.

KRUTIKHOVSKAYA, Z. A. & PASHKEVICH, I. K., 1970. Results of investigations of the relationship between magnetic and gravity fields with the tectonics of the Ukraine Shield. *Geophys. Symp. 38*, Naukova dumka, Kiev.

——, —— & SIMONENKO, T. N., 1973. Magnetic anomalies of Precambrian shields and some problems of their geological interpretation. *Can. J. Earth Sci.*, **10**, 629–636.

MOSHKIN, V. N., *et al.*, 1972. A computor-based descriptor-type information retrieval system for geology. *24th Internat. Geol. Congr.*, **16**, 169–172.

MOSSAKOVSKII, A. A., 1972. Palaeozoic orogenic volcanism of Eurasia (Principal complexes and the tectonic factors in their distributions). *Geotectonics (transl.)*, 14–22.

MURATOV, M. V., 1972. Main structural elements of the crust on continents, their interrelations and age. *24th Internat. Geol. Congr.*, **3**, 71–78.

NALIVKIN, V. D., 1963. Graben-like depressions in the east of the Russian Platform. *Soviet Geol.*, No. 1, 53–59.

——, 1972. Relations between the type of major tectonic structures and the density of oil and gas reserves. *24th Internat. Geol. Congr.*, **5**, 168–171.

PALEI, I. P., 1963. Main tectonic features of the Baltic Shield. *In:* Problems of regional tectonics of Eurasia. *Trudy Akad. Nauk*, **92**, 11–34.

PARFENOV, L. M., 1970. The tectonics of the Precambrian of Eurasia. *Geol. i Geofiz.* No. 8, 12–23.

PEIVE, A. V., 1945. Deep faults in geosynclinal regions. *Izv. Akad. Nauk. S.S.S.R., ser. geol.*, No. 5.

——, 1969. Oceanic crust in the geological past, *Geotektonika*, No. 2.

——, *et al.*, 1971. The oceans and geosynclinal processes. *Dokl. Akad. Nauk S.S.S.R.*, **196**, No. 3.

——, 1972, Problems of intracontinental geosynclines. *24th Internat. Geol. Congr.* **3**, 486–493.

RAABEN, M. E., 1969. Columnar stromatolites and late Precambrian stratigraphy. *Am. J. Sci.*, **267**, 1–18.

——, 1971. The Upper Riphean as a subdivision of the general stratigraphic time-scale. Moscow, 45pp.

RAST, N., 1962. (Ed.) *Applied Geophysics U.S.S.R.* Pergamon Press, Oxford, London, 429pp.

——, 1971. Isotope dating in the U.S.S.R. – an essay review in: The Phanerozoic time-scale – a supplement. *Lond. Geol. Soc. Special Publication*, **5**, 39–49.

SALOP, L. I., 1964. Geochronology of Precambrian and some features of the early stage of the Earth's geological development. *21st Internat. Geol. Congr. Dokl. Sov. Geol.* Nedra. Moscow.

——, 1972. A unified stratigraphic scale of the Precambrian. *24th Internat. Geol. Congr.*, **1**, 253–259.

SALOP, L. I. & SHEINMANN, YU. M., 1969. Tectonic history and structures of platforms and shields. *Tectonophysics*, **7**, 565–597.

SEMENENKO, N. P., 1972. Precambrian geochronology and problems. *Internat. Geology Reviews*, **14**, 947–953.

——, SCHERBAK, N. P., & BARTINSKII, E. N., 1972. Geochronology, stratigraphy and tectonic structure of the Ukrainian Shield. *24th Internat. Geol. Congr.*, **1**, 363–370.

SEMIKHATOV, M. A., 1972. On the general Precambrian stratigraphic scale. *24th Internat. Geol. Congr.*, **1**, 273–277.

——, & KOMAR, V. A., 1965. On the practical use of species of columnar stromatolites in the interregional correlation of the Rhiphean deposits. *Dokl. Akad. Nauk S.S.S.R.*, **165**, 1383–1386.

SHATSKII, N. S., 1946. Greater Donbas and the Wichita System. Comparative tectonics of ancient platforms. *Izv. Akad. Nauk S.S.R.*, *ser. geol.*, No. 6.

——, 1964. *Collected works*, II, Nauka, Moscow.

SOBOLEV, N. V., 1972. Petrology of xenoliths in Kimberlitic pipes and indications of their abyssal origin. *24th Internat. Geol. Congr.*, **2**, 297–302.

SOKOLOV, B. S., 1972. The Vendian Stage in Earth history, *24th Internat. Geol. Congr.* **1**, 78–84.

SOLLOGUB, V. B., *et al.*, 1967. Deep crustal structure of the East Carpathians and adjacent regions of the Ukraine from D.S.S. *In: Geophysical Investigations of the Earth's crust in south-eastern Europe*. Nauka, Moscow.

SUVOROV, A. I., 1968. *The structural laws and the formation of deep faults*. Nauka, Moscow, 316pp.

SYCHEV, P. M., 1969. The gravity anomalies and causes of vertical crust motion in the Asia – to – Pacific transition zone. *Geotektonika*, No. 1, 13–25.

TUYEZOV, I. K., 1969. Geophysical researches on the Far East region of Circumpacific belt. *Sakhalin Trudy Akad. Nauk S.S.S.R.*, No. 29.

USTIYEV, Ye. K., 1963. Problems of volcanism and plutonism. Volcano-plutonic formations. *Izv. Akad. Nauk S.S.S.R.*, *ser. geol.*, No. 12.

——, 1970. Relations between volcanism and plutonism at different stages of the tectonomagmatic cycle. *In:* Eds. G. Newall and N. Rast. *Mechanism of Igneous Intrusion*, Gallery Press, Liverpool, 372pp.

——, 1971. Basic concepts and terms of the theory of magmatic formations. *Internat. Geol. Reviews*, **13**, 95–110.

VASHCHILOV, Yu. Ya., 1963. Deep faults of the south of the Yana-Kolyma fold zone and the Okhotsk-chaun volcanic belt and their role in the formation of granite intrusions and the construction of structures (geophysical data). *Soviet Geol.*, No. 4.

VASSIL'KOVSKII, N. P., *et al.*, 1971. Japan Sea relict of an ocean (abstract). *In: Island arcs and marginal seas* (abstract in English). Tokai Univ. Press, Tokyo, 311pp.

VASSOYEVICH, N. B., *et al.* 1972. Sedimentary basins. *24th Internat. Geol. Congr.*, **5**, 187–194.

YANSHIN, A. L., *et al.*, 1966. *The tectonics of Eurasia*. Nauka, Moscow, 487pp.

ZONENSHAIN, L. P., 1972. Similarities in the evolution of geosynclines of different types. *24th Internat. Geol. Congr.*, **3**, 494–502.

Part I

INTRODUCTION

1

THE INVESTIGATION OF THE
GEOLOGICAL STRUCTURE
OF THE U.S.S.R.

PRE-REVOLUTIONARY PERIOD

Tens, possibly, hundreds of thousands of years ago Stone-Age man took the first step in the search for useful deposits and in the utilization of mineral wealth. He knew where to find chips of flint, and knew some of its properties. It is possible that he also knew salt-bearing sand or clay with rock-salt encrustations, which gave him great pleasure when he licked it.

Once the later Stone-Age men started to use stone for building homes and protective barriers the knowledge and use of the rocks became more widespread. During this period coloured stones and minerals were first used as decorations and soft, compact rocks were used for the most primitive sculptures. Coloured and friable rocks were used as pigments for rock pictures, some of which show amazing vitality and accuracy. All these deposits, which were the first ones exploited by our forefathers, were non-metallic ores except for some pigments. With the beginning of the bronze age (many thousands of years ago) in addition to the non-metallic mineral deposits man started discovering metallic ores, first copper and then iron ores. At the same time the extraction and melting of the precious metals—gold and silver—and of the then no less precious tin and lead had begun.

After the metallic ores, but still many thousands of years ago man started using mineral fuels. Initially the fires which they found at the points of emergence of the mineral gas were found useful in the preparation of food. Then petroleum and coal started being used in the south and the peat and brown coal in the north.

Many thousands of years ago in Egypt, the Near Eastern states, in China and Korea the first hypotheses of the structure of the earth, the earth crust and the causes of formation of volcanoes, mountains and seas evolved. At the same time the significance of fossil animals and plants was correctly but not widely understood.

By the 16th and 17th centuries in Russia the mining industry was already flourishing. Numerous 'rudoznatsy' (the then so-called ore-knowers) fearlessly penetrated the most distant corners of the Urals and even Siberia. They collected the first geological data. However, in those days the clergy were all powerful and ruthlessly suppressed the use of books for disseminating knowledge, except religious knowledge. The most interesting data of the rudozantsys have reached us only as a few manuscripts, reports, petitions and instructions. One instruction on the search and extraction of saline solutions is amazing in its detail of technical terminology.

In the time of Peter the Great, non-religious books first appeared, and thereafter the output of books steadily increased. Some of the early geological works are so valuable that they are still referred to now. One can quote *The description of the Kamchatka area* by S. P. Krashenninikov (1755), *Diary of the Expeditions in 1768–1771* by I. I. Lepekhin, investigations of R. Pallas in Siberia, Altai and Crimea and of G. Shurovskii in the Urals (1841) and Altai (1846) as well as a series of researches on the 'Moscow basin' (1866–1867).

Amongst the theoretical investigations the work entitled *On the Layers of the Earth* (1783) by M. Lomonosov, that astonishing Russian genius, is outstanding.

In 1775 Moscow University, the first Russian institution of higher education, was founded, and in 1773 the Mining Institute in St Petersburg was opened. Many of the greatest Russian geologists came from within their walls. Moscow University with its branch of the Moscow Geological Surveying Institute together with the Leningrad Mining Institute still maintain their leading roles. Nowadays there have been added many other first class educational institutions, where geologists are being trained.

At the end of the 18th and the beginning of the 19th century geology did not exist as an independent science; only mineralogy—in a completely different sense than at present—was being taught in the institutions of higher education. This type of mineralogy included petrography, geology, palaeontology and the study of useful mineral deposits.

It was only after 1812 that geology (then called 'geognosy') and palaeontology (then called 'oryctology') were separated from mineralogy. The study of useful mineral deposits was included in geognosy. At the same time the first specifically geological (geognostic) investigations began. These investigations were mainly concerned with the regions in which mineral deposits were being mined, but soon investigations spread over large areas in Altai, the Urals, Donyets Basin, the north of the Russian Platform and its central parts.

The first journals containing papers on geology appeared at this time. These journals were the *Mining Journal* (*Gornyi Zhurnal*), which started publication in 1825 and the Memoirs (Zapiski) of the Mineralogical Society (from 1830) and the University of Moscow.

The Tsarist government, which always lacked confidence in its native scientists and treated them with contempt, decided to invite foreign scientists to collect together the growing volume of geological knowledge of Russia. Fortunately outstanding geologists, namely the great English geologist Roderick Murchison and the foremost French palaeontologist Eduard Verneuil, were invited. For three years they travelled over the Russian Platform and the Urals. They had at their disposal huge collections of geological and palaeontological materials collected by Russian geologists.

In 1845 the monograph *Geology of Russia* appeared, the first volume of which (geology) was by Murchison in English and the second (palaeontology) was by Verneuil in French. The first volume was immediately translated into Russian and together with appendices by the Russian geologists was published in the *Mining Journal*.

Both volumes in their time were outstanding monographs. It is sufficient to point out that the Permian System was established in the first volume and the Devonian

System was first recognized in Russia in the second volume. Now the first volume has only a historical significance, but the second volume is still widely used by Soviet palaeontologists as a primary source.

With every decade the Russian mining industry developed further, and this brought about a considerable growth of geological investigation. The work of the Moscow geologists was especially important. At that time the government and the centre of cultural life was in St Petersburg, while Moscow was a backwater. Nevertheless scientists grouped round the Moscow University carried on with very important scientific work in various branches of knowledge, including geology. Thanks to the researches of G. Shurovskii, Sh. Rul'ye, G. Troutshol'd, P. Semenov, V. Meller and other investigators the Moscow area and the surrounding provinces became the most carefully studied region of Russia.

The majority of research carried out between 1840–80, however, has a spasmodic, incidental character and involved small areas. Thus arose the necessity for planned, prolonged investigations, which would illuminate large industrially important areas, for example the central parts and the south of the Russian Platform and the Urals.

In 1881 the Geological Committee was founded and given the task of preparing a geological map of European Russia to the scale of 10 verst to 1 inch (1:420 000); this map became known as 'desyativerstka' (ten verst map). The first director of the Geological Committee was G. P. Gel'merson who was soon succeeded by A. P. Karpinskii, the future president of the Academy of Sciences of the U.S.S.R. After Karpinskii, F. N. Chernyshev was the director. The initial staff of the Committee included only eight geologists.

The area, intended for mapping, would have covered 141 ten-verst sheets. Four to six years of field work was to be spent on each sheet. A simple calculation indicates that the survey was not planned realistically; many sheets were in fact completed by the Soviet geologists, while some areas remained unsurveyed.

The research work of the Geological Committee during the Pre-Revolutionary times was outstanding and created a deserved reputation for Russian geologists. Outstanding publications were F. N. Chernyshev's monograph on the Urals and Timan, S. N. Nikitin's monograph on the central provinces and N. N. Sokolov's monograph on the southern part of European Russia.

Research carried out at the Mining Institute and in the universities was also of importance. For instance there was the work of I. V. Mushketov and G. D. Romanovskii on Central Asia, of A. P. Pavlov on the southern part of the Russian Platform, of N. I. Sinstov on the Tertiary deposits of the southern provinces, of N. A. Golovinskii on general geological problems, of P. A. Kratov on sub-Ural regions, and many others.

The beginning of the 20th century was characterized by a further development of the mining industry which, thanks to the building of new railways, spread further and further to the south and east. Naturally geological investigations also increased.

A feature of the time was the major regional surveys based on the industrial regions. Geological surveys of the Donyets, and later of the Kuznetsk, Basins were carried out under the guidance of that prominent geologist L. I. Latugin and serve as brilliant examples of such investigations. The investigation of the Karaganda Basin followed, conducted by one of the best coal geologists, A. A. Gapaev. In the Caucasus detailed geological surveys of the most important oil-bearing regions began; these surveys were headed by I. M. Gubkin in Maikop, by D. V. Golubyat-nikov in Baku, and by K. P. Kalitskii in Central Asia. In the Urals the platinum deposits were studied in detail by N. K. Vysotskii, while A. N. Zavaritskii did brilliant work on the iron and copper orefields. In eastern and western Siberia there were major expeditionary researches on the study of gold deposits directed by V. A. Obruchev and A. P. Gerasimov.

Geology then penetrated eastwards and in Siberia (Tomsk) a notable school of geologists came into existence under the leadership of V. A. Obruchev and M. A. Usov. This school has contributed enormously to our knowledge of the geological structure of western Siberia, the Kuznetsk Basin, the Altai and central Kazakhstan.

By 1914 the Geological Committee had 58 geologists. The number of institutions of higher education was also on the increase, and new chairs of historical geology, geology of ore-desposits and crystallography were being established in the universities. In the University of Moscow, the school of geologists was established under the leadership of one of the greatest Russian palaeontologist stratigraphers, A. P. Pavlov and his disciple A. D. Arkhangel'skii. This school has done much valuable research on methods, and also has contributed largely to our understanding of the geology of the Russian Platform and of the Caucasus. The Kazan school of geology, headed by A. A. Shtukenberg, A. P. Krotov, M. E. Noinskii and A. V. Nyechayev, concentrated on investigations in the Volga region and the Cis-Urals. Its investigations on the Permian deposits are especially important. The Yekaterinoslav (Dnyepropetrovsk) Mining Institute, largely due to N. I. Lebedev, the holder of the chair of geology, expanded and improved our knowledge of the Donyets Basin.

Summing up the achievements of Pre-Revolutionary Russian geology it should be noted that while the European part had been studied relatively thoroughly, this did not apply to the Asiatic regions. A geological map of the European region was published by the Geological Committee in 1892 and again in 1915. These maps were of high quality and as good as the best foreign maps. The international map of Europe included European Russia with the Urals and the Caucasus, which were all surveyed by Russian geologists.

However in Asiatic Russia only isolated traverses and expeditionary investigations had been carried out. Large scale geological surveys had rarely been attempted and for large areas there were no maps of any kind. Thus in 1917 when the Geological Committee issued a geological map of Asiatic Russia, it had a very curious appearance. The largest part of it was entirely white and another part had only narrow coloured strips and only in places in the south were there small completely ornamented areas. Thus no general map of the Russian territory existed at that time. Such a map was accomplished only in 1937, after prolonged and meticulous work by Soviet geologists.

SOVIET PERIOD

Nowadays the Ministry of Geology and Preservation of Mineral Wealth, instead of the Geological Committee with its limited staff, guides geological investigation, surveys and exploration. The Ministry includes, apart from the All Union Geological Institute (VSEGEI, the former Geological Committee) a number of other

A. P. KARPINSKII
(1847–1936)

F. N. CHERNYSHEV
(1856–1914)

V. A. OBRUCHEV (1863–1956)

M. A. USOV (1883–1939)

A. A. BORISYAK
(1872–1944)

N. G. KASSIN
(1885–1949)

A. D. ARKHANGEL' SKII (1879–1940)

scientific research institutes; it also includes various geological departments situated in numerous republics and provinces of the Soviet Union. The total staff of these geological institutions numbers many thousands, thus there has been an incredible growth. In addition, the Academy of Sciences of the U.S.S.R. has a large geological institute, which in itself is larger than all the geological institutions of Tsarist Russia in its final period. Almost all the Academies in the Republics and many of the affiliated branches of the Academy of Sciences of the U.S.S.R. have geological institutes. Large geological establishments are supported by the Ministries of Oil, Non-ferrous Metallurgy, Ferrous Metallurgy and Coal, while many other Ministries have smaller geological departments. The reorganization of the industrial management and the establishment of the Economic Soviets has even further linked geology to the mining industry, and thus has had a considerable effect on the development of geology.

At present the Soviet Union has instead of a few dozen, as in Pre-Revolutionary times, tens of thousands of specialists at its disposal. This has only been achieved by the exceptional growth of the institutions of higher education, where the future geological personnel is being trained. The universities and the institutes, which existed before the October Revolution, greatly increased their intake of students, while new professional chairs were established in specialist branches of geology. The University of Moscow not only radically increased its size, but formed a new educational institution, the Moscow Geological Survey Institute (MGRI).

When the author entered the Mining Institute in Leningrad (1906) it specialized only in geological surveying with an annual intake of 20–25 students. Now it has three departments—geological surveying, oil, and geophysics—with an annual intake of 250–300 students.

Many geological institutions for training geologists have been established. The most recent development is the establishment of national geological staff who are trained and work in their homelands.

MAJOR TRENDS IN SOVIET GEOLOGY

There are three main areas of activity: (*i*) the search for useful mineral deposits; (*ii*) the construction of geological maps; (*iii*) systematic research.

Of these the search for useful mineral deposits is most important to Soviet geologists. Many new raw materials have been discovered which were unknown under the Tsarist government. Amongst these some are of considerable economic importance, as, for instance, bauxite, which is the main aluminium ore. Tens of thousands of new deposits of the known minerals, such as iron ore, coal and oil have been found. What is most important is that there has been a radical change in the geographical spread of these deposits. Formerly they were all concentrated in a few provinces—mainly in the Urals and the southern part of European U.S.S.R. At present the search is continuing for deposits of the most useful minerals in all the principal parts of our huge motherland—the Far East, Western Siberia, Kazakhstan, Central Asia and the Caucasus. This is making it unnecessary to transport these materials long distances by rail and creates favourable conditions for the independent development of the furthest parts of the U.S.S.R.

During the first three five-year plans the exploration and research were undertaken by the Geological Committee. However the growth of the national economy was so fast that the Committee, despite its size, was not able to cope with all the problems which arose. Consequently the Central Geological Survey Department (GGRU) was established, which was later transformed into the Committee for Geological Affairs attached to the Soviet of Peoples' Komissars (the Cabinet) of the U.S.S.R. The Geological Committee was then regrouped into the Central Scientific Geological Survey Research Institute (TsNIGRI), but what is more important is that, in a number of industrial centres and in the capitals of the Republics, local geological survey departments were opened, which at first were called the Trusts and later the Offices (Upravleniya). These were directly subordinate to the Committee for Geological Affairs and they were empowered to conduct basic geological surveys. This reorganization has been most successful.

I. M. Gubkin, a great Soviet oil geologist, had a leading role in making these changes. We also should mention his assistant A. A. Blokhin, whose insistence and faith in his prognosis resulted in the exploitation of the Ishimbai oil-deposit.

I. I. Malyshev, a Ural geologist and specialist on titaniferous ore, was the next head of the Committee for Geological Affairs. In his time the scope of operations used in the solution of geological and survey problems expanded further and in 1946 the Committee for Geological Affairs was reconstituted into the Ministry of Geology, U.S.S.R. The tasks carried out by TsNIGRI were altered and it was renamed the All Union Geological Institute (VSEGEI). The Geological Survey 'offices', were transformed into specialist geological departments which were further developed. At present a number of the geological departments—for instance that for the Urals, and that for Kazakhstan—have geological staffs of considerable size. In Kazakhstan a Ministry of Geology was initiated.

There are a very large number of Soviet geologists working on mineral deposits. We can only mention some of the most outstanding. S. S. Smirnov and Yu. A. Bilibin would be counted as such in ore geology. Both have contributed much to our knowledge of useful deposits in the eastern provinces of the U.S.S.R. The brilliant development of Siberia owes a lot to them.

A. E. Fersman used to surprise one by the variety of his research. He has done much useful work on mineral deposits (mainly non-metallic), such as apatites of Khibina, precious stones of the Urals, sulphur from Karakum, pegmatites, etc. A. N. Zavaritskii did some quite valuable research on the metalliferous ore deposits of the Urals. In the field of coal—particularly of the Donyets Basin—the work of P. I. Stepanov is important. The monographs of D. V. Golubyatnikov on the Caucasian oil are outstanding in their great accuracy and detail. The basis of the study of the oil deposits in Uzbekistan and Turkmenistan was laid by the brilliant research of K. P. Kalitskii.

Construction of geological maps. This is a rather important operation, but it must be subordinate to and connected with the exploration and search for useful deposits. Geological survey should be conducted not independently, but as a necessary basis for exploratory, and very frequently for prospecting operations. The geological map shows the structure of a region, its stratigraphy, and tectonics, and not infrequently its lithology. Only by knowing the geological structure of a region is it possible to start explorations correctly. Thus a detailed geological map of a small region must be a necessary concomitant of exploration. Formerly the main stratigraphic subdivisions, outcrops of igneous rocks, faults and the main elements of structure used to be shown on a geological map. Nowadays much other valuable information is also given. The age of the intrusions and of young extrusive rocks is indicated in colour or by a letter code. Tectonic data are made more accurate and interconnected. Fold axes, their culminations and depressions, and structure contours of the flexures are also shown. The age of the faults is sometimes indicated in colour or by means of symbols. An indication of the lithological composition is another very useful addition.

The main Soviet geological map may be complemented by additional sheets showing hydrogeological data, geomorphology, traverses and useful mineral deposits. Recently lithofacies maps have also been prepared. These show the lithological composition of the rocks and the position of the main facies. Such maps are usually available for the oil-bearing regions. The formulation of lithofacies maps is an important achievement of the Soviet geologists who use these industrially, as distinct from foreign geologists who refer to or use them only in connection with small-scale projects.

The next important achievement of the Soviet geologists was the mapping of the whole Soviet Union. The first map, to a scale of 1:5 000 000, was published in 1937 for the 17th International Geological Congress. The second, on a larger scale and in greater detail, was the 1:2 500 000 map printed in 1940. The third, which was considerably simplified, but included the latest data, was published in 1951 on a scale of 1:7 500 000. The year 1955 saw the publication of the second edition on the simplified map on the scale of 1:5 000 000, which was the first map in which all the blank spots (the unknown areas) were filled in. In 1956 a second map on the scale of 1:2 500 000 was published. It is better than all the previous maps in respect of detail. up-to-date nature, and the volume of material represented. The present author has been the principal editor of all the simplified maps of the U.S.S.R.

The construction of each map, to the scale of 1:2 500 000, is a major and complex effort, demanding the participation of a large number of geologists and many geological cartographers. The history of the construction of geological maps is interesting and complicated. It is interesting since we observe how the work carried out by one, albeit highly-qualified central organization, grows and spreads over many smaller departments, distributed throughout the territory of the U.S.S.R.

A. P. Gerasimov, the outstanding Soviet survey geologist, was particularly active in the organization of geological surveying and his instructions, advice and criticism were always most helpful. After decentralization the survey was carried out by many regional geologists. Amongst these one should mention N. G. Kassin who spent his whole working life on the investigation of Kazakhstan. The decentralization of the geological survey is an improvement but at the same time it is a complication. Its organization is more complex and the establishment of uniform legends, stratigraphy and presentation become all the more important.

In addition to the Ministry of Geology, surveying is conducted by many other geological institutions. For instance large-scale surveys were carried out by the Ministry of the Petroleum Industry in connection with the fact that the oil-bearing horizons often have very wide distributions. In general, oil geologists have made many important contributions to the knowledge of the geological structure of our country.

Large areas have been surveyed by ore geologists, working in regions of extensive ore mineralization. In such regions intrusive and extrusive rocks are often widespread. The ore geologists have prepared detailed maps of the ore-bearing regions, have studied the structure and tectonics of the intrusive massifs and have investigated the ages and relationships of the lodes.

All our coal-fields have been surveyed by coal geologists. Their maps have details of lithological characters, outcrops of sandstones, limestones and coals. The tectonic structures and especially the fault zones are mapped in great detail.

Geological surveys are also conducted by salt geologists, by those studying non-metallic deposits, by geologists seeking constructional materials, by hydrogeologists and by engineering geologists. In our country each major building project is preceded by the preparation of a geological map of the area.

Another important feature of the last few five-year plans has been the beginning of an entirely new type of geological survey—the aerogeological survey—which employs photographs taken from aeroplanes in conjunction with ground observations. This type of surveying makes the construction of geological maps, especially in the well-exposed regions, much faster and more accurate. An even newer aspect of geological surveying is the preparation of the geological maps of the sea floor. This involves observations from aeroplanes and ships in conjunction with ground and submarine investigations.

Some of the most valuable geological material is provided by geophysical research and deep drilling. Geophysical maps and maps of sub-surface structures together with the geological maps reveal the interior structure of the earth's crust.

Systematic investigations. These investigations are sometimes considered as secondary to the main aim of exploration-prospecting and survey. In reality systematic investigations build up a theoretical scientific basis, without which exploration and surveying would be incomplete and sometimes erroneous. The strength of Soviet geology lies in the harmonious interrelationship between these three types of geological investigations.

During the Soviet period the geological sciences have developed rapidly. Some sciences, which previously were studied by a few specialists, now have special scientific research institutes, as for instance crystallography, geochemistry, vulcanology, and cryopedology. A large number of new sciences have separated from the 'older' ones. Such for instance are tectonics and geochemistry. Many other branches of geology are entirely new and were not even mentioned in the Pre-Revolutionary literature, for instance the concepts of facies, the study of caustophytoliths, coal petrology, mineragraphy, luminescence analysis, the study of metamorphism, etc.

The annual number of systematic research projects has, of course, increased enormously and is at present counted in thousands. They are carried out by almost all institutions concerned with research exploration and surveying, but the leading role rests with the Institute and laboratories of the Academy of Sciences of the U.S.S.R. An important complex of projects was carried out by the scientific institutes (VSEGEI, VIMS) of the Ministry of Geology, of the former Ministry of Oil Industry (VNIGRI, VNIGNI) and numerous central scientific research institutes of oil corporations and trusts. Year by year the geological institutes of the Academies of Sciences of the constituent Republics and of branches of the Academy of Sciences of the U.S.S.R. grow larger.

The research projects carried out in U.S.S.R. both in their numbers and significance are on a higher level than such projects in any capitalist country. In many instances, although not in all, Soviet geologists are foremost in their field.

International Geological Congresses represent particular forms of review of geological achievement. Such congresses have been held every four years for the last 80 years, and geologists from nearly all countries attend. One of these congresses was held in Russia in 1897 and was most successful. Another was held in Moscow in 1937. It was even more successful than the 1897 congress and completely established the international position of Soviet geology. The descriptions (excursion guides) of those regions of the Soviet Union where excursions were held have not lost their value even now. All subsequent congresses have been attended by Soviet delegations, although these were relatively small. The number of papers presented by Russian geologists was also not large.

As a result of the decision of the Central Committee of the Communist Party of the Soviet Union and on the instructions of N. S. Khrushchev, a National Committee of Soviet geologists was established to strengthen international relations. The Committee meets in Moscow and is attached to the Geological Geographical section of the Academy of Sciences, U.S.S.R. The Committee made extensive preparations for the 1960 Congress in Scandinavia (Copenhagen) and Soviet geologists read more than four hundred papers which are printed in the 21st series of the Congress Transactions. Soviet geologists participated even more fully in the 1964 Congress in India.

In the first five years after the October Revolution the tempo and magnitude of scientific research decreased for a short while. Yet soon a development of science began. From the third five-year period onwards the pre-Revolutionary names were left behind and Soviet science progressed irrevocably.

The older scientists, who had with sincerity joined the side of the revolutionaries, played an important role in this development. A. P. Karpinskii, of course, was outstanding. F. Yu. Levinson-Lessing and his colleague D. S. Belyankin initiated the wonderful school of Soviet petrologists. Their assistants and students now occupy prominent positions in this field. V. I. Vernadskii and A. E. Fersman founded a large group of geochemists. The work of A. N. Zavaritskii on ore deposits, petrology and vulcanology has a considerable importance, and he also has trained a number of students. A series of major investigations by Soviet scientists on the petrology of sedimentary rocks included the outstandingly original and important work of V. P. Baturin. A. D. Arkhangel'skii played a leading role in the study of the conditions of formation of sedimentary rocks. He has trained many young men amongst whom we now find a number of important geologists. In tectonics the work of N. S. Shatskii, under whose guidance the first tectonic maps of U.S.S.R. were prepared, has an important place.

In the sphere of stratigraphy and palaeontology the investigations of A. A. Borisyak are exceptionally important. He introduced into Soviet geology the study of facies and palaeo-ecology, and established the biological evolutionary approach to palaeontology.

The numerous papers and the expeditions of V. A. Obruchev have an exceptional significance in the geology of Siberia. He, together with his pupil M. A. Usov, formed the Tomsk Siberian school of geologists, which has played a leading role in the study of the geology and of the useful mineral deposits of western Siberia. The investigations of V. N. Veber on the geology and palaeontology of Central Asia and adjacent provinces are outstanding. Amongst the investigations on the stratigraphy of the Russian Platform the work of A. N. Mazarovich has a major significance. He also did research work in general geology and wrote the first textbook on the geology of the U.S.S.R.

This has been only a short review of the achievements of the Pre-Revolutionary and Soviet geologists. We have mentioned only the most important names, and the most important projects, but there are many other names and projects worthy of mention if space had permitted.

2

GENERAL PROBLEMS

In this chapter the following problems, which are of great importance in the study of the geology of U.S.S.R., are considered: (*i*) geosynclines and continental masses; (*ii*) magmatism and tectonics; (*iii*) sedimentation and tectonics; (*iv*) geological regionalization of the U.S.S.R.; (*v*) Stratigraphical subdivisons; (*vi*) facies changes.

GEOSYNCLINES AND CONTINENTAL MASSES

Geosynclines and continental masses (platforms) are diametrically opposite regions which are closely associated, have intensive influence on each other, and are transitional into each other in terms of their development.

GEOSYNCLINE. This is a region of continuous and considerable sagging and a contemporaneous accumulation of enormous thicknesses of sedimentary and volcanic deposits. During the second phase of its existence a geosyncline as a result of folding emerges from the region of accumulation into the region of erosion, thus becoming a high-mountain country with a system of folded mountain ranges. At the end of this phase the geosyncline becomes a part of the continental mass. Erosion begins with the rise of the mountain chains, and eventually converts the ranges into a platform.

In the first phase a geosyncline is a marine, mainly off-shore region with islands. Its floor is a site of accumulation of enormously thick clastic (terrigenous), organic and effusive (volcanic) formations. Sagging and accumulation take place at the same time and they mutually off-set one another. Because of this, despite the accumulation of many kilometres of formations the depth and physico-geographical characteristics of the seas do not change greatly over long periods of time.

At times in some parts of the region instead of sagging the ground rises. Instead of sea, land appears in the form of islands, peninsulas, deltas, and archipelagoes. Sometimes parts of the geosyncline become entirely dry land.

In certain definite zones, at the bottom of the sea, on the islands and along the sea-shore numerous volcanoes, sometimes several kilometres in height, come into existence. These volcanoes are surrounded by a belt of accumulation of volcanic tuffs. The width of such belts reaches tens of kilometres. In the fissures and feeding channels of such volcanoes, the sub-volcanic intrusive rocks solidify and ore deposits are formed.

In the second phases the fold-generating movements begin. At first they progress at the same time as sedimentation, thus only comparatively slightly affecting the dominant marine regime. As folding intensifies, individual structures or complexes rise above sea level in the form of islands and peninsulas and eventually as major land ridges. The area of the sea begins to shrink and large parts of it are cut off and change into inland seas, which progressively decrease in size and eventually are completely filled up.

During the final stages of folding all the geosyncline, or a part of it adjacent to the continent, becomes a young mountain fold region, involving a system of mountain ranges, stretching along the strike of the geosyncline. Sea, which might have existed for several tens of millions of years, gives way to a highland area.

Where the fold-generating processes achieve the greatest intensity, which is in the deeper parts of the structures, intrusive masses—mainly of granites—originate. The size of such intrusive massifs not infrequently reaches tens and hundreds of kilometres. They are characteristic of the zones of intensive very powerful folding. In the zones of weak folding they are absent.

With the rise of the mountain chains erosion begins, and the transportation of the resultant products to the adjacent sub-montane lowlands and marine basins starts. During this period special sedimentary deposits such as molasse, Nagelfluh, flysch, greywacke ('pepper sandstones'), aspidic shales, and fan conglomerates (fanglomerates), are deposited. Such complexes are characteristic of periods of folding in the same way as are intrusions.

In the third phase of the existence of the geosyncline the fold movements weaken and afterwards cease completely. At the same time the rate of erosion grows until it exceeds the rate of uplift. As a result the relief of the young mountain range, considerable at first, is lowered, and after the ending of uplift is rapidly worn down to a land of low relief (peneplain, dolocline). Thus the geosyncline becomes a platform.

PLATFORM. This is a formerly folded area, which has been so compacted and rigid that later sedimentary formations resting on it are subject to little or no folding. Such formations are almost horizontal. If, in places, they are folded then such folds are gentle and of large wave length, or if sharp then they are flexures of very local distribution.

It is important that the horizontal strata over a platform are not metamorphosed, even where folds are developed in them. Moreover, their thickness is much less than in the geosynclines. Another distinguishing feature of these foundations is the absence of intrusions of granite massifs (Fig. 1).

The transition from geosynclines to platforms, has been mentioned above. The platforms in their turn change into geosynclines when a part of them sags, perhaps along faults, to below sea level and become a marine basin. As an example of such

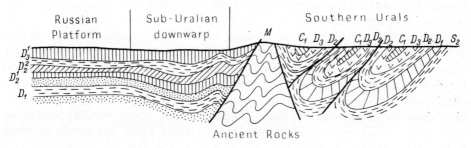

FIG. 1. A schematic cross-section across a platform, a marginal downwarp and a geosyncline, constructed on examples of the Russian Platform, the sub-Urals downwarp and the Southern Urals.

a change we can quote the present-day Indian Ocean, which was formed in place of the Gondwana Platform at the beginning of the Palaeogene. Its shores, with numerous coral reefs and volcanoes, represent a typical present-day geosyncline.

Geosynclines and platforms may be of widely varying ages. The geological age of a geosyncline is defined by the folding which converts the marine basin into the mountainous country. Thus, for instance, the Uralian Geosyncline is typically Hercynian,* the Grampian Geosyncline is Caledonian and the Mediterranean is Alpine. The age of the platform is defined by the folding from which it originated. For instance the eastern Kazakhstan is a post-Hercynian (epi-Hercynian) platform. Sometimes platforms are named after the formations which have accumulated on them. The Russian Platform is a post-Baikalian or Palaeozoic platform. Mangyshlak and Ustyurt are parts of a Neogenic platform, since within their boundaries all the deposits, beginning from Palaeogene and older, have been folded and metamorphosed, this has happened particularly forcefully to the Jurassic and the Triassic. Only the Neogene is almost unfolded and unmetamorphosed.

GEOSYNCLINAL AND PLATFORM DEPOSITS

The deposits of geosynclinal type have the following diagnostic features: (*i*) great thickness; (*ii*) complex folding; (*iii*) considerable metamorphism; (*iv*) synorogenic intrusions; and (*v*) geosynclinal suites.

The thickness of a System is measured in thousands of metres, of a Division in many hundreds and even thousands of metres, and the thickness of Stages in hundreds of metres.

All the deposits are compressed into systems of folds possessing definite strikes continuous for many hundreds and even thousands of kilometres. The folds are complex, compressed, not infrequently isoclinal or fan-like. They are broken up into a series of scales pushed one upon the other. The formation of the folds is always accompanied by metamorphism, which is not infrequently considerable. During the phases of particularly strong fold movements synorogenic (syn-folding) intrusions are generated inside the folds and often reach huge size (many tens and hundreds of kilometres along the strike).

Geosynclinal formations are complexes of deposits, found only in geosynclinal regions. Many of them are connected with the orogenic (mountain-forming) movements, while others are connected with major depressions. To the first group belong molasse, flysch, aspidic clays, graywackes, while to the second belong major reef massifs, siliceous and cherty formations, great thicknesses of rock salt and gypsum, thick coal bearing formations.

Molasse. This consists of the products of erosion of rising mountain chains. It is deposited at their front on submontane platforms or at the bottom of the marginal sea. Conglomerates, which alone are sometimes designated as molasses, are deposited near the chain. It must be remembered that not only the coarse conglomerates (*Nagelfluh*) enter the composition of molasse, but also sands and clays deposited away from the chains. Fauna is rarely found. Bedding is variable and can be of current or normal kind. Total thickness varies, but can be as much as 6–8km.

Flysch. This is also a product of erosion of rising ranges. It is closely associated

* The term Hercynian is synonymous with Variscan or Variscian.

with molasse and can be contemporaneous with it. Thin, regular and rhythmical bedding of flysch is characteristic. The other features are as in molasse.

Aspidic Shales. These shales are similar to flysch, but are distinguished by the absence of, or indistinct, rhythmic bedding. The degree of metamorphism is higher than in flysch.

Graywackes. These are greenish-greyish or brownish sandstones, shales and marls, deposited in near-shore regions of the sea. They are distinguished by a lesser or greater admixture of vulcanogenic materials. The colour of the sandgrains resembles that of ground pepper and consequently they are not infrequently called peppery sandstones. They are regularly bedded with marine fauna and lesser thickness of a few hundred metres.

Marginal downwarps. These are commonly situated between platforms and geosynclines. They are most frequently considered as parts of the geosyncline. The deposits of the downwarps are diagnosed by: (*i*) great thickness; (*ii*) weak, but distinctly orientated folds stretching parallel to the strike; (*iii*) weak metamorphism; (iv) absence of synorogenic intrusions; (v) specific types of formations. To the latter belong the massifs of calcareous reefs of 600–800m in thickness then thick layers of salts, and even thicker (10–12km) coal-bearing sequences. Great thicknesses of these formations are determined by rapid sinking of the basement, which reaches amplitudes of 3–12km.

Miogeosyncline and Eugeosyncline. Not infrequently the part of the geosyncline adjacent to the marginal downwarp or the platform is called a miogeosyncline: a 'median' or 'intermediate' geosyncline. The miogeosynclinal deposits occupy an intermediate position between the platform and the eugeosynclinal deposits. They have a great thickness and are compressed into distinctly linear folds, broken up by thrusts and normal faults, but are diagnostically less metamorphosed, more simply folded and what is most important lack entirely or have only weakly developed volcanic and intrusive rocks, as compared with eugeosynclinal deposits.

A eugeosyncline, a 'true geosyncline', occupies the central position. Its deposits possess all the typically geosynclinal features—great thickness, high metamorphism, complex folding and a great development of volcanics and intrusions.

In the Uralian geosyncline the western flank represents the miogeosyncline. The central part and the eastern flank belong to the eugeosyncline. It is interesting that the base of the Uralian miogeosyncline is underlain by the comparatively shallow crystalline basement.

The deposits of platform type are characterized by the following features: (*i*) small thickness; (*ii*) at most simple brachianticlinal folds; (*iii*) weak or non-existent metamorphism; (iv) absence of synorogenic intrusions; (v) platform formations. The thickness of a System is several hundred metres, of a Division from a few tens of metres to 100–300m, while a Stage is measured in tens of metres. Just occasionally these figures may be larger. The platform folds are simple with gently sloping limbs, which are often almost unnoticeable to ordinary observations. They have no linear orientation.

Diagnostic formations of platform type are few. Thus for instance it is possible to quote chemical deposits, dolomites, gypsum and rock salt, which despite their small thickness are distributed over great areas. Condensed deposits, such as beds of glauconitic sandstones with phosphorite pebbles, are only found in platforms. Despite their thickness of tens of centimetres and more rarely of 1–2m, they replace laterally deposits of normal sandstones and clays which are tens of metres thick.

The commonest rock-types of the platform deposits are, however, encountered also in downwarps and in geosynclines, differing only by their thickness being many times less.

With regard to the useful mineral deposits, geosynclines, and the downwarps are equally rich and important, and must be studied with equal attention. Sometimes an opinion is put forward that platforms are poorer in metallic ores than geosynclines. This is clearly incorrect. On platforms, within their folded basement we find the whole complex of useful mineral deposits of geosynclinal type, while in the cover the whole complex of platform type. In geosynclines only the geosynclinal complex is developed, although at their margins the platform complex is also always present. Geosynclinal and platform deposits differ sharply from each other. This is clearly seen on any geological map even if it is generalized. Each region—the geosyncline, the downwarp and the platform—is associated with diagnostic complexes of useful mineral deposits, but it is impossible to say which of these complexes is more or less valuable. It is useless and sometimes harmful to enter such comparisons. The three types of regions should be studied with equal attention and in equal detail. Each of the regions provides new discoveries of equal value and importance. The new Belgorod and Kola iron ores of the Russian Platform are just as important as the new Sarbai-Sokolovsk ores of the Urals, while nothing resembling the diamonds of the Siberian Platform has been found in geosynclines.

GEOSYNCLINES AND PLATFORMS OF THE U.S.S.R.

The outstanding Soviet geologists—Borissyak, Arkhangel'skii, Shatskii, Strakhov have studied and distinguished geosynclines and platforms over the whole territory of the U.S.S.R. An exception is provided only by the regions of development of the Precambrian, where this is incomplete since geosynclines cannot be easily distinguished owing to strong metamorphism and deformation of these formations. The oldest platforms are detected in the Lower Proterozoic of the Finno-Karelian Massif, north of the lake Onega and in the north-western termination of the Ukrainian Massif. The term Onegian can be proposed for the former platform and the term Ovruchian for the latter platform. Within the Onegian and the Ovruchian platforms the Lower Proterozoic of platform type lies almost horizontally or is folded into gentle flexures, is weakly metomorphosed, is thin and does not contain intrusive massifs (Fig. 9). Around those two platforms, and unquestionably existing in other platforms, there are situated geosynclines, which, however, have not yet been accurately delineated.

The Lower Palaeozoic geosynclines and platforms are formed as a result of Baikalian Orogeny, which has occurred at the end of the Upper Proterozoic, thus obliterating all the geosynclines which existed between the Upper Proterozoic platforms. These platforms thus now merge into each other, forming the basement of the Russian Platform.

In the east the Upper Proterozoic platforms of Anabara, Aldano-Vitim and Eastern Sayan merge into each other. They are supplemented by fold mountains, which had arisen in place of the Yenisei Massif. Thus the basic part of the Siberian Platform came into being.

To the north of the Russian and Siberian platforms was situated the Northern Geosyncline and between them the Uralian Geosycline, to the south of the Russian

Platform the Mediterranean, to the south of the Siberian Platform the Angara and to the east of it the Pacific Ocean (the Verkhoyanskian) geosynclines (Fig. 2).

Mesozoic geosynclines and platforms are characterized by even greater merging of platforms and decrease of their number and of the number of geosynclines. The Hercynian Orogeny caused the filling of the Northern, the Ural and Angara geosynclines. As a result a new huge platform which is called the Angara Platform, Angara Continent or in short Angarida came into existence. In the southern hemisphere there was another huge continent equivalent to Angarida (Fig. 2), called Gondwana. Between them was situated the Mediterranean Geosyncline involving a sea called Tethys. To the east of Angarida there was the Pacific Ocean Geosyncline and to the north of it the Northern Geosyncline, which in comparison with the Palaeozoic had changed its position (Fig. 2).

The Cimmerian folding which occurred from the Upper Trias onwards and included the Valanginian did not introduce radical rearrangements in position of geosynclines and platforms. Numerous mountain chains formed by it adhered to the eastern and southern margins of Angarida, thus increasing its size. Other chains came into being away from the shore as archipelagos of rocky islands, peninsulas and large land ridges. After the erosion of these chains an extensive but relatively narrow platform zone was formed. Within the limits of this zone, and lying with sharp unconformity over the folded and metamorphosed Jurassic and Triassic, are found the platform-type deposits of the Upper Cretaceous and Palaeogene, which are thin, predominantly continental, unmetamorphosed and weakly folded.

Similar post-Cimmerian platforms have developed within the Donyets Basin and Mangyshlak, where the folded and metamorphosed Jurassic are covered unconformably by the weakly completely unfolded and unmetamorphosed Upper-Cretaceous and Palaeogene. The post-Cimmerian platform reached an even larger size on the

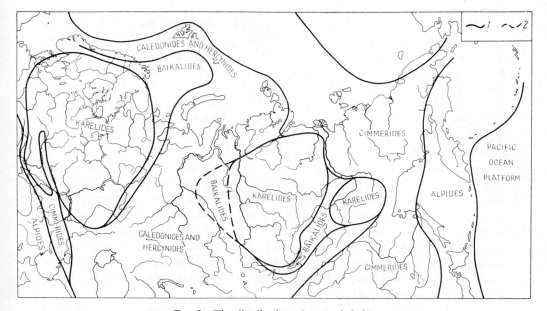

FIG. 2.   The distribution of orogenic belts.
1 — boundary to orogenic belts;   2 — suggested boundary to orogenic belts.

south-eastern margin of Angarida, thus stretching from Eastern Transbaikalia to the Sea of Okhotsk and even further to the east. Over this great distance the Jurassic and the Triassic are strongly deformed, folded into complex folds, so strongly metamorphosed that formerly in places they were considered Precambrian. These are intruded by the massifs of Cimmerian granites. Over the eroded tops of these granites lie the unconformable Upper Cretaceous and Palaeogene deposits, which are mainly continental, weakly folded or completely unfolded, unmetamorphosed, and thin.

The Alpine Orogeny, which had started in the Upper Cretaceous and is still active, has produced most of the high snow-covered ranges of the present day. These ranges such as the Pamir, Hindukush and Himalaya, followed the complete filling of the eastern part of the Mediterranean Geosyncline, thus forming the new continent, Eurasia. The western part of the Mediterranean Geosyncline is still un-filled, but undoubtedly after the termination of the Alpine Orogeny mountain chains will arise in place of the Mediterranean, the Adriatic and the Aegean Seas and will unite Eurasia and Africa. Those Alpine mountain chains, which were found during the initial phases of folding were quickly eroded away and reformed into platforms over which now lie the almost horizontal Neogene deposits which are thin and unmetamorphosed. Similar relationships are evident in Mangyshlak, Ustyurt and Transunguska Hills (Highlands).

MAGMATISM AND TECTONICS

Intrusive activity is causally connected with fold movements, and is contemporaneous with them. Effusive activity on the other hand reaches the greater development during the times between folding, but is also quite strong during epochs of folding. It may be unrelated to folding, and induced by deep fissures which can occur not infrequently independently of folding.

The mechanism of intrusion even now provokes arguments but it is definite that some intrusions are emplaced during the periods of greatest compression and folding. Thus for instance in the Pamir and in the Caucasus during the Cimmerian folding huge granite massifs came into being reaching often tens of kilometres across, intruding and metamorphosing Jurassic and earlier deposits. In the eastern Urals and in Karaganda the Cimmerian Orogeny is also evident, folding the same Jurassic sediments, but with no granites and showing only simple folds and weak meta-morphism. This gives us a right to claim that in the Caucasus and the Pamir the Cimmerian Orogeny was strong, while in the Urals and Karaganda it was weak.

Owing to the connection with folding the age of the granites is defined with respect to it. Sometimes granites and other intrusive massifs are defined as Old Palaeozoic or Precambrian, sometimes they are defined as Young Mesozoic and Cainozoic. Lately it has become possible to make such definitions more accurate and to distinguish the Caledonian, Hercynian, Cimmerian and Alpine granites. Rarely is it possible to determine the age of the granites even more accurately and to associate them with a definite phase. In Eastern Siberia the Kolymian granites are found associated with the Kolymian phase of the Cimmerian Orogeny, which happened during the Valanginian epoch.

Sometimes authors are afraid to associate granites with orogenies and call them Lower Palaeozoic, Middle Palaeozoic, Upper Palaeozoic, Mesozoic, Cainozoic.

Such caution is unfounded, since in substance the Lower Palaeozoic granites are the Older Caledonian, the Middle Palaeozoic are the Younger Caledonian, the Upper Palaeozoic are Hercynian, the Mesozoic and Cimmerian and the Cainozoic are Alpine. A greater achievement is the more accurate age determination of intrusions according to their system or stage; for instance the Upper Carboniferous intrusions which metamorphose the Middle Carboniferous, but are unconformably overlain by the Permian or the Ordovician granites which cut across the Upper Cambrian, but are unconformably covered by the Silurian.

At present when constructing any geological maps the determination of the age of intrusive massifs is obligatory. Without this the map cannot be considered complete. Such a determination is especially important since the intrusive massifs of various ages are associated with different ore mineralizations and metallogeny. The determination of the age of intrusive granites during the geological survey is a very important problem. Usually it is solved by considering the relationships to the surrounding formations. The intrusive massif is younger than those deposits amongst which it is situated but older than those which lie on its eroded surface and in the basal sands and conglomerates of which there are found grains and pebbles of rocks forming the intrusive massif. Consequently to determine the age of the intrusions the most important problem is a detailed examination of all the contacts of the massif under investigation. If they are covered, as happens very often, they must be exposed. The study of xenoliths—fragments of sedimentary rocks included in the intrusive mass—is no less important. Most frequently it is possible to determine the age of those rocks amongst which lies the intrusion and against which it has an active contact. This is valuable since it establishes the oldest possible geological age. The massif cannot be older than these rocks. A granite massif situated amongst the Jurassic deposits and metamorphosing the Middle Jurassic coal-bearing sandstones into crystalline schists and gneisses cannot be either Palaeozoic or Precambrian. This is a typical young massif of Cimmerian or Alpine age. It is more difficult to establish the upper limit to the age and to decide whether the massif is Cimmerian or Alpine. For this it is necessary to find deposits transgressive on the eroded surface of the granite and containing grains or pebbles of the granite. This is rarely achieved and consequently indirect data have to be used, such as the characteristics of those deposits which are younger than the deposits containing granites. If such deposits (in the present instance the Upper Cretaceous and Palaeogene) are thin, unmetamorphosed and weakly deformed, then they cannot contain within themselves intrusive granites and the age of the granite discussed is Cimmerian. This picture is observed in the post-Cimmerian platforms of the Eastern Transbaikal and Cisamuria, where the youngest granites are Cimmerian. Not infrequently these granites yield active contacts with the Upper Jurassic which determines their Lower Cretaceous, Kolymian age.

For the eastern Transbaikalia and similar provinces it is very difficult to distinguish the Cimmerian Intrusions from Variscan, and especially so when the intrusion is surrounded by the Lower Palaeozoic sediments with an active contact against them. Here the problem can only be solved if there are present transgressive Jurassic or Triassic sediments overlying the investigated massifs. Then the massifs are clearly Variscan. In a number of instances the problem cannot be solved and the age of the intrusions is determined as Hercynian–Cimmerian.

In the north-eastern Kazakhstan in the north and west not infrequently relationships diagrammatically shown in Fig. 3 are encountered. A granite massif is intruded

into ancient shales and sandstones, metamorphosing them into hornfelses. On it lie
sandstones or conglomerates with grains or pebbles of granite, which towards the
top are transitional into Lower Carboniferous or Upper Devonian marine sediments.
Such massifs are usually defined as Caledonian and are associated with Caledonian
folding which has widely and intensely affected these regions. Such a definition
would be correct only if an Ordovician fauna is found in the older metamorphic
rocks. If a Cambrian fauna is found then the granites may be Salairian related to
the Salairian phase of folding, which happened in the Upper Cambrian. If, however,
these older formations are found to be Upper Proterozoic then the granites could
also be Upper Proterozoic or Baikalian.

In the Urals up to 1930 all the intrusive massifs were considered Hercynian since
all the ancient metamorphic formations (Suite M) were relegated to the Lower
Devonian. Detailed geological surveying, which started in 1930 under the leadership
of the author immediately made the age of these formations more precise having
demonstrated that they are partly Proterozoic and partly Lower Palaeozoic
(Ordovician). This correction of the stratigraphy has permitted an introduction of
precision into the age of the intrusive massifs. The work of the Soviet petrologists,
amongst whom V. M. Sergiyevskii is most outstanding, has shown that the westernmost massifs, situated on the western slope amongst the Proterozoic slates, dolomites
and quartzites, are Proterozoic in age.

In the northern Urals on one of these massifs transgressive Ordovician sediments
were found. The next zone of massifs is situated further east on the eastern slope
of the Urals, but nearer to the watershed. These massifs are most frequently ultrabasic (serpentinites, amphibolites, dunites), granites are rarer and alkaline rocks are
very rare. V. M. Sergiyevskii and other geologists have proved that all these massifs
are of Caledonian age and on them in a number of places lie transgressive, Devonian
sediments.

Further to the east, on the eastern slope, there is a belt of main granite intrusions,
which stretch continuously for many tens of kilometres one after the other. The total
length of the belt (all the Urals) is more than 2000km. Granites apart, rarer ultrabasic
intrusions are encountered. The age of the main intrusion is proved by the fact that
they lie amongst the Devonian and Lower Carboniferous deposits, metamorphosing

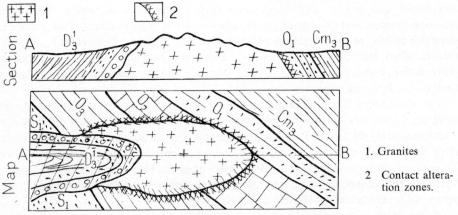

1. Granites

2 Contact altera-
   tion zones.

Fig. 3. Diagrammatic relationships of Caledonian granites in Central
Kazakhstan (section and map).

them, and are transgressively overlain by Mesozoic deposits. These are typical Hercynian massifs. The detailed study of the Hercynian granites has shown that they in turn vary in age. Some were formed at the beginning of folding in the Middle and Upper Carboniferous; others originated at the end of folding, in the Upper Permian and Lower Trias. Since Permian and Upper Carboniferous deposits are not known from the eastern slope of the Urals, the age of the Variscan intrusions is provisionally determined according to their participation in fold structures.

The early Hercynian granites are involved in the initial phases of folding. Consequently these granites participate in folding, are somewhat stretched out along their trend and have altered their original shape under the influence of compression during folding. This compression not only alters the outlines, the shape of the massifs, but also metamorphoses their rocks. Mineral grains are elongated parallel to the structural trend. Rocks alter *en masse*, and concentrate as irregular belts stretching along the strike. The granite starts acquiring gneissose features, which is especially noticeable at the marginal, peripheral parts of the massifs.

The late Hercynian granites originated during the last phases of folding, when folding was ending, and therefore did not affect them. Massifs of this age do not participate in fold structures, but cut across them and retain their original 'rounded' cross-cutting contacts. They are entirely, or almost entirely, internally unaltered and their rocks retain their original texture or as it said are 'fresh'.

The fresh, unaltered, post-Hercynian granites are also known from other provinces than the Urals. The mistakes made in the original dating of the allegedly Mesozoic granites in the northern mountain chains of Central Asia and Semipalatinsk region are methodologically interesting. In these regions granite massifs are intruded into the Upper Palaeozoic, are not involved in structures, are fresh, and have 'rounded' cross-cutting contacts. They were included in the Mesozoic on the basis of the absence of alteration associated with Variscid folding, but their relationships to the Mesozoic, particularly Jurassic sediments were unknown. Later investigations demonstrated that in a number of instances transgressive Jurassic deposits lie on these massifs, thus determining their age as late Hercynian.

In Central Asia in the process of studying the late Hercynian intrusions N. P. Vasil'kovskii (1952) established a very interesting fact. In a unique instance there was an active contact of such an intrusive against the Lower Triassic continental deposits. This leads us to consider that the last Hercynian intrusions and the corresponding folding occurred in the Lower Trias. The interesting conclusion is that this alters the normally accepted notion of the duration of the Hercynian Orogeny. Later the same worker confirmed his observation in many other regions.

Another methodological mistake made in dating young synorogenic intrusive massifs is that they are confused with sub-volcanic intrusions which fill the feeding pipes and fissures of the Mesozoic extrusives. Such bodies are sometimes hundreds of metres across and occasionally are larger and if not completely investigated can be considered as synorogenic intrusive massifs. The latter are easily distinguished by following diagnostic features: (*i*) much larger size of tens and even hundreds kilometres across; (*ii*) irregular outlines; (*iii*) absence of connection with volcanics or sills; (*iv*) situation amongst intensely folded, metamorphosed formations.

In the last decades the various methods of absolute dating of rocks, and especially granites, have been developing rapidly. The results obtained are very important and attract considerable attention (Polkanov and Gerling, 1960; Starik, 1961). With the aid of a number of methods the age of the granites is determined in millions of years. The figures obtained correspond well with the dates of the enclosing sediments, but it is not the figures, sometimes open to arguments, which are most important, but the fact that we now have a potentially incontrovertible method of relative dating of granite intrusions. We can now say definitely which granites are older and which are younger. In many regions this has an outstanding significance since it establishes the date of the mineralization.

Volcanic activity is, of course, happening to this day and has been studied much more widely than intrusive activity. The Academy of Sciences, U.S.S.R., has a special laboratory of vulcanology, which studies modern volcanoes of the U.S.S.R. Nevertheless there are still many obscure and controversial problems about the mechanism of vulcanicity. Vulcanicity is not always connected with folding and occurs equally during orogenies and in the periods between them, when it even reaches the widest development. We can name intrusive massifs after the orogeny with which they are associated, but volcanics cannot be called Hercynian, Alpine, etc. Volcanics thus bear the names of those formations amongst which they lie, for instance: Devonian volcanics, Upper Palaeozoic basalts, Ordovician porphyrites, etc. The more detailed the dating of the suite or formation containing the volcanics the better. A certain lava flow can be called Devonian, but it is more useful if it can be said whether it is the Lower, Middle or Upper Devonian, and even more if it can be attributed to a definite stage. Geological maps show volcanic rocks by the same colour as the sediments amongst which they lie, but the colour is additionally symbolized by overprinted symbols. An exception is made only for younger volcanics covering large areas of the present-day surface, as for instance in Kamchatka, Sikhote-Alin', Armenia, etc.; depending on their composition, they have special colours on the maps.

Sills. These are peculiar features completely associated with volcanics and having no relationship to the synorogenic intrusions. Their dating is not infrequently complicated since they can occur in any bedded formation transgressed by a feeding pipe of a volcano. Usually sills originate in the highest formations cut by the feeding pipe and on which lie the flows of the volcanic, thus the sills are coloured as the volcanics, but have a special symbol.

On the Siberian Platform there are widely distributed Siberian Traps—fissure eruptions of Triassic and partly Permian age. They are coloured as the Permian formations but are especially symbolized. There are contemporaneous sills associated with traps, which have the same composition. Sometimes such sills are found amongst Silurian and Ordovician limestones and other rocks. In such cases they are left indistinguished or coloured as Silurian, both of which are incorrect. It is best to indicate all sills by a special colour, which differs from the colours of intrusive massifs, as it is done on the 1955 general map of the U.S.S.R. with respect to the Siberian Traps.

SEDIMENTATION AND TECTONICS

In the accumulation of sediments tectonic movements play an important, frequently a leading, determinative role. Their significance cannot be overestimated. Tectonic elevation accelerates river currents over huge areas. Thanks to this the grain-size of the sediments alters, they become coarser, and the formerly deposited clays are followed by sands and sands by conglomerates. Tectonic movements considerably alter the relative sea level. With the rise of the sea level (submergence of the continent) depths are increased by tens, sometimes hundreds and in individual cases thousands of metres. On the sea floor great changes occur in the distribution of sediments over enormous marine basins. Where conglomerates were depositing first sands, then sandy clays and clays get deposited. Marine deposits change into terrestrial ones further from the basin. Lastly even greater elevations change regions of sedimentation into regions of erosion and sedimentation stops. This produces large pauses of different duration in geological successions.

The influences of orogenies is even greater. As has already been said they create in the place of previously existing marine basins high fold-mountain chains composed largely of sediments deposited on the bottom of such basins. A radical change in geography occurs which sharply influences not only the magnitude of sedimentation, but its distribution. To orogenies are related peculiar sedimentary complexes such as molasse, flysch, graywacke which are the distinctive products of rising mountain chains and which were deposited at the foot of such chains. These complexes have been already discussed, and an account in greater detail is found in *The Concept of Facies* (Nalivkin, 1957). Tens and hundreds of examples can be brought forward to illustrate the great importance of tectonic processes in the formation of sediments. Thus it is completely natural that many geologists—Soviet as well as foreign—have 'made a complete loop' and started suggesting that the formation of sediments and their characteristics can be completely explained by reference to tectonic movements and tectonics alone. This opinion constitutes a considerable methodological error since it forgets one of the basic premises of dialectical materialism, namely that a phenomenon cannot be understood if isolated from other associated phenomena. Sedimentation occurs not in a vacuum, but in a geographical province possessing its own climatic conditions and populated by its own fauna and flora. Thus it is influenced not only by tectonic movements, but by climatic conditions, fauna, flora and a series of other factors, and particularly by chemical processes (geochemistry). The influence of climatic factors is especially significant. Rains and droughts, periodically succeeding each other, influence the action of rivers no less, and perhaps much faster, than tectonic movements. They can cause periodic change in composition and thickness of sediments. Recently the fall in the level of the Caspian Sea has been more than 2m, causing the formation of a new shoreline and the disappearance of the bays Komsomolyets and Kaidak. This was accompanied by large variations in sedimentation, caused by the lessening of water in the River Volga, owing to the decrease in precipitation over the Russian Platform, which is a climatic fact.

It has been calculated that to form the Quaternary ice sheet the necessary water involved the lowering of the oceanic level by 40m. Such a lowering caused a very large change in sedimentation throughout the world. The basic cause was climatic cooling, with an increase in the quantity of sediments, again climatic factors. Many other examples of the influence of climatic changes on conditions of sedimentation can be quoted.

The influence of animals and plants on sedimentation is less than the influence of climatic factors, but is nevertheless considerable. Huge formations of reef limestone forming entire mountain chains are made by animals and calcareous algae. Plant remains accumulating *in situ* in large quantities produce thick coal seams and import a peculiar character to the coal-bearing formations which may reach several kilometres in thickness. Even more widely distributed are the oil or gas-bearing formations also related to the accumulation of organic substance, but of different composition.

It is important to point out that on all occasions the climatic factors or activity of organisms are not sufficient. Simultaneous tectonic movements occur as well. During the formation of the Quaternary ice apart from the dominant climatic factors, tectonic elevations and depressions also played a role. In the formation of reef massifs organisms play a leading role, but their activity must be accompanied by tectonic movements (depressions). In all large provinces of sedimentation several factors interplay simultaneously and only a study of their interrelationships enables an accurate understanding of the development of such a complicated process as sedimentation.

Let us now consider bedding—a problem of general character. Here again there are at least two standpoints. Quite a number of geologists consider that bedding is associated solely with tectonic movements consisting of periodic rising and falling of land masses. Others, including the author, claim that a series of factors, in particular climatic variations, enter into the origin of stratification, while tectonic movements have almost no significance. In the first place it is important to distinguish the origin of a formation from the origin of bedding within it. In the origin of the formation tectonic movements play a very important, commonly a leading role, but they have almost no relation to the origin of bedding within this formation. Conversely climatic conditions and a series of other factors have a small influence on the accumulation of the formation, but condition the origin of its bedding. This is obvious from the classic example of the Quaternary laminated clays deposited on the bottom of large lakes situated at the margin of an ice sheet. If the transport of material into the lake were perpetual and the bottom of the lake was lowered owing to tectonic movements then a formation of sandy clays without any bedding would have accumulated. However seasonal climatic fluctuations cause an unsteady, intermittent transport of various sediments into the lake and thus the formation of bedding. In the summer when ice is being melted fast, ice streams bring to the lake coarser sandy material, in the autumn melting weakens and the lake receives fine-grained clays. In the winter melting stops and there is a pause in sedimentation, during which the surface of the previously deposited material is rendered compact and the bedding-plane, which produces bedding, is formed. Then again melting begins, again a sandy lamination is deposited; in the autumn it is again succeeded by the clay lamination, in the winter it gets consolidated and a new bedding plane gets formed. Thus a multilayer of two laminations corresponds to every year. Such multilayers are known as rhythms. The origin of rhythmic bedding is considered in detail in the author's book *The Concept of Facies* in which many other factors causing bedding in sedimentary rocks are also discovered.

The example of bedding shows once more that, despite the power of tectonic movements, they cannot be regarded as a unique cause of sedimentation. In this complex process several phenomena always take place. Tectonic movements participate in a large number of cases directly or indirectly, but not always. For instance in the formation of the aeolian sands in desert or thick accumulations of dust (loess) near the mountains surrounding a desert, tectonic movements are irrelevant although the submontane regions are formed tectonically.

Only a full study of all the interconnected phenomena can give us an accurate characterization of sedimentation and the possibility of a correct determination of the conditions of formation of this or other formation developed with the enormous territory of our homeland.

GEOLOGICAL REGIONALIZATION OF THE U.S.S.R.

The basic concept which conditions regionalization is the tectonics,* but this apart a number of other concepts more or less closely connected with it have to be considered. Amongst the latter the properties of sediments, magmatism and useful deposits are especially important. Thus the regionalization advanced below can be called geological. In essence it is general.

The territory of the U.S.S.R. is subdivided into three parts: (*i*) Precambrian geosynclines; (*ii*) Palaeozoic geosynclines; (*iii*) Mesozoic geosynclines.

Precambrian Geosynclines. These are regions in which the last historical manifestations of folding accompanied by the formation of intrusive massifs happened in the Precambrian. Here only the Precambrian is intensively folded, metamorphosed and intruded. Palaeozoic, Mesozoic and Cainozoic are almost horizontal or weakly folded, not metamorphosed and have no intrusive massifs. Their thickness compared to the Precambrian is small. Amongst the facies, contemporaneously with the deposits of widely developed epicontinental seas there are also dominantly continental deposits. Magmatism is represented only by the development of Precambrian intrusive massifs. The Palaeozoic, Mesozoic and Cainozoic contain only volcanic formations and subvolcanic intrusive rocks of feed-pipes and fissures. Useful mineral deposits are characterized by a widespread development of Palaeozoic, and in places even Upper Proterozoic oil deposits and of mineral fields of Precambrian age.

Palaeozoic Geosynclines. These are regions where the last strong folding happened in the Palaeozoic. Here the Precambrian and the Palaeozoic are intensively folded, metamorphosed and intruded by massifs. The Mesozoic and Cainozoic are nearly horizontal or weakly folded, almost or entirely unmetamorphosed and have no intrusive massifs. Amongst the Mesozoic and Cainozoic facies these are widely developed continental deposits. All marine deposits belong to epicontinental seas and their thickness is relatively small. Magmatism is represented by a widespread distribution of the Caledonian and Variscan intrusive massifs. Precambrian intrusive massifs are rarer, while Mesozoic and the Cainozoic massifs do not exist. In the Mesozoic and Cainozoic there are only volcanic and intrusive rocks filling feed-pipes or fissures. Useful deposits are characterized by a widespread development of the Palaeozoic and Mesozoic deposits. Oil-fields of various ages are present.

Mezo-Cainozoic Geosynclines. These are diagnosed by a strong development of all orogenies—Precambrian, Palaeozoic, Mesozoic and Cainozoic. In their Alpine zone, folding continues at the present time.

Here all deposits including the Palaeogene are strongly folded, metamorphosed and include intrusive massifs. Only the Neogene and Quaternary deposits are moderately folded or almost not folded, weakly metamorphosed and do not include intrusive massifs. Amongst the Mesozoic and Palaeogene facies marine deposits of very great thickness are widely developed and in places contain large quantities of volcanics. Magmatism is diagnosed by the development of intrusive massifs of all

---

* Lately the term 'geotectonics' is often used. This can be only regarded as a tendency to use pompous terms. Geotectonics differs only in that it has three letters more.

ages, including Alpine. Young volcanics are widely developed and the development of the Neogene and Quaternary flood basalts is characteristic. Useful deposits are represented by a wide distribution of Tertiary oil deposits and young mineralization associated with the Cimmerian and Alpine intrusions. Sometimes in the Meso-Cainozoic geosynclines an outer Alpine and an inner Cimmerian zones are distinguished.

The position of Precambrian, Palaeozoic and Meso-Cainozoic geosynclines is evident on any general map and it is shown diagrammatically in Fig. 2.

With respect to the names of geosynclines and their borders, there is so far no generally accepted scheme and there are certain disagreements. Thus, for instance, the Donyets Basin is grouped with Precambrian geosynclines (Russian Platform), while Mangyshlak, Ustyurt and Karakum are grouped with Palaeozoic geosynclines. In this book the Donyets Basin, Mangyshlak, Ustyurt and Karakum are all grouped with the outer zone of the Mediterranean Geosyncline. The largest divergences exist with regard to the Western Siberian Depression. Many geologists consider it as a Palaeozoic geosyncline. The Tomsk school of geologists headed by M. K. Korovin allow the existence of a special Palaeozoic continent 'Tobolia' in its western part, so that it is structurally contiguous with the Siberian Platform, but is separated from it by a zone of Palaeozoic geosyncline. In the present book it is considered that Tobolia is connected to the Siberian Platform and is a promontory from it, and that a considerable part of the Western Siberian Depression is related to the Precambrian geosynclines. The basis for these standpoints will be given later when describing the corresponding regions. At present it must be said that such disagreements are inevitable and not lasting. Fresh factual material will make it possible to come to a common opinion. It is unquestionable that deep drilling and geophysical research, which has been widely expanded in Western Siberia, will help in the elucidation of the structure of the deeper zones.

STRATIGRAPHIC SUBDIVISIONS

The largest sedimentary subdivision in the earth's crust is a group, to which corresponds an interval of time called an era. A group is divided into systems, to which correspond periods of geological time. Systems are divided into divisions, of which there are two or three in each system. In isolated cases sub-systems are erected, for example the Carboniferous System which, at a special international meeting in Heerlen, Holland (1958), was divided into Dinantian and Silesian sub-systems. To the Silesian sub-System (shortened Silesian) belong the middle and upper divisions. The Dinantian sub-System (Dinantian) consists of a single lower division. The time corresponding to the divison or sub-system is an epoch. Divisions are sub-divided into stages, and each stage was formed during an age. Stages are divided into zones (horizons). More rarely sub-stages, which include several zones (horizons), are distinguished. 'Time' is sometimes used for the duration of a horizon as well as a sub-stage.

It must be pointed out that the generally accepted and obligatory scheme of subdivision involves groups, systems, divisions, and stages. The distinction of subdivisions and sub-stages is not obligatory. There are no general rules for sub-dividing stages, but such are commonly, especially in the northern and central provinces, called horizons. In the south the term 'zone' is more widespread. For the same subdivisions, not uncommonly, terms of rather wide meaning—'suite' and 'beds'—are used. Thus it is recommended to divide a stage into horizons or zones, but it is completely permissible to subdivide it into beds or suites, especially when their palaeontological characteristics have not been established.

It is generally accepted and obligatory to use the terms era and period as defined, but 'epoch' is also used more loosely. Thus in common usage not only the Upper Devonian epoch, but also the Frasnian 'epoch' and even the Semilukian 'epoch', are used, the term 'epoch' signifying intervals of time corresponding to divisions, stages and even horizons. The term 'age' is similarly used for small subdivisions, e.g. horizons and zones, as well as for a stage, to which it should be equated.

All the subdivisions introduced here are considered obligatory, and are widely distributed. Groups and systems have world-wide distribution. Divisions and stages are distributed over whole continents. Horizons and zones are also sometimes observed over whole continents but commonly the region of their development was more limited, being for instance the Russian Platform, the Urals, the Far East.

Apart from the obligatory terms there are terms of free usage. They are employed when the application of the obligatory terms becomes difficult or impossible. Most commonly this is caused by the absence of palaeontological characterization or because subdivisions include parts of several systems, divisions or stages. [Such usage is most widespread in the Precambrian.—Ed.]

The largest subdivision is a *series* consisting of several suites. The volume of a series varies, but is always large. Sometimes a series includes parts of two systems and may correspond in bulk to a system. More frequently a series includes parts of divisions, and sometimes only parts of stages.

Suite is a widespread term for sequences of medium volume. It sometimes corresponds to a division or even a sub-system, but is more frequently roughly equivalent to a stage. Most frequently parts of a suite are called beds or horizons. The term *pachka* (member) is used less frequently and it conveys the notion of a part of a true formation rather than of a suite. A suite may include parts of division. even of different systems, but only when it is not subdivided into smaller parts. As these parts become individualized they are called suites, while the former suite becomes a series. Thus the same stratigraphical subdivision may first be called a suite and subsequently a series. Most often a suite approximately corresponds to a stage.

Beds is an analogue of horizon or zone, being a relatively small subdivision rarely more than several metres or tens of metres thick. Its distribution varies, and is usually limited, being within a single geological region or a part of it. It is used when there are insufficient palaeontological characteristics to distinguish a horizon or a zone.

Bed is a fractional, primary stratigraphical subdivision. It is diagnostic by its homogeneity, considerable extent and almost parallel boundary surfaces (bedding planes). It normally cannot be divided into smaller parts. When however, an intercalation divides it into two or three parts then each part is also called a 'bed'. Thus it may be said that a clay bed is divided by a limestone intercalation into two clay beds. Thickness varies reaching rarely up to tens of metres.

Intercalation is a thin bed, sometimes of vanishing thickness, of an order of millimetres or centimetres. When a bed containing an intercalation is thick (many metres), then the intercalation can be of a considerable thickness up to a few metres. An intercalation only exists within a bed or a seam.

Lens is a lensoid bed of small extent with converging bedding planes. A lens wedges out over a short distance. Its width is not large being of an order of several hundreds metres and is often less. A lens which has a width of the order of millimetres or a few centimetres and is approximately equally thick is called a *lentil*.

Obligatory subdivisions and free subdivisions are defined in terms of their lithology, fauna and flora. There is a third group of subdivisions which are only defined lithologically, not involving fauna and flora. The third group includes the following subdivisions: formation, tolshcha,* member, seam.

Formation is in thickness and areally a large and sometimes a very large subdivision which is often measured in kilometres across the strike and in hundreds or thousands of kilometres along it. Its composition and correspondingly its terminology is peculiar to itself. In the Precambrian a formation is the largest subdivision corresponding to a group or a system, e.g. Gneissic Formation, Quartzite Formation. Sometimes they have geographical names, e.g. Keivian Formation, Karelian Formation. In the Palaeozoic and upwards formations are distinguished according to lithology, content of a particular mineral or useful mineral deposit: Flysch Formation, Saline Formation. Formations may be subdivisible into series or suites.

Tolshcha is also a formation but of a smaller thickness and distribution. Tolshchas are most frequently distinguished by their content of useful deposits, e.g. coal-bearing tolshcha, saliferous tolshcha, copper-bearing tolshcha. They are frequently named after a mineral or a rock, e.g. dolomite tolshcha, andalusite tolshcha, graywacke tolshcha, flysch tolshcha. A tolshcha may roughly correspond to a system or division and is divided into suites and beds.

Member is a small subdivision in terms of thickness and spread. It corresponds to a horizon, zone or beds, consisting of seams.

Seam has the character of a bed. It commonly consists of a useful deposit, e.g. coal seam, iron-ore seam, a seam of oil sand, but this is not obligatory. Not infrequently reference is made to a seam of sand, of limestone, of siltstone. A seam can be divided into separate parts, which are also called seams.

Relationships of Subdivisions

I. Obligatory	II. Free Usage	III. Limited Usage
Group System	Series	Formation
Order System	Suite	Tolshcha (formation)
Horizon (zone)	Beds	Member
Bed	Bed	Seam

The names of stratigraphical subdivisions have formerly been very variable and sometimes incidental, especially in the case of small subdivisions. Geographical names were used most frequently: thus Dinantian Division, Bashkirian Stage, Voronezh Horizon. Not uncommon are names connected with the lithological composition: Coal-bearing Formation, Dolomite Horizon, Diatomite Member. Often one meets names associated with organic remains: Rudistid Horizon, Coralline Limestone Formation, Nummulitic Member. For zones names after the main fossils are

---

Translator's Note: There is no precise English equivalent for this term and throughout the volume it has been translated as 'formation'.

widely current: *Gephyroceras* zone, *Belemnitella americana* zone. More rarely names are given in honour of geologists who study them: Kassin Beds, Rusakov Beds. There are also names dependent on the external features: Flagstone Horizon, Fragmentary Beds, Mottled Marls Horizon, Domanik.

For the sake of uniformity the use of only geographical names is now recommended. This suggestion has become widespread and almost obligatory. Geographical names can be after towns, rivers, lakes and other geographical elements as well as after the people living or tribes that had lived in a particular locality. Geographical names can be given not only after modern names, but after the ancient ones as well. For instance the term Riphaean formation has acquired its names after the word 'Riphae' which is the ancient name for the Urals.

In making up geographical terms the names are most frequently used in an adjectival sense: Moskovian Stage, Baikalian Series. Yet sometimes they are used without a change: Rakvere Suite, Sakareuli Stage.

It is recommended to choose short names, easily pronounced and spelled in Russian and Latin alphabets. Terms such as Kukisvumchorrskian, Fershainpennuazskian, Chokhataurskian, Yaupiebalgskian, etc. should be avoided. If the name consists of two or more words the name can be formed after one of these: thus instead of 'Karabogazgolian' one can use 'Bogazian'.

In distinguishing a new stratigraphic subdivision it is imperative to establish the stratotype, namely a section in which the new subdivision is most typical. Further detailed instructions are given in the book *Stratigraphical Classification and Terminology* (Gosgeoltekhizdat, Moscow, 1958).

FACIES CHANGES

Geological survey reports and sheet descriptions of geological maps, as well as stratigraphical papers are not uncommonly accompanied by general or type successions. The general succession of an area includes the basic features of sediments observed in various regions while their thicknesses are quoted as averages. Thus the general succession is one that does not exist in nature and is therefore artificial and invented. The main shortcoming of the general succession rests in the fact that they cannot include facies changes existing in the given area. Facies changes as a rule are considerable, may occur rapidly and over short distances, when the most diverse features such as lithological composition, fauna and flora, colour of rocks, structure of rocks, bedding and most commonly thickness, alter. Changes in lithological composition occur constantly. Pure coarse quartz sandstone passes laterally into argillaceous, finer and polymictic rocks. It can pass into salt, which in its turn becomes clay. Clay becomes calcareous, changes into marl and marl changes into limestone. Lithological changes are so numerous and diverse that they cannot all be recounted.

Parallel with lithological changes are fauna changes. Large thick-walled sessile forms are characteristic of reef and unbedded limestones. In marls these forms are replaced by immobile or mobile benthonic, small and relatively thin-walled species. During a change of the shoreline where sands pass upwards into sandy clays and then pure clays the faunal content changes sharply. The most significant changes occur during the change over from marine to lagoonal and then continental deposits. Lastly even in the layers of uniform lithological composition, for instance in bedded detrital limestones or dark shales, the fauna of one section may differ considerably from the fauna of another, separated by only a few kilometres.

The external features of the rocks, their colour and structure also undergo considerable changes, even over short distances. The well-known Lower Cambrian Blue Clay of the Baltic countries becomes red or brown in the Central provinces. Porous sandstones, saturated with oil, can, over a few kilometres change into compact, cemented sandstones without any signs of oil.

Bedding changes as well as other features. Sometimes thin rhythmic bedding persists over enormous areas, as for instance in flysch downwarps. In other cases massive reef limestones in a few hundred metres pass laterally into distinctly and regularly bedded marls. The most considerable and rapid changes occur in the thickness of individual beds and stratigraphical subdivisions such as horizons, suites, stages, division and systems. The scale of variations of thickness is considerable. Thin seams and horizons vary rapidly, but within narrow limits. Thicknesses of coal seams and coal-bearing formations have been particularly comprehensively studied. It has been established that they change variably. Sometimes a coal seam maintains its thickness for many tens of kilometres, but sometimes, as near washouts, the thickness of coal varies from zero to tens of metres over a distance of tens and hundreds of metres.

Frequently when the thickness of a stratigraphical subdivision is quoted, its variations are given: for instance 20–60m or 100–1500m. Often it is written: 'about 30m', signifying that the thickness can be several metres more or less than 30m, but if it falls to 20m one should write '20–30m'. When it is said: 'thickness up to 300m', this means that the thickness can exceed 300m, only by a few metres, at most 10–20m but it can be reduced to 0. Thus when in this book thicknesses of horizons, beds, suites, etc. are quoted they must be considered as average or as those which are most commonly encountered. In reality in different sections these figures are almost certainly lesser or greater. A complete wedging out of seams is also a common, sometimes widespread, phenomenon. Then thicknesses are given as 0–200m, 0–1500m or 'from 300 to 0', or 'complete wedging out'.

Summarizing, it can be said that not uncommonly changes of successions within an area are so variable and considerable that to express them in one general succession, irrespective of any introduced complications, is impossible. In such cases general successions are harmful since they distort reality. Consequently they have not been included in the following text. Instead of a general succession, tables comparing the most typical and the most exceptional sections should be given. Even such tables do not always reflect facies changes, but the data given in them are nevertheless much more complete and do not distort reality as this is done in general successions.

ABSOLUTE AGE

'Absolute ages' are those determined in units, tens, hundreds, thousands and millions of years. Previously determinations of absolute age were made comparatively rarely and had a local or episodic character. They were made on the basis of the speed of sedimentation or rythmic sedimentation. An example is the age determination of the Recent, Anthropogenic deposits by the rhythmic layering of the post-glacial lake sediments. Long and detailed study of this layering has shown that nearly 20 000 years has passed from the time of the last glaciation of the Russian Platform. Calculation of the annual layers in graptolitic slates of Scotland has shown that 700 000 years was necessary for their accumulation. Other examples can be quoted, but their number is small and the resultant figures have little application. In geological practice sedimentational methods have remained almost unused.

A few decades ago chemists and physicists studying radioactive breakdown established an outstandingly important fact—that radioactive breakdown occurs at a constant speed and can therefore, serve as a time index. This fact was discovered independently by Pierre Curie and Rutherford. In the Soviet Union the most distinguished geochemist V. I. Vernadskii was first to appreciate this fact, and thereafter until his death led all the relevant researches. The first determinations were made by the lead and helium methods in 1925 and 1926. Systematic development started in 1936 when a special commission headed by V. I. Vernadskii and including V. G. Khlopin and I. E. Starik was formed. Over the last thirty years the investigation of the radioactive breakdown has considerably progressed and what is most important acquired a wide application in geological practice. Nowadays in many cases the use of atomic methods is indispensable. This work is now done in the U.S.S.R., not by isolated investigators, as was the case not long ago, but by many laboratories with large numbers of workers.

The previously known lead and helium methods apart, the argon, strontium, ionium and carbon methods are widely used at present. The first three methods enable us to determine the absolute age of ancient deposits in tens of thousands, millions and thousands of millions of years. The carbon method based on the study of the isotope C_{14} permits the determination of tens, hundreds and thousands of years. Thus it has found a great application in archaeology and the investigation of the anthropogenic deposits. All these methods and the geochronological data are given in the very important and interesting book by I. E. Starik entitled *Atomic geochronology* (1961).

To determine the age of the sedimentary rocks, the loss of radiogenic argon by glauconite has recently been widely used. Since glauconite is quite often encountered in sedimentary rocks in large quantities, age-determinations using this method have been rapidly developed. It is described in the papers of N. I. Polevaya and her colleagues (1960).

Geologists at first looked upon the radioactive methods of age determination with certain reservations, which soon started disappearing. At present, as has been said, these methods have acquired a general application. It must be said that they have justified themselves and have opened wide possibilities for geologists. Clearly, as in any new method there are a number of ambiguities, divergencies and errors, but these do not affect the great significance of these methods.

In our geological literature absolute dates are every year becoming more common. Thus the data taken from I. E. Starik's book (1961) are quoted overleaf.

There are several geochronological scales of this type. In England one has been worked out by Holmes (1960), in the U.S.A. by Kulp (1960). They all approximate to each other, which makes them very convincing. The small existing divergencies show only that they have not all acquired a final form and that certain deviations or errors of dating are possible.

A large number of determinations are quoted in the papers included in the volume *Determination of the absolute age in pre-Quaternary formations* (Reports of Soviet Geologists. 21st International Geol. Congress, Moscow, 1960).

Geochronological Scale

Period, Epoch	Time at lower limit in millions of years	Duration in millions of years
Pliocene	10	–
Miocene	25	15
Oligocene	40	15
Eocene and Palaeocene	70 (70±2)	30
Upper Cretaceous	100	30
Lower Cretaceous	140 (135±5)	40
Jurassic	185 (180±5)	45
Trias	225 (225±5)	40
Permian	270 (270±5)	45
Carboniferous	320 (350±10)	50
Devonian	400 (400±10)	80
Silurian	420 (440±10)	20
Ordovician	480 (500±15)	60
Cambrian	570 (600±15)	90
Upper Proterozoic	1100–1200	500–600
Lower Proterozoic	1800–1900	700
Upper Archaeozoic	2600–2700	800
Lower Archaeozoic	3400–3500	800

Remark.—The figures in brackets are according to A. Holmes (1960).

Part II

PRECAMBRIAN GEOSYNCLINES

3

RUSSIAN PLATFORM

GENERAL SUMMARY

The Russian Platform is the best studied individual region of the Soviet Union. It represents the eastern part of the Baltic Massif. Its Precambrian basement consists of diverse structural parts. Where the basement consists of Archean or Lower Proterozoic gneisses or schists it is most compact and is conventionally called 'crystalline'. In places Archean platforms are overlain by the Proterozoic of platform type, which is weakly deformed, weakly metamorphosed and thin. In such a case the term 'crystalline' in inappropriate and the term 'consolidated' should be substituted. In the latter instance the boundary between the Precambrian and the Palaeozoic is often transitional, especially if it is determined by geophysical methods. The surface of the Precambrian basement is considerably subdivided despite its general planar aspect. In some regions the basement lies at depths greater than 2000m, thus below some modern oceanic floors, while in others there are elevations up to 1500m. Thus the present relief on the surface of the basement reach 3000–3500m or more.

The Palaeozoic fills the depressions in the Precambrian basement and consequently varies in thickness from 0 to 2000–3000m; it is almost horizontal and may show sharp angular unconformities against the Precambrian. The areas of the Precambrian of platform type, where unconformities are insignificant or absent, are exceptional. The Palaeozoic is unmetamorphosed (blue Cambrian clays of the Baltic countries are almost indistinguishable from the blue Quaternary clays) but is consolidated. Another diagnostic feature of the Palaeozoic is a widespread development of carbonates—limestones and dolomites—which reach several hundreds of metres in thickness and are traced over thousands of kilometres.

The Mesozoic is transgressive, but without an angular unconformity, over the Palaeozoic and has not been subdivided in detail. The thickness of the Mesozoic is much less, and does not exceed several hundred metres. It consists predominantly of clays and sands, while limestones are rare and thin. The Mesozoic is strongly eroded.

The Cainozoic consists of Palaeogene, Neogene and Quaternary deposits. The Palaeogene and the Neogene are not widely developed and after erosion have remained as separate patches and belts. Their thickness is measured in tens of metres, and they consists of sandy-muddy rocks, rarer marls and gaizes.* The Quaternary deposits, on the other hand, form a thin cover (a few or a few tens of metres), but are distributed over huge areas. Various continental facies predominate amongst these.

BOUNDARIES. The problem of the boundaries of the Russian Platform may

* The term 'gaize' is used for siliceous claystones.

seem simple. In reality it is a very complex problem, which in certain regions has not yet been solved. The northern boundary serves as an example. It is commonly drawn along the north shore of the Kola Peninsula. Such a boundary cannot be considered to be geologically significant. It is quite possible that a greater part of the floor of the Barents Sea has the same structure as the Timan-Pechora region, and this is confirmed topographically. Thus the boundary of the platform should be traced some distance to the north, somewhere on the bottom of the Barents Sea and possibly at the edge of the shelf.

Another vexed problem is whether Timan is a part of the platform or of the geosyncline. Many of the followers of A. D. Arkhangel'skii consider that it belongs to the Russian Platform, but the no less numerous followers of A. A. Borisyak attribute Timan to the Uralian Geosyncline. The investigations of the last few years, particularly by deep drillings, have shown that the truth as it often happens, lies in between. Timan cannot be a part of the Precambrian Russian Platform since Baikalian and possibly Caledonian orogenies were strong within its limits. These orogenies did not affect the platform. The structural character of Timan is also peculiar and within isolated downwarps there are great thicknesses of the Devonian (up to 3000m). On the other hand Timan cannot be made a part of the Urals. The Urals became a platform as a result of the Hercynian Orogeny, whereas the Timan-Pechora Province resulted from Baikalian and Caledonian orogenies. The Lower Permian foredeep or downwarp lies at the foot of the Urals separating them from the Timan-Pechora Platform. To be precise the Timan-Pechora Province should be considered as a post-Baikalian platform and must be considered with the Russian

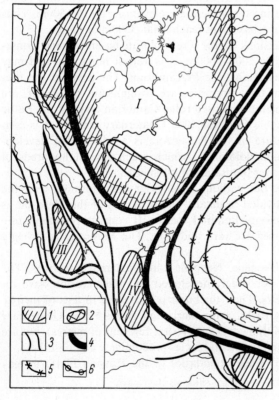

Fig. 4.

The boundary of the Russian Platform in the South-east.

1. Platforms and shields:
      I — the Russian Platform,
      II — the Ukrainian Platform,
      III — the South Caspian Platform,
      IV — the Karakum Platform,
      V — the Tarim Platform.

2. The Yaikian Massif.

3. Alpine and Cimmeridian structures.

4. Hercynian structures.

5. Caledonian structures.

6. Baikalian structures.

Platform, but as an independent, tectonically different region, as has been shown by N. S. Shatskii in his latest map. Consequently the boundary to the north-east of the Russian Platform should be drawn somewhat west of Timan and south of the outcrops of the folded Upper Proterozoic of the Kanin and Rybachii peninsulas.

The eastern boundary to the Russian Platform is drawn at its junction with the Urals Geosyncline. However, between the platform and the Urals there is situated the sub-Urals marginal downwarp, 50–60km or more in width. At present the downwarp is considered together with the geosyncline and the platform boundary is drawn at the edge of its western slope. The marginal downwarp can be traced as far as Orenburg. It is most probable that it continues south to the latitude of Aktyubinsk and then bends to the west. Bypassing a little north of Astrakhan, it margins the Yaikian crystalline shield on its southern side and merges with the Donyets-Dnyepropetrovsk downwarp. The outlines of the Russian Platform are shown in Fig. 2.

From Rostov-upon-Don to the west the Russian Platform is bound by the young Meso-Cainozoic Mediterranean Geosyncline. What happens between Aktyubinsk and Rostov-upon-Don is not completely known. Boreholes and geophysical observations show that structures of the same type as in the Donyets Basin are developed, these being formed by the Hercynian Orogeny and complicated by the Cimmerian folding and weak initial phases of the Alpine Orogeny. Structures similar to the Donyets Basin are taken to belong to the outer zone of the Mediterranean Geosyncline. Thus in the region of Aktyubinsk and southwards the Russian Platform, the Urals Geosyncline and the outer zone of the Mediterranean Geosyncline come into contact. The form of the contact has not been completely studied, but the most likely scheme is shown on Fig. 4.

OROGRAPHY

The orography of the Russian Platform is determined by the interaction of tectonic movements and processes of erosion and sedimentation.

Tectonic movements have produced block elevations and depressions involving large areas. Against the general background of a slow epeirogenic rise of all the northern part of the platform, separate regions rose or were depressed separately. Fractures bounding these regions have definite constant directions and have been given the name of 'the Karpinskii lines' after A. P. Karpinskii who first noticed them (Fig. 5). These lines determine the outlines of the Kola Peninsula, White Sea, Onega and Ladoga lakes, the northern river valleys and many other finer elements of the relief.

The differences in the amount of uplift of the separate blocks caused differences in their relief. Those blocks which have risen most, formed highlands, some reaching considerable height as for instance the Khibina-tundra (mountains) which reach a height of 1230m. Other uplands reach 700–900m, while most of the blocks are lower and some are under water forming the White Sea, the Onega and Lagoda lakes and the Bay of Finland.

However, it would be a serious mistake if we were to consider that the relief of our north-west is conditioned by tectonic movements alone. Erosion and sedimentation have a great significance. Any rise happens simultaneously with increased erosion, while the depressed areas become provinces of sedimentation. The largest of these provinces is in the region of the Boreal Transgression. Fig. 5 shows its southern boundary, but all our northern seas such as the Barents, the White and the Kara Seas belong to it. The deposits of these seas form terraces, which are obvious features of the relief.

For the north and even the centre of the Russian Platform the Quaternary Glaciation and the associated processes have an important significance. An incredible ice sheet which reached the thickness of 3–4km slowly moved off the Scandinavian and Kola Peninsula highlands and in the process of doing so smoothed, polished and in places eroded the underlying deposits. Such smoothed and polished rocks are not uncommon in the north-west of the platform where they are known as 'curly rocks' and roches moutonnées. The bottom relief of lake Lagoda, which is up to 90m in the north and very shallow in the south, has been generated by the ice moving over it.

The ice-sheet not only smoothed, gouged, and formed the negative forms of relief, but it also often produced positive forms, such as elevations, ridges and hills.

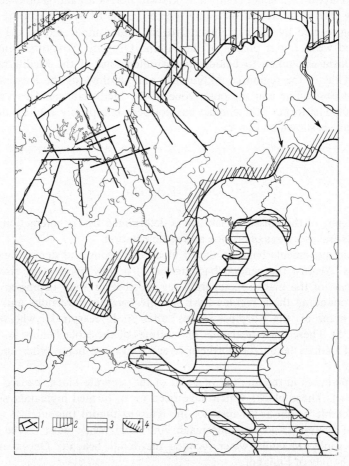

FIG. 5. Certain structural features of the Russian Platform.

1 — Karpinskii lines; 2 — the Boreal Transgression; 3 — the Akchagylian sea; 4 — the limit of maximum glaciation, arrows indicating the direction of ice movement.

Amongst these the most important are belts of terminal moraines causing a peculiar landscape and occurring as a series of irregular ridges and hills with depressions often filled by lakes (Fig. 5).

Tectonic movements, erosion and deposition apart, the formation of the relief of its Russian Platform has been influenced by the hardness and stability of rocks in the process of erosion. For instance the regions where compact, thick limestones belonging to the Visean Stage of the Middle Carboniferous are widespread are always mountainous (e.g. Valdaiian Uplands). Conversely where the soft easily weathered sandy and clayey, coal-bearing members are developed, the relief is marked by the presence of valleys, depressions and marshes.

The relief features of the deposits of Permian gypsum are peculiar and are widely developed on the eastern margin of the platform. It might seem that the easily soluble gypsum should develop numerous karst depressions and holes. In reality, however, the massive beds of gypsum are very stable and always form hills and ridges of not infrequently tens of metres in height. Depressions and kettle holes are only formed where the inhomogeneous, friable gypsiferous clayey or sandy rocks are widespread.

GEOGRAPHICAL ELEMENTS. Since a geographical map constitutes the base-map on which geology is portrayed every geologist must be a good geographer. The geographical elements of the Russian Platform are widely known and their description need not be repeated here. They must be known in sufficient detail. Thus such uplands as Obshchii Syrt, Ergeni and such rivers as Vyatka, Sura, Daugava and such provincial centres of mineral industry as Kirovsk, Krivoi Rog, Nikopol, Boksitogorsk, Oktyabrsk, Ishimbai should be known. The scholars geographical map on the scale of 1:5 000 000 (published 1950) serves as a useful aid in the study of geographical elements.

STRATIGRAPHY

Within the Russian Platform the Precambrian is most widely developed. Palaeozoic, Mesozoic and Cainozoic are also widespread.

PRECAMBRIAN

DISTRIBUTION. The Precambrian deposits form almost all the basement of the Russian Platform.* On the surface they outcrop over a much smaller area, thus forming the Baltic Massif, the eastern part of which, within the boundaries of the U.S.S.R., consists of the Kola Peninsula and Karelian A.S.S.R. as well as the Ukrainian (Azov-Podolia) Massif. The Voronezh Massif, which along the river Don sometimes outcrops at the surface, and the Byelorussian-Litva Massif (Poless-kian Ridge), occur at a small distance (100–150m) below the surface. The Yaikian Massif in the south-east of the platform lies deep. In a number of places the Precambrian has been found in deep boreholes; occurring, for instance, at the depth of 1648m in Moscow, at 920m in Staraya Rus', at 2750m in Kotlas. The greatest depth to the Precambrian is of the order of 3000–4000m.

LITHOLOGY. The Precambrian includes a great variety of rocks of sedimentary, intrusive and volcanic origin. Its most characteristic feature is the extensive

* Precambrian regions are called massifs, but the term 'shield' is no less widespread.

development of crystalline and metamorphic rocks, but it is mistaken to consider that all the Precambrian consists of such rocks. In the Archeozoic (Archean) the crystalline schists and gneisses predominate, but other metamorphic rocks are also encountered. In the Lower Proterozoic schists and gneisses are usual, but various weakly metamorphic rocks predominate, while slightly metamorphosed rocks are also encountered. In the Upper Proterozoic of geosynclinal type the slightly metamorphosed rocks of sedimentary intrusive and volcanic origin predominate, transformed into phyllites, quartzites, meta-limestones, with granites and intrusives.

In the Proterozoic of platform type (Jotnian Formation) comparatively weakly metamorphosed sedimentary rocks, such as slates, quartzites, effusives and their tuffs are present. Metamorphism is so weak that there has been a long argument about whether the Shokshinian, Ovruchian and Terian quartzites belong to the Devonian or not. The fact that over large areas within the Precambrian basement rocks are developed which, by their degree of metamorphism resemble the Lower Palaeozoic types, has a considerable importance for the geophysical methods of investigations.

THICKNESS.   The very strong deformation and metamorphism of the Archean so complicates its structure that the evaluation of its thickness is only possible in isolated instances. These calculations show that the thickness of the principal subdivisions of the Archean has to be measured in kilometres, while the total thickness in a few tens of kilometres. Violent structural deformation of the Lower Proterozoic permits an accurate evaluation of the thickness of only a few comparatively small subdivisions. Even here, figures thus obtained vary from many hundreds of metres to several kilometres. The total thickness is unknown, but is probably of the order of several tens of kilometres. The Upper Proterozoic of platform type yields accurate thicknesses, which are comparatively small, being of the order of several hundreds of metres.

FAUNA AND FLORA.   A most major palaeontological discovery is the finding of plant spores in the Upper and Lower Proterozoic (Fig. 6). Such spores can originate only in the air, since algae do not have them. At the same time there are no known terrestrial plant remains. It is clear that the plants, to which the discovered spores belonged, were aerial-aquatic, the greatest part of them was in the water, while the tops were above it in the air.

The spores are found mainly in muddy deposits, which have not recrystallized. The degree of metamorphism has no importance, since spores are preserved even in phyllites and schists. The Upper Proterozoic spores differ from the Lower Proterozoic and Lower Cambrian spores, thus permitting accurate age determinations. The spores of the Sinian complex are particularly characteristic. These spores at last present a possibility of palaeontological characterization of the ancient formations in which there is no fauna, while the flora is only represented by calcareous algae.

The biostratigraphical significance of the calcareous algae has provoked a lively discussion. Some palaeobotanists insist that the Upper Proterozoic algae, which are often found in profusion, cannot be distinguished from the Cambrian or Ordovician algae. Others point out distinct differences and consider that some genera, as for instance *Newlandia*, existed only in the Proterozoic. In either case the collection of calcareous algae must be considered indispensable to the field work.

In France Radiolaria, Foraminifera and animals similar to trilobites have been found in the Precambrian, but within the Precambrian of the U.S.S.R. as yet no animal remains are known, although such undoubtedly exist. The Lower Cambrian

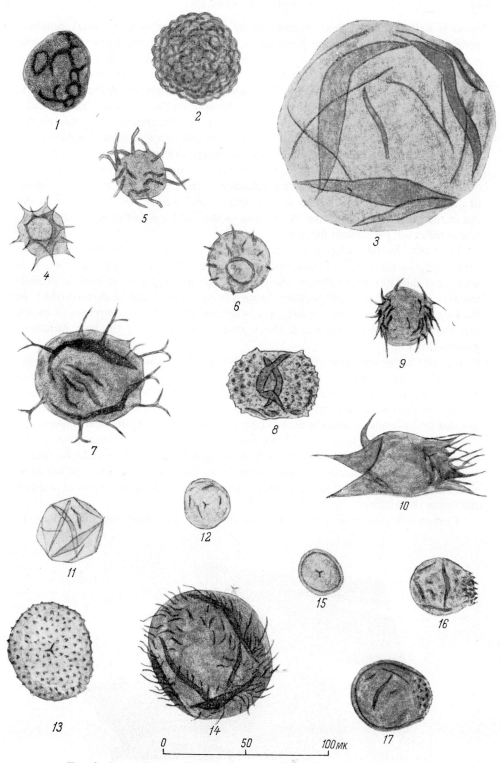

FIG. 6. Precambrian and Cambrian spores (from: B. V. Timofeyev, 1959).

Unicellular algae and zigospores: 1–3 Leio-sphaeridacea (Cambrian and Ordovician); 4–7 Hystrichosphaeridacea (Cambrian and Ordovician); 8–10 Diacrodiacea (Cambrian). *Oligo-triletate spores:* 11–15 Sphaeroligotriletacea (Proterozoic, Cambrian); 16 and 17 Ooidacea (Cambrian).

fauna is well developed. It is represented by the highly-evolved trilobites, brachiopods and gasteropods and the presence of these groups in the Upper Proterozoic
is thought necessary. It is possible that in the Precambrian these animals lacked hard
skeletons. This, however, does not exclude the possibility of their preservation and
the necessity of searching for them. The existence of a rich life in the Lower Proterozoic seas is proved by the discovery of schungite on the northern shore of Lake
Onega. Schungite is a black hard rock, which in appearance resembles earthy
graphite and is similar to it in composition. It represents strongly metamorphosed oil-
shales formed by the accumulation of huge quantities of microscopic organisms.

The presence of rich life in the Lower Proterozoic proves in its turn the presence
of life in the underlying deposits, thus reflected in their names—Archeozoic instead
of the previously adopted Archean. The suffix 'zoic' means life. Archeozoic is the
era of the most ancient life, the Proterozoic is the era of primitive life, the Palaeozoic is the era of the ancient life, etc.

FACIES. In the Archeozoic of the Russian Platform the distinction of facies is
almost impossible, since all these rocks are altered into schists and gneisses. It is
with great difficulty that amongst the gneisses, the paragneisses, formed from sedimentary rocks, and the metagneisses, formed from granites, can be distinguished. In
the Proterozoic it is already possible to distinguish all the basic complexes of facies,
although in the Lower Proterozoic this is often complicated by the high metamorphism of rocks.

Clastic facies—conglomerates, breccias, gravels, sandstones and clays—are
clearly developed. Amongst these, basal deposits such as conglomerates, and sandstones and rarer breccias and clays have a special significance. They allow the subdivision of the Proterozoic into smaller parts, especially if they are accompanied by
angular unconformities and the presence of the underlying deposits near it. Amongst
the peculiar facies let us note the Shokshinian and Ovruchian Jotnian quartz-sandstones, which are reddish, pinky-brown, dense quartzite-like homogeneous ringers
which are indistinctly bedded or false-bedded, representing the deposits of the near-
shore terrestrial valleys. They are widely worked as a first-class material used in
the facing of buildings.

Carbonates also have a wide distribution—both as limestones and dolomites.

FIG. 7.   Jaspilitic iron-ore from Krivoi Rog. *Photograph of a specimen
kept in the Chernyshev Central Geological Museum, Leningrad.*

They are homogeneous, bedded, rather thick, almost unfossiliferous or containing calcareous algae. There are no skeletal parts of animals in them and most of the carbonates represent chemical deposits or the result of algal or bacterial activity. In either instance carbonate rocks are related to very shallow marine basins, shallows, bay and straits.

Particularly peculiar are the jaspilite iron formations typically developed at Krivoi Rog. These are a few tens or hundreds of metres thick and consist of a rapid alternation of thin laminations of siliceous rocks (jasper) and iron ores, and are commonly deformed and folded (Fig. 7). These are the deposits of large enclosed stagnant basins.

PALAEOGEOGRAPHY. Continental massifs and marine geosynclinal regions can be distinctly recognized in the Lower Proterozoic. Compared with the Palaeozoic they are smaller and differ from the later ones by the low relief of the continents and extensive development of shallow, enclosed, quiet marine basins. The low, near shore valleys had large dimensions. On the surface of such valleys and anywhere else on the continents there were no rivers or fresh-water lakes and highly evolved plants did not exist. Consequently there are no coal-bearing formations and no coal. The occasionally observed carbonaceous sub-members represent strongly metamorphosed marine bituminous rocks.

Regional Subdivision

Baltic Massif

Within the Soviet part of the Baltic Massif there are found all the principal subdivisions of the Precambrian. In its subdivision at present the leading position is occupied by the determinations of the absolute age (Polkanov and Gerling, 1960; Starik, 1961).

Karelia

ARCHEOZOIC. The Archeozoic deposits are comparatively poorly developed. They are represented by the Gneissose Formation, consisting of biotite and biotite-garnet gneisses, amphibolites, migmatites, schists and in places marbles. A. A. Polkanov and E. K. Gerling (1960) have proposed dividing it into three large complexes: Katarchean (3100–3500 million years), Saamian (2100–2870 million years) and Byelomorian (1950–2100 million years). Altering these dates in accordance with I. E. Starik (1961) one can provisionally include the Katarchean (2700–3500 million years) in the Lower Archeozoic while the Saamian (2100–2700 million years) and Byelomorian (1900–2100 million years) are in the Upper Archeozoic.

KATARCHEAN has been preserved as isolated blocks amongst the Saamids. It is composed of gneisses and granites.

UPPER ARCHEOZOAN is comparatively widespread. It is composed of two complexes (formations)—Saamian and Byelomorian. The Saamian Formation is composed of gneisses, granitoids, ortho-amphibolites, migmatites and diorites. The Byelomorian Formation is composed of gneisses and granulites and is cut by pegmatites.

LOWER PROTEROZOIC (1200–1900 million years). The Lower Proterozoic has the widest distribution. Its lowest subdivision, the Lower Karelian Formation, according to K. O. Krats (1958) consists of four suites: the Ladogian, the Tikshezerian, the Gimolskian and Parandovian which pass into each other in various zones. They are composed of schists, gneisses, quartzites, amphibolites and altered lavas. The age of the Ladoga Suite is 1630–1820 million years. Above it lies unconformably the Upper Karelian Formation, which begins with basal conglomerates and sandstones with pebbles and grains of Archeozoic rocks. It consists of two suites—the Segozerian and the Onegian—separated by an unconformity. The formation consists of more or less metamorphosed deposits such as schists, quartzites, *Collenia*-bearing dolomites, schungite shales, diabases and porphyrites. It is possible that the second iron-bearing formation widely developed in Karelia (1770–1820 million years) belongs to the Karelides.

The Suisarian and Jotnian complexes lie on the Karelian deposits with a sharp, angular unconformity, outcropping on the western shore of Lake Onega. As against the ancient formations the Jotnian complex is weakly folded and has large, gentle folds. It is composed of weakly metamorphosed sedimentary rocks amongst which the Shokshinian Sandstones, which were formerly thought to be Devonian are most widely known. The sandstones have a diagnostic pink and red colour, they are not infrequently false-bedded and ripple-marked. They are in places underlain by grey quartzites which have a sharp, angular unconformity against the Lower Proterozoic, or are gradually transitional into the volcanic Suisarian Formation (see p. 153). In the upper part of the Jotnian there are thin horizons of phyllitic schists.

The Jotnian complex is intruded by the Rapakivi granites, the absolute age of which is 1400–1600 million years. This determines the ages of the Jotnian and Suisarian complexes as the Lower Proterozoic. Only a short time back they were attributed to the Upper Proterozoic (see p. 152).

UPPER PROTEROZOIC (from 570–620 to 1100–1200 million years). It is represented by weakly altered sandstones and clays lying on the Jotnian complex and under the Lower Cambrian, and it is not widespread.

A borehole to the south of Svir′ (Kharitonov, 1955) has shown under the Anthropogenic deposits the brightly-coloured Devonian (400m thick) and under it blue clays (64m), which are laminaritic towards the bottom and are underlain by almost unmetamorphosed compact sandstones and siltstones of Gdovian type. This makes the presence of the Valdaiian complex unquestionable. Under the Gdovian Sandstones lies a peculiar interbedded formation consisting of grey-greenish shales, siltstones and sandstones. The latter vary from being friable to compact and quartzose, always polymictic and arkosic. The lower part of the interbedded formation is strongly metamorphosed and is composed of feldspathic quartzites and phyllites. Judging by the degree of metamorphism and their pink colour the quartzites resemble Shokshinian quartzitic sandstones. The borehole stops in the pink quartzites.

The Svir′ borehole shows clearly that the Valdaiian complex lies on the interbedded formation, which gets more compact downwards and is transitional into the Jotnian. A similar transition of the Lower Sinian into compact quartzitic pink sandstones is also observed in the boreholes to the south-west of the platform (Orsha).

Kola Peninsula

ARCHEOZOIC. It is developed in the northern and central parts of the peninsula, and is represented by the Saamian Formation consisting of gneisses, amphibolites, granites and pegmatites. The iron-ore gneisses of Olenegorsk, in the central part of the peninsula (Polkanov and Gerling, 1960), belong to it.

LOWER PROTEROZOIC. According to L. Ya. Kharitonov (1958), one of the leading specialists on the geology of Proterozoic, one can distinguish in the Lower Proterozoic three suites—the Tundra, the Keivy and Imandra-Varsuga—as well as a series of intrusions of gabbro, peridotite and microcline granites.

The Tundra Suite is developed in the Monchegora region and at the eastern termination of the peninsula. It is composed of the schists of various tundras (mountains) such as schistose amphibolites and biotite, garnet, staurolite and other gneisses. It is no less than several kilometres thick.

The Keivy Suite forms the mountain Keivy in the central part of the eastern half of the Kola Peninsula and occurs over a rather large territory. The suite consists of staurolite mica schists, quartzites with partings of carbonate rocks, kyanite schists and biotite-garnet gneisses. The kyanite schists have an economic importance. The total thickness is about 2–3km.

The Imandra-Varzuga Suite (undifferentiated Proterozoic) represents a complex of metamorphosed sedimentary and volcanic rocks of great thickness and different ages. Its lower part belongs to the Lower Proterozoic while its upper part to the Upper Proterozoic. It occurs as a wide belt from the region of Kirovsk (lake Imandra) along the eastern part of the peninsula (river Varzuga). The suite is composed of basal conglomerates at the bottom, with dolomitized marbles, quartzites, sandstones and arkoses above, various slates and phyllites higher up and diabases and associated tuffs, keratophyres and quartz porphyries at the top. In the Monchegora region there are nickeliferous intrusives in the suites. The total thickness of the suite is 4km. Calcareous algae are encountered in the carbonate rocks. This makes it possible that a part of the Imandra-Varzuga Suite belongs to the Upper Proterozoic. Its structural style (Fig. 40) involves simple folds.

UPPER PROTEROZOIC. The Upper Proterozoic as in Karelia consists of two complexes—the lower and the upper. The upper part of the Imandra-Varzuga Suite belongs to the lower. The upper complex consists of suites which can be correlated with the Sinian complex of the Kanin Peninsula and the Timan. These suites are the Pechenga-Kuchin suites, the suite of the Rybachii Peninsula (Hyperborean) and the Ter′ Suite (and also Cape Tur′yev Suite).

The Ter′ and the Cape Tur′yev suites outcrop on the southern shore of the Kola Peninsula. The Ter′ Suite occupies a relatively large area, while the Cape Tur′yev Suite makes two small exposures. On Cape Tur′yev bedded greenish-blue sandstones outcrop, underlain by a basal conglomerate (3–5m), containing pebbles of alkaline basalts and syenites (Telyachii island). The sandstones are strongly folded and are thin. On the southern (Ter′) shore of the Kola Peninsula there is a series of exposures of red arkosic bedded sandstone lying almost horizontally, dipping 5–10°. They are underlain by basal conglomerates and become finer to the top where they have intercalations of slates. Ter′ sandstones are typical continental deposits, they are false-bedded and have desiccation cracks, ripple marks but no fauna. Thickness is unknown, but in the exposures it is of the order of a few tens of metres. The age of the Ter′ Suite has provoked lengthy and lively arguments. Some referred it to the Devonian, others to the Jotnian Formation, correlating it with the Shokshinian Sandstones. In 1957 B. V. Timofeyev found spores of the Upper Proterozoic (Sinian) type in the Ter′ Suite. This has determined their age.

Similar arkosic red and white sandstones, with conglomerates at the bottom and lying horizontally on the Archeozoic were found in 1949 on the north-eastern shore of the Kola Peninsula in the lower part of the Iokan′ga. This shows that the distribution of the red Upper Proterozoic sandstones has been formerly more widespread. They were probably developed over the whole of the peninsula; they have been removed by erosion.

If the arkosic sandstones have a typically platform character the other two suites are geosynclinal in type.

The Pechenga-Kuchin Suite is found at the north-western tip of the Kola Peninsula. It is transected by the Pechenga valley and in the east reaches the uplands of Kuchin-tundra. It is a thick (3–5km), complicatedly-folded, metamorphic, volcanic-sedimentary formation. It consists of phyllites, dolomites, carbonaceous slates, quartzitic sandstones, sericite slates and altered diabases, spilites and porphyries. Previously the suite was regarded as Proterozoic, but in 1948 badly preserved organic (?) remains were found in a dolomitized limestone. Amongst them bryozoans, crinoids and tabulates were determined, which has caused a temporary relegation of the suite to the Ordovician. Subsequent collections, however, did not confirm this determination. The re-examination of the collection by the Norwegian geologist Olaf Holtedahl, who is one of the foremost specialists on Scandinavian geology, has shown that what was considered as bryozoans, tabulates and crinoids were inorganic structures, also found in the Proterozoic of Norway. At the same time L. Ya. Kharitonov has found in the Pechenga-Kuchin Suite tillite-like conglomeratic members identical to those of the Rybachii Peninsula. This has given a possibility of correlating these two suites and the dating of the former suite as Proterozoic. This conclusion is confirmed by the resultant figures of absolute age determinations being 1720 and 1780 million years.

The Rybachii Peninsula and Kil'din Island Suite (north-western shore of the Kola Peninsula) represents peculiar sediments, which somewhat resemble the Jotnian Formation since they have simple and gentle folding and are weakly metamorphosed (Fig. 39). These deposits have a characteristically great thickness (up to 5–6km), regular bedding and widespread development of dolomites and muddy limestones. They contain calcareous algae (*Gymnosolen*) related to *Collenia*. They are distinguished as a formation called Hyperborean (Transnorthern). The absolute age of these rocks varies from 715 to 1030 million years.

The majority of geologists, including the author, consider the Hyperborean Formation as Upper Proterozoic. An examination of the specimens of its rocks has shown its identity with the carbonate rocks of the Upper Proterozoic of the Western Urals (Min'yarian and Katavian). Other geologists correlate the Hyperborean Formation with the Sinian Formation in Asia and consider it as Eocambrian, a Palaeozoic System, lying between the Cambrian and the Proterozoic. The establishment of such a system is not sufficiently definite and in this book the Hyperborean and Sinian Formations will be included in the Upper Proterozoic, together with the carbonate beds with *Collenia* which are found amongst the ancient metamorphic rocks of the Kanin Peninsula of Timan and the Polyudov Ridge. In general the determination of the accurate age of these rocks represents one of the most important problems facing Soviet stratigraphers.

Problems of age determination

The above description of the Precambrian section of the Kola Peninsula and Karelia has been done in the main according to the monograph of L. Ya. Kharitonov (1958). A few months later the review of K. O. Krats (1958) was published. The main subdivisions and their succession is identical in both publications, but there are a number of differences in their conclusions. Kharitonov considers that the Rybachii Peninsula and Pechenga suites are the same in age, while the view of Krats is that the former suite is younger. Moreover, Krats includes the Suisarian, the Segozerian, the Onegian and the Pechenga suites (he calls them series) in the Middle Proterozoic. The Precambrian successions claimed by N. G. Sudovikov (1937) and A. A. Polkanov (1947) differ even more.

Such divergences are unavoidable since the establishment of Precambrian successions involve special difficulties owing to the almost entire absence of fauna. Only calcareous algae and spores can be used in age determination. The spores are especially important, but palynology (the science which studies them) is as yet a young science, claiming few specialists in this sphere and consequently having few data at its command. The determinations of absolute age are widely used, but this method is also new and the number of determinations is limited. The use of this method is obligatory, but one must remember that single determinations cannot be considered final and especially when they conflict with geological data. Sometimes absolute age determinations made on numerous specimens from different regions and by means of varying methods force one to change the generally accepted geological diagnoses. The Jotnian Formation, consisting of the Shokshinian and Ovruchian Quartzites cut by Rapakivi granites, serves as an example. Geologists refer it to the Upper Proterozoic. The absolute age determinations on the Rapakivi Granite give an age of 1400–1600 million years, namely figures approaching the age of Lower Proterozoic suites: the Keivy Suite being 1570 million years, the Paranda Suite being 1560–1690 million years, the Ladoga Suite being 1650–1710 million years. At the same time the age of the Cambrian is 480–620 million years. Thus the Rapakivi Granite and consequently the Jotnian and the underlying formations must be grouped with the Lower Proterozoic. Geologists still continue grouping the Jotnian and the Rapakivi as the Upper Proterozoic, thinking that the chemists have made a mistake. However, the probability of this mistake becomes less with every year. Similar ages have been obtained by other laboratories for the Jotnian (Ovruchian Quartzites) and the Rapakivi of the Ukraine Massif. Lastly they are confirmed by three different methods. Evidently geologists will have to review their opinions and to group the Jotnian and the Rapakivi with the Lower Proterozoic. A. A. Polkanov and E. K. Gerling (1960) accept the age of the Rapakivi as 1620–1640 million years.

The determination of the age of the Jotnian Formation has shown how difficult it is to attribute a certain suite to one or other major subdivision of the Precambrian. Not infrequently a suite was attributed at different times to the Archeozoic or Lower Proterozoic. No less difficult is the separation of the Lower from the Upper Proterozoic or even of the latter from the Cambrian. The inclusion of the Hyperborean Formation of the Kola Peninsula (Rybachii Peninsula) to the Proterozoic does not provoke any disagreement. However, identical formations in the Timan and the western slope of the Urals are attributed by a number of investigators to the Cambrian. In this case the spores may give an answer, but in subdividing the Archeozoic palaeontology is so far helpless and only the absolute age determination can be used. Consequently at present in studying Precambrian stratigraphy the first step is to distinguish suites and series of suites. Each suite must be completely characterized so that its separation would not provoke doubts and would be accessible to any geologist. It must possess an accurate petrographic description. All the organic remains must be diligently studied and a palynologic analysis is imperative for the clayey formations and members. The knowledge of the strike-variations within a suite and its transition into other suites is necessary. Basal deposits such as conglomerates, sandstones, breccias and sometimes clays and limestones must be studied in equally great detail. Insufficient attention is being paid to the uppermost deposits, the significance of which is no less than that of the basal deposits.

Most important data are also obtained in studying tectonic relationships and particularly in studying the position of a particular suite in major tectonic units. In all this detailed mapping provides a great help. A whole series of problems cannot be solved without a detailed geological map.

The attribution of a suite to one or other major subdivision, as for instance Proterozoic or Archeozoic or Lower or Upper Proterozoic, presents difficulties with our present knowledge, but in making geological maps it is necessary and must always be done, and the degree of certainty of this particular identification must be indicated. In any case the study and separation of suites is of primary importance. Then follows the grouping of the suites into series and only then the inclusion of the series into this or that major stratigraphic subdivision. In solving the last problem disagreements are unavoidable and natural.

Ukraine Massif

The Ukraine Massif (Azov-Podolia Shield) is the second large region of Precambrian outcrops. The three basic subdivisions—the Archeozoic, the Lower Proterozoic and the Upper Proterozoic—are clearly distinguished.

ARCHEOZOIC. The Archeozoic is everywhere present, but is especially well-developed in the north-eastern part of the massif and forms the southern boundary to the Donyets Basin.

Earlier, according to the data of one of the greatest Ukrainian geologists—V. I. Luchitskii—it was considered that the Dnyepr Series is at the base, where it consists of biotite, plagioclase and other gneisses and amphibolites, and is intruded by granodiorites followed by aplites and pegmatites and has associated widely developed migmatites. The Dnyepr Suite is analogous to the Svionian Formation of the Scandinavian Massif and is included in the Archeozoic. It is overlain by Teterevo-Bug Series, which consists of paragneisses, biotite, sillimanite and graphitic schists, quartzites, crystalline limestones and amphibolites. In the Cis-Azov region it has an associated Korsak Formation of ferruginous quartzites. The Teterevo-Bug Series is intruded by charnockitic granites and diorites and by grey microcline granites. Judging by its high grade of metamorphism and development of charnockitic intrusive rocks the Teterevo-Bug Series is Archeozoic and belongs to its upper division.

Subsequent detailed investigations have shown that the succession proposed by Luchitskii is not always realized and relationships are considerably more complex and are not always clear. Consequently Yu. I. Polovinkina (1953), after spending decades on the study of the Ukrainian shield suggested that the Archeozoic should be grouped altogether as a single 'gneissic' formation to be named the 'Ancient-Gneiss Formation'. N. P. Semenenko (1953), the leading Ukrainian geologist, considers it possible to subdivide the gneissic formation into a Lower Archeozoic Bug-Dnyepr Series and an Upper Archeozoic Ingul-Ingulets Series. He includes the Saksagan (Krivoi Rog) ferruginous series, included by Polovinkina and many other geologists in the Lower Proterozoic, in the Upper Proterozoic. There are at present no data which would allow an undoubted subdivision of the Gneissose Formation into two divisions. The problem demands a very detailed study. It is more correct to group the Krivoi Rog Series with the Lower Proterozoic. In its lithological composition and grade of metamorphism it differs sharply from the Gneissose Formation and is separated from the latter by angular unconformities, basal conglomerates and sharply different strike. The Krivoi Rog Series has been traced as an independent complex towards the north as far as Kursk. In other regions of U.S.S.R. similar ferruginous formations are also attributed to the Lower Proterozoic.

LOWER PROTEROZOIC. It is developed only in the central part of the shield where it is represented by the Krivoi Rog (Saksagan) Series, widespread in the Krivoi Rog region. It includes jaspilitic iron ores which for tens of years provided the raw material for the heavy industry of the south. Its succession and tectonics are discussed on p. 132. The Ovruchian Series is developed in the north-western part of the massif. It is represented by folded, weakly metamorphosed sediments. Amongst them is the worked Ovruchian Sandstone which is very similar to the Shokshinian Sandstone. The Ovruchian Series is analogous to the Jotnian Formation. It is covered by the Polesskian complex, included in the Upper Proterozoic (see p. 60).

The regional researches of A. P. Vinogradov and co-authors and N. P. Semenenko and co-authors were published in 1960, based on numerous determinations of the absolute age. The schemes for subdividing the Precambrian, which are quoted in these works, basically correspond to each other, which makes them convincing. They are very similar to the scheme of subdivision of the Precambrian of the Baltic Massif (Polkanov and Gerling, 1960). Let us quote the scheme of Semenenko.

Lower Archeozoic (Katarchean). 2700–3000 million years or more. Separate blocks amongst the Dnyepr migmatites. The oldest gneisses of the Dnyepr region.

Upper Archeozoic. 1900–2700 million years. Dnyepr cycle—the Cis-Dnyepr migmatites of 2300–2700 million years. Podolian cycle—Chudnovo-Berdichev Granites 2100–2300 million years. Bug cycle—charnockites and monzonites, Kirovgrad and Zhitomir granites and migmatites, 1900–2100 million years.

Lower Proterozoic. 1100–1900 million years. Krivoi Rog cycle—Saksagan and Metabasite Series with jaspilites, 1600–1900 million years. Volynian cycle—Ovruchian Suite of meta-sediments and meta-volcanics, 1300–1600 million years. Korostenian cycle—massifs of post-orogenic granites and gabbros with xenoliths of Ovruchian rocks, 1100–1300 million years.

Upper Proterozoic. 550–1100 million years. Riphaean cycle—muddy, sandy and volcanic deposits underlying the Supralaminaritic Sandstone, the age of which is 570 million years. The age of the cycle is 570–630 million years. Older Riphaean formations, which have been discovered in the Russian Platform, are not found in the Ukrainian Massif.

Vinogradov and co-authors (1960a) give an age of 2100–2200 million years for the ferruginous series as well as for the analogous formation of the Kursk Magnetic Anomaly. Semenenko insists on an age of 1600–1900 million years. This disagreement can be explained by technical reasons, but it is even more possible that two different ferruginous jaspilitic formations exist, one in the Upper Archeozoic and the other in Lower Proterozoic as it is observed in the Baltic Massif. It should be pointed out that the Ovruchian Suite incontrovertibly belongs to the Lower Proterozoic.

Voronezh Massif

The Voronezh Massif, consisting of Precambrian rocks, is situated to the north of the Donyets Basin. The greatest part of the massif is covered by the Palaeozoic and Mesozoic and it lies at depths of 100–150m, but in the east it approaches the surface and in several places in the valley of the River Don it is exposed in a few small hills consisting of Upper Archeozoic (2080 million years) granites, as for instance near the town Pavlovsk. The Voronezh Massif is diagnosed in a number of boreholes and has been studied in greatest detail in the region of the Kursk Magnetic Anomaly and Byelgorod deposits. Here its structure is basically the same as in Krivoi Rog. The Upper Archeozoic and the Lower Proterozoic deposits form a number of complex folds. The Upper Archeozoic includes ferruginous sandstones. As distinct from the Krivoi Rog region, the region of the Kursk Magnetic Anomaly shows a considerably more intense rock-metamorphism and also the absence of typical jaspilites. The ferruginous quartzites of the Kursk Magnetic Anomaly contain from 25 to 45 per cent of iron, with 33 per cent as an average. Their age is not less than 2200 million years. The distribution and the shape of the magnetic anomalies associated with the iron ore deposits in the regions of the Kursk Magnetic Anomaly, the Byelgorod and Kursk deposits are shown in Fig. 8.

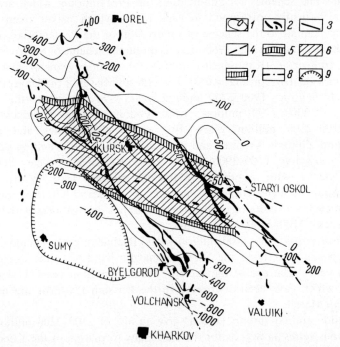

Fig. 8. Structural sketch-map of the Precambrian basement in the region of the Kursk Magnetic Anomaly. The map was constructed by N. G. Shmidt (1957) on the basis of geophysical data.

1 — depth isolines of the crystalline basement;   2 — the iron ore complex: 3 — the axial trace of the Kursk Magnetic Anomaly fold belt;   4 — the axial trace of the Voronezh Elevation;   5 — the distribution of Devonian rocks in the cover;   6 — the distribution of Jurassic rocks in the cover; 7 — the distribution of Carboniferous rocks in the cover;   8 — faults;   9 — fault zones intruded by basic magmas.

Yaikian Massif

The Yaikian Massif is situated to the east of the Voronezh Massif (see p. 136). Owing to the depth the rocks composing it are as yet unknown.

Byelorussian-Lithuanian Massif

The Byelorussian-Lithuanian Massif (Polesskian Ridge) is a region of shallow Precambrian situated to the north of Pripet and west of Minsk. It is a gentle ridge-like rise which divides the Moscow and Polish-Lithuanian depressions and stretches south to north.

Boreholes sunk here have discovered Precambrian at depths varying from 150 to 550m (Fig. 9). The massif consists basically of Precambrian, but on it in places lie the analogues of Gdovian Beds, being quartzose and arkosic sandstones, which are covered by the Laminaritic and blue Lower Cambrian clays. Previously these sandstones and clays were considered to be Devonian.

In the southern part of the Byelorussian-Lithuanian Massif in the Pripet valley is situated the Luninets outcrop, within which the Precambrian granites are found at the depth of about 20m under Quaternary deposits. This outcrop is bound to the south by a strong depression, which is from 1000 to 2600m, in depth and is a continuation of the Donyets-Dnyepr Depression. The Precambrian age of the Byelorussian-Lithuanian Massif is determined as the Lower Proterozoic being 1200–1700 million years (Vinogradov, 1960b). The map (Fig. 10) shows the relief of the Byelorussian-Lithuanian Massif, the Luninets exposure (Mikashvichi), the Donyets-Dnyepr Depression and the north of the Ukrainian Shield.

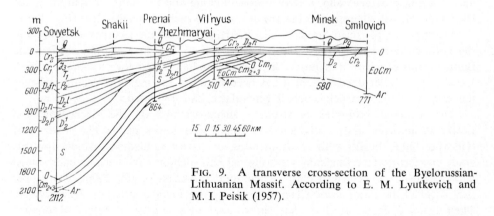

FIG. 9. A transverse cross-section of the Byelorussian-Lithuanian Massif. According to E. M. Lyutkevich and M. I. Peisik (1957).

Volga-Kama Massif

The Volga-Kama Massif has complicated outlines (Fig. 11) and consists of the Tokmovo and the Tatar Arches and Kotel'nich Rise, which continues towards the Timan Rise. The depth of the basement over the Tokmovo Arch is 800–1200m, over the Tatar Arch it is 1500–1700m. In the surrounding depressions the depth reaches 2700–3000m (Fig. 11).

It is interesting that over its whole area the Volga-Kama Massif is composed of Lower Proterozoic migmatites, injection gneisses which are plagioclasic, biotitic, microclinized and partly hornblendic. These migmatites have been found in boreholes in different points of the Tokmovo and Tatar arches and in the Kotel'nich region. The area occupied by them is almost equal to the area of the Ukrainian Massif. Their age is 1300–1700 million years.

Sinian complex

The term 'Sinian complex' had been adopted in 1956 by the Central Stratigraphic Committee to designate a particularly diagnostic complex of deposits underlying the palaeontologically characterized Lower Cambrian and overlying the metamorphosed Proterozoic. The differentiation of this complex is a considerable achievement of the Soviet geologists. It is found in all the Precambrian and Palaeozoic geosynclines.

The Sinian complex was first distinguished in China in the region of Pekin. Consequently it acquired the name 'Sinian' which means Chinese. Much later N. S. Shatskii (1952) has proposed the term 'Riphaean Group' for the similar deposits of the Russian Platform; the term meaning 'Uralian'. Even later the terms 'Eocambrian' and 'Late Precambrian' were employed in the Baltic Massif and the term 'Baikalian complex' for the region of Baikal and a series of other places. Until the accurate diagnosis of the boundaries of the Sinian complex in different places the use of these terms is completely permissible. Its distribution is very wide and includes in fact all the slopes of the rises and all the depressions of the Russian Platform (Fig. 11). There are two types of Sinian deposits in the Russian Platform— the platform type and the geosynclinal type.

PLATFORM SINIAN. It is developed in the central part of the Russian Platform, in the Leningrad Province, around the Baltic Sea, in the Moscow Province and neighbouring provinces and in Byelorussia, Podolia and the northern margin of the Ukrainian Massif where it overlies the Jotnian (Ovruchian Sandstones) (Fig. 12).

The type for platform Sinian is the Valdaiian (Vendian) complex developed in the north-eastern part of the Central Depression. It was first distinguished in the Baltic Countries where it acquired the name 'Vendian' (after the 'Vendian' tribes). Later in even fuller development it has been found in the region of Valdai, whence the term 'Valdaiian', which is now in general use, has originated.

The Valdaiian complex is strongly unconformable on the Archeozoic and Lower Proterozoic and is divided into upper and lower parts. The lower part (Gdovian Suite) begins with conglomerates, or coarse sandstones, which upwards grade into fine-grained sandstones transitional into siltstones and become interbedded with greenish and brownish clays. Their total thickness is 100–200m. The upper part is called the Laminaritic Clays (after the problematical structure *Laminarites*). Their detailed chemical study has shown that they consist of foliae of organic substance formed owing to the accumulation of plankton at the bottom of the sea. The thickness of the Laminaritic Clays in the Baltic Countries reaches 140m. Estonian geologists (Stratigraphical Reviews . . . 1958) have suggested the term Kotlin Suite for them.

The Laminaritic Clays are overlain by the Baltic complex—Supralaminaritic medium and coarse-grained sandstones and clays (30–70m), which are upwards transitional into the 'Blue' Clay with the Lower Cambrian fauna. In the Valdaiian complex the fauna is absent and only spores of a particular type have been found.

There are three points of view regarding the age of the Valdaiian complex. Some geologists headed by Shatskii, relying on the absence of the Lower Cambrian fauna, consider the whole of the Riphaean Group as Upper Proterozoic. Others led by B. S. Sokolov (1953) consider that the Valdaiian complex belongs to a special Palaeozoic System, lying under the Cambrian and named Sinian by the Chinese geologists and Eocambrian by the Swedish geologists. A third group of geologists relegates the Valdaiian complex to the lower horizons of the Lower Cambrian. Fauna and flora solves the problem. The study of the spores done by B. V. Timofeyev

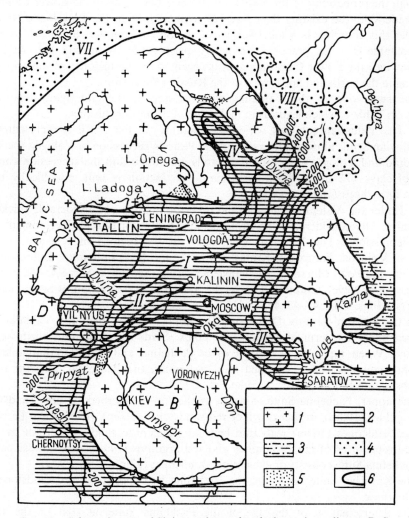

FIG. 12.   Schematic map of Sinian rocks on the platform. According to B. S. Sokolov, 1953.

1 — ancient shields: A — the Baltic Shield with Pishian (D) and Kola (E) massifs, B — the Ukraine-Voronezh Shield, C — the Volga-Kama Shield; 2 — Valdaiian Series; 3 — Serdobskian Series; 4 — undifferentiated Sinian at the edge of the platform; 5 — Jotnian (the sands of Ovruch, Shoksha and the Ter' coast); 6 — isopachytes of the Upper Sinian (Valdaiian) sedimentary cover. I — the Moscow syneclise in end-Sinian times; II — the Gorodok-Orzhian downwarp; III — the Ryazan-Saratov downwarp; IV — the Byelomorian downwarp; V — the Sulchona-Vychegorian downwarp; VI — the Volyn-Podolian edge of the platform; VII — the Caledonian marginal downwarp; VIII — the Kanin-Timan downwarp.

(1959), one of the best specialists in this subject, has shown that the spore contents of the Valdaiian and Baltic complexes is different (Fig. 6).

Of 16 groups of spores two were found only in the Valdaiian Complex, seven only in the Baltic and seven are insufficiently plentiful and diverse and so far have not been found in the Valdaiian complex. In addition the fauna of the upper part of the Baltic complex is characteristic of the lowest horizon of the Lower Cambrian, while the Valdaiian complex lies considerably lower. All this forces us to accept that although the probability of the Valdaiian complex belonging to the Lower Cambrian is considerable, the existing data force us to include it in the Proterozoic, thus to consider this complex as a local variety of the Sinian complex. Since this problem is disputable it is in practice completely permissible to include the Valdaiian complex into the Lower Cambrian as it is accepted by the majority of the geologists working in the north and north-west of the Russian Platform.

The succession of the Valdaiian complex and of the underlying deposits is shown on Table 1, which leads us to make the following conclusions.

1. The Valdaiian complex varies considerably. In Estonia it is represented by the Gdovian Sands alone. Conversely, in the Pestovo succession in Vologda (between Novgorod and Vologda), the Gdovian Sands are absent and the succession consists only of clays 172–217m thick. Its accurate distinction is only possible by using spores, or perhaps *Laminarites*.

2. Under the Valdaiian complex lie other complexes of considerable thickness. and universally considered to be Proterozoic. Their most typical development is in four depressions: The Principal Depression in the regions of Kresttsy, Valdai, Pestovo, Vologda, Soligalich, Kotlas, Yarensk; the Ryazan-Saratov Depression of Redkino, Moscow, Morsovo, Pachelma, Kaverino, Serdobsk, Saratov; the Vitebsk Depression of Nevel, Gorodok, Orsha; the Volga-Ural Depression of Severo'kamsk, Bavly (see Figs. 9, 12, 41, 43, 45).

In the Ryazan'-Saratov Depression the deposits older than the Valdaiian are grouped under the term of Serdobsk complex (Serdobsk is situated north-west of Saratov). When typically developed as in Pachelma and Kaverino boreholes the Serdobsk complex can be divided into two suites, the Pachelma or *sensu stricto* Serdobsk Suite and the Kaverino Suite.

The upper, Pachelma Suite, consists at the top of interbedded grey sandstones and argillites (190m), further down of red sandstones (160m). Below them lie black, compact siltstones (160m) with thin beds of glauconitic sands, underlain by thin, banded repetitions of grey dolomites and red argillites (140m). The lowest part of the Pachelma Suite consists of glauconitic sandstones (32m).

The Kaverino Suite lies under the Pachelma and consists of red sandstones up to 800m thick.

The determination of the age of the Pachelma Suite, as that of the whole Serdobsk complex is methodologically interesting. In order to establish the Lower Cambrian age of the rocks it is imperative to find over it the palaeontologically characterized Middle Cambrian, while to prove the Sinian age of the formation it is necessary to find the fossiliferous Lower Cambrian overlying it. Neither condition exists in the Ryazan'-Saratov Depression. The Lower Palaeozoic fossiliferous deposits are entirely absent and the oldest fauna-bearing sediments are the Middle Devonian. At the same time a number of investigators confidently consider the Pachelma Suite as the Lower Cambrian and the Kaverino Suite as the Upper Proterozoic. This confidence is of course premature and without foundation.

What is to be seen if we move from the north-west? In Valdai and Pestovo (Table 1) the quite typical Middle (Izhorian Beds) and Lower Cambrian and the Valdaiian complex underlie the marine Ordovician. In Redkino, 200–250km further south, Povarovo and Moscow, the marine Ordovician is absent and the Baltic and Valdaiian complexes, underlying the Middle Devonian, start losing their usual lithological nature and in Moscow, for instance, the 'blue' clays are brown or red. Further still to the south-east in Mosolovo under the quite typical Baltic and Valdaiian complexes (but now devoid of worms and *Laminarites*) lie the thick sandstones and clays, absent to the north, while nearby in Ryazhsk the Baltic complex disappears and only the nominal Laminaritic complex remains. A little to the south in Pachelma, Kaverino and Serdobsk the Valdaiian complex also disappears and all the formations underlying the Middle Devonian change into the peculiar Pachelma Suite. There is no basis for correlating the Pachelma Suite with the Baltic or Valdaiian complexes. These complexes could correlate with the Kaverino Suite or be, even more probably, entirely absent. These are the interesting conclusions which are obtained after the borehole analysis. The question must be considered unresolved and in order to solve it completely a fundamental and intensive study of the spores and microfauna is necessary. At present the upper part of the Pachelma Suite, i.e. the upper interbedded and red formations which are 560m in thickness, should be considered as Lower Devonian or Silurian. The lower part of the suite, i.e. the argillaceous and dolomitic formations, is correlated with the Lower Bavlinian Suite and is included in the Sinian complex. It is probable that the Kaverino Suite is of the same age. This is confirmed by the absolute age of 726–753 million years (Polevaya *et al.*, 1960).

In the Volga-Ural Depression, under the Middle Devonian marine deposits and the basal Takatinian-Eifelian Formation of continental origin, lie the Upper Bavlinian and Lower Bavlinian suites. They differ sharply in age. The Upper Bavlinian Suite is of Lower Devonian or Silurian age, whereas the Lower Bavlinian Suite is included in the Sinian complex. Its age is 1290 million years. The suites also differ sharply in their sedimentology. The Upper Bavlinian Suite consists of grey and red interbedded sandstones, siltstones and mudstones of 150–750m in thickness. These are typical sediments of near shore desert lowlands, and are very similar to the Old Red Sandstone of our western provinces and the Western Europe. The Lower Bavlinian Suite differs in its development of dolomites consisting of fine interbedding (ribbon banding) of grey and reddish dolomites with red and greenish dense clays (argillites) of 50–80 to 300m in thickness. In some regions diabasic lavas and sills are developed. Let us quote the Lower Bavlinian succession in boreholes at Bavly (south of Bugul'ma). From bottom to top there are up to 28m of lower red sandstones, up to 82m of dolomite-terrigenous red series, up to 285m of upper red sandstones. Above this, after a break, lies the Upper Bavlinian Suite, consisting of interbedded grey sandstones, argillites and siltstones reaching 140m. The lower red sandstones lie on the crystalline basement.

In other regions of the Tatarian Arch the succession retains its continuity despite the slight variations in thickness. In places the Lower Bavlinian Suite is absent.

In the western provinces of the Russian Platform the Sinian deposits are widely developed (Table 1, Fig. 12). They have been studied by the Byelorussian (Stefanenko and Makhnach; Makhnach *et. al.*, 1957, Makhnach 1958) and Ukrainian geologists (Shul'ga, 1952). A detailed review of the abundant data has been made by E. P. Bruns (1957) and V. K. Golubtsov and A. S. Makhnach (1961). The Valdaiian complex apart, E. P. Bruns distinguishes two older complexes—the Volynian and the Polesskian. Bruns points out that the division of the Valdaiian complex into Laminaritic Clays and Gdovian Sands is not realized in the west since the sandy and clayey formation have a complicated sequence. The Valdaiian complex is divisible into the Lower Valdaiian and Upper Valdaiian Beds. Both begin with a sandstone member and terminate by a clayey member, forming thus two sedimentary mesorhythms.

The Volynian complex is basically composed of lavas and tuffs with subordinate tuffogenous and sedimentary deposits. At the bottom on an eroded surface lie coarse

TABLE 1.

Upper Proterozoic and Cambrian Successions on the Russian Platform

		Southern slope of the Baltic Massif				
Strati-graphic Division		Tallin	Vykhma	Siverskaya	Porkhov	Valdai
Cover		Izhorian grey sands and sandstones with clay beds (10m).	Izhorian, soft pale quartz sands (21m).	Izhorian pale quartz sand-stones (24m).	Izhorian sands with beds of clays (33m).	Izhorian Beds. Complex inter-bedding of sandy and muddy clays with sandstones (117m).
Baltic Complex	Blue clay	*Eophyte*-bearing quartz sandstones (12m). Greenish clays (40m).	*Eophyte*-bearing sands (22m). Greenish clays with worm burrows (33m).	Bluish clays with worm burrows (122m).	At the top, a bed of white kaolinitic clay, followed down by greenish, compact clays, which be-comes sandy towards the bottom (52m).	At the top, pure, pyrite-bearing clays, Lower down, sandy clays with beds of sandstones (218m).
Baltic Complex	Superlaminaritic sandstones	At the top, pale fine grained, in places, glauconitic sand-stones. At the bottom, coarse sands with pebbles and boulders (77m).	At the top, pale fine grained glauconitic sands. At the bottom, coarse sands with pebbles and boulders (30m).	Fine grained sands (3m).	At the top, sandy clays with beds of sandstone. At the bottom, grey sand-stones (33m).	Grey sandy clays. At the bottom, a sequence of sand-stones (55m).
Valdaiian Complex	Laminaritic clays			Laminaritic clays with beds of sands (130m).	Grey bedded sandy clays with laminar-ites (97m).	Grey clays with laminarites (30m).
Valdaiian Complex	Gdovian sandstones			Fine grained, pale sands at the top, and coarse, boulder-bearing sands at the bottom (62m).	At the top, sands with beds of shales. At the bottom, coarse sands and sandstones (54m).	At the top, a se-quence (20m) of sandstones, transi-tional down into laminaritic clays. There are five rhythms.
	Basement	Archeozoic.	Archeozoic.	Archeozoic.	Archeozoic.	At the bottom, coarse sandstones. At the top, sands and even clays. Their older, Ser-dobskian age is probable, but the sequence has not been penetrated through (256m).

sandstones with fragments of quartzitic sandstones and granitoids of 20–25m in thickness. Above these lie basic tuffs 175–200m thick and above the tuffs basaltic lavas with intercalations of tuffites and sandy-clayey red beds. The thickness of the lavas varies considerably, from 0 to 160m. In places basalts (diabases) are in flows of up to 20–30m in thickness. They are worked in quarries. The centre of volcanic activity is situated in the region of the town Korbin. Away from it the succession changes: the lavas wedge out first and then the tuffs. The Volynian complex is closely related to the Valdaiian and is considered contemporaneous (Volyno-Valdai Series).

TABLE 1—continued

Main Depression				
Pestovo	Vologda	Soligalich (E. M. Lyutkevich, 1957)	Kotlas	Yarensk
Izhorian Beds. Interbedding of sandy clays with sequences of sandstones (137m).	Izhorian Beds. Grey sandstones with *Obolus*-bearing beds of clays (19m).	Izhorian Beds (?). Barren sandstones, sands and clays (73m).	Upper Devonian sandstones and clays (131m).	Upper Devonian (26m).
Pure greenish clays with a sequence of kaolinitic clays towards the top (39m).	Pure sandy clays, with beds of sandstone at the bottom (140m).	Greenish and red sandy and pure clays (143m).	Thick sandy clays with a sequence of sandstones in the middle (277m).	Green and brown sandy clays (89m).
Sandy clays (29m).	Sandstones and sands with beds of clays (57m).	Sandstones with beds of clays (72m).	Unrecognized.	Sandy clays (41m).
Homogeneous, grey sandy clays, silts and sandstones. A thin bed of coarse sandstone at the bottom (172m).	A thick sequence of sandstones, with clay beds towards the bottom. Spores of Valdaiian complex. (217m unpenetrated through).	Greenish and brown clays, some pure and some sandy and have beds of sandstone (95m).	Green and red clays with sandstones at the top (252m).	A rather thick sequence of sandy clays (452m).
		At the top, sands associated with Laminaritic Beds (40m).	At the top, thick sandy and pure clays with beds of sandstone lower down. At the bottom, thick sands and sandstones (845m).	At the top, a sequence of sandstones which are probably Gdovian.
Archeozoic.	—	Interbedded sands, sandstones and clays. It is possible that these rocks belong to the Serdobskian complex (336m unpenetrated through).	Unpenetrated through. A major part of the sequence is undoubtedly Serdobskian.	At first clays and sands and then thick sandstones. It is probable that they belong to the Serdobskian complex (166m unpenetrated through).

The Polesskian complex or the Byelorussian Series (Makhnach *et al.*, 1957) consists of a thick succession of red sandstones, lying directly on the crystalline basement and being 75 to 600m thick. The sandstones are quartzitic or quartzo-feldspathic and are usually current bedded. In the north they are coarser and thinner (up to 380m) and are called the Orshaian Suite. In the south they are finer, with frequent intercalations of red and brown argillaceous and silty rocks with clay balls. They reach 500–600m and are called the Pinksian Suite. Not infrequently the sand-stones of the lower parts of the Polesskian complex are considerably compacted and acquire the appearance of quartzite sandstones or are very similar to, if not identical with, the Ovruchian Sandstones developed even further south (Fig. 12). The inter-relationships of the Polesskian and Ovruchian quartzite-sandstones are as yet not clear. Borehole sections of Volynian and Polesskian complexes are given in Table 1.

The similarity between the Polesskian and Serdobskian complexes is noticeable both with respect to the lithological composition and to the stratigraphical position. This is a new demonstration of the Sinian age of the Kaverino Suite which is the lower suite of the Serdobskian complex.

On the south-western slope of the Ukrainian Massif analogous Precambrian rocks are also developed. As is seen in Fig. 7 they are direct strike continuations of the Valdaiian, Volynian and Polesskian complexes. These rocks have been described by Ukrainian geologists (Shul'ga, 1952; Krasheninnikova, 1956). The succession basically retains its character found in the north. Its upper part consists of a two-fold repetition of sandstones and greenish clays corresponding to the Valdaiian complex. Below them is the volcanic formation being the Volynian complex, and at the base lie the thick red sandstones, being the Polesskian complex. It is interesting that southwards the lower complexes change sharply. On the Dnyestr, east of Khmel'nitsk (formerly Kamenets-Padol'sk) the Polesskian complex is not present and lavas de-crease in thickness down to 70m. Only the Valdaiian complex is developed. To the south of the Dnyestr in the Moldavian and Odessa provinces the Upper Proterozoic and Cambrian deposits, normally underlying marine Silurian, are in places found in boreholes. They lie horizontally, are moderately metamorphosed, have a thickness of 400–600m and are composed of a rhythmic repetition of clays and sandstone formations. The absence of fauna renders the age determination difficult, but the lower clayey member reminds one of the Laminaritic Clays and it contains spores of Sinian age. This makes the development of the Valdaiian complex incontro-vertible. It is possible that the upper formations are Lower Cambrian. In places at the base of the succession red beds are found, which probably belong to the Polesskian complex.

Finally the Sinian complex is developed at the north of the platform, near Archangelsk, on the southern (Ter') shore of the Kola Peninsula, to the south of Svir' on the Karelian Isthmus to the north of Leningrad. Near Archangelsk the Valdaiian complex, included in the Cambrian, and the underlying Nenokskian Suite of red compact sandstones are developed. The Nenokskian Suite by its stratigraphical position and lithological content is identical to the Polesskian complex of the west of the platform. The red sandstones of the southern shore of the Kola Peninsula which have been called the Ter' Suite (Fig. 12), are also identical to it. The formations which to the south of Svir' (see p. 46) underlie the Valdaiian complex are also red beds. On the Karelian Isthmus under the Valdaiian complex there is developed a thin formation of grey and red sandstones containing members of volcanic rocks. It can be provisionally correlated with the Volynian complex.

Here we terminate the review of the most ancient platform type rocks the Russian Platform which overlie the crystalline basement. They are very interesting and variable. The formations have only been distinguished some years ago and their stratigraphy naturally still involves many controversial problems. To solve these problems it is more correct to group all these formations under the title of 'the Sinian complex' and to include them in the Upper Proterozoic.

GEOSYNCLINAL SINIAN. The geosynclinal Sinian is folded, metamorphosed and reaches huge thicknesses of several thousands of metres. It is developed along the northern and eastern margins of the Russian Platform.

In the north it includes the Hyperborean Formation of the Rybachii Peninsula (see p. 48), a greater part of the metamorphic formations of the Kanin Peninsula of the Timan, Polyudov Ridge and the eastern slope of the Urals.

The relief of the Precambrian Basement

The relief of the basement consisting of hard Archeozoic and Lower Proterozoic rocks is characterized by the existence of large elevations and depressions. The differences in height reach 4500–5000m. This naturally has had a great significance in determining the distribution of sediments and their facies variation. In the depressions a more complete and thicker succession is noticed, than over the elevations. Many suites are only present in the depressions and are absent over the rises. It is interesting that the relief at this stage also influences the Upper Proterozoic and Lower Palaeozoic deposits. This shows that it was basically formed at the end of the Lower Proterozoic. Obviously from those times onwards the relief has undergone many changes, but these changes did not alter the overall pattern of elevations and depressions. As examples we can quote (a) the Donyets Depression which was formed in the Middle Devonian and separated the Ukrainian Shield from the Voronezh Massif, and (b) the Kola Peninsula, which in the Upper Proterozoic was a high fold range with island peaks trending towards the southern part of the Kanin Peninsula and towards the region to the south-west of the present-day Timan. In the Timan area all this time there was a sea lapping the mountain ranges, which existed to the west of it.

The present-day relief of the basement is characterized by the development of two uplands belts—the northern and the southern—linked by the Byelorussian-Lithuanian and the Volga-Kama rises (Figs. 10 and 11). Between the two belts is situated the very large Central or Main Depression (synecline, trough) with irregularly rounded outlines. The uplands used to be margined from the outside by the western (Lithuanian-Polish-Podolian), southern (Mediterranean), eastern (Volga-Urals) and northern depressions. The Central Depression connects to the outer depressions by four narrow and deep troughs—the Latvian, the Pripetian, the Pachelmian (Ryazan-Saratov) and the Yarenskian. The Voronezh and Ukraine uplands are divided by a narrow and deep Dnyepr-Donyets trough.

The rises and depressions so far enumerated have determined the accumulation and the types of the Upper Proterozoic and Palaeozoic and also to a considerable extent of the Mesozoic and Cainozoic deposits.

TABLE 2.

Successions of Middle and Upper Cambrian and Ordovician Strata

Strati-graphic Divisions	Horizons (Beds)	Estonian, Vykhma	Siverskaya	Porkhov	Valdai	Vologda
Cover		Silurian, Porkunian pale coralline limestones (30m).	Devonian, Parovskian clays and marls (36m).	Narovian Beds (100m).	Narovian Beds (143m).	Narovian Beds (174m).
Upper Ordovician	Saaremyisa (Likholm).	Pale, compact and marly limestones (35m).	Missing	Missing	Missing	Missing
Upper Ordovician	Rakvere (Wesenberg).	Pale compact banded limestones and clays.				
Upper Ordovician	Vasalemma.	Cistoid-bearing limestones and dolomites.				
Middle Ordovician	Keila (Kegelian).	Yellowish dolomites.			Dolomites and limestones (19m).	
Middle Ordovician	Iykhvi (Iyevian).	Grey limestones and dolomites. The thickness of the Iykhvi-Rakvere Beds is 28m.		Dolomitic limestone (26m).		
Middle Ordovician	Shundorovian (Gubkian).	Pale dolomitized limestones with sponges (11m).		Muddy limestone with sponges (13m).	Greenish dolomites with sponges (35m).	
Middle Ordovician	Idavere (Itferian).	Greenish, muddy limestones and marls with beds of kukersite (12m).	Yellowish dolomites with beds of limestones (15m).	Marls and a muddy limestone (57m).	Interbedded limestones, dolomites and marls (87m). Idavere and Kukruse fauna.	At the top, muddy limestones; at the bottom, dolomites (60m).
Middle Ordovician	Kukruse (Kukersian).	Grey muddy limestones with beds of Kukersite (18m).	Grey muddy limestones (17m).	Muddy limestones with beds of Kukersite (20m).		
Middle Ordovician	Tallin (Echinosphaeritic).	Grey, bedded limestones with a pisolithic bed at the bottom (10m).	Grey, bedded limestones (13m).	Grey, bedded limestones; oolitic towards the bottom (24m).	Muddy limestones and marls (17m).	Grey limestones and dolomites with beds of clays (67m).
Middle Ordovician	Kunda (Orthoceratitic).	Variegated muddy limestones with pisolithic beds (7m).	Blotchy, muddy limestones (5m).	Muddy, grey limestones (10m).	Marls, clays and rarer limestones (30m).	Grey-green limestones and clays (33m).
Middle Ordovician	Volkhovian (Glauconitic limestone).	Variegated dolomitic limestones with glauconite (4m).	Muddy limestones with glauconite (4.5m).	Muddy limestones with glauconite (8m).	Marls and clays (21m).	Grey-green clays with graptolites and beds of limestones (30m).

TABLE 2—continued

Pestovo (West of Vologda)	Lyubim (South of Vologda)	Plavinyas	Vilnyus	Sovyetsk	Kaliningrad
Narovian Beds (113m).	Narovian Beds (85m).	Porkunian limestones.	Porkunian limestones.	Black, graptolitic Lower Silurian shales.	Black Llandoverian shales.
Missing	Missing	Absent.	Grey, muddy limestones (42m).	Upper part is eroded away. Pale-grey limestones (18m).	Grey limestones with beds of sandstone and dolomite (45m).
		Pale-grey, pseudo-brecciated limestones (10m).	Grey muddy limestones (8m).		
		Unrecognized	Missing	Bedded, grey limestones (20m).	Bedded, grey limestones with beds of black, graptolitic shales (27m).
Dolomitized limestones (130m).				—	
At the bottom, a sequence of Ifterian Beds (39m).				—	
	At the top, clays and dolomites; lower down, bedded clays and sandstones. At the base, sandstones (181m).		Grey muddy limestones with dolomitic beds (10m).	—	—
				—	—
	Grey clays with beds of limestones and dolomites (69m).	Grey limestones and marls (21m).	Grey limestones (3m).	—	—
Variegated marls with clays at the bottom (71m).	Greenish clays with beds of limestones (57m).	Compact limestones, which are muddy at the bottom (10m).	Greenish, muddy limestones (3m).	Grey, bedded limestones (18m).	—
Marls, clays and limestones (12m).	Towards the top, limestones and marls with greenish clays towards the bottom (37m).	Glauconitic limestones (5m).	Dolomites with pisolithic grains (2m).	Muddy limestones with glauconite (2m).	At the top, brown dolomitized limestones; while at the bottom, sandstones of varying coarseness (68m).

PALAEOZOIC

There is no generally accepted subdivision of the Palaeozoic at present and most diverse variations are encountered in literature. In the present book the following scheme, which has a considerable support, is used:

Upper Palaeozoic.
  Permian, Upper Carboniferous, Middle Carboniferous.
Middle Palaeozoic.
  Lower Carboniferous, Devonian, Silurian.
Lower Palaeozoic.
  Ordovician, Cambrian.

This scheme is based on the occurrences of orogenies. The strongest manifestations of the Caledonian Orogeny have occurred at the end of the Ordovician and separate the Lower and Middle Palaeozoic. The Middle Palaeozoic is an epoch of weak manifestations of orogenies. The Upper Palaeozoic is an epoch of strong manifestations of the Hercynian Orogeny.

Lower Palaeozoic

The Lower Palaeozoic includes the Cambrian and the Ordovician.

DISTRIBUTION. Formerly it used to be thought that the Lower Palaeozoic has a limited distribution over the Russian Platform, existing in its north-western and south-western parts. It is now possible to prove a widespread distribution of the Lower Palaeozoic in not only the northern but also the central part of the Russian Platform.

Deep drilling has shown that Cambrian and Ordovician seas had penetrated eastwards to the region of Vologda and southwards over the Valdai to the Kalinin Province and to the south-west into the Vale of Dugava. Behind the shorelines of these large epicontinental seas there must have been adjacent, near-shore plains with their deposits. In fact, the detailed study of the lower continental sandy-clayey

TABLE 2—continued

Strati-graphic Divisions	Horizons (Beds)	Estonian, Vykhma	Siverskaya	Porkhov	Valdai	Vologda
Lowr Ordovician	Leetse (Glauconitic limestone).	Glauconitic, calcareous sandstone (2m).	Glauconitic sandstone (2m).	Glauconitic sandstone (5m).	Glauconitic sands and sandstones (10m).	Glauconitic sandstones (4m).
Lowr Ordovician	Pakerort *Dictyonema* Beds.	*Dictyonema* shales (1.5m). *Obolus* sandstone (6m).	*Dictyonema* shales (1m). *Obolus* sandstone (5m).	*Dictyonema* shales (1m). *Obolus* sands (12m).	*Dictyonema* shales (2m). *Obolus* sand (12m).	Black bituminous clays (2m). Grey sands and clays (100m).
Middle Cambrian (?)	Izhorian Beds (Fucoid).	Grey sandstones and soft quartz sands (21m).	Grey sandstones with beds of clays (24m).	Grey quartz sand with a bed of clays (33m).	Sandstones, sands and clays (117m).	Grey sandstones and greenish clays (19m).

formations, developed in the northern and central parts of the platform, has shown that lithologically they differ considerably from the overlying Devonian sediments. The study of the spores has shown their Cambrian age. It is now possible to consider proven a wide distribution not only of marine, but also of continental Lower Palaeozoic deposits. This is largely due to the work of B. S. Sokolov (1952, 1953).

In the south-west on the platforms, in Podolia, the area occupied by the Lower Palaeozoic remains unchanged, but a greater part of the deposits previously considered Ordovician is now included in the Cambrian.

SUBDIVISIONS AND LITHOLOGY. In the case of the marine Cambrian and Ordovician these have been studied in detail and are shown on Tables 1 and 2.

The Valdaiian complex described above is considered by many investigators as the lowest complex of the Cambrian deposits. In practice this point of view can be considered as generally accepted. However, on the basis of a series of deductions other geologists consider the Valdaiian complex to be older and associate it with the Sinian or Riphaean complexes. This, as has already been said, is more correct.

The Baltic Complex, which includes the Supralaminaritic Grey Sands, sandstones, clays and the Blue Clay is universally accepted as Lower Cambrian. One has to agree with the suggestion made by A. I. Krivtsov (1955) to the effect that the term 'Blue Clay' should be changed into 'Leningrad Beds' since around Leningrad the Blue Clay is most fully exposed and found in boreholes. Incidentally the Blue Clay is never a true blue, it is pale bluish or greenish. Estonian geologists (Stratigraphical Reviews. . .1958) have suggested that the Supralaminaritic Sandstones should be called the Lomonosov Suite, while the Blue Clay should be called the Lontovas Suite and the *Eophyton* sandstone lying on it the *Piritas* Suite. The term 'Leningrad Suite', however, has priority.

Izhorian (Tiskresian, formerly Fucoid) sands and sandstones are grouped by many investigators with the Middle Cambrian. However, the detailed investigations of the last few years (Davydova, 1961) have shown that the Fucoid Sandstones are gradational into *Eophyton* Sandstone and it is possible that it is more correct to

TABLE 2—continued

Pestovo (West of Vologda)	Lyubim (South of Vologda	Plavinyas	Vilnyus	Sovyetsk	Kaliningrad
Greenish and brown clays with graptolites (28m).		Brown glauconitic sandstones (2m).	Clays with glauconite (1m).		
Black clays and grey sandstones (20m). *Obolus* sands (2m).	Not found.	Not found.	Not found.	*Obolus* sandstone (79m).	At the top, brown dolomitized limestones. At the bottom, sandstones of differing coarseness (68m).
Grey sandstones and clays.			Sandstones and clays (24m).	Sandstones with beds of clays. (50m).	Not penetrated into.

include them in the Lower Cambrian. The absence of the fauna in the Fucoid Sandstones complicates the issue.

The *Pakerort Horizon* includes *Obolus* Sandstones and *Dictyonema* Shales which were previously included in the Upper Cambrian. Some 20–25 years ago on the basis of its Tremadocian fauna it was included in the Lower Ordovician and this point of view is now generally accepted. Lately, however the question of including the Pakerort Horizon in the Upper Cambrian has been raised again. This problem is being at present discussed, especially since there is a suggestion that only the *Obolus* Beds should be put in the Upper Cambrian, while the *Dictyonema* Beds should be left in the Tremadoc; however, the two are interbedded.

The *Ordovician* (formerly Lower Silurian) is divided into three divisions. These divisions are subdivided into stages and horzons (Table 3).

The limits of the English stages do not always correspond with our horizons.

TABLE 3.

Subdivision of the Ordovician

Gt. Britain	Russian Platform (T. N. Alikhova, 1960)			Estonia	
Stages	Divisions	Stages	Horizons	Division	Series
Ashgillian	Upper	Ashgillian	Pirguián	Upper	Khar'yuskian
Caradoc		Plusskian	Wormsian		
			Nabalian		
			Wesenbergian		
	Middle	Iyevian	Kegelian	Middle	Viruskian
			Khrevitskian		
			Shundorovian		
Llandeilo		Llandeilian	Itferian		
			Kukersian		
			Tallinian		
Llanvirn					
Arenig	Lower	Arenigian	Kundian	Lower	Elandian
			Volkhovian		
			Myaekyulian		
Tremadoc		Tre-madocian	Pakerortian		

This circumstance not only complicates the situation, but not infrequently renders the applications of the English subdivisions to the Russian Platform impossible. The Estonian geologists (Stratigraphical Review...1958) have completely adandoned them (Table 3). For some horizons they advance more detailed subdivisions. For instance the Saaremiisa and the Tallin each are further divided into three horizons.

Earlier on, the horizons (then stages) were named after the German geographical localities (Wesenberg, Lyckholm, Kukers) or after the lithological or palaeontological features—thus, Echinosphaeritic, Orthoceratic, Glauconitic. The Estonian and Russian geologists have proposed new Estonian and Russian geographical names, which should be used. No one nowadays refers to Vilnius as Vilna or to Tbilisi as Tiflis and the old palaeontological terms also rapidly disappear from current usage. Consequently one cannot call Rakvere 'Wesenberg' and Keila 'Vegel'.

TABLE 3—continued

(P. M. Mannil, 1958)		North-western regions		
Stage	Horizon	Division	Stage	Horizon
—	Pirgy	Upper	Ashgillian	Saaremyisa (Likholm)
	Wormsi		Caradocian	
	Nabala			
	Rakvere			Rakvere (Wesenberg)
—	Vasalemma (Keila)	Middle		Keila (Kegelian)
	Iykhvi			Iykhvi (Iyevian)
	Iykhvi			Shundorovian (Gubkian)
	Idavere		Llandelian	Idavere (Itferian)
	Kukruse			Kukruse (Kuressian)
	Ukhaku			Tallin
	Lasname			
	Aseri			
Ontika	Kunda	Lower	Arenigian	Kunda (Orthoceratitic)
	Toila			Toila (Volkhovian)
	Leetse			Leetse (Glauconitic) sandstone
Iru	Pakerort		Tremadocian	Pakerortian (*Dictyonema-Obolus*)

It is evident from Table 2 that the lithological characters of the basic sub-divisions of the Ordovician are well represented throughout the area of distribution of the Ordovician Sea. Amongst the facies the pisolitic beds and kukersite deserve attention. The pisolitic beds or, more correctly, individual intercalations consist of accumulations of small phosphorite grains which have the size and form of lentils. Kukersite or the Gdovian Shale is an inflammable shale of brown or orange colour, very light weight, which catches fire from a match and is replete with the remains of marine organisms. It consists of the accumulations of microcopic unicellular algae. This shale is being worked in the Estonian SSR and Pskov province (Gdov).

The Baltic Ordovician succession is considered classical. It is described in textbooks on historical geology. Sections revealed by the new boreholes in the northern and central regions are given in Table 2. The best exposures of the Ordovician and Cambrian occur in cliffs, which are called 'glints' (Fig. 13) found on the southern shore of the Baltic Sea in Estonia.

FAUNA AND FLORA. The fauna is of a normal North Atlantic type and includes all the forms described in the textbook on historical geology. The flora is character-

FIG. 13.

Glint. Cliffs on the southern shore of the Bay of Finland (near Pakerort). *Photograph by K. Myuyuriseppe.*

1. Izhorian Tiskressian Sandstone.

2. *Obolus* Sandstone.

3. *Dictyonema* Shales (an uneven surface).

4. Glauconitic Sandstone.

5. Glauconitic Limestone.

istic, remains of unicellular algae being found most frequently. Their accumulations form intercalations of inflammable shales—*Dictyonema* Shales and kukersite. The study of the spores, a new group of plant remains, has produced most important results. This study has demonstrated the possibility of age determination by using spores, since both the Cambrian and the Ordovician contain sharply differing populations of spores.

PALAEOGEOGRAPHY. The Cambrian and Ordovician marine sediments are deposits of large epicontinental, comparatively shallow basins, which resemble the present-day Baltic Sea together with its bays. The Lower Cambrian sandstones and coarse, clastic deposits are the sediments of a near-shore, low-lying plain and (in the upper part) of a beach with the sea adjacent to it. The Blue Clay was deposited in a shallow stagnant bay of the same type as the modern Bay of Finland, but of larger size (it reached Vologda), sharper outlines and lacking the inflowing rivers. The Upper Izhorian Sandstones member points towards an upward movement, causing the shallowing of the sea and sometimes its complete regression and the formation of near-shore plains.

The carbonate formations (Ordovician) have originated owing to yet another, even more considerable depression. The sea reached to the east of Vologda up to the northern belt of uplands, it reached its largest size and developed currents which caused the formation of glauconite and phosphorite. It was populated by a very rich and diverse fauna.

At the end of the Ordovician an upward movement started, the sea became shallower and the currents were interfered with. Large shallow water shoals and bays were formed in which dolomites and dolomitized limestones were deposited. At the very end of the Ordovician the upward movements caused, with the exception of the Baltic countries, a complete regression of the sea from the Russian Platform.

The palaeogeography of the large low-lying regions situated to the north and south of the marine basins is peculiar. It involved huge plains of gentle inclination towards the sea. Their almost flat surface was only broken by dry, wide and irregular valleys of temporary streams. These streams originated after every powerful rainfall. During the rainy parts of the year they used to become short-living rivers. There were no permanently flowing rivers in the Lower Palaeozoic. The temporary streams deposited sand and gravel over the surfaces of the plains. In places near the shore, lakes and lagoons existed on these plains, the lagoons being connected with the sea. Finely laminated, fine grained clays and marls, less frequently gypsum and even less so rock salt, were deposited in the lakes and lagoons.

Even a short time back all the Lower Palaeozoic red beds were included in the Middle Devonian, but the analysis of spores has demonstrated the presence amongst them of a Cambrian population, identical to those found in marine sediments. It is quite probable that Ordovician forms will be found in the upper parts of the red beds. The dating of the red beds is not yet fully settled. The region of the development of marine and plain deposits was bounded by the rises of the Precambrian basment, which have already been described (p. 61).

Middle Palaeozoic

The Middle Palaeozoic includes the Silurian (formerly the Upper Silurian), the Devonian and the Lower Carboniferous deposits, which preceded the manifestations

TABLE 4.

Silurian Succession in Baltic countries and Podolia

Sub-stage			Beds	Estonia	Vykhma	Plavinyas
Cover			Beds.	Narovian Beds.	Anthropogenic deposits.	Variegated sands and clays.
Ludlow	Upper		—	Eroded away.	Missing	Missing
	Middle		Ochesaare.	Grey limestones with beds of marls and a marine fauna (10m).		
			Kauga-tuma.	Grey crinoidal limestones with brachiopods (15m).		
			Paadla.	Grey and brown dolomitized limestones (No less than 10m).		
	Lower		Kaarma.	Bedded limestones and dolomites with eurypterids (65m).		
Wenlock	Upper		Yagarakhu.	Bedded and massive stromatoporoid and coralline limestones (>20m).		Clays and marls with graptolites. Some are bituminous and others are bedded (77m).
	Lower		Yani.	Bluish marls and dolomites with a marine fauna (up to 46m).		Marls and clays with beds of limestones and a fauna (70m).
Llandovery	Upper		Adavere.	Muddy limestones and dolomites, with brachiopods (about 30m).	Grey and yellow dolomites with cherts (16m).	Bituminous shales and limestones with graptolites (5m).
			Raikkyula.	Grey and yellowish limestones and dolomites (30m).	Dark dolomites with beds of limestones (46m).	Green siltstones, grey limestones and clays; with marine faunas (6m).
			Tamsalu (Borealitic).	Marls and dolomites with banks of pentamerids (15–20m).	Grey limestones and green marls (27m).	Muddy limestones; graptolitic at the top and brachiopod-bearing and crinoidal towards the bottom (44m).
	Lower		Yuru (Ierdenian).	Variegated marls and limestones (13–19m).	Variegated marls.	
			Porkuni (Borkholm).	Brownish and white compact limestones (15m).	Pale limestones (20m).	Limestones. Brown clays with pebbles (12m).
Basement to Silurian strata.			—	Ordovician limestones.	Ordovician limestones.	Ordovician limestones.

Table 4—continued

Vilnyus	Sovyetsk	Kaliningrad	Olesko (near Lvov)	Podolia (O. I. Nikiforova, 1954)
Narovian Beds.	Lower Devonian red beds.	Upper Permian, gypsum and carbonate rocks.	Lower Devonian.	Lower Devonian.
	Upper Ludlow. Greenish clays and marls with beds of limestone (85m).	Thick grey clays with rare marls (377m).	Chortkovian Horizon of dark clays with beds of sandstone, and *Lingula* (75m).	Chortkovian grey shales with beds of limestones and sandstones, and a marine fauna.
	Grey, thickly bedded limestones (10m).	Grey clays, towards the bottom are interbedded with sandstones and limestones (168m).	Borshchovian Horizon of dark-grey shales with beds of limestone and a marine fauna (148m).	Borshchovian grey marls and shales with brachiopods.
	Greenish clays and limestones.			
Missing	Grey marls and graptolitic clays (270m). Grey compact limestones (40m).		Skalian Horizon of dark shales with limestone bands at the top, and limestones at the bottom. There is a marine fauna (208m).	Grey marls with brachiopods. Dark bituminous limestones with stromatoporoids. Grey, bedded limestones and shales. Thickly-jointed dolomites.
	Dark clays with limestone bands (72m).	Calcareous clays with limestone bands (217m).		
	Clays, marls and bituminous limestones (89m).	Clays with limestone bands (96m).	Malinovetsian Horizon of dark limestones, which are muddy and compacts and have beds of normal limestones and ostracods (133m).	Malinovetsian grey, granular limestones with a fauna. Ust'yevian yellow flaggy dolomites with eurypterids. Mukshian grey compact limestones with a rich fauna.
	Marls and laminated clays with a marine fauna (42m).	Interbedded, alternating clays and limestones (42m).		
Dolomitized limestones and marls with a Llandoverian undifferentiated fauna. Adavero Porkuni beds are 65m.	Black shales with greenish marls and beds of grey, muddy limestones; Upper and Lower Llandovery graptolites (14m).	Dark clays (18m).	Kitaigorodian Horizon of dark-grey, muddy, in places dolomitized limestones with a marine fauna (83m).	Kitaigorodian Horizon of grey and greenish, nodular limestones. At the top, banks of pentamerids. Brachiopod faunas. Restevo Beds of flaggy fossiliferous limestones and shales.
Ordovician limestones.	Ordovician limestones.	Ordovician limestones.	Ordovician sandstones.	Ordovician sandstones.

of the Hercynian Orogeny in the geosynclines. The Middle Palaeozoic epoch is sharply divisible into two parts. During the first part all or almost all the platform was dry, and not infrequently an upland area subject to erosion. During the second part owing to submergence a considerable part of the platform was covered by the sea. The first part includes the Silurian and the Lower Devonian, the second includes the Middle and Upper Devonian and the Lower Carboniferous.

Silurian

The Silurian marine deposits have a limited distribution; their continental parts have not been distinguished and it is possible that together with the Lower Devonian they are absent from the larger part of the platform.

The marine Silurian deposits are developed in the Baltic countries in Podolie and in the northern Timan.

The Silurian of the Baltic countries is exposed near the shore of the Baltic Sea and on the islands Saarema and Khiuma. It consists of carbonate and clay-carbonate rocks, which are well-bedded, grey or greenish in colour and thin, being up to 100m. In the open sea facies (limestones and dolomites) the fauna is normal, diverse, mainly consisting of corals and brachiopods; while in the lagoonal facies, represented by thinly laminated, fine grained marls and clays, the fauna is peculiar and impoverished, characterized by a large number of gigantostracans and in particular *Eurypterus*. The Silurian of the Baltic Countries is a direct continuation of the Swedish (Gotlandian) and English Silurian. These deposits belong to an epicontinental sea which connected to an open sea lying to the west. Only its eastern margin was situated within the frontiers of the U.S.S.R. Such a sea should be surrounded by a near-shore plain over which red beds may get deposited. These red beds have not been preserved, unless the lower part of the Baltic Devonian red beds belongs to them. It is much more probable that they were eroded away during the Lower Devonian, when all the Russian Platform was a region of erosion. The Silurian of the Podolia represents deposits of a shallow-water sea, which penetrate to the south-western marine of the Ukrainian Massif. These deposits consist of repetitions of limestones and muddy rocks of total thickness of 500–650m. The fauna is exceptionally rich and diverse and is of the southern type. Along the rivers Silurian exposures form high cliffs (Fig. 14). The subdivisions and lithology of the Baltic and Podolian Silurian are given in Table 4.

Devonian

The Devonian sediments have an exceptionally great economic significance since with them are associated major oil deposits—situated in a region between the Volga and the Urals—as well as thick deposits of various salts.

DISTRIBUTION. The Devonian sediments are developed over the whole Russian Platform with the exception of the uplands of the northern and southern belts. During the Frasnian epoch the southern belt becomes smaller owing to the penetration of the sea into the Donyets Basin. In the north (the Arkhangelsk Province) the deposits previously considered as the Middle Devonian are now included in the Cambrian (Lower Palaeozoic).

SUBDIVISIONS. The upward movements which started at the end of the Ordovician and continued in the Silurian reach their greatest intensity in the Lower Devonian. At this time almost the whole platform was an elevated region, thus a region of erosion. Only in the east within the limits of Second Baku was there a low-lying, near-shore plain margined by the Ural Sea. On this plain were deposited continental sands and clays which have been called the 'Upper Bavlinian Suite'.

During the Eifelian epoch the upward movements become replaced by downward movements and the sea at first penetrated into the Eastern Depression and then southwards from the Volga-Kama uplands (Fig. 11) into the central provinces. The greatest depth of the sea was reached during the Lower Frasnian epoch, since in the Upper Frasnian epoch the northern part of the platform started rising again and the sea withdrew to the south. At the end of the Famennian epoch it shallowed rapidly and broke up into a series of shoals and lagoons. The repeated and changing elevations and depressions had caused numerous alternations of marine deposits and of lagoonal and terrestrial sediments. Consequently Devonian successions of different regions differ. This variability has become even clearer after studying the cores of numerous deep boreholes, which passed through the Devonian deposits. Thus an urgent necessity arose to correlate and connect all these sections. This task was carried out by an All-Union Symposium organized by the All Union Institute of Petroleum (VNIGRI) in 1951. The decisions of this conference were published as a special pamphlet (Decisions of the All Union Conference. . .1951) and constitute the first scheme subdividing the Devonian of the Russian Platform and of the western

FIG. 14. Bedded Silurian limestones forming the cliff at Kamenets-Podolsk, with a medieval castle on the cliff. *Photograph by O. I. Nikifirova.*

TABLE 5.

Devonian Stratigraphic Correlations on the Russian Platform

Stage	Sub-stage		Western Provinces	North-western Provinces	
			1960	1951	1960
			Horizons	Horizons	
Upper Devonian — Famennian — Upper			Nadletizhian Letizhian Skervelian Ketmrovian	Nadbilovian	Nadbilovian
			Zhagarian	Bilovian	Bilovian
			Shvetian Murian Akmenian	—	Nadchimayevian
Famennian — Lower			Kurshaiian Kokishkissian	Nadchimayevian Chimayevian	Chimayevian Nadsmotinsko-Lovatian
Frasnian — Upper			Kruoiian Pakruoiian (Amulian) Stipinaiian (Bauskian) Orgskian (Parmushisian)	Nadsmotinsko-Lovatian Smo'tinsko-Lovatian Nadsnezhian Snezhian	Smotinsko-Lovatian Nadsnezhian Snezhian
Frasnian — Middle			Daugavian	Buregian Ilmenian Svinorodian Upper Shelonian	Buregian Ilmenian Svinorodian Upper Shelonian
Frasnian — Lower			Salaspilian	Lower Shelonian	Lower Shelonian
			Plyavinasian	Chudovian Pskovian Snyetogorian (?)	Chudovian Pskovian Snyetogorian (?)
			Amatian Gauiian	Podsnyetogorian (Oredezhian)	Podsnyetogorian
Middle Devonian — Givetian			Salatsian Narovian Pyarnusian	Tartuian Narovian Pyarnusian (?)	Tartuian Narovian
Middle Devonian — Eifelian			Kemerian	Missing	Pyarnusian
Lower Devonian — —			Stonishkaiian	Missing	Missing

TABLE 5—continued

Central provinces		Eastern provinces, the Volga-Ural region	
1951	1960	1951	1960
Horizons		Horizons	
Dankovo-Lebedyankian (including: Ozersko-Khovanian)	Dankovo-Lebedyankian (excluding: Ozersko-Khovanian)	Dankovo-Lebedyankian (including: Ozersko-Khovanian)	Dankovo-Lebedyankian (excluding Ozersko-Khovanian)
Yeletsian Zadonskian	Yeletsian Zadonskian	Zadonsko-Yeletsian	Yeletsian Zadonskian / Makarovian
Livenian Yevlanovian Voronezhian	Livenian Yevlanovian Voronezhian	Yevlanovsko-Livenian Voronezhian / Barmian Askynian	Livenian Yevlanovian Voronezhian / Undifferentiated
Petinian (?) Semilukian	Petinian (Alatyrian) Semilukian Rudkinian	Mendymian Domanikian	Petinian (Alatyrian) Semilukian / Mendymian Domanikovian
			Sargayevian
Upper Shchigrovian Lower Shchigrovian	Khvorastanian Kikinskian Archedinskian Yastrebovskian	Sargayevian Kynovskian Pashiiskian / Shugovovian	Kynovskian Pashiiskian
Starooskolian Mosolovian (?) Pyarnusian (?)	Starooskolian Vorob'yevkian Ol'khovian	Upper Givetian sub-stage (Ardatovkian Beds) Lower Givetian sub-stage	Mullinskian Starooskolian Vorob'yevkian Chernoyarian (Afoninian)
Missing	Mosolovian Morsovian Ryazhskian Novobassian	Takatinian	Moslovian Morsovian / Biian Calceolic Vyazovian Takatinian
Missing	Missing	Upper Bavlinian Suite (?)	Kazanlian / —

slope of the Urals (Table 5). Fresh stratigraphical material accumulates very rapidly. Thus in 1960 another conference was organized dealing with the Palaeozoic stratigraphy of the Volga-Urals and adjacent areas. The conference took place in Moscow in VNIGNI. The scheme accepted there differs from the 1951 scheme. The comparison is given in Table 5.

LITHOLOGY AND FACIES. Successions of the eastern, central and western provinces of the Russian Platform are different. In the east marine formations, sometimes relatively deep water and often rich in organic bituminous substances, are developed. The dark bituminous Middle Devonian limestones and bituminous shale-sandstones of Domanik and Shugurovian suites of the Frasnian age serve as examples. In the east between the Volga and the Urals are situated all the major oil deposits. In the central provinces marine deposits are also widespread, but they are of shallower origin, lack bituminous rocks, and have few oil deposits. The margin of the marine basins lay in the western provinces. Consequently, the red bed deposits, which are the sediments of the near-shore deserts, and dolomite and gypsum rocks, which are the sediments of shallows and lagoons (Fig. 15) are widely developed. In the east marine limestones and marls are widespread.

Eastern provinces. These are basically situated between the Volga and the Urals and are consequently named the Volga-Ural Province. Owing to the rich deposits of

FIG. 15.   Upper Devonian limestones, dolomites and marls on the river Velikaya, near Pskov. *Photograph by F. R. Gekker.*

oil the name 'The Second Baku' was given to this region, but is not used much now.

The Devonian succession begins from the Upper Bavlinian Suite of the Lower Devonian, possibly Silurian, age. In its stratigraphical position and lithological composition it is identical to the Old Red Sandstone of the Baltic countries and England. The Upper Bavlinian Suite underlies the Middle Devonian and overlies the Lower Palaeozoic. It consists of thick grey and red sandstones, clays and sometimes marls which constantly alternate but altogether make up a very homogeneous formation. The thickness of the suite varies. In places it is absent, while in others it reaches 200–300m. Near the Urals its thickness increases considerably and in Bashkiria (Salikhovo) it reaches 746m, while in Perm (Krasnokamsk and Severokamsk) it is 930–950m. No fauna has been found in it. The flora is represented by psilophytes and spores. According to E. F. Chirkova-Zalesskaya (1957) the psilophytes are of Lower Devonian age and are represented by several genera and are found over a wide area in Krasnokamsk, Polasn′ (near Perm), in Tataria and the Saratov region. The spores determined by B. V. Timofeyev are represented by Lower Devonian genera. The flora determines the age of the Upper Bavlinian Suite as the Lower Devonian, but the possibility that its lower part belongs to the Silurian cannot be excluded.

The Upper Bavlinian Suite is found in a majority of the regions of the Volga-Ural Province. The investigation of the Upper Bavlinian and similar formations was considerably influenced by the work of the Bashkirian geologist K. R. Timergazin. He correlated the Upper Bavlinian Suite with the Upper Proterozoic, considering it of the same age as the Ashian Suite. The succession of the Upper Bavlinian Suite is illustrated by the section of the Bavly-Tuimazy district, being from bottom to top:

Lower member of interbedded grey sandstones, siltstones and argillites (up to 145m).

Speckled quartz-felspathic (arkosic) sandstones (60–100m).

Upper member of interbedded grey sandstones, siltstones and argillites (80–100m).

In the Perm′ (Krasnokamsk, Severokamsk) region the succession alters somewhat. The lower grey member, retaining its character, reaches 470–490m in thickness. Over it lies a formation of dark grey sandstones, siltstones and argillites, up to 120m thick. The upper member of interbedded sandstones, siltstones and argillites differs by its variegated colour and great thickness of up to 340m.

The Lower Devonian age of the Kazanlinian Suite, found in boreholes near the village Kazanly in the Saratov Province is not in question. The age is determined by the Lower Devonian armoured fish (*Porolepis*) identical to those of Podolia and Baltic countries. The Suite consists of red sandstones and argillites, several tens of metres in thickness.

The Middle Devonian starts with basal sandstones and clays, which are continental at the bottom and contain psilophytes and are interbedded with marine limestones higher up. Their total thickness varies from 20 to 100m. Nearer to the Urals the sandstones are called the Takatinian Suite and are identical to the same suite in Western Urals. Further away from the Urals they change their lithological composition, become more clayey and are not called Takatinian any more. Judging by the spores and psilophytes the Takatinian Suite belongs to the Eifelian Stage. Higher on it gradually changes into clayey and then pure, bedded, dark, bituminous limestones with a marine brachiopod and coral fauna. These dark limestones are subdivided into several suites, identical to those in the Urals. The lower limestones used to be included in the Lower Givetian sub-Stage, the higher ones in the Upper Givetian. Recently the study of the fauna and the flora has shown that those suites, which were formerly considered as Lower Givetian, have to be included in the Eifelian Stage, while those which were considered as Upper Givetian correspond to the Givetian Stage. This opinion has been adopted at the Conference on Palaeozoic Stratigraphy, held in Moscow in 1959. Eifelian and Givetian limestones reach considerable thickness (up to 100m), have a wide distribution and a high bitumen content, which renders them an important oil-producing formation.

The Frasnian Stage (400–500m) commences with the basal Pashiiskian Beds, consisting of terrestrial sands, sandstones and clays with psilophyte flora; they reach several metres or tens of metres in thickness. Intercalations of friable, porous sandstones and sands frequently saturated with oil represent an important oil-bearing horizon. Towards the top the Pashiiskian Suite is gradually transitional into a thick marine formation of clays, marls and limestones. Amongst these a special attention should be paid to the Middle Frasnian Domanik and the facially similar Frasnian Shugurovian Beds. The Domanik is a formation consisting of thinly bedded clayey, siliceous and carbonate-bearing rocks characterized by the high content of organic substance (bitumens) and by a peculiar fauna. Of the fauna the goniatites (*Timanites* and *Gephyroceras*), the planktonic pteropods (*Styliolina*), the nautiloids (Orthoceratidae) and certain genera of pelecypods *(Buchiola, Pterochaenia, Ontaria)* are particularly characteristic. Amongst brachiopods rhynchonellids belonging to the genus *Liorhynchus* are encountered. In life pelecypods and liorhynchids adhered to algae, and their large number points towards great and thick growths of the algae. The Domanik facies was formed in sub-marine depressions, away from the shore-lines. The depressions were stagnant and overgrown with algae. The Domanik facies provides the main oil-producing formation. On the platform they are only encountered in the Frasnian Stage, while in the Urals they are also found in the Givetian and Famennian stages.

Other types of facies, contain 'normal' fauna consisting of brachiopods, pelecypods and gastropods, rugose corals tabulates and nautiloids are rarer; tribolites are very rare. The microfauna—foraminifera and ostracods—is abundant. In determining the age of cores from boreholes the microfauna has a great significance, and a large number of palaeontological-stratigraphical works have recently been concerned with it. The analysis of spores is being used progressively more frequently and especially so in determining the age of red beds. The study of the fish remains provides much information. Here the researches of D. V. Obruchev have a great significance.

The Famennian Stage is distinguished by the fact that the upward movements, which started in the north at the beginning of the Upper Frasnian epoch, reached considerable intensity in the Upper Famennian epoch. Consequently the sea narrowed down, spreading out laterally, and became much shallower, almost completely shoaling up and causing the accumulation of chemically deposited dolomites produced by algal and bacterial activity over the shoals. Along the northern shore of the sea a huge lagoon was formed, in which anhydrite and in places salt accumulated. In parts of the sea, where conditions were normal, limestones with a normal brachiopod fauna were deposited. Dolomitic shoals were populated by a less varied, but also typically marine, fauna which now crowds certain intercalations. In the highly saline lagoons lived only crustaceans—mainly *Estherias*—which also lived in saline basins.

The thickness of the Famennian dolomites and limestones varies considerably, reaching in places 200–300 and even 400m. Amongst the particular facies complexes one should mention the terrigenous complex developed in western Bashkiria and in south-eastern Tataria. It has a rather large number of members and horizons. The sands are not uncommonly oil-bearing and consequently the complex as a whole has a great industrial significance. Its structure is shown on Fig. 16. The Roman numerals I, II, III, IV and letters with numerical indices such as DI, DII, DIII, DIV, DV denote the main oil-bearing horizons. The thickness of the Pashiiskian Suite is here 20–40m. The distribution of the facies and their lithological characteristics are given in the *Atlas of Geological Facies Maps*, composed by the geological oil organizations and edited by V. D. Nalivkin. This atlas represents a major achievement. Its formulation was only possible owing to widespread and numerous geological and exploration projects carried out by the Soviet oil institutions. Numerous data have been included in the collection of papers entitled 'Oil and Gas Content of the Ural-Volga Province' (Moscow 1956) and in the paper by V. N. Tikhii (1957).

Central provinces. In the central provinces the study of the Devonian began over 120 years ago and the first palaeontological work of the German geologist Leopold von Buch was printed in 1840. The great monograph of the outstanding French palaeontologist Eduard Verneuil appeared in 1845 as the second volume of the capital treatise *Geology of Russia* written by Murchison, Verneuil and Keyserling. At the end of the 19th century the leading investigations of P. N. Venyukov and F. N. Chernyshev and of many others were published, but it was not until the Soviet period that in the search for oil stratigraphic and palaeontological research was especially widely developed. The number of projects is counted in hundreds. The basis on which the succession was established was laid by B. P. Markovskii, A. I. Lyashenko and the author. Fig. 18 shows the great detail of its subdivisions, since it involves eighteen horizons. Such detail is rarely encountered in world literature.

It must be pointed out that sandy-clayey deposits are distinctly predominant in the lower part of the succession, in the Middle Devonian and in the Lower Frasnian sub-Stage. Starting from the Lower Frasnian sub-Stage the carbonate deposits become dominant, with limestones, marls and calcareous muds in the Frasnian and dolomites and limestones in the Famennian Stage. In more northerly regions chemical sediments and especially anhydrites play an important part.

The three sedimentary mesorhythms are also interesting. The first (Eifelian) starts with basal Ryazhskian and Novobassovian Beds, which are the deposits of the nearshore plains, then follow the Morsovian Beds which are lagoonal chemical deposits, and the rhythm ends with marine Mosolovian clays and limestones. The second mesorhythm is Givetian, which starts from Ol'khovskian sands and clays, the deposits of the plains, and terminates with Vorob'evian and Starooskolian clays and limestones. The third mesorhythm is Lower Frasnian. It begins with Yastrebovian sandstones and terminates with Semilukian clays and limestones. There are indications of a fourth mesorhythm, which starts with Petinian sandstones (Alatyr Horizon) and ends with Yevlanovo and Livenovan limestones (Fig. 18).

The changes in the thickness of the succession on traversing from the centre of the Moscow Depression to its margins should also be noted. This is especially distinct on moving northwards. In the central parts of the depressions the thickness of the Devonian is greatest; in Archeda (north-west from Volgograd) it is over 2000m, in Serdobsk 940m, in Mosolovo 1017m, in Lyubim 923m, in Vologda almost 800m. At the margins of the depressions the thickness is considerably less: in Konosh (north of Vologda) 252m, in Kotlas 131 and in Yarensk 26m. Over the arches the thickness reaches intermediate dimensions as for example in Tomkovo 722m, near Penza (Prudy) 573m, in Ul'yanovsk 546m. The changes of the thickness can be clearly seen in Fig. 17. The changes in the composition and thickness of separate horizons can be seen in Table 6 in which are given the sections of the main standard boreholes. The position of the boreholes is shown in Fig. 17. The correlation of the sections allows one to draw the following basic conclusions.

Eifelian Stage. Continental, alluvial red beds (Narovian Horizon) predominate. In the south there are developed lagoonal facies in the form of gypsiferous dolomites and limestones with lenses of gypsum and rock-salt (up to 50m thick). Marine deposits have a limited distribution in the south-west. The inclusion of the Narovian Horizon in the Eifelian Stage or even its original separation provokes arguments. A number of specalists including D. V. Obruchev and B. P. Markovskii consider that this horizon and its marine analogues belong to the Givetian Stage.

Givetian Stage. In the west the red beds (Tartu Horizon) which are deposits of near-shore plains, predominate. Marine facies (Starooskolian Beds) are developed in the south, centre and east of the platform. The chemical deposits are almost absent.

Frasnian Stage. Most widely developed, it reaches a great thickness and is very variable. The marine formation of Upper Frasnian age, developed throughout the platform with the exception of the extreme north, is most typical. The formation consists of greenish and grey muds, limestones and marls 200–250m thick. It contains a very rich marine fauna. In the north of the platform the transgression of the top of the Lower Frasnian Stage had the greatest spread while in the south the Upper Frasnian transgression was widespread. The Lower Frasnian red bed formation of up to 250m in thickness is also widespread.

Famennian Stage. The sea shrank during this stage. Marine limestones and dolomites were developed in the northern part of the platform. At the margins of the sea the lagoons were well developed; these are the regions of accumulation of chemical deposits such as dolomite, limestone, gypsum. In the north there are widespread red beds—deposits of near-shore plains. Simplifying the relationships it can be said that the Devonian succession of the central regions consists of three formations: the Lower red beds being terrigenous; the Middle being greenish, grey marine limestones and clays, the Upper being halogenous red beds (Fig. 19). In a number of regions the section is complicated by the appearance of members with marine fauna amongst the red beds.

THE SALINE (HALOGENOUS) FORMATIONS. The sections (Fig. 19) exhibit the existence of two halogenous, dolomite–salt–gypsum formations. One of them is of Givetian age, while the other is Upper Famennian. Their formation is associated with the origin of lagoonal zones, of enclosed bays and of bitter saline near-shore lakes.

The gypsum of the upper formation, which is exposed near the Moscow Coal Basin is worked at a number of deposits.

Western provinces. The margin to the Devonian Sea was in the north-western and western provinces (Main Devonian Field), where the seas penetrated from the east. The red beds are widely developed and are the deposits of near-shore plains.

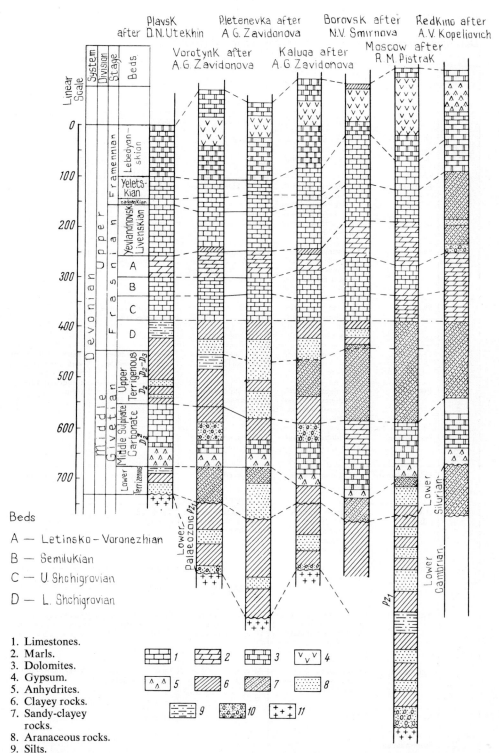

Beds

A — Letinsko–Voronezhian

B — Semilukian

C — U. Shchigrovian

D — L. Shchigrovian

1. Limestones.
2. Marls.
3. Dolomites.
4. Gypsum.
5. Anhydrites.
6. Clayey rocks.
7. Sandy-clayey rocks.
8. Aranaceous rocks.
9. Silts.
10. Conglomerates.
11. Crystalline basement.

FIG. 19. A correlation scheme of Devonian and Lower Palaeozoic strata in central provinces. According to A. G. Zavidonova (1951).

TABLE 7.

Devonian Successions in the western part of the Russian Platform

Stages and sub-stages	Beds, horizons	Pskov-Novgorod	Eastern		
			Porokov	Nevel	Gorodok
Cover	—	Anthropogenic deposits.	Anthroponic deposits.	Anthropogenic deposits.	Anthropogenic deposits.
Upper Frasnian and Famennian	Upper variegated Formation.	Variegated sands and clays with beds of marls (Up to 100m).	Missing.	Missing.	Missing.
Middle Frasnian	Buregian.	Variegated, dolomitized limestones, with marine faunas (15m).	Pale, blotchy limestones (9m).	Variegated fossiliferous dolomitized limestones (14m).	
Middle Frasnian	Ilmenian.	Clays with sandstone bands (20m).	Grey clays (15m).	Grey marls and clays (18m).	
Middle Frasnian	Svinorodian.	Limestones (multi-coloured) with marls and clays at the top; fauna is marine (15–17m).	Limestones. At the bottom, gypsum and marls; fossiliferous (25m).	Grey dolomitized limestomes (10m).	
Lower Frasian	Shelonian.	Greenish limestones, dolomites and marls. At the bottom, clays and gypsum (40m).	Unrecognized.	Greenish clays and dolomites (19m).	
Lower Frasian	Chudovian.	Grey limestones and marls; with marine fossils (6–20m).	Grey marls with faunas (11m).	Grey dolomitized limestones with a fauna (9m).	Yellowish dolomites with a poor fauna of Snetogorian to Chudovian ages (46m).
Lower Frasian	Pskovian.	Grey thickly bedded limestones and dolomites, with a fauna (20–25m).	Grey dolomitized limestones with a fauna (17m).	Grey dolomitized limestones with a fauna (16m).	
Lower Frasian	Snetogorian.	Grey dolomites and marls (3–12m).	Grey sandy clays (8m.)	Grey marls and clays (9m).	
Lower Frasian	Podsnetogorian (Amatian).	White quartz sands (6m).	Grey clays (9m).	Grey sandstones (18m).	Grey sandstones and clays (21m).
Lower Frasian	Oredezhian (Gauian).	Staritsian red beds on top (15–25m). Yashcherskian white quartz sands below (30m).	Red sandstones with beds of clay and fish (60m).	Thick red sands (110m).	Grey sands with clay bands (87m).
Givetian	Tartuian (Lugian).	Red sands and clays (Up to 200m).	Above, red clays and sands; below, sandstones (135m).	Red sandy clays with beds of sands (114m).	Above, sands and clays; below, grey sands (112m).
Eifelian	Narovian.	Variegated marls and clays. Grey marls, dolomites and clays (90–200m).	Above, marls and clays; below, a bed of gypsum (100m).	Grey marls, clays and dolomites (93m).	Grey marls and dolomites with beds of clays and gypsum (176m).
Eifelian	Pyarnusian.	Grey quartz sandstones with *Trochiliscus* (5–70m).	Missing.	Basal sandstones and clays (32m).	Grey sands (3m).
Eifelian	Kemerian.	Missing.		Missing.	Missing.
Lower Devonian	Stonishkaiian.				
Basement.	—	Silurian limestones.	Silurian dolomites.	Cambrian blue clay.	Cambrian blue clay.

TABLE 7—continued

Regions		Lithuanian-Podolian Depression		
Orsha	Kostyukovichi	Plavinyas	Sovyetsk (Stonishkyai)	Olesko (Lvov)
Anthropogenic deposits.	Anthropogenic deposits.	Anthropogenic deposits.	Permian sandstones, gypsum and clays.	Tournaisian shales and limestones.
Missing.	Missing.	Missing.	Missing.	Grey and brown, fossiliferous limestones and dolomites (174m).
		Yellowish dolomites (13m).		At the top, barren, alternating brown limestones and dolomites. In the middle, interbedded limestones, dolomites and clays; with a marine fauna. At the bottom, grey and green compact, calcareous and barren clays and argillites (35m).
		Grey dolomites and marls (15m).		
Cavernous dolomites with a fauna of Pskovian to Chudovian ages (38m).		Grey and brown dolomites with a marine fauna (6m).	Greenish marls, clays and dolomites with a marine fauna (38m).	
		Grey muddy dolomites with a marine fauna (20m).		
		Clays (above) and grey sandstones (below) (6m).	Marls and dolomites (10m).	
Above, grey clays; below, sandstones (64m).		White quartz sands with clays at the top (37m).	Grey clays and marls (20m).	
Grey sandstones and clays (51m).		Red sandstones and clays (60m).	Red sandy clays (50m).	
Red sandstones and clays (112m).	Sandstones and clays. Red sands at the bottom (94m).	Salatsian Suite. Grey clays, sands and siltstones (125m).	Red sandstones with sequences of clays (63m).	Interstratified shales and limestones with rare marine fossils. At the bottom sandstones and clays with gypsum (180m).
Above, grey sandstones and clays; below, marls, dolomites and a sequence of gypsum (142m).	Above, variegated rocks; below, grey marls, dolomites and gypsum (181m).	Greenish marls and dolomites. Gypsum at the bottom (129m).	Grey clays. At the bottom, marls and dolomites (143m).	
Grey sandstone (2m).	Grey sandstone with beds of clay at the top (29m).	Grey and greenish sands and sandstones with beds of clay (124m).	Grey sandstones and sandy clays (24m).	
Missing.	Missing.	Missing.	Kemerian clays and sandstones (76m).	Old Red Sandstone with fish (271m).
			Brown clays and sandstones (100m).	
Cambrian sands.	Cambrian sands.	Silurian.	Downtonian sandstones.	Silurian shales.

In the west of the Russian Platform, according to the data of the Latvian palaeontologist-stratigrapher P. P. Liepinsh (1955), the discovery of the Lower Devonian deposits, the age of which is determined by the finding of the armoured fish *Pteraspis* amongst the lower parts of the red beds, is especially noticeable. The overlying grey sandstones and clays contain spores of Eifelian age (Sovetsk; Table 7).

The whole of the Givetian Stage is represented by red beds with members of lacustrine thinly-bedded marls. The fish remains are abundant and diverse and amongst them the genera *Pterichtys* and *Asterolepis* are characteristic. Estherians (*Estheria membranacea* Pacht.) are often encountered in marls and clays.

The lower part of the Lower Frasnian sub-Stage is still represented by red beds with estherians and ostracods, but the genus *Bothriolepis* is characteristic amongst the fish. The red beds grade into clays and limestones of Pskov-Chudovian Beds possessing rich and variable marine fauna of *Ladogia meyendorfi* Vern. and *Lamellispirifer muralis* Vern. Marine deposits also form all the beds of the Middle Frasnian sub-Stage, but in the Upper Frasnian and Famennian stages the red beds predominate again and in places have lenses of gypsum. In the northern regions only the red beds are developed, while south of the lake Il'men' thin members of sandy-clayey and marly rocks with marine faunas are found amongst them and are known as the Upper Frasnian Smotinian-Lovatian Beds and Famennian Chimaevian and Bilovian Beds.

In the Latvian S.S.R., Lithuanian S.S.R. and the north of Byelorussian S.S.R. all the marine facies consist almost entirely of dolomites with somewhat impoverished, normal, marine fauna. Sections are shown in Table 7. In the south of Byelorussia and in the Pripet Forest, the region representing the north-western termination of the Dneypr-Donyets Depression, according to the Byelorussian geologist A. V. Fursenko (1953, 1957) marine and lagoonal deposits of Frasnian and Famennian age reach 1500–2000m in thickness. The salt massifs, which are analogous to the salt domes of more westerly regions, have a considerable thickness. The discovered marine fauna indicates that the part of the section, which underlies the salt, consists of repetitive clays, marls, limestones and sandstones and includes all horizons starting from the Upper Shchigrovian Beds and terminating by Zadonsk-Yeletskian Beds, thus including all the Frasnian and the lower part of the Famennian stages. The upper part of the Famennian Stage consists of red sandy-clayey deposits amongst which lie thick massifs of rock-salt and gypsum. The Famennian age of the saline formation is confirmed by the analysis of spores. The Devonian of the Pripet Depression is described in the section on the Mediterranean Geosyncline.

In the south-west in the Polish-Podolia Depression to the east of Lvov the thickness of the Devonian reaches 1200m and is it represented by three divisions (Table 7, Olesko). The Lower Devonian consists of red beds with armoured fish and is analogous to the Old Red Sandstone of Western Europe. Their thickness reaches 400m, but is often less, being 271m at Olesko. These are deposits of nearshore plains, lakes and land. The Middle Devonian is represented by a repetition of marine limestones and marls with lagoonal dolomites, gypsum and continental red beds. Its thickness is up to 180m. The Upper Devonian consists mainly of marine deposits and especially so in the Famennian Stage. In places it consists of just limestones and marls of up to 700m thick, but more frequently it has a variable composition (limestones, clays, sandstones) and is less in thickness. Above it lie the dark limestones of the Lower Tournaisian (equivalents of Yetrenskian Beds).

Along the northern margin of the Donyets Depression the occurrence of thin limestones with *Spirifer* ex gr. *anossofi* Vern. of Upper Frasnian age is very important. They are found in the salt domes of Romny, Slavyansk, Petrovsk, indicating a considerable transgression of the Upper Frasnian sea towards the south over the Voronezh Massif and the Ukrainian Shield. This transgression is the first stage in the formation of the main part of the Dnyepr-Donyets Depression. The same Upper Frasnian sea, containing the same fauna, penetrated from north to south to the Yaikian Massif. This is also indicated by the thin limestones with *Spirifer* ex. gr. *anossofi* Vern. found in the cap breccia of the Baskunchak salt dome.

PALAEOGEOGRAPHY. During the Lower Devonian epoch almost all of the Russian Platform was an elevated region of erosion. To the east of the Volga-Kama Uplands there was a low plain, which was the region of accumulation of the Upper Bavlinian Suite. An identical plain stretched to the west of the Byelorussian-Lithuanian Massif.

During the Eifelian epoch the sea covered all the eastern plain and penetrated along the depression to the south of the Volga-Kama Uplands into the Moscow (Central) Depression forming an enclosed basin, called the Moscow Sea.

During the Givetian epoch after a short-lived regression the Moscow Sea again re-established its former size. The older idea that it was directly connected with the seas of Western Europe has not been confirmed by the latest data.

During the Lower Frasnian epoch the marine transgression reached the greatest extent, but the sea was still a large enclosed inter-continental basin. In the Middle Frasnian epoch its size remained the same, but it receded slightly in the north-east of the platform. During the Upper Frasnian epoch the sea regressed considerably in the north, but was transgressive in Donbas, to the south.

During the Lower Famennian epoch the extent of the sea decreased in the south and in the Upper Famennian epoch it decreased even more both in the north and in the south. Large dolomitic shoals originated in the middle of the sea and a huge bitter salt lagoon came into existence in the north.

Lower Carboniferous

The Lower Carboniferous is the last epoch of the Middle Palaeozoic. It began with a rapid, considerable and short-lived elevation of the whole platform, causing the deposition of a coal-bearing formation, which includes coal deposits (Moscow and Kama coal basins), bauxites (Tikhvin, Onega), fire clays (Borovichi) and oil deposits ('Second Baku').

DISTRIBUTION. The regions between the Volga and the Urals, as well as the Moscow Depression, show the Lower Carboniferous. The latter depression in the south passed into the sea of the Donyets Basin, while the former depression in the region of Astrakhan was connected to the seas of the Mediterranean Geosyncline. Moreover, the Lower Carboniferous is fully developed in the Lvov-Volyn Depression.

SUBDIVISIONS. The unified standard scheme of subdivisions which was adopted at the All Union Conference at the All Union Institute of Petroleum (VNIGRI) in 1951 and published in the proceedings of the conference is generally accepted. The scheme is shown in the Table 8 supplemented, as decided by the 1960 Conference, by the Ozersko-Khovanian Horizon. In the 1960 Conference held in Moscow (VNIGRI) on the subject of the Palaeozoic stratigraphy of the Volga-Urals region the scheme of subdivisions was altered and amended.

Namurian Stage. The greatest differences of opinion are provoked by the boundaries of the Namurian Stage and its subdivisions. It has been proposed to divide the Namurian Stage into two sub-stages, but the position and the content of the Upper Namurian sub-Stage is not clear. A special conference on the content and position of the Namurian Stage was called in 1956, but it did not arrive at an agreed solution. At present many geologists propose to include the Namurian Stage in the Middle Carboniferous. Then the Middle Carboniferous must include three stages—the Namurian, the Bashkirian and the Moscovian. This is the most correct suggestion, but as yet has not been officially accepted. The reason for this indeterminacy is the gradual transition of the Lower Carboniferous into the Middle Carboniferous, observed in many regions. The pale grey bedded limestones of the the Visean Stage are succeeded by the identical limestones of the Namurian Stage,

TABLE 8.

Lower Carboniferous Successions of the Russian Platform

(Localities shown in Fig. 24)

Stage	Sub-stage	Horizon	Main		
			Valdai	Pestovo	Vologda
Cover	—	—	Anthropogenic sediments.	Vereiian sands and clays.	Vereiian sands and clays (24m).
Nam-urian	—	Protvinian.	Missing.	Pale limestones (18m).	Pink and grey dolomites (21m).
Visean	Serpukhovian	Steshevian and Tarussian.		Grey, fossiliferous silicified and dolomitized limestones. Grey dolomitized limestones with beds of clays at the bottom (59m).	Pale, sometimes silicified dolomites (26m). Grey dolomites and limestones (19m).
	Okaian	Mikhailovkian and Aleksinian.	Grey, sandy clays (13m).	Sandy clays, with bands of limestone (22m).	Grey and dark-grey limestones and dolomites; at the bottom, muddy and gypsiferous (24m).
	Yasnopoly-anian	Tulian, Coal-bearing and Bobrovkian.	Grey sandstones and clays (12m).	Grey sands, sandstones and clays (23m).	Missing.
Tournaisian	Cherny-shinian	Kizelian and Cherepetian.	Variegated sands and clays (2m).	Missing.	
	Likhvinian	Upian and Malyevkian.	Missing.		
		Ozersko-Khovanian.	—	Dolomites (20m).	Marls (73m).
Basement	—	—	Nadbilovian red beds (44m).	Nadbilovian red beds (44m).	Nadbilovian red beds (33m).

which towards the top are succeeded by the Bashkirian limestones of the same type and belonging to the Middle Carboniferous. The fauna throughout all these limestones changes gradually and imperceptibly. In the lower part of the Namurian stage there are many Visean forms, and it has been called the 'Protvinian Beds'. In the upper part of the Namurian there are many Middle Carboniferous forms, but this part has not been distinguished as a separate subdivision. Despite the presence of the Visean and Middle Carboniferous forms, the Namurian fauna has many peculiar, specific forms which impart to it a special character. This phenomenon is observed not only in the U.S.S.R. but also abroad, where the Namurian Stage has been distinguished a long time ago and belongs to the Middle (Upper) Carboniferous.

In addition, in separating major subdivisions, as in this case of the Lower and Middle Carboniferous, breaks in sedimentation, accompanied by sharp changes in lithological composition, have a great significance. Such a sharp change is in fact observed. Amongst the light grey pure limestones a member consisting of red and variable sandy-clayey rocks appears and is distinguished in the Russian Platform

TABLE 8—continued

Depression		Eastern Depression Krasnokamsk	Edge of Voronezh Massif—Serpukhov
Soligalich	Shar'ya		
Vereiian clays (38m).	Vereiian clays (46m).	Kashirian limestones (30m).	Vereiian clays and sands (21m).
Pale dolomites (8m).	Pale shales (13m). Pale limestones (24m).	Namurian, Serpukhovian and Okaian sub-stages. At the top, white saccharoidal crystalline carbonates.	Missing.
Pale-grey; at the top, gypsiferous dolomites (20m).	Pale-grey bedded limestones, sometimes silicified (30m).		Dark, muddy carbonaceous limestones (18m). Bluish bedded limestones (20m).
Grey and dark-grey dolomites (63m).	Grey dolomitized and normal limestones (34m).	Bedded, at the bottom, grey limestones and yellowish dolomites (cavernous). Some horizons are oil-bearing (259m).	Pale-grey hard limestones with beds of clays (31m). Grey, hard limestones (19m).
Grey bedded limestones. At the bottom, clays and sandstones (28m).	Grey sandstones with a limestone band at the top and clay bands at the bottom (31m).	Dark-grey and grey sandstones and carbonaceous clays with coal seams (47m).	Grey clays (21m). Grey carbonaceous clays and siltstones (12m).
Missing.	Missing.	Brown and grey, bedded limestones, often with algae and bands of clays (100m).	Brownish, muddy carbonaceous limestones. Grey laminated limestones and clays (27m). Ozersko-Khovanian dolomites (12m).
Dankovo-Lebedyankian clays (135m).	Dankovo-Lebedyankian dolomites and gypsum (210m).	Famennian dolomites.	Dankovo-Lebedyankian dolomites and gypsum (146m).

under the term 'Vereian Horizon' or even the 'Vereian Stage'. As is seen in Table 8 in the central provinces the Vereian Horizon lies directly on Protvinian Beds (Lower Namurian) and the boundary between the Lower and Middle Carboniferous coincides with the break in sedimentation or a sharp change of sediments. The situation is more complex in the Volga-Urals Province where the Bashkirian Stage of the Middle Carboniferous rests conformably on the Upper Namurian. Here as well as in the Urals, Tian-Shian', and many other provinces the break and sharp change in sedimentation occurs within the Middle Carboniferous and not at its boundary with the Lower Carboniferous. This is a very important event and complicates the problem even more. Thus not only the lower, but also the upper boundary of the Namurian becomes ambiguous.

TABLE 8—continued

| Stage | Sub-stage | Horizon | Tokmovian | |
			Mosolovo	Tokmovo
Cover	—	—	Vereiian clays and sands (24m).	Vereiian clays and sands (19m).
Namurian	—	Protvinian.	Missing.	Missing.
Visean	Serpukhovian.	Steshevian and Tarussian.		
	Okaian.	Mikhailovkian and Aleksinian.	White granular limestones with *Striatifera* (12m). White foraminiferal limestones (30m).	White hard limestone with banks of stratiferids (6m). White brachiopod limestones (22m).
	Yasnopolyanian	Tulian, Coal-bearing and Bobrikovkian.	Limestones and clays (21m). Grey siltstones and carbonaceous clays (5m). Missing beds.	At the top, pale limestones (20m). Lower down, white sands and clays (6m).
Tournaisian	Chernyshinian.	Kizelian and Cherepetian.	Grey and bluish limestones and clays (13m).	Missing.
	Likhvinian.	Upian and Malyevkian.	Grey limestones (14m).	
		Ozersko-Khovanian.		
Basement	—	—	Dolomites, limestones and clays (146m).	Dankovo-Lebedyankian dolomite and gypsum (60m).

What, then, is the practical solution? The only correct solution is to make meticulous collections of the fauna and particularly of the microfauna, leave it to be determined by the specialists and to draw the boundaries between the Visean and Namurian, between the Namurian and the Bashkirian and the Lower Namurian and the Upper Namurian according to their judgement. The second solution, which is simpler but incorrect, is to draw the boundary between the Lower and Middle Carboniferous on the basis of Vereian and other beds, associated with a break and a sharp change in sedimentation. It is easier to establish such a boundary in the field, but this method violates the basic principle of stratigraphy, that boundaries should be drawn according to faunal changes; faunal changes in the Middle Carboniferous groups appear considerably prior to the change in sedimentation. The third solution—radical, but the least correct one—is to dispose of the Namurian

TABLE 8—continued

Massif Ulyanovsk	Eastern Depression Buldar	Ryazan-Saratov depression, Archeda	Lvov-Volyn basin composite succession
Bashkirian limestones (33m).	Bashkirian limestones (20–50m).	Bashkirian sands and clay (122m).	Bashkirian Stage. Sandstones and clays (150m).
Pale, saccaroidal limestones (31m).	Pale limestones and dolomites (110–125m).	Grey organogenic limestones and marls (123m).	Limestones and clays (50m). Sandy-clayey, towards the top, coal-bearing strata (130–215m).
Pale dolomites and limestones (53m).

Pale, hard limestones with bands of dolomite and calcareous algae (100m). | Grey and yellowish dolomites (80–100m). | Grey, bedded, muddy limestones (106m). | Vizantian Stage—Ivanichian zone of argillites with subordinate bands of sandstones and limestones and fresh-water and marine faunas (100m). Poritskian zone of sandstones and clays with coal-seams and limestones (100m). Ustl'uzhian zone of bedded limestones with foraminifera (40m). Vladimirian zone of limestones, sandstones, clays and coals (80m). Yakhtorovskian zone with sandstones with clays and coal seams (60m). |
| At the top, limestones and dolomites (40m). At the bottom, carbonaceous clays with siltstones (28m). | Dark clays and marls with a marine fauna (30–35m). Thick sandstone sequence with clays at the bottom (Up to 160m). | Dark limestones and clays (66m). Dark sandstones and clays (20m). | Bussian zone. Alternation of limestones, clays and sandstones; Foraminiferal fauna being present (75m). Oleskovskian zone of dark limestones with brachiopods (45m). |
| Kizelian strata are missing. Grey hard bedded limestones (23m).

Grey, flaggy limestones with beds of green clays (17m). | A thick formation of bedded limestones and dolomites (170m). | Grey organogenic and flaggy limestones (73m).

Grey and greenish marls, clays and limestones (36m). | Tournaisian Stage. The Upper Suite consists of variegated sandy-clayey rocks with limestones towards the top (20–55m). The Middle Suite consists of variegated dolomites with beds of gypsum (40m). |
| Dolomites. | — | Black marls (66m). | The Lower Suite—grey limestones, sandstones and dark siltstones; d'Etreungt fauna (90m). |
| Dankovo-Lebedyankian dolomites, gypsum and limestone. | Frasnian limestones. | — | Upper Devonian. |

Stage altogether by including the Lower Namurian sub-Stage in the Visean and the Upper Namurian sub-Stage in the Middle Carboniferous. The boundary then will lie between the Lower Namurian and Upper Namurian sub-stages. This suggestion is incorrect for the following reasons.

1. The boundary between Namurian sub-stages is the most difficult to establish. In the field it cannot be established at all.

2. All the difficult problems remain unsolved. The attachment of the Lower Namurian sub-Stage to the Visean does not make this subdivision disappear, and the boundary between them is still difficult to trace. The same applies to the boundary between the Upper Namurian sub-Stage and Bashkirian Stage. In this way the third solution does not ease anything.

3. The Namurian Stage is in general use not only in the U.S.S.R., but in world stratigraphy and in order to liquidate it a resolution of an International Congress or of a special international conference is necessary. This, together with all other decisions which violate schemes accepted by world practice, is undesirable. For the Soviet Union this is not permissible since in industrial work it does not help at all. The question as yet remains unresolved, and in practice only the first solution can be accepted and particularly so when composing geological maps, since prior to an All Union Conference the Namurian should be included in the Lower Carboniferous.

LITHOLOGY AND FACIES. *Tournaisian limestones and clays.* The upward movements of the Upper Famennian epoch were succeeded at the beginning of the Tournaisian epoch by slight downward movements. Lower Tournaisian dolomites, limestones and marls of the Ozersko-Khovanian Beds of 10–30m in thickness were deposited in the Moscow Basin. Above them lies clays, marls and muddy limestones of the Malyevkian-Murayevnian Beds and Upian Beds containing marine, brachiopod-ostracod fauna. They are overlain by the Upper Tournaisian Chernyshinian limestones with a richer and variable coral-brachiopod fauna. The total thickness of the Tournaisian reaches 30–50m. It should, however, be borne in mind that this classical succession has a limited distribution and a short distance from the Moscow Basin undergoes changes, which are not uncommonly considerable. To the north the thickness of the Tournaisian decreases, and to the north of Borovichi it is absent. Over the Tokmovo Arch (see Fig. 25) there is no Tournaisian either. The succession also varies rapidly in the Volga-Ural Province where rather massive dark clays and muddy sandstones are widely developed. Thus, for instance, in Tataria the Tournaisian Stage consists in places of compact clays (argillites), siltstones, and clayey limestones, which are dark, bituminous and of great thickness (up to 200–300m in Kama-Kinel' Depression). In the Lvov-Volyn coal-field the Tournaisian Stage is also peculiar (Table 8). It starts with grey limestones, sandstones and dark siltstones reaching about 90m and containing a Strunian fauna and flora. Above it lie the variegated lagoonal dolomites and gypsum (up to 40m), lacking any fauna. The upper part begins with continental, variegated, sandy-clayey beds with plant remains; at the top limestone intercalations appear. The thickness of this suite is 20–55m, thus the Tournaisian as a whole reaches 150–200m. The variability of the Tournaisian succession is characteristic, and points to the differential variability of movements associated with its accumulation (Table 8).

The study of the Lower Carboniferous sections in the Volga-Ural Province has shown that in a number of regions, and in particular in the Kama-Kinel' Depression, there are three horizons named from below upwards: Ikchigolian, Rakovkian and Malinovian, and which separate the Tournaisian deposits and the coal-bearing sequence. The Ikchigolian and Rakovkian horizons are developed inside and outside the Kama-Kinel' Depression. They consist of grey, bedded limestones of 60–100m in thickness. The Ikchigolian and Rakovkian horizons contain a rich and variable fauna, identical with that of the Kizelian Horizon of the western slope of the Urals and are of the same age. The Malinovian Horizon (Formation) is developed in the Kama-Kinel' Depression. It underlies the coal-bearing formation and overlies the Rakovkian limestones, consisting of argillites and siltstones of 200–300m thickness, and containing a rich and diverse marine fauna.

In the 1960 Conference on the Palaeozoic stratigraphy of the Volga-Ural Province and in the subsequent session of the Central Information Committee on Stratigraphy the Ikchigolian, Rakovkian and Malinovian horizons were accepted as contemporaneous to the Kizelian Horizon and were included in the Lower Visean, while the Bobrikovkian and Tulian horizons were referred to the Middle Visean. This decision became binding in all geological investigations. Nevertheless in substance this is incorrect and in any case controversial. The faunas of the Ikchigolian, Rakovkian and Malinovian horizons together with the fauna of the Kizelian are closely related to the Upper Tournaisian fauna and differ considerably from the fauna of the coal-bearing formations and their limestone analogues developed in the Urals. A number of Soviet and foreign geologists consider the Kizelian fauna to be Upper Tournaisian. The same point of view is accepted by the author and is employed in this text-book.

The Coal-bearing (Bobrikovkian) Formation of the Lower Visean age lies on different horizons of Tournaisian clays and limestones and sometimes directly on the Devonian. In the west it is preceded by a break, while in the east clays and muddy limestones belonging to the Kizelian Beds gradually pass up into the clays of the Coal-bearing Formation. The Coal-bearing Formation is a most diverse complex of facies including terrestrial, lacustrine, fluviatile, lagoonal and marine, which are of great significance in mining. It is developed into the Moscow coal basin, the Kama basin, Selizhar and Borovichi coal deposits; it contains most valuable fire clays (Borovichi etc.), the Tikhvin, Onega, and Moscow bauxites and bauxitic rocks. In the east, in Kama-Ural Province, sands and rubbly sandstones are in places saturated with oil and are coal-bearing horizons. The Coal-bearing Formation consists of dark brownish and grey sandy-clayey rocks with lensoid beds of coal, clays and bauxites. Its thickness varies from 25–30m in the Moscow Basin to 10m in the north-west and up to 100m in the east (Table 8).

The greater part of the sands and clays lack organic remains, and represent deposits of large, low-lying and flat near-shore plains. These plains were crossed by relatively small, gently-flowing rivers whose beds received deposits of current-bedded sandstones. At the mouths of these rivers deltas were formed in the sea, and within these deltas the Coal-bearing Formation had the greatest thickness. The Coal-bearing Formation of the Moscow Basin represents such a delta of very large size. The Selizhar Deposit, the Borovichi and Tikhvin Basins are also deltas, but of smaller dimensions. On the surface of the deltas and the near-shore plains in the proximity of the shore-line large marshes and lakes existed. The former were sites of accumulation of great masses of plant remains forming the present-day coal seams. At the bottom of the lakes very fine muds were deposited, which were enriched in alumina and produced the subsequent bauxites, bauxitic rocks and fire clays. Enrichment by iron oxides caused the formation of seams of ferruginous rocks and in places of lacustrine iron ore.

Small downward movements were sufficient to allow the sea to penetrate into the low-lying surface of the deltas and the plains and to form the clayey-marly rocks with marine fauna. Such members are found in the west, in the Moscow Basin, but especially in the east. It cannot be considered that their presence proves a marine origin for the whole of the Coal-bearing Formation; since such an opinion is naïve and erroneous.

The Lower Carboniferous is the first epoch in the life of the Earth when the true terrestrial plants (lepidophytic flora) became dominant, and it is the first epoch when economic coal seams originated. The limestone intercalations in the upper part of the Coal-bearing Formation become larger, then they become as thick as the sandstones, clays and coals, and finally the whole formation becomes entirely formed of limestones. The Coal-bearing Formation thus passes gradually into a carbonate formation. The transitional formation is known as the Tulian, and together with the Coal-bearing Formation it constitutes the Lower Visean (Yasnopolyanian) sub-Stage. In the south-west of the platform it consists entirely of limestones and clays and the Coal-bearing Formation is absent.

The problem or the origin of the huge masses of clastic material-sands and clays—forming the Coal-bearing Formation is of importance in deciding its origin. Why did they not exist before or later? The answer can be only that during this epoch there were upward movements of the uplands of the Northern and Southern Belts. The upward movement of the uplands and the reverse in the depressions could have been the cause of the origin of the Coal-bearing Formation happening contemporaneously with the appearance of the terrestrial lepidophyte flora and the intensification of the activity of rivers.

The discovery of the new Kama coal basin is interesting and important. This basin follows an elongated and devious depression for which V. M. Pozner (1955) has suggested the name Kama-Kinel' (Figs. 20 and 21). This depression has been already traced over a distance of more than 500km.

Within it the thickness of the terrigenous formation increases considerably. If in the surrounding regions it varies from 10 to 30m, then in the depression it is up to 100–200m, and in the south even more than 400m. Investigations have shown that only the upper part of the terrigenous formation is coal-bearing. The whole formation has acquired the name 'Sarailinian' and its top belongs to the Visean age, its middle part to the Tournaisian and its greater lower part to the Famennian as is shown in the cross-section, Fig. 21.

The Coal-bearing Formation is overlain by the Middle Visean (Okaian) limestones. The study of the fauna has shown that the lower part of the Coal-bearing Formation here belongs to the Upper Tournaisian. In the north it includes coal seams of 2–3m thick. Its upper part contains a Lower Visean (Yasnopolyanian) flora and is equivalent to the Bobrikovkian (Coal-bearing) and Tulian horizons. It has the main coal seams of considerable thickness. According to N. I. Markovskii (1956), in various districts of Tataria coal seams of 2·5, 6 and 7m, and in the east over 15m, have been found by drillings to the depth of about 1000m. The thickness of the coal seams is inconstant. The Kama coal-field belongs to the concealed type. and does not outcrop at the surface. Its economic importance has not been determined in detail, but it is unquestionable. Its exploitation by subterranean gasification of coal is possible.

The Carbonate Formation of the Middle Visean (Okaian) and Upper Visean (Serpukhovian) sub-stages and of the Namurian Stage consists of limestones, dolomitized limestones, and more rarely dolomites reaching 150–250m in thickness (Fig. 22). Together with the overlying carbonates of the Middle and Upper Carbon-

Fig. 20.

The distribution of the Sarailinian and analogous formations in the Kama-Kinel' Depression.

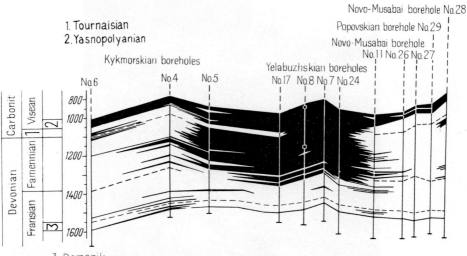

Fig. 21. A cross-section of the Sarailinian Formation in the lower reaches of the Kama (Yelabuga). Terrigenous rocks are shown in black. According to A. I. Kleshchev and others (1957).

iferous, which are also thick (300–400m) it is traced over thousands of kilometres, and forms one of the largest massifs of carbonate rocks known in the history of the Earth. The quantity of lime in it is so great that even a small part of it can satisfy the demands of all Soviet industry (Table 8). Porous intercalations in the dolomites of the Namurian Stage in certain districts of the Volga-Ural Province serve as oil-bearing horizons.

Carbonate rocks are deposits of a shallow sea, not exceeding 100–200m in depth and most commonly being only a few tens of metres. They were found by the accumulation of skeletons of animals and plants and the products of their disintegration, and also as a result of bacterial and chemical processes. The negligible amount of terrigenous materials points to the low-lying conditions of the adjacent land and an almost complete absence of rivers capable of carrying clay and sand into the sea.

The north of the Russian Platform during the Lower Carboniferous epoch was occupied by uplands. Thus in the northern direction seas are replaced by a near-shore alluvial platform, and consequently marine rocks get thinner, pass into sandy-clayey deposits and finally disappear (Kotlas). Further to the north the Lower Carboniferous red beds also wedge out and the Middle Carboniferous lies on the Frasnian red beds and the Valdaiian complex.

As an example of wedging out of the carbonate rocks one can select the section at Onega Bauxite Region (upper reaches of the River Onega). Here all the Lower

FIG. 22.   Visean limestones on the river Msta. *Photograph by R. F. Gekker.*

Carboniferous, lying between the Middle Carboniferous limestones and Frasnian red beds, is only 100m thick and is represented by sandstones and clays with the intercalations of Serpukhovian and Okaian limestones in its upper part. The lower part, corresponding to the Yasnopolyanian Stage, consists of variegated and black clays with lenses of bauxites and pisolitic iron ores. The thickness of the bauxite formation is 35–40m.

The Visean and Namurian stages of the Lvov-Volyn coal basin are peculiar. This basin is situated in the Galician-Volynian (Podolia) Depression, situated on the north-western flank of the Ukrainian Shield. Here the Lower Carboniferous has been traced from Lvov to the north, as far as Vladimir Volynskii. Its succession differs sharply from the succession in the Moscow and Donyets basins and resembles the successions of the Western European basins. The first diagnostic feature of the Lvov-Volyn Lower Carboniferous is that it lacks the Lower Visean Coal-bearing Formation. It is replaced by dark bedded limestones with *Gigantoproductus* towards the top. Its second diagnostic feature is the development of a coal-bearing formation (300–400m thick) which has the age of the Upper Visean and Namurian. The third peculiarity is the large number of limestone intercalations in the lower part of the Bashkirian Stage. The greatest thickness of the Lvov-Volyn Carboniferous is 750–950m, thus differing from the Donyets Basin where the thickness of the Carboniferous reaches 8000m.

PALAEOGEOGRAPHY. The palaeogeography of the Tournaisian epoch is very complex and has not been entirely clarified. In the western and central provinces there existed a shallow intercontinental sea margined by uplands on the north, west and south. It included the region of the Donyets Basin. In the west the sea had an indented shore-line with large number of bays, peninsulas and islands. It is possible that there was a frequently broken connection with the Lvov-Volyn Basin (Golubtsov and Makhnach, 1961). In the east it was open and deep, passing into the sea of the Urals Geosyncline.

During the Lower Visean (Coal-bearing) epoch this sea withdrew and dried up, thus producing a large, low-lying plain, covered by numerous lakes and marshes and transected by numerous small rivers. Nearer to the sea shore there were marshy lagoons. The sea was continuous to the south of the platform, in the Donyets Basin and in the south-eastern provinces. At times the sea penetrated into the interior of the plain, into the central and north-eastern provinces. The uplands of the Northern and the Southern Belts rose and became progressively eroded. The products of their erosion are carried out by the rivers to the plain and into the sea, forming sequences of fine-grained sandy-clayey material, characteristic of the Coal-bearing Formation. In the times when carbonate formations were deposited, a shallow, warm, tropical sea again penetrated the larger part of the platform, whose slow and lengthy downward movement allowed the accumulation of many hundreds of metres of limestones and dolomites. On the land the uplands were gradually smoothed down to the level of the plain, rivers disappeared and the transport of terrigenous material almost completely stopped.

Upper Palaeozoic

The Upper Palaeozoic is widely distributed and includes variable deposits, not infrequently of economic importance.

DISTRIBUTION. In the region of the Russian Platform it exists everywhere, with the exception of the uplands of the Scandinavian Massif and Ukrainian Shield, which together with the Byelorussian-Lithuanian Massif formed a western barrier to the Upper Palaeozoic seas and plains.

SUBDIVISIONS. The subdivision of the Upper Palaeozoic has for a long time been the object of discussion and argument. At a number of conferences it has been found possible to solve the main problems and to adopt new subdivisions. A scheme of subdividing the Middle and Upper Carboniferous has been accepted in an All Union Conference organized by the All Union Institute of Petroleum (VNIGRI). The decisions of the Conference were published in 1951 (Table 9). In 1950 the same institute had published 'The decisions of the Permian Conference' which contained a unified scheme of the stratigraphy of Permian deposits (Table 10).

Previously such unified, generalized schemes were formulated by single geologists. In the Soviet period the volume of factual data has become so great that a single person cannot deal with it. Generalized schemes and maps, and regional monographs such as *Geology of the U.S.S.R.*, are produced by teams of geologists, which are sometimes very numerous. Collective effort is one of the most important features of the Soviet scientific work. The latest generalized schemes of Carboniferous and Permian stratigraphy have a considerable detailed division of the succession as a new feature. Thus, new Lower Permian stages (Asselian and Sakmarian) have been distinguished and the Lower Permian now includes four stages— Asselian, Sakmarian, Artinskian and Kungurian. The Ufian Stage is included in the Upper Permian.

The separation of numerous Upper Palaeozoic subdivisions is done on the basis of the distribution of all animal groups, such as Foraminifera, corals, brachiopods, goniatites, ostracods, but lately fusulinids have occupied the first place amongst these. In this respect much has been contributed by D. M. Rauser-Chernousova and her students, working throughout the Soviet Union.

The boundary between the Carboniferous and the Permian is a controversial topic, which attracts great attention. To solve it all the animal groups were investigated, but especially so the fusulinids and the ammonites (goniatites), which were most contributive. Formerly the Permo-Carboniferous boundary was drawn at the bottom of the Artinskian Stage. The study of the fusulinids and of the ammonites has shown that, in deposits previously attributed to the top of the Upper Carboniferous, they have a Permian aspect and are represented by the same genera as in the Artinskian Stage. These rocks were then separated into a special stage—the Sakmarian—and included in the Lower Permian. The boundary was then drawn above the *Schwagerina* Beds, as was done in the 'Decisions of the Permian Conference (1950)'. Soon the detailed study of the *Schwagerina* Beds showed that the first representatives of the Permian fusulinids, ammonites and rugose corals appear in them. Thus subsequently the boundary was drawn under the *Schwagerina* Beds, as has been done in the discussions of the 1951 Conference. This decision has been reiterated by the 1960 Conference and is obligatory. At this conference it was suggested, and later confirmed by the Central Stratigraphical Committee, that the *Schwagerina* Beds (Horizon) should become a special stage—the Asselian—after the river Asseli in the Southern Urals (Table 9). It has also been decided to distinguish the *Pseudofusulina* Beds (Horizon) into a distinct Orenburgian Stage. The Permo-Carboniferous Boundary is drawn between these two stages.

The *Schwagerina* and *Pseudofusulina* horizons were promoted into stages on the suggestion of the leading specialist on the Upper Palaeozoic—V. E. Ruzhentsev. The basis for this rests upon the large volume of the subdivisions and the peculiar characteristic faunas. The change in the names has been adopted since it has now been decided not to name the subdivisions after any animal group and to change the old palaeontological names by the new geographical names. On the basis of the evolution of fusulinids the foremost expert on them, D. M. Rauser-Chernousova (1940), suggested drawing the Carboniferous-Permian boundary between the Tastubian and Sterlitamakian Beds. Later on in 1956 she accepted the unified 1950 scheme. Lastly, two leading Uralian Geologists I. I. Gorskii and V. D. Nalivkin have pointed out that in the Urals the most natural and convenient boundary lies under the Sakmarian Stage. These four points of view are reflected in Table 11.

TABLE 9. *Unified Stratigraphic Succession of the Middle and Upper Carboniferous of the Russian Platform*

1951 conference			1961 conference			
Division	Stage	Horizon or zone	Division	Stage	Sub-stage	Horizon or zone
Upper	Gzhelian	Pseudofusulinid	Upper	Orenburgian	—	Pseudofusulinid
		Zone of *Triticites jigulensis*		Gzhelian	—	Zone of *Triticites jigulensis*
		Zone of *T. stuckenbergi*				Zone of *T. stuckenbergi*
	Kasimovian	Zone with *T. ardicus* and *T. acutus*		Kasimovian	—	Zone with *T. ardicus* and *T. acutus*
		Zone of *T. montiparus*				Zone of *T. montiparus*
		Zone of *Protricitites*				Krutovskian
Middle	Moscovian	Myachkovian	Middle	Moscovian	Upper Moscovian	Myachkovian
		Podolian				Podolian
		Kashirian			Lower Moscovian	Kashirian
		Vereiian				Vereiian
	Bashkirian	Upper Bashkirian		Bashkirian	Upper Bashkirian	Melekeskian
						Chermanshanian
		Lower Bashkirian			Lower Bashkirian	Prikamian
						Severo-keltmian

TABLE 10. *Unified Stratigraphy of the Permian on the Russian Platform*

1950 Conference

Division	Stage	Sub-stage	Horizon or zone
Upper Permian	Tatarian	Upper Tatarian	—
		Lower Tatarian	—
	Kazanian	Upper Kazanian	—
		Lower Kazanian	—
	Ufian Suite	—	—
Lower Permian	Kungurian	—	Solikamskian / Filippovkian
	Artinskian	Upper Artinskian	—
		Lower Artinskian	—
	Sakmarian	Upper Sakmarian	—
		Lower Sakmarian	—
	Uralian	Upper Uralian	Zone of *Schwagerina constans*
		Lower Uralian	Zone of *Sch. vulgaris*

1960 Conference

Division	Stage	Sub-stage	Horizon or suite
Upper Permian	Tatarian	Upper Tatarian	Nefedovskian / Bykovkian
		Middle Tatarian	Severodvinian
		Lower Tatarian	Sukhonian / Nizhneustian
	Kazanian	Upper Kazanian	—
		Lower Kazanian	—
	Ufian	—	Sheshminian / Solikamskian
Lower Permian	Kungurian	—	—
	Artinskian	Upper Artinskian	Filippovkian
		Lower Artinskian	Iksian
	Sakmarian	Upper Sakmarian	Sterlitamakian
		Lower Sakmarian	Tastubian
	Asselian	—	Kokhanian / Nizhnesaksian / Batrakovskian

The first point of view has been accepted at the All Union Conference and is widely used in drafting geological maps. In the Volga-Ural Province, however, oil-geologists adopt the second scheme, since in this province there is a sharp change in lithology above the *Schwagerina* Beds, which is very convenient in comparing cross-sections. The third point of view is not widespread. The fourth point of view is only used by the geologists working in the Urals.

The subdivisions and Upper Palaeozoic lithologies in the main regions of the Russian Platform are given in Table 12.

LITHOLOGY AND FACIES. The Lower Bashkirian Stage of the Middle Carboniferous consists of the same kind of limestones as the Namurian Stage and is its direct continuation. The lower boundary of the Middle Carboniferous (Bashkirian Stage) is drawn according to the faunal change and in the first place on the appearance of the first fusulinids (*Profusulinella, Pseudostafella*) and first choristids (*Choristites bisulcatiformis* Semich).

The first break in sedimentation which occurs everywhere, but which was short-lived and thus did not upset the further accumulation of carbonates, is included in the bottom of the Moscovian Stage of the Middle Carboniferous. It is reflected in the deposition of the Vereiian Formation which is a thin member, normally several metres in thickness and consisting of variegated, reddish, bluish, pinkish clays and sands (Table 12). They have no fauna at the base, but contain plant remains, while towards their top appear intercalations with marine fauna.

During Vereiian times a portion of the western part of the Russian Platform became land and as a result of the weathering of the previously deposited limestones variegated clays were formed. In the east the marine regime evidently was not interrupted and the bright variegated colour of the deposited clays and sands depended

TABLE 11.   *Comparisons of suggestions for placing Permo-Carboniferous boundary*

Stage	Beds	According to N. P. Gerasimov, 1937; V. E. Ruzhentsev, 1946, 1951 and 1960 conference correlation tables	According to D. L. Stepanov, 1947 and unified 1950 scheme	D. M. Rauzer-Chernousova, 1940	I. I. Gorskii, 1948; V. D. Nalivkin, 1949
Artinskian	Sargian Irginian Burtseukian				Permian
Sakmarian	Sterlitamakian			Permian	Carboniferous
Sakmarian	Tastubian		Permian	Carboniferous	
Asselian	Schwagerinid	Permian	Carboniferous		
Orenburgian	Pseudofusulinid	Carboniferous			

on the primary colour of the rocks which formed the Vereiian lands or more probably on the crust of weathering of such rocks. Thus in the west the origin of the
Vereiian Formation depends on the formation of the land, while in the east it depends only on the conditions of transport of the bright variegated products of land-
weathering into the sea.

The overlying Middle Carboniferous horizons—Kashirian, Podolskian, Myach-
kovian and also Kasimovian, Gzhelian and Orenburgian Upper Carboniferous stages
consist of carbonates being 300–400m in thickness. In some provinces carbonate
sedimentation continues through to the Lower Permian. The deposition of the
pure carbonates can only occur in the absence of transport of terrigenous material.
The appearance of the large number of dolomite members, especially at the top of
the formation, points towards the formation of extensive shallows. This is confirmed
by the small Artinskian limestone reef massifs in the Kuibyshev Trans-Volga
districts.

It is interesting that the formation of carbonates in the Middle and Upper
Carboniferous occurred in the centre of the sea and was contemporaneous with the
accumulation of thousands of metres of sands and clays along its margins (Donyets
Basin). Similar relationships existed along all the Middle and Southern Urals. Here
again thick, shallow-water, bedded carbonates of the Middle and Upper Carboniferous and Lower Permian ages are gradually succeeded by the Upper Palaeozoic
molasse formations, many kilometres thick, containing sandstones, conglomerates
and marls (western slope of the Urals). Such an adjacent accumulation of hundreds
of metres of carbonates and clastic terrigenous rocks of several thousands of metres
along the margins of the same basin is a very interesting phenomenon.

Asselian, Sakmarian and Artinskian deposits, in comparison with the underlying Upper Carboniferous limestones and dolomites, are distinctive in the variations of their lithology. In the Volga-Ural Province, the Ishimbai district, which is
partly within the eastern boundary of the Ufa Plateau and partly to the north of
it, had suitable conditions for the formation of massive reef limestones (400–600m
thick) produced by downward movements. To the west of these, bedded limestones
were deposited, which are often dolomitized and silicified. Further to the west, from
the Tuymazinsk region to western Tataria and the Kuibyshev region, dolomite-
anhydrite beds up to 100–150m thick are widespread. To the east of the Volga the
marine Lower Permian is absent and islands or peninsulas were situated there. In
the Moscow Depression the peculiar carbonate Shustovo-Denyatino Beds (15m), containing *Schwagerina* and brachiopods of Lower Permian type and pelecypods of
Upper Permian type, appear again. Lastly in the Northern provinces the Lower
Permian is represented by terrestrial red beds above and a homogeneous formation
and thick beds of gypsum, anhydrites and dolomites below (200–350m). Such a
variability of facies as conditioned by the upward movements which had begun
to affect the whole platform (Table 12).

During the Kungurian epoch the upward movements continued, and the chemical
deposits of the bitter-saline lagoons and lakes, consisting of gypsum, anhydrite and
rock-salt, and the red beds, deposited on the plains surrounding the lagoons, acquired
an enormous spread. Facies (with marine faunas) such as marls and dolomitized
limestones were not widespread. A classical section of these is known at the village
of Polazn, near Perm. The marine Kungurian fauna shows the first appearance of
Kazanian forms such as *Spirifer regulatus* Kut. side by side with the predominant
Artinskian forms. The other Kungurian facies do not contain macrofauna, while

their microfauna and microflora have not been properly studied. The Kungurian age of the deposits is usually established, as of old, by lithological features such as the appearance of the chemical deposits. Thus the modern, Kungurian stratigraphy is on a very elementary level. Undoubtedly detailed study of the microfauna and microflora will show that many deposits which are now considered as Kungurian are in reality Artinskian, Ufian and Kazanian. In the central provinces the Kungurian Stage is not differentiated at all (Table 12).

The Ufian Stage is another formation the stratigraphy of which is not at all clear and the identity of which provokes doubts. In the central regions it is absent, but between the Volga and the Urals it is developed from place to place as red beds, variegated clays, sands and marls of up to 100–150m thick. This formation usually lacks any fauna and only in places, in lacustrine clays and marls, freshwater lamellibranches *(Anthracosia)* and ostracods are found. Nearer to the Urals its thickness increases and in the foothills of the Urals it is 1000–1500m thick. This indicates that in the Ufian times the Urals were mountain ranges. Their products of erosion went into the formation of the Ufian Stage.

The Ufian Stage consists of continental deposits. Until now it is not known to which marine deposits it corresponds. This is a problem. The age of these deposits is determined by their position above the Kungurian chemical deposits and by them being covered by the muddy limestones of the Kazanian Stage. At present, as it has been decided, they are included in the Upper Permian.

The Kazanian Stage, or *Zechstein* ('building stone' in translation) is a name given to a formation of greenish-greyish muddy, bedded limestones, marls and clays of 40 to 100m thick and containing a brackish water fauna (Fig. 23).

The bedded limestones are commonly broken into slabs and are used as foundation stones or for other building purposes, from which arises the old popular German term the *Zechstein*, which was given to the rocks of the Kazanian Stage in Western Europe. In exposures the compact greenish and grey limestones and marls belonging to the Kazanian Stage sharply differ from the friable variegated clays, sandstones or gypsum of the Kungurian Stage, from the red beds of the Ufian Suite or the still higher red beds of the Tatarian Stage. The distribution of the Kazanian Stage is peculiar. Its outcrops represent an enclosed rounded area, which in its form and size resembles the modern Caspian Sea (Fig. 99).

Earlier on, straits were suggested to connect the Kazanian sea and the Northern and Southern oceans. At present this hypothesis is abandoned since the fauna over the territory of the Kazanian sea has been found to be uniform. The similarity to the Caspian is also confirmed by the character of the fauna, which seems to be of salt-water type. The fauna consists of a very large number of types belonging to a few strongly variable species which are normally incorrectly considered as groups of species. Then the absence of animals such as echinoderms, cephalopods, colonial corals and trilobites which are typical of seas with normal salinity is also rather characteristic. These features indicate an abnormally low salinity of the Kazanian sea owing to it being enclosed and fed by fresh-water rivers, flowing off the young, high Urals. It is known that the Caspian Sea also has a low salinity of 1·5 per cent instead of 3·5 per cent of the open ocean, and this is caused by the influx of the fresh water of the Volga, the Urals, the Terek and the Kura. The problem of the origin of the Kazan sea and its fauna is interesting. Recent work by the Soviet palaeontologists has shown that the main Kazanian forms are encountered amongst the Kungurian faunas of the Kama region. It is unquestionable that the Kazan sea

was a remnant, enclosed, relict sea, which was an offspring of an open Lower Permian
sea in the same way as the Caspian Sea is a relict sea inheriting the reservoirs
and the faunas of the pre-existing Pliocene basins.

Text-books of historical geology describe in detail yet another enclosed sea of
Kazanian age, situated in the western Europe. The bedded muddy limestones which
were deposited in it have acquired the name *Zechstein*. In its form this is an inner
sea, which in terms of sediments and other characteristics was very similar to the
Kazanian of the Russian Platform, but was not identical. Its north-eastern part was
situated within the Baltic countries. It was margined by a series of bitter-salt
lagoons. The marine and lagoonal deposits are described in this book in the section
'Baltic Region' (p. 178).

The Kazanian Stage is divided into two sub-stages: the lower with a mainly
brachiopod fauna and the upper with a mainly pelecypod fauna. As has been
rightly shown by M. S. Shvetsov such faunal differences have local characters and
depend on the freshness of the water. In other provinces the pelecypod faunas can
be encountered in the lower part of the section, while brachiopod faunas occur in
the upper part. Furthermore in other provinces the fauna throughout the section
may consist of either pelecypods or brachiopods only. This is an example of a
frequently encountered stratigraphical mistake. Having established the sequence in
one region it is automatically applied to other, sometimes distant, regions. The
facies character of formations changes rapidly and considerable mistakes are often
committed owing to the fact that this important phenomenon has not been sufficiently considered. In the Orenburg region the Kazanian section differs considerably
from the section at Kazan. Owing to the proximity to the shore-line the rocks are
enriched in terrigenous material (Fig. 25). Lagoons with thick deposits of salt come
into existence. In the east the Kazanian sea did not reach the latitude of Perm

FIG. 23.   Kazanian muddy limestones at Vandova mountain in the lower
reaches of the Kama. *Photograph in the Geological Museum.*

and a wide near-shore plain separated it from the young Ural mountains. A similar plain bordered the Kazanian sea to the north-east and south-east. In the west no deposits of such a plain have remained. It is possible that they have been removed by erosion during the Mesozoic. It is also possible that they are included in the Tatarian Stage, which cannot be distinguished lithologically. The eastern near-shore Kazanian plain was a region of deposition of important red beds which reach tens of metres and more in thickness. It can be seen in the magnificent exposures of the high precipitous shores of the River Kama that these beds down the river change into marine greenish-grey and grey clays, marls and limestones. The red beds contemporaneous with the Upper Kazanian marine deposits are included in the Belebeyevskian Suite.

The efforts of the Kazan school of geology has a great importance in the study of Kazanian and in general Upper Palaeozoic deposits. Amongst its pre-Revolutionary representatives the names of A. V. Nyechayev, P. I. Krotov, A. A. Shtukenberg can be selected. After the Revolution M. E. Noinskii and his assistants V. A. Cherdyntsev and E. I. Tikhvinskaya have done a lot. They are all professors in the Kazan University. Amongst the Leningrad scientists B. K. Likharev has contributed much towards the study of the Permian deposits. The monograph by N. N. Forsha (1955) is especially valuable for the Ufian Suite and the Kazanian Stage. The investigations of A. N. Mazarovich—a professor at the University of Moscow—are important. On the subject of the Permian and Carboniferous of the North the paper by A. I. Zoricheva (1956) deserves attention. The Permian of the Baltic region is described by A. I. Vala (Vala et al., 1959).

Tatarian Stage. The continuing upward movements led to a complete drying out of the Kazanian sea. In its place originated a wide enclosed depression with a flat floor on which were situated numerous, both fresh-water and saline, lakes. On the surface of the plain under hot, dry, desert climatic conditions a complex of variable, rapidly changing, but basically uniform sediments was deposited. The uniformity of the complex was expressed in its small-scale variability. The rocks are bright, rapidly changing their colour, but in the main reddish, brownish and pink. Sandstones predominate and are at times coarse and badly sorted temporary streams deposits with lenses of gravel and conglomerate, while at times they are

FIG. 24. Tatarian strata on the river Vyatka. A typical landscape in the Kama district. *Photograph in the Geological Museum.*

fine or medium grained, well sorted unbedded or poorly bedded eolian wind deposits. In other instances the sandstones are medium or coarsely grained gravel-bearing, lensoid, current bedded deposits of rivers and temporary streams or fine grained finely and regularly laminated lacustrine deposits. There are numerous sandy, badly sorted muds, which are falsely or indistinctly and regularly-bedded deposits of temporary streams. Less frequently there are encountered regularly-bedded, gypsum-bearing or salt-bearing clays, with not infrequent lensoid beds of pure gypsum or rock-salt amongst them. These are the deposits of bitter-salt lakes. Even rarer are the pure, calcareous clays, marls and pure fine-grained limestone members, of up to 10–15m in thickness, which are the deposits of fresh-water lakes. These deposits laterally constantly change from one into the other, but collectively retain a uniform character (Figs. 24, 26).

The fauna is also characteristic. In the terrestrial facies it is represented by reptiles and amphibians. In lacustrine deposits fish, fresh-water lamellibranchs such as *Anthracosia*, ostracods and plant remains are quite common. In places plant remains are concentrated in large numbers forming coal and carbonaceous seams. In river deposits there are rare bones of herbivorous and carnivorous reptiles and rarer amphibians.

The total thickness of the Tatarian Stage is variable. It is, however, relatively small and does not exceed a few hundreds of metres, while its distribution is wide and peculiar. The Tatarian Stage is only developed in a huge enclosed depression which was previously occupied by the Kazanian sea. On maps the area of its distribution has the shape of a large rounded basin of irregular outline. The facial composition and the history of the Tatarian Stage leave no doubt that it represents a typical complex of deposits of a large desert. When the present author first advanced this point of view there were contradictions. It was asked: how could it be a desert if

FIG. 25.　Sandy, Kazanian limestones in the Orenburg region.
Photograph by V. D. Nalivkin.

at its centre there was a large fresh-water lake, and there were rivers crossing the 'desert', on the shores of which lived swamp-dwelling herbivorous reptiles (*Pariasauria*), which were destroyed by even larger carnivorous (*Inostrancevias*)? These objections were due to ignorance of what happens in the modern deserts. In the middle of the huge Gobi desert there are several large fresh-water lakes including the Lob-Nor lake. Deserts are traversed by large rivers such as the Amu Dar'ya and the Syr Dar'ya and their shore-lines have in places impenetrable growths of reeds reaching 4–5m in height and in which herds of wild boars live and which until recently were preyed on by tigers.

At present the Tatarian Stage is divided into two sub-stages and a large number of horizons of local significance (Table 12).

Vetluzhian Stage. Even before the October Revolution the well-known geologist S. N. Nikitin had included all the red bed formation lying above the Kazanian Stage into the Tatarian Stage. The investigations of the Soviet geologists and palaeontologists have shown that the upper part of this formation includes the Lower Triassic fauna (dinosaurs, fresh-water phyllopods and ostracods. A new flora appears in this upper part and in a number of provinces to the lower part was established. Consequently, the upper part of the Tatarian Stage was included in the Lower Trias under the term Vetluzhian Stage. Lithologically the Tatarian and the Vetluzhian stages are very near to each other and they are identical in distribution. They are a single formation bringing to the close the Palaeozoic cycle of sedimentation. The analogues of the Tatarian and Vetluzhian stages, also as red beds, are developed in the southern Baltic countries (p. 179).

FIG. 26. Exposures of Kazanian strata in the basin of the river Sok to the north-east of Kuibyshev. Hard bands are formed by beds of sandstones amongst clays. *Photograph by V. D. Nalivkin.*

MESOZOIC AND CAINOZOIC

On the basis of sedimentation the Mesozoic and Cainozoic can be divided into four epochs: (*i*) the Trias and the Lower Jurassic; (*ii*) Middle Jurassic, the Upper Jurassic and the Lower Cretaceous; (*iii*) The Upper Cretaceous and the Palaeogene; (*iv*) the Neogene and the Anthropogene (Quaternary deposits).

The Trias and the Lower Jurassic

The Trias and the Lower Jurassic, like the Silurian and the Lower Devonian, is an epoch of a major and general uplift of the Russian Platform. The platform consisted of an elevated region—an area of erosion. In the Lower Trias of the north and east of the Russian Platform there are developed continental, desert deposits of the Vetluzhian Stage, consisting of red and bright coloured beds, sandy, clayey, and pebbly deposits containing bones of Lower Triassic dinosaurs and fresh-water crustaceans. The thickness of the Vetluzhian Stage in places reaches considerable dimensions. In the Kirov Province it is 270m, in Aktyubinsk Province it is 200m and usually varies between 40 and 100m. The study of the lithology and biostratigraphy has allowed the division of the Vetluzhian Stage in these and other provinces into a series of suites and horizons.

Substantial biostratigraphical data have been obtained by studying the vertebrate bones, which are comparatively frequently encountered in sandstone members. They have enabled V. G. Ochev to subdivide the Vetluzhian Stage into three suites and to prove the absence of the Middle Trias.

The marine Lower Trias is found only in the Volga-Emba region. This is a thin member of dark and grey clay and marls with the long recognized fauna of the Mount Bogdo, near the lake Baskunchak, which has been separated into the Baskunchakian (Bogdinian) Suite (Fig. 27). Pale clays and marly rocks containing

FIG. 27. Mt. Bogdo near Lake Baskunchak. Dark rocks to the right are Permotriassic red beds, while pale rocks are marine Lower Trias pale and dark clays and marls. *Photograph by S. P. Rykov.*

Lower Triassic pelecypods have also been found in the Inder Mountains near the lake Inder in the lower reaches of the river Ural. The Lower Trias sea had penetrated into the Volga-Emba region from the Mediterranean Geosyncline (Mangyshlak), to the south.

Within the Russian Platform the Middle Triassic and Lower Jurassic marine deposits are unknown. The marine Lower Jurassic is developed in the Donyets Basin where alternating continental and marine deposits are a part of it. In the Volga-Emba Province during this epoch huge thicknesses of sandy clayey continental sediments many hundred metres thick, bright at the base and brown and dark above, have been deposited. Similarly thick formations, bright at the bottom, dark and brown above, and not infrequently coal-bearing, were deposited in Mangyshlak and Ustyurt (Tuar-Kyr). These are the products of erosion of the Russian Platform transported from its highlands and deposited over vast near-shore plains. In age they belong to the whole interval beginning with the Middle Trias and ending with the Middle Jurassic. Certain interesting data on the Lower Trias of the central and northern provinces have been brought forth by the work of D. L. Frukht (1959). She has described a series of sections of the normally developed Vetluzhian sediments. These are brown and red, brightly coloured clays and sandstones with rapid lateral variations and overall thickness of 100–120 metres.

In addition in the region to the north of Gorki there are developed two peculiar formations: the 'Crumpled' and the Conglomeratic-Brecciated. Both are developed in neighbouring regions to the south of the normal Vetluzhian Stage, replace it, and are covered by the horizontally lying Jurassic and underlain by the Upper Tatarian and older formations. The Conglomeratic-Brecciated Formation consists of brown, cherry-red and dark clays with fragments of gneisses and other rocks of varying ages (Cambrian, Devonian, Lower Permian). Its thickness varies up to 180m. It is considered to be an accumulation of the fragments of the rocks, that formed a neighbouring elevated ridge, which basically consisted of Precambrian gneisses.

The 'Crumpled' Formation is even more peculiar. It is exposed in the region of the villages Pychezh and Katunka and was previously considered a Quaternary deposit which infilled an overdeepened basin. According to D. L. Frukht (1959) the Jurassic rocks horizontally overlie this formation, which can be demonstrated in a series of boreholes over an area of some 4000km². The 'Crumpled' Formation consists principally of sandy-clayey, bedded rocks of Tatarian Stage, very strongly deformed and crumpled—hence the name. Amongst the rocks of the Tatarian Stage there are fragments and boulders of limestones of the Upper Carboniferous, Lower Permian and Kazanian stages and also blocks of gypsum and anhydrite of evidently Lower Permian age. Formerly the 'Crumpled' Formation was considered as a glacial Quaternary deposit. If it is covered by the Jurassic then its Quaternary age is obsolete and its relegation to the Vetluzhian becomes more appropriate, although the conditions of its formation remain an absolute problem (see p. 149).

The Middle and Upper Jurassic and the Lower Cretaceous

The Middle and Upper Jurassic and the Lower Cretaceous is an epoch of the most extensive Mesozoic transgression, which spread over the greater part of the platform from the west. The marine basin, thus formed, connected the Northern Ocean with the southern Tethys and had the form of an extensive intercontinental sea or a channel. Such seas do not exist at the present time.

SUBDIVISIONS. The Jurassic and Lower Cretaceous sections of the Russian

RUSSIAN PLATFORM

TABLE 13.

Jurassic and Lower Cretaceous Successions of the Russian Platform and the Dnyepr-Donyets Depression

Stages		Sovyetsk (E. M. Lyutkevich, 1957)	Mogilev (Korenevskaya) (N. T. Sazonov, 1957)	Chernigov (I. Yu. Lapkin and B. P. Sterlin, 1957)	Reiserovo and the south of the Dnyepr Depression (N. T. Sazonov, 1957)
	Albian	Basal strata of the Upper Cretaceous transgression.	Basal sands with pebbles at the bottom.	Basal sandstones of the Upper Cretaceous transgression (38m).	Cenomanian and Upper Albian basal sandstones with pebbles (65m).
	Aptian	Missing.	Missing.	Missing.	Missing.
	Hauterivian and Barremian				
	Valanginian				
	Volgian			Continental variegated clays with beds of sand (43m).	Red muddy sands and clays of continental origin (22m).
	Kimmeridgian			Lower Kimmeridgian clays with a marine fauna (20m).	
	Oxfordian	Black siltstones and clays with a rich marine fauna (70m).	Above, clays; below, bedded limestone. A marine fauna is present (37m).	Greenish clays with ammonites and foraminiferids; marls at the bottom (70m).	Calcareous clays with sequences of limestones and sandstones and a marine fauna (145m).
	Callovian	Above, clays with a bed of oolitic marl; below, black clays with beds of sand. Clays are carbonaceous (35m).	Marine, calcareous clays (43m). Continental sandstones with beds of marl and coal seams (44m).	Grey, bluish clays with ammonites. Basal sandstones at the bottom (30m).	Marine, grey clays (19m). Continental brown clays with brown coal seams and, at the bottom, sandstones (45m).
	Middle and Lower Jurassic	Black clays; barren and with beds of siltstones (28m).	Bathonian and Bajocian continental sands and clays with beds of brown coals and a flora. At the bottom a basal sandstone (90m).	Marine Bathonian and Bajocian clays (51m). Underlain by basal sandstones (20m).	Bathonian-Upper Bajocian, marine clays and sands (55m). Lower Bajocian-Upper Liassic continental clays and sands with coal seams and a flora (53m).

TABLE 13—continued

Kupyansk (I. Yu. Lapkin and B. P. Sterlin, 1957)	Donbas	Saratov (Yelshanka) (N. T. Sazonov, 1957)	Ul'yanovsk (N. T. Sazonov, 1957)	Penza, Prudy (M. F. Filippova, 1957; N. T. Sazonov, 1957)
Basal sandstones with grey mud balls (18m).	Basal, Upper Albian sandstones.	Grey, sandy clays (17m). A bed of phosphorite (1.5m). Alternating black clays and sands of Albian and Aptian ages (About 90m).	Clays and sands with a phosphorite layer (16m).	Basal sands and clays (28m). Sandy clays (30m).
Missing.	Missing.		Interbedded, clays, siliceous marls and sandstones with a marine fauna (82m).	Clays with sequences of bituminous shales at the bottom (30m).
		Black clays with concretions and belemnites (30m).	Upper Barremian glauconitic sands and dark clays (44m). Lower Barremian and Hauterivian black, compacted clays with ammonites (*Simbirskites*) (66m).	Thick, dark clays with beds of siltstones (82m). Black, compacted clays with concretions and *Simbirskites* (23m).
		A condensed phosphorite-glauconite bed with *Aucella* (1.5–2m).	Glauconitic sands with phosphorites (0.8m).	Condensed phosphoritic bed (0.2–0.4m).
Variegated continental clays and sandstones (64m).	Barren, continental sandstones (Up to 60m).	A phosphorite glauconite bed (0.25m).	Upper Volgian glauconitic sands with phosphorites. Lower Volgian glauconitic sands with phosphorites. Below, clays and bituminous shales (16m).	Phosphoritic derived pebbles in glauconite sand or gypsiferous clays, with *Aucella* (0.5m).
Red and grey marly clays (24m).	White, oolitic limestones with *Nerinea* (About 20m).		Dark-grey, but at the top, pale-grey clays with Kimmeridgian ammonites (43m).	Grey marls and clays with a Kimmeridgian fauna at the top and an Oxfordian fauna at the bottom (26m).
Coralline and algal limestones (14m). Pale, oolitic and brachiopod-bearing; at the bottom, conglomeratic Oxfordian to Middle Callovian strata (34m).	Coralline and oolitic limestones. Marls and sandy limestones (40m).	Pale grey clays with ammonites (20m).	Dark-grey calcareous clays with pyrites (40m).	
Lower Callovian dark clays with brown coal seams (22m).	Calcareous sandstones with a marine fauna. Sandstones and clays with plant remains (60–70m).	Dark and pale Callovian clays with ammonites (60m). In the middle, oil-shales (5m).	Callovian sands (1.5m).	Above, dark grey clays and marls with ammonites (11m). Below, dark noncalcareous clays, silts and sands (21m).
Tuffogenous sandstones and carbonaceous clays (31m). Clays and sandstones with Bathonian and Bajocian ammonites (156m). Upper Lias clays and sands (28m).	Sandstones and clays with marine faunas of Bathonian, Bajocian, Aalenian and Upper Lias ages. Sandstones with a flora (150–550m). Middle and Lower Lias variegated beds.	Dark, Bajocian clays and sands (35m). Upper Bajocian, thick clays with a marine fauna and sandy beds at the top (130m). Basal, land sands and clays (15m).	Basal sandstones and clays (14m).	Bathonian grey cross-bedded sands (17m). Bajocian dark sands and clays, with ammonites at the top and a flora at the bottom (16m).

TABLE 13—continued

Stages	Moscow (N. T. Sazonov, 1957)	The river Vyatka (E. M. Lyutkevich, 1957)	Pechora
Albian	Upper Albian not found. Middle Albian consists of sands with ammonites.	Micaceous sands with beds of black clays. (Up to 30m).	Not proved. Sands with plant remains and in places concretions with a Lower Aptian fauna.
Aptian	White quartz sandstones with a flora (Klinian Albian and Aptian).		
Hauterivian and Barremian	Ferruginous sands, and to the north dark clays with *Simbirskites*.	Hard, Hauterivian-Barremian micaceous clays. At the top, grey and brown sands. At the bottom, the main (1m) phosphorite bed (35m).	Dark clays with concretions. At the top, Barremian *Simbirskites*; at the bottom, Hauterivian *Aucella*.
Valanginian	Sands and phosphorites with *Polyptychites*. The Ryazan Horizon of glauconitic sands with phosphorites and *Aucella*.	Oolitic, ferruginous sandstones. At the top, dark clays. There are phosphorites and *Aucella* (2.5–3m).	Phosphoritic sands with *Polyptychites*. Sandy clays with ammonites.
Volgian	Upper Volgian—glauconitic sands with phosphorites (4.5m). Lower Volgian—dark clay with virgatites (6.5m). Sands (1m).	Upper Volgian sands, clays and oil shales (25m). Grey clays, glauconitic sands and oil shales (Up to 20m).	Upper Volgian unrecognized. Dark clays with oil shales and *Aucella*.
Kimmeridgian	A condensed bed of glauconitic sand with phosphorites (3m).	A condensed bed of clays, glauconitic sand and phosphorites.	A condensed horizon. Clays with sandstone bands, with Kimmeridgian and Oxfordian marine faunas.
Oxfordian	Upper Oxfordian-black clays with ammonites (10m). Lower Oxfordian—grey clays with phosphorites and *Cardioceras* (15m).	Marls and clays with phosphorites and a marine fauna (15–20m).	
Callovian	Upper Callovian—clays with oolites and ammonites (1–2m). Middle Callovian—sands with ammonites. Basal conglomerates and sands of Lower Callovian (0–9m), Bathonian (0–27m). In places both are partly or completely missing.	Dark clays, sandy clays and sands. At the bottom, sands and pebble beds. (Up to 50m).	Upper and Middle Callovian clays with a marine fauna. Lower Callovian sands with ammonites and beds of clays.
Middle and Lower Jurassic	Not found.	Not found.	Basal, barren sands.

TABLE 13—continued

Novouzensk (Ya. S. Eventov, 1956)	Astrakhan (Ya. S. Eventov, 1956)	Artezian, the lower Kuma (Ya. S. Eventov, 1956)	Southern Emba, Aznagul (G. Aizenshtadt, 1956)
Above, dark clays; below, dark clays and sandstones with plant remains (284m).	Sandy clays. Towards the bottom, sandstones and gravel stones (115m).	At the top, coarse sandstones; in the middle, shales; at the bottom, sandstones (290m).	Sandstones with beds of clays and plant remains (426m).
Dark compacted clays, in places sandy and with beds of sandstones. A marine fauna is present (179m).	Dark-grey silty clays (25m).	Alternating dark clays and sandstones, with a marine fauna and a basal conglomerate (163m).	Alternating suites of dark clays, sands and sandstones (144m).
Grey micaceous clays, sometimes sandy, with beds of siltstones and containing foraminiferids (121m).	Dark-grey sandstones at the top of the preceding sequence.	Missing.	Variegated clays and sands with plant remains (360m). Sandy-clayey dark suite (11m). Clays with lamellibranchs (50m). Greenish, fossiliferous sands and limestones and marls (39m).
Dark-grey muddy and glauconitic sandstones, alternating with muddy limestones. A marine *Aucella* fauna is present (119m).	Main part of the sequence is a formation of dark-grey clays with beds of sandstones (91m).		
Upper Volgian unrecognized. Top, Lower Volgian limestones and clays (142m), passing down into clays (95m) with beds of oil shale.	No Upper Volgian. Lower Volgian clays, sandstones and marls with *aucella* (49m).	Provisionally Upper Jurassic. At the top, a sequence of grey oolitic limestones passing down into grey dolomites with plant remains. At the bottom, grey quartz sandstones. No fauna found (135m).	No Upper Volgian. Lower Volgian limestones and marls with *Virgatites* (128m).
Unrecognized, but is probably represented by a condensed bed of glauconitic sand with phosphorites (0.5–1.0m).	A condensed bed of glauconitic sands with phosphorites.		Unrecognized.
Dark-grey micaceous clays with pebbles of phosphorites and a marine fauna (102m).			Only Lower Oxfordian present, consisting of pale marls and dark clays and limestones. Marine faunas found (22m).
Upper and Lower Callovian dark clays. Middle Callovian marls with a marine fauna (80m).	Dark clays with beds of sandstones and limestones and a marine fauna. (61m).		Upper Callovian—dark clays (24m). Middle Callovian—greenish siltstones (27m). Lower Callovian—clays with plants (60m).
Dark Bathonian clays with marine fossils (137m). Grey Bajocian sandstones and shales with a flora (164m).	Bathonian clays and Bajocian sands and sandstones. Alternating flora and fauna-bearing carbonaceous sequences (132m).	Dark-grey carbonaceous shales and grey sandstones, with plant remains. 540m have been penetrated.	Alternating clay and sand sequences, with plant remains. Coals (497m). Liassic (?) sandstones (134m).

Platform are classical and generally known. They are given in Table 13 together with the additional new borehole sections and the sections of the Dnyepr, the Donyets and the Astrakhan depressions (the latter for comparison). A comprehensive paper on the Lower Cretaceous of the central provinces of the platform is given by I. G. Sazonova (1958).

LITHOLOGY AND FACIES.   At the bottom of the Jurassic there is very often a basal member consisting of sandstones and clays, and at its base, occasional boulder and pebble conglomerates. The member is generally barren, but rarely contains plant remains and represents the deposits of near-shore plains. Its brown or grey colour, in contradistinction to the Permian and Triassic red beds, indicates a wet climate and high rainfall.

The deposits of the Jurassic and Lower Cretaceous seas differ sharply and are, it can be claimed, diametrically opposite to the preceding deposits of the Upper Palaeozoic seas. Whereas the latter mainly consist of carbonates, are not infrequently represented by thick pure limestones and dolomites and are characterized by their negligible amounts of terrigenous matter, the deposits of the Jurassic-Cretaceous sea on the contrary consist almost entirely of terrigenous material. This is mainly argillaceous, while carbonates are poorly developed or occur in small quantities forming thin marly beds and the pure limestones and dolomites are absent. The total thickness of these sediments is considerably less (Fig. 28). Huge quantities of fine grained terrigenous matter can appear only as a result of its transportation from the land by rivers. The intensified development of the river system can be explained by sharp changes of climate, which is hot, dry, lacking in humidity in the Upper Palaeozoic and cool, humid rainy in the Jurassic and Lower Cretaceous.

The widespread distribution of rather thick (up to 20–40m), pure, plastic, dark,

FIG. 28. Upper Jurassic strata in the Kamushka quarry near Moscow. At the bottom are pale and dark bedded Upper Carboniferous limestones, which are overlain by dark, Upper Jurassic (Oxfordian) clays. At the bottom of the Upper Jurassic sequence there is a thin horizon of basal Middle Callovian sandstones and clays, which are not seen on the photograph, but are exposed where geologists are standing.

black and bluish clays with an ammonite fauna point to a similarly widespread distribution of stationary, sometimes stagnant, waters. Under the conditions of an almost enclosed sea or straits such conditions are natural.

In some epochs, for example those of the Volgian stages a direct and wide connection between the Northern and the Southern oceans was established. Such a connection always initiates the appearance of fast, strong currents, which quickly cause considerable changes in the composition of sediments. The so-called 'condensed beds' appear, which consist of thin (0·5–1·5m) beds of peculiar coarse glauconite-phosphorite rocks consisting of phosphorite nodules and grains with a greater or lesser admixture of glauconite. Sometimes the whole deposit consists of glauconite sand. There are always considerable amounts of animal shells, which are commonly rolled and broken. These beds are called 'condensed' since in other regions clay members, which are tens of times thicker (20–25m) correspond to them. The accumulations of phosphorite concretions in certain regions are of commercial significance.

It is interesting that in the same parts of the stratigraphic column, i.e. almost together with the phosphorite-glauconite beds, there are encountered beds of oil shales, which require diametrically opposite conditions, namely of stagnant basins with immobile water. The study of the oil shales has shown that the stagnant basins in which they were deposited were situated at a distance from the seashore and represented depressions on the sea bottom. They were, therefore, called the black mud depressions or halistases. These depressions were filled by algae and inhabited by a rich characteristic fauna of principally ammonites and lamellibranchs. In their origin and the presence of large amounts of organic matter the Volgian oil shales resemble the Devonian Domanik. In a number of regions the oil shales are the object of industrial development.

The Jurassic and the Lower Cretaceous sections not infrequently show breaks in sedimentation. Previously these breaks were explained in terms of tectonic movements, the elevation of the sea bottom and the formation of land. Nowadays such breaks are explained in terms of erosive and transporting activity of strong bottom currents. Much more rarely the breaks are in fact connected with the oscillations of the shoreline or the formation of islands.

In the upper parts of the Lower Cretaceous successions there are developed mantles of clayey-sandy members with occasional, local plant remains. These complete the marine series of deposits, being the sediments of a near-shore platform, which was formed after the withdrawal of the sea. The character of such near-shore plains can be best judged from the peculiarities of the present-day Cis-Caspian Depression. A peculiar facies which is notable by its wide distribution is termed the *Aucella* Beds. This depositional member varies from clayey to coarse clastic sediments associated with the Jurassic-Cretaceous boundary and characterized by the abundance of a peculiar group of pelecypods (*Aucella*) and ammonites (*Polyptychites* and *Virgatites*). The upper part of the beds is especially typical and is included into the lower parts of the Lower Cretaceous, the Ryazan Horizon and Valanginian Stage—but the lower part of the beds includes the Volgian stages. The *Aucella* Beds are developed in England, the Scandinavian shores, the Russian Platform, over the whole of the Arctic and Eastern Siberia up to the river Amur. It is considered as a typical deposit of a northern, cold, boreal ocean.

The Upper Jurassic and Lower Cretaceous sediments constitute an enigma of the Jurassic and Lower Cretaceous stratigraphy of the Russian Platform. The seas of these epochs are studied in detail and their outlines are shown on the map. The seas, however are not suspended in the air, they are surrounded by shorelines and, under platform conditions, by widespread near-shore plains. On these plains there should have been deposited sediments conditioned by the humid, rainy climate and consisting of brown, dark and grey sandy-clayey often carbonaceous and coal-bearing formations. These types of deposits are not present in our stratigraphic column either because they have been eroded away or because we have not as yet learned to distinguish them. Although it is difficult to say which reason is right, it is likely that the second is more probable.

PALAEOGEOGRAPHY. During the Lower Jurassic epoch all the Russian Platform constituted an elevated continent. At its southern edge it had a wide near-shore plain. Under the tropical conditions of this plain, along its stretch from the Donyets Basin to Ustyurt, uniformly coloured sandy-clayey formations were being deposited; while in the marshes and lagoons huge masses of vegetation remains —the future coals—were accumulating. Within the Donyets basin this plain was at times invaded by the sea which deposited clayey-sandy rocks with ammonites.

During the Middle Jurassic epoch the eastern part of the platform began to get depressed and the sea penetrated further from the south and less from the north. It is interesting that in the south the sea is confined to the Vale of Volga and in the north to the Vale of Pechora, as if already outlining these valleys in Mesozoic times. Further to the north and west in a number of regions a near-shore plain was formed, over which basal pebble conglomerates were laid down, which during the Callovian were succeeded by marine deposits.

During the Callovian-Kimmeridgian epoch the depression of the platform and the corresponding area of the sea reaches a large size. One of its important features was a large shallow embayment, which penetrated far to the west reaching the Baltic Sea. Its deposits are developed in the Baltic area. During the Volgian epoch the sea retracted considerably and the large western embayment disappeared. The platform underwent several elevations and depressions, which caused fast and considerable oscillations of the shoreline and corresponding changes in facies. It is possible that in part these oscillations are only apparent, while the breaks which substantiate them have, as pointed out above, been caused by strong bottom currents. During the Lower Cretaceous epoch the sea retained approximately the same outlines as in the Volgian. In the Albian and Aptian the northern part of the platform rose and the sea changed into a near-shore plain over which the final sandy-clayey formations with plant remains were deposited to terminate a rhythm of sedimentation.

The Upper Cretaceous and the Palaeogene

During the Upper Cretaceous and the Palaeogene the sea sharply changed its outlines and became elongated east–west instead of north–south. The composition of the sediments changed as well and the quantity of carbonates increased sharply. Characteristic sediments, formerly not found, such as chalk and gaizes appeared.

DISTRIBUTION. The Upper Cretaceous deposits are developed over all the southern part of the platform, and in the south merged into the Upper Cretaceous of the Mediterranean Geosyncline. Similarly, they pass directly into the contemporaneous deposits of the Western Europe to the west and of Central Asia to the south-west. They have been preserved over a small area in the central provinces, to the north-east of Moscow (Fig. 29). The Palaeogene is found together with the Upper Cretaceous and is closely associated with it. Only in the central provinces is it almost entirely absent. The Upper Cretaceous and Palaeogene are deposits of one and the same sea and are parts of a single macrorhythm of sedimentation.

Upper Cretaceous

SUBDIVISION. The classical Upper Cretaceous sections of the Russian Platform are extensively described in text-books on historical geology. Some of them and a series of new sections from standard boreholes are shown in Table 14.

LITHOLOGY AND FACIES. The study of the lithology and facies of the Upper Cretaceous deposits was greatly influenced by a monograph (1910) written by A. D. Arkhangel'skii. This monograph was in its time a record of a highly original investigation. It was the first to make the principal generalized assessment of facies distribution, namely the predominance of terrigenous sandy near-shore deposits in the north, and of fine carbonate marls, chalk and clays in the south, away from the shore. The facies distribution (and subdivisions into epochs) has been studied on the basis of modern data and in greater detail by V. N. Sobolevskaya (1951), and for

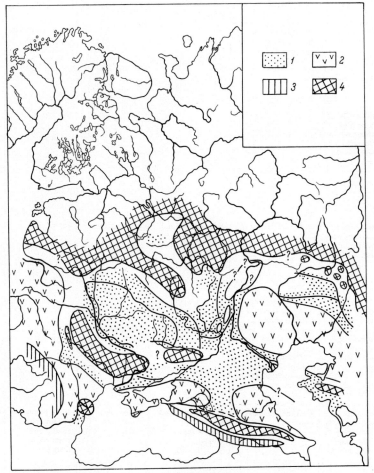

FIG. 29. The Cenomanian palaeogeography of the Russian Platform. According to V. N. Sobolevskaya (1951).

1 — quartz and glauconitic sands; 2 — clays and marls; 3 — flysch deposits; 4 — land subjected to erosion.

the central provinces by O. V. Flerova and A. D. Gurova (1958). Their lithological-facial sketch-maps provide substantial data for palaeogeographic purposes. These maps clearly show the distribution of the main sedimentary types. The sub-Moscow embayment, the large islands over the Ukrainian crystalline massif and the Hercynian-Cimmerian fold structures of the Donyets Ridge should be pointed out.

TABLE 14.

Upper Cretaceous Successions in the south of the Russian Platform

Stage	Sovyetsk (E. M. Lyut-kevich, 1957)	Chernigov	Smeloye, Romny (I. Yu. Lapkin, 1957)	Kupyansk	Penza, Prudy (M. F. Filippova and others, 1957)	Ul'yanovsk (M. F. Filippova and others, 1957)
Danian	Unrecognized.	Unrecognized.	White chalk with a fauna ranging Campanian to Maastrichtian (151m).	White chalk with a Maas-trichtian-Cam-panian fauna (240m).	Chalk-like marls (46m).	Unrecognized.
Maastrichtian	White chalk (20m).	White chalk with Maastrichtian, Cam-panian and San-tonian faunas. About 70m thick.			Marls and white chalk (31m).	Marls and white chalk (11m).
Campanian	White marls with beds of sandstones and containing foraminiferids (35m).					
Santonian	Glauconitic sandstones with beds of marl (22m).		White chalk and chalk-like marls with faunas of all stages (300m).	White chalk (238m).		Above, marls; below, greenish clays (33m).
Turonian-Coniacian	Dark-grey and greenish clays with beds of sandstones and foraminiferids (28m).	Chalk-like marls with faunas of both stages (About 70m thick). The total thick-ness of Turonian to Maastrichtian strata is 144m.		White chalk with beds of marls (79m).		Sandy clay (5m).
Cenomanian	Glauconitic sands with beds of dark clays and a horizon of phosphorites (17m).	Glauconitic sands (4m).	Glauconitic sand-stones with a marine fauna (48m).	Glauconitic sandstones (4m).	Basal formation of clays and sands, with plant remains (28m).	Basal formation of clays and sands (16m).
Albian	Upper Albian—sands and sand-stones (18m).	Continental basal sands (38m).	Basal sands and clays with lignite (24m).	Basal sand-stones (18m).		

An important addition to V. N. Sobolevskaya's ideas should be made. It is the fact that the Upper Cretaceous transgressions began not in the Cenomanian, but in the Upper Albian as a spread of continental sands, gravels and clays with plant remains. This basal formation is not infrequently transgressive on the Jurassic and even the Palaeozoic, as for instance in the Donyets-Dnyepr Depression and the sub-Caucasian plain. Upwards it gradually merges into Cenomanian sediments and the boundary between them is often arbitrary.

TABLE 14—continued

Kamyshin	Novouzensk (Ya. S. Eventov, 1956)	Novo-Kazanka (Malyi Uzen') (Ya. S. Eventov, 1956)	Astrakhan (Ya. S. Eventov, 1956)	Artezian, Lower Kuma (Ya. S. Eventov, 1956)	Southern Emba Aznagul (G. Aizenshtadt, 1956)
Unrecognized.	Gaizes and clays (57m).	Pale limestones (36).	Missing.	Clays, marls and limestones (33m).	Above, red marls; below, chalk and white marls (14m).
Above, sands; below, clays and marls with *Belemnitella*.	Above, muddy limestones; below, dark clays. Foraminifera are present (284m).	White limestones with a greenish tinge.	Unrecognized.	Limestones with a bed of clay near the top (77m).	White chalk with bands of green marl (123m).
Glauconitic sandstones, gaizes, black clays with *Belemnitella*.	Grey and pale-grey limestones with bands of clays and foraminiferids (232m).		White chalk-like limestones and marls with Foraminifera of Santonian and Campanian ages, at the top. At the bottom, Turonian inoceramids are found (251m).	Above, marls; below, limestones (102m).	White chalk, alternating with beds of green marl (112m).
Above, black clay and gaizes; below, marl with sponges and a phosphorite bed.	Above, sandstones and limestones; below, dark micaceous clays (99m).			Pale marls and limestones with Foraminifera. The rocks are provisionally considered as Santonian-Cenomanian. At the base there are conglomerates (42m).	White chalk and marls (34m).
White chalk-like marl with *Inoceramus*. White chalk and marls. At the bottom, sands with phosphorite pebbles.	Sandstones and alternating clays, with rare foraminiferids. Provisionally attributed to the Coniacian and the Turonian (21m).				Dark, greenish, hard silty marls (83m).
Micaceous sands with phosphorites and a marine fauna.	A part of Albian clays may be Cenomanian.	Unpenetrated.	Black micaceous clays with plant remains (107m).		Sandstones, dark clays and marls (66m).
Sands and sandy clays.	Dark clays with beds of sandstone (92m).	Unpenetrated.	Dark-grey clays with a marine fauna. Below are sandstones and gravelites (115m).	Coarse sandstones.	Thick sandstone sequence with clay members and plant remains (266m).

It should be pointed out that faunal studies by specialists allow a more detailed subdivision of successions (Table 14). In the Donyets Basin the Cenomanian and the Campanian, on the basis of foraminifera are divided into three zones.

The Cenomanian deposits sharply differ from the succeeding formations by a vast distribution of sandy facies, composed sometimes only of quartz sands, sometimes of quartz sands with glauconite and phosphorites.

In the Volga-Emba Basin they are replaced by clays, and in the west by marls and clays (Fig. 29). This feature can be explained by the intensified transport of sand from the northern part of the platform which rose during the Cenomanian epoch. In places the concentration of concretions and grains of phosphorite acquires an economic significance since by coalescing together they form entire phosphorite beds known as the 'Kursk lode'.

The Turonian and the Coniacian deposits (Fig. 30) are characterized by the vast distribution of white chalk, which covers the whole southern part of the Russian Platform, reaching a thickness of 750m near Kharkov. To the north the chalk is enriched with terrigenous matter and in the east, nearer to the Volga it is sandy, while in the west nearer to the upper reaches of the Dnyepr it is clayey and transitional into marl and siliceous clay (opoka). Lastly in the far north, near the former shore-line, sands and clays (Fig. 29) are deposited.

The condition of formation of the white chalk have been the subject of prolonged discourses. Some geologists considered it a deep-water sediment—an analogue of a deep-water *Globigerina* ooze; while others thought it to be a shelf sediment produced there and on the continental slope. At present the majority of investigators accept the latter point of view and chalk is considered a pseudoabyssal deposit. This is

FIG. 30. Upper Cretaceous strata on the Volga at Nizhnyaya Barmovka, to the south of Kamyshin. At the bottom lie chalky marls with *Inoceramus* (Turonion-Comiacian). At the top lie interbedded black clays and pale Santonian gaizes. *Photograph by N. S. Morozova.*

supported by the presence of large numbers of huge and thick-walled *Inocerami* and oysters in the chalk.

During the succeeding Sanatonian, Campanian and Maastrichtian epochs the quantity of the carbonate decreases. The white chalk continued in the west, but over the rest of the area marls were deposited. In the north, as before, clays, gaizes and sands were laid down. The Danian Stage of the Russian Platform provokes arguments. It is sometimes held that it is absent and that instead of it at the boundary between the Cretaceous and the Palaeogene in a number of places glauconitic sands with phosphorites indicate a break. P. L. Bezrukov (1936) shows a wide distribution of the Danian deposits found where the Maastrichtian is present and is similar to it in composition. The proof of the widespread development of the Danian Stage lessens considerably the importance of the break, formerly postulated at the boundary between the Cretaceous and the Palaeogene. In particular this is true of the eastern half of the platform where the deep-water Danian facies is directly succeeded by the deep-water Palaeocene. The glauconitic sands with phosphatic pebbles, which are developed at the boundary, were produced by bottom currents and do not necessitate the formation of land. The absence of a break in the central parts of the sea does not exclude movements of its shore-lines. In the north the sea receded to the south and the embayment, which in the Upper Cretaceous existed to the north-east of Moscow, disappeared; although the embayment along the Vale of the Volga remained and in the north almost reached the mouth of the Kama. Basically the outlines and characteristics of the Palaeogene seas were the same as those of the Upper Cretaceous seas.

The Palaeogene

SUBDIVISIONS. The succession and subdivisions of the Palaeogene are widely known, but considerable changes and additions have been introduced on making them more detailed. This is especially true in the upper and lower part of the succession. Certain of the subdivisions have fallen out of use such as, for instance, the Saratov Stage, which is partly replaced by the Kamyshinian Stage. The succession on the Russian Platform proper and in the adjacent Donyets-Dnyepr and Astrakhan downwarps to the south are closely related and are included in a single table (15), which also shows the successions found in a number of deep, standard boreholes.

LITHOLOGY AND FAUNA. The lithological and facies composition of the Palaeogene deposits is variable and characteristic. Its basic characters are as follows:

1. A sharp decrease in the carbonate fraction in comparison with the Upper Chalk is observed. Marls and muddy limestones are comparatively rare and thin.

2. Large quantities of siliceous rocks, such as siliceous clays (opoki) and diatomites, and glauconite-phosphorite facies are present.

3. The variability of the marine facies, ranging from bathyal clays and glauconitic sands with single corals and *Pleurotomaria* to the beach deposits of Nikopol (oolitic, rich, manganese ores) is notable.

4. The occurrence of near-shore plains, represented by the sandy-clayey rocks with plant remains, is common. Not infrequently these rocks are coal-bearing and commercially valuable, as for instance is the Dnyepr brown coal basin of Buchakian age. Brown coals are also encountered in the Kharkov and Poltava stages. In places, for instance the Nikopol region, the coal-bearing formations are underlain by bauxitic rocks of Buchakian age.

The proximity of land in the north and south-west (the Ukrainian Shield) caused rapid and considerable facies variations from south to north. Conversely, in the east–west direction the facies are constant over large distances. The facies variations have not been yet adequately studied and there is no complete lithological and facies information available. For example, it is considered that the Kiev Stage consists of marine clays and marls. This is only true for the deepest parts of the sea. To the north in Byelorussia the Kiev marls and clays pass into shallow-water muddy and pure sands which are in places glauconitic. Further still the marine sands change into continental sandy-clayey and partly carbonaceous formations. A short list of lithological characters of the main subdivisions is given in Table 15.

The Palaeogene is widely distributed and represented by various facies complexes. The deepest water complex is found in the Southern Ukraine and Balashov Province where it is called the Sumskian Suite, and also in the Volgograd Province where it is known as the Syzran Stage. These gaizes, clays and glauconitic sands are conformable on the Upper Cretaceous. The thickness of the gaizes is 70–90m. The near-shore, partly terrestrial and partly marine coarse rocks (mainly sands) are called the Kanev Suite or Stage. In addition there is a series of facies occupying an intermediate position between the Sumskian and the Kanev suites.

The Lower and Middle Eocene is even more variable and brown coal formations are extensively developed in it. In the west a complex of facies is predominant and is called the Buchakian Stage. It is characterized by rapidly alternating types with large quantities of sands and sandstones (Fig. 31). To the east and south of the Volga several almost entirely clayey suites are present.

The Upper Eocene is more homogeneous. From Chermigov to Emba relatively deep-water clays and marls of the Kiev Stage are predominant. The Lower Oligocene differs again by its multiplicity of facies. In addition to marine deep-water (bathyal) glauconitic sands and clays, more shallow marine sands with phosphorites are widely developed. Both facies are grouped together in the Kharkov Stage. In the east the clays, under other names, are again dominant and coal-bearing formations are

FIG. 31.   Palaeogene sands and sandstones on the right shore of the Volga, near Aktinov in the Kamyshin region. *Photograph by S. P. Rykov.*

not infrequent. The ore-bearing, green Nikopol clays and deposits of oolitic manganese ores are important. The Upper Oligocene and Miocene in the west shows a gradual upward transition of marine sandy and clayey deposits into similar deposits, but with terrestrial and fresh-water fauna. The whole complex of facies is known as the Poltava Suite or Stage. To the south and east a peculiar Maikop Suite consisting of clays with beds of sand, and almost entirely unfossiliferous, is developed. It will be discussed in detail as a part of the Mediterranean Geosyncline, where it is widely developed.

The siliceous rocks, especially gaizes (siliceous clays) but also diatomites, are characteristic. Formerly the accumulation of siliceous rocks was considered to be dependent on siliceous skeletal remains of plants and animals such as diatomaceous algae, radiolaria and lithistids. In modern literature the deposition of siliceous rocks is related to the deposition of silica produced by volcanic action. For many siliceous rocks, such as Devonian agates and siliceous shales this is a correct explanation. In the Palaeogene, however, there is no volcanic activity to produce the gaizes or diatomites and the action of organisms is evident and unquestionable.

The Neogene, possibly in places Anthropogene, age of the Variegated Clays, which previously were considered as Oligocene, is now generally accepted. The Clays are distinguished as a stratigraphical subdivision.

The sediments, previously known as the Poltava Stage, were after a detailed study found to have different ages and origins from one place to another. In parts of the central region of the Dnyepr-Donyets Depression the Poltava sands gradationally succeed those of the Kharkov Stage and contain a marine fauna. Their Oligocene age is not in doubt. To the south of Kharkov, in the lower part of the Potlava sands a rich flora and fresh-water fish bones of Aquitainian age have been found, and it is known that the lower part of the Aquitainian is Upper Oligocene, while the upper part is Lower Miocene. In places continental facies of the Poltava Stage contain a Miocene flora, but the most remarkable is the discovery of the marine Middle Miocene fauna at a number of localities, while at the top even a Lower Sarmatian fauna is found. This proves the Miocene age of a larger part, and in places of the whole, of the variegated and pale sands belonging to the Poltava Stage. Thus, the Poltava Stage is represented by a formation (10–30m, sometimes up to 60m thick) of pale and variable sands with thin laminae, which towards the base develop brown and greenish clay members, and which contain in places a marine fauna, and in others flora and coals. These are typical deposits of a near-shore plain which may be overrun by the sea or be a marshy or sandy beach. The greater part of its deposits, and commonly all of them, are included in the Miocene, but in places their lower parts are definitely Oligocene. This is a boundary formation and its age is similar to the Aquitainian and the Maikop.

The Upper Cretaceous and Palaeogene seas have been well investigated. To the north of them there was land, about which we know very little. To this land belonged not only the northern half of the Russian Platform, but also a large area recently drowned under the Barents Sea. This is indicated by the remains of Palaeogene palms found in Spitzbergen. No characteristic features of this huge northern land with southern flora are known at present, since plant-bearing Palaeogene sands and clays are only found in a few widely separated localities (Baltic countries, Mezen), while the Upper Cretaceous sands and clays are entirely unknown. It is impossible to state how this should be explained. Even if it is accepted that a greater part of the land was a region of erosion and nothing was deposited on it then large near-shore plains still should have existed at the margin of the sea and sediments, which have not been found, should have been deposited over them.

PALAEOGEOGRAPHY. The distribution of the seas and of the principal Cenomanian sedimentary types is given in Fig. 29. During Santonian and Turonian-Coniacian times relatively deep regions (400m–600m depth) were formed at the southern part of the platform. They reflected the inflexion of the shelf and the upper part of the continental slope (Fig. 29). Similar depths continued in the Palaeogene. With this inflexion, where characteristically strong bottom currents often existed, there are often associated glauconitic sands with phosphorites and a deep-water fauna with *Pleurotomaria* and simple corals. A deep-sea belt is characteristic feature of the Upper Cretaceous–Palaeogene sea and is not found in the preceding intercontinental seas.

Neogene and Anthropogene

Formerly the subdivisions Primary, Secondary, Tertiary and Quaternary were generally employed. At present the Primary deposits are known as Palaeozoic, the Secondary as Mesozoic and the Tertiary and Quaternary as Cenozoic (Cainozoic). The terms 'Primary' and 'Secondary' have long been discarded, while the term 'Tertiary' is rarer and will fall out of use entirely since Palaeogene and Neogene are now considered as independent systems. The term 'Quaternary' is still very widely used. However, with the disappearance of the terms 'Primary', 'Secondary' and 'Tertiary' this is not sensible and remains as a relict of the past, which does not deserve preservation. The term 'Anthropogene' (the age of the Man), which is analogous to the Palaeogene and the Neogene, has been suggested instead of the Quaternary. Thus the Cenozoic can be subdivided into three systems: the Palaeogene, the Neogene and the Anthropogene. The term deserves a wide usage.

The Neogene and the Anthropogene of the Russian Platform are noticeable for the predominance of continental conditions. Neogene seas spread only over its southern edge. Only the Akchagylian sea penetrated from the south to its eastern part as far as the middle reaches of the Kama. The wide region of the Barents, Kara and the White seas was land.

During the Anthropogene the southern seas withdrew still further, but in the north a large Boreal (Northern) Transgression occurred. The Northern Sea advanced and receded several times. At these times the widest extent of these seas was a little larger than at present. This is proved by the occurrence of the Middle and Upper Anthropogenic deposits of sands and clays with marine fauna on the continent at some distance from the present-day sea-shore. The modern Barents, Kara and White seas should be considered as a part of the land mass, which has been temporarily lowered below the sea-level. This is obvious on the general geological maps which show that the depth of these seas does not exceed 400m and is usually much less.

DISTRIBUTION. The distribution of the Anthropogenic and the Pliocene deposits is vast. They are developed over the whole platform, covering other sediments. The cover has many gaps, although they are small, since they are only found where more ancient formations are exposed, which in the main happens along river valleys.

The distribution and variety of Anthropogenic deposits is so large that a special branch of geological sciences, Anthropogenic (Quaternary) Geology, has been allocated to them.

The Neogene sediments are closely associated with the Anthropogene deposits. The Miocene and the Pliocene are strongly eroded and have been preserved over comparatively small areas in the south of the platform, the north being an area of erosion. The distribution of the Miocene and the Pliocene is shown on regional maps. The distribution of the Anthropogene is not normally shown on geological maps unless it is of a very great importance and special maps are prepared for these deposits.

SUBDIVISIONS. The subdivisions of the Miocene and Pliocene are widely known and are shown on Table 16.

The subdivisions of the Anthropogene are numerous and diverse. Some are based on lithology, others on the origin, still others on the fauna and flora. It is widely agreed to divide the Anthropogene into four divisions: (1) The Recent; (2) The Upper Anthropogenic (the Neo-Anthropogenic); (3) The Middle Anthropogenic; (4) The Lower Anthropogenic (Palaeoanthropogenic). These subdivisions have been proposed by the major expert on the Anthropogene S. A. Yakovlev (1956, 1958) and have been adopted on geological maps. At the beginning of 1960 the Central Stratigraphical Committee decided to consider the Holocene and the Pleistocene as divisions. The Pleistocene is subdivided into three stages, while the Pleistocene and Holocene stages are further divided into horizons of which there is a different number from region to region reaching sometimes 16 to 18. The Holocene represents

TABLE 16. *Neogene Successions in the south of the Russian Platform*

Division		Stage	Southern Ukraine	Northern Cisazovia	Apsheronian Peninsula
Pliocene	Upper	Apsheronian Akchagylian	Fresh-water deposits Taman Beds	Continental deposits	Apsheron Akchagyl
	Middle	Kuyalnitskian Cimmeridian	Kuyalnitskian Beds Marine strata		Continental Productive Formation
	Lower		Continental deposits		Marine deposits
			Marine Lower Pontian	Marine Lower Pontian	
Miocene	Upper	Meotic Sarmatian	Meotis Sarmat	Meotis Sarmat	Diatomitic Beds
		Konkian	Konkian Beds	Konkian Beds	
		Karaganian	Karaganian Beds	Karaganian Beds	
	Middle	Chokrakian	Chokrakian Beds	Chokrakian Beds	
		Tarkhanian	Tomakovkian Beds	?	
	Lower	Sakaraulian	Marine Lower Miocene	?	Upper Maikop

the beginning of a new division which explains its small thickness, which does not exceed that of a stage or even horizon. In every one of the stages genetic types of deposits can be distinguished: glacial, fluvioglacial, alluvial, lacustrine, marine, eolian, volcanic, chemical, eluvial, deluvial, eluvial-deluvial and proluvial (ambiguous origin). The type of the deposits is put in front of the name of the subdivision (stage).

The problem of the boundary between the Anthropogenic and the Neogenic deposits is of some complexity. In the marine deposits of the Caspian Depression the boundary has usually been placed at the bottom of the Baku Stage, though some 25–30 years back it used to be placed at the top of this stage. However, at present there are good reasons why it should be traced at the base of the Apsheronian Stage. It has been shown that the Apsheronian deposits within the continent pass into marine terraces which in turn pass into glacial deposits. The inclusion of the Apsheronian Stage in the Anthropogene and in particular in its lower division was considered the latest achievement of Anthropogenic geology, yet it soon appeared that the Akchagylian marine deposits are as closely connected with the glacial deposits of the Caucasian Mountains as the Apsheronian. Thus the Akchagyl has also been included in the Anthropogene.

Later deposits associated with glaciation in the supra-Sarmatian formations of the southern slope of the Caucasus and Kopet-Dag, of Meotian and Pontian age, were established. This indicates that if the Anthropogene boundary is to be drawn on the basis of glaciation then it has to be placed at the base of the Pliocene. V. I. Gromov on the basis of the evidence of the appearance of first hominoids, horses, oxen and elephants arrived at the same conclusion. He suggested that the Anthropogene system should be divided into three divisions: the Pliocene, the Pleistocene and the Holocene. This suggestion deserves attention.

There are other points of view about the position of the boundary between the Neogene and the Anthropogene. Adherents to one extreme doctrine deny entirely the existence of the continental, Quaternary Glaciation. The Central Stratigraphic Committee and the Editorial Committee of Geological Maps of the U.S.S.R. drew the boundary at the base of the Baku Stage and its equivalents. Where within the continents boundaries are indistinct the sediments are mapped as the 'undifferentiated Lower Anthropogenic and the Upper Pliocene'.

LITHOLOGY AND FACIES. The lithology and the facies of the Neogene and Anthropogene are exceptionally variable. Amongst the marine deposits there is a rather full representation of intercontinental and epicontinental seas including the Northern, Boreal, and the Southern, Caspian, types. The Akchagylian deposits should especially be noted. This sea during the Upper Pliocene not only filled the whole Caspian Depression, but reached far into the platform along the valleys of the Volga, the Kama and the Byelaya as far as the river Ufa (Fig. 6), The Akchagylian deposits consist of sands and clays with marine fauna typified by *Cardium* (*C. dombra*) and *Mactra*. Not infrequently this marine fauna is succeeded by an equivalent fresh-water fauna of *Unio* and *Planorbis*.

Such faunas and deposits are described in a monograph by A. P. Pavlov, one of the foremost Russian palaeontologist-stratigraphers, and elsewhere (Kirsanov, 1957). The deposits underlying the marine Akchagylian are interesting. They are widespread along the valleys of the Volga, the Kama, the Vyatka and the Byelaya and have been encountered in numerous boreholes. They consist of alluvial, lacustrine and terrestrial deposits, of grey, dark, banded clays; grey, current-bedded sands and silts, of sandy clays and beds of gravel. Their thickness varies from 30–75m in the valley of the Vyatka, to 100–150m in the valley of the Belaya, and to 200–250m in the vale of the Kama. The fauna is fresh-water, but sometimes beds with a marine fauna of Akchagylian type are encountered. The deposits are called Lower Akchagylian, sub-Akchagylian and Kinelian. They represent the infilling of the ancient river valleys of the Palaeo-Kama, the Palaeo-Volga and the Palaeo-Byelaya with the sediments of an ingressive sea and its fresh-water bays similar to the Bay of Finland. Their most probable age is Lower Akchagylian.

The older Miocene and Palaeogene continental deposits are developed along the Urals between the Ufa and Orenburg. Their upper part forms a coal-bearing formation of the South Urals brown coal basin in which seams of coal reach up to 105m. Their succession is shown in Table 15 (Southern Bashkiria).

In the north of the Russian Platform there are well developed terraces of sand, gravel and clay with a northern, marine fauna which are the deposits of the Upper Anthropogenic Boreal Transgression. This transgression still remains as the Barents, the Kara and the White Sea (Fig. 5). Amongst the unusually variable, continental deposits there are many identical to the present-day terrestrial, lacustrine, lagoonal, marshy and fluvial deposits. There are, however, two complexes of facies belonging to the Anthropogene only. These are the deposits of the continental ice sheets and the associated marginal, fluvioglacial deposits and the loessic loams and loess. In the regions of entire glaciation the ground moraines—the boulder clay of variable composition and colour (Fig. 32)—is most widespread. Its small thickness of the order of a few metres is very different from the gigantic thickness of the ice sheet, which reached many hundreds or even thousands of metres. Where the ice front was stationary, belts of terminal moraines originated as accumulations and high hills made of clastic fragments. These moraines, one after another, stretch over many hundreds of kilometres and possess a characteristic morainic landscape. Water currents under the ice and on its surface produced their own associated features such as eskers and kames. To the south, from the edge of the ice sheet, stretches a wide area of loess and loessic loams (see: *The study of Facies* by the author, 1957). In the sections of glacial and interglacial deposits usually one layer of boulder clay is encountered; less frequently two layers and in isolated cases a third layer can be seen. The correlation between successions in different areas has not as yet been finalized and consequently there are still most diverse views about the number of glaciations and their stages. Some Quaternary geologists consider that only one glaciation, with several stages, has occurred, each stage representing an oscillation in the position of the edge of the ice sheet. Others accept five glaciations. More frequently three glaciations are accepted, but this point of view is not proven.

FIG. 32. An exposure of boulder clay.

The names of various glaciations also vary, S. A. Yakovlev (1956) does not accept geographical names and uses age terms (Table 17). The geographic terms, however, are most widely used and the names of the main glaciations are generally accepted, as for instance are the Dneypr glaciation (maximal), the Moscovian, the Valdaiian and the Okaian (oldest). Nevertheless, it should be borne in mind that disagreements exist. For instance the oldest glaciation is sometimes called the Okaian and sometimes the Likhvinian. The interglacial at the beginning of the middle division is called either Likhvinian or Penultimate, or Likhvinian-Dnyeprovian or the Middle Interglacial. Here an authoritative meeting must decide. A comparison can be made between the sections in the central provinces and the succession of A. I. Moskvitin (1956), who is a leading expert on the Anthropogene (Table 17). A comparison shows that they vary depending on the point of view of the author. Moskvitin considers the Okaian glaciation as Akchagylian, the Yakovlev as Apsheronian, while many geologists consider it as Lower Anthropogenic. The last age is accepted in geological maps.

The thicknesses of Anthropogenic deposits, discovered in standard boreholes, are as follows: in the sub-Timan Lower Omza 13m, Zelenets 32m; in the north-west, Vologda 52m, Lyubim 13m, Soligalich 6m, Konosha (to the north of Vologda) 84m; in the north-east, Sovyetsk 20m, Valdai 55m, Daugavpils 200m; in Byelorussia, Korenevskaya 50m, Minsk 130m. In the Central provinces the thickness is greater, but still does not exceed 20–30m being 10m at Chernigov and 24m at Petrovsk.

TABLE 17.   *Suggested Schemes of Sequences of Glaciation*

S. A. Yakovlev, 1956	A. I. Moskvitin, 1956	A. V. Smirnov and D. N. Utyekhim, 1946
Present-day post-glacial	Holocene	Holocene
Post-glacial 4th Newer glacial 4th Newer interglacial	Ostashkovian glaciation Mologo-Sheksinskian inter-glacial	Valdaiian glaciation Kalinin-Valdaiian inter-glacial epoch
3rd Newer glacial 3rd Newer interglacial 2nd Newer glacial 2nd Newer interglacial 1st Newer glacial 1st Newer interglacial	Main phase of Kalinin glaciation Upper Volgian interstadial First phase of Kalinin glaci-ation Mikulinskian warm inter-glacial	Kalinin glaciation Dnyepr-Kalinin interglacial epoch
Middle glacial Middle interglacial	Moscow glaciation Odintsov interglacial Dnyepr maximal glaciation Likhvinian interglacial	Moscow glaciation Dnyepr-Moscow interglacial epoch Dnyepr glaciation Likhvinian-Dnyepr inter-glacial epoch
2nd Older glacial 2nd Older interglacial 1st Older glacial 1st Older interglacial Ancient glaciation (Apsheronian) Pre-glacial	Apsheronian glaciation Pre-Apsheronian interglacial Akchagylian (Okaian) glaciation	Likhvinian glacial epoch Pre-glacial epoch Apsheron Akchagyl

These figures suggest that thicknesses of 50–80m are very large, 20–30m are large and 10–20m are usual. Table 17 shows the 10–15 subdivisions of the Anthropogene. In the central provinces ten epochs are recognized which reach 15–20m. Thus each epoch is 1·5–2m thick. Even if this calculation does not always correspond to reality the thickness of the basic subdivisions does not exceed 5–10m and is only rarely greater. Compared to other systems these thicknesses are much less and point to the short duration of the epochs. Absolute age determinations confirm this. Figures obtained for the duration of the Holocene are 10–20 thousand years, for the Upper Anthropogenic epoch 100–110 thousand years, for the Middle Anthropogenic 240 thousand years and for the Lower Anthropogenic 100–110 thousand years (Yakovlev 1956). The duration of other geological periods has been established to be on average about 30 million years, which is not comparable to those of the Anthropogenic period. Thus the Anthropogene should really be considered as an epoch, but owing to its individuality, its detailed investigation, economic importance and vast distribution, at the present time the Anthropogenic deposits are considered an independent system.

The Holocene, the last epoch of the Anthropogene, is known in greatest detail in the north-west (Baltic Countries), where it is manifested by the oscillations of lake and sea level. K. K. Orviku (1958) subdivides it as follows:

Late Holocene. Mydian epoch: marine; present climate; pine, birch, spruce.

Middle Holocene. Limnean epoch: vast fresh-water lake; spruce. Littorina epoch: climate warmer than the present; broad-leaved plants and lime-tree.

Early Holocene. Ancillic epoch: vast fresh-water lake; climate like present; pine, alder. Eolean epoch: sea-channel connecting oceans. Cold climate; birch predominates.

Upper Pleistocene. Baltic ice-lake epoch: cold climate; sub-arctic conditions.

Palaeogeography. The elevation of the Russian Platform at the end of the Oligocene and beginning of the Miocene ended with a complete regression of the sea. Its surface became a flat land, very similar to the present day; the principal difference consisting in its considerably larger size. In the north it spread up to the edge of the shelf. Spitzbergen, Franz-Josef Land and Severnaya Zemlya were uplands situated at the edge of the land.

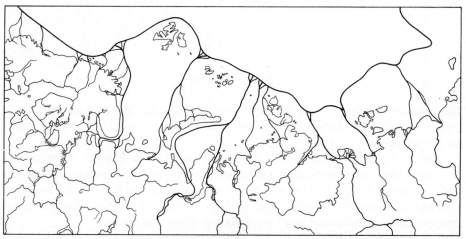

FIG. 33. River system in Upper Anthropogene times. The thick line indicates the shore.

Novaya Zemlya was a rather high, rocky range resembling the present-day Polar Urals. Along its eastern foothills was situated the river Ob', which reached the ocean to the east of the Franz Josef Land. Its valley is even now detected along the bottom of the Kara Sea. The Pechora and its tributaries flowing off the Novaya Zemlya range (Fig. 33) slowly meandered over a vast plain, which is now the bottom of the Barents Sea. The Severnaya Dvina, which flowed on the bottom of the White Sea, also flowed over the above-mentioned plain and bending towards the north-west skirted the Kola range, before entering the Atlantic Ocean.

At the end of the Pleistocene (Akchagylian epoch) the amount of water in the Caspian Basin increased sharply; the sea filled the entire depression (Fig. 5) and penetrated far to the north, the rest of the platform including its northern part remaining dry land. At the beginning of the Upper Anthropogenic epoch the northern part of the platform became depressed below the level of the ocean. It is probable that this depression was caused by the formation of a continental ice sheet, which depressed the Scandinavian Massif and other northern uplands. The sea then penetrated into the platform giving rise to the Boreal Transgression. At the time of the most intensive depression the Atlantic and Northern Ice Oceans were connected by wide, open channels. Then the general rise, which is still continuing, caused the formation of the Bay of Finland, the Ladoga and Onega lakes and the White Sea in place of the last (Mydian) channel. In the North the greater part of the transgessive sea is still present as the White, the Barents and the Kara seas.

The Miocene climate was warm and humid over the whole of the platform. Even in the north it was approximately the same as now in the Ukraine and along the north shores of the Black Sea. Everywhere there were forests, vast marshes, lakes and numerous rivers. The Pliocene climate gradually became colder; the the southern forests gave place to huge steppes with herds of fast *Hypparions* (relations of the present-day horses), of antelopes, camels, slowly moving *Mastodon* and extremely fast *Elasmotheriidae*. At the beginning of the Anthropogene the cold became most intense and with intervals remained so during the whole period of the Anthropogene. Together with a considerable rise of the northern part of the platform it was the cause of the formation of the continental ice sheet. The history of glaciation is as yet incompletely known, thus causing numerous discussions. The maximum extent of glaciation is shown in Fig. 5.

TECTONICS

STRUCTURES

The structures of the Russian Platform are easily divided into two groups: (1) the Precambrian geosynclinal and (2) the Proterozoic, Palaeozoic and Meso-Cenozoic of the platform type.

The Precambrian geosynclinal structures are characteristically complex and deformed, they have associated intrusive massifs and their rocks are strongly metamorphosed, while the Archeozoic ones have been recrystallized. The Proterozoic platform structures are simple, weakly deformed and have very few associated intrusive massifs, while their rocks are weakly metamorphosed.

The Palaeozoic structures are all of platform type—simple, weakly deformed, and without associated igneous massifs. They are commonly structures of the cover,

unaffected by tectonic movements. Their rocks are not metamorphosed, although they
are diagenised.

The Mesozoic and Cenozoic structures are the simplest and least deformed.
They commonly show their primary dip and their rocks are uncompacted and friable.

Precambrian Structures

The Precambrian structures occur on three tectonic levels, that of the Archeozoic,
Lower Proterozoic and Upper Proterozoic.

ARCHEOZOIC STRUCTURES.   In the Karelian A.S.S.R. the lower part of the
Archeozoic is made up of the Byelomorian Formation consisting of various gneisses
such as biotite, two-mica, and kyanite-gneisses, and amphibolites, also marbles, and

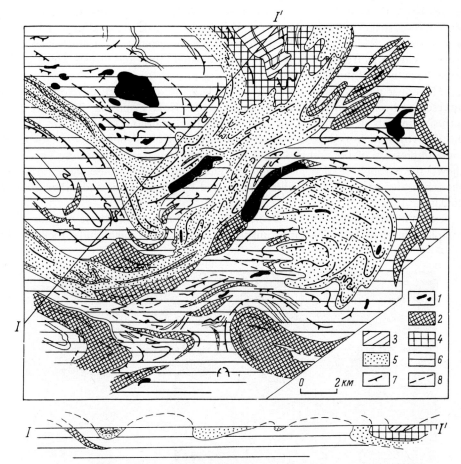

FIG. 34. A schematic geological map of one district of North Karelian
Byelomorides. According to V. L. Duk, K. A. Shurkin and N. I. Yaskevich
(1960).

1 — ultrabasic and basic rocks (drusites);   2 — oldest orthoamphibolites;
3 — hyperaluminous gneisses of the Kaitatundrovian Suite;   4 — plagioclase
biotite gneisses of the Knyazhegubian Suite;   5 — kyanite-garnet and garnet-
biotite gneisses of the Chupinskian Suite;   6 — biotitic gneisses with bands
of kyanite gneisses and bodies of orthoamphibolites of the Loukhian Suite;
7 — strike lines;   8 — faults.

micaceous and kyanite-bearing schists. The whole of this formation is strongly dislocated. There are mainly isoclinal, overturned, upright folds cut by thrust planes and later normal faults. Detailed mapping shows their extreme complexity (Fig. 34). The limbs of the large folds are complicated by minor folds and the latter by foliation (Fig. 35). The structures are complicated by syngenetic intrusions of granodiorites, associated with various migmatites (gneisses, lit par lit intrusions).

In the Kola Peninsula according to A. A. Polkanov (1937, 1947) the Lower Archeozoic consists of two gneiss complexes, the garnet-biotite and the biotite gneiss, both of which also include various migmatites. The extremely complicated structures form various fold belts such as Saamides, which were formed during the Saamian Orogeny in the middle of the Archeozoic. The Saamides form two belts: the northern in the north-west part of the peninsula where it enters from Norway and the southern forming the southern part of the peninsula in the north of Karelia, whence it stretches to Finland. It is a system of complex folds partly isoclinal with a more or less constant strike, which to the east of Kandalaksha is nearly east–west. The most complex minor folding is developed around Archean granites. At a much later time the Archean structures were broken by many normal faults and thrusts which determine the outlines of the White Sea.

The upper part of the Archeozoic in the U.S.S.R. is not well developed and is represented by the Ferruginous Formation of the Kola Peninsula. There are also intrusive massifs developed at the time of the Byelomorian Orogeny. The map of the Kirovogorian iron deposit (western part of the Kola Peninsula, Fig. 36) gives some impression of the complexity of the Upper Archeozoic structures. This map shows the strongly folded, ferruginous quartzites, which consequently rapidly change in thickness and are further complicated by granite veins and dykes of later diabases.

LOWER PROTEROZOIC STRUCTURES. The Lower Proterozoic structures are complex, have widely developed intrusive massifs, and high-grade rock meta-

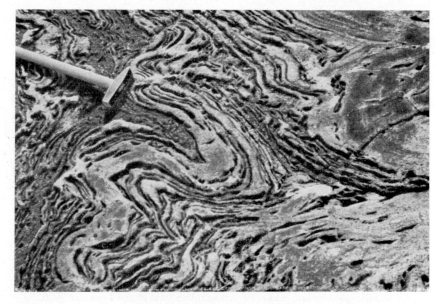

FIG. 35. Archeozoic gneisses of Karelia.

morphism giving rise in places to greenstones and amphibolites. In comparison with the Archeozoic, large, fundamental structures also broken by thrusts, the Lower Proterozoic structures have much more definite outlines and their trend is indicated on geological maps. The secondary and minor folding is much weaker than in the Archeozoic. Finally foliation is practically absent and is only observed in plastic clay slates. In the Karelian A.S.S.R. the Proterozoic is represented by the Karelian (Tatulian) Formation consisting of Segozerian and Onegian suites, which are separated by a fold-episode and an unconformity.

The Segozerian Suite, in the region of the lake Onega, consists of a continental

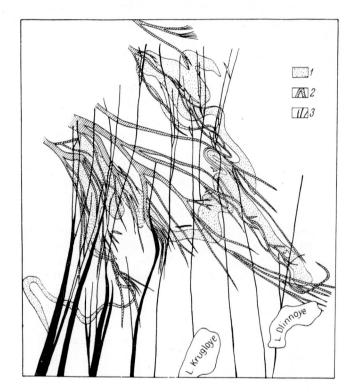

1. Ferruginous quartzites.

2. Pegmatites.

3. Diabases.
   White country rocks are gneisses.

FIG. 36.

Map of Krivoi Rog iron deposits. According to L. Ya. Kharitonov (1958) (E. A. Gedovius).

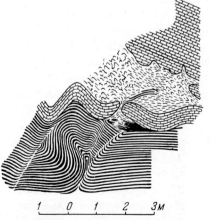

FIG. 37.

Minor folds in schungites occurring in a Schungian quarry. According to N. G. Sudovikov (1937).

1. Schungite.

2. Dolomite.

3. Siliceous schungitic shales.

formation of current-bedded sandstones and quartzites with intercalations of sericitic pelites and thick diabases and porphyrites overlain by a dolomite member. The total thickness is 700m.

The Onegian Suite consists at the base of limestones and dolomites, overlain by clayey, siliceous and shungitic slates of 300m in thickness, as well as predominant diabases. The slates are deformed into long, compressed, tight and sometimes dome-shaped folds trending towards the north-west and commonly broken up by thrust-faults. At Schunga the deposits of schungite show intensive minor folds of the schungite slates, the folds being disharmonic to the overlying limestones (Fig. 37).

Between lake Onega and lake Segozero the Lower Proterozoic is strongly metamorphosed into a green-schist facies and consists of actinolitic, sericitic and schloritic schists as well as dolomites, diabases and porphyrites, transected by granite intrusions. These rocks show complex folds, which are broken and thrust upon each other in several thrust slices.

The Lower Proterozoic structures of the Ukrainian Shield have been studied in detail in the region of Krivoi Rog since they are associated with one of the largest deposits of jaspilitic iron ore. Here within the rocks of the Gneissose (Bugian) Series, attributed to the Archeozoic, are wedged tight, long, narrow folds of the Lower Proterozoic Krivoi Rog Series.

The region has been described by N. P. Semenenko (1951, 1953), Yu. I. Polovinkina (1953) and A. P. Nikol'skii (1953). The Gneissose Series consists of gneisses, granites and green schists, which towards the south-west are very strongly deformed. The Krivoi Rog Series, which is deformed and broken up to the south-east, is divided into three parts. The lower division of basal arkoses, sandstones and conglomerates is uncomformable on various horizons of the more highly metamorphosed Gneissose Series. Phyllites and quartzose schists follow. The middle division starts with talcose and actinolitic schists amongst which there are found barren and ferruginous hornstones. The schists are overlain by ferruginous hornstone and jaspilites forming an important ore-bearing horizon. The upper division includes the Supra-ore Formation, consisting of alterations of hornstone and chloritic and pelitic schists and sandstones. Still higher there are carbonaceous and quartz-mica schists, which have dolomitic intercalations at the top. The succession is completed by a thick series of quartz-mica schists which at the base may have intercalations of talcose and ferruginous hornstones.

The thickness of the Krivoi Rog Series is not less than 1000m. Its succession, according to Nikol'skii (1953) is shown in Fig. 38. Owing to strong folding separate horizons are in places squeezed out, or thickened to several times their original thickness. Repeated folding causes the ore-bearing layer, which is 50m on average, to reach 1500m to the south of the Ingulets mine. In the region of Krivoi Rog the series is folded into a single, sharply defined major 'Fundamental Krivoi Rog Syncline'. It is no more than 5–6km wide and stretches over a distance of 60km. It is strongly compressed, is in places isoclinal and overturned to the east. Its flanks in particular the eastern, are complicated by a large number of secondary smaller folds and thrusts, producing very complex structures. Large compressed Lower Proterozoic folds, which pass into each other along the strike can be traced over hundreds of kilometres in the Baltic Massif as well as in the south of the platform.

Proterozoic Platform-type Structures consist of the rocks belonging to the Jotnian Formation. They are well developed along the south-west shore of the Onega Lake and stretch towards the north-west into Finland. The Shokshinian

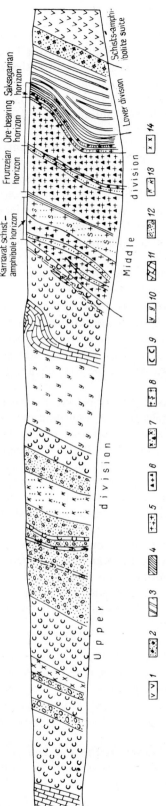

FIG. 38. Section across the Krivoi Rog series. According to A. P. Nikol'skii (1953).

1 — amphibolites;   2 — quartzites of the arkosic horizon;   3 — phyllites;   tite schists;   9 — sericite-quartz schists;   10 — graphitic   schists;   11 —
4 — talc   schists;   5 — chlorite-biotite   schists;   6 — martito-magnetitic   marbles;   12 — conglomerates;   13 — quartzites of the upper division;   14
hornfelses;   7 — hydrothermally   altered   hornfelses;   8 — amphibole-bio-   —gneisses.

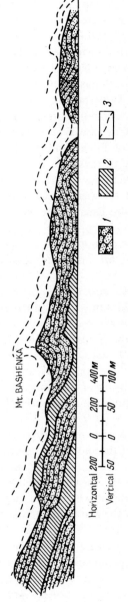

FIG. 39. Folds in the Hyperborean Formation on the Rybachii Peninsula.
According to D. D. Tenner and L. Ya. Kharitonova (1958).

1 — arkosic sandstones with beds of shales;   2 — pelitic sericite slates with
sandstone bands;   3 — zones of deformation.

reddish quartz-sandstones form a very gentle syncline with dips of 10°–15° and north-westerly strike. At its margins the dip sometimes increases, reaching 75°. The syncline is complicated by younger step faults, to which the formation of the lake Onega is related. In the Ovruch region at the north-west margin of the Ukrainian Shield, the Jotnian Ovruchian Sandstones are analogous to the Shokshinian and lie almost horizontally unconformable on the eroded Lower Proterozoic and Archeozoic structures. The red and greenish sandstones of the Kola Peninsula (Ter' Suite, Turiev Nos Suite etc.) are also almost horizontal or gently folded (5°–15° dip).

The Upper Proterozoic geosynclinal structures involve the Hyperborean Forma-tion (Riphaean) developed in the north-west of the Kola Peninsula, on the Rybachii Peninsula and Kil'din island. In the north-west it stretches out into Norway and in the south-east into the Kanin Peninsula, Timan and the western slope of the Urals. The Hyperborean Formation consists of a thick series of red sandstones which have conglomerates, shales and bedded dolomites, limestones and marls. These rocks, as shown by direct comparisons, are identical with those of the Upper Proterozoic of Timan and the Western Urals (Min'yarian and Til'merdakian suites). Analogously to these the Hyperborean Formation is relatively strongly folded into regional structures. The Hyperborean structures are large, elongated folds stretching over tens of kilometres and several kilometres wide (Fig. 34). Dips vary up to in-verted, but are usually low, not exceeding 10°–30°. The folds are often broken by thrusts of small amplitude. Folding is sometimes accompanied by intrusive massifs. As examples of more complex, but still rather simple, structures may be mentioned those folding the Imandra-Varzuga Suite in the eastern part of the Kola Peninsula (Fig. 40).

The structures of the unexposed part of the Russian Platform have not been much studied. Only in its north-eastern and eastern margins is certain information available, which enables one to explain the Proterozoic structures of the Kanin Peninsula, the Timan and the eastern slope of the Urals. The Proterozoic here includes thick quartzites and conglomerates. The study of the constituent grains and pebbles of these rocks has shown that they are products of erosion of mountain chains consisting of metamorphic rocks and large granite intrusives which were in existence within the Russian Platform to the west of the Urals and the south-east of Timan.

Since this erosion occurred at the beginning of the Upper Proterozoic, then the formation of the mountain chains should be attributed to the Karelidian Orogeny. Thus M. M. Gran', a leading specialist on the Precambrian of the Urals, has called these chains the Eastern Karelides in the Palaeo-Ural and correlated them with the Karelides proper, which were mountain chains of the Scandinavian Shield and were produced by the Karelidian Orogeny.

OROGENIES. The review of the structures shows that there were several Pre-cambrian orogenies, causing the formation of angular unconformities, transgressive oversteps, differences in the degree of metamorphism and occurrence of intrusive massifs. Four main orogenies are distinct, but there were probably more. The four orogenies were the Lower Archeozoic or Saamian, the Upper Archeozoic or Byelomorian, the Lower Proterozoic or Karelidian and the Upper Proterozoic or Baikalian.

The Saamian Orogeny occurred at the end of the Lower Archeozoic and pro-duced most complex structures, high metamorphism and formation of the most ancient intrusive granitoid and gneiss-granitoid massifs.

The Byelomorian Orogeny within the Soviet part of the Scandinavian Massif has caused the intrusion of massifs and the complex folding of the Ferruginous Suite.

The Karelidian Orogeny is one of the strongest Precambrian orogenies and occurred at the end of the Lower Proterozoic and the beginning of the Upper Proterozoic. It has caused the formation of the Karelidian chain which margins the Russian Platform in the north, north-east and east. The Karelides increased considerably the area of the Lower Proterozoic platforms and caused them to merge into a single unit of the main part of the present-day Russian Platform. Moreover, the Karelidian Orogeny has caused sharp angular unconformities and transgressions at the boundary between the Lower and the Upper Proterozoic, has caused high metamorphism of the Lower Proterozoic and has caused the formation of various intrusive massifs. The same orogeny in many regions has conditioned unconformities between the Palaeozoic and the Lower Proterozoic and the Archeozoic.

The Upper Proterozoic or Baikalian Orogeny, the latter name having been proposed by N. S. Shatskii, was much weaker than the Karelidian Orogeny, which had already converted a larger part of the Russian Platform into a platform. Thus the central provinces do not show any signs of Baikalian Orogeny, while to the north and east it has formed simple, gentle, brachyanticlinal folds, similar to those in the Palaeozoic of the Volga-Ural Region. Only at the very margin of the platform, in the north and east, the Baikalian Orogeny has formed new fold ranges and has welded them to the Karelides. In the north this fold zone is now in the main covered by the sea, but the outcrops of Hyperborean in the Rybachii Peninsula and Kil'din island belong to its margins. The central part of the fold-belt, which is more deformed and metamorphosed, is found in the Kanin Peninsula, Timan and Western Urals.

Palaeozoic Structures

General Characteristics

The Palaeozoic structures form potential oil-fields and in many localities contain oil and gas deposits. This important fact has attracted much attention and many of these structures, especially in the Volga-Ural region, have been studied in detail, and an extensive literature has been devoted to them. The review papers of N. S. Shatskii (1945, 1946), who is a major Soviet specialist on tectonics, are leading in this respect. The work of V. V. Byelousov (1944, 1948), another leading tectonician, who has introduced the practice of drawing isopachytes (lines of equal thicknesses of sediments), are also important. Of general papers by the oil-geologists those of A. A. Bakirov (1948, 1951) should be noted.

The abundant and valuable factual data on the structures of the Volga-Ural region is presented in the book *Volga-Ural Oil Region, Tectonics* compiled under the leadership of V. D. Nalivkin (1956). Some new data on the tectonics of the central provinces have been included in the chapter on Tectonics in the memoir by P. G. Suvorov and others (1957). The memoir *Geological Structure of the U.S.S.R., Tectonics*, vol. 3 (1958), which has been compiled by VSEGEI is also important. The structure of the U.S.S.R. should be considered in conjunction with the 'Tectonic map of the U.S.S.R. and adjacent countries' (1956 and 1959) and the *Explanatory Comments* (1957) edited under the leadership of N. S. Shatskii.

A characteristic feature of the Palaeozoic structures is their platform character.

They do not form either fold zones or belts, and have no definite linear orientation. In a majority of instances Palaeozoic structures have a small dip (not exceeding 1–3°), which is commonly measured in minutes. Such angles of dip cannot be easily detected by the naked eye. The mapping of such structures involves instrumental surveying when accurate instrument readings are used rather than the normal geological mapping. In their size, form and origin the Palaeozoic structures are quite variable and include gigantic syneclises and anteclises, which are several thousands of kilometres across. Other structures involve large thicknesses of Palaeozoic deposits many hundreds and even thousands of metres thick, which in age correspond to the whole of the Palaeozoic (Cambrian to Permian) or a major part of it.

Geophysical observations and deep drilling have shown an unexpected and a very important fact that structures in the lower deposits (e.g. Devonian) differ from those in the upper deposits on the surface, which are Permian. The differences are reflected in shapes, sizes and positions of crestal, central points. Sometimes the crests of anticlines are displaced sideways by several kilometres (Fig. 46).

Shatskii (1945) has suggested that the platform structures should be divided into three groups (orders). Thus structures of the first order include the fundamental structures: the 'syneclises' and 'anteclises' as well as the marginal downwarps; the structures of the second order which were called plakanticlinalia (Shatskii) included all others beginning with tectonic ridges and ending with flexures, while the structures of the third order are joints, faults, graben.

The oil-geologists who worked in the Volga-Ural region accept a somewhat different three-fold subdivision. The first order corresponds to that suggested by Shatskii. The second order includes major elongated upfolds, tectonic ridges, of several kilometres in length. The third order includes all the small isolated domes (brachyanticlines), which are situated on the structures of the first or second order. They serve as main oil reservoirs. This classification is adopted here.

First Order Structures

The first order structures can be really defined as regional since they involve large areas. Before Shatskii's syneclises and anteclises they were called *muldes* and *anticlinals*. The regional structures are not numerous, but they are large extending over hundreds and even thousands of kilometres.

In the area the northern belt of large upfolds (anteclises) is a determinative structure. In the west it begins from the Baltic Massif, continues as the Kanin Elevation, then the Timan Elevation (Anteclise) to the south of which separated by a minor saddle lies the Volga-Ural Elevation (Anteclise). According to the oil-geologists the latter anteclise consists of Tokmovian, Tatarian and Bashkirian arches and the Vyatka Ridge (zone) (Fig. 12). A southern belt lies in the juxtaposition to the northern belt. It starts in the east as the Yaikian Massif which is now traversed by the river Ural (Yaik). Thereafter follow the Ukrainian Shield and the Voronezh and Byelorussian Massif, already described.

The Yaikian Massif is situated in the northern part of the Ciscaspian Depression and lies at a great depth. Previously it was considered that all the south-eastern corner of the Russian Platform consists of a very large and deep depression, within which the base of the Kungurian (Upper Permian) lies at a depth of 7–9km. The latest geophysical investigations have shown that the structure of the depression is much more complex. In its northern part there are situated two large gentle elevations: the Khobdinskian and the Aralsorskian (Fig. 5). These elevations form a massif which rises 2–3km above the bottom of the depression (Slepakova, 1961). Some time ago the present author has established that fragments of the Upper Frasnian limestones, found in the cap breccia of the Baskunchak salt dome were identical to the fragments from the breccia of the Roman, Slavyansk, Isachkov and Petrovsk domes situated on the southern margin of the Voronezh Massif. Consequently a suggestion arose that the Baskunchak salt dome was also placed at the margin of some massif analogous to the Voronezh Massif and is its continuation. This massif the author called the Yaikian after the river Ural, which was formerly called Yaik. The Yaikian Massif has not been as yet completely investigated and the southern part of the Ciscaspian Depression has not been much studied either. It is probable that the depression is a part of the massif which is here deeply buried. The Voronezh Massif is margined in the south by the Dnyepr-Donyets Depression which is directly continuous with the Astrakhan Depression. The Astrakhan Depression in turn is marginal to the southern part of the Yaikian Massif, which partly lies to the south of the southern belt of elevation. To the east the Astrakhan Depression merges into the South Emba Elevation.

The southern and the northern belts together outline the position of huge depressions (syneclises). The belts form a ring, within which there is situated the Central (Moscow, Middle Russian, Main) Depression (Figs. 12, 40). From the west the ring of elevations is paralleled by the Western Depression, and from the east by the Eastern Depression. These two depressions connect with the Central Depression by the Latvian Trough and the Ryazan-Saratov (Pachelma) Trough.

The Central Depression is a large low area which is surrounded by uplands. On a geological map the Central Depression has a shape of a large syncline (mulde, syneclise) closing to the south-west and gradually opening and widening to the north-east. From geological maps it is clear how the structures of Timan and the Kanin Peninsula connect with the Kola Peninsula and enclose the depression in the north. A cross-section of the depression is given in Fig. 41 and shows the infilling deposits.

Deep boreholes have shown that throughout the whole of the Lower Palaeozoic, Silurian, and Devonian the Central Depression was bound on the east by the Volga-Ural Elevation (Kotyel'nich, Glazov), dividing it from the Eastern Depression (Krasnokamsk, Fig. 41). From the Famennian onwards the Volga-Ural Elevation ceased to exist and sedimentation continued homogeneously over the central part of the platform. This is clear from the structure of the coal-bearing series. During this epoch the Central and the Eastern depressions merged, giving rise to a single, huge depression connected to the Uralian Sea. In the Upper Carboniferous the Urals became dry land, forming an eastern margin to the depression, which again became entirely surrounded. This depression later had within itself the Kazanian Sea and thereafter the Tatarian and Vetluzhian deserts, which were huge enclosed low plains.

The Central (Middle Russian) and Eastern depressions retained a low relief despite tectonic movements which transformed them into truly tectonic structures. These movements were mainly negative (downward) and more rarely positive (upward). Downward movements were slow, balancing the sedimentation and retaining more or less similar palaeogeographical conditions. These movements were on a large scale and have led to the formation of sedimentary rocks of 3000m and more in thickness. Consequently along the northern and western margins of the Middle Russian Depression there arose large concentric normal faults and fractures. Along the cracks rose the basaltic magma and formed flood basalts and sills of diabases (basalts). These diabases are now found in a number of localities from Timan to Chernigov.

The upward movements were fast, of short duration and small extent. They transformed shallow marine basins in the depressions into lagoons, near-shore plains,

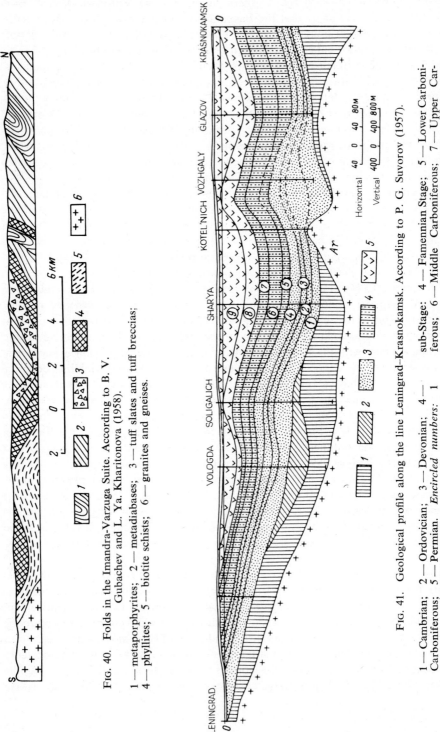

FIG. 40. Folds in the Imandra-Varzuga Suite. According to B. V. Gubachev and L. Ya. Kharitonova (1958).

1 — metaporphyrites; 2 — metadiabases; 3 — tuff slates and tuff breccias; 4 — phyllites; 5 — biotite schists; 6 — granites and gneisses.

FIG. 41. Geological profile along the line Leningrad–Krasnokamsk. According to P. G. Suvorov (1957).

1 — Cambrian; 2 — Ordovician; 3 — Devonian; 4 — sub-Stage: 4 — Famennian Stage; 5 — Lower Carboniferous; 6 — Middle Carboniferous; 7 — Upper Carboniferous; 8 — Lower Permian; 9 — Upper Permian. Carboniferous; 5 — Permian. *Encircled numbers:* 1 — Givetian and Eifelian stages; 2 — Lower Frasnian and middle Frasnian sub-stages; 3 — Upper Frasnian

and even low elevations (almost flat). Consequently within the marine deposits there are breaks and red and variegated rocks, while rarely there are pale (millstone) basal sandstones, gravels and clays. Such basal rocks contain intercalations of sands and friable sandstones which in the east of the platform are oil-reservoirs. Such non-conformable members are known at the base of the Middle Devonian (Takatinian Suite), at the base of the Upper Devonian (Pashiiskian Suite), at the base of the Upper Frasnian (Petinian Suite), at the base of the Visean (Coal-bearing Suite), and in the Middle Carboniferous (Vereiian Suite). The small duration of the breaks, and the small thicknesses of the continental (basal) formations, show that the positive movements were shortlived.

The Eastern Depression is the most important oil-bearing province. Consequently it has been studied in detail by means of geophysical methods and cut by many thousands of deep boreholes. These researches have enabled us to divide the Eastern Depression into a series of structures of first and second order (Fig. 42). In the north there is the large Pechora Depression. It is separated by a small South Timan pitch depression from the equally large Upper Kama or Glazovskaya Trough, to the south of which there are situated the Bashkirian, Tatarian, and Tokmovo arches. To the south-west of these lie the Melekasskaya and the Ulyanovsk-Saratov depressions, which are separated from the south-eastern slope of the platform by the Don-Myedvedits, tectonic, Ridge and the Pugachev Arch. The south-eastern slope of the platform continues into the Ciscaspian Depression. As has been pointed out, it is probable that in the central part of this depression lies the Yaikian Massif, which has the Volga-Emba salt-dome region adjacent to it to the south.

The Western Depression is divided into the Baltic, Polish-Lithuanian, and Podolian Syneclises, which to the south are connected with the Black Sea Depression (Fig. 12).

A most important, fundamental feature of the regional structures is the character of the tectonic movements which have produced them. These movements were directed sometimes downwards and sometimes upwards and affected at times the whole platform, and at other times considerable parts of it, inducing epeirogenic, block-displacements. The movements were often associated with faults, which

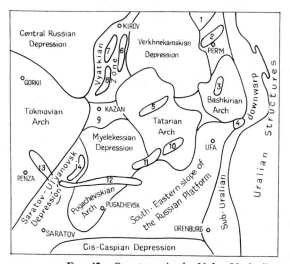

1. The Kama Arch.
2. The Krasnokama Elevation (tectonic ridge).
3. The Chernushinskii Elevation.
4. The Karatauian structures.
5. The Grakhanian Elevation.
6. The Urzhumian Elevation.
7. The Kukarian Elevation.
8. The Shurgian Elevation.
9. The Kazan Saddle.
10. The Tuimarzinskian Elevation.
11. The Baituganian Elevation.
12. The Zhigulev Elevation.
13. The Borlinskian Elevation.
14. The Sursko-Mokshinian Elevation.

Fig. 42. Structures in the Volga-Ural oil province. According to V. D. Nalivkin.

broke up the Precambrian basement as well as the Palaeozoic cover. Epeirogenic movements were not associated with orogenic movements in geosynclines and occurred independently of them. The only detected connection is that during the Caledonian and Variscan orogenies the platform suffered the most significant upward movements, thus becoming an elevated landmass, lasting a long time and becoming an area of erosion. This occurred during the Silurian and Lower Devonian epochs and again in the Permian and Lower Triassic epochs.

Second Order Structures

Structures of the second order are numerous, variable and widely distributed. Most of them are of intermediate size and several hundred kilometres across. Amongst them there are the following types: (1) arches; (2) tectonic ridges; (3) mega-flexures; (4) depressions.

ARCHES. Arches, to oil-geologists, are large gently convex elevations with rounded outlines, reaching 200–400km. So far four have been mapped and are the Tokmovian, Tatarian, Bashkirian, and Pugachevian arches. These arches are often considered as structures of the first order.

The Tokmovian Arch is a large structure situated to the south of the vale of Volga, between Gorki and Kazan, and in its central part lies the valley of the Sura. Within the arch the Precambrian basement, which nearby is situated at depths of up to 2000m, reaches up to depths of only 700–1000m. The dips of its flanks are 15–20 min. Both Bavlinian suites are absent over the arch. This means that at that time the Tokmovian Arch was an elevation and an area of erosion. Beginning from the Eifelian epoch onwards, the Tokmovian Arch became an area of sedimentation and a typical huge envelope structure. It has rounded to oblong outlines and a cross-section of 300–400km (Fig. 42). The Tokmovian Arch is a part of an even larger elevation called the Volga-Ural. The arch, however, also has smaller structures, including the Alatyrian Ridge as parts of it. A cross-section across the Ryazan' Saratov Depression, the Tokmovian Arch, and the Melekessian Depression is given in Fig. 43.

The Tatarian Arch reaches 400km in length, 200km across, and 200–1000m in height. The dips on its flanks are 12–15 min. The arch begins south of Kama, includes parts of Tataria, and continues to the north of Kirov. It is also a part of the Volga-Ural Elevation and possesses all the peculiarities of the Tokmovian Arch. The Tatarian Arch has a large number of oil-bearing structures.

The Bashkirian Arch is of smaller dimensions, being up to 220km long, up to 160km wide, and 600–1000m high. It is situated to the north-east of the town of Ufa and includes the middle reaches of the river Ufa and the Ufa, tectonic, Ridge. Within the arch the Bavlinian Suite and the Givetian Stage are well developed and the arch at those times was a low-lying area, which was transgressed by the Middle Devonian and Frasnian seas. Structures of second and third order are associated with the Bashkirian Arch.

Like the Tokmovian and Tatarian arches the Bashkirian Arch is a typical supratenuous cover structure and is not related to orogenic processes. Thus it is different from the Ufian tectonic Ridge, which is a typical marginal anticline trending parallel to the Uralian strike and produced during the Hercynian Orogeny.

The Pugachev (Zhigulev-Pugachev) Ridge is situated to the south of the Zhigulev Ridge (Fig. 42). The ridge is areally 160 by 120km, and over 800m high. It has subsidiary structures of third order, many of which are oil-bearing.

RIDGES. This term is applied to large, elongated, gently convex, vaguely defined structures, which are up to some hundred kilometres long and have flanks dipping from tens of minutes to 3°. In the central part of the platform structures such as Oka-Tsninian, Alatyrian, and Vyatkian ridges serve as examples.

On geological maps the Oka-Tsninian Ridge is clearly identifiable from the outcrops of the Carboniferous and Lower Permian rocks surrounded by the Upper Permian deposits. The ridge is first detected in the south (basin of the river Tsna) where the Visean limestones outcrop, while further to the north the Middle and Upper Carboniferous limestones and the Shustovo-Denyatinian (Lower Permian)

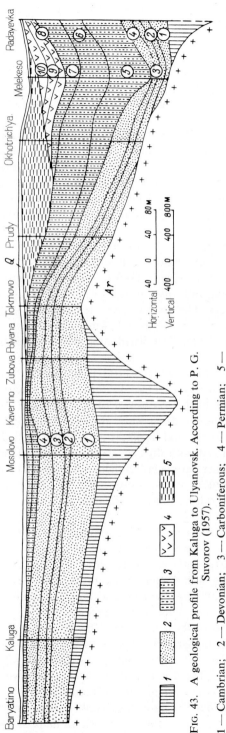

FIG. 43. A geological profile from Kaluga to Ulyanovsk. According to P. G. Suvorov (1957).

1 — Cambrian; 2 — Devonian; 3 — Carboniferous; 4 — Permian; 5 — Mesozoic, Palaeogene, and Neogene. *Encircled figures:* 1 — Givetian; 2 — Shchigrovian and Semilukian beds; 3 — Voronezh, Yevlanovian and Livonian beds; 4 — Famennian; 5 — Lower Carboniferous; 6 — Middle Carboniferous; 7 — Upper Carboniferous; 8 — Lower Permian; 9 — Kazhanian; 10 — Tatarian.

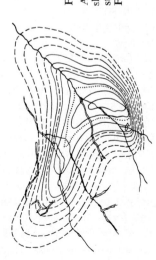

FIG 44.

An anticline with a steep southern slope and a flat crest in Saratov structures. According to S. F. Fedorov (1950).

marls are traced to the north of the river Oka to the neighbourhood of the town of Murom. The length of the ridge is about 250km, while its width reaches 50–60km. Dips are low, rarely exceeding 3° and most frequently measured in tens of minutes. Deep-drilling has revealed an unexpected and a very important feature: although the structure is most distinct in the Carboniferous and Permian rocks it is not present in the Devonian rocks. The Devonian deposits are almost horizontal and are not flexed.

The Alatyrian Ridge is situated to the east of the Oka-Tsinian Ridge, around the town of Alatyr and also has a north–south elongation. It is indicated by the Carboniferous and Permian outcrops as an even gentler and vague ridge-like structure. It is not obvious in the Devonian.

The Vyatkian Ridge (zone) (Vyatka-Kama Ridge) begins in the valley of the river Kama, where the Vyatka joins it. The ridge continues to the north, slightly veering to the east for 360km, and crosses the latitude of the town of Kirov. Its width is up to 150km. The ridge consists of a series of anticlines, which follow each other and are sometimes continuous. These structures are identified from the outcrops of the Kazanian Stage surrounded by the rocks of the Tatarian Stage. They are very gentle, large-scale, and of irregular outlines. At depth these structures can be traced at the top boundary of the Famennian Stage, but do not affect the Pashiiskian Formation, while the top surface of the crystalline basement is actually depressed.

In the eastern part of the platform the term ridge is used by the oil-geologists for elongated large structures 100–150km long and with variable outlines, which can be traced down to the Precambrian basement. These structures often originate due to displacements (normal and reverse faults) in the basement, which cause flexure type folds and sometimes associated normal faults. The distribution of the ridges is shown on Fig. 42.

Zhigulevskian Ridge.   This structure, which in its central part includes the Zhigulevskian Mountains, is a typical flexure-type ridge. In the east it reaches Mukhanov on the Kinel', in the west it ends to the south-west of Syzran', near the town of Kuznyetsk. The ridge is 360km long, 10–20km wide, and 700–800m high. The northern limb is steep with dips of up to 80° and in places is probably faulted. The southern limb is wide and gently inclined with dips of about 40 min. Along the axial part of the Zhigulevskian Ridge there are small anticlines trending along its strike and reaching from 2×4 to 3×6km in dimensions. Similar structures are known from the southern limb of the ridge. Many of the structures are oil-bearing.

The Don-Medvyeditsian (Saratov) Ridge consists of a series of flexure-type, box-like, irregularly bent, horse-shoe shaped structures consisting of Palaeozoic and Mesozoic rocks. Their limbs are not infrequently steep, reaching 45–55°, their arches are gentle or slightly cross-folded (Fig. 44). Similar structures are also developed to the south in the region of the juncture of the Don and the Medvyeditsa. Of such structures the Archedinskian and the Don domes are well known.

The Ufian Ridge is situated in the region of the Ufa valley. The ridge is sometimes called the Ufa Anticlinal. Its modern topographic expression consists of the Ufa plateau, which is a relatively low, flat, very picturesque, wooded and wild upland made of Upper Permian limestones. The Ufian Ridge stretches east–west and forms the eastern margin of the Bashkirian Arch. The western flank of the ridge is gentle with dips of a few tens of minutes. The eastern flank is also gentle, but dips are somewhat steeper, reaching up to 3–4°. It forms the western flank of the sub-Uralian Depression. The Ufian Ridge is not associated with fracture of the base-ment and is consequently not a flexure-type structure. It is more similar to the central ridges (e.g. Oka-Tsinian) and like them is only distinct amongst the Upper Palaeozic rocks. At the top of the Devonian it is almost unnoticeable.

The numerous small and medium-size ridges, situated to the north of the lati-tude of the town of Kuibyshev, are called collectively the Volga-Ural ridges. Near the Zhigulevskian Ridge are situated the Borlin and the Mordovian Ridge. To the north-west of them are the Elkhosko-Borovskian, Sernovodsko-Shugurovian, Baitu-ganskian, and Tuimazinskian ridges. On the Kama there are the Grokhansko-Yelabuzhian and Kransnokamsk-Polaznenskian ridges and a number of others are known.

The length of the ridges does not exceed 150–180km and is often less, while their width reaches 30–50km. The Tuimazinskian Ridge can be taken as an example: its length is 90–100km, its width 20–30km, its height 100–150m. The dips on its south-eastern flank are 2–6° and on the north-western flank 30–40 min. All the ridges, as well as the associated structures of third order, are oil-bearing.

The Sukhonian Ridge stretches from Kotlas to the town of Tot'ma along the valley of the Sukhona in a south-westerly direction. Further on in line with it is a series of relatively small structures, including the Soligalich, the Lyubimskian, the Vologda, which ends near the town of Rybinsk.

The Loknovian ridge-like Elevation to the west of Pskov is 150km long, 30–40km wide and about 400–500m high (Fig. 45). It is not obvious on the surface and has been discovered in boreholes. The Upper Devonian lies horizontally, but all the older formations have been greatly affected by the Loknovian Elevation, which served as a barrier to the Silurian sea.

FLEXURES AND MONOCLINES. Typical flexures, being bent horizontal for-mations, are rare. They are evidently most typical around the Ciscaspian (Yaikian) Depression, where the map shows the major Volgograd, Tokarev, and Buzulukian flexures.

The so-called steps (monoclines) traceable in the Palaeozoic formations of the north-western regions and other areas are similar to flexures. The large step between Staraya Rus' and Kresttsy reaches 900m in height and stretches from south to north.

TROUGHS (DOWNWARPS) AND DEPRESSIONS. These are the terms given by oil-geologists to the negative structures of second order. The troughs correspond in form and dimensions to arches, and depressions to ridges. On tectonic maps both types of structures are called downwarps. Minor depressions connecting elevations are called saddles. For instance the Birskian Saddle lies between the Bashkirian and Tatarian arches, and the Kazanian Saddle between the Tatarian and Tokmovian arches. The main troughs (downwarps), such as the Melekessian Trough, are shown on Fig. 42.

The Melekessian Trough is situated between the Tokmovian, Tatarian, and Pugachevian arches and is similar to them in shape and dimensions. It is a rounded trough some 250km long, 200km wide and 400–600m deep. The Melekessian Trough differs from the arches in that it has below it ancient Bavlinian suites, which are absent from the arches. As on the arches, ridges, depressions, and anticlinal and synclinal structures of third order are developed in the trough.

The Stavropolian Depression is of a similar size to the Zhigulevian Ridge, to which it is connected, and lies to the north of it (Fig. 42). It has, however, steeper flanks than the Zhigulev Ridge and its total length is 250km, width is 20–40km, and depth is 700–800m.

Third Order Structures

Structures of the third order are most frequently called anticlinals, brachyanti-clinals, or simply structures. They are found everywhere, their number reaches hundreds, and their shape varies from irregular to rounded or elongated. Their size varies from, say 2–20km in width and 30km in length to more frequently 5–10km in length and 3–7km in width. Their amplitude is small, being usually some tens of

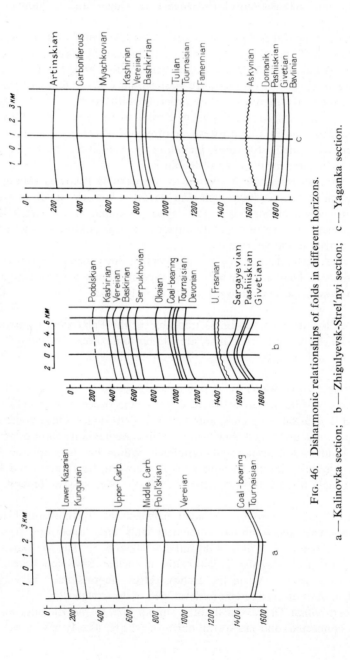

Fig. 46. Disharmonic relationships of folds in different horizons.

a — Kalinovka section; b — Zhigulyevsk-Strel'nyi section; c — Yaganka section.

metres (20–50), reaching sometimes 200m. These structures are of great industrial importance, localizing the main reservoirs of oil. Third order structures have often an important peculiarity, demonstrated by Soviet geologists, insofar as their shape changes from the upper layers to the lower layers. These inconsistencies are of several types, of which four are most important.

1. Elevations proved in upper horizons disappear in the lower ones and sometimes become downwarps (Fig. 46).
2. Elevations proved in the lower, Devonian horizons are absent from the Carboniferous or Permian.
3. The apex of an elevation may change its position from layer to layer (Fig. 46c).
4. The amplitude of the elevation may change from horizon to horizon.

Another important peculiarity of these structures is their dependence on the structures of the first and second order on which they are situated. On the steep and narrow ridges, as for instance the Don-Myedveditsian Ridge, the third order structures have an elongated form, a considerable height (50–200m), steep flanks (up to 50°), but small other dimensions (5–10 by 2–3km). On gentle ridges and arches the third order structures have more rounded outlines, gentle flanks (40 min.–4°), small height (20–70m), and large other dimensions (10–20 by 5–15km). An example of the first type is shown in Fig. 47a. This is the Syzran' structure, lying on the narrow and high Zhigulev Ridge. A structure of the second type is shown in Fig. 47b. This is the Polazn structure situated over the gentle Krasnokamian Ridge.

FAULTS. A peculiarity of platforms on all scales is the small number of faults. Large numbers of faults have been conjectured, but detailed investigations have shown that there are very few faults. In the Volga-Ural region many tens of structures have been studied in detail, without finding a fault.

There are certainly faults in the crystalline basement and in large numbers, reaching sometimes downthrows of up to 1500m, but they rarely penetrate the Palaeozoic cover, which normally shows flexures and monoclines and flexure-like ridges of second order above the faults in the basement.

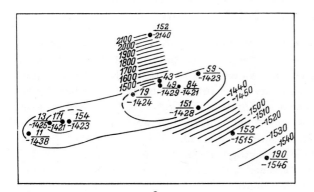
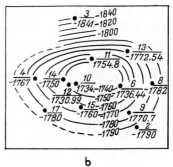

a   b

FIG 47.   The form of third order structures.

a — Syzran' structure drawn on the top surface of the basement;   b — Polazn structure drawn on the top surface of the Kashirian Horizon. Black points are boreholes, numerators indicating their number and denominators the depth of the marker horizon.

Mesozoic and Cainozoic Structures

It should be pointed out that Mesozoic and Cainozoic structures are much less widespread and less numerous than the Palaeozoic structures. This is explained by the smaller area of Mesozoic and Cainozoic rocks, the lesser knowledge about these formations and the lesser deformation affecting them, especially in the central and northern provinces.

In the regional structures the Lower Trias (Vetluzhian Stage) completely fills the lowermost part of the Central Depression (Mid-Russian Syneclise), thus terminating its existence. Later on in the history of the platform the depression becomes either a plain or an elevated region.

Further south a large elevation, elongated east–west, appears. This elevation is transected by the valley of the Volga and as a result in Kazan, at the base of the section exposed on the steep eastern bank, the lower part of the Kazanian Stage appears. Lower down the river it is overlain by the marls and clays of the upper part of the Kazanian and thereafter by the Tatarian Stage. Still further down the Upper Jurassic clays and sands appear and at Ulyanovsk are succeeded by the black Lower Cretaceous muds, which are succeeded by the Upper Cretaceous and near the Samara bend by the Palaeogene. Such a continuous series of outcrops indicates the southern flank of a large gentle elevation (dome) or a northern flank of a large depression. This depression, which is elongated east–west, started forming at the beginning of the Upper Cretaceous epoch, determining the extent of the Upper Cretaceous and later on the Palaeogene seas. It filled up at the end of the Oligocene and during the Neogene became an elevation.

The youngest regional structure is the Quaternary Boreal Depression, which was filled by the Upper Quaternary Boreal Sea. At present the depression is filled by the Barents, Kara, and White seas.

The Mesozoic and Palaeogene play a considerable role in building the southern ridges from Zhigulev onwards. They almost completely form the cover of the Don-Myedveditsian Ridge.

TECTONIC MOVEMENTS DURING THE PALAEOZOIC, MESOZOIC, AND CAINOZOIC

The main difference between these movements and those in the Precambrian is the absence of strong orogenies. A number of investigators, probably mistakenly, deny any signs of orogenic movements. Numerous structures at the eastern margin of the Russian Platform affect Permian and Carboniferous deposits, which strike parallel to the general strike of Hercynian folds in the Urals. The Ufian Ridge serves as an example. At the southern margin of the platform east–west structures originating during the Mesozoic and Palaeogene are common. They are parallel to Cimmerian and Alpine structures of the Mediterranean Geosyncline.

At any rate, even if orogenic movements do affect the margins of the platform they are weak, do not form complex linear folds, and have no metamorphism or igneous intrusions.

The encroachment of orogenies onto the margins of the platform is confirmed by the fact that platform structures are gradational into geosynclinal structures without any definite boundary. The structure of the eastern margin of the Russian Platform is succeeded by the structures of the sub-Uralian downwarp, where structures are still large and indistinct, but have somewhat steeper limbs. Then there are the structures of the Western Urals, which are still large, but clearly distinct and with high angles of dip. It is significant that the metamorphic state of the platform and the Western Urals (in the miogeosyncline) is almost the same and facies are also identical. Volcanics and intrusions are absent. Only in the Eastern Urals complex (in the eugeosyncline), do intensely deformed, frequently isoclinal structures appear. The metamorphic grade here is much higher, rocks passing into green schists. Volcanics and intrusives become widespread.

The basic, most widely developed type of tectonic movements on platforms (including the Russian) are epeirogenic oscillatory movements. They are responsible for many structures including the flexure-like ridges. The elevations and depressions thus produced vary in amplitude and lateral extent. Often they extend over large parts of the platform, while at other times they affect areas which are a few tens of kilometres across.

In their study of epeirogenic movements Soviet geologists stand in the forefront of the world, since they have developed the method of isopachytes, which employs the principle that the extent and magnitude of these movements are estimated from the thickness of deposits. Isopachytes are lines of equal thickness of sediments accumulated during a definite epoch. Thus isopachytes are also lines delimiting areas of equal sinking. Widespread application of isopachytes has demonstrated differential movements in different regions of the platform. The oscillatory movements of the Russian Platform have been described in hundreds of publications. Amongst these the first place is occupied by the papers of A. P. Karpinskii (1887, 1894), which have been frequently republished in the Soviet period. In these papers Karpinskii demonstrated clearly and convincingly the great importance of epeirogenic movements during the history of the Russian Platform. He was tens of years in advance of his contemporaries and only now do we understand the significance of his research.

Pseudotectonic Structures

This term is given to structures which are not normally associated with tectonic movements. They are frequently encountered and have variable forms and origins. They are often complex, non-metamorphosed, and local. If somewhere in a ravine or steep river cliff several small-scale, locally developed, rather complex and broken-up folds, affecting non-metamorphosed rocks, are found, then one can be certain that they are pseudotectonic structures. Most pseudotectonic structures are associated with landslips, glaciations, gypsum decollement, and collapse.

Landslip structures are encountered on the cliffs of modern and ancient river-valleys. Modern landslips are easily recognized and they can be easily proved. Ancient landslips in river-valleys, which have since changed their course and were filled by other Anthropogenic deposits, lose their connection with river cliffs. Consequently their identification is much harder and requires geomorphological study of the whole region.

Glacial structures are usually of two types, compressional and transported. Compressional structures originate when a thick ice sheet moving forward encounters obstructions such as a steep river or lake cliff. If the cliff is composed of plastic rocks, such as clays or sands, then they are deformed into small-scale, complex folds, which are commonly overturned and broken. This occurs in Lower

Cambrian blue clays, the *Dictyonema* shales and the *Obolus* sands on the shores of Popovka near Pavlovsk in exposures well known to every geology student from Leningrad. When the cliff consists of compacted stable rocks then they do not get deformed, but broken into large blocks which move upon each other. Such thrusts are seen amongst the Upper Devonian limestones near the township of Burega on the southern cliff-shore of the lake Ilmen' (Fig. 48).

Transportational structures are variable. They are formed when the edge of the ice sheet meets an obstruction, penetrates it and lifts a torn block to its surface and transports it for hundreds of kilometres. During transportation the rocks bend, forming small, gentle folds. The structures formed in an erratic of Lower Cambrian and Ordovician and situated in the valley of the Lovat' near the town of Kholm serves as an example. Other transportational structures are formed in deep kettle holes,

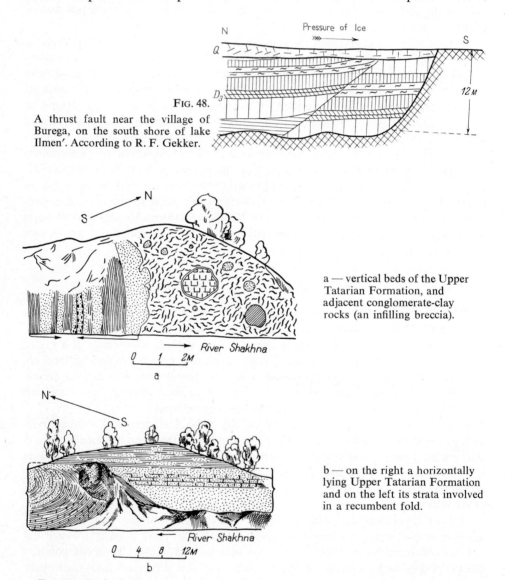

FIG. 48.
A thrust fault near the village of Burega, on the south shore of lake Ilmen'. According to R. F. Gekker.

a — vertical beds of the Upper Tatarian Formation, and adjacent conglomerate-clay rocks (an infilling breccia).

b — on the right a horizontally lying Upper Tatarian Formation and on the left its strata involved in a recumbent fold.

FIG. 49.   Puchezh structures on the river Shakhna. According to E. A. Kudinova (1939).

excavated by ice and filled by different boulders transported by ice. Such structures were sometimes considered as complex folds, as has been the case near the Mount Mishina near the town of Gdov, where there is a chaotic accumulation of boulders of gneisses, Cambrian rocks, and Ordovician limestones up to 100m in total.

Glacial structures in the region of Puchezh, Katunki, and Chkalovsk (Vasilevo) to the north of Gorki have been investigated in detail by E. A. Kudinskii (1939). For a distance of 10km typical fill-in structures alternate with folds. The bottom of one kettle hole is 100m lower than the level of the Volga. The hole is filled by boulders of Permian and Carboniferous limestones with clayey cement. The fold structures are complex, but unmetamorphosed and relatively small. Fig. 49 shows the contact between vertical Permian clays and marls with the fill-in breccia.

Fig. 50 clearly shows horizontal Upper Tatarian clays and marls locally folded into a small overturned fold. Such structures and others have been interpreted in different ways. The majority of geologists consider them either the products of true or salt tectonics. The first explanation cannot be accepted since such violent tectonic folding is always accompanied by a considerable degree of metamorphism. The second explanation is possible, but must be rejected since drilling failed to show any underlying salt, but instead revealed Permian deposits. A group of geologists have suggested a glacial origin of the Puchezh structures, which is a correct explanation, since these structures are identical to glacial structures of the Leningrad region. All these structures, as well as those found in Cambrian and Ordovician of Vyshnii Volochek, which were formerly considered to be tectonic, caused by the Caledonian Orogeny. It is interesting that some distinguished geologists support this contention. Only drilling and general mining operations have proved their glacial origin.

In this connection, a few years ago an event occurred which produced a surprise only possible in geology. A paper by D. L. Frukht (1959) appeared, reporting

FIG. 50. Structures near the village of Tetyusha on the Volga. The folds affect Tatarian red beds. *Photograph by Yu. E. Koreshkova (1952).*

that in some boreholes folded Puchezh and Katunki formations were found to be overlain by Jurassic rocks. This record requires verification. If it is confirmed then the Anthropogenic age and glacial origin of the folded formation must be rejected. Then the problem arises how such a folded formation has originated, if it is covered by the horizontally lying Jurassic rocks? Frukht has remained silent on this subject. This may be cautious, but from the productive point of view incorrect. It is imperative to have a positive point of view based on the large volume of collected data. Such opinions are always valuable.

It should be pointed out that the explanation of the genesis of the folded strata is not easy. It is clear, however, that such folding of continental formations can only be achieved by moving the strata. What agency caused the movement and how it occurred is not obvious. Glacial agency is inapplicable since there is no evidence of Triassic glaciation. Movement by diapiric action is again unlikely, since then the degree of deformation should be much more intense and involve brecciation. Movement by sliding (landslips) seem the most likely, and modern landships do produce such structures. Limited extent, relatively small thickness, lithology, and the existence, at the time, of a nearby range of hills confirm the theory of sliding, but a fuller explanation will have to be left for the future.

Gypsum structures originate at the edges of massifs of anyhydrite when it changes into gypsum and increases in volume. As a result, rocks adjacent to the massif experience strong pressure and get folded into small, complex folds or are brecciated. The folds develop over distances of no more than a few hundred metres and pass into horizontally lying formations.

Collapse structures originate as a result of collapse of the roofs of underground caverns. The overlying rocks on collapsing form flexures and box-like folds complicated by normal faults. If the caverns were deep the structures are broken in a complicated way. Modern collapses often originate in coal basins, as for instance in the Donyets Basin, when the ground collapses above the extracted seams of coal. This involves a serious hazard to the ground buildings and railway lines. Thus Soviet mining engineers employ special scientific techniques under the title of 'cover control'.

In conclusion it must be pointed out that Soviet geologists have made great achievements in studying the Russian Platform, and it has been studied much more fully than any other platform. Nevertheless there is still much room for regional, production, and theoretical research. Over a whole range of problems one has to pass from theoretical, often academic constructions to concrete field work, supplemented by complex, borehole exploration and geophysical undertakings.

MAGMATISM

Magmatic activity within the Russian Platform differs sharply on transition from the Precambrian to the younger formations. In the Precambrian, magmatism was of geosynclinal type, while later on it had platform character.

PRECAMBRIAN MAGMATISM IN THE NORTH OF THE PLATFORM

A review of magmatism on the Russian Platform is found in the monograph entitled: *Geological Structure of the U.S.S.R.*, vol. 2, *Magmatism* (1958). In this monograph many new and important data are introduced, but the quoted ages for the Russian Platform are not always unquestionable. For Karelia and the Kola Peninsula the information is supplied by T. V. Bilibina, N. A. Volotovskaya, N. A. Yeliseyev, and D. F. Murashev, while the data on the Ukraine Shield are summarized by Yu. I. Polovinkina, on the basement of the platform by L. A. Vardanyants and on the Palaeozoic formations by Z. G. Ushakova.

In the Precambrian Era magmatic activity reaches an extraordinary intensity and spread and is particularly widespread in the Archeozoic and Lower Proterozoic, while in the Upper Proterozoic it weakens. Precambrian magmatism can be divided into a number of cycles. Each cycle consists of three phases: the effusive phase, the phase of major intrusions, and the phase of minor intrusions. All the phases are consecutive. Magmatic cycles are closely related to orogenies. The effusive phase precedes orogenies, or more rarely is contemporaneous. The intrusive phase occurs at the same time as the peaks of orogenies, while the phase of minor intrusions coincides with the latest manifestations of folding. Thus the number of magmatic cycles corresponds to the number of orogenies. The violence of orogenic movements determines the magnitude of the coincident volcanic cycle. Weak periods of folding are not accompanied by intrusive activity.

Saamian magmatic cycle. It is virtually impossible to distinguish the effusive rocks which preceded the Saamian Orogeny, since such rocks are recrystallized and converted into gneisses and crystalline schists. The phase of major intrusions, how-ever, is well developed and involves huge and variable massifs. N. G. Sudovikov (1939) and A. A. Polkanov (1947) have called them the first group of intrusions, while previously they were referred to as post-Svionian. In Karelia, in the west, granodiorites (oligoclase granites) occupy large areas. In addition there are small outcrops of hypersthene diorites and gabbro-amphibolites. In the Kola Peninsula the oligoclase granite gneisses (first group) form huge plutons, which are involved in folding. In addition there are intrusive massifs of gneissic hypersthene diorites and gabbro-amphibolites (amphibolite gneisses). All these intrusions are strongly metamorphosed, deformed and rendered gneissose. Minor intrusions are repre-sented by pegmatites and aplites, associated with the granodiorites.

Characteristic features of all the Saamian intrusions are widespread zones of mig-matization and injections into the surrounding gneisses. A. A. Polkanov and E. K. Gerling (1960) have determined the age of the Saamian granites as 1700–2600 m.y. They have also distinguished the Katarchean granites with an age of 1840–3250 m.y. The latter granites are not widespread.

The Byelomorian cycle commences with Upper Archeozoic effusives and is not represented in the U.S.S.R. Then follow large intrusive massifs, called Svecofennian and Post-Bothnian which are classified under the second group of intrusions. They were formed during the Byelomorian Orogeny.

According to N. G. Sudovikov (1939), in Karelia the Byelomorian, Post-Bothnian granites are more widespread than the Saamian granites. The former groups are found in western Karelia, on the western shore of lake Onega and the shores of the White Sea. These granites are involved in fold-movements and have thick migmatic aureoles. Petrographically they have much microline, although there are also numerous granodiorites.

A well developed phase of minor intrusions reaches the shores of the White Sea where in the region of the Chupin fjord pegmatites have a commercial importance.

On the Kola Peninsula, according to A. A. Polkanov (1947), Byelomorian folds involve a synkinematic granulitic pluton in Lapland, which is about 20 000km² in total area and of which the eastern part is within the U.S.S.R. In the granulites differentiation from basic to acid magmas is apparent. The cycle is completed by the intrusions of granodiorite and microcline granites, which form the enormous Murmansk pluton on the shores of the eastern Murman', and also other plutons in the southern and western parts of the Kola Peninsula.

The Karelian cycle is completely represented by three phases. The effusive phase has metadiabases (spilites) of Segozero, which are interbedded with the Segozerian quartzites. With the spilites are associated the gabbro-diabases representing the feeding channels. On the Zaonezh'ye Peninsula there are also diabases associated with dolomites and schungitic slates which can be considered partly as thick sills of subvolcanic type and partly as lava flows. The Karelian cycle is well developed in the southern part of Karelian A.S.S.R. where it is represented mainly by effusions, Rapakivi-type intrusions and rarer sills and dykes.

Volcanic activity begins in the so-called Suisarskian volcanic complex (c.f. Sudovikov, 1937). It has previously been studied by a leading Soviet petrographer, Yu. F. Levinson-Lessing, who named it as the Olonyets Diabase Formation. Suisarsian rocks cut the shungitic slates and lie upon the latter. The volcanic rocks are covered by Shokshinian sandstones.

According to Sudovikov (1937) the Suisarsian volcanic complex includes diabases, various porphyrites, mandelsteins, aphanites, pillow lavas, tuffs, and volcanic agglomerates. The pillow lavas with pillows of up to 0·5–1·0m and tuffs are especially widespread, while mandelsteins and porphyrites are quite common. The lavas have a high Na_2O content and are widely albitized. Thus the Suisarsian complex represents typical spilites, it is only weakly metamorphosed, and pyroxene is almost entirely preserved.

The Suisarsian effusives are followed by the Rapakivi-type granites. The term 'rapakivi' (rotton stone) is based on the fact that on weathering the Rapakivi granite breaks down into coarse gravel and sometimes into pebbles, since it consists of large grains (feldspars of several centimetres in diameter). The Rapakivi massifs are developed on the north-eastern shore of the lake Ladoga and to the south of Viburg (Great Viburg Massif). The granites are used as building stone. Thus all the shores of the Neva in Leningrad are paved by this rock. The Rapakivi granites are found in many regions of the U.S.S.R. and are always of Upper Proterozoic age.

An interesting situation arose in connection with the determination of the age of Rapakivi granites. Geological data suggested that these granites were formed in the middle of the Upper Proterozoic (600–800 m.y.). When, however, the age was checked by geochemical methods, then surprisingly they were found to be 1200–1600 m.y. and thus belonging to the Lower Proterozoic. Geologists immediately accused chemists of errors, but verification by numerous methods did not change the result. Consequently geological conclusions have to be revised to include these granites in the Lower Proterozoic.

On the Kola Peninsula the Lower Proterozoic contains well developed metamorphosed lavas, represented by a complex of schistose amphibolites.

The major intrusion phase is represented by the Karelian massifs (third group of intrusions). In Karelia they are identified in a number of regions and consist of granodiorites and granites, which intersect the Karelian Formation. The granites have much microcline and hornblende, and are dark pink with large phenocrysts of microcline. Intrusions are large. On the Kola Peninsula such intrusions are even more strongly developed. The acid intrusions are gneissose microcline granites and porphyritic granites and are associated with widespread migmatization. The minor intrusions phase is represented by pegmatites and basic sheets which cut the Archeozoic and the Karelian intrusions.

The Baikalian cycle on the Kola Peninsula begins in the Upper Proterozoic as lavas and diabase sills in the Imandra-Varzuga Suite. Younger diabases and porphyrites, spilites, tuffs and tuffites metamorphosed to green-schist facies belong to the Pechenga-Kuchin Suite. These are of great thickness, forming belts of 8–15km wide amongst steeply dipping strata. The basic and ultabasic rocks of the Pechenga-Kuchin Suite belong to the intrusions of Baikalian age. These intrusions have an associated copper-nickel mineralization of great practical importance. The intrusions are syngenetic, with the formation of diabasic lavas and sills, and are the youngest cross-cutting sheets. Since the ultrabasic rocks lie amongst the Pechenga-Kuchin rocks they used to be considered as Caledonian. Now that the Upper Proterozoic age of the suite is established, the intrusions must also be considered as Upper Proterozoic.

According to L. Ya. Kharitonova (1955, 1958) the numerous minor ultrabasic and alkaline intrusions of the Kola Peninsula also belong to the Upper Proterozoic. These intrusions form two north-westerly trending belts. The first is situated in the central part of the peninsula and includes the Pechenga-Kuchin intrusions, the alkaline massifs of Keivy, the middle reaches of the Ponoi, White Tundra, and many others. This belt was called the 'ophiolitic belt' by A. A. Polkanov (1937). The second, southern belt includes the massifs of Kovdor, Africand, Lesnaya, and Ozernaya Varaka, Kanozer, Salnye Tundry, and others of the south-western part of the Kola Peninsula.

The youngest Upper Proterozoic intrusive-effusive complex is of gabbro-diabases on the western shore of the lake Onega. The gabbro-diabases here are within the Shokshinian quartz-sandstones. Usually these bodies of gabbro-diabase are regarded as sills, but the presence of porphyrites and aphanites amongst them (as in the Suisarsian complex) leads one to accept the presence of lavas amongst the intrusions. Such associations are typical amongst the diabases of the platform. On the shores of the lake Onega the gabbro-diabases are cut by mainly vertical veins of fine-grained aplites and coarse pegmatites. These rocks are considered to be residual products of gabbro-diabase crystallization and have been intruded during and after the consolidation of the latter. It is interesting that evidently similar 'granite' veins have been found, by boring, in the Sinian complex of the Karelian Peninsula in the region of Vuoksa. Here these veins cut across red, compact sandstones with layers of conglomerate and dolomites lying under Gdovian sandstones and above the crystalline basement. Similar veins were found in a borehole at Orsha and again amongst the red Sinian. Sinian red beds all along the Vetrenoi Poyas are, as shown by boreholes, metamorphosed by porphyrites.

The Upper Proterozoic diabases are encountered in boreholes of the Archangelsk region, at Kresttsy (to the south-east of Novgorod), and in the region of Kaluga (diabasic tuffs). At Kresttsy the Gdovian sandstones are underlain by thick tuffites with diabase sills. In the Volga-Ural oil province the Upper Proterozoic gabbro-diabases are found in a number of boreholes. They reach thicknesses of 4–165m and form sills and dykes of the Lower Bavlinian Suite. The Upper Proterozoic diabases (basalts) are widely distributed in the south-west of the Russian Platform where they compose the Volynian complex, described already (p. 56).

Diabases, considered as sheets, are also found in the Hyperborean Formation of the south-western part of the Kola Peninsula.

PRECAMBRIAN MAGMATISM IN THE SOUTH OF THE PLATFORM

The sequence and content of magmatic cycles in the south of the Russian Platform are very similar to those in the north. The cycles have been studied in detail by leading Ukrainian petrographers N. P. Semenenko (1951), Semenenko and others (1960), and Tu. I. Polovinkina (1953).

The Saamian cycle begins with granodioritic intrusions and ends with aplitic granites, aplites, and pegmatites. There are also associated migmatites. The cycle is widely represented throughout the Ukrainian Massif and is encountered in the Voronezh Massif. The rocks of the cycle have been called the Kirovgrad complex by Polovinkina.

The Byelomorian cycle is also distributed throughout the Ukrainian Massif and at a number of points in the Voronezh Massif, such as the granite outcrops on the river Don at Boguchar. In the Ukrainian Massif the intrusions of the Byelomorian cycle intrude the Teterev-Bug Series. The intrusions consist of grey biotite-microcline granites and charnokitic granites, syenites, and diorites, with common migmatites, aplites, and pegmatites. The green-schist rocks represented by amphibolites are evidently Upper Archeozoic and were originally diabase lavas. The rocks of this cycle are units of the Dnyepr-Tokov complex.

The Karelian cycle starts with labradorite-gabbro intrusions and ends with Rapakivi granites which intrude the former. The granites are younger than the Jotnian Ovruchian sandstones. The rocks of this cycle form part of the Korostenian complex, the age of which has already been given (p. 45).

The Baikalian cycle.   The youngest Precambrian rocks, as in the north, are basic effusives (lavas and sills). They are widely distributed in Byelorussia and Podolia. Surrounding the Ukrainian shield from north and north-west as a continuous belt, they nearly reach the Dnyestr (Fig. 17). The effusives and tuffs form parts of the Volynian complex which underlies the Valdaiian complex (p. 56). The thickness of the complex reaches 400m. Basaltic (diabasic) lavas of 20–30m outcrop at the surface. They have good columnar joints, are very tough and have been worked.

Diabasic sills and lavas are widely developed amongst Sinian dolomitic formations of the western slope of the Southern Urals. Evidently identical diabases have been found within the shale-dolomite formation of the Lower Bavlinian Beds in the Kotlas borehole (lower reaches of the Byelaya). In general the Precambrian cycles of the north and south of the Russian Platform strongly resemble each other.

PALAEOZOIC MAGMATISM

The economically important alkaline rocks of the Kola Peninsula are attributed to the Palaeozoic magmatic cycles. Accurate determination of their age is at present difficult owing to inadequate information about Palaeozoic sediments and especially their fauna and flora. The latter is only known from one locality at Lovozero Massif, where there have been found xenoliths of amphibolized slates with remains of psilophyte flora (*Psygmophyllum*, *Archeopteris*, *Rachipteris*), which according to A. N. Krishtafovich belongs to the Upper Devonian. If these identifications are accepted then the alkaline massifs should be included in the Hercynian magmatic cycle. However, the preservation of the plant remains in xenoliths is bad and a psilophyte flora is also encountered in the Silurian. This permits to include the alkaline rocks to the Caledonian magmatic cycle, which is well developed in the Grampian Geosyncline closely related to the Kola Peninsula. This conclusion is confirmed by their absolute ages being 300–500 m.y.

A. A. Polkanov (1960) indicates that there are three epochs of formation of alkaline rocks. The main intrusions are the youngest (300–500 m.y.), the preceding epoch is Lower Proterozoic (1500–1700 m.y.), and the oldest is Upper Archeozoic or Upper Saamian (2420–2490 m.y.). The main belt of alkaline intrusives is situated in the central part of the Kola Peninsula. With it are associated the largest massifs, including the Khibina-tundra (Fig. 51) which reaches 1000–1200m in height, and the Lovozero Tundra Massif, which contains the aforementioned xenoliths. The structural situation of the massifs is complex. The Khibina Massif on one side is adjacent to Archeozoic gneisses and on the other to Proterozoic pillow lavas, quartzites and schists. From the west it is in contact with hornfelsed slates of possible Palaeozoic (Devonian?) age. The massif has rounded outlines suggesting that it was formed during the end-phases of an orogeny. The Massif of Khibina-tundra reaches 40km across, is cut by stream valleys and has a very characteristic, almost flat top (Fig. 52).

In addition to the main belt, small massifs and outcrops of alkaline rocks are known to the north and south of it. The southern belt passes near Kandalaksha and Tur'yev Point on the north-western shore of the White Sea.

The Khibina pluton (Massif), like many others, is a complete body, formed during several phases. In the early phases alkaline syenites predominate while in the subsequent phases nepheline syenites are common and lastly the alkaline granites appear (Fig. 53). According to Polkanov (1937) titaniferous magnetite ores are related to syenites, the apatites are related to nepheline syenites, and the mica to pegmatites.

The alkaline rocks of Khibina and Lovozero are considered to be the youngest igneous rocks of the Kola Peninsula and their absolute age is about 300 million years.

Alkaline intrusions belonging to the Hercynian and perhaps Mesozoic cycle are developed near the Azov Sea in the Ukrainian Shield. The rocks here consist of alkaline granites and syenites, nepheline syenites and their minor intrusive analogues (monchiquites, camptonites, etc.). There are only very rare igneous rocks amongst the Palaeozoic formations of the Russian Platform and there are no synorogenic intrusions, which can be expected because of weak manifestations of orogenies.

The Palaeozoic igneous rocks are represented by diabases, identified as basalts or palaeobasalts. These are dark, compact, heavy, fine-grained rocks, which are black or dark green or brownish and occurring in conformable sheets, which are sometimes intrusions, possessing active upper and lower contacts, but may be lavas having only a lower active contact. Pipe-like and irregular bodies representing feeding channels are much rarer and consist of peculiar igneous rocks. These rocks are not uncommonly associated with sulphide mineralization and they may contain diamonds.

The age of the lavas is determined with reference to the deposits amongst which they lie. The age of the sills is also thus determined, although strictly speaking this is incorrect since rising diabase magmas can intrude any formation. In the Timan both the effusive and intrusive diabases have been studied in great detail and have been found in boreholes as well as on the surface. Here and on Novaya Zemlya and the Kanin Peninsula they are of Lower Frasnian age. There are no known outcrops of diabases over the Russian Platform, but they have been discovered in boreholes. Diabases have been found in Soligalich where they occur as black basalts, some 15m thick, amongst Narovian red beds. Diabase lavas and sills have also been found in

FIG. 51. Khibina-tundra Massif in the valley of the Byelaya. In the distance is lake Imandra. *Photograph by V. D. Nalivkin.*

FIG. 52. Central part of the Khibina-tundra Massif, where its flat top surface is well seen. *Photograph by V. D. Nalivkin.*

deep boreholes in the Tatarian A.S.S.R., where they are present amongst the Upper Bavlinian (Lower Devonian) Suite. Younger lavas of Lower Frasnian age are also found here. In Kazaklar, for instance, a lava flow is 45m thick. It lies at the top of the Pashiiskian Suite, as do the diabases found in a borehole near the village of Radayevka, Kuibyshev Province. Devonian diabases are also known in the Perm Province and Bashkirian A.S.S.R.

Far to the south, along the northern boundary of the Dnyepr-Donyets Basin, Devonian diabases have been found in a number of boreholes. The thickest section

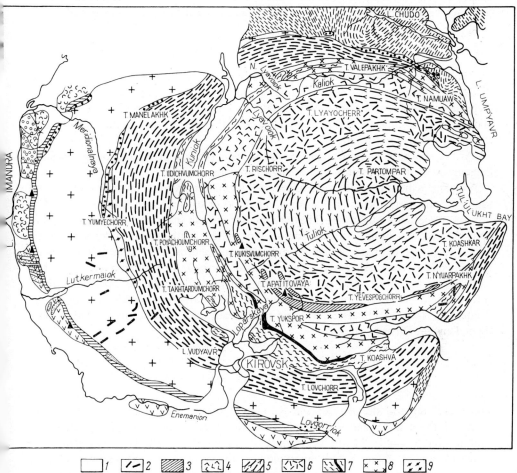

FIG. 53. The structure of the Khibina-tundra Massif. According to E. N. Volodin, adopted by L. Ya. Kharitonov (1958).

1 — Quaternary deposits; 2 — young minor intrusions; 3 — fine grained nephiline syenites; 4 — medium grained aegirine nepheline syenites; 5 — granitoid foyaites; 6 — massive foyaites; 7 — ijolite-urtites, apatite bearing rocks; 8 — rischorrites; 9 — alkaline syenite porphyries; 10 — trachytoid khibinites; 11 — massive khibinites; 12 — alkaline and nepheline syenites; 13 — Palaeozoic slates and hornfelses; 14 — Proterozoic hornfelses; 15 — quartz gabbro-diabases; 16 — metagabbro-diabases; 17 — pillow lavas, green schists and tuffogenic-sedimentary rocks; 18 — Archean gneisses.

has been revealed in the Chernigov borehole (Table 6), where basalts, diabases, porphyrites, tuffs, and tuff-breccias reach an enormous thickness of 455m and together with basal and covering sedimentary tuffogenous rocks form a thickness of 842m. This corresponds to a complete volcano. Fragments of diabases are found in the cap-breccias of a number of salt-domes of Ukraine, as for instance at Romny.

In all the enumerated regions diabases and basalt lavas and sills are developed over confined, relatively small areas. In none of these is there any connection between faulting and volcanicity, which indicates that, unlike the Siberian traps, which are associated with fissures, the Russian traps (diabases) are related to central volcanoes. One such centre, for example, can be identified near Korbin (Volynian complex).

MESOZOIC AND CAINOZOIC MAGMATISM

Mesozoic and Cainozoic effusives are only found at the margins of the Russian Platform. To the east of the Pechora Depression at the upper reaches of the Usa there are comparatively widespread basalts of Upper Cretaceous (?) age. Mesozoic and possibly younger basalts are widely distributed on the Arctic islands and especially Franz-Josef Land.

At the southern margin of the platform, near the sea of Azov, young, possibly Palaeogene basalts, andesites, and trachytes are encountered.

Volcanic ashes are widespread amongst the Quaternary and Neogene rocks of the southern part of the Russian Platform. They occur as fine-grain, rubbly sands and silts and sometimes mudstones. They are grey or greyish in colour. The beds are thin, up to 1·5–2m, but usually of the order of tens or few centimetres. Under a lens they appear to consist of sharp or slightly rounded glass shards. They stick to the tongue. In many cases even experienced geologists mistake them for ordinary sediments, but with certain familiarity and willingness to find them they can be easily identified. According to N. N. Karlov (1957) they are found amongst loessic mudstones of Quaternary age and Pliocene clays and mudstones, and in places in the Palaeogene (Don, Dnyepr, Dnyestr, etc.).

USEFUL MINERAL DEPOSITS

The useful mineral deposits of the Russian Platform are numerous and diverse. Many of these have national significance and the Nikopol manganese deposits, Khibina apatites, and oil from the Volga-Urals region are of international importance. It should be pointed out that although many of the deposits are well known and have been systematically exploited over several hundred years (salt and iron-ores), and although geological investigations have been carried out over the last two hundred years, Soviet geologists continue finding new deposits. Some of these such as the Devonian oil of the Volga-Urals Province are enormous. It follows that within the Russian Platform, which seems to be well explored, one can expect a whole series of discoveries of great scientific and industrial significance.

METALLIFEROUS ORES

These are numerous, diverse and sometimes large scale deposits. They are divided into two groups—geosynclinal and platform type. The former are found in the

Precambrian and are either magmatic or sedimentary, the latter are found in younger formations and are almost entirely sedimentary.

Ferrous Metals

The deposits of ferrous metals are numerous and sometimes of large magnitude.

Iron. Three principal groups can be distinguished. The first includes the deposits of Krivoi Rog, the Kursk magnetic anomaly and Byelgorod. It has a great industrial importance. The ores are of sedimentary origin, of Lower Proterozoic age, and strongly altered. The stratigraphy and tectonics of Krivoi Rog (p. 131) and Byelgorod (p. 52) have already been discussed.

The second group includes the Kola Peninsula deposits. Ores are of the same type as in Krivoi Rog and are represented by banded magnetite and haematite schists and quartzites, but of a more ancient age. These rocks are associated with biotitic Archeozoic gneisses. The Cherepovets metal factory, which is the forerunner of metallurgy in the north-west, was opened in 1955 to process the open-cast ferruginous quartzites of Olyenogosk and Eno-Kovdor deposits.

The third group differs strongly from the two preceding groups. It includes the Vyksunskian (near Murom), the Tula, the Lipyets and the Khopersk deposits. All the ores are sedimentary; the first three are lacustrine Jurassic and the last is marine Cretaceous. There are many such deposits and consequently, despite the small size of the individual ore-bodies, they have collectively the same resources as the Krivoi Rog deposit in 1938. The ores are accessible and the long exploited deposits supply the metallurgical establishments of the central provinces.

Manganese. The largest (Nikopol) manganese region lies on the right bank of the river Dnyepr, between Krivoi Rog and Donbas. The irregular surface of Precambrian gneisses and granites is covered by Oligocene sands and clays, amongst which lies the ore-bearing horizon. The sands and clays are overlain by Sarmatian sands and marls. The ore-bearing horizon (1·5–2m and rarely up to 4m thick) consists of light coloured clays rich in psilomelane, pyrolusite, and manganite concretions and oolites and represents near-shore marine deposits. The Nikopol ores supply the metal industry in the south of the U.S.S.R. and are also exported.

The Great Tokmak deposit, recently discovered, is in direct continuity, and lies to the west of the Nikopol deposit. It includes the greatest manganese deposit in the U.S.S.R. (Syelin, 1962) and is situated on the left bank of the river Dnyepr in the Zaporozhy'e Province of the Ukraine (Gryazov and Syelin, 1959). The fauna of the ore-bearing horizon is identical to the fauna of the Khadumian Horizon in the Caucasus and belongs to the Lower Oligocene.

Nickel. In the Kola Peninsula there are a number of copper-nickel sulphide ore-deposits associated with ultrabasic intrusions (peridotites) of the Baikalian magmatic cycle. The deposits of Pechenga, Monche-tundra, and Volch'ya Tundra are of this type.

Titanium. Minerals containing titanium, zirconium, rutile and ilmenite occur within the Neogene marine sands. They represent disintegration products of Precambrian rocks and were concentrated in marine beach sands by wave action. Enriched sands occur in the Poltavan and Sarmatian stages in Ukraine.

Non-Ferrous Metals

Non-ferrous metalliferous ores are much less significant in their distribution and quantity.

Copper. There are both magmatic and sedimentary copper deposits. The magmatic deposits are associated with the peridotites of Monche-tundra and Volch'ya Tundra. Thus the recently built town Monchegorsk is a metallurgical centre of the copper deposits of the Kola Peninsula.

The sedimentary copper deposits are represented by marine sandstones of Upper Permian age, situated at the western margin of the platform. They form a zone stretching from Perm to Orenburg. The near-surface deposits have already been worked out, but there is a possibility of finding further deposits at depth. It is of interest that the copper-bearing sandstones were the cause of the establishment of the first mining college—the Leningrad Mining Institute. Bashkirian mine-owners, in whose lands the deposits lay, petitioned Catherine the Second 'that on the basis of their experience they learn and become versed in mining economy, but in order to establish their craft and to strengthen the economy would like to beg for a school of mining.' The necessity for this was already felt by the advanced sections of contemporary society, and the petition served as a last push, as a result of which the Mining Institute was inaugurated in 1773. A whole series of engineers graduating in the Institute took an active part in the development of the deposits and smelting of the ore.

Aluminium. The most important aluminium-ore (bauxite) is represented by the Tikhvin and Olonyets bauxite basins, by the deposits of the Ukraine and smaller deposits on the southern margin of the Moscow basin. In the Tikhvin basin the bauxites are found amongst the Lower Visean lacustrine beds constituting the lower part of the coal-bearing section. These beds consist of clays, fire-clays, bauxitic rocks and bauxites overlying the Upper Devonian red beds. The lakes in which the bauxites formed were situated in wide valleys merging into a near-shore plain adjacent to the Visean sea. The Olonyets deposits are of the same type.

Nepheline. Nepheline serves as a raw material for extracting aluminium. Its deposits amongst the alkaline rocks of the Kola Peninsula are enormous. Alumina apart, the industrial method involved gives by-products of potash, calcinated soda and high-grade cement.

Rare and Dispersed Elements. Zirconium, niobium, tantalum, helium, and others belong to the dispersed elements. They have diverse valuable applications and properties and are part and parcel of modern technology.

FUELS

Fuels are widespread and diverse. Supplemented by the Donyets and Pechora coal basins the deposits of fuels completely satisfy the needs of our highly-developed industry. The deposits include black and brown coals, lignite and peat, oil and gas.

Coals

The Lower Carboniferous coals have the widest distribution and resources and are intensively worked in the Moscow basin. Remains of lepidophytes, ferns, and cordaites accumulated in large amounts in large near-shore swamps, thus producing coal seams. The seams have an average thickness varying from tenths of a metre to two metres, and occasionally reach up to 7·5 metres. The size of the swamps has determined the extent of the seams which rarely exceeds tens of kilometres. Deposition in the near-shore area, under a regime involving the rapid influx of sands and clays has been the cause of the frequently observed contamination of the seams.

This of course spoils the quality of the deposit, but their convenient situation with respect to the central industrial provinces is such that they are actively worked.

The Moscow basin produced in 1956 a quantity of $41 \cdot 7 \times 10^6$ tons and its resources to the depth of 200m are $17 \cdot 5 \times 10^9$ tons of which $16 \cdot 6 \times 10^9$ tons have been proved. In its importance the basin stands in the third place after Donbas and Kuzbas. In the last few years the margin of the known coal has been pushed to the west and east. New deposits such as those of Dorobug, Kaluga, Kozyel, Koroblino (south of Ryazan') have been explored and are being exploited. The propected resources have been multiplied by ten.

In Tataria, Bashkiria, and neighbouring regions, after the last war, borings for oil disclosed one to three coal seams of $0 \cdot 3$ to $0 \cdot 23$m thick amongst the Carboniferous coal-bearing suite. The deposits have a lensoid form, small extent, and lie at the depth of over 1000m. These are being here called the 'Kama coal basin' (p. 92). Geological resources of this basin are over 10×10^9 tons, but they are not exploited.

Relatively recently the industrial significance of the Lvov-Volyn basin has been demonstrated. Shafts have been constructed already and in 1956 production reached $0 \cdot 5 \times 10^6$ tons. The coal seams lie amongst a sandy-clayey Upper Namurian Suite. The proved resources are $1 \cdot 3 \times 10^9$ tons. There are 3–8 seams of $0 \cdot 5$–$3 \cdot 5$m in thickness.

The Upper Carboniferous and Lower Permian coals developed in the Donyets and Pechora basins are unknown in the central part of the Russian Platform. The Mesozoic coals, which are widely developed in Western Siberia and Siberian Platform, are almost unknown in the Russian Platform.

Palaeogene and Neogene coals and lignites are found in the Ukraine. To the south of the Dnyepr the Dnyepr brown-coal basin is situated amongst the beds of the Buchakian Stage. The 1–3 seams reach 20m and their total resources are $3 \cdot 3 \times 10^9$ tons. Production in 1956 was $9 \cdot 8 \times 10^6$ tons.

Peat and Lignite

During the Upper Anthropogenic epoch the areas of the melted ice-sheets became the site of boulder clay (bottom moraine). This sheet of clay was ideal for the formation of widespread bogs. Huge bogs became the loci of concentration of plant remains, and exist even now. The peat which was produced as a result is 6–8m, or even more, thick. Owing to the extent of the bogs the areas of peat are enormous and so are the resources. Certain turf complexes are still boggy, while others are dry and tree-covered. In a number of regions, especially say near Moscow and Leningrad, peat is extensively worked and is important as fuel.

In places amongst the Anthropogenic and Neogenic deposits beds and lenses of lignite—a compacted and altered peat—are found.

Petroleum

The discovery of major oil deposits on the Russian Platform is a considerable achievement of Soviet geologists and has a world significance. The initial discovery was made by two geologists. One of them was old and experienced (P. I. Preobrazhenskii) while the other quite young had just begun his scientific career (A. A. Blokhin). They were also greatly helped by one of the most distinguished oil-geologists (I. M. Gubkin). The industrial importance of the first major deposit (at Ishimbai) was proved by Blokhin in 1932. At that time oil on the platform was a subject of investigation by only a few geologists, while now there are thousands thus occupied. Around the new deposit new towns such as Ishimbai and Oktyabirsk with tens of thousands of population have grown and new factories and railways have been built. The tempo and dimensions of this expansion have exceeded anything known previously in the oil industry.

Two types of deposits are known here—those found in reefs and those in anticlines. The reef deposits were found first in 1924 near the Upper Chusovaya townships on the river Chusovaya. In 1932 the first commercial oil was struck at Ishimbai. Here oil lies in huge limestone reef massifs of Lower Permian age and is associated with porous, so-called sieve dolomites. The reef massifs reach 600–800m in height and 2–7m² in area and occur on the eastern margin of the platform where it is transitional into the Uralian marginal downwarp. Their largest group is found to the south of the Ufa near the towns Styerlitamak and Ishimbai. The top of some of the massifs has been uncovered by erosion and the massifs then outcrop as small individual mountains known as *shikhan*. They are representatives of the hidden massifs covered by Kungurian salt and gypsum and containing oil and gas (Fig. 54).

The second, anticlinical, type is much more widely developed and accounts for the majority of deposits. The oil lies in the arches of the anticlines described in the section on tectonics, and is concentrated in the porous beds. Three main types of such porous beds are known—those with primary porosity, secondary porosity, and breccias. Amongst the first group, sands and rubbly sandstones are important and are commonly found amongst basal and sandy-clayey formations belonging to the Devonian and the Carboniferous (Pashiiskian Suite, Vereiian Suite). Amongst the second group porous, cavernous and sieve dolomites and rarely limestones predominate. The reservoirs of the third group are the rarest and are found on the flexed parts of anticlines and other folds, where the brittle limestone brecciates in the process of folding, and the porous breccias become sites of accumulation of oil and gas.

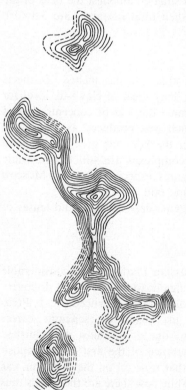

Commercial oil is found in beds of different ages ranging from the Upper Permian to the Bavlinian Suite, and in different places different horizons are oil-bearing. The sands and porous sandstones of the Vereiian Suite of the coal-bearing formations and the basal beds of the Upper and Middle Devonian are particularly important. In other deposits porous, Lower Permian dolomites and limestones have a great significance.

Fig. 54.

Under surface relief of Ishimbaian reef limestones according to A. A. Trofimuk.

In Bashkiria (Ovanesov, 1961) 85 per cent of all the oil is obtained from Devonian sandstones, 10 per cent from Lower Permian reef-limestones and 5 per cent from Carboniferous sandstones. It is significant that the number of commercial horizons within the Devonian and Carboniferous limestones (not only reefs) grows each year.

In the Perm region the relationships are different. There the Visean coal-bearing suite contains up to 73 per cent of explored deposits and the Devonian has only 3·2 per cent.

Upper Chusovaya Townships. The first reef deposit was found by Preobrazhenskii in 1924. Oil lies in porous dolomite beds amongst Lower Permian limestone reefs. The deposit is now exhausted.

Ishimbai Deposits. Of the same type as above, but larger. The zone of the oil-bearing reefs extends along the sub-Uralian downwarp from the latitude of the town Ufa to the south of the town Styerlitamak. A second type of deposits in the north-east begins with Krasnokamsk deposits on either side of the Kama, to the north and south of Perm. The oil here is found in several horizons, from Kungurian dolomites down to Devonian sands. Near the town Bugul'ma are the Tuimazinskian deposits found in dome-type structures of various sizes. The oil-bearing strata are the same as before.

Tatarian Deposits, which were recently found, occur in Southern Tataria to the south of Kama. The main oil-bearing horizons are associated with the Devonian.

Syzran-Kuibyshev Deposits occur in anticlinal structures on the Zhigulev Ridge. To the south of these lie the Buguruslanskian deposits in which the oil is found in the Upper Permian. To the south-west of Saratov there is a group of Saratov deposits where the reservoir rocks of breccia type were found amongst the Lower Carboniferous limestones.

Recently oil was found in the Orenburg sub-Urals. All these deposits are parts of the vast area known as the 'Volga-Ural region'. The name 'Second Baku', formerly applied to it, is now rejected since the 'Second Baku' is considerably more productive than the 'first Baku'.

High-pressure oil has been discovered in the Komi A.S.S.R., in the Pechora Depression, near Timan, and in the area of the western Tebuk (60km east of Ukhta). According to A. Ya. Krems (1961) this deposit is rich and high-pressure and is associated with several thick beds. It has relatively large (for the Timan-Pechora region) industrial resources of light, high-grade oil. In addition oil deposits are found to the north of Perm (Maikor), which renders all the region between Timan, Kama, and Ural suitable for investigation.

Oil seepages were also found in boreholes in the Palaeozoic of the Dnyepr-Donyets region and in the Pripet Depression. The structure of these depressions will be further discussed.

It is interesting that the Kaliningrad borehole has yielded porous quartz-sandstones of Upper and Middle Cambrian age (49m thick) with a bed of 3m thick uniformly saturated by oil (Dikenshtein, 1957).

Gas

For the Russian Platform gas is a new type of fuel, the utilization of which is expanding. The best example of this is served by the gigantic pipe line from Saratov to Moscow. At present the network of pipe lines has reached a considerable size and is expanding annually.

As a rule gas is associated with oil deposits. Gas-bearing oil deposits have accumulation of gas at the top, and on being tapped by a borehole this gas surges upwards carrying the oil as well, throwing it up as fountains tens of metres high, as is the case with the Ishimbai deposit. In some regions the gas is found on its own and is the object of exploitation. The Korobkovo deposit (near Saratov) is of unique importance in this respect.

Gas deposits are known along the southern margin of the platform all the way to the Western Ukraine. To the south of Kharkov a new important gas deposit has been found at Shebelinka, which led to the building of the Shebelinka-Kharkov-Kursk-Orel-Bryansk-Moscow pipe line. Gas is present in the Lower Trias, Lower Permian, and Upper Carboniferous deposits of the north-western section of the Donyets Depression. In addition deposits of gas have been found within this depression to the south and west of Poltava in the Lower Carboniferous, Trias and Middle Jurassic. Here the Lower and Middle Carboniferous oil has also been established. The largest deposits of gas have been found in the Volga-Ural Province. In most cases gas is found together with oil and they can be called oil-gas deposits. There are also some purely gas deposits which are in the main situated on the margins of the regions with oil-gas development.

To the south the gas-bearing zone starts from Volgagrad and continues along the Volga, via Kamyshin, to Saratov. From Saratov it bends to the east and is traced as far as Orenburg. The gas deposits have been described in a volume entitled *Gas resources in the U.S.S.R.* (1959).

Oil-Shales

It is interesting that the distribution of the oil-shales and oil do not coincide. Oil is only found in the Middle and Upper Palaeozoic, while in the Lower Palaeozoic and Mesozoic it is almost entirely unknown. The oil-shales, however, are developed in the Lower Palaeozoic and Mesozoic and are much less widespread in the Middle and Upper Palaeozoic. The existence of the oil-shales as such does not involve the formation of oil. They must be buried to the depth of over 1000–1500m under the condition of elevated pressure and temperature in order to produce liquid and gaseous distillates. It is possible that the parent rocks of oil are not only oil-shales, but all the rocks with higher than average content of fats and proteins, including bituminous limestones. The main factor thus is not the concentration of organic matter, but its total amount and conditions suitable for its distillation.

Precambrian oil-shales are unknown, although schungite is probably a metamorphic representative of the oil-shales of Lower Proterozoic age. Lower Palaeozoic oil-shales are widely found in the Baltic countries and are extensively exploited. Two horizons are known: the Lower Ordovician *Dictyonema* shales and the Upper Ordovician, Estonian kukersite. The *Dictyonema* shales are not suitable as a fuel owing to the high ash content. The Estonian or Gdovian shales lie amongst the limestones and dolomites of the Kukruse (Kukersian) Horizon of the Ordovician. They were discovered as far back as 1910 near the locality Kukruse (Kukers) in Estonia and have an areal extent of about 300km² and thickness of up to 3m. From Estonia they extend into the Leningrad Province where they wedge out In the Gdov region their thickness decreases to 1·8m and near Vermansk to 0·6–0·75m and further to the east they disappear altogether.

The Estonian (Gdovian) shales consist of kukersite, which is a yellowish-reddish light shale full of remains of diverse and rich marine fauna and involving mainly large numbers of unicellular algae with fatty and protein-rich protoplasm. The rock originated in large muddy depressions on the sea bottom away from the shore. It is widely worked as fuel and a source of gas. The gas is used both in Leningrad and Tallin.

The Middle Palaeozoic oil-shales are the Domanik Beds, and Domanik-type facies of Middle and Upper Devonian and Lower Carboniferous age. The term Domanik Beds originates from the stream Domanik in the Timan where they were first discovered; they consist of thinly-bedded bituminous limestones and shales, with a peculiar goniatite and pteropod fauna, reaching up 40–60m and more. The age of the Domanik is Middle Frasnian, and although exposed in the Timan and western flank of the Urals, it is only found in boreholes in the Russian Platform.

Domanik-type facies are lithologically and faunally similar to Domanik, but contain other types of goniatites belonging to the Givetian (Infradomanik), the Lower Frasnian (Shugurovian), the Famennian and the Tournaisian. The Domanik and Domanik-type facies are considered as the parent rocks of the Volga-Ural oil. To the west of the Volga they are replaced by clays. The Domanik is not used as fuel since it contains a large number of clayey and carbonate beds. The Upper Palaeozoic oil-shales are not known over the platform, but are found in the Artinskian deposits of the western slope of the Urals (river Yurezan'). The Mesozoic oil-shales are associated with marine Upper Jurassic and Lower Cretaceous deposits. The largest deposits are known from the upper reaches of the Vyatka and the Kama and from the eastern shore of the Volga, near the village Kashpur, below Syzran'. The Kashpur shales belong to the Lower Volgian. Their thickness is about 3m and total thickness of the worked horizons reaches 1·6m. The shales are black and lie amongst grey clays, which are sometimes bituminous and contain a marine fauna. Like kukersite they were formed in submarine muddy depressions.

Cainozoic oil-shales of the platform are lacustrine and are known from the Ukraine, where they occur together with brown coals.

NON-METALLIFEROUS DEPOSITS

There are many different deposits of non-metalliferous materials and sometimes they are of world significance. The non-metalliferous deposits can be divided into two large groups, the rock materials and the deposited materials, the first group involving magmatogenic and the second group sedimentary processes.

MAGMATOGENIC RAW MATERIALS. These are represented by the following raw materials found in the Precambrian. *Apatite and Nepheline* are found in alkaline massifs of the Kola Peninsula and they are particularly numerous in the Khibinian Massif (Khibina-tundra) near which a new town—Kirovsk—has been built (Fig. 55).

Apatite is an acid phosphatic salt of calcium with some admixture of fluorides and, rarely, chlorides of calcium and sodium. The P_2O_5 content reaches 41–42 per cent, thus is much higher than in sedimentary phosphates. Thus apatite is a valuable material for preparation of fertilizers and in particular of superphosphate. The apatite occurs in lenses, the most important having a length of 4km and a thickness of up to 200m. The total resources of apatite are over 2×10^9 tons and exceeds the resources of all the western nations by a factor of 2·5.

Nepheline is a mineral which formerly was used in small quantities in the glass industry. Its high alumina content has suggested its use as a raw material for aluminium. The problem has been successfully solved and nepheline is intensively exploited. The resources of nepheline are unlimited since it forms the bulk of the alkaline complexes.

Feldspar, mica, and quartz enter the composition of the Archeozoic pegmatites and aplites. They are particularly widely developed in the northern part of Karelian A.S.S.R., near the shores of the White Sea and in the region of the Chupinskian fjord to the south of Kandalaksha.

Precious, coloured, and technical minerals are also found in the Precambrian. Deposits of topaz and labradorite are worked in the north-western part of the Ukrainian shield. Deposits of talc are known in the southern part of Karelian A.S.S.R.

SEDIMENTARY NON-METALLIFEROUS RAW MATERIALS. Sedimentary non-metalliferous materials are found in the Palaeozoic, Mesozoic, and Cenozoic. The deposits are diverse, numerous, and are found over the whole territory of the Russian Platform, often reaching considerable size and having all-union importance.

Carbonates—limestones and dolomites are essentially Palaeozoic and have been used for many thousands of years. Moscow is known as 'white stoned' since many buildings, even before the Tatar invasion, were built of flaggy Middle Carboniferous limestones.

Clays. Clays are no less ancient useful raw material than the carbonates. Clay pottery was made many thousand years ago. Amongst clays there are three principal varieties: plastic clays, fire clays, and kaolinites.

Plastic clays are most widely distributed and are worked in thousands of quarries. They are widely used in the brick industry. Plastic clays were formed in lakes and less frequently in marine embayments. The Anthropogenic varved clays serve as an example.

Fire clays have frequently a low plasticity and are then called 'dry' clays. Flint clays have a high silica percentage and are compact and non-plastic. The

FIG. 55. The Khibina-tundra. Apatite quarry above Kirovsk.

Palaeogene Chiasovar deposit in the Ukraine and the Lower Carboniferous Borovichi deposits can be quoted as examples as having all-union importance by supplying the ceramic industry. The fire clays most frequently form part of marine or shoreline deposits.

Kaolinites, formed from the weathering of felspathic rocks, are widely distributed on the Ukrainian crystalline massif where more than 150 deposits are known and many are worked. The thickness of the kaolinite blanket reaches 10m and more. Eighty per cent of Soviet kaolinite comes from the Ukraine. Kaolinite has the same properties as clays and is widely used in the paper industry.

Salt. Some of the largest salt deposits are situated on the margins of the platform and the sub-Ural Depression in Verkhynekamsk and Il'yetsk. Immeasurable resources of salt are present in the salt domes of the Volga-Emba Province and Slavyano-Artyemovsk deposits of the Donyets Basin. Within the platform the salt is, and has been obtained for several hundred years, from salt solutions, but large deposits are unknown.

Gypsum and Anhydrite. There are many known deposits of gypsum and anhydrite, many of which are large and are being worked. The Lower Palaeozoic deposits are known from the Baltic and Arkhangel'sk regions and Devonian deposits stretch from Leningrad Province, through Pskov, to Minsk and they are also worked in the Moscow basin. Over a large area in the north-east gypsum and anhydrite are associated with Permian deposits. It should be pointed out that large deposits of this age are found in Zvoz (the north Dvina). Gorkii, Kama, Ufa, and Kuibyshev. Palaeogene and Neogene gypsum is found in the Ukraine and Podolia.

Sulphur. The main deposits of native sulphur are concentrated in the Ukrainian S.S.R. The sulphur is of Neogene age and originated as a result of reduction of gypsum by bitumens. The deposits are situated along the south-western margin of the Ukrainian crystalline massif.

Phosphorites. Hundreds of deposits of diverse geological ages are found in the Russian Platform. Thus the resources are enormous, but the P_2O_5 content is everywhere low. Thus the economic significance of these deposits is less than that of the Khibina apatites and Karataris bedded phosphorites. Nevertheless many deposits of the platform are successfully worked. They belong to the Volgian and Valanginian stages as at Yegor'yevo (near Moscow), Vyatka-Kama, and Kineshemka, to the Cenomanian in the southern Russian Depression, to the Senonian as at Aktyubinsk.

The peculiar Lower Ordovician deposits in Estonia and Leningrad Province represent accumulations of large numbers of obolid shells, lying in beds amongst Tremadocian sandstones. The shells have economic amounts of P_2O_5. The deposits have been surveyed and are ready for exploitation.

Sands, Sandstones and Quartzites. Deposits of building sand, moulding sands, abrasive and glass sands are numerous and belong to diverse geological ages. Sandstones and quartzites are more rarely used and are mainly Proterozoic. The Shokshinian and Ovruchian quartzitic sandstones are particularly well known.

REGIONAL DESCRIPTION

Here three regions will be considered, the Volga-Emba region of salt tectonics, the Baltic, and the Timan-Pechora regions.

VOLGA-EMBA REGION

General Description

The region is characterized by a widespread regional devolpment of salt-domes.

BORDERS. The eastern margin of the region is drawn along the foot of Mugodzhar, the western along the Volga. The salt-domes are, also, undoubtedly developed to the west of the Volga towards Yegreni and Donbas. The southern border is drawn to the south of Emba along the Caspian shore, although it can be supposed that salt-domes are also developed below the sea. The northern border is arbitrary, since there is a gradual transition into the Volga-Ural Province. The boundary is normally drawn at the last salt-domes situated on the tectonic step stretching from Ural'sk to Sol'-Ilyetsk.

The relief is flat and has been conditioned by the numerous transgressions of Neogene and Anthropogene seas. The uniformity of the plain is only broken by small mountains and hills, which despite their low height (tens of metres) are seen for long distances. These hills are tops of salt-domes. In the west they are rarer and higher as for instance: Mount Bogdo on the lake Baskunchak (Fig. 27), Mount Chapchachi, and the Indera mountains on the shores of the lake Indera. It is interesting that the bitter-salt lakes of Elton, Baskunchak, Indera and Chelkar lie directly over the tops of salt massifs and the lake salt is derived from the saline ground water. In the east the number of hills is large, reaching 300, but they are lower (Fig. 56).

Stratigraphy

Pre-Devonian deposits are unknown.

DEVONIAN. Flaggy greenish-greyish muddy limestones with *Spirifer* (*Theodossia*) ex gr. *anossofi* Vern. were found near lake Baskunchak as fragments in the

FIG. 56.   On the side of the Kai Kashi salt dome. In the distance is the vast Ciscaspian Depression.

cap breccia overlying the salt massif and formed when the massif moved up. These limestones are identical to the Yevlanovian limestones found in cap breccias of salt-plugs in the northern part of the Donbas and the region of Romny. Devonian deposits were also pierced by the Zhanasu borehole sited at the southern border of Emba basin. The Devonian deposits of Uralian type are of considerable thickness, are somewhat metamorphosed, and consist of limestones and shales. The borehole encountered only the upper and possibly a part of the middle division.

CARBONIFEROUS. The Carboniferous has been also encountered in the Zhanasu borehole and consists of more than 500m of limestones of Upper, Middle, and Lower Carboniferous age. Moreover, flags and fragments of Upper Carboniferous limestones with fusulinids have been discovered to the south of lake Baskunchak in the same cap-breccia which has yielded Devonian fragments.

LOWER PERMIAN DEPOSITS. The Lower Permian deposits have been palaeontologically identified and are developed on the western slope of the mount Mugodzhar, where they reach 2000–2500m. These rocks have been encountered in boreholes at Astrakhan and Tygurakchan to the south of Emba. The Sakmarian and Artinskian stages consist of clays, limestones, and dolomites some 238m thick. The Kungurian gypsiferous formation consists of clays, sandstones, and anhydrites reaching 280m in thickness. In other regions the Kungurian varies from 200 to 2500m.

The salt-domes, by analogy with the western slopes of the Urals, are normally considered as Kungurian. In the eastern region the Lower Permian age of the salt has been proved on the basis of spores. In the western regions the problem is more complex and the cap-breccia involves fragments of Upper Carboniferous and Upper Devonian limestones. Consequently the salt cannot be Lower Permian. The accurate age has not been determined, but the development of two salt-bearing horizons of Lower Permian and Middle Devonian ages is possible.

UPPER PERMIAN AND TRIAS. There are two types of deposits above the salt massifs. The first type is of limited distribution along the northern and eastern margins of Volga-Emba region. It is characterized by the red beds (500m) of the Ufian Stage, overlain by Kazanian red sandstones (up to 800m) with intercalations of grey marls and limestones, which in the north yield a marine fauna. Above these lie Tatarian red beds (900m) and the Lower Trias red beds (300m). The total thickness reaches 2500m. The second type is developed over the rest of the extensive Volga-Emba Province. It is characterized by the so-called Permo-Trias lying over the salt. The Permo-Trias is a uniform, almost unfossiliferous succession of red beds reaching 2000m in thickness. The detailed prolonged investigation by E. I. Sokolova has enabled her to distinguish the following parts of the Permo-Trias succession.

Upper Trias.   Variegated clays and sandstones with conglomerates towards the base (200m).
Lower Trias.   Baskunchak stage consisting of grey clays with marine fauna (up to 80m).
Vetluzhian Stage. Variegated sandy-clayey formation with conglomeratic sandstones towards the base (250m).
Tatarian Stage.   Red sandstones and clays (600m).
Kazanian Stage.   Red clays (500m).
Kungurian salt and gypsum at the bottom.

MARINE LOWER TRIAS. In the higher horizons of the variegated continental formation there is a thin (tens of metres) series of beds of grey and dark clays containing Lower Triassic fauna with *Dorycranites bogdoanus* and *Mytilus dalai-lamae* These beds are found on Mount Bogdo near lake Baskunchak and are established as the Baskunchak Suite.

In the Indera mountains, amongst variegated rocks, there is a group of light marls and clayey limestones with lamellibranchs (*Myophoria*), probably also of Lower Triassic age. These outcrops represent the deposits of an extensive embayment, which penetrated into the Ural-Emba Province from the south, having been a branch of the Tethys.

CONTINENTAL JURASSIC overlies the continental Upper Trias. It is 300–600m thick and consists of sandstones and clays of dark grey and brown colour and contains plant remains. Its lower part is Lower Jurassic while the upper part is Middle Jurassic, which is coal- and oil-bearing and is sometimes called the Dossorian Suite. Amongst it there are encountered horizons of coal, which are worked at Aktyubinsk. Sands, which are also found in it, are in places compact and barren, but elsewhere are porous and oil-bearing. The oil is obtained near a number of salt-domes, it is primary and has originated from bituminous Middle Jurassic rocks.

UPPER JURASSIC AND LOWER CRETACEOUS. Deposits are of the same type and have the same fauna as on the Russian Platform, but are thicker and reach 1000–1500m. The marine transgression begins at the end of the Middle Jurassic as there are Bathonian marine sediments. The Barremian is represented by continental variegated sands and clays (up to 360m), which are commonly oil-bearing. The discovery of limestones with a Volgian fauna at Orenburg and of a Lower Cretaceous *Nerina*-bearing fauna at Baskunchak is of interest.

THE UPPER CRETACEOUS AND PALAEOGENE are directly continuous from the Russian Platform and are similar throughout both lithologically and faunally, but are thicker reaching 800–1200m. Clays, limestones, and chalk are characteristic of the Upper Cretaceous, while clays and gaizes are developed in the Palaeogene. Santonian phosphorites, worked near Aktyubinsk, have an industrial importance.

THE NEOGENE AND ANTHROPOGENE are widespread and reach 300–600m in thickness. Lithologically they consist of repetitions of marine, fresh-water, and terrestrial deposits consisting of friable sands and clays.

Amongst marine transgressions of the Neogene the Akchagylian transgression, which is evident throughout the Ciscaspian Province with deposits overlying salt domes, is the most important. It is also evident along the valleys of the Volga, the Kama, and the Byelaya. The succeeding, Apsheronian transgression is only evident in the Ciscaspian region. Anthropogenic transgressions begin with the Greater Baku transgressions developed throughout the Volga-Emba Province, and succeeded by the Khozarian transgression in the south-western part of the Ciscaspian depression. The latest Khvalynian transgression was more widespread than the Khozarian transgression and is developed over the whole of the Ciscaspian region.

These transgressions have been caused by the increase in the effluence of waters brought in by the Volga, the Terek, and the Kura. Undoubtedly the heightened water supply of the Volga, which has been especially important, was determined by the changes in the melting of ice sheet over the platform, as the territory occupied by the continental ice sheet increased so did the water supply. It is possible that the ice sheet was also the cause of the Akchagylian transgression since otherwise it is inexplicable, as the Akchagylian sea was already self-contained and its size determined by the water supply of the rivers.

Tectonics

The structure of the Volga-Emba region is rather peculiar and is characterized by a widespread development of salt-domes. The Anthropogene and Neogene lie horizontally and are only rarely faulted. The Palaeogene and the Upper Cretaceous cover large areas and show distinct folds of platform type, which are large, ovoid, gentle and imperceptible. All these deposits are pierced by numerous salt plugs of which, according to the latest data, there are more than 900. On geological maps the plugs are visible along the outcrops of Lower Cretaceous, Trias, and Permian. The sub-salt formations lie at depths of 5–7km and in places 9–10km (Slepakova, 1961).

The salt-domes of the region are diverse and have been widely studied. The structures arose as a result of cores of Devonian and Lower Permian salt breaking through the anticlinal covers. Salt massifs in the process of rising raise the overlying deposits, tear them apart and displace them as large blocks of different outlines. The massifs have different shapes varying from bulbous lenses to high domes and stock-like bodies of great height. Normally they consist of rock salt, although there are rare examples of potash plugs. In the upper parts of the massifs gypsum is common and a cap-rock of several tens of metres thick usually marks their tops.

On upward movement a cap breccia, formed as a result of frictional breakdown of overlying rocks, originates. The breccia includes huge fragments of most diverse deposits. The cap breccia of the Baskunchak dome contains Upper Devonian and Upper Carboniferous fragments. It is possible that Mount Bogdo is a block of Permian and Trias rocks elevated to the surface from a depth of not less than 1000–1500m (Fig. 27). The thickness of the cap-breccia varies from a few metres to several hundreds of metres. The empty spaces in the breccia are sometimes filled by petroleum.

The depth of the salt-domes is enormous and so far not one dome has been pierced through by drilling, although boreholes penetrated some domes down to depths of 2000m. Geophysical data (Slepakova, 1961) suggest that the domes vary from 2 to 7km in height and are sometimes even higher. In horizontal cross-sections domes are also rather large, reaching 3–4×6–8km across, although most frequently they are 2–4×3–5km. The structure of a dome is shown in Fig. 57, which is based on the example of the Dossor dome. The salt core, which lifts the overlying block of Permian, Triassic, and Jurassic strata, has a gently convex cross-section. Amongst the Middle Jurassic Dossorian Suite there are four horizons of oil-bearing sands bounded by a fault. Along the fault the salt is in contact with the Lower or even Upper Cretaceous. In other domes salt reaches the surface.

The structure of the Dossor salt-dome is also obvious on the map (Fig. 58). S. T. Mironov distinguishes three types amongst the domes of the Emba region. The first type (Makatian) is characterized by the almost unbreached dome and it is rare. The second type (Dossorian) has one entire half, while the other is broken up into blocks. The third type (Imankarian) has a dome completely broken into a smaller number of blocks (Fig. 58). Recently over ten types of domes have been recognized (Isenshtadt, 1956).

The cause of the formation of salt-domes rests in the extraordinary plasticity of salt. Weak manifestations of folding which in the sandy-clayey deposits produce the hardly noticeable platform structures cause significant changes with the lensoid salt deposits. The salt becomes plastic, mobile, and moves towards the central part of the lens where it penetrates upwards along joints, thus bending and rupturing the overlying formations. The total mass of salt does not change, but its form alters, the lensoid mass with a flat top surface decreases in diameter and becomes dome-shaped. The domes of the Emba region (Fig. 58) are not very convex and salt merely raises the overlying formations, rupturing them only rarely. Near the town Romny in the Ukraine, on the other hand, the salt massif becomes stock-like and breaks through the overlying formations, reaching the surface (Fig. 244). The domes of the lakes

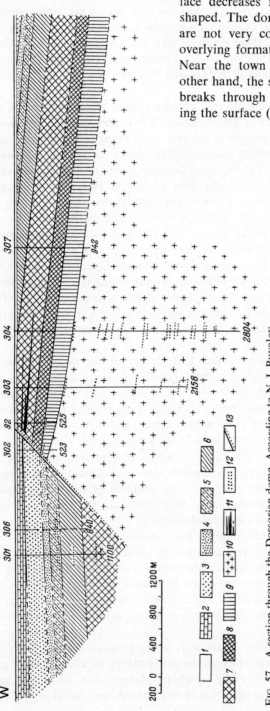

Fig. 57. A section through the Dossorian dome. According to N. I. Buyalov, adopted by V. A. Sel'skii (1936).

1. Recent.
2. Senonian and Turonian.
3. Cenomanian.
4. Albian.
5. Aptian.
6. Neocomian.
7. Upper and Middle Jurassic.
8. Lower Jurassic.
9. Permo-Trias.
10. Permian salt.
11. Petroliferous sands.
12. Oil seepage.
13. Fault.

Baskunchak and Inder probably also have stock-like forms and the lakes rest over their tops which have reached the surface.

The formation of salt-domes is a long and complicated process. Each orogeny and its phases occurring in adjacent geosynclines (the Mediterranean in this case) is reflected in salt deposits, causing further variations in their form and complicating the structure of the overlying sediments. In the Emba domes the faults produced cut across the Upper Cretaceous deposits and are transgressively covered by the un-deformed Akchagylian, thus the domes stopped forming at the end of the Palaeo-gene. This is supported by the fact that the whole of the Volga-Emba region from the beginning of the Neogene has become a stable platform.

Magmatism

There are no magmatic rocks known in the Volga-Emba district.

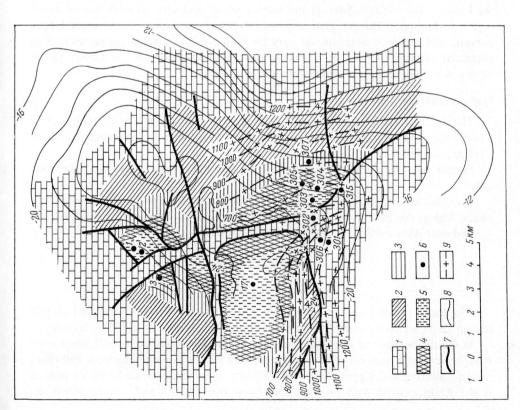

FIG. 58. Map of the Dossor dome. According to Sel'skii (1936). 1. Senonian and Turonian. 2. Cenomanian and Albian. 3. Aptian. 4. Neocomian. 5. Jurassic. 6. Boreholes. 7. Faults. 8. Isogammas. 9. Depth isolines of the salt from seismic data.

Useful Mineral Deposits

The mineral deposits are peculiar and considerable, although there are no metalliferous ores, but oil is of national and coal of local importance.

Oil Deposits

Oil deposits are associated with salt-domes. Oil-bearing horizons are numerous and are encountered from the Permian to the Anthropogene. The quantity of oil is determined by the porosity of the reservoir rocks. Amongst the Dossorian Suite the sandy horizons are most important as reservoir rocks. Those domes in which the reservoir rocks are porous are oil-bearing, but if the sands are compacted the oil is absent. The parent strata of the oil are not established and it is possible that they exist amongst the unpenetrated Devonian sediments, as is the case in the Volga-Ural province. In the west of the Volga-Emba province the oil is almost always accompanied by gas, but purely gaseous accumulations have not as yet been discovered.

The fact that not all the resources of the Volga-Emba province have been tapped is shown by the recent discovery of new deposits. For instance the major Kenkyak deposit, which is situated to the south of Aktyubinsk near the station Mortuk on the Tashkent railway, has oil within the Permian and Lower Triassic red beds. Oil has been found at Barankul and Prorva situated 400km southwest of Mortuk station and to the south of the Emba. Here the oil is present in the Middle Jurassic sandstones.

Until comparatively recently the Middle Jurassic oil was a unique feature of the Emba basin, but now it has been found in Mangyshlak and on the other side of the Caspian Sea in Ozek-Suat, in the sub-Caucasus and very recently indeed in the region of Bokhara. This suggests that the Middle Jurassic and the younger Upper Jurassic and Lower Cretaceous oil may be found in all adjacent provinces and in particular on the Ustyurt, in Karakum and in the foothills of the Gissar range. Mesozoic oil has very great significance and prospects.

Non-Metallic Deposits

These are of All-Union importance. There are great quantities of rock salt of which only a small part is being exploited in lakes such as Baskunchak, Elton, Chelkar, and Inder.

Potash salts are found in some domes as groups of strata amongst the rock salt.

Phosphorites are encountered in the Upper Jurassic, Lower and Upper Cretaceous, and in the Palaeogene. The Upper Cretaceous (Santonian) phosphorites are worked near Aktyubinsk and have great significance.

BALTIC REGION

General Description

The Baltic region forms an indivisible part of the Russian Platform, but despite having all its main characteristics has also certain diagnostically different features.

It should be pointed out that the quite full and variable section of the Lower Palaeozoic and Sinian complex is a type-section for the Russian Platform and other regions. Silurian and Upper Devonian are fully and characteristically developed, as is the Anthropogene with its large and numerous erratics and overdeepened depressions. The useful mineral deposits are relatively poor, there are no oil or coal or gas deposits. Only in the Estonian S.S.R. and Leningrad region is there a large deposit of oil shales. The Baltic region includes the Estonian S.S.R., the Latvian S.S.R. and Lithuanian S.S.R., the Kaliningrad, Leningrad and Pskov regions.

Stratigraphy

PRECAMBRIAN. This is exposed in the northern part of the Karelian isthmus. Here the huge Viborg Massif of Proterozoic granites and the overlying grey sandstones of Upper Proterozoic age can be examined, although the latter are often masked by the Anthropogene deposits and are only found exposed in isolated localities. The sandstones are found in boreholes of Vuoksa where at the base, overlying the gneisses, are brown and grey sandstones (87m) which towards the top are interbedded with argillites (30m) and covered by a bed of white dolomite (2·5m). Overlying these are are red sandstones (75m) with conglomerates (1m) at the base, belonging to the Gdovian Beds. The latter fact is demonstrated in a borehole south of Svir' where the red beds of the Upper and Middle Devonian (400m) are underlain by typical blue clay (64m), the lower part of which consists of Laminaritic clays. Below these there are the almost unmetamorphosed sandstones of Gdovian type. At the base there is a bedded formation of greenish and grey argillaceous, silty and sandy rocks. Towards the top the formation is less metamorphosed, while towards the base it is more metamorphosed into feldspathic quartzites and phyllites and is similar to the underlying Shokshinian quartzites at the bottom of the borehole. Thus between the Valdaiian complex and Jotnian there is the so-called Svirian complex. The latter is identical to the Nenokian Suite of the Arkhangel'sk region, or the Serdobskian or Polessian complexes of the south of the platform. The Valdaiian complex, as has already been said, consists of Gdovian sandstones and sands (up to 60–70m) and Laminaritic clays (Kotlinian Suite up to 130m). The Valdaiian complex is typically only developed in the east Leningrad and Pskov regions. In West Estonia, Latvia and Lithuania the Laminaritic clays are replaced by sandstones throughout the complex.

The crystalline basement is usually formed of Archean gneisses and granite gneisses. The region of Ikhva in Estonia (to the west of Narva) shows large magnetic anomalies. Boring has demonstrated that amongst the gneisses, a metagabbro and charnockites there is a group of magnetite quartzites, similar to those of the Kola Peninsula.

CAMBRIAN. The Upper and possibly Middle Cambrian are absent and the Pakerort (*Obolus*) Tremadocian beds rest directly on the Izhorian (Tiskresian or Fucoidal) Middle Cambrian (?) sandstones, which are 20–40m thick and have clayey, in places bituminous, intercalations. The Lower Cambrian is composed of *Eophyton* sandstones (Piritas Suite, 5–30m) at the top, the sandstones being in places replaced by blue clay (Leningrad Suite), while below lie the Supralaminaritic sands (Lomonosov Suite of 2–35m). The Lower Cambrian is known as the Baltic complex. Frequently the underlying Valdaiian complex is also grouped with the Lower Cambrian, but owing to almost complete and universal absence of a Cambrian fauna it is best considered as Upper Proterozoic.

In the west (Polish-Lithuanian Lowlands) all the Cambrian is represented by sandstones included either in the lower or middle divisions. Comparisons with Sweden suggest that the latter point of view is correct.

ORDOVICIAN.   Ordovician successions are shown in Table 2. In practice there are still more detailed subdivisions. As a result the limestones which are some 170–210m in thickness (80m in the south-west) are divided into 26 horizons. The detailed work has been done by the Estonian geologists.

Ordovician deposits have a distinctive uniformity over a large area. However, as a result of local rises in the sea floor in some regions certain horizons wedge out. Over the Lokno Rise the Ordovician is much thinner and is in places absent (Fig. 45).

SILURIAN.   The Ordovician only locally wedges out against the Lokno Rise, spreading out again to the east, while the Silurian stops against the rise and is completely absent to the east of the meridian passing through the Chudskoye lake and the Lokno Rise. It seems most probable that the Silurian of the area was eroded during the Lower Devonian continental regime.

Silurian successions are shown on Table 3. In the north the Silurian consists mainly of carbonate rocks of 100–130m in thickness. In the south-west at Plyavinas (south of Riga), at Sovyetsk and Kaliningrad the clayey and marly rocks predominate and are quite often graptolitic, while limestones are rare. At the same time the thickness of the formations increases reaching over 200m at Plyavinas, 506m at Sovyetsk and 936m at Kaliningrad.

DEVONIAN.   The Devonian is quite fully developed. Between Pskov and the Ilmen' lake there is exposed a complete, typical section of marine Devonian of the north-west type (Table 17). Here in the upper, Variegated Suite, beds with marine fauna are completely absent, but south of lake Ilmen' on the river Lovat there are three groups of marine strata that have been called Smot-Lovat, Chimaevo and Bilovo Beds (Table 18). Further to the south the thickness of the marine group increases, they completely replace the red beds and the succession becomes normal as in central provinces (Table 16).

Moving to the west, in the valley of the Western Dvina (Daugava) in northern Latvia and Lithuania the Middle Devonian as before is composed of sandstones (Fig. 59) while the Upper Devonian succession remains mainly marine, although lithology changes sharply owing to an almost complete changeover from limestones to dolomites which are exposed in the river cliffs (Fig. 60). The dolomites are yellowish-grey, brownish, often barren, although certain beds have concentrations of fauna. These beds are numerous and permit the subdivision of dolomites into a number of horizons (Table 18).

Still further to the south-west the Devonian succession becomes thinner. In Plyavinas all the horizons above the Svinorodian Beds are absent, in Sovyetsk the Chudovian Beds are absent and in Kaliningrad and outside the U.S.S.R. in Leba all the Devonian is absent. Successions found in the Polish-Lithuanian Depression (Table 7) show red lower Devonian beds with armoured fish, not found to the east. The Devonian of the region has been studied by I. A. Dalinkevicius (Lithuanian), S. I. Zheiba (1960), also a Lithuanian, and P. P. Liepinsh (1954, 1955), a Latvian geologist. Rarely the Devonian is folded into gentle platform folds (Fig. 61).

CARBONIFEROUS AND LOWER PERMIAN.   These are absent in the Baltic region, which is at variance with the situation in the Central Regions of the Russian Platform. In the west the Devonian is covered by the Upper Permian, which is formed in a depression cut in the Devonian and even Silurian beds. Further to the west the Devonian is covered by Anthropogenic deposits.

The Lower Carboniferous starts far to the east in the Valdai region (Fig. 45). Its absence to the west is probably explained by Mesozoic and Cainozoic erosion.

Fig. 59.

White and red Gauian sandstones (uppermost Givetian) on the river Gauya in Latvia.

Photograph by V. A. Gravitis (1954).

FIG. 60. Famennian dolomites on the river Venta in Latvia.
Photograph by P. P. Liepinsh.

UPPER PERMIAN AND TRIAS. The Upper Permian and Lower Triassic (Vyetluzhian) deposits form a characteristic complex which is sharply distinct from the underlying Devonian and Silurian sediments and also from the overlying Rhaetic and Jurassic. The Vyetluzhian deposits are only widespread in the Polish-Lithuanian Depression and to the west in Poland and Germany.

The Upper Permian is exposed in the north, in Latvia, where it consists of a thin (1–2m) basal conglomerate, and 20–40m of overlying limestones and dolomites with marine fauna typical of the German *Zechstein*. To the south Permian has been detected in many boreholes. In Sovyetsk the succession changes and basal conglomerates and sands (1–1·5m) are overlain by a thin group of limestones (4–7–11m) covered by anhydrites (40–70m). In the Kaliningrad borehole the Silurian is overlain by a group of carbonate beds (15–20m) with oil shales, and above the groups are anhydrites (65m), a rock-salt formation (170m), and again anhydrites (60m). Lithuanian geologists (Vala and others, 1959) have studied the Upper Permian and Trias in great detail. At its base the Karpenyaisian Suite overlies Devonian or Silurian. The suite consists of dark and brown bedded limestones, dolomites and marls, which are sometimes silicified. At the bottom the suite has a thin group of sandstones and conglomerates. The suite varies from 5 to 60m, but at a number of localities it wedges out entirely. Its fauna is marine, of *Zechstein*-type, and includes mainly western European types such as *Productus horridus*, *Spirifer alatus*, *Strophalosia*, and also bryozoans and ostracods.

Above follows the Myenchyaisian Suite, which is in places absent. This suite reaches 20–80m and in Kaliningrad up to 323m owing to the appearance of rock salt and anhydrite. The suite includes limestones and dolomites as well as locally dominant rock-salt and anhydrite. The impoverished lamellibranch fauna indicates less marine conditions of the basin with bitter-salt lagoons and lakes adjacent to it.

FIG. 61. The western flank of the gentle Plyavinas structure, involving Frasnian dolomites in Plyavinas region, Latvia. *Photograph by O. V. Birzgalis* (1939).

TABLE 18. *Upper Devonian Successions of the Baltic countries*
(According to S. I. Zheiba, 1960; P. P. Liepinsh, 1954)

Stage	Sub-stage	Horizon Lithuania	Horizon Latvia	Horizon North-western provinces
Famennian	Upper	Nadletizhian	Natsasian	Nadbilovian
			Paplavian	
		Letizhian	Letizhian	
		Shkervelian	Shkervelian	
		Ketlerian	Ketlerian	
		Zhagarian	Kapsedian	Bilovian
			Zhagarian	
		Shvetian	Shvetian	Nadchimayevian
		Murian	Murian	
		Akmenian	Akmenian	
	Lower	Kurshaskian	Kurshaskian	Chimayevian
		Ionishkian	Ionishkian	
		Kruoian	Kruoian	Nadsmotinsko-Lovatian
		Pakruoian	Amulian	
Frasnian	Upper	Stipinaiian	Bauskian	Smotinsko-Lovatian
		Pamushisian	Orgskian	Nadsnezhian Snezhian
	Middle	Istrian	Daugavian	Buregian Ilmenian Svinorodian
				Upper Shelonian
	Lower	Tatulian	Salaspilsian	Lower Shelonian
		Kupishian	Plyavinasian	Chudovian
		Suozian		Pskovian
		Ierian		Snetogorian
		Shventoian	Amatian	Podsnetogorian
			Gauian	
Givetian	—	Upninkaiian	Salatsian	Tartuian

On the limestones and gypsum of the aforementioned two suites lies a thick
continental formation of red beds. Its lower part (Suduvian Suite) of 6–96m in
thickness consists of brown gypsiferous clays and marls and in parts of variegated
clays and sands. Its flora is Upper Permian. The middle part is a variegated formation of Lower Trias age. It consists of green and brown clays and sands with intercalations of lacustrine limestones and marls. The clays are sometimes gypsiferous
and the fauna is fresh-water or salt-water consisting of *Estheria* and *Darwinula*
(ostracod), and plant remains are not infrequent. The formation is divided into
Nyamunian (15–84m) and the overlying Palangian (15–115m) suites. The upper
part of the continental formation is thin (5–80m) and consists of pale variegated
kaolin clays and sands containing remains of Rhaetic-Liassic flora, and collectively
known as the Taurage Suite.

The Middle Trias and most of the Upper Trias are absent and the continental
formation underlies Middle Jurassic sediments. Lithuanian geologists correlate the
Kaipenyaisian Suite with the Ufian Stage, the Menchyassian Suite with the Kazanian
Stage, the Suduvian Suite with the Tatarian Stage, and the Nyamunian and Palangian
Suites with the Vetluzhian Stage. This is probably incorrect, and local names of
various subdivisions should be used. It is in fact possible that correlation should be
made not with the east but with the west which is more closely related to the Baltic
region.

The analysis of abundant, but monotonous, fauna has permitted three important
conclusions. Firstly, the Upper Permian sea of the Baltic region was a salt saturated
brackish basin on the margin of the Western European Zechstein Sea and was
itself margined by numerous, large lagoons. Secondly there was no connection
between the Baltic region and Kazanian sea of the eastern part of the Russian Platform. Thirdly, the age of the fauna is mainly Upper Permian (Kazanian), although at
its base the succession yields some older Kungurian forms. It is probable that the
fauna corresponds to all the three subdivisions of the German and Polish Permian.

The Lower Trias lies on the eroded surface of the carbonate formation or on
anhydrites. The Trias is represented by continental red beds with *Estheria*, ostracods, ganoid fishes and fresh-water algae that lived in fresh-water and saline lakes.
The *Estheria* are the same as those in the Vetluzhian deposits of the east of the
Russian Platform, which enabled the correlation of the red beds with the Lower
Trias. The fauna, however, has not been sufficiently studied and the ostracods are represented by the genus *Darwinula*, which is found in the Tatarian Stage. Thus a
possibility exists that the lower horizons of the red beds should be included in the
Tatarian Stage. The red beds are 60m in the north, in the south in Sovyetsk they
are 216m, while in Kaliningrad they are 422m thick. In Kaliningrad a group of grey
marls and sills with peculiar Vetluzhian *Estheria* and fish reaches 22m and occurs
in the middle of the succession. This permits a conclusion that a full Trias succession
is here developed, with the lower red beds being the Lower Trias, the grey group
corresponding to the Middle Trias (a shelly limestone) while the upper red beds belong to the Upper Trias. This conclusion is confirmed by the fact that in Poland to
the south of Kaliningrad, and near the town of Pish, a borehole penetrated the tripartite succession of the normal Germanotype Triassic.

The disposition of the Upper Permian and Lower Trias is shown in Fig. 62, the
Upper Permian filling a depression in the Devonian and Silurian sediments. The
Lower Trias overlies the Permian essentially conformably, the Jurassic being un-
conformable on the Trias.

JURASSIC. Jurassic is only developed in the south-west and mainly in the Polish-Lithuanian Depression. There are only Middle and Upper Jurassic rocks present, the Lower Jurassic being absent. The Jurassic consists of dark grey and brown clays and sandstones with marine ammonite faunas. At outcrop the thickness of the Jurassic is 10–30m, while to the south-west in boreholes it is much greater reaching 130m in Sovyetsk and about 250m in Kaliningrad.

The Middle Jurassic in Sovyetsk consists of black clays and marls and is 28m thick, while in Kaliningrad 86m of sandstones are found. At a number of localities the Middle Jurassic is absent and the succession consists of Callovian and Oxfordian. It is possible that in these localities the Middle Jurassic is represented by an unfossiliferous group of continental grey and white sands and clays with lenses of brown coal. The marine Callovian is most widely distributed and consists of grey and yellowish sands with beds of sandy limestones of 12–30–61m in thickness. The Oxfordian consists of black and brown clays and silts with a marine fauna. In Sovyetsk it reaches 70m, but is usually less, while in Kaliningrad thick sandstones of up to 145m are present. The Kimmeridgian is developed in places, but more often is absent. The Volgian Stage also is absent, as well as almost the whole of the Lower Cretaceous, since in these times the region was dry land.

UPPER CRETACEOUS. The Upper Cretaceous transgression is in evidence throughout the Polish-Lithuanian depression and its eastern margin. The transgression affects the southern half of the Byelorussian Massif and southern regions of the Russian Platform. During the epoch when the transgression occurred the western and eastern seas were again connected giving rise to a single, large, southern basin which merged into the Tethys. The northern margin of the Upper Cretaceous sea continues from the mouth of the river Neman towards Kaunas, Vilnius, Minsk,

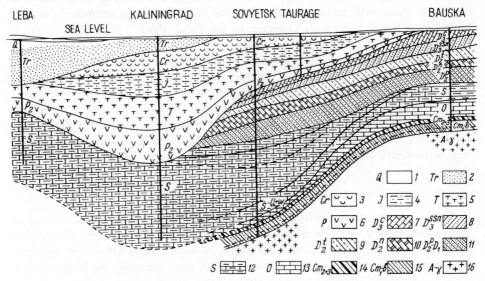

FIG. 62. Section — Leba (Poland) — Kaliningrad — Bauska — Lokno. The initial part of the section given in Fig. 45.

11 — Anthropogene; 2 — Tertiary; 3 — Cretaceous; 4 — Jurassic; 5 — Trias; 6 — Permian; 7 — Carbonate Formation; 8 — Lower Red Formation; 9 — Tartu Beds; 19 — Narovian Beds; 11 — Pyarnu Beds and Lower Devonian; 12 — Silurian; 13 — Ordovician; 14 — Middle and Upper Cambrian; 15 — undifferentiated Cambrian; 16 — granite gneisses.

Smolensk, and Moscow. As in the rest of the Russian Platform, the transgression began in the Upper Albian, above which all stages with the exception of the Danian are recognized. In the north, in Lithuania, near the margin of the sea, terrigenous clastic deposits of 100–140m in thickness predominate. Thus, for instance, in the borehole at Sovyetsk the Albian (18m) consists of glauconitic sandstones with a basal bed of phosphatic pebbles and beds of black shales. The Cenomanian (16m) in the same borehole also consists of glauconitic sandstones with clay laminae and a phosphatic horizon. The Turonian and Coniacian (29m) are represented by undifferentiated dark clays with beds of sandstone. In the Santonian (22m), amongst the sandstones and clays, the first chalk-like marls appear. In the Campanian (35m) the chalk-like marls are interbedded with glauconitic sandstones and lastly the Maastrichtian (Kaunas, 20m) consists wholly of chalk.

To the south the thickness of the Upper Cretaceous increases rapidly, reaching 200m in Kaliningrad, 313m in Polessk, 347m in Lidsbark and 800m in Lublin (Poland). At the same time the clastic content of the sediments decreases and the quantity of the chalk increases, appearing in the Campanian, and even in the Turonian and Coniacian-Santonian the white chalk-like marls predominate and the glauconitic sands with layers of phosphorites and clays are developed only in the Cenomanian and Albian. The described succession is observed from the Vilnaus-Minsk line to the south.

PALAEOGENE. The Palaeogene sea receded to the south in comparison to the Upper Cretaceous. Its margin stretched from Kaliningrad to Grodno, Minsk, Gomel and further east. As distinct from the Upper Cretaceous the Palaeogene consists entirely of clastics, which in the Palaeocene and Eocene are especially rich in glauconite. In the Eocene there are near-shore continental grey clays and sands with beds of lignite and primary accumulations of amber. The Oligocene also consists of glauconitic marine sands with some beds containing secondarily derived fragments of amber, which are collected on the Baltic shore. The total thickness of the Palaeogene in Kaliningrad is over 100m.

NEOGENE. The Neogene consists of continental deposits of clastics some tens of metres in thickness (Lithuania). The Miocene contains in places lenses of lignite and is subdivided on the basis of spores and pollen.

ANTHROPOGENE. This is widely developed and is represented by all the stages, reaching in total a considerable thickness of tens of metres and in kettle-holes it is much thicker (up to 250m). The erratics in it are widely distributed reaching huge dimensions. Erratics of chalk are sometimes so large that they provide whole quarries.

Tectonics, Magmatism, and Useful Mineral Deposits

These features are not widely developed in the Baltic region and have been already mentioned in the general review of the Russian Platform. The oil-shales of Estonia and the amber of Lithuania are worthy of note, and the *Obolus* phosphorites of Estonia are also of interest. Clays, sands, marls, and limestones are being worked.

TIMAN, KANIN PENINSULA, AND PECHORA DEPRESSION

General Description

The Timan-Pechora region, which includes the Kanin Peninsula, is a post-Baikalian platform lying between the Russian Platform and the Urals. Since the

Palaeozoic found within it is of platform type, the region is considered together with the Russian Platform rather than with the Urals. Undoubtedly the post-Baikalian Platform continues to the north from this region and forms the bottom of the Barents Sea. This is indicated by the sub-marine relief and Palaeozoic fragments in Quaternary moraines. The lithology and the faunal content of these fragments are identical to the Palaeozoic of the Timan-Pechora region.

Timan is a low and wide mountain range, dissected by rivers, and covered by forests. The peaks reach 300–450m (Chetlasski Kamen' Mountain is 461m). To the south the range passes into the Polyudovian range of the Urals.

The Pechora region is fairly elevated and forested to the south while in the north it becomes a low Bolshaya Zemlya type of tundra with occasional elevations of up to 200–250m, amongst which the Chernyshev range, which reaches heights of up to 200m and is long and narrow, is worthy of mention. The range is named after the great geologist F. N. Chernyshev who was one of the first investigators of Timan.

In the south of the region separate elevations are known as 'parmas'. The south-eastern continuation of Timan has the anticlinal elevation denoted on geological maps and known as Dzhezhim-Parma. All the elevations of Timan consist of hard and metamorphosed Proterozoic rocks, as is the case with the median (Seredinnyi) range of the Kanin Peninsula.

The well-developed river system is related to the Pechora and includes the tributaries of the Usa, Izhma, and Tsyl'ma. One of the tributaries of the Izhma is the Ukhta which together with the Domanik crosses the Ukhta oil-region. The western slope of Timan is drained mainly by subsidiaries of the Mezyen' and in the south by the Vychegda.

Stratigraphy

The region involves the geosynclinal Upper Proterozoic and platform-type Palaeozoic, Mesozoic and Cainozoic. The Upper Proterozoic has a great thickness, is strongly folded and metamorphosed, and in this differs from the corresponding formations of the Russian Platform where they have typical platform features including small thickness, absence of metamorphism, and very weak folding.

UPPER PROTEROZOIC. At the base of the succession lies the Chetlassian Suite consisting of banded phyllitic sericite-quartz slates with rare current-bedded quartzite beds (2000–2500m). Absolute age—1130 m.y.

Dzhezhimian Suite. In the south at Dzhezhim-Parma there are red arkosic quartzite sandstones with beds of gravel conglomerates towards the bottom and reaching 1000m. In the north there are 300m of red quartzites with a basal conglomerate. The suite is transgressive over the Chetlassian Suite and is 640 m.y. old.

Bobrovskian Suite. This consists of dark, brownish and greenish quartz-sericite slates which are carbonaceous at the bottom and have beds of quartzites at the top. The suite reaches 1000–2000m and is 540 m.y. old.

Bystrukhinian (Bystrinian) Suite (on the river Bystrukha). The suite consists of dark and light meta-dolomites and limestones which are not uncommonly banded. Algal limestones and dolomites with *Collenia, Gymnosolen,* etc. are widely developed and reach over 1000m.

Oselkovian Suite. This consists of dark grey and steel-grey quartz mica slates (up to 1100m) and is sometimes considered as the upper part of the Bystrukhinian Suite.

The Chetlassian Suite is normally included in the Upper Proterozoic. The last three suites, by analogy with the Urals, are sometimes considered to be Cambrian. They do, however, lack Cambrian faunas, which implies their Upper Proterozoic age (Sinian, Riphaean), which has already been fully substantiated.

All the Upper Proterozoic suites have been described in detail by V. A. Kalyuzhnyi (1959) and O. A. Solntsev (1959).

In the Kanin Peninsula the Upper Proterozoic lies in two belts. The greater of the two is the northern, of 12km in width and forming the Kanin Kamen' ridge. Here the thinly bedded schists and phyllites (2500m) with beds of quartzite and marble lie at the base, while sandstones and quartzites (1500m) with beds of schist lie at the top. The absolute age of the phyllites is 480–620 m.y. The second, southern, narrower belt shows dark and yellow meta-dolomites with calcareous algae of *Collenia* and *Gymnosolen*; the beds (1000m) being analogous to the Bystrukhinian Suite of Timan.

Figure 63 shows the relationships of the Proterozoic formations in south-west Timan, where the strongly folded Proterozoic with intrusive granites is unconformably overlain by the Middle Palaeozoic and the later horizontal Devonian. Formerly the Upper Proterozoic rocks of Timan and the Kanin Peninsula were denoted as schists (M) of undetermined age. To the west these formations are found on the Rabachii Peninsula and Kil'din island, while to the east they are exposed in the Dzhezhim-Parma anticline and Polyudov ridge and continue to the western slopes of the Urals where they were relegated to the Cambrian, but not on a sufficient basis.

A number of investigators who simultaneously studied the geology of the Urals (M. I. Garan', B. M. Keller and N. G. Chochia, 1955) relate all the suites to the Proterozoic and suggest the following correlation of local suites.

Middle Urals	Timan	South Urals, Karatan
Churochnian	Oselkovian	—
Niz'vanian	Bystrukhinian	Min'yarian
Deminskian	Bobrovkian	Inzerian
—	—	Katavian
Rassolinian	Dzhezhimian	Zil'merdakian
—	Chetlassian	Avzyanian

A more detailed correlation of these and other analogous suites is given by N. G. Chochia (1955). The Churochnian Suite was in his monograph included into the Cambrian, but in 1956 Sinian spores were found in it by B. V. Timofeyev. In Karatau the Ashinian Suite of the Sinian complex overlies the Min'yarian Suite and it is possible that the former corresponds to the Oselkovian and Churochnian suites.

CAMBRIAN. There are no Cambrian formations proved by fossil evidence, in Timan, the Kanin Peninsula, the Polyudovian ridge or the western slope of the Northern and Middle Urals. Within the Pechora Depression boreholes did not encounter any Cambrian deposits bearing an appropriate fauna. Metamorphosed rocks and in particular *Collenia* bearing limestones and dolomites have been attributed to the Cambrian by many workers in Timan, but as has already been mentioned the absence of Cambrian fauna from these formations forces their relegation to the Upper Proterozoic.

ORDOVICIAN. There is no palaeontologically proved Ordovician in Timan or the Kanin Peninsula or in borehole cores of the Pechora Depression.

It is, thus, possible to conclude that during Cambrian times not only the present belts of the Upper Proterozoic but extensive tracts of land to the east, including the Pechora Depression and the western slope of the Urals, formed land. In the Ordovician the eastern provinces were invaded by sea, but the belts of the Proterozoic still remained as elevated land.

In the Middle Palaeozoic the eastern part continued to sink, and in the Silurian the sea penetrated Northern Timan. In the Lower Devonian there was another regression and the sea receded to the present western slope of the Urals. In the Eifelian and Givetian epochs the land sank again and in the Lower Frasnian the sea advanced from the east covering the whole of Timan up to the belts of Proterozoic ridges. These ridges and the regions to the west of them remained land all the time.

The Silurian is exposed on the shores of Northern Timan over a small area. The basal sands are overlain by clays and marls, and higher up there are greenish argillaceous limestones of the Llandovery Stage, which contain large pentamerids (*Pentamerus oblongus*), *Favosites*, *Leperditia*, etc. and reach 25m. Above these lie dolomites and limestones (25–30m) with an obscure brachiopod fauna. The Silurian deposits are weakly folded and are almost non-metamorphosed.

The Silurian is better developed to the east in the Chernyshev range. Here Silurian sediments consist of limestones which sometimes stand 'on their head' (Fig. 64).

In the Pechora Depression the standard borehole at Nizhnyaya Omra (south-east of the Ukhta) has revealed the Izhma-Omra Formation of sandstones and clays with dolomites and anhydrites at the top (total 752m, Table 1), underlying the Middle Devonian. The identified corals indicate that the obscure rhynchonellid-coral fauna, similar to that of Northern Timan in the dolomites belongs to the Silurian. The Izhma-Omra Formation is developed throughout the Izhma-Pechora zone where Devonian is present, and also to the east of it. To the west in the Eastern and Western Timan zones it is absent and the Devonian lies directly on metamorphic

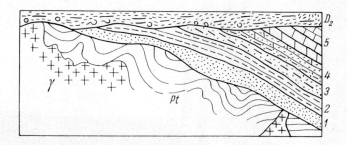

FIG. 63. The relationships of the ancient rocks of Timan. According to V. A. Kalyuzhnyi (1959).

Pt — metamorphic rocks and ancient granites; γ — ancient granites; 1–5 — the Izhma-Omrinian complex: 1 — the Sediol'skian Suite of red and white, cross-bedded sandstones; 2–4 — the Nibelian Suite of argillites, sandstones and siltstones; 5 — Vas'kerskian Suite of limestones and dolomites, with beds of argillites; D₂ — Nizhnechib'yuian (Lower Chib'yuian) Suite of sandstones and conglomerates.

rocks. Silurian sediments are also found on the eastern shore of the Kanin Peninsula, where they are of the same type as in Timan.

DEVONIAN.   The Devonian of Southern Timan and the Pechora Depression has oil and gas and has consequently been well studied and penetrated by boreholes.

The investigation of the Devonian began over 100 years ago. The Ukhta succession was described by Keyserling, and the fauna which he collected by Verneuil, in the well-known monograph *The Geology of Russia* published in 1845. Of the subsequent research the work of F. I. Chernyshev (1889–1890) is outstanding. In Soviet times N. N. Tikhonovich has done much and the latest information on biostratigraphy has been supplied by A. I. Lyashenko (1956) and on the Devonian geology by T. I. Kushnareva (1959) and V. A. Kalyuzhnyi and K. P. Ivanova (1959).

The Devonian is widely distributed throughout Timan and the Kanin Peninsula as well as the Pechora Depression and the western slope of the Urals. Throughout this huge area it is almost horizontal and little altered and has platform characters. It differs from the Devonian of the Russian Platform by somewhat larger thicknesses (1000–1200m and in the Pechora Depression up to 2500–3000m) and by the fact that it is folded into a number of large gentle anticlines obvious even on generalized geological maps. The main subdivisions of the Devonian are given in Table 19.

At the bottom of the succession, on the metamorphic rocks of the Izhma-Omra complex, rests the Chib'yuian Suite. It consists of sandstones and gravel-stones with groups of clays reaching a total thickness of 15–90m, over which at the Ukhta lie tuffites with diabases and shales of total thickness of 50–60m. The Chib'yuian Suite belongs to the Eifelian Stage (on ostracods), although formerly it was included in the

FIG. 64.   Vertical Silurian limestones of the Shar-Yu river in the Chernyshev range. *Photograph by A. A. Chernov (1956).*

Givetian. The suite is divided into lower and upper and in the east there is also a Suprachib'yuian sub-Suite. The lower sub-suite has an oil-bearing sandstone member. Overlying the Chib'yuian Suite is the main oil-bearing Pashiiskian or Nyefteiolian Suite. It consists of sands and sandstones with plant remains, *Estheria* and fishes and represents continental deposits and a commencement of a new marine transgression. The suite was formerly considered as Lower Frasnian, but at present opinions are expressed that its lower part belongs to the Givetian Stage. Its thickness varies from 5–7 to 20–30m.

The overlying Kynovian Suite has a passage zone with the Pashiiskian Suite. The Kynovian Suite consists of clays with limestones towards the top. The limestones contain a marine, principally brachiopod, fauna with the characteristic spiriferid *Cyrtospirifer murchisonianus* which is now called *Uchtospirifer*. The thickness of the suite on the Ukhta is 60–75m and to the east 25–35m. The Lower Frasnian sub-Stage is completed by the Sagaevian Suite of grey marls, clays, and limestones with a marine, brachiopod fauna. The suite reaches 50m.

The Middle Frasnian sub-Stage consists of two suites: the Domanik and the Mendym. The Domanik is considered as a parent suite for the oil. It is described (p. 328) later on and reaches 50–70m on the Ukhta. The Mendym involves grey marls and limestones with goniatites and brachiopods and on the Ukhta is 40–45m thick.

The Upper Frasnian sub-Stage is marine at the bottom and lagoonal and gypsiferous at the top. It is 500–550m thick on the Ukhta. The Famennian Stage is marine (marls and limestones 160m thick). The Devonian successions are shown in Table 19.

Rapid and important changes in the succession are characteristic for either an east–west transect, or especially for a north–south section. The east–west section cuts across five zones (Fig. 65). The most westerly or western Timan zone has only continental red beds of mainly Frasnian age (250–400m) in the north while thin marine facies appear as well in the south. In Dzhezhim-Parma the Upper Devonian is only 150m thick (Table 19). The eastern Timan zone (Fig. 65) has a dominantly marine facies (1000–1200m thick) and the Givetian Stage rests on metamorphic rocks. The Izhma-Pechora zone is again a mainly marine, but thinner (600–800m) succession, which overlies the thick (750m) Izhma-Omra complex already described. The fourth, Kamenka-Pechora zone differs from others by its immense thickness of up to 2800–3000m, and consists mainly of marine desposits. Such a great thickness can only originate in a rapidly sinking region such as a marginal downwarp, which has probably been the case, since the fifth zone (the western Urals) shows distinct and strong folding, considerable metamorphism and a complete succession consisting of all three divisions in mainly marine facies (Table 19), of geosynclinal type and related to the Uralian Geosyncline.

From south to north the succession varies even more considerably as is obvious in the successions of Middle and Northern Timan (Table 19). The variations are expressed in the diminution of the number and thickness of marine suites and the appearance of large thicknesses of continental formations with plant remains. The upper part of the Frasnian Stage and all the Famennian Stage is absent. In Northern Timan on the river Volonga the variegated formation of Frasnian age shows coal-bearing groups of strata with the oldest known coals, which reach 1–2m in thickness, and have amongst them the sapropelic as well as humic varieties.

An important feature of Devonian deposits is the presence of oil in them.

Analogously to the Volga-Ural region the main oil-bearing sands are associated with the Pashiiskian (Nyefteolian) Suite, but are also encountered in Lower Frasnian and even Lower Chib'yuian and Suprachib'yuian sub-stages. The Domanik, which is considered as the parent formation, is widely developed (Fig. 66). Another important feature of the Devonian is the wide development of basaltic (diabasic) lavas.

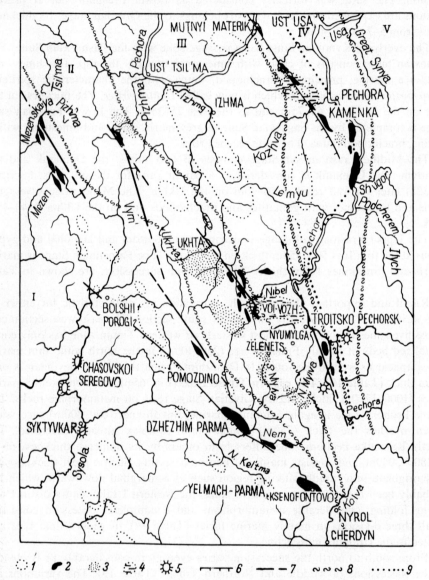

FIG. 65. Structure of the Timan-Pechora region. According to O. A. Solntsev (1957).

1 — structures of unknown nature; 2 — folds related to faults; 3 — growth structures in the Devonian; 4 — growth structures in the Carboniferous; 5 — salt domes; 6 — post-Permian monoclinal flexures; 7 — faults; 8 — boundaries of tectonic regions; 9 — western margin of the Permian Cisuralian downwarp. I — Western Timan zone; II — Eastern Timan zone; III — Izhma-Pechora zone; IV — Kamenka-Pechora zone; V — Western Uralian zone.

These lie within the Givetian and Lower Frasnian terrigenous formations. The lavas are frequently considered as Upper Givetian, but in Northern Timan there are indications that they are found not only at the bottom of the Frasnian Stage, but also towards the top of it. The sub-volcanic intrusions, also present, have an associated dispersed mineralization.

The Upper Devonian of the Kanin Peninsula is represented by continental red beds of Frasnian age, similar to those found in Northern Timan.

LOWER CARBONIFEROUS. Regressions causing a complete or partial absence of the Famennian Stage in Western Timan continued into the Lower Carboniferous, thus the Middle Carboniferous often rests on Frasnian. In Eastern Timan and the Pechora Depression, however, the Lower Carboniferous is well developed and is represented by three stages, which are most fully represented in the Izhma-Pechora zone and to the east of it. In these localities clays and sandstones are common in the Tournaisian, while the Visean and Namurian consist of limestones and dolomites. The total thickness of the Lower Carboniferous reaches 150–275m.

Amongst the rubbly sandstones and sands of the Tournaisian Stage to the south-east of Timan there have been discoveries of oil-shales (Jebol) and oil. The oil has also been found amongst porous Middle and Carboniferous limestones. Visean limestones have working mines of asphalts; as a result, the Lower Carboniferous has lately been extensively investigated (V. A. Raznitsyn, 1959). Raznitsyn points out that the Lower Carboniferous succession changes considerably from east to west. In Western Timan, on the Ukhta, the Lower Carboniferous is absent. To the south of the Ukhta region its thickness is 55m while in the Izhma-Pechora zone it is 150–275m and to the east of the Pechora 700–800m. On the Pechora Ridge the Lower Carboniferous is only 250–350m. Visean sandstones bear oil. According

FIG. 66. Almost vertical Domanic strata on the river Shar Yu, in the Chernyshev range. *Photograph by A. A. Chernov (1956)*.

to N. N. Lapina the Bol'shaya Zemlya tundra has the Uralian type of Lower Carboniferous, which is 550m thick.

MIDDLE AND UPPER CARBONIFEROUS. These are deposits belonging to a transgression which penetrated far to the west, covering Upper Proterozoic elevations and causing a direct connection between the Russian Platform and Pechora Depression seas. The transgression began in the Moscovian with resultant deposits of basal clastics. The Bashkirian Stage where present, and the closely associated Namurian, are represented by similar rocks consisting of often silicified limestones of 10–60m in thickness.

The Moscovian Stage and the Upper Carboniferous constitute a single formation of carbonate rocks (limestones, dolomites, marls) with rare intercalations of clays and sandstones. In the Upper Carboniferous silicification and development of caverns is quite commonly observed. The fauna is marine, abundant and variable, with choristids and fusulinids being characteristic. The thickness of the Moscovian varies from 40 to 200m, and of the Upper Carboniferous from 20 to 130m.

PERMIAN. During the Permian the western and eastern successions of Timan become again sharply different. To the east Permian deposits have a limited distribution and a Uralian type of succession. Upper Carboniferous marine limestones and dolomites are succeeded by the similar limestones of Sakmarian (220–230m) and Artinskian (133–200m) stages. Kungurian deposits are represented by a mixed facies of continental and lagoonal origin (160–1250m), and consist of gypsum and dolomites. Towards the sub-Uralian downwarp Kungurian deposits become coal-bearing, as they are in the Pechora basin. The Upper Permian is everywhere continental, terrigenous, grey or brown, and reaches 270–1250m in thickness. To the east it is coal-bearing. Desert and red bed facies are absent from the Permian.

To the east of Timan the Permian, where developed, is succeeded by red Lower Trias (Vetluzhian) beds and is of platform character. The Sakmarian Stage where present consists of limestones and dolomites some tens of metres in thickness. The Artinskian and Kungurian stages are represented by a lagoonal formation of anhydrites, gypsum, rock salt, and dolomites with sandstones and mudstone groups. The thickness of the two stages varies, sometimes reaching 250m and more. These chemical and lagoonal deposits are overlain by the red beds of the Ufian Stage which at Seregovo reach 170–250m, but are normally less.

The Kazanian Stage is widely developed, but is rarely exposed. In the south, near Dzhezhim Parma, it consists of continental near-shore or fresh-water flags (Soli-kamian type) and red beds. To the north typical marine formations appear (clays, marls, argillaceous limestones), but even these are rich in clastics brought in from Timan. The fauna is the usual one for the Kazanian, while thickness varies, reaching sometimes 200–300m.

The Tatarian Stage is widely developed, reaching 350m. It consists of red clays, sandstones and marls of continental origin, interspersed by rare beds of lacustrine limestones.

LOWER TRIAS. The Vetluzhian Stage has a wide distribution to the west of Timan. It consists of red continental deposits with not infrequent fluviatile and lacustrine facies. It is 100–150m thick. To the east of Timan the Trias is recognized in the Pechora coal basin.

JURASSIC AND LOWER CRETACEOUS. Both have a limited distribution and are transgressive over older rocks. The Jurassic succession begins with basal continental sands with clay layers and pebble beds and also fragments of wood. Their age is Bathonian, and in south-eastern Timan they are 150m thick, but normally are much thinner, being no more than 10–15m. The overlying marine Upper Jurassic sediments are best developed in the Izhma-Pechora zone where they reach 150m. Amongst the latter deposits dark grey clays predominate although beds of sandstones—often glauconitic—are not uncommon. Common ammonites and belemnites indicate Callovian, Oxfordian, Kimmeridgian, and Lower Volgian stages. The last contains groups of oil shales which in quality and quantity equal the type Volgian ones.

The Lower Cretaceous is only developed to the west of Timan. It begins with Valanginian sandstones and black clays, above which lie compact Hauterivian-Barremian micaceous clays several tens of metres in thickness. The succession is completed by Aptian grey and yellow sands with beds of clay amongst which continental facies are predominant. The total thickness of the Lower Cretaceous is up to 100–150m, and its top is often eroded away.

Upper Cretaceous, Palaeogene, and Neogene have not been established in the region, but their ultimate finding as small outliers of continental sands and clays which have escaped erosion is likely.

ANTHROPOGENE. The Anthropogene consists mainly of moraines, fluvioglacial, fluviatile, and lacustrine deposits. Along the coast there are terraces of the Boreal Transgression. The thickness of the Anthropogene varies depending on geomorphology. In areas of accumulation such as zones of terminal moraines, lake depressions and river valleys its thickness reaches 100m and more, while normally it is of the order of tens of metres and sometimes, over elevations, it is of the order of a few metres.

Tectonics

Tectonic structures of the Timan-Pechora region are interesting in being intermediate between the structures of the Russian Platform and the Urals. The Russian Platform is essentially post-Karelian, having been formed after the Karelidic Orogeny at the end of the Lower Proterozoic, while the Timan-Pechora region is post-Baikalian and the Urals post-Hercynian parts of the platform. Over the Russian Platform the Proterozoic (Sinian and Jotnian) is of platform-type, in Timan it is geosynclinal. The Middle and Upper Palaeozoic in Timan are of platform-type while in the Urals they are geosynclinal. These relationships determine the structures of the Timan-Pechora region. Its metamorphic suites (Oselkovian and Bystrinian, etc.) are intensely folded forming numerous, complex often isoclinal folds broken up by thrust planes into nappes. The folds are cut by granites, alkaline rocks and other intrusions which are synorogenic.

Structures in which the Middle Palaeozoic, and in particular the Devonian, is involved are large gentle, sometimes box-shaped folds. Dips reach 20°–25°, but are usually much less while the strike is constantly north-westerly (Timan trend). Near the Urals they become uraloid (north–south), in trend, while in the north at the Chernyshev range they trend north-east. The position and form of the most important folds are shown in Fig. 65. Folds with such orientations are not present in the Russian Platform proper, but approximately these directions can be discerned in the Palaeozoic of the Siberian Platform to the north-west of lake Baikal (Fig. 137). In the Upper Palaeozoic, separated from the Middle Palaeozoic by unconformities and breaks, the folds are even more simple and gentle.

The Jurassic and Lower Cretaceous are not folded and lie transgressively and unconformably on various divisions of the Palaeozoic. The younger rocks sometimes show sharp and complex folds unaccompanied by metamorphism. Such folds are known as pseudotectonic, and are caused probably by ancient landslides, although glacial action is not excluded. Two salt-domes, Seregovian and Chasobian (Fig. 65), are known. At Seregovo boreholes through the middle of the dome have shown over 600m of salt.

In the east the Pechora Depression is marginal to the sub-Uralian downwarp of Permian age, which contains the Pechora coal basin with Permian deposits reaching several kilometres. This downwarp forms a part of the Uralian fold-province and will be discussed later.

The thickness of the Devonian in the Kamenka-Pechora zone (Fig. 65), where it reaches 2·5–3km, should be noted. This suggests the existence of a downwarp more ancient than the sub-Uralian, but situated further to the west.

The origin of Palaeozoic folding is debatable. Considering the platform character of the region, the folds are explained sometimes in terms of adjustment to the faults in the crystalline basement, and sometimes in terms of unequal sedimentation during the Devonian and Carboniferous. Certain geologists working in Timan (Solntsev, 1957) especially adhere to the latter point of view (Fig. 65). To the south in the Volga-Ural province a number of tectonic ridges and flexures are associated with structures in the crystalline basement, but their form, strike and style differs sharply from the brachyanticlines of Timan. Thus the explanation for the latter in terms of unequal accumulation of sediments is more likely since similar supratenuous folds exist in the Volga-Ural region (numerous tectonic ridges, arches and downfolds), but they are considerably larger in size than those in Timan, and what is more have no constant trend. Most structures in Timan including the 'growth' brachyanticlines (Fig. 63) have a regular trend.

We have already noted the similarity of structures in Timan with those of the Lower Palaeozoics of the Lena basin where they are interpreted as the results of the Caledonian Orogeny affecting the platform. It is also likely that the Timan structures have been produced by the Hercynian Orogeny affecting a semi-stable platform, causing faults and folds independently of each other.

Large scale faults are also characteristic of the Timan-Pechora region. The faults are mainly normal and have a considerable extent and throw. The larger ones are shown in Fig. 65. Within the limits of Timan the faults are deeply penetrative and have been channels for the penetration of basaltic magma, a feature which is observed from the sea-shore to the southern end of the ridge. The effusive activity has been accurately dated as at the end of the Givetian and beginning of the Frasnian epochs, which also dates the faulting. Although the faults are usually considered as normal their true identity is unknown since they have not been completely investigated, and it is possible that some of them are reversed. In the Volga-Ural region such faults are rare.

Magmatism

Two complexes of eruptive rocks are known. The first is Upper Proterozoic (Baikalian) in age, and is encountered in metamorphic rocks of Northern Timan and Kanin sea-shore. It is represented by small intrusions of granites, gabbro, and alkaline rocks (nepheline syenites). The alkaline rocks of this type are not present in the Middle Palaeozoic and cannot be attributed to the Hercynian Orogeny, which is a useful guide for the age of alkaline intrusions of the Kola Peninsula. At any rate such intrusions cannot be younger than Caledonian and are most likely to be Upper Proterozoic or Baikalian. Outcrop of such intrusions are shown on generalized geological maps.

The second complex of eruptive rocks is Devonian and is represented by continental basalts and diabases. Similar rocks (p. 153) have been discussed in describing magmatism of the Russian Platform, where they were produced during both Upper Proterozoic and Devonian times. Rocks of this type are widely developed in platform regions and are absent in geosynclines. In Timan basalts, or as they are often called, diabases, are dark, almost black or brown, compact medium-grained rocks. They form lava flows of great extent but small (2–30m) thickness, which occur amongst marine or more rarely continental Devonian deposits. The number of such flows varies from 4 to 13 and sometimes they rest directly on each other, although often there are intervening sandstones and clays. In the Ukhta region their total thickness is 50–80m, but in Northern Timan it is 240m. The belt of the development of basalts is traced from Southern Timan (Ukhta region) to the north coast for a distance of 600km, but the width of the belt does not exceed the order of 30–40km. The basalts are present in Eastern Timan (Fig. 63) and are absent from Western Timan and to the east of it or the Upper Izhma zone. It can be suggested that the basalts were extruded from deep fractures which margin metamorphic massifs to the east in the south and to the west in the north. The eruption of basalts is commonly accompanied by intrusions and sub-volcanic complexes which fill magmatic channels. In Timan such channels have not been found, but their presence cannot be doubted. It is possible that the dispersed vein mineralization proved in Middle Timan is associated with such sub-volcanic intrusions.

The age of the eruption is debated and there have probably been several episodes. In the Ukhta region a tuffaceous suite consisting of tuffs, basaltic lavas, clays and conglomerates of total thickness varying from 50 to 80m lies over the Chib'yuian (Givetian) Suite and is overlain by the Pashiiskian (sand-clay) Suite of Lower Frasnian age. Correspondingly such effusives can be either Upper Givetian or Lower Frasnian. In Southern Timan the basalts are most often considered as Upper Givetian since they are overlain by the Pashiiskian Suite, which is at the bottom of the Lower Frasnian. In Northern Timan conversely the volcanic rocks are attributed to the Lower Frasnian and individual lavas are found even higher up in this stage.

Useful Mineral Deposits

Metalliferous Ores

In Middle Timan a disseminated polymetallic mineralization has been known for a long time. Ore minerals occur in veins occurring amongst Palaeozoic country rocks, but none so far found are of economic importance.

Fuels

Petroleum and gas deposits are associated with the sands of Givetian and Lower Frasnian age. Signs of the presence of oil on the Ukhta were known before the Revolution, but industrially important deposits were found by Soviet geologists. The southern part of the Pechora Depression, where it is near to Timan, is particularly promising. Here high-pressure oil fountains were obtained.

Seams of sapropelic and humic coals reaching 1–2m and found in Frasnian deposits of the western slope in Northern Timan (Volonga river) have been investigated but were not found to be industrially profitable. The Pechora coal basin is important, but is more related to the Uralian Geosyncline. Oil shales are widely developed, are of high quality and are found both in the Upper Devonian and Upper Jurassic. So far they have not been used in industry.

Non-metallic Deposits

Non-metallic deposits are variable and include: barytes, Iceland spar, gypsum, rock salt, and building stone, but so far are of local significance only.

Ending the description of the Timan-Pechora region, it should be emphasized that despite great achievements in studying its southern part much of it still remains incompletely investigated. Its northern parts are not well known and further research involving deep boring and geophysical exploration will give significant scientific and industrial results.

4

SIBERIAN PLATFORM

GENERAL DESCRIPTION

The Siberian Platform is the second region in which the last strong manifestation of folding occurred in the Precambrian. It shares a number of characters with the Russian Platform, but also differs strongly, mainly with respect to its stratigraphy.

So far many of its parts remain unstudied. Recently investigations in the Siberian Platform have been intensified, and new data accrued from year to year, thus permitting the compilation of a general geological map (scale 1:500 000), which in its detail approaches that of the Russian Platform. The Siberian map has been edited by T. N. Spizharskii who has spent a number of years studying the region and its northern provinces. The achievements of the Siberian geologists are manifest by comparing geological maps of the U.S.S.R. to the scale of 1:500 000 (1937 and 1955) with that to the scale of 1:250 000 edited in 1940 and 1956.

The northern part of the Siberian Platform is included in the Arctic region, which under the leadership of F. G. Markov has been mapped on the scale of 1:250 000. The map is accompanied by a detailed explanatory memoir by a series of authors.

BOUNDARIES. The Siberian Platform is outlined depending on the structure and the position of the Palaeozoic and Mesozoic rocks. Those regions are included in the platform when the Mesozoic and the Palaeozoic are of platform type, lying almost horizontally or forming gentle rounded folds (tectonic ridges or brachyanti-clines). The deposits are almost unmetamorphosed, thin, and do not involve syn-orogenic intrusions. Normally the platform is bounded by marginal downwarps, which are adjacent to geosynclines, thus allowing the drawing of accurate boundaries to the platform (Fig. 2). Between the Anabara and the Yenisei the Palaeozoic is almost horizontal. In the west it forms large, gentle, ridge-like structures, which are not accompanied by metamorphism or intrusive massifs. The thicknesses of separate systems are small. The succession has many breaks, but transgressions and breaks are not accompanied by angular unconformities.

To the east there is the large scale North-Siberian or Khatangian Depression, being a typical marginal downwarp. Further to the north there is the no less typical Palaeozoic Taimyr Geosyncline which is very similar to the Urals and involves strongly compressed, strongly metamorphosed, elongated folds, cut by a large number of granitic intrusions. Yet the comparison of the Palaeozoic of Noril'sk, Kureika and Fat'yanikha illustrates the platform character of the latter localities. All this allows the drawing of the northern boundary of the Siberian Platform along the Khatanga Depression. To the north-east of the Khatanga, the Anabara and up to the river Lena, and in the east along the Lena, the Aldan and the Maya the Siberian

Platform is bounded not by the Palaeozoic Northern Geosyncline, but by the Meso-zoic Verkhoyansk branch of the Pacific Ocean Geosyncline. Consequently the boundary is drawn in terms of the Mesozoic (Jurassic, Cretaceous) structures.

Between the Khatanga and the estuary of the Lena and further to the south and west of the Lena near to the Precambrian massifs the Jurassic and Cretaeous are almost horizontal, forming large, irregular, gentle flexures. The rocks are not meta-morphosed, lack granite intrusions, are relatively thin, and are typically of platform type. To the east of the Lena, despite the fact that the degree of metamorphism and the structural aspect remain the same, the thicknesses increase; for instance the Lower Cretaceous alone reaches 1500–2000m and more. This indicates the presence of the Cisverkhoyansk marginal downwarp. Indeed further to the east there is a typical geosyncline where the Jurassic and the Lower Cretaceous are very thick, and are strongly deformed into complex folds of a definite trend. The rocks here are heavily metamorphosed and involve numerous intrusive massifs. Here the margin of the platform can be arbitrarily drawn from the mouth of the Khatanga and the Anabara to the mouth of the Lena and southwards along it. The boundary con-tinues thereafter along the Aldan, thus margining the area of undislocated platform Cambrian and Jurassic deposits. Bending round the basin of the Maya, a subsidiary of the Aldan, the boundary lies along the northern slope of the Jugjur range, and bending towards the south-west along the Stanovoy range it lies to the south of the Aldan-Vitim Precambrian Massif. Then the boundary bends round the southern part of the Patomian Highlands and continues along the eastern shore of lake Baikal to its termination, thereafter continuing through the middle part of Eastern Sayan to the Yenisei.

All the way from the Jugjur to the Yenisei the Siberian Platform is bounded by the Angara Geosyncline, which is affected by the Caledonian Orogeny. The inclusion of the southern part of the Patomian Highlands into the zone of Caledonian Orogeny is based on the discovery of strongly deformed archeocyathid and trilobite-bearing limestones at the southern part of these highlands and near the join of the Muya and the Vitim.

According to one point of view, which has many aherents, the Cisbaikalian region is in the zone of Caledonian Orogeny. Nevertheless, the latest palaeontological investigations have demonstrated that over the greater part of the northern Patom-ian Highlands Cambrian rocks have a platform character, since they are thin, unmetamorphosed and unaffected by synorogenic intrusions. Consequently, the boundary of the platform does not go round the Cisbaikal region, but can be traced from the river Olekma to the south-west towards the southern termination of lake Baikal (Fig. 2). In the north-eastern part of the Russian Platform the Baikalids of Timan and the Bol'shaya Zemlya tundra are considered as parts of the platform, whereas in the south-east of the Siberian Platform identical Baikalids are mis-takenly not considered as parts of the Siberian Platform.

It is interesting that two diametrically opposite points of view were employed in compiling the two best tectonic maps of the Siberian Platform. The 1957 map (1:500 000) edited by Shatskii groups all fold regions, including the Caledonides and the Hercynides with the Siberian Platform, which is hardly correct. The 1958 map (1:250 000), edited by Spizharskii, on the other hand excludes all the fold regions, including the Baikalids, from the platform. It is probably most correct, as suggested by L. I. Salop (1958 b), to divide the disputed region into two zones—that of the Baikalids and that of the Caledonides and Hercynides—and to consider the former as a part of the Siberian Platform and the latter as a part of the Angara Geosyncline (Fig. 2).

In Eastern Sayan the boundary separates the Precambrian contribution of the basement to the platform from the region of the folded and metamorphosed Lower and Middle Palaeozoic. The western boundary of the platform is arbitrarily drawn somewhat to the west of the Yenisei. The problem of the West-Siberian Lowlands will be considered later. At this stage it suffices to state that its north-eastern part probably represents the continuation of the Siberian Platform (Fig. 2).

If the diversity of opinions about the boundaries of the Russian Platform is great, it is even greater with respect to the Siberian Platform. The western boundary, for which there are insufficient data, and even the northern boundary, which is most clear, provoke lively arguments. For instance the Yenisei structures developed in the region of Norilsk, along the rivers Kureika and Fatyanikha, and shown on general geological maps, are disputed. A. D. Arkhangel'skii (1941) considers them as Hercynian folds. It is, however, sufficient to compare these structures with those of the Taimyr Peninsula to become confident of their platform character. This conclusion is confirmed by the fact that between the Yenisei and the Taimyr structures there is a typical downwarp. The form and situation of the Yenisei structures is identical to the structures of the Ufa Plateau in the east of the Russian Platform. There is no room here to sort out all the debatable problems, since a special research monograph would be needed, but everyone studying the geology of the Siberian Platform should be aware that its boundaries are taken differently by different workers.

OROGRAPHY

The Siberian Platform, like the Russian Platform, represents a typically flat, almost planar, region. Its surface consists of a system of wide almost flat watersheds divided by wide and deep river valleys. The heights of the watersheds reach 500–1000m. The north-western part of the platform is elevated and is called the Central Siberian Plateau, which exceeds 500m and in the north to the west of Norilsk rises into the Putoran Mountains (Fig. 80) reaching 1500–2000m (Mount Kamen' is 2037m. The plateau is formed by the Siberian Traps.

The eastern part of the plateau, conversely, loses height down to 500m as it is formed of the softer Mesozoic and Cainozoic rocks. In the basin of the Vilyui, the middle reaches of the Lena and the lower reaches of the Aldan, there is a large Central-Yakutian Depression with heights less than 200m. The large North-Siberian Depression, which lies to the north of the platform, is situated over Mesozoic and Cainozoic rocks.

There are no very high mountain ranges, but along the southern margin of the platform there are a series of ridges of ancient rocks reaching 2500–3000m in height. Along the Okhotskian Sea there is the Jugjur range, succeeded to the west by the Stanovoy range, succeeded in turn by the Yablonovoy range. The Baikalian range has heights over 2500m, and to the south-west there is Eastern Sayan which reaches up to 3500m. The Precambrian massifs owing to their hardness form elevations, as for instance within the Aldan-Vitim Massif separate peaks reach 2000m. The Yenisei Massif is considerably lower, being a ridge with the greatest height of 1122m. The Anabara Massif is even lower with heights varying from 500 to

850m. These variations in height are connected with young block-elevations which have been strongest in the south.

Enormous areas of the platform are covered by unbroken, virgin forest (taiga). Even smaller hills (sopkas) are covered by the forest. Rarely the sopkas have no forest cover and stand out as bare rocks, in which case they are known as gol'tsi (singular: golets). The river system is well developed and there are numerous large, full rivers. The population is small and concentrated in the southern part of the platform, near to the railway and along the river valleys.

STRATIGRAPHY

The Precambrian stratigraphy of the Siberian Platform is very similar to that of the Russian Platform, but the Palaeozoic stratigraphy differs sharply. The Lower Palaeozoic and Silurian of the Russian Platform are comparatively poorly developed while in the Siberian Platform these formations occur very widely and achieve a great thickness. The Devonian and Carboniferous on the other hand are here almost absent while over the Russian Platform they are widely developed. The Permian is equally well developed in both platforms, but over the Siberian Platform the Permian is almost exclusively continental and is covered by the Siberian Traps, which are Lower Triassic in age and are characteristic of the Siberian Platform. In the Mesozoic and Jurassic and Lower Cretaceous coal-measures are typical of the Siberian and absent in the Russian Platform. Conversely the Upper Cretaceous-Palaeogene sea, which covered the south of the Russian Platform, was absent from the Siberian Platform. The almost complete absence of the Palaeogene and Neogene is very interesting. Amongst the Anthropogenic deposits the less widespread development of boulder clays and other glacial and fluvioglacial deposits such as terminal moraines, kames and eskers is worthy of note.

PRECAMBRIAN

The Precambrian deposits form the basement to the platform, but are exposed in four massifs—the Anabara, the Aldan-Vitim, the Eastern Sayan, and the Yenisei—as well as in three inliers, which are: the Olenek, the Turukhanka, and the Chadobets, all situated to the east of the Yenisei Massif.

The structure of the massifs and the inliers differs. In the Anabara Massif, Archeozoic gneisses and schists predominate. In the Aldan-Vitim and the Eastern Sayan Massifs a a full Precambrian succession is developed. In the Yenisei Massif the Proterozoic predominates over the Archeozoic, while in the inliers only the Proterozoic is exposed.

Everywhere the Cambrian overlies the Precambrian, often with an angular unconformity. In rarer cases the Cambrian is conformable on the Proterozoic. In these instances arguments arise as to the position of the boundary and certain formations are considered as Cambrian by some and as Precambrian by others. Lately the investigation of the Precambrian of the Siberian Platform has been considerably advanced but is still behind the similar studies in the Russian Platform and especially so where the Archeozoic is concerned. A review of data is made in volume I of *The Geological Structure of the U.S.S.R.* (1958).

Archeozoic

THE ANABARA MASSIF. The Anabara Massif is a large highland area of considerable size (60 000km²) and is situated in the higher reaches of the Anabara. The massif consists of the oldest Archean crystalline gneisses and schists, amongst which plagioclase-quartz-hypersthene gneisses are especially abundant (50 per cent). The complex is similar to that of Karelia and the Kola Peninsula. Crystalline

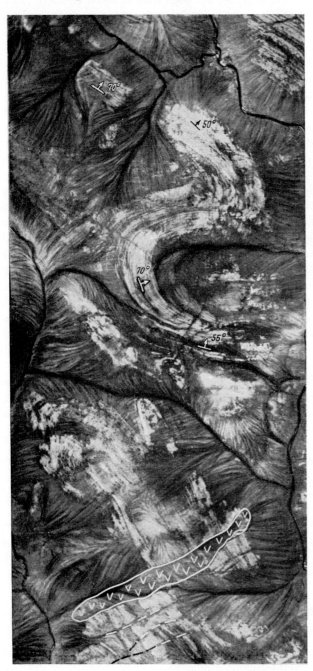

FIG. 67.

Archeozoic rocks of the Anabara Massif.

Complex folds in gneisses.

Black lines are rivers.

Fan-like features are ravines.

V=symbol for a gabbro-norite dyke.

Dashed lines are dykes of diabase.

Aerial photograph by Institute of Arctic Geology.

schists are intruded by Archeozoic granites with associated aplitic and quartzose veins, and also have widespread migmatite zones. Granites (5–7 per cent) are not very abundant and commonly are alaskitic. Basic intrusives are even less widespread and form small lenses along the strike of the country rocks. The basic rocks are composed of pyroxenites and peridotites and are younger than the granites. All rocks are cut by diabases.

The Archeozoic of the Anabara Massif has been studied by G. G. Moor (1940), who called it the Charnockite Series corresponding to the oldest Archeozoic formations of the Scandinavian Massif, the Ukrainian Shield, the Aldan Massif, and other regions. Moor has subdivided these rocks into three series (Fig. 67).

According to M. I. Rabkin (1956) the Anabara Massif consists of Archeozoic hypersthene gneisses, basic crystalline schists, quartzites, marbles, paragneisses and migmatites, reaching the total thickness of about 20km (arbitrary). The rocks are intruded by Archeozoic granitoid and anorthosite bodies. Moreover, along joints peridotites and pyroxenites intrude the Proterozoic rocks. The Archeozoic rocks are strongly deformed into a series of isoclinal folds with a north-westerly trend and are unconformably overlain by the Sinian complex. Rabkin divided the Archeozoic into four series.

THE ALDAN MASSIF. In its southern part the massif consists of Upper Archeozoic rocks (Fig. 68) with charnockitic gneisses being dominant and alaskitic granites common. According to D. S. Korzhinskii (1936) three suites can be recognized here. The lower 'Iengrian' Suite consists mainly of pyroxene-plagioclase schists, amphibolites and hypersthene gneisses, but there are also thick quartzitic rocks. The second suite of 'Charnockitic Gneisses' has widely developed hypersthenic rocks. The third series—Dzheltulinian—consists of schists, gneisses, amphibolites and granulites. There are also not infrequent lenses of marble and there are disseminated flakes of graphite in schists (Fig. 68). Korzhinskii considers that the schists were formed as a result of the alteration of both intrusive rocks (charnockitic schists) and sedimentary rocks (granulites, silimanite gneisses). Quartzites, marble, and graphite are also of sedimentary origin.

All rocks were affected by the intensive action of Archeozoic alaskitic granites. Later Yu. K. Dzevanovskii (1961), one of the prominent specialists on the geology of Eastern Siberia, made a detailed study and expanded the data and conclusions made by Korzhinskii. Dzevanovskii suggested a so-called Timptonian Suite, which as well as the Dzheltulinian Suite he divided into a number of subdivisions.

The thickness of the Archeozoic has not been calculated, but it is huge and of the order of 14–17km. The detailed study of the Iengrian Series has shown that its middle, quartzitic suite alone is 3500m thick. The upper carbonate Fedorovskian Suite of the series contains useful mineral deposits, including phlogopite.

The subdivision of the Archeozoic in the south-eastern part of the Aldan Massif (Uchuro-Maiskian region) as proposed by Dzevanovskii (1960, 1961) is as follows (bottom to top):

Iengrian Series

Chekcheian Suite—Biotite, amphibolitic and pyroxenic gneisses with layers of crystalline schists, amphibolites and quartzites (up to 800m).

Upper Aldanian Suite—Quartzites with beds of sillimanite-, cordierite- and biotite-garnet gneisses and ferruginous quartzites (3000–3500m).

Fedorovskian Suite—Amphibolitic and amphibole-pyroxenic gneisses with layers of biotite-garnet gneisses and quartzites at the bottom. Towards the top marbles, diopsidic rocks with phlogopite and beds of quartzite (no less than 1500m).

Timptonian Series

Lower Sunnaginian Suite—Hypersthenic and diopsidic schists with beds of charnockitic gneisses and amphibolites (no less than 500m).

Ugenian Suite—Hypersthenic plagiogneisses with beds of granulite (800–1000m).

Upper Sunnaginian Suite—Charnockitic gneisses with rare amphibolites (up to 3000m).

Kyurikanian Suite—Interlayered biotite-garnet schists and biotite-hypersthene gneisses and amphibolites; about 1500m. Includes Sutmian Suite?

FIG. 68.   Archeozoic rocks of the Aldan Massif.
Photograph by V. G. Tarasova.

Dzheltulinian Series

Sutamian Suite—At its base the suite has deformed conglomerates and quartzites overlain by diopsidic schists, biotite-garnet gneisses with beds of marble. Towards the top the gneisses predominate (no less than 1500m).

The above succession has been widely applied, but various parts of it had to be altered surprisingly rapidly. In 1958 the Olekmian Series consisting of four suites (total 2600m) was recognized at the base of the succession. In 1960 it was rejected, but at the top of the succession Khaikanian and Lurikanian suites (3000m) were added. In 1961 their recognition was considered to be baseless. Above the Fedorovskian Suite a new Idzhakian Suite was distinguished, but later rejected. Such alterations of the Archeozoic succession are unavoidable and natural, but their frequency is too fast and is caused by the inadequacy of radiogenic age determinations.

Such detailed stratigraphic schemes have also been formulated for other regions of the Aldan Massif and regions of the Siberian Platform. A large number of local suites have been proposed, ensuring detailed geological mapping of these areas; their correlation, however, is a problem for the future.

VITIM MASSIF AND PATOMIAN HIGHLANDS. The Archeozoic of this region has not been subdivided. As a representative example the Mamian, crystalline, Suite, exposed in the lower reaches of the Vitim and the valley of the Mama, can be quoted. It consists of crystalline schists and gneisses, the schists being micaceous, graphitic, staurolitic and garnet-bearing. The suite includes also marbles and quartzites and the well-known Mamian mica deposits. The total thickness of the suite is 8–10km. In other regions, as for instance in the Patomian Highlands, Lower and Upper Archeozoic are distinguished, the former consisting of ortho-gneisses while the latter are para-gneisses, marbles and schists.

BAIKAL REGION. This is of the same type as the Archeozoic of the last region and has been studied by E. V. Pavlovskii (1939). It is divided into three parts: the Lower Archeozoic consisting of gneisses and garnet-mica schists interbedded with quartzites, the Middle Archeozoic consisting of marbles and schists and the Upper Archeozoic consists of thick marbles (Fig. 69) with beds of quartzites and quartz schists. The total thickness is about 10km, while other investigators quote larger thickness up to 23km. The Baikalian Archeozoic, according to L. I. Salop (1958, a, b), is constructed from four massifs within a Proterozoic fold belt. To the south is the Baikalian Massif, then the Yuzhnomuian, Severomuian, and well to the north the Charian Massif (valley of the Chara). The latter massif shows a wide development of magnetite quartzites with beds reaching up to 30m. The carbonate (micaceous) formation of the South Baikal region includes well known phlogopite deposits (Slyudyanka) as well as deposits of lazurite, spinel, coloured marbles, graphite-rich gneisses, pegmatites, and apatite-diopside rocks (Salop, 1958 a).

EASTERN SAYAN. Again the local Archeozoic is divided into three: the Lower Archeozoic (10km) of gneisses, schists, and amphibolites, the Middle Archeozoic (10km) of gneisses, schists, and dolomites, and the Upper Archeozoic (3km) of fine-grained gneisses, schists and marbles. The Archeozoic here is intruded by alaskitic and porphyritic granites, amphibolites and quartz-diorites of pre-Proterozoic age.

YENISEIAN RIDGE. The ridge is principally formed of Proterozoic rocks, but at its southern end it has Archeozoic formations divided into lower and upper divisions. The Lower Archeozoic consists of schists, garnet and hypersthene gneisses, granulites and pyroxenites. The Upper Archeozoic is composed of variable rocks such as schists, amphibolites, gneisses, para-amphibolites, feldspathic quartzites, dolomites, and silicified marbles. The Archeozoic is intruded by granites.

By comparing the Archeozoic successions of the Siberian Platform two types can be distinguished. The first type is similar to the succession in the Kola Peninsula and Karelian A.S.S.R. and differs only by the fact that the Lower Archeozoic is developed. The Upper Archeozoic is absent and is probably represented by intrusions. The Anabara and Aldan Massifs show this type of Archeozoic. The second type is encountered in the Baikal region, Eastern Sayan and Yenisei Ridge. Here marbles, quartzites and schists, absent in the regions of the first type, are present and are accompanied by various intrusions. These formations represent the Upper Archeozoic. The analogues of the Lower Archeozoic are either absent or are represented over limited areas. Calculations of thickness have only a general significance since the structures are very complex and faulted. It can only be said that the thickness of the Archeozoic reaches tens of kilometres.

FIG. 69. White, Archeozoic marbles on the shore of lake Baikal, Cape Sagan (white) on the Ol'khon island. *Photograph by V. V. Lomakin.*

Proterozoic

AREAL DISTRIBUTION. The Proterozoic is very widespread and in places reaches a great thickness exceeding 20km. The boundaries of the Proterozoic and its stratigraphy provoke animated arguments. Its upper boundary is especially strongly debated. It is only clear where folded Upper Proterozoic is unconformably overlain by unfolded Lower Cambrian. Such areas exist in the south as a belt surrounding ancient platforms. Normally the non-folded, platform-type Upper Proterozoic is conformably or almost conformably overlain by a similar Lower Cambrian. In such cases the identification of the Upper Proterozoic and its separation from the Lower Cambrian presents great difficulties. Generally accepted differences have not as yet been established and the boundary is drawn in accordance with local relationships or the opinion of some geologist. The only undeniable diagnostic feature is the presence of the Lower Cambrian fauna. Only those underlying unfossiliferous formations which are closely allied and form a general sedimentary macro-rhythm with the palaeontologically diagnosed Cambrian sediments can be included in the Lower Cambrian. To the Lower Proterozoic should be relegated all those rocks which are separated from the Lower Cambrian by a break and angular unconformities, or are included in another macro-rhythm of sedimentation which does not contain a Lower Cambrian fauna or flora. The degree of metamorphism is also an important diagnostic character.

In the ancient rocks of Siberia, often below the Lower Cambrian with archeocyathid limestones, lies a thick formation of limestones with calcareous algae and groups of shales and sandstones. This formation does not contain a Cambrian fauna, is separated from the Lower Cambrian, and constiutes a specific macro-rhythm of sedimentation. Lithologically this formation is very similar to the Lower Cambrian and the break does not always exist and consequently many geologists consider it as Lower Cambrian, although on the basis of the above-mentioned evidence it is more correct to include it in the Upper Proterozoic. This point of view is accepted here and by others including V. A. Obruchev—the father of Siberian geology.

In 1956 at a conference on the stratigraphy of the Siberian geology it was decided to separate the complex of deposits underlying the Lower Cambrian, but resting on the Proterozoic with an unconformity, as the Sinian System, which is widely used in the Chinese Peoples' Republic for similar rocks. Later on at a meeting of the Central Committee for Stratigraphy it was unanimously decided to accept it as a single stratigraphic unit. With regard to the name and its stratigraphical role opinions were almost equally divided. By a small majority the term 'Sinian complex' (Sn) was accepted, as well as its position between the Palaeozoic and Proterozoic. The term 'Sinian' has been suggested earlier than the widely used term Riphaean and consequently has a priority. As to its position between the Palaeozoic and Proterozoic, this is only provisional, and has been proposed to underline its uncertain age and the necessity to solve this problem. As has already been said it is here included in the Upper Proterozoic.

The discovery of rather varying spores in the Sinian complex is very important. They differ from the even more variable and numerous Lower Cambrian spores and from the rare uniform Lower Proterozoic spores. Detailed investigation of spores belonging to the Sinian complex of the Siberian Platform and China has shown that they are almost identical. This allows us to claim a single age for these deposits. The study of the ancient spores was pioneered by S. N. Naumova and lately B. V.

Trofimov has contributed much. A valuable review has been produced by T. N. Spizharskii (1958).

TYPE SUCCESSIONS. There are two differing types of successions, platform and geosynclinal. The Lower Proterozoic is, in both, folded and metamorphosed. In platform-type successions the Upper Proterozoic is entirely, or only in its upper part, almost horizontal, weakly folded, non-metamorphosed and thin. In geosynclinal successions it is folded, thick and metamorphosed and is commonly characterized by the development of thick carbonate formations and synorogenic granite intrusions.

The platform-type Proterozoic is developed in the Anabara and Aldan Massifs and the Olenek inlier. In the Anabara Massif no Lower Proterozoic has been found. On the Archeozoic, already described, there lies in the south-eastern part of the massif a thin, unconformable formation of quartzites, phyllitic and talc-chlorite schists and layers of limestones. On this lie formations, normally considered as Sinian, which begin with basal conglomerates and sandstones, which are compact, quartzose and brick red, and sometimes have a ferruginous matrix. Over the latter rocks lie pink quartzites and sandstones which have an arkosic or calcareous matrix and are transitional towards the top into siliceous, in places dolomitized, limestones, sometimes with calcareous algae, and dolomites, argillites and sandstones. Higher up lie the archeocyathid limestones of the Lower Cambrian. The thickness of the algal limestones and dolomites (Billyakhian Suite) is 1200m, and on the southern flank it is 200m; while the pink and red sandstones and quartzites (Mukunian Suite) are 600–750m thick and on the southern flank 300–500m thick.

The Olenek inlier on the lower reaches of the Olenek, and between it and the Lena, has the Upper Proterozoic phyllitic formation at the base of the succession, which consists of cleaved sandstones and siltstones, phyllites, slates and sericite-quartz slates, in total 1600–2000m. The rocks are folded into tight, steep folds of north-westerly trend and are covered by the almost horizontal Sinian and Lower Cambrian. The Sinian complex is about 700m thick and it begins with basal conglomerates and sandstones, overlain by algal limestones and dolomites with partings of sandstones with characteristic spores.

In the Aldan Massif along the eastern and northern margins the Archeozoic is unconformably overlain by a thick formation of mainly carbonate rocks of the Lower and Middle Cambrian. In places in its lower parts are encountered Lower Cambrian archeocyathids. Sometimes the lower horizons are barren or algal and underlie archeocyathid limestones. It is possible that such lower horizons as in the Anabara Massif are atributable to the Sinian (Upper Proterozoic). On the flanks of the Stanovoy range and in the basins of the Olekma and its tributary the Chara a full Proterozoic section is present. It begins with strongly folded and metamorphosed Lower Proterozoic, overlain by the weakly folded and metamorphosed Upper Proterozoic Sinian complex.

The next province where the Lower Cambrian and possibly the Sinian of platform type lie transgressively on the Archean is the north-easterly flank of Eastern Sayan and its continuation to the north, where it is overstepped by the Cambrian as for instance in the upper reaches of the Anabara. In outcrops and boreholes there are found almost horizontal Sinian and Lower Cambrian red-beds resting on the Archeozoic.

The second (geosynclinal) type of Proterozoic succession is developed around lake Baikal, Western Sayan and in the Yenisei range, in other words in the fold belt marginal to the southern projection of the Archeozoic basement to the platform. A characteristic feature of the geosynclinal Proterozoic is a full, very thick succession reaching over 20km and consisting of folded and metamorphosed formations, which have intrusive massifs. The Lower Proterozoic is strongly deformed and metamorphosed by the Karelidian Orogeny. The Upper Proterozoic often consists of two divisions and is always folded and metamorphosed, but to a lesser extent than the Lower Proterozoic. The Lower Cambrian is weakly folded and metamorphosed and lies on the Sinian either conformably or slightly unconformably.

In the Baikal region according to E. V. Pavlovskii (1939), who spent years in studying it, the Upper Archeozoic is overlain by the Lower Proterozoic (quartz schists) and the Middle Proterozoic (greenish and lilac phyllites and quartzites). The Upper Proterozoic is known as the Baikalian complex, consisting of three suites: the lower one (Goloustenian) of quartzites, arkoses and shales of 1·5–2km total thickness, the middle one (Uluntuiskian) of algal limestones, dolomites, phyllites (400–600m), and the upper one (Kachergatian, 500–2000m) is again mainly clastic with arkoses, quartzites and phyllitic slates. The total thickness of the Proterozoic is 4–6km.

The age of the Baikalian complex has been argued about for a long time. Some investigators considered it as Lower Cambrian and others as Upper Proterozoic. The spores found by B. V. Timofeyev ended the discussion since they were found to be Upper Proterozoic (Sinian). Thus the complex can be correlated with the Sinian complex of Eastern Siberia and China. The Baikalian complex, the Lower Proterozoic and the Archeozoic have been described by I. P. Karasev (1959), B. M. Frolov (1961) and I. K. Korolyuk (1962).

The Proterozoic fold belt lies to the south of the Archeozoic basement and according to Salop (1958a, 1958b), two zones can be distinguished in it: the outer near the basement and the inner more separated from the basement. The outer zone has thick clastics (conglomerates, sandstones and quartzites). In the west they are known as the Mama-Bodaibinian Series (9–13km), in the east as the Udokanian Series (9–11km). The quartz sandstones of the Udokanian Series have deposits of muscovite pegmatites and the gold sands of the Lena region.

The inner zone consists of diverse pyroclastic and effusive rocks, interbedded with sedimentary sandy and shaley and carbonate rocks forming the Muian Series (12–15km).

Eastern Sayan has a Proterozoic succession, which differs but little along the strike. In the Lower and Middle Proterozoic carbonate formations and phyllites predominate. In the Upper Proterozoic there are phyllites, quartzite and limestones. In places, for instance in the Khamar-Daban range, the Proterozoic consists entirely of limestones up to 1·5–2km thick. In the basin of the river Onot the Proterozoic includes ferruginous quartzites and siliceous slates similar to the jaspilite formations of Krivoi Rog. The total thickness of the Proterozoic is calculated according to different principles as either 8–10km or 19–23km.

The Yenisei Ridge, which is essentially formed by Proterozoic rocks, is a continuation of Eastern Sayan. The thickness of the Proterozoic of the Yenisei Ridge is enormous, the lower division being 3km, the middle division being 7–8km, and the upper about 9km.

The lower division is represented by a strongly altered metamorphic complex consisting of micaceous, hornblendic, black graphitic and quartzose schists, cut by granites, which are of the same age as the Karelian granites of the Russian Platform.

The middle division rests unconformably over the lower and consists of less meta-morphosed rocks, including: phyllitic schists, slates, sandstones, and limestones; all of which are thinly bedded and thus resemble flysch. In places algal limestones and dolomites are developed. Over this formation lies the unconformable upper division, which has a platform character, but is distinctly folded and reaches a thickness of about 9km. It is weakly metamorphosed, as also is the unconformably overlying Lower Cambrian (Kirichenko, 1958, also Fig. 70). In the lower reaches of the Angara, Obruchev divides the Upper Proterozoic into eleven suites of a total thickness of 9km. Amongst these rocks, shales and calcareous shales and sandstones predomin-ate, but in places there are quite common formations of limestones and dolomites, which are sometimes bituminous. In the upper part of the section lies the Kras-noyarskian Suite (600m) consisting of red quartzites, siliceous slates and jaspilitic haematite ores, similar to those at Krivoi Rog. The suite is overlain by some 2000m of shales and algal limestones. The shales and limestones also outcrop further to the north along the Yenisei between Turukhansk and Igarka in the cores of gentle structures and are unconformably covered by the Lower Cambrian (Fig. 7).

Geologists of the Krasnoyarsk Geological Department (Lesgaft, 1958) have suggested the following scheme for the subdivision of the Proterozoic in the Yenisei Ridge (from bottom to top):

Penchenginian Suite. Lies on the Archeozoic and consists of marbles, amphi-bolites, mica, and amphibole schists and quartzites (1200–1500m).

FIG. 70. An unconformity between Lower Cambrian strata (in the cliff) and almost vertical Sinian rocks in the Kamenka river bed. Fine grained sandy-shaly rocks with occasional conglomerates are seen at the bottom of Cambrian strata. *Photograph taken in the Yenisei range by V. I. Dragunov.*

*Udereian Suite. Uniform grey slates and phyllites in places passing into quartz-*mica and quartz-amphibolite schists (1200–1300m).

Pogoryuian Suite. Fine slates and sandstones, which are current-bedded. There is much pyrite (600–800m).

Kartochka Suite. Variegated, bedded marls and limestones. The thickness is variable (100–500m).

Alad'inian Suite. Grey dolomites with veins and lenses of magnetite and large, commercially important bodies of magnesite. The suite shows a local development (300–400m).

Potoskuikian Suite. Banded red and green slates and dolomites (700–800m).

Kirgiteian Suite. Composition is variable, consisting of sandy slates, various carbonate rocks and rare sandstones 1200–1400m).

Nizhneangarian Suite. At the bottom there are haematitic gravelites, sandstones and slates (100m). At the top there are variegated slates and current-bedded sandstones (500m). It is developed in the east. The magnetitic gravelites and quartzites in places form economic deposits.

Dashkain Suite. Grey and greenish, bedded marls and limestones. The thickness varies up to 1500m.

The above succession in detail and with respect to thickness varies considerably from the successions suggested by G. I. Kirichenko, S. V. Obruchev, V. I. Dragunov and later on by S. P. Mikutskii (1961). Clearly successions in different regions, and opinions of different geologists, have not been sufficiently interrelated.

The description of all these suggested schemes is given in the 'Trudy' of the meeting on the stratigraphy of Siberia (1958) and in *The Geological Structure of the U.S.S.R.*, vol .1 (1958). The 'Trudy' have a paper by Timofeyev in which he points out that all the suites starting from Udereian contain spores of Sinian age, which are almost identical to those from the Goloustenian, Uluntuian and Kachergatian suites. Consequently, Kirichenko (1958) considers that all the suites higher than Penchenginian are Sinian in age. At the same time he points out that the middle part of the succession, and in particular the Udereian Suite, show different grades of metamorphism varying from barely perceptible phyllitization to conversion into crystalline schists. This depends on the position in tectonic zones or proximity to granites. Analogous deposits, also attributed to the Sinian, are found in the Chadobets inlier to the east of the Yenisei Ridge and in the Turukhan inlier to the north of it (Mikutskii, 1961).

The Proterozoic of the Siberian Platform is similar to the Proterozoic of the Russian Platform. It is divided into three parts. The lower division corresponds to the Karelian Formation and is Lower Proterozoic. The two upper divisions are Upper Proterozoic. The lower of the two is analogous to the Hoglandian and Jotnian formations. The upper division is the Sinian complex underlying the Lower Cambrian. The Sinian is equivalent to the Hyperborean Formation of the north-west of the Kola Peninsula or the Ter' Suite of its southern shore of the Valdaiian complex and its analogues in the Russian Platform.

FACIES AND PALAEOGEOGRAPHY.   Marine, continental, volcanic and intrusive rocks are developed. Complex tectonics and limits of outcrops do not allow the drawing of boundaries to these facies. Even in the Upper Proterozoic it is a problem for the future.

Algal limestones and dolomites, often reaching a hundred metres in thickness, are well developed. Calcareous algae are rock-forming not only in seas with normal salinity, but also in basins with elevated or depressed salinity. It should be remembered that calcareous algae are also encountered in the Cambrian and the Ordovician.

Jaspilitic, siliceous-haematite formations quite characteristic for the Precambrian, are found in the Yenisei Ridge, in Eastern Sayan and the Baikal region. They are associated with the Proterozoic, but in the Kola Peninsula and in northern Karelia they are also known in the Archeozoic. Structurally and in their origin these formations are identical to the jaspilites of Krivoi Rog.

Bituminous, carbonaceous and graphitic slates, sandstones, limestones and dolomites are often found in the Proterozoic and are also known in the Archeozoic. They are analogous to the schungites of lake Onega and are rocks enriched in planktonic micro-organisms. These accumulations produce bitumens including liquid oil, while metamorphism converts the bitumens into carbonaceous and graphitic rocks, which do not have any relationship to the true coal. The first coals occurred only in the Devonian.

FAUNA AND FLORA. Remains of animal organisms are as yet unknown. Calcareous algae are often encountered, but their use in stratigraphy is difficult since many genera, as for instance *Collenia* are found in the Cambrian and Ordovician as well as the Proterozoic.

The study of the spores has given substantial results, the spores having been abundantly found in argillaceous rocks of the Sinian complex. This study has shown that these spores differ from the spores found in the Lower Cambrian deposits as well as from the spores which lately have been found in the metamorphosed Proterozoic rocks, including phyllites.

UPPER PROTEROZOIC GLACIATION. In the Upper Proterozoic deposits of the Yenisei Ridge, in a number of regions, peculiar breccia-conglomerates, simulating

FIG. 71. Tillite-like, Sinian rocks on the right shore of the river Lena, above the mouth of the Greater Patom. Rounded boulders in the rock are not found in true tillites. *Photograph by G. F. Langersgauzen.*

tillites, have been found. The discovery of these rocks has suggested an Upper Proterozoic Glaciation (Churakov, 1941) over the Siberian Platform. The data quoted, however, do not permit one to consider this suggestion as proven.

The suggested tillites are represented by a dark grey or greenish unbedded, compact, calcareous, sandy or argillaceous rock, which has throughout its mass angular or weakly rounded boulders and fragments of diverse size and composition. Many boulders have smoothed surfaces and rarely there are glacial striae. The total thickness of these beds reaches 800m. The nature of the top surface of the underlying rocks is unknown (Fig. 71). This together with their great thickness suggests a more probable conclusion that these rocks represent alluvial cones of temporary streams. Their glacial origin, however, cannot be excluded. A. N. Churakov (1941) considers them as marine glacial deposits implying washed and redeposited morainic material.

More recently it has been suggested (Grigoriev & Semikhatov, 1958) that the so-called tillites are slide conglomerates in the Lower Cambrian sea. This, however, is not proved in the absence of the Lower Cambrian fauna.

PALAEOZOIC

Lower Palaeozoic

As distinct from the Russian Platform, the Lower Palaeozoic is very widely developed—much more so than any other deposits. Its thickness is several thousand metres and in places it is almost completely horizontal, while elsewhere it is folded into large broad folds of platform type. Consequently the metamorphism is weak and there are no synorogenic intrusive massifs. In places there are sills (traps) of diabases of a younger age. Areally and in its thickness the Cambrian rocks predominate. The Ordovician and the closely associated Silurian are of subordinate significance, although as is shown on synoptic maps they are also quite widely distributed.

Cambrian

The Cambrian is of a great economic importance since it contains numerous, large salt deposits, as yet confined for Eastern Siberia and the Far East. In addition it contains bauxites, oil, gas and other useful mineral deposits.

DISTRIBUTION. In the north-east the Cambrian outcrops over a large area around the Anabara Massif and the Olenek inlier. The best sections are in the upper reaches of the Olenek. To the south the Cambrian disappears under the Jurassic. It is probable that the Cambrian of the Olenek inlier is continuous into that around the Aldan Massif, thus being a continuous outcrop which begins in the east in the valleys of the Aldan and its tributary Maya and then through the valleys of the Olekma and Chara passes into the middle reaches of the Lena.

To the west the Cambrian continues towards the upper reaches of the Lena into the valleys of the Ilim and the Angara, surrounding the Precambrian of the Eastern Sayana and the Yenisei range. To the north its inliers continue along the Yenisei valley and into the valleys of the Tunguskas. Here the Cambrian has yet again a large outcrop known as the Yenisei-Lena Field.

SUBDIVISIONS. The large area of distribution of the Cambrian ensures a substantial difference in successions. The comparison of these successions has not as yet been completed and many details sometimes principal features are not clear.

Thus there are many diverse view-points as for instance about the position of the lower boundary, of the upper boundary, of the boundary between the middle and lower divisions and even about the existence of the Middle Cambrian as such. The different view-points on the Cambrian stratigraphy in the southern part of the Siberian Platform are reflected in the papers in the book *Geology and Oil and Gas in Eastern Siberia* (1959). The leading paper in the valuable new material is by I. P. Karasev and it is supported by borehole sections. In the south-east of the Siberian Platform—basins of the Aldan, Amga and the Middle Lena—the Lower and Middle Cambrian has a large and diverse fauna, the study of which has allowed N. E. Chernysheva (1955) to divide it into stages. From the bottom to top the Lower Cambrian is divided into the Aldanian and the Lenian stages and the Middle Cambrian into the Amgian and Maian stages. The Upper Cambrian remains unsubdivided. All the stages are defined in terms of trilobites and their correlations are given in Table 20.

LITHOLOGY AND FACIES. Two kinds of successions are recognized—the platform and the downwarp types. The geosynclinal type is developed in the adjacent Angara Geosyncline.

The Platform type shows a general distribution with regional variation, but is everywhere thin, of the order of several hundreds of metres, is almost horizontal, and is weakly metamorphosed. Salt deposits are either absent or thin. Amongst the facies the most widely developed are those of shallow epicontinental seas consisting of bedded limestones, marls, dolomites and clays with marine fauna (Fig. 72). The widespread development of archeocyathid limestones is characteristic of the Lower Cambrian. There are also commonly encountered deposits found in low-lying plains consisting of red and variegated sandy and argillaceous deposits lacking marine fauna. Amongst these, lacustrine and lagoonal facies of grey, greenish and dark, thinly-bedded clays, marls and dolomites are common, while gypsiferous and saline limestones with almost black bituminous, organic, carbonate rich rocks, which are parent rocks for oil and gas, are much rarer.

Along the northern and western foothills of the Anabara Massif there are encountered alluvial cone deposits consisting of thick, homogeneous, irregularly-bedded, rapidly altering conglomerates and sandstones, which are red, variegated or grey, and are almost entirely barren.

As an example of platform successions one can select the sections along the Middle Lena and the Aldan and Amga. On the Lena near the mouth of the Sinei the section begins with the Lower Cambrian, variegated, mainly red reef-limestones (20m) with archeocyathids. Upwards the limestones change into similar variegated limestones, marls and clays (total thickness 120m) with clastics, and a marine fauna of archeocyathids, brachiopods and pteropods. The red beds are typical of near-shore marine deposits formed at depths not exceeding tens of metres.

Higher up lies a formation of partly pale limestones with beds of dolomites and red clays at the top. These clays are gypsiferous and saline, are in places dark, layered, bituminous and alternated with bituminous paper shales (total thickness 300m). The bituminous rocks possess a rich fauna of trilobites (*Protolenus, Microdiscus*) and brachiopods (*Acrothele, Obolella*). The pale limestones are shallow-water marine deposits changing towards the top into a lagoonal, saline and gypsiferous formation of 50m in thickness. The dark bituminous limestones with shales and trilobites are the deposits of a large muddy basin of 100–200m in depth. They are typical parent deposits of petroleum (Fig. 73).

TABLE 20. *Cambrian Successions of the Siberian Platform*

Division	Stage	The Anabara Massif, Southern slope and Eastern regions	The Aldan Massif (According to N. E. Chernysheva)	The Middle Lena	Irkutsk amphitheatre (According to Ya. K. Pisarchik)	The Yenisei Massif	Turukhanian region (According to S. P. Mikutskii)
Upper	—	Marine and lagoonal deposits. Above, variegated rocks with gypsum; below, limestones and dolomites (Up to 500m).	Isolated small outcrops of marine limestones and marls.	Variegated suite with algal limestones with a marine fauna (100–200m). Cupriferous red beds (250–400m).	Verkholenian Suite of variegated sandstones, silts and marls with spores (350–500m).	Balaganskian Suite of red beds—sandstones, siltstones and marls, without a fauna (200m).	Grey and variegated marls, dolomites and limestones, with a rich fauna. In the north —800–1000m thick; in the south—150–250m thick.
Middle	Maian	Above, greenish muddy limestones (300–600m); below, variegated muddy limestones and sandstones (100–300m).	Above, bedded limestones with a marine fauna; below, thick algal limestones (Up to 1400m).	Grey and dark-grey bedded limestones and marls which are quite frequently dolomitized and have a poor marine fauna. In the west, 360m thick; in the east, 1100m thick.	Interval.	Interval.	In the north, in the basin of Khantaika thick grey limestones and greenish marls with a rich marine fauna of the top of the Angian stage (1100m).
Middle	Amgian	Bituminous shales and limestones with trilobites (20–25m).	Pale, massive archeocyathid limestones, passing laterally into bedded marls (Up to 800m).	—	Interval. Litvintsovian Suite of dolomites and limestones with trilobites (150m).	Interval. Zeledeyevian Suite of pale limestones, marls and sandstones (80–90m).	In the south, variegated limestones and dolomites, lacking a fauna (100–250m).

TABLE 20—continued

Lower							
Lenian	Bituminous shales and limestones (10–15m). In places replaced by greenish limestones (50–60m).	Grey limestones, dolomites and marls. In places sequences of bituminous shales and archeocyathid limestones (800m).	Saline and gypsiferous sequence (50m). Dark bituminous limestones and shales; in places pale limestones (300m).	Angara Suite of variegated rocks (250–450m). Byelian Suite of dolomites, anhydrites and salt (500–600m). Usol'skian, saline, Suite (600–1000m).	Agalevian Suite of grey limestones, marls and sandstones (250m). Kliminian Suite of flaggy limestones and dolomites (300m).	Kontinskian Suite of pale dolomites with algal seams. Above, lenses of limestones with trilobites and brachiopods (1200–1300m).
Aldanian	Emyaksinian Suite of variegated, muddy limestones with archeocyathids and trilobites (100–200m). Manykaiskian Suite of grey, muddy limestones with variegated sandstones and conglomerates at the base (150m).	Variegated marls and limestones with trilobites (30–100m). Yudomian limestones and dolomites with Collenia (250–550m).	Variegated limestones and shales with a marine fauna (120m). Red, massive, archeocyathid limestones (20m).	Motian Suite of alternating sandstones and anhydrites with dolomites (340–475m). Ushakovkian Suite (Tyretskian) of alternating sandstones, clays and silts (Up to 1000m).	Red sandstones and argillites with beds of dolomites (1000m).	Platonovskian Suite of barren pale dolomites and shales. Basal sandstones are present (200–300m).

FIG. 72.   Attractive cliffs of horizontal Upper Cambrian limestones on the river Olenek. *Photograph by V. Ya. Kaban'kov.*

FIG. 73.   Lower Cambrian limestones of the 'Pyanyi Byk' cliff in the middle reaches of the Lena. The limestones are dark, bedded and bituminous at the bottom and pale and massive at the top. *Photograph by P. I. Glushinskii (1941).*

The Middle Cambrian consists of grey, dark and light bedded limestones which are commonly dolomitized and contain a rare marine fauna. Then total thickness is 360m in the west and 1110m in the east. These are normal, marine shallow-water deposits.

The Upper Cambrian is again of variable composition. In the west it begins with a formation of interbedded red clays and calcareous limestones with gypsum and indications of current action. The total thickness is 400m and these are the deposits of a near-shore plain. The top of the Upper Cambrian consists of pale, marine limestones (200m), which are sandy and glauconitic and in places oolitic. In the east there are limestones and marls (total thickness 150m) with trilobites. These deposits are covered by brown calcareous lensoid sandstones (120m). Higher up there are quite thick dolomites and limestones of the Lower Ordovician.

On the Aldan and the Amga, according to N. E. Chernysheva (1955), at the base of the Lower Cambrian there is the Yudomian Suite (250–550m), consisting of yellow massive dolomites and limestones containing *Collenia*. The finding of a peculiar trilobite (*Judomia*) has forced the inclusion of this formation in the Lower Cambrian (Aldan Stage) although it resembles the Upper Proterozoic (Sinian). The question arises of whether the trilobite (*Judomia*) could be Upper Proterozoic, but there is no answer to this problem.

The Yudomian Suite is overlain by a variegated suite of Aldanian Stage. The fauna includes trilobites, brachiopods, archeocyathids and pteropods. The finding of Olenellidae accurately dates these as Lower Cambrian. The variegated suite consists of reddish and yellowish limestones and marls and has a thickness of 30–100m.

The Lenian Stage includes a number of suites differing in different regions. It consists of limestones, dolomites and marls of different lithological composition, including dark oil shales, bituminous limestones and dolomites, of a total thickness of about 800m. The fauna is relatively uniform, trilobite and brachiopod bearing, and in places (the Amga) there are archeocyathids.

The Middle Cambrian (Amgian and Maian stages) are also divided into a number of suites. At the bottom there are pale, in places reef, limestones, which along the strike change into bedded marls and limestones. Above them lie thick (1400m), barren, massive algal limestones (Ust′milian Suite), and above these are limestones and marbles with trilobite and brachiopod fauna. The total thickness of the Middle Cambrian varies considerably and in the Upper reaches of the Aldan and on the Amga the massive limestones are not well developed and reach only 350–500m. Near the confluence of the Maya into the Aldan these limestones develop a great thickness of 2300m.

No less great an area is occupied by the limestones and dolomites, which underlie the Lower Cambrian red beds. These carbonate rocks are known as the 'Tolbinian Suite' and at the river Tolba (confluent of Lena) they have liquid oil in uneconomic quantities. The Tolbinian Suite (300–1000m) is analogous to the Yudomian and used to be included in the Cambrian. Recently, however, Zhuravleva *et al.* (1961) has advanced a point of view that its lower part is Sinian. This opinion deserves attention.

The Cambrian deposits of the northern flank of the Aldan Massif, and their oil, are described in a series of papers which appeared in the book *Geological and Geophysical data on the geology of Yakutian A.S.S.R.* (1959) and in the work of A. K. Bobrov (1960).

The Upper Cambrian in the north of the platform consists of various limestones and dolomites some 600m thick, and with trilobite faunas. In the south there is the Verkholenian Suite, which is continental, variegated and gypsum bearing. It is 350–500m thick.

The downwarp type deposits have a great thickness of thousands of metres. At the same time the thicknesses of salt increase to hundreds of metres, but the general succession remains as in the platform type.

As an example of the succession may be noted the Cambrian deposits in the reference borehole in Taseyevo, some 200km to the north-east of Krasnoyarsk. The bottom of the hole (2195m deep) is in dark dolomites with gypsum, and limestones and groups of clays. These rocks are in the Usolkian, Lower Cambrian, Suite penetrated for 295m. Over them lie the Belian Suite (548m) which is salt-bearing at the bottom (285m) with beds of salt up to 15m, and marine (263m) at the top, consisting of limestones with clays with a trilobite-fauna. Overlying it is the saline Bulaian Suite (364m) of dolomites and limestones with thin beds of salt. The Lower Cambrian section is completed by the Angarian Suite (180m) consisting of rock-salt at the bottom and dolomites and limestones at the top.

The Middle Cambrian is absent and Upper Cambrian (764m) consists of red, terrigenous deposits, which are mainly siltstones with beds of sandstone and gypsum. The Cambrian is overstepped by the Jurassic. Thus the Taseyevo succession lacks the Middle Cambrian, is thick (over 2200m) and has saliniferous carbonate rocks in the Lower Cambrian.

In the lower reaches of the Angara, near the Yenisei range, there is an identical succession. Formerly, however, the limestones and dolomites were considered as Middle Cambrian, while now they are thought of as Lower Cambrian and the Middle Cambrian is missing. This is a general, but not a proved, opinion.

The total thickness of Cambrian sediments is great (2000–2500m), which in conjunction with their development near a mountain range permits one to consider them as the deposits of a sub-montaine downwarp (foredeep) which separated the Sayan and the Yenisei fold-belt from the Siberian Platform. The Taseyevo borehole is situated in the western part of the sub-Sayan Cambrian downwarp. In its eastern part there is the standard borehole at Byelaya, to the south of Cheremykhov. It cuts only the thick Lower Cambrian (over 1900m). At the bottom lies the Ushakovkian (Tyretkian) Suite (290m) of variegated beds with dark carbonaceous sandstones at the top (58m). Then follows the Motian Suite (447m) consisting of variegated sandstones at the bottom and gypsum and dolomites at the top. It is followed by the Usolian Suite (596m) of dolomites, anhydrites and rock-salt. The succeeding Belian Suite (489m) in addition to dolomites, limestones and gypsum, has a limestone bed with trilobites and the Bulaian Suite (90m) is of dark dolomites. At the top of the borehole there are 9m of Anthropogenic deposits.

The sub-montaine downwarp at lake Baikal bends to the north-east. The succession in the Baikal (Angara-Lena) downwarp is well illustrated by the standard borehole at Thigalovo (upper reaches of the Lena). It is similar to the previous section and has at the bottom the Motian Suite (442m), of dark carbonaceous sandstones and clays below and dolomites and gypsum towards the top. Overlying the Motian Suite is the Usolian Suite (1417m) of interbedded rock-salt and subordinate beds of dolomite and anhydrite. This is followed by the Byelian Suite (550m) of gypsiferous dolomites and limestones, then by the Bulaian Suite (182m) of homogeneous dark dolomites, then by the Angarian Suite (117m) of grey dolomites, limestones and marls, and the section ends with 45m of Anthropogenic deposits. The succession in the borehole has been described by Sulimov (1961).

In the review paper on data from standard boreholes (Kondrat'yeva, 1960) their succession is interpreted somewhat differently. The Ushakovkian and Motian suites are included in the Sinian since they lack a Cambrian fauna and flora. The lower carbonate part of the Usolian Suite is separated into the Tolbinian Suite, which is not sufficiently defined. There are also other differences. Further, there is a notable opinion (Grigoryev, 1956) according to which the lower part of the Lower Cambrian Aldan Stage has the structure of flysch, both along the eastern edge of the Yenisei Massif and within the Irkutsk amphitheatre (Ushakovkian Suite), and further to the north along the western margin of the Baikal range. Z. A. Kondrat'yeva (1960) holds the same point of view. In sections up to 300m thick there are found inter-bedded beds of sandstone (2–7m) with groups (30m) of rhythmic alternations of sandstones, siltstones and argillites. Each rhythm (10–50cm) begins with a fine-grained, grey sandstone with sole-markings followed by greenish siltstones, followed by dark calcareous clays. The flysch formation of the Yenisei Massif is overlain by thick, red sandstones and conglomerates of molasse type; in the Baikal region the red-beds of the Motian Suite correspond to these.

If the flysch and the molasse types of the Lower Cambrian at the southern margin of the platform are confirmed then during this epoch fold movements must have occurred in the adjacent Angara Geosyncline. This is interesting and important. So far folding and mountain building is known only from the Sinian period at the end of the Upper Proterozoic. Thus either this folding continued in the Lower Cambrian or the flysch and molasse types of deposits belong not to the Lower Cambrian, but to the Upper Proterozoic. The latter point of view has already been expressed and is quite likely.

Geosynclinal successions are developed in the Baikal region, in the southern part of the Eastern Sayans and in the Angara Geosyncline. Here Cambrian deposits are not only enormously thick, but are intensively folded and metamorphosed.

The Cambrian of the upper reaches of the Vitim serves as an example. Such Cambrian deposits belong to the Angara Geosyncline and not to the Siberian Platform.

Of the distinct, separate facies characteristic of the Cambrian the archeocyathid limestones and bituminous and saline formations can be considered.

Archeocyathid limestones. These are bioherms, or in other words irregular, rounded mounds formed from the accumulations of animals which grew over each other on the bottom of the sea. If they reached, as is possible, the surface of the sea then they became true reefs. Thus 'reef-like limestones' can be accepted as their proper description. Such limestones are unbedded or poorly bedded, massive, compact, often pale coloured, or pink or bluish, or even dark. On the platform they are thin, being a few tens of metres. However, by repetition within a succession they can account for several hundreds of metres. In downwarps, with prolonged depression of the sea bottom, the thickness of these limestone massifs reaches hundreds of metres (Fig. 135).

Bituminous limestones and shales are not uncommon in the Lower Cambrian of the Siberian Platform. They are usually dark coloured, distinctly and thinly bedded, fine-grained and thin, varying from metres to a few tens of metres. A fauna consisting of trilobites, brachiopods and a pteropod (*Hyolites*) is found in them. The bituminous rocks are the deposits of enclosed stagnant basins, rich in plankton (unicellular animals and plants) and thus are rich in organic matter. They are parent deposits of petroleum and gave rise to the oil found in the Cambrian dolomites of the Tolba, a tributary of the Lena, and in other regions.

Saline and gypsiferous formations are widely distributed in the Lower Cambrian of the southern part of the Siberian Platform. They represent the deposits of large near-shore plains. Where the plains were near to the sea lagoons originated, in which the rock-salt was deposited. Under the conditions of continuous, prolonged sinking of the plains the salt acquired great thicknesses of hundreds of metres (Fig. 70). Thinner formations could have been formed in bitter salt lakes.

FAUNA AND FLORA. The Cambrian fauna of the Siberian Platform differs strongly from that of the Russian Platform. It is much richer and more varied, and is widely represented by all the main groups such as archeocyathids, trilobites and brachiopods.

Archeocyathids and the accompanying calcareous algae were the first reef-builders in the history of the earth. They are commonly responsible for light coloured massive reef limestones, many tens of metres thick, and take part in the origin of thick bedded limestones, but in this respect the archeocyathids are less important than the calcareous algae. The investigations of the last few years have shown that the archeocyathid limestones are associated mainly with the Lower Cambrian, are rare in the Middle Cambrian and non-existent in the Upper Cambrian. Certain well-known specialists, however, including A. G. Vologdin, do not accept this point of view and consider, as formerly, that the greater part of archeocyathid limestones is Middle Cambrian in age.

The trilobites are most numerous and are often found in all marine facies, except the reef limestones. They are most common and variable in argillaceous and bedded carbonate rocks deposited in quiet and stagnant waters. The very diverse trilobites belong mainly to undescribed genera and their Cambrian age in the field is most easily recognized by their dissimilarity to the well-known Ordovician genera.

Cambrian brachiopods are also peculiar, usually being small, horny, and similar to Lingula (*Lingulella*), or are hood-shaped (*Acrothele*).

The flora consists mainly of algae, and microscopic planktonic algae accumulated in large numbers. Remains of land flora are unknown. The argillaceous and fine-grained carbonate rocks have peculiar spores of plants the top of which projected above water.

PALAEOGEOGRAPHY. At the end of the Sinian epoch the greater, western part of the platform rose and became land. At the western margin of the uplifted part, fold-movements took place causing angular unconformities. Further to the east the break was not accompanied by angular unconformities. The lesser, eastern part of the platform did not move and the algal Sinian limestones were gradually succeeded by the archeocyathid Cambrian limestones.

At the beginning of the Cambrian a slow lowering of the land occurred and the southern part of the platform became wide, low, flat near-shore plains. In the south-west it passed into the foothills of a mountain chain situated in the Eastern Sayans. In the south-east and east the plain was bound by the sea and there was a belt of lagoons and bitter salt lakes. Over the surface of the near-shore plain more or less thick deposits of diverse red beds accumulated. Continued lowering at the end of the Lower Cambrian converted the near-shore plain into a shallow sea with a series of low islands and numerous stagnant regions where bituminous, bedded limestones containing trilobites and brachiopods were deposited.

In the Middle Cambrian the general sinking continued, but at different speeds. At first the sinking was faster than deposition and almost the whole platform became a continental sea of intermediate depths. Then subsidence became balanced by deposition, and the sea showed little change in depth, leading to the accumulation of several kilometres of thick carbonates. At the end of the Middle Cambrian the movements became weaker and the sea became shallower, with dolomites being deposited over the shallows.

In the Upper Cambrian in some regions the subsidence was succeeded by an elevation and the Upper Cambrian is missing. In other regions the depression continued, which is indicated by the relatively thick deposits, but since the sedimentation was faster a near-shore plain was formed in place of the sea, where red and variegated clastics were deposited. Short-time increases in the rate of subsidence caused occasional ingressions of the sea causing the deposition of thin groups of clay-carbonate rocks with marine fauna.

In the north of the platform, where the subsidence had been considerable and uniform, the sea remained.

The above history of the Cambrian of the Siberian Platform is based on relegating the upper part of the archeocyathid limestones to the Middle Cambrian. If one accepts a more modern controversial point of view according to which the lower part of such limestones belongs to the Lower Cambrian then the Middle Cambrian is missing from the great majority of stratigraphic sections and an interval in sedimentation must have occurred which in places continued into the Upper Cambrian. At this time the greater part of the platform became an elevated land and an area of erosion.

Ordovician

The Ordovician is widespread and is closely associated with the even more widespread Cambrian. The Ordovician is, however, thinner and more uniform. The red clastic deposits of near-shore plains predominate, although there are not infrequent grey and greenish groups of marine carbonates.

DISTRIBUTION.   The spread is considerable as can be seen in synoptic maps. The Precambrian massifs such as the Anabara, the Aldan and the Baikal, and the Eastern Sayans were areas of denudation. The rest of the platform remained an area of sedimentation of land and sea deposits.

SUBDIVISIONS.   The Ordovician stratigraphy of the Siberian Platform has been less studied than that of the Russian Platform. There has been no general palaeontological research, and there are too few stratigraphical successions which have been established in detail. Only fairly recently an expedition of the Institute of Geology at the Academy of Sciences has succeeded in gathering data comparable to those from the Russian Platform. The expedition was led by V. V. Menner. Another expedition led by O. I. Nikiforova—one of the foremost Soviet specialists on the fauna and stratigraphy of the Silurian and Ordovician of Siberia—under the auspices of the All Union Geological Institute (VSEGEI) of the Ministry of Geology has been particularly successful. The stratigraphic scheme (Table 21) shown here is based on the data gathered by O. I. Nikiforova (1955) and O. N. Andreyeva (Nikiforova and Andreyeva, 1961). They suggest that the Lower Ordovician should be divided into two stages, the Ust'kutian and Chunkian; the Middle Ordovician into Krivolukian and Mangaseiian; and the Upper Ordovician into Dolborian and the 'uppermost Ordovician'. Owing to the lack of precision about the distribution and volume of these stages they should be called suites as has been done formerly.

The tracing of the lower boundary to the Ordovician presents difficulties since it lies within a thick series of red sandstones, clays, marls and limestones which are poorly fossiliferous. This series has a horizon characterized by *Obolus* and *Finkelnburgia*. Formerly these brachiopods were considered to be Cambrian. Nowadays, however, in the Russian Platform they are referred to the lowermost Ordovician Ust'kutian (Tremadocian) and the lower boundary of the Ordovician is drawn at the base of this stage.

TABLE 21.

Ordovician Successions of the Siberian Platform (According to O. I. Nikifirova, 1955

Division	Suite	North-west of the platform. Rivers Kureika and Khantanka	River Moiyero in the upper reaches of the Vilyui and Olenek	River Morkoka to the north of the Vilyui
Upper	Dolborian	In the north (Noril'sk); graptolitic shales and gypsiferous marls. In the south (Khantaika): grey shales and limestones with brachiopods (60–100m).	Grey marls and limestones with a rich fauna (25m).	Dark, bituminous clays with lenses of marine limestones, conglomerates and gypsum (10–25m).
Upper	Mangazeian	Sandstones and dolomites with beds of limestones with phosphorites and brachiopods (60–80m).	Marls and shales with intercalations of sandstones and phosphorites and a marine fauna (35m).	Above, clays and marls with beds of algal limestones; below, gypsiferous clays, marls and limestones, which are bituminous and carry brachiopods (90–100m).
Middle	Krivolukian	In the north: grey shales and limestones with a marine fauna (100–150m). In the south: grey and white barren quartz-sandstones (10–40m).	Muddy limestones, clays and dolomites with phosphorites and a marine fauna (50m).	
Middle	Chunian	Above, sandstones and limestones with *Angarella*; below, limestones and dolomites (350–450m).	Dolomites interbedded with clays and sandstones, and at the bottom with limestones. A marine fauna with *Angarella* is present (150–250m).	Above, reddish-brown dolomites, clays and sandstones with gypsum; below, algal limestones with *Angarella* (Oldondian Suite) (80m). Below, thick dolomites which are algal,
Lower	Ust'kutian	Grey and variegated dolomites, marls and limestones, with calcareous algae and trilobites (430–520m).	Dolomites and limestones. Above, with beds of sandstone and gypsum; below, algal and conglomeratic. A marine fauna is present (120m).	bituminous and gypsiferous; at the base, conglomeratic dolomites (Morkokinskian Suite) (160m).

In the south, along the Lena and Angara, not only the Ust'kutian, but a considerable part of the Verkholenian (Balaganian) Suite, if not all of it, were included in the Ordovician. On the Vilyui the Nyuisian Stage is included in the Tremadoc.

The upper boundary of the Ordovician is drawn at the base of beds with a Llandoverian fauna. In brachiopod bearing facies this fauna is distinguished by numerous smooth pentamerids (*Pentamerus oblongus*, *P. schmidti*). Amongst the graptolitic facies it is characterized by Llandoverian forms.

LITHOLOGY AND FACIES. The lithology of the Ordovician sediments is exceptionally variable owing to the preponderance of near-shore deposits—red beds

, P. Mikutskii, 1961)

| Podkamennaya (Stony) Tunguska | Southern part of the platform | | Suite |
	The Middle Lena (near Polovinka)	River Ilim	
Limestones and marls with a rich fauna; at the base, sandstones (40m).	Barren, red sandstones, clays and marls, often gypsiferous (100–150m).	Bratskian Suite of red sandstones and barren clays with gypsiferous layers (100–150m).	Makarovskian
Marls, rarer argillites, with beds of phosphorites and limestones with a rich marine fauna (40–50m).	Red and green argillites with beds of limestones and sandstones with brachiopods and trilobites (30–50m).	Mamyrian Suite of variegated sandstones with beds of marls carrying a brackish fauna (50–80m).	
Red sandstones, clays and shell beds (25m).	Greenish sandstones and clays with phosphorites and shell beds. At the bottom, red sandstones and clays (70–80m).	The Transitional Horizon of variegated clays with beds of sandstones and dolomites (40m).	Krivolukian
Above, reddish, Baikitian sandstones with *Angarella* (0–100m); below, pale limestones and dolomites with beds of sandstone with *Angarella* (100–160 m).	Greenish limestones with a marine fauna and beds of clay, gypsum and sandstone (50m).	Above, variegated barren clays with beds of sandstones; in the middle, algal limestones; below, variegated sandstones with beds of algal limestones (About 100m).	Ust'kutian
Above, Proletarskian Suite of dolomites and often oolitic limestones; below, Turamian Suite of algal limestones with oboloids (100m).	Dolomites with beds of algal limestones and shales with a marine fauna (100–150m).		

and variegated clastics. Amongst them there are found more or less considerable
series of marine limestones, marls and clays, with trilobites, tabulates and brachiopods. Lagoonal and lacustrine sediments are quite common and consist of thinly
bedded and fine-grained carbonate argillite rocks, which are often saline and
gypsiferous.

In individual cases the enrichment in organic matter as a result of plankton concentration in lagoons and stagnant marine depressions can be observed. Nearer to
mountainous elevations consisting of the Precambrian, the red beds lose horizons
with marine fauna and become the sediments of the submontaine zone—coarse
sandstones with conglomeratic groups. Their thickness is small, and pebbles of
Middle Cambrian marls and limestones are found in them. This circumstance
indicates that the elevations, which were being eroded and which were situated on
the sites of Eastern Sayan, the Baikal region and the Patomian Highlands, were not
high and involved Middle Cambrian, as well as Upper Proterozoic limestones. In
the south the Ordovician includes the upper part of the Balaganian (Verkholenian)
Suite, the Ust'kutian, Krivolukian (Mamyrian and Bratian), and the Makarovian
suites. The Chun'kian Suite is not recognized.

The section in the region between the Lena and the Ilim can be selected as an
example. Over a series of barren interbedded red and green sandstones and marls
of the upper part of the Verkholenian Suite (70m) lie conformably the similar rocks
of the Variegated Horizon of the Ust'kutian Suite (100–150m). Nearer to the Lena,
i.e. towards the old mountains, sandstones predominate, while near the Ilim the
limestones do so. On the Ilim there is a group of algal limestones and the brachiopod
Angarella has been found. The upper sandy carbonate horizons (150–220m) consists
of compact limestones and dolomites, which are dark grey or brown, not uncommonly sandy and with sandstone and flat-pebble conglomeratic members containing sandstone and dolomite pebbles. The horizon forms high cliffs along rivers.
It commonly yields brachiopods (*Angarella*) and gastropods of the Chunian Suite,
which is not distinguished here. Above this horizon is the transgressive Mamyrian
Suite of barren interbedded variegated sandstones and marls of total thickness up
to 150m, and above this lie the red coarse, ferruginous sandstones of the Bratian
Suite (up to 150m). A similar section is developed to the north in the middle reaches
of the Lena and near Kirensk. The section ends with the Makarovian Suite (100–
200m).

To the north-west along the lower Angara, the Ordovician lithology remains the
same, but the fauna becomes more abundant and is most varied in the marly members, while in the red sandstones *Angarella* is found.

With respect to the conditions of formation of the Ordovician in the south of
the platform it can be said that the complex coexistence of terrestrial (near-shore
plains) lagoonal and marine facies is characteristic for it. In the south the continental
facies is found, while in the north-west and north, in the upper part of the section,
the marine facies predominate.

In the northern part of the platform the marine facies become even more predominant, although there are terrestrial facies as well indicating temporary regressions (Table 21).

On the Podkamennaya Tunguska the Ordovician begins with grey and yellowish
limestones and dolomites with frequent layers of algae and oolitic limestones and
rare red or white sandstone members which have yielded *Angarella;* the total thickness of rocks is about 150m. Higher in the succession follow red and green clays,

marls and sandstones with limestones and dolomite members, reaching in total some 300m.

Along the lower Tunguska and the cream-coloured Cambrian limestones are conformably overlain by limestones alternating with marls and shales, which are in places thinly bedded, bituminous and yield trilobites. In the middle part they have various colours, from grey to red or yellowish and in total reach about 500m. Further to the north along the Kureika and Sukhaya Tunguska an analogous group of alternating limestones, dolomites, marls and shales with sandstone members has variegated red and green colours (200–500m).

In more northerly sections the thickness increases, as can be expected for a downwarp zone. For instance in the region of Norilsk according to V. V. Menner the Verkholenian Suite of the Upper Cambrian is overlain by limestones and marls, some 500m thick, covered by massive, thickly bedded limestones (350m). They are overlain by shales and marls (200m) with shales towards the top (50m). Thus the total thickness of the Ordovician reaches 1100m.

FAUNA AND FLORA. The fauna is rich and diverse, but has not been sufficiently studied and modern palaeontological monographs are almost non-existent. The group *Angarella* has been described in detail by B. P. Astakin (1932). The brachiopods belonging to it are inarticulate, and large with individual valves resembling low cones. They lived in a near-shore zone, as indicated by their presence in sands and their co-existence with lingulids. It is possible that like lingulids they lived in brackish and even fresh waters.

The Bryozoa are described by V. P. Nekhoroshev (1960), while the brachiopods by O. I. Nikiforova and O. N. Andreyeva (1961). The field atlas of principal forms edited by Nikiforova is of great help.

Amongst the flora the last widespread development of the calcareous algae *Collenia* is characteristic. These algae appear in the Upper Proterozoic, continue through the Cambrian, and die out at the end of the Ordovician. Like the modern calcareous algae they could exist not only in marine but in saline and bitter-saline conditions. Thus the discovery of calcareous algae does not necessarily indicate normal marine conditions.

PALAEOGEOGRAPHY. During the Ordovician the Siberian Platform was a vast plain, which in the north-east and along the southern border was margined by low hills of folded Precambrian. The hills did not have a river system as was the case in the Cambrian. Temporary streams, formed after rainfalls, carried only fine grained and sandy clastic material with occasional flat unrounded pebbles. A hot tropical climate, involving the alternation of dry and wet seasons, imparted a variegated bright colour to the rocks.

Wide plains, especially in the south, were often lifeless sun-burnt land. On their surface temporary streams deposited variegated sandy and clayey sediments. Strong prevailing winds developed sand storms, greatly contributing to the formation of deposits. Such winds, in particular, picked from the shore grains of glauconite and carried them far in to the interior of the continent mixing them with the red sands. To the north and west the plain was bordered by the sea and consequently even small oscillations caused far-reaching ingressions of the sea with a consequent deposition of carbonates with marine faunas.

The northern part of the platform was below sea-level and only rarely rose as land. To the south however, continental conditions prevailed and the sea only ingressed here for relatively short periods of time.

Middle Paleozoic

Silurian

The Silurian deposits are closely associated with the Ordovician, are litho-logically similar, and have the same general faunal characters. The Silurian rocks are not equally widely distributed and are mainly found in the northern part of the platform, where they consist of various mainly marine facies, although there are

FIG. 74. Bedded Ludlovian limestones and marls on the river Gorbiyachin, to the north of the river Kureika. The dip reflects the position on the eastern slope of the Turukhanian Elevation. *Photograph by V. I. Dragunov.*

also not infrequent lagoonal and near-shore deposits. The Llandoverian and Wenlockian stages are widely developed, while the Ludlovian Stage is only found in
certain northern localities since the southern part of the platform at these times was
elevated as dry land and became an area of erosion.

DISTRIBUTION. The Silurian is found in the region of Norilsk, the lower
reaches of Podkamennaya Tunguska, around the Turukhan Massif, the western margin of the Anabara Massif, in the middle reaches of the Lena and in the upper
reaches of Vilyui. It is undoubtedly developed in intervening areas, but in the south
of the platform only the Llandoverian Stage is found.

SUBDIVISIONS. Detailed work on the Silurian is only now beginning and
Table 22 is based on the data supplied by Nikiforova (1955) and Nikiforova and
Andreyeva (1961).

LITHOLOGY AND FACIES. Lithologically the Silurian is very similar to the
Ordovician and includes marine (dominant), lagoonal and terrestrial deposits. The
Llandoverian consists of limestones, which are often muddy and contain groups of
shales. Its thickness varies from 20 to 225m. At its base there is commonly a group
of graptolitic shales, up to 100m thick. In the south of the platform there is a thin
group of dolomites with a poor marine fauna.

The Wenlockian consists of limestones, dolomites and marls, reaching 70–300m
in thickness. The limestones are variable often with tabulates and stromatoporoids
or algae (Fig. 74), and are sometimes massive with banks of *Megalomus* (a thick-
shelled lamellibranch). The dolomites and marls form groups and individual beds.
The stage is subdivided into two sub-stages.

The Ludlovian Stage (200–300m) consists at the base of the deposits of a receding sea, lagoons and near-shore plains while towards the top it has deposits of
bitter-saline lagoons, lakes and near-shore plains while marine deposits are absent.
Thus at the bottom there are faunally-poor, bedded limestones, marls and dolomites
with groups of red and gypsiferous rocks. Towards the top there are variegated and
red sandstones and clays with marls bearing ostracods and *Eurypterus*. Amongst
the separate facies tabulate and stromatoporoid limestones and graptolitic shales
should be noted. These limestones contain *Multisolenia*, *Favosites*, *Syringopora*
and stromatoporoids as rock-forming organisms. There are also frequent, coexisting brachiopods and molluscs and the limestones have sometimes a massive,
reef-like structure. These limestones are analogous to the Ordovician algal limestones, which as lithologies are absent in the Silurian. The Silurian limestones are
usually grey or greenish, relatively thin (a few tens of metres) and are well-bedded.
They are, however, often repeated in the succession and in total reach hundreds of
metres (Fig. 75).

The graptolitic shales are black, platy or laminated shales rich in organic matter.
Graptolites occur as white impressions on bedding planes. In geosynclinal regions
they reach hundreds of metres, but in the Siberian Platform only a few tens of
metres. Their origins are diverse and over the platform they represent deposits of enclosed lagoons and stagnant bays of shelf-depressions more or less away from the
shore. If the graptolitic shales are found to pass into barren continental sandstones
or red beds then they are lagoonal or bay deposits. If, however, they are surrounded
by normal marine sediments and pass into them laterally then they were formed in
the marine shelf depressions.

FAUNA AND FLORA. The fauna is abundant and variable with brachiopods
(spiriferids and atrypids) being the most important forms, although these are also

TABLE 22. *Silurian Successions in the Siberian Platform* (According to O. I. Nikifirova and O. N. Andreyeva, 1961)

Stage	North-western part of the platform. On the river Kureika and the Norilsk region	River Moiyero, in the upper reaches of the Vilyui and the Olenek	River Morkoka, to the north of the Vilyui	Middle reaches of the Vilyui	Podkamennaya (Stony) Tunguska
Ludlovian	Gypsiferous red beds (120–140m). Limestones and dolomites with beds of algal limestones. A fauna of Stromatopora and brachiopods is present (50–180m).	Gypsiferous, red, variegated clays and marls, with ostracods (130m). Dolomites, marls and gypsum. At the bottom, algal limestones (125m).	—	—	Limestones with *Protathyris* and dolomites with *Eurypterus* (10m).
Wenlockian	In the south, massive and bedded stromatoporoid limestones with *Megalomus* (70m). In the north, interbedded limestones, marls and argillites, with rugose corals and brachiopods (140–190m).	Thick sequence of stromatoporoid and algal limestones with marl and clay sequences. Tabulates, stromatoporoids and rarer brachiopods and rugose corals are present (300m).	Interbedded limestones, clays and marls, often bituminous. A marine fauna is present (150m).		
Llandoverian	Above, grey brachiopod-bearing limestones (100–200m); below, black, graptolitic shales with *Climacograptus*, *Rastrites* and other graptolites (20–110m).	Organogenic and muddy limestones with sequences of marls and rarer dolomites. A rich and variable fauna is present; at the bottom, a suite of dark graptolitic calcareous shales (170–210m).	Bedded limestones with bands of marls and shales; at the bottom, bituminous, brachiopod limestones (225m).	Meikian Suite of a combined Wenlock Llandovery age. Bedded dolomites with *Eurypterus* and limestones with brachiopods and tabulates (50–60m). Graptolitic shales (10–20m).	Kochumdekian Suite of a combined Wenlock Llandovery age. Above, dolomites, passing down into brachiopod-coral (tabulates) limestones. At the base graptolitic shales underlain by sandstones and conglomerates (80–100m).

many trilobites (*Encrinurus*), nautiloids and gastropods. The large, smooth penta-merids (*P. oblongus, P. schmidti*), often occurring in banks are characteristic of the Llandovery.

No terrestrial flora has been found and algae of *Collenia* type are absent, although other calcareous algae are not infrequent.

PALAEOGEOGRAPHY. The palaeogeography is basically similar to the Ordo-vician, the basic difference being that the southern part of the platform was land, which was elevated in the extreme south and flat further northwards. The plain received only barren red sands and clays, which are even now commonly considered as the top Ordovician. Only during the Llandovery the sea for a short time pene-trated to the south.

In the north-western part of the platform the marine regime continued. The seas were shallow and continental and often receded, giving place to near-shore plains and lagoons and then transgressed again. In the Upper Ludlow the sea finally re-gressed even here.

Devonian

The Devonian rocks are closely associated with the Silurian, form con-tinuous successions in the same regions and consists of the same facies complexes —marine carbonate muds, and red beds of near-shore plains.

DISTRIBUTION AND FACIES. In the north-west of the platform the Devonian rocks are relatively fully developed, although they occupy small areas as inliers

FIG. 75. Flaggy limestones at the bottom and massive dolomites at the top, belonging to the Llandoverian Meikian Suite, outcrop on the river Vilyui. *Photograph by N. N. Tazikhin.*

within the Tunguskian Formation. According to V. V. Menner (1961) the fullest succession is in the region of Norilsk.

Lower Devonian. At the base lies the barren Zubovian Suite (210–220m) of variegated sandstones, argillites and gypsum. The succeeding suite is Kureikian (75–85m) consisting of red beds, argillites and muddy limestones. Armoured and bony fish (*Porolepis, Pteraspis*) occur; in the opinion of D. V. Obruchev these have a Siegenian character and belong to the lower part of the Lower Devonian. The Razvedochninskian Suite (120m) of grey argillites and sandstones also contains *Porolepis* and *Pteraspis* as well as marine brachiopods.

Middle Devonian. The Mantourovian Suite (>300m) consists of red beds with armoured fish (*Angarichthys hyperboreus*). The Yuktinian Suite (30m) of grey muddy limestones with brachiopod fauna of Dzhaltulinian type belongs to the Givetian Stage.

Upper Devonian. The Nakakhozian Suite (30–60m) is a red gypsiferous formation. The Pastichnian Suite (30m) consists of grey dolomites and argillites. The Kalargonian Suite (110m) of grey bedded limestones and dolomites with brachiopod fauna of Uralian type is of Frasnian age. The Famennian Stage has not been faunally recognized, but a red bed formation underlying Tournaisian limestones evidently belongs to it.

In the valleys of the Kureika and Bakhta the succession is of the same type and only differs in details. On the Bakhta the Yuktinian Suite (30m) is especially fossiliferous.

In the south between Kansk and Krasnoyarsk the red beds overlie the deformed Cambrian and are subdivided into several suites. Four of the lowermost suites are 1000m and more in thickness and belong to the Givetian Stage since they contain a Middle Devonian flora. In the Krasnoyarsk region the lower suite (Assaf'yevkian) has well-developed thick labradoritic and diabasic porphyrites. The thicknesses of individual Givetian suites and the succession within the stage varies considerably.

With respect to the stratigraphy of the Devonian red beds there are considerable disagreements caused by the insufficient palaeontological characteristics. Many suites are recognized by some and categorically denied by others. The disagreements are particularly great with respect to the lower part of the succession, which underlies the Givetian Stage. This part of the succession is sometimes considered to be Givetian and at others as Lower Devonian. The latter point of view adopted by local workers seems to be better established (Gorbachev, 1961).

The overlying suites belong to the Frasnian Stage since a number of bony and armoured fish including *Holoptychus* and *Bothriolepis* are found within them, as well as phyllopods and plant remains of Upper Devonian type. The total thickness of the Frasnian Stage varies from 300 to 1100m. Its highest part (Krasnogor'yevian) consists of greenish-grey sandstones and argillites and has Famennian spores. Its thickness is 200–350m.

The similarities between fresh-water fish of southern Siberia and the Baltic region is interesting and suggests a former connection between these continents. No less interesting is the occurrence of red beds in the middle reaches of the Vilyui between the Ordovician and Jurassic. The complex of spores and pollen within these have an Upper Devonian-Tournaisian character. Upper Devonian spores and pollen were also found in the region of the Kempendyaian salt-domes in red beds which were formerly considered as Cambrian (Fig. 70). Here a barren Kempendyaian Suite lies at the base and is several hundreds of metres thick. The overlying Kurunguryakhkian Suite of red beds (300–350m) has thin beds of marl with spores and pollen. It is overlain by the Upper Permian.

Summing up it is obvious that Devonian successions of the western and southern part of the Siberian Platform consist of red beds amongst which there is a marl group with fossiliferous Givetian shales. The Lower and Upper Devonian lack marine faunas and represent deposits of near-shore plains and lagoons. The Devonian successions of the north-eastern part of the platform are of an entirely different type. Here red beds are almost unknown and there is no Lower Devonian. The successions begin with Givetian transgressive sediments overlying Silurian and Lower Palaeozoic deposits. The widespread distribution of marine Upper Devonian deposits of European and Uralian type is characteristic. In the basin of the Kotui, a right-hand tributary of the Khatanga, and the neighbouring regions the Middle Devonian is exposed as outcrops of yellowish and brown limestones overlain by sills of Siberian traps. The thickness of the limestones is 25–60m. The fauna consists of brachiopods and rugose corals of Givetian age. In this region there are also Upper Devonian marine deposits represented by brown and dark grey bedded limestones with Frasnian brachiopods (*Cyrtospirifer*, *Atrypa*). The limestones are 30–50m thick, and are overlain by the traps. The succession is similar to that at Norilsk.

Another region where marine Givetian and Upper Devonian limestones are developed is situated on the watershed between the Kotui and the Olenek. The dark and brownish Givetian limestones overlie Ludlovian limestones and are covered by diabase sills. The Givetian limestones are up to 80m thick, they possess a rich and diverse fauna of brachiopods, pelecypods, gastropods, rugosa and nautiloids. The main form is *Stringocephalus*. The presence of the Upper Devonian species indicated the possibility of the upper horizons of the succession to belong to the Frasnian Stage.

In the estuary of the Khatanga (Port Nordvik), amongst the cap breccias of the Yurung-Tumus salt-domes, there have been found fragments of limestones with Givetian and Frasnian fauna. Some fragments are of bituminous Domanik-type limestones which are of Frasnian age. Boreholes penetrating Tournaisian limestones have shown a succession (190m thick) of dark muddy limestones with Famennian foraminifera. In the middle reaches of the Lower Tunguska and Podkamennaya Tunguska, in the valleys of the Uchama and Djatulia, dark limestones with rich and diverse brachiopod fauna have been found. At first these were thought to be Lower Carboniferous, but identifications carried out by the present author have shown that they are Givetian of the same type as the fauna of the higher reaches of the Olenek. The faunal composition is however, more similar to the southern *Spirifer cheehiel* fauna.

PALAEOGEOGRAPHY. The isolated faunal finds described above do not permit detailed palaeogeographical reconstructions. The only more or less determined feature is that marine facies are found in the north-east and deposits of near-shore plains in the east and south, thus indicating the existence of land on the site of the Western Siberian Lowlands. All this is confirmed by successions of the western Urals. Clearly highlands existed on the site of the Sayans and Baltic countries. The almost entire absence of the marine Lower Devonian is explained by the Siberian Platform being dry land during this epoch.

In the Middle Devonian two marine transgressions penetrated this land. The first marine incursion moved from the south-east and brought in the southern

Spirifer cheehiel fauna. The second incursion occurred later and penetrated from the north-east. The resulting sea was populated by a fauna of different type, including the northern stringocephalids. Distinct differences between the two fauna suggest that the transgressive seas did not connect.

In the Upper Devonian in the south-west the sea regressed beyond the Kuznetsk Basin and the platform became a dry land covered by numerous lakes and lagoons populated by armoured fish and *Estheria*. The northern sea continued to exist.

Lower Carboniferous

The distribution of the Lower Carboniferous deposits is even more restricted than that of the Devonian. They are known in the northern part of the platform in the regions of Norilsk, the Kureika, the Fat'yanikha, the Kotui and the lower reaches of the Khatanga and the Olenek.

In the west, the Yenisei region, the Tournaisian Stage is conformable on the Upper Devonian. In the north (Norilsk, the Kureika) the Tournaisian consists of interbedded sandstones, limestones and marls, which are 80–180m thick and contain a marine coral-brachiopod fauna (*Spirifer tornacensis* Kon.). In the south (the Bakhta, the Fat'yatnikha) marine deposits are replaced by continental grey and white quartz sandstones and arkoses, which are barren and reach 30–50m. In the west Lower Visean limestones are locally developed.

The Namurian Stage (Apsekanian Suite) consists of greenish grey and yellowish sandstones and siltstones with plant remains and indeterminable marine fauna. Their thickness is 100–120m. Overlying them are dark carbonaceous sandstones and shales with coal seams. These rocks are at the base of the Coal-bearing Upper Palaeozoic Formation. They are known as the Katian Suite. The flora found in them determines their age as Middle (possibly also Upper) Carboniferous (Fig. 76).

In the basin of the Kotui at the eastern edge of the Changoda lava field there is an isolated inlier of brownish limestones (30–60m) with *Spirifer tornacensis* and other Tournaisian brachiopods. To the north of the Anabara Massif a few small

Fig. 76.   A residual hill of Lower Carboniferous limestones lying on Upper Devonian marls with a dolerite sill (black rocks to the right of the hill) at the bottom. *Photograph taken in the Norilsk region by V. I. Dragunov.*

inliers of limestones with marine Tournaisian fauna, outcrop from underneath the Tunguskian formations.

The most north-eastern inlier of the Lower Carboniferous is situated in the lower reaches of the Olenek in the proximity of Proterozoic inliers. Here at the base of the Carboniferous lie sandstones and muds with beds of gypsum, all being barren and some 80m thick. They are overlain by grey and brown bedded brachiopod-bearing, possibly Tournaisian limestones yielding *Spirifer* ex gr. *tornacensis* and reaching 60m in thickness.

The study of the flora and fauna of the lower part of the Tunguskian coal-bearing formation has shown that they are identical to the Ostrogorian and Balak-honkian suites of the Kuznetsk basin. In particular the flora found in the Apsekanian Suite of the coal-bearing formation of the Norilsk region is identical to the Ostro-gorian flora. This suite is placed in the Namurian and possibly also in the Upper Visean and its analogues are distributed throughout the western part of the Siberian Platform. In the south it is replaced by the Tushanian Suite.

Upper Palaeozoic

Marine Middle Carboniferous and Upper Carboniferous and Permian deposits have not been found on the Siberian Platform. Their development as thin groups of muddy and carbonate rocks at the base of the coal-bearing formation is possible, since such are widely distributed in the geosynclinal regions, which margin the plat-form and are especially abundant in the east.

Continental Upper Palaeozoic and Lower Triassic deposits occupy large areas and are divided into two complexes—the coal-bearing and the tuffaceous. According to A. N. Krishtofovich the Continental Upper Palaeozoic is grouped as the 'Tunguskian Suite' and the Continental Mesozoic as the 'Baikalian Suite', both being parts of the Angara Series. Nowadays the latter two terms are not used and the 'Tunguskian Suite' is also falling out of usage. In the future this term will be re-stricted to the coal-bearing formation.

Tunguskian Coal-bearing Formation

The Tunguskian Formation belongs to the multiple-age type of coal-bearing formations such as are found in the Kuznetsk Basin and Karaganda.

In the Kuznetsk basin the whole of the coal-bearing formation had formerly been considered as Permian, but later its upper part was proved to be Jurassic and Triassic. Still later its lower part—the Ostrogian and Balakhonian suites—was demonstrated to be Carboniferous. In Karaganda a lower Carboniferous age is generally accepted for the lower suites. An Upper Palaeozoic age is possible for the upper suites and the uppermost horizons have proved to be Jurassic.

Similarly the Tunguskian Formation was formerly thought to be essentially Permian with a possibility of its lowermost horizons being Carboniferous. Corre-spondingly its age was thought to be Upper Carboniferous and Permian. Later on it was proved that the upper horizons of the volcanic formation are Triassic and the question arose of whether the whole of the Tunguskian Formation should not be Permian.

Nowadays we know that in the east of the Siberian Platform in places the whole of the Coal-bearing Formation, and in others its lower suites, are Carboniferous and correspond to the Ostrogian and Balakhonian suites. In the east, over a large area, the Coal-bearing Formation or its upper suites faunally or florally correspond to the Kol'chuginian Suite (Upper Permian) of the Kuznetsk basin. The highest parts of the formation belong to the Triassic and modern data indicate that in the south-east the upper parts of the Tunguskian Suite are Jurassic.

The Coal-bearing Formation of the Tunguska basin, like that of the Kuznetsk basin, can be referred to Carboniferous, analogous to the Ostrogian and Balakhonian; Permian, analogous to the Kol'chuginian; and Jurassic. The coal-bearing formation of the Kuznetsk basin is, however, geosynclinal, folded, metamorphosed and occupies a large area; while that of the Tunguska basin is of platform type, non-folded or weakly folded, weakly metamorphosed and is considerably thinner.

DISTRIBUTION. It is vast and the Tunguska formation covers almost the whole western part of the Siberian Platform. In the south the formation is found on the western slope of the Yenisei Massif in the Kan region. In the south-east it is distributed over the upper reaches of the Vilyui and is adjacent to the Jurassic deposits of the Vilyui Depression. Individual outcrops of continental Upper Permian rocks (Sorosskian Suite) are found in the region of the Kempendyai salt-domes situated in the middle reaches of the Vilyui.

SUBDIVISIONS. In those regions where the Coal-bearing Suite is widely studied it is divided into a number of divisions of local significance. Thus, for instance, in the southern regions amongst the productive part of the succession three suites (from below upwards) are distinguished: the sandstone-shale, Karapchankian Suite up to 300m thick, the tuffoidal Badarmanian Suite of 150m, and the Katian coal-bearing sandstone-shale Suite of 100m. In the Lower Tunguska region the following suites (from below upwards) were distinguished: the Anakitian, the Bugarikhtinian, the Burusovian and the Korvunchanian. However, even such a good observer as V. P. Teben'kov (Teben'kov, Gantman and Einor 1939) considers that only three suites—the Bugarikhtinian, the Ayaklinkian, and the Tuffoidal—can be distinguished in the Lower Tunguska. These correspond to the Productid, Transitional and Tuffogenous suites previously distinguished by S. V. Obruchev. The Ayakalinkian (Transitional) Suite is developed in all successions and is evidently a lateral equivalent of the Tuffoidal Suite.

The work on the flora and the study of the successions have permitted E. S. Rasskazova (1958) to distinguish the following suites. The oldest or Tuskanian Suite is present only in the south of the platform and consists of greenish-grey sandstones and clays (40–50m) with Ostrogian flora belonging to the Upper Visean-Namurian. The succeeding Katian Suite is present everywhere, is Middle Carboniferous in age (50–150m), and consists of rapid alternations of argillites, sandstones, coals and carbonaceous shales. It is an analogue of the Lower Balakhonian Suite of the Kuznetsk basin.

The Burguklinian Suite (up to 400m) is only developed in the north-west, belongs to the Lower Permian and consists of grey argillites, sandstones and coals. It is an analogue of the Upper Balakhonian Suite.

The Noginskian (100–150m), Pelyatkian (300m) and Degalinian suites of the Upper Permian to the east and south are replaced by the Stryelkinkian Suite (70–90m) and consists of grey and dark slates and sandstones with seams of coal.

In the north-west in places there is developed the Korvunchanian Suite with Lower Triassic flora. It is absent in places and the Upper Permian is overlain by the Tuffogenous Suite. In the east there are the Jurassic coal-bearing deposits known as the Chaikian Suite (120m).

It should be pointed out that the similarity with the Kuznetsk basin is great and the main subdivisions have identical floras and ages. The Kuznetsk basin has, however, a thicker succession, which is much more deformed.

LITHOLOGY AND FACIES. In the north-west of the platform, in the region of the Kureika, the Fat'yanikha and Norilsk, the lower part of the Tunguskian Formation considered as the Carboniferous is well developed. It is lithologically diverse. At its base lies a group of dark limestones, marls and shales with marine fauna found around the Kureika and Norilsk. This group is overlain by coarse arkosic sandstones, a few metres thick, which are succeeded by alternations of dark carbonaceous shales with coal seams, muddy sandstones and arkoses with pebble beds. The thickness of the series is 150m in the Norilsk region, not less than 500m on the Kureika and over 120m on the Fat'yanikha. Higher up lie effusive rocks, which in Norilsk begin with tuffites (10m) and are succeeded by the Lower Lavas (250m), Upper Tuffites (20m) and the Upper Lavas (250m).

In the north-east of the platform, between the lower reaches of the Khatanga and Lena, Permian rocks have been studied in detail by the personnel of the Institute of the Geology of the Arctics. The Lower Permian consists of sandstones, siltstones and muddy rocks showing rhythms, which often begin in the marine deposits and end with coal seams. Marine microfauna suggests an Artinskian and Kungurian age of the suites.

The Upper Permian consists of two groups: the lower (terrigenous) and the upper—mainly vulcanogenic. The lower formation is coal-bearing and consists of rhythmically alternating siltstones and sandstones. At the bottom of the formation muddy rocks, and at the top shaley rocks predominate. Rich macrofauna and microfauna indicate the Kazanian age of the formation. The upper (volcanogenic) formation is divided into the lower-Tuffaceous and the upper-Tuffolava suites. The Tuffaceous Suite includes sandstones and clays with Tatarian fauna, while the Tuffolava Suite in its lower part also has seams of clay with Tatarian ostracods. The latter fact is important since it indicates that Siberian Traps belong not only to the Triassic, but also to the Upper Permian.

As an example the succession at the lower reaches of the Anabara can be quoted. On the Middle Cambrian limestones and dolomites lies the Lower Permian (200–250m). It begins with basal coal-bearing clays and sands overlain by a group of sandstones with plant remains. Higher up there is an alternation of dark clays, silts and sands and the succession is completed by an upper group of sandstones with plant remains.

The Upper Permian is divided into two suites. The lower, Coal-bearing Suite (300–350m) commences with a group of sandstones, which is followed by rapid alternations of sands, silts and clays, with coal seams (0·4–0·7m). The upper suite is of tuffolavas (140m), begins with alternate sands and tuffs, and terminates with lavas, diabase sills and tuffs with beds of sands and clays. Above this suite lie the variegated rocks of the Upper Trias. A little to the west in the valley of the Popigai the succession is the same, but the thickness of the Lower Permian is 300–350m, of the Coal-bearing Suite 200–250m (coal-seams up to 1·5m) and of the Tuffolava Suite 200m. Moreover the succession is completed by a Lava Suite (200m) consisting of basalt flows with flat and more rarely columnar or spheroidal jointing (Gramberg, 1958).

In the central part of the Tunguska basin, on the Lower Tunguska and further to the south, the proximity of the sea cannot be recognized since marine groups at the base of the succession are absent. Here all the formations are continental, of a younger age and are lithologically different. Sandstones and white, grey and yellow sands predominate and include pure, arkosic, fine-grained and coarse varieties. There are also conglomerates at the base and within the formation, while clays and shales, which are grey or dark and carbonaceous and contain coals and sphaerosiderites, play a subordinate role. The total thickness of the formation is 400–500m and more.

In the extreme south-west between the Kan and the Biryusa the Coal-bearing Formation is only 70m thick. It begins with a bed of conglomerate (10m) with angular fragments, which is overlain by white sandstones (30m), which in turn are succeeded by clays (30m) with sandstone layers and coal seams. The flora has a Balakhonian character. The basal breccia found in the valleys of the left tributaries of the Lena—the Nyuia and the Zherba—in the Angara-Ilim region is of some interest. It lies at the base of the Coal-bearing Suite, reaches 30m in thickness, and is composed of angular fragments in a sandy matrix. Quartzites and pink granites reaching 1–3m in diameter are most common amongst the fragments. These fragments are sometimes considered as the products of the basal moraine of a glacier which descended from the Vitim-Patomian Massif. This is a likely explanation, but demands a palaeogeographical verification. It should be remembered that such a tillite should lie on a polished surface of the underlying rocks. This is a necessary condition to prove the morainic origin of the breccia.

The inlier of the Upper Permian-Lower Triassic deposits (Sorosskian Suite) in the valley of the Vilyui lies amongst the Kempendyanian salt-domes. Here the red beds of Upper Devonian are overlain, and Jurassic rocks are underlain, by grey marls, stromatolitic limestones, sandstones and argillites of tens of metres in thickness and containing ganoid fish of Upper Permian-Lower Triassic Age. It is also possible that some 70m of grey sandstones and sands found in the middle reaches of the Vilyui are also Upper Permian. These rocks also lie between the Devonian red beds and the Jurassic and also contain fish scales and teeth.

Both these inliers lie a relatively short distance away from the main Permian outcrop and may connect with them under the intervening Jurassic cover.

FAUNA AND FLORA.   The fauna is relatively rare and each discovery is worthy of notice. In all the known cases the fauna is fresh-water and lacustrine, and amongst it lamellibranchs predominate resembling modern Unionidae, but of a smaller size and more primitive construction. The genus *Anthracomya* found on the Podkamennaya Tunguska serves as an example. Crustaceans and in particular *Estheria* are not uncommon.

The flora is diverse and common and basically resembles that of the Kuznetsk Basin, but also includes a number of peculiar forms. The flora of the lower part of the Coal-bearing Formation in the region of the Kureika and Norilsk and of the west of the platform has not been described in monographs. The flora from the upper, most widespread, part of the formation has been studied by G. P. Radchenko (1940), a foremost specialist on Palaeozoic and Mesozoic floras. According to him it is restricted although it is represented by no less than 110 species, since predominantly three groups—Cordaites, *Calamites* and ferns are represented. Amongst the Cordaites the genus *Noeggerathiopsis*, represented by 16 species, is most common and determines the Upper Palaeozoic (Permian) age of the flora. In addition to the above mentioned plants Cycadophytes, Ginkgoales, Conifera and other Gymnosperma of undetermined order are found, but in small and subordinate quantities.

PALAEOGEOGRAPHY. The total span of time when the Tunguskian Formation was accumulating was long, beginning in the Lower Carboniferous and ending
with the Upper Permian. During this time the western half of the Siberian Platform
experienced a number of considerable depressions. Such depressions were associated
with fractures of great extent and width, through which enormous amounts of
basic magma poured out as lavas, while some fragmental material was thrown
out as tuffs and tuffites. Low but large volcanoes, especially during the Permian
and the beginning of the Triassic, were important features of the relief. The
volcanoes were continuous in lines and interrupted the monotony of the vast plain
covered by lakes and lagoons. This plain continued far beyond the Western-
Siberian Lowlands and was margined in the west by the young Urals and in the east
by the heights of the Anabara and Aldan massifs, which at the time formed parts
of a continuous ridge. Highlands also margined the Tunguskian Plain to the south,
while to the north lay a cold sea populated by a peculiar boreal fauna. The sea was
situated outside the boundaries of the Siberian Platform and only lapped against its
border.

Tuffogenous (Korvunchanian) Suite

The Tuffogenous Suite, which was formerly considered as the upper part of the
Tunguska Suite, is nowadays considered as a separate suite lying on the Tunguskian
Coal-bearing Formation. The Tuffogenous Suite is mainly Lower Triassic, but also
extends into the Upper Permian.

At its base the Tuffogenous Suite is transitional into the Coal-bearing Suite. Not
infrequently the transitional horizons are separated into independent suites as is, for
instance, the Aayakinian Suite of the Lower Tunguska. Towards the top the Tuffogenous Suite is gradational into the Lava Suite (see p. 265). The composition of the
Tuffogenous Suite is distinguished by the preponderance of beds of volcanic tuffs
and tuffites, which are interbedded with subordinate sandstones and shales with
plant remains. The general thickness of the suite varies and in places it is altogether
absent. Around the Lower Tunguska the thickness is 200–500m. The shales of the
upper part of the suite yielded Lower Triasic *Estheria*. The volcanic tuffs of the
suite are described in the section on magmatism (p. 267).

Around the Lower Tunguska the tuffs in place suddenly become less abundant
and the Tuffogenous Suite passes laterally into Korvunchanian (Lower Triassic)
Suite (see below). Consequently the Tuffogenous Suite should also be called
Korvunchanian.

In the north-east of the platform, between the lower reaches of the Khatanga and
the Lena, as has already been mentioned the whole Tuffogenous Suite is included in
the Tatarian Stage. Tatarian ostracods were found in beds of argillites even at the top
of the lava part of the suite. The differences in the age of tuffogenous suites
indicate that in the northern part of the platform volcanic activity began and ended
earlier than in its central part.

MESOZOIC

Within the Siberian Platform Mesozoic formations are just as widespread as
they are over the Russian Platform. The Siberian Mesozoic rocks, however, as
distinct from those of the Russian Platform, contain coal-bearing deposits often
reaching hundreds of metres. Marine deposits are widely found in the north-east

and east, in the basins of the Vilyui and the Lena, and in the north near the Olenek, the Anabara, the Khatanga and the Yenisei.

Triassic

Continental deposits are developed in the western part of the platform and the upper part of the Effusive (Tuffogenous) Formation forms a part of them.

DISTRIBUTION AND FACIES. In the middle reaches of the Tunguska the Korvunchanian Suite is included in the Lower Triassic. The suite consists of sandstones and clays with beds of tuffite. The thickness of the suite varies from 200 to 650m. The age of the Korvunchanian Suite is based on the Triassic flora and *Estheria* found in it and analogous to those of the Vyetluzhia Stage of the Russian Platform. Recently a skeleton of a *dicynodont*—a fast-moving reptile with two very large teeth—has been found in it. The reptile was of the size of a lion and lived in savannas with scattered bushes. The Korvunchanian (Tuffogenous) Suite is succeeded by the Putaranian (Lava) Suite.

The marine Triassic is developed only along the north-eastern margin of the platform, around the Lena and the Olenek. It consists of massive black shales (180–250m), with Lower Triassic ammonites (*Olenekites*). The shales are followed up by the Middle Triassic shales and sandstones (400–500m). Similar shales are also widely developed in the Verkhoyansk Geosyncline. From the lower reaches of the Olenek the Triassic passes towards the Chekanovskian Range and towards the seaboard, where Lower Triassic sandstones and shales with *Olenekites* are succeeded by red and green laminated Middle Triassic mudstones and the Upper Triassic dark shales with *Halobia*. To the west marine deposits are absent and on the Anabara there are only Rhaetic and Liassic terrestrial deposits.

PALAEOGEOGRAPHY. The palaeogeography is relatively simple since virtually the whole of the Siberian Platform was land. Its eastern half—the Anabara Massif, the Vilui Depression and the Aldan-Vitim Massif—were parts of an elevated region of erosion, with the southern and western margins determined by the massifs of Precambrian and Lower Palaeogoic rocks. The greater, northern part was a low plain—a region of sedimentation. The absence of coal-bearing rocks from this region suggests a dry steppe or even desert climate. The sediments consists dominantly of fluviatile deposits of temporary rivers and lakes formed in a steppe or desert. There are also aeolian sands.

In the plain, along huge faults, stretched rows of low volcanoes—which were centres of effusion of basic, mobile lava spreading over large areas. To the east and north-east the Siberian continent of Triassic times was bordered by a sea, which occasionally penetrated, but not far, into its interior.

The products of continental erosion accumulated in the sea as thick black aspidic shales and silts and more rarely sandstones. The absence of coarse-grained deposits indicates the low relief of the continent and the absence of mountain ranges. The relief and general character of topography resembled the present-day topography of the Cis-Urals region and the Central Kazakhstan, but of course a difference can be detected in the presence of volcanoes and local lava fields during Triassic times.

Jurassic and Cretaceous

Of all the Mesozoic deposits those of the Jurassic have the widest distribution and are almost everywhere associated with Lower Cretaceous sediments. Their areal distribution is immense and in the basins of the Vilyui, the Lena and the Aldan alone is larger than the whole of the Caucasus and Transcaucasia. The Jurassic and the Lower Cretaceous reach 3000–4000m in thickness and have yielded large deposits of coals and gas. Important stratigraphical information on the Jurassic and Cretaceous can be found in the *Transactions of the National Conference on the Stratigraphy of Siberia* (1957) and *The Geology of the Soviet Arctic* (1957), from which the successions quoted here have been extracted. Later on further details have been added in the 'Collection of papers' of the Yakutsk Conference of 1961.

DISTRIBUTION.   The initial distribution of Jurassic and Cretaceous rocks was wider, and probably much wider, than that of the present day. These deposits have suffered erosion and are now found in five isolated areas: (1) the north of the platforms; (2) the Vilyui-Aldan area; (3) the isolated outcrops of the Aldan-Vitim Massif; (4) the Irkutsk-Kana coal-bearing region; (5) the isolated outliers on the Tunguska Formation.

In the north the Jurassic occurs as a narrow belt stretching from the Olenek through Anabara towards the lower reaches of the Khatanga and the region of the port Nordvik. It is then traced up to the Yenisei to the west and the Jurassic is developed throughout the whole of the Northern-Siberian Depression.

In the lower reaches of the Lena the Jurassic is almost completely hidden under the Lower Cretaceous. To the south of the Aldan estuary it crops out and occupies the enormous area of the Vilyui Depression, the middle reaches of the Lena, and stretches far southwards along the valleys of the Aldan and the Amga. The Lower Cretaceous is also present.

In the Aldan-Vitim Massif small outliers of the Jurassic have been preserved in the east, in the upper reaches of the Aldan and the tributaries of the Maya. Here the Jurassic, in terms of its thickness and the presence of coal, resembles that of the Irkutsk basin. On the Olekma and to the west of it the Jurassic and the closely associated Lower Cretaceous have only been preserved in small depressions, surrounded by Precambrian and Palaeozoic massifs. There are, however, many tens of such depressions, the largest of which is situated in the valley of the Selenga.

To the north-west along the Angara the Jurassic forms the large Irkutsk (Cheremkhov) coal basin. Further to the west Jurassic rocks have been eroded away only to reappear again forming the Kana coal basin.

It was formerly considered that over the area occupied by the Tunguska Formation the youngest deposits are Lower Triassic as is, for instance, the Korvunchanian Suite of the Lower Tunguska. Investigations, carried out in recent years, have shown that in places there are also continental Jurassic sediments similar to those found in the Kuznetsk and Karaganda basins. The Jurassic rocks form a large outcrop to the west of the region of the township Yerboyachen situated in the upper part of Lower Tunguska.

SUBDIVISIONS, LITHOLOGY AND FACIES.   In the north-east of the platform, along the lower reaches of the Olenek, the Anabara and the Khatanga and near to the seaboard, a more or less complete development of marine Jurassic can be observed and related to the three main divisions. At the base, in places, there is a basal conglomerate with pebbles of gabbros and diabases. The conglomerate grades upwards into sandstones with plant remains which can be attributed to the Rhaetic and Lias. Overlying this is the Lower Jurassic (Lias) of 250–350m in thickness with a variable marine fauna of mainly ammonites and belemnites (*Amaltheus*).

The Middle Jurassic is peculiar. It starts with the Aalenian Stage (190m) consisting of flaggy sandstones with *Pseudomontis* (*Arctotis linaensis* and *Inoceramus retrorsus*). Higher up follow thick sands (200–250m), sandstones and clays, in some places grey and brown and in others very dark and pyritiferous. They have a dominant marine fauna with *Inoceramus retrorsus*, but in some groups there are also found plant remains. All these rocks are attributed to the Bajocian and Bathonian.

The Upper Jurassic is in places very thin (only 20–40m) and consists of condensed sequences of glauconitic-phosphoritic beds, amongst which the thicknesses of the whole stages are reduced to 1·5–2·0m. The fauna suggests the presence of the Callovian (*Cadoceras*), the Lower Oxfordian (*Cardioceras*) and the *Aucella*-bearing Upper Oxfordian, Kimmeridgian and Lower Volgian (*Aucella pallasi*) stages. The Upper Volgian Stage is not recognized. Where it is fully developed, the Upper Jurassic is 500–600m thick and consists of dark clays and marls.

Higher up follow the thick, dark and light grey clays and sands of the *Aucella*-bearing Valanginian, which also has polyptychitids and reaches 200m in thickness. This is followed by the coal-bearing formation of grey and yellow sands and dark clays with coals (total thickness: 800–900m).

The succession in the Lena coal basin according to A. I. Gusev (1958) and V. M. Lazurkin (1957), who spent many years studying the Arctic region, shows a wide development of coal measures. The succession begins with a lower series consisting of marine clastic deposits with a marine fauna (Middle Lias to Middle Jurassic). The series is 400–450m thick. The succeeding Yakutian Series also consists of clastics of Bathonian to Lower Volgian age. These rocks to the north (near Zhigansk) have beds with marine faunas, but to the south are entirely continental and coal-bearing and reach 300–500m. The Jurassic Coal-bearing Formation is divided into two series: the Middle Jurassic (Bailykian) and the Upper Jurassic (Chechumian or Chechomian). The latter series is divided into two suites: the Dzhaskoian and the Sytochinian. The Dzhaskoian (in the north the Byelikanskian) Suite is continental (20–60m thick) and belongs to the top of the Oxfordian and Kimmeridgian. It is transgressively overlain by the Sytochinian sands and clays, which belong to the Lower and Upper Volgian stages. The basal conglomerates of the suite yield diamonds. (Collection of papers of the Yakutsk conference, 1961.)

The Lena Series (Valanginian-Lower Albian) is the main coal-bearing series. At the mouth of the river Anabara it is marine and in the region of Bulun (on the Lena) it is marine at the bottom and coal-bearing at the top; while further to the south it is entirely coal-bearing. In the Bulun region it is 2000–2400–3000m thick. The overlying Olenek Series belongs to the Upper Cretaceous macrorhythm, which begins with the Upper Albian and Cenomanian sandstones, with groups of coal-bearing shales at the top. In the lower reaches of the Olenek the series is 1300–1500m thick. Along the Lena it is developed in the south, and to the north of the junction with the Vilyui the Lena and the Olenek Series are divided into a number of suites.

The upper, Vilyuian Series consists of current-bedded sands with beds of gravellites and lenses of clays with lignites. Its lower part (700m) is Turonian and the upper part (200m) is Senonian (Danian Stage). The series is only developed in the south.

Large thicknesses of the coal-bearing suites are explained by the fact that they have been laid down in a sub-montaine downwarp, which sank as the deposits were being sedimented. In the more southerly region information was provided by V. A. Vakhromeyev (1958), who is a leading specialist on the flora and stratigraphy of the Mesozoics in Soviet Asia. The succession which he has constructed has the same sequence, except that he proposes a 'Sangarian Series', which includes the Lena and the Olenek Series of Gusev. The Upper Cretaceous continental deposits he still groups as the Vilyui Series.

The Vilyui-Aldan basin is closely related to the Lena basin and lies to the south of the latter. Its stratigraphy has been described by Vakhromeyev (1958) and Zabalduev (1959) and Ivanov (Ivanov and Lyufanov 1961). The marine deposits are only developed in the lower part of the succession, thus corresponding to the Upper Lias and Lower Dogger. In the region of the Aldan these are found far to the south. The marine bands divide the Jurassic succession into three parts: the lower-fresh-water, the middle-marine, and the upper-coal-bearing (Fig. 77).

The lower part begins with basal conglomerates and sandstones unconformably lying on different Palaeozoic horizons. The total thickness is up to 100m (Fig. 77). The middle part is marine and consists of greenish and brownish sandstones and

FIG. 77.   Lower Lias continental sands and gravels of the Angara-Vilyui Mesozoic downwarp outcrop along the river Tetere, a tributary of the Podkamennaya (stony) Tunguska. *Photograph by N. N. Tazikhin (1952).*

dark shales (about 100m in total). At the bottom there have been found Middle and Upper Liassic ammonites, belemnites and lamellibranchs, while at the top occurs *Pseudomonotis (Arctotis) lenaensis*, which indicates the Aalenian (lower Middle Jurassic) Stage.

The Upper Coal-bearing Series (Sidorov and Slastenov, 1961) consists of grey, yellowish and brownish sands and sandstones alternating with groups of dark, carbonaceous clays with brown coal seams. In the middle part of the Vilyui near the mouth of the Markha the thickness of the coal-bearing formation is 900–1050m. Its age is Upper Jurassic (Chochumian Series) and Lower Cretaceous (Sangarian

TABLE 23. *Jurassic and Cretaceous Successions, based on Stratigraphic (reference) Boreholes*
(Data of the Reference Drilling Department of VNIGRI, 1957)

Period	Vilyuisk	Bakhinoi (200km north of the mouth of the Vilyui)	Namtsy (slightly to the north of Yakutsk)
Cainozoic	Anthropogenic deposits. (26m).	Anthropogenic deposits. (25m).	Anthropogenic deposits. (31m). Grey, Pliocene sands and clays, which are unconsolidated and without coals; Pollen is present. (109m).
Upper Cretaceous. Vilyuiskian Series	Sands and sandstones, with beds of clays without coals. (627m).	Unconsolidated, grey sands and clays without coals. (321m).	Grey and greenish coal-less, soft sands and clays. (272m).
Lower Cretaceous. Sangarian Series	Upper part of the coal-bearing sequence—sands with clays and coal seams; above, sandstones and conglomerates. (1000m).	Coal-bearing sequence of grey sandstones, silts and clays with brown coal seams. A current of gas was detected. (955m).	Grey and white soft sands and clays with brown coals. (928m). Grey compacted sandstones and clays with coals. (567m).
Upper Jurassic. Chechumian Series	Lower part of the coal-bearing sequence—sands and sandstones with sequences of clays, coal seams and beds of limestones. (1040m).	Coal-bearing sequence of grey sandstones and clays with brown coals, spores and pollen. (505m).	Grey clays with beds of sands and coal seams. (791m).
Middle Jurassic. Bailykian Series	Rhythmically-bedded sandstones, siltstones and clays. (243m).	Grey siltstones and argillites with beds of sandstones, in placeous coal-bearing. (278m).	Grey and greenish sandstones and clays without coals. A current of gas was detected. (334m).

Series). Near Vilyuisk and to the east of it the thickness of the formation increases to 2500m owing to the thickening of the Sangarian Series and the appearance of Upper Cretaceous (Vilyuisk Series) cross-bedded sands and sandstones with, in places, groups of carbonaceous clays, which can be subdivided into the Timerdyakhian and Lindenian suites. The Jurassic and Cretaceous succession is shown in Table 23, where the sections encountered in three boreholes are compared.

The tripartite Jurassic continues far to the south-east up to the valley of the Yudoma—a right tributary of the Maya. Here the lower part of the section has all the characters of typical deposits at the bottom of a range situated to the east. This has been described in detail by Yu. K. Dzevanovskii (1956). The marine group becomes thinner and has only an Aalenian fauna. The coal-bearing strata are also thin.

In the third region of development of Jurassic deposits, beginning with the upper reaches of the Maya and the Aldan and up to the valley of the Selenga, marine deposits are absent and the Jurassic is represented by continental deposits in general encompassed by the term 'Jurassic Coal Formation'.

South-Yakutian coal basin. The Jurassic Coal Formation has its greatest areal spread and thickness between the upper reaches of the Olekma, the Aldan and the Timpston. It forms a recently discovered South-Yakutian coal basin. At the base lie the intensively folded and metamorphosed Lower Proterozoic and Archeozoic rocks (Fig. 78). Over these rocks lies the non-folded platform type Cambrian (200m) with sandstones and conglomerates at the base and marls and archeocyathid limestones towards the top. The platform character of the Cambrian determines the existences of the whole basin within the Siberian Platform. On the Cambrian lies the disconformable coal-bearing Jurassic. It consists of four suites reaching 850m in thickness. In isolated areas there are also two other suites, thus increasing the total thickness of the Jurassic to 1300m. Each suite represents a rhythm of sedimentation, which begins with coarse rocks (sandstones) succeeded by silts and clays and ending with coal-bearing rocks. The coals are of good quality and reach thicknesses of over 4m. They are laterally continuous and therefore the resources are considerable. The coal-bearing formation lies almost horizontally and is weakly metamorphosed (Fig. 79). The South-Yakutian basin has been studied in detail by an expedition organized by the Laboratory of Coal Geology, Siberian Academy of Sciences, led by V. V. Mokrinskii (1956, 1958).

Considerably greater thicknesses (up to 3100m) are suggested by the geological survey of the Yakutian district (Bredikhin, 1961). At the bottom lies the Yukhtian Suite (Lower Jurassic?) consisting of sandstones and gravels without coal reaching 320–350m. Over them lies the Middle Jurassic Duraian Suite of sandstones and argillites with coal seams (350–400m). To the Upper Jurassic belongs the thickest Gorkitian Suite (800–1000m) of sandstones and silts with coal seams. To the Lower Cretaceous belong three suites without coal: the Kholodnikanskian gravel-bearing sandstones and conglomerates (450–500m), the Updytkanian conglomerates sandstones and silts (400m), and lastly, the Karaulovkian effusive suite of plagioporphyries, quartz porphyries, felsites and tuffs (up to 500m).

To the west, in the valleys of the Olekma, the Vitim and the Chara and their tributaries, up to the valley of the Selenga, the coal-bearing Jurassic is preserved as small outliers, but is basically the same. In places it is succeeded by compact limestones or marlstones with impressions of fish, *Estheria* and water-fleas typical of lacustrine deposits. These are known as the 'Turga' Horizon, formerly considered as Upper Jurassic and nowadays as Lower Cretaceous.

The Irkutsk and Kana Jurassic coal basins are situated along the northern foothills of the Eastern Sayan. They are transected by a railway and the valley of the Angara, and have been intensively exploited, studied in detail and described in hundreds of papers. The eastern part, known as the Irkutsk (Cheremkhov) coal basin, has been described in the important paper by Yu. A. Zemchuzhnikov, while the western Kana basin has been studied by M. K. Korovin.

The succession in the Irkutsk basin is well investigated and uniform. It is divided into three suites. The lowermost suite (Zalarinian) is not coal-bearing

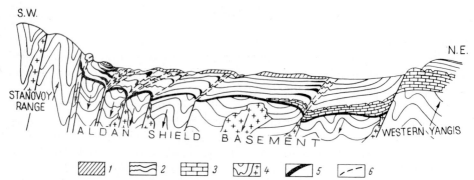

FIG. 78. The coal-bearing complex of the Southern Yakutian coal basin. The cross-section through the depression is according to V. V. Mokrinskii (1956).

1 — Anthropogenic cover; 2 — coal-bearing Jurassic strata; 3 — Cambrian; 4 — Precambrian; 5 — boundary between structural storeys; 6 — faults.

FIG. 79. Jurassic Coal-bearing Formation in the Southern Yakutian coal basin. Thrusting occurred from left to right. *Photograph by G. S. Makarychev.*

(0–120m) and consists of sandstones and conglomerates. The middle suite, Cherem-khovian, is coal-bearing (40–350m) and consists of grey, brownish and black sands, clays and coal-bearing shales. Its age is debatable and is probably Lower Jurassic. The coal seams reach 10–12m, but wedge out quite rapidly. Sapropelic coals are quite common. The upper suite, Cissayanian, has no coal (100–150m) and also consists of sands and clays.

A recent paper about the Irkutsk basin is by Yu. P. Deyev (Trudy. . . . , 1957) who considers the whole of the Coal-bearing Formation as Middle Jurassic. His information on facies changes is of some interest. The platform type of the succession is only 130–150m thick and consists exclusively of fine-grained sediments. This type of succession is characteristic of the central part of the area of sedimentation, away from the mountain chains. The second (Sayanian) type originated much closer to the mountains. The thickness of this type of succession reaches 720m, and at the base lies a group of conglomerates and sandstones (110m), and conglomerates are also encountered higher up. The coals reach a considerable thickness. There are three other types of succession characteristic of the foothills of Jurassic ranges. There are no coals in them and coarse-grained deposits predominate. One type (Baikalian) consists solely of conglomerates (400–800m).

Kana-Achinsk coal basin is about 650km long and about 100km wide. The coal-bearing formation is 120–1100m thick. The Jurassic, here, is normally almost horizontal, weakly metamorphosed and the coals are brown. Only in the Sayan-Partizan region—nearer to the Sayans—the Jurassic is deformed into an asymmetric anticline with dips of 8°–12° and even 45°–90° on its flanks. The coals here are black and bituminous.

According to the latest data (Aksarin 1957), in the Kana basin the Jurassic lies on an eroded surface of Palaeozoic rocks and can be divided into three suites. The lower, Pereyaslavkian (Partizanskian), is 85–180m thick, commences with conglomerates (10m), followed by sand and clays, which at the top of the suite bear coal seams. The suite belongs to the Lower Jurassic. The middle suite (Kamalinskian,=Sayanian+Ivanockian) is 170–320m thick, and begins with sandstones, overlain by greenish and dark argillites and sandstones with coal seams. The upper suite—Borodinskian—is 220m thick, and starts with pale sandstones followed by interbedded argillites and siltstones and rarer sands. Both middle and upper suites are Middle Jurassic. The top of the upper suite is highly productive, one of its coal seams (the Borodinskii) is over 40m thick.

In the central part of the Kana-Achinsk field the Middle Jurassic coal-bearing formation gradually passes into a variegated formation of 90–300m. The study of the flora and fauna in the latter formation has shown that its age is between Upper Jurassic and Barremian and that it correlates with the Ilekian Suite of the Western Siberian Lowlands (Neustruyeva, 1961).

In the west the Kana basin is almost directly transitional into the Chulyn-Yenisei basin, which has a similar stratigraphy and coals and is nowadays considered as a part of a single Kana-Achinsk basin.

The Jurassic of the Tunguska coal basin has only been recently discovered. As in the Kuznetsk basin there are remanent patches of the Jurassic continental coal-bearing sediments on the Tunguska Formation. A region to the west of the middle part of the Lower Tunguska, near to the township Yerbogachen, serves as an example. Here grey and brown sands with a Jurassic fauna and flora (*Czekanowskia* and *Cladophlebis*) are exposed.

According to Kirichenko and Tuganova (1955), as well as other geologists, remains of Jurassic continental deposits can be found not only near Yerbogachen, but throughout the area between the Vilyui and the Irkutsk and Kana coal-fields. To such deposits belong a majority if not all of the so-called 'Watershed Gravels' the age of which has been very disputable. The study of such gravels in the region of Terere, to the south-west of Yerbogachen, has shown that they consist of brown, in places ferruginous, cross-bedded, friable sandstones and conglomerates consisting of unsorted, badly rounded pebbles, the thickness of which is unknown, but which in river cliffs reach several tens of metres. The analysis of their spores and pollen has shown that they are Middle Jurassic.

Similar sandstones and conglomerates (60–80m) with Jurassic spores and pollen were found in the valley of the Chona, to the north-east of Yerbogachen. Here in the upper part of the section a bed of shells of the lamellibranch *Tancredia* indicates an Upper Liassic age. Even more recent investigations (Odintsev, 1961) have shown that the sandstones and conglomerates form the base of the Jurassic coal-bearing formation, and are overlain by sandstones and argillites with coal seams. The total thickness of the Jurassic is no less than 150m and the area of its distribution is greater than the Kana basin. Odintsev has suggested the term 'Murian coal-field' after the river Muru where the Jurassic is best developed. The discovery of such a large coal-field shows how badly the Siberian Platform is known and how much can yet be discovered in its vast area.

The Jurassic age of the Watershed Gravels in the southern part of the platform does not exclude the possibility of a different age of such gravels in other regions. Near the junction of the Aldan and the Lena similar deposits have yielded a fresh-water fauna of Miocene and Pliocene age.

PALAEOGEOGRAPHY. The palaeogeographical trends of the Upper Palaeozoic continue into the Mesozoic Era. The general prolonged downward movement of the basement led to the accumulation of thick terrigenous, generally coal-bearing deposits. The greater part of the Siberian Platform remained land.

During the Triassic, at first the palaeogeography remained the same as in the Permian. The western half of the platform represented a plain with lakes and swamps. Later on the latter becomes less frequent and the coal-forming conditions less common. At the same time volcanic activity intensified, becoming very widespread. The complete absence of terrestrial Middle Triassic sediments indicates a short-lasting elevation of land turning the platform into a region of erosion. At the end of the Triassic during the Rhaetic epoch the palaeogeography changed and the eastern half of the platform, which hitherto remained elevated, suffered a downward movement and a greater part of it became a near-shore plain, which in the north and east was margined by sea. On this plain were deposited Rhaetic-Liassic conglomerates, sandstones and clays, which were brought in from the west by temporary streams and weakly developed rivers.

In the Middle Lias the downward movements intensified and the sea entered the platform covering its northern and western borders and penetrating into its central region along the Lena-Vilyui Depression. With minor oscillations, the sea remained here throughout the Middle and Upper Liassic. During the Aalenian epoch the platform sea reached its maximum extent and in the south-east advanced into the region of the valley of the Maya and its right side tributaries. The sea also advanced further along the Vilyui and the Lena. This marine advance was a precursor of a fast and considerable regression. The whole eastern half of the platform was converted into a near-shore continental plain and throughout the Middle and Upper Jurassic remained a fairly elevated dry land, indicated by thin sediments and rare coal-bearing strata found over it.

At the beginning of the Cretaceous (Valanginian epoch) subsidence took place and accelerated, and the sea again penetrated into the eastern part of the platform, but not nearly as far as during the Lias-Aalenian transgression. The marine fauna with *Polyptychites* and other ammonites as well as numerous aucellids is widely distributed along the Vilyui and is found near the town Vilyuisk. The Valanginian transgression was short-lived. The sea regressed again, and a low near-shore plain, then became the site of large swamps and lakes. The well-developed river system brought in large quantities of sand and clay, causing the formation of a thick coal-bearing Lower Cretaceous series of rocks.

The western part of the platform in the Jurassic and Lower Cretaceous epochs was a hilly plain, with lakes and lagoonal swamps in its lower parts becoming sites of future coal-bearing formations. In the south-east and south there were low mountain chains—sources of clastic sediments.

The Upper Cretaceous sea, which penetrated from the south and drowned the huge area of the Western Siberian Lowlands, was margined to the north by Cimmerian Highlands. The sea formed an embayment along the Khatanga Depression towards the east along the north-western part of the Siberian Platform. The embayment, however, did not penetrate into the interior of the platform. To the east of the embayment there was a large plain, on the surface of which was deposited the thick Upper Cretaceous series of the Bulunskian deposits at the mouth of the Lena, port Tiksa and the Vilyui coal-field. The rest of the platform was a region of erosion.

CAINOZOIC

The area occupied by the Cainozoic deposits is characteristically small. The continental Palaeogene and Neogene deposits are not widespread and the corresponding marine deposits are altogether absent. The Anthropogene, although widely distributed, is relatively thin and does not cover the older formations below it.

The Palaeogene

The continental Palaeogene deposits are closely associated with the Upper Cretaceous. The Neogene is no less closely associated with the Lower Anthropogenic deposits.

As examples of disputed Palaeogene deposits there are the whitish-grey and yellowish cross-bedded sandstones in the famous Chirimii cliff on the Lena, near the town Zhigansk. These deposits yield pieces of wood, concretions of sphaerosiderite, and lenses of clays. The clays and concretions yield a mixed Tertiary-Upper Cretaceous flora. The abundance of the swamp plant *Taxodium* has suggested the inclusion of these sandstones, together with those on the Bakhanoi cliff and the island Agrafena, in the Miocene. Later on Krishtofovich, a major Soviet palaeobotanist, attributed these rocks to the Palaeogene, and as such these rocks are marked on the 1:5 000 000 map, but owing to the indeterminacy of the flora these sandstones are shown on the 1:2 500 000 map as Upper Cretaceous. Very similar sands and sandstones with remains of coniferous plants and amber are found along the lower reaches of the Anabara and Khatanga.

In the south-western part of the Vitim Highlands in the valleys of the tributaries of the Vitim there are white and grey sands with black clays reaching a total thickness of 20m. The analysis of the micro-flora has indicated the presence of the pollen of coniferous trees and more rarely of angiosperms of Palaeogene age.

The Neogene

The continental Neogene is more widely distributed. In the lower reaches of the Yenisei, in the basin of the Lesser Kheta there are lacustrine sandy clays with vivianite and fresh-water lamellibranchs. In the lower reaches of the Aldan and along the Lena below the Aldan, the watersheds of the left side tributaries and those of the right side of the Aldan have outcrops of sandy gravels and clayey deposits, varying from brown to yellow. These underlie the Anthropogene. These deposits are thick (up to 900m) and according to N. A. Ignatchenko (Trudy.... 1958) and Zabaluyeva (1959) are divided into two suites: the lower coal-bearing (450–700m) and the upper coal-free (150–200m).

The Lower Suite (Tandinskian) is basically of Miocene age and begins with a formation of sands and gravels (150–400m) without coals, which is overlain by identical sands but with groups of clays and silts with seams and lenses of brown coals (up to 300m in total). The Upper Suite (Bayaginskian) is of Pliocene age (determined by pollen analysis) and begins with grey and greenish sands overlain by coal-free silts and clays.

A somewhat different stratigraphic relationship is adduced for the Ust'-Aldan Depression by G. F. Lungersgauzen. He considers the Tandinskian Suite as Upper Oligocene, and the middle part of the sequence he separates as the Nama Suite (Miocene). To the Pliocene he attributes the formation found on Mount Mamont and the Tabaginskian Suite.

The Neogene on the left bank of the Aldan and the Lena lies almost horizontally, dipping to the north at $\frac{1}{3}°–\frac{1}{2}°$, but on the right shore of the Aldan and in the Cis-Verkhoyansk downwarp the Neogene forms large folds with dips of $30°–59°$ which, however, soon flatten out to $1°–2°$.

The coal-bearing Miocene formation contains 10–15 coal seams of 1–26m in thickness. The coals are of low quality. The seams are continuous for tens of kilometres, which implies large resources in this coal-field.

It is quite possible that the Neogene continental deposits are also developed on the shores of lake Baikal and in analogous depressions to the north-east and west of it. The accumulation of Neogene deposits occurred simultaneously with the sinking of the bottom of the Baikal graben, and their thickness as a result reaches 1400–2200m. In the region of the delta of the Selenga and to the west the facies are variable and rapidly changing. In the west near to the Khamar-Daban range there is a great thickness (800–1000m) of conglomerates and coarse sandstones, representing a typical sub-montaine alluvial cone. Somewhat to the east there is a formation of rapidly and irregularly alternating clays and sands of green, brown and dark coal-bearing banded aspect, representing the deposits of swampy shores of the bays of lake Baikal and nearby lakes. Further to the east the clays and marls are overlain by cross-bedded sands, which are coarse, with lenses of brown coals, and reach 200m in thickness. These are the deposits of a river delta. The dark, compact clays and sandy clays smell of petroleum and are even saturated with it.

The sections of two standard boreholes in the valley of the Barguzin—to the north-east of lake Baikal—and in the valley of the Tunka—to the south-west of lake Baikal—can be quoted as examples. The Barguzin borehole ended in Proterozoic gneisses at a depth of 1401–1414m; these are overlain by the Middle Pliocene (323m) consisting of dark sands and clays which are quite often coal-bearing. Their age is determined by pollen. The Upper Pliocene is even thicker (441m) and consists of brown and greenish sands and clays with pollen. The Anthropogenic deposits, however, reach the greatest thickness (637m) and consist of sands, sandstones and gravellites. The Tunka borehole shows a large number of basaltic lava flows. At its base between 2117m and 715m lies the Miocene. At first there is a basaltic lava flow of 13m in thickness. It is followed by micaceous sands, silts and argillites (343m) and then another basaltic flow (26m) is succeeded by an alternation of basalts and sands with a group of clays with lignite (670m). Towards the top there are grey sands and clays (350m) with plant remains. The Pliocene (297m) consists of grey sands with layers of clays and lignites and there are basaltic flows. The Anthropogenic deposits (418m) are grey sands and clays with groups of basalts.

The Cainozoic deposits of the eastern slope of the Yenisei ridge have been studied in detail where the ridge is crossed by the Angara. These deposits have accumulations of lacustrine bauxite. At the bottom of the succession, on the ancient rocks, lies a sandy-clayey coal-bearing formation (60–70m) of Upper Jurassic and Lower Cretaceous age. Over it lies a bauxitic formation (100–150m) of variegated rocks with seams of lacustrine hydrargillitic bauxites. The study of the flora, and especially of the pollen, has proved the Eocene age of the bauxite formation. It is covered by coal-bearing dark Oligocene clays with the pollen of chestnut, beech and hornbeam. These clays are succeeded by Miocene sands and clays and then by Pliocene sediments. The total thickness of the upper part of the section is 120m. The pollen succession indicates a gradual oncoming of cold conditions. The chestnut and beech forests of Oligocene become replaced by alder and birch woods of Miocene age and then by wooded steppes with a northern flora of Pliocene age.

The Cainozoic lacustrine fauna has been studied by Martinson (1956). According to him, during the Upper Cretaceous epoch in the Central Asia there were several huge fresh-water lakes (in north-eastern China and Mongolia). Their large sand beaches had a swampy rich tropical vegetation populated by wandering dinosaurs, huge alligators and turtles.

In the Palaeogene these basins decreased in size and broke up into a series of smaller and larger lakes. The climate was still warm, and the shores of the Palaeogene Baikal had lotus growing, as on the shores of the present-day Nile.

During the Neogene the cold began to set in, while the lakes continued contracting and acquired the present-day dimensions. The huge reservoirs of Central Asia became enclosed small bitter-salt and rarely fresh-water lakes. The fauna, which migrated gradually, finally populated lake Baikal and those lakes which existed on the surface of alluvial plains of the Siberian Platform and in tectonic, graben-like depressions, amongst the Palaeozoic and Precambrian massifs.

The Anthropogene

The Anthropogene deposits are variable, widely distributed and are sometimes thick. Their various facies resemble the Anthropogenic deposits of the Russian Platform. In the north glacial deposits are characteristic, but lacustrine, lagoonal fluviatile and various land deposits are also frequently encountered. Successions found in various regions are described in the 'Trudy' of the National Conference on the Stratigraphy of Siberia (1958) and the transactions of papers read at the conference in Yakutsk (1961).

Marine Interglacial Deposits are developed principally in the North-Siberian Depression, spreading to the south of the Igarka (along the Yenisei valley) and into other regions of the northern margin of the platform. These deposits consist of gravels, gravels and sands, sands and also blue plastic clays. A large number of lamellibranch and gastropod shells are found in such marine sediments. A great majority of these animals live nowadays in the Arctic Ocean, but certain warmer climate forms found in the Anthropogenic sediments are not present there today. This fact indicates that the Anthropogenic sea was warmer than the present-day ocean. The northern aspect of the fauna has, nevertheless, forced the term 'boreal' being applied to the marine transgression. The thickness of the boreal deposits is usually 20–40m, but more ancient marine deposits will be later described (p. 298) under the heading of the Yamal'skian Series.

Glacial Deposits are most commonly represented by boulder clays (moraines) of several metres in thickness and consisting of unsorted sandy clays, lacking bedding and containing a greater or lesser number of unrounded, unsorted and diverse fragments of different rocks fluvioglacial deposits are not uncommon, and re-present washed moraines, thus showing an indistinct bedding, some sorting and rounded fragments and pebbles. Sometimes in trough-like valleys (cirques) there are remanent terminal moraines, which cross such valleys as irregular ridges of fragments in a sandy-clayey matrix. The upper trough-like terminations of these valleys (cirques) serve as evidence of ancient glaciation. The actual cirques are often present, but cirques on their own are more rarely preserved (Fig. 80).

Investigations of the glacial deposits and relief has shown that the deposits are most widely developed on the Taimyr Peninsula. This is intelligible since even at the present time the neighbouring Severnaya Zemlya is glaciated. The glacial deposits of the Norilsk region (the Putoran Mountains, up to 2000m high; Fig. 80) are also well developed, but the Anabara Massif (up to 900m) does not show them. In the Norilsk region and the Taimyr Peninsula two glacial epochs have been recognized: the earlier (Zyryankian), involved a widespread glaciation, is followed by the inter-glacial Boreal Transgression and is succeeded by a later (Sartanian) glacial epoch

FIG. 80.   Glacial morphology in the Putoran Mts. of the Norilsk district.
Photograph by A. A. Mezhvilk.

characterized by a restricted valley glaciation, the evidence for which is found in upper parts of highland valleys.

In other regions the first glaciation occurred at the end of the Lower Anthropogene, had only a limited development and has not acquired a general term. It is, however, commonly known as Torskian. The second glaciation, which occurred in the Middle Anthropogene had a great spread and is known as Samarovian. A third glaciation, at the end of the Middle Anthropogene, is known as Tazian, while the fourth (Zyryankian) and the fifth (Sartanian) belong to the Upper Anthropogene. The Sartanian glaciation was only local and is sometimes included in the Recent, while the Tazian epoch is not recognized everywhere.

In the east the Anthropogenic glaciation is well represented in the Verkhoyansk range where glaciers advanced well away from the range and especially so to the south. The centres of glaciation are found in the upper reaches of the Vilyui and the Northern Baikal Highlands, but the magnitude and type of this glaciation is not clear and there are various explanations of it. A particular feature of the Anthropogenic deposits in the north of the platform is the presence of lenses and masses of ice buried in the sediments and reaching 10–15m in thickness. Such accumulations of ice yield carcasses of frozen extinct animals (mammoth, woolly rhinoceros) proving that the ice is ancient. The origin of the ice, nevertheless is debatable. Formerly it was considered as residual from the ice sheet or from the fields of hardened snow (so-called firns). At present the ice is explained as infilling of cracks which originated on the surface of ancient river valleys since very often the ice occurs as columns and wall-like features and not as horizontal layers.

Lake and Bog Deposits consist of bedded clays, marls and peat found here and there.

Deluvial Deposits. The severe northern climate of the Siberian Platform caused general and vigorous manifestations of freezing. The surface of the platform is a region characterized by permafrost, and continuous and deep freezing and thawing breaks up the structure of rocks and forms a blanket of deluvial deposits, which either remain *in situ* or move down the slopes.

Alluvial Deposits consist of fluvial, lacustrine and bog sediments and locally fluvio-glacial deposits, which as a whole have a wide distribution, but which on geological maps are often shown where they obscure the underlying rocks. The alluvial deposits occur at the surface of the great alluvial plains (the Yeniseian, the North-Siberian, the Lenian and the Vilyuian). The deposits usually consist of medium and fine grained, more rarely coarse, clastic sediments such as sands, clays, silts, gravels and marls. Their thickness varies from a few to many tens of metres, and their age is established on the basis of mammoth bones, woolly rhinoceros, northern elk, musk deer and many other sub-polar animals.

The alluvial deposits are thickest in the Baikal rift; in the Irkutsk valley to the west of the southern termination of lake Baikal the Anthropogenic sands and clays reach 500–600m. River terraces reach huge sizes and number as many as 15–18. Alluvial gold and other economically important minerals (Yu. A. Bilibin; 1938), as well as diamonds are found here.

SUBDIVISIONS. The subdivisions of the Anthropogene, until recently almost completely unsubdivided, has been of late successful and in this respect is as advanced as the similar work over the Russian Platform. During the 1958 conference a four-fold subdivision of the Anthropogene has been adopted for the Siberian Platform. The Lower Anthropogene includes the pre-glacial alluvial sands, gravels and clays lying in the depressions of the ancient relief. The Middle Anthropogene consists of the boulder clays of the maximal glaciation and the interglacial deposits with leaves of oak, *Corylus*, *Tilia*, *Ulmus* and remains of warm climate animals. On the Angara, terraces of 18–35m and 40–80m correspond to this epoch. The Upper Anthropogene is the epoch of the last glaciation and maximal freezing, when in the lower reaches of the Angara the flora was the same as now in the lower reaches of the Indigirka and the Kolyma. There were herds of mammoth, woolly rhinoceros, and along rivers settlements of early Arctic hunters (Solutrean Stage of the older Stone Age). The Recent or Holocene includes all the post-glacial deposits, dominantly alluvial and deluvial. The flora and fauna (badger and red deer) indicate a climate similar to or even warmer than the present.

The four major subdivisions are again divided into 16 minor ones, which points to the detailed nature of the present-day research on the Anthropogene. A new important contribution to the stratigraphy is by E. A. Vangengeim (1961) who has also described the mammals and various plants and contributed good photographs. The details of the Anthropogenic successions of the Yakutian A.S.S.R. is given in the Transactions of papers presented at the Yakutsk conference (1961).

The Anthropogenic sediments of the North-Siberian (Taimyr) Depression have been studied in detail. According to V. N. Saks (*The Geology of the Soviet Arctic Region*, 1957) the following succession is recognized.

The lower subdivision is only recognized in boreholes and is represented by gravels and sands (12–18m), representing washed moraine of the oldest glaciation. The middle subdivision consists of periglacial, mainly marine deposits of sands and clays (70m), which are regularly bedded and contain a marine fauna. The upper part of the subdivision is represented by the deposits of the maximal glaciation. These deposits are present everywhere in the depression indicating a wide distribution of the ice sheet. Boulder clays (typical moraine) or washed fluvioglacial sands and gravels predominate. The total thickness of the glacial deposits is small (12–25m). The maximal (Samarovian) glaciation is correlated with the Dnyepr glaciation of the Russian Platform.

The upper subdivision is itself divided into three suites. At the bottom lie the interglacial marine sediments. These are the deposits of the second or Boreal Transgression, they begin with terrestrial and near-shore sands with beds of gravel and pebbles, reaching in total 25–75m. Higher up lie typical marine facies (Sanchungovkian Horizon) consisting of clays and silts with beds of sand (up to 80m in total). These marine deposits contain a rich and varied marine fauna of Northern (Boreal) type. The upper horizon (Karantsevian) consists of sands and silts with a poorer and more cold-water marine fauna.

The second suite consists of glacial deposits of the third (Zyryankian) glaciation with sands interbedded with boulder clays forming a succession of 100–130m in thickness and of widespread distribution. These are most commonly fluvioglacial, but typical moraines are represented by boulder clays.

The third suite normally rests on the second and consists of lacustrine and fluviatile sands and clays with a fresh-water fauna. At the top there are often thick peats (4–6m).

The Recent covers all the older formations as a continuous unbroken mantle of comparatively small thickness (10–15m). It includes the most widely distributed fluviatile (terrace) deposits, but also lacustrine and marine and even terrestrial sediments, the latter having originated on wide, flat watersheds.

TECTONICS

STRUCTURES

The structures of the Siberian Platform are divided into the same three groups as the structures of the Russian Platform, thus: the Precambrian structures, the Palaeozoic structures, and the Meso-Cainozoic structures. The principal character-istics of these structures are again similar for both platforms, although the Palaeozoic structures are in places distinguished by somewhat stronger folding.

The Precambrian Structures

Three groups of Precambrian structures are recognized—the Archeozoic, the Lower Proterozoic and the Upper Proterozoic.

The Archeozoic structures are found in the Anabara and Aldan massifs, in the northern part of Eastern Sayan, and the southern part of the Yenisei Massif, and are complex. The structures of the Anabara Massif are represented by tight, steep, not infrequently isoclinal folds with a north-westerly trend. Where granites are intruded zones of intense deformation are produced. The Archeozoic is overlain by the platform-type Sinian with a sharp, angular unconformity. The structures of the Aldan Massif are very complex and cannot be characterized in detail. They have a generally east–west, somewhat north-westerly (Sayanid) trend. In the Archeozoic structures of the Baikal region the trend is north-eastern or Baikalid, but in Eastern Sayan the north-western (Sayanid) trend is again predominant. The structures are everywhere strongly deformed and dislocated.

The structures of the Yenisei Massif are found in its southern part in the Archeozoic crystalline block. They represent continuations of the structures in Eastern Sayan and consist of a complex system of minor folds, orientated in a north-westerly direction and transected by a large number of acid intrusions.

THE STRUCTURES OF THE LOWER PROTEROZOIC are similar to those of the Archeozoic, but are more uniform and less dislocated. The trend is more con-stant and can be easily traced. In the Anabara Massif the Proterozoic is almost entirely unknown, and it has not been detected in the Olenek inlier either. In the Aldan Massif the Proterozoic is absent, but is developed along the southern and western margins and its structures border the Archeozoic Massif. Further to the west in the basin of the Vitim the structures acquire a north-north-western strike with deviations towards an east–west or north–south strike. In the Baikal region large anticlinoria and synclinoria with a typical Baikalian, north-eastern trend and a complex structure are developed. Their inner complexity is due to minor folds, which are broken up and displaced by numerous intrusions.

In Khamar-Daban and Eastern Sayan the Lower Proterozoic structures margin the Archeozoic ones in the east and south. They are distinguished by great com-plexity, including isoclinal, overturned and faulted folds, which have been pushed on each other as a series of scales. The folds are cut by intrusions, which add to the general complexity.

In the Yenisei range the structures are as complex as in the Sayans. The whole of the Lower Proterozoic is strongly folded into tight folds, which are recognized with difficulty owing to the masking effect of the schistosity. The dominant trend is north–south and then into north-eastern. Consequently the Lower Proterozoic folds refold the Archeozoic ones. The north-eastern fold trend coincides with the same trend in Western Sayan, which is here adjacent to Eastern Sayan. The Yenisei folds are shown in Fig. 81.

THE UPPER PROTEROZOIC STRUCTURES in most places have a platform character and consequently differ from the Lower Proterozoic structures. The Upper Proterozoic folds are gentle, brachyanticlinal, have a definite trend, and are complicated by faults and secondary folds. The weak folding, as normally, is not accompanied by major synorogenic intrusions and the metamorphism is weak. It is important to point out that the Upper Proterozoic folds are similar to those affecting the Palaeozoic. Despite the unconformity between the Palaeozoic and the Upper Proterozoic the folds are identical in character and trend, thus the folded Cambrian shows folds which are similar to those in the Upper Proterozoic, but are even simpler in character.

Platform folds in the Upper Proterozoic exist at the margins of the Anabara and Aldan massifs, along the northern boundary of the Olekeno-Vitim Massif, the northern boundary of Eastern Sayan, the margins of the Yenisei Massif and in the Turukhanian and Olenek inliers.

Geosynclinal folds of the same type as in the Lower Proterozoic are found in the Patomian Highlands and in the Baikal region. The folds are tight, complicated by intrusive massifs and dynamically metamorphosed.

Palaeozoic Structures

The structures of the Lower Palaeozoic are different from those in the Russian Platform. In a number of regions they are well developed, involve high dips and were thought to have been affected by the Caledonian or even Hercynian orogenies. The investigations carried out in the last few years have shown that manifestations of the Caledonian and possibly Hercynian Orogeny are restricted to the Baikal region and the Patomian Highlands. In the south these manifestations are seen in the Khamar-Daban and southern part of Eastern Sayan. On the Siberian Platform proper they are absent. The folds of the Siberian Platform are non-geosynclinal and resemble the Upper Palaeozoic structures of eastern Russia and have similar dips and regularity of trends. The presence of saline formations complicated these folds even further and in places there are typical salt-domes as for instance in Kempendyai and the region of Zhigalov (Fig. 85).

Bearing in mind the existence of zones with more intense folds, it should be remembered that over the rest of the huge area the Lower Palaeozoic is almost

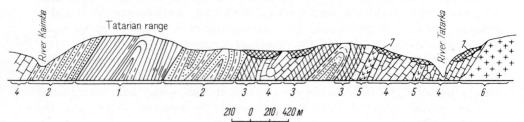

FIG. 81. Structures of the Yenisei Ridge along the river Tatarka, a tributary of the Angara. According to E. Shchukina and G. I. Petrov (1936).

1 — phyllites; 2 — quartzites; 3 — amphibole mica schists; 4 — marbles;
5 — amphibolites;   6 — boundaries;   7 — bauxitic Palaeogene formation.

horizontal or forms very large and gentle structures, like those on the Russian Platform. Consequently the Lower Palaeozoic is widely developed as can be seen on the general geological maps.

Around the Anabara Massif the Cambrian and Ordovician are almost horizontal or form large, gentle, platform-type structures. The Cambrian and the underlying Sinian rest with a sharp unconformity on intensely deformed and recrystallized Archeozoic. To the north and east of the Aldan Archeozoic Massif thick Sinian and Lower Cambrian limestones lie horizontally on Archeozoic schists and gneisses.

Over the western half of the Siberian Platform around the Tunguska and Angara the Lower Palaeozoic is very widespread, and is almost horizontal forming very large, gentle flexures. To the west of the Baikal Patomian elevation there is the large Angara-Lena downwarp (Fig. 82), which starts from the upper reaches of the Angara and lies along the upper Lena, over the upper reaches of the Lower Tunguska, and the valleys of the Peledui and the Nyuia. Beyond it is the Vilyuian depression filled by Jurassic strata.

Behind the Angara-Lena downwarp lies the Katanga Anteclise, stretching from the Yenisei Massif to the valley of the Chona (a right tributary of the Vilyui), and occupied by Cambrian sediments. To the north, between the Katanga Anteclise and the Anabara Massif, lies yet another enormous depression (the Tunguska Syneclise). Along its axis, from east to west, the Cambrian, Ordovician, Silurian, Devonian and near the Kureika the Lower Carboniferous gradually appear. The greater part of the Coal-bearing and Tuffogenous formations are exposed within the Tunguska Syneclise. On the flanks of these great regional structures are situated the secondary, sharper and smaller structures. These become most intense around the North-Western Depression and along the Angara-Lena downwarp and especially within the latter. They are also well developed along the eastern margin of the Yenisei Massif and the Turukhanian inlier (Fig. 74). All such folds are platform-type and the constituent strata are not metamorphosed. The folds are similar to the structures of the eastern part of the Russian Platform, but are more intense. Their origin is associated with the waning manifestations of the Caledonian Orogeny and block movements of the Precambian basement. The participation of salt beds in the formation of such structures (Fig. 83) is demonstrated by A. A. Ivanov (1956). The Motian Suite at the base of the section of the Zhigalev structure lies horizontally. In the overlying saline series, however, owing to the distribution of the salt the fold becomes progressively more apparent. In other words the salt intensifies the flat almost unnoticeable folds produced by the Caledonian movements in normal strata.

The structures of the Middle and Upper Palaeozoic are flatter and simpler than those of the Lower Palaeozoic. The Middle Palaeozoic structures of gentle, platform-type are seen in the regions of the Kureika and Norilsk and can be traced on small scale geological maps. The Upper Palaeozoic part of the Tunguska Formation is almost horizontal and forms very large, hardly noticeable elevations and depressions. Faults with throws of up to 400m are frequent, stretch over long distances and break up the Tunguska Formation and the traps into isolated blocks.

Fault Zones. The study of the tectonics and magmatism of the Siberian Platform has permitted I. I. Krasnov and V. L. Masaitis (1955) to identify a very interesting feature which, they think, demands a special investigation. They have shown that the region of development of the Tunguska Coal-bearing Formation and the Siberian Traps (Mid-Siberian Plateau) represents a huge region of downward movement. This region is bounded on the north-west, north-east and south-east by fault zones

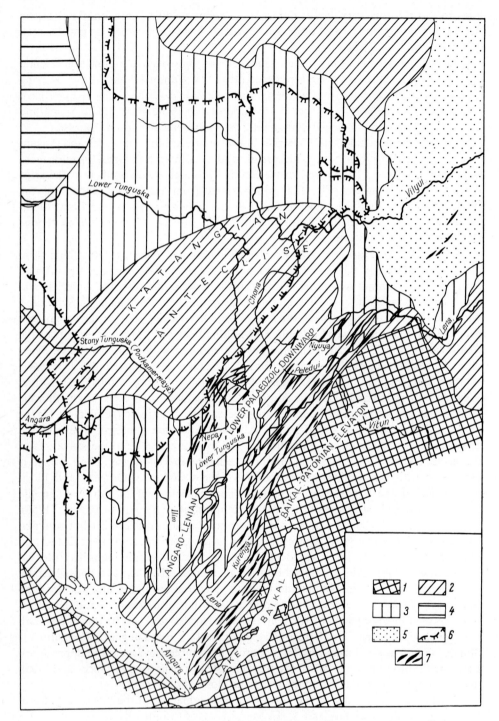

FIG. 82. Structures in the southern part of the Siberian Platform. According
to D. A. Tugolesov (1952).

1 — Precambrian; 2 — Lower Palaeozoic; 3 — Silurian; 4 — Devonian;
5 — Jurassic; 6 — boundary to the Tunguska facies; 7 — anticlinal structures.

formed in places of maximal flexure. The fault zones are 50–150m wide and thousands of kilometres long and within them joints, normal faults and intrusions representing channels for effusive activity are extremely common. The zones are

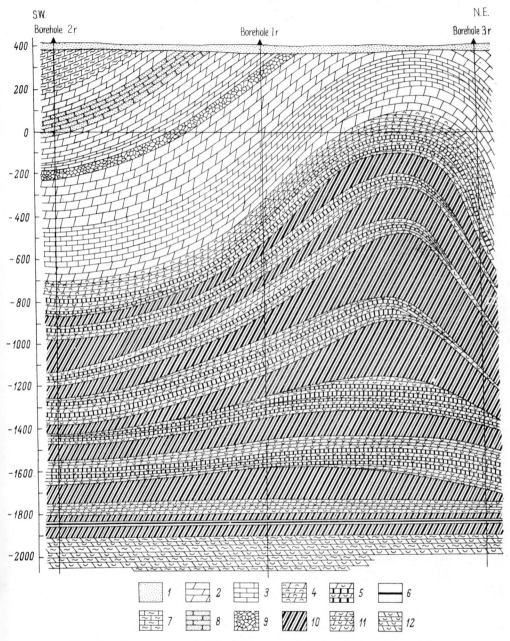

FIG. 83.   Section across the Zhigalov structure, in the upper reaches of the Lena. According to A. A. Ivanov (1956).

1 — Anthropogene clays and sands; 2 — Upper Cambrian marls and Lower Cambrian; 3 — dolomites; 4 — sandstones; 5 — limestones; 6 — breccia; 7 — gypsiferous dolomites; 8 — rock salt; 9 — salt and anhydrite; 10 — anhydrite-dolomite; 11 — a diabase; 12 — Motian Suite of dolomite-anhydrite and marls.

associated with positive magnetic anomalies. The south-western zone has been provisionally named as the 'Angara-Katanga' but the name 'Angara-Yenisei' is better. The south-eastern zone is shown as the 'Angara-Vilyui Zone' and the north-western one as the 'Vilyui-Kotui Zone' (Fig. 84).

Mesozoic and Cainozoic Structures

The Mesozoic and Cainozoic structures are even less intense than the Upper Palaeozoic structures. The Mesozoic and Cainozoic are as a rule horizontal, although there are three exceptions: (*a*) the marginal folds to the Verkhoyansk downwarp, (*b*) the salt domes, and (*c*) the block faults at the southern margin of the platform. In no case is there any metamorphism.

The marginal folds are developed in the valley of the Lena along the north-eastern border of the platform and within the Verkhoyansk sub-montaine downwarp. The folds involve the Jurassic, the Cretaceous and the Palaeogene, and have been best studied in the coal-fields. For instance in Sangar the dips reach 30° to 40° and in Zhigansk there are gentle folds. The Bulunsk coal-field also has well-developed folds and they are gentle and individualized as in other regions.

The salt domes are most typical in the basin of Kempendyai—the right tributary of the Vilyui. The Kempendyai domes are indicated on general geological maps, which show inliers of the Cambrian and Ordovician rocks surrounded by the Jurassic. As on the Russian Platform the domes are faulted into separate blocks. The centres of the domes show salt plugs of Lower Cambrian (?) age. The plug is often surrounded by blocks of Ordovician red beds and limestones, and also Lower Jurassic deposits. The Kempendyai, Kygyl-Tus and Kyundyai domes are tens of kilometres across, and show up as elevations of relief, reaching 120m in height. The microflora of the deposits, which were formerly considered as Cambrian, has suggested that many of them are Devonian. Thus the Kempendyai red beds would be

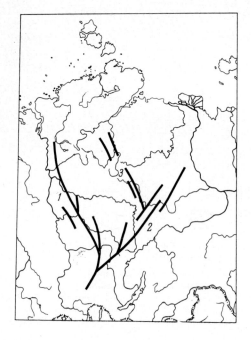

Fig. 84.

Fault zones on the margins of the Tunguska Depression. According to I. I. Krasnov and V. L. Masaitis (1955).

1. The Angara-Yenisei zone.
2. The Angara-Vilyui zone.
3. The Vilyui-Kotui zone.

probably Devonian and the salt Ordovician, but a few years ago new, convincing evidence has indicated its Lower Cambrian age (Lyutkevich, 1959).

The second group of the salt domes is situated far to the north, near the mouth of the Khatanga and the region of the port Nordvik. Here the best known salt dome is the Solyanaya Sopka (Urung Tumus) found before the Revolution and investigated in detail by Soviet geologists. There are also signs of petroleum near it. The age of the salt has not been determined, but amongst the cap blocks Upper Devonian (Domanik) rocks have been identified. The domes pierce Jurassic and Lower Cretaceous rocks.

Thrust-faults are developed along the southern margin of the Siberian Platform, and particularly along the northern foothills of the Stanovoy range and Eastern Sayan. They originated during the Neogene and Anthropogene movements, causing the elevation of Precambrian and Lower Palaeozoic blocks and massifs. These elevations were responsible for the present-day relief of the Stanovoy and Yablonovoy ranges, of Khamar-Daban and Eastern Sayan. Contemporaneous depressions caused the Baikal and other troughs. The rise of the blocks caused the folding of the adjacent Jurassic and lower Cretaceous into complex, often overturned and faulted structures. Along the plane of movement crush breccias are often observed, but there is no metamorphism and the zone of movement is narrow, varying from a few hundred metres to 2 kilometres.

The best known is the Cisangaran thrust block on which many papers have been published. The fault zone lies across the Angara, some 6km from its origin, and continues far to the west for hundreds of kilometres showing as a devious line on the map (Fig. 85). The thrust was from the south and its plane dips 20–30°, and its structure in cross-section is shown in Fig. 86. The magnitude of the movement is of the order of several hundreds of metres. In places it is divided into a series of parallel thrust planes, elsewhere it is interrupted and the Jurassic rests normally on the Precambrian. Similar block thrusts are widely distributed in Eurasia and have been described in detail from Central Asia.

A second zone of block thrusts is situated to the north of the Stanovoi range. Here there are commonly east–west areas of coal-bearing Jurassic. These areas are bounded to the south by the block thrusts where the Precambrian overrides the Jurassic along relatively steeply inclined thrust planes. The Jurassic nearby is strongly deformed, but the zone of deformation is normally no more than 1–2km across. The thrusts are normally marginal to the depressions filled by coal-bearing deposits. There are also associated normal faults which break up such deposits into separate blocks. This is especially evident in the South Yakutian basin.

It is interesting that until recently the Cisangaran and Stanovoi thrusts were considered as zones of Mesozoic folding and the thrusts as gigantic nappes. Such mistaken views disregarded the limited nature of the structures, the absence of any associated metamorphism and of synorogenic intrusions. True zones of folding are many tens or hundreds of kilometres wide, and involve strongly metamorphosed rocks and large intrusive massifs, which are absent from the area under consideration. Similar errors, however, were committed in other regions, such as Central Asia and the western slope of the Urals. These errors were first demonstrated as such by V. A. Obruchev (1938).

THE BAIKAL RIFT VALLEY. The youngest tectonic structures are the huge and vertically important rift depressions, which have caused the formation of lake Baikal, the Barguzin and Irkut valleys and the lake Kosgol depression in Mongolia.

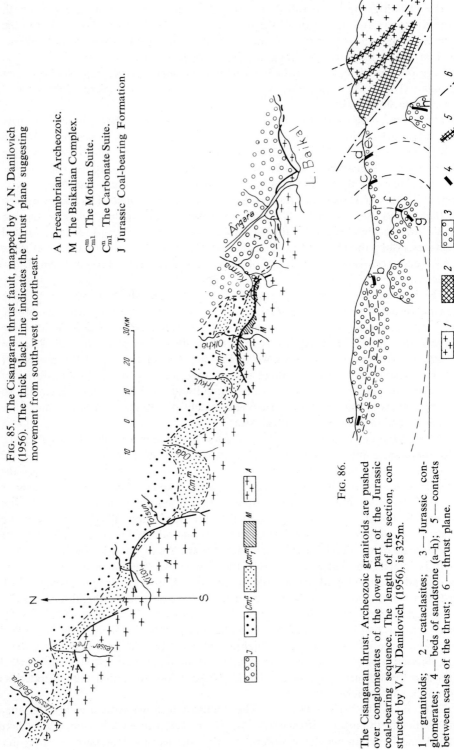

FIG. 85. The Cisangaran thrust fault, mapped by V. N. Danilovich (1956). The thick black line indicates the thrust plane suggesting movement from south-west to north-east.

A Precambrian, Archeozoic.
M The Baikalian Complex.
C_{m1}^m The Motian Suite.
$C_{m1}^{ n}$ The Carbonate Suite.
J Jurassic Coal-bearing Formation.

FIG. 86.

The Cisangaran thrust. Archeozoic granitoids are pushed over conglomerates of the lower part of the Jurassic coal-bearing sequence. The length of the section, constructed by V. N. Danilovich (1956), is 325m.

1 — granitoids;   2 — cataclasites;   3 — Jurassic conglomerates;   4 — beds of sandstone (a-h);   5 — contacts between scales of the thrust;   6 — thrust plane.

These structures are in part contemporaneous with, and in part younger than, the Cisangaran thrusts. This is clear from the fact that the shore-line of the Baikal cuts across the line of thrusting. The Baikal rifting is still continuing and is associated with high seismic activity. Earthquakes, which are sometimes severe, occur no less frequently than 2–3 times a year. Repeated levellings have shown that movements reach 1–2cm/yr (1–2m in a century) and some regions move up, while others move down. The depression of the bottom of lake Baikal started in the Neogene, since boreholes have yielded over 1200m of Neogenic and Anthropogenic sediments in the delta of the Selenga. The Baikal at its deepest is 1741m; it is 680km long and 60–80km wide and it is important to remember that the rift was accompanied by the rise of the mountain blocks around it. In its dimensions the rift approaches the African rift with the large East African lakes, the Red Sea and the Dead Sea in Palestine.

THE HISTORY OF TECTONIC MOVEMENTS

In the Archeozoic an orogeny at the end of the Lower, and another one at the end of the Upper, Archeozoic are most clear. The oldest orogeny is equivalent to the Saamian of the Russian Platform and it is apparent in the Anabara and Aldan massifs, where the Upper Proterozoic (Sinian) and Lower Palaeozoic rest unconformably on the Lower Archeozoic. In the Yenisei range again the Upper and Lower Archeozoic are sharply different.

The second orogeny is equivalent to the Byelomorian and is identified everywhere where the Lower Proterozoic overlies the Archeozoic, as for instance in Eastern Sayan and the Yenisei range. The Archeozoic and Lower Proterozoic are always separated by an angular unconformity, different degrees of metamorphism and different character of sediments. With the Byelomorian Orogeny there are always associated large and small post-Archeozoic intrusions.

In the Proterozoic a third, Karelidian Orogeny is also widely represented, causing an angular unconformity between the Lower and Upper Proterozoic, which have differing degrees of metamorphism, and has associated numerous intrusive massifs.

The fourth or Baikalian Orogeny, which occurred at the end of the Upper Proterozoic, is much weaker and its structures are often of platform-type, the metamorphism is weak and the associated unconformities are slight. In many regions there is no unconformity between the Sinian and the Cambrian and no evidence for the Baikalian Orogeny. This orogeny is strongly manifest only in the Baikal region and the Patomian Highlands, where the Sinian deposits are strongly metamorphosed, strongly folded, and include Upper Proterozoic intrusive massifs. In the Yenisei Massif and the Turukhanian inliers the Upper Proterozoic is deformed clearly, but weakly. The folds are large and comparatively simple. The metamorphism is weak and granite intrusions are absent. At the same time the Upper Proterozoic is everywhere clearly folded, somewhat more metamorphosed than the Cambrian, and is separated from the Cambrian by an unconformity. The intensity of folding here is almost the same as in the Hyperborean (Sinian) rocks of the Rybachii Peninsula, Kil'din island, and the north-west Kola Peninsula.

The Caledonian Orogeny did not greatly affect the Siberian Platform. Even in the Angara-Lena depression, where the folds are best developed and in places are complicated by salt tectonics, they are of platform-type and are not accompanied by metamorphism or intrusive massifs. Diagnostic features of the Caledonian Orogeny, such as the following, are absent: (*a*) angular unconformities between the Cambrian and the Ordovician, the Ordovician and Silurian, the Silurian and the Middle Devonian; (*b*) difference in metamorphism between the Lower and Middle Palaeozoic; (*c*) Caledonian granites. In the literature however, there are suggestions of manifestations of the Caledonian Orogeny in the Siberian Platform. These suggestions are based on an insufficient analysis of structures. In Transbaikalia and the southern part of Eastern Sayan the Cambrian and the Ordovician are intensely disturbed and metamorphosed, suggesting effects of the Caledonian Orogeny despite the vagueness of Cambrian-Devonion relationships and the absence of Caledonian intrusions. However, this area belongs to the Angara Geosyncline.

The Hercynian Orogeny is also not manifested over the Siberian Platform. The structures in the Kureika region at the north-western margin of the platform are often considered as Hercynian, but this is wrong, since the folds are simple and large, their rocks not metamorphosed and there are no intrusive rocks. These folds have nothing in common with the typical geosynclinal structures of the Taimyr Peninsula.

The Cimmerian Orogeny cannot be traced over the Siberian Platform, but is typical of the geosynclinal areas which margin the platform to the north-east, east and south-east. In the South Yakutian coal-fields near the Precambrian blocks there are complex minor folds in non-metamorphosed rocks. Such structures are sometimes considered as geosynclinal associated with the Cimmerian (Alpine)* Orogeny, but this is incorrect, since the block thrusts and the associated deformations in many cases are of a later age.

The Alpine Orogeny is absent from the Siberian Platform and the associated fold belts, thus the Neogene and Anthropogene are not folded and only show block faulting.

The oscillatory movements of the platform reached an unusual magnitude. The total thickness of the sediments forming the platform, arrived at by summing up the maximal thicknesses of individual formations, is of the order of 30–60km, determining the major depressions. Precambrian oscillatory movements occurred principally between the various Precambrian epochs, but also occurred during orogenies. The total thickness of the Precambrian is enormous and for the Patomian Highlands and the Baikal region reaches 40–45km.

The Palaeozoic upward and downward movements are also considerable both in area and amplitude. The Lower Palaeozoic downward movements were especially vast, involving the whole of the Siberian Platform. Their amplitude can be estimated from the total thickness of 2000–3000m of sediments. In the Middle Palaeozoic, however, slow continuous upward movements were the rule, which caused the platform to be an elevated land mass despite continuous erosion over its area. The great thicknesses of the Upper Palaeozoic (1000–2000m) originated as a result of alternations of upward with downward movements. Upward movements were predominant in some regions, while downward movements were predominant in other adjacent areas.

The Mesozoic oscillatory movements up to the Middle Triassic were the same as in the Upper Palaeozoic and lavas, tuffs and interbedded terrigenous sediments

---

* Several geologists consider the Cimmerian Orogeny as a part of the Alpine Orogeny, using the same term for both. This procedure involves unification of structures with diverse ages and distributions and is not acceptable.

were laid down. In the second half of the Triassic the whole Siberian Platform with the exception of its north-eastern margin became land. The Jurassic and the Lower Cretaceous show distinct oscillations causing marine transgressions and subsequent formation of near-shore plains (regions of accumulation of coal). Such oscillations were especially noticeable in the eastern part of the platform, the western part remaining all the time a region of erosion. The greatest downward movements, of the order of 1500–2500m, were observed in the Cisverkhoyansk downwarp.

The Cainozoic is distinguished by a definite predominance of slow and lengthy upward movements making the platform a region of erosion and accumulation of thin continental deposits. Only the northern and eastern regions were exceptional. There, in a narrow zone, marine, boreal sediments were formed, while the eastern margin shows thick Cretaceous and Palaeogene sediments.

The Baikalian Depression (rift) was formed by the Baikalian downward movements, which were also responsible for a number of other troughs to the north-east and south-east of it. The history of these movements is not as yet completely clear. It should be remembered that the rift depressions were accompanied by the elevations of marginal massifs, thus forming them into highland regions. Then the actual downward movement was a lengthy process involving decelerations and even pauses. The process began in the Miocene and is still active, causing strong earthquakes. However, with the exception of such troughs, the Siberian Platform is at present a region of upward movement and erosion.

MAGMATISM

In the Precambrian, magmatic activity in the Siberian Platform was very similar to that in the Russian Platform and is typically geosynclinal. The subsequent magmatism in both platforms is non-geosynclinal, but shows a different history in each. In the Russian Platform, Proterozoic, Lower Palaeozoic and Devonian basaltic lavas and sills are widely distributed, but are absent in the Siberian Platform. On the latter, Permian and Triassic effusions of basaltic magma (Siberian Traps), and young Neogenic and Anthropogenic basalts associated with the Baikal Rift, are present, but do not exist in the Russian Platform. The differences in the magmatism over the two platforms can be explained in terms of the history of downward movements. Over the Russian Platform the most significant downward movements occurred in the Devonian, while in Siberia they were Permian or early Triassic, Jurassic and Neogene-Anthropogenic. Jurassic effusives on the Siberian Platform were undoubtedly present, but as yet have not been sufficiently separated from the Siberian Traps.

Precambrian Magmatism

The Precambrian magmatism is represented by four cycles. The Saamian magmatism is most widespread and is encountered everywhere where the Archeozoic is present. The Byelomorian and Karelidian magmatisms are again widely developed, while the Baikalian (Upper Proterozoic) magmatism is only local and is encountered in the Cisbaikalian folds zones and Eastern Sayan. Each cycle shows all the phases including the effusive rocks, the major and minor intrusions.

The Saamian magmatic cycle reaches its greatest development in the Anabara and Aldan massifs, where it is dominant. It is also represented in other parts of the Archeozoic. The Lower Archeozoic lavas are recrystallized and converted into crystalline schists and gneisses. The Saamian intrusions are most widely distributed and characteristic.

In the Anabara Massif the so-called 'Charnockitic Series' is especially widespread and in particular has predominant quartz-hypersthene diorites and derivatives such as pyroxenites, peridotites and leuco-diorites. The rocks are usually altered into plagioclase-quartz-hypersthene gneisses, and there are also hypersthenic granites. The intrusions are associated with widely distributed migmatitic rocks which are identical to the migmatites of Karelia and the Kola Peninsula. The Charnockitic Series apart, there are also microcline granites, which are deformed to gneisses. The granites form lenses of 5–7km^2 in area. They are also associated with migmatites and accompanied by aplites and quartz veins. The granites are younger than the Charnockitic Series and are possibly Svecofennidian. In the Aldan Massif and the Stanovoy range according to D. S. Korzhinskii there is also a Charnockitic Series of the same type as in the Anabara Massif, consisting mainly of hypersthenic gneisses. Pyroxenic amphibolites and recrystallized utrabasic rocks also occur, but are rarer. However, in the Aldan Massif there is a widespread development of granitic intrusions of vast size. Amongst the granites two types can be distinguished: (*a*) the older, probably Saamian grey and brownish-grey biotite or biotite-amphibole gneisses and (*b*) the younger flesh-red granites of Byelomorian cycle. There are also abundant migmatites.

In the Olekmo-Vitim and Patomian Highlands in Cisbaikalia and Eastern Sayan the nature of the Archeozoic magmatism is the same as in the Aldan Massif. At least two magmatic cycles can be distinguished: (*a*) the first, represented by orthogneisses produced from acid, basic and ultrabasic rocks and (*b*) the second, characterized by massive granite intrusions, analogous to the Byelomorian cycle.

Lastly in the Yenisei ridge there are also signs of two magmatic cycles. The first cycle is considered as Saamian and is represented by charnockitic rocks forming concordant and rarely discordant bodies. The 'charnockites' according to Yu. A. Kuznetsov are rather peculiar rocks. Compositionally they are similar to granites, granite-syenites, syeno-diorites and quartz-norites, but differ by their granoblastic texture and the presence of garnet and hypersthene as main mafic minerals, and by the absence of contact alteration in the adjacent rocks. The Charnockitic Series is characteristic of the Lower Archeozoic throughout the world, including India, Canada and Scandinavia.

The Byelomorian magmatic cycle in the Siberian Platform is attributed to the Upper Archeozoic and begins with metamorphosed lavas, followed by major intrusions consisting of microcline granites, granitoids and granulites and finally by minor intrusions, pegmatites and aplites.

Siberian geologists do not recognize the Saamian and Byelomorian cycles, and do not use these terms even when the presence of two Archeozoic magmatic cycles is undoubted. This is because the study of the Siberian Archeozoic was conducted independently from the study of the Scandinavian Precambrian by different geologists. Since, however, these terms are used for the Precambrian of the Russian Platform it is proper to use them for the identical features of the Siberian Platform. Upper Archeozoic lavas certainly exist but have not been sufficiently described.

The large intrusions, on the other hand, have been studied and established in all those regions where the Archeozoic is present. In the Anabara Massif they are not widespread and consist of microcline granite, forming small lensoid bodies of 5–7km^2 in area. They have associated migmatites, aplites and quartz veins.

In the Aldan Massif and Stanovoi range, in the Olekmo-Vitim highlands, Cisbaikalia, Eastern Sayan and the Yenisei range, the great Byelomorian intrusions are widely developed forming large massifs some tens of kilometres across. The massifs involve different rock types including acid, basic and ultrabasic. The flesh-red microcline-bearing medium and coarse grained granites (alaskites) are particularly characteristic. In the eastern part of the Aldan Massif it has been demonstrated that the Byelomorian intrusions started with red and pink biotitic and biotite-amphibole granites, forming the main part of the massifs. They are transected by typical alaskitic microcline granites which are again red or pale. The pale type is found in small bodies associated with pegmatites and orthotectites. The alaskitic and biotitic granites are found in the Olekmo-Vitim highlands in Cisbaikalia and in Eastern Sayan. The Yenisei range has the great Tarakian Massif of grey porphyritic, gneissose granites. The massif is cut by pegmatites and aplites.

The widespread development of the Upper Archeozoic intrusions is explained by the vast and severe manifestation of the Byelomorian Orogeny causing the widespread angular unconformities between the Archeozoic and the Proterozoic. Across the unconformity the grade of metamorphism is different.

The Karelian magmatic cycle is again widespread and powerfully represented. It is of the same general age as the unconformity between the Lower and Upper Proterozoic. All this depends on the Karelidian Orogeny at the end of the Lower Proterozoic. The most important peculiarity of the Karelidian cycle is the vast distribution of lavas and tuffs. The lavas were porphyries or porphyroids and are commonly metamorphosed into green schists and porphyroid slates. Thick intrusions of diabase are common. In the Anabara Massif the Proterozoic is almost absent and consequently the diabase intrusions which cut all the Archeozoic rocks are attributed to the Karelidian cycle. In the Aldan Massif the Karelidian cycle is absent, but in the western part of the Stanovoi range there are diabases and porphyritic lavas, which are in places metamorphosed into slates.

To the effusive phase belong the green schists, diabases and porphyrites of the Baikalian highlands, the porphyritic schists and porphyroids of Western Transbaikalia, the porphyritic tuffs and tuffogeneous schist of Eastern Sayan, and the actinolitic schists, formed at the expense of sills or lavas of diabase, of the Yenisei range. The magmatic channels of these occur as uralitized diabase dykes cutting the Upper Proterozoic Tarakian granites.

The phase of the major and the associated minor intrusions is much more widespread. The intrusions are only absent in the Anabara and Aldan massifs, but are commonly encountered in the Stanovoi range, where the major intrusions are known as the 'Older Stanovoi'. The intrusions consist of grey and whitish biotitic granites with dominant oligoclase. In the zones of intense isoclinal folding the intrusive bodies are compressed and have become foliated, fine-grained gneissose and sometimes granite-gneissic rocks. In the regions of more moderate folding the intrusions have rounded outlines, and are medium grained and sometimes porphyritic. They are associated with biotitic and muscovitic pegmatites and aplites.

In the basins of the Olekma, the Vitim and Barguzin the Older Stanovoi granites are widespread, but are coarse-grained and sometimes porphyritic, with biotite and hornblende, and not uncommonly have many associated varieties from syenites to alkaline syenites on the one hand, to diorites and gabbro-diorites on the other. Similar granites are found in Cisbaikalia and Eastern Sayan, but their age is considered to be different and is Upper Proterozoic in Sayan.

In the Yenisei range the Lower Proterozoic granites are widely distributed and are accompanied by migmatization; the last phase of the orogeny has red porphyritic granites as the main concomitant intrusions.

The above examples do not exhaust in the least the great variety of the Karelidian intrusive rocks. Along the southern margin of the Siberian Massif they have a wide distribution forming very large massifs.

The Upper Proterozoic Baikalian magmatic cycle is only developed in the regions of strong Baikalian folding, at the end of the Upper Proterozoic and where the Upper Proterozoic and the Cambrian are separated by an angular unconformity and differing grades of metamorphism. Cisbaikalia, Eastern Sayan and the Yenisei range are such areas. The extrusive phase is represented by thick porphyries, porphyrites, tuffs and diabases which are more strongly differentiated than the Lower Proterozoic igneous rocks.

The phases of the large and minor intrusions are well developed and granites predominate. In Eastern Sayan and Khamar-Daban the granites are micaceous, while in the Yenisei range they are similar to the Lower Proterozoic granites, being grey or pink, microcline-bearing and sometimes porphyritic accompanied by numerous varieties.

PALAEOZOIC MAGMATISM

The Caledonian magmatic cycle is fully developed, but only in the regions of strong Caledonian folding and deformation of the Cambrian and Ordovician. This is observed only in the Angara Geosyncline. Within the Siberian Platform proper, the Caledonian Orogeny was absent or very weak and consequently there are no synorogenic intrusions.

The Hercynian volcanic cycle of geosynclinal type is absent from the Siberian Platform. Its manifestations are restricted to the Hercynian fold belts, which margin the platform from north and south. In the Siberian Platform the traps belong to it. The Siberian Traps represent one of the most amazing events in the history of the Siberian Platform. The rocks of the traps are dark grey or black and cover an area of 1·5 million km². They have long been shown on geological maps and were clearly outlined. The area covered by the traps is of the order of magnitude of the largest basaltic fields of the earth's crust.

The thickness of the traps is not considerable being in most exposures of the order of 100–200m, but in isolated localities increasing to 800–1000m. Initially the total thickness was probably up to 2000m, or even more, but the traps have been subject to considerable erosion and their upper part has been eroded away. The traps are of two types—effusive and intrusive. The effusive facies show transitions from basalts and dolerite-basalts to tuffs, tuffites and tuffogenous sandstones. The intrusive facies are more variable, ranging from fine-grained aphanites (dolerite-porphyrites) to medium-grained dolerites and coarse gabbros, gabbro-dolerites and dolerite-pegmatites. There are also rare hypersthenic varieties and the accompanying ultrabasic derivatives and various differentiates (alkaline veins, diabase-pegmatites).

A typical trap rock is holocrystalline and dark, consisting mainly of labradorite, clinopyroxene and olivine. The hypersthenic varieties are found in the Yenisei part of the platform, where their ultrabasic derivatives (picrites) are accompanied by an important mineralization.

The effusive varieties occur in huge lava flows, traced over tens of kilometres and interbedded with tuffs. It is important to point out that the Tuffogenous Suite consists mainly of tuffs, reaching 600m in thickness, and consisting of pyroclastic products which indicate the explosive nature of volcanism. The explosions as well as the eruptions involving lava occurred along very large cracks. The lavas form widespread sheets and their outcrops commonly show columnar jointing (Fig. 87).

The intrusive types in the main are represented by sills which have penetrated the Cambrian, Ordovician, Silurian, Carboniferous and Permian deposits parallel to the bedding planes. The sills sometimes cause the formation of broad, flat water-sheds between rivers, stretching up to 150km and showing steep cliffs of tens of metres where they are adjacent to rivers (Fig. 88). In isolated cases the thickness of the sills reaches 350m. Other intrusive bodies such as dykes, veins, stocks, necks, and laccolites are also found. In part they form bodies unexposed on the surface but acting as feeding channels. Some bodies are large (Fig. 89); some dyes reaching 1–1·5km in width and continuing for 40–70km. Usually, however, the igneous bodies are smaller, varying from a few metres to tens of metres across.

In the lower reaches of the Kutui (a right tributary of the Khatanga) there is a series of pipe-like intrusive bodies (Gulinskian Intrusion). The intrusions consist of alakaline rocks and their ultrabasic derivatives. Phlogopite and titan-magnetite deposits are associated with these rocks.

Fig. 87. Columnal joints in traps of the Norilsk region. The height of the columns at the bottom of the section is 2–10 m. *Photograph by A. A. Mezhvilk (1955).*

FIG. 88. A trap sill (dark with columnar joints) intruded into pale, Silurian limestones on the river Angara, near the Aplinian rapids. *Photograph by N. S. Malich.*

FIG. 89.

A pipe-like body of Siberian Traps (intrusive dolerites) crossed by the river Kondormo The body has irregular rounded outlines and appears darker than the surrounding rocks. *Aerial photograph by N. S. Malich.*

These are also interesting and economically important explosion pipes, which are vertical, almost pipe-like bodies with circular or oval cross-sections on the surface, the diameter of which varies from tens to a hundred metres and more (Fig. 90). The pipes are filled with kimberlite—a greenish or bluish rock with fragments of effusive rocks and sediments in it. The Siberian kimberlites are compositionally identical with the celebrated South African kimberlites and like them contain diamonds. These pipes represent volcanic necks originating by explosion and the infilling rock is the ultrabasic rock with brecciated structure.

The age of the Siberian Traps is a complex and difficult problem, which has provoked lively disputes. Only the lavas and tuffs lying amongst palaeontologically datable deposits, or containing lenses and beds of such deposits, can be dated with a greater or lesser accuracy. In the region of the Kureika, in the north-west of the platform, the layers of tuffites have yielded an Upper Carboniferous fauna. Between the Khatanga and the Lena the Tuffogenous and the Lava Suite contain Upper Permian ostracods. In some regions beds of clay in the upper horizons of the Lava Suite have Lower Triassic *Estheria*. Finally in the basin of the Vilyui the traps are covered by the Lower Jurassic (Rhaetic-Jurassic) continental deposits.

It is more difficult to determine the age of the intrusive facies (sills, dykes and necks). They are younger than the rocks amongst which they lie or transgress. The sills are known from all the members of the succession, beginning with the Cambrian and finishing with the Permian and even possibly Jurassic. The range of the age of the dykes, which cut across not only all the Palaeozoic deposits but the Jurassic as well, is even wider. It should also be pointed out that in the Cis-Baikalian zone there are also Neogenic and Anthropogenic basalts and the associated dykes cut not only Jurassic, but also Cretaceous and Palaeogene deposits.

At present a great majority of the investigators consider that the Siberian Traps range from the Upper Carboniferous to the Lower Triassic. The following sequence of events can be traced. In the Upper Carboniferous and Permian the accumulation of the Tuffogenous Suite with occasional lavas (in the north-west) took place. Most of the sill-like intrusions were then emplaced at the end of the Permian. At the end of the Permian and the beginning of the Triassic (the third epoch) the main lavas poured out of fissures. It is possible that in a number of places the formation of the tuffs, the extrusion of the lavas and the intrusion of the sills occurred concurrently. It is also possible that the age of the complex varies from place to place (from the north-west to the central parts). It is unquestionable that in the central region the Tuffogenous Formation preceded the lavas and this is seen in cross-sections and exposures.

Since all suites should have geographical names the Tuffogenous Suite has been united with the Korvunchanian and the Lava Suite was called the Puteranian (Fig. 80).

The extrusion of large masses of basaltic magma could have been caused only in two ways: (a) by the increased pressure in the basaltic layer, and (b) by the formation of fissures. Both phenomena occur under conditions of major block elevations and depressions. It is probable that apart from the depressions of the western part of the platform, the principal cause of the extrusion has been the depression of that part of the platform which at present forms the northern part of the Western-Siberian Depression and the adjacent Palaeozoic fold belts. The depression of the massifs has been the cause of the rise of huge masses of the basaltic magma.

Mesozoic and Cainozoic Magmatism

The Cimmerian magmatic cycle, like the Hercynian, did not develop completely in the Siberian Platform. The effusive phase is represented in the central region of the platform and along its eastern and southern margin, but is not widespread. In the central region, for instance the basin of the Vilyui, there are basaltic dykes cutting the Jurassic, but true Jurassic lavas and tuffs have not yet been found. Along the eastern margin of the Aldan shield there are porphyrites and tuffs, some 150m thick, which are cut by the Lower Cretaceous granites. Such isolated indications emphasize the limited occurrence of the undoubted Jurassic effusions.

The intrusions have been proved in the south-eastern part of the platform, from the basin of the Vitim to the basin of the Aldan. The intrusions include small massifs and dykes of light-grey and pink biotitic leucogranites, granite-aplites and quartz-syenite-porphyries. The intrusions are transgressive to the Jurassic and are overlain by the possibly Upper Cretaceous effusive rocks. This suggests a Cimmerian and more accurately Kolymian age for the granites. The associated pegmatites and spessartites are relatively rare. The Cimmerian intrusions are synorogenic and associated with young Mesozoic faulting.

The Alpine magmatic cycle is fairly widely developed, but has a typically platform character. In the northern part of the platform there are more indications of the Mesozoic (Cretaceous) effusive rocks (the Popigai graben on the river Popigai). In the south the Alpine cycle is represented by two complexes of considerable importance—the Aldan alkaline complex and the Baikal basalts.

The Aldan alkaline complex consists of alkaline and sub-alkaline rocks varying from the typical nepheline-rich varieties to syenites, syenite-porphyries, granite-porphyries, and aegirine-granites. They form small hypabyssal intrusions, sills, laccoliths, dykes and small necks. The sills vary from 2m to 80m in thickness and can be traced continuously up to 10–12km. Amongst the Cambrian limestones, amongst which they are most frequently found, they form skarn zones of up to 8–10m in thickness. The age of the alkaline rocks is Cainozoic since they cut across the Coal-bearing Formation and the Cimmerian granites. The alkaline rocks are widespread from the township of Nel'kan in the east to the Olekma, and their outcrops are shown on the general map in the upper reaches of the Maya and to the south of the town Aldan. The alkaline rocks are typical platform formations analogous to the Siberian Traps and Baikalian basalts. They represent the deeper part of a huge effusive complex, the effusive parts of which have been removed by erosion. Like the traps, the alkaline rocks are associated with the zones of large, deep fractures.

Recently there have been discoveries of pegmatitic veins with very large crystals of biotite and feldspar. The determination of the absolute age of these has given a Lower Cretaceous (107–135 m.y.) age. This forces a reconsideration of the age of the alkaline rocks.

The Baikalian basalts are developed to the west of the Aldan alkaline province and are associated with very large fractures along which the rise of the Cisbaikalian ranges has occurred.

According to V. A. Obruchev (1938) the Baikalian basalts are widely distributed along the southern margin of the Siberian Platform. The rocks are dark-grey, black, greenish-black, or brownish-black, fine-grained and massive. The upper parts of the lava flows show vesicular and amygdaloidal structures and columnar jointing is common. Pyroclastic varieties such as breccias and tuffs, which sometimes form low conical volcanoes with craters up to 1km wide, and with heights of 100–150m, are rarer. Compositionally the basalts are variable. Most commonly they are olivine-bearing, but olivine-free basalts are also common and there are also olivine diabases similar to the Siberian Traps. Nepheline basalts and limburgites, which are similar to the rocks of the Aldan province, point to the possible connection of these two volcanic formations, but are relatively rare.

The thickness of the Siberian basalts is normally of the order of a few tens of metres, but thicknesses of the order of 120–150m are encountered. There are lava fields and flows of great continuity, stretching for many tens and even hundreds of kilometres, and low volcanoes are rarer. The greater part of the effusive activity occurred through fissures, and belongs to two epochs—the Neogene (Miocene) and the Anthropogene.

The most westerly region of the development of the Baikalian basalts is in Eastern Sayan, in the upper reaches of the Oka. Here the earlier, plateau basalts form the tops of mountains which reach 2000m in height, while the later lavas fill valleys, having been extruded from a small volcano. Fissures filled by basalt are commonly encountered en route along the Circumbaikalian railway. The basalts reach a considerable thickness in the valleys of the Dzhida and the Selenga. Again, the basalts here are of two generations. The plateau basalts form the tops of hills of up to 1500–1600m, and their sheets reach 140m in thickness. The basalts of the second generation are lava flows filling the valleys and reaching 20–70m in thickness. The lavas emerged from fissures and more rarely emerged out of volcanoes.

A very large development of basalts exists in the Vitim highlands to the west of the Vitim. On the surface of the basalts small volcanoes remain and two of them are known as Mt. Obruchev and Mt. Mushketov.

The most easterly outcrop of basalts is found in the upper reaches of the Chara, to the west of the Olekma. To the east of the Olekma the basalts are unknown, while the alkaline rocks are predominant. It is interesting that the zone of the development of the basalts is not associated with the Baikal graben or the Kosogol and Barguzin troughs, but transects the fault zone in the Tunkin depression. In the east the basalt zone lies to the south of the zone of trough faulting, while in the west it lies to the north of it. This is interpreted in terms of the extrusion occurring from open fissures, while the faults of the Baikalian rift are everywhere sealed because of pressure across them. Thus there is no basalt on the shores of lake Baikal (general map).

USEFUL MINERAL DEPOSITS

If the Russian Platform, which has been studied well, still continues to yield new economic discoveries to the Soviet geologists, the much less investigated Siberian Platform presents even more favourable prospects. This has been demonstrated comparatively recently. Soviet geologists have found new, often large deposits of iron ore, polymetallic ores, gold, bauxite, diamonds, etc.

METALLIFEROUS ORES

Metalliferous ores of the Siberian Platform are abundant and variable.

Ferrous metals

IRON ORES.   Iron ores of different origins and ages are common. The Angara-Ilim deposit has been known for a long time. The deposit, at the junction of the Angara and the Ilim, consists mainly of hydrothemal magnetite associated with the Siberian Traps. The ore most commonly occurs in veins which can be either simple or branching and occurring both in the traps and in the Lower Palaeozoic deposits. The thickness of the veins sometimes reaches 20m and there are also stocks and nodes. The estimated reserves are up to 600 million tons. Deposits of this type are also found in association with the Siberian Traps on their western margin and up to the region of Norilsk in the north.

Proterozoic and Archeozoic jaspilitic ores form the second group of deposits. The ferruginous quartzites are found in Eastern Sayan, in the Yenisei Massif, in the Aldan region and in Cisbaikalia. There are now some exploited deposits, as for instance the Lower Angara deposit, which is one of the largest in Eastern U.S.S.R. The widespread development of the Proterozoic along the southern margin of the platform provides favourable prospects for further search for the ores of this type. The Lower Angara deposit occurs in the so-called Angara-Pit iron-ore basin, which has 1000 million tons of known and 5000 million of estimated reserves.

An important iron-ore district in Eastern Siberia is the South Aldan region where large reserves of good magnetite ore have been surveyed. The region is virtually adjacent to the Aldan (South Yakutian) coal basin, which has deposits of coking coal, thus permitting the establishment of a new coal-metal industrial centre (Antropov, 1959).

The Tunguska Formation has deposits of siderite. One such deposit—near the village Dvoryets on the Angara—was worked as far back as the eighteenth century.

NICKEL has been found in sulphidic ores associated with intrusive varieties of traps in the north-west of the platform of Norilsk.

TITANIUM is found in the region of the traps, in alluvial concentrates of ilmenite. The titaniferous ores, associated with the Gulinskian ultrabasic intrusion in the lower reaches of the Kutui, are of some interest.

SULPHIDES of copper, nickel, platinum, silver, and other metals occur in the intrusive rocks filling the feeding channels of the Siberian Traps. They are typically developed in the Norilsk region, but are also found to the south in the sub-Yenisei zone of the traps. The ore occurs in lenses, nests and veins in gabbro-diabases.

COBALT.   The main deposits of cobalt are found in the sulphide ores of the Norilsk region and to the south of it.

Non-Ferrous metals

The discoveries of numerous new deposits of these metals constitutes an important achievement of Soviet geologists.

CUPRIFEROUS SANDSTONES are associated with the Cambrian and Ordovician red beds, and reach their greatest development in the Ust'kut region in the middle reaches of the Lena, where they are traced over a distance of 500km. Despite the thin nature of the copper-bearing horizons and their low mineral content the cupriferous sandstones, owing to their considerable reserves, attract attention.

BAUXITES, lacustrine and of Cainozoic age, have been found to the south-east of the Yenisei range. The ore lies in the middle part of continental deposits which were formerly considered as Upper Jurassic or Lower Cretaceous. Now they are proved to be Eocene and lie over the Lower Cretaceous and the Precambrian (Fig. 82).

Noble metals

GOLD. The deposits have been worked since Pre-Revolutionary times. Soviet geologists have enlarged the size of the previously known deposits and have found new alluvial and vein gold. The relationship between mineralization and intrusions of different ages has been established. The gold in the youngest intrusions of the Aldan alkaline province is interesting. The best known gold-bearing regions are the Aldan (near the town Aldan), the Tommot area to the north of Aldan, the Bodaibinian on the Vitim, and the North Yeniseian in the northern part of the Yenisei range.

PLATINUM is found in sulphides associated with sub-volcanic intrusions of Siberian Traps in Norilsk.

FUELS

Fuels are abundant and are represented mainly by coal. The known oil deposits, gas deposits and oil shales are small, but search for them is continuing.

COAL AND BROWN COAL. Coals and brown coals are concentrated in five horizons—the Carboniferous, the Permian, the Lower and Middle Jurassic, the Upper Jurassic, and the Lower Cretaceous and Neogene. Along the north-eastern margin there are coals of Upper Cretaceous age. Black coals are developed in the west of the platform and are restricted to the lower horizons of the Tunguska Formation, underlain by marine deposits. Permian coals are most widespread and lie in the main part of the Tunguska Formation, which occupies the central part of the Siberian Plateau. The area of the Tunguska coal-field is huge and calculations in 1956 suggested reserves of up to 1516×10^9 tons of which 429×10^9 tons lies down to 300m. Owing to the absence of detailed surveys the proved reserves are only $2 \cdot 4 \times 10^9$ tons. The coal is only worked in the Noril'sk region, the basin as a whole being a reserve for the future.

Jurassic coal is developed in the Irkutsk and Kana-Achinskian basins. The former often contains sapropelic coals rich in volatiles suitable for distillation. The basin is traversed by a railway and is intensively worked. Geological reserves of the basin are 67×10^9 tons and the proved reserves are $21 \cdot 7 \times 10^9$ tons.

In the Kana-Achinskian basin in the east (the Kana basin) the Jurassic is represented by a coal-bearing formation of 40–200m in thickness. It lies on the Devonian and Lower Palaeozoic. In the west (the Achinskian basin), and more precisely in the Chulyin-Yenisei trough, the Jurassic reaches 600m and at Chulym 1200–1300m, and the proportion of coal also increases, the total number of coal seams being 44 and the thickness of some of them 40–75m. The Lower and Middle Jurassic Coal-bearing Formation is overlain by continental, and in the north by marine, Upper Jurassic and Lower Cretaceous sediments. In places (Chulym) the Jurassic Coal-bearing Series is underlain by the Upper Palaeozoic Coal-bearing Series with seams of black coal. The latter is similar to the Tunguska Formation, has not got much coal, is relatively thin (400–1000m), unmetamorphosed and almost horizontal.

Geological reserves of the basin are huge—1208×10^9 tons, while the proved deposits reach $61 \cdot 4 \times 10^9$ tons.

The coals of the eastern part of the platform were formerly considered as Upper Jurassic. At present it has been demonstrated that a greater part of them is Lower Cretaceous and that in places the coal-bearing formation lies on the Valanginian *Aucella* beds. A minor part of the coals is undoubtedly Upper Jurassic. The Jurassic coals are mainly found in the west, and the Lower Cretaceous coals are abundant in the east. The latter are very widespread, stretching from the Aldan and the Maya in the south to the estuary of the Lena in the north. Thanks to the fact that the Coal-bearing Formation over the whole of this area is in a sub-montaine downwarp, it is not uncommonly up to 2–3km thick, has a large number of thick coals of high quality, and is folded into large gentle folds. The Lower Cretaceous deposits are relatively unexplored and their geological resources are 2400×10^9 tons, thus exceeding the resources of the Tunguska basin. Owing to their inaccessibility the deposits are not much worked, thus being in the nature of a reserve.

The Lena Coal-field. This includes a number of deposits of black and brown coal, all those found between the Sangara deposit near Yakutsk, which is also included, and the lower reaches of the Lena. The number of coal seams reaches 20–25, but they are thin, being about $1 \cdot 5$–$2 \cdot 0$m.

The Vilyui Coal-field. This coal-field includes all the deposits found in the western part of the Vilyui depression, such as the Markha deposit, the Vilyui deposit and the Kempendyai deposit. According to Strugov (1956) the succession in the coal-field is as shown in Fig. 75. The economic seams are found in the upper suite and reach 10–30m in thickness. The seams are almost horizontal and are often exploited in galleries. A provisional assessment of these brown coals suggests that their reserves are the greatest in the U.S.S.R.

The Southern Yakutian (Aldan) Jurassic Coal-field is the most recently discovered deposit. The rocks are continental Upper Jurassic and Lower Cretaceous, resembling the Lena deposits. The coal-field has a series of high quality seams of black coal of working thickness, varying generally up to $0 \cdot 4$–$3 \cdot 0$m and sometimes up to 10–18m and developed over a very large area (Fig. 78). The coal-field also includes major deposits of coking coal of high quality. The geological reserves of the coal-field are assessed in tens of billions of tons ($55 \cdot 7 \times 10^9$ tons according to Mokrinskii, 1958).

Neogene coals are found in the Baikal rift valley where the thickness of alluvial, lacustrine and bog deposits (clays and sands) reaches up to 1000m. Seams of brown coals and lignites of workable thickness are known in the delta of the Selenga and in the Barguzin Trough. The economic importance of the Neogene coals is being determined. They are mentioned by Florensov (1956) who has studied the trough filled by Mesozoic and Cainozoic deposits. A coal-field of brown Neogene coals has been recently found in the region of the lower reaches of the Aldan (see p. 246).

The Siberian Platform as a whole has exceptionally rich reserves of coal, much exceeding any other regions of the U.S.S.R. Owing to their inaccessibility, however, the deposits are not worked out, thus forming a large reserve for the future.

OIL-SHALES. These have not been much studied although two types—a lacustrine and a marine—have been distinguished. The lacustrine oil-shales are found amongst the Jurassic and Lower Cretaceous deposits. In the Cheremkhov (Irkutsk) basin they are closely associated with sapropelic coals, being their ultimate stage when the sapropelic matter is most predominant and the cells least common.

The marine oil-shales are found in the Lower Cambrian of the Aldan Massif, in association with black, bituminous limestones. The Lower Cambrian oil-shales are even more widespread in the lower reaches of the Olenek where the reserves are possibly about 100×10^9 tons (Vakar 1957).

PETROLEUM. The finding of petroleum is one of the most important problems facing Siberian geologists. There are favourable indications. Considerable accumulations of organic matter are known in the Cambrian limestones and dolomites, Upper Devonian carbonate-shale rocks and Jurassic and Lower Cretaceous lacustrine deposits. There is also a possibility of discovering other bituminous formations.

The Lower and Middle Cambrian carbonates have a high content of organic material, found in various localities, leading to a greater or less degree of bituminous development in rocks, which is especially apparent in the south-east on the northern flanks of the Aldan Massif. Here oil-shales are found, while in the valley of the Tolba (a right tributary of the Lena) small amounts of oil are found in porous Lower Cambrian dolomites. In the Irkutsk basin, in the valley of the Osa, oil has been found in a borehole which penetrated the Lower Cambrian Motian Sandstones. More recently economic deposits of oil have been found in a borehole sunk on the left shore of the Lena in the Ust'Kut region, near the village Verkhnemarkovo, at a depth of 2164m amongst the Lower Cambrian Usolian sandstones.

In the Upper Devonian deposits of the northern margin of the platform there are bituminous facies like the Domanik of the Russian Platform. Indications of oil are found in the salt domes of Port Nordvik, the domes forming suitable structures for accumulation of oil.

The Mesozoic lacustrine deposits of some regions, mainly in the south of the platform, include found oil-shales, rendering the discovery of petroleum possible. Economic accumulations of oil in Jurassic continental deposits are found near the border of the U.S.S.R. and Chinese Jungaria and in North-East China. In the U.S.S.R. oil has been found in the lower part and the delta of the Selenga, which enters lake Baikal from the east, but this oil is not economically significant

GAS. A large deposit of gas has been found in the Ust'Port region, the lower part of the Yenisei. Recently large accumulations of gas were found over a large area in the valley of the Lena and the lower reaches of the Vilyui in the region of the Cisverkhoyansk sub-montaine downwarp. Near Zhigansk, 300km north of the mouth of the Vilyui, the Lower Cretaceous continental and marine deposits have several gas-bearing horizons. At the mouth of the Vilyui a gas fountain arose from the Lower and Middle Jurassic sediments and was accompanied by a small quantity of petroleum. Finally, near Yakutsk, 300km south of the mouth of the Vilyui, the Lower Jurassic deposits have also yielded gas.

The Taas-Tumus (Ust'Vilyui) fold offers the best prospects; here powerful gas fountains produced $1.5 \times 10^6 m^3$ in 24 hours. The gas-bearing formation is 46m thick and according to Lyutkevitch (1959) has reserves of $4 \times 10^{10} m^3$. In 1960, in the same region, a second gas deposit with small admixture of oil was found (Tikhomirov, 1961).

Considerable gas streams have been discovered in the Motian (Lower Cambrian) Suite of the Irkutsk region.

The gas prospects of the Siberian Platform are favourable, but much remains to be studied, although a considerable volume of research into the gas content of the Cambrian rocks has been done over the last thirty years.

The Devonian oil of the north-eastern part of the platform has not been studied, although it is of the same kind as the Devonian oil of the Russian Platform and has unquestionably good prospects. Information on the geological structure and oil and gas content of the Cambrian and Mesozoic rocks of Eastern Siberia has been collected in 'Data on the geology and oil of the Yakutian A.S.S.R.' (1959). Similar material is to be found in the book: *The geology and the oil and gas of Eastern Siberia* (1959).

NON-METALLIC DEPOSITS

The non-metallic deposits are diverse and considerable and have often an All-Union or even international significance.

RAW MATERIALS. Three mica-bearing regions are most considerable, the first two being the regions of Slyudyanka, the southern termination of lake Baikal, and the Maina to the north of the lake. There are deposits of phlogopite and muscovite, associated with the Upper Proterozoic pegmatites which are often for inadequate reasons considered as Caledonian. The third mica-bearing region has been recently found and is known as the Aldan phlogopite region, where this mica is found amongst Archeozoic carbonates and especially the marbles of the Dzheltulinian Suite. Large-scale deposits of first-rate phlogopite permit one to consider the Aldan Massif as a major phlogopite-bearing region and the largest in the U.S.S.R. The U.S.S.R. has the richest phlogopite deposits in the world.

Large-scale deposits of mica, found in the Precambrian pegmatites, are known from the northern flank of the Eastern Sayans (Biryus) and at the southern extremity of the Yenisei Massif (Barga).

In the lower reaches of the Kotui, a tributary of the Khatanga, there is the large Gulin (Sabyda-Gulin) phlogopite deposit, associated with a sub-volcanic intrusion of dunites, peridotites and alkaline rocks of Lower Triassic (?) age.

Diamonds. The discovery of primary diamonds is one of the most important geological discoveries since the second world war. At first small quantities of diamonds were found in sediments, which led to the determination of their concentrations. Detailed studies of these indicated that explosion pipes (Fig. 90) represent their sources, the pipes having formed during the Permo-Triassic volcanicity, although some pipes may be Jurassic. Their structure is identical with the Kimberley pipe of South Africa, with which the largest diamond deposit in the world is associated. There are four diamontiferous areas in Siberia—the Vilyui, the Olenek (on the middle reaches of the Olenek), the Muna, and the Aldan regions.

The Kimberlitic pipes of the Vilyui region so far have been most fully investigated. The richest of the pipes are 'The Peace' and 'The Successful' (Mir, and Udachnaya). The richest alluvial deposits are near them (Antropov, 1959). Owing to these discoveries the U.S.S.R. at present possesses a dependable reserve of diamonds which are so necessary in the development of national economy.

Iceland Spar. The traps of the Vilyui and the Lower Tunguska regions are known to have some of the richest deposits of Iceland Spar in the U.S.S.R.

Graphite. The graphite deposits also have an All-Union importance and are of two types—magmatic and metamorphic. The first type is found amongst the Archeozoic deposits at a number of localities along the southern margin of the platform. The best known deposit of this type is the Botogolian, which is situated in the Eastern Sayans, 250km to the west of Cheremkhov. The succession of Archeozoic crystalline schists and marbles is intruded by a massif of nepheline syenite which forms the Botogolian mountain. Graphite flakes of up to 1mm across are concentrated and form irregular nests and stocks in the central part of the mountain. The stocks are up to 25×50m.

The metamorphic graphite is widespread in the north-west of the platform—in the valleys of the Kureika, Lower Tunguska, Bakhta and Fatyanikha. The coal seams here are metamorphosed into graphite as a result of action of the intrusive traps. On the Kureika the thickness of graphite is on average 15m. The total reserves are enormous.

Corundum and kyanite are found in the Archeozoic plagioclase gneisses intruded and altered by granites. The most important is the Chainytian deposit in the northern foothills of the Stanovoy Range. Red corundum-bearing and bright green kyanitiferous rocks are found within muscovite-chlorite lenses. The origin of corundum evidently lies in the regional metamorphism of the sedimentary rocks, which included bodies of bauxite. The rocks of the formation were altered into gneisses and the bauxites into corundum rocks.

Precious and semi-precious stones are found at a number of localities in the Eastern Sayans, Cisbaikalia and Transbaikalia, and after the Ural deposits are most important. Some minerals are associated with pegmatites of Archeozoic

FIG. 90. Explosion pipe 'Zarnitsa' situated on the crest of an anticline of Cambrian limestone. The oval outcrop of the body is crossed by straight traverse lines. The body is 500×560m and is overgrown by forest. *Aerial photograph by V. M. Barygin (1958).*

and Lower Proterozoic age, as are for instance, topaz, tourmaline, vorob'yevite, lazurite and spinel. Amongst other semi-precious stones of a different genesis nephrite and agalmatolite should be mentioned. Rock crystal is found in the Precambrian.

SEDIMENTARY NON-METALLIC DEPOSITS.   These deposits are quite considerable, and amongst them common salt and phosphorites are most important. In addition gypsum and anhydrite, fire clays (kaolin) and various building materials are being worked.

Rock salt is widely distributed. The Siberian Platform is one of the most important saliferous regions of the earth. At present the salt assures the local needs, but its reserves are so great that it can satisfy the whole Far East. The problem is that of transport. The salt deposits of the Siberian Platform are up to the present the most easterly in the U.S.S.R. These deposits are divided into two groups—the southern and the northern. The southern group is associated with the Lower Cambrian red beds and is widely distributed from the upper reaches of the Angara, along the whole length of the Lena and up to the salt domes of the Vilyui basin. The most important deposit (Usolskian) is on the Angara, some 60km below Irkutsk. Formerly it has been considered as the largest, but in the last few years a series of larger deposits, some hundreds metres thick, have been found.

A great thickness of salt is found in the salt domes of Kempendyai, Kygyl-Tus, Kyundyai (in the Vilyui basin) where hills of up to 120m with in places precipitous slopes are formed. Here the salt was first considered as Lower Cambrian, then Ordovician, and now again as Lower Cambrian (Lyutkevich, 1959).

The northern group of deposits is situated along the northern margin of the Siberian Platform, where it is found in the salt domes of the Port Nordvik region. The best known of these is the Solyanaya Sopka (Tus-Takh). The age of the salt is debatable. Some investigators consider it as Cambrian, but it is more probably either Lower Devonian or Eifelian. In the cap breccia of the massif the oldest fragments are Middle or Upper Devonian, while Cambrian, Ordovician and Silurian rocks are absent.

PHOSPHORITES.   These are important as beds found amongst the metamorphic Sinian deposits, of the type seen at Karatau and Central Kazakhstan. Economic concentrations of phophatic pebbles and grains are found in the Ordovician.

KAOLIN.   The deposits of kaolin are associated with Jurassic rocks and are of economic importance in the Irkutsk region.

REGIONAL DESCRIPTION

CISBAIKALIA AND WESTERN TRANSBAIKALIA

General Account

The Cisbaikalia and Western Transbaikalia constitute a peculiar elevated region, which has provoked more heated discussion than any other region in Siberia. One group considered the region as a typical Precambrian platform like the Aldan Massif, while another group suggest that it is a region of manifestation of strong Caledonian folding or a typical Caledonian geosyncline. The adherents to the second group consider all the micaceous granites and pegmatites as Caledonian. According to the first point of view the region became a platform in the Upper Proterozoic, after the Karelidian Orogeny. According to the second standpoint the region became a platform only in the Middle Palaeozoic after the Caledonian Orogeny. Facts suggest a compromise point of view, with the western and northern parts of the region being attributed to the Baikalids of the Siberian Platform and the western and southern regions to the Caledonides of the Angara Geosyncline (Fig. 2). The Karelidian zone has the Upper Proterozoic and Cambrian of platform type. In the zone of the Baikalids the Upper Proterozoic is geosynclinal, but the Cambrian is platform-type and the zone belongs to the Siberian Platform. Lastly in the Caledonian zone the Upper Proterozoic and Cambrian are geosynclinal, which relegates it to the Angara Geosyncline.

Stratigraphy and Palaeogeography

The Archeozoic of the Cisbaikalia and Western Transbaikalia is of the same type as in the regions of the Olekma and Aldan, the Stanovoi range and the Eastern Sayans. The Archeozoic has already been described in the section on the Siberian Platform; suffice to say that the succession is complete and very thick and that both the Lower and Upper Archeozoic are developed. The Proterozoic is also rather fully developed and reaches a great thickness of tens of kilometres.

The Lower Proterozoic is everywhere metamorphosed, homogeneous, and has different names in different places. The Proterozoic of the Baikal Mountains has been described by L. I. Salop (1958a). The facies changes in the Lower Proterozoic (Muian Series, Fig. 91) are interesting. In the south the series begins with the Samokutian Suite (sm) of basal sandstones and conglomerates (200–500m). It is overlain by pale and dark crystalline limestones (450m) of the Bulundinian Suite (bl). In the centre the thicknesses of both suites increase sharply and above the Bulundinian Suite (1200m) a volcanic complex (Kilyanian Suite, kl), of 9–10km in thickness, appears. This is overlain by the Delyunian Suite (dl) of variegated sandstones and conglomerates (500m). In the north the succession changes again and the Bulundinian and Kilyanian suites are replaced by the corresponding Chayangrinian Suite (2500m) of crystalline limestones with beds of shales and lavas. Higher up lies the Dzhalagunian Suite (d) which consists below of carbonaceous slates, meta-volcanics and limestones (600–1000m), in the middle of dominant carbonaceous slates and phyllites (800–1200m), and at the top of the black carbonaceous slates and phyllites alternating with pale quartzites (>600m). Further to the east the Muian Series passes into the Bodaibinian Series and then into the Udokanian Series. All this stresses the variability of the Lower Proterozoic successions.

The Upper Proterozoic, unlike that of the central part of the platform, has a geosynclinal character, it is thick, strongly deformed and folded, strongly metamorphosed, and includes large synorogenic intrusive massifs, with the accompanying minor intrusions and various veins which are often mica-bearing. In Cisbaikalia the Upper Proterozoic is represented by the 'Baikalian complex' already described Its lower, Goloustenian, and upper, Kachergatian, suites (each reaching 1·5–2km) consist of terrigenous clastic rocks such as sandstones, arkoses and shales. These rocks could only have originated at the expense of the erosion of some elevated region, which was situated comparatively near and which included granitic massifs. The arkoses are formed as a result of erosion of granitic massifs situated near to the region of deposition. Such elevated regions are normally fold mountains of Lower

Proterozoic and Archeozoic to the east of Cisbaikalia, or in Western Transbaikalia. Thus in the Upper Proterozoic Western Transbaikalia ceased to be a region of sedimentation (sea or coastal plain) and became a region of fold mountains.

Along the strike, especially northwards, the nature of the Goloustenian and Kachergatian suites changes considerably. The carbonate formations become widespread and the total thickness decreases. Nevertheless sandy groups are found everywhere and often are very thick, suggesting sedimentation from the east, and in the north (nearer to the Lena) from the south.

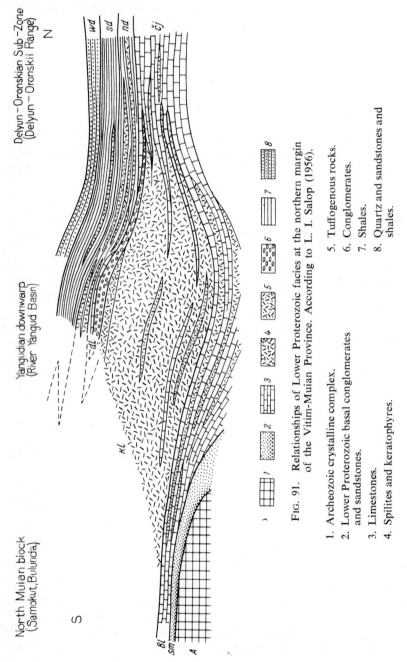

FIG. 91. Relationships of Lower Proterozoic facies at the northern margin of the Vitim-Muian Province. According to L. I. Salop (1956).

1. Archeozoic crystalline complex.
2. Lower Proterozoic basal conglomerates and sandstones.
3. Limestones.
4. Spilites and keratophyres.
5. Tuffogenous rocks.
6. Conglomerates.
7. Shales.
8. Quartz and sandstones and shales.

The existence of a high mountainous region to the east and south of the zone of the Baikalian complex is shown especially clearly by the so-called Conglomeratic (Balaganakhian) Suite. The suite first appears at the northern end of lake Baikal and is traced to the lower reaches of the Greater Patom. The suite varies considerably in its content and thickness, but the existence of conglomerates, sandstones and rarer shales is everywhere typical for it. In some localities the suite consists entirely of conglomerates, while elsewhere sandstones predominate. In still other localities beds of sometimes bituminous limestones appear. The thickness of the suite varies from several hundred metres to 5000–6000m. The Conglomeratic Suite in the east terminates against ancient porphyries and granites and in the west it is covered by the Baikalian complex. The conditions of the formation of the suite are correctly compared to the molasse of the Alps and to Neogenic and Anthropogenic conglomerates at the foot of Central Asiatic mountains. The Conglomeratic Suite undoubtedly proves the existence of a considerable mountainous land to the east and south of it.

The Upper Proterozoic of the Baikal mountainous region has been studied in detail by L. I. Salop (1958a). According to him to the north of lake Baikal the Baikalian complex passes laterally into the Patomian Complex (Series), which lies on the Lower Proterozoic Bodaibinian Series and begins with the aforementioned Balaganakhian (Conglomeratic) Suite (3000–6000m). It is overlain by the middle part of the complex consisting of four suites: the Mariinskian, the Dzhemkukanian, the Barakunian and the Valyukhtinian, which altogether are 4400–5000m thick. The middle part of the complex consists of dark, carbonaceous shales and limestones with groups of sandstones and siltstones. The upper part of the complex (1800–2100m) includes the Zhdinian and the Chenchinian suites and consists of greenish and lilac marls, slates and limestones which are aphanitic towards the top. Higher up, and conformably, lie sandstones and conglomerates with a marine, Lower Cambrian fauna. Frequently the upper part of the Upper Proterozoic (Sinian) is separated as the Zhuinian Series (up to 4000–5000m)—Fig. 92.

The total thickness of the Upper Proterozoic in the Patomian Highlands is enormous (9000–13 000m), indicating the geosynclinal conditions of its deposition. During the Baikalian Orogeny these rocks were folded into a young mountain region, the type area of the Baikalids. The position of this zone is shown in Fig. 2. In the south the Baikalids are margined by Caledonides of the Angara Geosyncline.

Certain adherents of the Caledonian Geosyncline attribute the Conglomeratic Suite together with the Baikalian complex to the Lower Cambrian. The error of this point of view has already been commented upon, but if it is accepted the existence of the Caledonian Geosyncline can be denied even more completely. During the Cambrian, all Caledonian geosynclines were regions of deposition in mainly marine basins. The presence of mountain ranges—zones of erosion—precludes the existence of a geosyncline.

The Cambrian deposits, their composition and stratigraphy confirm the absence of a Caledonian Geosyncline.

Along the western and northern margins of the Cisbaikalia and the Patomian Highlands there are Cambrian deposits represented by the same tripartite succession as in the adjacent Siberian Platform. The Lower Cambrian is red and saliferous, the Middle Cambrian is carbonate-bearing and the Upper is again red.

In the south between the source of the Angara and the upper reaches of the Lena at the base of the Lower Cambrian there is a peculiar Ushakovkian Suite (500m) of greywackes and conglomerates. These deposits are typical of the foothills region of the mountains rising in the central part of Cisbaikalia. The Ushakovkian Suite is overlain by the Motian Suite, which is found, despite interruptions, along the whole of Cisbakalia and the Patomian Highlands in the north. The suite consists of red sands and clays with lenses and beds of gypsum and rock salt, being typical of near-shore plains with salt-lakes and lagoons where the gypsum and salt had accumulated. To the west the plain joined the sea and to the east passed into a high area of erosion. This indicates that in the Lower Cambrian the central part of the region was also elevated as it has been in the Upper Proterozoic. However, what is most important is that all the three subdivisions of the Cambrian are considered unanimously by all investigators as platform deposits, on the basis of their thickness, lithology, and degree of metamorphism. Even in the Patomian Highlands these features are found. What is, therefore, the nature of the Caledonian Geosyncline which in the main consists of platform Cambrian? It should also be added that in the central part of Cisbaikalia the Cambrian and the Ordovician are entirely absent. The latter is explained by the fact that this region was a region of erosion. In the north-east the Cambrian retains its platform character which is the same as in the basins of the Olekma, the Lena, and the Aldan. To the north-east of Barguzin, for instance in the region of Proterozoic development there are small outcrops of

FIG. 92.   Goloustenian quartzites in the Baikal range.
Photograph by L. I. Salop.

Lower Cambrian (450m) represented by conglomerates (at the bottom), sandstones and limestones with trilobites. The Cambrian is almost horizontal and overlies the Upper Proterozoic with an angular unconformity. The Cambrian is almost unmetamorphosed and lacks intrusions.

Cambrian of geosynclinal type is found to the east in the lower reaches of the Muya, the valley of its tributary—the Kydymit—and to the west along the Turka. Here the Cambrian is the same as in the southern part of the Eastern Sayans, along the Dzhida and Eastern Transbaikalia. The Cambrian is folded into complex folds, is metamorphosed and intruded by granites, and lies unconformably of the Proterozoic. Its lower part consists of a conglomeratic-sandy-shaley suite of conglomerates, arkosic sandstones, chlorite schists and phyllites. Over this is a group of limestones, which are white, fine grained and rich in archeocyathids. The upper part consists of greenish sandy slates, sandstones and limestones. The total thickness of the Cambrian is great and reaches 4000m. It is intruded by granitoids, varying from granites and granodiorites to syenites, which are accompanied by various vein rocks.

The Ordovician is only developed in the western and northern parts of the region and lies conformably on the Cambrian. The Ordovician is of platform type and has already been described (p. 221).

The Silurian is developed along the margins and is conformable on the Ordovician (p. 225). Within the fold belt it has not as yet been discovered, but its presence is unquestionable since it is developed in the Eastern Sayans and the Eastern Transbaikalia.

The Devonian and Lower Carboniferous are only found in the extreme south, in the basin of the Dzhida. Both are geosynclinal, folded and metamorphosed slates and limestones with a marine fauna. So far only few individual outcrops are known.

The Upper Palaeozoic has not been palaeontologically dated, but it is possible that the brown arkosic sandstones of the upper reaches of the Vitim belong to it.

The Mesozoic is everywhere of platform type and is preserved in a number of limited regions, which are grabens, surrounded by the Precambrian. The Trias and the Lower Jurassic are unknown, while the Middle Jurassic is widely distributed. In the northern zone of the grabens (the Dzhida, the Tarbagatai and the Bukachach) it is represented by coal-bearing continental deposits, which normally begin as conglomerates and sandstones and continue upwards into sandstones and clays with black coal seams. The thickness of these beds is about 100–150m, but in places reaches 500m and more. At the base, under the conglomerates, there is often an effusive-tuffogenous suite.

The Lower Cretaceous is either conformable on the Jurassic or unconformable on the Precambrian. Most commonly it commences with the Turginian Suite of lacustrine deposits, which are whitish and grey marls and shales with bedding planes full of fresh-water Estheria, insects and fish. Apart from the marls there are sands with clays, which are sometimes coal-bearing. Higher up lies the Dayan (productive) Suite, which is coal-bearing and consists of sandstones and argillites with lenses and seams of brown coals. Its thickness is 200–1000m. Upper Cretaceous sediments are not known, but the presence of their continental facies in isolated troughs is unquestionable. This is indicated by the recent discovery of Palaeogene facies, which were not formerly known.

The Cainozoic has been preserved in isolated depressions and consists of sandy-clayey rocks with plant remains and a fresh-water fauna. The thickness of the

Cainozoic is small. The Palaeogene is found in the upper reaches of the Vitim (the Kydymite basin) and consists of grey and yellowish sands with beds of black shales, some 20–25m thick. Pollen analysis has demonstrated the presence of conifers and deciduous trees and spores of ferns which are most probably Palaeogene in age. The Neogene is distributed much more widely and is represented by different facies. Most commonly lacustrine and fluviatile deposits such as marls, sands and clays with a fresh-water fauna are encountered. In the lower reaches and the delta of the Selenga river they reach a great thickness of 1000–1200m and have already been described (p. 246).

The Anthropogene of Cisbaikalia and Western Transbaikalia consists of a variable complex of facies covering a large area. It has been described in detail by Obruchev (1938), who distinguishes pre-glacial and alluvial deposits, moraines and fluvioglacial deposits of two epochs and also various post-glacial deposits. The main wealth of the region—the Lena goldfields (basins of the Baodaibo, the Zhuya and others)—are associated with it.

Tectonics and Magmatism

The tectonics and magmatism in the Archeozoic and Lower Proterozoic are the same as for the rest of the southern part of the Siberian Platform, beginning from the Aldan Massif and the Stanovoi range and continuing into Cisbaikalia and ultimately the Eastern Sayans and the Yenisei Ridge. All these large provinces form a single belt of the Sayano-Stanovoi belt of Precambrian rocks.

In the Upper Proterozoic this unity is broken and two large platforms come into existence—the Angara-Yenisei and the Olekmo-Aldan, separated by the Cisbaikalian Upper Proterozoic Geosyncline. Within the platforms the Upper Proterozoic (Baikalian complex) has a platform nature. Within the geosyncline it has a geosynclinal character, is intensely folded, metamorphosed and includes granite intrusions. To the south of these three tectonic regions is the Angara Geosyncline (Fig. 2) in which the folding involves not only the Upper Proterozoic but also the Lower and Middle Palaeozoic. The Cisbaikalian Upper Proterozoic Geosyncline, like the Donyets Geosyncline, lies between two platforms of relatively small sizes. In the south this geosyncline widens and merges into the Palaeozoic Angara Geosyncline. At the end of the Proterozoic the Cisbaikalian Geosyncline became filled, was converted into a platform and united with the Angara-Yenisei and Olekmo-Aldan platforms thus forming a single Siberian Platform.

The Caledonian Orogeny did not occur in the Siberian Platform, but it has been vigorous along its southern margin and the southern half of the Eastern Sayans in Khamar-Daban and the southern part of Western Transbaikalia. At the end of the Palaeozoic this belt became a platform, adhered to the Siberian Platform and increased its dimensions.

In the Mesozoic on the Siberian Platform the Cimmeridian Orogeny was not manifest, but large-scale faulting caused the rise of Palaeozoic and Precambrian blocks, which formed regions of erosion, while other blocks became troughs in which the Jurassic and Lower Cretaceous sediments accumulated.

In the Neogene and Anthropogene the block movements become very large scale, conditioning the present relief of high mountains and deep troughs, one of which is filled by the waters of lake Baikal. The tectonic movements were accompanied by concurrent magmatic activity, and the folding in geosynclinal regions was associated with large and small intrusions. Block movements often cause the injection of basaltic sills and the Neogene and Anthropogene block movements are accompanied by widespread effusions of basalts.

Useful Mineral Deposits

The useful mineral deposits of Cisbaikalia and Western Transbaikalia are numerous and diverse and have already been described. Gold and mica are of prime importance and indications of the development of Proterozoic jaspilitic iron-ores are of interest. The copper deposits are considerable. One of them has recently been discovered in the Udokan region of Chita Province.

Cisbaikalia and Western Transbaikalia represent a structurally very difficult region, where detailed studies have only begun.

5

WESTERN SIBERIAN LOWLANDS

GENERAL DESCRIPTION

The Western Siberian Depression is one of the largest depressions in the world. The greater part of these lowlands is covered by bogs and the taiga, while the largest system of marshes in the U.S.S.R. is situated here and is known as Vasyugan'ye. In the north the taiga is transitory into tundra of the coastal parts of the Arctic Ocean and in the south it passes into steppes (Barabinian and Kulundinian), which are now being converted into agricultural land.

The great river Ob' and its tributaries drain the Western Siberian Lowlands. The valley of the Ob' in its lower reaches has a great width of 50–70km, and is covered by the meanders of the river and ox-bow lakes. The adjacent tundra has innumerable small lakes (Fig. 93). Almost the whole of the surface of the Lowlands is covered by a sheet of Anthropogenic deposits. Only in a few isolated localities are there Neogenic and Palaeogenic inliers (Fig. 94). Around the margins of the Lowlands, nearer to its Palaeozoic frame, areas of Neogene, Palaeogene, and Cretaceous reach a great size. The Palaeozoic is nowhere exposed in the middle of the Lowlands, but skirts it from the west, south and east.

Thick (3500m and more) Neogene, Palaeogene and Mesozoic rocks underlie the Anthropogene. The basement consists of the Palaeozoic and Precambrian. The Precambrian is everywhere folded, although the Sinian is in places probably almost horizontal as it is in the Siberian Platform. The region of the Precambrian Platform spreads probably into the north-eastern part of the Lowlands. In the western and south-western regions the Palaeozoic and the Sinian are folded as a result of the Caledonian and Hercynian Orogeny. The structure of the basement is a fundamental problem to be solved by Soviet geologists and geophysicists, while the existing suggestions have a provisional character.

STRATIGRAPHY

PRECAMBRIAN AND PALAEOZOIC

The Precambrian and the Palaeozoic are little known since they are not exposed and occur only in deep boreholes placed comparatively near to the Palaeozoic frame. The Precambrian is found in the region of Berezov where it is represented by gneisses, granites and talc-chlorite schists. The Lower and Middle Palaeozoic are most frequently encountered. They do not normally contain any organic remains, which renders the accurate determination of their age and even the separation of Lower and Middle Palaeozoic impossible. In isolated localities Devonian spores and a Lower Carboniferous fauna have been found. In the west and south of the Lowlands the Lower and Middle Palaeozoic are geosynclinal and consist of metamorphosed and dislocated limestones, slates, sandstones, quartzites, porphyries and

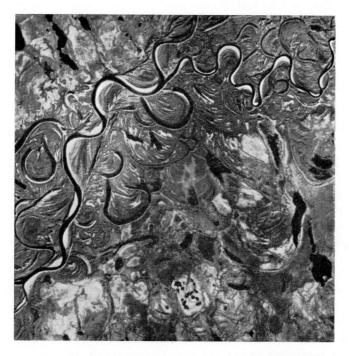

FIG. 93. An aerial photograph of the Ob' valley in its lower part. Numerous meanders and residual ox-bow lakes are obvious. Black patches are lakes on the surface of the tundra.

FIG. 94. Malyi Atlym exposures of Palaeogene strata on the river Ob'.
Photograph by V. D. Nalivkin.

tuffs. In the east of the Lowlands platform-type Cambrian dolomites and Devonian sandstones which are weakly deformed or horizontal, have been found.

The Upper Palaeozoic is quite widely developed and is represented by the Turinian, effusive-sedimentary, often coal-bearing, Formation, which is similar to the coal-bearing and tuffogenous formations of the Siberian Platform. For the large central part of the Lowlands the nature of the basement is almost unknown and its structure is determined from geophysical data.

PRECAMBRIAN. This provisionally includes the gneisses, granites, and chlorite-talc schists found at depths of 1300–1800m in the region of Berezov, in the lower reaches of the Ob′. The biotite gneisses found at a depth of 2600m in a borehole at Nagino are very interesting and are almost in the centre of the Lowlands. Near Berezov, adjacent to the gneisses, there is folded and metamorphosed Palaeozoic, indicating that the Precambrian is found here in the geosynclinal Palaeozoic zone. The Precambrian at Nagino probably occurs in a similar situation.

LOWER AND MIDDLE PALAEOZOIC. In connection with the search for oil and gas in a region between the Urals and the lower reaches of the Ob′ and the Irtysh, the Tyumenian Geological Survey has sunk a large number of exploratory boreholes. Tens of these reached the basement and often intersected metamorphosed and folded Palaeozoic deposits. Slates and quartz-schists and sandstones predominate and quite frequently dark meta-limestones and various volcanic rock (porphyrites, porphyries) and their tuffs are encountered. In a number of boreholes granites of Palaeozoic type are developed. In Berezov a Caledonian absolute age of gneisses has been recorded. Organic remains are found in a borehole to the south of Tobolsk implying marine Lower Carboniferous. In the southern part of the Lowlands the Middle Palaeozoic is found in Barabinsk at the depth of 2200m. It consists of slates interbedded with sandstones and porphyritic tuffs. Spores and radiolarians indicate an Upper Devonian age.

In the south-east of the Lowlands to the north-west of Achnisk the Byelogorian borehole has penetrated one of the best Carboniferous sections, reaching a total thickness of 445m. It is represented by interbedded sandstones, siltstones and argillites, tuffs and rarely limestones towards the bottom. Its fauna is diverse and indicates the Middle and Lower Carboniferous age for these rocks.

The weakly folded Devonian sandstones and argillites, which are in places coal-bearing and have spores, are found in Kolpashev and Mariinsk. Their weak metamorphism (argillites, silts) engenders a doubt about their geosynclinal status and they are more similar to the Devonian of the south-west of the Siberian Platform. In all other cases the age remains indeterminate and even the separation of the Lower and Middle Palaeozoic is impossible.

The platform-type Lower and Middle Palaeozoic has been found in stratigraphical boreholes on the rivers Yeloguya and Kasa, some 100–150km west of the Yenisei. On the Kasa these rocks are red beds: sandstones, argillites and siltstones, which are in places gypsiferous, 932m thick, and have been included in the Devonian. On the Yeloguya there are 418m thick dolomites, limestones and silts of Cambrian age.

UPPER PALAEOZOIC. In the region of Tyumen′ and to the south of Tobolsk there is a peculiar and interesting formation, exceeding 500–800m in thickness and consisting of dark coal-bearing and siliceous shales, red sandstone, conglomerates and shales. In its upper part there is a flow of olivine basalt and at the base an intrusion of olivine gabbro-diabase. Its distribution over a considerable distance indicates that the formation is almost horizontal and shows similarity to the upper part of the Tunguska Coal-bearing Formation and the Siberian Traps. The age of these formations is proved palaeontologically.

It is interesting that the palaeontologically dated Permian deposits, and the lavas which cover them, and the sills at the base of the Coal-bearing Formation, have been found in boreholes of the Chelyabinsk basin and to the north of it in the Bulanash-Yelkinsk coal-field. The Upper Palaeozoic coal-bearing formation (400–1000m) with economic coal seams has been found underlying the Jurassic in boreholes sunk within the Chulym Jurassic coal-field—near to the Kuznetsk Basin. In the north of the Lowlands the traps have been found to the west of the Yenisei valley. These data indicate the possibility of a considerable distribution of the Permian Tunguska Coal-bearing Formation and traps within the Lowlands. Geophysical data indicate that narrow long north–south structures of Uralian type continue to the east of the Urals, to the valley of the Tobol and lower Ob′. Further on they are displaced by irregular, rounded structures of Eastern Kazakhstan type. The geophysical data are important in understanding the structure of the basement of the Western Siberian Lowlands. The interpretation of these data varies and allows of differing reconstructions.

N. N. Rostovtsev (1956), who for many years has been in charge of the investigations made in the Lowlands by VSEGEI, has reviewed the geophysical and other data. Considerable material can be found in collections of papers by VNIGRI (Drobyshev and Kazarinov, 1958).

MESOZOIC AND CAINOZOIC

The Mesozoic and Cainozoic of the Western Siberian Lowlands, the eastern slope of the Urals and the Turgaian Depression is a single, homogeneous complex of mainly terrigenous sandy-clayey deposits. The area occupied by these deposits is very large and approximates to a square with 1600km sides. The thickness of the deposits is on average 2–3km, but in the north is 4–6km, thus the volume is some $(5-7\cdot5) \times 10^6 km^3$, which is an astonishing figure. It is sufficient to point out that it is equivalent to 20–30 Caucasian ranges, or in other words at least 20 of such ranges must be eroded away in order to fill the Lowlands with its Mesozoic and Cainozoic sediments. Evidently the ranges and highlands surrounding the Western Siberian Lowlands were elevated and eroded away several times.

The detailed investigation of the Mesozoic and Cainozoic of the Western Siberian Lowlands and the adjacent areas started a comparatively short time ago. In connection with the search for gas, oil, coal and iron the research has been greatly developed, and hundreds of specialists have collaborated. The achievements in the fields of stratigraphy are examples of the most important contributions by the Soviet stratigraphers and palaeontologists. In addition to the work of N. N. Rostovtsev and his colleagues (1956, 1957), V. P. Kazarinov (1958), D. V. Drobyshev, and Kazarinov (1958), V. N. Saks and co-authors (1957) in the north, and of V. I. Bodylevskii (1944) on the marine Jurassic and Cretaceous, the following collections of papers can be mentioned: 'Geology and oil in the west of the Western Siberian Lowlands' (1959), 'Stratigraphy of the Mesozoic and Palaeozoic of the Western Siberian Lowlands' (1957), and 'Proceedings (Trudy) of the Regional Conference on the Western Siberian Lowlands'.

The Meso-Cainozoic sediments are divided into the following complexes:

1. the Lower Trias, Rhaetic-Lias and Middle Jurassic, which are continental;
2. the Upper Jurassic and Lower Cretaceous, which are marine;

3. the Lower and Upper Cretaceous—continental;
4. the Upper Cretaceous-Palaeogene—marine;
5. the Neogene.
6. the Anthropogene.

Age	Lar'yak	Pokur
Quaternary	Sands and clays 24m	Alluvium 24m
Neogene	Continental grey sands and clays with plants (243m)	Pliocene: Yellow sands (40m) Miocene: grey sands and clays with plants 168m.
Upper Oligocene	Not recognized	Continental grey sands and sandstones. Plants (35m)
Lower Oligocene	Greenish clays with diatoms (67m)	Cheganskian Suite of light gaizes and clays with foraminifera (341m)
Eocene	Gaizes, clays and foraminifera (96m)	
Palaeocene	Dark clays and glauconitic sandstones (101m)	Dark compact fossiliferous clays (72m)
Campanian and Maestrichtian	Light clays and marls with a marine fauna (117m)	Grey and greenish, siliceous clays and marls with a marine fauna (120m)
Coniacian-Santonian	Grey, gaizes-like clays and glauconitic sandstone (275m)	Grey and greenish gaizes and clays with a marine fauna (133m)
Turonian	Greenish sandstones and sandy limestones with a fauna (55m)	Greenish clays with phosphorites —a marine fauna. Condensed sequence of 5m.
Cenomanian	Dark coal-bearing clays and sandstones. Pollen (429m)	Pokurian Suite of greenish and grey sandstones and clays with amber near the top and coals near the bottom (702m)
Aptian-Albian	Sandstones and clays with a flora (591m)	
Hauterivian and Barremian	Terrestrial and fresh-water sands and clays (295m)	Grey sandstones and clays with coals—continental (516m)
Valanginian	Dark clays with a marine fauna	Dark clays with sandstones towards the top and a marine fauna (194m)
Upper Jurassic	Dark clays with a marine fauna (287m)	Not penetrated
Middle Jurassic	A coal-bearing formation. Not penetrated to the bottom	Not penetrated

These complexes change considerably along the margins of the Lowlands, but in its central part are relatively uniform, reaching an overall thickness of 2500–3000m and more. Their composition can be assessed from the sections of two stratigraphic boreholes. The Pokur borehole is situated on the Ob', near the mouth of the

Vakha, which Lar'yak is situated on the Vakha, some 250km to the east of Pokur.

Substantial differences between various borehole sections and the large number of investigations by many organizations caused the proliferation of suites, which were sometimes unified into series. The correlation of these suites demanded a great effort by N. N. Rostovtsev (1957) and his colleagues. His scheme, with some additions and modifications, is shown in Table 24.

The detailed sections are given in descriptions of standard stratigraphical boreholes, in the series known as *The Standard Boreholes of the U.S.S.R.*, published by VNIGRI. The study of certain of these sections has shown that the boundaries given in Table 24 are often arbitrary and approximate and in different regions their position can be different.

LOWER TRIAS. These deposits form the upper part of the Coal-bearing-effusive Formation found in the Siberian Platform and the Western Siberian Lowlands (Turinian Series, Table 24). The lower part of the formation is Upper Permian, while the upper part on the basis of spores is Lower Trias, and possibly Middle Trias (p. 345).

RHAETIC-LIAS COAL FORMATION (Chelyabinskian Series). This formation is widely found around the margin of the Western Siberian Lowlands, but is preserved in basins of a small size. It is everywhere clearly, but weakly, folded into large, gentle flexures broken by thrusts and normal faults. At the marginal thrusts against the Palaeozoic the formation is deformed into small and complex folds unaccompanied by metamorphism. Within the Lowlands it is the same, but in a number of regions it is absent. Where the Rhaetic Liassic formation is preserved, it is represented by sandy-clayey coal-bearing rocks of varying thicknesses and not infrequently involves volcanic formations.

In the Omsk region the Rhaetic-Liassic Formation is 420m thick and includes a sheet of basic volcanic rocks, while in the region of Barabinsk it is only 17m thick. Here at the base lie green sandy clays (2m) probably representing a mantle of weathering. The constituent grey and brown argillites contain thin layers of felsites and quartz porphyries. The beds dip at 5°–8°. The Rhaetic-Liassic Coal-bearing Formation is developed on the eastern slope of the Urals and the Turgai Depression. In the Chelyabinsk Basin its thickness reaches 4000m, it is divided into four suites and has been studied in detail (Kareva, 1959). In the Turgai Depression there is a large coal-field (Kushmurun) which has also been studied in detail (p. 345).

The marine Rhaetic-Liassic deposits are only known from the extreme northeast of the Lowlands, in the Ust' Yenisei trough. They consist of dark argillites and silts with beds of sandstone with conglomerates at the bottom (400m). A Middle and Upper Lias fauna consisting of lamellibranchs has been found and indicates the brackish conditions of the marine embayment in which it lived, while the open sea existed to the east in the Khatanga Depression.

MIDDLE AND UPPER JURASSIC (continental). The Middle Jurassic (Tyumenian Suite) is similar to the Lias in its constitution and extent. Over the greater part of the Lowlands it is sandy and clayey, continental and often coal-bearing. In the west the suite is relatively thin (100–200m), but is considerable in the east (400–600m).

In the Ust' Yenisei trough are found marine clays and sandstones with Aalenian and Bajocian faunas, but the Bathonian is again continental. In the south-east the Jurassic Coal-bearing Formation crops out forming the Chulym-Yenisei coal-field, which is now considered as the western part of the single Kana-Achinsk Basin.

TABLE 24.

Mesozoic and Cainozoic subdivisions in the Western Siberian Lowlands (1960)

Stratigraphic subdivisions				Turgai Depression	Western Part	Central Part	Eastern Part	
Gp.	System	Division	Stage				Narym-Kolpashev Region	Maksimkin-Yar and Chulyma Region
Cainozoic	Neogene	Pliocene.	Upper.	—	—	—	Burlinian Suite (50–80m).	—
			Middle.		Kustanaiian Suite (30m).	Pavlodarian Suite (60–90m).		Miocene sands with beds of clays and pebble conglomerates (110m).
		Miocene.	Lower.		Zhilandian Suite (20–60m).			
			Upper.					
			Middle.		Aral'skian Suite (20–30m).			Missing.
			Lower.					
		Oligocene.	Upper.		Chagraiskian Suite (30–50m).		—	
			Middle.		Chiliktian Suite (60m). Kutanbulakian Suite (0–50m).	Turtaskian Suite (40m). Novomikhailovskian Suite. Atlymian Suite (50–70m).	Yurkovskian Formation.	
			Lower.	Cheganian Suite (20–280m).				
		Eocene.	Upper.	Saksaulian Suite (Up to 100m).		Lyulinvorian Suite (95–245m).		Symian Suite (150m).
			Middle and Lower.	Tasaranian Suite (Up to 280m).				
	Palaeogene	Palaeocene.	—	Missing.		Talitsian Suite (30–125m).		Kasian Suite (240–330m).
		Upper.	Danian.	Tyulyusaiian Suite (10–30m).	Gan'kinian Suite (10–230m).	Gan'kinian Suite (10–230m).	Kolpashevian Suite.	Simonovian Suite (260m).
Mesozoic	Cretaceous		Maastrichtian.	Zhuravlevkian Suite (40m).				
			Campanian.	Eginsaiian Suite (50m).	Berezovian Suite (35–260m).	Slavgorodian Suite (50–170m). Ipatovian Suite (5–100m).		
			Santonian.					
			Coniacian.					
			Turonian.	Ayatian Suite (25–100m).	Kuznetsovian Suite (20–65m).			
			Cenomanian.	Shetirgizian Suite (30–50m).	Uvatian Suite (270–315m).	Pokurian Suite (380–1100m).		

Era	Period	Epoch	Stage					
Mesozoic	Cretaceous	Lower	Albian. Aptian.	Taldykian Suite (60m).	Khanty-Mansian Suite (100–260m).	Pokurian Suite (380–1100m).		Kiian Suite (50–270m).
			Barremian.	Missing.	Vikulovian Suite (72–288m). — Leushinskian Suite (180–445m). | Kiyalian Suite (60–650m).	Vartian Suite (560m).	Kiyalian Suite (230–650m).	Ilekian Suite (600–745m).
			Hauterivian. Valanginian.		Tarian Suite (50–200m).			
	Jurassic	Upper	Volgian.		Kulomzinskian Suite (120–200m).		Tebiskian Suite (245m).	Tyazhinskian Suite (105m).
			Kimmeridgian.		Deminskian Suite (50–70m).	Achimovskian Formation		
			Oxfordian.		Abalakian Suite (65m).	Maryanovkian Suite (70–150m).		Itatian Suite (540m).
			Callovian.		Vogulinskian Formation	Barabinskian Formation (15m).		
		Middle	Bathonian.	Ubaganian Series (600–700m).	Tatarian Suite (160m).	Tatarian Suite (120m).	Tyumenian Suite (135–410m).	Makarovian Suite (200m).
			Bajocian.		Pokrovskian Suite (25–280m).	Tyumenian Suite (50m).		Omskian Suite (50m).
			Aalenian.			Omskian Suite (180m).	Togurskian Suite (0–120m).	—
		Lower	—	Chelyabinskian Series (1500m and more).				—
	Trias	Upper	—		Missing.	Missing.	Missing.	—
		Middle	—	Missing.				
		Lower	—	Turinian Series (800m).			—	—

The thickness of the Coal-bearing Formation in boreholes is 300–500m. It consists of dominant sandstones, amongst which lie, mainly in the middle and upper part of the section, seams of brown coals, reaching 14–18m and even 25–75m in thickness. Over the Coal-bearing Formation are variegated beds which are almost free from plant remains.

The depression is the only region where all the Upper Jurassic is represented by continental deposits—coal-bearing at the bottom and variegated at the top.

In a number of the regions of the Lowlands the Middle Jurassic is separated by breaks and unconformities from the Upper Jurassic. With the Middle Jurassic ends the first Mesozoic epoch of continental regime, while the succeeding Upper Jurassic and Valanginian are marine.

MARINE UPPER JURASSIC AND LOWER CRETACEOUS. These sediments are found over the whole area of the Western Siberian Lowlands and in places even overstep its margins. The sea transgressed along a wide front from the north covering the Lowlands well to the south of the railway. This was an open epi-continental sea of 200–400m in depth, evidently without islands and resembling the present-day Sea of Barents, but being somewhat larger. In the west it impinged against the eastern Urals and in the south it reached the slopes of the Palaeozoic highlands, not penetrating into the Kulundinian steppe, but in the east reaching to the slopes of the Precambrian highlands at the valley of the Yenisei.

Lithologically and faunally the Western Siberian marine Upper Jurassic and Lower Cretaceous are similar to the contemporaneous deposits of the Russian Platform. The basic distinction is the much lesser development of the condensed glauconite-phoshorite beds which are characteristic of the Russian Platform. This is explained by the fact that the Middle Russian Sea was in the nature of wide straits with strong currents, while the Western Siberian Sea was a bay, which was connected only to the north to an open ocean and had much slower currents. Amongst the Upper Jurassic and Lower Cretaceous Western Siberian sediments argillaceous deposits, normally as dark argillites, predominate. At the bottom of the succession there are also subordinate sandy and silty rocks. Total thicknesses are considerable; thus in Tyumen' (the west) the marine Upper Jurassic and Lower Cretaceous are 198m; in the south, at Barabinsk, they are 256m; in Omsk 298m; in the Kolpashev region (the middle Ob') 86m thick. In the central area the thickness is greater, being 277m in the region of Tara and 166m in Berezov, in the north-west. None of these figures include the continental Lower Cretaceous.

Sinking of deep boreholes in the Berezov district and around the lower Irtysh (Uvat) has led to the discovery of a clay formation 700m thick (Uvat), with oil shales, amongst the Upper Jurassic and Lower Cretaceous deposits. The oil shales (Maryanovian Suite) are probably the parent deposits of gas and oil discovered in the gentle anticlines of the Berezov district and to the south of it. The Berezov gas-deposit has been connected by a pipeline with Sverdlovsk and other towns in the Eastern Urals. The borehole sections are quoted by N. N. Rostovtsev (1956). In 1960 commercial oil was found near the village Sahim on the upper Konda, some 200km to the east of Ivdel' in a distant and isolated region of the Tyumen' province. The oil and gas are found in the sandy Vogulinian Formation, which lies at the base of the clays. The fauna (mainly Foraminifera) shows that the marine transgression in the Western Siberian Lowlands begins in the Callovian, continues as a sea in the Oxfordian and reaches the greatest depth (thickest clays) in the Kimmeridgian. In the Lower Volgian and Upper Volgian the marine regime was variable and so are the facies. In the south of the Lowlands the marine succession ends in the Valanginian by beds of the same age as the *Aucella* Beds in the north, but with a different fauna. The higher Lower Cretaceous horizons show red beds. In the north (Berezov) and lower Irtysh there are no red beds and the Lower Cretaceous deposits pass into the marine Upper Cretaceous.

CONTINENTAL UPPER AND JURASSIC LOWER CRETACEOUS. The constituent rocks consist of diverse sandy-clayey, often coal-bearing, deposits with bauxites and iron ores, and are found at the periphery of the Lowlands.

LOWER AND UPPER (CONTINENTAL) CRETACEOUS. These sediments were formed in the Hauterivian, Barremian, Aptian-Albian, and Cenomanian. This is the second Mesozoic epoch of a dominant continental regime. Almost all of the Western Siberian Lowlands were a plain, at times dry, when the red beds accumulated, and at others wet, when coals were formed. In the Hauterivian-Barremian a brackish embayment existed in the north-west and opened to the north (Saks, 1960). At the end of the Aptian and the beginning of the Albian the western part of the Lowlands sank and the sea penetrated to the south almost as far as the Turgai Depression. The Upper Albian-Cenomanian again became an epoch of a receding sea. It is interesting that the existence of the land was accompanied by lengthy and considerable downward movements. Consequently, sandy-clayey formations of a great thickness (up to 1200–1500m) came into existence. The successions in different regions of the Western Siberian Lowlands are diverse and consequently a number of suites and series of local significance have been distinguished (Table 24). The Hauterivian-Barremian red beds and Aptian-Albian coals have the widest distribution.

The Lower Cretaceous red beds (Variegated Formation, the Ilekian Formation) consist of variegated clays with beds of sometimes glauconitic sandstones. These clastic rocks are continental near-shore sediments of Hauterivian-Barremian age, varying in thickness from 60 to 700m. Their formation indicates a rise of land causing the appearance of continental deposits. The near-shore plain, where they accumulated, was nevertheless sinking slowly and continuously which allowed the accumulation of 500m of sediments produced under uniform palaeogeographic conditions.

The Lower Cretaceous coal-bearing rocks of Aptian-Albian age lie on the variegated clays, and consist of interbedded sandstones, silts and clays with seams of brown coal. In the east this group of sediments reaches a considerable thickness (280–300m), but its thickness decreases to 150–200m in the centre and 70–140m in the west. In the south-east the continental deposits are represented by grey or greenish sands and clays which are variegated towards the top. The continental deposits correspond to the Cenomanian and pass downwards into the Albian-Aptian coal-bearing rocks. This causes the formation of a single complex (Pokurian Suite) which reaches an enormous thickness of 600–800m and at Kolpashev as much as 1120m.

UPPER CRETACEOUS AND PALAEOGENE MARINE SEDIMENTS. The marine Upper Cretaceous is widely found over the whole of the Western Siberian Lowlands, although in the south-eastern part it passes laterally into continental deposits. In the north-east the marine deposits (Turonian-Maastrichtian) are found in the mouth of the Yenisei and in the north they are found in the basins of the Taz and the Pur and the lower Ob' (Fig. 95). Further to the south the outcrops of the marine Upper Cretaceous form a wide belt along the eastern slope of the Urals. Boreholes have shown them over the Lowlands to the meridian of Novosibirsk. As distinct from the Jurassic the Lower Cretaceous sea transgressed not only from the north, but also from the south through the Turgai and Orskian straits. The depression started in the Cenomanian, causing the formation of widespread near-shore plains with terrestrial, lacustrine, lagoonal and in places marine deposits. The first main transgression, drowning almost the whole of the Western Siberian Lowlands, occurred in the Turonian, forming in many places greenish and dark laminated shales, some 8–60m thick. Higher up lie sediments of Coniacian, Santonian, Campanian and Maastrichtian stages. The sediments are grey, greenish and dark calcareous and sandy clays with beds of gaizes (siliceous clays) and occasionally glauconitic sandstones. The fauna is rich and variable and most often with Foraminifera. The thickness varies. In the west, in Tyumen', it is 160m, in the south-east (Kolpashev) it is 210m, while in the centre (Khanty-Mansiisk) 265m, and in Berezov 223m.

PALAEOGENE. In the Maastrichtian in the north of the Western Siberian Lowlands the sea started to recede and in the Danian-Palaeocene it ceased to be connected with the Arctic Ocean. The Upper Cretaceous straits were succeeded by the Palaeogene enclosed sea, analogous to the present-day Baltic. Its dimensions gradually became smaller and it disappeared altogether at the beginning of the Oligocene. The sea was replaced by a vast plain with fresh-water and saline lakes. The Palaeogene sea was connected to the open ocean only in the south.

The Danian Stage has not been recognized in the Lowlands and the deposits overlying the Maastrichtian are referred to as Danian-Palaeocene, or more frequently Palaeocene, which is one way of getting round a complicated problem, although it is an obviously erroneous approach, since one cannot arbitrarily remove an internationally recognized stage.

The Palaeocene, Eocene, and Oligocene are divided into different suites (Table 25). The Palaeocene is conformable on the Upper Cretaceous, although there is a break along the eastern slope of the Urals. Manganese deposits are found associated with this break. In the west, nearer to the Urals the Palaeocene is usually represented by argillites, which are often siliceous. The Eocene consists of gaizes with diatomites towards the top. The Lower Oligocene is represented by greenish-grey clays with siderite nodules and fish teeth. The Middle and Upper Oligocene consist of continental sands. The Palaeogene fauna is relatively poor, with common Radiolaria, Foraminifera, lamellibranchs, and shark teeth. A similar succession is found in the north, on the Tazov Peninsula. The Upper Chalk is overlain by the Lower Eocene (Serovian Suite; 75m) of grey and laminated gaizes and clays. Overlying this is the Irbitian Suite (50–70m), belonging to the Middle and Upper Eocene and consisting of diatomaceous clays and diatomites. The Cheganian Suite (140m) of greenish, compact, laminated clays with shark teeth and lamellibranchs follows and belongs to the Lower Oligocene, while the Middle Oligocene (Atlymian Suite; 40m) is continental and consists of sands and clays with plant remains. The suite is covered by kaolinized sands (up to 150m) and the Anthropogene (Andreyev and Byelorusova, 1961).

In the east the siliceous clays are succeeded by ordinary clays and argillites, and sands and pebble-conglomerates become abundant. Nearer to elevations of Palaeozoic and Precambrian massifs the marine deposits are replaced by continental deposits of vast alluvial plains with coals and bauxites. The thickness of the marine Palaeogene in Cisuralia varies from 160 to 450m, being near Omsk 140–360m, in the east (Kolpashev) some 65–265m and in the north, near to the mouth of the Irtysh, 320–540m.

The succession of the Western Siberian Palaeogene can be correlated with the Turgai Depression, since they are deposits of the same marine basin. Nevertheless the ages of some suites are debatable. For instance, some consider the Cheganian Suite as Upper Eocene in age, considering that the continental regime starts in the

TABLE 25. *Correlation of Palaeogene Successions in the Western Siberian Lowlands with those of the Turgai Depression*

Stage	Western Siberian Lowlands (N. N. Rostovtsev, 1956)	Turgai Depression (V. V. Lavrov, 1957)	Northern and central parts of the Lowlands (S. B. Shatskii, 1956)	Southern part of the Lowlands (I. G. Zaltsman, 1956)
Upper and Middle Oligocene	Bottom part of the Irtyshian Series—Nekrasovian Suite of continental sands and clays. (50–200m).	Indicoterian Suite of pale sands and clays with bones of *Indicoterium*. Age: Middle Oligocene.	Irtyshian Suite of Neogene and Upper and Middle Oligocene age. Above, variegated clays and sands; below, grey and dark grey lacustrine-lagoonal and fluviatile clays and sands with brown coals. (100–250m).	Irtyshian Suite. Lower part—Nekrasovian Suite (Middle and Upper Oligocene, Lower Miocene) of lagoonal (marshy) and fluviatile coal-bearing and variegated clays and sands. (90–250m).
Lower Oligocene	Tavdian Suite of greenish, pyritous clays with beds of sands and a marine fauna. In places plant remains and lenses of brown coals. (60–150m).	Cheganian Suite of greenish, laminated, pyritous clays with marly concretions. The strata have a rich marine fauna and are sandy at the top. (40–100m).	Cheganian Suite (Upper Eocene and Lower Oligocene) of greenish, laminated clays with marine Foraminifera. (120–170m).	Cheganian Suite. In the north: marine greenish clays; southward, in Kulunda grey clays with lignite. Further south is missing. (80–170m).
Eocene	Makushimian Suite of clays and gaizes, varying grey to green. A marine fauna is present. Thickness: from 60–80m. up to 250–350m. Upper two-thirds consist of clays and gaizes of Eocene age. Lower third consists of Palaeocene and Danian strata with not infrequent glauconitic sands.	Saksaulian Suite. In the north: pale gaizes and silts, sometimes sandy. (50–60m). In the south: sands and silicified sandstones (sometimes muddy) and clays. Thickness: from 22–30 up to 120m.	Lyulinvorian Suite (Lower and Middle Eocene) of sandy gaizes-like clays with beds of glauconitic sands and a marine fauna. (190–230m).	Lyulinvorian Suite (Eocene) of marine gaizes-like clays; to the south of Kulunda pass laterally into continental deposits. (20–90m).
Palaeocene		Tasaranian Suite—at the base glauconitic sands with phosphorites; above, grey siliceous clays. A marine fauna is present. Thickness: 60–70m, but in places up to 240m.	Talitsian Suite (Palaeocene and Danian) of dark sandy clays with marine foraminiferids. (110–130m).	Klyuchian Suite (Palaeocene-Danian) of continental dark-grey clays and quartz sands occurs together with the Lyulinvorian Suite. (40–115m).

Lower Oligocene. The ages suggested in Table 25 are more probable. After the 1956 conference the study of the continental deposits from the Middle Oligocene upwards has advanced considerably. Instead of the Nekrasovian Suite (Table 25) three new suites have been distinguished and three suites were recognized in the Neogene. Consequently in the south-western part of the Lowlands and the Turgai Depression the following succession has been adopted (Decisions . . . 1959).

Middle Oligocene { Atlymian Horizon
{ Chiliktinian Horizon
Upper Oligocene—Znamenian Horizon
Lower and Middle Miocene—Aralian Horizon
Upper Miocene and Lower Pliocene—Pavlodarskian Horizon
Middle and Upper (?) Pliocene—Kustanaian Horizon

The Atlymian Horizon (50–70m) consists of pale quartz-sands with beds of clays and lignite towards the top. The Chiliktinian (Novomikhailovskian) horizon (80–100m) consists of greenish-grey and brown clays with seams of lignite.

The Znamenian (Byeshchenlian or Turtassian) Horizon (30–50m) includes the Chagraian and Naurzumian suites. Its lower part probably belongs to the top of the Middle Oligocene. In places white and grey clays and ferruginous sands are present.

NEOGENE. In the west of the Western Siberian Lowlands it begins with the Aralian Horizon (Lower and Middle Miocene).

Aralian Horizon—Greenish clays (20–70m) with marls and iron nodules. The clays are in places gypsiferous.

Pavlodarskian Horizon (Upper Miocene and Lower Pliocene). Grey and brown clays with beds of sand; a rich terrestrial fauna with *Hipparion*. Total thickness 30–80m.

Kustanaian Horizon (30m—Middle and possibly Upper Pliocene). Greenish and grey clays, marls and sands with a fresh-water fauna.

The Neogene is widely distributed in the central part of the Western Siberian Lowlands. The Miocene is divided into two suites: the lower (50–70m), consisting of light grey, sandy, laminated sediments, and the upper (80–100m) of interbedded clays, seat earths and sands with characteristic lenses of brown coal and lignite.

The Pliocene, which is up to 65m thick, consists at the bottom of brownish-grey, sandy-clayey rocks, overlain by bluish-grey loams, sandy clays and sands, which are capped by variegated clays with numerous calcareous concretions. The Neogene is normally hidden by the Anthropogene, but in places, as for instance the middle Tyma, a right tributary of the Ob', it is found in river-cliffs (yars) as brown clays with seams of brown coal. The leaf remains and pollen indicate their Miocene age.

ANTHROPOGENE. The Anthropogene consists of variable deposits of continental origin. Only in the extreme north along the sea-shore are there deposits of the Boreal Transgression, being sands and clays with a marine fauna. In the north approximately down to the 60° latitude there are widely developed glacial deposits and the boundary of the glaciation is shown on maps. The deposits include ground and terminal moraines, kames, and fluvioglacial sediments. The marine ground moraine represented by loams is particularly common. The study of the boulders has shown that the ice sheet descended off the Northern and Arctic Urals, Taimyr, and the Norilsk Uplands (Putoran Mts., Fig. 80). The central part of the northern Lowlands was free from the ice sheet.

In the north and north-east current-bedded sands are widespread, representing the deposits of deltas, and the lower Ob' and Yenisei, which often changed their direction. Thus, for instance, the Ob' entered the Kara sea via the Baidarats inlet and the Yenisei via the Gydanian inlet. At certain times the Ob' and the Yenisei joined each other in their lower reaches. The thickness of the alluvial deposits on the watersheds was 5–20m and in valleys up to 250m, and in the north 400m.

South of the boundary to the glaciation, the Anthropogenic deposits are diverse and in places as much as 100m thick. The deposits involve sands, loams, clays, peats, and rarer gravels, thus being the products of rivers, lakes, large marshes, and flat watersheds. The great development of the impermeable loams is the cause of the existence of the present-day marshes. A notable factor in the formation of the soils is the uninterrupted forest—the taiga. The glacial structures, identical to those of the Russian Platform (p. 148), are of some interest. These occur as separate hills, rising above the Anthropogene sediments and consisting of Mesozoic and Palaeogene rocks. The Mt. Samarova, near Khanty-Mansiisk, consisting of deformed Palaeogene gaizes and the occurrences of the Jurassic on the Yugan serve as examples. They were at first considered to be tectonic structures, but boreholes have shown Anthropogenic boulder clays (moraines) underlying them.

The subdivisions of the Anthropogene have been lately worked out in some detail, which was achieved by means of boreholes. A clear scheme was established at the Regional Conference on the unified stratigraphy of Siberia. In the papers presented at this conference and published in the 'Trudy' (1957) successions in the different parts of the Lowlands have been published. Such successions show differences from place to place, but the general succession is, as a rule, sustained and has been adopted in the 'Decisions' (1959), as follows (in ascending order):

Lower Anthropogene (Lower Pleistocene)
1. Pre-glacial deposits.
2. Sediments of the Yarian Glaciation correlated with the Okaian Glaciation of the Russian Platform—morainic clays, primary or redeposited.

Middle Anthropogene (Middle Pleistocene)
3. Interglacial (Tobol'skian) lacustrine and alluvial deposits.
4. Moraines of the maximal Samarovian Glaciation (the Dnyepr Glaciation).
5. Interglacial (Shirtaiian) deposits.
6. Boulder clays and sands of the Tazovian Glaciation (Moscovian Glaciation).

Upper Anthropogene (Upper Pleistocene)
7. Interglacial Kazantsevian, Sanchugovkian and Messovian deposits.
8. Zyr'yankian moraine, either primary or redeposited.
9. Karginskian interglacial alluvium and peat.
10. Sartanian Glaciation. In the extreme north there are loams and further south the alluvium.

Recent (Holocene)
11. Peat epochs of warmer climate. Eolian sands.
12. A number of terraces. Contemporary flora and fauna.

This is a generalized succession and it is not anywhere fully developed. In southern regions the Sartanian Glaciation is not developed, then further south other glaciations are not manifest, so that in the extreme south of the Lowlands there are no glacial or marine sediments.

Conversely, in the north the marine deposits are dominant and the moraine deposits are subordinate and represented by marine-glacial sediments. Numerous boreholes (Reinin, 1961) show a thick Yamalian Series in overdeepened valleys. The series consists of clays and sands with morainic marine-glacial boulder clays. It is Lower and Middle Pleistocene in age and has a marine fauna of Foraminifera, diatoms and lamellibranchs, and is divided into three suites, being from below upwards, the Poluian (50–60m), Kazymian (50–60m), and Salekhardian (up to 200–220m). The marine transgression reaches its maximum in Salekhardian times, when sea-level was 150m above the present and almost the whole of the northern Lowlands was covered by sea. From the Upper Pleistocene and in the Holocene the sea recedes and alluvial, lacustrine, and glacial deposits become just as widespread as the marine. Since the Salekhardian Suite has a marine fauna characteristic of the Sanchugovkian deposits it is possible that a part of the Salekhardian Suite belongs to the Upper Pleistocene.

In individual regions detailed investigations show that although the Anthropogene sediments have been successfully investigated there are still considerable problems remaining. For instance the correlation of regions of widespread glaciation with the regions of partial glaciation or the unglaciated regions still has to be done.

TECTONICS

Mesozoic and Cainozoic Structure

The structure of the Mesozoic and Cainozoic is evident from the sections prepared by the Boreholes Branch of VNIGRI. The first section is arranged across the central part of the Lowlands, from the lower Ob', to the middle Yenisei. Malyi Atlym is situated somewhat to the north of the mouth of the Irtysh, Pokur is near the mouth of the Vakha, Lar'yak is on the middle Vakha, the Yelogui is a river some 150m west of the confluence of the Bakhta and the Yenisei. The section is based on four standard (stratigraphic) boreholes (Fig. 95).

The next section (Pudino-Kas-Yenisei) cuts the south-eastern half of the Lowlands, 350km to the south of the former (Fig. 96).

The third section (Maksimkin-Yar-Chulym) is placed in the south-eastern corner of the Lowlands, within the Chulym-Yenisei Trough (Fig. 97). Vertical dimensions are greatly exaggerated and dips at angles of less than 1° rather than the apparent 60° are encountered. All deposits are almost horizontal which is a surprising feature of the Meso-Cainozoic Lowlands. The second feature of the Lowlands, obvious in the cross-sections, is a surprising constancy of thicknesses over large areas. It is accompanied by the generally observed sharp lessening of the thicknesses towards the margins of the Lowlands, and this is accompanied by a change from marine to continental facies. In Yelogui, Kas and Byelogorka there is only the continental Meso-Cainozoic.

The third feature of the Meso-Cainozoic rocks is the upward reduction of dips. For instance the Jurassic deposits often have dips of 3°–5° on the flanks of the tectonic structures, while the Palaeogene horizons are measured as minutes, or rarely up to 1°. Thus the Western Siberian Lowlands have the form of a huge shallow cup with an irregular bottom. At its edges and bottom there is a whole host of quite large gentle elevations and depressions (Fig. 98). These elevations and depressions, often and especially over young Hercynian parts of the basement, are elongated, thus inheriting the tectonic forms existent in the folded basement.

Amongst such structures the North-Sos'vinskian Arch is separated from the Urals by the Lyapinskian Depression. Among other large structures the Khanty-Mansiisk Trough, which includes the middle Ob' and the lower Irtysh, should be mentioned, since it is one of the deepest troughs in the Western Siberian Lowlands. There are also the Ust' Yenisei Trough, the Nazym Trough and the Ket'-Tym Trough, which is in the south-east of the Lowlands and is filled by continental deposits. Lastly the Nizhnevartovian Arch is being at present investigated in the search for oil.

The aforementioned large structures apart, which are several hundred kilometres

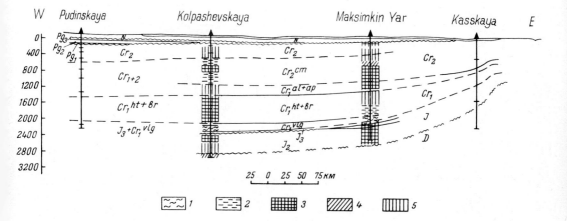

FIG. 96. Section across the south-eastern part of the Western Siberian Lowlands.

1 — marine deposits; 2 — marine shallow water deposits; 3 — alternating marine and continental deposits; 4 — lagoonal deposits; 5 — continental deposits.

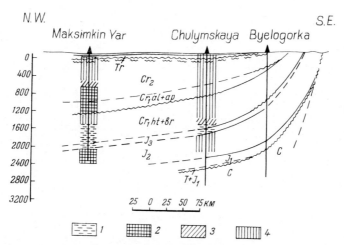

FIG. 97. Longitudinal section through the south-eastern part of the Western Siberian Lowlands.

1 — marine deposits; 2 — alternating marine and continental deposits; 3 — lagoonal deposits; 4 — continental deposits.

long, smaller normally rounded domes of 5–10km in diameter are also known. Gas and oil deposits are associated with these. There are also long tectonic ridges of 70–150km and more in length and not more than 7–20km wide.

THE HISTORY OF THE TECTONIC MOVEMENTS

The history of the development of the Western Siberian Lowlands is determined by two factors: tectonic movements, and contemporaneous sedimentation. Until the end of the Permian period the Lowlands did not exist and in its place there were folded Precambrian and Palaeozoic provinces identical to the adjacent Hercynian fold regions of the Urals and Taimyr; to the Caledonian regions of Kuzbas and

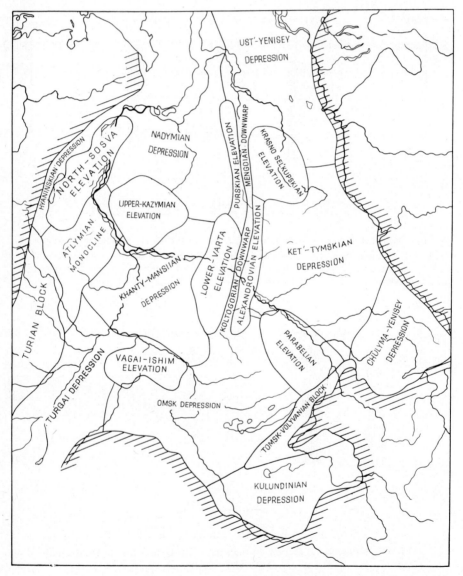

Fig. 98. The distribution of elevations and depressions of the basement in the Western Siberian Lowlands. According to V. D. Nalivkin (1960).

the Kazakh steppes; and to the Precambrian regions of the Siberian Platform. Consequently, the basement of the Lowlands should be considered as a complex structure, which includes all types of folded regions.

The most probable structure of the basement to the Lowlands has the following elements. The north-eastern part is a Precambrian platform formed during the Baikalian Orogeny (Baikalids). Before this orogeny it consisted of two parts: the more ancient western Lower Proterozoic platform and the eastern part parallel to the valley of the Yenisei, which was an Upper Proterozoic geosyncline. The Upper Proterozoic, Baikalian Orogeny affected this geosyncline and welded the western platform to the Siberian Platform. Some geologists think that the Upper Proterozoic geosyncline continues to exist in the Lower Palaeozoic and was affected only by the Caledonian Orogeny. This opinion cannot be considered correct since the structures formed by the Upper Proterozoic (Sinian) are typical geosynclinal structures and their formation was accompanied by a considerable metamorphism and intrusive massifs. The unconformably overlying Cambrian is in places unfolded and in others forms large, flat platform structures, not accompanied by metamorphism and unaffected by granitic intrusions. The folds affecting the Devonian and Lower Carboniferous of the Kureika and adjacent regions cannot be considered as geosynclinal either. These are also typical platform structures.

The Caledonian Orogeny caused the enlargement of the Precambrian Platform in the north, west and south. The Hercynian Orogeny enlarged the platform further still by adding new fold belts, although in places the older Caledonian fold belts have been reworked. After the Hercynian Orogeny all the basement to the Western Siberian Lowlands became land which connected the Russian to the Siberian continents.

At the end of the Permian and the beginning of the Trias important events caused the initial formation of the Western Siberian Lowlands. Along the present boundaries to the Lowlands large faults or series of faults occurred and along these faults the whole region of the Lowlands began a downward movement. The movement was accompanied by mass-fissure eruptions of the Siberian Traps. Their existence has been demonstrated in the north-east, the west and the south of the Lowlands. To these traps belong the Tyumen' and Chelyabinsk lavas and sills.

The subsequent Meso-Cainozoic history of the Western Siberian Lowlands can be divided into five epochs—three continental and two marine.

The first epoch includes the Upper Trias, Lower and Middle Jurassic and is a period of accumulation of coal-bearing rocks at first accompanied by volcanicity.

The second epoch (Upper Jurassic and Valanginian) involves a large marine transgression which covered practically the whole of the area of the Lowlands. The transgression was caused by an acceleration of downward movements.

The third epoch (upper Lower Cretaceous) commenced with a sharp upward movement, causing an almost complete regression of the sea and the formation of a vast plain where variegated and coal-bearing formations accumulated. The great thickness of these deposits is associated with a slow downward movement, which balanced the accumulation of sediments, brought in and distributed by numerous rivers.

The fourth epoch (Upper Cretaceous-Palaeogene) was an epoch of a new transgression of sea arriving from the south, and caused by an acceleration of downward movements. The south-eastern part of the Lowlands remained above sea-level and continental deposits continued to accumulate there.

The fifth epoch (end of Oligocene, Neogene, and Anthropogene) was governed by a continental regime, which continues at present. Only in the extreme north the Anthropogenic movements provoked the Boreal Transgression which still exists as the Kara Sea.

As a result of tectonic movements and a continuous supply of the products of erosion a great thickness of friable rocks, which are horizontal and almost un-metamorphosed, has accumulated. This mass represents a complete cover some 2000–3500m thick over an area of 1500–1800km wide and 2000–2500km long. The volume of these rocks is $(5–7·5) \times 10^6 km^3$. The accumulation of these sediments was permitted by the balance between movements and deposition. This kind of down-ward movement is widely known in the history of the Earth's crust. The result is that at the bottom of a sea some 30–50m deep some 300–400m of sediments get accumulated without any substantial variations in the depth of water.

The history of the Western Siberian Lowlands in the Mesozoic and Cainozoic is a typical example of prolonged balanced downward movements, interrupted by short-lived and fast upward and downward movements causing sharp changes in palaeogeographic conditions, but not affecting the accumulation of sediments.

USEFUL MINERAL DEPOSITS

The useful mineral deposits are divided into two groups: those found in the basement and those in the cover. It can be assumed that the rich resources of the basement are very great and are of the same type as the mineral deposits of the Eastern Urals, the North-Eastern Kazhakhstan and the west of the Siberian Platform. At present, however, such deposits are almost unknown in the Lowlands even where the base-ment lies at a shallow depth. Only Permian coals can be mentioned, which have been found in the Achinsk region underlying the Jurassic Coal-bearing Formation. The Permian coals are almost identical to those of the Kuznetsk Basin and the Siberian Platform.

The useful mineral deposits of the Meso-Cainozoic cover are also comparatively little studied since they are not exposed at the surface. However, what little is known suggests their diversity, abundance and value.

METALLIFEROUS ORES

Ferruginous metals (iron and manganese) form especially important deposits. The deposits of oolitic iron-ores are so great that it is more correct to refer to them as ore-fields. The first of these fields—Kustanaian (Ayatian) and Lisakovian—were found after the last world war in the north-western part of the Turgai Depression. The ores are sedimentary and lie amongst Upper Cretaceous, marine, near-shore deposits. Their origin is identical to the Kerch ore-field, although the latter ores are of Pliocene age. Another ore-field (Kolpashevian) is situated near the village Kol-pashev on the Ob', to the north of Tomsk. Its age is again Upper Cretaceous and the ores are of the same type and origin as in the Ayatian field. The often quoted re-sources of the Ayatian field are estimated as 12×10^9 tons ($3·6 \times 10^9$ tons of work-able ore). The Kustanaian and the Kolpashevian fields are situated at the opposite edges of the south of the Western Siberian Lowlands. It is quite possible that the ore will also be found in between.

Lately a so-called Western Siberian ore-field in the eastern parts of the Lowlands has been established. The field has a north–south length of 1500km, from the Trans-Siberian road to the Yenisei inlet. At first the Kolpashev deposit and then a whole lot of others were discovered. For instance, the so-called Bokchar deposit is situated some 250km north of Novosibirsk. The iron-ore is found at four levels: The Lower (Narymian) is Campanian-Cogniacian in age (86 m.y.), the second (Kolpashevian) is Maastrichtian in age (84 m.y.), the third (Chigorinian) is Danian in age (72 m.y.) and the fourth (Bakcharian) is Palaeocene (56 m.y.) in age. All ores are sedimentary, associated with a near-shore province of Upper Cretaceous and Palaeogene seas (Klyarovskii, Dmitriev and others, 1961).

Manganese deposits have been known for a long time and occur as oolitic ores amongst the Palaeocene near-shore marine sediments. The ores are found on the eastern slope of the Urals, to the north of Ivdel'. The largest deposits are often considered as being within the Ural province, but in fact these deposits are marginal Palaeogene sediments of the Western Siberian Lowlands. Undoubtedly similar deposits can exist in other regions and margins of Palaeogene seas.

Bauxite deposits are known from the Lower Cretaceous lacustrine sediments on the eastern slope of the Urals and in the centre of the Turgai Depression. Deposits of a similar type occurring under identical conditions exist in other regions of the Western Siberian Lowlands, indicating the possibility of finding new deposits of bauxite.

Along the south-eastern margins of the Lowlands there are known bauxites of the same type, but associated with Palaeogene continental deposits. Such a wide distribution of bauxites widens the possibility of finding new deposits in lacustrine and marshy deposits of Jurassic, Upper Cretaceous and Neogene age as well.

FUELS

The Upper Triassic and Jurassic coals are important and are being actively worked along the margins of the Western Siberian Lowlands: the eastern slope of the Urals, the Turgai Depression and the Achinsk region. Small deposits of Cretaceous, Palaeogene and Neogene lignite are even more numerous. In the north-west of the Lowlands, along the lower Ob', oil shales and gas have been found in the Upper Cretaceous and Upper Jurassic marine sediments. The deposits of oil shales in Berezov are large. The gas is concentrated in basal sandstones (1–37m) of the Upper Jurassic. One hundred kilometres to the south-west of Berezov in the Tyumen' region the Igrim gas-deposit has been found. It is much larger than that of the Berezov region. From 1953 to 1961 ten gas deposits have been found in the Berezov-Igrim regions (Prospects . . . 1961).

The industrial oil-field found in 1960 on the Konda (left tributary of Irtysh), near the village Shaim, at some 200km east of Ivdel', has an exceptional importance. Together with the Berezov gas deposits it suggests a vast gas-oil field, which has even been called the third Baku. In 1961, yet another oil deposit (Megionian) has been found in the Surgutian region on the Ob'. This increases the size of the gas-oil region further still.

NON-METALLIC DEPOSITS

The deposits are variable, numerous, of local significance. In conjunction they are valuable, conditioning the construction of various branches of industry which is being widely developed in the south of the Lowlands.

Part III

PALAEOZOIC GEOSYNCLINES

6

THE URALIAN GEOSYNCLINE

INTRODUCTION

Those geosynclines which in the Palaeozoic were converted from marine basins into fold-ranges are known as Palaeozoic geosynclines. These regions have been widespread, of long duration as marine basins, and show strong manifestations of the Hercynian and Caledonian Orogenies. As a result the following features have been developed:

1. The juxtaposition of folded Palaeozoic and non-folded or weakly folded Mesozoic and Cainozoic sediments.
2. Strong unconformities between the Palaeozoic and the Mesozoic.
3. Sharp differences in the nature and grade of metamorphism of the Palaeozoic and Mesozoic.
4. The formation of large synorogenic Palaeozoic intrusions and the absence of such intrusions in the Meso-Cainozoic.
5. Preponderance of Palaeozoic ore-genesis.

In the territory of the U.S.S.R. the following Palaeozoids are recognized: (1) the Uralian Geosyncline; (2) the western part of the Arctic Geosyncline; (3) the Angara Geosyncline—Central Kazakhstan, Altai, Kuzbas, the Sayans and Southern Transbaikalia; (4) northern and central ranges of Middle Asia.

The Urals are a typical Palaeozoic geosyncline, intensively deformed by the Hercynian Orogeny and showing a great development of Hercynian intrusives and Hercynian ore-genesis. At the end of the Palaeozoic the Urals became a folded mountain range and the subsequent Mesozoic and Cainozoic seas only penetrated to its foothills.

OROGRAPHY

RELIEF. In terms of their relief the Urals are divided into three north–south trending regions: Cisuralia, the Urals Mts., Transuralia (Fig. 99).

Cisuralia. This is a geographic region continuous with the Russian Platform and has certain common features with it. Its typical feature is the existence of ridges, highlands and separate mountains consisting of Lower Permian and Upper Carboniferous limestones and the Kungurian gypsum.

The Ural Mts. stretch north–south and represent a system of isolated, narrow, low, craggy or forest-covered ridges and ranges situated in parallel and separated by wide and flat ancient longitudinal valleys. The transverse valleys are young, narrow, deep, with precipitous sides which impart to them a canyon-like appearance (Fig. 100). In the Southern Urals the number of parallel ridges reaches 6–8 and their average heights vary from 900 to 1200m while separate peaks reach 1638m (Mt Yaman-Tau) and 1586m (Mt. Iremel'). It is interesting that the watershed ridge, Ural-Tau (tau=mountain), is lower (1000m) than the ridges to the west of it, which are cut by the transverse valleys. This is explained by recent block movements and will be discussed.

From the Karatau ridge (Black Mts.) the strike of the western ridges turns sharply to the north-east and then to the north-west. In the process the ranges are deflected against the Ufa Plateau, a part of the Russian Platform. The ranges bending round the plateau are known as the Ufa amphitheatre.

The Central Urals begin from latitude 55°N, passing through the Chelyabinsk and Zlatoust and continue to the edge of Timan where the low Polyudov Ridge deviates from the Urals. Along latitude 60°N the central part of the Central Urals is the lowest and consists of 2–3 ridges so weakly developed that the passage through by rail from Europe to Asia is almost unnoticeable. The average heights here are no more than 600–750m. This low region is not extensive and further to the north the

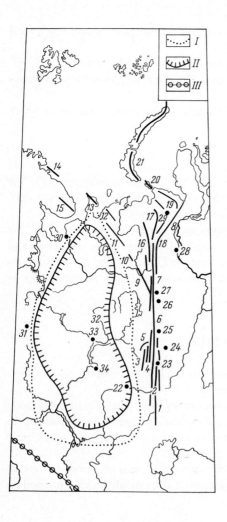

I. Boundary of the Tatarian Depression.
II. Boundary of the Kazanian Sea.
III. Boundary of the Tethys.

1 — the Mugodzhars; 2 — the watershed crest in the Southern Urals known as the Ural-Tau; 3 — the Alatau range; 4 — the Zilmerdak range; 5 — the Karatau range; 6 — the Central Urals; 7 — the Northern Urals; 8 — the Arctic Urals; 9 — the Polyudov Ridge; 10 — the Dzhemim-Parma range; 11 — Central Timan; 12 — Northern Timan; 13 — the Kanin Peninsula; 14 — the Rybachii Peninsula; 15 — the Kola Peninsula and the river Ponoy; 16— the Pechora Ridge; 17 — the Chernyshev range; 18 — the Sablya Massif; 19 — the Pai-Khoi; 20 — the island of Vaigach; 21 — Novaya Zemlya; 22 — Orenburg; 23 — Magnitogorsk; 24 — Chelyabinsk; 25 — Sverdlovsk; 26 — Serov; 27 — Ivdel'; 28 — Berezov; 29 — Vorkuta; 30 — Archangelsk; 31 — Moscow; 32 — river Vyatka; 33 — Kazan; 34 — Kuibyshev.

Fig. 99.

The Orography of the Urals and Timan. The Kazanian Sea and the Tatarian Depression are outlined.

ridges gain height, their number increases to 4–5 and the height reaches 1200–1500m.

The Northern Urals begin from the Polyudov Ridge and continue northwards to the Sablya Massif, which is the highest in the Urals with peaks of 1894m (Mt. Narodnaya) and 1425m (Mt. Sablya). The Northern Urals again form a system of longitudinal ridges and ranges which in 3–4 parallel rows pass into each other.

The Arctic Urals begin at the Sablya Massif and the strike turns sharply to the north-east towards Taimyr. The main features of the Northern Urals is the development of modern and ancient glaciations. The present-day glaciation has a relict nature. The ancient glaciation was of larger dimensions and has left behind large cirques.

In the southern part of the Arctic Urals and to the west of it there is a low branch, known as the Chernyshev Ridge, which continues to Cap Sin'kin and the island of Dolgii. In the northern part the Pai-Khoi range, which continues to the

FIG. 100.   A canyon-like valley of the Zilim in the Southern Urals.
Photograph by V. D. Nalivkin.

island of Vaigach and Novaya Zemlya, also branches away from the Urals. The axial part of the Urals does not, contrary to former views, bend into Pai-Khoi, but continues to the north-east, disappearing at the Baidarats inlet under a cover of Mesozoic and Cainozoic sediments of the Yamal Peninsula to a depth of no less than 6km.

Transuralia is a region of Palaeozoic and Precambrian outcrops and is situated to the east of the Ural Mts. In the Southern Urals this zone is quite wide (200–250km) and is a hilly region. Its homogeneity is disturbed by higher elevations of granites and other massive rocks and also by wide flat valleys of slow, small rivers. In places there are large numbers of lakes, which are usually small and irregular in outline. In the Central Urals the width of Transuralia is much smaller and the Palaeozoic outcrops are covered by an unbroken cover of Tertiary and Anthropogenic deposits of the Western Siberian Lowlands. In the Northern and Arctic Urals Transuralia is almost absent and the Lowlands come directly against the eastern slope of the mountains.

Landscapes. The change in the landscapes from south to north is of some interest. The southern termination of Mugodzhar is a typical semi-desert and even the mountains have no forest. To the north the semi-desert zone is succeeded by steppes and pastures, the mountains acquire trees which gradually become a continuous forest. Lastly the forest develops on the slopes and ultimately Cisuralia, the Ural Mts. and Transuralia are entirely covered by a continuous and virgin forest, which gradually passes into the Siberian taiga. The forest zone covers the greater part of the Central and Northern Urals. Still further to the north in the Arctic Urals the forest becomes more and more open, the mountains, bare and craggy. On the slopes and foothills small woods of dwarfed, distorted trees appear and the zone of forests yields to the zone of tundra which continues to the shore of the Arctic Ocean, or more correctly, to the seas of Barents and Kara.

RIVER SYSTEM. The river system is well developed and the rivers are numerous and of comparatively small size. On the western slope they are fuller and in the mountains have rapids and waterfalls, while on the eastern slope there is much less precipitation and the rivers are quieter. In the south, in Transuralia, in the summer, they tend to dry up. The rivers of the western slopes are related to three major rivers: the Pechora, the Kama, and the Ural.

In the basin of the Pechora the Usa is the largest river which drains the Arctic Urals, while to the south of it are the Shygor and the Ilych.

In the basin of the Kama the Vishera, the Kos'yeva (running across the Kizel coal-field) and the Chusovaya—the main river in the Central Urals—are large. Further to the south the rivers are small tributaries of the Byelaya and amongst these the Ufa (Bashkirian A.S.S.R.), the Sim, the Inzer, the Zigan can be mentioned. The Byelaya rises on Yaman-Tau, then flows to the south, sharply turns to the west, cutting across the Urals and appears in Cisuralia as a full river which turns to the north and joins the Kama. The river Ural starts not far from the source of the Byelaya, also flows first to the south and then cuts the Urals, acquires the waters of the tributary Sakmara and turns to the south, towards the Caspian Sea. The western slope of the Mugodzhar is drained by the Emba, the eastern slope by the Irgiz.

The greater part of the rivers on the eastern slope of the Southern and Central Urals belongs to the system of the Tobol, while in the Northern Urals the rivers are parts of the system of the Sos'va.

THE FORMATION OF THE RELIEF IN THE URALS. The principal factor in the formation of the relief is the sharp intensification of river erosion during the post-Akchagylian epoch. In the Arctic Urals the action of the large Anthropogenic glaciers was also important.

The intensification of river erosion is reflected in deep down-cutting (up to 200m and more) of the present-day river valleys, especially of the transverse rivers (Fig. 100) into the ancient peneplain surface with its typically clayey crust of weathering. The down-cutting can be demonstrated by means of remains of river terraces at levels up to 120m above the present courses of rivers, by the youthful profile of the valleys, the absence of meanders and by a much greater height of the western ridges in comparison with the watershed ridge of Ural-Tau.

These phenomena indicate young block faulting of Palaeozoic massifs in the regions to the west of the watershed. Movements of the order of 600–800m (the maximum) occurred in the Southern and Northern Urals, but were not active in the lower, Central Ural region. Thus the present-day relief of the Urals represents an ancient pre-Akchagylian peneplain which has been broken and elevated, thus rejuvenated, as a result of young block movements and intensified river erosion, and in the Arctic Urals, glacial erosion. The block movement occurred at the end of the Pliocene and in the Anthropogene.

STRATIGRAPHY

An exceptionally full development of fairly thick Palaeozoic deposits of geosynclinal type with a rich and diverse fauna and flora is a characteristic feature of Uralian stratigraphy. Mesozoic and Cainozoic deposits are widespread at its margins, penetrating into its centre only along the valley of the Ural, between the Urals and the Mugodzhar.

METAMORPHIC ROCKS

In the axial part of the Ural Mts. and especially along its watershed metamorphic and crystalline rocks are widely developed. The complex tectonics and absence of fossils makes their subdivision and dating difficult. Consequently, formerly they were known as the 'Metamorphic Formation' and were denoted by the letter 'M'. In the last few years the Ural geologists have succeeded in subdividing and dating this formation. Not everything, however, is as yet clear and the same rocks in the Southern Urals that are marked Cambrian and Proterozoic as are marked as Palaeozoic in the Northern and Arctic Urals. At present both points of view can be accepted as working hypotheses, which need further investigation.

Comparison with the Russian and Siberian platforms indicates that the metamorphic rocks include palaeontologically dated Lower Palaeozoic, but in small amounts. The greater part of the formation is Upper Proterozoic and in particular the Sinian (Riphaean) complex. A small fraction of the formation belongs to the Lower Proterozoic or even Archean.

HISTORY OF INVESTIGATIONS. In the middle of the last century all the ancient formations in the Urals were attributed to the Silurian System, owing to the discovery of occasional rare Silurian fossils. In 1882 the newly organized Geological Committee began the first systematic regional survey of the Urals on a 'ten verst' scale (1:420 000). The survey employed a number of leading geologists including the great Russian geologist F. N. Chernyshev who was at that time a young graduate of the Mining Institute. Soon, other faunas were found in the ancient formations, subjected to detailed studies and in a monograph by Chernyshev, it was demonstrated that these faunas were not Silurian, but Devonian. This unexpected discovery caused a change of views and all the ancient rocks became accepted as Lower Devonian even when they were unfossiliferous and metamorphosed.

The 'Devonian' idea continued up to 1930 when a regional survey on twice the former scale was attempted. The greater detail of this work caused within the first few years the discovery of new faunas, including Silurian, Ordovician and Cambrian. The re-interpretation of older results indicated that many of the fossils were Silurian rather than Devonian. Then it was demonstrated that the palaeontologically dated Cambrian and Ordovician, were unconformable on some older, more metamorphosed rocks. These latter could only be Proterozoic and were shown pink on geological maps. A greater part of the metamorphic rocks thus was included in the Precambrian.

The latest, most detailed surveys have verified in the Southern Urals the existence of the Middle and Lower Palaeozoic and Upper and Lower Proterozoic. In the Central Urals an even more complicated situation has been found. It has transpired that the metamorphic rocks of the watershed, when traced to the west, become less metamorphosed and have Ordovician fossils, although lower horizons of these rocks have been arbitrarily assigned to the Precambrian. Thus it was found that in the Central Urals all the metamorphic rocks are Lower Palaeozoic. The cross-sections of the Northern and Arctic Urals is very similar to that of the Southern Urals and together with the Lower Palaeozoic it has a widely developed Proterozoic. As a result of all these investigations it has been established that the ancient rocks have a complex structure and include Devonian, Silurian, Ordovician, Cambrian and Upper and Lower Proterozoic rocks.

PRECAMBRIAN

Archeozoic

The Upper and Lower Proterozoic and possibly the Archeozoic have been so far established. There is no definite evidence of the presence of the Archeozoic demonstrably of a higher metamorphism than say the overlying Proterozoic. However, the vast development of the Proterozoic renders the presence of the Archeozoic possible. It is possible that the Taratashian Suite described below is Archeozoic.

Proterozoic

In the Southern Urals the Ordovician fossil-dated sediments are transgressive over barren older rocks of great thickness. Many workers consider the upper suites of the latter as Cambrian and the lower as Proterozoic. M. I. Garan', who has studied these rocks for many years in detail, regards them as entirely Precambrian and this point of view is probably correct.

Let us consider the succession suggested by Garan' on the basis of his investigations in the northern part of the Southern Urals, between the Bakal and Kusinskian factories. It is in ascending order:

Taratashian Suite. Metamorphic schists, micaceous and ferruginous quartzites (jaspilites, haematite and magnetite quartzites), orthogneisses and paragneisses, migmatites, amphibolites, schistose effusives and tuffs (about 2000m).

Aian Suite (1900–2500m). Polymictic conglomerates at the bottom with pebbles of Taratashian Suite. Arkosic quartzites with sandy slates and slates towards the top.

Satkian Suite (2250–2400m). Diverse dolomites which are siliceous and sandy below, at the top pure and bituminous. There are also associated deposits of magnesite of the Satkian region. Calcareous algae have been found.

Bakalian Suite (1100–1200m). At the bottom slates and sandy slates, At the top amongst the clay slates, two formations of carbonate rocks (limestones, dolomites and siderites), each about 100m thick. Some of the iron ores of Bakal, the best in the Urals, are associated with these formations. There is also an horizon with magnesite.

Zigalginian Suite (200–250m, 600–1500m in the south). At the bottom a bed of quartz, quartzite and slate pebble conglomerate. This is overlain by quartzites and quartzite sandstones forming the top ridges of Zigal'ga, Suka, Shuida, Yvak and others. Above these are quartzite sandstones with black carbonaceous slates.

Zigazino-Komarovskian Suite (650m). Diverse slates (clay-, sandy and quartzose), sandstones and dolomites. Brown iron ores are found.

Avzyanian Suite (1000–2000m). Thick formations of dolomitized limestones interbedded with thinner formations of mica-chlorite-quartz sandstones and slates. Iron ore deposits of Katav-Ivanovskian and the iron ores and magnesites of the Byeloretskian regions.

Zil'merdakian Suite (800–900m). Arkosic sandstones and conglomerates at the bottom, sandy slates and slates with bedded quartzite sandstones at the top. Forms the summit of the Zil'merdak range.

Katavian Suite (400m). Variegated bedded, so called banded, marls and limestones. Calcareous algae *Collenia buriatica* Masl., *Conaphyton lituus* Masl. Forms large cliffs (Fig. 101).

Inzerian Suite (400–1300m). Fine and homogeneously grained quartzitic and arkosic, sometimes calcareous, sandy rocks with glauconite in places.

Min'yarian Suite (800m). Limestones and grey dolomites which are sometimes siliceous and oolitic. Some varieties of carbonates contain the alga *Collenia buriatica* Masl.

Higher up lies the Ashaian Suite (1000–1400m), which is Upper Proterozoic and will be described later (p. 315). The ancient formations of more southerly regions have been studied in detail by O. P. Goryainova, E. A. Fal'kova and K. A. L'vov (1939, 1956). They arbitrarily assign all the suites lower than the Zil'merdakian to the Proterozoic and the Zil'merdakian and subsequent suites to the Cambro-Ordovician. Their stratigraphic sequence in general coincides with that formulated by M. I. Garan', but it differs by the greater details of the succession and in certain other features. The two suggestions are at present most widely used. The general maps record the succession of Goryainova, Fal'kova and L'vov. The lower suites are Precambrian (Proterozoic) and the upper five Lower Palaeozoic. At present their inclusion in the Upper Proterozoic (Sinian, Riphaean) has been proved. On the tectonic map of Bashkiria (by A. I. Olli and V. A. Romanov, 1959) all suites beginning with the Ashaian and up to Zigal'ginian are attributed to the Riphaean structural level. More ancient suites, including the Aian have been included by them (1960) into the Upper Proterozoic.

The problem of the age of the ancient rocks in the Southern Urals is exceptionally interesting and important, but cannot be easily solved. In the opinion of the
present author the point of view, represented by Garan' and Olli, that all the suites
are Precambrian, is more correct.

One of the most significant changes in our concept of the geological structure of
the eastern slope of the Southern Urals has been introduced by Soviet geologists and
involves the metamorphic rocks of the Tobol-Ural watershed and more easterly
regions. Until quite recently they were considered as Lower Carboniferous. Now
the presence of the fossil-dated Lower Palaeozoic (Ordovician) strata amongst these
rocks has been established. These strata overlie still older Proterozoic and Archeozoic metamorphic and crystalline rocks. The lower complex within these rocks consists of paragneisses and orthogneisses, mica schists, amphibolite schists, quartz-
schists and other less deformed rocks. This complex is assigned to the Lower
Proterozoic and Archeozoic. The upper complex consists of less metamorphosed
rocks, which are coal-bearing, sericitic, tuffogenous slates, quartzites and cleaved
volcanic rocks. It is partly Upper Proterozoic and partly Lower Proterozoic. Owing
to the strong manifestations of the Caledonian Orogeny various Palaeozoic horizons from Ordovician and Silurian to the Visean rest on the metamorphic rocks
with a strong angular unconformity.

The new data indicate that the Tobol-Ural watershed and all the regions to the
east of it, including the Turgai Depression, are identical in their structure with
Central Kazakhstan and differ considerably from the Urals. This is a point of view
first expressed of N. G. Kassin, a major specialist on the geology of Kazakhstan.

FIG. 101. Variegated limestones and marls of the Katavian Suite forming
the 'Krasnyi Kamen' cliff on the river Sim, below the mouth of the Min'yar
in the Southern Urals. *Photograph by F. N. Chernyshev (1879).*

In the lower part (the Central Urals) of the mountains the ancient Proterozoic rocks are absent.

In the northern part of the Central Urals (the upper Kos'va and neighbouring rivers) and in the Northern Urals, including the Sablya Massif, the ancient rocks are again widespread and consist of a series of suites, analogous to those of the Southern Urals and reaching a thickness of 10–12km. The lower suites are included in the Proterozoic.

Summarizing, it can be said that in the opinion of most geologists, all the suites underlying the Zil'merdakian of the Southern Urals, and of its analogues in other regions, are included in the Proterozoic. In the Southern Urals these will be the Taratashian, the Aian, the Satkinian and Bakalian, the Zigal'ginian, the Komarovo-Zigazinskian and the Avzyanian suites, of a total thickness of no less than 7–9km.

The characteristic features indicating their Proterozoic age are: (1) jaspilitic iron ores and ferruginous quartzites as in Krivoi Rog and Kursk Magnetic Anomaly; (2) thick dolomitic formations and associated large deposits of magnesite as in the Proterozoic of Canada, Korea and North-Eastern China; (3) intrusions of Rapakivi-type. The organic remains are in the main represented by calcareous algae of *Collenia* type.

Ashaian Suite (series). This is the youngest Precambrian Suite, of a highly debatable age. The suite is found on the western slope of the Urals and in the north it spreads over the eastern slope as well. Throughout it is a uniform succession of terrigenous continental deposits, consisting of greenish-grey, steel-grey and reddish sandstones with clay slates, which are finely rhythmical and flyschoid. The organic remains are primitive spores, but macroscopically it is barren. The total thickness of the suite is 1000–1600m and it can be subdivided into four sub-suites: (1) Uryukian, consisting of coarse arkoses, gravel sands and conglomerates (35–100m); (2) Basian, of fine, rhythmically-bedded sandstones, silts and argillites, which are greenish and reddish (300–850m); (3) Kukkaraukian, of red and brown conglomerates and sandstones, showing distinct differential weathering (50–270m); (4) Ziganian, of massive greenish and reddish sandstones, silts and argillites with groups of silts with clay-balls and ripple-marks (350–450m). The Ziganian and Basian sub-suites have a typical flyschoid character, the bottoms of their bedding planes show sole-marks and problematic structures typical of flysch. All these rocks are weakly metamorphosed. The age of these rocks is of interest. Before the Ashaian Suite was distinguished its strata were considered by Chernyshev as Lower Devonian, since it lay below the marine Upper Devonian. This argument was accepted in 1931, when the suite was differentiated, and the sandstones underlying the Takatinian Suite were also included in the Ashaian Suite. Subsequently this was demonstrated to be an error. Under the Takatinian Suite in various regions lies rocks belonging to the Lower Devonian, Silurian and Ordovician, which also underlie the Ashaian Suite. In one of these suites in 1948, C. I. Domrachev found remains of psilophytes and attributed it together with the Ashaian Suite to the Lower Devonian, and the present author for some time held the same point of view. In 1934 L'vov and Olli included the Ashaian Suite in the Ordovician-Upper Cambrian, having established that the marine Upper Ordovician lies on it. C. N. Naumova, E. V. Chibrikova and B. M. Keller included it, on the basis of spores, in the Cambrian. In 1956 K. A. L'vov, on the basis of its transition into the marine Lower Ordovician, observed by him in the Northern Urals, stated categorically that the suite is Lower Ordovician. N. G. Chochia did not accept the transition and included the suite into the Cambrian, and this opinion lasted for some time, since there were no palaeonto-logical data. At the same time B. V. Timofeyev determined the spores from the Ashaian Suite as Lower Devonian. This together with the discovery of psilophytes has led me to express an opinion that two suites, one Middle Palaeozoic (Lower Devonian-Upper Silurian) and the other Lower Palaeozoic (Cambrian), were in-volved.

In 1945 N. S. Shatskii included the Ashaian Suite in the Riphaean System, pro-posed by him as the uppermost Precambrian, equivalent to the Sparagmites of Scandinavia. In 1959 the present author in company with a group of other geologists made a special, detailed study of the Ashaian Suite in the basin of the Sikaza, in the Southern Urals. It was established that it completely lacked Devonian fauna and flora and was under the marine Ordovician, Silurian and Lower Devonian, which are transgressive over it. The absence of the Palaeozoic fauna and flora and its position under the Ordovician made me inclined to include the suite into the Upper Pro-terozoic (Sinian, Riphaean), as it was done by Shatskii. At the same time Olli has independently arrived at the same conclusion, while K. R. Timergazin considered such an age as the most probable, but that the upper part of the suite may be Cam-brian. The visit to the classical exposures of the Sparagmite System in Norway, where it is transgressively overlain by the Lower Cambrian, has convinced me of the complete identity of the Ashaian Suite and the Sparagmites, and finally established its Precambrian Upper Proterozoic age. Nevertheless there still are adherents of the Devonian and Cambrian age of the suite. It is significant that the age determinations done in 1961 on the basis of the palaeomagnetism have given an Ordovician age. Such differences of opinion about the age of a group of rocks are rather rare even for the Precambrian.

PALAEOZOIC

Lower Palaeozoic

In the Urals the Palaeozoic rocks which have a great industrial significance and are widespread, are divided into three parts: the Lower, the Middle, and the Upper.

Before the Revolution Lower Palaeozoic rocks were almost unknown in the Urals and the opinion was often expressed that they were altogether absent. Thus the achievements of the Soviet geologists are very considerable since they demonstrated that such rocks are present throughout the Urals, from Mugodzhar to Pai-Khoi. Owing to the metamorphism of the Lower Palaeozoic deposits, discoveries of faunas are rare, and in many regions the Lower Palaeozoic has not been separated from the Precambrian and its succession has not been sufficiently studied.

Cambrian

The palaeontologically ascertained Cambrian rocks have only been found in one region, the basin of the Sakmara in the south of the Southern Urals. The rocks here are pale, dark and pinkish massive archeocyathid limestones of Lower Cambrian age. The archeocyathids are numerous and include *Archeocyathus* and *Rhabdo-cyathus*. In addition the brachiopod *Kutorgina perrugata* Walc. has also been found and there are often encountered algae such as *Epiphyton*, *Renalcis*, *Chabakovia*, and others. The archeocyathid limestones occur as separate blocks and have obscure stratigraphic relationships against the surrounding igneous and sandy rocks of the Tereklinian Suite.

The information on the Min'yarian archeocyathids of the Southern Urals is based on poor material and consequently requires a further examination. The same applies to the information supplied by L'vov (1939) from the Northern Urals. On the Kola Peninsula and in Norway fossils resembling archeocyathids are found in Sinian sediments. The Zil'merdakian Suite of the Southern Urals is arbitrarily assigned to the Lower Cambrian and the Katavian, the Inzerian and the Min'yarian and their analogues in the Northern Urals to the Middle Cambrian. This is based on the fact that the Min'yarian Suite rests on the Ashaian Suite and on the presence of *Collenia*, *Eophyton* and *Cenophyton* in the Katavian and Min'yarian sediments. These fossils are also found in the Middle Cambrian limestones of Siberia. However, not only the Cambrian but also the Proterozoic can lie under the Ordovician, and similar algae are found in the Proterozoic of North America.

The Cambrian sediments are a little, but not much, better dated in the Northern and Arctic Urals. Here L'vov found, in the marble limestones of the Shchokur'inian Suite, the calcareous algae *Collenia* and *Osagia* and also *Archeocyathus*, hyolite molluscs and remains of brachiopod (?) shells. The insufficient determinacy of such organic remains may mean that they are either Cambrian or Ordovician. In the last few years the Uralian Ordovician has yielded life-remains very similar to the archeocyathids. What, however, is definite is that the Shchokur'inian Suite cannot belong to the Upper Proterozoic.

Thus owing to the impossibility of an unambiguous palaeontological dating the Cambrian age, ascribed to many ancient rocks of the Urals and the Timan, has to be considered as provisional. This is emphasized by the fact that in the north of the Arctic Urals they are overlain by quartzitic sandstones and slates of up to 1000–1500m thick at the bottom of which Upper Cambrian brachiopods (*Billingsella*) have been found by A. M. Ivanova (1958).

Ordovician

DISTRIBUTION. As distinct from the Cambrian, the Ordovician fauna has been found in tens of regions along the Urals (from the Mugodzhar to the Pai-Khoi). Nevertheless the area of the Ordovician outcrops is relatively small since they occur as a series of narrow belts. The Ordovician is mainly represented by normal marine sediments of great thickness and of geosynclinal type.

SUBDIVISIONS AND LITHOLOGY. The Ordovician according to A. N. Ivanov (Ivanov & Myagkova, 1950a) is divided into stages recognised in Europe—Tremadoc, Arenig, Llandeilo and Caradoc. Yet in the Urals there is no single section where all the four stages can be palaeontologically dated. Normally the typical fauna of one or very rarely two stages is found, but even then it can only be correlated with difficulty. Consequently, as a rule, local divisions are used in different regions, and what is worse, by different authors.

The relatively high degree of metamorphism of these rocks and the paucity of faunas complicates the Ordovician stratigraphy. The correlation of the faunas permits the establishment in various regions of the Urals, of the four stages represented in marine deposits. Let us consider separate areas.

The Ordovician sections of the western slope of the Central Urals have been studied by Ivanov and Myagkova (1950a, 1950b), who divided it into three subdivisions. The Lower Ordovician consists of pale, micaceous quartzites (500–800m) lacking organic remains and transitional into phyllites (150–200m). The Middle Ordovician is the thickest and consists of the following formations. At the bottom lie the phyllites and sericite-quartz and chlorite-quartz schists (500m). Some lime-stones occurring at the bottom of this succession contain *Collenia, Praesyringopora* and *Palaeofavasites*. Higher up follow interbedded phyllites and limestones (200–250m) with nautiloids and tabulates of Middle Ordovician age. These rocks are over-lain by marls (150m) with Middle Ordovician brachiopods and bryozoans. Higher up lie dolomites (200m) with a poor trilobite-brachiopod fauna. The dolomites are capped by dark-grey compact limestones (250m) with a rich Caradoc fauna mark-ing the beginning of the Middle Ordovician (Ivanov & Myagkova, 1950a). At the very top there are limestones (200m) and calcareous sandstones (100m) with stro-phomenids. Above this succession and separated from it by an angular unconformity follow the basal conglomerates and sandstones of the Silurian. Certain investigators consider the micaceous quartzites at the bottom of the succession and a part of the phyllites to be of a more ancient age. Thus, for instance L'vov begins the section in the Kolvo-Visherskian region with the Tel'possian Suite of conglomerates, sand-stones (calcareous towards the top) with crinoid ossicles and brachiopods (*Finkel-burgia, Angarella*), of total thickness of 150–600m. Overlying this is the Khydeian Suite (300–600m), consisting of sericite-chlorite quartz slates and calcareous slates with beds of dolomites and limestones with peculiar tabulates (*Tetradium*). The section ends with a carbonate Schugorian Suite (700–800m) of dolomites and lime-stones with a rich fauna of *Megalaspis* and *Asaphus*. Comparing it to the sections in the Central Urals it is obvious that, the micaceous quartzites apart, the two successions coincide and over thin basal conglomerates and sandstones lie slates covered by thick carbonates, dark, bedded limestones and dolomites with a Caradoc fauna. The absence of volcanic rock is important.

The same features apply to the western slope of the Southern Urals. Here at the Upper Byelaya are sandstones and conglomerates at the base, overlain by fine-grained sandstones and greenish-grey shales, resembling the Artinskian Stage, but containing Middle Ordovician orthids, strophomenids and trilobites. The section is completed by pale quartzitic Caradoc sandstones with a rich fauna of brachiopods, trilobites and cystoids. The Ordovician lies transgressively on Proterozoic sediments and its total thickness is more than 1000m. In the south of the Urals-Sakmara-Ora region—the succession is of the same type and the lower arkosic sandstones yield a Tremadoc fauna of *Obolus, Acrothyra*, orthids and the trilobite *Euloma*.

The eastern slope and the western slope at the Ufa amphitheatre show a different type of succession. The rocks here are in general more metamorphosed and contain widespread meta-volcanic rocks. On the western slope at the Ufa amphitheatre (Nizhneserginsk), the Silurian is underlain by maroon and brown phyllitic slates, quartzitic sandstones, tuffs and spilites, which are thick and strongly deformed. The fauna of tabulates suggests the presence of the Caradoc. To the east of the eastern slope of the Southern Urals Ordovician sediments are widespread, and consist of metamorphosed slates and volcanic rocks as well as frequent quartzite sandstones. The thickness is considerable, but owing to the complexity of tectonics has not been worked out. In the Brendensk region trilobites (*Euloma*) and brachiopods of Lower Ordovician age have been found amongst slates, phyllites, quartzites and sandstones of a total thickness of about 500m. The Ordovician is transgressively covered by various horizons of the Middle Palaeozoic, beginning from the Silurian.

PALAEOGEOGRAPHY AND FACIES. The Ordovician of the Urals is typically geosynclinal and marine and its sediments were deposited in a regime involving continuous downward movement. In the east there are manifestations of volcanic activity, including the characteristic spilites, which were submarine. Parallel with the dominantly marine limestones and shales there are sandy and quartzitic formations of near-shore or even terrestrial deposition.

Amongst the different facies the thick greenish-grey graywackes (pepper sandstones), which are different from the Artinskian pepper sandstones, can be noted. Both types, however, can be compared with the Alpine flysch or partly with molasse. This suggests that the Ordovician rocks are synorogenic with the Caledonian Orogeny.

Other important facies are pale and completely white or pink coarse arkosic sandstones, overlain by granular sandstones with marine faunas. These rocks represent the initial deposits of marine or terrestrial sediments.

Middle Palaeozoic

In the Urals the Middle Palaeozoic (Silurian, Devonian, and Lower Carboniferous) represents a single complex of deposits which are gradational into each other, have a common history and tectonics, a common magmatism and similar useful mineral deposits. The Middle Palaeozoic is sharply separated from the Lower Palaeozoic, the former having been deformed by the Caledonian and the latter by the Hercynian orogenies.

The Middle Palaeozoic successions vary laterally and two general types can be distinguished: the western Uralian and the eastern Uralian. The former has three characteristic features: (1) the marine regime was established from the Eifelian onwards; (2) the folds are relatively simple; (3) there are no volcanic rocks. The eastern Uralian type on the other hand shows the evidence for the marine regime throughout, the folds are complex and highly faulted, and metamorphism is in places considerable, resulting in green schist facies in the Central Urals. Furthermore the succession includes volcanic rocks which in places are of some importance.

The western type is basically found on the western slope and the eastern on the eastern slope. Only in the Southern Urals—in the Ufa amphitheatre and to the south—is the eastern type found on the western slope as well.

Silurian

The Silurian is represented by its three stages. In the Llandovery and Wenlock the sandstones and shales predominate and there are widespread volcanic rocks. In the Ludlovian, limestones including reefs are predominant. On the eastern slope and central region from Pai-Khoi to Mugodzhar the Silurian is everywhere present, but on the western slope it is often absent or replaced by barren continental sediments.

SUBDIVISIONS. The Silurian is subdivided into three stages: the Llandoverian, the Wenlockian and the Ludlovian, the latter being divided into the lower and the upper. The Upper Ludlovian was, formerly, sometimes called Downtonian, but this term is not now used since the Downtonian of England consists of lagoonal facies contemporaneous with the marine Upper Ludlovian. The leading work on the stratigraphy and palaeontology of the Urals is by A. N. Khodalevich (1939, 1949, 1951) one of the major stratigraphers of the Urals.

The Silurian succession of the eastern slope of the Northern Urals has, in connection with the bauxite deposits, been studied in greatest detail; Table 26 follows S. M. Andronov and A. N. Khodalevich.

LITHOLOGY AND FACIES. As a result of the movements associated with the Caledonian Orogeny there were mountain chains to the east of the Uralian Sea, which existed throughout the first half of the Silurian. The products of erosion of these chains were introduced into this sea as sands and muds and as near-shore conglomerates. Consequently, for the Llandoverian and Wenlockian epochs thick sandstones and shales are characteristic for the Urals as a whole. The contemporaneous downward movements provoked a widespread volcanic action resulting in thick lavas and tuffs. All this has conditioned a peculiar nature of sedimentation, in so far as thick clastic-volcanic complexes were found. Only in isolated localities thin groups of dark muddy limestones with coral-brachiopod faunas came into existence.

Amongst the individual facies the graptolitic shales, which are dark, calcareous, fine-grained and with many white graptolites, deserve attention. The thickness of these sediments varies from tens to a few hundred metres.

Beginning from the Upper Wenlockian the supply of the clastic material into the Uralian Sea became sharply reduced and its supply from the west stopped altogether. Thus the quantity of the clastic sediments diminished and carbonate formations (mainly limestones) up to 2000m in thickness accumulated. Only along the eastern margin of the sea, near to the western Siberian continent, the deposition of clastic sediments interbedded with carbonate-mud deposits continued.

During the Upper Ludlovian along the western seaboard there was a barrier reef separated from the shore by a zone of lagoon. The reef consists of massive pale limestones with, in places, a very rich fauna. The precursors of the reef massifs were calcareous banks formed by mass accumulation of similar organic remains,

TABLE 26. *Silurian rocks of the Eastern Slope of the Northern Urals*
(According to S. M. Andronov and A. N. Khodalevich, 1956)

Subdivision	Description of rocks	Thickness
Gedinnian Stage	Pale, massive limestones with a mixed Silurian-Devonian fauna of *Plectatrypa marginalis* Dalm. and *Karpinskia vagranensis* Khod.	100–200m.
Upper Ludlow	Pale massive limestones passing laterally into a volcanic-terrigenous formation; limestones have a rich and variable brachiopod-coral fauna.	300–700m.
Lower Ludlow Colongian Suite Bankian Horizon	Dark, bedded limestones with bands of shales and sandstones, passing in the east into a volcanogenic formation. In places limestone banks of huge *Conchidium vogulicum* Vern.	100–300m.
Voskresenkian Suite Striatian Horizon	Red massive limestones, in places consisting of *Brooksinia striata* Eichw. (a pentamerid).	200–300m.
Wenlock Pokrovkian Suite	Volcanic-sedimentary strata with thick sheets of basic lavas and beds of brachiopod-bearing limestones.	>600m.
Llandovery	Found in places, consisting of graptolitic shales and brachiopod-bearing (*Pentamerous oblongus* Sow.) limestones. Unconformity.	—
Ordovician	Metamorphic rocks.	—

lying side by side, or stuck to each other. Most frequently such banks were formed by thick-walled pentamerids, but more occasionally they were formed by large, thick strophomenids and huge lamellibranchs (*Megalomus*).

The bedded pale or dark limestones with a diverse fauna consisting mainly of brachiopods but with frequent trilobites, ostracods, gastropods and lamellibranchs, are especially widely distributed. The dark, muddy limestones have common coral-tabulate facies where the rock-forming organisms were tabulates and especially the genus *Halysites*, but there are also strophomenids and rugose corals. These rocks are very thick and individual suites reach several hundred metres.

FAUNA AND FLORA.   The flora is represented only by calcareous algae, common in reef facies. In near-shore continental conditions a psilophyte flora grew, but its representatives have not been found as yet in the Urals. The fauna is rich and diverse and is of a typical Middle Palaeozoic type with a preponderance of brachiopods and coelenterates (tabulates, Rugosa and stromatoporoids). The Arthropoda such as trilobites and ostracods and molluscs such as nautiloids, gastropods and lamellibranchs have a considerable significance. Other groups are rarer, although crinoid ossicles are sometimes rock-forming.

PALAEOGEOGRAPHY.   During the Silurian the Ural Geosyncline was a large straits-like sea elongated in the north–south direction. The sea was one and a half or twice as wide as the present area of the Silurian which has been folded into complex folds.

The western edge of the sea lay along the western boundary of the Urals, but its precise eastern edge is not known. It was probably somewhere in the region of the valley of the Ob'. Along its western and the eastern shores there were large near-shore plains where red beds accumulated. Behind these plains were low uplands of the Russian Platform and marginal chains of the Siberian Platform. The margin of the Siberian Platform was then considerably to the west of its present position. The sea was generally shallow but there is a possibility of the presence of deep downwarps in which argillaceous, calcareous and siliceous sediments or all three types contemporaneously were being deposited and the fauna was almost absent. In its eastern part there were a series of craggy, volcanic islands appearing above the water and being destroyed by the waves.

In the first half of the period corresponding to the Llandoverian and Wenlockian epochs the Russian and Siberian uplands were higher and were being intensively eroded by rain storms, which produced powerful but ephemeral streams. The streams brought in large amounts of clastic materials into the sea where they accumulated as shales and sandstones. The large amounts of sand and mud thus deposited rendered the conditions for animal life almost impossible especially if these animals were lime secreting. Thus limestones are generally rare and thin. In the second half of the period—Lower and Upper Ludlovian—the rise of the continents was interrupted, erosion ceased to be active, and the supply of clastic material drastically diminished. This produced conditions favourable for the development of lime-secreting animals. Limestones and dolomites started being deposited and reached great thicknesses towards the end of the epoch.

The western shore of the Uralian Sea was a vast water-free desert which was gradually sinking. These conditions were favourable for the formation of barrier reefs, continuous without interruption from the Pai-Khoi to the Mugodzhar. There were large submarine banks of communities of massive brachiopods and lamellibranchs.

TABLE 27B.

Lower and Middle Devonian Successions on the Southern slope of the Urals

Stage	Sub-stage	Suite	The Southern Urals		The Karatau range		
			River Askyn	River Inzer	River Asha	River Minyar	River Chusovay
Pashiiskian Suite			Brown, ferruginous clay (0.25m).	Pale quartz sandstones (2m).	Variegated clays (3m).	Pale sandstones (8m).	Sandy-clayey sequence (1–40m).
Givetian	Upper	Cheslavkian	Dark limestones with *Stringocephalus* (16.5m).	Pale, massive limestones with *Stringocephalus* (13.5m).	Yellowish limestones with a marine fauna (4m).	Grey, thickly-bedded limestones with a rich fauna (7m).	Grey, bedded limestones with *Stringocephalus* (20–30m).
		Chusovian			Black and greenish sandstones with a psilophytic flora (15–20m).	Grey sandstones with a band (4m) of bituminous sandstone (12m).	Yellow sandstones and shales with beds of limestone. A fauna and a flora are present (1.5–8m).
	Lower	Afoninian			Missing.	Missing.	Dark, bedded bituminous limestones and shales, with goniatites (up to 25m).
Eifelian	Upper	Biian	Dark, bituminous limestones (25m).	Dark, bedded limestones with a marine fauna (3–8m).			Dark, thickly-bedded limestones, in places bituminous (0–40m).
		Calceolitic		Yellowish marls and shales with a marine fauna with *Calceola* (7m).			Dark, bedded limestones with a marine fauna (20–30m).
	Lower	Vyazovian		Dark, bituminous sandstones (3m). Limestones with ostracods (3–5m).			Grey, bedded marls and limestones with *Calceola* (20m).
		Takatinian and Vanyashkinian	Yellowish and white quartz sandstones (30–40m).	Greenish shales and sandstones (15m). Pale sandstones and clays (15–20m).			Pale, arkosic sandstones, which are muddy at the top (15m).
Basement.			Ashaian Suite.	Ashaian Suite (1500m).	Ashaian Suite (600m).	Ashaian Suite (600m).	Ashaian Suite.

TABLE 27B—continued

	The Central Urals		The Northern Urals		
River Kos'va	River Niz'va, a tributary of Kolva	River Mutikha, in the Vishera basin	River Akchim, a tributary of the Vishera	River Vishera
Sandstones and clays (0–45m).	Grey quartzites.	Bedded limestones and shales.	Grey sandstones.	Grey sandstones (50–60m).
Missing.	Missing.	Missing.	Grey, bedded sandstones.	Shales and sandstones (3m).
			Sandstones and shales (4m).	Limestones and shales (6m).
Dark, bituminous limestones and oil shales with a fauna of Domanik type (up to 25m).			Unrecognized.	Dark, bedded limestones and shales of Domanik type (33m).
Dark, dolomitized limestones with pentamerids (80m).		Pale, crinoidal limestones (80m).	Dark, bedded, bituminous limestones (80m).	Sequences of sandstones and limestones (38m). Dark, bedded bituminous limestones, with a marine fauna (42m).
Dark, bedded limestones with ostracods (15–30m).				
Bedded limestones; towards the bottom, shales, with *Calceola* (120m).			Dark shales; limestones towards the top.	Above, limestones (3m); below, shales (12m).
Pale and yellowish arkosic sandstones (120m).		Pale, coarse sandstones (150m).	Pale-grey sandstones. Unpenetrated (160m).	Pale quartzose sandstones. Unpenetrated (10m).
Ashaian Suite.	Proterozoic, Nizvian dolomites.	Silurian dolomites.	—	—

Along the eastern shore, although the influx of the classic sediments was reduced it nevertheless did not completely stop and sandstones and shales were ultimately formed. The vulcanicity continued as before, but the volcanic chains shifted somewhat to the east. The facies became very variable.

Devonian

Devonian sediments were widely distributed throughout the Urals—both on the western and eastern flanks. They reached a great thickness (2000–3000m), are very variable and possess a rich fauna. The Devonian succession of the Urals in terms of their completeness and the details of investigations and richness of fauna are classical and are a standard of comparison not only in relation to the other regions of the U.S.S.R., but also for Western Europe and the U.S.A.

The Devonian of the Urals was established over 100 years ago. Murchison in his travels in 1843 proved their considerable spread, but assigned a considerable part to the Silurian. In the 1880's F. N. Chernyshev had shown this to be a mistake, but fell into another pitfall by ascribing to the Devonian a whole series of barren formations. After his work almost all of the Urals were shown on maps as Devonian. Only the detailed geological survey which began in 1925 and is still continuing has permitted clarification and has shown that a number of suites, formerly considered as Devonian, are really Silurian, Ordovician and Cambrian. Even so the area occupied by the Devonian is quite considerable.

The number of papers on the Devonian of the Urals is large and can be counted in hundreds. Amongst the latest palaeostratigraphical researches the following can be mentioned. On the Upper Devonian there are the papers of B. P. Markovskii and A. K. Krylova. On the Middle Devonian the work of A. P. Tyazheva and on the Middle and Lower Devonian the monograph by A. N. Khodalevich should be noted. In addition there are monographs on Rugosa by N. Ya. Spasskii, on goniatites by A. K. Nalivkina and on trilobites by Z. A. Maximova.

DISTRIBUTION.   The Devonian is widespread from the Pai-Khoi to the Mugodzhars. Owing to its complex tectonics the Devonian outcrops appear as narrow belts elongated along the strike. In the Central Urals nearer to the central part the Devonian deposits are heavily metamorphosed and turned into green schists.

SUBDIVISIONS.   The sediments of the western slope are most fully investigated since they were important for the understanding of the oil-bearing Devonian rocks in the Volga-Ural region. The successions are given in Tables 27A and B.

The successions of the eastern slope have been less fully studied and only in the Northern Urals and especially in the bauxite-bearing regions, have full and detailed successions been established. These have shown the presence of all the six stages, represented mainly by marine deposits. Examples of the successions on the eastern slope are given in Table 28.

In carbonate rocks the faunal boundary between the Silurian and the Devonian can only be drawn with difficulty. The previously considered diagnostic Devonian fossil *Karpinskia*, although widespread in it, has also got species in the Upper Silurian and Eifelian. Thus *Karpinskia* is a facies fossil of Hercynian limestones to be described and not just of Lower Devonian. The discovery of *Karpinskia* in dark bedded limestones of the Vyazovaian Suite of the Eifelian Stage of the valley of the Ik (S. Urals) is of interest.

In sandy-shaley and effusive, almost barren, rocks the boundary between the

TABLE 28.

Devonian Successions of the eastern slope of the Urals

(According to S. M. Andonov and A. N. Khodalevich)

The Northern Urals		The Southern Urals
Western Zone	Eastern Zone	
Famennian Stage is not recognized. Frasnian Stage—above, pale-grey massive limestones with brachiopods (*Hypothyridina cuboides* Sow. (100–300m); below, sandstones and shales with beds of brachiopod limestones (0–100m).	Famennian Stage—sandstones, shales and conglomerates. In places bands with marine faunas (> 300m). Frasnian Stage—pale, massive limestones with *Hypothyridina cuboides* Sow.	Famennian Stage: Zalairian Series, bottom of the Kiian Suite—shales and siliceous shales with lenses of Clymenid limestones (300–400m). Bakeshevskian Suite of green sandstones, argillites and siliceous shales (400–500m). Karantauian Suite of green sandstones, gravellites and shales (10–500m). Frasnian Stage: Koltubanian Series—above, limestones (in places massive) and shales (0–100m); in the middle, jaspers and siliceous shales (50–100m); at the base, sandstones and conglomerates (10–50m).
Givetian Stage—pale massive or dark, bedded limestones with a marine *Stringocephalus* fauna (100–300m). Dark bedded amphiporous limestones with *Conchidiella* (100–200m). A bauxite-bearing formation.	Givetien Stage—grey massive limestones with *Stringocephalus* (No more than 100m). Dark, bedded limestones with a bauxite-bearing or sandy formation at the base (0–100m).	Givetian Stage: Aushkulian Suite of grey, bedded and massive limestones, which in places wedge out. A rich brachiopod fauna with *Stringocephalus* is present (0–150m). Askarovskian Suite of greenish, bedded sandstones and argillites, similar to the Zalairian Suite (150–200m). Tatlybayevskian Suite of conglomerates, tuffaceous sandstones, tuffs and siliceous tuffites (25–300m).
Eifelian Stage—pale massive limestones with pentamerids (0–200m). Dark, flaggy limestones of domanik type (0–300m). Pale massive limestones with *Spirifer superbus* Eichw. (10–300m). Dark, bedded, amphiporous limestones (30–75m). A bauxitic formation.	Eifelian Stage—pale, massive limestones with pentamerids (0–200m). Dark, flaggy limestones of domanik type with *Buchiola* and pteropods; dark, carbonaceous shales, rhythmically alternating with sandstones and bituminous limestones. The sequence is flysch-like (100–200m).	Eifelian Stage: Kalmykovian Suite of thick tuffs and tuffites with *Calceola sandalina* L. (350–400m). Kusimovskian Suite, red jaspers (30–50m). Khasanian Suite of basal conglomerates and sandstones, with grey, bedded limestones towards the top (25–50m).
Koblenzian Stage—limestones with brachiopods. Gedinnian Stage—pale, massive limestones with *Karpinskia vagramensis* Khod. (100–200m). Upper Ludlow—pale, massive limestones with brachiopods (300–700m).	Lower Devonian and Ludlow are not separated, being represented by a thick volcanogenic formation.	Lower Devonian. Tanalykian Suite of basal conglomerates and sandstones with limestones above carrying Koblenzian brachiopods and rugose corals (0–75m). Gedinnian Stage has not been recognized. Upper* Devonian. Bugulygyrian Suite of thick red jaspers (50–150m). Irendykian Suite of porphyrites, albitophyres spilites and tuffs (500–1000m).

NOTE—A. N. Khodalevich assigns the dark limestones with *Conchidiella* not to the Lower Givetian but to the Upper Eifelian.

* Original gives Upper Devonian whereas probably Lower Devonian is intended [N. Rast: editor].

Devonian and the Silurian is arbitrary depending on some constant recognizable horizon.

The boundaries between the divisions and stages are drawn depending on faunas and when the fauna is poor then the boundaries are arbitrary. On the western slope the rhythmic quality of sediments helps the positioning of the boundaries. Each rhythm starts with basal sandstones and is succeeded by clays, shales and bedded limestones. This is clear from the succession at the river Min'yar (Table 27A). Boundaries are drawn at the base of the sandstones.

The boundary between the Givetian and Eifelian has provoked arguments. Formerly this boundary on the western slope was drawn at the base of dark, bedded bituminous limestones (Biian, Calceolitic and Vyazovian Beds) and the underlying muddy formation (Vanyashkinian Beds). The underlying pale quartz sandstones (Takatinian Beds) were considered to be Eifelian. Later on a detailed study of the fauna has shown that a great majority of animals in the dark limestones are Eifelian. As a result the 1956 stratigraphic conference in Sverdlovsk decided to draw the boundary above the limestones with *Cochidiella* and below the limestones with *Stringocephalus*, that is at the former junction between the Upper Givetian and the Lower Givetian. This decision has been approved by the Regional Geological Committee and should be used in applied geology.

The boundary between the Devonian and the Carboniferous is most commonly within homogeneous formations and does not coincide with sedimentary rhythms. Consequently, it is drawn on faunal changes and the appearance of the first Tournaisian groups such as *Spirifer tornacensis* Kon. and *Productus niger* Goss. amongst the brachiopods. The positioning of this boundary is often difficult.

LITHOLOGY AND FACIES. In the Devonian System there are three complexes of deposits: the marine generally distributed type; the deposits of near-shore plains found on the western slope; the effusive facies typical of the eastern slope.

The marine Devonian rocks are diverse and thick. On the western slope they appear in the Eifelian while on the eastern slope they occur throughout. Their composition is also different. On the western slope, limestones predominate (Fig. 102) and in places there are dolomites reaching 450–500m in a total succession of 700–800m. It should be pointed out that there are two peculiarities—a rapid change in successions and a distinct sedimentary rhythm. The change in successions depends on the type of facies and thickness of deposits and occurs continuously over short distances. In the Southern Urals some tens of kilometres along the strike is sufficient for a group of siliceous, thinly-bedded and fine-grained goniatite-bearing shales and limestones, say 30–40m thick, to change into barren dolomites of 450–500m in thickness. Such changes have been noticed also in the Central Urals and have been described by oil-geologists such as Domrachev (1952), N. G. Chochia and K. I. Andrianova (1952). Fig. 103 shows indistinctly bedded Famennian dolomites of great thickness.

The sedimentary rhythms (see above) are obvious and continue for hundreds of kilometres and sometimes along the whole of the Urals. These features are bounded by their lateral extent. As is obvious from the successions (Table 27A) the number of the rhythms reaches four in the Middle Devonian and two in the Upper Devonian. The basal sandy-shaley deposits of the rhythm are representatives of deposition in the near-shore plains and include useful mineral deposits, which were laid down in lakes and lagoons and marshes on the surface of these plains. In the Pashiiskian Suite at the bottom of the Lower Frasnian rhythm, bauxites, bog iron ores and

FIG. 102. Upper Devonian limestones near Ust' Katav factory on the river Katav in the Southern Urals, showing typical relief of the western slope in the Southern Urals.

FIG. 103. Famennian, yellowish, faintly bedded dolomites at Kamen' Olenii on the river Chusovaya in the Central Urals. *Photograph by G. A. Smirnov.*

fire-clays have been found. Somewhat to the west on the Russian Platform the Pashiiskian Suite is an important oil-bearing horizon. In the Orlovian Suite, which lies at the bottom of the Upper Frasnian rhythm there are in the Southern Urals bauxites and fire-clays, but in the Central Urals the suite is not present.

On the eastern slope the variability of succession is even greater. As well as limestones of great thickness (up to 1000–1500m) there are widely developed formations of sandstones and shales, which reach a thickness of many hundred metres. The diversity is even more enhanced by the thick groups of lavas and tuffs. In the Central Urals, in their axial part, the Devonian deposits are intensively metamorphosed into green schists and marbles. In the Northern Urals three bauxite horizons appear amongst the Eifelian and two horizons amongst the Givetian limestones. The total thickness of the Devonian on the eastern slope is 2000–3000m and sometimes is more. The sedimentary rhythms are well developed, but not everywhere and on the eastern slope they are not typical.

Amongst the individual facies the following will be discussed: bituminous limestones, Domanik facies, cherts and siliceous shales, Hercynian sandstones, near-shore plain deposits.

Bituminous limestones. These are bedded, black or dark-brown rocks with a strong smell of oil and are enriched in a bituminous, organic substance. They are tens of metres thick and spread over hundreds of kilometres. Thus despite the small percentage, the total amount of organic material in them is huge, and as a result they can be considered as important parent rocks for oil. These rocks have a rather sparse coral-brachiopod fauna, with quite frequent ostracod-bearing limestones, full of ostracod shells. According to the conditions of formation these are typical marine sediments of regions with relatively quiet waters, i.e. depths of over 50–60m or of isolated embayments.

Domanik facies. These are groups and suites of bituminous, thinly-bedded rocks with a peculiar pseudopelagic fauna, characterized by a large number of goniatites (*Gephyroceras, Timanites, Tornoceras*), pteropods (*Styliolina*), pseudopelagic lamellibranchs (*Buchiola, Pterochaeria*) which adhered to algae, and of almost smooth brachiopods *(Liorhynchus)* (Fig. 66). The rocks of this facies are typically enriched in bitumens so that rocks like oil shales may occur, but normally they are bituminous limestones and shales, although there are siliceous shales and limestones. The Domanik facies are considered as oil-generating formations and with them is associated the Volga-Ural oil and the oil of the western slope of the Urals. The thickness of the Domanik facies is relatively small (up to 30–40m) and more rarely up to 60–90m. Their continuity is considerable and is sometimes measured in hundreds of kilometres.

As regards their age, the facies vary. The typical Domanik is Middle Frasnian and the name is derived from the river Domanik, which is a tributary of the Ukhta (the Timan) where it is 90m thick. The Domanik is found in an area beginning from the Timan, Novaya Zemlya and Pai-Khoi and outcrops throughout the Urals up to the transverse part of the Byelaya. On the eastern slope it is unknown. The Infradomanik is a group of rocks resembling Domanik and with a fauna of the same type, but of Middle Devonian, Givetian age. It is less widely distributed than the Domanik, but is also found on the eastern slope of the Urals. The Upper Frasnian and Famennian Domanik facies are encountered on the western slope of the Central and Northern Urals and in particular in the Kizel coal-field, where their thickness is over 200m. The areas of development of the Domanik have rounded closed outlines, corresponding to the stagnant muddy depressions in the sea-bottom in which they accumulated. Faunal features (the abundance of suspended forms) suggests that these depressions were filled by algae.

Cherts and siliceous shales are widely distributed throughout the eastern slope of the Urals forming groups of several hundred metres thick. These often show concentrations of manganese compounds of industrial importance. The manganese imparts red and pink hues to rocks. The siliceous shales and cherts have a high content of silicon and represent the deposits of shallow parts of the sea—the lagoons, the bays and the straits. The silicon concentrates at the expense of accumulation of siliceous shells of radiolarians and is also introduced volcanically.

Hercynian limestones. Thick, massive or indistinctly-bedded pale limestones forming considerable cliffs are sharply distinct from the surrounding shales. These are reef massifs, parts of the barrier reefs, which stretched along the seashore to the east of them. They have been observed from Novaya Zemlya and Pai-Khoi along the Urals to the Mugodzhars. They are generally developed on the eastern slope, but in places pass on to the western and central parts of the Urals. The formation of the Hercynian limestones began in the Upper Ludlovian, which is confirmed by the discovery of a Silurian fauna in their lower horizons. A greater part of them belongs to the Lower Devonian, while the upper horizons belong to the Eifelian. The representatives of the genus *Karpinskia* are found throughout the formation from bottom to top. The Hercynian limestones are of a high quality and inexhaustible resources (Fig. 104).

The Near-shore deposits were formed in the Lower Devonian and the Eifelian, on flat plains which on the western slope of the Urals margined the sea. In the Middle Devonian these plains were submerged under the sea, which penetrated far to the west into the Russian Platform. Here the succession of the near-shore plain-deposits is represented by white and grey arkoses, coarse sandstones of tens of metres thick and distinguished as the Takatinian Suite of Lower Eifelian age (Fig. 105).

On the western slope of the Southern Urals the outcrops of the peculiar Zalairian Suite, which is basically Famennian, cover a large area. The suite consists of a rhythmic alternation of groups of greenish-grey graywackes, argillites, beds of limestones and siliceous shales. There are few organic remains. The formation is similar to flysch and is attributed to the near-shore plain environment. It has been described by B. M. Keller (1949).

PALAEOGEOGRAPHY. During the Lower Devonian epoch the outlines and the general character of the sea and land were the same as in the Upper Ludlovian epoch. The Ural Geosyncline was a wide straits-like sea. In the west the sea was margined by a plain, which extended far into the Russian Platform, as far as the Volga-Kama Uplands. The plain was a desert, lacking continuous water streams. This created favourable conditions for the development of reefs along it, as far back as the Silurian. In the east there was a high continent and the streams carried into the sea large amounts of clastic material. On the shore and in the sea near to the shore there were numerous volcanoes.

At the beginning of the Middle Devonian the Russian Platform began to subside and the sea transgressed on the western slope of the Urals, which during the Eifelian epoch were covered, with the exception of small islands, one of which existed on the place of the present-day Karatau. During the same epoch the sea penetrated into the platform thus submerging the whole of the near-shore plain. The huge sea, which in

330

Fig. 104.

A cliff of Hercynian limestones on the river Greater Ik, on the western slope of the Southern Urals.

Photograph by L. S. Librovich

Fig. 105
Lower Eifelian, Takatinian sandstones in the upper reaches of the river Ilych, in the Northern Urals.

Photograph by L. G. Kondain.

the west was epicontinental, was geosynclinal in the east. Large stagnant muddy depressions originated on the sea-floor, were overgrown by the algae and became the sites of accumulation of Domanik facies. The barrier-reefs subsided and stopped their growth and black, bituminous bedded limestones were deposited over them. In other regions with more favourable conditions reef massifs came into existence, but as individual structures resembling the present-day table reefs and atolls of small size and irregular distribution. Over the greater part of the Urals the supply of the terrigenous material almost stopped, causing the abundant growth of lime-secreting organisms and the deposition of hundreds of metres of limestones.

On the eastern slope the palaeogeographical conditions remain without change. Owing to the balanced subsidence thousands of metres of homogeneous deposits accumulated. These deposits consist of interbedded sandstones, siliceous and clayey shales with subsidiary limestones. Vulcanicity was widely developed. The sea was shallow and had a complicated shore-line. The shores were flat, but elevated and affected by marine erosion.

Lower Carboniferous

The Lower Carboniferous is essentially identical to the Devonian, differing sharply from the Upper Carboniferous. Its characteristic distinction from the Devonian is a wide distribution of coal-bearing formations with economic coal seams which were absent from the Devonian and more ancient deposits.

DISTRIBUTION. Throughout the Urals and both slopes one finds the Lower Carboniferous. On the eastern slope of the Southern Urals as far back as 1930 almost all barren coal-bearing formations were attributed to the Lower Carboniferous. The work of the last decade has shown that these formations are metamorphosed, bituminous graptolitic rocks of Ordovician age. In the central, more elevated regions of the Urals the Lower Carboniferous has been eroded away.

SUBDIVISIONS. The Lower Carboniferous limestones known as the 'Mountain Limestone' and the Coal-bearing Formation were recognized in the Urals over 120 years ago, before Murchison's travels. He has made a wrong deduction that the Coal-bearing Formation lies above the Mountain Limestone, but his authority was great and even later investigators repeated the same error.

Only in the middle eighties of the last century has the work of the Geological Committee and especially of A. A. Krasnopol'skii established the true position of the Coal-bearing Formation, suggesting the following divisions for the Lower Carboniferous:

1. Limestones with *Productus giganteus* Mart.
2. Coal-bearing Formation.
3. Limestones with *Productus mesolobus* Phill.

This succession was adopted for the next forty years. Only at the beginning of the 1920's K. I. Lisitsin and after him N. I. Lebedev have demonstrated the presence of the Tournaisian and Visean in the Urals. In 1926 the present author established this fact independently and since then the two stages have been generally adopted. At the end of the thirties it was demonstrated that a Namurian fauna is present at the top of Visean beds and they are therefore as in Western Europe, Middle Carboniferous. Thus the Carboniferous of the Urals is divided into: Tournaisian, Visean and Namurian. The Coal-bearing Suite is now considered as a facies of Lower

Visean age on the eastern slope and possibly partly Tournaisian age on the western slope.

By comparing and compiling the results of research by many geologists in the last few years the present author has formulated the following Lower Carboniferous succession on the western slope of the Urals:

Middle Carboniferous

1. Bashkirian Stage. Limestones with the first choristids.

Lower Carboniferous

1. Namurian Stage. Limestones with the last *Striatifera*.

2. Visean Stage.
   Upper Visean sub-Stage. Pale dolomitized limestones, saccharoidal and almost barren with rare *Striatifera*.

   Middle Visean sub-Stage. Dark, thick, bedded, compact limestones with typical gigantoproductids and *Striatifera*.

   Lower Visean sub-Stage. In the south these are dark and grey, bedded limestones with first *Lithostrotion* and small, atypical *Gigantella*. In the north the Coal-bearing Formation. In places there are pale, massive limestones with a very rich and variable fauna.

3. Tournaisian Stage.
   Upper Tournaisian sub-Stage.
   Upper Horizon of grey, bedded limestones with gigantic *Chonetes*

Fig. 106. An outcrop of Strunian limestone (0·6m) on the river Ryauzyak on the western slope of the Southern Urals. Members of the geological society of the Pioneers Palace in Leningrad are collecting fossils. *Photograph by D. V. Nalivkin (1940).*

(*Davisiella*) *comoides* Sow., *Productus* (*Plicatifera*) *sublaevis* Kon. and large *Caninia*.
Lower Horizon of different facies, mainly limestones.

Lower Tournaisian sub-Stage.
Upper Horizon of limestones with *Spirifer* (*Paulonia*) *medius* Leb.

Lower Horizon (Strunian) of limestones with *Productus niger* Goss., *Spirifer ternacensis* Kon., *Syringothyris uralensis* Nol., *Cymaclymenia* sp., *Wocklumenia* sp., *Phacops accipitrinus* Phill. (Fig. 106).

Below this lie the Upper Famennian limestones with brachiopods (*Cyrtiopsis*) or clymeniids (*Laevigites*).

At the 1951 conference on the unified stratigraphic Carboniferous succession on the Russian Platform and the western slope of the Urals and at the Sverdlovsk 1956 conference on the unification of stratigraphical succession in the Urals (Theses of Papers, 1956) the following succession has been adopted for the Urals:

Lower Permian
    Sakmarian Stage

Schwagerina Horizon

Upper Carboniferous
    Gzhelian Stage
    Kassimovian Stage
Middle Carboniferous
    Moscovian Stage
        Myachkovian Horizon
        Poldolian Horizon
        Kashirian Horizon
        Vereiian Horizon
    Bashkirian Stage

Lower Carboniferous
    Namurian Stage
        Protvian Horizon
    Visean Stage
        Serpukhovian sub-Stage
        Steshevian Horizon
        Tarussian Horizon
        Okaian sub-Stage
        Venevian Horizon
        Mikhailovkian Horizon
        Aleksinian Horizon
        Yasnopolyanian sub-Stage
        Tulian Horizon
        Coal-bearing Horizon
    Tournaisian Stage
        Chernyshinian sub-Stage

TABLE 29.

Lower Carboniferous Successions in the Urals

Stage	Sub-stage	The Southern Urals			The Central
		River Ik	River Zilim	River Min'yar	River Chusovaya
Bashkirian		Black marls, argillites with limestones, sandstones and goniatites. Fusulinids and choristids are found (400–600m).	Pale, bedded limestones with fusulinids and choristids (80m).	Brown, bedded limestones with choristids (20–40m).	Missing.
Namurian		Dark limestones with goniatites and brachiopods (150m).	Pale limestones with *Striatifera* (100m).	Grey, bedded limestones.	
Visean	Upper	Dark, bedded limestones, in places pseudo-brecciated and cherty. Goniatites are found (200m).	White saccaroidal limestones with *Striatifera*.	White, massive dolomitized limestones.	White saccharoidal, vaguely-bedded limestones with *Striatifera* (up to 200m).
Visean	Middle	Marls, argillites and cherty limestones. Foraminifera are present (450–650m).	Grey, bedded limestones with *Gigantoproductus*.	Grey, bedded limestones with brachiopods and rugose corals.	Grey, bedded, cherty limestones with *Gigantoproductus* and other marine fauna (150–200m).
Visean	Lower	Argillites and sandstones with beds of limestones and cherts (400m).	Dark, bedded limestones with *Gigantoproductus jasvensis*. A sequence of sandstones and shales.	Dark, bedded brachiopod limestones with *Gigantoproductus jasvensis*.	Sandstones, shales and marls with a marine fauna (40m). A coal-bearing sequence (80m).
Tournaisian	Upper	Argillites and sandstones (250m).	Grey limestones with *Davisiella*. White, crinoidal limestones with brachiopods.	Missing.	Grey, bedded limestones and shales with a Kizelian fauna. In places strata are eroded away (0–40m).
Tournaisian	Lower	Argillites and siliceous shales with beds of limestones. Foraminifera are present (300m).	Grey limestones with *Spirifer medius*. Dark limestones with d'Etrœungt faunas.		Limestones with *Spirifer medius*. Dark limestones with a brachiopod fauna (200m).
Famennian	Upper	Argillites with beds of sandstones. A fauna and a flora are present (1 000m).	Grey limestones with *Liorhynchus ursus*.		Dark cherty limestones and shales with goniatites (30–80m).

TABLE 29—continued

Urals	The Northern Urals	The eastern slope of the Urals		
River Kos'va	River Vishera	Magnitogorsk	Kamensk	River Sos'va
Bluish-grey clastic foraminiferal, bedded limestones with cherts and choristids (60–100m).	Pale limestones and dolomites (25–50m). Dark aphanitic limestones (0–10m).	Conglomerates, sandstones and shales with beds of limestones (200m).	Above, pale and dark limestones; below, pseudobreccias (120–130m).	Not exposed.
Grey, bedded limestones with the first choristids and *Striatifera* (100–180m).	White and grey dolomites and limestones with numerous foraminiferids (40–70m).	Shartymian limestones with brachiopods and goniatites (200–250m).	Conglomerates, sandstones and cherty limestones with a marine fauna (300–450m).	
Pale, vaguely bedded limestones with *Striatifera* (200–250m).	White dolomitized limestones with *Striatifera* (120m).	Pale indistinctly bedded limestones with brachiopods (150–200m).	Pale, massive or bedded limestones with *Striatifera* (300–400m).	
Grey, bedded limestones with a rich fauna (up to 200m).	Grey, bedded limestones with *Gigantoproductus* (240m).	Dark-grey and grey bedded limestones and locally volcanic rocks (400–500m).	Grey limestones with *Gigantoproductus* (400–500m).	
Sandstones, shales and limestones with a marine fauna (20–80m). Shales, sandstones and coals (140–200m).	Thinly bedded limestones with *Gigantoproductus maximus* (60–80m). A coal-bearing sequence without workable coals (0–60m).	Dark limestones; in places, volcanic rocks (150–200m). Spilites, porphyrites and tuffs (150–250m).	Shales, sandstones and tuffs, with a marine fauna. A coal-bearing sequence. In places there are volcanic rocks (150–500m).	
Pale limestones with *Davisiella* (up to 40m). Grey, bedded limestones and shales (up to 60m).	Hard, cherty limestones with a Kizelian fauna (up to 100m).	Grey limestones with *Davisiella*; volcanic and tuffogenous rocks and sandstones (100–300m).	Grey, bedded limestones.	Grey, sandy limestones, shales and tuffs; below, diabases and tuffs. Kizelian fauna is present (280m).
Dark limestones with cherts and *Spirifer medius*. Grey compacted limestones with a d'Etrœungt fauna (20 m).	Dark, bedded limestones with *Spirifer medius*. Greenish marls and shales (40–50m).	Basic and acid volcanic rocks (600–700m). In places shales and sandstones.	Dark, bedded limestones with a fauna. In places, sandstones with a d'Etrœungt fauna.	Yellowish limestones and volcanic rocks (140m). Diabases and tuffs (100m).
Flaggy, dark, bituminous limestones with goniatites (200m).	Yellowish dolomitized, oolitic limestones (30m).	Shales, siliceous shales and limestones with *Clymenia*.	Dark shales and sandstones with ostracods.	Shales and volcanic rocks with beds of limestones with *Clymenia*.

Kizelovian Horizon
Cherepetian Horizon
Likhvinian sub-Stage
Upaian Horizon
Malevkian Horizon

Upper Devonian
  Famennian Stage
    Upper Famennian sub-Stage

Both successions in principle coincide, but the first is more correct since it ob-
viates the use of Central Russian terms in the Urals, where the lithology and the
faunas are very different. For instance in the Urals there is no Malevkian Horizon,
but there are dark compact limestones with Strunian fauna. Likewise in the Urals
there are no Upaian and Cherepetian faunas. There are also differences in Namurian
and Visean faunas. Consequently the employment of Central Russian terms in the
Urals can only be recommended where the faunas are identical.

LITHOLOGY AND FACIES. As in the Devonian the western and eastern successions differ considerably. Great differences in the western successions are also
noticed from south to north. The successions of the eastern slope are more constant
and vary a little from west to east (Table 29).

On the western slope there are thick, bedded, occasionally massive limestones
with a rich and variable brachiopod-coral fauna with occasional ostracods or
goniatites (Fig. 107). Their uniformity is disturbed by the Lower Visean coal-bearing
deposits and groups of sandstones and shales in the Tournaisian.

On the eastern slope some sections show limestones, but more frequently sandstones and clays, which are often coal-bearing, and lavas and tuffs are widely
developed. The reef limestones in the Lower Visean of the Southern Urals are interesting and have been described by A. A. Sultanayev (Fig. 108).

Of the individual facies let us select the coal-bearing groups. These are uniform,
brownish, dark or grey bedded sandstones with groups of muddy sandstones and
clays of subordinate thickness. Amongst the clays there are commonly carbonaceous
types and coal seams. The composition of the sandstones is variable. On the western
slope, in the Kizelian basin, they are uniform, fine-grained and consist almost exclusively of quartz grains. On the eastern slope the sandstones are more variable
and include arkoses and also beds of very coarse rocks. In Kizel their thickness is up
to 350m and on the eastern slope it is above 1000m.

FAUNA AND FLORA. The fauna is of the same type as the Devonian and includes the same groups. Its most characteristic features is the flourishing of produc-
tids, which reach a very large size (30–40cm) as gigantoproductids.

The variations and changes of the flora are important, since the lepidophyte
types become widespread. This flora consists of large tree-like plants such as *Sigil-
laria*, *Calamites* and ferns. The plants grew on land, but near to the sea-shore. For
the first time in geological history plant remains accumulate in marshes in such
huge quantities that they gave rise to the coal seams.

PALAEOGEOGRAPHY. The palaeogeography in the Lower Carboniferous was
the same as in the second half of the Devonian. A single, although important, distinction had been the appearance of permanent rivers of considerable length. Where
these rivers entered the sea, deltas originated, which in places (Kizelian basin) have
had a typical conical, fan-like shape.

FIG. 107

Middle Visean limestones at Kamen'
Vinokurennyi on the river Chusovaya
in the Central Urals. *Photograph by
G. A. Smirnov.*

FIG. 108.

Middle Visean limestone reef massif on the
river Ui to the north of Magnitogorsk. The
flat hilly relief is typical of Transuralia. *Photograph by A. A. Sultanayev.*

Upper Palaeozoic

The Upper Palaeozoic of the Urals forms an international standard succession, with which successions throughout the world are compared. Its classical sections have become known as a result of the work of A. P. Karpinskii (1874, 1891) and F. N. Chernyshev (1889, 1902) and of many Soviet geologists amongst whom D. L. Stepanov (1941, 1948, 1951), D. M. Rauzer-Chernousova (1937, 1940, 1949) and V. D. Nalivkin (1949, 1950) can be singled out.

DISTRIBUTION. On the western slope and in Cisuralia the Upper Palaeozoic is widespread. All the divisions and stages are represented by different facies and reach a great thickness (2000–3000m and more). In the central part the Upper Palaeozoic is absent. On the eastern slope it is developed as a facially different, thinner and incomplete succession. The marine fauna here is only Middle Carboniferous. The barren sandstones and the conglomerates which overlie the Middle Carboniferous can be considered as Upper Carboniferous. Marine Permian is entirely absent. The published notices of the discovery of marine Lower Permian have not been confirmed and all the claimed outcrops turned out to be Middle Carboniferous.

SUBDIVISIONS. The subdivisions are the same as in the Russian Platform (see

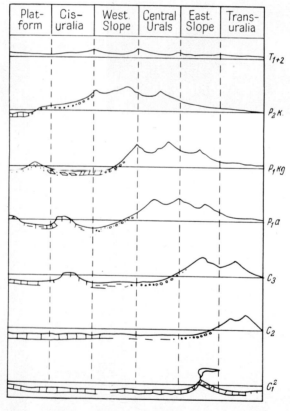

FIG. 109. Schematic demonstration of the advance of the Hercynian Orogeny in the Urals. Horizontal lines represent sea level. Reconstruction by D. V. Nalivkin (1952).

p. 97). The Middle Carboniferous is divided into Bashkirian and Moscovian stages and the Upper Carboniferous into Kassimovian, Gzhelian and Orenburgian, the Lower Permian into Asselian, Sakmarian, Artinskian and Kungurian, while the Upper Permian is usually represented only by the Ufian Stage, although in the Orenburg region and to the south of it deposits of Kazanian and Tatarian stages appear as well.

The position of the lower boundary of the Upper Palaeozoic is determined on the same basis as on the Russian Platform (p. 96). The upper boundary is found under the unconformable Mesozoic.

LITHOLOGY AND FACIES, PALAEOGEOGRAPHY. The lithology, the facies, the palaeogeography and the Hercynian Orogeny are directly associated with each other. The Hercynian Orogeny conditioned sharp changes in palaeogeography, which affected the facies and lithologies.

The first manifestations of the Hercynian Orogeny began in the first half of the Middle Carboniferous epoch to the east of Transuralia, causing the formation of high mountains the products of the erosion of which formed sediments to the west of them. In subsequent epochs the fold-belt moved progressively to the west, causing the formation of mountains and marginal sediments being formed in successively more western regions (Fig. 109), and resulting in the following succession of events.

Middle Carboniferous

The Siberian Massif produced by the Precambrian and Caledonian orogenies during the Bashkirian epoch started gradually growing towards the south-west. Its margin, which lay approximately where the present-day valleys of the Ob' and Irtish are, encroached on and deformed the Devonian and Lower Carboniferous deposits. The first high fold mountains come into existence (the first Hercynides). Their erosion occurred contemporaneously with their elevation and the products of erosion were deposited on the eastern slope of the Urals as thick clastic sediments with subordinate units of limestones and dolomites with choristids and fusulinids. Homogeneous, thick, Lower Carboniferous and Namurian limestones and dolomites are suddenly superseded by the sandstones and shales of the Middle Carboniferous. The boundary between the two groups is sharp, involves a faunal change in lithology (Table 30).

On the western slope of the Urals there was no folding during the Bashkirian, and Namurian limestones are gradationally succeeded by the Bashkirian limestones. The boundary between the Lower and Middle Carboniferous is not sharp and can only be determined faunally, on the basis of mass-appearance of first choristids and fusulinids.

In the Moscovian the margin of the Siberian continent moved westwards and the resultant fold-belts involve progressively more western regions, approaching the eastern slope of the Urals. Here the deposits become coarser and conglomerates start being deposited. Beds with marine faunas are rarer and at the end of the epoch disappear altogether, and are succeeded by beds with plant remains, which at one time were mistakenly determined as Lower Permian.

To the east of the western slope, sandstones and shales appear for the first time and in places completely replace the limestones, which occur only as beds of marl with quite common goniatites. To the west of the western slope in Cisuralia the clastic deposits do not penetrate and there the pure, bedded limestones are continuous (Fig. 110).

FIG. 110. Visean limestones on the river Ai in the Southern Urals.
Photograph by V. D. Nalivkin.

FIG. 111. Artinskian sandstones and marls standing vertically. In the centre
of the exposure is a slump sheet. *Photograph by V. D. Nalivkin (1945) from
the river Chul'pan on the western slope of the Southern Urals.*

'Artinskian' pepper sandstones. On the western slope of the Urals, in the Upper Palaeozoic, there are widespread 'pepper' sandstones which in colour and appearance resemble ground pepper. These sandstones are interbedded with dark and grey shales and marls and in conjunction with these are many hundreds of metres in thickness (Fig. 111). The sandstones almost always have a marine fauna amongst which fusulinids and goniatites are typical and brachiopods and plant remains are quite common. The study of their petrography has shown that these sandstones are typical graywackes, i.e. sandstones with grains of volcanic rocks.

Before the Revolution and in the first decades after it all the pepper sandstones were considered Artinskian and such they were commonly called, since they were found above the Upper Carboniferous limestones. Soviet geologists, however, have shown the contemporaneity of these 'Artinskian' sandstones and 'Upper Carboniferous' limestones. In various horizons of the sandstones different faunas were found. At the bottom the fauna is Middle and Upper Carboniferous; in the middle the fauna is Sakmarian and Artinskian, while at the top it is Kungurian. Thus the 'pepper' sandstones represent a facies formation ranging in age from the Middle Carboniferous to the top of the Kungurian. In more westerly regions the massive and bedded limestones are contemporaneous with the sandstones. The limestones were formerly considered as Upper Carboniferous, but were found to range from the Middle Carboniferous to the Kungurian. The relationships between the sandy and carbonate facies have been described in detail by V. D. Nalivkin (1949, 1950).

Upper Carboniferous

The movement of the Siberian Massif continued to the west and fold-belts came into existence on the western slope of the Urals and at their western flanks were deposited conglomerates and breccias similar to tillites. There are no macroscopic faunas in the coarse deposits and they are provisionally considered as Upper Carboniferous. On the western slope the facies are the same as in the Middle Carboniferous, but in its eastern regions sandstones and shales are dominant, although in the west they give place to limestones and dolomites.

In the Upper Carboniferous the first indications of the sub-Urals marginal downwarp appeared. In the eastern part of the downwarp sandstones and even conglomerates were deposited; in its central part clays and marls are found and in its western part limestones with reefs were formed.

Formerly the Upper Carboniferous was divided into two stages: Kasimovian and Gzhelian. In 1960 the Stratigraphic Committee adopted a third—Orenburgian—Stage, which corresponds to the *Pseudofusulina* Horizon. In the Lower Permian the Asselian Stage (previously *Schwagerina* Horizon) was recognized. Thus the succession of stages is:

Lower Permian
     Kungurian
     Artinskian
     Sakmarian
     Asselian

Upper Carboniferous
     Orengurgian
     Gzhelian
     Kasimovian

The detailed investigation of fusulinids has enabled the subdivision of the Lower Permian stages in Cisuralia as follows:

Kungurian Stage
Solikamskian Horizon
Irenian Horizon
Sarsian or Filippovskian Horizon

Artinskian Stage
Sylvenian Horizon
Saranaian Horizon
Burtsevo-Irginkian Horizon

Sakmarian Stage
Sterlitamakian Horizon
Tastubian Horizon
Lower Sakmarian sub-Stage

Each of these horizons is characterized by a definite assemblage of fusulinids, traceable throughout Cisuralia.

Lower Permian

Artinskian and Sakmarian epochs. These are the epochs of the maximal development of the Hercynian Orogeny. The whole of the eastern slope and the Central Urals became high, snow and ice covered young mountains, such as are the present-day Alps and Caucasus. These Hercynian mountains have not lasted. Their magnitude can only be judged from the thick conglomerates and sandstones of up to several hundred and sometimes 2000–3000m. The thickness of conglomerates suggests that the mountains were no less than 5000–6000m high. These mountains were situated near to the zone of accumulation of the clastic deposits and their position is estibated to be where the present Central Urals are. The composition of the conglomerates indicates that the boulders were transported by full and turbulent rivers such as the present-day Terek. The existence of such rivers depends on snow and ice capping of the mountains. Thus there has been a Lower Permian glaciation.

Artinskian and Sakmarian conglomerates are typical alluvial cones. To the west, away from the mountains their thickness rapidly decreases, the size of the pebbles diminishes and the number of sandy beds increases. The conglomerates, thus, pass into normal pepper sandstones and these in turn pass into clays and marls. In the deepest part of the downwarp, black plastic clays of small thickness (40–60m) accumulated. On the western slope of the downwarp the clays rapidly pass into marls and limestones. The latter are normally bedded, fragmentary and with fusulinids, brachiopods and bryozoa. Reef massifs are relatively rare amongst them.

A comparison of the distribution of conglomeratic formations and reef massifs (Senchenko, 1960) has led to the discovery of an interesting and important feature: where the conglomerates are found very thick reefs are absent. This is explained by the fact that conglomerates are related to rivers and therefore to the influx of fresh water, which kills the reef-builders.

Reef Facies reach their greatest development in the Sakmarian and Artinskian epoch, but they started in the Upper Carboniferous and disappeared in the Kungurian. The reef massifs have been described in detail by V. D. Nalivkin (1949, 1950) and A. A. Trofimuk (1950), while valuable data have been given by D. M.

Rauzer-Chernousova (1949). The reefs have attracted attention owing to the fact that they contain major oil and gas deposits and especially so in the Ishimbai region. There are hundreds of papers on these features, and they have been studied by the 'Permian' excursion of the 17th International Geological Congress (1937) and their description has been quoted in the guide for this excursion (southern route). Normally the massifs are buried under younger Kungurian and Ufian deposits and are found only in boreholes, but the number of such boreholes is large and in places the actual massifs (shikhans) are exposed and their form and structure is well known.

The reef massifs have an irregularly conical form (Fig. 112) with one or two apices. Their height is 300–500m and sometimes up to 600–800m while the diameter fluctuates from 2 to 20km. Normally the massifs occur individually, follow each other irregularly along the slope of the sub-montaine downwarp, or on rarer occasions unite into elongated ridges.

In a reef massif, bedding is absent or indistinct and the study of the faunal horizons has shown that they are almost horizontal and consist of porous, so-called sieve or sponge dolomites. The reef limestones and dolomites are quite pure. They are covered by black, muddy limestones, which have been deposited under conditions in which the reef-building organisms were killed off. Still higher follow the red beds of the Ufian Suite. The massifs form a belt some tens of kilometres wide, but which stretches along the Urals from the Polyudov Ridge (beginning of Timan) up to the Mugodzhars (Fig. 113).

Kungurian Epoch. The growth of the Siberian continent towards the west decreased, and at the end of the epoch ceased altogether. The fold mountains became of lesser height and the associated foothill zone of smaller dimensions; the thickness of the fan conglomerates and sandstones decreased near the mountains, there was a belt of small alluvial cones transitional into the sandstones of 'Artinskian' type. Then there was a belt of clays and marls corresponding to the axial part of the downwarp. To the west there was the region of accumulation of carbonates which were mainly dolomites, but sometimes limestones. The Kungurian Stage thus shows two closely related features; the sharp decrease in the size of the marine basins, and the development of bitter-salt lagoons in their place. On the bottom of such lagoons common salt, gypsum and rarely potassium salts were deposited. A Kungurian coal-bearing formation, which developed in the Pechora coal-field, is also of some considerable interest.

The succession in the Pechora coal-field is as follows. At the base lie Lower Carboniferous limestones and over them in the southern part of the fields are the *Schwagerina* limestones of the Asselian Stage. Over the rest, and greater part, of the basin Asselian and Sakmarian rocks are not recognized and are possibly absent. The various horizons of the Lower Carboniferous are directly overlain by a coal-bearing formation; its lower part is Artinskian and is distinguished as the Yunyakhian Suite (800–1900m). It consists of sandstones and argillites with beds of limestones and marine faunas. Higher up lies the Vorkutian Suite (Kungurian) with a peculiar fauna and flora. The suite consists of continental sandstones and argillites. At the bottom the suite has beds with marine faunas, but at the top it is entirely fresh-water. The suite is everywhere coal-bearing and contains the best coals in the coal-field. Its thickness varies, reaching 2400m. The Upper Permian begins with the Pechora Suite (1000–3500m), which is the upper coal-bearing formation with coals of great thickness, but poor quality. It consists entirely of continental

FIG. 112. Tastuba (a rocky hill) reef massif of Lower Permian limestones. *Photograph by V. D. Nalivkin (1943).*

FIG. 113. Sakmarian reef massifs on the river Vishera, in the Northern Urals. *Photograph by V. D. Nalivkin (1948).*

deposits such as sandstones and argillites, with beds of marls. Nearer to the Urals there are coarse sandstones and conglomerates. In the north the suite is replaced by the Paemboiian or the Yangareiian Suite. The succession is completed by a supra-coal-bearing formation (1200m) of red beds and brown sandy-clayey rocks without coal. In places there are thick sandstones and conglomerates. In the north these are replaced by the so-called Kheiyaginian Suite (1800m) of the same composition, but according to the latest information this suite and its analogues belong to the Trias.

The deposits with marine faunas form only thin groups and in the Central and Northern Urals have a subordinate significance. Only in the Arctic Urals and the Arctic in general do they have a wide distribution and development. To the south lagoonal sediments (salts) are widespread and thick. This is caused by a continuous depression of the sub-montaine downwarp within which the lagoons had formed.

Upper Permian

The fold-generating movements were very weak or non-existent. The elevation of the western slope and Cisuralia caused a regression of the sea and drying out of lagoons. Large sub-montaine plains originated on which the hot desert conditions caused the deposition of Ufian red beds, which are generally thin (100–200m), but in the Southern Urals, near to the mountains, reach 1500–2000m. The clastic material of the red beds was derived from the, by then, low and eroded Urals Mts. Permanent rivers were absent or rare and the transport of sediments took place by sheets of water which uniformly spread the sediments (sands and clays) over the large plain, forming in the north the Solikamskian limestones and in the south, near Orenburg, formations of Kazanian limestones and clays with rock-salt being deposited near the sea-shore. No analogues of Vetluzhian or Tatarian stages have so far been recognized, but their presence in the Southern Urals is undoubted. In other regions upper horizons of Ufian are possibly of Vetluzhian and Tatarian age.

The mountain-forming processes ceased operating at the end of the Upper Permian and beginning of the Trias and the Urals became converted into almost a flat peneplain. On the eastern slope of the Southern Urals the plain gradually subsided and became covered by a thick Upper Permian-Lower Trias effusive sedimentary formation. This formation at Kopeisk (Chelyabinsk region) has a maximum thickness of 1700m, penetrated by a borehole. At the bottom on metamorphosed pelites, lie basal sandstones and conglomerates (22m), followed by basalts and tuffs (555m), followed by interbedded basalts and sandstones and shales (442m). In the latter suite there are Upper Permian plant remains. Still higher lies a tuffogenous suite (377m) of conglomerate tuffs, sandstone tuffs and silts. This suite is overlain by a clayey-silty suite of 300m with *Estheria*, spores and Lower Triassic pollen. Higher still is a sandy, pebbly coarse clastic suite with a Rhaetic flora marking the beginning of the Rhaetic-Jurassic coal-bearing Chelyabinsk Formation (Malyavkina and Kareva, 1956; Kareva, 1958, 1950). A similar effusive-sedimentary Upper Permian-Lower Trias Formation has been found by drilling in the Tyumen region and to the west near the town Ishim and Kushmurun in the Turgai Depression (or 600km away), and in the north in the Bulanash-Elkin region.

The effusive-sedimentary formation has been called 'the Turinian Series'. It is analogous to the traps of the Siberian Platform and the Kol'chuginian and Mal'tsevian suites of Kuzbas. The development of the Turinian Series at such great distance from Kuzbas and the Siberian Platform is very interesting. The Trap Formation is typically continental suggesting that in the Upper Permian all the Western Siberian Lowlands and the Urals became a platform.

MESOZOIC AND CAINOZOIC

As the Palaeozoic rocks of the Urals were typically geosynclinal, so the Mesozoic and Cainozoic were typically platform formations. Only Rhaetic-Liassic rocks reach more than 1000m in thickness, are weakly metamorphosed and deformed into gentle folds. The rest of the Mesozoic and Cainozoic are thin, unmetamorphosed and unfolded, and are known as rubbly rocks (unconsolidated rocks). Local violent disturbances are pseudotectonic. The Mesozoic can be divided as follows:

1. Rhaetic-Liassic Coal-bearing Formation.
2. Jurassic continental deposits.
3. Upper Jurassic and Lower Cretaceous marine deposits.
4. Continental Cretaceous and Palaeogene.
5. Marine Upper Cretaceous and Palaeogene.
6. Continental Neogene and Anthropogene.
7. Marine Neogene and Anthropogene.

These are the same divisions as in the Russian Platform and Western Siberian Lowlands, and have the same features, fauna and flora. These formations are mainly developed in the Ural foothills and are almost absent from the true Urals.

The Rhaetic-Liassic Coal-bearing Formation is economically significant and coals are extracted in the Chelyabinsk, Veselovo, Bogoslavsk and Volchansk deposits. It is only developed on the eastern slope and consists of terrestrial, lacustrine and lagoonal deposits of brown and grey sandstones and clays which are often carbonaceous, thick and possess enormous coal seams. The Coal-bearing Formation used to be called the 'Chelyabinskian Series'. It lies on the Upper Permian-Lower Triassic Turinian Series, already described. The age of the Chelyabinskian Series is considered as Upper Trias (Rhaetic), but its great thickness (up to 4000m) and its flora indicate that the Upper Suite of the series is Liassic. Formerly it was thought that the Chelyabinsk Series consisted of two suites: the lower—Chulyakian conglomeratic (500m) and the upper—Korkinian, coal-bearing (1500m). A detailed investigation of spores (Malyavkina and Kareva, 1956) and of stratigraphy (Kareva, 1958, 1959) has shown that the series consists of four suites (from bottom to top); Glubokovkian, Kopeikian, Karierian and Sugoyakian. Each suite is a mesorhythm of sedimentation and consists of two sub-suites—the lower (conglomeratic-sandy) and the upper (sandy-clayey with coals). The conglomeratic sub-suite represents typical deposits of the foothills and there were, therefore, mountains with emerging alluvial cones. The coal-bearing sub-suites are deposits of plains covered by marshes and dissected by slow-flowing rivers. The four-fold repetition of the suites indicates a four-fold repetition of tectonic movements which have produced the mountain chains. The epochs of tectonic movements are interrupted by the epochs of quiescence when the large, marsh-covered plains originated.

In the Chelyabinsk coal-field all the four suites with the sub-suites are found: the Slubokovkian consisting of conglomerates (up to 450m) and coal-bearing horizons (up to 1000m), the Kopeikian of conglomerates (250–300m) and with coal seams (900–1700m); the Karyerian (300m of conglomerates and 1200m of coal-bearing) has a coal seam, the second lowest, which is 100m thick; the Sugoyakian (150m altogether). In the Bulanash-Elkin field there are only the three lower suites of a total thickness of up to 2000m. In the north in the Bogoslovian, Veselovian and Volchanskian basins there is only the Karyerian Suite with thicknesses of 220, 350 and 500m.

Owing to the initial phases of the Cimmeridian Orogeny the coal-bearing strata are folded into simple and gentle folds, which are broken up by normal faults and thrusts. These strata have been preserved in grabens of much later origin. During the formation of such narrow and long grabens the margins of the massifs were deformed, but only over short distances and without any metamorphism of the rocks.

The Jurassic continental deposits are developed both on the western and eastern slopes and consist of sands and clays which are often white or brightly coloured reaching small thicknesses of ten of metres or rarely more. There are few organic remains except for leaf-impressions which sometimes render the rocks carbonaceous. The lacustrine deposits of Jurassic age often contain useful mineral deposits, such as iron ores, fire-clays, and seams of brown coal and bauxite.

The Cretaceous continental deposits are also widely distributed throughout the Urals and lithologically are indistinguishable from the Jurassic. Their age is determined palaeobotanically. On the eastern slope of the Central Urals lacustrine Lower Cretaceous rocks have several deposits of gibbsitic bauxite of industrial significance. In the Southern Urals these are accompanied by nodular iron ores.

The Marine Upper Jurassic and Lower Cretaceous deposits have been laid down in a sea which advanced from the north over the Siberian Lowlands and spread up to the eastern slope of the Polar and Northern Urals. These deposits consist mainly of black clays and sands, which are normally thin, fossiliferous and not infrequently glauconitic. On the western slope of the Southern Urals, near Orenburg and along the Mugodzhars, there are marine Upper Jurassic and Lower Cretaceous deposits of the type found over the Russian Platform although the Uralian deposits contain pale limestones with a Lower Volgian fauna, which is not found on the platform.

The succession on the eastern slope of the Northern Urals (Trudy . . . 1957) serves as an example of Jurassic and Lower Cretaceous deposits. At the base of the succession, the Palaeozoic is covered by a crust of weathering consisting of variegated kaolinite clays (10–20m). Higher up lie the Upper Jurassic coal-bearing strata (150–200m), known as the Obian Series and divided into two or three suites. The series, of Callovian and Lower Oxfordian age, consists of conglomerates and sandstones at the bottom changing upwards into dark clays and sands with workable brown coals. The series is covered by a marine formation (180–250m), which is Upper Oxfordian-Barremian in age and can be divided into a series of suites based on marine faunas. The rocks of the series are grey, compact mudstones and siltstones and in the middle it has a rather characteristic and continuous group (10–20m) of ferruginous, glauconitic sandstones (Fedorovskian Suite) of Volgian and Lower Valanginian age. The lower group of clays is Upper Oxfordian and Kimmeridgian and the upper is of Middle and Upper Valanginian and Hauterivian-Barremian ages. The succession ends by a continental coal-bearing formation with a seam of brown coal. The thickness of the formation varies from 30–150m, and it belongs to the Barremian-Aptian (Table 24).

The Marine Upper Cretaceous and Palaeogene deposits on the eastern slope are widely distributed from the Arctic Urals, where they pass over to the western slope into the basin of the Usa, and end at the Mugodzhars. In the south they join through the Turgai Depression and the valley of the Urals to the deposits of the Russian Platform and Central Asia. Throughout their extent the marine Upper Cretaceous and Palaeogene deposits pass laterally into the analogous deposits of the Western Siberian Lowlands, where they are widespread, but hidden under the cover of younger sediments. Along the slope of the Urals they outcrop continuously, and lie transgressively and horizontally upon the older rocks.

The Upper Cretaceous and Palaeogene of the eastern slope of the Urals are deposits of a marginal zone of a vast enclosed sea which covered almost the whole of the area of the Western Siberian Lowlands (see p. 293 and Table 24). These deposits consist of sandy and clayey rocks with widespread glauconitic sands and gaizes. Their total thickness reaches 200–400m, and apart from the effects of local pseudo-tectonics they remain entirely horizontal.

The history of the Upper Cretaceous and Palaeogene basins is complex, has not been completely worked out and is therefore debatable. There is evidence for no less than three transgressions: the Turonian, the Santonian and the Maastrichtian-Palaeocene. Each of these transgressions involved extensive erosion of the pre-existing sediments, thus complicating palaeogeographic reconstructions. Four channels, in particular, are being disputed. The first connected the Santonian Sea of the eastern slope to a bay, which in the Northern Urals existed in the valley of the Usa. The existence of the channel was claimed by A. A. Chernov and denied by A. D. Arkhangelskii, who considered that the Usa embayment was the terminal end of a narrow bay, which lay along the valleys of the Kama and Pechora and connected to the Central Russian Sea. Such a long, narrow bay is very improbable and the point of view of A. A. Chernov is based on better evidence. The second channel is suggested along the Khatanga Depression. It was, supposedly, connecting the Western Siberian Turonian Sea to the Arctic Ocean. The suggestion is unlikely, since in all the regions further to the north and north-east the Upper Cretaceous and Palaeogene are represented by continental deposits. The third and fourth channels were to the south (Orian and Turgaiian) and are generally accepted for younger seas. Boring has demonstrated that the Turgaiian channel lacks the deposits of the Cenomanian Sea, which is explained by some by the non-existence of a channel, but is probably due to subsequent erosion. The succession on the eastern slope of the Southern Urals (Trudy . . ., 1957, Sigov) is as follows:

Khantymansiian Suite (Albian-Cenomanian)—grey clays with beds of sand (30–100m); almost barren. These are terrestrial near-shore deposits.

Kuznetsovkian Suite (Turonian)—greenish-grey silts (150–180m). In the west nearer to the land there are beds of glauconitic sands with phosphorite pebbles. In places the Mugaiian iron-ore formation replaces the suite. The formation is widely developed in the Southern Urals (Ayatian deposits) and Northern Urals (Marsyata). It consists of ferruginous sandstones with beds of oolitic ore.

Slavgorodian Suite (Santonian-Coniacian)—greenish argillites, and glauconitic sands, silts and gaizes (total of up to 130m) in the south. The suite has a marine fauna with *Pteria tenuicostata*.

Gany'kinkian Suite (Campanian-Maastrichtian). At the base there are iron ores with glauconitic sands with phosphorites. Higher up there are clays, marls and sandstones, with beds of gaizes (total up to 80m).

Talitskian Suite (Danian-Palaeocene). Manganiferous strata (Polunochnoe deposit) lie at the base (up to 34m) with diatomites or argillites (Marsyatkian sub-Suite, 150m) in the middle. The upper Ivdelian sub-Suite, which is Upper Palaeocene and up to 200m thick, consists of dark gaizes and argillites with Foraminifera.

Lyulinvortian Suite (Eocene). The gaizes and glauconitic sandstones (up to 80m), at the bottom, are followed by diatomites and then muddy sandstones with a total thickness of up to 150m. The fauna includes shark teeth and foraminifera and is sometimes divided into the Tasaranian and Akchatian suites.

Cheganian Suite (Lower Oligocene)—greenish paper shales (up to 80m); good marker horizon with shark teeth and lamellibranchs.

Turgaiian Suite (Middle and Upper Oligocene)—continental sandy-clayey deposits (100–130m).

Aralian Suite (Lower? Miocene). Variegated, green and dark-grey clays (30m).

Kustanaiian Suite (Middle and Upper Pliocene) of grey and brown sands and marly clays with a fresh-water fauna (30m). This succession basically corresponds to the successions of the Western Siberian Lowlands and the Turgai Depression (p. 295).

On the western slope of the Urals marine Upper Cretaceous and Palaeogene are developed only in the extreme south, in the regions of Orenburg and the Mugodzhars. To the north only the Santonian sands with *Pteria tenuicostata* are found in the region of the Ufa and Krasnoufimsk. To the north of the latter locality the marine Upper Cretaceous and Palaeogene are absent. In the south the Upper Cretaceous sections exhibit chalk, marls and nummulitic sandstones (in the extreme south of the Mugodzhars) and indicate an open and warm sea, which connected with the seas of the Crimea, and Caucasus and Central Asia. Another feature of the western slope is the marine Palaeogene succession. Almost all of the Eocene and Oligocene consist of continental sediments. The Urals and the Mugodzhars as such were land masses even during the Santonian when the sea reached its greatest extent. The marine Upper Cretaceous has deposits of phosphorites, chalk and glauconite. In the Palaeocene of the eastern slope of the Northern Urals there have been found manganese ores (Polunochnoye) and in the south there are known deposits of phosphorites, diatomites, clays and sands.

Marine Neogene and Anthropogene. These formations, as distinct from the Upper Cretaceous and Palaeogene, have a negligible distribution. The marine Neogene of the eastern slope is absent and on the western includes the Akchagylian sediments which reach Cisuralia in the Ufa region (p. 122).

The marine Anthropogene is represented by the deposits of the Boreal Transgression, which are widespread on the northern shore of the Russian Platform and the Western Siberian Lowlands (p. 298). These deposits are developed in the Pai-Khoi and Arctic Urals regions and consist of sands, clays and more rarely gravels and shell-gravels no more than 20–40m thick and commonly less. These sediments form river and shore terraces. They have a boreal fauna of lamellibranchs and gastropods. In the east of the Arctic Urals, near the shore, there is the Yamalian Series (up to 200m) which is marine (p. 298).

The Continental Neogene and Anthropogene are found everywhere. They cover especially vast areas in sub-montaine plains, along the valleys of the present-day and former rivers, which penetrate far into the range, but in isolated areas remain as elevated, formerly peneplained districts which at present form quite high, flat ranges of hills.

The thickness of the continental Neogene and Anthropogene is variable and normally does not exceed a few tens of metres, but in isolated depressions reaches 400–500m. The lithology is relatively homogeneous, being sands and clays and their mixtures. Coarse-grained rocks such as gravels, boulder deposits and breccias are rarer and amongst the latter, in the Arctic and Northern Urals, there are frequent ground moraines. Further to the south stone-rivers and sheets are developed as a result of the fragments of ancient quartzites sliding down on slopes and river valleys and filling them. Fluviatile and lacustrine deposits are widespread and the former on the eastern slope often contain alluvial gold and platinum and, on the western slope, diamonds.

Along the western slope of the Urals from Ufa to the Sakmara there is a belt of narrow depressions filled by Mesozoic, Palaeogene and Neogene rocks. On the Oligocene and Lower Miocene sands and clays (50–60m), which are in places coal-bearing, lies a Lower Miocene coal-bearing formation (150–200m) of grey clays and brown coals, the latter as much as 140m thick. Higher up lie Middle and Upper Miocene clays and sands (60m). The Pliocene (200m) consists of sands, clays and gravels. The Anthropogene loams (20m) top this peculiar and important succession.

The Mesozoic and Cainozoic of the eastern slope is directly continuous with the Mesozoic and Cainozoic of the Western Siberian Lowlands (Tables 24, 25). A majority of the suites found in the Lowlands can also be identified on the eastern slope of the Urals. The proximity of the mountains, however, is reflected in their lithologies, which often involve the use of a series of local names.

TECTONICS

The Urals represent a typical Palaeozoic geosyncline, the basic features of which are determined by the Hercynian and to a lesser extent Caledonian and Precambrian orogenies. On the eastern slope and in the extreme south of the Mugodzhars respectively the initial phases of the Cimmeridian Orogeny and the middle phases of the Alpine Orogeny are manifest.

STRUCTURES

Four structural stages are diagnosed: (1) Archeozoic and Lower Proterozoic, (2) Upper Proterozoic and Palaeozoic, (3) Rhaetic-Liassic, (4) Meso-Cainozoic. The first and the third groups have a very limited distribution and the whole of the Urals are in principle formed by the structures of the second and fourth storeys.

Archeozoic and Lower Proterozoic Structures

In the Southern Urals these structures are only exposed in the region where the Taratashian Suite is found, while on the western slope they are found to the north-west of Zlatoust. The structures resemble those of Krivoi Rog, but are more complex and show microfolds in jaspilites interbedded with the iron ores. Gneisses and schists are highly deformed (Olli & Romanov, 1959). In the Northern and Central Urals the Archeozoic and Lower Proterozoic are unknown and probably unexposed. In the Arctic Urals the Lower Proterozoic are rarely exposed and then strongly dislocated.

Upper Proterozoic and Palaeozoic Structures

These structures affect the greatest part of the exposed Ural Geosyncline, and are overlain by Meso-Cainozoic structures. The Proterozoic and Palaeozoic structures vary, depending on constituent rocks, their age and situation with respect to compression.

Three groups of structures can be identified on the basis of their age: (1) Baikalo-Caledonian, affecting the Upper Proterozoic and Lower Palaeozoic rocks; (2) Early Hercynian, affecting the Middle and Upper Palaeozoic; (3) Late Hercynian, affecting the Upper Permian and Lower Trias.

The Upper Proterozoic has several stratigraphic breaks, as well as the breaks between the Upper Proterozoic and Lower Palaeozoic, between the Upper and Middle Proterozoic and the Middle and Lower Proterozoic. On the western slope these breaks are not associated with angular unconformities and as a consequence the structures are the same throughout and the strata are involved in identical large-scale tectonics and show the same grade of metamorphism (Fig. 114).

There are many sections showing monoclinal successions of beds dipping uniformly in one direction. The sections involve Upper Proterozoic, Devonian, Carboniferous and Lower Permian strata. These relationships indicate an almost complete absence of the effects of Baikalian and Caledonian orogenies. Different relationships are found in the extreme west and on the eastern slope. In the extreme west, beginning from the Timan and the Polyudov Ridge, the Middle Devonian is transgressive over the Lower Palaeozoic and Sinian, which are more deformed and metamorphosed. This is also observed in the easterly inliers of the Southern Urals, where the Lower Carboniferous is strongly unconformable on the ancient rocks. On the eastern slope, acid and ultrabasic Caledonian intrusions are well represented, indicating distinct and powerful manifestations of the Caledonian Orogeny.

Lithological composition, as everywhere, has a strong influence on the structures. Thick massive limestones, dolomites and quartzites, even where strongly compressed, as on the eastern slope, fold into relatively simple, large anticlines and synclines. These structures are not complex even when they are dislocated and dragged on the underlying pelitic rocks. Conversely the thinly bedded plastic rocks get folded into numerous small and complex folds even where compression has been relatively weak. As a result disharmonic folds originate between the strongly folded pelites and the weakly folded, overlying limestones.

The form of the structures is influenced most strongly by the direction of compression, which determines their position. During the Hercynian Orogeny—the main one in the Urals—the pressure was from the east, as a result of the movement of the Siberian Massif. Consequently, in the Eastern Urals the folds are much more complex, more deformed and include the greatest number of intrusive massifs. In the central part of the Urals the folds are simpler and the Caledonian structures predominate over the Hercynian. Lastly, to the west of the western slope, the folds are simplest and the intrusions are almost absent or are of Baikalian age. The metamorphism conforms to the degree of folding, being most violent in the east and the least vigorous in the west.

The zone of *green rocks* deserves some attention and occurs on the eastern slope of the Central Urals within which the complexity of structures and metamorphism of the Palaeozoic rocks is of a high degree. The sedimentary sandstones and shales become converted into crystalline garnet-mica schists, mica-kyanite schists and even into paragneisses. The limestones of the green zone are recrystallized into marbles. Lenses of bauxite and bauxitic rocks have been converted into corundum-bearing varieties. The lavas and tuffs have changed into green-schists with secondary minerals such as chlorite, epidote, sericite, albite. At the same time the primary structures of the rocks have disappeared on conversion into quartz-sericite schists, chlorite-albite schists, and chlorite schists, all more or less pyritized. Here they are chalco-pyritic. Acid rocks, poor in alkaline earths, are often changed into berezites and rocks rich in alkaline earths into listvinites. The ultrabasic rocks are often serpentinized resulting in the formation of serpentine, asbestos and talc, as well as actinolitic and chloritic rocks. A great majority of the aforementioned metamorphic rocks are green, thus the term 'green-rock zone'. Laterally these rocks gradually pass into normal, unaltered rocks. Consequently the boundary of the zone is indistinct. In the north the zone begins to the north of Lower Tagil and in the south near Mias. Its width varies within the limits of tens of kilometres. On a geological map it is obvious that the zone reflects the outcrop of the crystalline basement of the Russian Platform in the Ufa Plateau. This outcrop causes a clear deflection of all Central Ural structures, which bend round it. At this projection the pressure and dynamic metamorphism reached the highest grade. This conclusion is confirmed by the fact that green rocks have suffered the most intensive cataclasis, faulting and deformation. Hydrothermal and other diffusive processes occurred at the same time as the dynamic metamorphism. The zone has a series of large deposits of useful minerals such as asbestos, gold, chalcopyrite ores, corundum, precious stones talc, etc.

The structures of the eastern slope show a considerable complexity even beyond the green-rocks zone, showing isoclinal, recumbent and fan-shaped folds. The folds are broken by numerous penecontemporaneous thrusts and transformed into scales (nappes). Even greater complications are introduced by numerous intrusive massifs of, sometimes, great size. All these important characters are evident from the general geological maps and the maps of the Urals. As an example of the section with isoclinal folds, scales and intrusive massifs, one can quote the section (Fig. 115) to the south of Magnitogorsk investigated by L. S. Librovich. The structures of the eastern part of the Central Urals can be seen in the valley of the Pyshma (Fig. 116) to the north of Kamensk where they are somewhat simpler, but still very complex. As examples of disharmonic

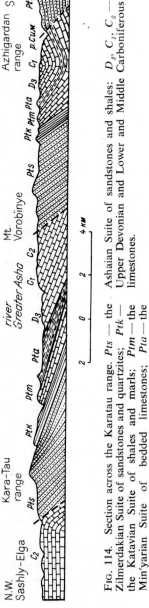

FIG. 114. Section across the Karatau range. Pts — the Ashaian Suite of sandstones and shales: D_3, C_1, C_2 — Zilmerdakian Suite of sandstones and quartzites; Ptk — Upper Devonian and Lower and Middle Carboniferous the Katavian Suite of shales and marls; Ptm — the limestones. Min'yarian Suite of bedded limestones; Pta — the

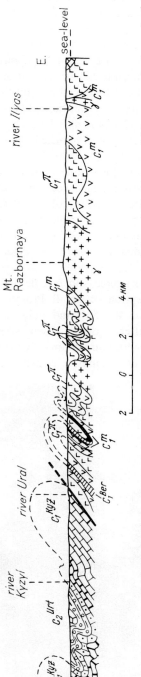

Fig. 115. Section across the Kizil region to the south of Magnitogorsk. According to L. S. Librovich (1944). C_2^{urt} — the Urtazymskian Suite of conglomerates, sandstones and beds of limestones; C_1^{Kyz} — the Kizilian Suite of limestones which are sandy towards the top; C_1^{Ber} — the Berezovkian Suite of limestones, tuffs and carbonaceous shales; C_1^m — diabases and porphyrites; C_1^π — porphyrites and felsites; γ — Hercynian granites.

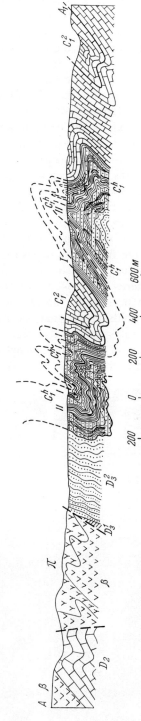

Fig. 116. Cross-section along the river Pyshma. According to I. I. Gorskii (1944). D_2 — Middle Devonian limestones; D_3^1 — Frasnian limestones; D_3^2 — Famennian and at the top Tournaisian shales, sandstones and limestones; C_1^h — the coal-bearing sequence; I, II — clays with coal seams; C_1^z — Visean and Namurian limestones; β — Devonian porphyrites and tuffs; π — Devonian albitophyres and tuffs.

folding one can point at the small, very complex and numerous structures of the coal-bearing formation.

The structures of the Urals have been studied in detail by I. I. Gorskii (1943) who has provided much information on its geology. Amongst many hundreds and thousands of structures, he had selected several gigantic, complex structures, hundreds of kilometres long and tens of kilometres wide (anticlinoria and synclinoria). On the eastern slope they are less obvious owing to the existence of large thrusts and intrusive massifs. The large Magnitogorsk synclinorium can be quoted as an example. It is situated to the south of Magnitogorsk and can be traced beyond Orsk. In its core is Carboniferous, surrounded by Devonian and Silurian. The synclinorium is about 500km long and about 100km wide. On its flanks there are tens of folds of second order, which are tens of kilometres long, and hundreds of third order structures of smaller magnitude.

The structures of the central region of the Urals are also quite complex and deformed and especially so in the zone of Caledonian intrusions, near the watershed ridge, but they are nevertheless simpler than the structures of the eastern slope. This is especially obvious in the regions of Upper Proterozoic and Lower Palaeozoic rocks, which include thick quartzites and dolomites, as for instance in the region of the Bakal iron deposit (Fig. 117). In the districts where the rocks are pelitic the folds are rather strong.

The structures of the western slope are much more simple and less deformed. The intrusive massifs are absent, with the sole exception of small Proterozoic plutons. The folds are regular, normally with dips of 40°–50° and only rarely greater. The folds follow the regional strike one after another. Their size is measured in tens of kilometres in length and breadth. The folds are commonly broken by thrusts and normal faults, but undisturbed folds are also common. Nearer to the

Fɪɢ. 117.   Bakal iron-ore region. Bakalian quartzites in the near distance.

central part there are in places large anticlinoria and synclinoria. For instance in the north of the Southern Urals, even a small scale general map to the scale of 1:7 500 000 shows a large anticlinorium with Proterozoic in its core, surrounded by the Cambrian, Silurian, Devonian, Carboniferous and Lower Permian. Its length is 400km and its width 75–150km. The structures of the western slope in the proximity of the petroliferous and coal districts have been studied and mapped in detail, such as the structures of the Kizel coal basin in the Central Urals. These structures are shown on the guide map of the Permian excursion (northern route) of the 17th International Geological Congress (1937). Fig. 118 shows a somewhat simplified southern part of it. The section through its northern part along the river Kos'va is shown on Fig. 119. The map shows the form of the structures and the principal

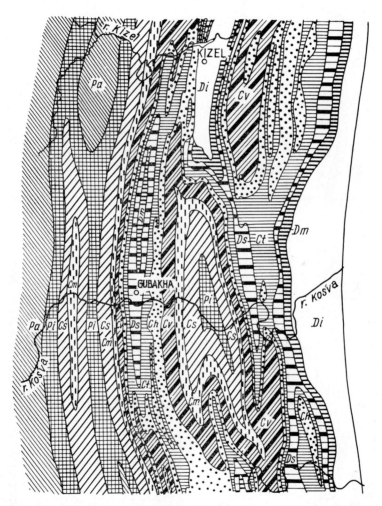

FIG. 118. Map of the central part of the Kizelian coal-field. According to
I. I. Gorski (1937).

Di — Lower Devonian;   Dm — Middle Devonian;   Ds — Upper Devonian;
Ct — Tournaisian;   Cv — Visean and Namurian;   Ch — the Coal-bearing
Formation;   Cm — Middle Carboniferous;   Cs — Upper Carboniferous;
Pi — Lower Permian;   Pa — Artinskian.

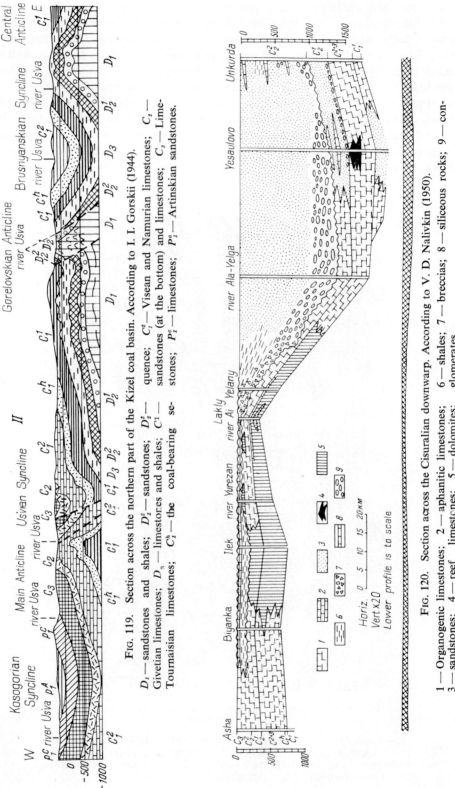

FIG. 119. Section across the northern part of the Kizel coal basin. According to I. I. Gorskii (1944).

D_1 — sandstones and shales; D_2^t — sandstones; D_2^g — Givetian limestones; D_3 — limestones and shales; C^1 — Tournaisian limestones; C_1^h — the coal-bearing sequence; C_1^q — Visean and Namurian limestones; C_2 — sandstones (at the bottom) and limestones; C_3 — Limestones; P_1^o — limestones; P_1^a — Artinskian sandstones.

FIG. 120. Section across the Cisuralian downwarp. According to V. D. Nalivkin (1950).

1 — Organogenic limestones; 2 — aphanitic limestones; 6 — shales; 7 — breccias; 8 — siliceous rocks; 9 — con-
3 — sandstones; 4 — reef limestones; 5 — dolomites; glomerates.

thrusts. The section shows a conformable sequence of strata, beginning from the Lower Devonian and ending with the Lower Permian.

The Sub-Urals Downwarp is a peculiar geological tectonic structure of great economic importance. The downwarp was discovered in the thirties and described in detail by V. D. Nalivkin (1949, 1950). The sub-Urals downwarp is situated between the Russian Platform and the Uralian fold-belt. The downwarp can be traced along the Urals, beginning from the Pechora basin and continuing to the Mugodzhars. From the Mugodzhars it continues to the south-west of Astrakhan around the Yaikian Massif (Russian Platform). Over its great length (about 1500km) the downwarp is a wide, rather flat syncline of 5–6km in depth and 60–80km wide. Thus the flanks dip no more than 1–3°, and usually tens of minutes. The exaggerated and unexaggerated profiles of the downwarp are shown in Fig. 120. The formation of the downwarp started in the Middle Carboniferous, at the same time as the initiation of the Hercynian Orogeny, and was completed in the Kungurian epoch by the formation of thick beds of gypsum, rock-salt and potassium salts. The facies distribution in the downwarp is shown on the same illustration. In the east, as has already been pointed out, in the Lower Permian coarse deposits, such as conglomerates and sandstones, predominate. Then follow sands and clays and the centre of the downwarp is filled with clays. Along the western edge of the Cisuralian downwarp, on the Russian Platform, there are reef massifs which are in places oil-bearing. The composition of the facies indicates that the downwarp most frequently was a deep marine basin and at times, especially during the Kungurian epoch, it was a zone of marshes and nearshore plains. The Urals miogeosyncline and eugeosyncline are typical and serve as examples of such tectonic structures established by N. S. Shatskii. The miogeosyncline of the Urals was where its present-day western slope is. Within it the structures are relatively simple, weakly deformed and not far displaced. Consequently igneous rocks are not widespread and often are absent, while the Palaeozoic is weakly metamorphosed. Geophysical investigations have shown that on the western slope the crystalline basement of the Russian Platform lies at a comparatively shallow depth. The Uralian eugeosyncline occupies the central part, the eastern slope and the adjacent part of the Western Siberian Lowlands. The eugeosyncline shows widespread Palaeozoic magmatism, complex shapes of tectonic structures and a high grade of Palaeozoic metamorphism.

Structure of the Rhaetic-Liassic Coal-bearing Formation

As distinct from the other, younger Mesozoic deposits the Rhaetic-Liassic coal-bearing formation of the eastern slope of the Urals is everywhere folded into gentle folds, sometimes accompanied by thrusts. The folds are regionally distributed, are not accompanied by metamorphism and can be regarded as manifestations of a weak orogenic movement. Since this orogeny occurred contemporaneously with the accumulation of sediments it is considered Rhaetic-Liassic and represents the initial phases of the Cimmeridian Orogeny. Similar folds are present in the Jurassic coal-bearing strata of the Angarids. They are known in the Northern, Central and Southern Urals, in Karaganda, Central Asia, and Kuznetsk Basin, Tuva and Transbaikalia. Important features of the Rhaetic-Liassic coal-bearing strata are normal faults and thrusts. These faults are of different ages and throws and the sense of downthrow. Some of the faults cut across the formation. The faults are relatively small-scale, a few kilometres long, have throws of a few tens of metres and are penecontemporaneous with folding. An example is the fault cutting across the coal-bearing formation of the Bogoslovian coal-field. A great majority of the faults are marginal and are contemporaneous with folding or later than it. Often repetition of movements can be detected on the faults. Sometimes the marginal faults are large, as for instance the Chelyabinsk coal-field which is margined by a fault 250–300km long and with a throw of 1500–3000m. Its fault-breccia is tens of metres thick. The thrust planes often show deflection of the adjacent layers into sometimes very complex structures which, however, rapidly disappear and are not accompanied by metamorphism.

Meso-Cainozoic Structures

The Meso-Cainozoic structures are typical platform features. The Mesozoic and Palaeogene deposits in places show large, gentle, almost unnoticeable brachy-anticlines such as are found in the Palaeozoic of the Russian Platform. Such structures on the western slope are of special interest since they may have gas and oil. In the extreme south at the southernmost termination of the Mugodzhars the Upper Cretaceous and Palaeogene deposits form in places more obvious brachyanticlinal folds such as the Chumka-Kul' (Kabanie) lake anticline, which is the continuation of the Mugodzhars and is obvious on the general maps.

Against the background of almost horizontal strata there are in places sharp folds, sometimes broken by faults. These folds occur over small areas, which are surrounded by almost horizontal beds of the same age as those involved in folding. These beds are of the same degree of lithification throughout and the folds must be due to pseudotectonic movements, such as landslips, ice dislocations and sinks. It is an error to consider them, as is sometimes done, as manifestations of Alpine folding.

The Neogene and Anthropogene beds are not involved in folding, are almost horizontal or show original dips, associated with deposition on slopes. Even in the south of the Mugodzhars these beds are transgressive over Mesozoic and Palaeogene folds. Nevertheless the Neogene and Anthropogene strata are affected by tectonic block movements, which affect regions of Palaeozoic and Precambrian rocks, uplifting them as horsts or more rarely causing the formations of depressions; thus producing the present-day relief. The Palaeozoic massifs in the process of these movements affect the overlying, almost horizontal, Mesozoic and Palaeozoic cover to the extent of 1500–1800m.

The dimensions of the young blocks are variable from gigantic massifs, hundreds of kilometres long and tens of kilometres wide, forming whole mountain ranges, to small horsts of several kilometres or even metres across. The blocks normally show almost vertical movements on vertical planes, but in places they are inclined and move on inclined surfaces. One such gently-inclined thrust has been established in a water-hole near Ivdel', on the eastern slope of the Northern Urals near the Lozvinian jetty and 20km to the north. Cross-sections established by drilling are shown in Fig. 121. The upper cross-section shows a displacement of 800–900m with the Devonian overlying the Palaeogene.

The lower cross-section shows the same upthrusted block, 20km to the north where the displacement is much less. The Palaeogene in the section is sometimes vertical and even overturned. Such deformation involves frequent minor folds, which have even been considered as the manifestations of the Alpine Orogeny. In fact this is not a regional deformation, but the effect of block movements. V. A. Obruchev often pointed out the significance of such structures in the Altais and the Tian Shian. The blocks were formed at the end of Neogene and Anthropogene periods. The blocks are elongated in the direction of the mountain ranges, but sometimes are transverse to the strike forming sharply delineated massifs. The Karatau Block in the Southern Urals and the Polyudov Ridge in the northern part of the Central Urals are examples of such transverse blocks. The blocks are the youngest structures in the Urals and are responsible for its relief. It is interesting to point out that the faults associated with the blocks were relatively shallow, extending downwards for several kilometres, which is confirmed in that they are never associated with outpourings of basalts, which is characteristic for other regions of young block movements (Baikal). The faults are, nevertheless, sufficiently deep to affect the circulation of underground water (Budanov, 1957). With the faults there are associated numerous springs, which are sometimes warm or hot, and also considerable currents of subterranean water found in boreholes. The small depths of the faults explains the weak seismicity over the Urals compared with other regions of young block faulting.

The young Anthropogenic platform-type structures are interesting and are found in the north of Transuralia. A ridge known as Muzhinskian Urals serves as an example. It is situated to the west of the Northern Muzha (lower Ob') and has been described by N. G. Chochia and his co-authors (1961). The range of Muzhinskian

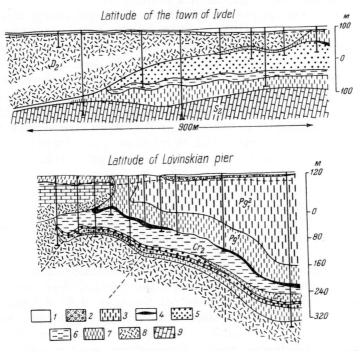

Fig. 121. Young block thrusts in the Ivdel region on the eastern slope of the Central Urals. According to N. V. Budanov (1957).

Anthropogenic sediments: 1 — loams; 2 — boulder clay. Palaeogene sediments; 3 — gaizes; 4 — carbonate deposits. Upper Cretaceous strata: 5 — quartz-glauconite sands; 6 — argillites. Palaeozoic strata; 7 — crust of weathering; 8 — Devonian porphyrites; 9 — Silurian limestones.

Urals is 150–170m high, about 40km long and 12–15km wide. It is visible locally and identifiable on geological maps, representing Upper Cretaceous inliers surrounded by the Anthropogene strata. Only the Upper Cretaceous (Coniacian-Campanian) is exposed, but the Upper and Lower Cretaceous up to 700–800m thick have been found in boreholes. The clays and gravellites of the Valanginian unconformably overlie the Palaeozoic basement. Tectonically the Muzhinskian Urals represent a gentle easterly inclined monocline with dips of $\frac{1}{2}°$ down to $1°$. The monocline is considered as a flank of a flat arched anticline, elongated parallel to the Urals. In fact it is a block structure associated with the fracture of the basement and the upward movement to the east of the fracture.

The History of Tectonic Movements

The history of the Urals Geosyncline, as of other Palaeozoic geosynclines, clearly shows three main epochs: (1) the epoch of accumulation, and of a mainly marine regime; (2) the epoch of the interplay of accumulation and erosion, and (3) the epoch of erosion and of a mainly continental regime.

The first epoch lasted until the Middle Carboniferous, and preceded the Hercynian Orogeny. The second epoch included the interval of time between the Middle Carboniferous to the Lower Trias. The third epoch continued from the Middle Trias to the present time, and succeeded the Hercynian Orogeny. Over the area of the Urals all orogenic movements can be recognized, but the Hercynian Orogeny is the most important, when a basin of sedimentation was converted into a mountainous district of erosion. Until recently it was thought that only the Hercynian Orogeny affected the Urals. Detailed studies have indicated that the true picture is more complex, and Precambrian, Caledonian, Hercynian, Cimmeridian and Alpine orogenies were found to have affected the mountains, although the intensities and areal extent of these orogenies were different. The Hercynian Orogeny was most intense and had the greatest influence, but the initial elements of Uralian structures were introduced by Precambrian orogenies and in places complicated by the Caledonian Orogeny. Furthermore, on the eastern slope Cimmeridian folds are superposed on the Hercynian, while in the south of the Mugodzhars weak, but clear manifestations of the Alpine Orogeny had occurred, although it was absent from the Urals proper.

PRECAMBRIAN OROGENIES. These occurred within the Precambrian and between the Precambrian and Lower Palaeozoic. In the south, in the valley of the Belaya, according to O. P. Goryainova and E. A. Falkova and Krauze, an angular unconformity separates the Ordovician (faunally dated) and the Precambrian. Further to the south the contact between the Lower Palaeozoic and the Precambrian is tectonic, but a stratigraphical break is suggested by the presence of grains and pebbles of Precambrian rocks in the Lower Palaeozoic sandstones and conglomerates.

The breaks within the Precambrian are accompanied by differences in the grade of metamorphism. In the Southern Urals the oldest Taratashian Suite is represented by gneisses and crystalline schists. The overlying suites show a much lesser degree of metamorphism, but are more metamorphosed than still higher Precambrian suites. The latest proof of Precambrian orogenies is the development of the ancient Precambrian granites amongst which the Rapakivi type, resembling those of Karelia and the Ukraine, are especially important. The Berdyaushian and Troitskian massifs of the western slope of the Urals consist of such granites.

The Salairian phase. The orogenic movements of the Upper Cambrian are known as the Salairian phase, resulting in an unconformity between the Ordovician and the Middle Cambrian. The following facts prove the existence of this phase.

1. In the region of Byeloretsk and Tirlyan (in the Southern Urals) sandstones and quartzites with an Ordovician fauna are transgressive over bedded, barren limestones, considered as Middle Cambrian.

2. On the western slope of the Southern Urals the Ashaian Suite is unconformable on limestones with *Collenia* of the Min'yarian Suite considered as the Middle Cambrian.

3. Amongst the Ordovician sediments clastic sandstones, quartzites, conglomerates and sandy shales are most widespread and form great thicknesses. Some of these sediments have a flysch character and suggest an elevated area of young mountains in the process of erosion.

The above points, if definite, would have been quite convincing, but there are doubts about them.

The pre-Ordovician folding of the Byeloretsk and Tirlyan regions is undoubted, but their date has not been established since the underlying limestones are not Cambrian, but Proterozoic.

The unconformities between the Ashaian and Min'yarian Suites again do not provide an accurate date, since the age of the latter suite is Proterozoic and not Middle Cambrian. The Ashaian Suite itself is also Upper Proterozoic.

The character and lithology of the Ordovician and the thickness of its strata are most substantial in this argument. These features indicate Caledonian movements, which however, occurred during the Ordovician, not where the sediments were accumulating, but in neighbouring areas, probably to the west in the region of the Western Siberian Lowlands. Thus at present there is no sufficient evidence for the Salairian phase and the claimed supporting features should be studied in detail. It is most probable that these features were caused either by the later Caledonian Orogeny or the earlier Baikalian Orogeny.

THE CALEDONIAN OROGENY. The problem of the Caledonian Orogeny is one of the most debated questions of the tectonics of the Urals. In the literature, there are two diametrically opposite points of view. The first claims that there is no sign of the Caledonian Orogeny in the Urals and the second claims that it is manifested throughout the Urals. In certain regions of the Urals the manifestations of the Caledonian Orogeny are undoubted, but these areas are marginal to the chain and over its main area signs of Caledonian Orogeny are absent. Stratigraphic breaks, unassociated with angular unconformities, must have been produced by epeirogenic movements. The undoubted evidence for the Caledonian Orogeny is found on the eastern edge of the eastern Urals, and the eastern slope of the Mugodzhars. Here, in a whole series of outcrops, Visean limestones are found to be transgressive over and unconformable on Lower Palaeozoic sediments. In the Dombarovskii region the Visean sediments are unconformable on ancient granites of possibly Caledonian age. Similar relationships are observed on the eastern slope of the Mugodzhars. A similar picture exists in the Central Kazhakhstan nearest to the Urals and the Mugodzhars. It is undoubted that these three regions are all parts of a continent constructed by the Caledonian Orogeny and that the continent was dry land in the Devonian and was only overrun by sea during the great Visean transgression. Fig. 4 shows the most probable outlines of this continent.

Another region of the manifestation of Caledonian Orogeny is Timan and the adjacent western slope of the Urals known as the Polyudov Ridge. In Timan, Givetian sandstones are transgressive over Silurian and Lower Palaeozoic sediments. In the Polyudov Ridge the Llandovery and the Middle Devonian (Takatinian Suite) are transgressive over Upper Proterozoic rocks. It is possible that these unconformities have been caused by the Baikalian rather than the Caledonian Orogeny.

In the Urals proper, unconformities corresponding to the Caledonian Orogeny are normally absent, although there are considerable stratigraphic breaks. On the western slope of the Southern Urals over a considerable distance there are numerous exposures of the contact between the Lower Devonian and the Middle Devonian on one hand and the Lower Palaeozoic on the other, but there are no angular unconformities. On the eastern slope of the Central and Northern Urals, Eifelian rocks overlie directly the higher horizons of the Silurian. Similar relationships have been found in tens of exploratory traverses in the North Urals bauxite regions. There is much evidence for erosion at the base of the Eifelian Stage, nevertheless it is everywhere conformable to the Silurian, which suggests epeirogenic movements as the cause of this erosion, and the signs of Caledonian activity in the Central Urals have to be denied any validity.

The above generally accepted conclusion, nevertheless, has to be revised. An ever increasing series of facts indicates that orogenic movements need not be accompanied by angular unconformities if folding and sedimentation occur contemporaneously. Caledonian acid and ultrabasic massifs in the central part of the Urals provide an important line of evidence for the Caledonian Orogeny in these mountains (p. 365). These massifs form a continuous series in a zone tens of kilometres across and many hundreds of kilometres along the watershed of the Northern and Central Urals and to the east of the watershed in the Southern Urals, transgressing on to the western slope (Krak Massif) in the valley of the Byelaya. To the west of the Caledonian intrusions there are only the Precambrian (Baikalian) massifs and to the east of it there is an even wider zone of Hercynian intrusions. The continuous regional development of large and diverse Caledonian intrusions throughout the central part of the Urals can only be explained by a regional development of the Caledonian Orogeny. In this case, this positive indication counts more than the negative indication of the absence of unconformities. It is possible that more detailed observations will demonstrate the existence of minor unconformities, as happened when the relationships of the Upper Cretaceous and Cainozoic deposits of the Fergana valley (p. 548) were being studied. All this suggests that the Caledonian Orogeny affected not only the margins of, but the Urals proper themselves. The Orogeny, however, was weaker than the Hercynian and did not cause the cessation of the geosyncline.

HERCYNIAN OROGENY. The Hercynian Orogeny was the main mountain-building episode in the Urals, it had caused most important palaeogeographic changes in the history of the Urals. Thus in place of marine basins, which existed for tens of millions of years, high mountains resembling the Caucasus or the Alps, came into existence. After the Hercynian Orogeny the Urals became a part of the platform, forming a large continent, and the central part of the mountains has remained land ever since. During the Hercynian Orogeny gigantic granite intrusions of the eastern slope and Transuralia were produced and contemporaneous mineral deposits were emplaced.

As has already been mentioned Hercynian structures are more complex on the eastern slope compared with the western slope of the Urals. The same asymmetry is observed in the distribution of the Variscid granites which are absent on the western slope. Lastly an analogous asymmetry existed in the succession from sea to land. In Transuralia this happened in the Middle Carboniferous. In Cisuralia throughout these times there existed a shallow sea and the land only appeared in the Upper Permian.

The above features also indicate the asymmetric nature of the Hercynian Orogeny. In the east it started in the Middle Carboniferous and was powerful, while in the west it started in the Lower and Upper Permian and was considerably weaker. All this is completely confirmed by the analysis of Upper Palaeozoic sediments of the western and eastern slopes of the Urals (p. 355).

The Phases of Hercynian Orogeny. The recognition of individual phases in the Urals is very difficult and in places impossible, since on the eastern slope only Middle Carboniferous sediments are found, while in the Central Urals Upper Palaeozoic strata are entirely absent. On the western slope the Orogeny begins only in the Lower Permian. Another highly important feature which is responsible for non-recognition of phases is the regional conformity of strata within the Upper Palaeozoic and on the boundary of the Upper and Middle Palaeozoic. In many sections throughout the Urals various Upper Palaeozoic horizons gradually pass upwards without angular unconformities, which circumstance has often been mentioned in literature. Indications of unconformities are infrequent and isolated and often demand a further verification. As an example of these one can quote the often described unconformities between the Artinskian sandy-clayey formations and Carboniferous limestones. This unconformity was claimed on the basis of the Artinskian sediments resting on different horizons of the Carboniferous, as far down as the Visean. A detailed study of the fauna has indicated that this interpretation is wrong. The fauna in the lower horizons of the 'Artinskian' sediments is typically Upper Carboniferous and in places Middle Carboniferous. In other regions, in the west, the Upper Carboniferous and the Lower Permian are presented by reef limestones transitional into each other. Thus the unconformity is a facies change of limestones by clastic deposits. These relationships are apparent by comparing the Upper Palaeozoic succession in different regions of the Urals.

A single unconformity, which is distinct throughout the eastern slope, the central regions and the western slope, is that between the Palaeozoic and the Lower Mesozoic (Upper Trias and Lower Jurassic). This unconformity is only indistinct in Cisuralia where Permian strata are almost horizontal. The unconformity is accompanied by differences in the grade of metamorphism of the Palaeozoic and Mesozoic, and by sharp distinctions between structures, by vast palaeogeographic changes and by granitic intrusions. All this indicates that although the individual phases of Hercynian Orogeny cannot be recognized, unquestionably it existed and was powerful.

CIMMERIDIAN OROGENY. The Cimmeridian Orogeny has regional manifestations throughout the eastern slope of the Urals, but it was weak, since although the Rhaetic-Jurassic coal-bearing strata are folded, the folds are simpler, relatively gentle and do not show complexities such as isoclinical or recumbent folds. The weakness of the orogeny is also confirmed by the insignificant grade of metamorphism of the Rhaetic-Jurassic coal-bearing strata by comparison with the locally exposed Lower Carboniferous rocks and their coals. Nor are there any Cimmeridian granites which are common in the Caucasus, the Pamirs and in Kolyma, where they cut across the Trias and the Lower Jurassic and are as common as the Hercynian granites of the Urals. The Cimmeridian Orogeny of the Urals was as distinct and weak as in Central Kazakhstan and Kuzbas.

Intensive deformation found over small areas in the neighbourhood of the thrusts is of some interest. The deformation is akin to that observed in Central Asia where Cainozoic sediments are overthrust by Palaeozoic massifs (p. 536). In both cases the explanation is the pressure of the massifs at the edges of depressions filled with Mesozoic and Palaeozoic sediments. With such young movements (Neogenic or Anthropogenic) intensive, but localized deformation is of local significance and is not associated by metamorphism. These young movements used to be considered as the evidence for strong Cimmeridian movements despite their local distribution and differences with the usual folds. The best refutation of this idea is the comparison of such structures with those in the Lower Mesozoic of the Caucasus and Kolyma, where the Cimmeridian Orogeny has been truly strong.

In the literature there can be found a fortunately rare opinion that the Ural structures are Cimmeridian rather than Hercynian. This opinion is based on an inadequate appreciation of quantitative factors affecting the Urals. With the holders of this opinion facts of regional, general validity are not sufficiently taken into account, while those facts which have a purely local significance such as the deformations in the Trias-Jurassic coal-bearing strata are assumed to be characteristic of the Urals as a whole.

The Cimmeridian Orogeny can be dated, since where it affects the Lower Mesozoic rocks the Upper Jurassic and Lower and Upper Cretaceous are horizontal. Thus the orogenic movements occurred contemporaneously with the deposition of Upper Triassic and Lower Jurassic sediments. The folds affecting the Lower Mesozoic rocks are only found on the eastern slope. In the central parts of the Southern Urals and on the western slope all Mesozoic sediments including the Triassic and the Jurassic are horizontal.

ALPINE OROGENY. The Alpine Orogeny only affected the southern Mugodzhars, where the Upper Cretaceous and Palaeogene strata are plicated into gentle, large, barely noticeable folds. The folds are found regionally, but there are fewer of them than of the Cimmeridian folds on the eastern slope. The folding occurred at the end of the Palaeogene, and the Neogene sediments are unaffected. In the Urals proper the Alpine Orogeny did not occur.

MAGMATISM

The magmatism in the Urals reflects completely its tectonics. The Palaeozoic magmatic cycles—Hercynian and Caledonian—are distinctly dominant. The Precambrian cycles are also fully developed, but are less widely distributed than the Palaeozoic. The Meso-Cainozoic cycles are only represented by effusive rocks which are relatively rare and cover small areas. The magmatism in the Urals and the associated ores have been studied by A. N. Zavaritskii and V. M. Sergievskii.

PRECAMBRIAN MAGMATIC CYCLES. The Lower Proterozoic (Archeozoic?) has a rather local distribution in the Southern Urals (Taratashian Suite), but has all the features of geosynclinal formations and therefore the Karelidian magmatic cycle can be expected. At present the granites and granite gneisses of the Taratash Massif in the Mugodzhars are considered as such where they are covered by metamorphic rocks.

The Upper Proterozoic is widely found occurring over large areas, but magmatic rocks are rare in it. In places near to the watershed strongly metamorphosed volcanic rocks have been found and to the west of it dykes and sills of diabase are found. The Upper Proterozoic intrusive massifs are small and infrequent. In the Southern Urals there is the Berdyaushian laccolite of Rapakivi-type granites and associated alkaline rocks. In the Northern Urals there is the Troitskian Massif of porphyritic granite with associated manganiferous iron ores. A part of the Taratash granite massif to the north of the Berdyaushian mass is probably Baikalian.

THE CALEDONIAN MAGMATIC CYCLE. The cycle is mainly developed in the central part of the Urals along their length and is much more widespread than the Upper Proterozoic cycle. The Caledonian cycle has not been sufficiently studied and the age of a whole series of magmatic complexes is not known, but then 25–30 years ago there was no concept of Caledonian magmatism in the Urals. The main achievement here is due to V. M. Sergievskii and other petrologists, but much has to be done in the future. The Caledonian cycle begins with the Lower Palaeozoic (Cambrian and Ordovician) widespread outpourings of acid and basic magmas. In the Northern Urals the effusive formation, lying at the bottom of the Serebryankian (Ordovician) Suite, serves as an example. Here amongst the cherry red and variegated tuffogenous slates lie porphyrites, tuffs, tuffolavas and amygdaloidal rocks. The volcanic rocks wedge out to the south. In the Southern Urals along the Orenburg-Orsk railway, under the graptolitic Llandovery slates and Silurian spilite-diabases and separated from these by an unconformity, lies an Ordovician volcanic series of quartz-albitophyres, felsites and mainly tuffs and compact felsitic tuffs. Their thickness reaches 700m. The volcanic rocks are stratigraphically discontinuous and are overlain by tuffites and tuffaceous sandstones of about 1000m thick, which are again overlain by the Llandovery graptolitic shales.

The major intrusive phase is again widespread. In the Northern and Central Urals it comprises a western zone of ultrabasic (peridotite, dunite) massifs, which are often situated near the watershed and are frequently platiniferous. In the Southern Urals the massifs are on the western slope and immediately to the east of them are plagioclase granites, which in the opinion of Zavaritskii (1939) are parts of the same magmatic complex. Sergievskii calls the ultrabasic rocks—the pre-Ludlow gabbro-peridotites—and the granites—the Ludlow plagiogranite complex. The Caledonian age of peculiar granite-gneiss massifs of the central part of the Central Urals and Southern Urals is less definite and they may be Baikalian.

In the Urals the unique alkaline complexes of the Ilmen' and Vishnevye Mountains deserve attention. The rocks consist mainly of nepheline syenites (miaskites) and are found on the eastern slope between Sverdlovsk and Cheliabinsk and near to Mias. Their age is unknown, but they are closely associated with granite gneisses and consequently a Caledonian age is probable, although Sergievskii considers them to be Hercynian (p. 367).

The Caledonian phase of minor intrusions is widespread and its rocks very variable. The pegmatites, most often found with the deep granite intrusions, are associated with gneissose rocks and are interesting, since with them are associated precious stones and industrially important minerals including: emerald, topaz, columbite, beryl. The deposits associated with the Ilmen' alkaline pegmatites are especially well known.

The main area of Caledonian magmatism is the central part of the Urals and it is also widely developed in the east of the eastern slope of the Southern Urals and the Mugodzhars, which have the same structure as the adjacent Central Kazakhstan. In a number of areas the granite massifs are unconformably covered by Devonian and Visean sediments. There are also some Lower Palaeozoic volcanic rocks, but the possibility of these being Baikalian cannot be excluded.

THE HERCYNIAN MAGMATIC CYCLE. Until fairly recently this cycle was thought to be the only one in the Urals. Soviet geologists have proved the presence of the Precambrian and Caledonian igneous rocks, but even now the Hercynian magmatic cycle, like the Hercynian Orogeny, is considered the main one. The Hercynian granite massifs are up to 200km long and have associated iron ores of Blagodat', Vysokaya and Magnitnaya.

The Hercynian cycle begins in Middle Palaeozoic times with volcanic effusions of various, widely distributed types, which have been studied in detail by V. M. Sergievskii (1948) and others (*Geological Structure of the U.S.S.R.*, 1958). Beginning in the Llandovery and ending in the Visean the volcanic effusives of the eastern slope of the Urals have a wide distribution and reach a great thickness of several kilometres. In many regions the sedimentary rocks lie amongst them as thin beds, or are completely absent. Only three phases of increasing effusive activity, divided by periods of quiescence, are noticed. The first phase began in the Llandovery, reached a maximum development in the Wenlock and Lower Ludlow and ended in the Upper Ludlow or the beginning of the Gedinnian. The volcanic centres lie in the westernmost zone, near to the watershed and include rocks of two types: (1) porphyrites (augite-plagioclase bearing basalts and basalt-andesites) accompanied by tuffs; (2) spilites and diabases, plagioclase porphyries and albitophyres in which fissure-eruption products were dominant and tuffs almost absent. The second phase encompassed the whole of the Middle Devonian. The centres of eruption were further to the east. Andesitic and dacitic rocks and tuffs predominated and bedded psammitic tuffs with mixed andesite and dacite fragments were especially dominant. Large lava flows were rare. Potash keratophyres of peculiar type are worthy of interest. The third phase began at the end of the Famennian and beginning of the Tournaisian, was most intense in the Tournaisian and died away in the Visean. The eruptive centres were then still further to the east and in the Southern Urals lie along the valley of the Ural. In the Lower Carboniferous rocks of two types are widely found: firstly the basic diabases, porphyrites and palaeobasalts, and secondly sodic porphyries. The olivine diabases and palaeo-basalts are widespread.

The Upper Palaeozoic effusions present a problem which as yet has not been solved. In all Palaeozoic geosynclines Upper Palaeozoic volcanic rocks are widespread, yet in the Urals neither Upper Palaeozoic lavas nor tuffs are known. Consequently, it is concluded that there was no Upper Palaeozoic vulcanism in the Urals. Actually it had occurred on the eastern slope, but the rocks have been eroded away. This hypothesis is demonstrated to be true since the Upper Carboniferous conglomerates have pebbles of very fresh andesites, trachytes and biotite-dacites, unknown amongst the solid rocks. The similarity of these rock types to the Upper Palaeozoic effusive rocks of Kazakhstan and Central Asia suggests their Upper Palaeozoic (Middle or Upper Carboniferous) age.

In the last few years the presence of even younger effusive rocks corresponding to the Siberian Upper Permian and Lower Triassic age has been demonstrated. These rocks have been found in boreholes in the region of Kushmurun, Tyumen', Chelyabinsk and the Bulanash-Yelkino Trough (300km to the north of Chelyabinsk) and consist of sills of olivine gabbro-diabases (150m thick) overlain by red beds (110m), which are in turn overlain by alternations of olivine palaeo-basalt lavas and carbonaceous clays (245m). All the rocks are almost horizontal at a depth of 1500–2000m. The absolute age determinations of the lavas of Yelkino and Kushmurun regions give 167 and 164 m.y., corresponding to the Lower Trias. As well as basalts there are liparites.

The Hercynian major intrusions (second magmatic phase) reach large dimensions of over 200km in length and up to 75–100km in width. These intrusions are elongated parallel to the tectonic strike and follow each other almost without interruption forming a zone of 1200km in length and 100–200 km in width. In the north and south the intrusions lie below the Cretaceous and Tertiary strata along the whole of the geosyncline. V. M. Sergievskii (*Geological Structure of the U.S.S.R.*, 1958) calls them the batholitic granitoid intrusives. Their age is determined by the fact that they exhibit active contacts against Lower Carboniferous limestones and are transgressively overlain by the coal-bearing, Rhaetic-Liassic strata or the marine Upper Jurassic. The composition of the intrusive rocks varies although various granites predominate. The Sverdlovsk Massif (Fig. 122) is composed of pale, whitish, yellowish and grey granites, consisting of an acid plagioclase, microcline and biotite. The granites are transitional into syenitic and dioritic facies and sometimes develop porphyritic varieties. In the Southern Urals, in the region of Chelyabinsk and Cochkar' the granites are microcline-bearing, biotitic or two mica granites, often accompanied by granodiorites. The Mt Magnitnaya granites are also associated with granodiorites or even quartz diorites, in which microcline is absent and is replaced by orthoclase and anorthoclase.

Syenitic massifs are rarer and the largest and best known amongst them are those of Mt Vysokaya and Mt Blagodat' in the Central Urals, where iron ore is found in association with these intrusions.

The ultrabasic massifs are relatively rare and small such as those between Alapaevsk and Kamensk, in the Central Urals where the rocks are of gabbro-peridotite type, being peridotites, pyroxenites, gabbros and serpentinites.

The alkaline complex at Ilmen' which includes nepheline syenites, mainly miaskites and alkaline syenites and alkaline soda-granites, is younger than the preceding massifs *(Geological Structure of the U.S.S.R.*, 1958).

Minor intrusions, which cut across the major intrusion, are widespread and variable. Pegmatites, aplites and granite-porphyries predominate and are accompanied by precious minerals, as for instance in the region of Murzinka and Shaitanka, to the south-west of Alpaevsk, and also the region of Adui, some 40km to the south of the latter.

CIMMERIDIAN AND ALPINE MAGMATIC CYCLES. Mesozoic and Cainozoic volcanic rocks are rare and have a limited distribution. Major intrusions into Meso-Cainozoic strata are entirely unknown, unlike the Meso-Cainozoic geosynclines of the Caucasus, the Pamirs and the Far East.

The youngest and most widespread volcanic rocks of the Urals are the basalts of Pechora edge and of Pai-Khoi, where they lie on dislocated Permian deposits as horizontal lavas associated with Upper Cretaceous (?) sediments. The largest area of basalts is found in the valley of the Usa, to the south of Vorkuta. The rocks are diagnostic, sometimes olivine-bearing with a matrix of brown, opaque glass with microporphyritic structure, but more rarely amygdaloidal. Their thickness does not exceed a few tens of metres and according to L. L. Khaitser they are Upper Permian. To the south of the Usa, on the western slope and in the central regions there are no known Meso-Cainozoic volcanic rocks. On the eastern slope, as for instance south of

Kushmurun, there are a number of indications of the development of Meso-Cainozoic volcanic and hypabyssal diabases and basalts. Undoubtedly young extrusive rocks are known from Central Kazakhstan, and their existence in the eastern parts of the eastern slope of the Southern Urals and the Mugodzhars is quite possible.

On the eastern slope in a number of regions diabase sheets, which cut the Carboniferous strata and the granites, have been found. The intrusive rocks are very fresh and resemble dolerites and basalts. Sometimes they are picritic and bear almost no plagioclase. These rocks are the youngest igneous rocks of the eastern slope.

FIG. 122.   Weathered Hercynian granites near Sverdlovsk (rock tents).

USEFUL MINERAL DEPOSITS

The Urals represent one of the most important mining regions of the U.S.S.R. Within the Urals there are found thousands of mineral deposits, of which many have a national and sometimes world significance. Amongst the latter, platinum, potassium salts, magnesite, asbestos and chromite can be mentioned. The significance of these deposits is enhanced owing to the economically favourable position of the Urals which have a well-developed network of communications and a large population.

Of the useful minerals, in the first place stand the metals and in particular iron, chromium, nickel, aluminium, platinum, and copper; in the second place stand non-metals such as potassium salts, magnesite, asbestos and limestones, while the fuels, although considerable in magnitude, are insufficient for the Urals.

ORES

Ferrous Metals

Iron deposits of the Urals are the most important amongst those of the ferrous metals. The deposits of iron fall into two groups: (1) deposits associated with magmatic rocks; (2) deposits associated with sedimentary rocks. Magmatic deposits occur on a larger scale. Mt Magnitnaya in the Southern Urals, for instance, has ore deposits of 325×10^6 tons and around it has grown the new major town Magnitogorsk. The ore is magnetite, lying at the contact between Lower Carboniferous limestones and porphyrites (Fig. 123).

The newly discovered contact magnetite deposits of Sokolovskoye, Sarbaiskoye and Kacharskoye have considerable magnitudes. These localities are situated in the Southern Urals in the Kustanaian Province. The exploitation of these deposits has already started and reserves are estimated as $2 \cdot 5 \times 10^9$ tons. In the Central Urals there are two mountains formed of iron ore and covered by mines. These are Mt Vysokaya and Mt Blagodat'. The former supplies the ore for a large metallurgical and machine-making town—Nizhnii (Lower) Tagyl which formerly was a small factory settlement and now is a major industrial centre of the Urals. Near Mt Blagodat' there is the large Kushvinskian factory. In both localities magnetite is the ore found at the contact between syenitic intrusions and shales, tuffs and limestones of Silurian and Lower Devonian age. The total deposits of iron are much more than at Mt Magnitnaya. The magnetite deposits of the more northerly Serovian and Ivdelian regions are identical in age and genesis. They are worked in second and third Severnyi, Pokrovskii, Ayerbakh, Vorontsov, and other mines. Large deposits of titaniferous magnetites are associated with rocks of gabbroic series and groups of diabases and include those in the Pervouralskian region near Sverdlovsk, in the Kusinskian region, in the Verkhnevisherian region (Kutimian deposits) and in the Ivdelian region (first Severnyi mine). The recently discovered Kochkanarian deposit is also large.

Among the sedimentary deposits the complex ores in the Proterozoic reach the largest size. The Bakal region which has a very high grade ore (turgite) is particularly important. The Primary ore, siderite, enters the composition of the carbonate-rich Bakalian Suite. The siderite is syngenetic with other carbonates (limestones and dolomites) of the suite. As a result of secondary changes brown ironstones (tur'ites) are produced. A second group of deposits (Zigazino-Komarovskian) is also Proterozoic and is of the same type. This group is found in the central part of the Southern Urals and provides the ore to the Byeloretskian plant.

The two groups of deposits are often considered as metasomatic or hydrothermal. The importance of these processes in the formation of tur'ites is undoubted, but the siderites are originally sedimentary ores.

There are also other iron ores, of differing origin, found amongst the Proterozoic rocks of the Southern Urals, of which the jaspilitic ores, identical with those of Krivoi Rog, are particularly important. These have been found by M. I. Garan' and G. A. Smirnov to the north of the Kusinskian factory where they occur amongst the Taratashian strata.

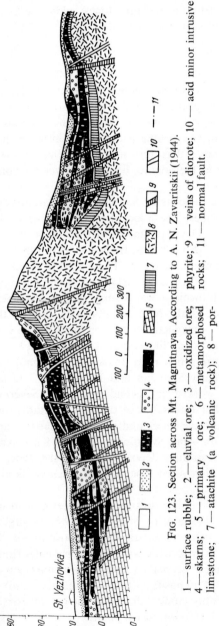

Fig. 123. Section across Mt. Magnitnaya. According to A. N. Zavaritskii (1944).

1 — surface rubble; 2 — eluvial ore; 3 — oxidized ore; phyrite; 9 — veins of diorote; 10 — acid minor intrusive 4 — skarns; 5 — primary ore; 6 — metamorphosed rocks; 11 — normal fault. limestone; 7 — atachite (a volcanic rock); 8 — por-

On the eastern slope of the Central Urals in the Nevyansk and Verkhne-Osyetsk regions there are numerous deposits of metasomatic ores of brown ironstone in the Devonian, Carboniferous and Silurian limestones and dolomites. On the western slope of the Urals beginning with the Pashiiskian mines and ending with the Mini'yarian factory there is an almost uninterrupted belt of small deposits of oolitic brown ironstones associated with the near-shore clastic sediments of Upper Devonian age. These ores were formed under the same conditions and in the same basins of deposition as the bauxites of the western slope. Similar brown ironstones are also found amongst the coal-bearing deposits of the western slope and some of them are being worked.

The following group of iron ores is found amongst the continental Jurassic and Cretaceous sediments. The initial heavy industry of the Urals was established on the basis of these deposits. The most widespread type of ore is found as lenses and nests of hard, red and brown ironstone lying on the Palaeozoic rocks, amongst variegated clays. These are lucustrine and marsh deposits with iron content between 35–65 per cent.

On the eastern slope the classical region of such ores is the Alpayevskian region. From here they continue south to the Kamensk and Bagaryak region. They are also known from the Bogoslovian region in the north and Orskian region in the south, as well as from a number of intermediate points. On the western slope the Jurassic iron ores are mainly found between the Vishera and the Vil'va and also on the Ufa Plateau. The Jurassic ores of the Khalil deposit in the Southern Urals, are compositionally somewhat different. They are produced as a result of the deposition of the products of weathering and erosion of serpentinite massifs, on the shores of Jurassic lakes. Consequently these ores always have chromium and nickel impurities, which improves their quality.

On the eastern slope of the Central Urals the Jurassic ironstones often contain an admixture of manganese. Similar manganiferous ironstones are known from the Verkhne-Neivinsk, Serebryansk and Bilimbai regions. Amongst the Upper Cretaceous deposits of the eastern slope, the oolitic iron ores are known from the rivers Ayat, Tagil (Mugaiskian deposits) and Sos'va. The ores are typical marine deposits. The primary ores are sideritic and their quality is often high. The exploration of the Ayatian and the adjacent Lisakovian deposits has shown high reserves of $3 \cdot 6 \times 10^9$ tons (Antropov, 1960). There are no known Cainozoic deposits.

The total number of iron ore deposits in the Urals reaches hundreds, of which many have been worked out, but the greatest part remains and consequently the Urals are still a major centre of iron ore. The post-Revolutionary finds have increased the resources by 15 times, while the yield has increased 20-fold.

Manganese, nickel, cobalt. The deposits of manganese are small but numerous, and are industrially valuable. The deposits are divided into Palaeozoic and Palaeogene.

The Palaeozoic ores are found on the eastern slope amongst cherts and siliceous shales and quartzites interbedded with porphyrites and tuffs. In deeper zones the ore is rhodonite, which in the oxidation zone is converted into psilomelane, wad, pyrolusite and other minerals. The largest deposits are found in the Sverdlovsk, Mias, Magnitogorsk and Baimak regions. There are also small isolated deposits of vein (psilomelane) and contact ore associated with granites, gneisses and syenites.

The Palaeogene ores are found at the base of the Cainozoic sediments on the eastern slope, where the deposits are Palaeocene marine sands and clays. The primary carbonate ores (oligonite) occur as lenses and beds amongst clays. As a result of oxidation pyrolusite, psilomenane and wad get formed. The largest deposits are found in the Northern Urals in the Serov and Ivdelo regions, where the deposits of Polunochnoye, Berezovskoye and Marsyata are the most important and their ores are used in the Urals factories (Antropov, 1960).

The nickel deposits are associated with serpentinites and disseminated sulphide (pentlandite) has been found in peridotites. The first group of deposits originated as a result of breakdown of the serpentinites and the subsequent concentration of nickel silicates mainly at the contact with limestones. The largest deposits are found in the Verkhne-Ufalei, Rezhevsk and Verkhne-Neivinsk regions. The second group includes the deposits with nickel and chromium-bearing brown ironstones, occurring as irregular accumulations in the zone of weathering on the surface of the serpentinite, pyroxenite and peridotite massifs. The Khalil deposit is the best known in the Southern Urals.

The work of the last few years has shown that the deposits of Khalil and siliceous type can be traced far to the south into the Aktyubinsk Province and the Mugodzhars. In this province and the Orsk-Khalil region rich ores have been found. At present the deposits of the Southern Urals constitute one of the largest nickeliferous regions in the U.S.S.R. and the world. These deposits have considerable reserves spread over large areas and shallow depth (5–20m) and are above the water table. All this produces favourable conditions for exploitation and reduces the cost of the ore. In the second five-year period, over 25 deposits have been found in the Aktyubinsk region and over 10 deposits in the Orsk-Khalil region. The nickel deposits in the Urals are second in importance in the U.S.S.R.

Deposits of cobalt are rare. There is a deposit in the Serovskian region where it is associated with contact iron ores. The primary ores are cobaltite and glaucodote. Recently new deposits have been found and the exploitation of cobalt is in progress. The Yelizavetinskoe deposit has the cobalt ores as such.

Chromite. The chromite yield in the Urals is not only the highest in the U.S.S.R., but throughout the world. The ore is found in association with serpentinites formed as a result of the alteration of peridotites and dunites. Depending on their utilization the ores are divided into high-quality, used for the manufacture of ferrochrome, and medium-quality and low-quality employed in the refractory and chemical industries. The main deposits of high grade ores are found in the Mugodzhars and on the eastern slope of the Southern Urals. Amongst the isolated deposits the Yuzhno-kempirsai in the Mugodzhars can be singled out. The deposits of medium and low quality are concentrated on the eastern slope of the Central Urals (Saranovskian deposit). There are many other deposits, but they are much smaller than the Yuzhnokempirsai and the Saranovskian. The former is particularly important in making alloys.

Non-ferrous Metals

The deposits of non-ferrous metals are numerous and are found throughout the Urals, but are normally small or medium-size. Large-scale deposits are rare, but altogether they are quite important in the development of non-ferrous metallurgy.

Aluminium. The reserves of aluminium in the Urals are the most important in the U.S.S.R., but the individual deposits are either small or medium-sized. Two large groups are recognized: the marine Devonian and the continental Lower Cretaceous. The Devonian deposits are most important since they have continuous ore-bearing strata with large reserves. These deposits are associated with intervals in the marine regime. Beds of bauxite lie at the base of a transgressive marine series normally amongst clastic sandstones, clays and conglomerates. The ore is almost always oolitic or pisolitic. Six horizons of bauxite have been established. On the eastern slope the main horizon lies at the base of the Eifelian limestones (Northern Urals bauxite field) in the Serovsk region. The horizons at the base of the Ludlovian limestones (Isovskian region) and of Givetian limestones (Bogoslavskian region) are of secondary importance. Small deposits have been found in the basin of the Chusovaya at the bottom of the Lower Frasnian limestones (Pashiiskian Horizon) and of Upper Givetian limestones (Ust'-Utkian Horizon). According to the latest data (Knyazeva, 1958) the Eifelian limestones of the eastern slope contain not one but three bauxite horizons and also two Givetian horizons. Thus the total number of horizons is 9. They are all associated with intervals in sedimentation of limestones. Certain horizons of bauxite have a marine fauna of tabulates and brachiopods. Other horizons (red bauxites) lack a marine fauna and are probably freshwater.

The Mesozoic deposits are situated on the eastern slope of the Central and Southern Urals, in the Kamensk, Rezhevsk and Alpayev regions. The bauxites are found amongst the Mesozoic continental deposits overlying the Palaeozoic and covered by the marine Senonian. The age of the bauxite-bearing strata is Lower Cretaceous.

Owing to the ease of exploitation and working of the ore (hydrargillite type) the Mesozoic deposits are intensely worked and constitute an important raw material for the aluminium factories of the Urals. Recently a new bauxite horizon has been found. The bauxites lie at the base of the Upper Trias coal-bearing formation (Chelyabinsk Series) found on the eastern slope of the Urals. The bauxites are hydrargillitic and are associated with the basal, variegated, continental and lacustrine group.

Copper. All the most important deposits are associated with the green-stone belt of Upper Palaeozoic age and found along the eastern slope of the Central Urals. In this belt there is the so-called copper-bearing or chalcopyritic tuff-slate zone where most deposits are concentrated. The mines and the copper-extracting factories form here a number of large concerns, such as Krasnouralian or the Northern in the north and the Kalatinian and Karabashian and Kirovgradian progressively to the south. In the South Urals there are the Tanalyk-Baimak group of deposits and the Blyavinskian chalcopyrite deposit.

The lensoid deposits of chalcopyrite lie in groups among the volcanic rocks. Formerly it was thought that these ores originated due to the activity of Hercynian intrusions. Later on A. N. Zavaritskii and V. M. Sergievskii advanced the suggestion that the ore is syngenetic with the enclosing strata and was formed well before the Hercynian intrusions. This is quite possible, since the overlying younger deposits contain grains and fragments of the chalcopyrite ores.

There are also much rarer copper deposits associated with acid intrusions, these include the Tur'inian copper mines, where the ore is found at the contact of granodiorites and Lower Devonian limestones, as well as the well-known malachite mine near the Polevskian factory where the ore is at the contact of the porphyritic granites and limestones.

Recently, in the southern regions of the Urals and in particular in Bashkiria, a number of new deposits has been found including the Sibai and the Buribai deposits where open working is possible. After Central Kazakhstan the Urals have the largest deposits of copper.

Titanium. Titanium is a metal of the future progressively acquiring great industrial significance. It is silvery, relatively light (S.G.=4·5), rather heat resistant, noncorrosive and is used in high-quality alloys. In the Urals there are both primary and alluvial deposits associated with basic igneous rocks. The main minerals are rutile, ilmenite and titanomagnetite.

Zinc. Zinc is often associated with the chalcopyrite ores of the Central Urals. It is especially abundant in the Karpushikhinskian and Kuznechikhinskian deposits.

Lead and Silver. Rarer Chalcopyrite ores are polymetallic and include copper, silver, gold, zinc and lead. The Blagodat' polymetallic quartz veins serve as examples of such ores.

Precious Metals

Precious metals of the Urals are of great importance. The deposits of platinum have long been among the largest in the world. Gold deposits although much less extensive are nevertheless significant.

Platinum is associated with ultrabasic gabbro-peridotites and dunites. These intrusions are situated on the eastern slope of the Northern and Central Urals and continue from south to north as somewhat discontinuous en échelon zones. The easternmost belt of Caledonian age is of greatest significance, and it follows the contact with the metamorphic suite, starting from the Upper Tagil region and further to the north. All the main centres of alluvial and primary platinum are in this zone. The deposits of pure platinum, found only in the Urals, are most important and have yielded the most platinum obtained so far. The primary deposits occur as segregations in dunite, or as ore-shoots in dunites and pyroxenites.

Gold. The deposits of gold are of different origins. Most of the economic deposits are associated with granites, granodiorites, quartz diorites, aplites and quartz porphyries. The gold is found in shoots and patches in quartz veins and as gold-bearing sulphides. Large gold deposits have been found in the Southern Urals. The gold-bearing alluvia have also a great industrial significance. The Anthropogenic alluvia apart, gold is found in Jurassic continental deposits and even in Artinskian conglomerates and sandstones of the western slope.

Rare Metals

The deposits of rare metals are numerous and although small their significance is ever increasing. The largest number of deposits is associated with acid volcanic rocks.

Tungsten found in scheelite, is found in the gold-bearing quartz and quartz-feldspar veins, associated with aplites, granite-porphyries and beresites. The most important deposit is at Gumbeika in the Southern Urals.

Arsenic is found in quartz veins with arsenopyrite, which occur in fold-bearing regions amongst the granite massifs of the Kochkar, Chelyabinsk, Dzhetygar, Mias and other regions.

Antimony (as mispickel) is discovered in the Verkhne-Isetian region in veins with quartz, mica, carbonates, cinnabar, realgar, pyrite, arsenopyrite, mispickel and bismuthinite. The mineralization is associated with the dykes of albite porphyries which cut hornblende granite-porhyries.

Mercury is found with mispickel, but cinnabar has also been found amongst the Lower Devonian limestones at their contact with the porphyrites of the Nizhne-Turinian and Ivdelian regions.

Iridium and Osmium are found together with platinum in the Nev'yansk, Sysert and Mias regions.

Vanadium is found in titaniferous magnetites.

RARE AND DISSEMINATED ELEMENTS. Such elements are of great significance in contemporary technology such as rocket aviation, atomic industry, radio-technology, television. Their properties and applications have been described by P. Ya. Antropov (1960). A series of new deposits of rare minerals, of high prospects, have been found in the Urals.

FUELS

Fuels are relatively sparse, but nevertheless are important in the economy of the Urals. Various coals—black, brown, anthracite and sapropelic—stand in the first place. The second position is occupied by oil and the third by peat. The yield of the fuels is increasing every year. Wood must not be forgotten either and numerous kilns producing high-quality metals use charcoal.

Coals

Devonian coals are not known and notes to this effect found in the literature are mistaken, since compact bitumen-like substances filling cracks in Domanik rocks were considered to be coals.

Lower Carboniferous coals are widely distributed and are of great industrial significance. Their deposits are found on the western and eastern slopes. The yield of the Kizel coal-field, with the towns Kizel and Gubakh at the centre of it, is most important. The field is situated on the western slope of the basins of the Kizel, the Kos'va and the Us'va. The coal-bearing strata are also found to the north in the valley of the Vishera, and to the south in the valley of the Chusovaya. The industrial significance of these regions is considerably less.

The coal-bearing strata are Lower Visean and are thickest (250–300m) in the Kizel basin (Gorodetskaya, 1948). To the north, south and west they gradually wedge out into marine deposits. In the Kizel coal-field there are three workable coals of a total thickness of 4·35m. The Simplicity and regularity of tectonic structure (Figs. 118, 119) allows the sinking of large pits. The coals are sulphurous and ashy (4–31 per cent) with an average ash-content of 20–25 per cent and 1·5–8 per cent of sulphur, and 28–50 per cent of volatiles. These are coking coals and can also be used as a source of chemicals. The proved reserves are $0·8 \times 10^9$ tons and the yield in 1956 was $11·4 \times 10^6$ tons.

On the eastern slope there are the following coal-fields: Alpaevsk-Kamensk field and Yegorshinskian part, Poltava-Bredinsk field, Dombarovsk field and in the south of the Mugodzhars the Bergchogur field. All these fields show complicated tectonics and are in places affected by granite intrusions. Consequently the Lower Visean and Tournaisian coal-bearing strata are deformed and metamorphosed. The coal seams are disturbed, broken up, compressed and altered. Thus there are few large pits, but small local pits are being used. Coals are coking or locally anthracitic and in places have been converted into graphite. The proved reserves are $0·2 \times 10^9$ tons and the yield in 1956 some 1 000 000 tons.

Permian coals are found only in the Pechora district, in the valley of the Usa and its tributaries, where they occur in the large Pechora coal-field. The largest number of coals of highest quality are found in Kungurian deposits (Vorkutian Series) of essentially continental origin (3000m thick). The Vorkutian deposit on the river Vorkuta is being intensively worked. Here on the western slope of the Arctic Urals there are found black high-grade coking coals. There are seven basic workable coals of considerable size. To the south on the right shore of the Kos-Yu there is the Inta-Kozhim coal-field, with gas coals. There are 8–17 seams varying in thickness from 0·6 to 2·83m. The Upper Permian coals of the Pechora basin (continental Pechora Series) are brown, 6–8m thick and are found between the Kos-Yu and the Usa. The total reserves of the Pechora basin down to the depth of 600m are $110·5 \times 10^9$ tons of which $7·0 \times 10^9$ tons have been proved. The yield in 1956 was $15·4 \times 10^6$ tons.

To the north of Vorkuta there is the newly discovered Silovskian coal-field, which has an Upper Permian (Kazanian) coal series (3300m thick), which is strongly folded and metamorphosed, and is known as Paemboiian Suite and lies on the Vorkutian Suite. It is probable, however, that the Paemboiian Suite is Lower Permian (Upper Kungurian).

Trias-Jurassic coals are found on the eastern and western slopes. The Trias-Jurassic continental strata with coals are thickest (1000m) on the eastern slope where they are Upper Trias (Rhaetic) and Lower Jurassic in age and have been regionally folded. The folds are large, simple, with low dips, but in places have been involved in dislocations. The simplicity and regularity of the folds, weak compaction of rocks and shallow coals aid their extraction in open-cast mines. The exploitation is also made easy by the great thickness of the seams which are often tens of metres thick (60m near Karpinsk). The coals are almost always brown, ashy with much pyrite and volatiles, thus they easily break apart and suffer auto-ignition. Nevertheless they are cheap, satisfactory and their yield is rapidly increasing. The proved deposits of the Chelyabinsk basin are $1·3 \times 10^9$ tons, of the Serovsk basin $0·5 \times 10^9$ tons, of the Bylanash-Elkino basin $0·12 \times 10^9$ tons. The yield in 1956 was 19·4, 18·6 and 1·0 million tons respectively. The exploitation in the Chelyabinsk region, where the productive measures are 1500m thick and contain 5–30 seams of 50–200m total thickness, is the most intensive. To the north in the Central Urals in the Bulanash-Elkino region the measures are about 450m thick and have ten coal seams. In the third major field (Serovskian or Bogoslavskian) near Karpinsk (Bogoslavsk) there are three coal seams of 5, 4·5 and 20m in thickness. The last seam is in places 60m thick and lies at a depth of only 100m. The Jurassic coals, which are almost horizontal, are in places found further to the north, as far as Obdorsk.

On the western slope, the Jurassic continental clays and sands in the area between Solikamsk and Ishimbai show in places thin seams of brown coal. To the south of Ishimbai to the Caspian Sea there are much larger deposits of brown Jurassic coals. The coals of this Ural-Caspian basin are used for local needs and are extracted, for instance in Sol'Iletsk and Aktyubinsk.

Cretaceous and Cainozoic coals are found in a series of regions on both slopes as thin lenses of brown and peaty coals of normally negligible economic importance. Nevertheless in Bashkiria, to the south of the Ufa, the Miocene continental karstic sediments have rather large deposits forming the South Urals brown-coal basin. The deposits continue into the Orenburg province. The existence of thick (up to 140m), horizontal seams is conducive to easy exploitation. The largest deposits are at Kuyurgaz and Babayevo, 100km to the south of Ishimbai (Malyutin, 1956).

The recently found Iboganskian coal-field is closely associated with the western slope of the Southern Urals. The few seams are very thick (30–50–70 and even 120m), horizontal and are being worked in open-cast. The measures are Jurassic. The proven reserves are $1·4 \times 10^9$ tons and the yield in 1956 was about 3×10^6 tons.

Petroleum and Gas

The Volga-Ural oil and gas province (p. 158) is situated between the Urals and the Volga. Most of its deposits are generally related to the Russian Platform. To Cisuralia belong only the Ishimbai and other analogous deposits, closely associated with the western margin of the Sub-Uralian downwarp and differing from the other deposits of the Volga-Ural province. These deposits have already been described (p. 160).

The oil and gas are found in porous 'sieve-like' and 'spongy' dolomitized lime-stones of the Lower Permian and Upper Carboniferous reefs: In the Ishimbai region the massifs lie at a depth of 480–670m under formations of rock salt and gypsum of Kungurian age and the red beds of the Ufian Formation. To the east the massifs (p. 96) are exposed forming single mounts known as 'shikhans' (Fig. 124).

In the Pechora basin, within the Urals, there are signs of oil in the form of small accumulations of liquid bitumens in the Lower Carboniferous sandstones and limestones. In the Kizel region liquid bitumens found in a borehole were filling cavities in corals and cracks in Visean limestones. The economic significance of these discoveries has not been followed up.

With the Urals are closely associated gas deposits of the Berezovo region, in the valley of the Ob′ and oil deposits in the Shaim region on the Konda—a right tributary of the Irtysh. These and other deposits nearby are within the Khanty-Mansiian National Region and lie in the Western Siberian Lowlands, but economically they belong to the Urals and a gas pipe from Berezov to the Urals is being planned. The oil and gas deposits are found in the Upper Jurassic and Lower Cretaceous sediments and are economically important (p. 303).

Oil shales

Oil shales and sapropelic coals can be raw materials for liquid fuels and have been found in several localities amongst the Devonian, the Permian and the Trias Jurassic strata of the eastern and western slopes.

Peat

Peat deposits are quite large. According to preliminary estimates the deposits of dry peat are 4×10^6 tons even if the Tobolskian North is not considered. The peat deposits are favourably situated near the main industrial centres such as Nizhnii-Tagil, Sverdlovsk and Verkhnekamsk, but they have not been much used.

The fuels on the eastern slope of the Southern Urals are nevertheless in relatively short supply and are being imported from Karaganda and even Kuzbas.

NON-METALLIFEROUS DEPOSITS

The non-metalliferous deposits of the Urals are numerous, diverse and their deposits colossal and sometimes of world importance.

Magnesite. The magnesite deposits of the Urals are very vast and have international significance as the raw material for high-refractories and magnesite cement. The Satkat deposit is a bed lying with the Proterozoic dolomites belonging to the Satkian Suite. The mineral has a mixed hydrothermal-metasomatic-sedimentary

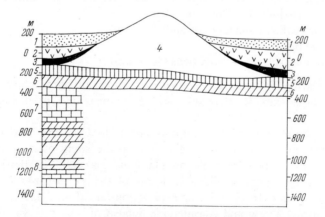

FIG. 124. Diagrammatic section of the Tra-Tau shikhan.

I — Ufian red beds; 2 — Kungurian gypsum and anhydrite; 3 — Upper Artinskian shales and marls adhering to the reef; 4 — the reef massif; 5 — Lower Sakmarian bedded limestones, underlying the reef; 6 — Upper Carboniferous dolomites and limestones; 7 — Middle Carboniferous-bedded limestones at the top and massive dolomites at the bottom; 8 — Lower Carboniferous-massive dolomites at the top and bedded limestones at the bottom.

character. Other deposits of the same type have been found in the Katav-Ivanovskian and the Bakal regions. There are reasons for believing that they will be found elsewhere in the Proterozoic carbonate suites of the Urals.

Asbestos. The deposits of asbestos are variable, numerous and often large. The asbestos yield of the Urals is the largest in the U.S.S.R. and is one of the largest in the world. The enormous Bazhenovskian deposit to the north-east of Sverdlovsk is of world importance. The largest reserves are of chrysotile associated with serpentinized ultrabasic rocks. The chrysotile is found in cracks and joints and its main deposits occur on the eastern slope of the Central Urals in association with the Bazhenovskian, Alpayevskian, Rezhevskian and Nev'yanskian peridotite intrusions. Large deposits are known in the Northern and Southern Urals.

In the last few years major deposits of anthophyllite have been found in the Sysert region. This mineral represents xenoliths of basic rocks reworked by hydrothermal emanations from the surrounding magma. P. M. Tatarinov has investigated the asbestos deposits. The recently-found blue asbestos is also important.

Diamonds. The study of diamond-bearing alluvia on the western slope of the Central Urals began in 1938. At present economic exploitation is in progress (Antropov, 1960). The diamonds are obtained from the western diamontiferous belt associated with the zone of carbonate and clastic rocks of the Middle and Upper Palaeozoic.

Potassium salts. On the western slope of the Central Urals there are the Solikamskian and the Verkhnekamskian potassium salt deposit which is the largest in the world and tens of times larger than the Stassfurt deposit. The succession in the Urals is as follows:

1. At the bottom lie Artinskian clays (40–50m) with goniatites and brachiopods.
2. Clay-anhydrite beds (about 380m).
3. Lower or underlying rock salt (260–400m).
4. Potassium salts divided into the lower, sylvinite (10–60m) zone and the upper, carnallite (45–90m) zone.
5. Top rock salt (1–70m and normally 25–30m).
6. Gypsum beds (8–85m).
7. Calcareous—clayey—marly beds (40–180m).

The salt-bearing horizons are Kungurian. There are large chemical concerns working on these salts in Berezinki and Solikamsk and producing potassium fertilizers.

Rock salt. The reserves of Permian rock salt in the Urals are inexhaustible. This has been shown in the Solikamsk region and to the south of it in the Ishimbari region where salt is unconformable on limestone massifs. The Iletskian deposit, associated with the margin of the salt domes of the Volga-Emba type, has been worked for a long time. In 1930 during exploration for oil near Shumkova (30km to the east of Kungur) a new and considerable deposit of rock salt was found. These deposits are Kungurian although the Iletskian salt is Kazanian.

Gypsum. The reserves of gypsum are again inexhaustible. There are hundreds of deposits on the western slope and the belt of gypsum massifs continues without interruption from Cherdyn in the north to Iletsk in the south. The gypsum is associated with Kungurian sediments. Those deposits which are near to railways are being worked as for instance in the Kungur region.

Limestones and dolomites. The reserves of limestones and dolomites are very large and especially so on the western slope. Hundreds of metres of limestones are found in a continuous belt along the Northern, Central and Southern Urals. They date from the Middle Devonian to the Lower Permian. In the central part of the Urals Lower Palaeozoic and Precambrian carbonates are widespread. On the eastern slope limestones are less abundant, forming individual massifs and belts, but their reserves are still considerable.

Cement. The Urals are completely self-sufficient with raw materials for cement and there are a number of cement factories.

Barytes. All barytes deposits are found in hydrothermal veins. The largest deposit in the U.S.S.R. (Medvedovkian) is situated near the village Medvedovka in the Zlatoust region. The deposit is a 4–8·5m thick vein, lying within slates and carbonates of Proterozoic age. The barytes is of high quality.

Graphite. The reserves of graphite are considerable and it is obtained at the Taikinskian deposit from where it is used in the largest factory in the U.S.S.R.

Marble. The deposits of marble and other facing-stone are large and are intensively worked.

Precious and semi-precious stones have been widely known for a long time and there are more than 55 varieties. In terms of their value, first are emeralds, followed by amethyst, topaz, beryl, demantoid, tourmaline and varieties of quartz. Of semi-precious stones, rhodonite, agates, aventurine, malachite, selenite and serpentine are important.

Fire clays are widely found in large deposits, mainly in continental Mesozoic and Cainozoic sediments, although some occur in the Lower Carboniferous coal-bearing strata and in the Devonian bauxitic suites.

Phosphorites. Large deposits are only known in the west of the Mugodzhars where there are Upper Cretaceous phosphorites in the Aktyubinsk region.

Piezoquartz. The largest deposits in the U.S.S.R. are found in the Arctic and Southern Urals.

Quartzite. High quality quartzite is found amongst the Proterozoic metamorphic rocks of the Central Urals. These deposits are important and are the largest in the U.S.S.R. In the Sverdlovsk Province there is the Mt Karaul'naya deposit worked for the Pervouralsk factory. In the Southern Urals there are other deposits including that at the Yurezanskian Ridge for which a factory at Katav-Ivanovskii is being built.

Building materials. There are over 200 large deposits of building materials. In addition there are deposits of abrasives kyanite, acid resisting rocks, roofing slates, feldspar, talc diatomite and moulding sands. New varieties of mineral deposits are zircon, lepidolite (for rubidium and caesium), dunite, serpentine, monothermite, marshallite (quartz dust), vermiculite and nepheline.

Only a few most important deposits of the Urals have been mentioned here to indicate its wealth in mineral resources.

7

WESTERN ARCTIC

GENERAL DESCRIPTION

The Western Arctic represents a typical Palaeozoic geosyncline. It has strongly folded and metamorphosed Palaeozoic and Precambrian strata intruded by synorogenic plutons. The Mesozoic and Cainozoic are of platform type and are either gently folded or horizontal, almost unmetamorphosed, and lack synorogenic intrusions. These features of the Western Arctic are different from the Eastern Arctic where there is a Meso-Cainozoic geosyncline.

BOUNDARIES. The eastern boundary lies to the west of Khatanga Bay and slices the western termination of Taimyr which is a region of geosynclinal Mesozoic. Further on it bends to the east of Novaya Zemlya and Franz-Josef Land and follows the Siberian shelf to the frontier. The southern boundary should follow the polar circle, but in fact the northern part of the Arctic Urals, the Pai Khois, the islands Vaigach, Novaya Zemlya and the peninsulas Yamal, Gydan and Taimyr are also included in the region as well as Severnaya Zemlya and the islands of the Kara Sea.

OROGRAPHY. The orography is characterized by the alternation of low and narrow mountain ranges with vast lowland plains. The greater, northern part of the Western Artic is a region of marine Boreal Trangression. Consequently the plains represent the bottom of shallow epicontinental seas and the ranges represent archipelagos of islands rising above the sea level. In the west there is the Barents sea plain with depths of no more than 200–400m, followed by the Kara Sea, in the south including the low peninsulas of Yamal and Gydan, the Lapp Sea on the other hand is entirely in the Eastern Arctic.

The Arctic Urals with peaks reaching 1500m bend towards the Pai-Khoi with heights of 1360m in the south and 467m in the north. The straits known as Yugorskii Shar and Karskie Vorota separate off the island of Vaigach. Further on follows Novaya Zemlya with peaks of 1000–1100m, which consists of two islands separated by the Matochkin Shar straits (Fig. 125). To the north of Novaya Zemlya lies the archipelago of Franz-Josef Land with a highest point of 735m. A second large archipelago—Severnaya Zemlya—is lower and its height varies from 200 to 300m. The Taimyr Peninsula is a Palaeozoic mountain range, but the highest Byrrang Mts in it reach 400–500m in the west and 1100–1200m in the east (Fig. 126). These mountains are transected by the valleys of the Pyasina and Taimyra, with which lake Taimyr is associated. Glaciation is widespread in Franz-Josef Land, Severnaya Zemlya and the northern island of Novaya Zemlya. The climate is rigorous and arctic. The population is small, and lowland plains are vast tundras. Transport is difficult and sometimes dogs have to be employed (Fig. 127).

STRATIGRAPHY

The mountainous and elevated areas consist of folded, metamorphosed Palaeozoic and Precambrian rocks. In the plains under the cover of Anthropogenic sediments lie almost horizontal Mesozoic and Cainozoic strata.

PRECAMBRIAN

The Precambrian is exposed in the Arctic Urals, in Taimyr and in Severnaya Zemlya. The Archeozoic has not been recognized. The Lower Proterozoic consists of gneisses, schists and amphibolites of over 7km in thickness. The Upper Proterozoic is divided into a number of suites and consists of phyllites, quartzites, marbles with *Collenia*, slates and volcanic rocks. The thickness of the Upper Proterozoic reaches 8km.

In the Arctic Urals there is a gneiss-amphibolite complex of muscovitic and two-mica granite gneisses interlayered with amphibolites. The plagiogranites and granites of the complex are converted into orthogneisses. The total thickness is unknown, but is over 1km. Conditionally the gneiss-amphibolite complex is included in the Proterozoic but it may be Archeozoic.

Overlying this is a suite (Neroveiian, Verkhneikharbeiskian) of carbonaceous, quartzose and phyllitic slates with marbles (total thickness 1000m). It is followed by a suite of greenish slates and quartzites with basal sandstones and conglomerates at the bottom and banded chloritic and sericitic slates at the top. Its total thickness is 2000m. The next suite—the Menvinian—is 800–1300m thick and consists of green chloritic and carbonaceous sandy slates, quartzites and marmorized limestones, with volcanic rocks at the top of the suite. There are no organic remains in these suites and they have been provisionally included in the Cambrian since Ordovician strata lie on them. This is also shown in synoptic maps. It is, however, more probable that the lower suites are Proterozoic.

In Taimyr the Proterozoic involves two complexes of crystalline and metamorphic rocks. The lower is no less than 7000m thick and is included in the Lower Proterozoic, but could equally be Archeozoic. It consists of gneisses and schists with deformed gabbroid intrusions, orthogneisses and granitoids, being often associated with wide zones of migmatites. The complex is divided into three series.

The Upper Proterozoic complex (up to 8000m thick) lies on the Lower Proterozoic with an unconformity, and includes many intrusions of meta-gabbro and cataclased granites. At the base there is the Pronchishchevian greenstone suite (1200–2000m) of green schists representing altered volcanic rocks and tuffs and also metamorphosed terrigenous sediments. There are also beds of metamorphosed dolomites and limestones. Higher up lies the Oct'yabrskian Suite (1000–1500m) mainly of coarse clastic sediments such as quartzites and conglomerates. The suite is succeeded by the more widespread and thicker (1700–2600m) phyllites of the Zhdanovian Suite with beds of dolomites, limestones and graphitic slates.

The succession ends with the Laptevian Suite (400–1200m) of local development. The suite consists of sandstones and conglomerates at the bottom, followed by metamorphosed sandstones with beds of marbles and capped by metamorphosed volcanic rocks.

It is not proper to include the above thick and complicated succession in the

complex as Lower Proterozoic, especially since the overlying Cambrian has no
Upper Proterozoic. It is quite possible that the lower part and maybe the whole
faunas and in some features resembles the Sinian (Upper Proterozoic).

FIG. 125.   The shore of the Mityushikha inlet on the northern island of the
Novaya Zemlya. The cliffs consist of Middle Devonian clastic strata.
Photograph by V. I. Bondarev (1955).

FIG. 126.   Central part of the Byrrang Mts. *Photograph by L. A. Chaika (1950).*

The Precambrian of Severnaya Zemlya is represented only by the Upper Proterozoic. The Upper Proterozoic here is lithologically similar to the Upper Proterozoic of Taimyr, but also consists of two formations—a more pelitic below and more psammitic above. The lower formation consists of green chlorite-sericite-quartz schists with graywacke-like meta-sandstones and green phyllites. At the top there is a group of dark graphitic slates. The total thickness of the formation is 2500m. The upper formation is conformable on the lower and consists of quartzose and polymictic meta-sandstones with thin beds of slates, graywackes and phyllites. At the top there are flows of meta-volcanic rocks. The thickness of the formation is 1500–2000m.

Higher up lie barren formations provisionally considered as Lower and Middle Cambrian. However, they may very well be Upper Proterozoic Sinian, while the so-called Upper Proterozoic may be Lower Proterozoic.

PALAEOZOIC

Lower Palaeozoic

The Lower Palaeozoic has a limited distribution and especially so the dated Cambrian. It should be remembered that many subdivisions of the Cambrian strata shown on maps have no fauna and are probably Upper Proterozoic.

CAMBRIAN. The maps of the Arctic Urals and Pai-Khoi show widespread Cambrian strata, but Cambrian faunas are found only at isolated localities in rocks which unconformably overlie the so-called Cambrian of the maps. Thus in the Arctic Urals the suggested Cambrian is unconformably overlain by variegated quartz-sandstones and shales (Miniseiian Suite) 1000m thick in the lower part of which an Upper Cambrian brachiopod *Billingsella* has been found.

FIG. 127. Dog transport on the slopes of the Byrrang Mts.
Photograph by V. A. Vakar.

On Novaya Zemlya there are Middle and Upper Cambrian deposits with trilobites. The sediments are found on and near to the shore of Matochkin Shar. In thin beds of siliceous shales *Paradoxides* were discovered. The Upper Cambrian (400m thick) strata consists of green chloritic and variegated quartz sandstones and phyllitic and chloritic slates, in which trilobites have been found.

In Taimyr various faunally barren sediments overlying the Proterozoic are included in the 'Cambrian'. In the east these strata consist of variegated sandstones, shales and limestones of a total thickness of 100m. In the west only the bottom part of the formation is variegated, while the top part consists of dark massive algal dolomites and limestones and the total thickness is no less than 1000m (possibly up to 2500m). These deposits are identical with the Sinian deposits of the Urals and Siberia and should be considered as Upper Proterozoic. The Middle Cambrian is found everywhere. In the west it is represented by shales, siltstones and beds of dolomite. To the east the carbonates thicken and on the river Taimyra *Collenia*-bearing limestones are widespread and thick (1900–2500m). The absence of a fauna makes their age provisional and again there are similarities to the Sinian. Lastly in eastern Taimyr variable thick (1300–1500m) limestones, which are often algal, are found. At their top they have yielded Middle Cambrian trilobites.

The Upper Cambrian is represented by limestones and dolomites with algae and beds of shale. The limestones are sometimes bituminous, but lack a fauna. Their thickness varies from 250–1600m, and their age is established with reference to the overlying limestones with an Ordovician fauna.

The Cambrian of Severnaya Zemlya is divisible into three, but owing to the absence of a fauna this is entirely provisional. The lower division begins with basal conglomerates and sandstones, followed by greenish and dark-grey phyllitic and slaty pelites, which are topped by sandstones. The total thickness is 1000–1100m. The middle group consists of variegated shales and marls (450–500m). The upper part of the succession consists of grey shales and sandstones, some 3000m thick. At the bottom of the upper group *Obolus*, and higher up orthids and crinoids of Ordovician type, have been discovered. If the *Obolus* beds are Tremadocian then the whole upper group is Ordovician.

ORDOVICIAN. Ordovician strata are widely found and dated. The fauna allows a three-fold subdivision and in places the recognition of finer subdivisions. In the Arctic Urals the top Cambrian and Lower Ordovician rocks (Miniseiian Suite) are unconformable on the Cambrian (?) or older rocks, and there is a conglomerate at the unconformity. These relationships are often explained by the manifestation of the Salairian phase of the Caledonian Orogeny, although there is a possibility of the break being due to manifestation of the Baikalian Orogeny at the end of the Upper Proterozoic.

The Miniseiian Suite consists of quartzites and phyllites with a marine fauna with *Billingsella*. The suite is over 1000m thick. The upper part of the Lower Ordovician and Upper Ordovician are represented by suites which replace each other laterally. The sedimentary Syangurian Suite consists of siliceous and phyllitic shales with groups of carbonates with a marine fauna and reaches 600–700m in thickness. The volcanic Ochenyrdian Suite consists of basic lavas such as porphyrites and microdiabases and their tuffs. More rarely there are albitophyres and felsites and sediments. The thickness of the complex is 1000–1300m. The Upper Ordovician phyllitic and siliceous shales with limestones are trilobite-bearing and up to 700m thick.

In Pai-Khoi the Ordovician covers large areas. At the bottom, as in the Urals, are sandstones and sandy limestones with a marine Lower Ordovician *Angarella*. Higher up follow the uniform grey and dark shales with limestones (800–900m). The marine fauna found at various horizons indicates the presence of the Middle and Upper Ordovician.

On Novaya Zemlya and Vaigach the Ordovician is well developed and thick (3000–4000m). The marine fauna indicates the presence of all its divisions. In the north of Novaya Zemlya the Upper Ordovician is either partly or completely eroded away under the unconformable Llandovery red beds. This suggests the manifestation of the first phases of the Caledonian Orogeny. The lithology is uniform and grey and black bedded, or more rarely massive, limestones predominate. Beds of shales and dolomites are quite frequent. On the northern island of Novaya Zemlya the Middle Ordovician has variegated sandstones. Sometimes thick (1400m) green chloritic slates, sandstones and lavas are placed in the Lower Ordovician, but the absence of a fauna renders this suggestion only provisional.

In Taimyr the Ordovician is exposed in narrow belts in the central part of the peninsula. It is represented by different facies, amongst which thick limestones and graptolitic shales are important. In the west there are bedded and massive limestones and dolomites varying from 500m to 1500m and containing thin beds of shales with rare corals. In the east sandy and clayey shales appear and thicken rapidly to 2400–3000m. Amongst these shales graptolitic assemblages suggest Tremadoc, Arenig, Llandeilo, and Caradoc stages. The graptolites are of Western European type.

On Severnaya Zemlya the already described *Obolus*-bearing shales and sandstones belong to the Lower Ordovician. These beds are unconformably overlain by limestones interbedded with sandstones and underlain by a basal conglomerate. The discovery of *Angarella* in these strata, which are 300–350m thick indicates that they are upper Lower Ordovician. Higher up lies a thick variegated formation (1600–1800m) of dolomites, marls and sandstones with gypsiferous layers towards the top. There is almost no fauna, which complicates the accurate determination of their age. The Ordovician succession ends with barren, dark and grey limestones, dolomites and sandstones of a total thickness of 250m. The thickness of the *Angarella*-bearing Ordovician is of the order of 2250–2500m, but together with the *Obolus* shales it is 5000–5500m.

Middle Palaeozoic

The Middle Palaeozoic is widespread and has a rich marine fauna. It is a thick and typical stratigraphic complex.

SILURIAN. The Silurian is quite widely found and is represented by mainly marine deposits of a considerable thickness. In the Arctic Urals, in the west, there are thick (up to 1000m) limestones and dolomites of Upper Silurian age with a brachiopod fauna. Nearer to the central part the limestones are replaced by shales and remain only in isolated beds. Amongst the black carbonaceous shales there are graptolitic groups indicating the existence of all the stages of the Silurian. On the eastern slope the appearance of volcanic formations at the bottom and the top of the succession is characteristic. The volcanic rocks are interlayered with the Llandoverian and Wenlockian shales. The Lower Ludlow stage consists mainly of richer fossiliferous, sometimes reef, limestones of marine origin.

In Pai-Khoi graptolitic, carbonaceous, clayey and siliceous shales (slates) predominate. The graptolites indicate the presence of all Silurian stages. In places shales
are replaced by limestones. On Novaya Zemlya and Vaigach island, there are also
thick limestones and shales. The Lower Silurian sometimes is represented by variegated sandstones, while in the Upper Silurian normally dark, muddy and bedded
limestones predominate with a marine fauna of smooth brachiopods (*Lissatrypa* and
Protathyris). The shales are often graptolite-bearing. The rich marine fauna allows
the distinction of the Llandoverian, the Wenlockian, and the Ludlovian stages. The
latter is divided into two sub-stages: the lower and the upper. The lower has various
pentamerids while the upper has typical *Protathyris didyma* and species of *Lissatrypa*. Volcanics are rare and the thickness is great. Llandoverian variegated beds
reach 500m in the south and up to 3000m in the north. The limestones are 1500–
2000m thick (Fig. 128).

In Taimyr the general lithology and fauna of the Silurian is the same as on
Novaya Zemlya and in the Urals and its thickness is of the order of 1000–2000m.
The Silurian, here, has been less adequately studied, since there are fewer fossiliferous horizons, which complicates the subdivision of the succession. In various
places there are different horizons which altogether suggest the presence of all the
stages. In the west there are dominant limestones, often with tabulates and stromatoporoids. In the north and especially north-east clastic sediments—shales and silts—
predominate. Here the sandy Caradoc limestones are conformably overlain by black
phyllitic shales (400–500m) with Llandoverian graptolites. Higher up there are

FIG. 128. River Pestsovaya on the northern island of Novaya Zemlya.
Silurian limestones for a typically high relief of the elevated part of Novaya
Zemlya. *Photograph by S. V. Cherkesova (1955).*

black sandy shales (300–500m) with Wenlock graptolitic shales. The succession ends with similar almost black siltstones (400–450m) with beds of shale and a poor Upper Silurian fauna. This is a typical shaly succession.

On Severnaya Zemlya the Silurian is no less widespread than the Ordovician, reaches 1600–1700m in thickness, and has a rich fauna. It is unconformable on the Ordovician and has a different structure. The Lower Silurian (Llandoverian and Wenlockian) begins with limestones (500m) of Llandoverian age and contains tabulates and pentamerids. The Wenlock (600m) also consists of limestones with beds of sandstones and dolomites. The fauna is diverse and marine. The Upper Silurian (Ludlovian), 600–650m thick, has a large number of sandstone layers and gypsiferous beds at the top. In the main it consists of grey and black, massive and bedded limestones with a marine fauna. The gypsiferous dolomites at the top have only ostracods *(Leperditia, Isochilina).*

DEVONIAN.   In the Central and Northern Urals three types of successions are found: the western, the central and the eastern. In the western successions marine sediments begin in the Eifelian or even Frasnian. Below them are the continental red beds corresponding to the Lower Devonian. Sometimes they are much thinner or absent and Eifelian strata rest with an angular unconformity on the Silurian and Ordovician. In the Northern Urals such a type of succession is absent, but is developed to the west in the Chernyshev range and the islands of Bolshoi Zelenets and Malyi Zelenets.

The central successions have all Devonian strata of marine origin. Such successions are encountered in the Arctic Urals and Pai-Khoi, Vaigach and the southern

FIG. 129.   Upper Devonian (Frasnian) limestones. Southern flat part of Novaya Zemlya. *Photograph by V. I. Bendarev (1954).*

island of Novaya Zemlya. The lower part of the Devonian (the Lower Devonian and the Eifelian) have isolated massifs (300–600m) of reef, Hercynian, pale massive limestones with the typical brachiopod *Karpinskia*. The massifs are surrounded by thick shales. The upper part of the Devonian (the Givetian and the Upper Devonian) consists of bedded carbonates (Fig. 129) and clastic sediments (600–1000m) and contain a rich marine fauna of a typical nature. In Pai-Khoi only this type of succession is present.

The eastern type of succession differs from the central by the appearance in some regions of volcanics which are sometimes rather thick. The sedimentary rocks, however, are the same. In Pai-Khoi the eastern type is evidently exposed at the bottom of Baidarats Bay. In Novaya Zemlya again only two types of successions are found and the third (the eastern) is exposed at the bottom of the Kara Sea. The western type is found on the western shore, mainly in the north of Novaya Zemlya. Here the Devonian rocks lie with an abrupt unconformity on Silurian and older rocks and the succession begins with a basal sandy-conglomeratic group overlain by limestones and shales of Middle and Upper Devonian age. The basal group often has lavas, sills and tuffs of basalt-diabase composition, and in such instances the thickness of the basal volcanic-sandy formation increases to 750–1000m (on the northern island). Since this formation is succeeded by the lower strata of the Upper Devonian, it is possible that it partly belongs to the Givetian. The basal formation is overlain by massive limestones with subordinate shales which are sometimes bituminous. The limestones contain a rich and diverse marine fauna of mainly brachiopods, which indicate the presence of all three substages of the Frasnian and two substages of the Famennian. The leading fossils are the same as on the Russian Platform. Frasnian limestones are 400–800m thick and Famennian limestones are about 400m thick. The goniatite-bearing Domanik facies should be noticed, since this is the same as in the Urals and Timan. The sediments are rich in organic matter.

The second type of Uralian succession is found in the east on Vaigach island and the southern and northern islands of Novaya Zemlya. The Lower, Middle, and Upper Devonian is present and is represented by marine facies. The Lower Devonian includes three complexes: the shaly-sandy, the dark bedded limestones and pale massive reef limestones (Hercynian limestones with *Karpinskia*). The thickness of the Lower Devonian is 400–600m. The Middle Devonian is represented by Eifelian and Givetian strata with a marine brachiopod-coral fauna. Grey and black bedded limestones and shales predominate, but there are also common quartz sandstones, quartzites, and conglomerates interbedded with shales. The thickness of the Middle Devonian is 400–700m. The Upper Devonian is the same as in the western type of successions in Timan, but differs by the absence of the basal volcanic-sandy formation. On Givetian limestones lies the similar Frasnian limestones and shales, which are overlain by Famennian limestones and dolomites. The total thickness of Upper Devonian strata is 400–600m.

The Devonian of Taimyr is very similar to that of Novaya Zemlya and is continuous with it. Formerly only successions of the western Timan type were known and involved the Givetian and higher stages, represented by limestones, resting unconformably on the Silurian. In the last few years in the southern and central regions of Taimyr central-type successions have been found involving a conformable passage of Ludlovian limestones into Lower Devonian limestones succeeded by Eifelian and Givetian limestones.

The Lower Devonian is found in the western termination of the Byrrang Mts and in the upper reaches of the Tareya and the Leniva and is represented by the Tareya Suite (600–700m). The lower part of the suite consists of dark limestones and bituminous shales with armoured fish, while the upper part consists of grey and greenish coral and brachiopod limestones, marls and shales. A rich and variable fauna has a Lower Devonian aspect. The Tareian Suite is overlain by lagoonal variegated shales, marls and sandstones with beds of gypsum (100–150m) which are followed by red, terrestrial sandstones and shales with groups of conglomerates. These are probably Middle Devonian. The Middle Devonian consists of diverse lithologies. Apart from the variegated, gypsiferous group (Tareian) and the red beds in the south-eastern termination of the Taimyr Peninsula and near to the mouth of Khatanga there is a gypsiferous formation (no less than 100m) which towards the top is transitional into Givetian limestones. This gypsiferous group correlates with the gypsum and salt formation with which the Port Nordvik salt domes on the other shore of the Khatanga are associated. The marine Eifelian sediments are so far only found in one region where they consist of brachiopod-bearing limestones. The marine Givetian, on the other hand, is widespread and its sediments are often unconformable on the Silurian. The basal conglomerates pass up into green and grey limestones and shales with *Stringocephalus* (150–200m).

The Upper Devonian often encountered, is 400–700m thick and consists of marine limestones, marls and shales. A rich and diverse fauna permits the identification of the Frasnian and Famennian, and also finer subdivisions of Ural type. Amongst characteristic facies the Domanik of the same type as in the Urals and with goniatites should be noticed. The Domanik may be the parent horizon of oil.

The Devonian of Severnaya Zemlya is unusual and its facies differ from the Taimyrian and Uralian Devonian facies, but are very similar to the Devonian of Spitzbergen and other northern Caledonides. The Devonian of Severnaya Zemlya is similar to the southern Caledonides: the Krasnoyar region and the Minusinskian Depression. A diagnostic feature of the Devonian in Severnaya Zemlya is the predominance of the Old Red Sandstone facies including red and variegated sandstones, siltstones, shales, marls, and dolomites. Various armoured fish are widely found. Gypsum and salt-bearing groups are common. There are also frequent groups of dark, grey limestones and dolomites with ostracods (*Isochilina, Leperditia*) representing the deposits of salt lakes and marshes. Limestones or marls with marine faunas occur, but are very rare. The age of these rocks is the same as in the Minusinskian Depression (Upper Eifelian and Upper Givetian). The investigation of the armoured fish, the lamellibranchs and the ostracods has shown that in Severnaya Zemlya there are three divisions of the Devonian. The Lower Devonian, a part of which is probably Upper Silurian, reaches 500–800m in thickness. The Middle Devonian is about 400m thick and has variable facies. The Upper Devonian is overlain by the Anthropogene strata and is folded into gentle folds.

LOWER CARBONIFEROUS. The Lower Carboniferous is closely associated with the Upper Devonian and is represented by the same facies of limestones with marine faunas. Usually the Tournaisian, the Visean, and the Namurian are well developed, although Visean rocks are most widespread and can be divided into three substages. The Tournaisian is less widely encountered and in some sections is entirely absent and the Visean is unconformable on older rocks. The Namurian is again less widespread than the Visean.

In the Arctic Urals and Pai-Khoi the Tournaisian is represented by brachiopod-bearing limestones (100m) or dark goniatite-bearing shales and limestones (50–60m).

The Visean almost everywhere consists of coral-brachiopod limestones, varying in thickness from 250–400m to 800–1000m. The Namurian is often represented by shales or limestones with brachiopods and foraminifera and is 50–100m thick. The Lower Carboniferous succession is identical to that of the Western Urals and differs from the succession on the eastern slope by the absence of volcanicity and a smaller content of clastic rocks. The succession in Novaya Zemlya and Vaigach has the same features. The Tournaisian consists of limestones and shales (200–300m) or is absent. The Visean limestone group (300–700m) is most widespread and is in places overlain by the brachiopod-bearing Namurian. A curious feature on the eastern shore of Novaya Zemlya is the absence of marine Lower Carboniferous. Here between the Devonian and the Permian is an almost barren group of quartz sandstones with thin limestones.

The Lower Carboniferous of Taimyr (600–1000m) again resembles that in the Urals and consists of limestones and shales with a rich marine fauna, with mainly Visean, occasional Tournaisian and rarely Namurian forms. In Severnaya Zemlya no Lower Carboniferous strata have been found.

Upper Palaeozoic

The Upper Palaeozoic was an epoch of intensive manifestations of the Hercynian Orogeny, resulting in the formation of high mountains, which as a consequence of erosion supplied material for thick clastic formations including conglomerates, sandstones and shales. Another feature of the Upper Palaeozoic of the Arctic is the small thickness of the Middle and Upper Carboniferous. The Lower Permian here rests directly on the Lower Carboniferous and older strata.

There are frequent coal seams in the Permian sediments of the Arctic Urals and Taimyr.

THE MIDDLE AND UPPER CARBONIFEROUS of the western slope of the Northern and Central Urals are represented by clastic deposits and massive limestones—sometimes reef-limestones. Limestones with rich marine faunas are particularly characteristic.

In the Arctic Urals and Pai-Khoi there are only limestones of Middle and Upper Carboniferous age. The limestones are less widespread, strongly eroded and occur as separate outcrops, where their thickness is 80–200m. In the majority of regions such limestones have been completely eroded away and Lower Permian conglomerates and sandstones rest directly on the Lower Carboniferous. The conglomerates include pebbles of Lower and Middle Carboniferous limestones.

On the island of Vaigach and Novaya Zemlya the number of isolated outcrops of Middle and Upper Carboniferous limestones is less and basal Permian rests not only on the Lower Carboniferous, but on older formations. Some geologists thought that the thick clastic deposits of the Lower Permian were Upper and Middle Carboniferous. This view so far has not been palaeontologically confirmed, although it is correct in the Central and Southern Urals. In Taimyr there are several isolated outcrops of limestones and clayey and siliceous shales, which are conformable on the Lower Carboniferous. In these outcrops brachiopods, corals and goniatites of Middle Carboniferous age have been found, although the rocks are sometimes considered to be Upper Carboniferous. The thickness of these rocks varies reaching 400–500m. Higher up there are clastic sediments, which are barren at the bottom, but contain a Lower Permian fauna at the top.

In Severnaya Zemlya the Upper Palaeozoic is unknown, but further to the east between the mouths of the Khatanga and the Lena the Upper Palaeozoic is widely known and is in general of Taimyr type.

PERMIAN. The economically important coal measures of the Permian are marine and continental, very thick and very widespread. This most peculiar complex of sediments, closely related to mountain building, has been produced as a result of the Hercynian Orogeny. In the Lower Permian there are frequent groups and suites with marine faunas, but the predominant continental deposits are either barren or include plant remains and in places coal seams. In the Upper Permian the groups with marine faunas are rare and continental, sometimes coal-bearing facies are dominant. The thickness of the Upper Permian is less than that of the Lower Permian.

In the Arctic Urals and Pai-Khoi the Permian occupies large areas and consists of a uniform thick formation of brown, grey and black clastic sediments, such as sandstones, siltstones, and shales with coal seams. It is interesting that in many sections at the base there are sandstones or conglomerates and green marls, with occasional pebbles. The marls have a Sakmarian goniatite-brachiopod fauna. Their thickness is only 3–12m and they pass upwards into the Gusian Suite, which is also of Sakmarian age and has at its bottom goniatites and at its top plant remains. The suite consists of black shales with sandstones and shales towards the top. Its total thickness is 200–500m. The Artinskian Stage includes two suites—Bel'kovskian and Talatian—which together are 850–1500m thick. The rocks are black and grey muddy sandstones interbedded with siltstones and sandstones and more rarely sandy limestones. The fauna is rich and variable, with predominant brachiopods, goniatites and lamellibranchs. The Lower Permian succession is completed by the Vorkutian Suite (1500–1600m), which is the main coal-bearing formation of the Pechora basin and consists of grey sandstones, argillites, siltstones, and pebble conglomerates. At its bottom there are thin beds with marine faunas, but in the main the suite consists of continental lacustrine, fluviatile and marsh deposits as well as land deposits with plant remains and as many as 100 seams of coal.

The Upper Permian is entirely continental. The lower suite (Paemoboian) is 3000m thick and has many coal seams of a workable thickness. The suite consists of black and grey siltstones, argillites, and sandstones with plants. The upper suite (Tal'mausian) is lithologically different and lacks coal seams. It has a thick (600m) group of conglomerates passing upwards into coarse sandstones and then siltstones. The total thickness of the suite is about 1000m. In Pai-Khoi there are fewer conglomerates and sandstones and shales predominate, for which a possibly excessive thickness of 3000m is given. In the upper suite there are two thin layers of basalts and a Lower Trias fish has been found. It is possible that the suite is Lower Trias in age.

In Novaya Zemlya the Permian is similar to that in the Urals, but lacks coal seams and coarse clastic rocks and is also thinner. The Lower Permian (1700m) strata are at the bottom sandy and include a number of marine bands with brachiopods, but there are also plant-bearing groups. The Upper Permian (300m) is only found in one region where it consists of green sandstones and shales with brackish-water lamellibranchs.

In Taimyr, on the southern part of the peninsula, the Permian is widespread. It is of the type found in the Urals and is divided into a lower and an upper division. A typical feature of the Taimyr Permian is the presence of lava flows and basic sills analogous to the Siberian Traps. Each division of the Permian represents an individual macro-rhythm of sedimentation beginning with marine and ending with continental, coal-bearing series, as happens in the Arctic Urals.

In the west (the Cisyenisei region) the Lower Permian starts with a formation of sandstones and siltstones of 1100–1200m in thickness and has a number of marine bands. The upper part of the succession consists of a coal-bearing formation (800–1100m) with workable coals. In the central part of the peninsula the lower part of the succession has limestones and dolomites (400–1200m). The limestones have a rich marine fauna. The carbonate rocks are overlain by a clastic group with local coals and tuffogenous groups of 600–800m in thickness. In the east the limestones are replaced by sandstones, siltstones and shales with beds of limestones with marine faunas. The upper part of this succession consists of a coal-bearing formation with workable coal seams. The total thickness of the Lower Permian is 2000m.

The Upper Permian of Taimyr like the Lower Permian starts with a suite with common marine bands and groups, but also includes coal-bearing groups. The upper part is continental and often coal-bearing. Thick limestones are absent from the Upper Permian and the fauna is found in calcareous sandstones. Sandstones and shales predominate throughout. The total thickness of the Upper Permian is 1200–1600m.

The traps, of Upper Permian and Lower Triassic age, are characteristic of Taimyr. The whole complex includes basic lavas and tuffs which have beds of clastic sediments with fresh-water faunas and plant remains. The complex varies in thickness from 100–150m to 800–1000m and its age has not been accurately determined. The fresh-water lamellibranchs are Upper Permian, but the flora is Triassic. Consequently the complex is considered to be Upper Permian and Lower Trias. To the east, between the Khatanga and the Lena, the traps are Upper Permian, while to the south on the Siberian Platform they are mainly Lower Triassic. The complex completes the Upper Palaeozoic sedimentary rhythm, which ends, as often elsewhere, in the Lower Trias.

MESOZOIC

The Mesozoic, which begins with the Upper Trias, is in comparison with the Palaeozoic much thinner and less widespread. It is weakly folded or horizontal and is of platform type. Only in the northern part of the Western Siberian Lowlands, and especially around the lower Yenisei, is the Mesozoic quite thick. At the eastern edge of the Taimyr the succession changes and acquires a geosynclinal aspect and the Mesozoic becomes in its lithology, thickness and tectonics identical to the Pacific Ocean Geosyncline. Thus the eastern edge of Taimyr and the shore between the Khatanga and the Lena should be included in the Pacific Ocean Geosyncline, which will be discussed later.

UPPER TRIAS AND LOWER AND MIDDLE JURASSIC. Coal-bearing formations of this age, which are found on the eastern slope of the Arctic Urals, have not been discovered in the Arctics proper, but their existence is probable. On the lower Yenisei, around Dudinka, synoptic maps show anticlinal structures involving the Cretaceous. Here boreholes have encountered the Lower and Middle Jurassic consisting of sands and clays (600–1000m). Marine suites, here, alternate with those in which plant remains and brown coal seams are found. An analysis of the fauna has shown that the Lower Jurassic sea penetrated from the east along the Khatanga

Depression as far as the lower Yenisei. Further to the west the marine Lower and Middle Jurassic is unknown and in many regions is absent. The discovery of marine Lower and Middle Jurassic deposits within the Western Siberian Lowlands is of great interest.

On Franz-Josef Land the Mesozoic succession begins with the Carnian Stage of the Upper Trias. On the island of Vil'chek, on the coast, there are outcrops of dark limestones and shales with a marine fauna of ammonites (*Sirenites*) and lamelli-branchs (*Halobia*). The limestones are covered by Upper Triassic and Liassic, continental sandstones, siltstones, and argillites with plant remains and of a total thickness of 300m. The Middle Jurassic (no less than 30m) consists of clays and siltstones. Beds with marine faunas suggest the Aalenian and Bathonian stages.

UPPER JURASSIC AND LOWER CRETACEOUS. The deposits of this age are widespread and have been found in outcrops as well as boreholes. Marine deposits with rich and diverse northern faunas are predominant, but in places there are continental coal-bearing sediments.

On the eastern slope of the Arctic Urals the Upper Jurassic and Lower Cretaceous are only found in boreholes sunk in the Salekhard region and to the south. The Upper Jurassic begins with a coal-bearing group (50m), and ends with grey marine clays (60m). The Lower Cretaceous begins with Valanginian clays (200m) with *Aucella*, ammonites and Foraminifera. Higher up there are grey and greenish clays and silts of Hauterivian and Barremian age, some 100m in thickness and representing an alternation of continental and lagoonal deposits. The succession ends with a coal-bearing formation (160m) of suggestedly Aptian-Albian age.

In Pai-Khoi and Novaya Zemlya there are no *in situ* Jurassic or Cretaceous rocks, but the quantity of boulders of sandstones with marine faunas is surprising. Their rich faunas suggest all ages from Bathonian to Aptian-Albian. It is possible that on the northern island of Novaya Zemlya there are *in situ* deposits which are hidden under ice, and this is supported by the abundance of the relevant boulders in moraines.

In Franz-Josef Land the Upper Jurassic and Lower Cretaceous sediments can be seen to emerge from under the ice. The Upper Jurassic is represented by dark clays and brown sandy limestones with concretions. The total thickness of these rocks is no less than 200m. A rich and variable fauna of ammonites and belemnites suggests the presence of Middle and Upper Callovian, Oxfordian, Kimmeridgian and Lower Volgian strata. Higher up there is the Lower Cretaceous volcanic-coal series of up to 200m. It consists of lavas and sills of basalts and dolerites (2–60m) separated by sands and carbonaceous clays with lenses of brown coals and plant impressions (Fig. 130).

On the islands of the Kara Sea there are Lower Cretaceous coal-bearing formations, which are thin and resemble the coal-bearing series of Franz-Josef Land.

In Severnaya Zemlya there are exposures of greenish clays and sandstones (15–20m) with Kimmeridgian ammonites and sandstones with Lower Cretaceous plant remains and greenish sandstones with *Aucella* of Valanginian age.

In Taimyr there are isolated exposures of Upper Jurassic and Lower Cretaceous strata in the central and southern parts of the peninsula. In places there are Callovian (150–190m) and Lower Oxfordian sands (50–70m). The dominant rocks are sandstones and clays with *Aucella*, reaching tens of metres in thickness and being of Volgian and Valanginian age. The upper part of the Lower Cretaceous is represented by a coal-bearing formation (up to 70–100m) with 1–3 brown coal seams.

Around the lower Yenisei, in boreholes sunk on the Malokhetian anticline, Upper Jurassic sands and clays rapidly thicken to 600m. Their marine fauna indicates the participation of all the stages, of which the strata of the Volgian Stage are particularly thick (340m). The Lower Cretaceous is also very thick (2000m) and its lower part (Valanginian-Hauterivian) reaches 500m, and consists of clays and sandstones with *Aucella* and ammonites. The upper part is coal-bearing, lacks a marine fauna and is divided into a number of suites; altogether some 1500m thick (*Geological structure* . . . 1958). The coal-bearing Lower Cretaceous formation is exposed in the valleys of the Bol'shaya Kheta and the Malaya Kheta.

UPPER CRETACEOUS. This is the epoch of a major transgression, which advanced from the south into the southern parts of the Western Arctic. In the northern part of the Arctic, in Franz-Josef Land, Novaya Zemlya and northern Taimyr the Upper Cretaceous sediments are unknown and probably were not deposited. In the south the Upper Cretaceous is found in the Arctic Urals, Pai-Khoi, northern parts of the Western Siberian Lowlands and southern margin of Taimyr and the Northern Siberian Lowlands.

On the western slope of the Arctic Urals, in the basin of the Usa, there are widespread glauconitic sands and gaizes, some 60–90m thick. Their marine fauna (*Pteria tenuicostata* Roem.) renders them Santonian in age. Santonian siliceous clays and marls with radiolarians, diatoms, and lamellibranchs (*Inoceramus cardissoides*) are found in a locality near the Saa-Yaga on the north-eastern sea-shore of Pai-Khoi.

In the northern part of the Western Siberian Lowlands the Upper Cretaceous is widespread, but is only exposed in isolated localities. Santonian marine clays and sands are most frequently encountered, but there are also marine Campanian-Maastrichtian deposits and Cenomanian continental sands. The Danian terrestrial quartz sands and kaolinite clays are typical and in the basin of the Pur occupy a

FIG. 130. Upper Jurassic on the island of Berghaus on Franz-Josef Land. The island is 370m high and consists of Upper Jurassic sandstones capped by a sill of dolorite. *Photograph by V. D. Dibner.*

large area. In the deep borehole at Berezov the Upper Cretaceous is about 500m thick.

Around the lower Yenisei the Upper Cretaceous is found over large areas and has also been encountered in boreholes. At its base there are continental, coal-bearing, Albian-Cenomanian deposits. They are overlain by marine Upper Cretaceous strata consisting of grey and greenish clays and siltstones with gaizes and sands. The total thickness is 700m. A rich and diverse fauna indicates the presence of the Lower Turonian, Upper Turonian, Coniacian, Santonian, and Maastrichtian.

CAINOZOIC

PALAEOGENE. The typically marine Palaeogene is widely found to the south, while in the western Arctic it is absent or represented by continental deposits. Near the 'Lower Yenisei' sands and clays with Palaeogene pollen are found on the flanks of the anticlines. A number of outcrops of gaizes with Eocene diatoms are found in the lower Nadym basin (south of the Ob' estuary). These are the sediments of a bay, which connected to a southern sea. The gaizes are in places overlain by brown sands and greenish clays with Lower Oligocene diatoms.

The Palaeogene sediments of the southern part of the Tazovskii Peninsula have the same lithology and origin. They represent a direct easterly continuation of the Nadymian Palaeogene. The Lower Eocene (75m) consists of gaizes and gaize-like clays with small lamellibranchs and diatoms. The Middle and Upper Eocene (50–70m) consist of greenish diatomaceous earths and diatomites. Its outcrops cover large areas on watersheds. The Lower Oligocene (up to 140m) is represented by greenish, bedded mudstones with shark teeth, lamellibranchs and plant remains. The mudstones resemble the clays of the Cheganian Suite. Higher up there are the continental deposits of the Middle Oligocene consisting of pinkish and white sands with nests of kaolin and beds of carbonaceous clays. The Upper Oligocene-Neogene is represented by a kaolinized formation (up to 150m) of sands and gravels with fragments of wood. Higher up is the Anthropogene (Yamalian Series). The Palaeogene is described by Yu. F. Andreyev and Zh. M. Byelorusova (1961) in the book *Papers on the Geology of the North . . .* (1960).

NEOGENE. It is represented by a few outcrops of continental loams, sands and clays (10–20m thick) with plant remains and pollen of Miocene and Pliocene age. The outcrops are found in the north of the Western Siberian Lowlands as for instance in the valley of the Pur.

ANTHROPOGENE. It is found everywhere and consists of deposits of differing origins. The most complete and thickest sections are found in the lower reaches of the Yenisei and the Ob' and in the regions between them where the thickness of Anthropogene sediments reaches 150–170m and more and is usually 40–60m. All four subdivisions are found (see p. 297). A review of the Anthropogene of the Soviet Arctics is given by V. N. Saks (1953).

TECTONICS

The Western Arctic represents a typical Palaeozoic, Hercynian fold province, where the Precambrian and Palaeozoic structure differs from the Meso-Cainozoic structures; the former being geosynclinal and the latter of platform type. The Pre-cambrian and Palaeozoic are divided into four main structural storeys, although more detailed distinctions are also possible. The Meso-Cainozoic is divided into two storeys; the first of the two includes the Upper Trias and the Lower and Middle Jurassic and has simple but well-developed folds. The second storey shows no fold-ing except in association with faults or pseudotectonic structures (landslides, glacial structures, chemical processes) and is completely unmetamorphosed.

STRUCTURES

Precambrian Structures

The Lower Proterozoic structures are found in the Western Taimyr gneiss com-plex, where they occur as relatively small isoclinal folds with almost vertical, some-times overturned, limbs. In the west they are east–west, while in the east they gradually acquire a north–south trend.

The Upper Proterozoic structures of Taimyr have the same trend as the Lower Proterozoic. On the Chelyuskin Cape they have an almost north–south or north–north-easterly trend. In the south of Severnaya Zemlya the strike is north–south and in the north of it the strike is north–north-westerly. Thus the Upper Proterozoic structures of Taimyr and Severnaya Zemlya form a large arc over 1000km long.

In Taimyr there are large rather tight anticlines and synclines, tens of kilometres across. These folds are complicated by numerous isoclinal folds. In Severnaya Zemlya all the Proterozoic is folded into a large anticlinorium with many folds of second and third order on its flanks.

Palaeozoic Structures

The Cambrian and Ordovician structures margin those of the Proterozoic and inherit its trend. In Taimyr they are brachyanticlinal, stretching 100–150km, but have relatively gentle limbs, especially so where limestones are abundant. The Lower Palaeozoic structures unlike the Precambrian lack minor folds of isoclinal style. As a rule the Lower Palaeozoic is unconformable on the Precambrian and has basal sandstones and conglomerates. In Severnaya Zemlya Cambrian and Ordo-vician strata form linear structures of the same type as the Proterozoic. These struc-tures are of the same dimensions and dips of from 25–30° to 60–80°. In those areas where the gypsum-bearing formations exist the folds have complex forms produced as a result of the hydration of anhydrites.

Middle Palaeozoic Structures. The Middle Palaeozoic structures of Taimyr and Severnaya Zemlya are distinctly different from the older structures. In Taimyr, Novaya Zemlya and Pai-Khoi the Middle Palaeozoic structures are typically geo-synclinal, complex and broken by numerous thrusts of the same strike as the regional trend. In Severnaya Zemlya the structures involving the Devonian are of platform type (Fig. 131) and occur as domes and basins of 4–16km across. Local dips vary from 5–10° to 15–20°. In arches and troughs of the folds the beds are almost horizontal. Differences in structural style are accompanied by lithological differences. Geosynclinal structures affect marine limestones and shales. In Severnaya Zemlya the continental beds of O.R.S. type—such as is found in Scotland and Scandinavia—are predominant.

These features force us to include Severnaya Zemlya, the southern part of the

Kara Sea and the northern margin of Taimyr into the Caledonides, while the south Taimyr, a major part of Novaya Zemlya, Pai-Khoi and the Arctic Urals belong to the typical Hercynides.

Upper Palaeozoic Structures. On Severnaya Zemlya the Upper Palaeozoic is absent. It is widespread in the region of Hercynides where it is thick, intruded by the Hercynian massifs and deformed into a system of numerous linear, more or less complex folds, cut by thrusts and similar to the folds of the western slope of the Urals and the Pechora coal-field.

The Mesozoic and Cainozoic structures are everywhere of typical platform type. They are large, gentle, almost unnoticeable brachyanticlines, such as the folds which affect the Cretaceous sediments of the lower reaches of the Yenisei (see synoptic maps). Most frequently the Meso-Cainozoic strata are not folded, and have an original dip, but since they are faulted they are never entirely horizontal. Angles of dip of tens of seconds to tens of minutes are not common.

THE HISTORY OF TECTONIC MOVEMENTS

The history of tectonic movements is almost the same as in the Urals and is especially so for the Hercynides. The strong and distinct signs of Caledonian Orogeny, of Precambrian orogenies and of the ultimate Hercynian Orogeny are characteristic of the Arctic. The Caledonian zone stretches from the north of Taimyr to Severnaya Zemlya and affects the bottom of the Kara Sea. Within it the last strong orogeny was the Caledonian, including a post-Silurian phase. The Devonian and younger deposits are of platform type.

MAGMATISM

Three magmatic regions can be distinguished. The first—Western Uralian— includes the western slope of the Arctic Urals, Pai-Khoi, and Novaya Zemlya. It shows a limited intrusive and extrusive activity. Lavas, sills and sub-volcanic basaltic intrusions formed at the end of the Devonian can be noted, and were succeeded by small granites, of suggested Hercynian age. These igneous rocks are found on the northern island of Novaya Zemlya. There are also young Upper Cretaceous lavas of the Arctic Urals and Pai-Khoi. The rocks of the latter locality may be Upper Permian.

The second region—Taimyr—shows a peculiar igneous activity. In it there are

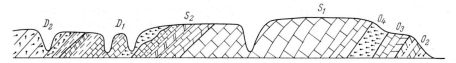

FIG. 131.   Structure of Devonian and Silurian strata of Severnaya Zemlya.
According to B. Kh. Yegiazarov (1954).

O_2 — variegated strata, gypsiferous towards the top; O_3 — limestones and dolomites; O_4 — pale sandstones of 100m in thickness; S_1 — Llandoverian Wenlockian thick (1100m) grey bedded limestones; S_2 — Ludlovian grey bedded limestones, which are dark towards the top, flaggy, gypsiferous and sandy; D_1 — sandstones passing up into dolomites and limestones; D_2 — red beds with armoured fish.

numerous and large ancient intrusions of Precambrian and Caledonian age. Ancient volcanic rocks, which are more or less metamorphosed and converted into amphibolites and greenstones, are widespread. Hercynian granites are found in some regions, but they are small and unlike the huge Hercynian granites of the eastern slope of the Urals. There is also the trap formation of Permo-Trias age, which is identical to the Siberian Traps. No younger volcanic rocks have been found.

The third region—Severnaya Zemlya—shows only the ancient Caledonian and Precambrian magmatism.

USEFUL MINERAL DEPOSITS

The inaccessibility and the small population of the Arctic region complicates exploratory and prospecting projects. Nevertheless, Soviet geologists have found several major deposits of fuels and non-metalliferous ores. These deposits form reserves for future industrial expansion of the Soviet Union. The Pechora coal-field even now is of national importance.

ORES

Many thousands of individual ore deposits are known from the Palaeozoic and Precambrian terrains. The mineralization varies from locality to locality, but no major deposits of economic importance have so far been found. For instance in Taimyr there is a belt of polymetallic mineralization. It lies in the central part of the Byrrang Mts and stretches for hundreds of kilometres. Numerous veins and ore showings are found in this belt and are associated with mineralized Permian breccias. In the western part of these mountains there is a long belt of realgar-cinnabar mineralization associated with faults affecting Silurian and Carboniferous rocks. Lastly there is the development of pyrrhotite associated with the traps.

FUELS

Coals

The coals are especially important. The major Pechora coal-field supplies coal to many regions of the Soviet Union. The Lower Permian, Vorkutian Suite of Kungurian age contains over 100 coal seams. The number of commercial coal seams is 30 and their total thickness is 30m. The coals vary in composition, but are usually ashy (15–20 per cent). A large number of seams has been found in the Paemboiian Upper Permian Suite. The Pechora basin was found by A. A. Chernov.

In Novaya Zemlya, despite the large area covered by Permian sediments, so far no coal has been found. In Taimyr the coals are again widespread and are found mainly in the Upper Permian. The coals are high quality and sometimes are coking coals. There are 40 workable horizons with some seams being 6–12m thick. The coal is worked for local needs. Geological reserves of the basin are 75.5×10^9 tons and the proved reserves are 0.7×10^9 tons. In the north-east of Taimyr, amongst the Lower Cretaceous deposits there are brown coals of 4–5m in thickness.

Petroleum and Gas

Considerable accumulations of gas are found in boreholes sunk to the Lower Cretaceous sands in the Malokhetian anticline (Yenisei folds) and the Berezovian anticline (lower Ob', near Berezov). Oil in non-commercial quantities has been obtained in both localities.

Non-metalliferous Deposits

The non-metalliferous deposits are numerous, diverse, but small. Only one major deposit of fluorite at Amderma (northern shore of Pai-Khoi), where the ore lies amongst Ordovician limestone, is worthy of mention.

The study of the Arctic was pursued by the Institute of Arctic Geology, under B. V. Tkachenko. The work of F. G. Markov on the Palaeozoic, of N. S. Voronets and V. I. Bodylevskii on Mesozoic faunas, of V. N. Saks on the Meso-Cainozoic and of V. A. Vakar and M. G. Ravich on older rocks and magmatic complexes should be mentioned. N. N. Uvrantsev has elucidated the structure of the Nordvik region and Novaya Zemlya and P. V. Vittenburg has done outstanding work on Vaigach and adjacent areas.

8

ANGARA GEOSYNCLINE

GENERAL DESCRIPTION

The Angara Geosyncline is represented by a vast mountainous tract, stretching from the Urals to Baikal and then to the Dzhugdzhurs. In the north it is bounded by the Western Siberian Lowlands and by the northern part of the Eastern Sayans and in the south it merges into the mountains of Central Asia, then its boundary disappears beyond the bounds of the U.S.S.R.

The geosyncline represents the southern part of Angarida, a continent for which the name was coined by A. A. Borisyak. After the Hercynian Orogeny the geosyncline became a part of the platform and has been called the 'Ural-Siberian post-Palaeozoic Platform' and 'the Ural-Siberian Platform with folded Palaeozoic basement'. Its typical feature is the strong manifestation of the Hercynian Orogeny and in some parts, of the Caledonian Orogeny. Such industrially important regions as Eastern Kazakhstan, the Altais and the Kuznetsk Basin are parts of this geosyncline area.

The region has been primarily investigated by the geologists of VSEGEI. The work began in the pre-Revolutionary period (A. K. Meister, P. I. Preobrazhenskii, and L. I. Lutugin), but has been greatly developed since. In Kazakhstan the investigations were led by N. G. Kassin with participation of M. P. Rusakov, who found the Kounrad and other deposits and I. S. Yagovkin who has done much for the understanding of Dzhezkazgan. V. P. Nekhoroshev spent his life studying the geology and the useful deposits of the Altais and V. I. Yavorskii has investigated the Kuznetsk Basin. In Cisbaikalia G. L. Padalka has been a leading geologist. Tens of other geologists, to be mentioned later, have participated in the work of VSEGEI.

In addition research done by the expeditions of the Geological Institute of the Academy of Sciences of the U.S.S.R. were of importance, and especially so in their contributions to the understanding of the tectonics of Kazakhstan. These expeditions were directed by N. S. Shatskii.

Considerable results also accrued from the work of the expedition of Moscow University, led by A. A. Bogdanov. Nowadays a group of geologists based at Alma-Ata and headed by K. I. Satpayev and R. A. Borukayev carry out the work in Kazakhstan. There are also groups of geologists based at Novosibirsk and Irkutsk and the Siberian branch of the Academy of Sciences, where the Institute of Geology is chaired by A. A. Trofimuk and includes academicians and corresponding academicians.

BOUNDARIES. In the north-west the Angara Geosyncline joins the Western Siberian Lowlands, the boundary being drawn along the edge of the Upper

Cretaceous-Palaeogene cover. In fact the geosyncline continues further to the north (Fig. 2). In the north-east the boundary lies in the central part of the Eastern Sayans, then bends to the north-east and continues along the eastern shore of lake Baikal, margining the southern part of the Patomian Highlands and the Aldan Massif from west and south, all the way to the edge of the Dzhugdzhur range (Fig. 2). In the south-east the geosyncline abuts against the Pacific Ocean Geosycline, Cisamuria and Eastern Transbaikalia. In the upper reaches of the Chikoi its south-eastern boundary disappears into Mongolia.

Both the north-eastern and the south-eastern boundaries are quite arbitrary and their detailed position is disputed. The type and tectonics of Cambrian and Jurassic strata are considered as the main diagnostic factors. The areas where the Cambrian and the Jurassic have a platform character, are considered as parts of the Siberian Platform. On the other hand where the Cambrian is geosynclinal and its rocks strongly folded, metamorphosed and intruded by granites, while the Jurassic is of platform type, the Angara Geosyncline is established. Where the Cambrian and the Jurassic are both geosynclinal the rocks constitute a part of the Pacific Ocean Geosyncline. In those areas where the Cambrian and the Jurassic are not well developed the drawing of the boundaries is arbitrary.

Again inside the U.S.S.R. the boundary of the Angara Geosyncline enters the Alakul Depression and through lake Balkhash and the valley of the Chu continues to the valley of the Syr-Dar'ya.

The western boundary follows the eastern edge of the eastern slope of the Urals. The whole Turgai Depression is within the Angara Geosyncline.

OROGRAPHY

RELIEF. The relief is conditioned by two main factors: the prolonged Meso-Cainozoic erosion and the young Pliocene and Anthropogene block movements. Where only the first factor has been active the relief consists of gentle hills approximating to a peneplain. This is mainly seen in the north-western provinces. In the south and east block movements of Palaeozoic massifs cause the formation of high mountains of the northern Tian Shians, the Altais and the Sayans. The block movements have also produced the Baikal trough and Teletskoe and Kosogol lakes of Mongolia.

The Turgai Depression. This is a large down-faulted area filled with horizontal Jurassic, Cretaceous, Palaeogene, and Neogene strata. It is surrounded by the Palaeozoic and Precambrian massifs of the Southern Urals and the Mugodzhars on one side and the Central Kazakhstan Massif on the other. The river Turgai flows through the depression. The country has no forest and consists of steppes in the north and semi-desert in the south. The semi-desert opens into the Aral basin.

North-Eastern or Central Kazakhstan. This used to be known as the Kirgiz steppe and now is sometimes called 'Sary Arka'. The region is characterized by flat, almost unnoticeable, watersheds and wide river valleys, which are out of equilibrium with the present-day rivers which do not carry much water and often dry out in the summer. In the north there is the wide, flat depression of Tengis (Teniz) salt-lake. There are small rounded hills (sopka) in the steppes, giving their name to the term 'small-sopka plain' (Figs. 137, 146). Sometimes they are quite high reaching 1300–1600m, forming true mountain ranges, as are for instance the Karakalinian Mts

(1339m). In the south the region is bounded by lake Balkhash and Ala-Kul' (Kul'=
a lake).

The *Kulundinskian and Barabinskian steppes* represent a vast low plain of
continuous steppes, which separates the Palaeozoic massifs of Central Kazakhstan
and Kuzbas. The region has a large number of freshwater and saline lakes and the
exposed rocks are Cainozoic continental sediments.

The Altais and the Tarbagatais. The Altais form a high mountainous region in
the centre of the Angara Geosyncline. The Altais are divided into the Ore Altais and
the Rocky Altais. The latter range is situated in the upper reaches of the Ob', has a
highly dissected relief and reaches 4000–4500m in height (Mt Byelukha is 4506m).
The Ore Altais are lower, are near to the valley of the Irtysh and are 2600–2800m
in height. Within the range there are several major mineral deposits. The Kalbinskian
range (1608m) lies within a large loop of the Irtysh, to the west of its valley.
The large depression of lake Zaisan (386m) separates the Altais and the Tarbagatais,
the latter being quite high mountains (2500–3000m).

Kuzbas and Salair. The Kuznetsk Basin and the Salair ridge form the next,
elevated, Palaeozoic Massif. Their northern parts have a small-sopka relief, but in
the south there are the Shoria Highlands (the Altais). A well-developed river system
and an almost unbroken forest (taiga) are typical of this region, which is a major
mining and metallurgical district.

Kuznetskian Alataus. The Kuznetskian Alataus are a region of forest-covered
mountains consisting of hard metamorphosed Lower Palaeozoic and Precambrian
rocks characterized by a high relief. To the east there are the much less metamorphosed Devonian and Carboniferous sediments, showing a lower relief and forming
the large Minusinskian Depression, which is traversed by the Yenisei.

Minusinskian Depression. This depression lies to the south-east of the Kuznetskian Alataus with the towns of Abakan and Minusinsk located in the middle of it.

Tuva Depression. The Tuva Depression lies to the south of the Minusinskian
Depression. In the north it is bounded by the Eastern Sayans (2500–2900m), in the
south by the Tannuol range (peaks of 2591m and 2930m). The bottom of the
depression lies near Kizyl at a level of 621m.

The Eastern Sayans and Western Transbaikalia is a large mountain region,
overgrown by forest and with a highly dissected relief (2900–3000m). The southern
branches of the Eastern Sayans apart, it includes the Khamar-Daban range (to the
south of Baikal) with heights of 2500–3000m, and the Yablonovoi range (1500–
2000m) and the Stanavoy range (2000–2500m). Between the ranges and uplands
there are depressions of often tectonic origin (parts of the Baikal Rift), which are
filled with Mesozoic and Cainozoic sediments and young basalts. The largest
depression is the Baikal Trough, followed by the Irkut, the Barguzin, the Upper
Angara and the lower Muya troughs. There are widespread basalts in the middle
reaches of the Vitim. Here there are the extinct, Obruchev (1085m) and Mushketov
(1012m) volcanoes.

HYDROGRAPHY.   There are many rivers and lakes in the region and the rivers
are often full. They belong to three major systems (the Irtysh, the Ob' and the
Yenisei). There are a few quiet rivers with little water in the dry steppes of
Kazakhstan. The largest of these are the Ishim and Sary-Su which enter the lower
Chu, which dries out in the summer. The lake Teniz depression is quite large.

The Irtysh rises outside the U.S.S.R., flows through lake Zaisan (Zaisan-Nor),
then turns sharply to the north-east round the Kalbinskii range and follows the

Irtysh zone of faults. Here it flows in a deep narrow valley thus enabling the construction of large hydro-electric stations. The basin of the Ob' includes the Altais and the Kuznetsk Basin, the latter being traversed by the river Tom'ya on which stands the town Tomsk. Lake Teletskoe lies in the Altais.

Between the Ob' and the Irtysh lies a vast region of steppes. The Kolundinskian lakes lie in its southern part, where it is known as the Kolundinskian steppe. The northern part of the region is known as the Barabinskian steppe, and like the southern region it has numerous freshwater and saline lakes and marshes, which lie to the north. Lake Chany is the largest lake in the northern part of the region.

The Yenisei, in the south, drains the Tuva Depression and to the north it traverses the Minusinskian Depression. The upper reaches of the Yenisei lie in the Western and Eastern Sayans where there is permanent snow and a widespread glaciation. Thus the Yenisei is one of the fullest rivers of the U.S.S.R.

Lake Baikal is the deepest lake in U.S.S.R. (1742m) and the river Selenga flows into it. To the east of it is the basin of the Vitim with many tributaries and the upper part of the Olekma basin.

STRATIGRAPHY

A distinct feature of the Angara Geosyncline is the presence of two main complexes: (1) the Palaeozoic and Precambrian, and (2) the Anthropogene and Neogene.

Palaeozoic and Precambrian rocks are strongly deformed and metamorphosed, dominantly marine, but partly continental sediments. The Proterozoic iron formations and the Carboniferous and Permian coal-measures are worthy of a mention. Palaeozoic and Precambrian strata form high regions, varying from small-sopka districts to mountainous areas. In lowland areas these rocks are buried under the cover of Meso-Cainozoic deposits.

The Anthropogene and Neogene cover is present in all lowland areas, but is also found in some mountainous districts. The strata of the cover are entirely continental, horizontal and unaltered.

Mesozoic and Palaeogene sediments also almost entirely lack marine facies, which can only be detected along the north-western, western and south-western boundaries of the geosyncline. Mesozoic and Palaeogene strata are thin and unfolded.

At the margins of the geosyncline continental sediments occur as a continuous cover, while patches of them are found in the internal depressions surrounded by Palaeozoic rocks. Coal-bearing formations of ages varying from the Trias to the Middle Miocene are typical.

ANCIENT FORMATIONS

In Siberia the ancient formations are just as widespread, or even more so, as in the Urals. The rocks are of a similar type, barren or containing only calcareous algae, which can be found equally in the Cambrian, Ordovician and Proterozoic and which are unsuitable as indications of the age. Only the rocks with Cambrian and Ordovician faunas can be accurately dated. The ages of other formations, forming whole mountain ranges, remain indeterminate or are decided on the basis of

personal views of this or that geologist. The older metamorphic rocks underlying
the palaeontologically dated Ordovician strata have been included in the Cambrian,
the Proterozoic, the Lower Palaeozoic or even the Ordovician. Only the palaeontologically dated formations, or their lateral equivalents, should be included in the
Ordovician or Cambrian.

The rocks which underlie the dated Lower Cambrian and which contain
Collenia-like algae are Upper Proterozoic and have been separated as the Sinian
complex. The subdivision of the older Precambrian rocks is done on the basis of
angular unconformities, variations in structural style, grade of metamorphism and
distribution of igneous massifs. 'Spore' analysis gives significant results.

ARCHEOZOIC

Archeozoic gneisses and schists are less widespread than the Lower Proterozoic
rocks. In many regions such gneisses or schists are either absent or have not been
separated from Upper Proterozoic formations. Archeozoic rocks are so deformed
and altered that no accurate estimation of their thickness can be made, but they
must be many kilometres or even tens of kilometres thick. Gneisses and schists
apart, there are also quartzites, marbles, and frequent granites which are more or
less altered, rendered into gneisses and associated with migmatites.

The age of the Archeozoic is determined on the basis of separation of its rocks
from Proterozoic formations by unconformities with basal conglomerates, and by
differences in structures, metamorphism and synorogenic intrusions. Absolute age
determinations are particularly significant.

It should be remembered that gneisses or schists as such can be not only
Archeozoic, but may be found in folded contact metamorphic zones of any age.
Such zones are rarely more than 10–12km wide and are closely associated with
intrusions. Consequently such zones, depending on the outlines of intrusive massifs,
which had caused them, soon disappear along the strike. The Archeozoic gneisses on
the other hand were produced regionally, have a particular stratigraphic position
and can be traced for many hundreds of kilometres.

Lower Proterozoic

The main diagnostic feature of Lower Proterozoic rocks is their position under
the Upper Proterozoic; Lower Proterozoic strata having been strongly deformed are
separated from the Upper Proterozoic by angular unconformities of regional extent
and possess intrusive massifs. The metalliferous ores also reflect the differences
in age.

The Lower Proterozoic consists of strongly deformed and metamorphosed rocks,
including slates and even schists. There are also numerous phyllites and quartzites,
which are sometimes micaceous. Volcanic rocks are common and have been metamorphosed into amphibolites and talc-chlorite schists. In places there are Proterozoic
gneisses and schists and numerous and various intrusive massifs. The thickness of
Lower Proterozoic rocks is great, reaching 7000–9500m. Normally there are no
organic remains, although the higher horizons (limestones) have primitive calcareous
algae. There are also quite frequent carbonaceous and graphitic rocks suggesting
the former existence of non-skeletal organisms.

Upper Proterozoic

To the Upper Proterozoic belong almost barren limestones, dolomites, slates, quartzites, and volcanic rocks. Their thickness varies from a few hundred metres to 3km. They lie unconformably under the palaeontologically dated Middle and Lower Cambrian strata. The only organic remains consist of calcareous algae and 'spores'.

The age of these rocks raises sharp disputes. Formerly, they were often thought to be Cambrian or even Ordovician, depending on the overlying formations. According to the decision of the National Committee the algal limestones and other associated sediments which underlie dated Lower Cambrian rocks should be included into the Sinian complex, without prejudice as to their being either Palaeozoic or Proterozoic. Other terms such as 'the Riphaean Formation' can also be used. In practice the Kazakhstan and Altaian geologists use neither term and consider these rocks either as Proterozoic or as Palaeozoic. The absence of a Lower Cambrian fauna tends to imply that they should be, more correctly, considered as Upper Proterozoic.

The Palaeozoic-Proterozoic boundary should be drawn at the base of the oldest, faunally dated, Lower Palaeozoic strata. If such strata are Ordovician or Middle Cambrian then other features such as unconformities, differences in metamorphic grade, intrusions, and so on should be used to establish the break. The Lower Cambrian, however, may pass downwards into the Upper Proterozoic (Sinian) without a stratigraphic break.

As an example of the Precambrian succession in the western part of the Angara Geosyncline we quote the succession in the Kokchetavskian region (E. D. Shlygin, 1956).

Upper Proterozoic

The Yerementauian Limestone-Volcanic Series (3500m). At the top lies the Zhel'tauian Suite of limestones interbedded with quartzites and cherts with flows of porphyrites (1500m). At the bottom lies the Tiesskian Suite of meta-volcanic rock (porphyrites and tuffs) with beds of limestones and cherts and porphyries (2000–3000m).

Unconformity.

The Akdymian Quartzite Series (3000–3500m). At the top there are quartzites and cherts and at the bottom, quartzitic slates, chlorite-sericite slates, quartz-graphite slates, siltstones, and sandstones.

Unconformity.

Lower Proterozoic

The Borovskian Series. Kotchetavskian Suite (1000m) of micaceous quartzites, graphitic slates, sericitic and ferruginous quartzites with beds of green schists and dolomites.

Yefimovskian Suite (1500–1700m) of mica schists, amphibolites, porphyritoids, quartzites, graphite-chlorite schists, phyllites, and lenses of marmorized limestones.

Unconformity.

Kuuspekskian Porphyroid Suite (2000m) consisting of porphyroids, porphyritoids, sericitic schists, amphibolites, and alkaline gneisses.

Unconformity.

Archean

The *Zerendinskian Gneiss-Amphibolite Series* (4500–5500m). Consists of orthogneisses, paragneisses, amphibolites, marbles, schists with sillimanite, cordierite, mica and tourmaline. It is divided into two suites.

The Kokchetavskian succession correlates with the neighbouring Ulutau region (Shtreis & Kolotukhina, 1948). Here the unconformity between the Graywacke (uppermost) Suite and the Kokchetavskian Suite is accompanied by granite intrusions. The thickness of the suites corresponding to the Kokchetavskian thus decrease to 800–1000m. Between the Yefimovskian and the Porphyroidal suites there is an unconformity and the thickness of the latter is reduced to 1300m. Otherwise, however, the two successions are identical.

For the eastern part of the Angara Geosyncline the succession of Gornaya Shoria (south of Kuzbas), as established by K. V. Radugin (1952), is quoted here.

Cambrian

Mrassian Formation. This lies at the bottom of the Cambrian and consists of shales, sandstones, conglomerates and reef limestones with archeocyathids.
Folding. Intrusion of diorites.

Upper Proterozoic (Algonkian)

Not less than 4300m thick.
Ust'-Anzakian Suite (1500m). Consists of black and siliceous shales, sandstones, breccias, marmorized limestones.
Unushkolian Suite (800m) of marmorized limestones, black cross-bedded shales with bands of conglomerates. Many *Collenia*-like algae in limestones with genus *Newlandia* at the bottom.
Tyzassian Suite (2000m) of grey, greenish, and black graywacke sandstones and shales, with rarer conglomerates.
Munzhaian Formation. In some sections the Upper Proterozoic starts with dark marbles, shales and sandstones, with dirty-grey basic diabase lavas.
The main fossil throughout the Upper Proterozoic is *Newlandia*.
Folding. Intrusion of granites.

Middle Proterozoic (Siberian)

Sagaian Formation (2000m). Consists mainly of dark, bedded, marmorized limestones. Towards the top there are black shales, with, in places, siliceous shales, sandstones and lavas.
Western Siberian Formation of bedded algal dolomites with black anthraconite marble at the top. There are numerous algae and stromatolites belonging to the genus *Algostroma*.
Kobyrzinian Formation (more than 1500m). Pale stromatolitic dolomites at the bottom and black anthraconitic marbles at the top.
All three formations form a characteristic, thick, algal carbonate group. Its relationships to the Lower Proterozoic are not clear and they probably overlie metamorphic green schists.

Lower Proterozoic

Slates (schists), quartzites, marbles, dolomites, porphyroids.

Archeozoic

Upper Group. Quartzites, diopside marbles and mica-schists.

Middle Group (Kuznetskian Formation). Banded amphibolites, schists with siliceous marbles and diorite gneisses at the top.

Lower Group. Paragneisses, injection gneisses and microcline granite-gneisses.

The Precambrian succession of Gornaya Shoria resembles that of the western part of the Eastern Sayans (Fig. 132), near to the Minusinskian Depression and that of the Precambrian in Kuznetskian Alataus. Here the Upper Proterozoic consists of slightly metamorphosed shales, sandstones, limestones, and lavas. These rocks underlie the palaeontologically dated Cambrian. Under it with a large break lies the Middle Proterozoic, consisting of phyllites and talc and other schists with dark graphitic marbles of great thickness and some conglomerates. The total thickness is more than 9000m. The Lower Proterozoic consists of amphibolites, talc-chlorite schists and green schists. Biotite and muscovite gneisses and gneissose amphibolites and marbles are included in the Archeozoic.

Thus the Upper Proterozoic here includes the Upper Proterozoic of Siberia and the Lower Proterozoic includes the Middle and Lower Proterozoic of Siberia.

In the eastern part of the Angara Geosyncline in the Stanovoi range the succession resembles that of the Aldan Massif. Proterozoic deposits with an absolute age of 730–985 m.y. predominate, forming the 'Ancient Stanovoi' complex of biotite and amphibole gneisses and gneissose granites. Amongst these occur gneisses and schists belonging to the Timptonian and Dzheltulinian Series with absolute ages of over 1200 m.y. and belonging to the Lower Proterozoic and Archeozoic. The lower, Archeozoic, series of the Aldan Massif (Iengrian) is not present in the Stanovoi range.

In the Patomian Highlands the Lower Proterozoic consists of phyllites, carbonaceous quartzites, sandstones and limestones of a total thickness of 11–13km. It is overlain by the Upper Proterozoic some 8–9km thick and represented by limestones, sandstones, and conglomerates.

In the Mama-Vitim region other investigators suggest much smaller thicknesses. On the Mamian Crystalline Suite (Archeozoic) rests the unconformable Bodaibinian, metamorphic, Suite (3km) consisting of calcareous, graphitic, and phyllitic slates, graywackes, quartzites, and conglomerates. Higher up lies the unconformable Zhuian, green and red, Suite (1–1·5km), which begins with basal sandstones and conglomerates, passing upwards into striped shales and quartz sandstones with beds of limestones. The suite resembles the Hyperborean Formation of the Kola Peninsula where it is also included in the Sinian. The Proterozoic is unconformably overlain by the folded and metamorphosed Lower Cambrian with archeocyathid and trilobite-bearing limestones.

PALAEOZOIC

Lower Palaeozoic

The Cambrian and the Ordovician are closely associated with each other and with the Precambrian, often forming a single, thick, strongly deformed and metamorphosed complex, consisting of sedimentary, volcanic and intrusive rocks (Fig. 133).

The fauna in the Lower Palaeozoic is sparse and is sometimes found only after prolonged search. This is because the rocks are metamorphosed and because in the Cambrian many animals had chitinous-calcareous carapaces and shells which do not get preserved. The paucity of the fauna complicates the dating of many horizons and formations. An important, but difficult work on dating has yet to be done.

Cambrian

Cambrian rocks are relatively rare, but with each year the discoveries of Cambrian fossils multiply. The lower boundary of the Cambrian is determined by the oldest Lower Cambrian fauna. All the barren formations below it, especially those with calcareous algae, should be included in the Proterozoic. It should also be remembered that the archeocyathid limestones, formerly considered as Middle Cambrian, are now regarded as Lower Cambrian. In some regions, but not everywhere, the boundary of the Cambrian is determined by a break associated with an angular unconformity, caused by the Upper Proterozoic Baikalian Orogeny. The boundary is drawn at the bottom of the beds with the oldest Tremadocian fauna. More rarely it corresponds with a break and unconformity produced by the Salairian phase of the Caledonian Orogeny.

SUBDIVISIONS. The Cambrian is divided into the lower, middle and upper. Finer subdivisions are entirely local (Table 31). In the west, in Central Kazakhstan, according to R. A. Borukaev (1955, Resolutions . . ., 1958) and his co-workers the Lower Cambrian starts with the Teleskolian, tuffaceous sedimentary, Suite overlain by the thick spillite-keratophyre

FIG. 132 Proterozoic rocks of the Eastern Sayans. (Figuristye Byelki). *Photograph by A. G. Vologdin.*

Boshchekhulian Suite. The total thickness of the Lower Cambrian is over 6000m (Fig. 134). Above this lies the uncomformable Middle Cambrian (up to 4000m). It begins with the Agyrekian, tuffogenous-sedimentary, Suite of 700m, followed by

FIG. 133. Cambrian rocks of the Eastern Sayans (Prorvinskie Byelki).
Photograph by A. G. Vologdin.

FIG. 134. Lower Cambrian quartzite and chert ridges in Cisbalkhashia. A typical small sopka semi-desert relief. *Aerial photograph by V. N. Moskaleva.*

the cherty-slaty Maidanian Suite (500–800m). Both suites have beds with trilobites and atrematous brachiopods. *Erbia sibirica* (trilobite) is typical of the Agyrekian Suite and *Dinesus* is typical of the Maidanian Suite. Sometimes the Agyrekian Suite is considered as a lower horizon of the Maidanian. The upper part of the Middle Cambrian consists of the Sasyksorian Suite (2500m) of sandstones and shales, divided into three horizons. The suite includes a number of bands with marine fauna, mainly of trilobites (*Agnostus Anomocare*).

There is a distinct unconformity between the Middle Cambrian and the Upper Cambrian strata, of which at least a third is missing from the succession. Higher up is the Lower Ordovician. The Upper Cambrian is represented by the lower part of the Tortkudukian Suite (300–2500m) of sediments and volcanic rocks divided into four palaeontologically distinct horizons, containing *Agnostus*, *Euloma* (at the top) and the atrematous brachiopod *Billingsella*. The upper part of the suite is Lower Ordovician (Nikitin, 1956). The Cambrian of Kazakhstan has been studied in considerable detail.

From the north-eastern Altais to the east, including Mount Shoria, Salairs, Kuznetskian Alataus and the Western and Eastern Sayans the Cambrian is thick and widespread.

To the Lower Cambrian belong thick effusive and tuffaceous rocks, barren sandstones and shales and rarely limestones and dolomites. It is possible that a part of these strata belongs to the Sinian complex since they are overlain by the archeocyathid limestones (Fig. 135). The lower part of the latter was formerly even thought

FIG. 135.   A reconstruction of Southern Siberian landscape in Lower
Cambrian (Motian) times. According to A. A. Predtechenskii (1960).

to belong to the upper part of the Lower Cambrian. Recently an opinion has become widely held that these limestones correspond to the Lower Cambrian in its entirety.

The detailed investigation of trilobites and archeocyathids has permitted the distinction of five horizons in the Lower Cambrian of the Sayan-Altai province, as follows (from bottom to top): the Bazaikhian, the Kameshkian, the Sanashtykgolian, the Solontsovian, the Obruchevian. These horizons correspond to the Sinyaian, the Tolbachanian, the Olekmian, the Ketemian, and the Yelanian of the Aldan Massif (Repina, Khomentovskii). Although the correlations seem to fit in the Aldan Massif, all these horizons are placed in the Lenian subdivisions of the Lower Cambrian, while in the Sayan-Altai province they are considered to range through the Lower Cambrian. Thus there is a serious divergence in our understanding of the limits of the Aldanian and Lenian stages. This is one of the problems to be solved by the National Committee.

The Middle Cambrian is more widespread and embraces various limestones and dolomites, including the archeocyathid varieties, as well as sandstones, shales, and rarer lavas. The Upper Cambrian is less fully developed. In some sections, owing to the effects of the Salairian phase it is entirely absent, while in other sections it is represented only by its higher horizons, which are closely associated with the

FIG. 136.   Archeocyathid limestones in the Eastern Sayans, on the river Kizyr. *Photograph by A. G. Vologdin.*

Ordovician (Table 31). A. F. Byelousov (Byelousov, Sennikov, 1960) has given detailed data on the Altais.

TABLE 31.

Cambrian and Ordovician Successions of the Angara Geosyncline

System	Division		Central Kazakhstan, the Bayanaulian region (According to R. A. Bokurayev, 1955; L. I. Borovikov, 1958)	Ore-Altais, Chingiz, Kalba (According to M. N. Bartseva, 1957; L. I. Borovikov, 1958)	North-eastern part of the Rocky Altais (According to V. P. Nyekhoroshev, 1958)		Salair (V. D. Fomichev, 1956, 1961)
Ordovician	Upper		Zharsorian Suite of porphyrites and tuffs with beds of sandstone and limestone (1500–3200m). Angrensorian Suite of sandstones, conglomerates, shales and limestones with trilobites (2000m).	Khankharian Suite of thick, coarse, variegated clastic strata with a marine fauna (500–2000m). Uskuchevskian Suite of green pepper shales and sandstones with orthids (1200m).	Kabian Suite—above, sandstones and conglomerates (1500m); in the middle, grey calcareous shales and sandstones with beds of limestones. A fauna is present (900m).		Above, limestones, with a Caradoc fauna of trilobites, brachiopods and corals, forming the Veberovian Horizon. Urian Suite of sandy-shaley rocks with tuffs and lavas. A fauna of trilobites, graptolites and *Lingula* is present. Basal conglomerates and sands (1200–1600m).
	Middle		Yerkebidaikian Suite of sandstones, silts, jaspers and limestones with a marine fauna (1500–2000m). Bel'suiskian (Arenig-Llanvirn) Suite of basic tuffs, sandstones, jasper and limestones with a marine fauna (5000–6500m). Upper part of the Tortkudukian Suite—volcanic rocks with beds of limestones carrying a Tremadocian fauna (600–3000m).		Savelovskian Suite of gravelites and conglomerates; below, sandstones with a fauna (850m).	Below, greenish-violet flysch-like formation (1800m).	
	Lower			Bugryshikhinian Suite of dark shales and sandstones with trilobites (600m).	Choiskian Suite of green and violet shales and sandstones; at the base, conglomerates. A Tremadoc fauna is present. Interval.	Gornoaltaiian Suite of green-violet flyschoid shales (2000–3000m).	
Cambrian	Upper		Lower part of the Tortkudukian Suite—limestones with trilobites; at the bottom, sandstones and conglomerates (300–2500m). Interval.	Variegated sandstones and shales with beds of limestones, and a rich fauna. Interval.	Kul'bichian Suite, of grey and brown conglomerates, limestones, shales and sandstones, with a marine fauna (200–500m).		Tolstochikhian Suite of carbonate rocks rich in Upper Cambrian trilobite fauna (400m). Interval.
	Middle		Sasyksorian Suite—above, shales; below, sandstones with beds of limestone. A marine fauna is present (1000–2500m). Maidanian Suite of variegated jaspers, sandstones and shales; at the base, conglomerates. A marine fauna is present (1000–2700m).	Sandstones, shales and limestones with a marine fauna (1600m).	Yelandian Suite of grey and variegated shales, quartzites and limestones with trilobites (300–2000m). Kaimian Suite of shales with beds of marble, sandstone and in places volcanic rocks. Trilobites and archeocyathids are present (up to 3500m).		Biryulian Suite—above, greenish-grey marine sandstones with beds of limestones; below, dominant red beds. A lot of volcanic rocks. Basal conglomerates and sandstones (2000–2500m).

Certain individual isolated outcrops of geosynclinal, Cambrian sediments of Western Transbaikalia are of interest. These suggest that the region is a part of the Angara Geosyncline. The most northerly outcrops are situated in the south of the Patomian Highlands, to the north of the valley of the Murya. According to L. I.

TABLE 31—continued

Kuznetskian Alataus (T. N. Alikhova, 1958; A. L. Dodin, 1958)	The Western Sayans and in Tuva (According to E. V. Vladimirskaya, 1958; A. L. Dodin, 1959)	The Eastern Sayans	Transbaikalia, the middle Vitim (L. I. Salop, 1954, 1961)
Izasian Suite of purple and grey porphyrites, tuffs, sandstones and argillites with trilobites (1000m).	In the Western Sayans, greenish and grey sandy-shaley deposits of several kilometres thickness; divided into the Lower (Kemterekian) and Upper (Shignetian) complexes. In Tuva, thick, coarse, sandy-shaley deposits with a rare, poor brachiopod, gastropod fauna (up to 4000m). In places sandstones with *Angarella*. Basal conglomerate is present.	No Ordovician strata are known.	A thick sequence of schists, phyllites and crystalline limestones; at the base, conglomerates. The age is provisional and the sequence may be Devonian.
Unknown.			
Kozhukhovskian Suite of argillites and sandstones with a Tremadoc fauna.			
Kitatskian Suite of carbonates and argillites (500m).	Upper Cambrian and the upper part of Middle Cambrian strata are missing.	Upper Cambrian and the upper part of Middle Cambrian strata are missing. In places the lower part of a sequence of greenish sandstones and shales is provisionally assigned to the Upper Cambrian. Kizirskian Suite of porphyrites, diabases and tuffs (1500–1800m). Dzhebian Suite of chloritic and other slates (schists); at the base, conglomerates (2000m). Unconformity.	Missing.
Interval.			
Taidonian Suite of sediments and volcanic rocks (1500m). Kanymian Suite of spilites with lenses of reef limestones and beds of siliceous sediments and sandstones (2500m).	Sandstones, shales, porphyrites and limestones with trilobites (1200–3000m).		Yangudian Formation of barren, bedded marls and limestones (300m). Dark finely-laminated limestones with trilobites (500m).

Salop (1954), who worked here for a long time, the lower part of the Lower Cambrian consists of thick phyllitic shales, sandstones and marls which are grey, greenish, buff, and violet in colour and reach a total thickness of 3000–4000m. In places there are lavas and tuffs (quartz porphyries and porphyrites) reaching 1500–2000m in thickness. The only fossils are calcareous algae, and consequently the age is indeterminate. It is probable that a major part of the succession belongs to the Sinian (Upper Proterozoic). The subsequent formation (Mamakanian), which is up to 1800m thick, starts with basal conglomerates and sandstones and higher up consists of phyllitic shales, marls, and dolomites. This suite of rocks is gradually transitional into limestones with the top Lower Cambrian and Middle Cambrian fauna and consequently correlates with the Aldan Stage of the Lower Cambrian.

The carbonate formation (Yangudian) reaches 1500–2000m in thickness and consists of grey and dark-grey massive and bedded limestones with beds and groups of dolomites. The massive varieties contain archeocyathids and calcareous algae, while the bedded varieties have also yielded trilobites and brachiopods. Its age is Lenian and Amgian. The rocks are folded (Fig. 136), although some geologists consider the folding to be related to local thrusts. According to them the palaeontologically dated Upper Cambrian strata of the Patomian Highlands are of platform type and only the Upper Proterozoic is geosynclinal. The next outcrops of folded and metamorphosed Lower Cambrian limestones with archeocyathids have been found to the south (the middle Vitim) in the valley of the Kydymit where they are undoubtedly of geosynclinal type (see p. 281). Even further south in the Uda and Dzhida basins according to V. G. Byelichenko and others (Theses . . ., Chita, 1961) the Cambrian, which is as much as 6–8km thick, is palaeontologically dated. At the bottom of it there are thick conglomerates and sandstones succeeded by shales, sandstones and spilite-keratophyre lavas. The upper part of the Lower Cambrian consists of shales, sandstones, lavas and limestones, the latter being either bedded or reefs with archeocyathids. The Middle Cambrian is not known. To the Upper Cambrian has been provisionally attributed a suite of terrigenous red beds, 1300–1400m thick. It is quite possible that the upper part of the suite of shales, sandstones and limestones is Middle Cambrian, especially since a Middle Cambrian fauna has

TABLE 31—continued

System	Division	Central Kazakhstan, the Bayanaulian region (According to R. A. Bokurayev, 1955; L. I. Borovikov, 1958)	Ore-Altais, Chingiz, Kalba (According to M. N. Bartseva, 1957; L. I. Borovikov, 1958)	North-eastern part of the Rocky Altais (According to V. P. Nyekhoroshev, 1958)	Salair (V. D. Fomichev 1956, 1961)
Cambrian	Lower	Boshchekulian Suite —above, spilites, keratophyres, albitophyres, porphyrites and tuffs (4000m). Teleskolian Suite of sandstones, porphyrites, tuffs, greywackes, jaspers and shales (2000m).	Spilite-keratophyre sequence—above, a band with trilobites. Sandstones, silts and limestones with archeocyathids and trilobites.	Kayanchian Suite of limestones and marbles with beds of shales; abundant archeocyathids. At the base, conglomerates (1000–2000m). Baratalian Suite (Upper Proterozoic). In places, spilites and keratophyres; elsewhere, thick limestones and dolomites, with calcareous algae and quartzites (2000–4000m).	Listvyankian Suite of archeocyathid limestones, white and massive above and black and bedded below (up to 1500m). Lukovian Suite of shales and sandstones, at the base, conglomerates and volcanic rocks (400–700m). Zolotukhian Suite of keratophyres and spilites with archeocyathids (2000m).

been found to the west, along the strike, in the Western and Eastern Sayans (Table 31).

LITHOLOGY, FACIES, AND PALAEOGEOGRAPHY. Shallow water marine sediments predominate. In the west they are represented by thick formations of sandstones and shales with beds and lenses of limestones and dolomites and minor groups of conglomerates, which are especially common in basal horizons.

The unbedded reef massifs built by archeocyathids and algae are rare and small. Some barren sandstones and shales represent the sediments of near-shore plains. Such plains were not vast and did not resemble those huge plains where the Cambrian red beds of the Russian Platform or the Siberian Platform accumulated. Thus the coast-lines had rocky outlines. The near-shore parts of the continents (islands and sea) saw the growth of numerous volcanoes ejecting masses of lava and tuff of spilite-keratophyre type.

The reconstruction of the Lower Cambrian landscape in Southern Siberia was attempted by A. A. Predtechenskii (1960), portraying the Siberian Platform and the adjacent Angara Geosyncline (Fig. 137). In the north and east there were large epicontinental seas: the Yakutian, the Irkutian, and the Tungusskian. To the west and south the seas were bounded by the Baikalid fold belts, which were situated where the present-day Stanovoy range, the Pash, and the Sayans are. The Patomian Highlands formed a peninsula and so did the Yenisei Ridge. The Turukhanian inliers, the Anabara Massif, and the Olenek inliers were islands or shallow. To the west of the Baikalids there was an open and deep geosynclinal sea. In the valley of the Vitim it formed a narrow and long inlet. Since the sea filled the Angara Geosyncline it can be referred to simply as the Angara Sea, which stretched to the west outside the limits of Fig. 137. Along the eastern shore of the sea there was a chain of volcanic and reef (archeocyathid) islands, partly shown in the reconstruction. In the north-west it was the Kuznetsk archipelago, followed by the Singilenskian and Dzhidinskian islands. Within Eastern Transbaikalia lay the Gazimurskian archipelago.

Amongst the various facies the archeocyathid and algal limestones of the eastern part of the Angara Geosyncline are especially worthy of consideration. There they are better developed than anywhere else in the U.S.S.R. The limestones form unbedded

TABLE 31—continued

Kuznetskian Alataus (T. N. Alikhova, 1958; A. L. Dodin, 1958)	The Western Sayans and in Tuva (According to E. V. Vladimirskaya, 1958; A. L. Dodin, 1959)	The Eastern Sayans	Transbaikalia, the middle Vitim (L. I. Salop, 1954, 1961)
Usian Suite of limestones and marbles with archeocyathids. In places shales, sandstones and volcanic rocks (1200m). Kondomian Suite of volcanic rocks and sediments (2000m). Bel'sian Suite of archeocyathid limestones. In places shales and spilites (3000m).	Volcanic-carbonate strata with at times dominant archeocyathid limestones (2000m) and at others a sequence of spilites and keratophyres (up to 9000m). In places there are greenish shales and sandstones with beds of limestones (3000m).	Boksonskian Series of marls and limestones with *Collenia* and archeocyathids (3000m). Sarkhovian Suite of spilites, keratophyres (1500m). Okian Suite of carbonates, in places being an entirely carbonate succession.	Grey and dark-grey massive and bedded limestones and dolomites with archeocyathids and *Collenia* (1500–2000m). Mamakanian Suite of barren, red sandstones, marls and dolomites; below, red sandstones and conglomerates (up to 1800m).

or poorly-bedded massifs of different sizes, sometimes a hundred metres high. The massifs consist of mainly pale limestones and dolomites, which vary from pink to white, pale-grey, bluish, and rarely dark-grey or black colour. Their smooth surfaces show numerous cross-sections of archeocyathids which were formerly thought to be badly preserved corals. The age of these limestones is of interest. Up to 1945 they were considered to be Middle Cambrian and only the oldest ones were thought to be top Lower Cambrian. At present the majority of these limestones are included in the Lower Cambrian and only a part of them is thought to be Middle Cambrian. The fine distinctions are made by specialists, but even they dispute which archeo-cyathids are Lower Cambrian and which are Middle Cambrian. The trilobites found in such limestones help in the diagnosis of their age. The archeocyathid massifs are surrounded by bedded fine-grained limestones and marls, which by merging with the massifs form thick carbonate groups, which stretch for hundreds of kilometres, resulting in whole ranges, as on the eastern slope of the Salairs and the Sayans. In the Eastern Sayans bauxites (Buksonskian deposit) are found in association with these massifs, just as the Northern Uralian bauxites are associated with the Silurian and Devonian reef limestones of the Urals.

At the end of the Middle Cambrian and in the Upper Cambrian epoch the

FIG. 137.   Cambrian strata on the river Upper Angara in the south of the Patomian Highlands. *Photograph by L. I. Salop.*

Salairian phase of the Caledonian Orogeny caused the elevation of a large landmass
—an area of erosion. Consequently a part of the Middle Cambrian and Upper
Cambrian sediments is often missing from the succession. In places there are thick
flyschoid shales. The topmost Cambrian and Ordovician sediments rest unconformably on the Middle Cambrian, Lower Cambrian and older rocks. Consequently,
where the Upper Cambrian is present, it consists dominantly of clastic sandstones,
conglomerates, and shales, and has lavas and tuffs.

Ordovician

Ordovician sediments are closely associated with Cambrian and Proterozoic
deposits and have a similar distribution and great thickness and lithology.

SUBDIVISIONS. The Ordovician is divided into lower, middle and upper, has
quite a peculiar fauna and consequently, although it can be correlated with European
Ordovician stages, the correlation is difficult and cannot be considered as having
been established. Thus the adoption of local subdivisions is more rational and is
often attempted. The lower boundary is placed at the bottom of the beds with
Tremadocian faunas such as *Obolus* and *Dictyonema* or the trilobites *Euloma* and
Niobe. The upper boundary is traced at the unconformity caused by the Caledonian
Orogeny (Taconic phase), which had occurred in the Upper Ordovician. Ordovician
successions are shown in Table 31.

LITHOLOGY AND FACIES. In the Ordovician as well as in the Cambrian,
marine facies predominate, while continental facies have not been studied
and remain almost unrecognized. Sandstones and shales predominate and there are
no archeocyathid limestones as in the Middle and Lower Cambrian. There are
almost no reef limestones, but much more conglomerate is present. Such a
predominance of clastic sediments is interpreted by the occurrence of mountain-
building movements, caused by the Caledonian Orogeny. The flyschoid rocks resemble the true flysch and are thinly and rhythmically bedded. Such rocks are found
in Kazakhstan ('green flysch') in the Altais and other regions. Graywackes are found
everywhere as thick, bedded, greenish-grey, and pepper sandstones similar to the
Artinskian sandstones of the Urals, and being reminiscent of the Alpine molasse.
Their fauna occurs as impressions of numerous orthids and strophomenids indicating
an Ordovician age.

Graptolitic shales are of some interest, are rhythmically bedded and may be
considered as orogenic facies.

FAUNA AND FLORA. The flora consists of calcareous algae, similar to the
Cambrian types, but less frequently found. No terrestrial plants are known.

The fauna differs from the Cambrian fauna and is characterized by forms with
thick calcareous shells, while Cambrian types have thin chitinous or chitino-cal-
careous shells. The brachiopods, lamellibranchs and gastropods show this feature
especially well. Amongst the brachiopods the orthids and the strophomenids are
dominant and *Angarella* makes its appearance. There are no archeocyathids, while
rugose and tabulate corals appear. Graptolites, beginning with the genus *Dictyonema*,
are numerous and variable.

PALAEOGEOGRAPHY. The details of the palaeogeography have not been
worked out, but the main feature has been the recurrent change from marine to land
conditions, although the former conditions were dominant. The presence of sub-
montaine downwarps does not provoke any doubts. There were, also, numerous
volcanoes.

MIDDLE PALAEOZOIC

The Middle Palaeozoic deposits are thick, folded, metamorphosed and invaded by Caledonian and Hercynian intrusions. Although marine facies predominate, continental sediments are also quite widespread and well represented. Two types of successions—Caledonian and Hercynian—can be clearly distinguished (Fig. 138).

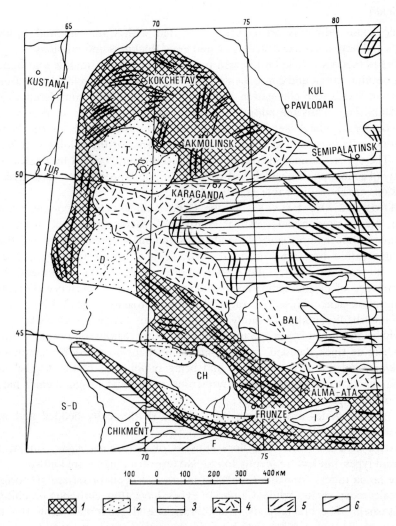

FIG. 138.   Caledonian and Hercynian structures of Central Kazakhstan.
According to A. A. Bogdanov (1959).

1 — Caledonian structures;   2 — major depression in the Caledonian basement filled by clastic Upper Palaeozoic sediments; 1 — Tenizian, D — Dzhezkazganian; 3 — Hercynian   structures;   4 — intervening   volcanic belt;   5 — strike of main structures;   6 — major depression in Caledonian and Hercynian basement filled by Mesozoic and Kainozoic sediments: Kul — Kulundinian, Tur — Turgaian, Ch — Chuian, F — Ferganian, Bal — Balkhashian, I — Issikkulian, S-D — Syr-Dar'yaian. The sketch map clearly indicates the position of the Kazakh macroisthmus. It consists of Caledonian structures, trending north-eastwards, beyond the limits of the map.

The Caledonian type is found in the areas affected by the last phases of the Caledonian Orogeny, which formed high ranges that existed at the end of the Silurian and in the Lower Devonian and Eifelian. Consequently, the corresponding sediments are missing from the succession and Siluriarn or Lower Palaeozoic strata are overlain by the Givetian or younger deposits.

The Hercynian type is found in the areas of weak effects of Caledonian Orogeny or a complete absence of such effects. The orogenic movements, which occurred in the middle of the Middle Palaeozoic are here not noticeable and the topmost Silurian and the whole of the Devonian are represented by marine sediments, such as Hercynian limestones.

The Caledonian type is found in the Western Sayans, the Minusinskian Depression, the Kuznetskian Alataus and in the northern and western parts of Central Kazakhstan and to the south-east in the Karataus, the Chu-Ilii Mountains and the north-eastern ranges of the Tian-Shians and north China.

The Kazakh macroisthmus. In all the regions just mentioned in the Lower Devonian there were high mountains stretching for hundreds of kilometres. At first the mountains were represented by isolated islands, such as the present-day Kuril Islands and Japan. Then the islands merged into each other and formed an entire large isthmus of mountains and volcanoes. Such isthmuses do not exist at present and can be given the term macroisthmuses.

The Kazakh macroisthmus was washed by the Western European Sea from the west and by the Siberian-China Sea from the east. Each of these seas had a different fauna, which confirms the existence of the isthmus (Fig. 138). In the Visean the isthmus was partly destroyed and large cross-channels were formed along which the Western European (Uralian) gigantoproductids penetrated into the Central Kazakhstan Sea.

Silurian

The Silurian is quite widespread and represented by different facies. In the provinces of the strongest and prolonged continuation of the later phases of Caledonian Orogeny Silurian strata are almost entirely missing since these regions were represented by dry land. Such regions existed over the area of the present-day Eastern Sayans, the Minusinskian Depression, the Kuznetskian Alataus and the northern and western parts of Central Kazakhstan. Only in some places there are thin groups with marine faunas. In other regions the Silurian succession is variable with breaks and consists of marine and land deposits. In the regions affected by weak late phases of Caledonian Orogeny the Silurian succession is complete and is represented by marine limestones and sub-marine lavas. Successions of this type are found in the southern part of the geosyncline—in the Salairs, the Altais, and Cisbalkhashia.

SUBDIVISIONS. The Silurian is divided into the Llandoverian (Llandovery), the Wenlockian (Wenlock), and the Ludlovian (Ludlow) stages. The first two stages are known as the Lower Silurian, while the Ludlovian, itself divided into lower and upper, is known as the Upper Silurian.

LITHOLOGY AND FACIES. In Eastern Kazakhstan, in the north, there are predominant coarsely clastic conglomerates and sandstones, while shales and limestones form thin beds, found only at the bottom of the succession. In the upper part of the succession there are the widely distributed terrestrial red sands and clays while at the very top there are thick lavas and tuffs. The succession in Central Kazakhstan serves as an example.

TABLE 32.

Devonian Successions in the Angara Geosyncline

Stage	Central Kazakhstan			The Ore-Altais (N. L. Bublichenko, 1956; V. A. Komar, 1957)	The Central Altais (According to I. I. Byelostotskii)
	Karaganda	The Bayanaulian region	North-eastern Cisbalkhashia (L. I. Kaplun and T. V. Rukavishnikova, 1958)		
Famennian	Sulciferian and Meisterian Beds of pale and dark limestones and silts. In places red Clymenic limestones (80–500m).	Sulciferian and Meisterian Beds of grey bedded limestones and marls with a brachiopod fauna (150–900m).	Unrecognized.	Pikhtovskian Suite of green, grey sandstones, gravelites, lavas and tuffs (600–1500m). Snegirevkian Suite of sandstones, albitophyres, tuffs. A marine fauna is present (500–1000m).	Cheremshankian Suite of sandstones, shales, limestones; and at the bottom, conglomerates. A marine fauna is present (500m).
Frasnian	Maian Beds—marine clastic deposits. In the north—red beds with a flora. In places volcanic rocks are present (100–1400m).	Maian Beds of continental red beds, rapidly varying in thickness. Beds of limestones with a marine fauna are present and in places there are volcanic rocks (80–2000m).	Maian Suite of continental red beds, with in places bands with marine faunas; in places there are volcanic rocks (200–900m).	Kamenevkian Suite of volcanic rocks, sandstones and shales; at the bottom, reef limestones (Gerikhovskian subSuite) (800m).	Malafeyevian Suite of shales and sandstones with beds of limestones. Above, a Frasnian fauna; below, a Givetian fauna (750m).
Givetian	A volcanic formation. Red sandstones, silts and conglomerates with flora in some beds and fauna in others (at the top (400m).	Akbastauian Suite of red beds with marine limestone bands or sandstone beds with floras. A sequence of volcanic rocks (300–600m).	Aldailian Suite of marine and continental sandstones and shales. Faunas and floras are found (400m).	Shipunovkian Suite of clastic and volcanic rocks, with beds of limestones with marine fossils at the top. Basal conglomerates are present (800–1000m).	Porphyrites, albitophyres, tuffs with beds of sandstones (1000m).
Eifelian	Albitophiric Formation of lower Middle Devonian and Lower Devonian ages. At the top, albitophyres, tuffs, tuffaceous sandstones and sandstones; below, porphyrites, tuffs and tuffites (1300–2000m). Beds of limestones in the upper part yielded a marine Eifelian fauna.	Kaidaulskian Suite of albitophyric lavas, lava breccias and tuffs; sequences of sandstones with floras and faunas are present (Lower and Middle Devonian) (2000–4500m). Above (Eifelian and top Koblenzian) the Suite sometimes has sandstones and tuffs with bands of limestones with a marine fauna.	Kazakhian Suite of brown tuffaceous sandstones and tuffites. Above, a bed with marine Eifelian fauna and flora (200m).	Talovkian Suite of acid volcanic rocks and tuffs (400–2000m). Berezovkian (Losikhian) Suite of limestones, shales and tuffs, with *Calceola* (600–1000m).	Partly bedded, partly reef massive limestones (400–1000m).

TABLE 32—continued

Altais	Kuzbas region			Eastern depressions	
The Eastern Altais	Salair (V. D. Fomichev, 1961)	The Barzasian region	Minusinskian (V. S. Meleshchenko, 1956; N. G. Chochia, 1956)	Central Tuva
Variegated, in places grey terrestrial sandstones and shales with a flora (1000–1500m).	Rassolkinian Suite of red beds with conglomerates at the base. There is a bed with a marine fauna (300–350m).	Podoninian Suite of red beds (30–150m). Kurundusian Suite of marine limestones and shales (30–120m).	Tubian Suite of sandy red beds with plants and fish (200–700m).	Dzhargian Suite of red sandstones and conglomerates with plant remains (200–1200m).
Variegated beds and volcanic rocks in places, replaced by black shales and silts with a marine fauna (1000–2000m).	Often missing. At the southern margin grey shales and limestones with a marine fauna are found.	Kel'besian Suite of marine limestones and shales (30–80m). Sergiyevskian Suite of red beds (30–150m). Vassinian Suite of marine shales and limestones (20–160m). Orlinskian Suite of red beds (80–180m).	Kokhaian Suite—red beds and grey, lagoonal marls (150–300m). Oidanovian Suite of red beds (300–700m).	Kokhaian Suite of variegated silts, marls and sandstones with armoured fish and phyllopods. Begredian Suite of red sandstones and silts, with a flora. phyllopods, fish (400–600m).
Beian Suite of limestones and shales, with a marine fauna (300–400m). In places it is replaced by dark shales (1000m). Volcanic-sedimentary formation of up to 2500–3500m.	Indospiriferian Horizon (Safonovian Suite) of shales and limestones (600m). Kerlegeshian Suite of thick shales and sandstones, with bands of limestones and lavas; marine fauna (800–1000m). Akarachkanian Suite of sandstones and basal conglomerates (400–500m).	Beian Suite. Cheehiel marine horizon (50m). Barzaskian Suite of dark, sometimes bituminous shales and lagoonal deposits (up to 300m).	Beian Suite of limestones, shales and sandstones (100–170m). Ilemorovian Suite of greenish sandstones and marls with a fresh-water fauna. Toltakovskian Suite of red beds (100–400m).	Uyukian Suite of red and grey sandstones, shales and limestones. Ilemorovian Suite of grey sandstones and silts (600m). Atakshilian Suite of lavas and red beds (1500–2000m).
Above, marls, limestones (in places) with a marine fauna (400m). Below, thick continental red beds and sequences of (volcanic rocks 2000–2500m).	Mamontovian Suite of shales, sandstones and limestones (350m). Shandian Suite of limestones and argillites (250–1500m). Salairkian Suite of argillites, sandstones and limestones (300–500m). At the base, sandstones and a bauxite-bearing Berdian Suite (0–100m).	Red, continental strata with psilophitic flora. Volcanic rocks.	Tashtypian Suite of grey limestones with brachiopods (350m). Tolochkovian Suite of thick red and volcanic rocks (up to 1500m).	Tashtypian Suite of grey limestones and marls with brachiopods (up to 300m). Saglian Suite of red beds and volcanic rocks. Psilophytes are present (700–2500m).

Lower Devonian. Sandstones, conglomerates, and shales with thin brachiopod-bearing (*Leptostrophia rotunda*) beds amongst the sandstones.

Ludlow. The top strata are green sandstones and limestones with brachiopods and local lavas and tuffs (Ainasuiian Horizon). The middle and lower part of the Ludlovian are similar to the above, but have occasional reef limestones and green siltstone resembling molasse (Akkanian Horizon). The fauna is diverse and in places rich in tabulates (*Halysites*), brachiopods (*Conchidium*) and trilobites (*Encrinurus*).

Wenlock. Thick barren sandstones with local lavas and fossiliferous limestones (total thickness up to 3000–4500m)—Zhumakian Horizon.

Llandovery. Embraces the grey-green sandstones, siltstones, conglomerates with beds of limestones. There are brachiopods and tabulates (Alpinian Horizon).

The total thickness of the succession is 3000–4000m and in the Chingiz range it is up to 6000–8000m. Further to the east marine deposits predominate. In the Rocky Altais and the Salairian ridge the lower part of the succession (Llandovery and Wenlock), is represented by thick (1000–2500m) shales with beds of sandstone and limestone. In the Salairs these strata are known as the Yurmanian Suite (1600m) and in the Altais as the Chinetikian Suite with graptolites and tabulates found amongst the shales. The upper part of the succession (Ludlow) consists of limestones and shales (400–1500m thick) with a rich and diverse fauna with brachiopods and corals. In the Altais these rocks are known as the Chagyrkian Suite and in the Salairs as the Tomskozavodian (Ostracodic) as suggested by Fomichev and Alekseyeva (1961).

In the north-east—Transbaikalia, the Eastern Sayans, the Kuznetskian Alataus—there are no Silurian strata; thus mountain ranges (the Caledonides) must have

TABLE 32—continued

					The
	Central Kazakhstan				
Stage	Karaganda	The Bayanaulian region	North-eastern Cisbalkhashia (L. I. Kaplun and T. V. Rukavishnikova, 1958)	The Ore-Altais (N. L. Bublichenko, 1956; V. A. Komar, 1957)	The Central Altais (According to I. I. Byelostotskii)
Koblenzian	Albitophyric Formation (lower Middle and Lower Devonian) (see above).	Kaudalian Suite (albitophyric) (see above).	Sardzhal'skian Suite of greenish and brown sandstones, tuffaceous sandstones and tuffs. Limestones with marine fauna and flora present.	Missing.	Baragashian Suite of shales and limestones with a marine fauna; at the bottom sandstones and conglomerates (up to 1500m).
Gedinnian			Pribalkhashian Suite of greenish and brown sandstones with beds of silt and towards the top tuffaceous sandstone. Beds with marine fauna and psilophytes are present (300–600m).		Obscure.

existed at that time. To the south of them continental red beds appear and in the Western Sayans and Tuva green limestones, shales and sandstones (up to 4000m thick) appear and have a rich marine fauna. These sediments were deposited in the same sea as existed in the area of the present-day Eastern Altais and the Salair range.

PALAEOGEOGRAPHY. In the Llandoverian epoch the sea was most extensive and as isolated embayments penetrated most of the regions of the geosyncline. In the Wenlockian epoch the sea retreated to the south and the Kazakh macroisthmus became distinct. In the Ludlovian epoch the sea receded still further to the south, into the central part of the geosyncline. In the northern half of the geosyncline the accumulation of terrestrial land deposits and thick lavas was in progress. In certain regions, which were at the time elevated, there are no top Silurian strata.

Devonian

The distribution of Devonian strata is considerable and they are found throughout the Angara Geosyncline. Their thickness is also considerable, although this varies from a few hundred metres to 8–10km.

SUBDIVISIONS. Western European stages of Gedinnian, Koblenzian, Eifelian, Givetian, Frasnian, and Famennian are adopted. The peculiar nature of the fauna, which is often different from that of the Urals and Western Europe, complicates the distinction of horizons and consequently local terms should be used (Table 32). The lower boundary is normally arbitrary. When the Lower Devonian is continental it is drawn above the last beds with Silurian fossils and is fixed on the basis of lithological features.

BLE 32—continued

Altais	Kuzbas region			Eastern depressions	
The Eastern Altais	Salair (V. D. Fomichev, 1961)	The Barzasian region	Minusinskian (V. S. Meleshchenko, 1956; N. G. Chochia, 1956)	Central Tuva
In places Lower Devonian strata are missing.	Nadkrekovian grey and white bedded limestones with *Karpinskia* and tabulates (up to 230m).	Lower Devonian strata are either missing or represented by the lower part of the red beds. The question is ambiguous.	Imekian Suite of grey and greenish sandstones and argillites (250m). Chilanian Suite of red sandstones, conglomerates and silts. At the top, sheets of porphyrites (1 500m).	Barykian Suite of grey and greenish silts, marls and sandstones, often tuffaceous (up to 3000m). B'yertdagian Suite of red sandstones, conglomerates and basic lavas. Psilophytes and armoured fish (900m).
	Krekovian Suite. Dark and grey tabulate and brachiopod-bearing limestones (350m).			

Where the Lower Devonian and Silurian are marine and transitional into each other the boundary is drawn on some faunal basis (e.g. north-eastern Cisbalkhashia).

The upper boundary is also arbitrary. On the basis of faunal changes it is drawn on the boundary between marine Famennian and Tournaisian stages, but in many sections either or both stages are represented by continental red beds. Under such conditions the boundary is drawn on the basis of lithology or under horizons with marine faunas. In the subdivision of the continental red beds armoured fish, ostracods, phyllopods, plants, and spores are used.

LITHOLOGY AND FACIES.   The facies are very diverse and change rapidly as the result of the Caledonian Orogeny. Where this orogeny has been powerful, as in the east, north, and south of the geosyncline Old Red Sandstone type of deposition predominated producing in places up to 8–10km of sediments (Fig. 139). Amongst these sediments and especially in their upper part there are beds and groups of rocks with marine faunas. In the east (the Eastern Sayans) there are no marine bands and in the Minusinskian Depression and adjacent districts the most widespread is the Givetian Group of muddy limestones with brachiopods and bryozoans. The main form here is the Chinese (*Spirifer cheechiel* Kon). The group is variously known as the Beiskian Suite, the Mart'yanovian Stage and the Cheehielovian Horizon (Fig. 140). A second marine horizon occurs at the top of the Eifelian Stage. In the Minusinskian Depression it is known as the Tashtypian Suite.

In the south of the central part of the geosyncline—the Salairs, the Ore and Rocky Altais—the marine facies (Fig. 141) predominate. In the Lower Devonian there are thick, massive, and poorly bedded reef limestones analogous to the Hercynian limestones of the Urals, but with a different fauna, resembling that of the North American Helderberg Limestone. In the Salairs at the bottom of these limestones lie rich deposits of bauxite which are sometimes altered into emery as a result of alteration at contacts with intrusives.

FIG. 139.   Middle Devonian sandstones in South-western Tuva.

Devonian facies of the Altais are very variable and in addition to those shown in Table 32, many others have been recognized. Amongst them the Korgonian Suite of Eifelian age is found in the Korgon, Tigeretskian and Kholzunskian ranges. The suite consists of a thick lava group (up to 3500m) with subordinate groups of tuffites, sandstones, and silts with marine faunas and in the middle of it is a horizon of commercial iron ore (Iron-ore deposits . . ., 1958–1959).

In the south-east of Central Kazakhstan there are also mainly marine limestones with rarer sandstones and shales (Fig. 142). Lavas of different ages and composition are widespread and often are 2000–4000m thick.

FIG. 140.   Red beds and the marine Beiskian Formation in the North Minusinskian Trough. In the near distance are Middle Devonian red beds overlain by pale marine marls and clays of the Beiskian Suite showing up as white coloured rocks to the left of the central part of the photograph. The suite is overlain by Upper Devonian red beds. *Photograph by E. A. Shneiderov.*

FIG. 141.   Middle Devonian marine limestones and shales of the Belgebashian Suite on the river Chuya in the Rocky (Mountainous) Altais.

In the west and south-west Devonian strata are either missing or are represented by continental red beds of, sometimes, great thickness; thus indicating the existence of the Kazakh macroisthmus.

The monographs by N. L. Bublichenko and L. L. Khalfina are concerned with the Devonian of the Altais and Kazakhstan; while that of Kuzbas and more easterly regions is treated by M. A. Rzhonsnitskaya and L. L. Khalfina.

FAUNA AND FLORA. The first terrestrial psilophyte flora appeared in the Silurian, reached its peak in the Devonian and then disappeared. In the Famennian epoch the psilophytes were already succeeded by lepidophytes. The representatives of the latter had a well developed woody substance and gave rise to the first coal seams.

The fauna is rich and variable with brachiopods, and especially spiriferids and rhynchonellids as the most dominant forms. There are also numerous rugose and tabulate corals, lamellibranchs, gastropods, bryozoans and ostracods. The trilobites, although already on the way to extinction, are still numerous. The first goniatites appear and are stratigraphically important. In lacustrine and lagoonal facies, amongst the red beds, there are found armoured fish, phyllopods and ostracods.

PALAEOGEOGRAPHY. In the north and east there was dry land surrounded by vast near-shore sun drenched plains with red beds as deposits, and surmounted by volcanoes. To the west was the Kazakh macroisthmus (Fig. 138), while to the east of the isthmus there was a warm tropical sea, with occasional coral reefs and volcanic islands.

Lower Carboniferous

Lower Carboniferous rocks are found in the same regions as the Devonian and are of great thickness. Coal-bearing formations with major national deposits, such as at Karaganda and Kuzbas (lower coal seams), have Lower Carboniferous coals.

FIG. 142. Upper Devonian and Lower Carboniferous sandstones and conglomerates on Mt. Kandyk-Tas, in Southern Kazakhstan. *Photograph by L. I. Borovikov.*

SUBDIVISIONS. The normal Western European-Uralian subdivisions are used and the Lower Carboniferous is divided into the Tournaisian, the Visean and the Namurian. The fauna differs from that in the Urals and is similar to the Chinese and North American. Consequently it is more correct to use the American stages of Kinderhook, Burlington and Keokuk and even more so to employ local terms, such as are already used for continental deposits. Successions of Central Kazakhstan are given in Table 33.

LITHOLOGY AND FACIES. Marine and continental facies are equally frequently encountered, while lavas are less common. The absence of thick conglomerates and sandstones is characteristic and indicates that there were no mountain ranges caused by mountain-building processes. Consequently, throughout the length of the geosyncline the Lower Carboniferous succession remains almost constant. In the lower horizons of the Tournaisian and Lower Visean stages, in the main only marine limestones and shales are found. These strata have a rich and diverse fauna with brachiopods, rugose corals and bryozoans. Goniatites are rarer, but not less important. The limestones are often many hundred metres thick (Fig. 143).

The upper part of the succession (Middle and Upper Visean and Namurian) is lithologically variable. In the west—Central Kazakhstan—there are in places marine deposits with Uralian forms of fossils, including gigantoproductids. Here there are also thick coal-bearing formations and other continental deposits. In the east marine facies are rare and land deposits predominate. The succession of the Minusinskian basin (Bogomazov, 1961) serves as an example:

Namurian Stage—Conglomeratic suite (200–300m).

FIG. 143. Shaley limestones and shales of a Lower Carboniferous age in Tarbagatai. Complex folds are interesting. *Photograph by M. M. Vasil'yevskii.*

TABLE 33.

Lower Carboniferous and Upper Devonian Strata of Central Kazakhstan
(According to M. S. Volkov, 1957)

Sub-stage or stage	The Ishim region; thickness = 170 to 590m average = 400m	The Dzhezkazgan region; t = 1030–6680m a = 3850m	The Karaganda region; t = 910–6425m a = 3660m	The Chingiz region; t = 4700–9990m a = 7350m
Middle Carboniferous.	Continental deposits.	Dzhezkazganian Suite of continental deposits with cupriferous sands.	Naddolinskian Suite of continental deposits.	Continental deposits.
Namurian.	Greenish sandstones and shales with a marine fauna. Red beds with plant remains (65–130m).	Lagoonal, greenish sandy-clayey deposits with a fauna of lamellibranchs predominant. Rarer marine clastic strata with rare beds of limestones occurs instead (250–560m).	Dolinskian and Nadkaragandian Suite. Continental sandy-clayey deposits; in places, coal-bearing (1000–2000m).	Continental and marine terrigenous deposits with a basal conglomerate (1000–2150m).
Upper Visean.	Upper Yagovkian Beds. Grey sandstones with rare bands of clays and limestones with marine faunas (30–170m).	Upper Yagovkinian Beds of mainly terrigenous, clastic rocks but sometimes also marine and sometimes lagoonal deposits (150–200m).	Karagandian Suite of grey and brown sandstones. Silts and argillites with workable coal seams. The main productive strata (250–300m).	Terrigenous—volcanic deposits, with a marine fauna (500–900m).
Middle Visean.	Lower Yagovkian Beds of grey and greenish sandstones and in places red beds (25–80m).	Lower Yagovkian Beds—homogeneous rocks of a mainly terrigenous type; sometimes rhythmically interbedded with limestones. A fauna of lamellibranchs is present (100–200m).		Homogeneous terrigenous—volcanic deposits with bands of limestones possessing a marine fauna (400–1000m).
Lower Visean.	Upper Ishimian Beds of greenish sandstones, shales and limestones, with a marine fauna (20–50m).	Upper Ishimian Beds of mainly greenish sandstones, silts and shales, but occasional beds of limestones exist. A lamellibranch fauna is present (75–700m).	Ashlyarikian Suite of interbedded marine and coal-bearing deposits with workable coals. Akkudukian Suite of mainly marine sandy-clayey deposits with sequences of coal-bearing deposits.	Sandstones, silts and shales with limestones possessing a marine fauna. In places coal-bearing sequences (400–1500m).

Upper Tournaisian.	Lower Ishimian Beds—greenish sandstones, shales and limestones, with a marine fauna.	Lower Ishimian Beds—variable limestones and greenish sandstones and siltstones, with marls. A marine fauna is present (100–500m).	Terektian Beds—siliceous marls with goniatites (250–1500m).	Greenish and grey sandstones, silts, shales and tuffs with beds of limestones carrying a marine fauna (400–1000m).
Middle Tournaisian.	Rusakovian Beds—grey limestones, shales and sandstones with a marine fauna. Basal sandstones and conglomerates are present (10–60m).	Rusakovian Beds—grey, silicified limestones and in places cherty marls, bedded sandstones and silts. A marine fauna is present (60–750m).	Rusakovian Beds—grey, bedded limestones with sandstones and shales in places. A marine fauna is present (20–200m).	Terrigenous, volcanic and argillaceous—cherty deposits with marine, fossiliferous bands (1000m).
Lower Tournaisian.	Missing.	Kassinian Beds of pale and dark bedded and massive limestones with a marine fauna (50–700m).	Kassinian Beds—pale limestones with a rich brachiopod fauna (40–250m).	Terrigenous—volcanic sequence with beds of limestones (1000–1500m).
d'Etreungt.		Dark bituminous limestones with a rare fauna (20–200m).	Posidonian Beds of brown and grey shales with cherty limestones and silts. In places brachiopod limestones are present (30–150m).	Unrecognized.
Famennian.		Calcaratusian and Sulciferian Beds—dark and pale, bedded limestones with a brachiopod fauna (150–900m).	Pale and dark limestones and silts with a marine fauna. In places there are Clymenic variegated siliceous limestones (80–500m).	Unrecognized.
Frasnian.		Continental red sandstones and conglomerates, rapidly varying in their thickness (80–1970m).	Maian Beds of marine, clastic (terrigenous) deposits with bands of limestones. In the north, red beds with a flora (Akbastauian Suite (100–1375m).	Continental red beds with basal conglomerates. In places bands with marine faunas (200–900m).
Basement.	Proterozoic.	Proterozoic and albitophyric formation.	Albitophyric formation.	Albitophyric formation.

Visean Stage—(480–1375m).

Bainovskian and Podsinskian suites (200–370m). Consist of interbedded tuffs, tuffites and sandstones with plants and fish.

Krivinskian, Solomenskian, Komarkovian and Sogrinskian suites of brown, red and green tuffs and tuffites, interbedded with siliceous limestones, argillites and sandstones. All the suites have analogous spores and pollen, but poor plant remains. The thickness is 240–790m.

Samokhvalian Suite (40–215m) of tuffs, tuffites and sandstones with in places basal sandstones and conglomerates. A good Lower Visean flora is found.

Tournaisian Stage—(80–600m).

Nadaltaiian Suite (20–190m) of grey, yellow and greenish tuffs, tuffites and sandstones with beds of limestones. The fossils are: ostracods, fish, plants and brachiopods (orthotetids).

Altaiian Suite (50–170m) consisting of red, maroon and grey tuffs and tuffites with rare beds of sandstones and limestones with fish.

Bystraynian Suite (10–235m) of green and grey bedded tuffs, tuffites, sandstones, limestones and dolomites. The fossils are: fish, ostracods, lamellibranchs (freshwater), plant remains and rare brachiopods (orthotetids).

The Tournaisian Stage, thus, consists of continental sediments with marine bands. The Visean and Namurian stages are represented by entirely continental, freshwater and terrestrial deposits. The abundance of tuffs and tuffites indicates the former presence of neighbouring volcanoes which were active throughout the Lower Carboniferous epoch.

Coal-bearing formations. The facies of these formations are most peculiar and industrially important. They differ from coal-bearing formations of the Lower Carboniferous of the Russian Platform and the Urals by much thicker and more frequent coals. The Coal-bearing Formation of the Urals is however Visean and in places Tournaisian and its coal seams are sparse and thin. In Karaganda and Ekibastuz the Coal-bearing Formation is Upper Visean and Namurian and its upper horizons are Middle Carboniferous in age. The coal seams are numerous and sometimes very thick, and especially so in Ekibastuz. The formation itself is also very thick, reaching 4000–4500m in Karaganda. The outcrops of this formation are not restricted to Karaganda and Ekibastuz, but occur in other areas, thus representing a vast near-shore plain. This plain was situated along the southern margin of the Western Siberian continent stretching from the Turgai Depression to the valley of the Irtysh. In the easterly deposits near Irtysh (Minusinskian Depression and Kuzbas) the accumulation of continental, in places coal-bearing, sediments (Ostrogian Suite) began in the Namurian, but most of the coal seams are Upper Palaeozoic.

The lithofacies of the Visean-Namurian coal-bearing formation are similar to the Uralian Lower Visean measures, with the dominant sediments of near-shore plains, vast low-lying deltas, rivers, floods, ox-bows, near-shore marshes, lakes and wide sandy watersheds. It is possible that there were 'mangrove swamps'. The strata although thick and broadly regular are constantly changing from relatively fine-grained sandy and muddy sediments to subordinate lacustrine marls and bog coals. Their colour is mainly dark grey, brown and lighter grey. The stratification varies from finely rhythmic to irregularly cross-bedded. There are plentiful plant remains and rarer phyllopods and ostracods, freshwater lamellibranchs and fish. The Karaganda succession is given in Table 34.

PALAEOGEOGRAPHY. The main feature is the position of the land in the north and sea in the south. The position of the shore-line is represented by the belt of coal-bearing strata which were laid down on the near-shore plain.

In the west the Kazakhstan and Ural seas were, as before, isolated by the Kazakh isthmus, as can be demonstrated by their faunal differences. At the beginning of the Visean epoch the isthmus was overrun by the sea and a migration of the fauna began. Thus for instance in Kazakhstan the first gigantoproductids appear, although specifically they are different from those found in the Urals. Another important group of Uralian Visean brachiopods belonging to the group of *Productus*

TABLE 34. *Coal-bearing Sequence of Karaganda*
(Accepted at a stratigraphic conference in 1958, in Alma-Ata)

Division	Stage	Lithological description
Upper		Shakhanian Suite of variegated argillites and silts with beds of sandstone; 3 coal seams. No fauna was found. (300–500m).
Middle		Tentekian (Naddolinskian) Suite of dark argillites with coal seams and considerable sequences of greenish sandstones, sometimes gravelites; 17 ashy coal seams. Phyllopods found in argillites. Horizons of tuffs and porphyries. (300–700m).
		Dolinskian Suite of dark, grey and bluish argillites and silts with beds of sandstones, 11 workable coal seams. Flora, phyllopods and ostracods are present. (400–500m).
Lower	Namurian	Nadkaragandian Suite of rapidly alternating grey and greenish argillites and sandstones with thin bands of limestones possessing a fresh-water fauna; no coals. In places the upper, sandstone, part is distinguished as a separate suite. (500–600m).
		Karagandian Suite of alternating, continental, coal-bearing sequences of silts, sandstones and argillites with beds of argillites and marls with marine faunas; 20 workable ashy coal seams. (750–800m).
	Visean	Akkudukian Suite of grey sandstones and dark argillites with beds of sandy limestones. Marine deposits with sequences of continental, coal-bearing strata. Unworkable coal seams. (600–700m).
		Terektinian Beds (Horizon) of variegated, cherty limestones, and shales with goniatites and brachiopods. The strata are only found in places. (0–100m).
	Tournaisian	Rusakovian Beds (Horizon) of grey and dark-grey, bedded, not infrequently silicified limestones with a marine fauna. In places the strata are replaced by shales and siliceous shales, marls and sandstones. (70–125m).
		Kassinian Beds (Horizon) of pale, bedded or massive limestones with a rich fauna. The limestones pass laterally into siliceous shales or a sandy-clayey sequence. (40–100m). Posidonian Beds (Sokurskian Horizon) of grey and dark marls and shales with brachiopods and *Posidonia*. In places the strata pass into limestones with brachiopods of d'Etroeungt age. (60–100m).

striatus is not found in Kazakhstan, but is known from Kuzbas. In the central part of the geosyncline the Tournaisian-Visean transgression to the north forms a bay between the present-day valleys of the Yenisei and the Ob'.

Upper Palaeozoic

Upper Palaeozoic strata are less widespread than the Middle Palaeozoic rocks. This is due to the occurrence of the Hercynian Orogeny in the Middle Carboniferous, which had caused the elevation of mountain chains. These mountains were advancing progressively further to the south into Central Asia. In the Upper Permian there was no sea in the Angara Geosyncline.

SUBDIVISIONS. Amongst marine deposits, where they are developed, Lower Carboniferous, Upper Carboniferous and Permian strata can be recognized. More detailed subdivisions can be erected locally. Continental formations are divided into local suites of often debatable age. This applies even to major coal-fields such as Kuzbas and Karaganda. The Balakhonkian Suite of Kuzbas, for instance, was formerly included in the Lower Permian. V. I. Yavorskii even now accepts this point of view. A majority of the investigators of Kuzbas on the basis of fauna and flora consider the upper part of the Balakhonkian Suite as Lower Permian and its lower part as Upper and Middle Carboniferous. The upper coal-bearing formation of Kuzbas (the Kolchuginian) may be Lower Permian, to be correlated with the Vorkutian Formation. A majority of coal-geologists, however, consider the Kolchuginian Suite as Upper Permian. The Kuzbas succession is given in Table 36.

In Karaganda the lowermost coal-bearing suite (the Karagandian) is regarded by some as Lower Carboniferous, which is supported by evidence, while others consider it as Middle Carboniferous.

Such divergences in views are mainly explained by the local character of the flora and fauna, which is considerably different from the fauna and flora of European provinces and in any case has not been adequately studied.

The boundary between the Lower and Upper Palaeozoic is marked by frequent unconformities and sharp lithological changes. The boundary between the Upper Permian and Lower Trias is not clear and its delineation is a problem for the future. In some regions Permian and Triassic rocks are entirely absent. Middle Trias strata have not been found anywhere and are probably entirely missing.

The top part of the Upper Trias (the Rhaetic) shows weakly folded and almost unmetamorphosed rocks overlying unconformably various horizons of Middle and Upper Palaeozoic, which involve more strongly deformed and metamorphosed rocks.

At the western margin of the geosyncline—Kalbinian range, the Ore Altais and the Tarbagatais—the Upper Palaeozoic stratigraphy has in the last few years suffered a considerable rearrangement. The investigation of the local fauna and flora has shown that the rocks which were formerly thought to be Lower Permian are in the main Lower Carboniferous (top Visean-Namurian) corresponding to the Ostrogian Suite of Kuzbas. The upper horizons contain a flora of the Mazurovian Suite and are Upper Carboniferous.

Higher up lies the Serzhikhinskian Suite (from 500–800m to 1600–1800m) of lavas and effusive-sedimentary rocks. It has no organic remains and is provisionally considered as Upper Carboniferous-Lower Permian. The Upper Permian and Lower Trias deposits are not known, but can be recognized in the southerly Kenderlyk coal-field, which is situated in the south-eastern part of lake Zaisan

FIG. 144. Upper Palaeozoic oil shales of the Kenderlyk deposit. Lenses of limestone lie within the shales. *Photograph by V. P. Nekhoroshev.*

FIG. 145. Triassic conglomerates of the Kenderlyk deposit in the region of lake Zaisan. Columnar weathering is typical. *Photograph by V. P. Nekhoroshev.*

Depression. At the base lies a thick volcanic group (porphyrites, tuffs and volcanic breccias) of Lower Carboniferous age. This is overlain by the first (lower) coal-bearing shaley suite of 750m in thickness. Its lower part begins with sandstones and conglomerates, followed by sandstones, lavas and shales with coal seams (2·0–2·5m each) and an Angaran *Pteridium* flora. The upper part consists of dark grey sandstones and shales with three groups of oil-shales which total up to about 40m thick (Fig. 144). The age of the lower suite was formerly thought to be Lower Permian, but nowadays a part of it is thought to be Namurian and Middle Carboniferous.

The second (upper) coal-bearing suite is 800–1000m thick and consists of sandstones and shales. Its upper part has 9 coal seams of a total thickness of 18m. It is included in the Upper Permian. Still higher lie the thick (800m) Triassic conglomerates (Fig. 145) overlain by the third coal-bearing formation (1000m) of Rhaetic-Liassic age and which has 25 seams of brown coal.

Upper Palaeozoic coal measures are also found in the Cisirtysh coal region (north of Semipalatinsk), in the Tuva Depression, and in a series of depressions belonging to the Minusinsk coal-field.

The Minusinsk succession is similar to that of Kuzbas, but is thinner (2000–3000m), has continental Visean strata and no Upper Permian. The Visean is overlain by the lower coal-bearing formation of the Namurian and Middle Carboniferous age. The upper coal-bearing formation belongs to the Upper Carboniferous and Lower Permian. The suggested successions of G. P. Radchenko, G. A. Ivanov and V. M. Bogomazov (1961) are shown here.

G. P. RADCHENKO	G. A. IVANOV	V. M. BOGOMAZOV
Lower Permian	*Upper Permian*	*Lower Permian*
Arshanovian Suite	Narylkovian Suite	Narylkovian Suite (600m)
Upper Carboniferous	Byeloyarian Suite (750m)	Byeloyarian Suite (300–500m)
Poberezhnaya Suite	Bezugol'naya Suite (70–150m)	Bezugol'naya Suite (70–150m)
Middle Carboniferous	*Lower Permian*	*Middle Carboniferous*
Khakasskian Suite	Chernoyarian Suite (350–400m)	Chernoyarian Suite (270–370m)
		Namurian
	Conglomeratic Suite (200m)	Conglomeratic Suite (200–300m)
Visean Stage	*Upper Carboniferous*	*Visean Stage*

At present the ages suggested by Radchenko and subdivisions advanced by G. A. Ivanov are accepted.

LITHOLOGY AND FACIES. The intensive manifestations of the Hercynian Orogeny produced young fold mountains in the Upper Palaeozoic. The mountain-building activity spread from north to south, filling the geosyncline and converting it into a part of the continent Angarida (Angaraland). The ranges were being eroded at the same time as they rose and the products of erosion were transported and de-

posited in seas and on plains. Consequently the marine Upper Palaeozoic deposits
are always rich in clastic materials. Pure limestones are rare and limestone reefs are
missing. On the other hand sandy and clayey limestones, shales, sandstones and
conglomerates are widespread. In Kazakhstan there are no conglomeratic or sandy
groups such as the Artinskian rocks of the Western Urals. This indicates the absence
of high snow-capped mountains with well-developed rivers. Kazakhstan mountains were no higher than 2000–3000m, and did not have a snow-cover or torrential
rivers. They were surrounded by desert plains and were like the Karatau range which
at present lies as a black range of mountains between the Syr Dar'ya and the Chu.
The desert nature of the Upper Palaeozoic Permian continent is supported by the
wide development of red and variegated sandstones and shales approaching in aspect
Ufian and Tatarian sandstones of the Russian Platform. They also contain copper-
bearing sandstones, with which the Dzhezkazgan deposit is associated.

The successions of the Dzhezkazgan area and the Tengiz Depression to the north
of it have been worked out in 1958 (Resolutions ... 1958):

Dzhezkazgan syncline (Mulde)	Tengiz Depression
Upper Permian	
Absent (?)	Shoptykulian Suite of grey and red silts, argillites and sandstones with beds of black freshwater limestones 300–900m.
Lower Permian	
Kingirian Suite of red beds and grey lacustrine marls (total 900–1000m).	Kiiminian Suite of red sandstones and argillites with beds of black limestones. Fossils: reptile bones. 200–1600m thick.
Zhidelisaiskian Suite of fine-grained red beds with cross bedding (up to 300m).	Kairaktinian Suite of rhythmically bed-ded red sandstones, argillites and limestones with freshwater fish and flora (200–1000m).
Upper Carboniferous and Upper Middle Carboniferous	
Dzhezkazganian Suite of red sandstones and argillites with beds of conglomer-ates and cupriferous sandstones. Total thickness: 400m. Erosion.	Vladimirovskian Suite of variegated, cross-bedded sandstones, argillites and conglomerates. It is 100–1500m thick and has a flora. Erosion.
Lower Middle Carboniferous and Namurian	
Toskudukian Suite of regressive deposits —continental and cupriferous at the top, and some beds with marine faunas at the bottom (total: 300–600m).	Kireiskian Suite of sandstones and shales with some beds of limestones with marine faunas (100–1000m). Sometimes (Kumpan, 1960) the Zhidelisaiskian Suite is considered Upper Carboniferous and the Kingirian Suite is correlated with the Kairaktinian.

The Middle and Upper Carboniferous contain groups of cupriferous sandstones. In the Tengiz Depression there are thick (3000–4000m), terrestrial and freshwater Permian red beds with horizons of lacustrine limestones.

On the surface of the large near-shore plains huge swamps and lagoons originated and became the sites of accumulation of plant remains which were subsequently converted into coals. The Upper Palaeozoic near-shore plains were situated in Cisirtyshia. The sea which was adjacent to these plains was on the site of the Kalbinskian range and stretched south across present-day lake Zaisan and the Tarbagatais. The conditions of formation of the Permian Kolchuginian coal-bearing formation of Kuzbas were different, since the nearest outcrops of marine Lower Permian rocks are 500km, and of Upper Permian rocks 1000km, away. Thus the Kolchuginian Suite was not formed on near-shore plains. The swamps which gave rise to this suite were situated in an enclosed depression similar to the present-day Aral Sea, lake Balkhash or Lob Nor. Conditions of the present-day lake Chad in Central Africa are probably analogous.

FAUNA AND FLORA. The Upper Palaeozoic flora and especially that of the Permian show the first coniferous trees, which suggests an environment away from the sea. The higher plants, thus, were growing in central parts of continents.

The fauna is poor and uniform. Marine forms are associated with marine transgressions, which penetrated the geosynclinal region from the south. The fauna is less variable than that of the open sea. The faunas of the Upper Carboniferous and Lower Permian cannot be easily distinguished, since they have only rare fusulinids which are so widespread in the Urals and Central Asia. Freshwater lamellibranchs, ostracods, phyllopods (*Estheria*) and fish are of great importance. The terrestrial vertebrates are very rare. L. A. Ragozin has described the lamellibranchs.

PALAEOGEOGRAPHY. In the Upper Palaeozoic the Angara Geosyncline has become mainly land, which was more elevated in the north and east and flat in the south and west. The land was a part of a large northern continent—Angarida. At the end of the Middle Carboniferous the sea transgressed furthest, covering the whole of Cisbalkhashia, but not reaching the Karaganda basin. The next, Upper Carboniferous, transgression occurred further to the east within the Dzhungarian Alataus and to the north of them. Formerly it was thought that the subsequent, Lower Permian, transgression occurred still further east over the Tarbagatais and Zaisan Depression penetrating into the Kalbinskian range, but the study of the fauna and flora has shown that this transgression was Namurian rather than Permian. In the east the Lower Permian sea had entered the Chikoi basin from the south. During the Upper Permian epoch all the Angara Geosyncline was land and continued to be so in the Trias and Jurassic periods.

The vulcanicity also weakened, but in the south-west certain volcanic groups of rocks are found. The climate was hot and tropical.

MESOZOIC AND CAINOZOIC

Mesozoic and Cainozoic rocks are closely related and consist basically of continental shale-sandstone sediments, which are friable, almost horizontal and thin. The Rhaetic-Jurassic Coal-bearing Formation is an exception, reaches over a thousand metres in thickness and is folded into gentle, but obvious folds which are faulted. while the rocks are metamorphosed.

Within the Palaeozoic and Precambrian massifs the Mesozoic and the Cainozoic occur as small patches in depressions and graben. Along the margins of the massifs the Neogene, Palaeogene and Cretaceous sediments are widespread and surround them. The complete absence of the Mesozoic from the Altais is interesting.

SUBDIVISIONS. The Lower Trias is normally not separated from the Upper Permian and what is called the Palaeozoic often includes the Lower Trias. The Middle Trias has not been found and is possibly missing—representing that epoch when the entire Angara Geosyncline was a highland area subject to erosion. A greater part of the Upper Trias is also missing and the Mesozoic succession begins with the Rhaetic.

The Mesozoic and Cainozoic can be divided into the following complexes: (1) Rhaetic-Jurassic coal-bearing formations; (2) Upper Jurassic-Lower Cretaceous continental deposits; (3) Upper Cretaceous-Palaeogene continental deposits; (4) Neogene-Anthropogene continental deposits; (5) Mesozoic and Palaeogene marine deposits. The boundaries between these complexes coincide with sharp lithological and floral changes, which can be traced without difficulty. On the other hand the boundaries between the Mesozoic and Cainozoic and between the Neogene and Anthropogene sediments occur within homogeneous complexes, are indistinct, and are arbitrarily drawn along some lithological horizon, which cannot be correlated from one area to another.

LITHOLOGY AND FACIES. The Rhaetic-Jurassic coal-bearing formations are found throughout the geosyncline but only in isolated small areas of different ages from west to east. In the west they are Rhaetic, while in Kazakhstan and Kuzbas they are Lower Jurassic and in the east Middle Jurassic. The lithology and general relationships are everywhere the same. The strata are relatively thick (up to 1500m). At the bottom they are the conglomerates and sandstones typical of areas adjacent to rising mountains. The upper, coal-bearing, part consists of sandstones and shales which are the deposits of flat intramontaine plains with marshes, lakes and rivers. The uniform, grey brownish and dark colours of the rocks and signs of abundant vegetation suggest a moist temperate climate (Fig. 146).

In the Turgai Depression the coal-bearing formation of the coal-field is 600–800m thick and fills several depressions in the Palaeozoic basement. The formation is Lower or Middle Jurassic and is known as the Uboganian Series, being divided into four suites. In Karaganda and Kuzbas the Jurassic coal-bearing formation is unconformable on Palaeozoic coal-measures, suggesting a repetition of favourable conditions for accumulation of coal. A rainy climate over large swamped depressions is suggested. The Jurassic succession in Karaganda, according to T. K. Kyshev, is as follows:

The Upper Coal-bearing Suite (Mikhailovkian) consists of argillites with occasional beds of sandstones and thick lenses of brown coal. The age is Upper Middle Jurassic and the thickness 80–150m.

The Upper Conglomeratic-Sandy Suite consists of sandstone with occasional beds of conglomerates and argillites with thin coal seams. The age is Upper Lias-Lower Middle Jurassic and the thickness is 100–200m.

The Lower Coal-bearing Suite (Dubovkian) of argillites, sandstones with coal seams. The age is Lower Liassic and the thickness 30–100m.

The Lower Conglomeratic-Sandy Suite of conglomerates and sandstones interbedded with argillites. The age is Rhaetic-Lower Liassic and the thickness 40–60m.

Around the upper Yenisei in the Tuva A.S.S.R., near the town Kyzil there is the considerable Ulukhemskian coal basin. The basal part of its succession has workable coals (3 to 9) of Upper Palaeozoic age, but industrially the higher Jurassic coal-bearing series is more important. It is 1500m thick and is divided into four suites.

The lower, Elegestian Suite (up to 120m) has about 70 per cent of sandstones and a Lower Jurassic flora. This suite is overlain by the Erbekian Suite (230–585m) with 3–5 workable coals, and consisting of sandstones and conglomerates (up to 200m) at the bottom and sandstones, argillites and coal seams at the top. The third or Saldamskian Suite (750m) is mainly fine-grained, has many thin seams and 2–3 workable coal seams. The uppermost or Bomskian Suite (315m) has no coals and consists of sandstones (67 per cent) and silts (31 per cent). The Erbekian and the Saldamskian suites have a Middle Jurassic flora. The Bomskian Suite is barren, lithologically different and is possibly not Jurassic (Losev, 1955).

In Transbaikalia the Jurassic and Cretaceous freshwater molluscs have been studied by G. G. Martinson (1956). He has proved the presence of several faunal complexes, and distinguished a series of stratigraphical horizons. The lowermost has been called the Bukachachinskian Suite, which is found in Eastern and Western Transbaikalia (Dzhidinian Depression), but is less well developed in the lake Gusino Depression and is missing in the Vitim and Aldan highlands. The next or Ulanganginian Suite is overlain by the Turginskian and Dainskian suites. The first two suites (Lower Cretaceous) are widespread throughout Western and Eastern Transbaikalia, while the third (Upper Cretaceous?) has not been well defined and is

FIG. 146. Jurassic coal-bearing strata, seam Nadezhdyi, in a quarry of the Saryadyr deposit in the Yerementau region of Central Kazakhstan. The gently dipping coal seam is in the middle of the photograph while in the distance typical small sopka topography can be noticed. *Photograph by A. S. Kumpin.*

only found in a few troughs of Eastern Transbaikalia. Martinson's work is interesting since it shows that even continental deposits can be subdivided on the basis of molluscs. Phyllopods, spores and plant remains are also useful in this respect.

Upper Jurassic-Lower Cretaceous continental sediments. In the west this is a relatively thin group of strata (50–60m to 100–150m) of sands, clays and, at the base, occasional conglomerates and breccias. The strata often lie on the weathered surface of the Palaeozoic. This weathered surface involves kaolinitic, striped clays, sands and even rare breccias. The clays and sands sometimes show the bedding and veining of Palaeozoic sediments. The age of the crust of weathering is provisionally accepted as Mesozoic.

The Lower Cretaceous, as on the eastern slope of the Urals, is characterized by a development of lacustrine and swamp deposits, which are variegated, bauxitic and black lignitic clays and variegated sands. In some regions of Kazakhstan and Cissalairia amongst the lacustrine and swamp clays there are lenses of gibbsitic bauxites and bauxitic rocks. The age of the Lower Cretaceous is determined on the basis of plant remains, pollen and freshwater molluscs. It is separated from the Rhaetic-Jurassic coal-bearing formations by an angular unconformity, is not metamorphosed, is much thinner, and lies horizontally.

In the east, in Western Transbaikalia the Upper Jurassic and Lower Cretaceous succession changes, its thickness increases and becomes coal-bearing in the Lower Cretaceous as well as in the Upper Jurassic. As an example the lake Gusino coalfield in the Selenga basin to the south of Ulan-Ude can be quoted. Here the coal-bearing formation is more than 2000m thick. It was formerly thought that its lower part is Jurassic. According to V. M. Skoblo and G. A. Dmitriev the formation as a whole is Lower Cretaceous, as is indicated by the mollusc-dinosaur fauna. Although this is possible it has not as yet been proved. The lake Gusino Coal-bearing Formation starts with a suite (150m) of conglomerates and sandstones at the bottom, and is succeeded by the coal-free silty-shaley suite (200m). The main part of the formation consists of a repetitive series of brown and grey silts, shales and sandstones, which are often arkosic. At the top the sandstones predominate and there are a few coal seams.

Upper Cretaceous-Palaeogene continental sediments. During this epoch the sea approached the Palaeozoic massifs, margining them from north and west (Lavrov, 1957). On transition into the massifs the marine deposits gradually change into continental near-shore deposits. The latter have no marine fauna and contain bones of land animals.

The sequence of the Upper Cretaceous, Palaeogene, Neogene and Anthropogene continental deposits is well exposed in the southern part of the Turgai Depression and Golodnaya Steppe (Betpak-Dala) to the east of it. K. V. Nikiforova (1960) has studied and described this sequence in detail.

The Upper Cretaceous is represented by grey and reddish sands and clays of a total thickness of about 100m. It is divided into three suites. The presence of broad-leaved subtropical plants and dinosaur bones indicates a warm and damp climate. The subsequent series includes a series of suites consisting of bedded clays and rarer sands with marine faunas. These are the Tasaranian, Saksaulian and Cheganian suites of Eocene and Lower Oligocene age. Their thickness varies from 60m to 150m and they have been described in the chapter on the Northern Siberian Lowlands. To these corresponds the continental Amangel'dynian bauxite formation found in the east.

The following continental series belongs to the Middle and Upper Oligocene and is known as Turgaiian. Formerly it was called Indricotherian owing to the discovery of the bones of gigantic hornless rhinoceros *Indricotherium* in it. The flora indicates a wet and warm climate of the type found in near-shore plains.

The Aralian Series (20–30m) belongs to the Lower and Middle Miocene and consists of several suites. Lacustrine clays and alluvial sands constitute the dominant lithology. The fauna is of forest type, tropical with rhinoceros, tortoise and mastodons being the main representatives.

The following faunal complex, which is widespread and well represented consists of *Hipparion* which lived in wooded steppes with a colder and dryer climate. The fauna with *Hipparion* appears in the Sarmatian, reaches a peak of development in the Maeotian and dies out in the Pontian. In Kazakhstan its equivalent is represented by the fauna of the Pavlodarian Suite consisting of sands and clays, with a total thickness of 120m.

The Aralian and Pavlodarian Series constitute the Neogene system. All the overlying deposits are Anthropogene. The Anthropogene is divided into Pliocene or Eopleistocene, Pleistocene and Holocene. The Pliocene is divided into Akchagylian, Apsheronian and Bakuian stages, while the Pleistocene consists of Khozarian and Khvalynian stages. The Holocene is not differentiated.

Specific faunal complexes are found in Pliocene stages, thus the Villafranchian (Khaprovian in the U.S.S.R.) is found in Akchagylian; Tamanian is found in Apsheronian, Tiraspolian is found in Bakuian. The Pleistocene Khazarian Stage has a Khazarian mammalian fauna and the Kvalynian Stage has the Upper Palaeolithic mammoth fauna. The Holocene fauna is similar to the present, but suggests warmer conditions. In the region under consideration the Akchagylian Stage embraces most of the Kenshagyrian Suite (7–8m) of red sandstones and gravels. The suite also includes the strata of Apsheronian age. The Bakuian Stage is represented by the Verkhnegobiian Suite of conglomerates (5m) and the third terrace of the river Chu. The Pliocene deposits suggest colder climate, but not glaciation. The Pleistocene, on the other hand, shows all the features suggesting a nearby glaciation and a sharp drop in temperature. To the Khazarian Stage belong the second terraces and to the Khvalynian the yellow-brown loams of the first terrace. The Holocene includes all the terrace deposits and other recent sediments formed under the conditions of cold arid climate. Nikiforova in her monograph gives the ages of Pliocene and Pleistocene sediments of other regions of the Angara Geosyncline, the Urals and the Western Siberian Lowlands. The subdivision of the Neogene and Anthropogene, as suggested by Nikiforova, has been worked out in detail by V. I. Gromov. This scheme is widely accepted, but has not been recommended at a national conference. Thus here the Pliocene is included in the Neogene, although it cannot be separated from Pleistocene and Holocene.

The continental Palaeogene embraces the rocks of the Semipalatinsk region. At the bottom of the succession, overlying the weathered Palaeozoic rocks, lie breccias, conglomerates and coarse sandstones which upwards change into light-grey quartz-bearing and brownish-yellow ferruginous siliceous sandstones and clays of a total thickness of about 100m. These strata are thought to be Eocene and they are overlain by barren red and greenish-grey, in places gypsiferous, clays of about 100m in thickness. These latter strata are provisionally included in the Oligocene. Both formations have been considered marine, but a complete absence of marine fauna in them proves their terrestrial origin. These are typical deposits of near-shore plains, inclined towards the north and with numerous bitter-salt lakes. The analogous, barren, continental variegated sands and clays can be found along the northern margin of the geosycline, between the Kuznetskian Alataus and the Yenisei. These rocks contain lenses of bauxites and bauxitic rocks of Palaeogene age. In the interior of the geosyncline the continental Upper Cretaceous and Palaeogene deposits occupy large areas of the lake Tengiz depression in Kazakhstan and lake Zaisan depression in the Altais and the valley of the Ilii. There are also small outcrops elsewhere. Amongst these rocks variegated clays, which may be gypsiferous or carbonaceous, predominate, but sands and cherty sandstones are also common. Their thickness is 40–50m and the absence of a fauna complicates their dating. For the Tengiz and Zaisan rocks the Palaeogene age is normally accepted, but the discovery of dinosaur bones in the analogous strata of the Dzhungarian Depression suggests a possible Upper Cretaceous age for their lower horizons. The continental origin of all these sediments is not in doubt. In separate depressions the thickness of the Palaeogene sharply increases to 800–1000m (see p. 530).

Neogene and Anthropogene continental sediments. This last complex of friable sediments lies horizontally and has a widespread distribution. Its succession has already been described (p. 439). On synoptic geological maps these strata are not shown. Their thickness normally does not exceed a few hundred metres, but can reach a few thousand, as in the deposits accumulated at the foot of young ranges (Kirgiz range—over 2000m) and in lake depressions (lake Zaisan—1190m).

The facies of these sediments are very similar to those of the present day with lacustrine deposits, bog sediments, fluviatile deposits and those of wide watersheds being often found. Near to mountain ranges loess and loessic sediments, gravels and sands of alluvial fans are common. In mountainous regions there are the products

FIG. 147.   Glaciated landscape of Mt. Ya. S. Edelshtein in the Eastern Sayans. Arrêtes, a corrie, corrie steps and a glacial lake are seen. *Photograph by A. G. Vologdin.*

of glacial and fluvioglacial erosion consisting of unsorted, poorly rounded, unbedded masses of fragments, pebbles and sands (Fig. 147).

As an example of the lowland successions, that near Pavlodar (Irtysh) recorded by Nikiforova can be quoted (Trudy . . ., 1957; Nikiforova, 1960). Here the Lower Oligocene green paper shales of the Cheganian Stage and near-shore marine origin are overlain by thick continental Middle and Upper Oligocene sediments (Turgaiian Series), divided into four suites. The series consists of grey, greenish and chocolate coloured clays interbedded with sands at the bottom, and sands of gravels with lenses of clays at the top. The series is overlain by the Lower Miocene grey and greenish calcareous clays and marls distinguished as the Aralian Suite (20–70m). This suite is widespread over the whole Northern Cisaralia and contains bones of Mastodon, rhinoceros, bears, *Hipparion*, and tortoises, which lived in forests interspersed with steppes with numerous lakes.

Higher up lies the Pavlodarian Suite (30–80m), which belongs to the Upper Miocene and Lower Oligocene. It is known to have rich accumulations of vertebrate bones. At the base of the Pavlodarian Suite lies a formation of fluviatile, current-bedded, white and rusty sands covered by grey unbedded fine grained sands with accumulations of bones. At the time of their formation these sediments were in the nature of quicksands, sucking in animals such as *Hipparion*, giraffes, gazelles and other herbivorous, forest-steppe creatures. Of carnivorous animals the bones of sabre-tooth tigers have been found. The bone-bearing sands are overlain by barren clays which are grey, bedded and lacustrine at the bottom and red and brown un-bedded, sandy and land deposits at the top.

The succession ends with the Middle Pliocene and Upper Pliocene and Anthropogene loams and sands amongst which there are found bones of *Elasmotherium* and horses of varieties found in steppes. The Miocene forest was succeeded by the steppe which still exists at present, and has been recently ploughed. The Neogene and Palaeogene succession of the intramontaine Zaisan Depression has been pierced by a stratgraphical reference borehole sunk 20km to the north of the town Zaisan (Vasilenko, 1956):

Anthropogene
    Depth=0–90m sands and clays.

Pliocene
    Reddish formation (Karabulakian Suite) with *Hypparion*. Depth=90–223m.
    Clay formation (Kalmakpaiian Suite). Depth=223–718m.
    The strata are grey sandy clays, silts and fine-grained sandstones and clayey silts with indistinct bedding and calcareous nodules, which have no organic remains. It is possible that these rocks represent fossil loess.

Miocene
    Sandy formation (Sarybulakian Suite). Depth=718–940m. The rocks are grey sandy clays, silts and fine-grained sandstones. At the top there is 60 per cent of sandy rocks and at the bottom 30 per cent. There are almost no organic remains.
    Pale-grey and greenish clays and silts (Akzharian Suite). Depth=940–1088m. There is an abundant freshwater micro-fauna of ostracods, diatoms and *chara*.
    Grey sands and clays (Nurinskian Suite). Depth=1088–1190m. Ostracods. fine bones of fish, spores and pollen have been found.

Oligocene

Green, bluish and red clays (Ashutasskian Suite). Depth=1190–1373m. The clays at the top are calcareous with fish and molluscs and Upper Oligocene plant remains at the bottom.

Variegated clays (Terektinian Suite). Depth=1373–1573m. Red, green and brown clays of this suite are gypsiferous in the middle and sandy at the bottom. At the base there are friable sandstones and gravels.

Middle Carboniferous (?)

Highly altered andesitic basalts.

Geophysical investigations have shown that in other regions of the middle of the depression the thickness of the Cainozoic reaches 1600m, but towards the margins the thickness decreases to 200–300m. Such great thicknesses although exceptional can be found in other depressions with a sinking bottom and considerable influx of terrigenous materials. This confirms the data of V. K. Vasilenko (1956), who has worked on the data from the Zaisan and Ilii depressions. The latter depression at the depth of 1200m has Upper Miocene strata, rendering the projected thickness of the Cainozoic as about 2000m.

It was proved later that in other parts of the Zaisan Depression the thickness of individual suites changes and a lower suite of Eocene-Palaeocene age (200m) appears at the bottom of the succession. The Anthropogene stratigraphy of Central and Southern Kazakhstan has been worked out by N. N. Kostenko (1960), who together with V. S. Bazhanov has correlated it with the Anthropogene deposits of Kazakhstan and analogous deposits of other Asiatic and European countries. The mammalian faunas have been described by B. S. Kozhamkulova (Materials . . ., 1960).

There follows the Anthropogene succession of Central and Southern Kazakhstan:

Present-day Stage (*Recent*)

Various deposits with a present-day fauna (Holocene). It is thin.

Upper Stage (*Riss and Wurm*)

Sediments of various origins and thickness. Near mountains there are young alluvial cones. In the mountains there are the last moraines and in the plains the first terrace. The fauna consists of mammoth, woolly rhinoceros and bison. At this time the last glaciation occurred in the mountains, while the plains acquired a colder climate.

Middle Stage (*Mindel-Riss and Riss*)

The mountains show moraines. Coarse clastic sediments and loess is found near to the mountains, and the plains show the second terrace. The fauna is Khazarian with bones of trogontheres, camels, bison, oxen, reindeer. The thickness of strata in the mountains is 180–200m and in the lowlands 20m.

Lower Stage (*Pre-Mindel and Mindel*)

Upper Horizon. In the mountains the greatest glaciation produced moraines and flavioglacial deposits. Near the mountains there is 400–500m of loess. In the plains the steppe with woods and terraces (up to 20m). The fauna is Koshkurganian (Tiraspolian) with the giant camel, *Elasmotherium*, saiga, bison and rhinoceros.

Lower Horizon (Akchagylian-Apsheronian). In the mountains the first glaciation occurs and molasse-type deposits 1000–1500m thick appear. In the plains the wooded steppes were widespread and terraces (30–40m) are found. The fauna is Iliian (Villafranchian), consisting of mastodons, mountain elephants, gazelles, ostrich, and two-horn rhinoceros. The climate was warm.

Neogene

Is represented by the deposits of plains and sub-mountain regions. The fauna is of hipparions of Lower Pliocene age.

The work of N. I. Kostenko is important and interesting. He, however, erroneously places episodes of dislocation and erosion at the boundary of successive epochs. This is wrong since such phenomena occur at the end of one or the beginning of another epoch.

The problem of successive glaciations in mountain regions remains unresolved. In the plains the continental ice-sheets advanced and receded and glacial epochs are distinguished by intervening interglacial sediments. In the mountains the problem is difficult since it is improbable that mountains of thousands of metres high were ever entirely free from snow and ice.

The Anthropogene of the Altais has been described in detail by E. N. Shchukina (1960). The correlation of various districts of the Altais (plains and mountains) amongst themselves and the rest of the U.S.S.R. is of some considerable importance. The comprehensive paper of L. D. Shorygina (1960) dealing with the Cainozoic of western Tuva, also deserves attention. Her photographs are especially valuable and the data on glacial deposits are interesting. These papers together with that by K. V. Nikiforova and others are published in the book, *Stratigraphy of Quaternary (Anthropogenic) deposits of the Asiatic part of the U.S.S.R. and their correlation with European deposits* (1960), edited by V. I. Gromov. This book has many references and correlation tables.

In the eastern part of the Angara Geosyncline the Upper Cretaceous and Palaeogene are continental, platform-type, limited in extent and have not been much studied. The Neogene and Anthropogene are more widespread, but again have not been widely investigated.

For instance in Western Transbaikalia the Buryatian geological office (Theses . . ., Chita 1961, Bazarov) has only relatively lately established the following sequence:

The Palaeogene—consisting of breccias, gravels, sands and loams which are thin. The age has been established by pollen analysis. The flora of oak, maple, chestnut, *Gingko* and bog ferns has a warm aspect.

The Neogene—basically consists of valley gravels and sands (80–100m) with a flora of coniferous trees like the present-day taiga. In places the Pliocene consisting of red gruss, clays and silts (20–30m) can be distinguished. The age is based on a *Hipparion*-bearing fauna.

The Lower and Middle Pleistocene consists of thick (150–200m) alluvial sands and silts, with a *Trogontherium*-bearing fauna at the bottom.

The Upper Pleistocene is represented by terraces (12 and 20m) of sandy gruss proluvial deposits and loessic loams (10–15m). The fauna of mammoth, woolly rhinoceros, horses and reindeer has a cold aspect.

The Holocene—consists of various deposits (30–40m) with a present-day fauna and fossil man.

The Anthropogene succession of Tunkin Depression (west of lake Baikal) is typical of a young intramontaine trough, by being thick and having glacial deposits associated with a group of basalts.

The Lower Pleistocene according to E. I. Raevskii (Theses . . . Chita, 1961) consists of thick (500m) sands and gravels at the bottom, overlain by a tuffogenous formation (80–100), succeeded by clays and sands (20m).

The Middle and Upper Pleistocene (500m) has four horizons of moraines interbedded with interglacial sands and clays with mammoth.

The Holocene consists of sands and clays of smaller thickness and a modern fauna.

Cainozoic strata of the Barguzin valley are also very thick (p. 246) *Marine Mesozoic and Palaeogene deposits.* These sediments are represented by two complexes: the Upper Jurassic-Lower Cretaceous and the Upper Cretaceous-Palaeogene. The first complex is only found in the north-west part of the geosyncline. The second complex is more widespread along the northern margin of the geosyncline, from the valley of the Ob' to the west, across the Turgai Depression (straits) and along the Chu valley stretching far to the south. The successions of these complexes have been given in the descriptions of the Western Siberian Lowlands where they are characterized by a great development (Table 25). Grey, dark-grey and greenish sands with marine faunas are mainly encountered, although there are also glauconitic facies and silicified clays and sandstones.

FAUNA AND FLORA. Amongst continental facies the often well preserved plant remains predominate. Their study has been advanced by A. N. Krishtofovich, V. D. Prinada, I. V. Palyubin, V. A. Vakhromeyev and other palaeobotanists, but much work remains to be done, especially on comparative biostratigraphy. Various horizons at present could be either Palaeogene or Lower Cretaceous, Middle Jurassic or Rhaetic etc. The investigation of the spores and pollen is of importance.

The fauna is represented most frequently by lamellibranchs, gastropods and ostracods, with rarer fish bones or even rarer land vertebrate bones.

PALAEOGEOGRAPHY. The strong manifestation of the Hercynian Orogeny at the end of the Permian caused the elevation of the geosyncline and the formation of the continent Angarida. In the Upper Permian and Lower Trias epochs quite high mountain ranges were separated by vast desert valleys. In the Middle Trias and Upper Trias epochs the surface of Angarida was gradually levelled into a small-sopka peneplain.

In the Rhaetic the first mountain-building movements of the Cimmeridian Orogeny occur in the west. At the foot of rising ranges conglomerates and sandstones accumulated forming large alluvial fans. At the end of this epoch the movements relaxed and the intermontaine valleys had a vast network of swamps connected by slowly-flowing rivers. The remaining larger part of the geosyncline remained a small-sopka plain.

In the Lower Jurassic the mountain-building movements spread their influence to the east, including the Kuzbas, and causing the growth of high ranges with a north-westerly strike. The valleys between these ranges were covered by rivers, swamps and lakes overgrown by a luxuriant vegetation. In the Middle Jurassic the same phenomena spread into the eastern part of the geosyncline and in the Upper Jurassic-Lower Cretaceous epoch into Western Siberia and the Far East.

The Upper Jurassic transgression, which advanced from the north, reached only the north-western margin of the geosyncline. Upper Cretaceous transgressions, advancing from the north, west and south-west penetrated much further into the geosyncline, at times overrunning the Turgai Depression. Palaeogene seas were even more widespread, but did not reach further than the margins of Palaeozoic massifs. The greater, central part of the old geosyncline remained dry from the Permian until now. This landmass can be called the young nucleus of Asia comparing it with the ancient nucleus situated to the east.

TECTONICS

The Angara Geosyncline is a typical Palaeozoic structure, showing strong manifestations of Precambrian and Palaeozoic orogenies. The Hercynian Orogeny caused its filling and conversion into a mountainous fold-region. Only the weak initial phases of the Cimmeridian Orogeny occurring in the Rhaetic and Lower Jurassic have affected this old geosyncline. Block fault-movements are typical of Neogene and Anthropogene times.

STRUCTURES

The following structural storeys can be distinguished: (1) Archeozoic and Lower Proterozoic; (2) Upper Proterozoic and Lower Palaeozoic; (3) Middle Palaeozoic; (4) Upper Palaeozoic and Lower Trias; (5) Rhaetic and Lower and Middle Jurassic; (6) Upper Jurassic, Cretaceous, and Cainozoic.

ARCHEOZOIC AND LOWER PROTEROZOIC. The Archeozoic and Lower Proterozoic strata are found in central cores of large anticlinoria, shown on the 1953 tectonic map edited by N. S. Shatskii. These structures are largest and most numerous in the east of the geosyncline, in the Kuznetskian Alataus and in the north and west of Central Kazakhstan. This structural storey is everywhere equally dislocated and consists of deformed and faulted isoclinal folds, complicated by granitic massifs. The rocks are strongly metamorphosed and often recrystallized and the description of isolated structures is difficult.

UPPER PROTEROZOIC AND LOWER PALAEOZOIC. The second structural storey also involves deformed and faulted folds and numerous intrusions. The grade of metamorphism is again high, but it is less than the metamorphism of the first structural storey,* which is always separated from the second by angular unconformities.

The thick calcareous formations of Upper Proterozoic and Lower Cambrian age are much less violently folded and are simpler, since these rocks do not react to fold-forming movements.

Within the second structural storey there are two unconformities. The first separates the Sinian complex from the Lower Cambrian. The unconformity is not everywhere present and is of relatively minor significance. The second unconformity is between the Lower Middle Cambrian on the one hand and the Upper Cambrian or Ordovician on the other. This unconformity is more pronounced and is found

---

* *Editor's footnote*—storey in the sense of *étage* in French.

almost everywhere. It was formerly thought that in Kazakhstan the second unconformity did not exist but recently it has been proved to be present. The unconformity is attributed to the Salairian, first phase of the Caledonian Orogeny. The
strike of the structures of the first storey is not clear. The structures of the second
storey on the other hand, although not everywhere observed, coincide in trend with
Caledonian structures.

MIDDLE PALAEOZOIC. The Middle Palaeozoic is everywhere strongly deformed, metamorphosed and intruded, but the deformation is less intense than in
the second structural storey. Isoclinal folds are rare and found only in argillaceous
rocks. Normal anticlines and synclines predominate, but are often disrupted into
separate scales, overriding each other. Dips are high and may be inverted. In addition to the thrusts synchronous with folding the structures are broken up by later
normal faults. N. G. Kassin has shown a good example of the structures involved
in the third storey, in Central Kazakhstan (Fig. 148).

In the areas of thick bedded formations the folds become more complex (Fig.
143). They are also more complicated in the south of the geosyncline, as for instance
in the Altais and Rocky Shoria where the Middle Palaeozoic is represented by the
greatest thickness of strata. Conversely in the east and north, where the lower part
of the Middle Palaeozoic succession is missing and Middle Devonian and even
Lower Carboniferous beds lie on the intensively deformed and compacted Lower
Palaeozoic, the Middle Palaeozoic folds are much simpler.

The Lower and the Middle Palaeozoic successions are almost always separated by
unconformities. There are no obvious unconformities within the Middle Palaeozoic
successions, especially in complete sections of Hercynian type. The Caledonian
type successions show the Middle Devonian clearly unconformable on the Silurian.

UPPER PALAEOZOIC AND LOWER TRIAS. The structures of the fourth
storey are as a rule simpler than those of the third storey, are accompanied by lesser
degree of metamorphism and a smaller number of intrusive massifs. In the northern
and western parts of Kazakhstan they are especially simple and cannot be easily
distinguished from the structures belonging to the Lower Mesozoic (Fig. 149).
Nevertheless, the structures of the thick coal-bearing formations, such as the Balakhonkian of Kuzbas (Fig. 160), are complex. They are also complex in the south of
central geosynclinal zone where the Upper Palaeozoic succession is thick. The same
applies to the Zaisan Depression and the Kalbinskian range (Fig. 150).

A peculiar group of structures associated with the Hercynian Orogeny is large
zones of deformation and cataclasis stretching for hundreds of kilometres and being
many kilometres, or even tens of kilometres wide. The first of these has been found
long ago and studied by V. P. Nekhoroshev and is known as the Irtysh zone. It is
over 500km long and trends parallel to the Irtysh, cutting across the Ore Altais. Its
position is obvious on geological maps. Another zone has recently been shown to
pass to the west of lake Balkhash. Again it is of large dimensions. Both zones are
composed of a whole system of faults which are either parallel to each other or
converge. Some of these fractures become channels of penetration of magmas or ore-
bearing solutions. In the Altais these zones of brecciation are associated with commercial mineralization. The faults margin separate blocks and lenses of rocks of
differing ages. In the Altais the age of such blocks varies from the Lower Palaeozoic
to the Upper Palaeozoic. The rocks are strongly deformed and brecciated and are
often metamorphosed into crystalline schists.

Recently a third (Uspenskian) zone of complex faults has been described by

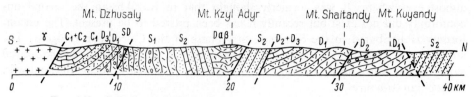

FIG. 148. Structures of Upper Palaeozoic rocks in the Bayan-Aula district
of Central Kazakhstan. According to N. G. Kassin (1941).

Suvorov (1961). It is situated in the region of the upper Sarysu, to the south of the
Uspenskian mine and has a south-westerly trend. It is almost parallel to, but somewhat different from the other deep faults.

It is thought that unconformities are most often found between the Lower and
Middle Carboniferous. Yet detailed investigations show that it is not so. A major
unconformity and associated lithological changes begin in the middle of the Middle
Carboniferous, consequent upon the deposition of Bashkirian strata. These strata
together with their choristids and a generally typical Middle Carboniferous fauna
are continuous downwards into strata belonging to the Namurian Stage.

There are sometimes local unconformities within the strata belonging to the
Upper Palaeozoic.

FIG. 149.

A brachyanticlinal fold of Middle and
Upper Carboniferous rocks of the
Teniz Depression in Central Kazakhstan. On the aerial photograph black
patches are lakes.

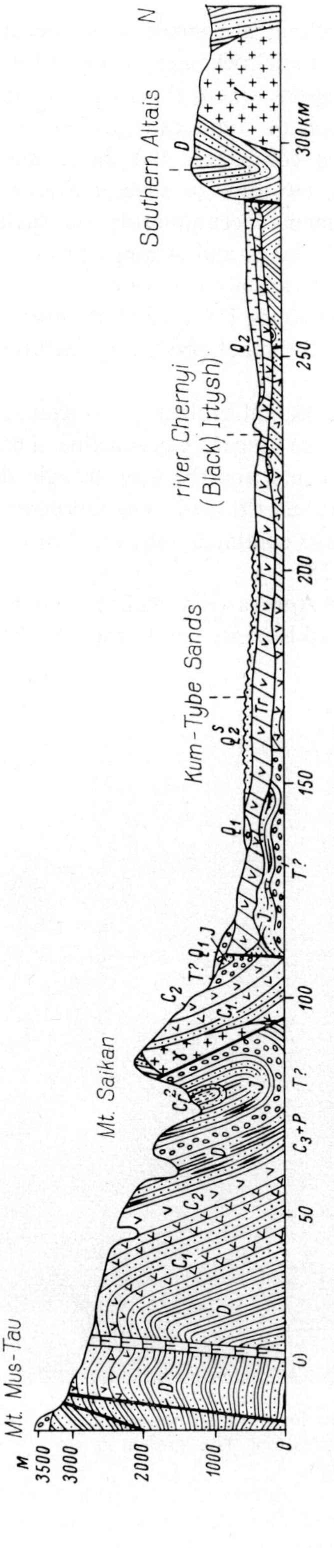

FIG. 150. A section across the Saur range, the Zaisan Depression and the southern Altais. According to V. P. Nekhoroshev (1941).

The unconformity, which is inaccurately referred to as that 'between the Palaeozoic and the Mesozoic', is found everywhere. The inaccuracy arises out of the fact that Upper Permian rocks are continental, almost barren sandy and clayey deposits. The study of their flora has shown that the upper horizons of these sediments are often Lower Triassic. Thus what geologists call 'the Mesozoic' in fact begins with the Upper Trias-Rhaetic. The whole of the Middle Trias and most of the Upper Trias are missing from the succession owing to the unconformity, which does not follow the boundary between the Permian and the Trias, but affects Triassic rocks.

RHAETIC, LOWER AND MIDDLE JURASSIC. The fifth structural storey is affected by a mild orogenesis. The constituent strata of the storey are always continental, often coal-bearing and are folded into regionally trending folds complicated by thrusts and normal faults. This proves the manifestation of the initial phases of the Cimmeridian Orogeny. The small dimensions of the folds (often 10–20°) without metamorphism and the absence of synchronous intrusive massifs indicate that the orogenic movements were weak. The unconformable Upper Jurassic and Lower Cretaceous sediments are flat and unmetamorphosed.

Rhaetic-Jurassic coal-bearing formations are often found over small areas. Only in Karaganda and the Kuzbas do they occupy relatively large areas, and the following have been investigated. In Kuzbas the Jurassic is affected by a series of gentle, small folds. Only in places do dips reach 45° and more normally they do not exceed 30° (Fig. 146).

UPPER JURASSIC, CRETACEOUS, AND CAINOZOIC. The last structural storey shows no regional folding, metamorphism, or synorogenic intrusive massifs. Sometimes deformed rocks and isolated folds of local development are detected. Such features are observed along fault lines, where the rising Palaeozoic massifs deform the adjacent soft sediments, producing complicated folds. Such folds disappear over a few kilometres and often over a much smaller distance

Local folds are also observed in ancient landslips and at the margins of gypsum massifs. There is also a possibility that isolated folds developed as glacial drag-folds.

All such folds, sometimes quite complex and faulted, are small scale, occur over small areas, pass out into undeformed strata and their rocks are unaffected by metamorphism. They occur as isolated features in regions where the rocks are in general horizontal. A completely horizontal disposition of strata can nowhere be seen, since continental deposits never accumulate on completely flat, horizontal surfaces. On alluvial fans, watersheds, inclined hills, etc. dips of several degrees reaching 20–30° are found. Consequently the sediments accumulating on such slopes have a primary dip of a few, or even 20–30°. In lakes and swamps primary dips are less (tens of minutes), but even here the sediments are not horizontal.

Young Faults. Throughout the geosyncline there are young Pliocene and Anthropogene faults (normal and reverse). They often affect the relief producing features such as Dzhungarian Gates (Fig. 151).

Baikalian Rift. The fault system, with which the Baikal graben is associated, continues to the south-west of it outside the bounds of the Angara Geosyncline. The valley of the Irkut filled by thick Anthropogenic sediments, and the deep trough of lake Kosogol situated in Mongolia, belong to the Baikalian rift zone. The Obruchev fault, which bounds the rift in the north and west, also continues into the Angara Geosyncline. The Baikal Rift has been described on p. 257.

CROSS-FOLDS. In the Urals and eastern part of Angara Geosyncline folds of different ages have the same trend. In the Urals the trend is north–south, while in the

FIG. 151. Dzhungurian Gates, representing a narrow long graben bound by young faults which have a clear topographic expression. The graben is several kilometres wide. *A drawing by A. N. Afonichev.*

Eastern and Western Sayans is is north-westerly. Younger folds inherit the trend of the older folds and as a consequence a so-called inherited folding originates.

In the central and western parts of the Angara Geosyncline a different situation arises. The older (Caledonian) and younger (Hercynian) folds are almost at right angles to each other. Cross-folds of this type have been especially fully studied in Central Kazakhstan where Caledonian folds have a north-easterly trend and Hercynian folds a south-westerly trend. The north-east trending folds affect the Lower Palaeozoic strata, while the north-west trending folds are found in the Devonian and Lower Carboniferous. A study of their relationships has shown that in those regions where north-easterly trending folds were formed these folds do not become unfolded by the later deformation, but become broken into blocks which rise or become depressed beneath the Middle Palaeozoic anticlines and synclines.

In the areas of well-developed Lower Palaeozoic and Precambrian, folds of north-eastern trend predominate. Such is the case in the north-eastern part of Central Kazakhstan. In the south of Central Kazakhstan where the Middle and Upper Palaeozoic sediments predominate the north-western trend is found (see Shatskii's tectonic map). The two main trends, however, are not unique. At the western margin of the Kazakh Steppe, in the Turgai Depression and the south-eastern margin of the Urals north–south structures, typical of the Urals, are found. This conclusion was first reached by N. G. Kassin. A tectonic map of Eastern Kazakhstan formulated by G. Ts. Medoyev in 1953 shows all the principal structures, their outlines and trends.

THE HISTORY OF TECTONIC MOVEMENTS

The main features of the area are caused by strong Precambrian and Palaeozoic orogenies, weak Cimmeridian Orogeny and considerable Neogenic and Anthropogenic block movements. There are no structures associated with the Alpine Orogeny.

Archeozoic

As everywhere the Archeozoic tectonics are complicated by a high degree of metamorphism and strong deformation. The outlines of Archeozoic continents and geosynclines remain as yet unsketched, thus not permitting the delineation of epeirogenic movements. The existence of thick Archeozoic marbles indicates prolonged, gradual movements during which the depth of the sea remained almost constant despite the rapid accumulation of carbonates.

There has been also an undoubted orogeny at the end of the Archeozoic, which has produced structures which are now unconformably overlain by the Lower Proterozoic. This orogeny was analogous to the Byelomorian of the Russian Platform.

Proterozoic

The Proterozoic is much more widespread and has been more investigated than the Archeozoic, but again the outlines of Proterozoic continents and geosynclines are not known. It is generally accepted that in the Lower Proterozoic the Angara Geosyncline was a single geosynclinal region, but this is not substantiated by evidence and it is more probable that within its bounds there were several small Archeozoic continental massifs bordered by relatively narrow geosynclines. The downward movements, in these basins, reached great magnitudes and lasted a long time. Figures of the order of 12–15km are often quoted. Great thicknesses (1500–2000m) of shallow-water, algal, bituminous limestones and dolomites could only have formed during slow balanced movements. There are no indications of contemporaneous mountain chains in the process of formation. Successions lack conglomerates or sandstones, which could have originated from such mountains. The suggestion that in the Lower Proterozoic and Archeozoic there were no mountain ranges deserves attention. Evidently high mountain chains originated at the end of the Proterozoic and have special characteristics.

The orogeny at the end of the Lower Proterozoic was powerful and very wide-spread. It can be called Karelidian after the analogous orogeny of the Russian Platform, since in the Urals and Kazakhstan it has not been given a special name. A number of investigators also distinguish a second Lower Proterozoic orogeny, which separates the lower volcanic group from the upper sedimentary. This orogeny remains unnamed and owing to its local manifestation and obscure age probably deservedly so.

The Baikalian Orogeny, which occurred at the end of the Upper Proterozoic, was widespread and powerful. It caused a great angular unconformity between the Proterozoic and the Cambrian, and also was associated with a fairly high degree of metamorphism and strong intrusive activity. This orogeny has been recognized in the northern and western parts of the geosyncline, but is not recognized in the south-east where thick carbonate strata of the Sinian complex are gradational into similarly thick Cambrian archeocyathid limestones. Some geologists consider both carbonate formations as Lower Cambrian, which is incorrect. Others adduce archeo-cyathid limestones to the Upper Proterozoic, which is even worse. The most rational approach, adopted by the third group, is to draw an arbitrary boundary under the lowest horizons with archeocyathids.

If the obscurity of the signs of the Baikalian Orogeny make stratigraphical studies difficult then, conversely, the shortcomings of stratigraphical investigations interfere with the recognition of this orogeny. If Upper Proterozoic strata are mistakenly included in the Lower Cambrian then manifestations of orogenic movements can be confused with the Salairian phase and the Karelidian Orogeny may be assumed to be Baikalian.

Palaeozoic

At the beginning of the Palaeozoic the whole area under consideration became a vast geosyncline, which in the north-west merged into the Uralian Geosyncline. In the north-west this geosyncline was adjacent to the Siberian Platform. The Caledonian Movements caused the southerly and westerly extension of the Siberian Platform, thus reducing the area of the Angara Geosyncline. The Hercynian Orogeny induced the filling of the Angara and Uralian geosynclines and connected the Baltic and Siberian continents into a single new continent of Angarida.

Palaeozoic orogenies produced considerable palaeogeographic changes, and caused the growth of tectonic structures and intrusive massifs, which are associated with the most widespread and important mineralizations.

CALEDONIAN OROGENY. The Caledonian Orogeny consists of three phases: the Salairian, the Taconic and the Erian, of which the first two were especially powerful. The Salairian phase is sometimes considered as a separate orogeny, but should be really considered as the first phase of the Caledonian Orogeny.

Salairian phase. This phase has acquired its name from the Salairian range where it was first recognized. It is demonstrated by the fact that the Upper Cambrian or Ordovician rests uncomformably and transgressively on older rocks, which is often seen in the Angara Geosyncline. Here successions with transitional Cambrian and Ordovician are rare and always some part of the succession is missing. The date of the phase is determined from the appearance of thick sequences (1000–1500m) of conglomerates and sandstones, which are typical sub-montaine deposits, in the succession. These conglomerates appear at the base of the Ordovician, and according to some investigators are partly Upper Cambrian. Consequently the Salairian phase is adduced to the beginning of the Ordovician and possibly the end of the Cambrian.

Taconic phase. This term refers to the movements which caused the uncomformity between the Ordovician and the Silurian. This unconformity is often associated with different structural trends across it and with intrusive activity. The movements occurred at the end of the Ordovician and possibly the beginning of the Silurian, were not contemporaneous from one place to another, and in places recurred several times.

The grey-green, pepper-graywacke sandstones of a few hundred metres in thickness suggest movements in the second half of the Ordovician. There are also flyschoid rocks which are rhythmically bedded. Both groups of rocks originated contemporaneously with orogenesis and mountain building. Molasse type deposits in thick formations appear in the succession at the beginning of the Silurian. These deposits again were contemporaneous with mountain building.

Erian phase. This phase is much weaker and in many regions cannot be recognized, since there Silurian and Devonian rocks are conformable. There are no contemporaneous rocks, which are normally associated with orogenic phases. The occurrence of the Erian phase is indicated by a widespread emergence which produced the Lower Devonian continent and the Kazakh isthmus. It is, however, possible that the emergence was caused by block movements.

The Caledonian Orogeny, in its totality, was the main orogeny in the south-east, north-west, and west of the Angara Geosyncline where it has produced great palaeogeographic changes, by producing new land-masses and a new structural trend. With the orogeny there are associated major, widespread intrusive massifs and some mineralization.

HERCYNIAN OROGENY. The movements began not at the boundary between the Lower and Middle Carboniferous, but at the end of the Bashkirian epoch. Thus the lower Middle Carboniferous is continuous with the Namurian, from which it is only distinguished faunally. Abrupt lithological changes and unconformities appear only after the Bashkirian. The first Hercynian orogenic movements are often called the Sudetic phase.

Sometimes the Hercynian Orogeny is thought to begin with the Devonian-Carboniferous unconformity, but this Bretonic phase did not occur either in the Angara Geosyncline or in the Urals of Central Asia. The observed depositional breaks, beds of conglomerates, breccias, and sandstones, were caused by epeirogenic movements and not by orogenic folding. As a result there are no widespread unconformities, changes in trends, effective metamorphism, or intrusive masses which could be associated with the Devonian-Carboniferous boundary.

In the Altais and the adjacent Rocky Shoria two local phases are recognized: the Tel'bessian (in the Givetian) and the Saurian (in the Middle Visean). These phases have been fully described by V. P. Nekhoroshev (1958).

Tel'bessian phase. This term refers to tectonic movements which had provoked the Givetian transgression and caused an abrupt change in sedimentation. Sometimes intrusions of ultrabasic, basic, and acid rocks are referred to this phase.

These movements, which cannot be denied, should not be called an orogenic phase, since they represent typical oscillatory movements, unassociated with angular unconformities, any metamorphism or synorogenic intrusions. The movements were of considerable magnitude and were associated with faults invaded by intrusions, which cannot be considered as synorogenic.

Saurian phase. This phase was recognized by Nekhoroshev (1958) in the Saurian range where it occurred in the Lower Visean epoch. The Saurian range is situated to the south from the Altais and is separated from this range by the Zaisan Depression. In the range the Lower Visean marine sediments are intruded by the granites of the Saurian complex, while the overlying continental sediments in places rest on the eroded surface of these granites and contain pebbles of them. The age of these sediments is in places determined as Upper Visean and consequently the movements are Middle Visean. These small granites, however, cannot be attributed to orogenic movements since: (1) there is no unconformity between the Lower and Upper Visean; (2) no differences in degree of metamorphism can be noticed between the Lower and Upper Visean; (3) there are no large synorogenic intrusions.

The Saurian movements thus are again oscillatory movements of large amplitude, but of small extent. The movements were accompanied by faults with intrusions along fault planes.

Sudetic phase. The Sudetic phase began in the Vereiian epoch and continued throughout the Middle Carboniferous, producing thick clastic formations which in the north and east were continental. This phase constitutes the most important and powerful manifestation of the Hercynian Orogeny, which affected the whole geosyncline and produced all major structures. The phase is accompanied by the largest and most widespread intrusions. The manifestations of this phase can be depicted in the cross-section across the Rocky Altais (Fig. 152), where the differences between the Upper and Middle Palaeozoic can be seen on the left of the diagram. After the Sudetian phase the whole Angara Geosyncline became land margining the continent Angarida. In the north and east it had a high relief, but in the rest of the region it was a plain tilted to the south and south-east. All later marine transgressions affected only this plain.

Uralian (Saalian) phase refers to those orogenic movements, which culminated in the Lower Permian. This phase after the Sudetic was the strongest and most widespread, especially affecting the southern part of the geosyncline.

As a result of the Uralian phase the Angara Geosyncline became a highland area into which the sea did not penetrate either in the Upper Permian or in the Mesozoic. The orogenic and mountain-building movements were accompanied by metamorphism and intrusion. The new 'Asiatic core' was formed at this time.

Pfalzian phase. The last manifestations of the Hercynian Orogeny occurred in the Upper Permian and Lower Trias and are collectively known as Pfalzian. These movements were much weaker than before. In the north and east the Pfalzian phase is almost undetectable and is most pronounced in the south of the geosyncline. In the Kuznetsk Basin this phase causes abrupt, regional, angular unconformities between the Upper Permian and Jurassic coal-bearing formations, as well as metamorphism and various structures. In the Semipalatinsk region the Semeitau intrusions, which invade Upper Palaeozoic strata, are associated with Pfalzian movements.

It should be remembered that some unconformities and breaks between the Mesozoic and the Palaeozoic are not Pfalzian, as it is sometimes imagined, but Uralian (Saalian). Only the unconformities separating the Upper Permian from the Mesozoic can be considered as Pfalzian. This phenomenon is rare since Upper Permian rocks have a limited distribution and are often missing altogether. In these regions the Uralian and even Sudetic movements may be responsible for the break between the Upper Palaeozoic and the Mesozoic.

It should also be borne in mind that the phases such as Sudetic, Uralian, and Pfalzian are somewhat arbitrary and cannot be adduced to narrow intervals of time or to individual unconformities. There are, in fact, many more local unconformities than phases. The phases thus embrace groups of movements, which occurred over long intervals of the Middle Carboniferous, Lower Permian, Upper Permian, and Lower Trias. Sometimes these phases are deliberately disregarded and terms such as 'fold-movements of the Middle Carboniferous times or 'the Upper Permian folding' are used. Such a terminology lacks emphasis on the periodicity of movements, which is stressed through the employment of the word 'phase'.

In sum total the Hercynian Orogeny has been most important in conditioning the palaeogeographical reconstruction of the Angara Geosyncline and by promoting the metamorphism of its rocks and the emplacement of intrusions and ore bodies. The orogeny has converted a mobile, geosynclinal zone into a part of a stationary and rigid continent.

OSCILLATORY PALAEOZOIC MOVEMENTS. Long and slow downward movements which were continuous over parts or even whole periods dominated throughout the Palaeozoic Era. During these movements the palaeogeographic conditions of the affected regions did not basically change, which indicates the balanced nature of such movements. Such balanced movements occurred on a very large scale of hundreds of metres and even kilometres. Thus the total thickness of Palaeozoic deposits reaches 10–15km.

The slow downward movements were interrupted by abrupt and short-lived oscillations, which disturbed the preceding palaeogeographic conditions. In particular

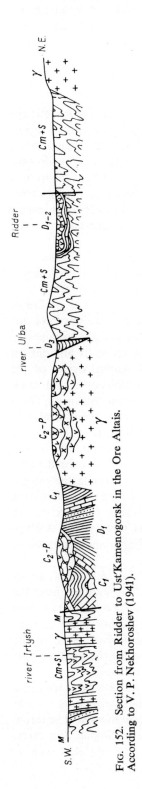

FIG. 152. Section from Ridder to Ust'Kamenogorsk in the Ore Altais. According to V. P. Nekhoroshev (1941).

rapid downward jerks initiated marine transgressions, as for instance the Givetian.

Large-scale downward movement often entailed large faults of hundreds and thousands of metres throw, hundreds of kilometres long and tens of kilometres deep. These faults often acted as channels of penetration of basaltic magma, giving rise to large intrusions.

Mesozoic and Cainozoic

CIMMERIDIAN OROGENY. The Cimmeridian Orogeny affects the Rhaetic and Jurassic coal-bearing formations. This orogeny produced simple and gentle, regionally distributed folds, which are often complicated by thrusts and normal faults. All these structures are most widespread in Karaganda and Kuzbas. The orogeny commenced in the Rhaetic with the growth of considerable mountains at the front of which thick conglomerates and sandstones underlie the coal-bearing formation proper. The orogeny ended in the Middle Jurassic, and Upper Jurassic rocks remain unfolded. In this area the orogeny was weak and was not accompanied by metamorphism or intrusive massifs.

BLOCK MOVEMENTS. Beginning from the Upper Jurassic onwards the oscillatory block movements were the only movements which affected the Angara Geosyncline.

The small thickness and the dominantly fine-grained composition of Upper Jurassic, Cretaceous, and Palaeogene clastic rocks indicate that at this time the upward and downward movements of the separate Palaeozoic and Precambrian massifs did not extend for more than a few hundred of metres vertically and did not spread over large areas. But in the Upper Pliocene and Anthropogene of the southern margin of the geosyncline great thicknesses of conglomerates and sandstones appear, reaching hundreds (even 1500–2000) of metres in thickness. These deposits continue for many hundreds and thousands of kilometres, lying parallel to the fronts of the present-day mountain chains. These deposits indicate that the mountains in question existed at the end of the Pliocene and continued to exist throughout the Anthropogene period. This situation could only have existed under the conditions of uninterrupted rise of the Palaeozoic massifs forming the ranges. The age of the sediments—mollasse formation—determines the date of the initiation and the duration of the movements. In the south of the Angara Geosyncline the movements began at the end of the Pliocene and continued through the Anthropogene. It is possible that in some places block movements began and continued in the Anthropogene, and may even continue at the present time.

Turkestan-Sayan belt of block faulting. A vast mountain region, which is elongated along the southern margin of the Angara Geosyncline, stretches from the valley of the Syr Dar'ya (town Turkestan) in the west of the Cisbaikalia ranges in the east and includes a part of Central Asia. The structure of the belt is the same as in the north of the geosyncline as it is obvious in synoptic geological maps. At the same time the northern part is a plain or small-sopka region, while the southern part is mountainous with snow-capped peaks and well-developed glaciation.

The problem posed by these facts has been solved by V. A. Obruchev, who in a number of papers has demonstrated that the mountain building is caused by vertical or almost vertical movements of the Palaeozoic and Precambrian massifs. The throw involved reaches 5–6km and in isolated instances is even more. Similar, but rarer, downward movements also occur, leading to the formation of deep lakes such as Baikal, Kosogol, Teletskian, and Issyk-Kul.

In many regions it has been established that the magnitude of the throw involved in block faulting varies from one place to another, often large massifs 200–300km long and more than tens of kilometres wide rise by about 5–6km, thus forming large mountain chains. These massifs arise in an *en échelon* pattern and collectively stretch for thousands of kilometres giving rise to the Tian Shians, the Altais and the Sayans. No less frequently blocks of smaller dimensions arise as isolated masses with marginal throws of a few hundred metres to 1000–2000km. Very frequently the larger massifs are broken up into systems of blocks with differing magnitudes of movement involved. Investigations conducted in ore-bearing areas have shown that smaller blocks of a few hundred metres to a few kilometres across show highly differential and diverse amounts of movement; and that the ore bodies affected by such movements become torn into large fragments, which sometimes rise towards the surface and at other times become depressed by several or tens of metres.

The movements of the small blocks are undoubted, while the movements of larger blocks again are generally accepted, but require verification. The movements of the largest massifs, in principle not different, are as yet not universally agreed upon; nevertheless their occurrence cannot be denied. The most important proof consists of regions of elevated plains, which impart to the summits of the ranges an almost horizontal appearance. Such elevated plains are often found in the Tian Shian Mts, forming good grazing land called 'syrtas'. Sometimes continental deposits become raised to the top of mountains, but in other regions marine Upper Cretaceous and Palaeogene sediments can be found. For instance to the south-east of Tashkent in the Chatkal Mts, near the Chapcham Pass, there is a plateau covered by the Palaeogene marine sediments of a few tens of metres in thickness. The plateau is situated at a height of 3000m, while in adjacent regions the same deposits lie at a level of a few hundred metres only. These relationships, which are observed in many regions, can only be explained by block-faulting.

The uplifted and down-thrown massifs, blocks and fragments are bounded by faults of different throw and extent. Movements along them sometimes produce catastrophic earthquakes, such as the 1911 earthquake at Alma-Ata (Vernensk) and those in the Baikal region. There are cases when young movements follow the old Palaeozoic and Mesozoic fissures and a new fault-breccia comes into contact with an old fault-breccia. The fault planes are always nearly vertical, but often with some inclination to this or that direction. Where the faults are inclined under the rising blocks, thrusts originate and in association with the larger thrusts there arise regions of intensive marginal folding of the overlying, softer rocks. The simplest of these zones of deformation have minor folds with a few tens of degrees of dip on their limbs. Where the deformation is more intense the folds become isoclinal, over-turned and separated by thrust planes. The most complex forms arise when all the folds are broken up into isolated wedges, which override each other like small-scale nappes. In isolated cases fragments of Palaeozoic strata get involved. Even the most deformed regions, however, differ from those affected by orogenic movements. The rocks adjacent to faults are not metamorphosed, and not intruded, by synorogenic intrusions. Furthermore, the zones of deformation associated with faults are narrow (a few kilometres) and the folds die out away from the faults, and in any case the thrusts and associated folds are clearly related to the margins of the Palaeozoic massifs. Conversely the finding of such deformation zones indicates block movements, since the minor folds produced cannot be attributed to the large gentle arches.

MAGMATISM

The dominance of the Palaeozoic and Precambrian volcanic rocks is characteristic for the region. Mesozoic and Cainozoic igneous rocks are rare, found in limited areas and consist of lavas and very occasionally the intrusions of feeding channels.

ARCHEOZOIC MAGMATIC CYCLES. These are fully developed, but only locally found, since Archeozoic rocks can only be found in isolated areas. The most complete data are available for Central Kazakhstan and the Rocky Shorias and adjacent regions.

In Kazakhstan the intrusions are represented by orthogneisses, metamorphosed and deformed granites, amphibolites, diorites, and gneissose granite porphyries. The lavas are metamorphosed into amphibolites and gneisses.

In the Rocky Shorias and adjacent regions there are Archeozoic ortho-amphibolites (altered diabases) and gneisses with porphyroblasts of labradorite representing altered labradorite porphyries. The existence of other strongly altered and recrystallized effusives cannot be doubted. Intrusive rocks are more widespread and diverse and include orthogneisses, migmatized gneisses, microcline gneisses, granites and amphibolites. In many regions Archeozoic magma cycles have not been investigated and the presence of several independent cycles can be suspected.

LOWER PROTEROZOIC MAGMATIC CYCLES. These cycles again have not been sufficiently studied. The Karelidian cycle which was current at the end of the Lower Proterozoic can be recognized. The Lower Proterozoic lavas are less metamorphosed than the Archeozoic, although their conversion into gneisses and schists can be observed in zones of igneous contacts. The lavas retain indications of original structures. The intrusions, which are always strongly deformed and altered, also retain some primary structures. The gneissification of granites is widely observed although, unlike that in the Archeozoic, it is incomplete.

In the west (Central Kazakhstan) the Lower Proterozoic succession begins with a thick porphyroid formation (up to 2000m) consisting of porphyries and albitophyres, which are altered into porphyroids. At the top of this group there are diabasic porphyrites and tuffs. In the Kokchetavskian region there are amphibolites and alkaline gneisses. The Karelidian Orogeny is accompanied by intrusions of granite gneisses and porphyroid granites.

In the Altais, in the Katun basin, the Turgundinian intrusive complex serves as an example of Proterozoic intrusions. Its rocks consist of flesh-red and pink porphyroid granites and their associates—adamellites, granodiorites, diorites, and alkaline granites such as alaskites and plagiogranite porphyries.

In the western part of the Eastern Sayans at the base of the Proterozoic there are amphibolized gabbros and basic lavas and greenstones. Intrusions of microcline granites, which are also present, are associated with the Karelidian Orogeny.

UPPER PROTEROZOIC MAGMATIC CYCLE. The Upper Proterozoic cycle is much less widespread than the Lower Proterozoic. Upper Proterozoic igneous rocks are in places included in the Lower Palaeozoic, since the boundary between the two eras is debatable. In the cycle lavas are common and intrusives are relatively rare. In the east—the Rocky Shorias and in the west of the Eastern Sayans—there are basic lavas (diabase porphyrites) which are included in the Munzhian Formation, which is analogous to the Spilite-Keratophyre Formation of Kazakhstan. Intrusions are rare, but include trondhjemites and diorites.

CALEDONIA MAGMATIC CYCLE. The rocks of this cycle are widely found and especially so in the regions of strong manifestation of the Caledonian Orogeny. Mineralization, often in commercial quantities, is associated with this cycle. The cycle, like the Caledonian Orogeny, can be divided into three phases (stages). The first, or Salairian phase includes the Lower and Middle Cambrian lavas followed by the major Upper Cambrian-Lower Ordovician intrusions and Lower Ordovician minor intrusions which cut the granites. The second—Taconic—phase starts with Ordovician lavas and ends with the minor and major intrusions of the Upper Ordovician and lowermost Silurian times. The third phase (Erian) includes Silurian and Lower Devonian lavas and late Silurian intrusions.

Sometimes the Caledonian cycle is divided into: the Lower Palaeozoic and the Middle Palaeozoic. The former includes the Salairian and Taconic phases and the latter the Erian phase. This division in the last few years has acquired some popularity, but cannot be considered to be correct, since magmatic cycles should be dated on the basis of tectonics rather than stratigraphy.

In the West-Central Kazakhstan all three phases are well represented. Many investigators, including R. A. Borukaev, include the Spilite-Keratophyre Formation, which is 2000–3500m thick, in the Salairian phase. The rocks of the formation consist of diabase porphyries, plagioclase porphyries, albitophyres, and associated tuffs and spilites. The intrusive rocks are mainly basic and ultrabasic and have a limited distribution. The Upper Proterozoic age of this formation is possible, since it lacks a Cambrian fauna and is separated from the dated Cambrian by unconformities and breaks. This formation apart the Cambrian and Ordovician have other thick lava groups. Lower Palaeozoic lavas are rich in iron, calcium and sodium and consist mainly of submarine flows, which produce sheets some 10–50m thick and hundreds of square kilometres in extent. The lavas are interbedded with sediments and are characteristically represented by pillow-lavas and amygdaloidal varieties. Terrestrial volcanoes, which must have reached up to 2–3km in height, are rarer. Their rocks include lava flows, tuffs, volcanic breccias, volcanic bombs, and lapilli.

The composition of the lavas is variable. In places andesite-basalts are most common, while elsewhere the lavas are more basic. At the end of the cycle, in the Silurian, there are dominant acid lavas and tuffs including acid porphyrites and albitophyres. As a rule the Lower Palaeozoic lavas are altered into greenstones. Their augite and hornblende are changed into secondary green amphiboles, epidote and chlorite. Chloritization is in general most widespread. Consequently the effusive rocks often have a green or grey-green colour; the original dark brown or black colour is rare.

The volcanic activity continued throughout the Silurian, but was most vigorous at the end of it and at the beginning of the Devonian. Siluro-Devonian lavas are widespread throughout Eastern Kazakhstan—from the Ulu-Tau to the Altais. They reach 2500–3000m in thickness and are known sometimes as the 'Albitophyre Group'. In a majority of cases the group consists of sub-aerial flows, but rarer submarine lavas are also encountered. The rocks are dark grey or brown, but may also be greenish. The group is sometimes thought to be Lower and Middle Devonian in age. Great accumulations of pyroclastic debris such as tuffs, tuff-breccias, and agglomerates are typical. These rocks are found intercalated with either massive or vesicular and amygdaloidal lavas. In places the actual volcanoes with their necks filled with porphyritic rocks are found. The volcanoes occupy an area of 15–20km² and consist of diabasic and dioritic porphyrites and the associated volcanic ejecta.

The composition of the lavas is very diverse with andesites and rarer basalts and dacites appearing first and being followed by black variable porphyrites, quartz-free albitophyres and andesitic and oligoclase-bearing lavas. This sequence is not precisely the same everywhere.

In the Rocky Shorias, to the south of Kuzbas and near to the Kuznetskian Alataus there is a lava group which has been named by M. A. Usov as Tel'bessian which is also accompanied by various intrusions of the Caledonian cycle. The Tel'bessian iron mineralization is associated with these intrusions. The group in its composition and conditions of formation is identical to the Siluro-Devonian group of Kazakhstan. The dominant rocks are porphyrites and their tuffs and tuffites, which have bombs and lapilli in them. In places, basic and acid effusive rocks are inter-bedded with sedimentary rocks, such as sandstones and shales.

The association of lavas with siliceous and cherty rocks has been stressed by N. G. Kassin. Thus silicification is attributed to the chemical precipitation of silica of volcanic origin, formed at the expense of decomposition of volcanic ash.

Intrusives. The variety and abundance of lavas are paralleled by the variety and abundance of major and minor intrusions. P. A. Borukayev has proved that there were no less than five epochs of intrusive activity. Amongst these epochs the Salairian, Upper Cambrian intrusions are most important and are followed in importance by the Upper Ordovician and Silurian-Devonian major and minor intrusions.

Compared to the Caledonian intrusions of the Urals those of Kazakhstan have more granites and less ultrabasic varieties, so characteristic of the Urals. As a result platinum, chromite, and nickel ores, which are associated with the ultrabasic intru-sions of the Urals are rare in Kazakhstan. The Salairian intrusions are not numerous and consist of granites, porphyries, and rare ultrabasic rocks. The Taconic intrusions are more widespread and are composed of granodiorites, quartz-diorites, granites, syenites and gabbros (Fig. 153).

The Siluro-Devonian intrusions are most widespread and are found almost

Fig. 153. Caledonian. Lower Palaeozoic granites at lake Buibenskoye in the Western Sayans. *Photograph by Ya. S. Edel'shtein.*

everywhere. Acid intrusions—mainly granites—form huge massifs, with often steep sides, but sometimes they are stock-like bodies. It is possible that the stock-like masses represent volcanic necks or their channels along which lavas ascended. The basic peridotites, pyroxenites, and gabbros always occur in small stocks. The acid rocks are normal granites, granodiorites, diorites, and monzonites. They are rich in mafics and are darker than their Hercynian counterparts, but the colour alone cannot be used as an indication of the age of intrusive rocks.

The major intrusions are cut by minor intrusions and various veins with both of which gold, copper, and iron mineralization is associated.

The Caledonian magmatic cycle is even better represented in the east, but is of the same general type as in the west. The Lower Palaeozoic lavas are found everywhere and are thickest (over 2000m) in the eastern parts, such as the Sayans, the Kuznetskian Alataus, the Minusinskian Depression, and in the Salairs. They are less widespread in the Altais. The end of the Cambrian and the beginning of the Ordovician saw the climax of volcanic activity. In the Lower Cambrian, when the archeocyathid limestones were developing, the volcanicity was weak. It should, however, be borne in mind that thick, so-called Lower Cambrian, volcanic groups may often be Upper Proterozoic, and are overlain by the archeocyathid limestones. Basic lavas such as porphyrites and albitophyres, as well as the corresponding tuffs, tuffites and tuff-breccias are most common, but acid quartz-bearing and quartz-free keratophyres are also quite abundant. The volcanic activity occurred in the near-shore part of the continent in a shallow-water sea. In places there were fissure-eruptions of diabase.

Major and minor intrusions are widespread often reaching large size, as for instance in the Kuznetskian Alataus and in the Sayans (Fig. 154). The three age-groups—the Salairian, the Taconic and the Erian—are distinguished and the former two groups predominate. In the Kuznetskian Alataus and the Mariinskaya Taiga granite intrusions are 60–100km across and have a well-known gold mineralization associated with them. In places there are gabbros, pyroxenites, and diorites. The Krasnoyarskian intrusions serve as an example of the late Caledonian granites, and have a variable composition. The alkaline syenites found in association with them form the well-known stacks (Stolby) on the shore of the Yenisei, opposite the town Krasnoyarsk (Fig. 155).

HERCYNIAN MAGMATIC CYCLE. The cycle is widespread and the central southern part of the Angara Geosyncline is the main one, but over the greater part of the geosyncline, in areas affected by the Caledonian Orogeny, is less important than the Caledonian cycle.

The regions where the Hercynian cycle is dominant consist of Eastern Cisbalkhashia, the Tarbagatais, the Kalbinskian range, the Altais and the Southern Salairs. The cycle is divided into several phases corresponding to the Hercynian Orogeny. These phases apply especially well to intrusives, and are less obvious with respect to lavas, which are sometimes contemporaneous with the intrusives.

The lavas signal the outburst of vulcanicity at the beginning of the Lower Devonian and the end of the Middle Devonian. Then the vulcanicity somewhat fades in the Upper Devonian and resurges at the end of the Lower Carboniferous. The lavas of this epoch are thick and are accompanied by siliceous shales and cherts. The distribution of the lavas is not limited to the region of dominant Hercynian vulcanicity, but they are also found in the Minusinskian Depression and in the Western Sayans. The acid varieties usually consist of thick porphyrites, porphyries,

FIG. 154

Buibinskian Lower Palaeozoic por-
phyritic granites of the Western
Sayans.
  Photograph by V. P. Ivanov.

FIG. 155

Alkaline syenites of Caledonian age
forming the 'Perya' rocks.
  *Photograph in Central Geological
Museum.*

and the corresponding tuffs and tuffites. The lavas are often interbedded with marine limestones and sandstones. The volcanic rocks erupted from large fissures and large contemporaneous intrusions are rare.

In the Upper Palaeozoic the vulcanicity wanes and centres of its activity are displaced to the south, but the intrusions become widespread and large. In the Middle and Upper Carboniferous the lavas appear only in the southern provinces and in the Tarbagatais, the Saurs, the Southern Altai and the Semipalatinsk region. The nature and composition of these lavas is similar to the Middle Palaeozoic types. The flows and sheets of andesinophyres and albitophyres, which are sometimes quartzose, alternate with tuffs and sediments such as sandstones, conglomerates and shales which are usually terrestrial but are rarely marine.

In the Permian and the beginning of the Mesozoic the vulcanicity of the southern part of the Angara Geosyncline weakens and in places dies out altogether. In the north outpourings of lavas similar to the Siberian Traps occurred at about the same time as the latter. The rocks are dark basalts and diabases occurring as flows and sills, the identification of which is sometimes difficult, rendering their age debatable and indeterminate. Judging from the well studied traps of the Siberian Platform the effusion of lavas and the emplacement of sills occurred in the Upper Carboniferous, Permian, Trias and Jurassic, differing from one region to another. In the Angara Geosyncline the traps are known in the extreme west, near lake Ubagan in the Turgai Depression, where they have been found in boreholes, and also in the Kalbinskian range, in the Semipalatinsk region, in Eastern Kazakhstan (the Yeremen' Tau) and lastly in the Kuznetsk Basin. In the last region M. A. Usov considers all the lavas and sills as post-Jurassic but there are reasons to believe that there are some Upper Palaeozoic traps as well.

The Effusive-sedimentary Formation contemporaneous with the Siberian Traps is known from the east of the geosyncline in the Eastern Sayans and Western Transbaikalia. In the regions of the rivers Dzhida and Khilok (tributaries of the Selenga) there is a thick group of basic and acid lavas and tuffs with groups of siltstones, sandstones and shales with plant remains. The study of the plant remains (Theses . . ., Chita, 1961, Kozubova) has shown that there are two suites present: (1) the Petropavlovskian consisting of Upper Permian basic lavas, and (2) the overlying Tamirian, consisting of acid lavas of the Lower Trias age.

The Hercynian (Upper Palaeozoic) intrusions have a wide distribution, but occur mainly where, as in the Altais, the Tarbagatais and Cisbalkhashia, the Hercynian Orogeny was strong. Three phases of intrusive activity are recognized: the Middle Carboniferous, the Lower Permian and the Upper Permian-Lower Triassic. The first of these phases is the most important and it embraces most of the granites and granodiorites of Eastern Kazakhstan and the Altais. The intrusions are often very large, mainly consist of biotite granite, but also involve other rock types. The massifs are cut by aplites, lamprophyres, syenite-porphyries and rarer pegmatites.

The Lower Permian intrusions are generally rare and smaller, but in Kalba and the Tarbagatais are represented by large batholiths (Fig. 156). Pink, red and pale-grey granites predominate. Peculiar alkaline intrusions, including nepheline syenites, are interesting and are found in the Ishim range, the Talassian Alataus and in the Ulu-Taus.

The Upper Permian-Lower Triassic intrusions are rare and are represented by the Semeitauian intrusions, situated to the north of Semipalatinsk. These intrusions vary in form, are small-scale and variable in composition. Their characteristic feature

FIG. 156

A batholith of Hercynian granites in the Tarbagatai range. Weathering and topography are typical of the range. *Photograph by M. M. Vasil'yevskii.*

FIG. 157

A dyke of porphyries intruding Upper Palaeozoic oil shales of the Saur range at Kenderlyk deposit in the region of lake Zaisan. *Photograph by G. P. Kleiman.*

is that they are not associated with Hercynian folds and have not been deformed, have rounded outlines and consist of fresh minerals. They penetrate the Lower Permian.

Minor intrusions cutting major intrusions and Upper Palaeozoic rocks are common.

CIMMERIDIAN AND ALPINE MAGMATIC CYCLES.   Rocks of these cycles are rare and are represented by lavas and even rarer sub-volcanic intrusions. The best example is the 'Melaphyre horseshoe' of Kuzbas, which is shown on geological maps. It consists of basalts and diabases of often alkaline affinity (essexites and anamesites). These rocks occur as sills and lavas; the sills lie within the Permian, Triassic and Jurassic marine deposits, while the lavas are found on the Trias or in the Jurassic. Their thickness is 30–40m in the south, while in the north the sills, lavas and intercalated marine sediments form a group of 1000m in thickness. V. A. Obruchev (1936) considers all these rocks as analogous to the Siberian Traps, of which they represent the Upper (Mesozoic) part.

The Cainozoic lavas are even rarer, although small groups of such lavas are, according to A. G. Gokoyev, found in several localities in Kazakhstan. They are also found in the Eastern Sayans and in Tuva where they are associated with young Pliocene and Anthropogene faults. The effusive rocks consist of basalts and pyroclasts overlying Neogene and Anthropogene rocks. The basalts are usually few tens of metres thick but rarely may reach 150–200m. In places small volcanoes of 600–750m in height have escaped erosion.

USEFUL MINERAL DEPOSITS

The Angara Geosyncline can be truly called the Soviet region of Mining. In the pre-Revolutionary era only a few polymetallic and gold deposits in the Altais, the Salairs and the Kuznetskian Alataus were worked. Coal was either unexploited or was used for local needs only. Only few geologists then visited the area amongst whom V. A. Obruchev, who started as a mining engineer, was outstanding. He was responsible for the formation of the Tomsk Geological School, attached to the local Polytechnic. Amongst its distinguished professors M. A. Usov, M. K. Korovin and V. A. Kuznetsov can be singled out.

The changes introduced by the Soviet regime are difficult to imagine. On the shore of lake Balkhash there is now a huge copper-smelting plant. In the middle of the formerly wild steppe the second plant at Dzhezkazgan was built adjacent to a new coal-mine at Baikomir. In the same area the manganese deposit at Dzhezda was discovered. In the centre of Kazakhstan a new coal-field with its principal town Karaganda grew up. To the east of it at Ekibastuz a coal-field with the thickest seams in the world has been surveyed. On the Irtysh a large hydro-electric station has been constructed supplying the industry with cheap electricity. The Kuznetsk Basin became a pride of our country with its large coal-pits, while to the south the Telbes iron ores have been found and in the town Novokuznetsk a metallurgical plant has been constructed.

The most important discoveries have been made in the Turgai Depression and along its margins. Here, underlying the Cretaceous and Palaeogene sediments, lie Palaeozoic sediments and intrusive rocks. They contain the Sokolovsk-Sarbai group of magmatic iron ores, which are already being worked. The same rocks contain the Dzhetygar asbestos deposit. The Jurassic of this area has the great Ubogan coal-field and the Lower Cretaceous the Amangeldin deposit of bauxites. Lastly in the Upper Cretaceous in the valley of the Ayat there is a large basin of oolitic iron ores. None of these deposits are exposed at the surface and their discovery depended upon geosphysical aids to surveying and on intelligent drilling. Much yet remains to be done by hundreds of geologists employed on work in the Angara Geosyncline.

ORE-MINERALS

The non-ferrous metal ores of the Angara Geosyncline and in particular the Kazakhstan copper and the polymetallic ores of the Ore Altais are of national importance. Gold is found in the north of Kazakhstan and in the Kuznetskian Alataus. Ferrous metals are less widespread, but iron and manganese deposits are used in Kuzbas metallurgy.

Non-ferrous metals

Copper. The richest deposits in the U.S.S.R. are found in Central Kazakhstan (Dzhezkazgan) of sedimentary-metasomatic origin. Its ores are found in the Upper Palaeozoic copper sandstones. The deposits of Kounrad and Boshche-Kul are of magmatic, disseminated type and the ore is found in porphyritic granodiorites of Hercynian age. Dzhezkazgan is situated near to the granite massif of Ulu-Tau, and includes over twenty different deposits of copper and many more smaller ore bodies. All of them are found amongst the sandy and shaley continental formation (over 900m) of Middle and Upper Carboniferous age and known as the Dzhezkazgan Suite. The mineralization is associated with a group of sediments of 200–300m in thickness and divided into seven horizons, consisting mainly of grey sandstones with calcareous-clayey cement.

The primary mineralization is disputed, but is most likely of sedimentary origin since the ore is restricted to distinct lithological and stratigraphic horizons. The primary ore has been affected by various metasomatic processes associated with the circulation of solutions along fault planes. There are no grounds for suspecting hydrothermal activity since Upper Palaeozoic vulcanicity had not affected this region. At the same time K. I. Satpayev considers the deposit to have been produced entirely metasomatically by circulating solutions

The Kounrad deposit is almost of the same size as the Dzhezkazgan deposit, but is of an entirely different origin. Its ore is disseminated in small quantities and is relatively poor. The deposit is a massif of Hercynian porphyritic granite intruded into Silurian sediments and lavas. The upper part of the massif and the adjacent contact altered lavas have been silicified into quartz-sericite rocks (secondary quartzites). The copper-bearing chalcopyrite and pyrite are disseminated more or less throughout the secondary quartzites, which are 200m thick. This deposit has been described by M. P. Russakov. The smelting plant is on the shore of lake Balkhash.

Boshche-Kul is a deposit of the same type and approximately the same size as at Kounrad and it is situated to the east of Tselinograd. The three deposits apart there are tens of other copper deposits. Sedimentary cupriferous sandstones have also been found in the continental Devonian and Silurian. The copper porphyritic ores are always associated with Hercynian and often late Hercynian intrusions, thus forming a series of medium-size and small deposits. Some have economic concentrations of gold. Small deposits also exist in zones of deformation and crushing and also in veins. The copper deposits of Central Kazakhstan are the first in the U.S.S.R.

Polymetallic Ores

The Ore Altais represent a major national centre of polymetallic ores. Smaller deposits are also known from the Salairian range. Even smaller ore-showings have been found in many localities of Central Kazakhstan, the Altais and the Sayans.

Silver-head-zinc ores of the Ore Altais have been known for thousands of years and the site of local workings for silver and lead was known in pre-Christian times. The first smelting plants were built here in the eighteenth century. Hydro-electrification of the district in the Soviet period has led to the realization of the idea of 'Greater Altai' and to modernizing the workings of this important district. Silver, lead and zinc apart the district has great reserves of pyrite and in places of ores with economic concentrations of gold and copper.

The origin of the Altaian deposits provokes arguments. V. P. Nekhoroshev, a leading specialist, and many others consider that two main fracture zones play an important role in localization of these deposits. One zone follows the valley of the Irtysh and the other is 70km to the east of it. These zones represent systems of complex faults which separate and then converge again. Upward and downward movements along these zones were by differing amounts thus causing the appearance of different deposits at the surface.

Other geologists deny the significance of the zones of deformation considering that the characters of the intrusive bodies are governing factors. Although a majority of larger deposits are associated with the zones of faulting, there are many other deposits which are not related to these zones. The reserves of polymetallic ores in the Ore Altais are very large. The Ridderovskian deposit is especially significant and Zyryanovskian, Zmeinogorskian, Sokolnian, and Zolotushinskian deposits are also known. The Altais have the greatest deposits of lead in the U.S.S.R.

Aluminium. The structures of the Angara and the Ural geosynclines are very similar and their bauxite deposits are of the same type.

Three types of deposits are known: (1) the lacustrine Mesozoic; (2) the marine Middle Palaeozoic; (3) the marine Lower Palaeozoic. The deposits of the first type are found over a large area from the Turgai Depression to the Yenisei. Red halloysitic bauxites occur as lenses amongst lacustrine Lower Cretaceous, Upper Cretaceous and Palaeogene sediments. The seven years plan envisages the opening of a large bauxite mine which will supply raw material to the projected Pavlodar aluminium plant. The bauxite deposits are known as Amangeldinian and their age is Upper Cretaceous-Palaeocene.

The second type of deposits are associated with Devonian limestones of the Salairian range with diaspore bauxites. In places these bauxites are invaded by Hercynian granites, lose their water and become converted into corundum.

The deposits of the third type are diasporic bauxites found amongst the algal Sinian (or Cambrian) limestones of the Eastern Sayans. The Buksonian (Boksonian) deposit serves as an example for this type.

The bauxites of Kazakhstan form the third largest reserves in the U.S.S.R. (Amiraslanov, 1957).

Precious Metals

During the Tsarist regime these were the most important worked deposits of the region. Now although their field has considerably increased it has fallen behind in comparison with the non-ferrous metals. Gold is most important and is associated mainly with Caledonian and even Upper Proterozoic granites. This has been proved in two gold-producing districts—the Stepnyakovskian in the north of Kazakhstan and the Kuznetskian in the Kuznetskian Alataus.

In the Ore Altais gold and silver is obtained from polymetallic ores. Platinum and associated metals are rare, but occur in small quantities. This is explained by the fact that ultrabasic intrusions are rare in the Angara Geosyncline and common in the Urals where these ores are found.

Ferrous Metals

The ferrous metals, unlike those of the Urals, are of secondary importance and were altogether unknown in pre-Revolutionary times.

Iron. Iron smelting plants were erected in Central Kazakhstan (at Temir Tau) despite the fact that the neighbouring Ata-Su and Konurlen deposits were inadequate and iron ore had to be transported from the Urals. After the recent discovery of the Ayatian sedimentary iron ores, which like the Kerch ores occur as a thick bed of oolitic rock, the situation changed, since the Ayatian deposits have reserves of over 12×10^9 tons. Nearby there is also the Lisakovskian deposit of the same nature and reserves of 7×10^9 tons, which is also convenient for exploitation.

Palaeozoic magnetite ores have been found at Sokolovka and Sarbai. These ores are of magmatic origin and resemble those at Magnitogorsk. The nearby Kacharian deposit is of exceptional quality and in its abundance of ore can be compared with Krivoi Rog, while its martensitic ores are the best in the U.S.S.R. These ores have also vanadium and cobalt impurities. Sulphidic ores also exist and have appreciable quantities of copper, zinc, nickel, cobalt and sometimes molybdenum.

The Kachar deposit is of skarn type. Here limestones and tuffogenous rocks belonging to the Lower Carboniferous are interbedded with quartz porphyries and orthophyres. The ores are formed as a result of the metasomatism of limestones and tuff. Nearby there are intrusions of porphyritic granites. The ore-bearing formation lies under a cover of Palaeogene and Upper Permian deposits of 150–200m in thickness. These deposits are already being worked and ensure the supply for the iron smelting industry of Kazakhstan. Reserves of iron ore in the Kustanaisian Province reach tens of thousands of million tons. In addition the large Karadzhal haematite deposit has been surveyed and is ready to supply the Karaganda smelting plant.

In Kuzbas where the iron ore still has to be imported the deposits of the Tel'bessian ore-field are of two types: (1) magmatic—contact-metasomatic; (2) sedimentary ferruginous quartzite of a Precambrian age. The Tel'bessian deposits are found in the Rocky Shorias, to the south of Kuzbas. The largest deposit is at Temir-Tau, where it is associated with late-Caledonian granites, which intrude the sandy-clayey continental rocks of a provisionally Silurian age. The main metasomatic ore-mineral here is magnetite. The Tel'bessian deposits played an important role in the development of the ferrous metallurgy in Western Siberia, but the deposits are not very large and new ore has to be found in the future. In this respect the discovery of the Krivoi Rog type of ferruginous quartzites has great prospects. The ferruginous quartzites (jaspilites) and the haematite-magnetite ores have been found in the Proterozoic of the Kuznetskian Alataus, in the south-eastern Tuva and in the Eastern Sayans. If the analogous deposits of the Yenisei range are added then the abundance of such ores becomes obvious.

Manganese. The deposits of manganese are numerous and are associated with siliceous shales, cherts and cherty limestones of an Upper Devonian age in Kazakhstan and an Ordovician age in the Kuznetskian Alataus. The ores are sedimentary, but the influx of manganese was caused by volcanic activity.

To the north-east of Kuzbas the Mazul deposit has been recently investigated, while in the south-eastern part of it the Usinskian deposit has been studied. Both are large and of sedimentary-metamorphic origin. The latter deposit, in particular, has an ore-bearing group of interbedded siliceous rocks and carbonate manganese ores with limestones, which have a Middle Cambrian fauna. The ore-bed is 10–20m thick and is widespread. The Mazul deposit is also found in a Cambrian siliceous-calcareous formation.

In Kazakhstan large deposits of manganese are known in Kara-Dzhal, Greater Ktai and Dzhezdy.

Cobalt. The Khovuaksian deposit in Tuva has the true cobalt ores of some economic interest.

Rare Metals

Tungsten. The deposits of tungsten are numerous, but small and are mainly found as tin-tungsten ores in the Kalbinskian range and the Ore Altais, where they are associated with Upper Hercynian intrusions. In Western Transbaikalia there is the Dzhida deposit found amongst the Lower Proterozoic metamorphic rocks intruded by a massif of gneissose granite-syenite. At the contact there is a stock of a younger granite with which the mineralization is associated. The ore is found in a series of quartz veins cutting across the metamorphic envelope and the granite. The main mineral is gubnerite and there are also alluvial accumulations.

In Kazakhstan vein and stockwork ores are common and in addition to tungsten contain tin and molybdenum, which allows of their working.

Molybdenum is found in Western Transbaikalia, in the middle reaches of the Chikoi, near the village Gutai, amongst Archeozoic gneisses and amphiboles. Molybdenite, here, occurs in quartz veins related to the younger granites.

In Kazakhstan molybdenum ore is found in Shalgiya, and in addition vein molybdenite and complex copper-molybdenum ores are known. The reserves of Kazakhstan are the second largest in the U.S.S.R. (Amiraslanov, 1957).

Tin. The deposits are relatively small.

Titanium. In the Kustanaisian region there are alluvial rutile-ilmenite ores.

Antimony. Quartz-antimonite veins cutting various rocks are known from Central Kazakhstan.

Mercury. Deposits are known in the Ore Altais.

Vanadium. According to the latest data of S. G. Ankinovich (1960) Central Kazakhstan is a vanadium-field, the metal being found in the carbonaceous-siliceous Cambrian shales.

FUELS

Back and Brown Coal

The coals of the region are of paramount importance in the U.S.S.R. Palaeozoic coals found in the Kuznetsk, Karaganda and Minusinskian basins are particularly important. Mesozoic coals are found in numerous dispersed fields and are of a secondary importance. Cainozoic lignites are only of local significance.

Palaeozoic coals are associated with the Carboniferous and Permian coal-bearing formations. The coals of the Karaganda and Ekibastuz basins and of the Balakhonkian Suite of Kuzbas are Carboniferous, while those of the Kol'chuginian Suite of Kuzbas, of the Kenderlyk deposit and of the Coal-bearing Formation of the Minusinskian basin are Permian.

Karaganda. This is the fourth largest coal-field in the U.S.S.R. It is situated in Central Kazakhstan and is connected by a railway with Alma-Ata, the Balkash plant (combine), Dzhezkazgan, Temyr Tau and the Urals. The succession is shown in Table 34 and is 5500m thick. The proved reserves reach 10.5×10^9 tons and geological reserves are 35×10^9 tons. The main coal seams are found in the Karaganda Suite (600–800m) which has an Upper Visean age. The seams reach 10–12m in thickness, but are commonly thinner. The basin is 100km long and 50km wide and is tectonically complex. The central part of the basin has a Jurassic coal-bearing formation as well as the Palaeozoic.

Ekibastuz. The thickness of coal seams in this field is unique in the U.S.S.R. There is a coal seam which including thin intercalations of shale is 150m thick, and must have been produced as a result of the accumulation of 450–600m of peat. Present-day peats, however are on average 6–8m thick, reaching 20–24m in isolated instances.

Ekibastuz is situated to the north-east of Karaganda, at a distance of 135km from Pavlodar. The field is a brachysyncline of 13km in length and 6km in width. The structure, thickness and age of the coal-bearing strata are the same as in Karaganda. Despite the limited size of the field the thickness of coals makes for considerable reserves of up to 10×10^9 tons.

Kuzbas. The geology of the Kuznetsk Basin will be described in detail later. The field is one of the largest in the U.S.S.R. and its geological reserves to the depth of 600m are 252×10^9 tons. The proved reserves are 34.5×10^9 tons, and the yield—second in the U.S.S.R.—is 63.8×10^9 tons.

There are three coal-bearing formations in the basin: (1) the Balakhonkian, Carboniferous and partly Lower Permian, Formation; (2) the Upper Permian Coal-bearing Formation, and (3) the Jurassic Coal-bearing Formation. The Balakhonkian Suite has the richest coals. It is 800–2700m thick and has 28 seams, varying from 0.75 to 18m in individual thickness and 78m total thickness. The Kolchuginian Formation and in particular the Yerunakovian Suite is 2600m thick and has 45 workable seams of a total thickness of 75m. The Lower Jurassic measures have not been sufficiently studied and are not being worked. There are 10 Jurassic, workable coals of a total thickness of 18m. Sapropelic coals, which are black and have a conchoidal fracture are of some interest.

Minusinskian Basin. This field is situated at the confluence of the Abakan and the Yenisei. The coal-field has several deposits connected by a recently built railway. The coal seams are both of Carboniferous and Permian age, but the latter are more numerous and thicker. Here seams of 3–8m and even 12–15m are encountered. The geological reserves of the field are 36.3×10^9 tons and the proved reserves are 2.1×10^9 tons. The yield in 1956 was 2.5×10^6 tons.

Kenderlyk Coal-field. This is one of the most important fields in Southern Kazakhstan. It has widespread thick (40m) oil-shales and its succession has been described on p. 432.

Mesozoic Coals. Jurassic continental coals occur in small deposits dispersed over a very large area. The largest Jurassic coal deposits are found in the Turgai Depression and in the Karaganda and Kuznetsk basins. The concealed coal-fields are particularly important. They are found where the Jurassic measures are hidden under a cover of Cretaceous and Cainozoic deposits. Recently the large Turgaiian (Ubagan) coal-field has been explored. It has a whole series of high-quality coals near to the surface. The total thickness of the Kushmurunian deposit reaches 80–100m and can be worked in open mines. The coal of this district is due for exploitation. The total reserves of the basin (35 × 10⁹ tons) are the same as in the Karaganda Basin. The basin has been described by I. I. Gorskii and N. I. Leonenok (1960).

In Tuva, to the east of Kyzl a new Ulukhemian coal-field has been found amongst Jurassic strata. Its reserves are approximately (6–10)×10⁹ tons and the seams are shallow and suitable for mining. The coals of this basin, if mixed with the 'lean' coals of Kuzbas, can be used in the production of the metallurgical coke of satisfactory quality. The succession has been described on p. 438.

Palaeogene and Neogene Coals. The continental Neogene and Palaeogene clastic rocks have frequent coal-bearing formations and groups. They are especially widespread in the western half of the geosyncline—in the Turgai Depression, the Teniz Depression, in the Kazakh steppes, Cisirtyshia and along the northern margin of the Altais. The brown coals are of three ages: Eocene, Middle and Upper Oligocene and Lower Miocene

Oil and Oil-Shales

Oil, in economic quantities, has not been found in the area of the Angara Geosyncline, although oil shales and sapropelic coals of high volatile content have been found in several regions. The main reasons for the absence of oil is the absence of suitable porous reservoir rocks and the complexity of structure.

Barzasian Oil-shales or sapropelic coals have been found in the Middle Devonian marine sediments of the Barzas valley, at the north-eastern edge of Kuzbas, where they occur as two beds of 0·5 and 3·6m thickness. The beds consist mainly of psilophyte and algal remains, with a high content of volatile matter. In the same region there are Middle Devonian dark oil-shales, which are 40m and even 56m thick, but contain much less volatiles.

The aforementioned Konderlykian oil-shales, associated with the Lower Permian marine transgression, are the second most important deposits after the Barzasian.

NON-METALLIC DEPOSITS

The non-metallic deposits are represented by many diverse deposits, which are almost always small and satisfy local demands. The deposit of abrasive material at Semiz-Bugu has had an All-Union significance.

Corundum and Andalusite. The Semiz-Bugu deposit in the central part of Kazakhstan, to the east of Karaganda, occurs where thick Silurian limestones and acid lavas are cut by a Hercynian granite. The contact zone has large bodies of andalusite with pure corundum or mixed corundum and muscovite in the middle of these bodies. In places corundum is found amongst marbles and has probably been formed at the expense of diasporic bauxite, similar to that found in the Salairian bauxite deposits. Economic accumulations of andalusite are also known from other localities in Kazakhstan.

Barytes. Veins of barytes are often associated with Hercynian granites and are in places of industrial importance, as for instance, in the central part of Kazakhstan, in the north of the Ore Altais and in the Salairs.

Asbestos. The deposits of chrysotile asbestos are the most important after those of the Urals. The deposits are associated with the serpentine belt of small ultra-basic massifs, which stretch one after another along the Sayanian strike. These deposits are shown on synoptic geological maps. The Aktovrat asbestos mine is associated with one such massif on Tuva. In the extreme west of the geosyncline at the boundary between the Southern Urals and the Turgai Depression a large asbestos deposit has recently been discovered near the village of Dzhetygar.

Iceland Spar. Deposits associated with crush zones and cavities in karst limestones are known in Tuva.

Phosphorite. The Lower Palaeozoic rocks, with which the Karatau deposit of bedded phosphorites is associated, continue into the western part of Central Kazakhstan. Consequently on the supposition that the phosphorite will also be continuous, they were looked for and found in 1954.

Rock Salt. Salt is obtained in a series of saline lakes of the Kulundinian steppe. The largest salt deposits occur in lake Tavolzhan, near to Pavlodar.

Mirabilite (Glauber's salt). This substance is again obtained in the Kulundinian lakes near to the station Kulunda.

Metallurgical Raw Materials. The widespread metallurgical industry of Kazakhstan, the Altais and Kuzbas is completely supplied by the local deposits of fluxes, limestones and quartzites.

REGIONAL DESCRIPTIONS

Within the Angara Geosyncline five regions with distinct structures, positions and economic importance can be recognized. The regions are: (1) Central Kazakhstan; (2) the Altais; (3) the Kuznetsk Basin; (4) Tuva and the Sayans; (5) the East including Transbaikalia.

CENTRAL KAZAKHSTAN

General Description

Central Kazakhstan is a large region of Palaeozoic rocks of small-sopka relief (Figs. 146, 158).

To the west, north and north-east this region is bounded by a plain with Neogene and Palaeogene deposits. In the south the boundary lies along the Balkhash depression and in the south-east an arbitrary boundary is drawn along the railway line.

Stratigraphy

Precambrian and Lower Palaeozoic. These ancient rocks are most widespread in the north (the Atbasarskian region) and in the west, around Mt Ulu-Tau. As is evident on geological maps, in the two above-mentioned areas the Precambrian and Lower Palaeozoic rocks are most extensive. Even the granites are Precambrian and Caledonian. Carboniferous and Devonian strata occur as merely isolated patches and are essentially continental in origin. Otherwise throughout the region the ancient rocks appear in the cores of anticlines as narrow belts and patches (Fig. 158). The

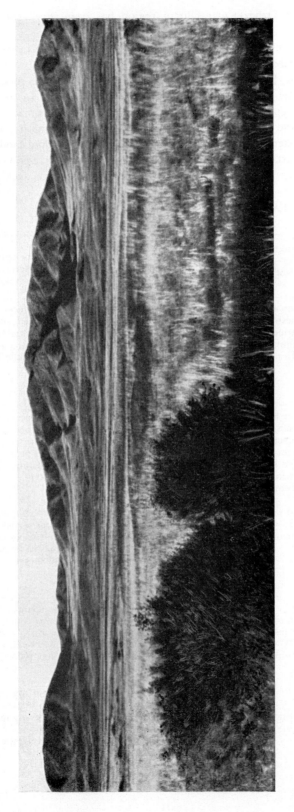

FIG. 158. Lower Cambrian quartzites with continental Neogene clays in forefront outcropping in the south-western part of the Karaganda region. *Photograph by V. D. Veznesenskii.*

Lower Palaeozoic of Dzhezkazgan-Ulutau region has been investigated by L. I. Borovikov (1955).

Recognized areas of Cambrian and Ordovician rocks are found in the south-east in the Chingiz and Cisbalkhash ranges.

Middle Palaeozoic. Silurian, Devonian and Lower Carboniferous rocks are found everywhere and in the south-east predominate over other rocks. In the south-east there are Hercynian types of succession with a marine Lower Devonian and Eifelian (Table 35). Only in the Givetian epoch does the marine transgression penetrate into the northern and western districts. The boundaries between sea and land determine the distribution of certain useful mineral deposits. The shallow regions of the sea, especially if, as then, they are near active volcanoes are sites for accumulation of

TABLE 35. *The Devonian of Kazakhstan*

Stage	Central Kazakhstan (According to B. I. Borsuk)		The Ore Altais (According to V. P. Nekhoroshev)
	North-Caledonian type	South-east —Hercynian type	
Fammenian	Grey and dark bedded limestones with a marine fauna. (200–600m). In places is missing.	Sulciferian Beds of limestones, sandstones and shales. (260–300m). Calcaratusian Beds of limestones and shales with a marine fauna. (250–300m).	Porphyrites and tuffs with beds of sand-stones and shales possessing a marine fauna. (400–800m).
Frasnian	Grey and red sand-stones, shales, con-glomerates and in places volcanic rocks. Plant remains are present. (2000–4000m).	Maian Beds of red and grey sandstones and shales, with bands of limestones and vol-canic rocks. (400–600m).	In places shales and limestones with a marine fauna; some-times lavas, tuffs and sandstones. (Up to 1000m).
Givetian	Red sandstones and silts, with plant remains.	Sandstones and vol-canic rocks with lenses of limestones with a Cheehiel fauna. (200–1000m). Red beds with a flora.	Predominant shales; rarer sandstones and limestones with marine faunas. (500–1000m).
Eifelian	Albitophyric Forma-tion of porphyrites, felsites and tuffs. (1000–1500m). In places is missing.	Green sandstones, shales and limestones, with a marine fauna. In places volcanic rocks are present. (400–1200m).	Sandstones and shales; rarer, limestones with marine fossils; in places volcanic rocks. (400–1000m).
Koblenzian and Gedinnian		Variegated sandstones and shales with a marine fauna. (250–400m). In places a sequence of porphyrites, tuffs and sandstones. (Up to 1200–1600m).	Sandstones and shales with bryozoans. In places are missing. Unrecognized.

siliceous and manganiferous rocks. Vast near-shore plains of a Lower Carboniferous age were areas of mass-growth of marsh plants, which has led to the formation of coals, which at present are largely eroded away, but are still found in tectonic basins. Such basins in a vast belt stretch from Atabasar, through the lake Tengiz Depression, into Karaganda towards the north-east into the Yeremen'tauian (Karzhantauian) deposits and Ekibastuz.

Upper Palaeozoic. There are virtually no Upper Palaeozoic rocks in the north, and where found they are continental, although at Dzhezkazgan such rocks occur over a large area and are thick. Near Semipalatinsk and to the north and south of it Upper Palaeozoic continental rocks are widespread and near the Irtysh coal-bearing, and at Semei Tau contain a complex of volcanic deposits.

Mesozoic and Cainozoic. Where the Palaeozoic strata are dominant, Mesozoic and Cainozoic rocks occur as thin continental facies preserved in small graben amongst the Palaeozoic massifs.

Along the western and northern boundaries of Central Kazakhstan there are extensive marine Upper Cretaceous and Palaeogene sediments, continuous with those of the Western Siberian Lowlands and the Turgai Depression. In the south-east within the Kenderlykian Depression there is the coal-bearing Trias. In the north-west the top horizons of the sedimentary-effusive Turinian Suite, of mainly Permian age, should be included in the Lower Trias, while the bottom of the essentially Jurassic Coal-bearing Formation is Rhaetic (Upper Trias). The continental Trias of the Kenderlykian Depression (The history of Lower Mesozoic accumulation of coal . . . 1961) has been studied and described in detail. The Lower Trias here is represented by rocks which mark the completion of the Upper Permian sedimentary rhythm. The rhythm begins with conglomerates and sandstones belonging to the Maichatskian Suite (150–310m), overlain by the coal-bearing Akkolkanian Suite (220–550m) consisting of silts and argillites with coal seams. The higher supra-coal group consists of lacustrine argillites and argillites with phyllopods and ostracods of a Lower Trias age.

The next sedimentary rhythm begins with conglomerates and sandstones of the Akhaltanian Suite (350–1000m) with Upper Trias insects and plants. Above this suite lies the Tologoiskian Suite (585–680m) of greenish and brown argillites and silts with coal seams. The freshwater fauna determines the age of the suite as Rhaetic-Liassic. The rhythm finishes with sandstones and argillites of the Taisuganian Suite (200–300m) with a Liassic flora. Higher up lie the continental gravels and sands of Upper Cretaceous and Palaeogene age.

There are also known outcrops of the Rhaetic-Liassic Coal-bearing Formation, exhibiting a slight folding and being a few hundred metres thick. The largest outcrops occur in the centre of the Karaganda Basin; at the eastern boundary of the geosyncline to the west of Pavlodar (Maikyuben'); at the western boundary; in the Turgai Depression and in the south-east in the Kenderlykian Depression. The available synoptic maps do not show any outcrops of the Cretaceous since they are so small that they cannot be indicated. Nevertheless, Lower Cretaceous variegated sands and clays with lenses of bauxite, fireclays and lignites are important. These rocks are most widespread along the western margin of Central Kazakhstan, near the edge of the Turgai Depression, in the upper reaches of the Ishim, near Akmolinsk and along the margins of lake Tengiz. Nowadays these sediments are included into a single Arkalykian Suite of Palaeogene age. The marine Upper Cretaceous is found along the northern and western margins of the Palaeozoic Massif. The marine

Palaeogene is exposed almost everywhere along the northern, north-eastern and western boundaries. Within the Palaeozoic Massif the Palaeogene and Neogene are represented exclusively by continental sediments.

The Anthropogene deposits are found everywhere, but only in the depressions occupied by lakes Tengiz, Balkhash, Ala-Kul and Zaisan, and at the foothills of young rising chains, such as the Kirgiz range, are these deposits of considerable thickness. The Anthropogene has been discussed by N. N. Kostenko (1960).

Tectonics

Tectonic structures of the region are conditioned by the Caledonian and Hercynian orogenies. Where the Hercynian Orogeny has been dominant a north-westerly structural trend is prevalent. In the west and north an east–west or north-east trend is produced by the Caledonian and Precambrian orogenies. To the north of Balkhash, Hercynian north-west trending folds cross the north-east trending Caledonian folds. Consequently a mosaic of trends is discerned on the synoptic maps.

The tectonic scheme of Central Kazakhstan (Fig. 138) has been constructed by A. A. Bogdanov (1959). The scheme shows the Caledonian zone trending to the south-east and forming the Kazakh isthmus. There is a belt of volcanic activity within it which was most vigorous. This belt separates the Hercynian zone from the Caledonian.

Notice must also be taken of the Upper Palaeozoic Depression such as the Tengiz and Dzhezkazgan structures, filled by Upper Palaeozoic terrigenous deposits. The Palaeozoic Massif is margined by young troughs such as the Kulundinian, the Turgai, the Chuian and the Balkhash depressions which are filled by mainly continental Mesozoic and Cainozoic sediments. The scheme also shows the main, large anticlinoria and synclinoria.

Magmatism

There is a sharp dominance of Palaeozoic and Precambrian intrusions. The large, obvious ultrabasic intrusions of the Urals are small here and on maps can only be found with difficulty, as for instance to the south-west of Pavlodar, in the Kalbynian range and to the north-east of lake Zaisan.

The Ishimian nepheline syenite mass, situated where the Ishim turns from an east–west to a north–south trend is peculiar and is provisionally thought to be Hercynian.

The granite massifs found in the north and west are Precambrian or Caledonian in age, while the massifs which occur in the south and east are mainly Hercynian. The early Hercynian massifs were intruded in Upper Devonian-Lower Carboniferous times. The youngest, Semeitauian intrusions to the north of Semipalatinsk intruded Upper Palaeozoic strata and are considered as Upper Permian or even Lower Triassic. For Central Kazakhstan the vast occurrence of Palaeozoic lavas, which have a total thickness of several kilometres, is characteristic. The most important volcanic suites are the Boshchekulian (Lower Cambrian) Suite of spilite-keratophyres (4600m) and the Upper Ordovician, Zharsorian Suite (300m), of albitophyres. Many other effusive formations sometimes have extensive outcrops and thicknesses (1500–2000m).

THE ALTAIS

General Description

The Altais are a highland area, with a relatively low north-eastern part known as the Ore Altais and bounded by the Irtysh valley. The larger ore-deposits are situated in the Ore Altais. The greater part of the highlands is known as the Rocky Altais, which to the north-east are margined by the Rocky Shorias, and in the east pass into the Western Sayans and the Tannu-Ola range. In the north the Altais are margined by hilly plains and steppes of the Altai edge, which is a region of virgin steppes subject to development.

The Ore Altais belong to the system of the Irtysh. The Rocky Altais are traversed by the Charysh, Katun' and Biya. In the upper reaches of the Katun' lies the high Katunian range with Mt Byelukha (4506m) as the highest peak. In the valley of the Biya lies the narrow and deep lake Teleskoye, which is typical graben with high beautiful cliffs bounding it.

Stratigraphy

The Altais is a region of folded and metamorphosed Palaeozoic and Precambrian rocks intruded by numerous Hercynian and rarer Caledonian and Precambrian granites. The region can be divided into two parts. In the west, in the Ore Altais Middle Palaeozoic strata are widespread, while only near to the Irtysh are there outliers of Upper Palaeozoic rocks. In the east on the other hand, the ancient metamorphosed Ordovician, Cambrian and Proterozoic formations are exposed, while the Middle Palaeozoic is much less extensive. Here Upper Palaeozoic and Mesozoic rocks have been completely eroded away.

Ancient Formations. The ancient formations of the Altais are of the same nature as in the rest of the Angara Geosyncline. They are thick, folded, metamorphosed and have not been adequately dated. Their stratigraphy cannot be considered to have been completely investigated. Many formations are of debatable ages and the Lower Cambrian and Upper Proterozoic rocks in particular are disputed. Middle Palaeozoic strata occur mainly in the west, the Ordovician in the centre and the Cambrian and Proterozoic dominantly in the east, forming a series of parallel belts.

Middle Palaeozoic. The strata of this age are mainly found in the west where their succession, of Hercynian type, is 6–8km thick and consists of mainly marine sediments.

The lower part of the Silurian is represented by graptolitic shales and groups of limestones, while its upper part consists of limestones and shales with tabulate corals and brachiopods. The very top of the Silurian is represented by terrestrial sandstones and conglomerates.

The Gedinnian is not identified and may be represented by a break, but Koblenzian and Eifelian limestones and shales with marine faunas have been found in several localities. The mainly marine Givetian, Frasnian and Famennian rocks are widespread (Table 35). The Devonian is gradational into the marine Lower Carboniferous. Devonian lavas with local iron ore horizons are thick.

To the east Middle Palaeozoic rocks become rarer and their thickness less and the Altaian succession is typically Caledonian in type.

Mesozoic. Mesozoic rocks are almost entirely eroded away. In the lower reaches of the Biya, where it merges with the Katun' there is a small area of continental Cretaceous sediments, which are thin and almost horizontal.

Cainozoic. The Neogene and Anthropogene are not separated. The Palaeogene is continental and its strata are found in the north-eastern and south-western margins of the Altais. In the Zaisan Depression they are 1600m thick.

The Anthropogene deposits are widespread and are represented by all varieties except the marine (Shchukina, 1960).

Tectonics

In the Western Altais both Caledonian and Precambrian movements have been recognized, but the Hercynian Orogeny must have been the strongest. Correspondingly most structures have a north-westerly trend.

In the Eastern Altais the Hercynian Orogeny has been weaker and the Caledonian has been stronger. Consequently both the north-easterly and north-westerly and westerly trends can be discerned. The westerly trends become even more dominant in the Rocky Shorias, the Western Sayans and the Tannu-Ola range.

The Ore Altais have large-scale deformation zones, accompanied by mineralization. The largest of these zones runs parallel to the Irtysh for a distance of about 500km (Fig. 152). The Zyryanovskian and Ust'-Kamenogorskian ore-regions are situated along this zone. Further to the north there is another zone of deformation with which the Leninogorskian (Ridder) ore-region is associated.

Magmatism

The vast extent and variability of magmatic rocks is one of the Altaian characteristics. Granites form the main intrusions and are of Precambrian and Caledonian ages in the east and of Hercynian in the west (Cisirtyshia). Although the intrusions belonging to all three Hercynian phases occur, the Upper Permian, possibly Lower Trias, intrusions similar to those of the Semeitauian region of Semipalatinsk are particularly diagnostic. Some geologists consider the polymetallic mineralization of the Ore Altais to be associated with these granites.

Useful Mineral Deposits

The Ore Altais represent the largest region of mining and refining of silver, lead and zinc in the U.S.S.R. Copper and gold are obtained as by-products. The powerful hydro-electric station on the Irtysh provides the necessary energy. Small tin-tungsten deposits are also found here. Locally important individual deposits of copper, iron and gold polymetallic ores are also found in the Ore Altais.

Certain non-metallic deposits such as ornamental stones—Altaian cherts and marbles—and semi-precious stones such as aquamarine and rose quartz are worked. No fuels have been found in the Altais, although to the south of Zaisan there is the Kenderlykian deposit of Permian coals and oil-shales.

Kuznetsk Basin

General Description

The Kuznetsk Basin is the second most important in the U.S.S.R. The enormous quantities of various coal apart, the basin and the surrounding ranges have many other deposits, amongst which iron ores are especially important. They are used in the Novokuznetsk smelting plant (combine). Thus in this formerly inaccessible region many new towns have been constructed around modern pits.

Unlike Donbas, Kuzbas is a low wooded area, surrounded by the Salarian range to the west, the Kuznetskian Alataus to the east and by the Rocky Shorias to the south. The administrative centre is Kemerovo, which is a provincial capital.

Stratigraphy

Mountain ranges and elevations surrounding the Kuzbas consist of Precambrian and Lower and Middle Palaeozoic strata. The large area of Devonian rocks to the north of the basin is worthy of mention. The Kuzbas itself consists of a thick (8–10km) coal-bearing formation. The lower horizons of these strata are Lower Carboniferous, then Lower and Upper Permian, Trias and lastly Lower Jurassic rocks. Anthropogenic deposits normally lie on these strata, while the Neogene, Palaeogene and Cretaceous continental sediments are only found around the outer margin of the mountains which surround the coal-bearing strata.

The Precambrian and Lower Palaeozoic successions are the same throughout the marginal uplands and have already been described (p. 403).

The Middle Palaeozoic in the west and north, where it is most widespread, is of Hercynian type, has a full succession and a great thickness: Ludlovian, Lower Devonian and Eifelian limestones, which sometimes may be reefs, are typical. In the south and east there are continental, often red, facies, with thin groups of marine shales and carbonates occurring as intercalations. The Givetian group is most widespread and with it are associated the oil-shales of the north-eastern Barzas

FIG. 159. The Kuznetsk coal basin; a typical topography along the river Tom'. *Photograph by V. I. Yavarskii.*

region. Devonian successions are given in Table 32 and a detailed description of the Middle and Upper Devonian has been done by Byel'skaya (1960).

The limestones and shales with marine Tournaisian and Visean fossils are divided into five suites. These rocks are overlain by the coal measures, which commence with the Ostrogian Suite.

The Coal-bearing Formation consists of clastic sediments reaching 9–10km in thickness and occupying the central, low region of the basin (Table 36). The systematic investigation of the formation began in 1931 under L. I. Lutugin. Amongst his colleagues, A. A. Gapayev and V. I. Yavorskii, who is still working on it, can be singled out. The first paper in the geology of Kuzbas had been written by Yavorskii and P. I. Butov in 1927. In 1940, volume 16 of *The Geology of the U.S.S.R.*, edited by Yavorskii, was published. It contained much data and generalizations. In 1954 at a conference in Kuzbas the correlation table given in Table 36 was adopted. Borehole data of considerable value can be found in a paper by V. S. M. Muromtsev (1956). In the north-east of Kuzbas the Leninskian, the Uskatian and the Il'inskian formations pass into sandstones of the contemporaneous Krasnoyarskian Formation.

The age of the Balakhonkian Suite cannot be as yet considered as finally

TABLE 36. *Stratigraphy of the Coal-bearing Formation in Kuzbas*
(Accepted at a stratigraphic conference in 1954)

Age	Suite	Sub-suite	Formation	Thickness m
Jurassic	Conglomeratic	—	—	700– 900
Lower Trias	Mal'tsevian	—	Upper Mal'tsevian (variegated)	300– 400
			Lower Mal'tsevian (tuffogenous)	280– 300
Upper Permian	Kol'chuginian	Upper Kol'chuginian (Yerunakovian)	Gramoteinian	1200–1400
			Leninskian	500– 700
		Lower Kol'chuginian	Uskatian	400– 500
			Il'inskian	600– 900
Upper-Lower Permian	Kuznyetskian	—	—	700– 800
Lower Permian and Upper and Middle Carboniferous	Balakhonkian	Upper Balakhonkian (Lower Permian)	Usyatskian Kemerovian	100– 170
			Ishanovo-Intermediate	160–1200
		Lower Balakhonkian (Middle and Upper Carboniferous)	Alykayevian	200– 600
			Mazurovian	300– 550
Namurian	Ostrogian	—	—	200– 600
Visean	—	—	—	—
Tournaisian	—	—	—	—

established. V. I. Yavorskii (1957) thinks it is Lower Permian, and the Ostrogian Suite Upper Carboniferous. At the 1954 conference the upper part of the former suite was also included in the Permian; but its lower part was referred to the Upper and Middle Carboniferous. At the same conference the Ostrogian Suite was relegated to the Lower Carboniferous (Namurian and Upper Visean).

The conditions which gave rise to the coal-bearing formation changed with its age. The Ostrogian Suite was undoubtedly deposited under near-shore conditions; at first in a shallow sea and afterwards on a near-shore swampy plain, thus resulting in the appearance of first coal seams.

The Balakhonkian (Upper Palaeozoic) Suite was deposited entirely on land, while the sea receded by some hundreds of kilometres from Kuzbas.

The Kol'chuginian Suite was deposited on the vast alluvial plains of a large enclosed region, separated from the sea by thousands of kilometres.

Abundance of Coal. The Ostrogian Suite has only a few thin coal seams lacking any economic importance. The Balakhonkian Suite is most important and has 30–34 workable seams, of which many are up to 10m thick and some reach 13–18m thick. Thereafter there is a break in coal accumulation, corresponding to the Kuznetskian, Il'inskian and Uskatian suites, which are altogether 2200m thick. These suites have no commercial coals and a part of them has been formerly called the Coal-free Suite.

The upper part of the Kol'chuginian Suite, which is nowadays called the Terunakovian sub-Suite, has a large number of workable seams of 0·5 to 7·25m in thickness. The succeeding Mal'tsevian Suite has no coals but there are a number of workable seams of different compositions in the overlying Jurassic strata. These coals, however, are not worked at present.

Tectonics

Structures belonging to the Precambrian, Caledonian, Hercynian and Cimmeridian orogenies can be found in Kuzbas. The Hercynian Orogeny, however, was here most important. The folds in the Balakhonian Suite, where it is marginal to the basin, are of some complexity (Fig. 160), but towards the centre of the basin the folds become simpler. In the Kol'chuginian Suite they are relatively gentle. The folds in the Jurassic rocks of the core of the basin are the simplest, since these rocks were only affected by the Cimmeridian movements.

Magmatism

The Precambrian and Upper and Middle Palaeozoic magmatism was the same as in other regions of the Angara Geosyncline. Rocks corresponding to the Siberian Traps are, however, peculiar to the Kuzbas and are represented by sills and lavas of mainly basaltic composition lying amongst the Mesozoic and Palaeozoic deposits. It is possible that as in the Siberian Platform the rocks of Upper Palaeozoic, Lower Trias and Jurassic ages are present. Their upper part forms the outcrop known as the 'melaphyre horseshoe' (see p. 465).

THE SAYANS AND TUVA

General Description

The mountainous tract of the Sayans and Tuva consists mainly of the outcrops of Precambrian and Palaeozoic rocks. The narrow and high mountain ranges of the

region are separated by wide, elevated plains, at heights of up to 800m. The
mountains consist of Precambrian and Lower Palaeozoic rocks while the rocks in
the valleys are Middle Palaeozoic and in places the continental facies of the Upper
Palaeozoic and the Jurassic.

To the north of the region there is the highest range of the Eastern Sayans with
peaks of 2500–3000m. To the west and south of it lies the industrially important,
large Minusinskian Depression, which is bounded to the south by the Western
Sayans with peaks of 2500–3000m. Behind the Western Sayans lies the Tuva

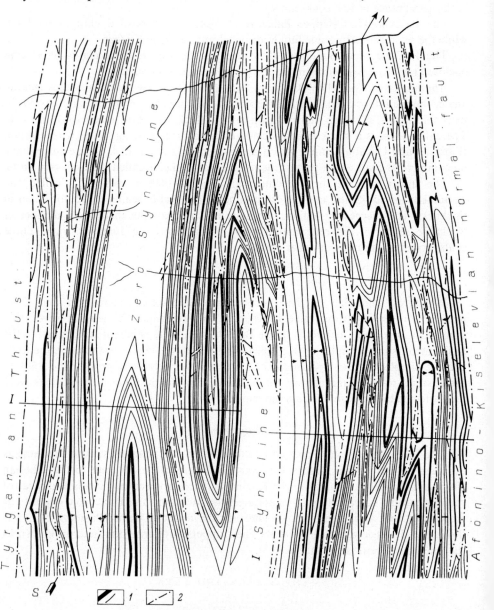

FIG. 160. Map of structures affecting the Balakhonkian Suite in the region
of the town Kiselevsk. According to V. I. Yavorskii (1957).

1 — coal seams;   2 — fault planes.

Depression with the central town Kysyl. To the south of the depression are the Tannu-Ola ranges with peaks of 2000–2500m and further to the south in the Mongolian Republic lies the salt lake Ubsu-Nur. Hereabouts is also the transverse deep and young graben-type depression of lake Kosogol (Khubsugol), which margins Tuva on the east.

Administratively the Eastern Sayans stretch through the Krasnoyarsk region, the Irkutsk Province and the Buryatian A.S.S.R.

Stratigraphy

The typical feature of the region is the dominance of Precambrian and Lower Palaeozoic rocks, forming all the mountain ranges. The less widespread Middle Palaeozoic and even rarer Upper Palaeozoic and Jurassic strata apart, there are Cainozoic sediments situated in small young graben. These have associated young volcanoes and their lavas and tuffs.

In studying the Precambrian it is imperative to remember that the northern part of the Eastern Sayans with their Archeozoic and Lower Proterozoic rocks constitute an extension of the basement to the Siberian Platform. Only the southern part of the Eastern Sayans belongs to the Angara Geosyncline. This part, as well as the branching Western Sayans, consists mainly of Upper Proterozoic, Cambrian and Ordovician strata, of a total thickness of over 10km. All these rocks are strongly folded and metamorphosed. There are too few fossil localities. which complicates their dating. Until a comparatively short time ago the greater part of the strata was considered as Ordovician. It has, however, been demonstrated that its structure is complex and that there is a thick geosynclinal Upper Proterozoic sequence, which is unconformably overlain by the Cambrian and the thick clastic and effusive rocks of the Ordovician. There are thick complexes of carbonates of which the lower is Upper Proterozoic and the unconformable upper complex is Lower Cambrian. Bauxites and bauxitic rocks (Boksonian deposit) are often found associated with the unconformity.

Middle Palaeozoic. The Middle Palaeozoic succession is of Caledonian type. The Silurian is represented by continental and marine sediments. The Lower Devonian and the Eifelian consist almost exclusively of red beds with rare beds of shales and limestones with marine faunas. The thin Givetian limestones and shales are extensive and possess a characteristic marine fauna. The Upper Devonian and Lower Carboniferous are again mainly continental with red and variegated sandstones and shales. Towards the Siberian Platform the number and thickness of marine bands decrease. The south of the Siberian Platform has only been invaded by the Givetian transgression. The Devonian succession (Table 32) of the Minusinskian Trough (Chochia *et al.*, 1956) has been studied in detail, since there were indications of oil (Teodorovich, 1950).

Upper Palaeozoic. The Upper Palaeozoic is only found in the central part of the Minusinskian Depression. It is represented by a coal-bearing formation, which is very similar to the upper part of the Kuzbas succession. The formation consists of a thick group of brownish-grey sands and shales, which are relatively simply folded and gently metamorphosed and include 7 to 40 workable coal-seams of 0·7 to 14m in thickness. The total reserves in the area are 36×10^9 tons.

Mesozoic. The Jurassic is found in the centre of the Tuva Depression, lies unconformably on the Palaeozoic, is weakly folded and represented by a coal-bearing formation. The age and structure of these rocks is identical to the coal-bearing Jurassic strata of Kuzbas. The coal-bearing rocks of Tuva occur in the Ulukhemian coal-field with proved reserves of 1.3×10^9 tons (Losev, 1955).

Cainozoic. The Cainozoic is peculiar and is represented by two complexes of deposits: lacustrine and volcanic. Both types of deposits are associated with graben, where gravels, sands and clays derived from surrounding mountains accumulated in the graben lakes. Formerly these sediments were thought to be Anthropogenic, but pollen analysis has shown that they are more probably of a Neogene or sometimes even of Oligocene age. Thus the younger lavas are also probably pre-Anthropogene.

Tectonics

The most important feature of the Sayans and Tuva is the dominant nature of the Caledonian Orogeny. The Karelidian and Baikalian orogenies also undoubtedly affected the region, but their effects have been masked by prolonged and powerful Caledonian movements. Even at the end of the Upper Cambrian the Caledonian (Salairian) movements were already important. The late Ordovician movements were just as strong and together with the Salairian deformation they are responsible for the main folds of the Eastern and Western Sayans and the eastern Tannu-Ola. The late Silurian orogenic episode was relatively weak, but after them the whole region was elevated above the sea level. Thus in the Devonian there were only short-lasting periods, marked by Devonian transgressions, when epeirogenic movements caused the penetration of the sea into the area.

The Hercynian Orogeny was weak and the Cimmeridian even weaker. Consequently the Upper Palaeozoic strata are folded into relatively simple folds, with rare Hercynian granites. In the Jurassic the folds are even gentler and there are no synorogenic intrusions.

The tectonic map of N. S. Shatskii shows an obvious arc of Caledonian structures, which in the west is truncated by the Hercynian folds.

The young and considerable block faults in the Eastern Sayans are tectonically important. These can be compared with the East African rift-zone and called the Baikalian Rift. The latter rift-zone begins in the Mongolian Peoples' Republic with lake Kosogol, traverses the Eastern Sayans, forms the Baikal Depression, and continues into the Barguzin Trough and the valley of the Vitim. Movements on faults belonging to the zone are still continuing, thus the region registers a high seismicity and earthquakes of 8th and 9th scale.

Magmatism

The magmatism is related to the tectonics and is conditioned by them. Thus the Caledonian and Proterozoic intrusions and lavas are most extensive and the Caledonian cycle is responsible for the main metallogenesis. The young magmatism related to the Baikalian Rift is of some interest.

The younger vulcanicity has been described by M. L. Lur'ye (1954) and before him by V. A. Obruchev (1938). There were two eruptive epochs. The first epoch of a Miocene or possibly Upper Oligocene age involved sheets of basalts, which erupted out of fissures and covered considerable areas. The sheets vary from 100 to 350m in thickness, but may be less. At the end of the Pliocene and in the Anthropogene these lavas were affected by faults and elevated to the height of 1500m and more, thus giving rise to plateaux. The second generation of basalts is related to individual volcanoes in the environment dominated by an appreciable relief, similar to the present day. The lavas flowed on the bottom of the present-day valleys while the volcanic centres were situated on the watersheds and flanks of ridges produced by young faults. The lavas are very extensive and are found in the eastern part of Tuva and the south-eastern part of the Eastern Sayans, as well as the Khamar Daban range, Western Transbaikalia and the adjacent areas of the Mongolian Peoples' Republic.

The Cainozoic basalts of Eastern Siberia occur in the interior of the continent and were not associated with marine basins. They thus resemble the Siberian Traps and the Devonian basalts of the Russian Platform.

EASTERN PART OF THE ANGARA GEOSYNCLINE

The eastern part of the Angara Geosyncline is situated to the east of lake Baikal and is parallel to the southern margin of the Siberian Platform. The region includes Western Transbaikalia, the Yablonovoi range, and systems of smaller ranges parallel to the latter. Further to the east is the Stanovoi range, which ends in the Dzhugdzhur range. The northern and the southern boundaries of the region are indistinct. In the north the Angara Geosyncline is bounded by the Upper Palaeozoic structures, which are unconformably overlain by the platform-type Cambrian un-intruded by the Caledonian massifs. Within the Angara Geosyncline Cambrian rocks are strongly folded, metamorphosed, contain lavas and are cut by the Caledonian intrusions. In the south the boundary is traced along the Cimmeridian structures of Eastern Transbaikalia, where Jurassic rocks are thick, folded, meta-morphosed and intruded by Cimmeridian granites. In the Angara Geosyncline Jurassic rocks are almost horizontal, unmetamorphosed and lack intrusions. The changeover from the Upper Proterozoic to Palaeozoic and from the Palaeozoic to Cimmeridian folds is gradual over tens of kilometres, rendering the boundary in-distinct.

The relief of the Yablonovoi range is similar to the Southern Urals. The range consists of narrow, parallel, north-west trending ridges of maximum heights of 1600–1700m. The strike valleys are wide and the transverse valleys are narrow. The rivers are full and are related to the upper Vitim and the upper Olekma. The similarity to the Urals is explained by the similarity of the geological structure, including the young Pliocene and Anthropogene faulting.

The Stanovoi range is similar to the Eastern Sayans, is up to 2500–2800m high and has no individual ridges. It also has been formed by the young block faulting. The Palaeozoic structures of the Stanovoi range in the east disappear under the Mesozoic structures of the Dzhugdzhur range, which belongs to the Pacific Ocean Geosyncline. The Stanovoi range is drained by the tributaries of the Amga, the Aldan, the Maya and in the south by the tributaries of the Amur and in particular by the Zeya.

The geology is characterized by the abundance of metamorphosed sedimentary and volcanic rocks of Palaeozoic and Proterozoic ages, and by Caledonian and Hercynian intrusions, which are very abundant. Owing to the strong metamorphism fossils are rarely found, thus the fossil localities in the folded and metamorphosed Cambrian rocks of the Upper Vitim are very important. The marine fauna here is abundant and diverse. In this and other regions there also have been discoveries of marine Ordovician and Silurian faunas. In the north the Middle Palaeozoic is of Caledonian type and consists mainly of red beds. In the southern zone a Hercynian type of succession with the mainly marine Devonian and Carboniferous is widespread. The Upper Palaeozoic is only found in tectonic depressions, as for instance in the Selenga region. It consists mainly of continental clastic rocks and lavas, but in places (Chikoi) there is the marine Lower Permian.

Mesozoic and Cainozoic rocks are only found in graben and are entirely continental. Jurassic and Lower Cretaceous rocks are often coal-bearing and hundreds of metres thick (Florensov, 1956). The Cainozoic is thinner and has the younger lavas found over large areas around the Upper Vitim and east of Barguzin. The Cainozoic of the Barguzin Trough consists of 636m of Anthropogene sands and shales and 777m of Pliocene sandstones, sands and shales, lying on Archeozoic gneisses.

Numerous intrusions are associated with ores, such as the well-known iron deposit (Petrovskian-Transbaikalian). There are also small deposits of rare metals and gold sands. There are also Jurassic and Lower Cretaceous black and brown coals (Gusinyi Ostrov coal-field). In the Udokan range (the middle Vitim) there is the large Udokan deposits of copper found in Precambrian rocks. Further search will no doubt expand the known reserves of this interesting, but isolated region of the Angara Geosyncline.

9

MIDDLE ASIATIC GEOSYNCLINES

GENERAL DESCRIPTION

The northern ranges of Central Asia are parts of the Angara Geosyncline, while the central ranges represent a continuation of the Urals Geosyncline. Lastly the southern ranges are parts of the Mediterranean Geosyncline. From a formal standpoint Central Asia should be divided between the three geosynclines but this would lead to the disappearance of one of the most interesting geographically and economically uniform regions of the U.S.S.R. and it has been left as a unit.

BOUNDARIES. The northern boundary is very clear and distinct. It is represented by a series of desert depressions, continuous into each other. To the north-east of the Caspian Sea the deserts lie to the north of Ustyurt and include the Aral Sea, Lower Chu, Balkhash and Ala-Kul depressions, ending against the mountain pass of Dzhungarian gates (Fig. 151). The other boundaries of the region coincide with the frontier.

Tectonic structures, which converge in Central Asia, diverge beyond its boundaries and curve round the massif of the Takla-Makan desert (the Tarim Massif), and this is obvious on Shatskii's tectonic map.

GEOLOGICAL REGIONS. Central Asia is divided into three systems of mountain ranges: the northern, the central and the southern. These ranges are also systems of arcuate structures and are at times referred to as ranges and at others as arcs.

The northern ranges continue from the western part of Central Kazakhstan and show a predominance of Precambrian and, especially, Palaeozoic folds. These ranges have widely developed marine Proterozoic and Lower Palaeozoic sediments and Precambrian and Caledonian intrusive massifs, with associated mineralization. The southern margin of the northern ranges follows the valley of Syr Dar'ya and of its main tributary—the Naryn—to the Khan-Tengri Massif.

The southern ranges belong to the Mediterranean Geosyncline showing all its characteristics, including the distinct manifestation of the Hercynian Orogeny and the subordinate effects of the Precambrian, Caledonian and Cimmeridian orogenies in the north and the Alpine Orogeny in the south. The Hercynian massifs of intrusive rocks are most common. The Middle Palaeozoic is most widely developed with a fairly full succession, consisting mainly of marine sediments. The southern boundary of the central ranges follows the lower reaches of the Amu Dar'ya, the vale of Zarafshan and the Alai valley.

The southern ranges belong to the Mediterranean Geosyncline showing all its features including powerful manifestations of the Caledonian and Cimmeridian orogenies and a widespread development of young intrusions and marine Mesozoic and Cainozoic sediments. The southern ranges include the Kopet-Dags, the Mangyshlak area, the Karakum desert, the ranges of southern Tadzhikistan and the Pamirs. All the southern ranges are convex northwards. Within them the Neogene is folded and there are Neogene oil accumulations.

The administrative subdivisions of the area are shown on Fig. 161.

OROGRAPHY

The relief of Central Asia involves mountains, hills and deserts. The mountains in the north consist of Precambrian and Palaeozoic rocks and in the south of Precambrian, Palaeozoic and Mesozoic rocks. The foothills are formed of moderately folded younger sediments of Upper Palaeozoic and Mesozoic ages in the north and of Cretaceous and Cainozoic ages in the south. The deserts are alluvial plains covered by almost horizontal Neogene and Anthropogene deposits, which are in places overlain by aeolian sands. Most of the cultivated land and towns are found in the foothills. The mountain ranges are numerous, often long and the highest in the U.S.S.R. The peak on the Akademia Nauk range is 7495m. The peak Pobeda in the Khan-Tengri range is 7489m and the Lenin peak in the Zalairian range is 7134m. There are many peaks of over 6500m.

The mountains of the northern and southern areas are of different origins. The northern ranges are young, having risen from a peneplain by block faulting, involving Palaeozoic and Precambrian rocks which were uplifted at the end of the Pliocene and in the Anthropogene epochs. Their structures are Caledonian or Hercynian. Since these mountains were parts of a plain their summits are also at present planar and are at levels of 4–5km. Such high-level areas form good pastures and are known as 'syrty'. There are no distinct peaks and the flat nature of the tops can be seen from a distance.

The southern ranges typical for the Pamir region are constructed of young folds, which came into existence in the Neogene and Anthropogene. These ranges are of the same type as the present-day Alps, reaching great heights and having numerous peaks rising above the surrounding mountains (Fig. 162). There are no syrty and the mountains are jagged. The folds are being produced even at present. This is demonstrated by the folding of the terraces of rivers and numerous strong earthquakes.

The central ranges are produced by block movements in the north and folding in the south. Their topography is most variable and complex (Fig. 163).

The foothills form a more or less narrow belt around the mountains. The foothills are often merged alluvial cones. Owing to the smooth relief, abundance of water in the rivers flowing out of the mountains and a widespread development of loessic soils these areas are good agricultural regions and have dense populations and large provincial centres: Alma-Ata (Kazakh S.S.R.); Frunze (Kirgiz S.S.R.); Tashkent (Uzbek S.S.R.); Dushanbe (Tadzhik S.S.R.) and Ashkhabad (Turkmen S.S.R.).

The zone of the foothills is especially wide in large intermontaine valleys of which the vale of Fergana is the largest and is of great economic importance. The valley of Zerafshan with the ancient cities of Samarkand and Bokhara is of importance.

The deserts also occupy large areas stretching along the northern boundary of

FIG. 162

Middle Palaeozoic strata of the Zerafshan range exhibiting a typical Alpine relief.
Photograph by V. R. Martyshev.

FIG. 163

A typical V-shaped erosion valley in Middle Devonian limestones. In the distance a series of transverse ridges diverging from the main Alai range.
Photograph by D. I. Mushketov in 1912.

Central Asia for a distance of 2500km. Their width in Karakum and Kyzyl Kum is 900km to 1200km. The most important feature of the deserts is the small annual precipitation of no more than 150–200mm. In the summer there is no rain at all and in the winter there is very little snow. Permanent rivers and freshwater lakes are almost entirely absent and only the lower reaches of the larger rivers such as the Amu Dar'ya, Syr Dar'ya, Chu and Ili cross them. The other feature of the desert is the blanket of aeolian sands. It is rather typical, but is not seen everywhere. Sometimes large areas have no sand and the desert is stony or clayey as is for instance the desert of Ust'yurt.

The deserts are generally plains with occasional low, flat and branching elevations. The largest of these elevations is the Ust'yurt plateau, the higher parts of which consist of compact horizontal Sarmatian limestones. The abrupt marginal cliffs of the plateau are known as chinks, which reach heights of 120m. To the south of the Ust'yurt the plateau is eroded by a former river, the dry valley of which is known as the Uzboi which connects the Sarykamyshian Depression (formerly a lake) with the Caspian Sea. To the south of the Uzboi the Sarmatian Highlands are eroded and known as the Zaunguzian Plateau, the southern slope of which is marginal to the dry valley of 'Unguz'.

Amongst the sands of the Kyzyl Kum desert rise a number of lifeless mountains of Palaeozoic rocks known as the Kyzyl Kum Highlands. The low fold range known as the Mangyshlak and its continuation of the Tuar-Kyr Highlands consist of Jurassic sediments and also desert deposits. The sands of the deserts are underlain by sandy-clayey river deposits, often with freshwater faunas. This indicates that formerly the deserts were river valleys and that their surface represents alluvial plains. The action of wind separates the coarser aeolian sands from the finer dust which is blown away to be deposited in the foothills as loess. The loess is agriculturally a very productive sediment.

Amongst our deserts the Karakum (Black Sands) is the largest and the Kyzyl Kum (Red Sands) is the second largest. Behind the Kara-Tau range there is the much smaller Moyun Kum (Camel Sands) and to the north of it is the semi-desert of Betpak-Dala (The Hungry Steppe). Owing to the persistence of D. I. Yakovlev a large artesian basin has been found under the Hungry Steppe and it has become a cultivated pasture region.

The rivers of Central Asia are numerous and full, since they are fed by permanent ice and snow of the mountains. In July the rivers are especially full since the ice and snow melt at the fastest rate. The rivers produce large alluvial cones as they emerge out of the mountains. The cones are often hundreds of metres high and consist of coarse, conglomerates, sands and gravels, which merge into more distant sands and silts. Owing to the fast currents of these rivers the pure clay rarely gets deposited. The largest river is the Amu Dar'ya which rises from the southern ranges. Pandzh and Surkhob are its tributaries (Fig. 161). The Syr Dar'ya is also large and is fed by the snow and ice of the central Tian Shian. It arises out of the confluence of the Naryn and the Kara Dar'ya. The Zarafshan, which rises in the Gissar range, has much water, which is entirely used up in the irrigation of Samarkand and Bokhara, and consequently the river does not reach the Amu Dar'ya. The Chu which drains the northern Tian Shian is much smaller and gets completely lost in the desert. The Ili—another large river of the northern Tian Shian—enters lake Balkhash. The Murgab and the Tedzhen are small rivers forming dry deltas in the southern part of the Karakum.

STRATIGRAPHY

All systems can be found in Central Asia and both their marine and continental facies are exceptionally well represented. In the northern ranges the Lower Palaeozoic and Precambrian sediments predominate; in the central ranges the Lower Palaeozoic rocks, and in the southern ranges the marine Trias, Jurassic, Cretaceous and Cainozoic, are widely developed.

Precambrian

Precambrian rocks are most extensive in the northern ranges, almost absent in the central ranges and are sometimes found in the southern ranges.

Archeozoic

The Archeozoic consists of highly metamorphosed thick gneisses (migmatites, gneissose granites), amphibolites and mica-schists found in the Transilian Alataus, the Kirgiz range and the Kandyk-Tas range. The Proterozoic rests on the Archeozoic unconformably and is in turn overlain by the fossil-dated Lower Palaeozoic.

The Archeozoic age of the lower part of the crystalline rocks in the south-western Pamirs (Vakhanskian Series) is proved by analogy with the identical migmatites, dolomite marbles and gneisses of Kashmir, where they underlie a thick Proterozoic formation, overlain by fossiliferous Cambrian rocks. In the Pamirs there are thick, bedded dolomite marbles, interbanded with great thicknesses of gneisses, crystalline schists and migmatites. The total thickness is at least several kilometres. Deposits of lapis lazuli and spinels, as well as other minerals, are found in association with marbles. Such mineral associations are also found in the Archeozoic of southern Cisbaikalia. The relationship of these rocks to Palaeozoic strata are unknown since all the observed contacts are mechanical. Nevertheless, the neighbouring Ordovician and Devonian rocks differ sharply in their grade of metamorphism and consist of limestones, slates and quartz sandstones typical of the Palaeozoic formations.

Archeozoic rocks maintain a uniform grade of metamorphism, which cannot be attributed to local contact effects. This permits the inclusion of all these rocks in the Precambrian, despite the expressed opinions that their age is Lower Palaeozoic. With the true palaeontologically-dated Lower Palaeozoic they have nothing in common. The Precambrian age of the crystalline rocks in the Pamirs is proved by the discovery of Proterozoic spores in them (Timofeyev, 1958).

Lower Proterozoic

Lower Proterozoic rocks have been observed in the northern ranges, where they are overlapped by Upper Proterozoic rocks, which are overlain by Lower Palaeozoic sediments. Lower Proterozoic strata are found in the Kirgiz range and in the Transilian Alataus, where they are divided into several suites. V. A. Nikolayev (1952) has combined these suites under the general term of 'the Lower Tian Shian complex'. In the west of the Kirgiz range a Spilitic Suite forms the top of the succession. It is underlain by the Kaindinian Suite of mica-chlorite schists and marbles. Below this suite are the garnet-chlorite mica schists, marbles, amphibolites and quartzites of the Nel'dinian Suite. At the bottom of the depression is the Makbalian Suite consisting of quartzites, quartz-mica-garnet schists, marbles and amphibolites. The total thickness of the complex is 4000–6000m. Lower Proterozoic spores have been found by B. V. Timofeyev (1958) in the Kaindinian Suite. In the central ranges no Proterozoic or Archeozoic rocks have been recognized.

In the southern ranges Lower Proterozoic rocks are found in the south-western Pamirs and in the Karategins. The top part of the Vakhanskian Series of the Pamirs is considered as Lower Proterozoic. In the Karategins the upper part of the analogous Garmian Series is correlated with it. These correlations are given in the conclusions of the Tashkent stratigraphic conference (Conclusions . . . 1959, Table 1).

Upper Proterozoic

In the Gissarian range and the Karategins the Upper Proterozoic includes the Barzanginian Series, which consists of sericite-chlorite schists, quartz-chlorite schists, graphitic schists, quartzites and hornfelses, reaching altogether 2000m. B. V. Timofeyev has found Upper Proterozoic spores in this series.

The Upper Proterozoic has also been identified in the northern ranges where it is represented by the Upper Tian Shian complex (Nikolayev, 1952). Nikolayev includes the metamorphosed slates and sandstones of the Karoiian Suite (3500m) of the Karatau in the Upper Complex. This suite is unconformably overlain by the Tamdynian limestones, which have a Cambrian fauna. He also includes the phyllitic slates and polymictic chlorite-sericite sandstones of the Lesser Karatau and the Talassian Alatau into the same complex. Owing to the complexity of their tectonics the thickness of the Upper Proterozoic is provisionally determined as 3–5km. Later on, the upper part of the Karoiian Suite has yielded beds of archeocyathid limestones. Thus only the greater, lower part of this suite can be considered as Proterozoic in age.

The Kakdzhotian Suite of the Karatau undoubtedly belongs to the Upper Proterozoic. This suite (2000–3000m) consists of greenish and steel-grey phyllites and cleaved sandstones.

After the publication of the synoptic paper by V. A. Nikolayev (1952) several other papers on Central Asia have appeared. As an example the dissertation of V. G. Korolev (1957) on the Precambrian of the Terskian-Alatau region and the neighbouring areas can be quoted. The succession proposed in this dissertation is as follows:

Lower Archeozoic (Aktyuzian Suite of 3000–5000m). The rocks are gneisses, amphibolites, marbles and quartzites, widely affected by migmatization.

Upper Archeozoic (Keminian Suite of 3000–5000m), includes mica schists, micaceous quartzites and subordinate marbles.

Lower Proterozoic (Sarybulakian Suite—5000m). Quartz-sericite and chlorite schists and marbles form the main part of the suite, but there are subordinate meta-poryhyries and porphyroids.

Upper Proterozoic. The Terskeiian Suite (3000m) consists of metamorphosed porphyrites, tuffs, chloritic slates and subordinate diabases and marbles. There are intrusions of quartz diorites and granitoids.

The Bol'shenarynian Suite (1500m) has acid lavas and tuffs at the bottom and slates, quartzites and marbles at the top.

An angular unconformity follows.

Sinian (Dzhetymian Suite—1500m). At the bottom tillite-like conglomerates and sandstones, carbonaceous shales and ferruginous quartzites. At the top variegated shales, sandstones and haematite-chert rocks. In the Dzhetym range there is a major deposit of iron ores, found in the ferruginous quartzites.

Lower Cambrian-Lower Ordovician (Shortorian Suite—500 to 800m). At the bottom there are tillite-like conglomerates and sandstones (50m) overlain by the vanadium- and molybdenum-bearing shales, coals and cherts. Higher still are Upper Cambrian limestones and Lower Ordovician fossiliferous shales.

Middle and Upper Ordovician (Dzhebaglinian Suite—200 to 250m). The suite consists of flyschoid, greenish, peppery sandstones and shales with fossiliferous marine bands and manganese-bearing siliceous groups.

In neighbouring areas other geologists distinguish other suites. Thus the Precambrian of the district is not sufficiently known. The discovery of jaspilitic iron ores is of some importance.

PALAEOZOIC

Lower Palaeozoic

The thick Lower Palaeozoic formations are mainly marine, folded, strongly metamorphosed and vary from thick limestones to thick sandstone-shale groups, some of which may contain continental deposits. Fossiliferous localities are relatively rare, which complicates the dating and differentiation from the Upper Proterozoic. Thus the study of the Lower Palaeozoic is at present an important problem facing Central Asiatic geologists.

DISTRIBUTION. The Lower Palaeozoic is widely represented in all northern ranges, in the Chu-Ilian Mountains, the Karatau, the Talasskian Alatau, the Kirgiz range, the Transilian Alatau, the Kyngei-Alatau and Terskei-Alatau, where it is shown on the maps as Ordovician and occasionally as Cambrian. In the central ranges Lower Palaeozoic is exposed much more rarely, occurring in uplifted tectonic structures. On the map of the U.S.S.R. to the scale of 1:7 500 000 Lower Palaeozoic rocks are shown as undifferentiated Silurian or Palaeozoic. In the southern ranges Ordovician strata are found over small areas and while the Cambrian is undoubtedly developed so far it has not been found.

Cambrian

Full Cambrian successions with all three subdivisions present are only known in the northern ranges where they are widespread.

In the Karatau range the eastern slope shows the thick (2000–3000m) Tamdinian limestones at the top of which an Ordovician fauna has been found and at the bottom of which rocks above a phosphorite-bearing group have yielded a fauna of Middle Cambrian trilobites. This group (75m), known as the Chulaktauian Suite, is mainly calcareous and overlies on the Karoiian Suite (1000m) of sandstones, shales and lavas. At the top of this suite beds of archeocyathid, Lower Cambrian limestones have been discovered. At the southern end of the Karatau the Cambrian succession begins with tillite-like breccias (Baikonurian Suite), overlain by the vanadium-bearing carbonaceous and siliceous shales (Kurumsakian Suite) overlain by Middle and Upper Cambrian limestones (200–400m), which in turn are covered by graptolitiferous Arenig shales. This succession has been observed far to the east (the Dzhetym range).

The range Kandyk-Tas, to the east of Frunze, shows the following succession:

1. at the base lie conglomerates (90m) with pebbles of syenites and porphyries;
2. quartzitic sandstones, phyllitic, chlorite-sericitic and aspidic shales with lenses of marble (500–700m);
3. grey and green arkosic and micaceous sandstones and quartzites (1000m);
4. porphyrite tuffs and lavas, tuffaceous sandstones and conglomerates and sandstones (550m), which are unfossiliferous and may be Lower or Middle Cambrian, although owing to the unconformity, which separates them from Upper Cambrian strata, they are more likely to be Upper Proterozoic in age;
5. unconformable dark-green sandy and clayey shales and often coarse sandstones, reaching 1000m and containing the Upper Cambrian brachiopod *Acrotreta*;
6. greenish-grey, fine-grained, muddy sandstones with dark violet shales, Ordovician, with *Asaphus* and *Illaenus*.

Similar formation of clastic metamorphosed sediments are widely found in the northern ranges, but no fauna has been found in them and it is probable that they are in the main of an Upper Proterozoic (Sinian) age.

In the central ranges Cambrian sediments occur in the Alai range, consisting of dark, bedded, siliceous and muddy fine-grained limestones with microscopic brachiopods (*Acrotreta*, *Lingulella*) and trilobites (*Agnostus*)

To the west, in the Turkestan range, Cambrian outcrops form a narrow, discontinuous zone. Within this zone there are dark and light grey, chloritic, alum-bearing and calcareous shales with groups of sandstones and limestones, with local developments of lavas. The thickness of these rocks is 2000–5000m. The fauna suggests that they are mainly Middle Cambrian, although in places there are archeo-cyathid strata of the Lower Cambrian.

The Cambrian succession of the Turkestan and Alai ranges has been most fully studied (Yaskovich, 1958; Decisions . . . 1959, Goryanov *et al.*, 1961) and palaeonto-logically dated. The Aldan Stage is not found, while the Lenian Stage is represented by the Altykolian Suite (625m) of limestones, shales, siliceous shales and sandstones with archeocyathids and trilobites. The Amgian Stage consists of two suites. The lower, Shodymirian Suite is 1600m thick, variable and consists of siliceous and normal shales, sandstones, lavas and limestones with a rich fauna of trilobites. The upper suite (Sulyuktinian—800m thick) consists of siliceous and normal shales with lenses of trilobite-bearing limestones. The Maian Stage is represented by the Ravutskian Suite (2700m) of shales, limestones and sandstones. A rich trilobite fauna permits its subdivision into two sub-suites. The Upper Cambrian strata consist of shales with lenses of trilobite-bearing limestones. The Upper Cambrian is often missing and Ordovician, Devonian, Carboniferous and even Jurassic sediments rest unconformably on rocks of Middle Cambrian age.

Far to the east, according to V. G. Korolev (1957), in the Dzhetym range (to the south of the Terskian-Alatau) the tilloid breccia is overlain by carbonaceous and siliceous shales of a Lower Cambrian age. The shales are succeeded by thick Middle and Upper Cambrian limestones.

It is possible that in the southern ranges and in particular in the Pamirs, certain thick shales, quartzites and limestones with Ordovician faunas at the top may involved Cambrian deposits at the bottom. Yet, so far, no Cambrian fauna have been found in the Pamirs.

Ordovician

The initial movements of the Caledonian Orogeny are known as the Taconic phase. They first occurred in the Ordovician period and are characteristic of it. The movements began at the beginning of the Ordovician and, with interruptions, continued to the end of it. Thick flysch-like and molasse-like deposits were produced by these movements. These sediments are often rhythmically bedded and are of graywacke type. There are widely developed lavas. Such sediments are widely developed in the north of the northern ranges, while in the southern ranges normal, marine limestones, shales and siliceous shales and sandstones (often quartzitic) are found.

The Ordovician is found over large areas in the north and is most fully developed in the western ranges, such as the Karatau, the Talassian Alatau and the Kirgiz range. Further to the east the Ordovician is found in all the ranges, but is less fully

Fig. 164. Massive Lower Palaeozoic limestones and quartzites of the Western Pamirs at the confluence of river Yazgulem and the Pandzh. *Photograph by D. V. Nalivkin (1915) showing an expedition fording the river.*

represented and its rocks occupy smaller areas. In the central ranges Ordovician strata are still rarer, with very few fossiliferous localities. Hence on the 1:7 500 000 map it is not separated from the Silurian. In the southern ranges Ordovician sediments have only been found in the Central Pamirs, in the Yazgulem range and in the region of Rang-Kul (Fig. 164).

The Ordovician successions have been given in the 'Conclusions' (1959, Table 3), and only the fullest are given here. In the south, in the Karatau, the following succession has been established by O. N. Khaletskaya (1958).

Upper Devonian (Tyul'kubashian Suite). Red sandstones and conglomerates are found.

Unconformity

Caradoc. Consists of greenish graywacke sandstones (600m) with silty bands, groups of lavas and orthids and strophomenids.

Llandeilo. The rocks are mainly shales with beds of sandstones, numerous trilobites (*Ampyx*) and rarer brachiopods. Total thickness of the strata is 450–500m.

Llanvirn. Consists of greenish argillaceous and calcareous slates (50m) with numerous graptolites (*Didymograptus*).

Tremadoc. At the top of the series there are shales with beds of limestones, 100–120m thick and with numerous brachiopods. In the middle of the series there are interbedded limestones and *Dictyonema* shales, reaching totally 250–300m. At the bottom there are massive, algae-bearing limestones and dolomites of 200–250m in thickness.

Thus the succession consists of carbonate-shale sediments at the bottom, graptolitic shales in the middle and brachiopod-trilobite bearing graywackes and shales at the top.

In the Kirgiz range there are two thick terrigenous series, arbitrarily assigned to the Middle and Upper Ordovician and having a thickness of 7000–10 000m. The lower series is known as the Linguloid and consists of interbedded sandstones and shales, which are reddish at the top and greenish-grey towards the bottom of the series. The fauna is found in thin bands and consists of lingulids and gastropods. The upper suite (Dzhartashian) also consists of shales and sandstones and is generally barren. Only in places are thin limestones with *Ampyx*, *Illaenus*, *Asaphus* and strophomenids found. The great thickness of these rocks is surprising.

Further to the east, in the Transilian Alatau there are thick groups of metamorphosed strata of phyllitic shales, sandstones and metamorphosed limestones, which in places have yielded *Remopleurides* and *Illaenus* and some brachiopods. Similar rocks are also found in the extreme east and in particular in the Dzhungarian Alatau (Decisions . . . 1959).

In the most northerly ranges the Ordovician succession is complete and is palaeontologically dated (D. I. Yakovlev). In the mountainous massifs of the Betpak-Dala steppe a bed of limestone in a succession of sandstones and shales has yielded a Tremadocian fauna, which has both the Cambrian-type *Agnostus* and Ordovician forms such as *Illaenus* and *Pliomera*.

In the Chu-Ilian Mountains (Fig. 165) there is a shale-sandstone suite, of some 1000m in thickness, in the lower limestone members of which a trilobite fauna of *Asaphidae*, *Illaenidae* and *Nieszkovskia* has been found. A higher sandy-conglomeratic formation is found in another region, where ancient granites are overlain by basal sandstones and conglomerates, which towards the top pass into a formation of argillaceous and calcareous graptolitic shales (100m) with various representatives of *Didymograptus* and *Climacograptus*. Some 200m above the base of the succession there is a limestone bed with asaphids lying amongst argillaceous shales. Still higher is a very thick formation of dark, barren shales with groups of sandstones and conglomerates. The first group of conglomerates is 100m thick and lies 600m above the base, while the second group (150m) is 2100m above the base of the succession. The second group is overlain by sandstones with a marine fauna (*Asaphidae, Illaenus, Scutellum, Lichas, Pliomera*). The very top of the succession 2500m above the base consists of shales with a bed of limestone containing strophomenids. In several other successions dark-grey Caradocian limestones with a variable brachiopod fauna are common.

In the Dzhetym range (south of Issyk-Kul) the Ordovician is represented by three divisions. The lower (200m) consists of black slates with atrematous brachiopods and graptolites. The middle and upper divisions (600–2000m) are represented by a thick formation of greenish sandstones and shales with a marine fauna (Korolev, 1957). In the central regions Ordovician faunas are rare. In the western part of the Turkestan range there are thick shales with beds of sandstones and occasional seams of graptolitic shales with *Didymograptus* and limestones with *Illaenus*. In the central Alai range the Cambrian is conformably overlain by Ordovician shales and limestones (Goryanov *et al.*, 1961). The trilobites suggest Tremadocian and Arenigian stages. Higher up there are limestones with Llandoverian tabulates. In the eastern part of the range grey-green, pepper-sandstones and graywackes with tabulates are typical.

FIG. 165. Shales of Ordovician age forming the rounded peneplained uplands of the Chu-Ilian Mts. *Photograph by L. I. Krys'kov (1957).*

In the southern ranges—the Central Pamirs and the Yazgulem range—thick quartzites, slates and marbles are found. At their top Upper Ordovician trilobites have been discovered, while at their bottom an Arenigian fauna is found. Upper Ordovician brachiopods are also found in shales and limestones of up to 1500m thick, occurring in the Rang-Kul district.

PALAEOGEOGRAPHY. The successions, so far quoted, show the variability of Lower Palaeozoic strata and the predominance of marine formations amongst them. No full palaeogeographic reconstruction can be made for Cambrian times, since the available data are very fragmentary.

For the Ordovician the available data are more complete and allow a deduction about the predominance of marine conditions and the accumulation of thick geosynclinal formations throughout Central Asia. In the second half of the period the northern part of the region saw the growth of young fold ranges, at the edge of which thick formations of rhythmically bedded sandstones and shales of flyschoid kind were deposited. In places sandstones alternated with conglomerates of molasse type and graywacke suites.

Middle Palaeozoic

The Middle Palaeozoic is widespread and its rocks which are variable, but especially in the south predominantly marine, are rather thick. In the northern ranges, the marine facies apart, there are red beds which are the deposits of near-shore plains. As a result of the Erian phase of the Caledonian Orogeny the succession in the northern ranges is of Caledonian type—thus Lower Devonian and Eifelian strata are either continental or missing. In the central or southern ranges the succession is of Hercynian type and the Devonian is represented by marine sediments and lavas (Fig. 166).

In the central ranges Middle Palaeozoic strata predominate in most structures; in the northern ranges they are less important than Palaeozoic and Precambrian rocks while in the southern ranges they occur as narrow belts in the cores of Mesozoic structures.

Ore deposits of Middle Palaeozoic age are numerous. In particular the antimony-mercury ores found in the Devonian limestones of the Alai range and the lead in the Lower Carboniferous limestones of certain northern ranges can be singled out.

Silurian

In the central ranges Silurian strata are widely found and all Silurian stages and divisions can be identified. The rocks consist mainly of limestones and shales (Fig. 167). In places there are thick groups of lavas, amongst which spilites (pillow lavas) are especially characteristic. The latest general papers on these rocks are by O. N. Khaletskaya (1958) and A. E. Dovzhikov (Dovzhikov et al., 1959; Decisions ... 1959).

In the northern ranges owing to the Caledonian Orogeny high fold-ranges were produced in the sea and in these areas the Silurian is missing and the Devonian is unconformably on the Ordovician. Around the ranges a belt of conglomerates and sandstones were laid down and passed laterally into fine-grained rhythmically-bedded flyschoid clays and marls and bedded limestones with rich faunas. In depressions the Silurian succession is thick and complete and in places involves volcanic rocks.

In the southern ranges Silurian rocks are found in cores of anticlinoria, where the strata consists of thick limestones and equally thick shales and sandstones. Here marine faunas are commonly encountered. In general the succession is the same as in the central ranges.

Fig. 166. Middle Palaeozoic limestones and shales in the Zerafshan range, with a geologists' camp at the bottom. The peak consists of Hercynian limestones. (See Fig. 168.) *Photograph by V. P. Martyshev.*

SUBDIVISIONS. The subdivisions are the same as in the Urals and Western Europe. The Lower Ludlovian and Upper Ludlovian stages are especially fully represented. The main successions are shown in Table 37 and in the 'Decisions' . . . (1959).

FAUNA AND FLORA. The fauna and flora are very similar to those of the Urals. Amongst the fauna, brachiopods (Nikiforova, 1937) and tabulates are most common and are sometimes rock-forming. Trilobites, gastropods, lamellibranchs and rugose corals are also normally found. The flora, consisting of psilophytes, is rare.

FACIES. The marine facies predominate and amongst them bedded, grey and dark-grey brachiopod limestones are most common and in addition to brachiopods have other fossils in them. Banks of pentamerids formed by the heavily-ribbed *Conchidium* and smooth pentamerids and *Brooksina* were characteristic of Silurian times. In many limestone massifs of hundreds of metres in thickness (the Hercynian limestones) the lower horizons have Ludlovian faunas. The limestones formed of tabulate corals are no less abundant and are usually dark, argillaceous, bedded and teeming with tabulates, rugose corals and other fossils. These limestones are interbedded with shales and pass laterally into them, but nevertheless often reach thicknesses of several hundred metres.

TABLE 37. *Silurian Successions in Central Asia*

Stage		Northern type: Turkestan, Alai and Kokshaalian ranges		Southern type: Zarafshan and Gissar ranges
Upper Ludlow		Pale, massive, more rarely bedded limestones with a rich brachiopod fauna; assigned to the Kunzhakian (Marginalian) Horizon (upper Upper Ludlow) on finding *Atrypa marginalis*. (150–200m).	Matchaian Suite	Grey, thinly-bedded brachopiod limestones. (80–100m). In places, pale massive reef limestones. (300–400m).
		Isfarian Horizon (lower Upper Ludlow)—grey bedded limestones with tabulates and brachiopods. (300–500m).		Pale, thinly-bedded limestones with rugose *Pholidophyllum*. (Pholidophyllian limestones). (400m).
Lower Ludlow		Dal'yanskian (Pentamerian) Horizon (Lower Ludlow)—grey, bedded limestones with banks of pentamerids and various tabulates. (800m). Merishkorskian Horizon—dark, bedded limestones with tabulates.	Pul'gonian Suite	Dark, almost black, bedded limestones and dolomites with horizons made of Amphipora. (Amphiporian limestones). (Up to 600m).
Upper Wenlock		Arkhakarian Suite of dark shales and sandstones with, in places, graptolitic bands and lenses, and suites of limestones with tabulates. (200–600m).		Grey and greenish, chloritic sericitic and calcareous slates with dark foliaceous limestones possessing tabulates and brachiopods. (500m).
Llandovery and Lower Wenlock		Syugetian Suite of interbedded dark shales and sandstones with sequences of basic lavas and tuffs. Graptolitic shales occur. (From 300–400m. to 2000–3000m).		Dark, homogeneous, thick shaley formation with beds of graptolitic shales (slates) and sequences of sandstones. (2000–4000m).

Shales are dominant in the lower part of the succession, and include graptolitic fine-grained, compact, calcareous and even sandy varieties (Fig. 167). The graptolites have been investigated by A. M. Obut (1958).

Devonian

Devonian strata are well developed, thick and mainly marine, but more rarely continental.

Fig. 167. Middle Palaeozoic rocks of the Kaklich-Uchar ridge near to the town Osha and forming a part of the Alai range. The ridge consists of Upper Ludlovian, Devonian and Lower Carboniferous limestones and shales. The flat, rounded hills at the foot of the ridge consist of Lower Ludlovian and Wenlockian graptolitic shales and sandstones. *Photograph by D. V. Nalivkin (1914).*

DISTRIBUTION. Devonian rocks are thickest and most widespread in the central ranges, where their succession is identical with that of the eastern slope of the Urals. All stages are present and are represented by marine formations with frequent and thick groups of volcanic rocks (Decisions . . . 1959). In the northern ranges Devonian strata are found over smaller areas and represent only the Middle and Upper Devonian as is the case in all the Caledonian types of successions (Sergun'kova, 1958). The sea advanced from south to north and in the most southerly localities basal red beds are overlain by Givetian and Upper Devonian limestones. Further to the north the succession begins with the Frasnian Stage while on the southern slope of the Karatau the Famennian Stage only is present. In the extreme north the marine Devonian is entirely missing (Table 38).

In the southern ranges all three Devonian divisions are represented by marine facies similar to those of the central ranges, but outcrops are relatively small, occurring in narrow belts.

SUBDIVISIONS. Devonian strata are divided in the same way as on the eastern slope of the Urals. Examples of full successions in the central ranges are given in Table 38. Here there are also quoted the successions of the northern ranges (Chatkalian range, the Karatau). In the Alai range the Devonian succession has been studied in greatest detail by G. A. Kaleda (1960) and G. S. Porshnyakova (Gorianov et al., 1961), and the following suites have been distinguished.

Lower Devonian

Akkulian Suite of conglomerates and sandstones with formations of shales and lenses of reef limestones. The suite shows a local development and reaches 100m in thickness.

Sandalian Suite. Limestones and dolomites, with *Karpinskia*, reach up to 100m.

Yashian Suite of interbedded tuffs, porphyrites, diabases, limestones and shales. Fossils include: brachiopods, tabulates and rugose corals. The thickness is 50–700m.

Lower-Middle Devonian

Almalykian Suite. Dark shales and sandstones with lenses of reef limestones have a fauna of tabulates and brachiopods of indeterminate Lower Devonian or Eifelian age. The suite is of a local development.

Aravanian Suite. A group of volcanic rocks with beds of siliceous shales and limestones. The fauna is of an indeterminate, but most likely Eifelian age. The thickness is 1000–2000m.

Kaindinian Suite. Tuffaceous sandstones, tuffs and shales with lenses of limestones. The fossils are: tabulates and amphipods. The thickness is 400–500m.

Dzhidalian Suite. This is a thick terrigenous formation with lenses of limestones. Some groups of sandstones are continental and have a flora. The fauna in the limestones is Lower Devonian or Eifelian and at the top Givetian in age. The thickness is 800m.

Middle Devonian

Katranian Suite of pale, massive, organic limestones with a variable marine fauna with Eifelian and Givetian forms. The thickness is 350m.

Yauruntusian Suite consists of grey, bedded limestones with Givetian tabulates and brachiopods. The thickness is 300m.

TABLE 38.

Devonian Successions of Central Asia

Stage	The Pamirs	The Gissar range	The Alai range	The Nura-Tau	The Chatkal range	The Karatau
Famennian	Dark-grey and reddish bedded limestones with a brachiopod fauna.	Dukdonian Formation of pale, massive limestones (Fig. 168). Faunally corresponds to the Upper Devonian and Givetian (400–800m). In places is replaced by the Pushnevatskian Suite of terrigenous and cherty rocks (500–2500m).	Grey, bedded limestones with Uralian-type brachiopods; *Cyrtiopsis* occurs (100–200m). In places, limestones with goniatites.	Continental sandstones and shales with psilophytes (300m). Lavas are present.	Dark, bedded limestones with *Spirifer* (300–400m).	Dark, massive, bedded limestones with *Spirifer* (Up to 2000m).
Frasnian	Effusive-sedimentary formation.		Pale, bedded and massive limestones with a marine fauna including *Hypothyridina euboides* (150–250m). In places, shales and sandstones.	Dark and pale, bedded limestones and shales with brachiopods (Up to 1200m).	Dark limestones and marls, with a marine fauna (200–300m).	Tyul'kubassian red continental sandstone strata (Up to 2000m).
Givetian	In places limestones with brachiopods, elsewhere, sandstones and shales.		Dark, bedded limestones and shales with *Stringocephalus* (200–600m).	Dark, bedded limestones, shales and sandstones. Above, stringocephalids; below, brachiopods, rugose corals and *Amphipora*; at the base—sandstones (900m).	Dark, lower down, red sandstones and shales. Towards the top beds with a marine fauna (*Stringocephalus*) (800m).	Missing.
Eifelian	Grey, bedded limestones with a marine fauna.	Magianian Suite of grey, bedded limestones with beds of shales (Up to 450m).	Pale, massive and bedded limestones with a marine fauna (100–250m).	Missing.	Missing.	
Coblen-zian	Massive, thick reef limestones; in places replaced by a volcanic-terrigenous formation.	Chimtargian Suite of thick, pale massive reef limestones corresponding to the Lower Devonian and the Ludlow. *Karpinskia* is present (700–1300m).	Pale, massive reef limestones with *Karpinskia* (250–400m).			
Gedin-nian			Manakian Beds—dark limestones with brachiopods (150–300m).			

Arpalykian Suite of dark, bedded dolomites of a total thickness of 400–800m, and with tabulates and strophomenids.

Kuruganian Suite. The suite (550m) shows a complex interdigitation of shales, sandstones, limestones and tuffs. The fauna is Givetian, with tabulates and brachiopods.

Kanian Suite. A metamorphosed shale-volcanic formation, with beds of tabulate-bearing Givetian limestones at the top. The thickness is considerable.

Middle-Upper Devonian

Boardinian Suite, consisting of dark, bedded limestones and dolomites with a uniform amphipod fauna, which is Givetian at the bottom and Frasnian at the top, the total thickness being 500m.

Upper Devonian

Kurchavaiian Suite of grey, bedded, fine-grained, aphanitic limestones (200m) with foraminifera and other fauna.

Adyrakouian Suite of grey, bedded limestones and dolomites with *Spirifer anossoi* Vern. and amphipods. The thickness is 350m.

Kshtutian Suite (up to 500m) of dark, bedded dolomites with *Cyrtospirifer archiaci* Vern., and reaching 500m in thickness.

It must be pointed out that despite the large number of recognized suites, they do not reflect the full lithological variability of Devonian formations. For instance the goniatite-bearing limestones of the Kichik-Alais remain undifferentiated. The Famennian, in addition to the dark, bedded limestones (Kshtutian Suite) is also represented by various other deposits, such as pale *Cyrtiopsis* limestones, which also contain other brachiopods. Many other examples can be quoted. Thus the subdivision of the local Devonian has just begun.

FAUNA AND FLORA. The fauna is very rich and variable, although goniatites are relatively rare. The fauna in its totality is of Uralian type, but there are occasional Caucasian and Himalayan forms. The flora is rare and in the northern ranges it is represented by psilophytes.

FACIES. The facies are exceptionally diverse, varying from reef limestones to red beds. Amongst the marine facies brachiopod-limestones and shales predominate. There are thick (400–600m) massive reef limestones, which are identical with the Hercynian limestones of the Urals (Figs. 166, 168).

Characteristic of the region are the very thick (up to 200m) black, bedded muddy limestones of the Karatau. These rocks have a rich, but uniform, brachiopod fauna of Famennian age. The limestones were found in the muddy, near-shore parts of the sea, the bottom of which was subject to fast, compensating downward movements. The limestones have been studied in detail and have been divided into four horizons and fifteen groups.

PALAEOGEOGRAPHY. A vast, shallow, tropical sea with volcanic and reef islands existed where at present there are the southern and central ranges. During slow and prolonged downward movements great thicknesses of limestones came into existence.

The northern ranges area during the Lower Devonian and Eifelian epochs was land, with a system of fold ranges, which were formed at the end of the Caledonian Orogeny. This land was part of the large Kazakh macroisthmus (Fig. 138) of

Central Kazakhstan. The land first appeared at the end of the Silurian, while during the Lower Devonian and Eifelian epochs it suffered an extensive erosion and a vast plain came into existence. On this plain red sandstones and clays were deposited. At the beginning of the Givetian the macroisthmus underwent a downward movement and the sea penetrated at first its southern margin and later, in Frasnian times still further northwards. In the Famennian epoch the sea reached the axial part of the Karatau which formed a barrier to Devonian and Tournaisian seas. The barrier was transgressed only in the Visean epoch.

Lower Carboniferous

The Lower Carboniferous is the last epoch of widespread marine development

FIG. 168. Sandy-shaley Lower and Middle Devonian and Tournaisian strata (at the bottom) and massive pale Ludlovian and Lower Devonian limestones (at the top), in the Zerafshan range. *Photograph by V. P. Martyshev (1957).*

in Central Asia. In the middle of the Middle Carboniferous the commencement of the Hercynian Orogeny caused the regression of the sea from most of the area of the northern ranges.

DISTRIBUTION.   Marine, Lower Carboniferous rocks occur throughout large areas of the central ranges and the southern part of the northern ranges. To the north of the axial region of the Karatau and Talassian Alatau Middle and Upper Visean and Namurian rocks have been sporadically found. Here certain barren sandstones and shales, which are considered as Upper Palaeozoic, may be in fact Lower Carboniferous. In the southern ranges Lower Carboniferous rocks are rare.

SUBDIVISIONS.   The subdivisions of the Lower Carboniferous are the same as in the Urals and Western Europe. Only in the northern ranges does the fauna resemble that of Kazakhstan, suggesting the existence of connecting straits through the southern Karatau and in the region of Son-Kul (Bensh, Sergun'kova and Solov'yeva, 1958; 'Decisions' . . . 1959). Comparative tables and sections are given in these 'Decisions' of the conference on the stratigraphy of Central Asia.

In the central ranges, for instance the Alais, two types of succession are of interest. In the troughs all the stages are represented by marine facies, while on the blocks the succession normally starts in the Middle Visean, which is disconformable on Devonian and older rocks.

The Tournaisian is most fully represented in the northern ranges where the dominant lithological type is variably bedded limestones, reaching 1000m in thickness. The Strunian Beds of the Lower Tournaisian and the massive limestones and dolomiate of the upper part of the Lower Tournaisian, as well as the grey crinoidal and Bryozoan limestones of the Upper Tournaisian, are well developed. In the central ranges the Tournaisian is represented by thick, barren limestones found in some troughs. Occasionally Tournaisian rocks are found in the southern ranges.

The Visean Stage is most widely represented, found in all the ranges and spread over vast areas. It is mainly represented by limestones which reach many hundred metres, up to 2000m, in thickness. Amongst their faunas gigantoproductids and striatiferids are especially common.

The Namurian Stage completes the Lower Carboniferous and is represented by different lithologies, such as dark, bedded limestones, quite common sandstones and shales, and occasional volcanic rocks. The fauna consists mainly of brachiopods and goniatites, and the total thickness of the rocks varies from 300–1000m. The most complete successions (Decisions . . . 1959) are given below.

NORTHERN RANGES		CENTRAL RANGES
Chatkal	*Son-Kul'*	*Alais, Oshian hills*
	Bashkirian Stage	
Limestones, sandstones and tuffs; *Choristites bisulcatiformis*; 200m	Calcareous conglomerate; 1000m	Dark, bedded limestones with *Choristites bisulcatiFormis*; 200m
	Namurian Stage	
Nauvalinskian Suite of dark, bedded limestones with brachiopods; 300m	Zanginian Suite of grey sandstones 185m; dark, bedded limestones	Ortotauian Suite of dark, bedded limestones with *Productus concinnus*; 100m

Upper Visean sub-Stage

Ugamian Suite of grey limestones with brachiopods (*Striatifera*); 550m	Zanginian Suite. Pale massive limestones with brachiopods; 150m	Charbakian Suite of grey limestones with striatiferids; 400m

Middle Visean sub-Stage

Aksuiian Suite of brownish bedded limestones and shales; 200m	Teshikian Suite of black shales and laminated cherts; 100–300m	Chil'mairamian Suite of coral, oolitic and bedded limestones with *Gigantoproductus giganteus* and rugose corals; 300m
Itelgiuian Suite of massive limestones with *Giganto-productus sarsimbai*; 250m	Itelgiuian Suite of massive and bedded limestones with *G. sarsimbai*; 100–350m	—

Lower Visean sub-Stage

Pskemskian Suite of grey limestone with flints; *Productus deruptus*; *Spirifer plenus*; 720m	Karakinian Suite of brown limestones and black shales with *Productus deruptus*; 200–1000m Dzhaprykian Suite of variegated muddy limestones with goniatites (*Pericyclus*); 10–50m	Grey bedded and massive limestones

Upper Tournaisian sub-Stage

Sargardonian Suite of pale, massive limestones; 300m	Akchetashian Suite of dark, flaggy limestones with flints; *Palaeochoristites*; 150–300m	Sasykkungurian Suite of pale, massive limestones (oolitic) with *Productus humerosus*; 450m
Kulas'inskian Suite of dark, bedded limestones with flints; 500m	Katunarykian Suite of pale saccharoidal limestones; *Spirifer desinuatus*; 100–150m	—
Koksuian Suite of lime-stones with brachiopods (*Palaeochoristites*); 100m		

Lower Tournaisian sub-Stage

Chavatikian Suite of bitu-minous, bedded limestones and dolomites; Foraminifera; 300–400m	Sonkulian Suite of thinly-bedded limestones with brachiopods (*Spirifer kickinensis*); 350–550m	Chil'ustunian Suite of detrital, bedded limestones with Foraminifera; 400m
Brichmullian Suite of thinly-bedded limestones and shales; 100m		

The Teshikian Suite of the Sonkulian succession is sometimes included in the Upper Visean, and the Zanginian Suite in the Namurian. In the successions of the Alai range the suites of the Oshian Hills are sometimes replaced by others of different lithologies and faunas (Goryanov *et al.*, 1961). These authors are guilty of a methodical error, which is unfortunately all too common in stratigraphical investigations. The suites which they employ completely correspond to stages; thus the Karadavanian Suite equals the Tournaisian Stage, the Peshkautian equals the Visean Stage, the Shuranian Suite equals the Namurian Stage. One should not use new strati-graphic terms if older equivalents are available. Even more serious is the employ-ment, by these authors, of a novel usage in terms of which several suites may be completely equivalent to a stage. Thus, both the Peshkautian Suite and the Pumian Suite are considered as equivalents of the Visean Stage. This suggests that the massive reef-limestones of Visean age would form a third suite, the terrigenous-effusive rocks a fourth suite, the cherty shales a fifth suite and the continental red beds a sixth suite. Thus the Visean Stage ceases to be a stratigraphic subdivision with a definite strata type and becomes a geochronological subdivision which can only be characterized by a large number of stratotypes.

FAUNA AND FLORA. The fauna is usually of Uralian type. Only in the northern ranges and in particular in the Karataus and near lake Son-Kul representatives of Kazakhstan faunas appear, as for instance *Productus deruptus* Rom. and *Spirifer plenus* Hall. The flora is rare since there are no coal-bearing formations.

FACIES AND PALAEOGEOGRAPHY. In Central Asia there are no Lower Carboniferous coal-bearing formations, so typical of the Urals. This is explained by the fact that the continental shoreline passed to the north of Central Asia in the Karaganda district. Central Asia was occupied by a sea with small, sometimes volcanic, islands.

Amongst the marine facies the various limestones are most common and are most often bedded, grey and black with numerous brachiopods and corals, but their fauna as a whole is variable and mixed. Crinoidal and oolitic limestones are not infrequent and the latter are much more common than in the Devonian and Silurian. The thickness of the limestones often is many hundreds of metres (Fig. 169).

The sandstones and shales are grey and brown, but are uncommon and thin. These features suggest that the supply of terrigenous material was not abundant, since the continent was far away.

Upper Palaeozoic

Upper Palaeozoic successions reflect the oncoming of the Hercynian Orogeny, which at first manifested itself strongly in the north of the present-day northern ranges. In this area, beginning with the middle of the Middle Carboniferous the sediments become continental. Marine transgressions penetrated into this region only in the Middle Carboniferous and were short-lived, depositing thin limestones and shales with marine faunas.

In the north of the northern ranges the Middle Carboniferous marine sediments occupy large areas, since here the first manifestations of the Hercynian Orogeny did not appear until the Lower Permian. Consequently Upper Permian and Triassic deposits of this area are continental.

In the southern ranges all the phases of the Hercynian Orogeny cause angular unconformities, but here these movements were weak and did not cause the conversion of the geosyncline into a platform. Thus marine deposition continued throughout the Upper Palaeozoic and Lower Trias. At the end of the Middle Carboniferous thick sandstones and shales, which in places are flyschoid and elsewhere coal-bearing, were deposited. Upper Palaeozoic rocks are found throughout Central Asia, but are nowhere outstandingly important. In the northern ranges they are less important than Lower Palaeozoic rocks; in the central ranges they are less significant than Middle Palaeozoic rocks and in the southern ranges than Mesozoic rocks.

Middle Carboniferous

SUBDIVISIONS. The Middle Carboniferous can be divided into two very different parts. The lower part, corresponding to the Bashkirian of the Russian Platform, is thin, being normally some tens of metres and rarely reaching 150–200m. It consists of marine deposits and often cannot be separated from the Namurian. The upper part corresponds to the Moscovian and is thick, being normally several hundred metres in the south and reaching 1500–2000m. It consists of terrigenous, mainly continental deposits, with groups of marine limestones, shales and sandstones at the bottom. In places there are thick groups of lavas. It is always preceded by breaks and angular unconformities. The fullest successions of the Middle and Upper Carboniferous are quoted overleaf (Conclusions . . . 1595, Tables 9 and 10).

DISTRIBUTION. The analogues of the Bashkirian Stage are found in the same areas where the Namurian Stage is present, and on maps they are shown as Lower Carboniferous. The upper part of the Middle Carboniferous is represented by sediments, which form the greater, lower part of the rocks shown as Upper Palaeozoic. The upper part of the Middle Carboniferous is represented by rocks found in the

FIG. 169. Lower Carboniferous limestones of the river Naryn, in the Kokirim-Tau range—a constituent of the Northern ranges. *Photograph by E. I. Zubtsov (1955).*

marginal part of the Dzhungarian Alataus, on the south-eastern slope of the Karataus and in the east of the Zalairian Alataus. A large area of Middle Carboniferous
sediments is found in the mountains to the east of Tashkent, where these rocks are
unconformable on older formations. In the central ranges, beginning with the
Kyzylkum uplands and ending with the Kokshaal range Middle Carboniferous
rocks are found everywhere, and they are also quite common in the southern ranges,
the Pamirs and Darvaz.

LITHOLOGY AND FACIES.   In the north there are abundant sandstones, which
are in places grey-green, of Artinskian type and interbedded with shales, lavas and
conglomerates, which sometimes reach great thicknesses of up to 300–500m. In the

	Southern Fergana	
Tashkent Region Kuraminian Mountains	Kara-Chatyr range	Gissar range

Upper Carboniferous

Kasimovian and Gzhelian stages

Oyaisaian Suite of liparite porphyrites, laparitic and dacitic tuffs and tuffolavas with sandstones, limestones and conglomerates at the bottom (thickness 500 to 2000m).	Dastarian Suite which is flyschoid and consists of siltstones, sandstones, lime-stones and conglomerates with sandstones and shales at the bottom. Fossils include: brachiopods, goniatites and pseudo-fusulinids (thickness 1000 to 1500m).	Gissarian Suite of shales, sandstones, conglomerates with limestone members and groups of lavas at the bottom. Foraminifera include *Triticites* and *Psuedofusulina*. Thickness is about 1000m).
Nadakian Suite (Upper). Sediments and volcanic rocks.	Uchbulakian Suite of con-glomerates, coarse sandstones and shales towards the top. Fossils include *Triticites*, (thickness – 1000 to 1800m).	
	Dzhilginsaian Suite of con-glomerates, sandstones, silts and lenses, and limestones. Foraminifera are found (thickness – 600m).	

Middle Carboniferous

Moscovian Stage

Nadakian Suite (Lower) of Dacites, quartz porphyrites, tuffs and sandstones.	Akterekian Suite of sand-stones, silts, lenses of lime-stones and conglomerates at the bottom (Upper Moscovian sub-Stage). Thickness – 1750m.	Sagdorian Suite of siliceous slates, sandstones, limestones tuff-breccias, lavas. Thickness – 500m.
Akchinian Suite of andesite-dacite and dacite lavas with corresponding tuffs. At the bottom there are con-glomerates. Thickness – 500–2000m.	Kalmakbulakian Suite of sandstones, conglomerates, silts and lenses of limestones (Lower Moscovian sub-Stage. Thickness – 1500 – 2000m.	Karatagian Suite of lavas, sandstones, shales and limestones. Fossils are represented by choristids. Thickness – 2200m.

| Sandstones, tuffites, limestones with goniatites. Trachytes, porphyrites. 50–200m. | Shuranian Suite (upper part), the analogue of the Bashkirian Stage. Sandstones, limestones. | Karagatian Suite (lower part). Lavas, sandstones, shales and limestones with choristids. 2,000m. |

Kuramian Mts	Kara-Chatyr	Gissars	Pamirs

Lower Trias

| Kyzylnurian Suite (Upper) of mainly volcanic rocks with plant remains. | Madygenian Suite (Upper) of sandstones and shales with floras. Thickness 200–250m. | Khanakian Suite (Upper) of sand- stones, tuffs and conglomerates with a Lower Triassic flora. (T. A. Sikstel). | Limestones and siliceous shales with lamellibranchs. |

Upper Permian

Pamirian Stage

| Kyzylnurian Suite (Lower) of liparites, liparitic tuffolavas. Thickness – up to 1000m. | Madygenian Suite (Lower). Coal-bearing and siliceous rocks with plants. Thickness 60m. | Khanakian Suite (Lower) of barren sandstones, argillites and conglomerates. | Karabelessian Suite of tuffaceous sandstones and limestones with small foraminifera. Thickness up to 150m. |

Murgabian Stage

| Ravatian Suite of lavas, tuffs, sandstones and conglomerates. Thickness up to 1000m. | Absent | Khanakian Suite (Lower) of sand- stones, lavas and gravellites. Thick- ness – 1225m. | Ganian Suite of lime- stones and siliceous shales, with fusulinids (*Sumatrina*). Thickness 15 to 150m. |

Lower Permian

Darvazian Stage

| Shurabian Suite of andesites, dacites, liparites, tuffs, beds of sandstones and conglomerates. At the bottom there are sandstones with plants and lavas. | Tuleikanian Suite of red sandstones and conglomerates. Thickness – 1200m. | Lyuchobian Suite of acid lavas, tuffs and sandstones. Thickness 1300–1500m. | Kubergandian Suite of limestones and shales with fusulinids (*Misellina*). Thickness 10 to 170m. |

Karachatyrian Stage

| | Karachatyrian Suite of limestones and shales with schwagerinids at the top (500m) and sandstones and silt- stones with rarer limestones at the bottom. The fossils at the bottom are fusulinids and the total thickness of the bottom sub-suite is 1000m. | | Shindinian Suite of lavas and tuffs (150m). Bazardarian Suite of thick, barren shales with beds of sand- stones. Thickness – 1000 to 1500m. |

Tashkent region the appearance of thick lavas (1500–2000m) is characteristic. In the central ranges the succession retains the same variegated terrigenous composition but groups with marine faunas have a wider distribution. In the southern ranges of the Pamirs thick successions of dark and grey, fine-grained, thinly bedded sandstones and shales are characteristic. These rocks have a rhythmic layering of flyschoid type. Their upper part belongs to the Upper Carboniferous, and their total thickness reaches 2000–5000m.

FAUNA AND FLORA. The fauna is rare and normally has choristids and fusulinids. Plant remains are very common, but normally badly preserved.

PALAEOGEOGRAPHY. During the Bashkirian epoch a warm, shallow sea spread almost everywhere in the region. In the Moscovian epoch the first indications of the Hercynian Orogeny appeared causing the rise of a series of young fold ranges. In the north, approximately along the line of the Karataus to the Transilian Alataus, the ranges were on land. Further to the south and especially in the central ranges young mountains rose from the sea forming mountainous, often volcanic, islands. In the Pamirs at the foot of the chain there was a vast submontaine plain where flysch-like formations accumulated.

Upper Carboniferous-Lower Permian

During these periods further Hercynian movements caused the formation of land not only in the northern but also in the central ranges areas.

DISTRIBUTION. Upper Carboniferous and Lower Permian rocks are found everywhere as Upper Palaeozoic terrigenous and volcanic formations.

SUBDIVISIONS. In the Middle Carboniferous the fauna and the lithological subdivisions are very similar to those in the Urals, while new, southern forms of animals appeared in the Upper Carboniferous and Lower Permian. This forces the distinction of local horizons worked out by A. D. Miklukho-Maklai (1958).

The fullest Permian succession in the northern ranges (Kuraminian Mts); central ranges (Kara-Chatyr range) and the southern ranges (the Gissar range and the south-eastern Pamirs) are given overleaf. The first three successions represent a continuation of the successions on p. 510).

LITHOLOGY AND FACIES. In the greater part of the northern region of the northern range there are sandy-shaley continental deposits reaching in places great thicknesses. Only in the extreme east in the frontier part of the Dzhungarian Alataus and the Transilian Alataus (the Ketmen' range) are there groups of limestones and shales, with marine Upper Carboniferous or Lower Permian faunas, interbedded with terrigenous sediments. The marine deposits reach their greatest development along the southern margin of the northern ranges and around the Vale of Fergana their thickness is over 4500m. The rocks consist of a flyschoid, sandy-shaley formation with beds of fusulinid-bearing limestones, and as a result can be divided into several faunal horizons. The marine deposits of the central ranges have a similar aspect. In the southern ranges their lower part consists of flyschoid strata while various marine sediments are predominant towards the top.

In the north-western Pamirs, according to M. K. Kalmykova (1958) the Upper Carboniferous is represented by limestones (about 200m) with marine faunas. The limestones are overlain by the Charymdarian Suite of flysch type, rhythmically alternating shales and sandstones with subordinate conglomerates. The suite is over 2000m thick (Fig. 170). Higher up follow the rather peculiar thick reef limestones (up to 800m), known as the Safetdarian Suite, which is absent in the north. These limestones have a very rich diverse marine fauna of Lower Permian age. Laterally the limestones pass into and are overlain by the Gundarian Suite of silts, sandstones and limestones with fusulinids.

FAUNA AND FLORA. The fauna is rich and diverse, but is rather peculiar and includes various Uralian forms (*Medlicottia, Helicoprion*) and a series of Indian and Indo-Chinese forms. The flora is found in continental deposits and is relatively poor.

PALAEOGEOGRAPHY. In the Upper Carboniferous epoch the shoreline moved to the south and at the end of the Lower Permian not only all the northern, but also the areas of the central ranges became land. The sea existed only in the area of the southern ranges.

The relief of the newly-formed land was mountainous resulting in the accumulation of thick sands, shales and even gravels derived from these young ranges. Along the shoreline there was a chain of volcanoes responsible for kilometre-thick lavas and tuffs. The young fold-mountains of the central ranges stretched towards the north-west into the region of the present-day Kyzylkum uplands and further to the north, thus joining the contemporaneous fold-belts of the Urals Geosyncline. The total length of these chains was no less than 3000–4000km long and tens, and in places hundreds, of kilometres wide. This mountain belt had a great palaeogeographical significance being a barrier to the Upper Permian, Triassic, Jurassic and Lower Cretaceous seas. Only in the Upper Cretaceous did the sea breach the barrier entering the Turgai Depression and the Aral Basin.

FIG. 170. Lower Permian shales and sandstones of the Charymdarian Suite in the north-eastern Pamirs (Darvaz), which are a part of the Southern ranges. *Photograph by S. I. Strelnikov (1958).*

Upper Permian-Lower Trias

The Upper Permian and the Lower Trias of the region cannot be separated from each other and constitute a single formation.

DISTRIBUTION. Upper Permian-Lower Triassic rocks are found in all the three systems of mountain ranges, but are relatively rare. In the northern and central ranges these rocks are represented by sandstones, conglomerates and shales with remains of peculiar plants. In the southern ranges there are mainly marine sediments. The Darvaz (north-western Pamirs) section serves as an example. The Safetdarian reef limestones are overlain and laterally pass into the Gundarian Suite of graywacke sandstones and silts, interbedded with fusulinid-bearing limestones. The suite passes into a thick, Lower Permian, barren group of violet sandstones and conglomerates (up to 2000m), forming the Iollikharian Suite, which is in turn unconformably overlain by gypsiferous sandstones of the Shakharsevian Suite (200m). The Permian part of the section ends with shales and limestones of the Chapsai Suite (500–1000m), which is conformably overlain by variegated, cross-bedded Lower Trias sandstones (up to 600m) with marine bands, with lamellibranchs and ammonites at the bottom. The Middle Trias is missing and the Upper Trias, which is entirely continental is associated with the Jurassic. The Upper Permian successions are given in the 'Decisions' (1959, Table 11), and some have been quoted here (p. 511).

LITHOLOGY AND FACIES. The lithology of the southern arcs is very variable and includes both limestones and red beds. In the southern Pamirs the Upper Permian and the Lower Trias are represented by marine limestones, while in the northern Pamirs only the bottom of the succession is marine, and the top is continental. In the central and northern ranges the succession is thin and consists of continental sands and shales. In places there are volcanic rocks.

FAUNA AND FLORA. The fauna is diverse, abundant and has a southern, Mediterranean and Himalayan character. Various ammonites and complex fusulinids are especially characteristic.

PALAEOGEOGRAPHY. To the north of the Hercynian fold belts, occupying the area of the central ranges, there was a vast plain with local swamps and lakes. To the south of the fold belts there was a large tropical sea with coral islands. In the north-west the sea stretched into the Northern Caucasus and in the south-east into the Karokorum-Himalayan region.

MESOZOIC AND CAINOZOIC

The Mesozoic and Cainozoic complex of sediments was conditioned by the position of the Hercynian mountains of the Urals and Central Asia. To the north and east of these mountains there are only continental platform deposits. To the south and west there are mainly marine, geosynclinal and platform sediments, although there are also contemporaneous continental coal-measures and red beds with layers of salt. There are also volcanic rocks and Cimmeridian granites.

Trias

The Lower Trias is continuous downwards into the Upper Permian. In the north the Middle Trias and the greater part of the Upper Trias is missing. In the south, both are present and are represented by marine facies of Mediterranean type. The top of the Upper Trias (Rhaetic) in the north represents the base of the

Jurassic coal-measures. In the Vale of Fergana and the Gissar range these measures have bauxites.

SUBDIVISIONS. As an example the South Pamirs succession can be quoted.

Lower Trias (30–80m) begins with basal conglomerates and sandstones, which higher up pass into flaggy, sandy limestones with numerous lamellibranchs (*Claraia*). The limestones are sometimes oolitic.

Middle Trias. The Anisian stage is represented by the upper part of thick massive limestones found on the southern shore of lake Rang-Kul. The fauna is abundant and variable.

The Ladinian Stage is represented in the South-eastern Pamirs by a cherty-calcareous suite (up to 200m) with *Daonella*. This suite passes upwards into the thick (300–1200m) Istykian Suite of sandstones and dark shales. The shales and the sandstones are interbedded with each other and with rare beds of limestones. The suite corresponds to the Upper Trias since its lower part has a Carnian fauna (*Halobia*) and its upper part a Norian fauna (*Halobia, Monotis*). The sands and shales of the suite in places pass laterally into limestones and dolomites forming the Ak-Tash Massif (Fig. 171). These limestones have yielded large, thick-shelled lamellibranchs (*Megalodon*), belonging to the Carnian fauna.

The Upper Trias is conformably overlain by Lower Jurassic conglomerates.

In the outer zone of the Mediterranean Geosyncline (Mangyshlak and Tuar-Kyr) Lower Triassic rocks consist of sandstones, shales and limestones of a total thickness of 800m, which on the basis of ceratitids are divided into five zones. The Middle Trias is represented by the Karaduanian red beds (500m) which have plant remains.

FIG. 171. An unconformity between bedded Callovian limestones at the top of the ridge and massive Aktashian, Upper Trias limestones in the main lower part of the ridge. *Photograph taken in the south-eastern Pamirs by S. I. Strelnikov (1958).*

The Upper Trias starts with a group of basal (40m) dark massive limestones and lamellibranch shell accumulates. Higher up follow dark shales, sandstones and limestones, which are collectively known as the Akmyshian Suite (1000m). The suite is conformably overlain by the Azmergenian Suite (200m) of black shales, bituminous limestones and flaggy sandstones. On the basis of their faunas both suites belong to the Carnian Stage, while rocks of the Norian and Rhaetic stages are absent.

LITHOLOGY AND FACIES. In the north—the Vale of Fergana—the continental Lower Trias is represented by a thin group of sandstones and shales of brown and grey colours. These rocks represent the sediments of an alluvial plain and are in parts fluviatile and lacustrine and are included in the upper part of the Madygenian Suite.

The Upper Trias (Rhaetic) begins with a group of basal sandstones and conglomerates, which lie unconformably on various Palaeozoic strata. Towards the top it gradually passes into Jurassic coal-bearing deposits, consisting of brown and grey variegated clays and sandstones, which are usually thin (tens of metres) and which at the bottom have greenish and reddish seams of bauxites and lacustrine bauxitic rocks. In the lower reaches of the Amu Dar'ya these rocks are known as the Takyrsaiian Suite (430m) in the Kugitang as the Kairakian Suite (20–60m), in the Yagnob region as the Ravatian Suite (30m), in the Transalaian range as the Mintekian (1000m) Suite, in southern Fergana as the Kamyshbashian Suite (20–60–200m), in northern Fergana as the Sarykamyshian Suite (70m) and in the Fergana range as the Kokkiian Suite (500m).

FAUNA AND FLORA. The fauna is of Mediterranean and Indian type, is very abundant and diverse. There are many ceratitids and ammonites and quite common lamellibranchs (*Halobia* and *Pseudomonotis*). Large thick-shelled *Megalodon* is typical of reef limestones. The flora is relatively poor and of Mesozoic type.

Jurassic

In the north the Juarassic is continental and coal-bearing and in the south includes both the coal-bearing Lower Jurassic and Upper Jurassic formations and thick reef limestones of a typically geosynclinal aspect.

DISTRIBUTION. The coal-bearing Jurassic of the northern ranges is found in a series of regions but occurs over small areas only. In the central regions the areas occupied by coal-bearing Jurassic rocks are large and the thickness of these strata increases rendering some of the coal seams productive and of economic importance, as for instance in the Uzgen field of the southern part of the Fergana range. The deposits of the southern part of the Vale of Fergana are also quite significant. In the southern ranges the coal-bearing Jurassic is found along the whole of the Mangyshlaks to the Pamirs, but only occurs over small areas. The largest outcrops are found in the Tuar-Kyrs and the Kugitangs.

The marine Jurassic is found throughout the southern ranges and especially so in the Kugitangs and the Pamirs.

SUBDIVISIONS. The detailed subdivisions recognized, have so far a local character (Table 39).

LITHOLOGY AND FACIES. The coal-bearing strata are of a normal type, consisting of rhythmically alternating sandstones, shales, marls and coals (Fig. 172). The overall colour is brownish, black and grey. The thickness of the strata is usually a few tens of metres, but in the Uzgen basin it reaches up to 2000–4000m. It is interesting that in the central Pamirs the coal-bearing formation changes its character, progressively losing the abundant plant remains and coal seams. The strata are still of the same colour, grain-size, rhythmic character etc., but become a true flysch. This is quite understandable since the Jurassic epoch is the time when the Cimmeridian Orogeny was at its height, and flysch is formed synorogenically.

The marine Jurassic is represented by various rocks, which in the Middle and Lower Jurassic are mainly terrigenous or volcanic, while in the Upper Jurassic they are carbonates. In the Upper Jurassic and Lower Cretaceous thick reef limestones were especially widespread, reaching thicknesses of up to 1800m and forming whole mountain ranges, which stretch for several thousands of kilometres. Such ranges are well represented in the Northern Caucasus, the Kopet-Dag range (Fig. 173), in the Gissar range and the Pamirs.

The limestones are pale-grey, compact, sometimes massive and sometimes bedded and characterized by an abundance of reef-forming algae, corals and thick-walled lamellibranchs (*Diceras*) and gastropods (*Nerinea*). These rocks are cut by modern rivers with resulting deep chasms and high cliffs. Superficially these rocks

TABLE 39. *Main types of Jurassic Successions in Central Asia*

Division	The Fergana range (south)	The Gissar range The Kugitangs	South-eastern Pamirs
Upper	Koshbulakian Suite (Upper Jurassic?) of red and green silts, passing down into cross-bedded sandstones. (150m).	Karabilian Suite (lower part) of fresh-water red beds. (150m). Gaurdakian Suite of Lower Titonian and Kimmeridgian salt and gypsum. (Up to 800m). Gissarian Suite of pale and dark, thick reef limestones (Oxfordian and Callovian). (Up to 750m).	A massive reef complex of massive and bedded limestones and dolomites. In places sequences and suites of grey, bedded marls are present. The fauna is of Titonian, Oxfordian, Catlovian and in places towards the top Lower Cretaceous. (Up to 1800m). Red, Callovian basal conglomerates.
Middle	A thin formation of coal-bearing strata. Zindanian (Middle Jurassic) Suite consists of sandstones and silts with coal seams. (350–550m).	Marine Upper Bathonian limestones, marls, shales and sandstones. (120m). Coal-bearing Lower Bathonian strata. (40m). Marine, Bajocian and Aalenian sandstones, and sandy and normal shales with ammonites. (100m).	Homogeneous, grey flaggy limestones and marls with Bathonian and Bajocian ammonites and lamellibranchs. (360–650m). In places Bathonian volcanic rocks—andesites, dacites, liparites, trachytes and tuffs—are present. (1500m).
Lower	Chaartashian Suite of conglomerates, sandstones and silts, with a flora. (130–450–700m). Tuyukian Suite of sandstones, silts, argillites and coal seams. (120–370m).	Coal-bearing strata consisting of interbedded sandstones and sandy and carbonaceous clays, with coal seams. A flora is present. (200m). Basal sandstones and gravelites. (100m).	Upper Liassic massive, pale limestones with ammonites. (150m). Middle and Lower Lias—above, alternating limestones, shales and volcanic rocks; below, sandstones and conglomerates. (300–500m).

resemble the Visean limestones of the Urals to such an extent that even I. V. Mushketov thought that in the Kugitangs they were Lower Carboniferous.

FAUNA AND FLORA. The peculiar freshwater, unionid lamellibranch (*Ferganoconcha*) was first described from the Jurassic coal-measures of the Vale of

FIG. 172. Jurassic coal-bearing strata in the region of Chakpak deposit of the Fergana range. *Photograph in Central Geological Museum.*

FIG. 173. Lower Cretaceous (Neocomian) bedded and massive reef limestones of the Kopet-Dag range, around Kyzyl-Arvat (Skobelevskian gorge). Geologists' automobile in forefront. *Photograph by M. P. Petrov.*

Fergana, while the flora of this group of rocks has been studied in detail by M. I. Brik. The marine fauna is southern, of Caucasian-Indian type and is both diverse and abundant.

PALAEOGEOGRAPHY. Towards the end of the Trias the whole northern part of Central Asia was a vast plain, being a part of the post-Hercynian Platform. During the Rhaetic epoch and in the Lower Jurassic orogenic processes begin, being partly block-forming and partly fold generating. New mountain ranges were dissected by wide valleys where coal-measures accumulated. These regions are found everywhere and consequently a conclusion can be drawn that in Lower and Middle Jurassic times the mountains and the valleys occupied almost the whole territory of Central Asia, including the area of the southern ranges.

At present no analogues of such palaeogeographical conditions are encountered; the present-day Western Transbaikalia or the Southern Urals are somewhat similar, in so far as in these regions low mountain ranges are divided by relatively narrow valleys where there are a few swamps suitable for the accumulation of plant remains. Furthermore the early Jurassic climate was warm and damp.

In the Upper Jurassic epoch the climate became still warmer and drier and coal accumulation stopped. In the north red beds started depositing, while in the south a tropical sea with coral reefs started spreading again. During the Malm epoch saline swamps replaced the reef limestones and were even more widespread in the Lower Cretaceous.

Lower Cretaceous

Lower Cretaceous deposits developed in the same areas as Upper Jurassic sediments were laid down, and are continuous from the latter. But in the Lower Cretaceous continental facies, such as red beds and thick saline and gypsiferous groups of strata, were more widespread. These continental conditions were caused by the last manifestations of the Cimmeridian Orogeny (Vinokurova, 1958). The sandstones are in places oil and gas bearing.

DISTRIBUTION, LITHOLOGY AND FACIES. In the northern and central ranges the Lower Cretaceous is represented by red beds of up to a few hundred metres in thickness. The red beds are exposed over a large area, in the lower reaches of the Syr Dar'ya, between Tashkent and Chimkent and at the margins of the Vale of Fergana (Fig. 174). In the southern ranges Lower Cretaceous rocks are more widespread and diverse. In the north-west and west it is found in the Mangyshlaks, the Balkhans and the Kopet-Dags, where marine sediments predominate, being represented by various limestones, shales, clays and sandstones with rich and variable faunas including diagnostic forms of ammonites and belemnites. The succession is quiet complete, and includes all the stages from the Valanginian to the Albian (Luppov, 1959, Vakhromeyev *et al.*, 1936).

To the south-east in the southern branches of the Gissar range (Kugitang) and the Tadzhik virgation (the Kulyaba region) the lower part of the Lower Cretaceous is quite peculiar, being a thick saline formation. The upper part consists of red beds with marine bands at the bottom. The saline formation is especially fully developed in Kugitang, where potassium salts are found amongst the rock salt. Here is also the Gaurdak sulphur deposit, which originated by the decomposition of gypsum by bitumens rising from the underlying Upper Jurassic limestones. Further to the east in the Pamirs there are more marine deposits which commence with thick limestones.

SUBDIVISIONS. The typical successions of the southern regions are shown in

Table 40. A synoptic chart of successions has been adopted by the 1958 stratigraphical conference ('Decisions' . . . 1959, Table 16).

PALAEOGEOGRAPHY. In the southern ranges a wide development of marine basins, varying from relatively open deep seas to sandy-shore basins (red beds with marine faunas) was typical. The bitter saline lagoons which were more local were even more characteristic. In other regions the red beds were continental, which indicates that near-shore plains were also in existence.

Upper Cretaceous and Palaeogene

This was an epoch of a major transgression. The Upper Cretaceous sea breached the barrier of the Hercynian structures entering the Aral basin, the lower reaches of the Chu and the Syr Dar'ya basin, including the Vale of Fergana.

Another feature of the Upper Cretaceous and Palaeogene is the almost complete absence of mountain-building movements and of corresponding coarse clastic sediments. The inactivity of mountain-building processes is explained by the weakness of the initial phases of the Alpine Orogeny.

Upper Cretaceous

Upper Cretaceous sediments are the deposits of the largest Mesozoic sea, which stretched from the south of the European part of the U.S.S.R. across the western half of Central Asia to China and India. The recently discovered oil deposits are found amongst Upper Cretaceous sediments.

FIG. 174. Lower Cretaceous red, continental sandstones of the river Tau, in the eastern part of the Alai range. An expedition caravan is on the bridge. *Photograph by D. I. Mushketov (1912).*

DISTRIBUTION. Upper Cretaceous sediments are exposed in all the structures of the southern ranges, where these rocks are mainly marine sediments. In the central ranges they are found almost everywhere in the foothills of the ranges and on the slopes of separate mountains. Up to the Karatau range and continuing to the Fergana range there are predominantly marine facies, while to the east of these ranges exclusively continental sediments of mainly red bed facies are developed. In the northern ranges, to the east of the Karataus, Upper Cretaceous red beds yield dinosaur bones and these are found in the foothills and in the valleys of the Chu and the Ili.

TABLE 40. *Lower Cretaceous Successions in Central Asia*

Stage	Platform type	Geosynclinal type	
	Bol'shoi Balkhan	The Kopet-Dag range	The Kugitang range
Albian	Sandstones, not infrequently glauconitic, rarer clays. Spherical concretions and a marine fauna are found. The sequence is divided into four zones. (300–450m).	Glauconite-quartz sandstones, sandy clays and clays with concretions. The sequence is divided into nine faunal zones. (600–1000m).	Grey clays with beds of shelly limestones. (150m). Aulatian Suite— above, limestones, clays and sandstones, with a marine fauna; below, dark clays and sandstones, with ammonites. (Up to 200m).
Aptian	Greenish glauconitic sandstones with beds of clays and marls, and a marine fauna. Three zones are recognized. (200–300m).	Greenish-grey glauconitic and quartz sandstones with abundant oysters and ammonites. The strata are divided into five zones. (550–900m).	Kaligrekskian Suite of greenish clays and oolitic and muddy limestones with a marine fauna. (80m). Red beds of the top part of the Okuzbulakian Suite.
Barremian	Silts, clays, sandstones and limestones with a marine fauna. (80m). Limestones. (125–140m).	Grey, bedded limestones and marls with beds of sandstone at the bottom. A marine fauna is present. (130–180m). Limestones with *Orbitolina* and other marine fossils. (200–350m).	Lower part of the red gypsiferous Okuzbulakian Suite. (90m). Lower down, grey clays, marls and sandstones with a marine fauna. At the bottom red beds and gypsum. (65m).
Hauterivian	Limestone with a sandstone member, and a marine fauna. (200–360m).	Massive reef, and bedded oolitic limestones with beds of silts and ammonites. (650–850m).	Kyzyltashian Suite of continental red clays, silts and sandstones. (130m).
Valanginian	Above, clays and sandstones; lower down, limestones and marls. A marine fauna with *Aucella* is present. (280m).	Black and grey limestones with beds of marls and sandstones. Ammonites and *Aucella* are present. (150m).	Almuradian red clays and gypsum, with a bed of dolomite carrying a marine fauna. (120m). Upper part of Karabilian freshwater clays and sands. (150m).

SUBDIVISIONS. Amongst the marine deposits the usual division into stages does not produce any difficulties. In the north-west, in the Mangyshlaks and the Tuar-Kyr range the stages are the same as in the Emba region and the Russian Platform in general. In the Korpet-Dags and the Gissar range the successions are similar to those in the Caucasus (Il'yin, 1959). In the Pamirs and in the Alai and Fergana valleys the successions show similarities to those of India and North Africa. Typical successions are shown in Table 41. A synoptic table of successions in Central Asia has been adopted at the stratigraphic conference in Tashkent in 1958 (Conclusions . . . 1959, Table 16).

LITHOLOGY AND FACIES. The lithology and facies are rather variable. In the region of marine deposition near the continental shore and islands there are terrigenous sediments of mainly sandstones and clays. Further offshore there are mainly carbonates including marls, marly limestones and compact limestones of pale-grey and white colour. The limestones are typical in the Mangyshlaks, where compact limestones form the whole of Mt Aktau (Fig. 175).

It is interesting that there is no chalk and no tropical reef limestones. Near the continental shore the sea transgressed two or three times, but in places existed right through the Upper Cretaceous epoch. Consequently marine suites are interbedded with, and underlain, or covered by continental sediments. Thus, for instance, to the north of Tashkent the succession begins with Cenomanian conglomerates and sandstones with dinosaur bones and tree trunks. The total thickness of the sediments reaches 40–60m. This group of terrestrial deposits is overlain by a marine formation of 70–140m, corresponding to the interval of Turonian-Campanian. The formation at the bottom consists of red clays and marls with *Placenticeras* and *Inoceramus labiatus*. The clays and marls are succeeded by variegated sandstones with *Trigonia* and then by pale limestones and marls with lamellibranchs. The succession ends with again continental, barren, yellow and red clays of 35–55m in thickness.

In the eastern regions there are only the deposits of alluvial plains, consisting of red and grey clays, sandstones and gravellites with plant remains and at the bottom with fragments of dinosaur bones. These rocks are often a few tens of metres (40–80m) thick and rarely reach 100–150m. Red and grey cross-bedded sandstones are typical while gravels and rarer breccias with fragments of dinosaur bones, which are more or less rounded and pieces of wood are common. The thickness of these rocks is 10–30m and less. These are typical deposits of dry rivers, which only had water after strong rainfalls.

FAUNA AND FLORA. The marine fauna is rich and variable and is northern in the north-west, North Caucasian in the south-west and mixed, mainly southern, in the south-east. Amongst the terrestrial forms the dinosaurs are interesting, although only single bones or bone-fragments are found.

PALAEOGEOGRAPHY. The main features of the district were the western sea and the eastern land which, despite the varying shoreline, remained there. There were no mountains in the Upper Cretaceous and the sea covered large areas at present occupied by the Chatkal Mts, Alaian and Transalaian ranges and the Pamirs and Sintsian ranges. Conversely the Tarim Depression was land, representing an elevated barrier, owing to which the sea could not penetrate further to the east.

Palaeogene

The Palaeogene sea continued on from the Upper Cretaceous sea even in the south where there are angular unconformities between Palaeogene and Upper

TABLE 41. *Upper Cretaceous Successions in Central Asia*

Stage	Platform type		Geosynclinal type	
	The Tuar-Kyr range	The Vale of Fergana	South-western Tadzhikistan	The central Kopet-Dag range
Danian	Organogenic-clastic limestones with echinoids. At the bottom calcareous sandstones. (20–30m).	Missing.	Above, sands with gypsum; below, grey clays and sandstones with echinoids. (90–100m).	Sandy limestones, marls and sandstones with echinoids. (250m).
Maastrichtian	White chalk with *Scaphites* and *Baculites*. (20m).	Nadradiolitic (Supraradiolitic) Suite of red gypsiferous clays with beds of sands. (100m).	Grey clays with layers of shell beds and oyster beds. *Scaphites* occurs at the top and oysters towards the bottom. (90–250m).	Bluish and greenish marls with *Scaphites* and *Inoceramus*. (120–190m).
Campanian	Marls with echinoids, *Inoceramus* and *Belemnitella*. (60–100m).	Radiolitic Suite of sandy limestones with rudistids, and a clay member. (30m).		Variegated marls with a marine fauna. (200–600m).
Santonian	Marls with echinoids.	Yalovach Suite of variegated sandstones and clays with beds of gypsum reaching 1.5m. Red sandstones. (100m).	Grey, calcareous clays with rare beds of oyster-shells. (200–300m).	Grey limestones, marls and clays with rare bands of red marls. A fauna of inoceramids is present. (250m).
Coniacian	Grey sandstones with *Inoceramus* and other fossils. (8m).	Red, banded sands of the Yalovach Suite with beds of variegated clays. Bones of dinosaurs and turtles are present. (20–50m).	Pale marls with abundant echinoids. (85m).	Marls and limestones with *Inoceramus* and ammonites. (5–20m).
Turonian	Grey marls, muddy limestones and clays with *Inoceramus*. (40–90m).	Thomasitic Suite of greenish clays and marls with *Thomasites*. Exogyra Suite of greenish clays and shell-beds. (30–40m).	Clays and oyster-shell beds. (45–115m). Variegated clays with gypsum. (50m). Clays, oyster-shell beds and limestones. (150–300m).	Marls with *Inoceramus* and marine echinoids. (10–40m).
Cenomanian	Greenish, glauconitic sandstones with phosphorites. Ammonites are present. (50–100m).	Changetian Suite (upper part) of red sandstones and conglomerates, with dinosaur bones. (270–340m).	Clays and limestones containing a marine fauna, interbedded with variegated clays with gypsum. (130–200m).	Above, glauconitic sandstones and silts; below, dark clays with beds of sandstones and limestone. A marine fauna is present. (170–350m).

FIG. 175. Mt. Aktau (white mountain) in the Mangyshlaks. The mountain consists of Upper Cretaceous pale-limestones and marls. The dip to the north is gentle. *Photograph by N. I. Andrusov (1909).*

Cretaceous rocks. The sediments of the Upper Cretaceous-Palaeogene sea represent a single macro-rhythm, which often starts with basal Cenomanian rocks and ends with Middle and Upper Oligocene rocks. In the Palaeogene this sea had an even greater extent than in the Upper Cretaceous. The commercial oil of the Vale of Fergana and the Termez region is found in Palaeogene rocks.

DISTRIBUTION. The distribution of the Palaeogene rocks is the same as in the Upper Cretaceous. The shoreline was also approximately the same, differing only by the fact that the bay in the lower reaches of the Chu was large and the Karataus appeared as a narrow, high peninsula. In the east amongst the continental deposits the distinction between the Palaeogene and the Neogene is not always possible and on the 1:7 500 000 map (1950) they are shown as undifferentiated Tertiary sediments.

SUBDIVISIONS, LITHOLOGY AND FACIES. In the north-west—the Mangyshlaks, the Tuar-Kyr and the Karakum, thin, uniform clay formations predominate (Fig. 176). Faunas found at isolated horizons suggest that these strata were deposited, in relatively deep parts of the shelf and possibly on the continental slope. To the east—in the Aral Basin—the succession retains the same character. Here nummulitic limestones of Middle Eocene age and dark clays are widespread, thin and lie horizontally. The clays have been palaeontologically characterized and several local subdivisions have been erected (Table 25). The fauna is again relatively deep-water, but from the middle of the Oligocene and onwards the lithology and the fauna change abruptly. The clays become sandy, beds of sandstone appear and the fauna becomes much more uniform, being characterized by the predominance of thick-

FIG. 176. Palaeogene green clays in the Kara-Kala Mts. in south-eastern Turkmenia, eroded into 'bad lands'. *Photograph by M. I. Petrov (1951).*

walled shallow-water forms. Higher up the succession has continental, fluviatile current-bedded sands and gravels interbedded with groups of fine-grained, sometimes coal-bearing or carbonaceous lacustrine and swampy clays and marls. The continental rocks are differentiated under the name of 'Aquitanian Suite'. Its top parts are Lower Miocene.

In the south-west in the Kopet-Dag marine, rather thick (up to 2000m) clays predominate. Towards the top they are continental or are represented by the Maikop facies (Table 42). In the south-east, for example the Vale of Fergana, the Palaeogene is associated with useful mineral deposits and particularly oil. Consequently its succession is well studied and subdivided (Table 42). Here carbonate rocks such as dolomites and limestones are typical and amongst the latter shelly varieties are quite common. In places large thick lamellibranchs (*Gryphaea esterhazi*) are so abundant that they are used as building stone for walls of local villages. Amongst the carbonates there are clays which sometimes have marine faunas and sometimes are gypsiferous, since they were lagoonal. The facies are very variable indicating the development of shallow-water, near-shore and deeper basins (Gekker *et al.*, 1960). The top of the succession is represented by a thick formation of variegated, maroon clays with sandstones of continental origin, which have not been completely studied. The sandstones and the dolomites of the succession often yield commercial quantities of oil. The Palaeogene sediments are weakly metamorphosed and folded into simple folds with normal and reversed faults (Mirkamalova, 1958). Such Palaeogene strata are found in the Alai range, the eastern part of the Transalaian range and continue to the south-east into the Yarkend and Sintsian regions.

In the Pamirs the Palaeogene is not well represented and in the western part of the Transalaian range it occurs as thick slates, showing a well-developed slaty cleavage, which is caused by the metamorphism associated with intense folding.

To the east of the line Chu-Ilian Mts-Fergana range only continental sediments are found. These sediments are typical deposits of vast plains, are relatively thin (tens of hundreds of metres), consist of mainly fine clays and sandstones and more rarely of sands and gravels, which are frequently current-bedded, red or variegated and sometimes are grey and brown. In places there are beds of gypsum and salt indicating at the former existence of bitter-salt lakes.

Palaeogene successions are given in the 'Decisions' (1959, Table 17).

FAUNA AND FLORA. The marine fauna is very rich and diverse. In the north it has a somewhat northerly character, owing evidently, to deposition taking place in deep cold water. Beside this sea, to the east of the Aral basin there was growing a rich tropical vegetation amongst which there were roaming numerous mammals including the gigantic *Indricotherium* which were the largest mammals ever. In the south the marine fauna was tropical.

The large number of oysters is characteristic and the region was an oysters' paradise. The gigantic *Gryphaea* apart, there are in places numerous large *Platygena asiatica* Rom., which in their form and size resemble plates and consequently the horizon with them was known as 'platy'. Foraminifera also have a great biostratigraphical significance and have been described in detail (Minakova, 1958).

PALAEOGEOGRAPHY. The most important feature of Palaeogenic relief was the absence of large mountains. In the west there was a vast sea with low, rocky islands and peninsulas, while in the east there were vast plains which in places became small-sopka country.

All this was quite different from the present-day, mountainous relief. Where at present there are snow-clad ranges there has been a tropical sea and boundless sun-drenched plains. In the place of the present-day high plateau of Ust'yurt the sea was at its deepest.

The climate was much warmer than at present, and where the Sea of Aral now lies there were large swamps and lakes, in which crocodiles lived and the shores of

TABLE 42. *Palaeogene Succession of the Kopet-Dag and the Vale of Fergana*

Division	Stage	The Kopet-Dag range	The Vale of Fergana
Oligocene	Upper	Maikopian Series (lower part)—dark clays with beds of sands. The series is developed in the west. (250–370m).	Mailisaiian Suite (lower part) of red clays and sandstones; gypsiferous and continental. (200–400m).
	Middle	Karagaudanian Suite of continental red beds, found in the east. (600m).	Sumsarian Suite of purple clays and sandstones, with oysters. (15–120m).
	Lower	Torymbeurian Suite of thick greenish and brownish clays with a rich Upper Eocene and Lower Oligocene marine fauna. (400–850m).	Khanabadian Suite of grey and green clays, with silts towards the top. A marine fauna and in places beds of gypsum are present. (30–70m).
Eocene	Upper		Isfarinian Suite of grey cherty clays with Radiolaria. (Up to 60m).
			Rishtanian Suite of greenish clays, sandstones and limestones, with oysters (*Platygena*). (30m).
		Koturian Suite of sandstones (muddy and calcareous) with a marine fauna of oysters. (80–200m).	Turkestanian Suite of green clays and marls with beds of limestones. A rich fauna of oysters is present. (30–100m).
		Kinderlian Suite of clays with Foraminifera. (200–500m).	
		Ezetian Suite of dark clays with fish scales and Foraminifera. (75–250m).	
	Middle	Oboiian Suite of pale marls and clays with Foraminifera. (50–250m).	Alaiian Suite of green clays and marls; above, limestones and dolomites. Oysters are present. (30–50m).
	Lower	Danatian Suite of variegated, sandy clays and marls, with Foraminifera. (100–220m).	Suzakian Suite of brown clays and marls with oysters and in places gypsiferous. (30–40m).
Palaeocene		Chaaldzhian Suite of greenish clays with arenaceous Foraminifera. (60m).	Bukharian Suite of gypsum, clays and limestones. Lagoonal and marine deposits. (50–100m).

TABLE 43.

Neogene and Anthropogene Successions in the western part of Central Asia

	Stage	Ciscaspian Lowlands, Turkmenia (W. E. Stepanaitys, 1958; P. V. Fedorov, 1955, 1960)	Transcaspia (A. G. Eberzin, 1958; S. V. Epshtein, 1958)
Holo-cene	—	Novocaspian deposits (Stage)—sands and clays with a present-day fauna (10–15m).	Novocaspian Transgression. Sands and clays with a present-day fauna.
Pleistocene	Upper	Khvalynian Stage—gravels, sands and clays (Up to 30–50m). Atelian, continental, Suite.	Khvalynian Transgression. Alternating marine and continental strata.
Pleistocene	Middle	Khazarian Stage—sands, clays and shell beds (Up to 100m). Urunduzhikian Horizon—sands and shell beds (10–20m).	Khazarian Transgression. Clays with shells. Pre-Khazarian Regression. sands and clays.
Pleistocene	Lower	Baku Stage—chocolate coloured clays and silts, with a marine fauna (150m). Tyurkyanian, continental, Suite.	Regression continues to a complete withdrawal of sea. Sands and clays.
Pliocene	Apshero-nian	Clays, sands and shell beds, with a marine fauna. Three sub-stages are recognized (Up to 800m).	Marine Regression, resulting in relict basins. Sands, clays and shell-beds of a marine fauna.
Pliocene	Akchagy-lian	Grey clays with beds of sands and a marine, lamellibranch fauna. Ostracods are found (40–400m).	Transgression, reaching the lower Amu Dar'ya. Sands and clays with a marine fauna.
Pliocene	Cheleke-nian	Red, continental clays with oil-bearing sands. A microfauna is present (2500m).	The whole of Transcaspian is land. In places thick red beds (Chelekenian Stage) have accumulated.
Pliocene	Pon-tian	Pale greenish and brown clays with a microfauna (600m).	Sands and clay with marine and lagoonal faunas.
Miocene	Upper	Meotic is not found. Sarmatian consists of grey clays and marls (100–150m).	Meotic Regression. Sands and gravels. Lower Sarmatian maximal Transgression. Clays and limestones.
Miocene	Middle	Konkian, Karaganian and Chokrakian horizons of grey clays and marls (200m).	All horizons present. In Chokrakian times a transgression begins. Sands and clays with a marine fauna.
Miocene	Lower	Unrecognized.	Burdigalian Stage—sands and clays. Aquitanian Stage—terrestrial sands and clays with plant and vertebrate remains.
Oligocene	Upper	Grey and greenish clays with Foraminifera (thin).	
Oligocene	Middle and Lower		

South-western Tadzhikistan (P. K. Chikhayev, 1956)	Foothills of the Pamirs and the Tian Shian (V. I. Popov, 1958)	Western Fergana Ak-Bel and Ak-Chop (N. A. Sadovskaya, 1958)
Kulyabian Suite of loessic loams, silts, sandstones and in places gravels, with a fresh-water fauna and plants. The age is not clear but may partly correspond to the beginning of the Anthropogene (200–500m).	Saryobian Suite of loess, gravels and present-day moraines.	Gravels, sands and clays of plateaux and terraces of the Syr Dar'ya. Some are dislocated and some are undislocated (40–80m).
	Kilimbian Suite of clearly folded, unconsolidated molasse deposits.	Soft sands and gravels (150m). Unconformity.
	Dzharidiridinskian Suite of bright-red clays and marls. At the bottom, grey conglomerates.	Grey sandstones and conglomerates with a cold-temperate flora (710m).
Polizakian Suite of ashy conglomerates and sandstones. In the west sands and clays (Up to 900m).	Dzhardarian Suite of grey-brown conglomerates and clays.	Variegated clays and sands (480m) and the top part of the Palevaya Suite (190m), the flora indicates cold-temperate dry conditions.
Karanakian Suite. Near mountains thick conglomerates; away from them yellowish sands and clay (450–1400m).	Vakhiinskian Suite of reddish-brownish sands and clays or conglomerates, with sub-tropical fauna and flora. At the base a sequence of grey-brown strata corresponding to a glaciation (Pont-meotic?). Unconformity.	Brown (Buraya) Suite and the lower part of the Palevaya Suite (150m). Sands and clays with a flora indicating dry-warm conditions, and Akchagylian ostracods. Brown sands and clays with a northern flora (50–70m).
Tavildarian Suite. Near the mountains (in Darvaz)—thick conglomerates. Away from mountains—pink sandstones and clays. The Suite is sometimes assigned to the Upper Miocene (600–1200m).		
Khingouian Suite of variegated purplish sandstones and clays. Near to mountains—conglomerates (Up to 1800m).	Surkhobian Suite of bright-red sandstones and clays, in places with gypsum.	Upper Gypsiferous Suite (690m). Middle Gypsiferous Suite (850m).
	Yezganian Suite of dark, red and purple clays and sandstones.	Lower Gypsiferous Suite (800m). Saline Suite (860m).
Childarian Suite of brick-red and grey sandstones and clays (400–850m). Kamalinskian Suite of sandstones (70–200m). Shurysaiian Suite of red clays, silts and sandstones (150m).	Childarian Suite of brick-red marls with gypsum, clays and sandstones. A flora is present of Upper Oligocene–Lower Miocene age (Up to 1000m). Shurysaiian and Sumsarian suites of continental, red, and marine clays.	Subsaline Suite (Podsolyenosnaya) of coarse, calcareous sandstones and conglomerates (1080m).
Sumsarian Suite of clays with oysters (50m).		Mailisaiian and Sumsarian suites of red beds.

which had rhinoceros and elephants as at present in Africa. To the south of this region the climate was much drier, possibly desert and instead of swamps there were bitter-salt lakes.

Neogene and Anthropogene

These periods were associated with strong mountain-building processes, which involved block-faulting in the north and east and folding in the south and west. The sea progressively retreated to the west and gradually acquired the present-day outlines. Sandstones and conglomerates, several kilometres thick and of molasse type, which are typical foothill deposits of high mountain chains, are widespread. Glacial deposits and loess are also typical.

DISTRIBUTION. The Neogene is widely developed in low desert areas. The whole of the Ust'yurt plateau and the Karakum desert consists of it. A major part of the Kyzyl Kum, the Moyun Kum and other deserts is also underlain by the Neogene. In the mountains, however, it has been almost completely eroded away. The Anthropogene is found everywhere and is very variable.

SUBDIVISIONS. In the west in the regions of marine basins the Neogene and Anthropogene successions have been subdivided in detail into all the main stages and horizons, shown in Table 43. In the east, however, where there are only continental sediments the differentiation has been much less detailed. Even the Plio-Pleistocene boundary is not always detected and in many regions the Pliocene and Lower Anthropogene are shown together. The molasse rocks, however, have been studied in greater detail than anywhere else in the world. For this investigation V. I. Popov, his students and colleagues can be congratulated (Table 43).

LITHOLOGY AND FACIES. In the north-west—the outer zone of the Mediterranean Geosyncline, the Mangyshlaks, the Ust'yurt, the Sarykamyshian and Aral depressions, the Karakum and the Kyzyl-Kum—the Neogene and the Anthropogene have a platform character. The rocks are horizontal, bedded sands and clays with beds and groups of limestones, shell-limestones and coarse clastic debris. The total thickness is small and does not exceed a few tens of metres. There are frequent alternations of marine and continental suites. The marine Miocene succession at the chink to the Ust'yurt is shown in Fig. 177, while continental Pliocene rocks of the Sarykamyshian Depression are shown in Fig. 178.

A second type of succession is observed in the north-eastern plains around the lower reaches of the Chu, Ili and other rivers which flow across the Balkhash and the Ala-Kul depressions. Here there are horizontal red sands and clays and less frequently fine gravels. The succession is thin reaching only a few tens of metres (Gramm, 1958).

A third type of succession, still of platform type, is observed in the northern ranges. Again only continental deposits, which are often red (Fig. 179) are developed, although there are also three new and peculiar types of facies complexes. The first complex of molasse deposits (hundreds of metres), consists of sandstones and conglomerates which stretch along the mountain chains for hundreds of kilometres. The second complex is of loess and loessic rocks and in places overlies the molasse sediments, but elsewhere is their lateral equivalent in the direction away from the mountains. Most centres of population are found in the areas covered by these deposits. The third complex consists of glacial and fluvial glacial Anthropogene and Pliocene sediments found within the mountain chains and more rarely adjacent to them. None of these sedimentary rocks have been orogenically deformed and only

Fig. 177

Chink—the marginal scarp of
the Ust'yurt plateau, formed of
Sarmatian limestones—responsible for the horizontal top
surface—and Lower Miocene
clays. In the forefront Oligocene and Eocene clays form
lower slopes. *Photograph by
V.N.I.G.R. museum (1949).*

Fig. 178

Alluvial (lacustrine) Pliocene
clays in the Kanga-Kyr cliff
situated in the Sarykamyshian
Depression. The clays show interesting rhythmic bedding.
*Photograph by M. P. Petrov
(1951).*

in places have been affected by block movements. V. V. Popov (1960) has produced a correlation table of Anthropogenic deposits.

In the intermontaine depressions, surrounded by the rising mountain chains there is a fourth type of deposit consisting of thick accumulations of Neogene and Anthropogene sediments, reaching up to 1500–2500m in thickness (Vasilenko, 1956, the river Ili Depression). The standard borehole sunk in the Ili Depression cut the following succession:

At 2756–2800m, at the bottom of the hole, there were Palaeozoic quartz porphyries followed by:

Upper Cretaceous-Palaeogene (91m). Variegated, barren, continental clays and sandstones.

Oligocene (about 167m) consisting of lilac conglomerates at the bottom and red clays towards the top.

Miocene (458m)—red and green clays and sandstones.

Lower Pliocene (1095m) consisting of variegated and dark clays at the bottom and grey sandstones at the top.

Upper Pliocene (445m)—sands with beds of clay.

Lower Anthropogene (100m)—grey sands with beds of clay.

Middle and Upper Anthropogene (400m)—grey and yellowish sandstones and gravels with beds of clay. All the above deposits are continental and their age is determined by pollen analysis.

The successions of the Anthropogenic sediments of Central Asia and Kazakhstan, their stratigraphy, lithology and other features have been discussed in numerous papers offered at the conference in Tashkent. Synoptic tables of sections and the text of the papers contain much valuable data, generalizations and conclusions. These papers have been included in the book—*Reports offered to the Middle Asia*

FIG. 179. Neogenic continental deposits in the valley of the river Naryn, which flows in the Central ranges. *Photograph by E. I. Zubtsev.*

and Kazakhstan conference on the investigation of the Quaternary period (Tashkent, 1960).

FAUNA AND FLORA. The fauna and flora are of typical southern Russian type.

PALAEOGEOGRAPHY. The palaeogeography is very interesting and variable. In the Lower Miocene and Middle Miocene epochs the sea penetrated into the eastern part of the Aral Depression. In the south, in the Central Pamirs and the Khorassan ranges (northern Iran) high fold ranges, representing the eastern part of the Mediterranean macroisthmus, were formed. This isthmus stretched from the Pamirs to the Alps separating the Sarmatian from the Mediterranean seas. The Sarmatian sea retreated to the centre of the Aral Depression. In the Meotic epoch the Ust'yurt and the Transunguzian plateau rose as a single massif, thus forming a barrier to the Akchagylian sea. In the south new fold-belts originated spreading progressively to the north. The sub-montaine zones of deposition also spread northwards.

At the end of the Pliocene, during the Apsheronian epoch, the Uzboi Depression originated and separated the Ust'yurt and the Transunguzian plateaux. This depression allowed the sea to penetrate into the Sarykamyshian Depression. In the Anthropogene period the Baku, the Khozar and the Khvalyn seas again retreated from the east and the size of the Caspian decreased abruptly.

The mountain chains which originated in the south of the Pamirs in the Middle Miocene spread progressively northward becoming particularly noticeable at the end of the Pliocene and in the Anthropogene epoch, when the present-day relief came into existence. Movements are even now in progress as is indicated by young structures and earthquakes. It should also be noted that the Anthropogenic glaciation reached a great extent; all the Pamirs and the Alai and the Gissar ranges were covered by snow and ice and glaciers descended into the foothills.

The melting of ice and snow caused a sudden increase in the supply of water to the Amu Dar'ya and the Syr Dar'ya. The Syr Dar'ya waters produced the Sea of Aral. The Amu Dar'ya at first flowed into the Caspian Sea, crossing the Karakum desert. Afterwards its course shifted to the east, filling the Sarykamyshian Depression. The excess of water in Sarakamyshian lake overflowed via the Uzboi into the Caspian Sea. At present the course of the Amu Dar'ya lies further to the east entering the Sea of Aral. The river itself has much less water than in the past.

TECTONICS

GENERAL CHARACTERS

Within Central Asia all orogenies beginning with the Byelomorian and ending with the Alpine can be recognized. The latter is still continuing, causing earthquakes. It is rather important that in various tectonic belts different orogenies had different intensities.

In the northern ranges the Caledonian and Precambrian orogenies were particularly important, producing structures of great extent. The Caledonian synorogenic granites are most widespread. Precambrian orogenies were obviously strong, as is shown by differences in degree of metamorphism of the Archeozoic, Lower Proterozoic, Upper Proterozoic and Lower Palaeozoic. Separating these strata are widespread unconformities, and associated synorogenic granites. Nevertheless areas occupied by the Precambrian are relatively small and in the northern ranges its structures are in the subordinate position. The Hercynian Orogeny is recognized ubiquitously, but its manifestations were weak and unaccompanied by synorogenic intrusions. The south-eastern (Dzhungarian Alataus) districts and the ranges round the Fergana valley are exceptional. The Cimmeridian Orogeny was again widespread, but was even weaker, lacking synorogenic granites. The Alpine Orogeny did not affect the northern part of the region, while the structures affecting the Cretaceous and the Cainozoic in the south of the region are of platform type. In many areas at the edges of the ranges the block movements of Palaeozoic massifs produced marginal deformation of Mesozoic and Cainozoic strata. Certain resulting structures are complex, but are not accompanied by metamorphism.

In the central ranges the Hercynian Orogeny was, as in the Urals, the strongest and structures affecting Middle and Upper Palaeozoic rocks and synorogenic intrusions occupy large areas. The Caledonian Orogeny even if it has occurred has been weak. Thus Lower Palaeozoic rocks are affected by the same structures as the Middle Palaeozoic rocks and have suffered an equal grade of metamorphism. Although between Lower Palaeozoic and Middle Palaeozoic rocks there are unconformities they are not particularly significant. Caledonian intrusions are unknown. Small areas of Precambrian rocks show Precambrian structures affected by a metamorphism which is of much higher grade than that affecting Palaeozoic rocks, thus indicating the violence of Precambrian orogenies accompanied by intrusive activity. Younger orogenies, such as Cimmeridian and Alpine can be identified throughout the central ranges, but were relatively weak and are not accompanied by synorogenic intrusions. Younger folds are nevertheless of great economic importance since they contain oil deposits.

In the southern ranges, as in the Caucasus, the Cimmeridian and the Alpine orogenies were most important. Folded Mesozoic and Cainozoic rocks are widespread, often complex, affected by a considerable metamorphism and invaded by synorogenic intrusions. In the northern part of the southern ranges (the Mangyshlaks, the Ust'yurt, the Tuar-Kyr and the Karakum) the Cimmeridian Orogeny is predominant, the Alpine Orogeny being weak and only the initial phases are represented. The Neogene rocks are not folded and the Sarmatian limestones represent the top of the great Ust'yurt plateau. The external zone is a typical Neogene platform, similar to the Stavropol plateau and Donbas. The former plateau is now called the Turanian Platform. In the internal zone all deposits, including Lower Anthropogene sediments, are compressed into distinct regional folds. The Alpine Orogeny was more powerful and consequently Palaeogene rocks are highly metamorphosed and intruded by synorogenic intrusions (in the Pamirs). The Cimmeridian Orogeny was also equally strong and its structures are associated with granite intrusions.

The Hercynian Orogeny was much weaker than in the central ranges, but its initial phases were quite strong, involved synorogenic intrusions and deposition of flysch-like and molasse-like sediments. The later phases of the Hercynian Orogeny are almost entirely wanting and marine Upper Permian strata are succeeded by marine Trias with only local unconformities and the boundary.

The Caledonian Orogeny was weak and in places did not occur at all. The degree of metamorphism of Lower Palaeozoic strata is the same as of Middle Palaeozoic strata and the same structures affect both. There are no known Caledonian intrusions. Precambrian orogenies undoubtedly occurred and were powerful and accompanied by synorogenic intrusions. The Precambrian (Archeozoic) has a sharply different metamorphism, as for instance in the south-western Pamirs, where gneisses, schists and marbles are intruded by ancient granites and are partly migmatized.

The outer zone of the Mediterranean Geosyncline is at present a subject of lively discussion. Here it is considered as a Neogene platform. It is sometimes considered to be an epi-Cimmeridian platform, but most commonly it is referred to as an epi-Hercynian platform. These differences of opinion are caused by the fact that only Palaeozoic and Lower Trias folds have a geosynclinal character. These folds are complex, their rocks considerably metamorphosed and accompanied by synorogenic intrusions. The younger folds are much simpler, weakly metamorphosed and lack synorogenic intrusions. These features impart a platform character to the folds as a result of which the district is considered as an epi-Hercynian platform.

A detailed study of the folds affecting Mesozoic and Cainozoic rocks has shown that the two sets of folds are of different kinds. The folds affecting Upper Trias, Jurassic and Lower Cretaceous rocks are much sharper and tighter than those found in platform situations. Furthermore they have a definite linear trend which is constant for hundreds and even thousands of kilometres. These features are related to an ubiquitous and distinct, though weak, manifestation of the Cimmeridian Orogeny. Thus this platform region should be known as epi-Cimmeridian rather than epi-Hercynian. Nevertheless the folds affecting Upper Cretaceous and Palaeozoic rocks, although they are gentler and simpler than those affecting the Mesozoic, are also distinct and have a linear trend. Thus the region should be known as the Neogene platform. In practice the three different names have been used since Mesozoic and Cainozoic structures have specific features. The most common and accurate term is the epi-Hercynian platform. The epi-Hercynian Turanian Platform is important since it has oil and gas. The platform continues southwards into Afghanistan.

STRUCTURES

Northern Ranges

The Precambrian structures of the northern ranges are complex consisting of thrust sheets, which have been pushed one over the other. The cleaved metamorphosed Lower Palaeozoic structures, of for instance the Karatau range, are no less complex. At the same time the thick carbonate Tamdinian Suite of the northeastern slope of the Karataus are deformed into simple large folds with relatively gentle flanks. Such a juxtaposition of complex, often isoclinal folds in closely bedded heterogeneous strata with simple large folds affecting thick homogeneous limestones is typical and has also been encountered elsewhere, outside Central Asia.

Middle and Upper Palaeozoic rocks of the north are folded into gentle, simple folds, but from the southern slope of the Karataus and further to the south the Upper and Middle Palaeozoic structures become complex and sometimes isoclinal. The folds consist of considerably metamorphosed rocks and are accompanied by granites of various ages, including those of Lower Triassic age, which conclude the Hercynian magmatic cycle. In places thick Upper Palaeozoic volcanic formations are involved in folding. In those regions where the Hercynian Orogeny has been strong the folds are often broken by faults into thrust sheets, wedges and blocks.

Lower Mesozoic rocks, and in particular the Jurassic Coal-bearing Formation, are always folded, but the folds are gentle and simple and are usually broken by later faults. The degree of metamorphism of the coal-bearing Jurassic is low and there are no synorogenic intrusions.

The Upper Mesozoic (Upper Jurassic and Cretaceous) and the Palaeogene of the north remain unfolded, lie almost horizontally or have a primary dip (6–8° or sometimes higher) and are often affected by small normal faults. In the south the Upper Mesozoic and the Palaeogene are folded into simple brachyanticlines complicated by normal faults and thrusts. Such faults are typical to the north of Tashkent, where they are responsible for a peculiar hilly region, known as 'chuli'.

The Neogene and the Anthropogene are only affected by block faults and associated fold deformation. Such fold structures are typical features of the northern ranges where they are widespread. Less commonly they are encountered in the central ranges.

Block Structures and Deformation. Towards Middle Pliocene times all the region of the northern ranges was a plain with gentle hills, but lacking any appreciable mountains. At the end of the Pliocene the southern half of this plain developed a large number of east–west and north–south fissures. Palaeozoic massifs bounded by these fissures began rising, sometimes quickly and at other times slowly. Such uplifts continued throughout the Anthropogenic epoch and in places occur even at present, causing earthquakes. These movements affected a zone some 100–200km wide and thousands of kilometres long. The zone passes eastwards into the Angara Geosyncline up to the Yablonovoi and Stanovoi ranges. Within Central Asia the zone includes the northern ranges and the area occupied by these ranges is the area of block movement while the height of the ranges represents the vertical magnitude of the movements reaching no less than 5–6km. The northern mountain ranges follow one another and are known as the Turkestan-Sayan belt of elevation (p. 456).

The block movements and associated belts of deformation have not been sufficiently investigated and there are almost no detailed investigations. This is a considerable drawback since block movements are widespread, variable and often affect sites of mineralization and determine the distribution and movement of underground water and streams. The causes of such block movements have not been sufficiently investigated either. An opinion is expressed that during lateral compression of almost horizontal plains truncating metamorphic strata of Palaeozoic and Precambrian age, wide gentle arches originate. As a result of concentration of stress at the more curved position of the arches the component rocks break up into isolated blocks, which then move independently at varying speeds. This view deserves attention, but other explanations are equally possible, as if for instance tracing of deep faults which do not involve the formation of such arches.

Central Ranges

Here Precambrian and Lower Palaeozoic rocks are rarely exposed in small outcrops and are everywhere intensely deformed, strongly metamorphosed and accompanied by synorogenic granite intrusions.

The Middle and Upper Palaeozoic are most widespread and occupy large areas. The structures affecting Middle and Upper Palaeozoic rocks are very complex and are broken by thrusts into wedges and slices, which show different amounts of displacement. The rocks are strongly metamorphosed and cut by Hercynian intrusions belonging to several phases including the Lower Triassic.

The fold-form here depends on lithology. Thick series of bedded sandstones and shales of Silurian and Upper Palaeozoic age are compressed into a series of small very tight, often isoclinal folds (Fig. 181) which are affected by faults and sometimes override each other. The equally thick reefs of Upper Silurian and Devonian age do not fold, but at the base become inclined in some direction. Sometimes limestone massifs are compressed amongst shales, when such shales flow around the carbonate masses, following their irregularities and forming compaction structures.

Faults and thrusts in general have the same trends occurring as linear structures. In the west the trend is north-westerly and eventually joins the trend in the Urals, while in the east the trend is east-westerly.

The Lower Mesozoic to the Upper Trias, the Lower and Middle Jurassic, the Bauxitic and coal-bearing formations consist of bedded sandstones and shales, of sometimes great thickness. These rocks are strongly deformed and folded into distinct, relatively small brachyfolds which are often transected by normal and reverse faults. Deformation and metamorphism is considerably less than that of Palaeozoic rocks, and there are no synorogenic intrusions.

Upper Mesozoic (Upper Jurassic or Cretaceous) and Palaeogene strata are even

FIG. 180. A young fault throwing dark Baku clays against pale Upper Anthropogene loams. *Aerial photograph of the Urundzhik area in western Turkmenia, taken by V. P. Miroshnichenko (1956).*

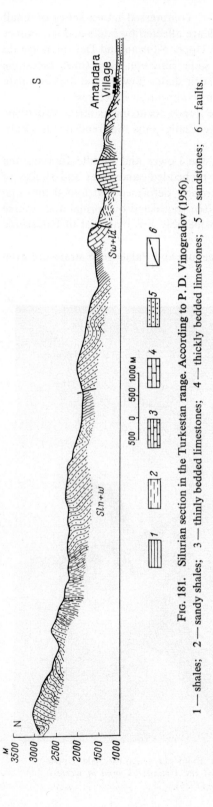

FIG. 181. Silurian section in the Turkestan range. According to P. D. Vinogradov (1956).

1 — shales;   2 — sandy shales;   3 — thinly bedded limestones;   4 — thickly bedded limestones;   5 — sandstones;   6 — faults.

more weakly folded forming brachy-anticlines of different sizes and rela-tively gentle limbs. The folds are often affected by normal and reverse faults. Sometimes Palaeozoic rocks are pushed over Mesozoic and Palaeozoic strata (Fig. 182).

Neogene and Anthropogene rocks are always folded, but the folds are very simple, gentle and least affected by faults. In the Vale of Fergana the folds, involving Neogene conglomerates and sandstones form a foothill belt of barren waterless hills, known as 'adyrs'. In places there are young, present-day anticlines the upper hori-zons of which consist of Upper Anthro-pogene deposits.

Southern Ranges

These ranges constitute a region of the most complex and variable fold and fault structures. In this respect the Pamirs are especially notable. In the cores of its anticlinoria lie Precam-brian and Lower Palaeozoic sediments as well as the fully represented Middle and Upper Palaeozoic strata. In various parts of the southern ranges the struc-tures are different.

In the north-west Mangyshlaks represent a large anticlinorium com-plicated by faults, which are both longitudinal and transverse and there are also smaller structures of second order of magnitude. In the core of the anticlinorium lie intensely folded and metamorphosed Permian and Lower Trias slates and sandstones, which are sometimes folded into isoclinal folds. These rocks are uncomformably over-lain by Upper Trias and Lower and Middle Jurassic coal-bearing and petro-liferous sandstones and shales of a great thickness. These rocks are again folded, but this time into regular secondary folds at the flanks of the anticlinorium. The marine Upper

Fig. 182. Snow-clad Palaeozoic massif in the upper reaches of the river Ak-Bura in the Alai range. The massif is overthrust on Palaeogene and Upper and Lower Cretaceous red beds, marls and clays showing as a pale patch at the foot of the massif. *Photograph by D. I. Mushketov (1909).*

Jurassic and Lower Cretaceous form a still higher unconformable group of sediments margining the anticlinorium, but having gentler dips. The next structural storey consists of Upper Cretaceous and Palaeogene strata lying at the outermost periphery of the anticlinorium and showing individual brachyfolds, which are large gentle and barely noticeable (Fig. 175). These folds are truncated by the horizontally lying Miocene, which forms the top of the Ust'yurt plateau (Fig. 176). The Pliocene and the Anthropogene are represented in terraces at levels much below the plateau and often cut into its escarpments. The major Tuar-Kyr anticlinorium and the folds which are adjacent to it in the south-east are constructed similarly to the Mangyshlaks. Identical structures are found under the Transunguz Plateau and the alluvial cover of the Karakum. On the Krasnovodsk Peninsula at depths of about 1000m folded Palaeozoic rocks have been penetrated by boreholes. These rocks are unconformably overlain by the gentle folded Upper Cretaceous and Palaeogene.

At the present time all these areas represent a Neogene platform which is known as the Turanian (p. 535). It is separated from the Neogene folds of western Turkmenia and the Kopet-Dags by a typical submontaine downwarp, which is almost 3000m deep.

In the structures of western Turkmenia, nearer to the Caspian Sea there are only Neogene sediments, including the oil-bearing red beds. These sediments are folded into a series of brachyanticlinal folds cut by normal faults (Figs. 180, 183). The faults are of diverse types and a majority are only of local significance having been developed within the confines of particular structures, but they are often numerous. Such faults break up individual structures into a series of blocks (Fig. 184). Relatively

Fig. 183. Monzhukla anticline, situated to the south of Mt. Nabit Dag in western Turkmenia. The anticline involves Akchagylian and Apsheronian red beds, is cut off by a large fault to the west and is broken up by smaller fractures. *Aerial photograph by V. P. Miroshinchenke (1952).*

few faults are of regional significance (Fig. 185) cutting across whole regions. With these faults there are associated mud volcanoes the centres of which are placed under thick Neocomian and Upper Jurassic limestones, thus lying at a depth of about 4000m. It is interesting that to the south, near to the river Artek, the trend of Neogenic structures changes from almost east–west to almost north–south. Geophysical data have shown that the Neogenic structures border an ancient shield which forms the southern Caspian Depression (Fig. 4).

The Tadzhik Virgation which lies to the south-east is a particular tectonic region, situated to the south of the Gissar range. It represents a system of fanning, divergent (virgating) folds affecting mainly Upper Cretaceous, Palaeogene and Neogene rocks. In cross section the relief consists of low ridges, separated by wide valleys. To the south these folds and ridges widen out, become gentle and plunge under a

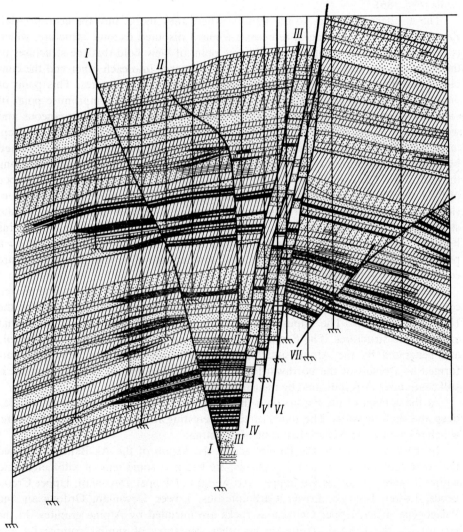

FIG. 184. Section across the Kum Dag fold in western Turkmenia. Numerous faults fragment the fold into separate blocks. According to T. B. Shvarts (1959).

I–VII — faults;   oil deposits shown in black.

northerly inclined alluvial cover. In the region of Termez some of the anticlines are oil-bearing and in the region of Badkhyz (Kushka) they are gas-bearing.

The southern boundary of the Tadzhik Virgation is determined by the Kugitangs. This range is an anticlinorium (in relief a low ridge) in the core of which lie Jurassic and even Palaeozoic strata. Massive reef limestones, forming high cliffs and escarpments, occur within Upper Jurassic sediments.

The Pamirs constitute a most peculiar tectonic region affected by strong and variable movements. For instance in the south-western cliff-wall of the peak Gorumda, in the Transalaian range, there are three recumbent folds, overlying each other and consisting of Neogene red beds. The height of the cliff is about 1000m.

The folds affecting the Mesozoic are the same as in the Greater Caucasus. Jurassic shales and bedded limestones form numerous structures complicated by faults (Fig. 186).

The northern boundary to the Pamirs, where they merge into the central (Alai, Zarafshan and Gissar) ranges, represents a much disputed tectonic structure, which is still debatable. Adherents of one extreme point of view hold that the structures of the Alais and the Pamirs are identical, gradually merge into each other and the contact between the two ranges does not involve any special structures. This point of view has in practice been abandoned. According to a second, compromise point of view, although the structures of the Pamirs and the Alais are contemporaneous and were formed in the Palaeozoic era, their contact is represented by a zone of deep, large and complex faults, which even now are sites of movements. This is indicated by numerous sometimes catastrophic earthquakes, the epicentres of which lie along the fault zone. The coincidence of the epicentres with the fault zone and the existence of the latter has been convincingly demonstrated by N. N. Leonov (1961), who considers that the Alai and Pamir structures are contemporaneous and that the arcuate structures of the Pamirs were formed in the Palaeozoic era. This opinion remains hypothetical since an impression is obtained that Leonov has been attracted by a preconceived idea and diligently collected data to prove it, thus his notions are controversial and unconvincing.

The followers of another extreme point of view which is most correct and proven consider the Alai structures to be basically ancient, Palaeozoic and generated by the Hercynian Orogeny. The Alpine Orogeny has merely complicated and somewhat altered these structures. The Pamir structures, however, are young, Meso-Cainozoic and generated by the Alpine and Cimmeridian orogenies. These structures were formed as a result of the northward push of the Indian Massif. This movement is still continuous as is indicated by the earthquakes.

At the contact of the Pamirs with the Alais there is thus a zone of most complex deep and shallow faults. The zone is a very interesting and important tectonic feature, which resembles the Akbaitalian zone of thrusting.

In the central part of the Pamirs across the region of the Ak-Baital watershed the Akbaitalian zone of thrusting passes. Its width is some tens of kilometres and within it there are successive nappes and wedges of Upper Devonian, Upper Cretaceous, Lower Jurassic, Lower Carboniferous, Lower Devonian, Ordovician and Palaeogene strata. Upper Cretaceous rocks are intruded by Alpine granites. In the territory of the U.S.S.R. there are no other structures of similar complexity and magnitude of movement. The origin of this zone of thrusting remains unexplained.

It is interesting that under the Akbaitalian zone there is another zone of deep earthquakes, the foci (epicentres) of which lie at depths of up to 300km and even

FIG. 185

Plane of movement of a large fault which has an east–west trend and lies to the north of Krasno-vodsk. Movements on the plane are still occurring causing catastrophic earth-quakes. *Photograph by V. D. Nalivkin (1955).*

FIG. 186. Folds in Jurassic limestones and shales of the Central Pamirs. To the left is the caravan of an expedition. At present the Pamir highway passes through this locality. *Photograph by D. V. Nalivkin (1915).*

more. The structures of the southern Pamirs crystalline Precambrian are peculiar. Here thick, massive, monolithic marbles and quartzites form large, simple folds with relatively gentle dips. The simplicity of these structures is at variance with the complexity of folds affecting Palaeozoic and Mesozoic rocks. This simplicity does not reflect the intensity of movements, which must have affected the marbles and quartzites. The detailed study, however, has revealed drag folds and possibly flowage folds resulting upon plastic movement within the strata. The form of these folds is suggestive of folding within salt formations (Fig. 187).

The tectonics of the Pamirs have not been as yet sufficiently investigated, but data so far available indicate an unusual degree of complexity and the presence of exceptional features as has been indicated by M. V. Muratov and I. V. Arkhipov (1961).

THE HISTORY OF TECTONIC MOVEMENTS

Precambrian

The existing data on the Precambrian are incomplete and disjointed and consequently a complete sequence of events cannot be established. However, it can be claimed that the Byelomorian, the Karelian and the Baikalian orogenies occurred widely and vigorously. The sequence of these events in the northern ranges is the same as in the Sayans and Kazakhstan. The great thickness of Precambrian strata is also analogous to that found in these regions, which proves that in Central Asia prolonged downward movements were accompanied by faults and igneous eruptions.

Lower Palaeozoic

THE CALEDONIAN OROGENY. The Sinian complex and the Lower and Middle Cambrian reach a great thickness of up to 10–12km and are represented by

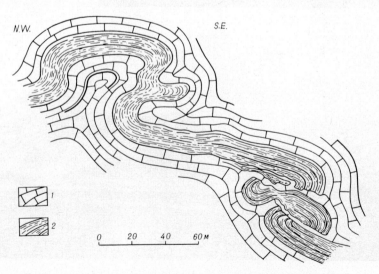

FIG. 187. Drag folds in Precambrian rocks of the south-western Pamirs. According to I. A. Khorev (1956).

1 — marbles; 2 — gneisses.

sedimentary and volcanic rocks. Their composition suggests the absence of orogenic movements and prolonged depression.

The Salairian Phase. The occurrence of this phase in the northern and possibly central ranges is probable, since some Upper Cambrian horizons are missing and there are unconformities between the Ordovician and older rocks. Yet the phase is not ubiquitously manifested. In the eastern Karataus the thickness of Tamdinian limestones and dolomites is over 2000m. These rocks towards the bottom yield Middle Cambrian fauna and towards the top an Ordovician fauna. There is no evidence of a break and thus no evidence for the Salairian phase. The Ordovician is usually represented by a thick series (several kilometres) of sedimentary and effusive rocks developed throughout the region. Prolonged downward movements continued, but regions of accumulation of sediments run parallel to the regions of erosion and rising fold ranges formed as a result of the Caledonian Orogeny. This situation is proved by the Upper and Middle Ordovician molasse found in the northern ranges, where greenish-grey 'pepper' graywackes are interbedded with shales and muddy limestones. Rhythmic bedding is widespread, which suggests an analogy with the flysch, but the frequent occurrence of marine faunas and rudaceous rocks indicates a closer similarity to the molasse. Both flysch and molasse occur at the front of rising fold-ranges and indicate the occurrence of folding.

Later Phases of the Caledonian Orogeny occurred in the Silurian times and affected the northern ranges. Here the Silurian succession is very variable, involves considerable non-sequences and thick groups of clastic rocks. The last phase of the Caledonian Orogeny caused the elevation of a greater part of the northern ranges to a level above the sea and joined it to the Kazakhstan macroisthmus. As a result a greater or lesser part of the Devonian is missing from the succession, while in the north sometimes there are no Devonian strata at all. The older rocks are there overlain by the Lower Carboniferous, in the Karataus by the Famennian, further south in the Chatkalian Mts by the Frasnian and in the extreme south (Kuraminian Mts) by the marine Givetian.

In the central ranges in many localities the Silurian and Devonian succession is of Hercynian type, complete, and is represented by marine sediments of continuously varying lithologies. This indicates that the last phases of the Caledonian Orogeny did not occur in the central ranges. The occurrence of the late phases in the southeast Fergana (Kuznetsov, 1960) is indicated by the presence of Ordovician graywackes and by unconformities between the Givetian Stage and the Silurian.

The Silurian and Devonian of the southern ranges (the Pamirs) are of the same type as in the central ranges, which consequently implies the general absence of the later phases of the Caledonian Orogeny in the south.

Middle Palaeozoic

The Upper Silurian is the epoch of the last manifestations of the Caledonian Orogeny, which in the north were particularly strong. Devonian and Lower Carboniferous times are distinguished by the absence of fold movements and the development of downward movements which were only at times interrupted by local upward movements.

The final series of Caledonian movements are sometimes collectively referred to as the Erian phase and continued throughout the Silurian epoch. As a result in

places mountain ranges grew causing breaks in the succession and also the formation of thick terrigenous sedimentary piles, which varied from argillaceous (flysch
type) to sandy and even conglomeratic (molasse type). The strongest and most widespread movements occurred in the Upper Ludlovian epoch, when the isolated island
ranges merged into a single, mountainous Kazakh macroisthmus. In the Lower
Devonian epoch all the northern ranges including the Karataus, the Chatkalian Mts,
the Son-Kul region and the Kungei-Alatau range, were a part of the macroisthmus.

In the central and southern ranges throughout the Silurian and Lower Devonian
epochs prolonged and considerable downwards movements caused the accumulation
of a thick sedimentary succession (up to 6–8km) of marine deposits. The existence of
locally developed continental groups and unconformities indicates at local short-
lived uplifts interrupting the continuity of sedimentation. Only locally is there
evidence for the occurrence of the Caledonian Orogeny, as in the south-eastern
Fergana.

In the Eifelian epoch the downward movements spread into the northern ranges
district and in the Givetian epoch these movements affect the whole region considered
here. During the Upper Devonian and Tournaisian epochs the subsidence continued
and the sea, advancing from south and west, occupied progressively larger areas in
the northern ranges. In the Visean and Namurian epochs the subsidence was at its
extreme and the sea breached the Karataus and the Talassian Alataus and advanced
north. As a result of the subsidence a thick pile of marine sediments reaching 4–5km,
and more, accumulated on the site of the northern ranges. In the region of the central
and southern ranges the downward movements continued and the total thickness of
marine deposits, including Lower Devonian and Silurian rocks, reached 10–13km,
which is a considerable quantity.

In the central regions at the end of the Devonian the subsidence was interrupted
by uplifts which in places involve considerable areas. As a result stratigraphical
breaks occur and in many local successions the Tournaisian is completely missing
and the Visean is unconformable on the Devonian and sometimes on older rocks. In
the neighbouring regions such uplifts did not occur and the whole succession, including the Tournaisian, is represented by marine sediments.

Upper Palaeozoic

The Upper Palaeozoic is distinguished by the ubiquitous Hercynian fold movements. At the same time there were also adjacent areas where downward movements
caused the accumulation of kilometres-thick terrigenous and volcanic formations.

HERCYNIAN FOLDING. As everywhere else the Hercynian Orogeny began at
the boundary between the Lower and Middle Carboniferous and in places at the
beginning of the Moscovian Stage. The Bashkirian Stage is continuous into the
Namurian Stage. The initial manifestations of the Hercynian Orogeny can be recognized by a sharp change in the lithology and the appearance of predominantly clastic
sediments and angular unconformities which involve the overlap of rocks of the
Moscovian Stage over various horizons of the Middle and even Lower Palaeozoic.

The initial phase of the Hercynian Orogeny, which occurred in the Middle
Carboniferous, was strong in all the ranges and was always accompanied by granite
intrusions. This phase was especially strong in the north of the northern ranges region where young mountain chains displaced the sea. To the south of the line
Karataus-Talassian-Alataus marine deposits alternate with continental sediments

indicating that there were oscillatory movements. The great thickness of deposits, however, indicates that subsidence was predominant.

The second phase of the Hercynian Orogeny occurred at the end of the Lower Permian. After this phase young mountain ranges grew in the southern region of the northern ranges and also in the central ranges, the sea retreating to the region of the southern ranges.

The third phase occurred at the end of the Upper Permian and the beginning of the Lower Trias, thus not entirely corresponding to the boundary between the Palaeozoic and the Mesozoic, but being slightly later. The third phase was weak in the north, comparatively strong in the central ranges and the southern part of the northern ranges and cannot be traced in the south of the southern ranges where the marine Permian is succeeded by the marine Lower Trias.

Mesozoic

Beginning with the Middle Trias and onwards the greater part of Central Asia —including the region of the central and northern ranges—became part of the platform. The geosynclinal regime continued only in the southern ranges. These features determine the further history of tectonics.

CIMMERIDIAN OROGENY.   In the platform regions the Cimmeridian Orogeny was weak, caused a weak metamorphism and is not accompanied by synorogenic granites. In the geosynclinal region the Cimmeridian Orogeny was powerful, conditioning the nature of structure and acompanied by a high-grade metamorphism and numerous, sometimes very large, intrusions.

In the northern ranges the mountain-building movements caused a formation of coal-bearing formations, initiated in the Upper Trias (Rhaetic) and reaching their maximum development in the Lower Jurassic. Often these movements (the rise of Palaeozoic massifs) are referred to the earliest, proto-Cimmeridian, phase of orogenic activity, but there is no evidence that this is the case. It is in fact possible that vertical rise of the massifs rather than folding was responsible for the mountain building. The folds which affect the coal-bearing strata also affect the overlying lacustrine Upper Jurassic sediments proving that the folds are late Cimmeridian in origin. The Upper Cretaceous and the Palaeogene consists of horizontally lying sediments.

In the central ranges the study of the Cimmeridian Orogeny is complicated by the superposition of the Alpine Orogeny. Nevertheless the complexity of folding affecting Jurassic rocks, their grade of metamorphism and the uncomformity between them and Upper Cretaceous strata leave no doubt that the Cimmeridian Orogeny was important although it was not accompanied by synorogenic intrusions.

In the southern ranges the situation is vastly different. Here the many-kilometres-thick Upper Triassic, Jurassic and Lower Cretaceous rocks are deformed into complex, sometimes isoclinal folds, the rocks being strongly metamorphosed and cut by Cimmeridian granites. The manifestation of the Cimmeridian Orogeny is especially clear within the outer, northern zone of this region where it borders the Hercynian platform, as for instance in the Mangyshlaks, the Tuar-Kyr and the Krasnovodsk district. Here Cimmeridian structures are unconformably overlain by folded Upper Cretaceous and Palaeogene rocks, and also by the horizontally lying Neogene and Anthropogene deposits. In the southern internal zone (the Caucasus, south-western Turkmenia and the Pamirs) the Alpine Orogeny has been very strong thus masking the effects of the Cimmeridian Orogeny. The analysis of structures indicates that even here the Cimmeridian Orogeny was intense and was accompanied by large synorogenic intrusions.

The proto-Cimmeridian phase (Lower Jurassic) occurred ubiquitously, but was weak and can only be recognized by gentle unconformities at the boundary between the Upper Jurassic (Callovian) and the older parts of the Jurassic succession. The younger Cimmeridian phase at the beginning of the Lower Cretaceous (Malm-Valanginian) was more intensive and the majority of the Jurassic, Cimmeridian granites are associated with it. Although this phase can be recognized ubiquitously it was most powerful in the internal, southern, zone and especially in the Pamirs.

Upper Cretaceous-Cainozoic

The Alpine epoch was the time of major mountain-building movements and considerable subsidence with which was associated the deposition of many kilometres of mainly terrigenous and also effusive rocks.

THE ALPINE OROGENY. This orogeny began in the Upper Cretaceous and still continues despite intermittent intervals. The orogenic intensity was at its height in the Pliocene and Anthropogene, when the highest mountains came into existence.

In the southern, first zone (southern ranges) the Alpine Orogeny was fully developed and together with the Cimmeridian Orogeny is responsible for the formation of the main tectonic structures. The initial phases did not cause large scale palaeogeographic changes, although the Laramide phase at the end of the Upper Cretaceous produced an unconformity between the Upper Cretaceous and the Palaeogene and the Savian phase at the end of the Oligocene was also important. The latter phase is responsible for the initiation of all the main mountain ranges and for the withdrawal of the sea from their confines. After an interruption in the orogenic activity, lasting almost the whole of the Miocene period, the newer phases, which are still continuing, began afresh. These powerful phases of mountain-building are responsible for the present-day relief, for the complex structures of the Upper Cretaceous and the Palaeogene and for a widespread metamorphism and associated granites.

In the second zone (central regions) all Alpine phases occur, but are relatively weak and are responsible for simple structures. Here the metamorphism is negligible and granites are absent.

In this region as for instance in the Vale of Fergana the so-called 'hidden' phases of orogenic activity are characteristic. The succession lacks any obvious unconformity and all the sediments beginning from the Upper Cretaceous and ending with the Anthropogene lie conformably. At the same time Upper Cretaceous rocks are steep and sometimes isoclinally folded, while Miocene rocks have much gentler dips and Anthropogene rocks are almost horizontal. Thus although the effect of the Alpine Orogeny is incontrovertible and there must have been isolated phases of it no obvious unconformity can be identified. This interesting situation used to be explained by continuous folding, but detailed observations indicate that this is more apparent than real. All the phases produce almost unnoticeable unconformities unaccompanied by breaks in sedimentation, thus remaining hidden.

The third zone includes the southern part of the northern ranges. Here the initial phases of the Alpine Orogeny were weak and produced only simple, gentle folds. The last phases did not occur at all and the Neogene and the Anthropogene are represented by almost horizontal rocks. Thus this Zone is a part of a Neogene platform.

The fourth zone encompasses a large, northern part of the northern ranges. Here the Alpine Orogeny is completely unrepresented and all the rocks beginning with the Upper Cretaceous remain unfolded.

By comparing the four zones it is clear that the Alpine Orogeny progressively weakened towards the north and north-east. This is an important phenomenon since it guides the distribution of useful mineral deposits and in particular of petroleum. The orogeny was the cause of large-scale palaeogeographic changes and in particular of the distribution of marine basins and mountain ranges.

UPLIFT, SUBSIDENCE AND BLOCK MOVEMENTS. Great thicknesses of Upper Cretaceous, Palaeogene, Neogene and sometimes Anthropogene rocks, some-times reaching several kilometres, bear evidence of prolonged and considerable periods of subsidence. These phenomena occur in geosynclinal zones, while in plat-form areas they are not important, but here uplifts associated with block faulting take place as for instance in the northern ranges. In the central ranges block move-ments can also be identified but of much smaller scale and occurring simultaneously with fold movements. In the southern ranges only fold movements occur and block movements are associated with relatively shallow faults, which only occasionally tap basalts thus becoming channels for the upward movement of basaltic magmas.

Young faults, which are especially obvious on aerial photographs are of some interest (Figs. 180, 183, 188).

FIG. 188. A young fault on the eastern Caspian coast, to the south of Nebit-Dag. The region to the west of the fault even a hundred years ago was described by N. I. Karelin, a well-known visiting geographer of that time, as a wide shallow-water bay. Aerial photographs such as this one show young tectonic structures particularly clearly. *Aerial photograph by V. P. Miroshichenko (1953).*

MAGMATISM

Magmatic cycles of Central Asia are just as numerous and diverse as local orogenies. The initiation and distribution of such cycles are conditioned tectonically.

In the northern ranges the ancient cycles are best represented in Precambrian and Caledonian strata. The Hercynian cycle was less widespread in its activity. Younger cycles are practically not represented. In the central ranges the rocks belonging to the Hercynian cycle predominate. Rocks belonging to more ancient cycles are fully developed but are not widespread. Younger cycles than the Hercynian are only represented by Jurassic and Neogene lavas of limited distribution.

In the southern ranges igneous rocks of the Cimmeridian cycle are best represented. The presence of Alpine intrusions which have only been slightly subjected to erosion is of importance. The Hercynian cycle is represented only by the rocks related to its initial phases. The Caledonian and Precambrian cycles although widely diagnosed have not been sufficiently investigated.

In general the dominant cycles are progressively younger to the south. V. A. Nikolayev has been a leading petrologist who has investigated the igneous rocks of Central Asia. Review papers on various provinces of the region can be found in the volume *Geological Structure of the U.S.S.R.*, vol. 2, *Magmatism* (1958), where the magmatism of the northern Tian Shians is given by E. N. Goretskaya, of the central and southern Tian Shians by E. D. Karpova, of the Pamirs by N. K. Morozenko and of Turkmenia by E. A. Khudobina.

PRECAMBRIAN MAGMATIC CYCLES. In the northern ranges the three main cycles: the Byelomorian, the Karelidian and the Baikalian are well developed. The Byelomorian magmatic cycle is most fully represented in the Chu-Ilian Mts and is associated with gneissose rocks. A part of these gneisses consists of altered granites, but there are also gabbro-gneisses and amphibolites. Most commonly the igneous rocks are so strongly altered that it is almost impossible to recognize their origin. The Karelidian and Baikalian cycles can be clearly distinguished, but with some difficulty. In the green metamorphic formation of the Chu-Ilian Mts there are numerous lava flows, sometimes composing the bulk of the succession. Amongst these lavas both acid and basic types (quartz albitophyres, diabases, porphyrites) and accompanying tuffs can be recognized. In places there are small bodies of intrusive basic and ultrabasic rocks. In the overlying sand-shale formation, normally attributed to the Upper Proterozoic and by some partly to the Lower Cambrian, there are thick (1000–2000m) porphyrites and diabases interbedded with tuffs and tuffaceous sandstones. In other areas the separation of Lower and Upper Proterozoic effusives has not as yet been achieved. As regards the ages of granite intrusions there are too few data available and they are normally denoted as Proterozoic or even Precambrian. For instance the Anderkenyn intrusion in the Chu-Ilian Mts cuts the gneisses and is transgressively overlain by Ordovician basal conglomerates with pebbles of plagiogranites of the intrusion. At a number of localities in the Terskian Alataus and the Kungei-Alataus Upper Proterozoic granitoids are covered by transgressive Cambrian strata.

Precambrian lavas and intrusives are widespread in the northern ranges, rare in the central ranges and found in the Pamirs amongst the Archeozoic Crystalline Formation (Fig. 189).

THE CALEDONIAN MAGMATIC CYCLE. This cycle is widespread and well represented in the northern ranges and its rocks are predominant in the north of these ranges, while in the south of the northern ranges Caledonian igneous rocks are as important as Hercynian. Volcanic activity occurred in the Cambrian, Ordovician and Silurian. In the Cambrian lavas are common, in the Ordovician they are rare and in the Silurian again common. In the Tian Shians porphyritic lavas and tuffs of an early Silurian age reach 500–800m. The end Silurian effusive rocks are even more widespread and thicker. Porphyritic and albitophyric lavas, tuffs and breccias are 1000–1500m thick. Intrusive rocks are also widespread and quite variable. Synorogenic granite massifs are often tens of kilometres across.

According to N. G. Kassin, the largest granite intrusion in the Tian Shians (S. and S.E. of Frunze) is of Caledonian age. This intrusion is over 200km long and 50–70km wide and can be identified in synoptic maps. The majority of large and compositionally variable intrusions in the Chu-Ilian Mts are also Caledonian. These intrusions involve light-grey, grey, red and pink granites, dark granodiorites and monzonites. There are also smaller basic bodies of gabbro-norites, gabbro-diabases and ultrabasic peridotites, dumites and serpentinites. Major intrusions are cut by minor porphyrites and aplites. In the south and east of the northern ranges there are widespread contiguous Caledonia and Hercynian intrusions and their separation often entails considerable difficulties. These intrusions have been described by E. N. Goretskaya (*Geological Structure of the U.S.S.R.*, vol. 2, 1958).

The age of the undoubtedly Caledonian intrusions can be demonstrated in detail

Fig. 189. A granite massif intruding Precambrian strata of the southwestern Pamirs, along the Pamir highway, at Toguz-Bulak. *Photograph by D. V. Nalivkin (1915).*

and it is known which ones are Salairian, Taconic and Erian. Late Caledonian (Erian) porphyritic granodiorites have been recognized in the Chu-Ilian Mts. These intrusions marmorize Ordovician limestones and are transgressed by Devonian red beds, which have pebbles of the granodiorites in the basal conglomerate and grains of them in the basal sandstones. In the same district there are also Salairian, Kurdaiian granodiorites intruding Cambrian rocks, pebbles of which are found in Ordovician rocks.

In the central and southern ranges the Caledonian magmatic cycle has a limited distribution and its rocks are uncommon, although they include various effusive, Cambrian Ordovician and Silurian rocks and certain intrusions surrounded by Lower Palaeozoic strata. Intrusive activity, as in the Urals, occurred during the initial phases of the Caledonian Orogeny.

THE HERCYNIAN MAGMATIC CYCLE.   The rocks of this cycle show an ubiquitous distribution and vary considerably.

In the north of the northern ranges Hercynian igneous rocks are rare, but in the south of these ranges the cycle is well represented by thick and widespread lavas of which there is a great number and which show a great diversity of composition. In the central ranges the Hercynian cycle is dominant and its rocks form the only well-developed and widespread group.

In the southern ranges, and especially in their northern zone, Hercynian igneous rocks are common and it is possible that the Hercynian cycle is dominant. In the central and southern zones of these ranges the Hercynian cycle is much less important than the dominant Cimmeridian cycle and only the initial phases of the Hercynian cycle are represented.

In the north-west of the northern ranges—the Karataus, the Betpak-Dalas and the Chu-Ilian Mts—Lower Carboniferous and Upper Palaeozoic lavas are common while intrusions are rare. In the north-east and the entire south of the northern ranges the Hercynian cycle is fully represented. Here Lower and Upper Palaeozoic effusives and two principal phases of intrusions (Lower Hercynian and Upper Hercynian) can be recognized. Lower Hercynian intrusions are emplaced in Lower Carboniferous strata, and their pebbles are found in Upper Palaeozoic rocks. The instrusions are normal, grey brotite-hornblende granites. Upper Hercynian intrusions are often pink and red granites which are sometimes alaskitic. These intrusions cut across Upper Palaeozoic rocks and are overlain by continental Upper Jurassic strata. Numerous minor intrusions accompany these granites.

There are also rarer intrusive rocks, as for instance alkaline, potassic shonkinites found in the western part of the Talassian range, where they occur in small massifs. Hercynian intrusions of the Kuraminskian Mts of the Tashkent region have been studied in greatest detail. According to N. P. Vasil'kovskii (1952), one of the foremost geologists in Central Asia, the relationships of these masses to the coal-bearing strata of various ages have helped in recognizing a series of intrusions ranging from the Middle Carboniferous to the Lower Triassic ages. The latter are particularly interesting being small granite intrusions with active contacts against sandstones and shales with an Upper Permian flora. N. P. Vasil'kovskii and before him two other investigators suggested a possibility of even younger Cimmeridian granites being present in the Tashkent region. Their evidence in this respect is not sufficient especially since the Cimmeridian Orogeny in this area is very weak. Yet the existence of sub-volcanic intrusive rocks, filling the pipes from which Jurassic lavas poured out, is quite possible and particularly so where Jurassic effusives are widespread.

In the central ranges the Hercynian cycle is dominant while other cycles are much less important. Acid and basic lavas and the associated tuffs and breccias are found amongst Devonian and Lower Carboniferous deposits, but are relatively uncommon. Upper Palaeozoic igneous rocks are much more common, where they are associated with sandy-shaley formations, forming both sills and lava flows. The sills are usually thin (tens of metres) but very continuous. Of acid rocks albitophyres and quartz albitophyres are most usual, while mainly augite diabases are the most frequently encountered basic igneous rocks.

Hercynian intrusions in practice account for all intrusions shown on the synoptic maps of the central ranges. Acid and intermediate intrusions predominate and include granites, diorites, granodiorites and syenites, although there are two small regions in the west (the Gissar and the Alai ranges) where alkaline intrusive rocks are well developed. In the east there is a small massif of basic rocks, but no ultra-basic rocks are known. In the latter respect Central Asiatic magmatism strongly differs from that in the Urals, where both Caledonian and Hercynian ultrabasic rocks are widespread.

Hercynian intrusions of the central ranges belong essentially to the later phases of the cycle and in the main have active contacts against Upper Palaeozoic strata. Hercynian intrusions and lavas have been described by E. D. Karpova (*Geological Structure of the U.S.S.R.*, vol. 2, 1958).

In the southern ranges the situation is entirely different. The Hercynian cycle is here less important than the Cimmeridian, although in the Northern Pamirs where Middle and Upper Palaeozoic strata outcrop the rocks belonging to the Hercynian cycle are widespread and consist of effusives of Middle and Upper Palaeozoic age and also of mainly acid intrusions.

The effusive rocks are mainly found in the western part of the Northern Pamirs and Darvaz (Kalaikhumb-Sauksai zone) and in a wide belt a hundred kilometres long they stretch across the Pamirs. N. K. Morozenko (*Geological Structure of the U.S.S.R.*, vol. 2, 1958) recognizes three effusive formations. The two lower formations lie within Silurian-Devonian strata and vary in thickness from 1000–3000m. The Pyandzh Suite consists of greenstones (porphyrites and tuffs). The Sauksai Suite consists of epidotic schists formed as a result of the metamorphism of basic lavas. The third suite is spilitic with mainly basic rocks varying from diabases, spilites, diorite porphyrites and their tuffs. Their age is Lower Carboniferous and their total thickness of up to 1000m. In addition thin groups of lavas and tuffs are known from Permian strata.

The intrusions are separated into two complexes—the Early Hercynian and the Late Hercynian. The former complex is emplaced amongst Silurian limestones and is covered by Lower Carboniferous rocks and may belong to the Caledonian cycle. The complex consists of numerous and diverse intrusions including ultrabasic, basic, granodioritic and granitic types. The Late Hercynian complex originated in the Middle Carboniferous and consists mainly of granitic types. The enormous Darvaz intrusion and a series of intrusions in the lake Kara-Kul district belong to this complex.

THE CIMMERIDIAN MAGMATIC CYCLE. In the northern and central ranges this cycle is represented by lavas. In the west, within the area of U.S.S.R. such lavas are very rare, thin and of a limited distribution. Amongst Jurassic rocks of lake Ala-Kul tuffs are known traversed by basaltic dykes. Outpourings of amygdaloidal olivine basalts are also known from Jurassic strata on the shores of lake Issyk-Kul.

In the east within the Chinese Peoples' Republic, in the Bogdo-Ola range Jurassic lavas cover large areas and are many hundred metres thick.

The problem of Jurassic synorogenic intrusions used to provoke and still provokes lively arguments. As has already been indicated some investigators think that there are such intrusions amongst Jurassic coal-bearing strata of the Tashkent region and also in the Chatkalian, Kuraminian, Alai and Turkestan ranges. Vasil′kovskii (1952) has gathered the fullest data pertaining to this question, but his facts do not prove the existence of synorogenic Cimmeridian intrusions in the Tashkent region, although such are found in the Pamirs. There are, however, pipe-like bodies and dykes which intrude Jurassic rocks of the Tashkent district. These intrusions represent typical sub-volcanic channels of Jurassic or even younger basalts. It must be pointed out that there are no known Jurassic lavas and tuffs in either the Tashkent region or in the Vale of Fergana, but there are indications of Neogene vulcanicity. Thus the intrusions may be of Alpine age.

In the southern ranges the Cimmeridian magmatic cycle is fully developed, its rocks are widespread and are the main magmatic rocks of the southern zone. In the northern zone—the Mangyshlaks, the Tuar-Kyrs, the Kugitangs, the Peter the 1st range and the Transalai range there are Jurassic lavas of limited extent, but there are no known Cimmeridian granites.

In the southern zone Lower and Middle Jurassic lavas are even more widespread and in many localities there are synorogenic Cimmeridian granites. The most westerly exposure of the latter is in Krasnovodsk region (Uffra Massif), but they are best represented in the central and southern Pamirs. In Krasnovodsk the granite forms a hilly peninsula, which juts out far into the Bay of Krasnovodsk (Fig. 190). The age of the granite and the associated igneous rocks is debatable. Some investigators think that they are Palaeozoic, while others consider them as Lower Miocene. These arguments are current since the contacts have not been sufficiently studied. A Palaeozoic age is improbable since Palaeozoic granites should intrude Palaeozoic sediments which are not present in this area. A Mesozoic (Lower Cretaceous) age is most likely (Fig. 191).

In the Central Pamirs—the valleys of the Bartang, the Kudara, the Pshart and the Murgab—granite massifs intrude the shales and sandstones belonging to the Middle and Lower Jurassic, but do not penetrate the overlying Upper Jurassic and Lower Cretaceous limestones. The Kudara intrusion, which has been described by V. A. Nikolayev, consists of pale, medium to fine grained biotite granites. Along its southern contact a thick formation (1·5–2km) of slates has been converted into garnetiferous and andalusite-bearing mica schists, while beds of limestone have changed into marbles. The granites are gneissose and have numerous apophyses of granite, aplite and pegmatite. The relationships between these granites and slates are the same as in the northern Caucasus. N. K. Morozenko (*Geological Structure of the U.S.S.R.*, vol. 2, 1958) distinguishes five Cimmeridian complexes: the Rushanian, the Ayazgulemian, the Koitezekian, the Pshartian and the Bazardinian. The two latter complexes are possibly Alpine. Granites predominate, but there are also basic rocks and various minor intrusions. Near the Koitezek Pass the granite massif outcrops over an area of 40×30km. Pshartian granites from massifs of 30×3km in area and possessing a general trend parallel to the prevalent strike. Haematite, quartz, molybdenite, wolframite and sulphides of zinc, copper and lead are associated with the Cimmeridian complexes.

ALPINE MAGMATIC CYCLE.　Rocks of this cycle are most widespread. Lavas

555

FIG. 190

The granite Uffra Massif near Krasnovodsk. The edge of the granite, faintly visible through the morning mist, penetrates deeply into the Krasnovodsk bay. *Photograph by V. D. Nalivkin (1955).*

FIG. 191

Southern part of the Uffra Massif near Krasnovodsk. *Photograph by V. D. Nalivkin (1955).*

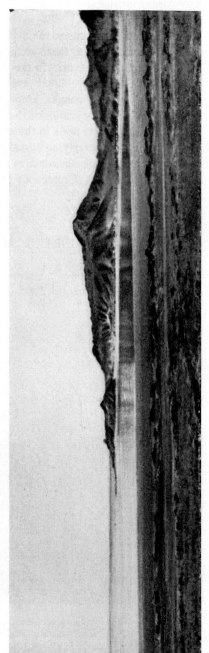

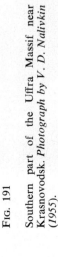

belonging to it are found ubiquitously but are often thin and are of a limited distribution. Typical synorogenic intrusions are only known from the Pamirs, to the north-east of Khorog, where they cut across Cimmeridian granites. In the northern and central ranges Alpine lavas are found in the Transiliian Alataus where the barren red beds of Upper Jurassic-Lower Cretaceous age have horizons of basalts. In the north of the Vale of Fergana Cretaceous sediments are intruded by andesite-basalt dykes. Basalts are also found in the Neogene deposits to the south of lake Chatyr-Kul. Alpine intrusions have been found in the Pamirs and possibly near Krasnovodsk (Fig. 191).

In the southern ranges the Alpine cycle is more fully developed. In the west (western Turkmenistan) there are beds of volcanic ash in the Pliocene and bentonite layers (altered volcanic ash) in the Palaeogene. In the extreme south of the U.S.S.R. near Kushka there are lavas and in places sub-volcanic intrusions of andesite, andesite-basalts and basalts amongst Eocene sediments of Badkhyz and Karabil (Fig. 192). Within the limits of the Tadzhik Virgation there are no Cainozoic lavas and likewise there are none in the northern Pamirs. In the south-eastern Pamirs there are lavas and tuffs overlying Jurassic limestones. The lavas consist of weakly altered biotite-pyroxene andesites, dacites and liparites. In the central and southern Pamirs in several areas the Cretaceous red beds are intruded by stock-like bodies of diabases and granitoids.

USEFUL MINERAL DEPOSITS

In Central Asia there are no mineral deposits of world importance, but there are many nationally important deposits of oil, coal, copper and lead ores, rare metals,

FIG. 192. Palaeogenic andesite basalts at Badzykh, near the town of Kushka in south-eastern Turkmenia. In the far distance is a 200m cliff of Pliocene red beds and to the left of it lake Er-Oilan-Duz. *Photograph by A. Lavrov.*

salts and sulphur. Hundreds of deposits of local significance are known and are commonly exploited.

In the pre-Revolutionary era the district was economically backward, but during the Soviet regime it has changed beyond recognition. Parallel with an active development of agriculture Socialist industry was built up on the basis of local mineral deposits. In this respect Central Asiatic geologists have played a significant role, by discovering each year yet newer and newer deposits.

ORES

Soviet geologists especially deserve credit for finding ores. When the present author in 1911 had started working in Central Asia (Vale of Fergana) he was told that 'Ores are rare and one should not waste time on them'. The continuous and sacrificial labour of Central Asiatic geologists has overturned this statement, which was based on a few rough traverses. A new antimony-mercury zone of deposits was found in the Vale of Fergana and near to it the copper deposit at Almalyk and several deposits have also been found in other regions of Central Asia.

Ferrous Metals

Despite the achievements of Soviet geologists, there still are unsolved problems. A series of iron ore and manganese deposits of various origins are known, but almost all of them are small or medium size and although the known reserves have been increased by tens of times, these deposits cannot be compared with the analogous deposits of the Urals. It should also be added that the rarity of ultrabasic rocks means that there are no nickel or chromium ores. Consequently further search for such ores is necessary.

The Precambrian jaspilitic iron ores present the best prospect and recently a large iron ore deposit has been found at Dzhetymtau, in the central Tian Shians.

Non-Ferrous Metals

The non-ferrous metals of Central Asia are some of the most important deposits in the U.S.S.R.

To the south of Tashkent there is the Almalyk copper deposit. The ores are disseminated and occur in vugs in an intrusive massif. The copper content is low, but the total reserves are large since the massif is large. Copper deposits of Central Asia stand in the third place in the U.S.S.R. Polymetallic ores of lead and zinc with admixtures of silver and cadmium are quite important.

In the west Karamazar (Kuraminian Mts) to the north of Leninabad the Kan Sai deposit, and in the Karataus the Achi-Sai deposit, are found amongst Middle Palaeozoic limestones. The ores from these deposits are worked in the Chimkent foundry. To the east of Frunze there is the Ak-Tyuz deposit, and in the Dzhungarian Alataus the Tekeli deposit, which again supply the same foundry. The lead deposits of Central Asia are second in the U.S.S.R. (Amiraslanov, 1957).

Bauxites and Bauxitic Rocks are found in several lacustrine deposits of Upper Triassic age, and lie at the bottom of the Jurassic Coal-bearing Formation. The deposits are small and the quality of the bauxite is low. There are also Middle Carboniferous bauxites, but again in small quantities. There is a good possibility that other Palaeozoic and Mesozoic deposits will also be found.

Precious Metals

Sparse alluvial gold associated with Palaeozoic intrusions is known from some localities and in particular in Darvaz (north-western Pamirs). Silver occurs in small quantities as an impurity of lead.

Rare Metals

The south Fergana antimony-mercury zone is particularly important. The mineralization is associated with breccias of young fault zones occurring in Devonian limestones. The minerals are cinnabar and antimonite with some fluorite and copper-bearing minerals. The best known deposits of mercury are at Khaidarkan, and of antimony at Kadamzhai. They are situated in the southern part of the Vale of Fergana in the foothills of the Alai range. The mineralization is Upper Palaeozoic in age. The mercury reserves of Central Asia are the most important in the U.S.S.R. (Amiraslanov, 1957).

Tin is found in Hercynian pegmatites of the Turkestan and Zerafshan ranges and also in polymetallic ores of the Ak-Tyuz deposit, which lies in the Kirgiz range, to the east of Frunze. Tungsten and arsenic have been discovered in the Karamazar region (Kuraminian Mts) in skarn-type deposits.

Indium, gallium and thallium occur as concentrated impurities amongst poly-metallic ores of Central Asia (Antropov, 1959).

FUELS

Fuels of Central Asia are of national importance and the reserves are constantly increasing, despite the fact that the deposits were known in pre-Revolutionary times. The increase in the reserves is taking place on account of exploration of old deposits as well as the discoverey of the new ones. Whole coal basins have been found including the Angrenian and the Uzgenian (South Fergana). Nevertheless oil is the most important fuel found and is intensively exploited. Coal satisfies local require-ments and recently there have been discoveries of rich deposits of gas.

Oil

Large oil provinces of various types have been discovered in recent years and it is possible that other deposits will also be found. In the last few years the significance of Central Asia with regard to oil and gas has been underlined (Denisevich, Diken-stein *et al.*, 1961).

The first, nationally most important province is known as the West Turkmenian (Turkmenneft), and stretches along the shore of the Caspian Sea from Krasnovodsk to Artek, as far as western Turkmenistan. Indications of oil have been known for a long time, but the discovery of commercial oil and the establishment of its great importance have been due to Soviet oil geologists and drillers. At present Turkmenia is a district which is just as significant as the Baku oil-fields.

It is interesting that the geological structures of the Baku and the Western Turkmenian oil provinces are almost identical. Thus the oil is found within the same type of anticlinal structures, which stretch across the Caspian Sea forming a sub-marine barrier obvious on maps. On both sides of the Caspian Sea the copper oil-bearing formation is represented by thick (1000m) continental deposits which have alternate clay and sand formations of an Upper Pliocene age. Within the Turkmenian

Soviet Socialist Republic these sediments are referred to as 'red beds' produced by a shallow delta of a river (the Praamu Dar'ya) which entered the sea from the east.

In Baku certain lower formations are also oil-bearing. Thus Upper Cretaceous sediments contain petroleum. It is undoubted that in Turkmenia such sediments are also oil-bearing, but here they have not been as yet drilled. The study of the mud volcano breccias has shown that the roots of these volcanoes lie amongst thick Upper Jurassic, bituminous limestones which are parent sediments of oil and gas. Such limestones are found both in the Caucasus and Turkmenia. In both regions the reservoir rocks are rubbly, relatively fine-grained sandstones. The structures are brachyanticlinal, relatively small (several kilometres) and possess quite steep flanks (10°–30°). The anticlines are often faulted. The upper horizons—Baku, Apsheronian, Akchagylian—involved in folding are compositionally similar in both regions.

Western Turkmenistan to the south of Krasnovodsk is a single petroliferous region which stretches along the Caspian Sea. The region can be subdivided into a series of isolated regions, which normally correspond with anticlinal structures. So far the Nebit-Dag and Cheleken (Fig. 193) oil-fields have the greatest industrial importance but oil has also been found in other regions. The Boya-Dag structure which is cut by the neck of an ancient mud volcano has hot springs and powerful streams of hot gas. These structures are found in a lifeless desert (Fig. 193). A strong oil fountain, arising from the bottom of the Caspian Sea is of some interest.

The second petroliferous province encompasses the Vale of Fergana and the Termez region known as Ferganian. This province differs from the Western

FIG. 193. The central elevated part of Cheleken structure consisting here of Pliocene strata. In the far distance is a low plain covered by Anthropogenic deposits. *Photograph by V. D. Nalivkin.*

Turkmenian by the fact that the oil-bearing horizons are found in Palaeogene strata, amongst which the porous dolomites of Palaeocene age, the porous Eocene lime-stones and the top Oligocene sands act as reservoir rocks. In places oil is found in the overlying Miocene continental formations—sandstones of the Bactrian stage (Gabril'yan, 1957; Simakov et al., 1957).

Accumulation of gas and liquid bitumens found in massive, bituminous, Upper Jurassic and Lower Cretaceous limestones are of interest. The bitumens react with the overlying beds of gypsum giving rise to sulphur deposits (Gaurdak). The Meso-zoic oil of the Vale of Fergana is very important.

One of the most important results of the research carried out in western Uzbeki-stan is the discovery of a third, Bukhara-Khivinian oil-gas province. The gas deposit at Gazli is the largest in the U.S.S.R. and Mesozoic strata are regionally petroliferous and gas-bearing (Zhukovskii et al., 1957; Dikenstein et al., 1959). According to the latest information (Gar'kovets et al., 1961) the commercial oil in the Gazli group of structures, to the north-west of Bukhara, and the Kagan and Mubarek groups of structures, to the south-east of Bukhara, is continuous for a total distance of 150km. The oil-bearing horizons are found in sandstones of the Lower Cretaceous, limestones of the Upper Jurassic and sandstones the Middle Jurassic.

The discovery of large accumulation of oil in the Mesozoic of the southern boundary of the Mangyshlaks and central Karakum is of considerable significance. These discoveries as well as the investigation of the mud volcanoes of western Turkmenistan permit a conclusion that the western part of Central Asia is a region of marine Lower Cretaceous and Jurassic and thus is highly likely to yield oil and gas. This is quite an important conclusion and promises high prospects to Soviet oil-geologists.

There is so far no oil found in Jurassic lacustrine sediments, but its existence is likely since such have been found in Chinese Dzhungaria. The parent rocks of this oil are Upper Jurassic oil-shales compacted into lacustrine sapropelit. The occur-rence of oil in lacustrine and overlying rubbly sandstones represents an important problem, which has been satisfactorily solved in the U.S.A. (Green River Suite) and the Chinese Peoples' Republic.

Oil-shales and bituminous shales have been found in the Palaeogene of Tadzhi-kistan and the Silurian of the Turkestan range (beds in graptolitic shales), but no commercial deposits are known.

Coal

Black and brown coals are abundant and sometimes form large deposits which are almost entirely of Jurassic age. No Lower Carboniferous coals are known, while Upper Palaeozoic (Middle Carboniferous) coals are only found in the Central Pamirs, but in small quantities.

Jurassic coals are ubiquitous and their largest deposits occur in depressions between the northern and the central ranges, as for instance in the Vale of Fergana and the Fergana range, which at the time of Jurassic sedimentation were a part of a vast plain. This plain was one and a half times wider than the present-day Vale of Fergana. The fold structures of the range of Fergana started growing in the last phases of the Cimmeridian Orogeny, at the beginning of the Lower Cretaceous. The range thus was a barrier to the Upper Cretaceous sea. The ultimate stage in the formation of the Fergana fold structures, as well as the folds in the Jurassic Coal-bearing Formation occurred much later, when Palaeogene, Neogene and Anthropogene sediments were affected. The formation of coal-bearing strata started at the end of the Upper Trias, reached its maximal development in the Lower Jurassic and drew to a close in the Middle Jurassic. Thus commonly these rocks are referred to as 'Jurassic'. During this epoch the areas of sedimentation were a series of inter-montaine valleys of diverse dimensions. One of these valleys—the northern—was situated to the north of Tashkent and was the site of accumulation of the deposits which are now called Chakpak (southern Karataus) and Keltemashat (near Chikment). In the area of the Kugitangs, the southern slope of the Gissar range and the Northern Pamirs there was a large near-shore plain which was margined by a southerly sea and abutted against northerly young low mountains, the erosion of which supplied the clastic fraction of the coal-bearing sediments. The largest plain, which was distant from the sea, was situated between Caledonian ranges, rising on the site of the present-day northern ranges, and Hercynian ranges of the present-day central ranges. This plain was elongated in an east–west direction from the Nuratau range to lake Issyk-Kul—a distance of over 1000km—and was 150–200km wide. It is probable that the plain was subdivided by subsidiary ridges with an east–west trend.

The three main plains were closely associated since their flora was the same and the climate wet and temperate. Over the surface of these plains flowed numerous rivers which entered the main rivers flowing in an east–west direction. Numerous lakes and swamps existed between the rivers, determining the position and magnitude of coal seams.

The dimensions and the shape of the northern plain have not as yet been established. The Chakpak and Keltemashat deposits originated in its central part. It is possible that this plain continued to the north along the Karataus, thus uniting with the region of the Baikonur deposit (near Dzhezkazgan) and then stretched further to the north, along the eastern slope of the Urals. It is in fact quite probable that it then continued to the east along the northern ranges, to join the large Kul'dzha deposit, situated in the Chinese Peoples' Republic. If these speculations would be verified, then the northern plain will be very large and continuous over several thousand kilometres.

The central plain is typically intermontaine, but judged by the present-day distribution of coal-bearing formations is the smallest, being only 1000km long. It is thus much smaller than the northern and southern plains. Along it there were a series of swamps which converged to form continuous marshy areas stretching for hundreds of kilometres. The Shurab and Kyzyl-Ki deposits in the south of the Vale of Fergana and the Issyk-Kul deposit were evidently related to individual swamps of this type. The recently discovered Uzgenian coal-field on the southern half of the Fergana range was a large continuous swamp. The central plain subsided continuously and as a result a great thickness of over 1000m of sediments accumulated. The last phase of the Cimmeridian Orogeny (Lower Cretaceous) has folded these rocks into complicated folds, which are especially tight in the Uzgenian basin. Consequently in this field metamorphism is most advanced and the coals are of the valuable true black variety.

The southern plain was at least as extensive and probably more so than the northern plain. It was in a direct connection with Jurassic seas which margined it from west and south. In the east and north it was bounded by highlands of Hercynian structures. This plain was a typical alluvial near-shore low area passing into a submontaine plain. Its elongation was the same as the shoreline of the continent

Angarida and thus several thousand kilometres long. Within Central Asia this province lies in the Mangyshlaks region, where its deposits include the Middle Jurassic Coal-bearing Suite and the more northerly Oil-bearing Suite, which is the continuation of the Dossorskian Suite of the Emba oil-field. Further on the coal-measures of the Tuar-Kyrs and the Greater Balkhan and Kugitang, the southern slope of the Gissar range (Ravat deposit), the Peter 1st range and the northern and central Pamirs belong to this province. The original plain then continued to the south-east into China (Sinkiang). The proximity of the sea to the uplands meant that the plain was often highly inclined, precluding the formation of swamps and accumulation of coal, such as in the central province. According to A. P. Nedzvetskii (1957)—one of the foremost geologists in Tadzhikistan—the Ravat deposit is the largest in Central Asia and some of its coals are of coking quality. The deposit lies in the valleys of the Yagnob and Fan Dar'ya.

The above facts do not exhaust completely the data on variations of Jurassic coal-bearing strata, but indicate their diversity and variability.

Non-Metallic Mineral Deposits

Non-metallic useful mineral deposits are numerous and variable. Deposits of hydrothermal fluorite found in Palaeozoic limestones are considerable and include the Aurakhmar deposit to the north-east of Tashkent, the Takob deposit in the Gissarian range to the north of Dushanbe and the Naugarzan deposit in the Karamazars (the largest in the U.S.S.R.). Numerous but small deposits of barytes and witherite are also known as for instance in Kopet-Dag (Karakalinian region).

Sulphur is found in the Gaurdak deposit in the Kugitangs where it is found as the result of reaction between Lower Cretaceous gypsum and bitumens arising from Upper Jurassic limestones.

There is also a known deposit (Sernyi Zavod) in the middle of the desert to the north of Ashkhabad. The sulphur at this locality has been produced by reactions between Palaeogene limestones and bitumens. The Shorsu and a series of other deposits are of such a type.

Phosphorites. Phosphorites are found in numerous and diverse deposits of which those in the Karataus are the largest in U.S.S.R.

Bedded marine phosphorites of a Middle Cambrian age are found at the base of the Lower Palaeozoic, Tamdinian limestones in the south-eastern Karataus. Cenomanian phosphorites similar to the Embian are known from western Mangyshlaks and the shore of the Sea of Aral. Palaeogene phosphorites are also found on the shore of the Sea of Aral along the western slope of the Karataus, then near Tashkent and in the Vale of Fergana.

Salts are widely found and are of various compositions. Reserves of common salt are inexhaustible. Near to Kulyab, in the east of the Tadzhik Depression, Upper Jurassic salt forms hills of up to 150m high, as is for instance the Khodzha Mumyn. Large deposits of salt are also known in the Kugitangs. In the valley of the Narya there are Neogene deposits, in western Turkmenistan sedimentary salt is obtained from lakes Kuuli and Dzhebel and in the Aral basin salt is obtained from bitter-salt lakes. Potassium salts have been found in the Kugitangs where they occur together with halite. The Karabogazgol Bay is a large salt-generating depression. Depending on the influx of water from the Caspian Sea various salts get deposited in it. Of these salts mirabilite ($Na_2 SO_4$, $10H_2O$) is especially valuable.

Gypsum is again widely found and especially so in association with Upper Jurassic salt and marine Upper Cretaceous and Palaeogene deposits. In Tadzhikistan there is also the large Gulisai deposit of celestine.

The region is also rich in various raw materials such as that used in producing cement, facing marble etc. Loess and loessic loams are widely used in local constructions.

REGIONAL DESCRIPTIONS

Of many differing districts and provinces of Central Asia we will consider only three: the Pamirs, the Vale of Fergana and the Karakum Desert. The Pamirs ('the roof of the world') is a young range of high mountains, the Vale of Fergana ('the pearl of Central Asia') is agriculturally and economically a most important province, which is an example of an intermontaine valley, while the Karakum is the largest desert in the U.S.S.R.

THE PAMIRS

General

The Pamirs constitute the highest plateau in the U.S.S.R., which has an average height of over 3000m. The plateau is the most elevated and eroded part of Alpine and Cimmeridian structures of the Asiatic portion of the Mediterranean Geosyncline. The tectonic position and geological structure of the Pamirs closely resemble the western part of the Great Caucasus range, as is obvious on synoptic geological maps. In both regions a complete Palaeozoic and partial Precambrian section is exposed in the core of a very large anticlinorium. Both cores are surrounded by strongly folded Triassic, Jurassic and Cretaceous sediments of principally marine origin. To the south of the Palaeozoic core of the Pamirs and separated from it by a zone of intensely deformed Mesozoic and Cainozoic rocks lie Precambrian massifs which show complex interrelationships. The elevation and consequent erosion in the Pamirs has been of a greater magnitude than in the Caucasus and consequently the Palaeozoic core is much wider and is surrounded by Meso-Cainozoic strata not in the Soviet Territory, but further west in Afghanistan, while in the east, Palaeozoic strata continue into the Kuen Lun Mts of China. According to its relief the Pamirs can be divided into two sharply different regions. The eastern region, populated by the Pamirian Kirgiz, is called the Pamirs, while the western region, populated by mountain Tadzhiks, is known as Badakhshan and its centre is the town Khorog. The western boundary to Badakhshan is the river Pandzh, which is one of the main tributaries of Amu Dar'ya. The eastern boundary to Badakhshan is indefinite, being a zone some tens of kilometres wide and is not populated by either Tadzhiks or Kirgiz people.

The Eastern Pamirs is a typically glaciated region and its main valleys are broad and U-shaped (Fig. 194). By merging into each other they form several large, flat depressions, tens of kilometres wide and occupied by glacial lakes such as Kara-Kul, Rang-Kul and Zor-Kul, which are wide and shallow. Everywhere there are remnants of terminal and lateral moraines, roches moutonnees and other indications of former glaciation. The watersheds are narrow and relatively low and often bear evidence of breaching glaciers. The present-day glaciation, owing to small precipitation, is negligible and as a consequence the Eastern Pamirs have a desert aspect.

FIG. 194. The Murgab valley in the south-eastern Pamirs. The wide desert valley has sparse vegetation and lies at a height of almost 4000 m amongst a present-day terrace. *Photograph by S. I. Strel'nikov (1958).*

FIG. 195

The Bartang valley in Badakhshan. In forefront are black Lower Jurassic slates and at the back Mesozoic volcanic rocks. *Photograph by P. P. Chuyenko.*

Great heights (valleys 3 to 4km, ranges 5–6km) render the climate harsh and cold with an average annual temperature below zero and permafrost is ubiquitous. This conjunction of desert conditions and permafrost is peculiar.

The Western Pamirs (Badakhshan) is a district of very deep gorges with surrounding slopes rising up to 3500m. At the bottom of these gorges with V-shaped profiles flow large rapid rivers which coalesce into the Pandzh. Amongst these rivers the Bartang (Fig. 195), the Yazgulem (Fig. 164) and the Vanch are especially important. The northern boundary to the Pamirs lies along the Kyzylsu, which flows along the Alai valley, which has a glacial character. Where the Transalaian range is followed by the Peter 1st range the environment of the Kyzylsu changes, its valley becomes narrow and deep, the profile becomes V-shaped, the current accelerates and after merging with the Muksu it is known as the Surkhab and further on as the Vakhsh. After the merging of the Vakhsh and the Pandzh the river is known as Amu Dar'ya.

The north-western part of the Pamirs, between the Vanch and the Surkhab, is known as Darvaz, which means 'the gates'. In Darvaz the river valleys are the deepest.

At the boundary between Darvaz and the Pamirs proper lies the Fedchenko glacier, which is the largest in the world. To the west of the glacier is the Akademiya Nauk range, which is the highest in the U.S.S.R. with a peak of 7495m high. The range is only 60km long and is a typical erosion range, striking across the dominant trend, while other ranges follow it.

Stratigraphy

In the Pamirs there are found all deposits beginning with the Archeozoic. Furthermore, the Neogene apart, all the systems are represented by mainly marine sediments; Neogene sediments are entirely continental. Unlike the other mountain arcs the Pamirs have large areas of Precambrian and Palaeozoic sediments and abundant marine deposits of Triassic, Jurassic and Lower Cretaceous ages, as well as a greater total thickness of geosynclinal sediments.

Archeozoic. The ancient crystalline metamorphic rocks of the south-western Pamirs are Archeozoic and consist of many kilometres of gneisses, mica schists and marbles which form mountain ranges of over 6000m in height. A comparison with the Karakorums and the Himalayas has demonstrated the identity and a direct connection between the crystalline rocks of the south-western Pamirs and the Archeozoic of these ranges. Recently B. V. Timofeyev (1958) has found Proterozoic spores in the paraschists (Goranian Suite) occurring at the bottom of the sequence. This suggests a possible Proterozoic, but a definitely Precambrian age of these rocks.

Proterozoic. The phyllites, massive quartzites and metamorphosed limestones of the Central Pamirs were originally included by the present author in the Proterozoic. Later on owing to the suggested correlation with palaeontologically dated Palaeozoic rocks the Central Pamirs formations were thought to be Lower and Middle Palaeozoic. The work of the last few years has shown that these rocks are laterally continuous into Proterozoic strata of the Kuen Luns where the latter are unconformably overlain by Lower Palaeozoic rocks.

Lower and Middle Palaeozoic rocks are the same as in the rest of the Mediterranean Geosyncline and in the central ranges. Marine Ordovician, Lower and Upper Silurian, and a full section of the Devonian and Lower Carboniferous have been discovered.

FIG. 196. A massif near the river Karasu, consisting of bedded Lower Trias limestones at the very top, dark tuffs under them and Upper Permian limestones and siliceous shales still further down. At the foot of the massif there are eroded Anthropogenic alluvial cones, and at the edge of the river a present-day terrace. *Photograph by S. I. Strel'nikov (1958).*

FIG. 197. Thick, rhythmically bedded, Middle Jurassic, flysch-like marls and limestones in the south-eastern Pamirs. *Photograph by S. I. Strel'nikov (1958).*

Upper Palaeozoic. The Upper Palaeozoic is peculiar and is represented by a thick, largely barren formation, which is in places carbonaceous, of sands and shales of Middle and Upper Carboniferous age. Lower Permian sediments are marine and are in places represented by reef limestones, while Upper Permian marine limestones have the same fauna as is found in the Northern Caucasus. The limestones are conformably overlain by sands and shales with marine Lower Trias faunas (Fig. 196). These rocks become red and barren upwards and represent typical sediments of a marine regression or in other words the deposits of a near-shore plain.

Mesozoic. The Mesozoic is typically geosynclinal of Mediterranean type, is several kilometres thick, has abundant marine remains and consists mainly of massive limestones and sandy and shaley rocks of flysch type. There are also coal-bearing or more rarely conglomeratic molasse-type strata with horizons of lavas. The Cimmeridian Orogeny left a strong imprint. It began in the Upper Trias and ended at the beginning of the Lower Cretaceous. Marine deposits are most abundantly found in the south-eastern part of the Pamirs. Upper Cretaceous deposits, which are closely associated with Palaeogene strata, are separated from underlying rocks.

Upper Cretaceous and Palaeogene. This is the epoch of a large-scale marine transgression, which overran the Northern Pamirs and the Kuen Lun, but stopped against the Tarim Massif, which is at present a depression. The deposits of the transgression represent a repetition of marine and lagoonal sediments, the latter being occasionally saliferous, gypsiferous and petroliferous (in China and the Tadzhik Depression).

Neogene and Anthropogene. During these epochs entirely continental deposits of great thickness accumulated; where molasse-type conglomerates were laid down the

FIG. 198. Relatively gentle, simple folds affecting thick Upper Jurassic limestones in the south-eastern Pamirs. *Photograph by S. I. Strel'nikov (1958).*

thickness reaches 8–9km. Glacial sediments are particularly abundant in the Anthropogene.

Tectonics

The tectonics of the district are very complex and interesting since all orogenies beginning with the Byelomorian and ending with the Alpine occurred here. Complex folding and a high grade of metamorphism is typical in Lower Cretaceous, Jurassic and older rocks. Jurassic strata of the Pamirs have similar styles of folds and grade of metamorphism to Carboniferous rocks of the Urals (Fig. 198).

Middle and Upper Palaeozoic and Mesozoic rocks, compressed between massive crystalline rocks of the Southern Pamirs and metamorphic strata of the Central Pamirs, are particularly strongly deformed. Consequently the so-called Akbaitalian zone of thrusting involves broken folds of all stratigraphic units, beginning with the Ordovician and ending with the Upper Cretaceous.

Magmatism

The magmatism is complex and diverse. Phases of intrusive and effusive activity overlap each other, which complicates the dating of synorogenic intrusions, the sills and the dykes. In the south of the Pamirs the age of igneous rocks varies from the Late Cimmeridian (Lower Cretaceous) to the Byelomorian (Upper Archeozoic). In the central part of the Pamirs many intrusions are differently dated as Hercynian or Cimmeridian. Despite these difficulties the widespread distribution of the Cimmeridian synorogenic granites in the central and south-eastern parts of the Pamirs is an established fact. No such intrusions are found in the central ranges, which conforms the inclusion of the Pamirs into the Mediterranean Geosyncline and the consequent similarity of magmatism in this region to that in the Caucasus. More detailed data for the Pamirs have already been quoted.

Useful Mineral Deposits

Even before the Revolution gold deposits were known in Darvaz. Such alluvial gold has been found by Soviet geologists in other regions. In addition iron and lead ore, coal, and piezo-optic quartz have also been found, but all these deposits are rather inaccessible.

VALE OF FERGANA

General

The Vale of Fergana is justly known as the 'pearl of Central Asia'. It is surrounded by high mountain ranges, which throughout the year are snow-capped. Large glaciers give rise to numerous full and turbulent rivers. The flanks of the ranges consist of Palaeozoic rocks gradually passing into foothills where brightly-coloured and variegated Cretaceous and Palaeogene sediments rival equally bright and variegated robes, carpets and hats of the Uzbek and Tadzhik populations. Hills and valleys of the foothills merge into the alluvial plain and disappear under screes and numerous alluvial cones which merge into each other. Mountain streams emerging in their turn are broken up into thousands of irrigation channels. Where water is present the hot sun helps in producing a luxuriant vegetation. The whole of the vale is surrounded by orchards, cotton plantations, rice paddies and brightly green lucerne and clover fields. Innumerable villages and towns continue one after another. The density of population is higher than in Belgium.

Yet not only is intensive and advanced agriculture the pride of the Vale of Fergana, for oil-fields and coal-fields, numerous mines of copper, mercury and antimony are also of great importance. At the same time extraction of gypsum, salt and raw materials of cement, marble and other building materials is taking place. The mining industry of the Vale of Fergana is of national importance and since the raw materials are available other types of industry have also been developed.

As the rivulets die out the green of the oases rapidly disappears and the ground becomes an ashy-grey, sun-drenched desert, which is in some places muddy, elsewhere pebbly, while sometimes it is covered by sand-dunes. In the middle of the desert, between low, silty but precipitous cliffs, flows the wide, rapid and muddy Syr Dar'ya, which is formed by the confluence of the Kara Dar'ya and the Naryn.

In the north the Vale of Fergana is bounded by the Chatkal range and its western projection known as the Kuraminian Mts (the Karamazars). To the south of the vale lie the snow-capped Alaian and Turkestan ranges. All these mountain belts are elongated parallel to the strike of the Palaeozoic structures, which form them. The Fergana range on the other hand lies across the trend and its southern part is formed by Jurassic structures. The majority of the oases with the main towns of Andizhan, Namangan, Kokand and Fergana are populated by the Uzbeks and belong to Uzbekistan. The western part including the Kuraminian Mts and the towns of Leninabad and Ura-Tyube, and also the adjacent part of the Turkestan range, is populated by the Tadzhiks and is known as Northern Tadzhikstan. All the foothills and mountain slopes with the towns Osh, Uzgen and Dzhalal-Abad belong to the Kirgiz Soviet Socialist Republic and have a mixed population. Kirgiz people form the main population of the mountains.

Geological Structure

Two concentric belts—mountainous and foothills—can be distinguished around the central part of the vale. The differences in the relief of these belts reflect their geological structure.

The mountainous belt includes Palaeozoic massifs. The only exceptional range is the southern part of the Fergana Mts which, as indicated on maps, consists of continental sediments of the Uzgen coal basin. Some marine Upper Cretaceous and Palaeogene sediments are preserved to the south-east of Osh, within the Alai range. These rocks suggest the former existence of a wide marine channel. In the Kuraminian Mts isolated plateau-like fragments at heights of up to 2000m have escaped erosion and represent remains of an ancient marine cover of Cretaceous and Palaeogene age. The rocks have been uplifted to this height by faulting. There are other small districts where Jurassic, Cretaceous and Palaeogene sediments are found, but all of these occupy very small areas, as compared with Palaeozoic strata.

Lower Palaeozoic. At the bottom of the Palaeozoic succession lies a thick shaley formation, consisting of dark argillites, with formations of siliceous shales and beds of sandstones and black limestones. In the latter in places (e.g. the Alai range) there has been found a fauna of tiny, to the naked eye hardly noticeable, trilobites (*Agnostus*) and brachiopods *(Acrotreta)* of Cambrian age. In higher horizons there is an Ordovician fauna mainly of graptolites. The strata are highly folded. No breaks or angular unconformities have been detected, thus there is no evidence for initial manifestations of the Caledonian Orogeny, but great thicknesses of clayey and sandy material indicate a proximate existence (possibly in the south) of a rising region in the process of erosion.

Middle Palaeozoic. The sediments belonging to the Middle Palaeozoic are the most widespread in the Alai and Turkestan ranges. Here the section is exceptionally complete and all the main subdivisions of the Silurian, Devonian and Lower Carboniferous are present as marine deposits. The lower part of the Silurian consists of thick shales and sandstones, which are rhythmically bedded, of flysch type and often graptolitic. These sediments indicate that the last phase of the Caledonian Orogeny was occurring nearby in more northerly provinces. This is supported by the nature of the succession in the Chatkalian and Kuraminian Mts, where it is highly variable and often consists of continental deposits. In places there are no Silurian deposits suggesting that the area was a region of erosion. Towards the end of the Silurian the Caledonian Orogeny caused the elevation of all the northern ranges, forming dry land, which existed throughout the Lower Devonian and the Eifelian epoch. The sea again re-advanced in the Givetian and thereafter gradually encroached northwards.

In the south the last phases of the Caledonian Orogeny did not occur and thick, massive Ludlovian limestones gradually pass upwards into similar Lower Devonian and Eifelian limestones as a result of which great thicknesses of Hercynian type limestones (Figs. 165, 167) reach 1500–2000m, thus forming whole mountains. Bedded limestones of sometimes great thickness, represent the Upper Devonian and the Lower Carboniferous (Fig. 166). The Devonian of the southern part of the Vale of Fergana has been described in detail by G. A. Kaleda (1960). In the Alai and Turkestan ranges there are sections in which Visean limestones lie with a break over various horizons of the Devonian and Silurian. This circumstance is explained by the former existence of islands which were only submerged during the Visean epoch.

Upper Palaeozoic. Upper Palaeozoic strata are widely distributed throughout the Vale of Fergana. The continuous Lower Carboniferous succession is captured by the Namurian which shows the signs of the beginning of the Hercynian Orogeny, first occurring in the middle of the Middle Carboniferous. Higher up follow unconformable limestones and in turn unconformable sandstones and shales with choristids of the Myachkovskian Stage. These rocks form the lowest strata of the enormous Upper Palaeozoic section, which begins with terrigenous rocks with subordinate thin and occasional beds of limestones. The lower part of the section has both marine and terrestrial sediments as well as thick volcanic formations. According to N. P. Vasil'-kovskii (1952) the thickness of the Minbulakian effusive suite, belonging to the top of the Middle Carboniferous of the Chatkalian Mts, reaches 3500m while the overlying Upper Carboniferous Aktashian volcanic complex is up to 5500m thick. Here the total thickness of Upper Palaeozoic rocks is more than 12 000m, since vulcanicity was widespread and volcanic outpourings occurred many times. The upper part of the Upper Palaeozoic includes not only the Upper Permian but probably the Lower Trias, represented by continental deposits with volcanic rocks. Intrusive activity began in the Middle Carboniferous when it was contemporaneous with initial folding. There are also Upper Carboniferous, Lower Permian and Lower Triassic intrusions. The latter are interesting since they confirm the ending of the Hercynian Orogeny in Triassic times.

Mesozoic. As a consequence of the Hercynian Orogeny the whole region of the Vale of Fergana became land which lasted a long time. The sea penetrated into this region only in the Upper Cretaceous and finally receded in the Oligocene.

The continental Lower Trias passes down into the Upper Permian. The Middle and a greater part of the Upper Trias are missing indicating the existence of an elevated land mass in the process of erosion. At the end of the Upper Trias, during the Rhaetic the new Mesozoic sedimentary cycle began with a thick coal-bearing formation (up to 2000–3000m) which succeeds a thin group of sandstones and shales containing (in the south) lenses of lacustrine bauxites and bauxitic rocks. The cycle ends with the development of barren red beds which are of an Upper Jurassic and Lower Cretaceous age and are several hundred metres thick (Fig. 174).

Mesozoic sediments everywhere overlie unconformably the Palaeozoic (and Lower Trias), and show simple folds. The folds are gentle and Jurassic coals have not suffered much metamorphism and remain brown, but in the southern part of the Vale of Fergana the folds are more complex and the black coals of a higher metamorphic rank are produced. The manifestations of the Cimmeridian Orogeny are weak and are not accompanied by intrusive activity. Jurassic lavas are unknown, but possible.

Cainozoic. The Palaeogene is represented by marine, the Neogene and Anthropogene by continental deposits. The Upper Cretaceous-Palaeogene sea covered the whole of the vale, with the exception of its western part where there has been a peninsula. As a result, in the latter area marine deposits pass into lagoonal, saline sediments and terrestrial red beds. The red beds form the beautiful, wild and curious hills of Akbel and Akchop (to the east of Leninabad) and reach a great thickness of 6500m.

Marine deposits are represented by grey, greenish and white clays, limestones and dolomites with abundant and variable faunas including masses of various oyster shells. At the bottom, in the middle and at the top there are formations of red sandy-clayey rocks which are in places gypsiferous and saliferous and which represent deposits of near-shore plains and of swampy lagoons on these plains (Gekker *et al.*, 1960). Porous dolomites and red sands in places yield oil which is exploited at a number of wells. Oil and gas react with gypsum giving rise to native sulphur, which is also being worked. Gypsum and rock salt are also being extracted. Certain muddy limestones are used as raw materials for cement.

Neogene and Anthropogene deposits are continuous and almost entirely continental. The boundary between the Neogene and the Anthropogene is often arbitrarily drawn at some diagnostic stratum. The combined Neogene-Anthropogene epoch is characterized by major mountain-building movements responsible for the present-day relief. The movements involved block faulting accompanied by weak folding. Fold movements are still continuous giving rise to the new anticlinal structures near Namagan as well as catastrophic earthquakes. The rising mountains were immediately subjected to erosion and the products of erosion were deposited as a molasse consisting of conglomerates and sandstones, which were studied in detail by V. I. Popov (1938 and later). The first uplift movements began at the end of the Oligocene when the Vale of Fergana became land and gradually acquired the present-day appearance. The recent tectonics of the south-eastern Fergana were described by Yu. A. Kuznetsov (1960).

THE KARAKUM

General

The Karakum is a translation from the Turkmenian 'black sands'. The term 'the Karakumy' is a Russified version and should not be used. The Karakum is the largest desert in the U.S.S.R. It stretches from the dry bed of the Uzboi to the frontier for 850km, and is 500km wide—from the foothills of the Kopet-Dag to the Amu Dar'ya. There is not a single freshwater lake or a river within it and precipitation is negligible. Drinking water is obtained from wells and a settled population is almost absent. The north-western part of the Karakum is low and represents a vast alluvial plain—the valley of the ancient Amu Dar'ya. The south-eastern part gradually rises. but is also a desert formed by the alluvial beds of the Murgab, the Tedzhen and the Balkha, all of which rise from the mountains of Afghanistan and Iran. The various districts of the high desert have local names such as the Karabil, the Badkhyz and the Obruchev steppe, where many years ago first investigations were carried out by V. A. Obruchev. The Tedzhan and the Murgab on reaching the Lower Karakum break up into dozens of branches forming peculiar land deltas, which are shown on maps. These deltas are nowadays flourishing oases and centres for growing high quality cotton. In the delta of the Murgab there are two large towns—Mary (the ancient Merv) and Baitamali—which have large numbers of cotton, cotton-oil, soap, canning and other factories. Until recently there has not been enough water for these oases and consequently the Great Karakum canal has been constructed, which branches off the Amu Dar'ya, near Kelif. The canal is a major project. It first follows an old tributary of the Amu Dar'ya known as the Kelif Uzboi, and then continues along an artificial channel across the sands.

All settlements are situated along the edge of the desert and include the towns Ashkhabad and Kyzyl-Arvat in the west and along the Amu Dar'ya Chardzhou, Khiva Turtkul, Urgench, Tashauz, and Nukus in the east. In the middle of the Karakum there is only the single township of Sernyi Zavod built in Soviet times, being the centre of extraction of sulphur and having a borehole which encountered oil.

In the north the Karakum is bounded by the plateau-like highlands of Ust'yurt with heights of 100–120m. To the south it abuts against the rocky ranges of the Tuar-Kyr consisting of Jurassic, Cretaceous and Palaeogene rocks. In the extreme north-west lies the high, lifeless rocky Bol'shoi Balkhan Massif. In the south-west is situated the low range of Kopet-Dag, which continues out of the U.S.S.R. At the foot of the Ust'yurt lies the dry valley of the Uzboi along which the water of the Sary-kamyshian lake once flowed into the Caspian Sea. At present in the place of this lake is situated a large depression into which formerly the Amu Dar'ya flowed and kept it brimful. At present the course of the Amu Dar'ya has shifted to the east and the lake has dried out. To the south-east of the Sarykamyshian Depression lies the low, but highly dissected Transunguzian plateau, which is a continuation of the Ust'yurt. Along its southern margin lie a series of dry valleys known as the Unguz.

The Karakum is the eastern, greater part of Turkmenistan which is a centre of breeding of Persian lamb.

Geological Structure

The Karakum is a typical Neogene platform. At its surface lie horizontal Neogene and Anthropogene sediments which are responsible for greater or lesser elevations bounded by normal faults. The Ust'yurt and the Transunguzian plateaux serve as examples of such elevations. Their surfaces consists of Sarmatian (the Ust'yurt) and Pliocene (Transunguzian plateau) sediments. The sub-Neogene sediments are folded. The folds affecting Palaeogene and Upper Cretaceous strata are simple, large and gentle, but become sharp and steep flanked in the Lower Cretaceous and Upper Jurassic. In the Middle and Lower Jurassic and Upper Trias the folds are even more individualized, steep and broken by normal faults; the metamorphism of the constituent beds is weak but noticeable. All four tectonic storeys are results of orogenic activity. The lowest (first) storey is affected by the last phases of the Hercynian Orogeny; the second storey is affected by the Early Cimmeridian phase, the third storey by the Young Cimmeridian phase and the fourth storey by the early Alpine phases. The late Alpine phases did not occur in the region and consequently a Neogene platform was formed.

The nature of the folds in each of the four storeys is clear in the Mangyshlaks and the Tuar-Kyr, since here erosion has revealed the beds which otherwise are hidden under the Karakum desert lying to the south-east.

The Neogene and Anthropogene sub-montaine downwarp begins a few tens of kilometres away from the Kopet-Dag range. The lower boundary of the Neogene is buckled into a deep down-fold rising at the Kopet-Dag to the surface and being involved in folding. The depth of the downwarp on geophysical data is about 3000m. A borehole sunk to the north of Ashkhabad traversed through 600m of Anthropogene conglomerates and shales but did not reach the base of the Anthropogene. The sub-montaine downwarp has been demonstrated and studied in the last few years and is of great interest, since the secondary anticlines on its flanks may be repositories of gas and even petroleum. To the north of the downwarp lies a Neogene (epi-Hercynian) platform forming the Karakums. Mesozoic and Palaeozoic folds of this platform have of late attracted attention since in the south of the Mangyshlaks they were found to contain oil and gas (Antropov, 1959). The work of a leading Soviet geophysicist Yu. N. Godin in this and other areas of the Turkmen Soviet Socialist Republic and other Central Asiatic republics has been of outstanding importance.

The platform sediments are divisible into three structural storeys: the first includes the Miocene and has been most uplifted, the second suffered much less uplift and includes the Pliocene, while the third, which has not undergone any block faulting begins with the Baku Stage and includes the Anthropogene. The Miocene is visible in the chinks (escarpments) of the Ust'yurt, on its surface and in the lower parts of the slopes of the Transunguzian plateau. The section is shown in Fig. 176. The lower part of the section consists of Lower and Middle Miocene and the lowermost Sarmatian deposits of clays with subordinate beds of sand. There is a rare marine fauna in these sediments. The Middle Sarmatian shelly limestones (20–30m) form the compact resistant top surface of the Ust'yurt and the precipitous top slopes of its chinks. To the south-west, behind the Uzboi, the limestones plunge under Pliocene sands and clays and then pass into gypsiferous clays and sands, which give rise to native sulphur. The Sarmatian sea covered the whole of the lower part of the Karakum and spread behind the region of the Sea of Aral.

At the beginning of the Pliocene all the northern part of the Karakum rose by some 40–60m and became a margin to the Akchagylian sea. The deposits of this sea appear as terraces adhering to the middle part of the Ust'yurt escarpment. Thus the Akchagylian sea was much less widespread than the Sarmatian. In the lower reaches of the Amu Dar'ya the former was represented by a mere embayment along the valley of the Uzboi.

At the end of the Pliocene and the beginning of the Anthropogene new block movements elevated the Ust'yurt escarpment and the Akchagylian terraces to their present level. The water level of the Caspian basins at these times rose and fell repeatedly as a result of variations in influx rather than tectonic movements. During one such period of large influx the Apsheronian sea penetrated along the valley of the Uzboi into the Sarykamyshian Depression, while on another similar occasion the Khvalynian sea filled the whole south-western part of the Karakum, flattening out its surface. The water level of the Khvalynian sea was 74m higher than the present-day Caspian, thus ensuring the wide spread of the former sea.

All Anthropogenic marine transgressions had a secondary significance in the history of the Karakum. The main factor which conditioned the formation of the Karakum surface was the shift in the position of the course of the Amu Dar'ya. In the Lower Anthropogene epoch the Amu Dar'ya had a pronounced westerly deviation slightly higher than the present-day Kelif and was flowing via the deltas of the Murgab and Tedzhen towards the base of the Kopet-Dag and then continued along the sub-montaine downwarp in the north-westerly direction, joining the proto-Caspian sea near Kyzyl-Arvat. The uplift of the Kopet-Dag has initiated the progressive shift of the river course to the east and in the Middle Anthropogene epoch the river was flowing at the foot of the Transunguzian plateau. The Unguz, which now consists of a series of unconnected valleys, even if it was not the old course of the Amu Dar'ya, must have been a part of its valley. The gradual eastward shift of the Amu Dar'ya was the cause of the alluvial plain, forming the Lower Karakum.

At the beginning of the Anthropogene epoch the Amu Dar'ya sharply changed its course and worked out its present bed which lies to the east of the Transunguzian plateau. Behind this plateau at one time there were subsidiary courses, which branched westwards, filling up the Sarykamyshian depression while the surplus of water continued along the Uzboi into the Caspian Sea. The number and size of the westerly courses gradually decreased, the water level in the Sarykamyshian Depression dropped and the Uzboi dried out. This happened at least a few thousands years ago since the shores of the Uzboi have no town-remains, but only indications of stone-age settlements. Even after this, however, some westerly branches of the Amu Dar'ya persisted and there was water in the Sarykamyshian Depression, since along its eastern shores and along the westerly branches there are numerous indications of towns and fortresses, which existed 2000 years ago and later. Nowadays the river branches, and the depression with its small central salt pan, are dry. The alluvial deposits of the Amu Dar'ya are therefore known as the Karakum Suite, which forms the base rock of the desert. After the disappearance of the river and until the present day strong winds winnowed the alluvial sediments producing proximal deposits of wind-blown sand and distal deposits of loess (clay and silt particles) against the distant mountains.

The present-day desert is an alluvial plain inclined to the north-west and covered by aeolian sands. The thickness of these sands is no more than 10–15m except in rare, occasional hills where it reaches 20–25m. Considerable areas of the desert have a clay-pebble rather than sandy cover. The surface of the desert develops during rainfalls freshwater lakes (takyrs) as well as bitter-saline shores both of which dry out in the summer and winter. It can be said that the placing of the desert is determined by the river course, its cause is the climatic change and the basic active factors are the activity of wind and rain water.

Part IV

MESO-CAINOZOIC
GEOSYNCLINES

FIG. 199. Mt. Byelalakaya in the Main Caucasian Range, above the Dombaiian valley. An expedition of geologists in the forefront. *Photograph by V. D. Nalivkin.*

10

THE MEDITERRANEAN GEOSYNCLINE

GENERAL DESCRIPTION

Two geosynclines—the Mediterranean and the Pacific Ocean—show strong manifestations of the Cimmeridian and Alpine orogenies. The regions occupied by the geosynclines have been sites of prolonged, continuous orogenic movements of young regional metamorphism, of the youngest synorogenic intrusions, the youngest mineralization and petroleum. Only a small part of the Mediterranean Geosyncline is within the territory of the U.S.S.R., since the geosyncline is one of the largest in the world. The Soviet part covers all the southern provinces of the European part of the U.S.S.R. and Central Asia, forming a mobile belt of highest, youngest and most interesting parts of the Soviet Union (Fig. 199).

BOUNDARIES. Only the northern boundary is geologically conditioned; all other boundaries are determined by the position of administrative boundaries and the state frontiers. The northern boundary is disputed and is not universally accepted, although the Carpathians, the Crimea, the Caucasus and Transcaucasia, the Kopet-Dag, Southern Tadzhikistan, the Kugitangs and the Pamirs undoubtedly belong to the geosyncline.

The problems of the northern regions are numerous and the most important ones are as follows:

1. Is the Ukrainian Shield a part of the platform or a nucleus of a geosynclinal province? The answer is that it is a part of a Precambrian platform, as indicated by almost horizontal Palaeozoic strata on its margins and by their low grade of metamorphism and their small thickness. Thus the platform character of Palaeozoic rocks is obvious.

2. Where is the junction between the Carpathians and the platform? The answer is that a sub-montaine downwarp separates these mountains from the platform. The existence of this boundary downwarp has been discovered by Soviet geologists. The boundary lies somewhat to the west of the Dnyestr and further to the south along the Prut. The exposures of Palaeozoic rocks in the Dobrudzha and the Zmein island have a geosynclinal character. From the estuary of the Danube onwards the downwarp trends first towards the south-east and then to the east along the southern margin of the Ukrainian Shield.

3. Does the Donyets Basin and the Dnyepr-Donyets downwarp in general represent a part of a platform or a part of a geosyncline? Some workers consider it to be a platform, since up to the Devonian it had been a part of the Russian Platform, while others attribute it to the Mediterranean Geosyncline since the Palaeozoic is of a geosynclinal type while regional folding affects the Jurassic and even the

Upper Cretaceous and the Palaeogene of this downwarp. A third answer, which is the most correct, is that the Dnyepr-Donyets downwarp is an independent tectonic region of a geosyncline type and is situated within a platform, being what N. S. Shatskii calls an 'aulacogene'.

4. What is the nature of the Ciscaucasian vast plain, separating the Caucasus and the Russian Platform? This question is related to the problem of the relationships amongst the Precambrian, the Palaeozoic and the Meso-Cainozoic geosynclinal provinces. The Precambrian province is represented by the Yaikian Massif forming the south-eastern corner of the Russian Platform and the sub-surface position of the Ukrainian Shield, which continues to the east almost as far as Sal'sk (Fig. 200). The Palaeozoic province forms the basement to the Ciscaucasian plain and includes folded and metamorphosed Middle and Upper Palaeozoic geosynclinal sediments. In the north (Fig. 200) these rocks are continuous from the Donyets Basin and form a low, sub-surface ridge, similar to the Donyets Ridge and known as the Karpinskii Ridge. A. P. Karpinskii who was one of the greatest Russian geologists, first indicated its existence. The Karpinskii Ridge has been traced in numerous boreholes as far as Astrakhan. Its cross section is shown in Fig. 246 and there is no doubt that it continues to the east, joining the Urals. To the south of the ridge Palaeozoic strata have been found in boreholes as far as the Caucasus, where they form the basement to a wide downwarp which is in the south bounded by the Greater Caucasus range (Fig. 200). The downwarp is sharply asymmetric and has narrow and relatively steep southern and wide, gentle and platform-like northern flanks. The downwarp is filled by Mesozoic and Cainozoic deposits.

Within Donbas, the Karpinskii Ridge, the Mangyshlaks and the Tuar-Kyr the Upper Trias, the Jurassic and the Lower Cretaceous are folded into large, gentle folds forming a wide arc traversing Palaeozoic structures (Fig. 4). This arc is considered as the outer zone of the Mediterranean Geosyncline. Its basement consists of thick, geosynclinal, strongly folded and metamorphosed Palaeozoic sediments and lavas. Thus this zone cannot be attached to the Russian Platform where Palaeozoic deposits are of platform type.

Many geologists consider the outer zone as a post-Hercynian platform, of the same type as the post-Hercynian, Uralian Platform. This point of view can be accepted since Mesozoic strata often have a platform character. Nevertheless, there are also clearly formed relatively tight folds, which have a continuous linear trend over large distances. Thus the linear structures of the Jurassic of Donbas are associated with the similar Karpinskii Ridge, the Mangyshlaks and the Tuar-Kyr. This feature permits the inclusion of the outer zone into the province of the Mediterranean Geosyncline rather than into the province of the Uralian Geosyncline. That part of the outer zone which forms Ciscaucasia is known as the Scythian Platform. As a consequence the margin of the Mediterranean Geosyncline bends round the Dnyepr-Donyets Depression and follows along the northern edge of the Karpinskii Ridge, the Mangyshlaks and the eastern edge of the Tuar-Kyr, as is shown in Fig. 4.

GEOLOGICAL REGIONS. Two zones can be distinguished within the Mediterranean Geosyncline—(1) the outer northern zone which is adjacent to older fold belts, and (2) the inner southern zone which is distal to such fold belts and is geosynclinal.

The outer zone includes regions with horizontal Neogene, weakly folded Palaeogene and Upper Cretaceous and distinctly folded Lower Cretaceous and Jurassic strata. To this zone belong the valley of the Dnyestr, the lowlands adjacent to the

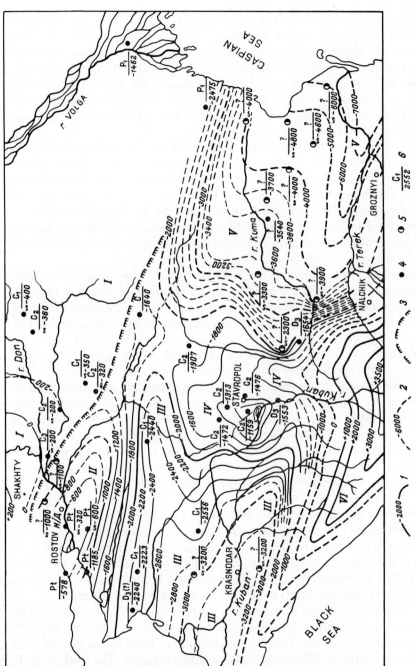

FIG. 200. The structure of Ciscaucasia. The contours indicate the depth of the top surface of Palaeozoic deposits. According to N. I. Tsibovskii (1956).
I — the eastern continuation of Donbas (Karpinskii Ridge): II — the top surface of Palaeozoic and Precambrian rocks; 2 — suggested depth termination of the Ukrainian Massif; III — the Kuban' downwarp; IV — isolines; 3 — the southern part of the Karpinskii Ridge; 4 — boreholes the Stavropol elevation; V — the eastern part of the downwarp (the Ter'- penetrating down to Palaeozoic strata; 5 — boreholes not reaching Kuma downwarp; Vi — the Caucasian range. 1 — depth isolines of the Palaeozoic strata; 6 — age and depth of penetrated deposits.

Black Sea and the Sea of Azov, the Odessa Bay, the Ciskubanian plain, the Stavropol plateau, the Nogai steppe and the valley of the Manych. Further east the zone includes the northern part of the Caspian Sea, the Mangyshlaks, the Tuar-Kyr, the Krasnovodsk plateau and the Lower Karakum.

The inner zone shows weakly but clearly folded Neogene and Anthropogene, strongly folded Palaeogene, Upper Cretaceous and older formations. The Cimmeridian and the Alpine orogenies affected the zone equally strongly and large anticlinoria form natural geological regions. Such anticlinoria form the Carpathians, the Crimean Mts, the Caucasus, the Lesser Caucasus, the Kopet-Dag and the Pamirs. The anticlinoria are separated by three regions of depressed fold axes: (1) the Danube delta and the Odessa gulf, (2) the Sea of Azov, the Kerch Peninsula and the Straits, (3) the central part of the Caspian Sea.

ADMINISTRATIVE SUBDIVISIONS. From west to east the region involves the following administrative subdivisions: the Ukrainian S.S.R., the Moldavian S.S.R., the Dagestan A.S.S.R., the Kabardinian A.S.S.R., the North-Osetian A.S.S.R., the Chechen-Ingush A.S.S.R., the Rostov Province, the Stavropol and Kransnodar regions, the Georgian, the Armenian and the Azerbaidzhan S.S.R.s.

GEOLOGICAL INVESTIGATIONS. The investigation of the Soviet part of the Mediterranean Geosyncline have proceeded unequally and most work has been done in the Caucasus.

The study of the Soviet Carpathians was begun by Polish geologists. Their work was of significance, but relatively limited. After the inclusion of a part of the Carpathians in the Soviet Ukraine the volume of work rose sharply owing to the search for and development of oil and gas and the inauguration of the Lvov coal-field. At present there is a literature of hundreds of publications, including the papers of A. A. Bogdanov (1949, 1950), O. S. Vyalov (1949), M. V. Muratov (1949), V. I. Slavin (1956) and the review paper on the Palaeozoic by G. Kh. Dikenstein (1953).

Owing to the absence of large deposits of useful materials the study of the Crimea proceeded relatively slowly. The recent search for oil and gas provided new data presented in the paper by G. Kh. Dikenstein *et al.* (1958). Valuable information has been supplied by M. V. Muratov (1949, 1955a, 1955b, 1960). Separate systems have been described by G. F. Veber, A. S. Moiseyeva, G. Ya. Krymgol'ts, V. F. Pchelintsev, A. G. Eberzin (1957) and S. A. Yakovlev (1955). There are also tectonic investigations by N. I. Andrusov (1893), M. V. Muratov (1949) and V. G. Korneyeva (1959).

The study of the Caucasus can be divided into three periods. Firstly in the second half of the 19th century there were unco-ordinated and unsystematic investigations of which those of G. Abikh, who spent over forty years can be singled out. The work of the staff of the Caucasus Mining Dept, who in 1908 published the first geological map with still a few blank areas, deserves comment. N. I. Andrusov, a leading palaeontologist-stratigrapher, began working at the end of the 19th century.

The second period embraces the first thirty years of the 20th century and can be justifiably called the 'epoch of the Geological Committee': when headed by A. P. Gerasimov its staff did a very thorough investigation. Thus the geological map issued by the committee in 1930 did not have any blank spaces. The work of V. P. Rengarten and K. N. Paffengolts is especially outstanding. Amongst other geologists of the Caucasus section one can name I. G. Kuznetsov, B. F. Meffert, A. L. Reingard and V. I. Robinson. The geologists of the Oil Division have also contributed a lot including: D. V. Golubyatnikov, I. M. Gubkin, M. V. Abramovich, N. I. Usheikin.

The work of the group of geologists headed by F. Yu. Levinson-Lessing and D. S. Belyankin and including P. I. Lebedev and S. S. Kuznetsov is of considerable importance.

The third period from the thirties to the present is the epoch of a previously unknown flourishing of Caucasian geology. The work of the VSEGEI (former Geological Committee) geologists stands foremost. K. N. Paffengolts in 1955 issued a geological map and in 1959 a short and compressed, but very full review paper entitled *The Geology of the Caucasus*, which can be recommended to all those interested in the Caucasus. Of VSEGEI geologists V. D. Golubyatnikov, I. F. Pustovalov, S. P. Solov'yev and N. I. Tsibuskii can be singled out.

The work of the Oil group of the Geological Committee was continued and developed by the All Union Oil Institute (VNIGRI). Of its staff much has been done by B. A. Alferov, D. B. Drobyshev, N. B. Vassoyevich, P. K. Ivanchuk and A. V. Ul'yanov. The work of Andrusov was continued and developed by palaeontologists such as A. G. Eberzin, V. P. Kolesnikov, B. P. Zhizhchenko, I. A. Korobkov, O. S. Vyalov, P. V. Fedorov. Amongst the investigators on the Palaeozoic the investigation of A. D. Miklukho-Maklai and on the Mesozoic that of T. V. Mordvilko are outstanding. The work of the professors of the University of Moscow and the Institute of Geological Surveying is quite considerable. The list of these savants includes: N. S. Shatskii, V. V. Byelousov, A. A. Bogdanov, M. V. Muratov, V. E. Khain as well as G. D. Afanas'yev, M. I. Varentsov, V. A. Grossgeim and V. I. Slavin. The main feature of the third period is the development of local, national geology, which is foremost especially where detailed investigations are concerned. Geologists of Azerbaidzan, Armenia, Georgia, Dagestan, Stavropol and Krasnodar regions and Rostov province have published thousands of papers, which form the basis of our knowledge of the Caucasus. Amongst the review books one can mention the multivolume *Geology of Azerbaidzhan* towards the compilation of which S. A. Azizbekov has contributed a great deal. There are so many local geologists that amongst the major ones only a few can be mentioned and they include the following:

Azerbaidzhan: M. V. Abramovich (oil geology), Sh. A. Azizbekov (regional geology), M. M. Aliev (Cretaceous), K. A. Ali-zade (stratigraphy), A. A. Ali-zade (general geology), V. A. Gorin (tectonics), M. A. Kashkai (ore-mineralization), S. A. Kovalevskaya (general geology), Sh. F. Mekhtiev (oil-geology), A. N. Solovkin (petrology), E. Sh. Shikhalibeili (regional geology), K. M. Sultanov (stratigraphy).

Armenia: D. A. Arakelyan (stratigraphy), A. T. Aslanyan (regional geology), L. A. Vardanyants (geology of the Caucasus), A. A. Gabrielyan (stratigraphy), I. G. Magak'yan (metallogeny), A. I. Mesropyan (general geology), S. S. Mkrtchyan (metallogeny), G. P. Tamrazyan (general geology).

Georgia: M. S. Abakeliya (geophysics), L. K. Gabuniya (vertebrates), P. D. Gamkrelidze (regional geology), A. I. Dzhanelidze (general geology). G. S. Dzotsenidze (petrography), G. M. Zaridze (magmatism), I. V. Kacharava (stratigraphy), A. D. Tskhakaya (seismology).

The study of the South Russian Depression began over one hundred years ago and its central part—the Donyets Basin—is one of the most fully studied territories in the U.S.S.R. After some sporadic investigation at the beginning of the 20th century the Geological Committee began systematic mapping, accompanied by theoretical memoirs. The mapping was directed by L. I. Lutugin and later P. I. Stepanov and carried out by S. V. Kumpan, B. F. Meffert and A. A. Snyatkov. This work provided a basis for the subsequent detailed investigation undertaken by many tens of geologists. Thus much new data were collected helping in adding to and amplifying the work of the Geological Committee. The discoveries of oil and gas in particular added to the volume of data. Considerable results have been obtained in studying the westerly continuation of the Donyets-Dnyepr and Pripet depressions and the eastern continuation of the Astrakhan Depression. The data obtained from deep boreholes and geophysical investigations are especially important. A result of such work is that the South Russian Depression was found to be an important oil province. On the Pripet Depression the work of N. I. Lebedev working in pre-Revolutionary times and D. V. Aizenberg, P. I. Shul'ga, G. F. Limgersgauzen, E. O. Novik, A. P. Rotaya, B. I. Chernyshev and M. V. Chirvinkaya can be singled out.

OROGRAPHY

The relief of the outer and inner zones of the geosyncline is very different, which is explained by the horizontal aspect of Neogenic and Anthropogenic sediments in the external zone and a lack of effects of orogenic folding or only a small scale of fault movements. Consequently the external zone over large distances of thousands of kilometres consists of weakly eroded, gently inclined plains. Some parts of the zone are elevated and the Stavropol plateau has heights ranging from 400–800km.

The inner zone conversely consists of a series of mountainous, sometimes high regions with heights of up to 5200–5600m and downwarped regions and graben, forming marine basins of sometimes a great depth as for instance the Black Sea (2100–2200m deep) and the south Caspian Depression, which is up to 980m deep. The relief involves differential heights of up to 8000m. This relief is the result of fold movements, which at present are accompanied by intensive erosion. Block movements are rarer, but are also responsible for variations in the relief and have produced the Black Sea and the submergence of one half of the Crimean anticlinorium.

The Eastern Carpathians are typical young fold mountains with a highly dissected topography, but the highest point Mt Goverl (2058m) implies a relatively low general elevation. The region of the Carpathians is relatively near to the Ukrainian shield, being separated from it by the valleys of the Dnyepr and the Prut, which correspond to a narrow external zone about 80–100km broad. To the east this zone widens rapidly and includes Bessarabia (in places over 300m high) and then the wide Cis-Black Sea Depression. Until recently this depression was a vast steppe. but is now an agricultural region where rotation crops are growing and the Kakhovskian hydro-electric station is active. Behind the shallow (14m) Sea of Azov lies the no less wide intensively worked Ciskuban' Depression, which through the Manych lowlands passes into the Nogai Steppe, which margins the Caspian Sea, and includes the vast delta of the Terek. The Stavropol Uplands separate the Nogai Steppe and the Ciskuban' Depression.

The Crimean Mts represent only one half of an anticlinorium, the other half of which lies at the bottom of the Black Sea. The height of these mountains is about 1500m, while the depth of the Black Sea is about 1500m which suggests a relative displacement of 3000m. The higher peaks of Crimea (Roman-Kosh, 1545m and Ai-Petri, 1233m) are precipitous and consist of thick Upper Jurassic limestones. In the core of the anticlinorium at sea level, there are Middle Jurassic shales, sandstones and volcanic rocks (Ayu-Dag) and a Lower Jurassic and Upper Triassic flysch formation—the Tavridian Suite.

The Greater Caucasus is the best known and most intensely studied mountainous region in the U.S.S.R. Its relief is the best example of low fold-movements causing the elevation of large segments of the Earth's crust into gigantic anticlinoria. Erosion sets in contemporaneously with the uplift, but since the uplift is faster than the erosion mountains come into existence. The influence of lithology on the relief can also be demonstrated in the Caucasus. Hard and massive rocks show up against the general relief as high ridges and independent ranges, while valleys originate over the outcrops of clays or friable and bedded rocks.

The watershed range corresponds to the crest of the anticlinorium where the uplift has been most, but even within this range the higher western part consists of compacted and metamorphosed Precambrian rocks and the individual high peaks such as Mt Dukh-Tau (5198m) are made of marbles and massive granites (Fig. 199). The highest peak in the Caucasus—Mt El'brus—is a young volcanic cone resting on the crest of a Precambrian ridge. Mt El'brus is 5633m high and has seen eruptions as recently as in the Upper Anthropogene epoch. Another volcano—Kazbek (5047m)— also situated in the watershed range has erupted even more recently. The crest of the Caucasian range in its western, higher part has a permanent snow and ice cover. Consequently the Kuban' and its tributaries the Malka and the Laba as well as the Terek system are full rivers while the Sulak and the Samur arising in the eastern Caucasus are much smaller and poorer rivers. All these rivers flow through deep gorges across the second range of the northern slope cutting through massive Upper Jurassic limestones which are the same as in Mt Chatyr-Dag and Ai-Petri of Crimea. The second range reaches 2500–2800m in height. The third range consists of hard, bedded Upper Cretaceous limestones and attains heights of 1000–1500m. Thus the northern slope of the Caucasus consists of three ranges, separated by depressions. The first range is known as the Upper Cretaceous, the second range as the Upper Jurassic and the third as the watershed.

The southern slope has a completely different relief since here there are no Upper Jurassic or Upper Cretaceous limestone formations. The slope is occupied by relatively homogeneous sandy and shaley formations and is consequently dissected by numerous and complex small valleys and transverse ridges. Of the latter the Kakhetinian range is most significant. Limited glaciation accounts for relatively quiet rivers. Thus despite their number all the rivers belong to two strike systems— the Kuru and the Rion—the valleys of which are separated by the small Dzirul Massif of Precambrian granites.

The Lesser Caucasus (Transcaucasia) constitute a completely dissected relatively low (3600–4000m) mountain region, situated to the south of the Kura and the Rion. The rocks of this region consists mainly of Jurassic, Cretaceous and Cainozoic lavas and sediments intruded by Cimmeridian and Alpine granites, with which the Dashkesan, the Zaglik, the Allaverdy and the Zangezuri mineralizations are associated.

The Eastern Lesser Caucasus form a mountainous tract reaching 3500–3900m in height, although the Neogene volcano of Aragats (Alagez) is 4095m. Here deep valleys alternate with mountain ranges and there are occasional plateau-like uplands formed by Anthropogenic basalt flows. To the north lies the large and high lake Savan and the southern boundary is marked by the wide valley of the Araks.

The Western Lesser Caucasus consist of alternate ranges, small plateaus, and river valleys. This is the region of young Cainozoic vulcanicity, with which are associated numerous mineral springs, such as at Borzhomi, Sairme etc.

The Kura Lowlands is a wide planar region, which is steppe-like at the margins, but is swampy in the centre and as it merges into the Kura delta.

The Apsheronian Peninsula is an important petroliferous region, which over a large part is sun-drenched desert, the uniform surface of which is only broken by cultivated districts around oil-fields.

STRATIGRAPHY

The most important feature of the region is the development of the marine Upper Permian and the preponderance of marine deposits amongst Triassic, Jurassic, Cretaceous, Palaeogene, Neogene and even Anthropogene sediments. This feature is used in distinguishing Meso-Cainozoic and Palaeozoic geosynclines.

PRECAMBRIAN

Precambrian strata are both exposed and encountered in the boreholes sunk in the Central Carpathians, in the south-eastern and southern parts of the Ukrainian Shield, in the Caucasus and Transcaucasia. In the Carpathians in the Rakhovka (Marmaroshian) crystalline massif there is the Byelopotokian Suite (480m) which is attributed to the Lower Proterozoic. The suite consists of quartz schists, para-gneisses and schists and is overlain by the Delovetskian Suite (1000m) consisting of mica-schists with layers of marbles, and sericite and chlorite schists and amphibo-lites. The latter suite is attributed to the Upper Proterozoic or Lower Palaeozoic and is covered by Upper Palaeozoic strata.

In the south-western and southern Ukrainian Shield Cambrian and Upper Pro-terozoic (Sinian) platform-type sediments (p. 50) are ubiquitously found in bore-holes. Sinian spores have been found in Ciscaucasian oil.

Precambrian strata and the associated ancient granites form over a consider-able distance the watershed ridge of the western part of the Greater Caucasus. These rocks are found at heights of over 3000m and owing to bad accessibility have not been extensively studied. According to A. P. Gerasimov (1939), who was one of the fore-most specialists on the geology of the Caucasus, the Precambrian can be divided into two lower divisions. The Lower Precambrian (Archeozoic) consists of gneisses, schists and marbles. The Upper Precambrian (Proterozoic) includes slates, phyllites and quartzites. The conclusions of A. P. Gerasimov were further amplified by V. P. Rengarten (*Geology of the U.S.S.R.*, 1941) in the Northern Caucasus and by K. N. Paffengol'ts (1948) in Armenia.

In the main ranges Rengarten recognized the following subdivisions:

A_1 Lower Precambrian (Archeozoic).
A^1_1 Mica schists, micaceous gneisses, marbles.
A^2_1 Amphibolites, biotite gneisses.
A^3_1 Orthogneisses, augen gneisses, schistose diorites.
A_2 Upper Precambrian (Proterozoic).
A^1_2 Mica schists, ferruginous quartzites and various slates of Baksan, Malka and Kuban'.
A^2_2 Chegemian Suite—quartzites and mica-chlorite schists.

The Precambrian age of these rocks is confirmed by the fact that pebbles and grains of these rocks are found in conglomerates and sandstones of Lower Palaeozoic age. In any case palaeontologically dated Lower Palaeozoic rocks are much less metamorphosed.

In Transcaucasia K. N. Paffengol′ts also divides the Precambrian in two, but this division is based on the grade of metamorphism and he does not think that it is always established. K. N. Paffengol′ts also indicates that the Upper Precambrian and the Lower Palaeozoic are often indistinguishable and consequently he groups their rocks together until a distinction can be established. P. A. Arakelyan (1957) has divided the Ancient Complex of Armenia into four suites of which the lowest, consisting of much migmatized mica schists, he considers as of Upper Proterozoic age, while the rest are included in the Lower Palaeozoic. No fauna has been discovered and consequently the age is tentative.

The Precambrian of the Dzirul Massif is represented by crystalline schists, gneisses, amphibolites and quartzites, and as yet has not been subdivided. The age of these rocks is determined from their relationship to the unconformably overlying Lower Cambrian archeocyathid limestones.

It is interesting to note that numerous and undoubtedly accurate radiometric age determinations of the older crystalline rocks and ancient granites have given an unexpected result since none of these rocks were found to be older than 400–350m.y. This is not only younger than the Precambrian, but also Cambrian (480–570m.y.) Ordovician (420–480m.y.) and even Silurian (400–420m.y.) (Starik, 1961; Rubinshtein, 1961). Some leading investigators took these figures to mean that there is no Lower Palaeozoic or Precambrian in the Caucasus. This conclusion was, however, premature since the interpretation of the new and important radiometric methods has not as yet been completely worked out. For instance in Czechoslovakia rocks of undeniably Lower Palaeozoic and Precambrian ages, which are faunaly determined, yield also radiometric ages of 400–350m.y. These ages were reinterpreted to be due to the impact of the Caledonian Orogeny and Hercynian Orogeny which altered isotopic constitution in rocks.

In other provinces of the U.S.S.R., Cambrian faunas have been found always in rocks yielding radiometric ages of 500–600m.y. and it is undoubted that the Cambrian of the Caucasus has the same age. The figures obtained in the Caucasus demand an appropriate interpretation of a geochemical type. To deny the existence of the Cambrian is inept since its presence has been demonstrated faunably, and consequently the Precambrian is also present since it underlies the Cambrian. The granites of the Dzirul Massif despite their apparent radiometric age of 330m.y. (Rubinshtein, 1961) are nevertheless Precambrian since they are unconformably overlain by Lower Cambrian archeocyathid limestones.

PALAEOZOIC

Lower Palaeozoic

Lower Palaeozoic strata are metamorphosed and have not been sufficiently investigated. Only few, isolated fossil localities are known. No Ordovician faunas have been found and there are only three localities where Cambrian faunas have been identified. Two of these localities are situated in the Northern Caucasus where in the valley of the Laba (Dzhentu, Fig. 201) archeocyathid limestones have been found and in the Malka valley black, Middle Cambrian, trilobite-bearing limestones have been detected. The third locality is in the Dzirul Massif where in the upper part of the section over Precambrian rocks lie metamorphosed slates, phyllites and sandstones with lenses of archeocyathid limestones. The fossiliferous rocks imply that a thick formation (no less than 2000m) of metamorphic slates and quartzites with lenses of limestones should be included in the Lower Palaeozoic. In the main range this formation is associated with Precambrian strata and is only exposed in the watershed ridge. In Transcaucasia it is again only known in association with the Precambrian.

Middle Palaeozoic

Rocks of this age are usually strongly metamorphosed, thick, have yielded few fossils and have not been sufficiently investigated. The existing palaeontological data do not permit a complete stratigraphical subdivision except in Armenia and the Nakhichevan A.S.S.R. where Middle and Upper Devonian and Lower Carboniferous successions have been well investigated.

Silurian

In the Main Caucasus range in the valley of the Malka relatively weakly metamorphosed quartzitic and arkosic sandstones and phyllitic slates of a total thickness of 1000m have been found amongst metamorphic strata of a Lower Jurassic age. In the upper part of the pre-Jurassic rocks a rich and variable fauna of Hercynian type has been found in intercalations and lenses of massive pale and dark limestones. The fauna consists in the main of Ludlovian forms, but some Lower Devonian species also have been found. All these rocks including quartzites and phyllitic slates, are considered as Silurian which at the base of the succession embraces 300m of volcanic rocks.

Devonian

The Devonian is found in the axial part of the Caucasus on the northern and southern slopes, in southern Armenia and the Nakhichevan A.S.S.R. In the axial

FIG. 201.   The Dzhentu Massif of Lower Palaeozoic limestones, in the basin of the river Lesser Laba. *Photograph by V. N. Robinson.*

part the Devonian is represented by metamorphosed slates with beds of dark, muddy limestones with homogeneous *Cyrtospirifer*-bearing brachiopod fauna belonging to the Frasnian. The slates are considered a part of the metamorphic formation of the Greater Caucasus. On the southern slope there are found identical dark slates with Frasnian brachiopod faunas. It is interesting that they were formerly considered to be Lower Jurassic on the basis of a much lower grade of metamorphism than that affecting older rocks in general. The succession on the northern slope is given on p. 591.

The low metamorphic state of Devonian pelites is confirmed by their low degree of alteration in Armenia, where they are found over a large area south of Erevan, and reach a thickness of 1700m. The succession begins with Middle Devonian (400–700m) limestones and shales with mainly coral faunas, amongst which a characteristic complex of Middle Devonian forms with *Calceola sandalina* Lam. is particularly significant.

The Upper Devonian consists of bedded limestones, marls, sandstones and shales. A rich and diverse brachiopod fauna indicates at the presence of the Frasnian (65m) and Famennian (900m) stages. The Devonian succession is shown in Table 44. The Famennian Stage is transitional into the Etroeungt Beds (Strunian) of the Lower Tournaisian overlain in turn by Tournaisian and Visean limestones. The Devonian of Armenia is compressed into a series of gentle relatively simple folds, which of course can be expected since the limestones are massive and thick (Fig. 202).

Upper Devonian limestones have been found in Azerbaidzhan, to the north of Nakhichevan. So far typical Lower Devonian deposits have been found neither in the Caucasus nor in Transcaucasia. In the Caucasus this is explained by the inadequate search for faunas and strong metamorphism precluding such work. In Transcaucasia (the Nakhichevan A.S.S.R.), in Velidag, a special borehole was sunk into outcropping rocks with the aim of discovering Lower Devonian strata. The borehole penetrated 1820m through homogeneous black slates and metamorphosed limestones, all almost entirely barren. Only in the uppermost 300m there were found Foraminifera of the type which on the Russian Platform are included in the Upper Devonian. How such Upper Devonian Foraminifera can occur under Eifelian sediments and what is the age of the underlying 1500m of black slates and limestones remained unclear. There are many possible solutions of the problem but it is most probable that the determination of the Foraminifera is incorrect and the slates should be included in the Eifelian Stage, the Lower Devonian and the Silurian as is shown in Table 44.

Judging from the situation in the provinces lying to the west and east of the Caucasus the development of the marine Lower Devonian in the range is undoubted, which indicates that later phases of the Caledonian Orogeny did not occur.

Lower Carboniferous

Lower Carboniferous rocks so far have only been found on the northern slopes of the main range, where they consist of coralline limestones belonging to the Tournaisian and Visean stages and are associated with phyllites and basal conglomerates. It is possible that as in the Donyets Basin the lower horizons of the coal-bearing formation are also Lower Carboniferous.

In Armenia the Lower Tournaisian is represented by the Etroeungt (Strunian) Stage characterized by a very rich and variable fauna. This stage (125m) is repre-

TABLE 44. *Devonian Succession in Armenia and the Nakhichevan A.S.S.R.*
(According to P. A. Arakelyan and M. S. Abramyan)

System	Stage	Sub-stage, Suite, Beds	Lithology	Thickness (m)	
Lower Carboniferous	—	d'Etroeungt Beds	Shales and limestones	125	
Upper Devonian	Famennian	Upper Famennian	Brownish sandstones and quartzite-like sandstones as well as dark shales and limestones with a brachiopod fauna. The strata are divided into three suites, being from top down— Gortunian, Tamamidzorian and Kadrlian.	210–350	480–730
Upper Devonian	Famennian	Lower Famennian	Dark and brownish shales, limestones and sandstones, with brachiopods (*Cyrtospirifer orbelianus*, Abich). Two suites are recognized— Ertichian and Noravanian.	270–380	480–730
Upper Devonian	Frasnian	Bagarsykhian	Yellowish sandstones and quartzite-like sandstones with beds of dark limestones and shales.	250–330	460–640
Upper Devonian	Frasnian	Chrakhanian	Dark limestones, black shales and brown quartzite-like sandstones with brachiopods (*Lamellispirifer*, *Cyrtospirifer*, *Atrypa*).	180–230	460–640
Upper Devonian	Frasnian	Danzikian	Grey, sandy limestones and reddish shales with brachiopods (*Cyrtospirifer*) and corals.	30– 80	
Middle Devonian	Givetian	Gumushlyugian	Grey and dark-grey limestones, in places with beds of black shales. A fauna of brachiopods (*Stringocephalus*) and rugose corals is present.	110–230	190–410
Middle Devonian	Givetian	Sadarakian	Limestones with beds of shales and sandstones, brachiopods (*Stringocephalus*) and rugose corals being present.	80–180	190–410
Middle Devonian	Eifelian	Arazdayanian	Dark limestones and purple shales with rugose corals (*Calceola*) and brachiopods. It is possible that the upper part (150–200m) of the underlying shales belongs to the Eifelian.	200–250	200–250 (?)
Lower Devonian	—	—	Under the Arazdayanian Suite lies a thick, barren, dark formation of shales and sandstones, with beds of limestones. The formation was found in a borehole (1710m) and is not exposed at the surface. It is possible that its upper part is Eifelian, its middle part is Lower Devonian and its lower part is Silurian.	1500 (?)	

sented by inter-bedded shales, limestones and sandstones. The Upper Tournaisian (65m) is represented by limestones overlain by Visean limestones. No Namurian strata have been found.

The Middle Carboniferous succession of the northern slope of the western Caucasus described by V. N. Robinson (1932), one of the leading specialists, serves as an example of such a succession in the Caucasus.

The Silurian sediments already described are overlain by the following strata.

Lower and lower Middle Devonian—First Suite consisting of grey phyllitic slates passing upwards into quartz porphyries, tuffs and tuffaceous sandstones. It is followed by phyllitic slates with beds of rarer limestones and more frequent sandstones. The suite is 1000–1500m thick.

Middle part of the Middle Devonian—Second Suite with cleaved conglomerates at the base, followed by tuffaceous sandstones with beds of limestones and still higher up variegated slates. The suite is 700m thick.

Givetian and Frasnian stages—Third Suite, which begins with shales, sandstones and limestones overlain by dark grey massive and bedded limestones with stromatoporoids, bryozoans, and calcareous algae of a most likely Givetian age. Towards the top occur variegated shales, tuffaceous sandstones and tuffs and at the very top dark grey bedded limestones of a Frasnian age and containing *Cyrtospirifer verneuli* Murch., *Camarotoechia* sp. and other brachiopods. The total thickness is 500m.

Famennian Stage—Fourth Suite (500m) of grey slates with sandstones, limestone and conglomerate intercalations.

FIG. 202. Devonian limestones of Southern Armenia.

It should be remembered that a fauna has been found only in the third suite, while the age of the older suites is provisional.

The Strunian Horizon includes grey bedded limestones with *Syringopora, Amplexus* and Foraminifera (*Endothyra*). The thickness is 150–200m. These limestones are associated with a formation similar to the Zalairian Formation of the Southern Urals, thus consisting of finely bedded, sometimes cleaved chlorite quartz slates with beds of metamorphosed limestones and groups of porphyrites. The thickness of the formation is up to 700m.

The Visean Stage (up to 1000m) consists of siliceous and pelitic slates with beds of white and dark grey limestones with *Lithostrotion.*

Higher up follow massive, barren formations of 2500–4000m in thickness and provisionally considered to belong to the Upper Visean and Namurian. The rocks of these strata vary from slates to sandstones and porphyrites. In the Teberda region there are slices of Middle Devonian limestones compressed amongst Jurassic slates.

In Ciscaucasia Middle and Upper Palaeozoic rocks have been found in numerous deep boreholes. The study of their cores has shown that in the south approximately on the latitude of Stavropol and to the south of it Middle and Upper Devonian and Lower Carboniferous deposits are predominant. In the north of Ciscaucasia only Upper Palaeozoic is found.

Middle Palaeozoic strata are folded and metamorphosed and are of the same composition as in the Northern Caucasus. Metamorphosed slates, sandstones and lavas are most common. Beds of limestones are rare and faunas are very rarely encountered. Amongst fossils microfaunas and spores are most common. These fossils indicate that the bored sediments belong to the Middle and Upper Devonian and Lower Carboniferous (Tsibovskii, 1956).

Upper Palaeozoic

Upper Palaeozoic strata of the Main range occur in the same localities as the Middle Palaeozoic. On the synoptic 1:7 500 000 map they are combined and shown as the undifferentiated Palaeozoic. On the northern slope exposures of Middle and Upper Carboniferous rocks continue from the upper reaches of the Malka to the upper reaches of the Laba. They are also known in southern Armenia and the Nakhichevan A.S.S.R. Upper Palaeozoic strata are everywhere unconformable on the Middle Palaeozoic.

Middle and Upper Carboniferous and Lower Permian

Lower Permian strata, often red beds and alternate marine and continental sediments, and are closely associated with Middle and Upper Carboniferous rocks. Marine Upper Permian sediments gradually, or only with a negligible break, pass into marine Lower Trias rocks, thus forming a single complex. This complex has a vast extent and has been found in the Carpathians, in Crimea, on the northern slope of the Caucasus in southern Armenia, in the Nakhichevan A.S.S.R. (Dzhulfa), in the Mangyshlaks, the Tuar-Kyr and the Pamirs.

In the Carpathians (the Rakhova crystalline Massif), the mica schists and marbles of the Lower Palaeozoic (?) Delovetskian Suite are overlain by phyllites and graphitic quartzites of the Kuzinskian Suite (200m), which are provisionally included in the Upper Palaeozoic.

In south-western Ukraine (the Odessa Province) to the north of the Danube, on the submerged northern slope of the Dobrudzha Massif, occur black silts and limestones with plant remains. These rocks (271m) are provisionally accepted as Upper Palaeozoic and are overlain by greenish Permian sandstones and conglomerates of 58m in thickness.

In Crimea according to Dikenstein and others (1958) Upper Palaeozoic folded and metamorphosed slates and limestones (up to 1600m) have been found to the north of the resort Saki on the Tarkhankutian Peninsula. In the Simferopol region the slates of the Tavridian Suite have yielded separate slices of limestones with Carboniferous and Lower Permian faunas. Carboniferous and Permian shales and limestones have also been found in boreholes.

Middle and Upper Carboniferous and Lower Permian strata of the northern slope of the Caucasus are grouped into two suites—the coal-bearing and the red beds —which occur between the valleys of the Malka and the Laba. The suites are analogous to similar formations in the Donyets Basin. The Coal-bearing Formation is 1000–1700m thick and has a basal conglomerate at its lower boundary which rests unconformably on the Lower Palaeozoic and Precambrian. The formation consists of almost non-metamorphosed sandstones and argillites with two or three groups of seams of workable coal. There are altogether 6–9 seams with thicknesses varying from 0·4–0·9m. In the lower horizons a Westphalian-type flora has been found indicating that the development of Middle Carboniferous deposition apart there is a possibility of finding Namurian and Visean strata, as is the case in the Donyets Basin. Higher horizons of the formation yielded a Stephanian flora; thus they belong to the Upper Carboniferous and in part to the Lower Permian. The Coal-bearing Formation is heavily dislocated by normal faults. It is exposed at heights of about 2000m in country with a highly dissected topography. The formation is overlain with a break by Permian red beds or by Lower Jurassic strata.

The Red Bed Formation consists of alternating groups of sandstones and conglomerates with subordinate intercalation of clays and silts, and its total thickness is 3000–4000m or even more. In the lower part of the formation there is a Lower Permian flora with *Walchia*. The formation as a whole represents a typical molasse consisting of deposits of a rising mountain range. The sandstones and conglomerates in places have alluvial gold deposits. This gold is concentrated as a result of erosion of ancient Precambrian granites.

In Ciscaucasia Upper Palaeozoic strata are widespread. To the north of the Manych valley the basement is almost entirely of Middle and Upper Carboniferous rocks to the south and of Permian sediments to the north (Dubinskii, 1956). Upper Palaeozoic strata, which here form the whole of the subterranean Karpinskii Ridge (p. 707), consist of thick, folded sandy and clayey formations. These formations represent a continuation of the Coal-bearing Formation of Donbas, are identical to it generally, but differ in their lack of commercial coals. The position of the Karpinskii Ridge is shown on Fig. 200. Rivers had a considerable bearing on the origin of the Coal-bearing and the Red Bed formations, and the question arises: where did these rivers flow and what was the nature of land over which they were flowing? Deep boreholes in Rostov, Zhdanov, Kayal, Novo-Minskaya and Peschanokopskaya have shown that in this region there are no Upper Palaeozoic rocks and Mesozoic and Cainozoic sediments rest directly on the Precambrian. It can be stated that the crystalline massifs, which were land in Upper Palaeozoic times, continued from Donbas to the south-east forming a peninsula. The rivers flowing on this peninsula brought in the sediments of the Coal-bearing and the Red Bed formations. The fine-grain nature and good sorting of the sediments indicate long courses of these rivers and the plain-like low topography of the peninsula. It is quite possible that the coal-bearing formations of the Northern Caucasus and Donbas were laid down on a single alluvial plain by a single river changing its course as it at present happens in China.

In the western part of the southern slope of the Greater Caucasus marine Permian sediments have recently been found. They consist of sandy-clayey deposits with beds of limestones. It is possible that along the southern slope of a high range, serving a source for molasse, there existed a narrow channel connecting seas which were in existence where the present Caspian and Black seas are.

In Transcaucasia there are known marine Lower Permian sediments of a total thickness reaching 350–450m. These sediments are found in the Nakhichevan A.S.S.R., in southern Armenia where in several localities they contain coral-fora-miniferal faunas. These massive grey and bedded limestones were previously known as *Stafella sphaerica* Beds. The same localities where Lower Permian strata are found have also Upper Permian beds of grey and dark grey limestones (300m), which are sandy and muddy towards the top and yied a marine fauna. The succession has been described by R. A. Arakelyan (1952).

Upper Permian

In the Carpathians, on the western slope over a small area, there are exposed barren variegated sandstones and conglomerates resembling the Alpine 'Verrucano'. These sediments are closely associated with the overlying Lower Trias marls and limestones with which they form a single formation. The sandstones and conglo-merates are provisionally considered as Upper Permian or Lower Trias.

In Crimea, at the foot of the Crimean Mts, lies the massive Tavridian Formation of flysch type consisting of alternations of shales and cherts. On the southern Crimean coast it has beds with Upper Triassic and Liassic faunas. In the Simferopol and other nearby regions in an analogous, evidently contemporaneous formation there have been found exotic blocks forming small hills of massive limestones con-taining Lower and Upper Permian fusulinids, ammonites, brachiopods and trilobites. What represents the Lower Trias is not clear, it is not clear either where are the terrigenous sandy-clayey Permian rocks, which should accompany the limestones. The problem is even more complex since one of the blocks has yielded a Lower Carboniferous fauna with *Lithostrotion*.

The relationships between the Upper Permian and the Lower Trias of Crimea are not clear, although in the Mangyshlaks, in the Donyets Basin and in the Tuar-Kyrs they form a single, indivisible formation. The formation consists of continental deposits similar to those of the Russian Platform. Only in the Lower Trias is there a thin group of clays with a marine fauna. In the southern, inner zone of the Mediter-ranean Geosyncline the Upper Permian and the Lower Trias are represented basic-ally by marine facies, occurring together and passing into each other either without a break as happens in Transcaucasia and the Pamirs, or being separated by a small break as happens in the Caucasus.

In the Ciscaucasian plain the Upper Permian and the Lower Trias participate in the structure of the basement. These sediments are the same as in the Donyets Basin, consisting of red, terrigenous thin series of strata (100–200m). On the Kar-pinskii Ridge and nearer to the Sea of Azov they are absent. In the Northern Caucasus the Upper Permian is represented by marine deposits with interesting reef limestones. These marine deposits are restricted to the upper reaches of the Kuban', the Greater and Lesser Labas and the adjacent areas; their thickness is 50–170m. Their succession begins (Miklukho-Maklai, 1956) (Fig. 203) with basal sandstones and conglomerates (8–27m) known as the Kutanian Horizon. It is followed by grey shales and bedded foraminiferal limestones (7–26m) called the Nikitinian Horizon, in turn succeeded by pale massive reef limestones (25–80m) of the Urushtenian Horizon. Laterally the reefs pass into platy marls and shales. The top of the succession has flaggy pale limestones (10–25m) with ammonites referred to as the Abagian Horizon. The reef limestones yield a very rich, variable fauna of southern type with fusulinids and peculiar attached brachiopods (*Richthofenia*), which are homeomorphs of corals. Upper Permian rocks lies on Lower Permian and more rarely on older strata and are overlain by basal, Lower Trias sandstones and conglomerates.

In Transcaucasia the best section of Permian and Trias sediments is found in the region of Dzhulfa (Nakhichevan A.S.S.R.), and has been described in detail by R. A. Arakelyan (1952). The lower part of the Upper Permian succession consists of limestones with inclusions of chert and a rich fauna of foraminifera, bryozoans, corals and brachiopods. Complex Upper Permian Foraminifera (*Polydiexodyina*) are characteristic. The upper part of the Upper Permian consists of variegated marls, alternating with carbonaceous sandy and muddy limestones. The total thickness of Upper Permian strata is 150–300m. Transition into the Lower Trias is gradual and the Trias is recognized by the appearance of ceratitids and pseudomonotids (*Claraia*).

Further interesting and important details of the succession are given by T. G. Il'yina (1962), who differentiates the upper part of the Upper Permian into the Dzhulfian Suite (40–50m) which consists of variegated marls and limestones with a rich and diverse fauna. The upper part of the suite was formerly included in the Trias since it contains the ammonite identified as *Otoceras*, but a detailed investigation has shown that it is really *Protoceras*. Thus as a result of reidentification the Mesozoic-Palaeozoic boundary was displaced.

The Dzhulfian Suite is conformably overlain by reddish limestones with Lower Trias ammonites *(Kashmirites, Paratirolites)* and other fossils. Higher up follow flaggy limestones with *Claraia*. The study of the calcareous faunas of the Upper Permian and Lower Trias has shown that they are related, while brachiopods and

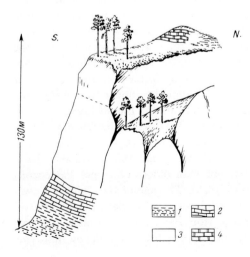

Fig. 203

Upper Permian strata of the Gefo Massif in the upper reaches of the Laba. According to K. V. Miklukho-Maklai (1956).

1. The Kutanian Horizon of sandstones and shales.
2. The Nikitinian Horizon of bedded limestones.
3. The Urushtenian Horizon of massive reef limestones.
4. The Abagian Horizon of flaggy limestones.

rugose corals are completely transitional from one epoch to the other. The discovery of Triassic rugose corals, which in all text-books are quoted as Palaeozoic, is particularly important. This fact again emphasizes the correct conclusion that the Lower Trias is the concluding epoch of the Palaeozoic system. T. G. Il'yina (1962) ends her contribution by a conclusion that Permian and Lower Triassic rugose corals gradually change into hexacorallids. The conclusion is not established and is incorrect since such a principal problem cannot be solved so simply on the basis of the limited material available. The available data suggest that rugose corals die out in the Lower Trias, while hexacorallids separate from them as a side branch.

MESOZOIC

Trias

Triassic sediments are not very widespread and occur in the same areas as Upper Palaeozoic strata. The top Trias sediments (Rhaetic) are completely associated with the Lower Jurassic forming a single complex of a vast distribution.

SUBDIVISIONS. The three main divisions are present and all stages occur, as is characteristic of the Alps. The Lower Trias usually consists of a single Scythian Stage, but in the southern Transcaucasia it is divided into Seisian and Campilian stages or beds. The Middle Trias is divided into the Anisian and the Ladinian stages and the Upper Trias into the Carnian, the Norian and the Rhaetic.

LITHOLOGY AND FACIES. In the Carpathians, within the main anticlinorium, a formation (10–250m) of red and green shales and sandstones with beds of cherts, limestones and volcanic rocks is attributed to the Lower Trias. At the base lie conglomerates which were formerly thought to be Upper Permian. Still higher is a formation of barren dolomites, limestones and marls (150–300m) with an Upper Trias fauna at the top; the lower part of the formation being considered as of a Middle Trias age. Compact conglomerates and quartzitic dislocated sandstones of Dobrudzha type Trias are found on the Zmein (Snake) island, to the east of the Danube delta. In Crimea no Lower Triassic sediments are known. The lowermost part of the Tavridian Formation is possibly Middle Triassic, but this is indicated by a single discovery of rather indefinite Middle Trias ammonites (Robinson, 1956). Upper Trias forms the main part of the lower section of the Tavridian Suite, which is a thick cleaved formation (4000–6000m) of siliceous and clay slates, bedded fine-grained sandstones and thin beds of dark limestones. The formation is very uniform, thinly-bedded, rhythmic, and poorly fossiliferous. The rocks which are strongly folded suggest flysch type of deposition and the contemporaneous occurrence of the proto-Cimmeridian orogenic phase. The lower part of the suite has yielded a homogeneous fauna with *Pseudomonotis caucasica*, belonging to the Norian Stage, while higher up the Rhaetic *Avicula contorta* has been found. The upper part of the suite has beds of black crinoidal limestones with a rich and variable Liassic fauna.

According to the latest data of A. I. Shalimov the Tavridian Suite reaches 1200m, has the characters of a flysch deposit, and belongs only to the Upper Trias. The overlying sandy-clayey formation he calls the 'Eskiordynian Suite', which is 700m thick. It is also of flysch type, but has horizons with blocks of Upper Palaeozoic, Triassic and Lower Liassic limestones (wildflysch). The age of this suite is Lower Jurassic and partly Lower Middle Jurassic. In places it is overlain by the effusive Al'minian Suite (500m) of a Middle Jurassic (top Bajocian, Bathonian) age. All these suites are strongly folded and are unconformably overlain by flat-lying

Upper Jurassic and Cretaceous strata. The Tavridian Suite has also been found in boreholes sunk east of the Sea of Azov in the region of Yeisk.

In the Northern Caucasus the Trias succession varies over relatively short distances. In the central region (River Laba) sandstones and conglomerates of a Lower Triassic age lie either with a small break on Upper Permian or on metamorphic rocks. Higher up follow massive and bedded limestones of the Scythian Stage, which contain ceratitids *(Meekoceras)* and pseudomonotids *(Claraia)*. The limestones are transitional into marls with Anisian ammonites, and the marls in turn pass into shales and sandstones of the Ladinian Stage with *Daonella* and the Carnian Stage with *Halobia*. Then there is a break (Fig. 204), followed by basal sandstones and conglomerates of the Norian Stage which pass up into shales and red and grey limestones with ammonites and *Pseudomonotis caucasica*. The succession ends with red massive Rhaetic limestones with brachiopods. The total thickness of the succession is 1200–1500m. To the west the Middle Trias disappears and Upper Trias strata rest directly on the Lower Trias, while still further to the west, in the valley of the Byelaya the Upper Trias is unconformable on the Permian (Fig. 205).

In southern Armenia, in the Vedi-Chai valley, the Lower Trias is represented by a formation (400m) which at the bottom consists of interbedded dark limestones and shales and at the top of grey, yellowish and reddish flaggy limestones (150–200m) with *Claraia*. Higher, along the valley, there is an isolated outcrop of Upper Trias. Here at the bottom there are shales and sandstones (10m) with lamellibranchs and ceratitids of a Norian age. Above them are Rhaetic coal-bearing beds consisting of shales, sandstones and lenses of coal. This formation is only partly preserved.

To the south in the Nakhichevan A.S.S.R. in the Dzhulfa region a classical internationally known section has variegated Upper Permian limestones gradationally succeeded by Seisian limestones (25–30m) which are barren at the bottom, but have ceratitids and *Claria* towards the top. Higher up lies a flysch-like formation (160–180m) of the Campilian Stage. The formation consists of barren, thinly bedded, rhythmically alternating muddy and flaggy limestones with fucoids. Still higher lie the grey oolitic limestones (25–40m) with *Pseudomonotis venetiana*.

The lower part of the Middle Trias (Anisian Stage) consists of variegated marly limestones (180–210m) with fine partings and pelecypods and gastropods. This part of the Trias is exposed not only in the Dzhulfa succession but also in the adjacent areas of the Nakhichevan A.S.S.R., as is also the last thick formation (500–1000m) of dark and grey cavernous dolomites and dolomitized limestones with gastropods and lamellibranchs. The age of this formation is Ladinian and Carnian and it is analogous to the Trias dolomites of Tyrol. Still higher are volcanic rocks of Liassic age. Owing to the absence of a fauna the age of its higher part is provisional.

FAUNA AND FLORA. The fauna is very rich and variable but as yet has not been described in monographs. Ammonites predominate and both ceratitids and ammonoids are important age indices. Numerous lamellibranchs are of a great importance and often of stratigraphic significance as are for instance *Claraia (Pseudomonotis clarai)*, *Halobia*, *Daonella*, *Pseudomonotis*. The genus *Megalodon* with a large very thick shell is common in reefs. There are numerous brachiopods and gastropods. The flora consists of calcareous algae, sometimes rock-forming.

PALAEOGEOGRAPHY. In the internal, southern zone of the Mediterranean Geosyncline a marine regime predominated; a warm, tropical sea with reefs followed directly upon a similar Upper Permian sea. In the external zone a continental regime

Fig. 204. An unconformity between Upper and Middle Triassic rocks
exposed in the basin of the river Laba. Thick bedded limestones at the top
of the mountain are Upper Triassic while Middle Triassic soft sandstones
and shales form rounded surface which does not show bedding. Middle
Triassic rocks continue into the forest. *Photograph by V. N. Robinson.*

Fig. 205

Massive Upper Triassic limestones forming the crest of Mt.
Tkhach, situated on the watershed between the rivers Byelaya
and Laba.
Photograph by V. N. Robinson.

was predominant and the sea penetrated into this region only locally and for short durations of time. In the Rhaetic epoch in several regions of both zones large near-shore swampy plains came into existence where coal accumulated.

Jurassic

The Jurassic system of the region is economically very important since all large coal deposits and a large number of non-metallic deposits are found within its strata. Upper Jurassic oil and gas deserve attention.

DISTRIBUTION. Jurassic rocks are widespread since they represent the deposits of the first system extensively uncovered by erosion in the mountainous regions. In the Carpathians of the U.S.S.R. it is not widely exposed, although Jurassic outcrops are present around the Marmarosh Massif and at the Carpathian escarpment. Boreholes have penetrated Jurassic sediments of southern Moldavia, to the north of the Danube, proving its existence under large areas. The three divisions of a total thickness of 3500m have been proved and are distinctly folded. The Crimean Mts consist mainly of Jurassic sediments of all three divisions. The massive Upper Jurassic reef limestones particularly influence the local relief. In the main Caucasian range Jurassic strata cover a particularly wide area. They girdle its western part, while in the eastern part which is less elevated and eroded they form its watershed (see synoptic maps). The north of the Lesser Caucasus is also covered by Jurassic rocks, as well as parts of the Nakhichevan A.S.S.R. and an area to the south-east of Erevan (Armenia).

SUBDIVISIONS. Jurassic successions of Crimea, the Caucasus and Trans-caucasia are of Alpine type and have all the features found in the Alps; the three divisions and all the main stages are present. The Lower Jurassic (Lias) is divided into lower, middle and upper. The Upper Lias is often known as Toarcian. The Middle Jurassic (Dogger) is divided into the Bajocian and Bathonian stages, but in some localities the underlying Aalenian Stage is also differentiated, although it is sometimes included in the Lias. There are, however, localities where Aalenian deposits are transitional into the Bajocian and have basal sandstones and con-glomerates and are even unconformable on the Lias. Elsewhere the Aalenian and the Toarcian are transitional with a break occurring above the Toarcian or being entirely absent.

The Upper Jurassic (Malm) is divided into Callovian, Oxfordian, Kimmeridgian and Tithonian (rather than Volgian), represented by marine limestones and marls with a rich southern fauna. In some successions the Lusitanian Stage, comprising the Oxfordian and the lower part of the Kimmeridgian, is recognized and is represented by carbonate facies, which are often massive reefs (see p. 606). The main Jurassic successions are given in Table 45.

LITHOLOGY AND FACIES. In the Carpathians Triassic limestones are over-lain by a suite of sandstones, shales, and spotted marls (50–60m) with a Lower Lias marine fauna. Higher up, separated by a break, lie belemnite-bearing limestones or marls (30–50m) of an Upper Lias or Bajocian age. These rocks are only found in the main anticlinorium. Middle Jurassic rocks are not well developed, but Upper Jurassic strata are widespread and are found almost in all zones, and although its facies are very variable various limestones and siliceous rocks with marine faunas predominate and reach a total thickness of 200–400m. The fauna indicates the presence of all stages, from the Bajocian up to and including the Tithonian (Slavin, 1956).

TABLE 45. *Jurassic Successions in the Caucasus*

Stage	The Northern Caucasus		
	Western	Central	Eastern
Overlying rocks.	Valanginian limestones. The flysch.	Valanginian limestones.	Valanginian limestones.
Titonian.	Bedded, variegated limestones with a rich fauna and flysch deposits in the west (400m).	In the east, limestones and marls; in the west, gypsiferous sandstones, shales and dolomites (100–200m).	Above, limestones and marls; below, lagoonal gypsum, marls and sandstones (150–700m).
Kimmeridgian and Oxfordian	Bedded and massive limestones (600m). In places variegated strata (300m). In the west sandy-shaley flysch deposits (500–600m).	Interbedded limestones and marls; massive limestones at the bottom (Up to 250m).	Bedded limestones and marls; sometimes bituminous (60–600m). Pale massive reef limestones (80–350m).
Callovian	Ferruginous sandstones with beds of oolitic limestones (15m).	Sandy limestones, sandstones and at the bottom conglomerates. In places brown coals (25–100m).	Above, oolitic limestones and marls with a fauna; below, sandstones and conglomerates (Up to 400m, but often less).
Bathonian.	Not recognized.	Grey shales with beds of sandstones. A marine fauna is present (500–600m).	Closely interbedded shales, sandstones and marls (200–450m).
Bajocian.	Grey shales with sphaerocideritids (150–400m).		Black and greenish shales with beds of sandstones. A marine fauna is present (200–900m).
Aalenian.	Upper, shales (300m). Lower, sandstones, shales and limestones (200–300m).	Ferruginous sandstones and argillites (50–120m).	Brown and brownish sandstones and shales; towards the bottom, coal-bearing (1500–2400m); above, marine (200–2000m).
Upper Lias (Toarcian)	Shales with beds of sandstones (Up to 800m). Basal sandstones (20–40m).	In the Kuban' basin volcanogenic rocks (400m) and shales (200m).	Dark, hard limestones with beds of sandstones. At the top a marine fauna (Up to 4000m).
Middle Lias	Shales and sandy shales with beds of sandstones. The strata are in places marine, with a fauna, and elsewhere coal-bearing (500m).	A coal-bearing formation of the Upper Kuban' and the Baksan (600–700m). Basal conglomerates.	Dark-grey, metamorphosed slates with ammonites (*Amaltheus*) at the top (1500–3000m).
Lower Lias	Basal sandstones with beds of clays. A marine fauna is present (50–170m).	Unrecognized.	
Underlying rocks.	Various deposits.	Palaeozoic.	Various deposits.

TABLE 45—continued

The Southern Caucasus		The Lesser Caucasus
Georgian Military Road	Abkhazia	Armenia and Azerbaijan
Valanginian limestones and marls.	Valanginian red beds.	Volcanic rocks. Sandstones.
Identical dark, bedded limestones and marls, almost devoid of fauna (250m).	Red, gypsiferous sandy-shaley formation (Up to 400m).	Volcanogenic-sedimentary formation with beds of limestones (Up to 500m).

A flysch formation divisible into several suites. Rhythmically repeated dark marls, calcareous sandstones and limestones. No fauna but there are problematic structures (1000–1200m).	In places, thick and variable limestones (400m).
	    In places a thick sequence of carbonate flysch.
		Massive and bedded limestones and marls with a reef fauna (200–400m).
		    In places the Kimmeridgian is represented by volcanic rocks.
		    Shaley-sandy and tuffaceous strata, with a marine fauna (From 70 to 400m).

Black shales, alternating with sandstone bands. Problematic flysch structures are present (No less than 1000m).	Coal-bearing formation of Tkvarcheli and Tkvibuli (150–200m).
	    Greenish sandstones (300m).
	    Dark shales (200m).
		Upper volcanogenic-sedimentary formation of porphyrites, tuffs and breccias; towards the top abundant tuffaceous sandstones with a Bathonian marine fauna (Up to 3500m).

	Green tuffaceous sandstones with ammonites (*Parkinsonia*). A thick, volcanic porphyrite sequence. In places a top horizon of quartz porphyries is present. Bands with marine Bajocian and Upper Lias faunas are found (1500–2000m).	Quartz-porphyries (plagiophyres). Beds of tuffogenous sandstones with an Aalenian fauna.
Dark shales of flysch type, alternating with sandstones (2500m).		Lower volcanogenic-sedimentary formation of porphyrites, tuffs and breccias. Beds of tuffogenous sandstones, with marine faunas of all Lias stages, occur. In places the formation is replaced by normal sediments, mainly sandstones and shales with marine faunas.
A volcanogenic suite of black shales, thin beds of quartzites, tuffs and lavas (1000m).	Volcanic, porphyrite suite with bands of shales. In places limestones (Up to 1000m).	
Quartzites and slates, at the bottom graphitic (450m). Basal sandstones and conglomerates (Up to 100m).	In places slates, elsewhere quartzites and graphitic shales (schists) (2000m).	
Palaeozoic and granites.	Palaeozoic.	Various deposits.

On the western and southern slopes of the Ukrainian Shield, as found in bore-holes, Jurassic strata are widespread. Nearer to the outcrops of Precambian the Jurassic is represented by thin sandy and clayey deposits (50–60m). Plant remains indicate an Upper Jurassic age for these strata. Away from Precambrian outcrops the succession thickens to 300m and marine and lagoonal deposits become predominant. At the top there are organic, Tithonian limestones (70m) while below them are sulphates and carbonates (200m) of Kimmeridgian and Oxfordian ages. The suc-cession has variegated Callovian sands and clays at the bottom (30m). Further away from the platform—in the downwarp—the succession rapidly thickens and Middle and Lower Jurassic strata appear. In Stryi Jurassic strata are about 900m thick, while in the south of Moldavia the Lias is 350–450m, the Middle Jurassic is 750–1600m and the Upper Jurassic is 300–1900m thick.

In the western part of the Cis-Black Sea Depression Jurassic sediments have a full succession, while in the Lower Danube sub-montaine downwarp their thickness according to geophysical data reaches 3000m. Here dark and pale grey marine shales and sandstones predominate although Tithonian sediments consist of thick red beds. From north to south the total thickness rapidly decreases and there is a change in facies. In the downwarp the Lias is represented by black, cleaved argillites with beds of sandstones, dolomites and limestones with Foraminifera. The thickness of the incomplete succession is 350–450m. Aalenian and Bajocian are not differentiated and consist of dark argillites with beds of foraminiferal sandstones, dolomites and limestones. The thickness of these rocks reaches 1300m in the south, but decreases to 5–30m in the north where grey clays and marls have an ordinary marine fauna. The Bathonian consists, in the south, of bedded limestones (500m) with inter-calations of clays and marls. In the west the limestones pass laterally into dark clays (460m) with *Posidonia*, while in the north sandstones (230–320m) with intercalations of shales are dominant.

The Upper Jurassic consists of laminated shales, varying from greenish to reddish and having thin beds of sandstones and limestones. The fauna is foraminiferal and the thickness is 170–400m in the north and 40–150m in the west. In the south Oxfordian strata are of the same type, contain Foraminifera and are 40–830m thick. In the north limestones (480–525m) predominate and are yellow and compact at the bottom, pale and oolitic at the top. The Kimmeridgian consists of limestones (20m) in the south and dolomites and clays (60–80m) in the north. Tithonian strata are peculiar with 60–130m of anhydrite and gypsum with intercalations of clays and dolomites at the bottom. Towards the top lie continental red and variegated sand-stones and clays (90–105m). In the north-west there are only red beds (285m) consisting of sands, clays and barren conglomerates.

In adjacent regions of the downwarp the succession changes slightly and Lias (350m) is represented by black calcareous argillites with beds of limestones and sandstones. A microfauna has been discovered in these rocks, and the sandstones are in places saturated with oil. The Aalenian and Bajocian (200–800m) also consist of dark argillites with beds of sandstones. The Bathonian (550–750m) is represented by grey limestones with intercalations of shales. The Callovian (70–400m) consists of dark clays. The Oxfordian (30–830m) is represented by grey clays with beds of limestone, the Kimmeridgian (30m) by coal-bearing strata and the Tithonian (150–600m) by gypsum and towards the top by clays.

In Crimea two distinct thick formations occur at the southern shore and condition its relief and consequently the position of many sanatoria and holiday centres. The upper formation represents a rocky calcareous ridge—the so-called Yaila—formed by Upper Jurassic reef limestones. Here individual mountains such as Ai-Petri (1233m), Roman-Kosh (1545m) and Chatyr-Dag (1525m) rise with steep escarpments above the lower flat part of southern Crimea which marks the outcrops of Middle and Lower Jurassic shales.

The mountains form a barrier against cold northerly winds and condition the warm healthy climate. The southern shore of Crimea is entirely overgrown by forest, is covered by orchards and vineyards drenched by warm southern sun, and consequently is one of the best holiday resorts of the Soviet Union. Nearer to the mountains amongst aromatic pine trees on Middle Jurassic sandstones lie mountain sanatoria, predominantly constructed for tubercular patients. Down the slope, on Lower Jurassic shales the forest becomes deciduous and passes into an uninterrupted near-shore belt of orchards and vineyards where again there are numerous sanatoria and holiday centres.

The Lower Jurassic is represented by a thick formation of dark and pale grey shales and cherts, interbedded with siltstones, fine-grained sandstones and limestones, all being rhythmically repeated and having few organic remains but numerous sole markings. The total thickness is no less than several thousands metres and all formations are intensively folded into small folds. These features leave no doubts that the Tavridian Formation has a flysch character, its upper part being of a Lower Jurassic age (p. 596). The flysch is associated with the ubiquitous older Cimmeridian orogenic phase. This phase produced mountains which were eroded away and supplied the sediments for the flysch which formed at the foot of the mountains. The formation of the flysch proceeded contemporaneously with the elevation and erosion of the mountains and manifestations of folding. The thickness of Lower Jurassic rocks is 50–1000m.

Middle Jurassic sediments are more diverse and rest with a break on the Lower Jurassic. Three types of deposits occur. First, alternations of marine clays, marls, limestones and more rarely sandstones. These rocks have a diverse fauna. The second sedimentary type consists of continental, often coal-bearing strata, which in the region of Beshui, to the south of Simferopol, are 1500m thick and include inconstant, strongly deformed seams and lenses of coal. Their age is Bajocian and possibly Toarcian. The third type is of volcanic deposits with rarer intrusions. The volcanic rocks in places occur instead of normal sediments. There is also the peculiar Bitakian Suite, which consists of rather thick (2000m) Bajocian conglomerates lying on the Tavridian Suite. It is developed near Simferopol.

The Upper Jurassic is typified by the development of thick reef limestones, which often unconformably overlie on older rocks. The succession begins with basal conglomerates and sandstones, which higher up pass into shales, sandstones and marls of the Callovian Stage (20–50m) which is succeeded by Oxfordian bedded marls, shales and limestones with ammonites of a mixed northern and southern aspect. Still higher lie pale and more rarely dark grey massive, generally vaguely, but in places well-bedded limestones with a southern fauna of single and colonial corals, hydroids, massive sessile lamellibranchs (*Diceras*) and thick-walled massive gastropods (*Nerinea*). The fauna is in the main Oxfordian but becomes Kimmeridgian towards the top. The thickness of the reef limestones reaches 700m, but is normally about 200–450m (Fig. 206). Thick massive reef limestones of Upper Jurassic age are often known as Lusitanian or simply as the Lusitan. The term refers to a definite facies complex of indeterminate age which includes both Oxfordian and

Kimmeridgian. Consequently the Lusitanian is not used in geological maps and there is no Lusitanian Stage.

To the east of Mt Demerdzha, near Alushta, the reef limestones are replaced by thick conglomerates (2000m) lying on Tavridian shales. Still further to the east, near Sudak, there is developed a Sudakian facies of grey, bedded marls and clays with corals, but even here reef massifs are encountered, surrounded by marls.

The Kimmeridgian and Tithonian are represented by various closely associated facies forming one single complex. In the west, near Balaklava, the Kimmeridgian and Tithonian consist of pale pink massive marmorized limestones of 300–400m in thickness. Polished slabs of these limestones are seen in the Moscow Underground Railway. Slightly to the east in the Baidar valley the limestones pass into flysch type slates and still further to the east they reappear again with massive and bedded varieties reaching altogether 400–450m. The limestones lie conformably on Oxfordian reef massifs, capping them throughout Yaila. To the east of Mt Karabi-Yaila the limestones disappear and are replaced by flysch type strata, lying on conglomerates or sands and shales. The flysch is represented by a thick laminated formation, consisting of rhythmically repeated thin units which are more sandy at the bottom and more argillaceous at the top. The colour of these formations is uniform, grey, dark-grey or brownish. The surfaces of sandstones show hieroglyphs or flysch features of unknown origin. The fauna is very rare and is normally absent. In places there are breccias consisting of angular fragments and sometimes large blocks of reef limestones, forming the so-called wildflysch, which is analogous to the wildflysch of the Caucasus and the Alps. The thickness of Upper Jurassic flysch is

FIG. 206. Upper Jurassic massive reef limestones passing on the left of the photograph into bedded limestones at Novyi Svyet, on the southern shore of Crimea. *Photograph by V. D. Nalivkin.*

variable, being 1100m near Theodosia, but reaching 1800m and even 2500m further to the west. Crimean flysch is a direct continuation of west Caucasian flysch. Further to the east in the central Caucasus the flysch is again replaced by reef limestones.

In the Caucasus Jurassic deposits are very widespread and thick (up to 15km) and consequently lithological and facies changes are particularly abundant. Here as in Crimea the Lower and Middle Jurassic are sharply separated from the Upper Jurassic. The change is lithological and is commonly associated with unconformities and breaks. The Lower and Middle Jurassic are typified by the great predominance of almost barren black shales of flysch character. The shales are commonly replaced by coal-bearing and locally volcanic rocks. In the Krasnodar region drilling has shown the presence of oil in Middle Jurassic sandstones. Limestones are encountered as subordinate beds and are commonly muddy or marly. Upper Jurassic strata, also thick but less so, form a sequence of carbonate rocks including pale reef limestones. Carbonate facies are most common in the north adjacent to the formerly existing land. In the south these rocks pass into flysch deposits and reefs are only found in isolated localities.

As an example one can adopt the correlation scheme of eastern Caucasian successions (Table 46) compiled by V. E. Khain (1950) who is one of the foremost specialists on the region. Table 46 shows that Tithonian limestones are found both to the north and to the south of the flysch. In Oxfordian and Kimmeridgian times the relationships were simpler as is evident in Fig. 207 but it should be remembered that the reef limestones are also found far to the south in certain parts

TABLE 46. *Facies variations of Middle and Upper Jurassic Strata in the South-eastern Caucasus*
(According to V. E. Khain, 1950)

	Southern Zone (Vandamian)	Central Zone (Dibrarian)	Northern Zone (Samur-Shakhdagian)
Titonian	Massive reef limestones.	Variegated marly-argillaceous flysch.	Massive limestones and dolomites. Variegated gypsiferous sandstones and clays.
Kimmeridgian	Missing.	Siliceous, calcareous argillaceous flysch.	Missing.
Oxfordian and Callovian		Siliceous, silty argillaceous flysch.	Pale, reef limestones.
Bathonian		Dark, silty argillaceous flysch.	Missing.
Bajocian	Volcanic porphyrite formation.	Thick sandstones with sequences of shales.	
Aalenian	Dark, muddy shales with rare beds of sandstone.	Dark silty-argillaceous flysch.	Dark argillites and sandstones with beds of conglomerates.

of eastern Transcaucasia. Conversely to the west of the Northern Caucasus along the Black Sea and along the southern slope as far as Tiflis the Tithonian is represented by flysch deposits. The main successions in the Northern and Southern Caucasus, the Lesser Caucasus and Southern Armenia are given in Table 45.

Of various facies complexes let us consider the following:

Aspidic Slates. A very thick (300m) formation of uniform dark grey highly altered Middle and Lower Lias slates represents this facies. The slates are closely interbedded with thin fine grained sandstones and siltstones. The strata are almost barren and Liassic faunas are very rare. The slates are used as roofing materials and as blackboards. The formation outcrops along the watershed of the Eastern Caucasus closely resembles flysch and is similarly associated with the Older Cimmeridian phase of folding. The main features and conditions of deposition of the formation are almost identical with the Tavridian Suite of Crimea. The age of the formation is also probably the same as the age of the Tavridian Suite, but no Upper Trias fauna has been found in the Caucasus.

Flysch. This has already been described (p. 604) and is identical with the Palaeogene flysch (p. 620).

Coal-bearing strata. These rocks are found amongst thick Lower and Middle Jurassic slates, and are closely associated with the flysch. In Dagestan the Aalenian coal-bearing formation lies on aspidic slates and is continuous with the latter. In western Georgia a Bathonian coal-bearing formation underlies the Upper Jurassic flysch. The lithology and other features of coal-bearing formations are described below (p. 683). They originated on near-shore plains.

Reef Massifs. Reef complexes of carbonate rocks are typical in the Upper Jurassic and are rarer in the Lower Cretaceous. Thick formations of massive pale pink and more rarely dark bituminous limestones and less frequent dolomites are traced as a thick unit 400–700m stretching hundreds of kilometres along the trend. Massive limestones are surrounded by bedded limestones and marls. The fauna is found in isolated accumulations where it is abundant and includes numerous corals. Thus there were typical barrier reefs margining the northerly land or isolated islands.

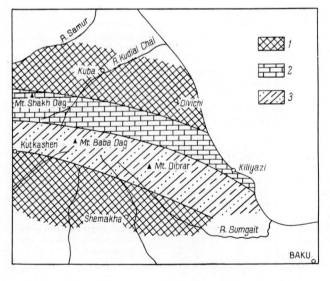

1. Regions of erosion.

2. Reef limestones.

3. Silty-clayey flysch.

Fig. 207.

Lusitanian facies according to V. M. Khain (1950).

Continental red beds. Continental red beds are widespread in the Upper Tithonian of Dagestan and western Georgia. Red sandstones and clays predominate and conglomerates and breccias as well as gypsiferous and dolomitic groups of strata are common. Red beds, like coal-bearing beds, were found on near-shore plains, covered by bitter salt lakes and swamps, but under dry desert conditions.

FAUNA AND FLORA. The fauna is very rich and variable. It includes mainly southern forms, such as *Phylloceras* amongst ammonites, *Diceras* amongst lamellibranchs and *Nerinea* amongst gastropods. Many of these creatures lived on the reefs or near to them. The flora is typically Eurasian (Angara type) and is characteristic of Jurassic coal measures of the Asiatic part of the U.S.S.R.

PALAEOGEOGRAPHY. The palaeogeography is very complex. The dominance of southern tropical marine basins with irregular shorelines and numerous volcanic, reef, and other islands of older rocks is typical. Coal-bearing strata and red beds indicate that there were vast near-shore plains The land on which these plains were situated was a northern continent and also groups of islands in more southerly provinces. The origin of flysch and aspidic deposits is debatable but they are most likely deposits accumulating on near-shore plains invaded by sea. Throughout the Lower and Middle Jurassic epochs the climate was wet and possibly temperate. In the Upper Jurassic epoch the climate became hot, tropical dry and typical of the desert.

The southern slope of the Greater Caucasus and the Lesser Caucasus were volcanic and numerous large volcanoes existed along huge faults. Frequent and numerous eruptions produced large amounts of pyroclastic material, covering large areas.

Cretaceous

The Cretaceous epoch is divided into Lower and Upper, each of which is of equal significance to Palaeogene and Neogene. Both the Lower and the Upper Cretaceous are composed of marine deposits, but along the northern margin of the geosyncline the Lower Cretaceous is often absent or is represented by continental deposits. The Upper Cretaceous is associated with the Upper Cretaceous transgression, which was the most widespread in the history of the Earth and resulted in ubiquitous marine sediments continuous into the Neogene. The recent discovery of commercial oil and gas deposits in Lower and Upper Cretaceous sediments of the Northern Caucasus and Crimea is very important.

DISTRIBUTION. Cretaceous sediments are widely found in the Carpathian, Crimean, Caucasian and Transcaucasian fold belts. Nevertheless the area covered by Cretaceous deposits is much smaller than that covered by Jurassic and Palaeogene sediments. In the northern part of the geosyncline Lower Cretaceous strata are almost missing, otherwise the succession is everywhere complete. In southern Moldavia Upper Albian rocks rest on the Tithonian. In the Cis-Black Sea downwarp the succession—a single sedimentation rhythm—begins with the Upper Albian (gaizes), which is succeeded by Upper Cretaceous sediments. In Ciscaucasia again the Lower Cretaceous is unrepresented and the sedimentation rhythm begins with Upper Albian sandstones and clays which are succeeded by a full Upper Cretaceous succession.

As has already been pointed out in the Mangyshlaks, the Tuar-Kyr, the Kopet-Dag and the Pamirs both the Lower and Upper Cretaceous is represented by mainly marine sediments including thick Lower Cretaceous reef limestones.

SUBDIVISIONS. The division of the Cretaceous into Lower and Upper being generally accepted, it is found that the last phases of the Cimmeridian Orogeny occurred in Lower Cretaceous and the first phases of the Alpine Orogeny in Upper Cretaceous times. The practice of differentiation of the Aptian and Albian as the Middle Cretaceous is now abandoned, while the recognized stages are: Valanginian, Hauterivian, Barremian, Aptian, Albian, Cenomanian, Turonian, Coniacian, Santonian, Campanian, Maastrichtian and Danian. The last is sometimes considered as a part of the Palaeogene while the Lower Valanginian is sometimes differentiated into the distinct Berriasian Stage. The grouping of the Valanginian, Hauterivian and Barremian into the single Neocomian Stage is done less and less frequently. The grouping of the Coniacian, Santonian, Campanian and Maastrichtian stages into the Emscherian and Senonian stages of the Mediterranean is not used. Each of the stages on the basis of its fauna is divided into two sub-stages.

LITHOLOGY AND FACIES. In the Soviet Carpathians the Lower and Upper Cretaceous is mainly represented by a thick flysch with all Cretaceous stages, with the exception of the Danian, being present. The Lower Cretaceous flysch (1000m) is found at the foot of the Marmarosh Massif composed of Precambrian and Palaeozoic rocks. At the beginning of the Albian the accumulation of flysch deposits was temporarily interrupted and Upper Albian and Cenomanian shallow water sandy facies (130m) appeared. These represent the beginning of a new sedimentation rhythm which continued, in some places without interval and elsewhere after a break, into the Palaeogene. The total thickness of the Cretaceous-Palaeogene flysch reached 4–5km and in the north 7km. Flysch strata are everywhere strongly folded and broken into nappes, 'skibas', which in translation means slices. The nappes are pushed over each other and as a result Cretaceous outcrops occur as long narrow belts. The Cretaceous flysch of the Carpathians is divided into several suites.

To the north of the Carpathians, away from rising mountain chains the quantity of the sandy and clayey matter rapidly decreased and consequently Lower Cretaceous sediments in places are completely missing. Thus the Upper Cretaceous commences with a thin clastic formation soon succeeded by marls and limestones and then by white chalk. The succession has a platform character and Upper Cretaceous and Palaeogene strata are gently folded and unconformably overstepped by the horizontal beds of Neogene.

Under the Palaeocene Yamnian Suite lies the Upper Cretaceous Stryian Suite, sometimes known as the *Inoceramus* Suite. Then after a break Albian and Aptian marls appear. Southwards the Borkutian and the Byelatissian suites correspond to the marls. In the north under Aptian strata lies the Spasskian Suite, while in the south under Byelotissian rocks there is the Rakhovian Suite, which is equivalent to the Barremian and Hauterivian, the Valanginian being absent.

The Cretaceous maintains this type of succession as far as southern Moldavia and even further to the east. In southern Moldavia Tithonian red beds are unconformably overlain by sands, sandstones, conglomerates and marls of total thickness of up to 150m and of an Upper Albian age. The Cenomanian (30–80m) consists of interbedded clays, sandstones, marls and limestones. In Turonian- Coniacian times greenish-grey marls (18m) with beds of limestones developed. The Santonian is represented by white chalk (176m), interbedded with marls. Finally the Campanian (130m) consists of pure white chalk, which is succeeded upwards by marls overlain by glauconitic marls of Palaeocene age.

In the Cis-Black Sea Depression Cretaceous deposits, as found in boreholes, begin with Upper Albian gaizes (15m) which are overlain by Cenomanian (70m) clays, marls and sandstones, followed by Turonian (50m) chalky marls and sandstones and Senonian (70m) chalky marls and limestones.

In Crimea Lower Cretaceous sediments, especially in the 'Foothill Ridge', are widespread and consist of various rapidly changing facies. Such changes are interpreted in terms of considerable tectonic movements, which occurred in the Lower Cretaceous epoch. A review of the Mesozoic and Cainozoic stratigraphy of Crimea is given by Muratov (1960). In the west, near Balaklava, Tithonian limestones are overlain by pebbly sandstones (20–40m) of Valanginian and Hauterivian ages. Still higher there are grey and yellowish clays (100m) with *Sphaerosiderites* of Barremian and Aptian age. The clays are overlain by Albian conglomerates, sandstones and green clays.

To the east, in the classic Biasola section, the Valanginian (3·5m) consists of sandstones and conglomerates, succeeded by Hauterivian muddy limestones and sandstones (80m) with an abundant fauna. A bed (2–3m) of yellowish muddy limestone has a particularly well preserved fauna of ammonites, including *Crioceras*. The limestones and sandstones are followed by thick (120m) grey bedded Aptian clays capped by a sandstone horizon (15m). Near Simferopol on the river Salgira the Albian is represented by the 'Salgirian Suite' (1500m), which is a typical flysch deposit. Even further to the east amongst clays and sandstones there are massive Urgonian rudistid limestones with *Requienia* and *Monopleura*. Lastly near Theodosia on the Upper Jurassic flysch lies a formation (100m) of clays, marls and limestones with Berriasian faunas, followed by thick (200m) olive-green Neocomian clays capped by red Aptian clays (50m). In the Crimean steppes Lower Cretaceous strata are 1600m thick. Upper Cretaceous strata of Crimea are exposed in a narrow belt of uniform lithology. The rocks are limestones and marls and sometimes are chalk-like. Their thickness is relatively small, varying from 60 to 500m. In the Crimean steppes borehole information indicates up to 1800m of Upper Cretaceous sediments. In the west in the well known sections of Bakchisarai, Albian serpulitic limestones are overlain by 5m of sandstone and then by 60m of white and greenish Cenomanian marls (60m). The Turonian is represented by white marls and *Inoceramus* limestones (45m). The Senonian and the Santonian (80m) again consist of marls, while higher up white chalk-like Campanian rubbly-marls (100m) make their appearance. The Maastrichtian is represented by marls (80m) at the bottom and sandy marls and marly sandstones (70m) at the top. The section ends with Danian sandstones and marls (6m) and white bryozoan limestones (30m) overlain by white limestones (18m) of Palaeocene age.

In the east, near Theodosia, the succession has a Caucasian aspect. Albian clays are succeeded by white and greenish Cenomanian and Turonian marls (43m), then by white platy marls (90m) of Lower Maastrichtian age producing a low (third) scarp. Still higher occur yellowish marls, limestones and clays (90m) of Upper Maastrichtian and Danian age. The succession ends by the 'Lysaya Gora Suite' (70–150m) of Upper Danian and Palaeocene age, which consists of almost barren bedded yellowish and platy marls of flysch type.

Within the vast Ciscaucasian plain (Scyptian Platform) which continues as far as the Manych valley, Cretaceous sediments have only been found in boreholes, in which Lower Cretaceous strata were found to be almost entirely missing, while a full Upper Cretaceous succession (200–1500m) begins with Upper Albian sandstones and clays. The Lower Cretaceous and Jurassic are developed only near the western

Caspian shore, including the Astrakhan region, and also near the shore of the Sea of Azov. In the Astrakhan region the Middle Jurassic, consisting of dark clays with beds of sandstones, is 400–600m thick. There are signs that it is oil and gas-bearing. The Upper Jurassic is represented by dolomites and silts of a total thickness of 135m. The Lower Cretaceous succession begins with Barremian basal conglomerates and sandstones (90m), which in the south (Ozek-Suat) contain commercial deposits of petroleum. The sandstones are overlain by Aptian (120–160m) dark clays, silts and sandstones and by Albian sandstones (275–500m). Large deposits of gas have been found in Aptian and Albian deposits (Promyslovoye). In the Krasnodar region Lower Cretaceous sandstones contain large deposits of oil and gas.

On the northern slope of the Main Caucasian range Cretaceous sediments occur as a narrow belt parallel to the strike. The general lithology of these sediments is similar to the Crimean Cretaceous.

LOWER CRETACEOUS.   Lower Cretaceous sediments of the northern slope are uniform and consist mainly of marls and limestones. To the west of Maikop they have a flysch character and reach a great thickness of 4000m. The Valanginian (800m) begins with a horizon of 'block conglomerates' (150m) analogous to wild-flysch, which is succeeded by pale marly clays and marls with abundantly fossiliferous localities. The Hauterivian also commences with coarse sandstones and conglomerates (60m), followed by the Lower Sideritic Suite consisting of black plastic clays with sphaerosiderites (up to 1000m). The overlying 1200m of Barremian strata are black sandy clays with beds of sandstones or in places dominant sandstones with occasional deposits of oil and gas. The 'Upper Sideritic Suite' (up to 1000m) is of Aptian and Albian ages and is of the same lithology as the Hauterivian.

In the Kislovodsk region flysch facies are replaced by carbonates. The Valanginian (100–120m) is represented by uniform yellowish and grey dolomitized limestones with a poor fauna. These limestones are followed by Hauterivian shelly limestones and sandy clays (30m). The Barremian (120m) is represented by yellow calcareous sandstones and oolitic sandy limestones. Lower Aptian strata (60m) begin with beds of conglomerates (3m) succeeded by bedded calcareous and muddy sandstones, while Upper Aptian strata (160m) consist of calcareous glauconitic sandstones with ammonites and lamellibranchs. These glauconitic sandstones continue into the Lower Albian (180m) becoming more muddy and friable towards the top. The Upper Albian (40m) is represented by black thinly bedded calcareous clays with a rich fauna of ammonites, belemnites and lamellibranchs. The succeeding Upper Cretaceous limestones form the next mountain range. Along the Georgian Military Road, to the south of Ordzhinikidze the succession is of the same type. The Valanginian (290m) consists of limestones with beds of marl at the bottom and compact oolitic limestones with a poor fauna at the top. The Hauterivian, the Barremian, the Aptian and the Albian are represented by a thick formation (870m) of sandstones, sandy marls and clays overlain by white, Upper Cretaceous limestones and marls (250m).

In central Dagestan the succession is the same and the total thickness reaches 1250m, while to the east towards Mt Shah-Dag the Valanginian, Hauterivian and Barremian are represented only by limestones and the total thickness decreases to 300m (Fig. 208).

The thicknesses of Lower Cretaceous strata are shown in Fig. 209 indicating the position of former Ciscaucasian land, of a mountainous island in the watershed region of the Greater Caucasus, and a Transcaucasian archipelago of islands.

A detailed study of faunas has allowed the introduction of details in the various successions. T. V. Mordvilko (1956) divides the Valanginian of the Northern Caucasus into two parts: the lower sub-stage (18–145m) of marls and the middle and upper sub-stages (up to 213m) of compact dolomitized limestones.

The Hauterivian is divided into two sub-stages and six zones. It consists of marls and clays with beds of limestones. The thickness varies from 6 to 300–400m.

The Barremian includes two sub-stages and two zones. The lower sub-stage (20–85m) embraces sandy limestones or sandy-clayey rocks while the upper (14–165m) consists of ferruginous sandstones and in the south of Dagestan of glauconitic sandstones and dark clays.

The Aptian has two sub-stages. The lower sub-stage of three zones (12–160m) has basal sandstones and conglomerates overlain by black clays with beds of glauconitic sandstones with spherical concretions and occasional beds of black clays. The fauna at the bottom is principally ammonitic and at the top pelecypodic.

The Albian is divided into three sub-stages of which the higher two are equivalent to the Upper Albian. The Lower Albian (30–235m) is differentiated into three zones and is mainly represented by glauconitic sandstones which coarsen and become

FIG. 208. Lower Cretaceous bedded limestone on the river Uruk (a tributary of the Terek) in Dagestan. *Photograph by F. Bubleinikov.*

cross-bedded towards the top. In Dagestan the sandstones are replaced by black clays. The Upper Albian consists of uniform black argillites, which are at the top interbedded with pale marls. The total thickness is 16–80m, but an abundant ammonite fauna allows the distinction of three zones in the middle sub-stage and two zones in the upper.

Lower Cretaceous strata of the southern slope of the Caucasus consist mainly of flysch deposits, especially well developed in the west and east—the regions of so-called 'flysch troughs'—where their thickness reaches 3000–4000m and more (Fig. 209). In western Georgia the flysch is replaced by calcareous facies such as the Urgonian Limestones. On the shore of the Black Sea, to the north-west of Sochi the Urgonian facies consist of pale, almost white thickly bedded or massive limestones, the Hauterivian and Barremian by grey marls with aptychi and higher up by variegated clays and marls, the Aptian and Albian by grey clays and sandstones. As is obvious (see Fig. 209) the total thickness increases to the north-west. To the south-east of Sochi the flysch is replaced by limestones of a total thickness of 1400m. The Valanginian (150m) consists of dark bituminous and yellow limestones, while the Hauterivian is represented by a thick, homogeneous formation (610m) of thickly bedded yellowish grey limestones with flint nodules. The Lower Barremian (250m) is represented by marls at the bottom and interbedded limestones and marls towards the top. Upper Barremian, so-called 'ringers' form a uniform almost barren formation (250m). Aptian marls (40m) are at the bottom finely laminated and have a rich fauna. The Albian (50–70m) is represented by compact limestones (10m) at the bottom and variegated marls at the top, becoming transitional into tuffogenous Cenomanian marls.

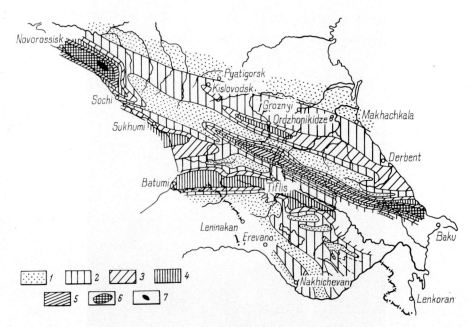

FIG. 209. Thicknesses of Lower Cretaceous strata according to V. E. Khain and L. N. Leont'yev.

1 — land; thickness: 2 — 0–500m; 3 — 500–1000m;
4 — 1000–2000m; 5 — 2000–3000m; 6 — 3000–4000m; 7 — over 4000m.

Calcareous formations continue to the south-east towards the region of Chiaturi and Tkvibuli which are situated to the west of the Dzirul Massif. Here the Valanginian, the Hauterivian and the Barremian (500m) are represented by a calcareous formation, the lower part of which is sandy and oolitic, while the upper major part consists of white medium bedded limestones known as the Urgonian rudistid facies. The Aptian consists of interbedded white marls and limestones (100m) succeeded by marls, sandy clays and muddy Albian limestones of 50–100m in total thickness. All the stages are characterized by marine faunas.

Urgonian Limestones. The limestones occur in the upper part of the Neocomian Calcareous Formation of Barremian age and have a thickness of 300–400m. The Urgonian facies consists of pale, almost white thickly bedded or massive limestones, which are in places dolomitized and are ubiquitously cavernous. In some localities a uniform, but abundant fauna is found and has typically a large number of rudistids (*Requienia, Monopleura*) indicating that the limestones were reef deposits originating at shallow depth and in violent currents. To the east of the Dzirul Massif carbonate facies including the Urgonian Limestones continue into the Lesser Caucasus and further to the south.

On the southern slope of the Greater Caucasus thick flysch deposits appear again. In the upper reaches of the Rion they are 2500m thick and consist of uniform, rhythmic repetitions of clays, marls and fine grained sandy rocks. They have almost no fauna, but a large number of problematic markings typical of flysch. In the region of the Georgian Military Road, to the north of Tiflis the thickness of the Upper Cretaceous flysch is about 2000m. Its lower part (the Valanginian) consists of alternating marls and limestones (250m) while the rest is composed of clays and sands.

In Kakhetia and Azerbaidzhan the Lower Cretaceous flysch is ubiquitously distributed and in the extreme east reaches a thickness of over 400m (Fig. 209). In Azerbaidzhan the Valanginian is unusual since it is composed of carbonate rocks (1500m) including marls, limestones and marly clays. The Hauterivian and Barremian are represented by a dark-grey silty and muddy flysch (1400m) with rare beds of limestones and marls. In the north of the zone there is situated the coarse wildflysch with huge blocks of Upper Jurassic limestones lying within grey Barremian clays. Aptian and Albian strata are much thinner (up to 400m) and show occurrences of variegated flysch sediments. Upper Albian and Cenomanian strata are often closely connected forming a single sedimentary rhythm.

The Lesser Caucasus in Valanginian and Hauterivian times was mainly an archipelago of islands as is shown on Fig. 209. A major part of Armenia also was land, which is indicated by the fact that the Lower Cretaceous is missing from Armenian successions and where at present the succession begins with the Aptian Kopalonosian Formation, which represents continental deposits, or with marine Upper Barremian limestones.

Aptian deposits such as limestones, volcanogenic strata and sandstones are more widely distributed. Upper Albian sandstones and marls (up to 200m) are also widespread and are closely associated with Upper Cretaceous sediments to which they are basal beds.

The full Lower Cretaceous succession is found in the Zangezur region of south-eastern Transcaucasia. Here porphyritic tuffs and tuff breccias (200m) are overlain by the organic thickly bedded Urgonian Limestones (200m) which have Upper Barremian rudistids. The Lower Aptian is represented by marls with beds of sandstones (130m) and the Upper Aptian by tuffogenous, coarse sandstones (200m) with beds of sandy ammonite-bearing limestones. Still higher lie Turonian deposits.

Lower Cretaceous deposits are widespread in the Kopet-Dag where the Neocomian is represented by massive, bedded limestones and the Aptian and Albian by shales, marls and sandstones. The Lower Cretaceous of the Kopet-Dag is the continuation of the Lower Cretaceous of the Northern Caucasus.

Upper Cretaceous. In the central and eastern part of the Northern Caucasus, hard Upper Cretaceous marls and limestones form the third tier, a high ridge of 800–1200m elevation, which stretches parallel to the Main Range. The rocks are pale grey, white and more rarely yellowish and greenish, pure and more rarely sandy limestones. The fauna, of which inoceramids are particularly characteristic, indicates that (the Danian, which is often absent, apart) all Cretaceous stages are present. The thickness of the calcareous formation is 250–500m, but in central Dagestan it is 1400m thick, which is accompanied by a considerable increase in the sandy fraction.

In the north the zone of carbonate rocks was marginal to land. It was also marginal to the southern Caucasian island, but to the west and east it soon passed into the zone of sedimentation of much thicker almost barren flysch deposits. In the Maikop region the marl-limestone facies of Upper Cretaceous age pass into carbonate flysch deposits known as limestone flysch, some suites of which are used as high quality raw material forcement, which is used by the Novorossiisk cement industry. The thickness of the flysch increases westwards and in the Novorossiisk region reaches 2000–3000m. The flysch is divided into several suites, differentiated on the basis of lithology and colour, thus—dark flysch on pale flysch. In these deposits the macrofauna is very rare and is represented by fragments of thick-walled inoceramids. The Upper Cretaceous flysch is found along the Black Sea shore and continues towards the southern slope of the Caucasus.

On the southern slope, at the shore of the Black Sea the lower part of the Cenomanian is composed of tuffogenous rocks and marls (100m) and its upper part —the Ananurian Horizon—by a suite of black and brownish thinly bedded siliceous sediments (phtanites with diatoms and radiolarians) of 40m in thickness. Still higher lies a thick formation of rhythmically bedded marls and limestones. Then over a distance of 200km there are no outcrops of Upper Cretaceous rocks and the next belt of Upper Cretaceous flysch begins in Southern Osetia, passes through the region of the Georgian Military Road and ends in the Kakhetian range. Again the Cenomanian begins with tuffogenous strata, overlain by black phtanites of the Ananurian Horizon and then by thick marl-limestone flysch deposits. The total thickness of Upper Cretaceous sediments is 1500–2000m. After another gap in outcrops Upper Cretaceous rocks are found in the Dibrarian (flysch) Depression, where they reach a thickness of 2000m.

In the Lesser Caucasus Upper Cretaceous strata are found ubiquitously, but in relatively small outcrops. Flysch facies are almost absent and instead normal marine sediments with rich faunas of all the stages are developed. The Cenomanian and Turonian consist of sedimentary volcanic rocks and in certain localities, as for instance along the north-eastern slope of the Lesser Caucasus, the development of Coniacian-Santonian volcanic rocks (500–1000m) is characteristic. Higher up there are thick bedded limestones (from 300 to 800m) of white and grey colour and Campanian and Maastrichtian ages. The limestones have a rich and diverse marine fauna.

FAUNA AND FLORA.   The fauna is exceptionally abundant, diverse and basically of southern type, being characteristic of the Alps and the Himalayas. Only in the external zone of the geosyncline, in Crimea and the Northern Caucasus, have occurrences of northern forms typical of the Russian Platform occasionally been recorded. Ammonites and belemnites are especially important. Amongst Berriasian ammonites the genus *Berriasella* is most typical and *Phylloceras*, *Lytoceras* and *Crioceras* also occur. Various forms of Hoplitidae are typical of the Albian and the Aptian. In the Upper Cretaceous the genera *Acantoceras* and *Pachydiscus* are stratigraphically most important. Various lamellibranchs were at the acme of their development during the Cretaceous period. In the Urgonian Limestones rudistids (*Requiemia* and *Monopleura*) were characteristic while in Upper Jurassic lime-stones of Transcaucasia *Radiolites* and associated genera are typical. Oysters and pectens are common, but inoceramids are the leading forms and are particularly common in marls, chalk and bedded limestones. Of individual forms *Aucellina gryphaeoides* Sav., characteristic of the Upper Albian, should be noted, since it determines the boundary between the Lower and the Upper Cretaceous and hails the commencement of the Upper Cretaceous sedimentary rhythm.

Of other groups Foraminifera are progressively acquiring a great biostratigraphic importance. Soviet micropalaeontologists concerned with their study are foremost in the world and have used them extensively in subdividing flysch deposits.

Terrestrial flora is rare, but calcareous algae are common and often rock-forming.

PALAEOGEOGRAPHY. Palaeogeography is variable in general and locally. Numerous local and regional uplifts and depressions caused rapid and short-lived changes, at times involving the formation of islands and peninsulas—new land areas—and at others producing deep-water depressions. The situation was even more complicated by volcanic eruptions in the south, giving rise to new islands. Barrier reefs and atolls originated as a result of biogenic activity of organisms, such as calcareous algae, corals and hydroids. Despite the variability the main geographic provinces remained relatively unchanged. Along the northern foothills of the Crimean ranges and the Main Caucasian range there was a region of accumulation of carbonate rocks. In the Lower Cretaceous this region often represented a vast barrier reef, analogous to the Great Reef of Australia. In the Upper Cretaceous epoch this region was a relatively deep sea. The sea was situated within the outer zone of the geosyncline, while in the Lower Cretaceous epoch much of this zone was land.

The watershed part of the range was an elongated and narrow, relatively low island similar to Sakhalin. The palaeogeography of the flysch downwarps is still not clear. The downwarps are usually considered as regions of deep sea, but it is more likely that they were wide valleys open to the sea and bounded by rising mountains. The bottoms of the valleys were progressively depressed, allowing for the accumula-tion of vast thicknesses of rocks and the development of volcanic, coral and ordinary islands separated by deeper or shallower straits.

CAINOZOIC

Palaeogene

Palaeogene strata constitute the second part of a major sedimentary macrorhythm which commenced at the beginning of Upper Cretaceous times. Consequently the areas of distribution of both groups of strata parallel each other and they have similar lithological and facial compositions, similar tectonics and palaeogeographies.

Palaeogene strata of the Caucasus, Crimea and the Carpathians are of considerable economic importance since they contain several major deposits of petroleum.

DISTRIBUTION. Palaeogene rocks are more widely distributed than the Upper Cretaceous. In the inner zone the Palaeogene is hidden under the Neogene and there are relatively few outcrops of the former. In the Soviet Carpathians Palaeogene sediments are more widespread than those of any other system. In Crimea Palaeogene strata occur in a narrow belt along the northern foothills, passing through the Kerch and Timan peninsulas and continuing on to the northern slope of the Caucasus as far as Baku, where the belt of Palaeogene strata crosses over to the southern slope. Further along the southern slope outcrops of Palaeogene rocks are sporadic and are well separated from each other. In Transcaucasia huge areas have lavas and sedimentary rocks of Palaeogene age and are more widespread than those of any other system.

SUBDIVISIONS. The Palaeogene normally is divided into the Palaeocene, the Eocene and the Oligocene. Only the position of the Danian is debatable, since lately a number of investigations have included it into the Palaeocene. As is shown in Table 47 illustrating the main successions, division into stages and suites is very complex and often has only a local character. The Maikop is the only exceptional stage, since it is found all the way from Bulgaria to Baku, and is a most important oil-bearing horizon. Sometimes Western European stages are used but not always and not everywhere.

LITHOLOGY AND FACIES. In the outer zone of the Mediterranean Geosyncline Palaeogene deposits have a platform character: they are weakly folded, almost unmetamorphosed and are thin, being of the order of a few hundred metres. In the west, in Ciscarpathia and Moldavia, as near the Black Sea, the deposits consist of marls and clays which are sometimes glauconitic. At the base and the top of the succession there are sands of Palaeocene and Upper Oligocene ages respectively (Table 47). The Ukrainian stages are used and include the Buchakian, the Kanevian, the Kievian, the Kharkovian and the Poltavian (see Table 15). The thickness is 300–500m.

In Ciscaucasia the southern part has foraminiferal clays overlain by the Maikop. The total thickness in the sub-montaine downwarp reaches 1600m. To the north the succession is of Ukrainian type and the thickness is less.

In the Carpathians the Palaeogene is represented by flysch deposits which have a great total thickness (4000–5000m), and are divided into several local suites (Table 47). Of these suites the lower (Yamnenian) which is oil-bearing, has the greatest economic importance. The suite consists of sandstones, which pass laterally into a rhythmically bedded formation of clays and sands (Fig. 210). Nummulites found in it determine its age as Palaeocene. The suite is overlain by sands and thinly bedded clays and sands of flysch type, again laterally transitional into each other (Fig. 210). These sediments are considered to be of an Eocene age and are sometimes combined under the term of 'Carpathian Series', but are most commonly divided: the Manyavian, the Vygodovian and the Bystritsian suites. A facies variant of the lower Bystritsian Suite is sometimes separated as the Popelian Suite, while the Manyavian and the Vygodovian suites are often united as the Vitvitskian Suite. Upper Eocene sandstones are in places oil-bearing and the parent oil-bearing suite is the bituminous Menilitovian, which consists of repetitions of brown, black and grey-greenish shales with fish remains and thin layers of fine grained, pale sandstones. At the bottom of the suite there are beds of siliceous rocks, shales and marls. The total thickness of the Menilitovian Suite is up to 1700–1900m and its age is Lower and Middle Oligocene, thus corresponding to the Lower Maikop of the Caucasus. The overlying Krosnian Suite is analogous to the Middle Maikop. In the north the upper part of the Krosnian Suite is known as the Polyanitsian Suite.

The flysch formation is uniform, but changes rapidly along the strike. Moreover it has not been well defined palaeontologically and as a result there are many variations of its subdivisions. The scheme shown in Table 47 is one of the most frequently employed but there are many others. Sometimes suites with identical names have different ages. For instance the Krosnian Suite or series is considered sometimes of Upper Oligocene age, sometimes of Lower and Upper Oligocene age and sometimes of Upper Oligocene and Lower Miocene age. In places it is equivalent to the Polyanitsian Suite, in places it embraces the equivalents of the Upper Menilitovian Suite as well, while in yet other places the whole of the latter is included. The latest reviews by O. S. Vyalov (1961) and V. G. Korneyeva (1959) differ considerably in their conclusions.

In Crimea carbonate rocks predominate in the west, while to the east they pass into muddy flysch deposits. Near Bakhchisarai the lower part of the succession (Table 47) consists of limestones and chalky marls (255m), which are succeeded by greenish and brown argillites (180m) with fish (*Lyrolepis caucasica*) and brown barren clays (500m). Amongst the limestones nummulitic varieties (35m) of Middle

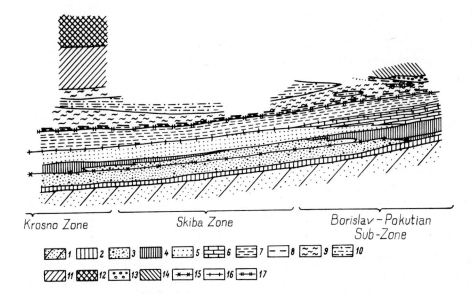

Krosno Zone   Skiba Zone   Borislav – Pokutian Sub-Zone

FIG. 210. Palaeogene flysch in the river Cheremosh basin of the Soviet Carpathians. According to Ya. O. Kul'chitskii (1957).

1 — Stryiian sediments; 2 — Yaremchanian Horizon; 3 — Yamnian sandstones; 4 — sandy-clayey flysch of Manyavian type; 5 — Vygodovian sandstones; 6 — siltstones, sandstones and marls; 7 — sandy-clayey flysch of Bystritsian type; 8 — hornfelsic horizon; 9 — Menilitovian shales; 10 — sandstones of Manyavian type; 11 — transitional beds; 12 — Krosnian deposits; 13 — conglomerates of the Polyanitsian Series; 14 — sandstones and argillites of the Polyanitsian Series; 15 — Eocene-Palaeocene boundary; 16 — Middle Eocene-Upper Eocene boundary; 17 — Oligocene-Eocene boundary.

TABLE 47.

Palaeogene Successions of the Mediterranean Geosyncline

Division			The Carpathians	Regions adjacent to the Black Sea	Western Crimea	The Kerch Peninsula
Lower Miocene			Strebnikian Suite (500–1300m). Vorotyshchian Suite (1000m).	Missing.	Missing.	Upper Maikopian—brown clays.
Oligocene	Upper		Polyanitsian Suite of grey shales and sandstones (500–800m). In the south, Krosnian Suite (500–1500m).	Greenish clays and gaizes with a marine fauna (up to 170m).		Middle Maikopian. Upper Kerleutian Beds—brown sandy clays with beds of sands. Lower Kerleutian Beds—brown and grey clays.
	Middle		Menilitiovian Suite. Upper Menilitovian Beds of black, bituminous shales with bands of hornfels (menilite) and sandstone. Fish is found (200–1000m).		Olive-green clays with a deep-water fauna. At the bottom, in places sandy and have beds of sandstone. Total thickness—up to 500m.	Lower Maikopian—brown and grey shales with ostracods and *Planorbella*.
	Lower		Lopyankian Beds—shales, sandstones, marls and dolomites (150–300m). Lower Menilitovian Beds of black and brown, sometimes bituminous shales, sandstones and menilites. Fish is found (250–400m).			Dyurmenian Horizon of grey shales with beds of sandstones. The thickness of all the Maikop is 1500–2000m.
Eocene	Upper		Bystritsian Suite—finely rhythmic flysch consisting of carbonaceous shales and fine grained sandstones (200–300m).	Brown and greenish marls, clays and sandstones.	Greenish and brown clays with fish (180m). Chalk-like, flaggy limestones and marls (100m).	Greenish, plastic clays with beds of fish-bearing marl (150–200m).
	Middle		Popelian Suite of sandstones and limestones with nummulites (40–150m).	Marls and limestones with nummulites.	White limestones with crabs (40m). Nummulitic limestones (35m).	Greenish calcareous clays (80m).
	Lower		Vygodovian Suite of massive sandstones with nummulites (up to 400m). Manyavian Suite of finely rhythmic flysch shales and sandstones.	Marls and limestones with nummulites. The thickness of Middle and Lower Eocene strata is 140m.	Greenish or brown clays with beds of nummulitic limestones. Oysters are found (40m).	Black clays with Foraminifera.
Palaeocene			Yamnian Suite of massive sandstones with beds of shales, with nummulites (50–300m).	White and glauconitic marls (12m).	Glauconitic, bluish marls (23m). White saccharoidal limestone (Montian Stage) (15m).	Grey marls, with beds of sandy marls (70–160m).

TABLE 47—continued

Stavropolia, River Kuban'	The Groznyi region	The Apsheron Peninsula	Western Georgia, Abkhazia and the Sukhumi region	Southern Armenia
Upper Maikopian. Ritsian Suite (300m). Ol'ginskian Suite (170m).	Upper Maikopian. Zuramakentian Horizon (150–200m).	Upper Maikopian. Zuramakentian Suite.	Upper Maikopian. Ketsakhurian Horizon (170m). Sakaraulian Horizon (230m).	Variegated, salt-bearing strata of Lower Miocene-Upper Oligocene age (750m).
Middle Maikopian. Karadzhalgian Suite of clays (80m). Zelenchukian Suite of sands (25m). The total thickness of Middle Maikopian strata—600m.	Middle Maikopian. Riki Suite (Septarian) of foliaceous shales with concretions (250–450m).	Middle Maikonian. Riki Suite (Septarian) of bedded clays with concretions towards the bottom. Bands of sandstone are present (200–250m).	Middle Maikopian—non-calcareous grey, foliaceous shales (65m).	
Lower Maikopian. Batalpashinskian Suite of foliaceous clays with lenses of dolomites and fish (100m).	Lower Maikopian. Miatlian and Mutsidakalian suites of grey, bedded clays with bands of sandstone (400–1100m).	Lower Maikopian—brown and variegated foliaceous clays and marls with fish and bituminous shales. The total thickness of the Maikop is up to 1000–1200m.	Lower Maikopian. Grey, foliaceous clays with fish scales and plant remains (50–60m). Khadumian Horizon of grey clays with pteropods (Planorbella) and ostracods. At the bottom there are sandstones (50m).	Shorakbyurgian (Shorbulakian) Suite; above, clays with lenses of coral-line limestone; below, tuffaceous sand-stones and clays with a marine fauna (Planorbella) (900m).
Khadumian Horizon of black and brown clays with beds of marls and sandstones. Planorbella is found (50–400m).	Khadumian Horizon of black and brown clays (100–300m).			
Byeloglinian and Kumian suites of clays and marls with Foraminifera and fish (40–70m).	Byeloglinian and Kumian suites of clays and marls with Foraminifera (70–90m).	Upper Kounian flysch—white silts and clays in rhythmic bedding (400–600m). Middle Kounian; brown, shaley, barren clays (200–400m).	Foraminiferal marls with Globigerina and Bolivina 15–50m). Brown foliaceous, fish-bearing marls with fish-scales (30m).	Rykusskian Suite of clays and sandstones with Foraminifera (600m).
Khadyzhenskian Suite of white and green marls with foraminiferids (120–160m). Kaluzhskian Suite of greenish marls) sometimes bituminous) with a marine fauna (200–400m). Kutaisskian Suite of rapidly repeated clays, marls and sandstones (150–180m).	Zelenaya (green) Suite of marls with nummulites. Pestrotsvetnaya (variegated) Suite of variegated clays with Globorotalia.	Lower Kounian; green-ish laminated clays, marls, siliceous argillites, tuffs and bentonites. The fauna includes Fora-minifera (Globorotalia) and fish) (100m).	Foraminiferal marls and clays with num-mulites. Pale, greenish fora-miniferal marls with Globorotalia (40m).	Nummulitic Suite. Flysch-rhythmic-ally repeated shales and fine-grained sandstones. Fora-minifera (Globoro-talia) are found (1500m).
Abazinkian Suite of marls and gaizes (35m). Goryachii Klyuch (Hot spring) Suite of sands and sandstones with beds of clay (30–50m). El'burganian Suite of compacted marls with a marine fauna (25m).		Sumgaitian Suite of bright-red, brown and dark-grey, towards the bottom, green clays and marls, with Foraminifera (Globorotalia) (100–200m).	Limestones with a marine fauna—white litothamnian and green glauco-nitic. Echinoids (Echinocorys) and nummulites are present (up to 30–60m).	

Eocene age are especially characteristic. The limestones consist entirely of nummulitic shells and represent a high quality building material from which the city of
Simferopol is built. Near Theodosia the whole succession consists of clayey and
sandy flysch deposits. The lower part (Eocene) of the succession contains a layer of
nummulitic limestones, while the upper part (Oligocene) is 1800m thick and is identical to the Maikop Suite of the Caucasus.

The Northern Caucasus show a great uniformity of the succession. Over a long
distance from Novorossiisk to Baku the Palaeogene is divided into two divisions.
The lower consists of the Palaeocene and Eocene and is represented by a thick
series of uniform clays formerly referred to as Foraminiferal Clays, but now divided
into a large number of smaller units, collectively known as the Foraminiferal Series
(Table 47). The thickness of the series generally does not exceed 500m and only in
the Novorossiisk region does it reach 1000m. The upper division (Oligocene) is represented by the Maikop Series or as it was formerly known—the Maikop Suite. Its
lower part is sometimes differentiated into the Khadumian Horizon.

The thickness of the Maikop Series increase away from the Caucasus. Thus in the
region of Grognyi it is 1000–2000m. The series is an important oil-bearing reservoir
and its lithology is described on p. 621.

In the southern Caucasus, in the west and the east deposition of mainly sandy-
clayey flysch continued in flysch downwarps. These sediments are of Lower Palaeogene age, do not exceed 1500–1700m in thickness and are described below. The
Upper Palaeogene is represented by the Maikop Series.

In the Lesser Caucasus the lithological composition of Palaeogene strata is as
diverse and variable as that of Upper Cretaceous sediments. The widespread development of thick effusive formations is particularly characteristic and as a result the
thickness of the succession reaches 4000–5000m. Intensive uplift of the Lesser
Caucasus in places caused the accumulation of great thicknesses of up to 3500m of
sands and clays with beds of conglomerates. Eocene nummulitic limestones are
widely distributed and in the south of Armenia upward movements gave rise to
Upper Oligocene and Lower Miocene saline rocks.

Flysch. Flysch deposits were first recognized in the Alps. The term was given by
Swiss stone-masons and means 'smooth stone' since the rock has smooth bedding
planes. The flysch implies a thick formation, often reaching several thousands of
metres, of uniform rhythmically repeating almost barren fine-clastic rocks. Depending on the original composition several varieties are distinguished. The calcareous
flysch consists of rhythmic repetitions of pelitic marls and limestones with subordinate amounts of terrigenous rocks. The sandy-clayey flysch basically consists of
repetitions of clays, silts and fine grained sandstones. The wild flysch has irregular
accumulations and lenses of coarse sands, gravel, breccias and even large blocks of
older, but local rocks in essentially fine grained matrix. Flysch deposits often show
group and individual suites of siliceous rocks, including spongolites and radiolarites and also bituminous formations. According to its colour the flysch is referred to
as white, brown and black. Its most diagnostic character is the continuous repetition
of individual rhythm varying in thickness from a few to tens of centimetres and even
a few metres. The structure of the rhythms has been discussed by N. B. Vassoyevich.
Flysch macrofauna and flora are very rare, while microfaunas, especially Foraminifera, are much more common.

The work of Soviet geologists on flysch and in particular of N. B. Vassoyevich
(1948, 1951) is of first class importance. Alpine flysch is essentially of Palaeogene

age, but Soviet geologists have demonstrated its development in Upper Triassic and Upper and Lower Palaeozoic times, thus in all mountain chains where folding occurred and associated deposits were produced. The nature of the transport of these sediments and the loci of their deposition provoke arguments. It is generally accepted, although not proved, that flysch deposits accumulated in deep oceanic depressions adjacent to mountain chains. The nature of transport is, however, not explained and not discussed. Rivers must have been main agencies of transport since otherwise it is difficult to envisage the accumulation of many kilometres of clastic sediments. There are no rivers in deep seas and consequently it is natural to consider flysch deposits as products of land deposition, accumulating on a near-shore alluvial plain at the foot of a high range. Wildflysch thus represents screes and landslides on the surface of such a plain. The flysch is closely associated with such complexes of facies as coal measures, oil-bearing formations, reefs and saline formations, none of which are deep water deposits. The final solution of the problem of flysch requires further investigations.

The Maikop Series. This is a depositional complex which is closely associated with flysch and is also of a disputed origin. Some consider the Maikop Series, like flysch, to be of a deep-water origin, while others think that it consists of terrestrial deposits of near-shore alluvial plains situated between rising mountain chains and sea. This is a point of view accepted by the author. The Maikop Series belongs to the Middle and Upper Oligocene and Lower Miocene and is divided into three parts. The Lower Maikop in the Maikop region consists of dark and pale grey clays with ostracods. The Middle Maikop is composed of dark-grey clays, some with fish remains while others are barren, non-calcareous, but with thick beds of oil-bearing quartz sands. The Upper Maikop is again composed of barren, non-calcareous, dark-grey clays. The successions of other regions show similar types of the Maikop. Its thickness varies from 100m to 1400–1700m and more. Its stratification is regular, but cross-bedding is sometimes observed and there is also evidence of wave action and desiccation marks. Organic remains are rare and consist mainly of fish and micro-organisms. In some very rare intercalations marine faunas are encountered.

Nummulitic limestones—represent a peculiar southern facies associated with the zone of wave action in Eocene seas. The limestones are a formation of grey, bedded or nearly massive rocks up to 30–40m thick and consisting almost entirely of num-mulite shells. The facies represents the deposits of a tropical beach completely covered by the shells brought in by the tidal surge.

Fauna and Flora. The fauna (including the Danian) differs sharply from the Upper Cretaceous by complete absence of ammonites, belemnites, inoceramids and rudistids. Other groups continue from the Upper Cretaceous. The flora has an upper Cretaceous aspect and consists of modern tropical and southern groups.

Palaeogeography. The palaeogeography is bascially the same as in the Upper Cretaceous epoch. The Caucasian island grows at the expense of the surrounding alluvial plains. The Transcaucasian archipelago also increased in terms of the number of constituent islands, many of which at the end of the Oligocene merged into surrounding alluvial plains and bitter-salt lakes and swamps originated. Volcanic activity reached a great intensity.

Neogene

The Neogene is the first part of a macrorhythm of sedimentation, which continued in the Anthropogene and even the Recent. During the Neogene Alpine orogenic movements reached a great intensity and the relief gradually acquired its present configuration. In the north of the region seas receded. A series of large oil and gas deposits are found in the Neogene.

DISTRIBUTION. The distribution is peculiar. In the north of the Mediterranean Geosyncline (the outer zone) Neogene rocks lie almost horizontally and have not been involved in folding. The Neogene, together with Anthropogene strata, form a very widespread but thin layer, which obscures the weakly folded rocks of Palaeogene and Upper Cretaceous age and older more ancient deposits.

Nearer to the Alpine fold structures Neogene sediments are involved in relatively gentle and simple brachyanticlines of, for instance, the Kerch and the Apsheron peninsulas. The zone of folded Neogene beds is narrow, but widens out in the Kurian Depression. In the Rion Depression and the south of Transcaucasia Neogene sediments are rarely found and form small outcrops.

SUBDIVISIONS. The Mediterranean Geosyncline is the region of the most complete, classical Neogene sequence of marine, brackish-water and continental sediments. The stratigraphy of this sequence is elucidated as a result of the work of thousands of geologists of tens of nationalities beginning with those from Indonesia, and Vietnam and ending with Moroccans and Portuguese. Many Soviet geologists of different nationalities contributed to this international effort both in pre- and post-Revolutionary periods. The work of outstanding scientists such as N. A. Sokolov, A. P. Pavlov, N. I. Andrusov and later of V. P. Kolyesnikov, A. A. Ali-zade, K. A. Ali-zade, A. G. Eberzin and B. P. Zhizhchenko was an important contribution to general investigations. The above-mentioned scientists have shown that Neogene strata of the U.S.S.R. have several sometimes universal features found on the territories of adjacent countries. Consequently stratigraphic subdivisions of Neogene rocks in the U.S.S.R. have rightly a number of local names, but such subdivisions are correlated with international subdivisions and in the first place with those employed in adjacent countries. The maintenance of such interrelationship is imperative and raises the quality of industrial and commercial conclusions, as for instance with regard to oil and gas. Thus the Sarmatian Stage everywhere corresponds to the Upper Miocene while the Middle Miocene is divided into the Konkian, the Karaganian, the Chokrakian and Tarkhanian horizons (beds). The Lower Miocene of the Carpathians and the Dnyestr region is divided into the Helvetian (above) and Burdigalian (below) stages. Continental Upper Oligocene and Lower Miocene sediments are often united under the term Aquitanian. In the Caucasus such a nomenclature is used only for marine Lower Miocene. Thus the Sakaraulian Horizon is at times equated with the Burdigalian and the Kotsakhurian Horizon with the Helvetian. The difficulty of equating Soviet with western European subdivisions is illustrated by the fact that the Helvetian Stage is sometimes considered as of Lower Miocene and sometimes as of Upper Miocene age. The first alternative is more widely accepted.

In Transcarpathia and Ciscarpathia the Neogene consists of flysch facies, which are thick, uniform and palaeontologically badly defined. As a result there are many schematic subdivisions. The schemes accepted in 1958 are shown in Table 48. Subsequently L. S. Pishvanova has carried out a major investigation of Foraminifera, as a result of which the previous scheme worked out by I. B. Pleshakov had to be modified. According to the new scheme (Vyalov et al., 1961) the Lower Sarmatian equivalents of Transcarpathia are divided into two suites—the Lukivian (upper) and the Dorobrativian—which in Ciscarpathia form a single Dashavian Suite. The

Upper Tortonian of Transcarpathia is divided into seven suites which from top downwards are: the Darolinian, the Peresnitsian, the Tachivian, the Shandrian, the Nankivian Tuffogenic, the Solotvian and the Tereblian. Formerly the Tereblian and the Solotvian suites were regarded as of Lower Miocene age, while now they are suggested to be at the top of the Middle Miocene. This is an important change. In the Lower Tortonian of Transcarpathia are included the Taraborian and the Novoselitsian suites and in Ciscaucasia the Bogorodchanian Suite. To the Lower Miocene (Burdigalian-Helvetian) of Transcaucasia belongs the Buralivian Suite and in Ciscarpathia the top part of the Lower Molasse.

The boundaries of the Neogene have provoked a lively discussion. Quite recently the Maikop was included in the Oligocene, while the Tarakhanian Horizon was considered to mark the base to the Miocene. The analysis of the microfauna has shown that the Upper Maikop is of a Lower Miocene age and the Tarakhanian Horizon is now thought to be the base of the Middle Miocene (Table 48). The Miocene-Pliocene boundary is normally drawn above the Meotic Stage. B. P. Zhizhchenko includes this stage in the Pliocene on the basis of the fact that faunally there are no common Sarmatian and Meotian forms, while several Meotian forms continue in the Pontian and become the prototypes of Pontian forms. This conclusion probably deserves serious attention, but as yet has not been confirmed by an authoritative meeting and consequently here the Meotian is included in the Miocene.

The Plio-Pleistocene boundary in the twenties used to be drawn above the Baku Stage. This was done on the basis of existing stratigraphic relationships. In the thirties on the basis of closely associated faunas the Baku Stage was included in the Lower Pleistocene. In the fifties it was proved that marine Apsheronian deposits pass into river terraces, which pass into glacial deposits. The Baku and the Apsheronian faunas are closely related and consequently on the 1955 synoptic map of the U.S.S.R. the Apsheronian was included into the Lower Anthropogene. Several investigators think that on the same grounds the Akchagylian also must be considered as a part of the Anthropogene. In 1956, however, at the All Union conference of specialists on the Palaeogene and Neogene it was decided to draw the boundary above Apsheronian, as is being done by oil geologists. This recommendation is accepted in the present text and should be used in practice.

Differences of opinion are also expressed on the subject of other boundaries within the Neogene. Such differences are unavoidable since the majority of subdivisions have a really local character and undergo rapid variations in facies which as yet remain largely uncorrelated. The main Pliocene sections are shown in Table 49.

LITHOLOGY AND FACIES. The most important lithological feature of Neogene sediments is a complete absence of thick carbonate rocks. The thickest reef-limestones (Podolian Toltras) are no more than a few tens of metres thick. There are also some bedded limestones, which are no thicker and amongst which shell limestones are widely developed.

Another equally important lithological feature is a widespread development of continental and saline sediments.

Marine sediments of normal salinity have a subordinate character and are encountered in the Lower and Middle Miocene and in the Anthropogene epoch (Tyrrhenian terrace). This is explained by the development of the Mediterranean macroisthmus (Fig. 213), which separated Upper Miocene and Pliocene seas from the open ocean, causing their subsequent desiccation accompanied by repeated elevation and erosion of mountain massifs. The latter condition led to the develop-

TABLE 48.

Miocene Successions in the Mediterranean Geosyncline

Stage horizon		Transcarpathia	Ciscarpathia	Black Sea region	The Kerch Peninsula	Stavropol Plateau
Upper Miocene	Lower Pontian	Lower part of the Gutinskian Suite.	Missing.	Yellow clays with beds of sands; shells are found (up to 100m).	Odessa Limestone (6–10m).	Clays and sands with a marine fauna.
	Meotic	Isian Suite of tuffs, tuffogenous sandstones and clays with a brackish fauna (200m)		Continental, variegated clays with beds of sands and gravels towards the base	Greenish clays and shell beds. In places small reef massifs with the Kerch Limestone at the bottom (80m).	Continental clays and sands. In the north, clays and marls with a marine fauna (50–160m).
	Sarmatian	Lipshinian Suite of clays, tuffs with brown coals and marine bands (400m). Vyshkovian Suite of clays with sequences of sandstones and tuffs. It has a marine fauna (700m).	Upper and Middle Sarmatian strata are missing; the Lower Sarmatian is represented by grey clays and sandstones with a marine fauna (300m).	Upper Sarmatian—greenish clays and sandstones (up to 150–180m). Middle Sarmatian—clays and limestones (up to 300–350m). Lower Sarmatian—limestones (40m).	Upper Sarmatian—bryozoan reefs, pale clays and diatomites (up to 200m). Middle Sarmatian—limestones and clays (up to 80m). Lower Sarmatian—clays (200–340m).	Marine clays and sands with a rich fauna. The three sub-stages are present (100–600m).
Middle Miocene	Konkian Horizon	Apshinskian Suite of a molasse type, with conglomerates, sandstones and clays. There are bands with marine faunas. Four sub-suites are recognized (2500m).	Pokutian (Galitsian) Suite of greenish clays with sequences of conglomerates and sandstones. At the bottom it is saliferous and gypsiferous (Tirasian Suite). Beds with marine and fresh-water faunas are present (500–1000m).	Tortonian grey limestones, passing down into sands and clays with brown coals; ultimately, basal conglomerates (10–50m).	Greenish clays, towards the top with a marine fauna and shell beds (100–140m).	Dark, deep-water clays with a marine fauna (15–50m). To the north and east sands appear, reaching 70–80m in thickness.
	Karaganian Horizon				Dark clays with *Spaniodontella* (up to 100m).	Clays, marls and sands with lenses of bryozoan limestones. *Spaniodontella* is present (40–310m).
	Chokrakian Horizon	Teresvian Suite of repeated conglomerates, sandstones and clays (molasse type). Beds with a marine fauna are found. Three sub-suites are distinguished (1500–2000m).	Ugerskian (Chaplian, Balichskian) Suite of sandstones, sands, clays and marls. At the bottom is the saline, Kalushian Horizon (200–1500m).	Missing.	Dark, deep-water clays with *Spirialis* and other marine fossils (up to 80m).	Pale and dark clays, marls, sands, shell-beds. Marine fauna with *Spirialis*. Variable thickness (0–700m).
	Tarkhanian Horizon				Dark clays and marls with a marine fauna (10–20m).	Developed only in the south. Dark-grey and brown (buff) clays and marls with a marine fauna (7–25m).

TABLE 48—continued

North shore of the Caspian, Chernyi Rynok	Shore of the Caspian, Central Yalam	The Apsheron Peninsula	Gori to the west of Tiflis stratigraphic (reference) borehole	Mokvi, Abkhazia	Erevan (Avan)
Missing.	Missing.	Brackish-water deposits.	Anthropogenic deposits (40m). Interval.	Marine deposits (up to 450m).	Basalts (61m).
		Meotis (Byurguian Horizon). Ashy-brown diatomitic, in places bituminous clays, marls and volcanic ash (up to 500m).	Dushetian Suite of molasse deposits—conglomerates with beds of sandstones and clays (655m).	Grey and dark grey clays with beds of sandstones. Ostracods are present (400m).	Zangian Suite of sandy-clayey deposits with shell beds. Fish and *Mactra* are present. Thickness in the Avan borehole is 15m, while in the river Rozdan gorge it is up to 800–1000m.

Grey and dark-grey clays with beds of sandstones. In the middle a sequence of marls. A fauna is present (290m).	Upper Sarmatian is missing.
	   Middle and Lower Sarmatian—grey, sandy clays with a marine fauna (250m).
		Sarmat (Akhudagian Horizon)—grey and white sandy clays with beds of fish-bearing marls (marine fauna) (90–500m).	Natskhorian Suite of molasse—bluish clays and sandstones (554m).
			   Middle and Lower Sarmatian—marine sandstones and clays (486m).
				Upper Sarmatian—an interval.
				   Middle and Lower Sarmatian—dark clays with a marine fauna (560m).
					Gypsiferous; towards the bottom, bituminous clays (58m).

Missing.	Missing.	Konkian (Baigushkaian) Horizon—ashy-grey foliaceous, diatomitic and fish-bearing shales with beds of volcanic ash (35–140m).	Grey, compacted sandstones with a marine fauna (57m).	Dark clays with beds of sandstones and a marine fauna (210m).	Upper Saliferous Formation of rock-salt with thin beds of clays, which towards the bottom are bituminous (365m).
Dark clays, interbedded with sands. A marine Karaganian fauna is present (190m).	Black clays with *Spaniodontella* and fish. Sequences of beds of sandstones are present (265m).	Karagan (Chikilchaiian Horizon)—brown clays with beds of dolomites and marls. Fish, ostracods and diatoms are present (40–100m).	Green and brown clays with beds of sandstones and a marine fauna (58m).	—	Plateau basaltic intrusion (200m).

Missing.	Dark clays with *Spirialis*. In the middle a sequence of sandstones (340m).	Chokrak dark clays with beds of dolomites and a marine *Spirialis* fauna.
		   Tarkhan (Kasumkentian Horizon)—dark and brown clays with beds of marl.
		   Chokrak and Tarkhan are sometimes combined as the Siyakian Horizon of a total thickness of 50–500m.
			Variegated marine clays and sandstones (54m).	Grey clays, sandstones and marls, with a basal conglomerate (175m).	Lower Saliferous Suite. Grey and brown saliferous clays with salt horizons (410m).

	Not recognized.		Not recognized.	Interval.	

ment of thick formations, coarse clastics including conglomerates and sandstones and representing typical molasse deposits at the periphery of mountain ranges.

In Transcarpathia and Transcaucasia deep faults of large extent gave rise to widespread vulcanicity. Here thick formations of lavas, tuffs and tuff breccias are interbedded with sedimentary deposits, complicating the facies complexes. Five basic facies complexes have been denoted. The first consists of deposits of normal seas connected to the oceans. Their sediments are only observed in the Lower and Middle Miocene and have a limited distribution. The second complex consists of deposits of enclosed more freshwater seas such as the Sarmatian and is more widespread. The sediments of alluvial plains at margins of internal seas are also quite extensive and constitute the third complex. Molasse sediments originating at the edge of mountains form the fourth complex of a more limited distribution. The fifth complex of effusive and mixed sedimentary volcanic deposits is found almost exclusively in the south.

The Soviet Carpathians involve all four main tectonostratigraphic zones: the margin of the platform, the sub-montaine downwarp, the main anticlinorium and the southern downwarp. The margin of the platform has an almost complete Miocene succession represented by muddy and carbonate marine deposits of small thickness. The Pliocene consists of continental sediments. The marginal downwarp alters its position and shape making abrupt incursions towards the north-east, thus widening and shallowing. By Pliocene times it was completely filled. In the central, axial part of the Carpathians the thickness of the sediments of the downwarp is enormous, Miocene deposits alone being 4000–5000m thick. Pliocene sediments, however, even here are continental and thin. The succession includes two saline and one coal bearing group of strata (Table 48). In the main anticlinorium Neogene sediments are not widely developed, but in the southern Alfoldian (in the west) and Transylvanian (in the east) downwarps they are again well represented and reach a great thickness of 5000–5500m. The Miocene consists of repetitions of marine, brackish, coal-bearing and saline deposits, which are in places replaced by thick volcanic complexes of true volcanoes. The Pliocene (up to 600m) is represented by freshwater and brackish sediments and volcanic formations (Table 49). The thickness of the Sarmatian alone is 1500m.

The Crimean and Cis-Black Sea Plain shows the repetition of Carpathian zones, except that the southern downwarp and one half of the main anticlinorium are

TABLE 48—continued

Stage horizon		Transcarpathia	Ciscarpathia	Black Sea region	The Kerch Peninsula	Stavropol Plateau
Lower Miocene		Tissenian Suite of clays and sand-stones with beds of conglomerate and tuffs. Seams of salt exist. A marine fauna is present (>2500m). Tereblinskian Suite—rock-salt with beds of clays (>250m).	Stebnikian Suite of variegated marly clays with beds of sands and tuffs. Foraminifera are present (500–1 300m). Vorotyshchian Suite of clays and sandstones with rock-salt and, at the top, conglomerates (900–2000m).	Missing.	Upper Maikopian—thick, uniform clays with sequences of sands. The fauna is poor (1200–1300m).	Dark, almost black clays of the Ritsian Suite of the Maikop (300m). Grey, greenish clays with a marine Ol'ginskian Suite with arenaceous Foraminifera (170m).

faulted down to the bottom of the Black Sea. The succession of the margin of the platform is the same as in the Carpathians (Table 48). The downwarp succession has been well investigated at its southern margin—northern Crimean foothills and Crimean steppes (Tarkhankutian and Kerch peninsulas). In this part the Lower Miocene is represented by clays and sands of the top of the Maikop Series and the Middle Miocene (400–500m) by clays and sands with intercalations of shell limestones. Sarmatian deposits (500–600m) described by N. I. Andrusov show different facies in the upper division, being bryozoan reefs from near Kerch and alluvial sands with *Hipparion* bones, near Sevastopol. The rich fauna of the latter has been described by A. A. Borisyak. The Meotian (80m) consists of marine sediments, which include the well-known 'Kerch Limestone' and the higher 'Odessa Limestone' of the Pontian Stage, which is the best building stone. The Pliocene is completely represented by Pontian (50m) marine or really brackish sediments, overlain by the Cimmerian Horizon (90m) which includes the well-known Kerch Iron Horizon, followed by the Kuyalnitsian Horizon (35m), the Taman Beds with an Akchagylian fauna and the Krasnokutian and Guriian Beds corresponding to the Apsheronian (Table 49).

Transcaucasia, the Caucasus and the Ciscaucasian Plain include the four Carpathian zones and show their features. The margin of the platform is well to the south of the valley of the Manych. Further to the south is the downwarp lying near the main range and divided into two by the Stavropol uplift; the northern part is wide and gently inclined, while the southern is narrow and steep.

The western part of the downwarp is characterized by the following data: to the south of Krasnovodsk a 2225m borehole did not penetrate through Middle Sarmatian strata, while in another nearby borehole the thickness of Anthropogene and Neogenic sediments is 2260m. This thickness to the north of Krasnodar falls to 1371m, further to the north to 625m and near Rostov to 92m. Over the Stavropol elevation the thickness of the Anthropogene and Neogene is 150–400m. In the eastern part of the downwarp near Kizlyar its thickness reaches 2400 and even 2700m. The greatest thickness is encountered on the Apsheron Peninsula. Here the thickness of the Miocene reaches almost 3000m, of the productive formation up to 3500m, of the Akchagylian and Apsheronian up to 1200m, and of the whole of Neogene up to 7500m.

The Miocene of the Northern Caucasus and Ciscaucasia is well developed and

TABLE 48—continued

North shore of the Caspian, Chernyi Rynok	Shore of the Caspian, Central Yalam	The Apsheron Peninsula	Gori to the west of Tiflis stratigraphic (reference) borehole	Mokvi, Abkhazia	Erevan (Avan)
Grey, compacted non-calcareous clays with sequences of sandstones. The thickness of the whole Maikop is >1430m.	Dark clays with thin beds of sandstone. The thickness of the whole Maikop is 135m.	Upper Maikopian (Zuramakentian Horizon)—dark, non-calcareous clays with siderite concretions towards the bottom (70–150m).	Sakaraulian Beds—pale arkosic sandstones and conglomerates (152m). Interval.	Kotsakhurian Horizon—muddy limestones and sandstones of a brackish water origin (170m). Sakaraulian Horizon—grey, marine sandstones and clays (230m).	Pestrotsvetnaya (variegated) Formation of variegated clays and sandstones with bands of gravels (465m).

TABLE 49.

Pliocene Successions of the Mediterranean Geosyncline

Stage, Sub-stage		Transcarpathia	Cis-Black Sea Plain		The Kerch Peninsula	The Stavropol Plateau	North-eastern shore of the Caspian Sea, Chernyi Rynok
Lower Anthropogene		Continental deposits.	Continental deposits.		Chauda Beds—sands, clays and limestones with a marine fauna.	Continental deposits.	Baku Stage of sands and clays. Total thickness of Anthropogenic deposits is 383m.
Apsheronian		Il'nitsian Suite of clays with beds of sandstones and tuffs; coal seams are also present. The fauna is of an indeterminate age (150m), yet it is Middle or Upper Pliocene and rests unconformably on the Gutinskian Suite.	Upper Poratian Beds—continental sands and clays with *Unio sturi.*	Levantian Deposits	Guriian Beds—sands, gravels and clays with a marine fauna and flora (10–15m).	Continental sands and clays (20–30m). In the east, marine deposits.	Dark clays with beds of sand and a marine fauna (647m).
Akchagylian			Lower Poratian Beds with continental deposits.		Tamanian Beds—sands, clays and shell beds with an Akchagylian fauna (20m).	Only in the east. Sands and clays with a marine fauna.	Grey clays with thin bands of sand. A marine fauna is present (154m).
Kuyalnitskian	Upper		Continental, Budnanskian Beds—cherry-red clays and loams, resting unconformably on Pontian strata (100m).		Kuyalnitsian Beds of sands and clays with a marine fauna (Up to 40m).	Upper Armavirian sub-Suite of repeated red, continental clays, sandstones and conglomerates. In the west the sub-suite replaces Akchagylian and Apsheronian.	Lower down, dark, Upper Sarmatian clays.
Kuyalnitskian	Lower						
Cimmeridian	Upper				Pantikapeian Beds—clays and sands with seams of ore (25m).	Lower Armavirian sub-Suite—red continental clays resting on the Sarmat (Up to 100m).	
Cimmeridian	Middle				Kamyshburunian Beds—a horizon of ore. Sands, clays and brown ironstones (50m).		
Cimmeridian	Lower				Azovian Beds—ferruginous sands and clays, with a marine fauna, and ore horizons (6–20m).		
Pontian	Upper	Gutinskian Suite of andesites, basalts, dacites and tuffs (600m).			Bosforian sub-Stage—shell-beds and clays (crags) (40m).	Sands and shell beds present only to the north of Stavropol (4–20m). To the south of Manych: bluish clays with a marine fauna (Up to 200m).	
Pontian	Middle				Novorossiiskian sub-Stage. Beds with *Congeria subrhomboidea*; shelly horizons (15m).		
Pontian	Lower		Lower Pontian, Odessa Limestone.		Lower Pontian—clays and limestones.		

TABLE 49—continued

Yalama—Central shore of the Caspian Sea	The Apsheron Peninsula	Daikent	Gori	Poti	Erevan
Anthropogenic, mainly continental deposits (460m).	Baku Stage. Turkyanian Horizon.	Total thickness of Anthropogenic deposits is 868m. Baku clays (188m).	Gravels and sands (4m).	Anthropogenic deposits (250m).	Anthropogenic, continental sands and clays (61m).
Grey sandy limestones and brackish-water clays (298m).	Apsheron—clays and sands with a marine fauna and limestones towards the top (500–1000m).	Grey and dark-grey clays with beds of sands and a marine fauna (1937m).	Dushetian Suite of molasse type—weakly cemented conglomerates with intercalations (beds) of yellowish and greenish sandstones and clays (691m). Interval.	Guriian Beds—homogeneous sandy clays with beds of normal clays and rarer gravels (362m). Interval.	Pliocene deposits are represented by basaltic lavas (60m). Lower down lies the Zangian Suite of sands and clays assigned to Miocene. The presence of pebbles at its base makes it possible to be Pliocene. A possibility is not excluded that the Upper Saliferous Formation is also Pliocene.
Grey and dark-grey clays; above, with Foraminifera; below, with ostracods (197m).	Akchagyl—dark, pure clays with a marine fauna (50–100m).	Grey, bedded, calcareous clays (336m).			
Productive Suite of continental, brown, sandy, calcareous clays and silts with an admixture of pebbles at the bottom (268m). Lower down, Middle Sarmatian clays and sands.	Productive Formation of continental, grey sands and clays. The upper part (Surakhanian and Balakhanian suites) has thick clays with horizons of petroliferous sands (1000–1100m). The middle part is a suite associated with an interval (break) and consists of coarse sands and gravels (100m).	Lower down follows the Middle Sarmatian.	Lower down follows Upper Sarmatian, Naukhorskian Suite of continental clays and sands (854m).	Bluish-grey clays; above, calcareous and fine grained; below, sandy (88m). Thick, pale sandy clays with sequences of fine grained clays with beds of sandstones (510m).	
	The lower part—clays with sequences of sands (400m). Total thickness is usually 1500–1600m, but reaches 4000m.				
	Babadzhanian Beds of grey clays with beds of sand.			Bluish-grey sandy and normal clays with beds of sandstones (695m).	
	Middle Pontian, homogeneous, dark clays.				
	Lower Pontian, dark clays with *Valencienesia*. Total thickness of Pontian strata is 50–200m.				

typical. The bottom part of it consists of dark, almost barren Upper Samatian clays which are most probably the deposits of near-shore plains. They are succeeded by the Tarkhanian Horizon which is a group of sandy marls with a peculiar marine fauna and a limited distribution.

The Chokrakian Horizon represents the deposits of a narrow marine basin elongated along the Caucasus. In the north there are argillaceous and muddy rocks, further south within the downwarp only clays with beds of marl were deposited, while in the extreme south, near the shore of the Caucasus the sediments are most diverse and include shell limestones, oolitic limestones, sandstones, silts and clays.

The Karaganian Horizon spreads far to the north and is lithologically similar to the Chokrakian, although differing faunally. In the north there are clays and sandstones, further south only clays and in the south, near the former shore, varying shallow-water sediments.

The Konkian Horizon consists of deposits of a marine basin, similar to, but of deeper-water type than, the Sarmatian.

The Sarmatian Stage consists of deposits of an entirely enclosed sea with depressed salinity and consequently a peculiar fauna (Fig. 213). In Upper Sarmatian times considerable upward movement of land caused the formation of the Stavropol Peninsula, which is adjacent to the Caucasian mainland. At the same time an isthmus originated, connecting the land of the Greater Caucasus and the Lesser Caucasus. Lithologically the Sarmatian Stage is characterized by dominant shales, but marls, bedded limestones and sandstones are also widely developed.

The Meotic Horizon indicates an even greater contraction of the sea and freshening of its water. Its fauna changes so much that the question arises as to whether it could be Pliocene. In many districts continental sandy-conglomeratic strata predominate. Amongst marine sediments sands are most common. The Miocene deposits of the Northern Caucasus and Ciscaucasia have been described in detail in a monograph by V. A. Grossgeim (1961).

Pliocene lithologies and facies are widely known and are tabulated in Table 49. The Productive (oil-bearing) Formation of the Apsheron Peninsula is especially peculiar. The formation represents the deposits of a delta, the surface of which was covered by large near-shore lakes. Thus the formation consists of repeated beds of clays and sands which are oil-bearing and water-bearing. The distribution and thicknesses of the formation are shown on Fig. 211. The sands have yielded and still yield large amounts of oil. In the region of the river Samur (Fig. 211) the productive strata pass into molasse deposits over 1000m thick.

Formerly the Productive Formation was only studied on land. At present it is also being investigated at the bottom of the sea surrounding the Apsheron Peninsula (Fig. 212). Submarine structures have been bored and produce much oil.

The regions of the Kura and Rion valleys are the sites of accumulation of unusually thick sediments, which from the beginning of the Neogene onwards reach 6000–7000m and from the Upper Cretaceous onwards 14 000–15 000m. In the Lower and Middle Miocene there are dominant, uniform almost barren clays, similar to the flysch. The most widespread marine diverse Sarmatian deposits are thick and consist mainly of clays, sandstones and shell limestones. The facies of Pliocene sediments are even more variable. The molasse developed near to the Caucasian range is most characteristic, and consist of thick, barren conglomerates interbedded with sandstones and loans, the latter yielding occasional terrestrial and freshwater faunas. In the Meotic times, or more accurately in an epoch between the Sarmatian and Akchagylian when the Shirakian Suite (1700m), or to the west of Tiflis the Dushetian Suite (2000m), was being deposited, the conglomerates were the best-developed sediments. The Productive Formation (3000–4000m) of the Apsheron Peninsula is a variant of the molasse, but is finer grained (sands and clays) on account of its distance from the mountain range.

The second epoch of molasse deposition began in Apsheronian times and continues until the present. Away from the range these deposits pass into deposits of near-shore plains and marine sediments. As an example of this is the Middle Apsheronian Kusarian Suite developed in the lower reaches of the Samur in juxtaposition with the Alazanian Suite (up to 3000m) of conglomerates and loams of an Upper Pliocene age, found in the upper reaches of the river Alazan, to the east of Tiflis.

The Lesser Caucasus and the Armenian Highlands from the beginning of the Miocene were in the main part of a landmass—the mountainous Mediterranean macroisthmus. At times a southern sea breached across it forming such seas, of

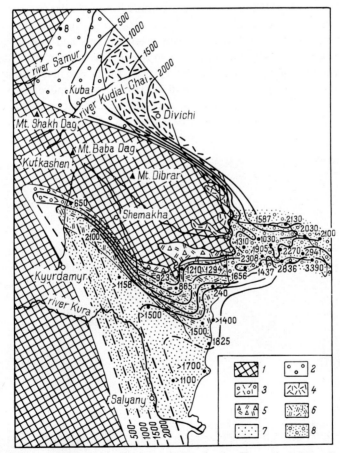

FIG. 211. Facies and thicknesses of the Productive Formation, According to V. E. Khain (1950).

1 — areas of erosion; 2 — Kubanian facies; 3 — Donguzdykian facies, with conglomerates; 4 — Kyzylburanian and Donguzdykian facies; 5 — Donguzdykian facies with breccias; 6 — Eastern Kabristanian and Lengebizian facies; 7 — Salyanian facies; 8 — Apsheronian facies.

normal salinity, as the Saksaulian (Lower Miocene) Georgian basin. The last, short-lived connection with the ocean occurred in the Meotic epoch, when groups of marine deposits are absent from this area, or are represented by saline and gypsiferous rocks—deposits of bitter-salt lakes and swamps—and by thin groups of limestones and clays with brackish or freshwater lacustrine faunas. Terrestrial vulcanicity became widespread during the Pliocene and gave rise to thick formations of lavas and tuffs.

FAUNA AND FLORA.   The fauna and flora is of the typically Mediterranean type.

PALAEOGEOGRAPHY.   The Palaeogeography is predictable, but complex. The predictability arises out of periodic enlargement of land and a consequent restriction of the area occupied by seas. The complexity arises out of continuous changes in the configuration of the shoreline accompanied by considerable displacements of facies.

The greatest uplift took place in the south of the Lesser Caucasus, where numerous islands, which existed in Palaeogene and Lower Miocene times merged to form a mountainous macroisthmus, which joined the Pamirs to the Alps (Fig. 213). The relief and other features of this isthmus were similar to the present-day Kamchatka, the Kuril Islands and Japan elevated so as to cause the emergence of the intervening straits. The Mediterranean macroisthmus had numerous volcanoes. In Upper Miocene times it completely separated the Sarmatian sea from the ocean thus rendering its waters fresh. The Caucasus, Crimea and the Carpathians stood out as islands within this sea. A renewed burst of upward movements occurred in the Meotian and Pontian epochs and as a result the Caspian and Black Sea basins separated and acquired their present-day outlines. A very important palaeogeo-

FIG. 212.   Pliocene outcrops at the bottom of the sea off the shore of the Apsheron Peninsula. The lighter part is land covered by sands; the dark part is sea with scarplets of limestones. Interpretation is according to V. V. Sharkov. *Aerial photograph in the Laboratory of Aerial Methods of the Academy of Sciences, U.S.S.R.*

graphical phenomenon was the Caucasian Main range glaciation and a possible glaciation in the Lesser Caucasus. This was formed during the post-Sarmatian epoch and can be demonstrated by the accumulation of thick conglomeratic formations such as the Shirakian and the Dushetian. Such formations are the alluvial cones of powerful mountain rivers emerging on to a plain. As the study of the present-day mountain ranges has demonstrated such rivers only originate if they are fed by permanent snows and glaciers. The Lower Pliocene or Upper Miocene glaciation is an undoubted fact. It is very important since it indicates that mountain glaciation is not characteristic solely of the Pleistocene epoch and cannot be used as a feature on which it is separated from the Pliocene. Thus the connection of Baku, Apsheronian and Akchagylian deposits with moraines and fluvioglacial deposits can only serve as a basis for including these formations in the Anthropogene if the Pliocene as a whole, as it has already been stated, is included in it.

Anthropogene

Anthropogene sediments are variable, widespread and in the lower reaches of the Kura are 1000m thick.

DISTRIBUTION. Marine Lower and Middle Anthropogenic deposits are developed in the same localities as Upper Pliocene strata. Continental facies are heavily eroded and have not been adequately studied or delineated. Upper Anthropogene deposits are not widely eroded and are ubiquitous and well investigated. They are not shown on synoptic maps, but special maps are prepared for them.

SUBDIVISIONS. The Plio-Pleistocene boundary is strongly debated as has already been pointed out (p. 623). It should be said that the latest investigations have shown that Akchagylian and even post-Sarmatian (Meotian and Pontian) sediments are connected to glacial deposits of mountain edges. Thus if logic is followed up completely the boundary to Anthropogenic deposits should be drawn under the Meotic and to include the whole Pliocene succession in the Anthropogene. This conclusion is also confirmed faunally and tectonically. As yet such a conclusion has

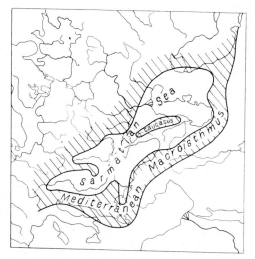

FIG. 213.   The Mediterranean macroisthmus in Sarmatian times.

not been accepted by a meeting of specialists and consequently the boundary is drawn above the Apsheronian.

At present the Anthropogene is divided into lower, middle, upper and Recent. Recent deposits embrace those of the coast dunes and marine terraces with *Cardium edule* and *Mytilaster*, as well as the first river terrace and the moraines and fluvio-glacial deposits of present-day glaciers. The correlation of Black Sea and Caspian sediments is given in Table 50. The correlation of continental sediments and those of alluvial plains and mountain chains represents a very difficult problem which is far from being solved.

LITHOLOGY AND FACIES. The lithological composition of marine sediments is relatively uniform. Sandy and pure clays predominate, although there are also sands and silts and shell-limestones which are sometimes lithified and sometimes are rubbly and form so-called faluns. Gravels and breccias are rare. Deposits of alluvial plains are the same as the present-day deposits of this type.

Such sediments consist of lenses of cross-bedded sandstones with laminations of fine gravel, being the deposits of river beds. The regularly bedded sands and clays

TABLE 50. *Correlation of Anthropogenic Successions*

Stage	Black Sea	Caspian Sea
Present-day	Present-day deposits. Older Chernomorian deposits with a fauna similar to the present-day. (1–2m to 20m).	Present-day deposits. Novocaspian Stage—terrace (9–10m high) covered with sands and shell-beds with *Cardium edule*.
Upper	Novoeuxinic deposits—sands, gravel and clays, with a freshwater fauna, found on the bottom of the Black Sea. (Up to 10–10m). Surozhskian Horizon (Alanian terrace). The first terrace of the sea of Azov—sands and clays. (6–10m). Karangatian Horizon (terrace, 10–45m) —sands and gravels with a rich Mediterranean fauna.	Khvalynian Stage. Novokhvalynian (Kemrudian) terrace, below the former terrace and below sea level. The sediments are sands and shell-beds, normally with pebbles. (10–15m). Older Khvalynian terrace of sands with pebbles and shells, shell-beds and rarer clays. (Up to 20m).
Middle	Uzunlarian Horizon (terrace, 30–40m). Sands and clays with a mixed Mediterranean-Caspian fauna. (5–10m). Older Euxinic deposits of sands, clays, shell-beds, with a brackish fauna. Terrace—60–80m.	Khazarian Stage—mixed sediments, including: gravels, sands, rarer clays and shell-beds. Two terraces—the older and the newer Khazarian—are recognized. (25–40m).
Lower	Baku Stage—sands, clays and pebble beds (tens of metres). Not widespread (Taman).	Gyurgianian Stage—sands, clays, gravels, locally shell-beds with marine fossils. Baku stage—brownish clays and sands with a marine fauna.
	Chaudian Beds—sands, clays and pebble beds, fully developed near Batum and in Tuapse and possess a marine fauna. (Terrace—100–150m). Sediments, from 15–20m to 100–150m thick.	Tyurkyanian Horizon of a thin sequence of clays with a freshwater fauna. Total thickness is 70–200–700m.

are almost barren being the sediments of flood-plains, spreading over very large areas. The regularly but thinly bedded, sorted calcareous strata with freshwater faunas are lacustrine deposits. Similar strata, but with brackish faunas, are deposits of lagoons and estuaries. Saline and gypsiferous formations are the deposits of bitter-salt lakes and swamps. Brown and reddish unbedded and inhomogeneous loams and clays, which are mostly loessic and even transitional into loess, are generally barren but contain rare terrestrial gastropods (*Helix*, *Pupa*) and bones. These sediments are the subaerial deposits of watersheds and their slopes, being aeolian and products of temporary rains.

Another important and peculiar facies complex, which has already been recorded from the Pliocene, is molasse or the deposit of alluvial cones, which existed on the plain at the edges of mountain chains. The molasse consists of great thicknesses of conglomerates and gravels with beds of clays and sands reaching hundreds of metres.

Moraines and river-sorted moraines are widely found in valleys of mountain rivers. The study of these moraines has shown the former existence of three epochs of glaciation. The Würm moraines are well preserved while the moraines of the preceding Riss glaciation are considerably eroded. Of the earlier Mindel and Günz glaciations only small remnants have been preserved and their investigation presents great difficulties. Some investigators even include Günz in the Pliocene which is quite probable. The pre-Akchagylian glaciation of the El'brus has been demonstrated by E. E. Milanovskii and I. V. Koronovskii.

The last facies complex consists of volcanogenic formations including lavas, tuffs, tuff-breccias and normally interbedded sediments. This complex is developed in the Lesser Caucasus and in various localities of the Greater Caucasus (El'brus and Kazbek) and is described later (p. 668).

PALAEOGEOGRAPHY AND FAUNA. The northern margin of the Mediterranean Geosyncline, beginning from the foothills of the Soviet Carpathians (to the northern shoreline of the Sea of Azov) and then to the valley of the Manych, was a low-lying wide near-shore plain to the south of which were marine basins. In the Lower Anthropogenic epoch this region had a warm climate and was probably a wooded steppe—a region of accumulation of red-brown clays which near the Azov Sea reach a thickness of 20–25m. Later on the climate became colder and boundless steppe succeeded the woods; the steppe continued until the present day and its surface became the site of accumulation of aeolian and atmospheric deposits such as loessic loams and loess to the thickness of 20m.

Further south there were wide, shallow, freshwater basins, which at times became sea-lakes populated by brackish *Didacna*, *Monodacna* and at others by a freshwater fauna of *Unio* and *Planorbis*. Such groups as *Dreicencia* and *Neritina* lived in either types of water but preferred brackish. Plains covered by sea-lakes and true lakes were also in existence to the south of Crimea and near the western foothills of the Main Caucasian range and the Lesser Caucasus. In the north-east the sea spread over the Manych Depression forming a considerable embayment. Crimea and the Caucasus were high mountain chains of probably a wider extent than at present. Their eastern part was covered by permanent snows and glaciers which in the middle of the Anthropogene were wider than at present. The Black Sea at that time did not exist.

In the middle of the Anthropogene period in the region of the present-day Black Sea a very curious phenomenon of the formation of Aegean-Black Sea rift took place. This rift is even larger than the Africa-Arabian rift. The formerly existing plain with brackish seas and freshwater lakes began to subside along extensive, but relatively shallow faults and the resulting depressions became submerged by the waters of the Mediterranean Sea. As a result the Black, the Aegean and the Marmora seas came into existence. One of the faults transected the Crimenan range and consequently its southern part was overrun by the sea. The south-western margin of the Caucasus was also faulted out. The initial movements at first gave rise to a basin with a mixed brackish and Mediterranean fauna, and its sediments are known as the Uzunlar Horizon (Table 50).

Continuous depression caused a further influx of marine water leading to the normal 3·5 per cent salinity and the appearance of a typical Mediterranean fauna with echinoids, which at present are not found in the Black Sea. A true, southern warm sea originated on the site of the former plain. The deposits of this sea, which were at one time known as those of the Tyrrhenian terrace, are now known as the Karangatian Horizon. The further history of the Black Sea is not entirely clear. It is sometimes thought that after the Karangatian sea the Novoeuxinian sea originated with its brackish fauna of the same type as that in the present-day Caspian Sea. It is, however, more likely that such faunal changes occurred only along the margins of the sea in embayments and lagoons. One such bay of the Karangatian sea penetrated into the Sea of Azov region and further to the north-east along the Manych Depression up to lake Manych-Gudilo (Surozhskian Horizon).

It should be pointed out that M. V. Muratov (1955a) considers the changeover from brackish to saline basins to have occurred without any rift faulting. The cause according to him was gentle undulating movements of large folds forming domes and synclinal structures. He also thinks that the Karangatian sea was first succeeded by the Novoeuxinian sea-lake with fresher water and then by the proto-Black Sea of normal salinity. During these changes there were two transgressions—the Karangatian and the Older Black Sea. This theory does not explain the absence of the southern half of the Crimean anticlinorium or the connection of the Karangatian sea with the Sea of Marmora and the Aegean. It is also difficult to understand how two marine transgressions could follow each other over such a short period of time.

The development of the Caspian in the Anthropogene occurred independently of the Black Sea. The Caspian was to a large extent conditioned by the regime of continental glaciation. Increased melting produced more river water for the Caspian causing a transgression, while the slow-down of melting produced a regression. The greatest trangression occurred in Middle Apsheronian times. In Upper Apsheronian times a regression occurred. A new transgression began in the Baku epoch and reached its greatest extent in the succeeding Gyurgyanian epoch. At the end of this epoch a new regression was again succeeded by the Khazarian transgression, of much narrower extent. After yet another regression an even more limited Khvalynian transgression occurred giving rise to the Neocaspian Sea, which was almost the same as the present-day Caspian. In the last thirty years the level of the Caspian has fallen by 2m causing changes in its shoreline. In Baku twelfth-century ruins, which until recently projected into the sea, are now completely on land. The use of the Volga for irrigation will lower the level of the Caspian still further, indicating how the intervention of man aids the considerable regression which is now occurring.

TECTONICS

There are thousands of papers in different languages written on the subject of the Mediterranean Geosyncline. The investigation of the Soviet part of the geosyncline stands high. Soviet geologists have demonstrated the absurdity of certain reconstructions which were and are pursued by some foreign geologists. The best-known misconception of such kind is the idea of gigantic displacement of folds and systems of folds in large sheets, *nappes de charriage*. The magnitude of displacement was considered of the order of hundreds of kilometres and especially complex and large nappes used to be envisaged in the Carpathians. The investigations of Soviet geologists have demonstrated their absence in this area. The tectonics of the Carpathians are truly complex but nappe displacements do not exceed 10–15km. This has been proved by M. V. Muratov (1949), A. A. Bogdanov, O. S. Vyalov, V. G. Korneyeva and a whole series of Ukrainian geologists.

The scheme of main tectonic structures and their relationships are shown in Fig. 214. In the west there are the Carpathians and in the east the Pamirs. The Russian Platform and the Ukrainian Shield are shown in crosses. Further to the south lies the region affected by the Hercynian Orogeny. This region to a large extent is a part of the external zone of the Mediterranean Geosyncline. The dash ornament shows those regions of the external zone which have simply and gently folded Mesozoic and Cainozoic rocks. Oblique lines are used for the folded strata of the Carpathians, Crimea, the Caucasus, the Kopet-Dag and the Pamirs. Black patches indicate Palaeozoic inliers. Lastly the obliquely dashed region further to the south is a zone of southern folds affected by young vulcanism.

Great achievements have also been made in the study of Caucasian tectonics. Regional and review papers by A. P. Gerasimov, V. P. Rengarten and K. N. Paffengol'ts (1959a, 1959b) are of especial importance. The monograph by V. V. Byelousov (1938–1940), who first indicated the significance of measuring and mapping of thicknesses in establishing tectonic movements, is also one of the leading publications. The work of V. E. Khain (Khain, 1950, 1954; Khain and Leont'yev, 1950) is important in the study of the eastern part of the Greater Caucasus.

STRUCTURES

The following structural storeys have been recognized.

1. Precambrian-Middle Palaeozoic
2. Upper Palaeozoic-Middle Trias
3. Upper Trias-Middle Jurassic
4. Upper Jurassic-Lower Cretaceous
5. Upper Cretaceous-Palaeogene
6. Neogene-Anthropogene.

The first storey in other geosynclines consists of four sub-storeys including those of the Archeozoic, Proterozoic, Lower Palaeozoic and Middle Palaeozoic. Their presence in the Mediterranean Geosyncline and particularly in the Carpathians and the Caucasus is possible, but so far they have not been differentiated since their outcrops are small and their stratigraphy has not been properly worked out.

The later storeys are also divided into sub-storeys, but their structures have relatively small points of difference. The nature of structures also varies according to their position in one of the following tectonic belts.

1. The Ciscaucasia or the sub-Caucasian Synclinorium
2. The Greater Caucasus (an anticlinorium)
3. The Rion-Kura Depression (a synclinorium)
4. The Lesser Caucasus (an anticlinorium)
5. The Erevan-Nakhichevan synclinorium.

Within these belts there are numerous secondary, but nevertheless large tectonic zones which are distinguished by different workers in different ways.

First Storey

Precambrian-Middle Palaeozoic strata are found in three regions: (1) the central Carpathian zone (the Marmaresh Massif), (2) in the western part of the Greater Caucasus and (3) the central part of the Lesser Caucasus anticlinorium, where they are exposed in the Dzirul and Kramian massifs and also in the Arzakent zone. The folds are everywhere complex, deformed and faulted and the rocks are often intensely metamorphosed and sometimes recrystallized. Consequently individual folds are difficult to recognize. The degree of deformation, metamorphism and faulting is so high that it is difficult to find any fauna. In the Greater and Lesser Caucasian anticlinoria faunas have been found, which has allowed a diagnosis of Devonian and Upper Palaeozoic metamorphosed strata. Devonian rocks are involved in very complex folds. Folds affecting Devonian and Lower Carboniferous strata are very distinct in the Erevan-Nakhichevan synclinorium. Here a formation of limestones (1700m) is folded into large simple flexures which are in places affected by normal faults (Fig. 202). The rocks are not highly metamorphosed and the fossils are excellently preserved.

In southern Ciscaucasia Upper Devonian and Lower Carboniferous sediments form a part of the basement. Again the metamorphism is much weaker than in the Caucasus and folds are evidently much simpler.

Second Storey

(Upper Palaeozoic-Middle Trias.) Owing to a weak manifestation of the Hercynian Orogeny and especially of its latest phases, Upper Palaeozoic strata are folded and metamorphosed only to the same extent as the Mesozoic and do not belong to metamorphic formations. On the northern slope of the western Caucasus the rocks of the second storey are fully developed over a small area. Although there is a break between the Trias and the Upper Palaeozoic it is insignificant and the same structures affect their formations, which are equally metamorphosed and dislocated (Figs. 203, 204).

In the Erevan-Nakhichevan synclinorium Permian deposits are transitional into Triassic and both are folded into simple structures and metamorphosed just as much as Jurassic sediments.

To the north of the Valley of the Manych as far as the Astrakhan-Kizlyar railway line—a district which forms the eastern continuation of Donbas—interesting, folded Upper Palaeozoic sediments have been found in boreholes. Similar rocks are also found in the Astrakhan region. Judging by dips, degree of metamorphism and evidently by the character of folding Upper Palaeozoic rocks of Ciscaucasia are identical with those of Donbas on the one hand and those of the western slope of the Southern Urals and the Mugodzhars on the other.

Third Storey

(Upper Trias-Middle Jurassic.) This structural storey is exposed over large areas and its rocks have been well investigated, since they contain large deposits of ores and coal. Extensive published material on its structures indicate that complex and small folds, which are usually tight and isoclinal, predominate. This feature arises out of the wide-spread development of thick, fine-

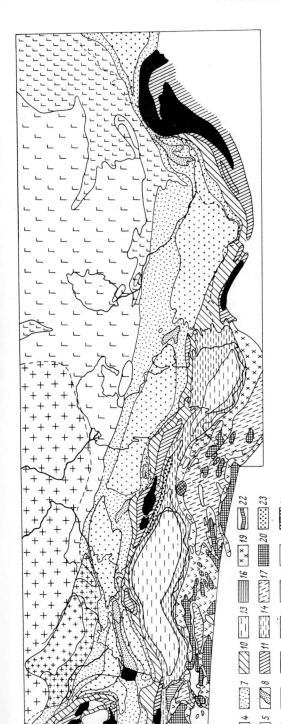

Fig. 214. Principal tectonic structures in the south of the U.S.S.R. According to M. V. Muratov (1949).

1–3 — the Precambrian platform; 4–5 — the Palaeozoic platform; 6–9 — the Neogene platform and the Neogene fold belt; 10–12 — a zone of mega-anticlines; Palaeozoic outcrops in black; 13–16 — inner depressions; 17–23 — the zone of inner structures; 24 — areas of Neogenic and Anthropogenic vulcanicity.

grained thinly and rhythmically bedded deposits, for which such folds are generally characteristic. The folds (Fig. 215) affecting the Trias and the Jurassic of Crimea serve as a good example.

The folds affecting Lower and Middle Jurassic shales of the southern slope of the Main Caucasian range in the region of the Georgian Military Road are of the same type (Fig. 216). Here the composite profile of V. P. Rengarten (Guides. . .1937) clearly shows complex folds continuing for tens of kilometres across the strike. Such folds are also detected in the overlying Upper Jurassic and Lower Cretaceous sediments and also in flysch-type slates. The profile shows only the main folds. In actuality they are much more numerous. The other feature of the area, large synorogenic intrusions trending parallel to the strike and dividing the rocks into a series of separate overriding scales (nappes) is also clear on the profile.

Nearer to the Black Sea in the Tkvibuli and Tkvarcheli coal-fields, Lower and Middle Jurassic rocks are strongly folded, but are simpler in style and more rounded since the thick series of sandstones are not sharply arched. Here again the whole formation of rocks is broken up into separate blocks overriding each other. The Lesser Caucasus anticlinorium has thick Jurassic effusive rocks, causing an even greater simplification of the fold style and the increase of the size of individual folds.

Fourth Storey

(Upper Jurassic and Lower Cretaceous.) A characteristic feature of the storey is the development of thick (400–700m) massive reef limestones forming the ridge (yaila) of the Crimean Mts and the second ridge of the Northern Caucasus. Again the thick unbedded limestones reacted strongly against the pressure, and as a result are folded into huge, simple folds with relatively gentle flanks. The folding is always accompanied by deformation causing secondary complex minor folds and thrusts. Where the limestones lie amongst shales, they have been broken up into separate massifs and blocks, lying now within the shales. The well-known belts of limestone blocks in the Carpathians are of this origin and form one of their characteristic features. The zone of the Carpathian blocks or klippen stretches for 500km from the Marmarosh Massif to Vienna and is from 1–2km to 20km wide. The zone includes most variable blocks, including massifs of 10–15km in length and 2–3km in width. Thus a narrow belt of rocky mountains, hills, peaks and cliffs stretches on and on. The number of individual blocks is very large, being in places tens and hundreds per kilometre. In the Soviet Carpathians to the north-west of the Marmarosh Massif there are two zones. The northern zone consists only of Upper Jurassic limestones, while the southern involves blocks of variable rocks, ranging in age from Liassic to Upper Jurassic. In both zones massive unbedded or indistinctly bedded reef limestones predominate. Formerly the zone of the klippen was considered as a front of a large *nappe de charriage*, but Soviet geologists have demonstrated the local association of the zones with broken, deformed anticlines.

In the Lesser Caucasus similar disharmonic relationships between shales and massive Urgonian reef limestones are also observed. Where the Upper Jurassic and Lower Cretaceous are represented by flysch formations as on the southern slope of the Caucasus the rocks show characteristic minor and complex folds also broken by thrusts.

The structures of the third and fourth storeys in the outer zone (see p. 697) are peculiar. They are hidden under a horizontal cover of Neogene and Anthropogene sediments. The widely developed borehole and geophysical explorations for oil and

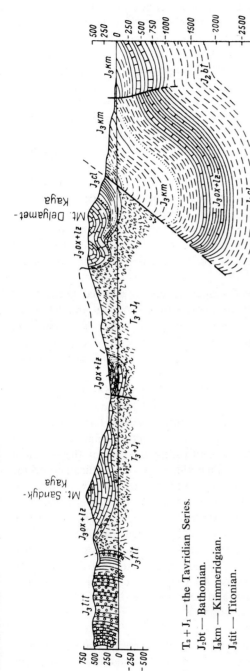

T$_3$+J$_1$ — the Tavridian Series.

J$_2$bt — Bathonian.

I$_3$km — Kimmeridgian.

J$_3$tit — Titonian.

Fig. 215. Complex folds affecting the Tavridian Suite and Jurassic strata of Crimea. According to D. S. Kizelal'ter and M. V. Muratov (1959).

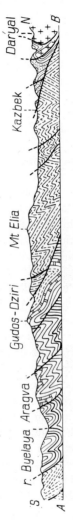

Fig. 216. Folds affecting Lower and Middle Jurassic rocks of the Main Caucasian Range, exposed along the Georgian Military Road. The section is 45 km long. According to V. P. Rengarten (1937).

gas have produced a large amount of new data, which is especially considerable for Ciscaucasia and the Black Sea region. These data have shown that in these provinces Mesozoic structures are nearly of platform type. In Ciscaucasia and the Black Sea region a rather gentle southerly regional dip is widely developed (Figs. 200, 218, 246). Against the background of this regional dip there are brachyanticlinal and brachysynclinal folds which are mostly of small amplitude and have gentle flanks. These folds follow each other along the trend sketched out by the folds of the inner zone. In Ciscaucasia they form several series trending parallel to the Caucasian anticlinorium.

The brachyanticlines represent rounded, closed folds elongated in an east–west direction. They are several kilometres to several tens of kilometres long, and the associated dips do not exceed a few degrees. These folds are quite often reservoirs of Mesozoic oil and gas, the larger of which are found in the Terek-Kuma plain, in Stavropol and the Krasnodar region.

The general monoclinal dip and the shape of the folds imparts to the Ciscaucasian Mesozoic rocks a platform character. Consequently Ciscaucasia as well as the whole of the outer zone is not infrequently considered as the Epi-Hercynian platform. This view is not entirely correct since typical platform folds are not related to orogenies and do not possess regional linear trends. As an example of the typical platform folds one can name those affecting the Palaeozoic of the Russian Platform.

The folds affecting Mesozoic rocks of Ciscaucasia have a regional linear trend parallel to that of the Caucasus. The influence of both the Alpine and the Cimmeridian orogenies is incontrovertible and consequently the whole of Ciscaucasia must be considered as a Mesozoic mobile zone, which only became part of the platform at the beginning of the Neogene. Thus the region is not an Epi-Hercynian, but an Epi-Cimmeridian platform.

The concept of the sub-Caucasian downwarps is considered from two points of view. According to one, already advanced, the downwarp is asymmetric and has a narrow and steep southern flank and a broad and gentle northern flank. Adherents of another standpoint consider the downwarp as a symmetric, relatively narrow structure with similar flanks and the greater gently dipping part of the northern flanks as the southern slope of the platform (Shcherik, 1958). It is difficult to decide between these two views, since the basic difference between them is how mobile and susceptible to folding is the northern flank of the downwarp; the slope of the platform would be less mobile and its folds independent of orogenies. The analysis of folds in the disputed area suggests that they have a linear trend parallel to younger fold belts, which can only be interpreted as the resultant effect of orogenies, implying a mobile character for the area and its inclusion in the marginal downwarp. It is probable that in Mesozoic times the whole of Ciscaucasia was a mobile zone intermediate in character between a platform and a geosyncline.

Fifth Storey

(Upper Cretaceous-Palaeogene.) In the outer zone of the Mediterranean Geosyncline the fifth storey is found ubiquitously, but is badly exposed. It is folded into large very gentle folds almost indistinguishable by the naked eye and which are detected in the Donyets Basin and in the Mangyshlaks. Only nearer to the Caucasus do these folds become tighter and steeper. The Stavropol plateau represents three gentle large folds affecting Palaeogene rocks and only the southernmost of the three shows dips of up to 20–30°. The Grozno oil region represents a system of folds elongated parallel to the Main range, since Miocene strata are also involved, these structures will be discussed later.

The Ciscaucasian downwarp which at present can be clearly delineated did not exist in Palaeogene times. The whole of the Ciscaucasian plain was at the end of the Palaeogene (Maikop) a very wide and shallow depression with steeper slopes at the edges thus resembling a tray. This pattern is established on the basis of the thickness of Palaeogene and Upper Cretaceous rocks. Along the northern margin of the depression, near Rostov their thickness is 410m, near Kayala (south of Rostov) it is 450m and near Astrakhan about 600m. Further to the south the thickness increases rapidly and a hundred kilometres south of Astrakhan it is 1450m and further to the south 1750m. To the south of Rostov the thickness of the fifth storey rocks is 1500–2000m and to the north of Stavropol it is 1800–2100m, while nearer to the Caucasus —on the northern flank of the downwarp smaller figures of about 1000m have been recorded.

In the Northern Caucasus carbonate rocks of Upper Cretaceous age and the overlying Palaeogene have a regional dip or form simple large folds. More rarely they are isoclinally folded and broken by thrusts and faults into separate slices (Fig. 217). In the west where flysch facies are developed main folds of this type trend parallel to the trend of the range and are broken up into separate blocks and scales (nappeslices). The flanks of the larger folds often develop secondary smaller and complex folding.

On the southern slope of the Main range in the region of development of flysch formations (Black Sea shore and to the north of Sochi) the main folds are elongated along the Main range watershed in a north-westerly direction. The folds are tight, elongated, always asymmetrical and verging, and even overturned, to the south, towards the sea. Their flanks are complicated by minor folds of second and higher orders. The main large thrusts are also verging to the south-west, towards the sea, and movement has been in this direction. Minor folds are also complicated by thrusts and faults. Another example can be selected from the central part of the Main range, along the Georgian Military Road (Fig. 216), where imbrication is even more clear, minor folds are equally complex and folds and thrusts are verging to the south.

The Lesser Caucasus anticlinorium is complicated by the appearance of thick volcanic and also widely developed carbonate formations. The presence of such rocks violates the regularity and homogeneity of trends and dips of the flysch formations. Here folds are variable, of differing shapes, and verging in different directions. In the Erevan-Nakhichevan synclinorium the folds are simpler and have a brachyanticlinal aspect.

Sixth Storey

(Neogene-Anthropogene.) The younger the strata of the sixth stage the simpler are the structures affecting them. The structures involving Lower and Middle Miocene rocks differ little from Palaeogene structures, while Upper Anthropogene sediments often possess the original depositional dip and lie effectively horizontally. Separate horizons are often divided from each other by breaks and distinct angular unconformities, but huge monoclines are also quite common and stretch for tens of kilometres along and many kilometres across the trend. The monoclines affect all the strata beginning from the Lower Miocene and ending with Upper Anthropogene sediments, succeeding each other without any intervening unconformities, while

Lower Miocene rocks are vertical and Upper Anthropogene deposits are almost horizontal as occurs in the Vale of Fergana. The reason for the formation of discontinuously bedded layers is clear, the interruptions being breaks in fold movements and the concentration of such breaks into distinct phases separated by intervals of immobility. The origin of the formation of conformably bedded strata is also clear representing an almost continuous sequence of orogenic movements occurring contemporaneously with sedimentation.

The outer zone of the Mediterranean Geosyncline throughout Neogene and Anthropogene times was a part of the platform which was called by M. V. Muratov the 'Scyptian'. The deposits of these epochs lie almost horizontally and are only locally broken by normal faults. Contemporaneously with the Neogene platform and to the south of it was a mobile fold zone which commenced at the northern slope of the Main range. Between the fold zone and the platform lay the submontaine downwarp. The downwarp, which was first formed in the Miocene, continues developing even at present. It did not much influence the relief, since it has been occupied by a succession of shallow water brackish marine basins and near-shore alluvial plains of the type that is there at present.

In the west the downwarp is situated to the north-east of the Carpathians. Here along its outer, northern edge, in a region of slow, but continuous downward movement grew reef massifs (toltras) of Tortonian and Sarmatian times. Further to the east the downwarp lies along the Cis-Black Sea plain, the steppe of Crimea and the Ciscaucasian plain. Further on it lies at the foot of the Kopet-Dag. As a result of the work of the greatest Soviet tectonician, N. S. Shatskii (1947), the downwarp has been the best studied in the Soviet Union.

The sub-Caucasian downwarp is a gigantic, gentle asymmetric syncline with a narrow and steep southern limb, which is in places complicated by secondary folds as in the Grozno oil region (Fig. 223, 224). The northern limb is wide and flat, with secondary folds of platform type. The depth of the downwarp from the bottom of Neogene strata reaches 2500–3000m. The axis of the downwarp is depressed in the west and east and its highest elevation occurs at the Stavopol plateau, where the downwarp is the shallowest and the total thickness of Anthropogene and Pliocene sediments is 150–400m, while Palaeozoic rocks lie under Upper Cretaceous (Albian) rocks, at a depth of 1000–1500m. The western part of the downwarp here is known as the Indolo-Kuban' marginal downwarp. Within this structure the top of the folded Palaeozoic lies at depths of over 5000m. The base to the Maikop (Upper Palaeogene) reaches depths of over 4500m, as is shown in Fig. 218. Deposits older than the Lower Cretaceous, have not been penetrated here.

The eastern part of the downwarp, known as the Terek-Caspian marginal downwarp, is even deeper, reaching 6000–7000m, filled not only by Cainozoic and Cretaceous sediments but also by Jurassic rocks. It is interesting that here the thickness of Anthropogene and Pliocene sediments reaches the unusual figure of 2500–2700m. K. N. Paffengol'ts (1959a) considers that the sub-Caucasian marginal downwarp originated in pre-Aptian times but its main structures arose in the Miocene period and were finally formed in the post-Pontian epoch. The present author agrees with this exposition.

The structure of the northern margin of the fold zone varies from district to district. In the Soviet Carpathians owing to the push of the Skiba zone, consisting of Upper Cretaceous and Palaeogene rocks. Miocene strata are deformed into complex folds, in places inverted (Fig. 238). In Crimea along the northern margin of the Yaila

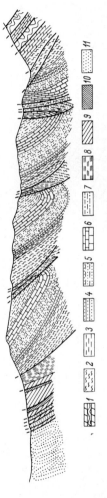

Fig. 217. Isoclinal scales in the south-eastern part of the Caucasian range. According to V. A. Grossgeim.

1 — Valanginian; 2 — Hauterivian; 3 — Upper Aptian and Middle Albian; 4 — Upper Albian; 5 — Cenomanian; 6 — Turonian-Coniacian; 7 — Santonian-Lower Campanian; 8 — Palaeocene; 9 — Lower and Middle Eocene; 10 — Upper Eocene; 11 — Oligocene-Lower Miocene.

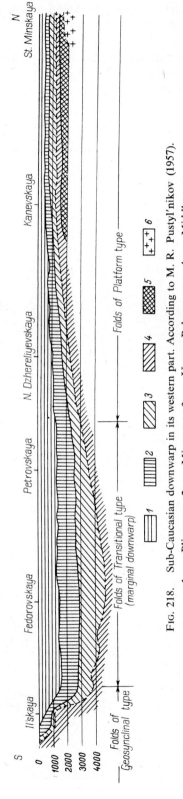

Fig. 218. Sub-Caucasian downwarp in its western part. According to M. R. Pustyl'nikov (1957).

1 — Pliocene; 2 — Miocene; 3 — Upper Palaeogene; 4 — Middle Palaeogene; 5 — Cretaceous; 6 — Palaeozoic.

range Neogene and Anthropogene rocks are often involved in a conformably bedded monocline which shows a gradual decrease of dip in younger deposits. The Neogene forms large relatively gentle brachyanticlinal folds, which are especially well represented on the Kerch Peninsula, where they have been studied by N. I. Andrusov (1893), who was a leading specialist on the Neogene and its faunas. The Kerch structures are oil and gas bearing and in them is found the major deposit of oolitic iron ore.

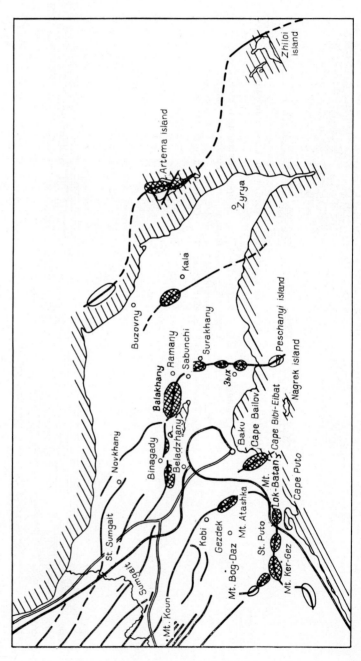

FIG. 219. Tectonic map of the Apsheron Peninsula with anticlines outlined as ovals (1937).

Identical structures are found where the axis of the Greater Caucasus anti-clinorium submerges on the Apsheron Peninsula. The main structures are shown in Fig. 219. It is clear that they have an almost north–south trend as if enclosing the anticlinorium. The brachyanticlines are broken up by a series of normal faults which are usually small with throws of several metres to tens of metres, but there are also faults with displacements of hundreds and even thousands of metres. Such faults penetrate down to the surface of Upper Jurassic, bituminous gas-bearing limestones. Gas emerging from these cracks bring up water, mud and rock fragments to the sur-face, giving rise to mud volcanoes characteristic of the Apsheron, the Kerch and the Taman peninsulas.

The nature of the Apsheronian brachyanticlines is illustrated in Fig. 220. Sub-surface isoclines sketched on the basis of hundreds of boreholes give a clear im-pression of the shape of the folds and the nature and extent of faulting. It should be noted that only the largest faults are shown, while their total number is very large and the terms 'broken plate' and 'broken cup' are not exaggerations (Fig. 221). Such a term has already been used in the case of the oil-bearing brachyanticlines of the Cheleken island (peninsula) in western Turkmenistan.

As an example of structures with the Palaeogene Maikop cores one can quote the Atashkinskian brachyanticline, situated to the west of Baku. Its cross-section is given in Fig. 222. This profile shows a core of Maikop clays broken by faults and standing vertically. These strata are overlain by Miocene clays which are also steep. They are overlain by thick clays with productive sands, which are in turn overlain by thin Akchagylian clays covered by very thick Apsheronian clays and Anthro-pogenic sediments. Throughout this massive, rather homogeneous succession of clays and sands there are no obvious unconformities, but while Oligocene and Miocene rocks are nearly vertical the newest strata are almost horizontal. This is yet another example of an almost uninterrupted sedimentation and contemporaneous folding.

The northern slope of the Main range is basically homoclinal Upper Cretaceous and Cainozoic sediments forming the southern slope of the sub-Caucasian downwarp. In the west the regional dip is almost undisturbed by secondary folds. In the central part, from the river Ardon meridian, the sheet dip is disturbed by several thrusts, verging to the north. In the eastern part the thrusts are accompanied by narrow, elongated folds.

The well studied and bored folds of the Grozno oil-field can serve as an example. As the cross-section on Fig. 223 shows they are basically similar to the folds of the Kerch and Apsheron peninsulas, by being tight and steep anticlines separated by wide shallower synclines. Anticlinal crests are often complicated by thrust-like faults causing crestal overturning (Figs. 223, 224).

The basic feature of Grozno folds is their linearity. While the ratio of length to breadth for Apsheronian folds is 2–3, the same ratio for Grozno folds is 7–8.

Wide almost plain-like depressions, situated to the south of the major Caucasian and Carpathian anticlinoria, have individualistic platform-type structures. This is explained by the fact that in these regions the main orogenic episode occurred in the Palaeogene, and in the beginning of the Neogene they become somewhat metamor-phosed and hardened massifs with great thicknesses of effusive rocks. Within the area of the massifs mainly downward movements occurred producing conditions favour-able for the accumulation of thick formations of sedimentary and volcanic rocks, reaching a total thickness of 4000–5000m. In some districts block upward movements produced flexures in Neogene strata. The flexures often pass into normal faults of a

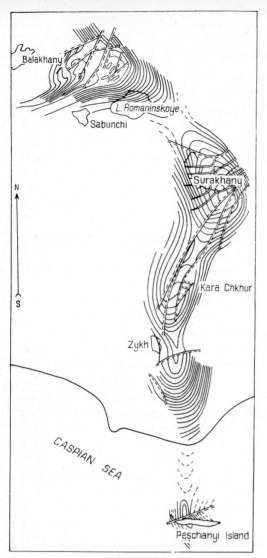

FIG. 220

An example of brachyanticlines at Romany, Surakhany-Zykh and Peschanyi island on the Apsheron Peninsula. Anticlines are broken up by faults (1937).

FIG. 221

A longitudinal section of the Su-
khanian fold showing a series of st
like faults (1937).

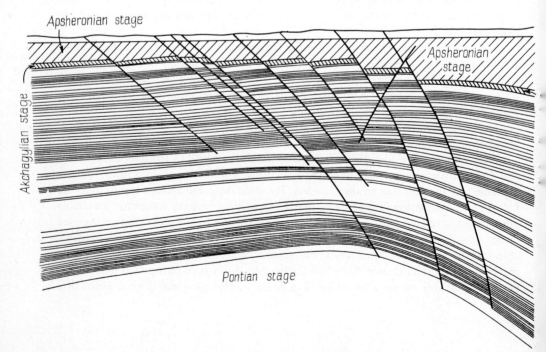

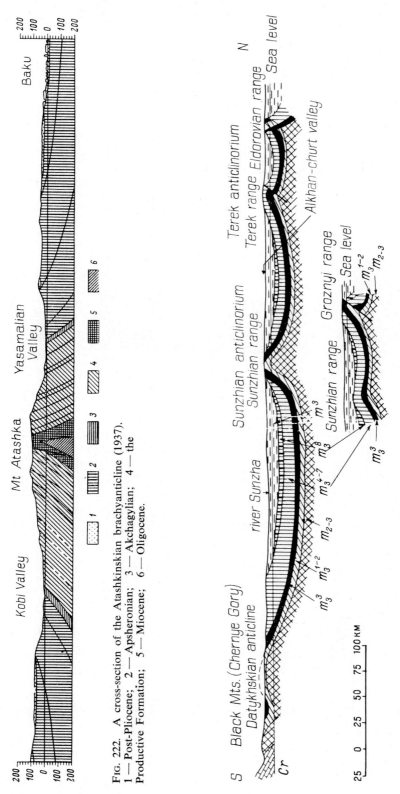

FIG. 222. A cross-section of the Atashkinskian brachyanticline (1937).
1 — Post-Pliocene; 2 — Apsheronian; 3 — Akchagylian; 4 — the
Productive Formation; 5 — Miocene; 6 — Oligocene.

FIG. 223. Section across the Groznyi oil region. According to B. A. Alferov (1937).

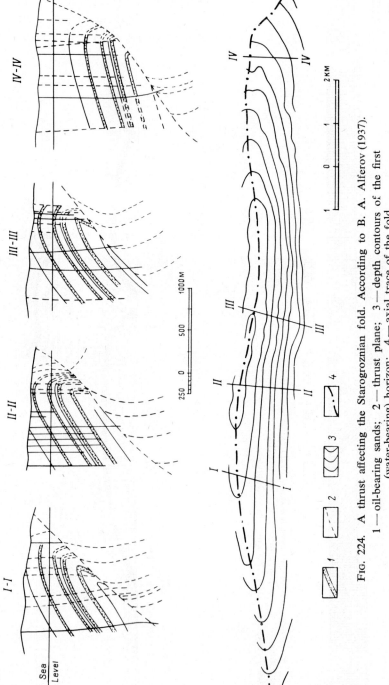

Fig. 224. A thrust affecting the Starogroznian fold. According to B. A. Alferov (1937).

1 — oil-bearing sands; 2 — thrust plane; 3 — depth contours of the first (water-bearing) horizon; 4 — axial trace of the fold.

great depth. These faults became the channels of penetration of basaltic lavas. At the same time local, but complex, structures originated near the faults since the compact Palaeogene and Cretaceous rocks in the process of upward movement were pushed against the less competent Neogene strata.

To the south of the main Carpathian anticlinorium there are two huge depressions—Alfoldian in the north and Transylvanian in the south. In the former Neogene and Anthropogene strata are over 4000m thick and are slightly inclined to the south-west. In the axial part of the adjacent narrow Solotvinian Depression there are gentle dome-like structures reflecting buried salt domes. The salt is of a Miocene age. In the Transylvanian Depression the thickness of Neogene and Anthropogene rocks is over 5000–5500m. Near to the edge of the depression they are gently inclined towards its centre. Nearer to the centre there is a ring-like zone within which Neogene beds are deformed into complex folds, sometimes of flexural type. In the central part of the depression Neogene strata are again almost horizontal, although in places there are gentle dome-like anticlines. These domes have gas and salt water and are probably produced by the underlying Lower Miocene salt domes. At the margins of the depression there are numerous Tortonian lithothamnian reefs, similar to the Podolian toltras developed at the margin of the northern submontaine downwarp.

The southern depression to the Crimean Mts at present forms the bottom of the Black Sea, which is distinctly horizontal. To the south of the Caucasian anticlinorium lie the Rion and the Kura depressions. Their structure is basically the same as the southern Carpathian depressions. In the Rion Depression as a result of Anthropogenic uplift Neogene strata are widely eroded, and their largest outcrop is in the lowest part of the depression near the sea. As an example of the almost horizontal Neogene cover one can quote the region to the east of Kutaisi, where Neogene, Palaeogene and Upper Cretaceous rocks overlie the Lower Palaeozoic granites of the Dzirul Massif. Further to the south Neogene rocks are again almost horizontal or form large gentle folds complicated by thrusts and faults. Near the edges of synclines the strata bend sharply upwards forming monoclinal flexures. Such marginal monoclinal flexures are quite common.

In the Kura Depression structures affecting Neogene and Anthropogene deposits are similar to those of the Rion Depression. There are also areas of almost horizontal Neogene deposits, which pass into zones of large, gentle brachyanticlines, locally broken up by thrusts and normal faults. There are again monoclinal flexures, which sometimes pass into box folds adhering to the blocks and grabens of the basement. The zone of Kura brachyanticlinal folds continues directly to the Apsheron Peninsula, where brachyanticlinal folds are most typical and are common in a small area (p. 646).

The structures of the western part of the Kura Depression to the north of Tiflis are shown in Fig. 225. The post-Sarmatian conglomeratic series (Dushetian Suite) is exposed at the foot of the Main range and is here strongly deformed into linear folds with isoclinal and inverted limbs. Further to the south nearer to the centre of the depression the folds become gentler and simpler. To the east the character of the structures remains the same. The formation of the folds occurred contemporaneously with epeirogenic downward movements of the Kura Depression. As a result the total thickness of Cainozoic rocks reached a great figure of over 9000m.

In the south of the Lesser Caucasus Neogene strata are continental, almost horizontal and folded into large, gentle folds. Where the folds are broken by thrusts local deformation and steep dips can be observed. Anthropogenic flood basalts, sometimes occupying large areas, are not affected by folding and lie almost horizontally, or follow the relief. In places they are also faulted.

In general, although Neogene and Anthropogene strata are in places affected by regional folding and complex structures, their structural style is much simpler than that of older rocks. The complexity of structure developed discontinuously, following the sequence of orogenic phases, which in the Mediterranean Geosyncline is very clear and instructive.

The History of Tectonic Movements

The history of tectonic movements in the Mediterranean Geosyncline has been traced more accurately and in greater detail than in any other geosyncline. Soviet geologists have had a leading role in these major and important investigations. The study of orogenies is considerably influenced by the method of their subdivision into phases. In the Caucasus not only all the phases recognized by H. Stille, but a number of others, have been recorded. In this respect the investigations of A. P. Gerasimov, V. P. Rengarten and K. N. Paffengol'ts are very significant. In the Greater Caucasus the method of isopachytes was first worked out by V. V. Byelousov and has since been applied throughout the Soviet Union. Other investigations have aready been mentioned.

Precambrian and Lower Palaeozoic

Complex and lengthy tectonic movements of Precambrian and Lower Palaeozoic ages have not been sufficiently investigated, since outcrops are limited and metamorphism is powerful. In the Lesser Caucasus the actual Precambrian or Lower Palaeozoic age of several formations has not as yet been established. Nevertheless the available data permit the drawing of the conclusion that the tectonic history of the Precambrian and Lower Palaeozoic era is not different from that in Palaeozoic geosynclines. The Byelomorian, Karelidian and Baikalian orogenies can be recog-

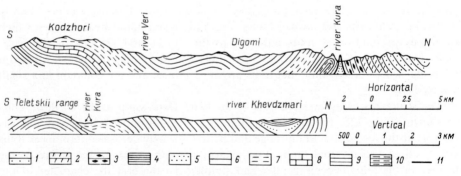

FIG. 225. Sections of the environs of Tiflis (Tbilisi). According to I. V. Kacharova (1937).

1 — conglomeratic series; 2 — Natskhorian (Upper Sarmatian) Series; 3 — Middle and Lower Sarmatian; 4 — Middle Miocene; 5 — Oligocene; 6 — Upper Eocene; 7 — Mamodavidian Series; 8 — Middle Eocene conglomerates; 9 — Middle Eocene; 10 — Lower Eocene; 11 — thrust.

nized in the Precambrian and the Caledonian Orogeny in the Lower Palaeozoic of the Caucasus.

The Caledonian Orogeny is especially obvious in the Lesser Caucasus where palaeontologically dated Cambrian and Devonian rocks show considerably differing degrees of metamorphism and structural forms. Cambrian rocks consist of phyllites, slates and meta-limestones folded into complex and highly dislocated structures. Devonian rocks consist of normal relatively unmetamorphosed limestones, shales and sandstones forming simple, slightly dislocated folds.

In the Greater Caucasus the initial Lower Palaeozoic phases can be proved by the sharp difference of metamorphic grade between Lower and Middle Palaeozoic strata. The black, Upper Devonian shales of the southern slope, in western Georgia, used to be considered as Lower Jurassic until the fauna was found. These rocks differ sharply from the nearby Cambrian strata of the Dzirul Massif. The discovery of pebbles of Lower Palaeozoic rocks in Lower Devonian (?) conglomerates of the northern slope of the Main range is also significant, since it suggests a break and an unconformity separating Lower and Middle Palaeozoic rocks. An important proof of the occurrence of the last phase of the Caledonian Orogeny derives from absolute age determinations, which record the oldest, undoubtedly Cambrian and Precambrian formations as 350–400 myr. This indicates that these rocks were tectonically reworked at the time of the Caledonian Orogeny.

Middle and Upper Palaeozoic

If the occurrence of initial Caledonian phases at the end of the Ordovician and possibly at the end of the Cambrian does not provoke any doubts; the situation with the final, end-Silurian phases is much more complex. In the Lesser Caucasus their occurrence is probable, since the Devonian succession begins with the Eifelian Stage and marine Lower Devonian strata are missing. The proof is still lacking because nowhere can one see the unconformity between the Eifelian and the Silurian. In the Greater Caucasus, conversely there is every reason to believe that, as in the Urals, the final phases of the Caledonian Orogeny did not occur. This is proved by the development of Hercynian limestones with a mixed Upper Ludlovian and Lower Devonian fauna. Similar limestones began to form in the Silurian and ended in the Lower Devonian which indicates that there was no break between the systems and the absence of a local manifestation of the Erian phase of the Caledonian Orogeny.

The Middle Palaeozoic epeirogenic movements separate the Caledonian Orogeny from the Hercynian. In the Erevan-Nakhichevan synclinorium they reached a great development and caused the accumulation of Middle Devonian and Upper Devonian and Lower Carboniferous sediments of over 2000m in thickness. The downward movements were balanced by the deposition of shallow-water marine sediments, and do not cause the deepening of the basin. Short-lived and small uplifts caused the appearance of barren clastic deposits of near-shore alluvial plains amongst marine limestones. In the Greater Caucasus even more lengthy downward movements probably occurred since thick Devonian and Lower Carboniferous strata encompass all the subdivisions of the systems.

Upper Palaeozoic epeirogenic movements began after the occurrence of the first phases of the Hercynian Orogeny and continued from the end of the Middle Carboniferous throughout Upper Carboniferous and Permian times, and in the Lesser Caucasus even extended in the Lower and Middle Trias. In the Greater Caucasus a minor, palaeogeographically insignificant break separates the Upper Permian from the Lower Trias. Here the total thickness of sediments accumulated during the epeirogenic movements is 5000–6000m. This indicates that downward movements were long and balanced, and were only interrupted by short breaks. It is interesting that in Middle and Upper Carboniferous and Lower Permian times the land as a near-shore plain was being depressed while in Upper Permian and Triassic times it was the subsiding bottom of a shallow-water coralline tropical sea. Thick (3000–4000m) Lower Permian red beds represent a typical molasse, being the deposits of the margins to mountain ranges. The accumulation of these rocks went hand in hand with the erosion and elevation of the mountain ranges situated in the north—the area of the Ciscaucasian plain—and in the south—the area of the Main Caucasian range. In the Lesser Caucasus beginning with the Lower Permian and ending with the Lower Trias there was a continuous accumulation of mainly marine deposits of 1500–2000m in thickness. These rocks imply continuous downward movements, but on a smaller scale than in the Greater Caucasus. The Hercynian Orogeny occurred ubiquitously but was relatively weak and only its initial phases, in Middle Carboniferous times, were quite powerful. These movements caused the formation of land and the Lower-Middle Carboniferous break. The initial phases were accompanied by the intrusion of granite massifs. Lower Permian molasse deposits indicate that orogenic and fold movements occurred during this epoch. These movements were the same as in the Urals and in the Tian Shian and were probably accompanied by granite intrusions, which have not been as yet identified. The last phases at the end of the Permian and beginning of the Trias were relatively weak, causing a break and an unconformity between Permian and Triassic strata. This break was very short-lived and was not accompanied by palaeogeographic changes. Marine Upper Permian strata were succeeded by marine Lower Triassic beds.

In the Lesser Caucasus the Hercynian Orogeny was weaker and the break which lasted the duration of the Middle and Upper Carboniferous, did not involve granite intrusions. Lower Permian molasse is absent indicating the inactivity of Lower Permian phases. Lastly the gradual transition from marine Upper Permian into Lower Trias rocks indicates that the last phase of the Hercynian Orogeny did not occur here.

Upper Trias, Jurassic and Lower Cretaceous

The interval of time encompassing Upper Trias, Jurassic and Lower Cretaceous represents an epoch of movements which in the Mediterranean Geosyncline were most important and are known as the Cimmeridian Orogeny. This orogeny was accompanied by intrusive and effusive activity and the formation of several major mineralizations.

The Cimmeridian Orogeny is completely and ubiquitously developed throughout the region. The orogeny began in the Upper Trias and ended in the Lower Cretaceous. There were two epochs when the orogeny was especially strong: (1) in Rhaetic and Lias times the Older Cimmeridian phase and (2) in Tithonian and Valanginian times the Younger Cimmeridian phase.

The evidence for the former phase rests in a thick accumulation of clastic rocks and for the second phase a sharp palaeogeographic change expressed in the formation of red beds and gypsiferous and saline suites. The investigation of the unconformities and local breaks permits the subdivisions of the two main phases into a series of other phases of shorter duration. In the Carpathians Trias and Jurassic deposits are badly exposed and have not been sufficiently investigated. From the top of the Trias and continuing into the Lias there was an epoch of accumulation of thick shaley formations of flysch type. The Tithonian is distinguished by its red beds which are lagoonal and gysiferous at the bottom and are about 300m thick and are at the top sandy-pebbly and continental (270m). In Lower Cretaceous times there was a wide-spread development of thick flysch formations indicating the effect of the Younger Cimmeridian phase.

In Crimea the Tavridian Suite of Upper Triassic and Liassic age is especially characteristic. This is a typical flysch formation of a great thickness, which has been formed as a result of the accumulation of sandy-clayey fine-grained products of erosion of a mountain chain situated at some distance away from the site of accumulation of the series and lying either to the north or to the south of this site. The accumulation of sediment and the rise and erosion of the mountain chain occurred simultaneously with the Older Cimmeridian orogenic phase.

The Middle Jurassic epoch is characterized by the weakening of orogenic forces and a sharp intensification of epeirogenic uplifts and subsidences, which were responsible for a great variability of facies, including the deposition of a thick (1500m) coal-bearing formation. The epeirogenic movements were accompanied by large longitudinal and transverse faults which served as guiding channels for thick and widely developed volcanic rocks. At the end of the Middle Jurassic elevations reached their maximum extent and in Lower Callovian times the whole of mountainous Crimea, of the present day, became land.

The Upper Jurassic begins either with Middle Callovian or with Oxfordian strata resting on various Middle Jurassic horizons. The sharp unconformity is explained by the aforementioned uplift. It is also possibly caused by the Adygeiian phase of the Cimmeridian Orogeny, but the former explanation is more probable. Upper Jurassic strata, especially from the beginning of Oxfordian onwards contain thick (2000m) conglomerates and sandstones of a typical molasse type, as well as thick flysch formations. Muddy formations similar to flysch were widely distributed in Lower Cretaceous times. All this leaves no doubt of considerable manifestations of Younger Cimmeridian orogenic phase, lasting throughout the Lusitanian and Kimmeridgian epochs and recurring again in the Lower Cretaceous. As shown by M. V. Muratov (1949), as a result of these movements two mountainous islands arose out of the sea to the south of the present-day sea-shore and supplied material for the molasse and the flysch.

The Younger Cimmeridian phase was responsible for the origin of the main Mesozoic structures of Crimea. In the Caucasus the Cimmeridian Orogeny was also very active and the Older and Younger phases are especially clearly distinguished and were separated by the Middle Jurassic epeirogenic epoch, which was accompanied by gigantic faults and an abruptly intensified volcanic activity.

In the Greater Caucasus the Cimmeridian Orogeny had the same features as in Crimea. Upper and Middle Trias strata are separated by an angular unconformity with basal sandstones and conglomerates occurring at the bottom of the former. The break is explained by a distinct orogenic phase for which A. P. Gerasimov has proposed the name Labinian. The movements occurred at the end of the Middle Trias and the beginning of the Upper Trias. It is most probable that they represent the initial Older Cimmeridian phase.

The Lias consists of very thick uniform slates—the so-called aspidic slates, which are many kilometres in thickness (ten kilometres in Dagestan) and form the watershed ridge of the eastern Caucasus. These slates are very similar to the Tavridian Series of Crimea and also indicate the occurrence of the Older Cimmeridian phase. Where the mountain chains were, which supplied the material of the slates, is as yet not clear, but they were most likely at the site of the Ciscaucasian plain. The rise of these mountain chains happened simultaneously with downward movements within the limits of the Main range, thus creating suitable conditions for the accumulation of a great thickness of fine grained clastic rocks.

The Middle Jurassic shows a great variability of facies; as well as various marine sediments there are coal-bearing strata and widely distributed volcanic rocks in places thick. All this indicates the ubiquitous nature of epeirogenic movements, uplifts and depressions, which were associated with deep faults. The movements were epeirogenic rather than orogenic since Middle Jurassic strata are involved in the same fold structures as Upper Jurassic strata. The last epeirogenic uplifts cause those unconformities and breaks which separate Middle and Upper Jurassic strata. It should be pointed out that another point of view is more commonly held, claiming that the unconformity between Middle and Upper Jurassic deposits is caused by the pre-Callovian (Adygeiian) phase of the Cimmeridian Orogeny. Considering the present state of knowledge this point can be maintained, but the Callovian Stage was preceded by the Bathonian and there are no facies complexes such as flysch or molasse of Bathonian age. The lack of such facies typical of orogenic stages forces one to doubt Bathonian folding.

The time problem of orogenic fold movements demands further investigations. The movements are conventionally dated on the basis of unconformities, and breaks. As has already been mentioned this method of approach can be used even now, but it should be supplemented by the study of sedimentation processes. Such additional considerations lead to fundamental changes of existing notions.

Upper Jurassic and Neocomian (Valanginian, Hauterivian and Barremian) strata show a distinctive distribution of basic facies complexes. On the northern slope thick carbonate formations are sharply predominant, while on the southern slope there are very thick shales. Such a distribution indicates that terrigenous components of flysch arrived from the south from some rising cordillera. The sinking of the flysch downwarp for several kilometres occurred simultaneously with the elevation of a young fold range under the impact of Younger Cimmeridian orogenic phases. In the Upper Jurassic epoch only fine-grained sediments were deposited and there were no molasse formations. This circumstance suggests considerable distance from the region of erosion. In the Neocomian, side by side with the flysch, there was a local deposition of conglomeratic molasse, which is explained by the approach of the region of erosion (young mountain chains) and the encroachment of Younger Cimmeridian orogenic phase.

At the same time as there were rising mountains in the south, in the north the region of accumulation of reefs and other limestones of Upper Jurassic-Neocomian age was undergoing a slow continuous depression also of the order of several kilometres.

In the Tithonian short-lived uplifts of the sea-floor produced a flat near-shore plain covered by bitter-salt swamps and lakes, where saline, gypsiferous and sulphur-bearing strata accumulated. This plain also underwent a slow sinking movement. In the Valanginian as a result of abrupt downward movements a shallow sea invaded the plain. Thereafter the bottom of the sea continued subsiding. Thus in Upper

Jurassic and Neocomian times long, continuous balanced movements were typical for the northern slope. These movements despite their great extent did not cause alterations in the palaeogeography.

In Tithonian-Valanginian times the downward movements were twice interrupted by an abrupt uplift and an abrupt depression. In the history of the earth such a combination of prolonged and abrupt movements is frequently encountered. As a rule these types of events are forerunners of orogenies. In this case the succeeding orogeny was Alpine, but before considering this topic let us examine the effect of the Cimmeridian Orogeny in the Lesser Caucasus.

In the Lesser Caucasus the clearly delineated flysch downwarps of long duration such as are typical of the Greater Caucasus and Crimea are absent. If during the epoch under consideration Crimea and the Greater Caucasus were mainly a region of sedimentation the Lesser Caucasus region was mainly the domain of erosion. In the latter region rising fold ranges were separated by marine straits and basins where diverse sedimentation took place. A great majority of mountain ranges and mountainous islands originated as a result of the Cimmeridian Orogeny, but at the same time there were in the same localities block movements and a concurrent vulcanicity. The vulcanicity can be detected in all the subdivisions of the Jurassic and Lower Cretaceous, which determines these epochs as times of formation of major faults and of block movements. Consequently there was a complex relationship between various tectonic movements which as yet has not been properly established.

Upper Cretaceous and Cainozoic

The Upper Cretaceous and Cainozoic represent an epoch of numerous and often considerable tectonic movements in the main contributing towards the Alpine Orogeny. At times epeirogenic movements also reached major proportions and are of great palaeogeographical significance as is for instance the Aegean-Black Sea rift which resulted in the formation of the Aegean, Marmora and Black seas, as well as the bisection of the Crimean anticlinorium. The epeirogenic movements were not accompanied by folding, but were extensive and reached a great amplitude such as is inherent in the depression of the Ciscaucasian plain and the Black Sea and the effects in the Lesser Caucasus. Such movements also affected the most diverse sections of the earth crust, including blocks of a hundred metres across and several metres in amplitude. Small-scale movements are badly investigated but often affect mineral deposits by breaking them up into separate blocks.

THE DEPRESSION OF THE CISCAUCASIAN PLAIN. Throughout almost the whole of the Lower Cretaceous the northern margin of the Mediterranean Geosyncline as a result of slow, balanced movements was land. At the end of the Albian as a result of an abrupt downward movement the uplift was interrupted and by the end of the Upper Albian epoch the whole of this region became submerged under the sea. The sea bottom continued sinking in a balanced way so that despite the accumulation of thousands of metres of sediments the facies did not change. The balanced depression continued, with minor breaks, throughout Upper Cretaceous, Palaeogene and Miocene times. Thus a thick sequence (3000–4000m) of horizontally lying, mainly marine sediments accumulated. The great depression was interrupted by a long epoch of uplift lasting throughout the Meotian and Pontian epochs and causing the first separation of the Black and Caspian seas. Later on depression returned since the thickness of Pliocene and Anthropogene strata reached 1000–1500m, but abrupt upward movements still caused the reduction in the size of the enclosed sea basins.

THE DEPRESSION OF THE BLACK SEA. This depression which occurred concurrently with the sinking of the Sea of Marmora and the Aegean Sea has already (p. 635) been described. It should be added that the movement was typically abrupt. The land surface was very rapidly, possibly catastrophically, succeeded by a deep (over 2000m) marine depression which has lasted until the present time. The investigation of its bottom sediments has demonstrated the suddenness of the depression. Bottom cores of several metres show that after the depression a thin sequence of deep-water deposits accumulated. There is no indication of shelf facies characteristic of depths of up to 200m. The deposits with a brackish fauna are succeeded without any transition by deep-water facies.

EPEIROGENIC MOVEMENTS IN THE LESSER CAUCASUS. A characteristic feature of the Lesser Caucasus, as compared with the Greater Caucasus, is a much wider development of young vulcanicity. In the Cenomanian, Palaeocene, Oligocene and Miocene a sudden intensification of vulcanicity occurred and beginning with the Akchagylian six epochs of powerful volcanic activity have been established. These events were caused by the intensification of deep faulting. The faults which were guiding channels for rising magma were accompanied by epeirogenic upward and downward block movements. The location of these movements and the outlines of the mobile massifs have already been shown; their dimensions and geographical and time distribution were considerable.

ALPINE OROGENY. The prolonged complete and powerful manifestation of the Alpine Orogeny, which began in the Albian epoch and is still continuing, is the most important feature of the Mediterranean Geosyncline. Throughout this long interval of time orogenic activity did not proceed continuously, at times becoming stronger and at other times becoming weaker or ceasing altogether. Twenty-one phases of intensified orogenic activity have been recognized in the Caucasus. Lately many phases have been subdivided into lesser divisions, as for instance the Austrian phase which has been divided into the Older Austrian, Newer Austrian and post-Austrian. This situation complicates the dating of separate phases and at times many investigators use similar terms for different phases. At present the naming of phases according to the time of their occurrence is acquiring a wide usage. Thus one refers to the pre-Cenomanian, pre-Akchagylian and pre-Upper Maikopian phases. Nevertheless the generally accepted terms are more useful.

Alpine orogenic phases have different distributions and intensity. The evidence for some is detected throughout the Caucasus and even beyond them, causing major unconformities, sedimentary breaks and palaeogeographic changes. Such phases are frequently accompanied by intrusive massifs. Other phases are only found in separate regions, do not produce major unconformities or palaeogeographic changes, are not accompanied by intrusions and cause only minor breaks. The former phases are called strong while the latter are called weak.

A review of Alpine orogenic phases in the Greater Caucasus and Lesser Caucasus has been prepared from the data of A. P. Gerasimov, L. A. Vardanyants and K. N. Paffengol'ts and follows here.

The first *Austrian* phase occurred at the end of the Lower Cretaceous in the Middle Albian (Older Austrian), at the end of the Albian (Newer Austrian) and in the Lower Turonian (post-Austrian).

The Older Austrian phase, the first main phase, is of a considerable strength and the unconformities caused by it are ubiquitous, while in the Greater Caucasus there are associated acid intrusions.

The Laramide phase occurred before the Danian Stage (Older Laramide) and in the Palaeocene (Newer Laramide). The Laramide phase is most important in the North American Rockies from where it has been named. It is also powerful in other regions of the Mediterranean Geosyncline but in the Caucasus it is weak and is only identified locally.

The Pyrenean phase (pre-Oligocene) is strong in the Lesser Caucasus where it is ubiquitous and is accompanied by acid and basic intrusions and a belt of ophiolites. In the Greater Caucasus it has been locally detected on the northern slope.

The Helvetic and Savian phases (Oligocene) are weak and have only been locally identified on the basis of breaks and unconformities within the Maikop Suite.

The Styrian phase (pre-Middle Miocene) is divided into the Older Styrian (pre-Tarkhanian) and Younger Styrian (pre-Chokrakian). The phase is ubiquitously manifested, but is only strong in the Lesser Caucasus, where it is accompanied by acid and basic intrusions and is the last phase producing fold structures.

The Attican phase (pre-Meotic) is the second main phase in the Caucasus and is powerful in the Greater Caucasus, where it caused major unconformities, palaeo-geographic changes and was accompanied by acid intrusions.

The Kabristanian phase (pre-Pontian) is closely connected with the Attican phase, representing its ultimate stages in the south-east of the Greater Caucasus.

The Eastern Caucasian phase (pre-Akchagylian) like all the subsequent phases has only been active mainly in the eastern part of the Greater Caucasus.

The Rhodanian phase (Rhonian, pre-Apsheronian) is weakly reflected in the east of the Greater Caucasus and in particular in the Kura Depression.

The Valakhian phase (pre-Baku) is the first Quaternary phase, being ubiquitous in the Greater Caucasus and is accompanied by acid intrusions. The phase together with other Anthropogenic phases—*Kalinkian* (pre-Mindel), *Kura* (pre-Riss), *Ismailian* (pre-Würm) and *Divichian* (pre-Recent)—constitutes an important diastrophic cycle, which is in fact the third main phase of the orogeny. Despite the fact that each of the Anthropogenic phases is weak and of a limited distribution, altogether they complete the formation of the Greater Caucasian fold structures and cause its relief.

The Lesser Caucasus during the Attican phase, i.e. at the end of the Miocene, re-presented a compacted rigid block, which subsequently only underwent epeirogenic elevations and depressions accompanied by deep faults and an intensive volcanic activity.

It should be remembered that in recognizing phases a full and accurate time of their initiation must be given. Little has as yet been done in this direction. Soon after H. Stille (1924) enunciated the main phases the times of their occurrence were quite mistakenly placed between successive epochs. For instance the Valakhian phase was said to have happened between the Apsheronian and the Baku epochs. Between these two epochs there is no interval of time and no orogenic phase could have occurred. Recently this has been taken into account and referred to the pre-Baku phase, but even though such a notion is more accurate it is indeterminate. The pre-Baku epoch may correspond to the Upper Apsheronian, but can also be both the Middle and Upper or even the whole Apsheronian. The problem thus arises at what time did this or other phases occur in the area under discussion.

The study of sedimentation helps in the solution of this, a not always easy problem. In the present case we know that relatively uniform, marine, muddy sediments are typical for the Lower and Middle Apsheronian. These indicate the absence of abrupt palaeogeographic changes. On the other hand the Upper Apsheronian differs in its considerable variability of facies and repetitions of marine, freshwater and terrestrial deposits, determining the time of the Valakhian phase as Upper Apsheronian.

Formations such as flysch and flyschoid suites and particularly molasse indicate the time of occurrence of phases, which is the time of elevation of fold belts. Certain phases were prolonged, lasting up to an epoch corresponding to the major division of a system, while others were short-lived, lasting only a part of an epoch. Formerly in assessing the duration of an orogeny the sequence of (1) sedimentation, (2) folding and (3) erosion of the folds was considered. In actual fact the three processes occur simultaneously in any particular epoch while the first two processes precede an orogenic epoch and the last two succeeded it. Sedimentation starts first and is soon joined by folds which appear at the bottom of the sea and are accompanied by sedimentation. Then when the folds reach the level of wave action all three processes of sedimentation folding and erosion proceed concurrently. Later when the folds rise above sea level the sedimentation stops, but formation of folds and their destruction proceed simultaneously. Lastly folding stops but erosion continues, which is theoretically possible, but in practice is unimportant since the duration of this stage is so short that on the scale of geological time it is insignificant.

The tectonics of the Mediterranean Geosyncline have been studied better than in any other. The result of this investigation shows how little we know about the process of folding.

Neotectonics and Seismicity. The Neogene and Anthropogene movements are known as neotectonics. There is an interesting review of them in a volume: *The Neotectonics of the U.S.S.R.* (1961). The seismicity and earthquakes are caused by faulting which occur at present. A large number of faults follow the already existing planes, which had come into existence in the Neogene and Anthropogene. Consequently neotectonics and seismicity are closely related. Observations on the largest earthquakes have shown that they cause displacements of no more than 4–6m. At the same time many faults are associated with displacements of the order of hundreds and even thousands of metres. They have doubtless originated as a result of repeated fractures sometimes over a very long period of time.

The distribution of Caucasian earthquakes is interesting. They are almost completely concentrated in the eastern part of the Main range, at its margins and mainly along the southern slope. In the Lesser Caucasus earthquakes are absent, since this area at present forms a stable monolithic block. Relatively few earthquakes occur in the western part of the Caucasus. The eastern part represents a mobile region, which is continuously moving and is affected by numerous faults. Many earthquake epicentres are situated along the sub-aqueous ridge which connects the Apsheron and the Krasnovdsk peninsulas. Thus the highly seismic regions of the eastern Caucasus and western Turkestan are closely associated.

MAGMATISM

The dependence of magmatism on tectonics, or more accurately the relationship between tectonics and vulcanicity, is exceptionally clear in the Mediterranean

Geosyncline. Each orogeny and even each main orogenic phase is accompanied by intrusions. The large number of orogenies and their main phases causes the unusual diversity of these intrusions. Epochs of epeirogenic movements separating or co-inciding with orogenic epochs are accompanied by vertical faults which penetrate down to the magmatic zone and serve as guiding channels of numerous volcanoes, which are often very large. The young volcanic activity reached particularly impressive dimensions in the southern regions of the geosyncline, namely Transcarpathia and Transcaucasia. Here folding occurred and finished earlier. From the Pliocene onwards these regions became almost platforms and almost all tectonic movements became expressed as block elevations or depressions accompanied by deep faults. It is this feature that causes the unusual development of various volcanic rocks, their tuffs and breccias.

In the Greater Caucasus, the Carpathians and the Crimean Mts on the other hand, Pliocene and Anthropogene strata are characterized by a maximum develop-ment of folding and elevation of mountain ranges. Consequently here the volcanic activity is limited, but young intrusions are widespread. Precambrian and Palaeozoic intrusions of all these regions are known as ancient or palaeointrusions, while Mesozoic and Cainozoic intrusions are known as young or neointrusions.

PRECAMBRIAN AND PALAEOZOIC MAGMATIC CYCLES. In connection with the limited distribution of ancient strata the rocks of the Precambrian and Palaeozoic magmatic cycles also show a limited development and have not been sufficiently investigated. The presence of all four magmatic cycles is undoubted, but the complete separation of Archeozoic, Proterozoic, Caledonian and Hercynian magmatisms is a problem for the future. Only Precambrian and Hercynian cycles have been in places differentiated with a sufficient confidence. The absence of areas or sections where both Lower Palaeozoic and Middle Palaeozoic rocks are exposed renders the identification of Caledonian intrusions only provisional.

In the Main range the relatively rare and highly altered augen-gneiss granites and amphibolites are the oldest. They are known as the Urushtenian complex and are typically found in the valleys of Urushten and Zelenchuk, which are the tributaries of the Kuban' on the northern slope of the Front range and also to the east and west of it (Fig. 226). Urushtenian serpentines and granites cut across the oldest metamorphic formation and their pebbles are found in conglomerates which are provisionally considered as Devonian, but may be older. The age of the complex is suggested as Caledonian, but the suggestion has not been proved. The regional gneissosity of the complex and the association of its rocks with ancient formations implies a probable Precambrian age. G. D. Afanas'yev (1955), who for many years worked on magmatic rocks of the Caucasus, has brought forward a series of radio-metric dates, varying within the limits of 280–380 myr, thus indicating a Lower Palaeozoic age. The significance of these dates must, however, be revised (p. 587), and the stratigraphic position of the complex is shown on Fig. 226.

The so-called 'grey' or 'central' granites are most widespread. They are traced for 400km as individual massifs continuous along the range, outcropping amongst metamorphic rocks (Fig. 226). The granites are diverse compositionally, and probably in their ages. They were formerly considered as Proterozoic since they cut meta-morphic rocks and some of their pebbles are found in Palaeozoic strata. Nowadays, however, these granites are thought to be Hercynian. G. D. Afanas'yev (1955) has based this opinion on the fact that the 'grey' granites intrude the Urushtenian complex and are older than Permian red beds, while their absolute age varies from

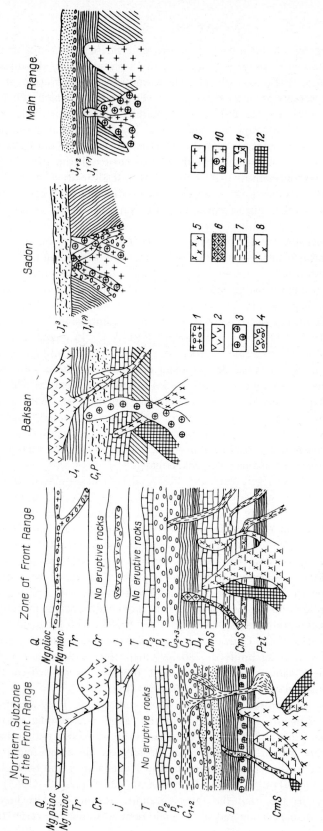

FIG. 226. Sequence of magmatic events in the Northern Caucasus. According to G. D. Afanas'yev (1955).

1 — post-Miocene andesites, andesite-basalts and dacites; 2 — post-Miocene liparites, trachyliparites and Pyatigorian extrusions; 3 — Tyrny Auz porphyritic granites; 4 — Jurassic andesitic and andesite-dacitic lavas; 5 — Lower and Middle Carboniferous granite porphyries of the Kuban' region; 6 — Lower and Middle Carboniferous minor intrusions of the Peredovoy range; 7 — Upper Palaeozoic intrusions and extrusions of palaeoliparites and palaeodacites; 8 — Upper Palaeozoic granodiorites, granites and alaskites; 9 — two mica granites of the southern sub-zone of the Main range; 10 — alaskites and porphyroid granites of the Main Range; 11 — granite gneisses and granite veins of the Urushtenian complex; 12 — serpentinites of the Urushtenian complex.

180 to 230 myr, although sometimes it reaches 300 myr. The last figure corresponds to the age of the Urushtenian complex.

It is more correct to consider the age of the 'central' or 'grey' granites as not established, but in a majority of cases not younger than Hercynian. The fact that they are of different ages is shown in Fig. 226, since some of the intrusions penetrate into the provisionally Lower Jurassic deposits. G. D. Afanas'yev (1960) gives absolute age determinations of 190–210 myr.

Central intrusive massifs are accompanied by pegmatites, aplites and hydro-thermal veins.

The Dar'yalian Massif serves as an example of 'grey granites'. The granites line the walls of the well-known Dar'yal gorge of the river Terek. The position of the massif is shown in Fig. 226. The massif is uncomfortably covered by the Lower Lias and clearly Mesozoic strata and its age is 125–195 myr.

In the Kuban' Basin at the very north of the region occupied by ancient strata 'red' or 'northern' granites emerge from underneath the overlying Liassic rocks. The granites intrude metamorphic formations and are transgressively covered by Lower Trias and Lias sediments. These intrusions are provisionally thought of as Caledonian but judging from their absolute age of 220–250 myr a Hercynian age is more likely. The significance of these figures must be reviewed.

The ultrabasic serpentinites of the upper reaches of the Laba and its tributaries are also found elsewhere and are considered to be Hercynian. The same age is attributed to quartz diorites, syenites, quartz porphyries and other intrusive rocks found in the central part of the western Caucasus where they intrude Lower and Middle Palaeozoic deposits.

Middle and Upper Palaeozoic lavas of a wide distribution also belong to the Hercynian cycle.

Dzirul Granite Massif. The massif is situated between the upper reaches of the Kura and the Rion and consequently between the Greater and the Lesser Caucasus. The massif is quite large, of up to 60km by 30km. Its surface forms a low plateau inclined to the south. In the south, north and east it is bounded by normal faults, but in the west it disappears under Cretaceous and Palaeogene sediments. In places it is transgressively overlain by Middle Jurassic and Lower Jurassic deposits. The massif consists of granites of two generations—the grey and the red—which are analogous to the grey and red granites of the Main range. In addition in the eastern part of the massif there are two ancient formations: (1) the Crystalline consisting of gneisses, schists and amphibolites, and (2) the Metamorphosed consisting of quartzites, slates and meta-limestones with archeocyathids. The Crystalline Form-ation is considered as Precambrian, since it is intruded by grey granites and its xenoliths are found in them. The Metamorphosed, Cambrian formation is cut by red granites which suggests their age as Caledonian or Hercynian. Associated with the red granites are aplites and pegmatites which also intrude the metamorphic rocks. The tectonic position of the Dzirul Massif is not clear. Some investigators (V. P. Rengarten) consider it as a horst-like basement block, of which the main part is a basement to the Rion and Kura Depression and which conditions the relatively simple structures of these depressions. The actual existence of such a block amongst the folded structures of the geosyncline is unlikely and demands further evidence. The other point of view advanced by K. N. Paffengol'ts is more likely. According to him the Dzirul Massif as well as the more southerly Somkhetian (Khramian) Massif, the Chatakhian inliers of ancient strata and other more distant inliers lying to the south-east are parts of the axial zone of major anticlines with a south-easterly trend.

Khramian granites. The term covers the Somkhetian Granite Massif and the Chatakhian outcrops found in the valley of the Khrami and its tributaries to the south-west of Tiflis. The rocks are grey and red medium to coarse grained micaceous, microcline-plagioclase granitoids. Near to them there are small inliers of micaceous gneisses and metamorphosed ancient strata, which may be Precambrian or Lower Palaeozoic. The Khramian granites have been even less investigated than the Dzirulian. Georgian geologists consider the former as part Caledonian—part Hercynian. There is no doubt that the keratophyric intrusions (granite porphyries) of the upper reaches of the Kuban' are Hercynian, since they intrude Middle Carboniferous strata and are unconformably covered by Lower Permian red beds.

In the Marmarosh Massif of the Soviet Carpathians there are centrally situated schists and gneisses, many of the latter being considered as deformed granites. Along the edges of the massif there are outcrops of less altered rocks (slates), the Palaeozoic age of which is quite probable. Amongst the slates are silts and dykes of porphyrites, of up to 500–1500m in thickness and stretching for 50km, as well as granitoid (granite porphyries) minor intrusions. A large syenitic intrusion is known in the south-west of the Marmarosh Massif on the territory of Rumanian Peoples' Republic.

CIMMERIDIAN MAGMATIC CYCLE. The Cimmeridian magmatic cycle is manifested ubiquitously and is of a wide distribution. The cycle is represented by lavas, major and minor intrusions and vein rocks. Its main diagnostic features are the occurrence of its rocks amongst Lower Jurassic and Middle Jurassic shales and sandstones and the unconformable relationship of Upper Cretaceous sediments to them.

There is no doubt that some Cretaceous intrusions lie amongst Palaeozoic strata, which must be penetrated first. The separation of such intrusions from truly Palaeozoic is a problem for the future. Another problem is the distinction of the Older Cimmeridian intrusions from the Younger Cimmeridian intrusions. The Older Cimmeridian intrusions can only be those with active contacts with Lower or Middle Jurassic strata and transgressively overlain by Upper Jurassic rocks, while pebbles or fragments of such intrusions are found in Callovian, Oxfordian and Lusitanian sandstones and conglomerates. Such interrelationships have been proved only in individual cases and consequently a greater part of the intrusive massifs lying amongst Jurassic sandstones must be considered simply as Cimmeridian. There is of course the possibility that some of them are Alpine.

In the Carpathians only Jurassic volcanics are known and in Transcarpathia, in the Muresh Mts they reach a great thickness, occupy a large area and consist of porphyrites, basalts, tuffs and tuff breccias. In places there are small basic intrusions being probably volcanic pipes. The intrusions sometimes lie, as in Crimea, under Tithonian limestones and sometimes under Lower Cretaceous conglomerates.

In Crimea, rocks of the Cimmeridian magmatic cycle are widely distributed along the south shore where they are sometimes responsible for considerable hills, as for instance Ayu-Dag ('Medved' Gora'), near Gurzuf (Fig. 227). All these rocks lie amongst Upper Jurassic (Callovian) sandstones and conglomerates and are associated with the Middle Jurassic. Small intrusive bodies cut across the Tavridian Series in places and represent guiding channels of Middle Jurassic volcanic rocks. Volcanic activity began in the Lower Bajocian and is found throughout the south shore. In the Bathonian epoch vulcanicity was local and occurred over a small area.

Most eruptions were submarine and originated in definite regions, giving rise to groups of volcanoes which in part rose above the sea level. The height of these volcanoes was no less than 1500–2000m. These figures are based on the thickness of volcanic rocks. Lavas consist of spilites, keratophyres and liparites and are accompanied by tuffs, tuff breccias and agglomerates.

The intrusions were of two types—subvolcanic and synorogenic. The former constitute stocks and dykes, being guiding channels of volcanic rocks. The latter reach large dimensions (several kilometres across) forming intrusive massifs and penetrating Jurassic strata during the pre-Callovian orogenic phase, which produced an unconformity between Middle and Upper Jurassic rocks. The pre-Callovian phase represents the final stage of the main Older Cimmeridian phase and occurred at the end of the Bathonian epoch. Thus Crimean intrusions present a rare example of definitely proved Older Cimmeridian intrusions.

In the Greater Caucasus Lower and Middle Jurassic volcanic rocks have a considerable distribution. An almost uninterrupted chain of volcanoes continued over the great distance from Black Sea to Kakhetia. The lavas and tuffs of these volcanoes form thick formations of many hundred metres. These formations are mainly acid varieties at the bottom and basic and intermediate at the top. Various porphyrites predominate and consequently the whole region of their development is known as 'the porphyrite belt'.

On the southern slope of the Main range the volcanic activity began in the Lias and continued into the Middle Jurassic and in some regions even into the Upper Jurassic. On the northern slope volcanic rocks are far less distributed in time and space and occur only in the Lias. In the Middle and Upper Jurassic and the Lower Cretaceous there are virtually no volcanic rocks.

FIG. 227. The intrusive massif of Ayu-Dag on the southern shore of Crimea. The town of Gorfuz is in the forefront.

Powerful manifestations of the Older and Younger Cimmeridian orogenic phases condition a considerable distribution of Cimmeridian intrusions. Such intrusive bodies are found amongst Jurassic and more recently identified intruding Palaeozoic and Precambrian deposits. Granite intrusions situated on the southern slope to the north-east of Sukhumi belong to the Older Cimmeridian phase. The intrusions intersect the ancient strata, the Lias and the Porphyritic Bajocian Suite, which is overlain transgressively by Lower Cretaceous rocks. The intrusions are represented by quartz diorites, granodiorites and granites. Intrusions of the same type are developed in the Dzirul Massif. On the northern slope a pre-Callovian intrusion is found in the region of the Sadonian mine where its rocks consist of granite-porphyries which intersect Upper Lias keratophyres.

The age of the Sadonian intrusion is debatable. A. G. Afanas'yev (1955) considers it Hercynian, but the absolute age figures quoted by him do not support this point of view. The Upper Lias keratophyres give an age of 70 myr and the alaskites and pegmatites which are younger than keratophyres give 120, 185, 190 and 250 myr. All this and the relevant geological relationships (Fig. 226) force one to consider these granites as Jurassic (Older Cimmeridian).

The Younger Cimmeridian intrusions of the north-west Caucasus intrude the Middle and Upper Jurassic and sometimes Barremian and Albian strata. Pebbles of the intrusions are found in Upper Cretaceous conglomerates. The intrusions consist of various complexes of acid to basic rocks.

The age of many other intrusions has not been accurately determined and is quoted as Jurassic (Cimmeridian). Lastly there are intrusions such as the Tyrnyauzian, which is considered by some investigators as Jurassic and by others as Lower Miocene (Alpine). Its radiometric age is 50–70 myr.

S. P. Solov'yev (1952), one of the leading specialists on Caucasian magmatism, has proved that minor intrusions such as pegmatites and aplites, which are widely distributed in Precambrian and Palaeozoic rocks, are very rare amongst Mesozoic strata, where quartz veins and quartz-carbonate veins are often observed.

A distinctive aspect is represented by the gabbro-diabase suite which denotes typical sub-volcanic formations, analogous to intrusive varieties of the Siberian traps. The rocks occur as dykes and sills consisting of basic rocks, most frequently similar to diabase and gabbro. The rocks are widely distributed throughout the northern slope, from the basin of the Kuban' to the basin of the Samur (Dagestan). Some are also found on the southern slope, in western Georgia, Kazbek and Kakhetia. The intrusions lie amongst Lower Jurassic and rarely Middle Jurassic (Aalenian) rocks. It is probable that they represent the sub-volcanic part of magmatic formations which came to the surface as lavas and tuffs of Lower and Middle Jurassic age. Many polymetallic and copper mineralizations are associated with the intrusions.

In the Lesser Caucasus the Cimmeridian cycle is also widely evident, but less so than the Alpine. Volcanic rocks of this region are basically associated with Lower and Middle Jurassic strata, but are also encountered amongst Upper Jurassic and rarely Aptian-Albian sediments. In the eastern part of the Lesser Caucasus anticlinorium volcanic rocks occupy especially large areas where they occur amongst Jurassic deposits. Here Middle and Lower Jurassic volcanic formations reach a thickness of 1500–2000m, and are accompanied by major ore deposits, situated to the west of Allaverdi, at Kadabek, Dashkesan, and to the east of Zangezura. Amongst sedimentary tuffs there are beds with marine faunas, while in the Nakhichevan region, although the Lias is still volcanogenic, almost the whole of the Middle Jurassic sequence is represented by sandstones and limestones with a marine fauna.

To illustrate the volcanogenic Jurassic succession let us quote the Kirovabad section of Kedabek and Dashkesan region, as given by K. N. Paffengol'ts.

At the base lies a thick (over 2000m) formation of porphyrites, tuff breccias, tuff conglomerates and other tuffaceous rocks. There are no fossiliferous horizons in it, but by analogy with the Dzirul region, where such horizons have been found, the Porphyritic Formation is provisionally considered as of a Liassic age. Higher up lies a suite of quartz porphyries of a variable (up to 500m) thickness, which in places wedges out completely but in general is found throughout eastern Transcaucasia. The age of the suite is Bajocian. The suite is stratified, representing a series of lavas, rapidly following each other. The suite is overlain by the Upper Porphyritic Formation of 800–1500m in thickness, which also consists of porphyrites, tuffs and tuffogenous rocks. Fossiliferous bands show that the greater part of the Upper Porphyritic Formation belongs to the Middle Jurassic and its upper part is in places Callovian or Oxfordian. Higher up lie Lusitanian limestones overlain by the Third Porphyritic Formation of up to 500m in thickness. The formation has no fossiliferous bands and is provisionally included in the Tithonian. The Kirovabad section is the most complete but its thickness decreases laterally and the Third Suite wedges out. The details of the section change much along the strike but the main two suites are continuous.

Cimmeridian intrusions lie amongst Jurassic deposits, having active contacts against them and being transgressively overlain by Lower and Middle Cretaceous deposits. Formerly it was thought that many intrusive bodies were of a Cimmeridian age. The investigations carried out in the last few years have shown that marginal parts of these massifs intrude Upper Cretaceous sediments and are consequently of an Eocene age. The relatively large (25×4km) Mekhmanian Massif in the Rocky Karabakh is undoubtedly Cimmeridian and is situated to the south of Kirovabad. The Younger Cimmeridian age of this massif is proved by the fact that the constituent granites intrude Middle Jurassic strata and their pebbles are found in Albian conglomerates.

THE ALPINE MAGMATIC CYCLE. The rocks of this cycle are also as widespread as those of the Cimmeridian, and occupy large areas in the Lesser Caucasus. In the Greater Caucasus, although they are frequently encountered, they cover a much smaller area. Effusions of differing ages—Cenomanian to Present—are most widespread, while intrusions are rare and small. The youngest of the exposed intrusions belong to the Lower Miocene.

In the Soviet Carpathians the Upper Cretaceous and the Palaeogene are represented by thick flysch formations, which always lack volcanic rocks. In the Neogene and the Anthropogene the accumulation of flysch deposits stopped and volcanic rocks became widespread. A volcanic chain originated along the south-eastern margin of the Carpathians, giving rise to the present-day Vygorlat-Gutinian range which stretches from Humenne to Czechoslovakia, through Uzhgorod to Hust. Here lavas flowed into valleys, covering large areas. Extrusion began in the Lower Miocene, but reached the greatest intensity in the Tortonian and Middle Sarmatian epochs and continued in Pliocene times. The first volcanic rocks were acid lavas and tuffs, and after an interval thick andesite-dacites and tuffs and tuff-breccias with sedimentary bands appeared. Pliocene andesite-basalts and andesite brought volcanic activity to a close. Volcanic rocks are interbedded with marine, saline and coal-bearing sediments and pass into them laterally. The total thickness including the sedimentary rocks reaches 4500–5000m. Volcanic rocks were formed in connection with deep, marginal faults at the periphery of the Alfoldian Depression.

In the Greater Caucasus the Alpine magmatic cycle is completely represented by volcanic deposits, major and minor intrusions, but the area occupied by the outcrops of these rocks is smaller than that covered by the rocks of the Cimmeridian cycle. Several mineral springs are associated with the former. Volcanic deposits are most widespread and range from Cenomanian to Upper Anthropogene. The mutual exclusion of volcanic and flysch formations is again distinctive. Where Upper Cretaceous and Palaeogene flysch sediments are present, volcanic rocks are absent, although there are rare thin sequences of tuffs representing ashes and other ejectamenta deposited at a long distance from extrusive centres. Consequently Upper Cretaceous and Palaeogene volcanics are absent from either the northern slope or flysch downwarps of the southern slope. A different picture can be sketched on the southern slope where flysch deposits are replaced by normal marine sediments which are often shallow water, as on the Black Sea shore, in south-western Georgia and south-western Azerbaidzhan. Here there is a frequently encountered Cenomanian (Cenomano-Albion) volcanic formation consisting of repetitive andesites, porphyrites, tuffogenous and sedimentary rocks, of a total thickness of up to 1000–1500m. In the north-east of the Lesser Caucasus there are volcanic deposits (porphyrites) of Coniacian-Santonian age (500–1000m).

In north-western Georgia a Lower Turonian so-called Mtavari Suite (400m) is distinguished. Its rocks consist of grains and fine pebbles (microconglomerates) of porphyritic rocks of dark-brown colour, cemented together by a calcite matrix, which includes numerous *Lithothamnium* algae. The suite is considered either as a primarily tuffogenous formation or as redeposited Jurassic porphyrites.

The next important phase of volcanic activity in the Greater Caucasus occurred at the end of the Pliocene and in the Anthropogene. Lavas of this phase concentrated in definite centres forming thick volcanoes. One of these volcanoes—El'brus—is the highest European volcano, while Kazbek—another volcano—although not quite as high is no less impressive. The third minor volcano is situated in the upper reaches of the Samur. The fourth volcanic centre is situated in the upper part of the Chegem valley, to the south-west of Nal'chik.

All these are shown on synoptic geological maps. With regard to Chegemian lavas an opinion has recently been expressed that their age is Oligocene, since they are invaded by Lower Miocene intrusions. A control borehole in Baksan has revealed a sedimentary-volcanic formation under the Miocene, which was previously considered as of Pliocene age. The problem is yet to be solved but the existence of an Oligocene volcanic formation is quite likely.

One of the great achievements of K. N. Paffengol'ts, a major petrographer of the Greater and Lesser Caucasus, was the demonstration of different ages of vulcanicity. On El'brus he has shown (Paffengol'ts, 1959b) that the older lavas (liparites) were dislocated, eroded and covered by the flows of younger lavas—dacites and andesites (Fig. 228). The existence of two epochs of volcanic activity has also been claimed by him for Kazbek. It is quite probable that two episodes of vulcanicity are of general occurrence for the whole of the northern slope of the Greater Caucasus and spread over into the Lesser Caucasus and in particular to the Aragats and southern Armenian volcanoes. Paffengol'ts considers the first episode as of Maikopian age, and Oligocene and early Miocene. The lavas and the associated intrusions which cut them are probably of the same age, the latter being Lower Miocene. Intrusive bodies concerned may have been guiding channels or sills. The sills have been noted by Paffengol'ts (1956).

The second epoch belongs undoubtedly to the Anthropogene and partly to the end of the Pliocene. Lavas start in the Akchagylian and finish at the beginning of the Holocene.

It is interesting that all the main volcanic centres are situated near to the Greater Caucasus watershed and are associated with young faults. This is because the Greater Caucasus here suffered the greatest uplift of block type and the consequent faults penetrated down to the depth of tens of kilometres as far as the basalt layer.

In the Greater Caucasus, Alpine intrusions are rare, occur over relatively small areas, and on the northern slope are associated with young faults. The intrusions of the Mineralovodian region have been investigated in greatest detail. They consist of almost white trachyliparites with a high content of sanidine. Their form is laccolithic and they have been emplaced as dome-like or conical bodies pushing up Maikop strata into which they are intruded into a corresponding shape. In their origin they are concordant injections of magma associated with a general rise of it along fractures. This is supported by the fact that some intrusions are dyke-shaped and fill actual fractures. Such intrusions have nothing in common with synorogenic varieties, since the former lie in weakly folded strata which in the main have a northerly sheet dip. Formerly the laccoliths were included in the Apsheronian epoch and considered as contemporaneous with El'brus and Kazbek effusions. Lately pebbles of trachyliparites have been found in proximate Meotic conglomerates. Consequently Paffengol'ts (1959b) includes the intrusions in the Lower Miocene-Upper Oligocene, thus suggesting their contemporaneity with the Maikop into which they are intruded. It is more likely, however that they are somewhat younger than

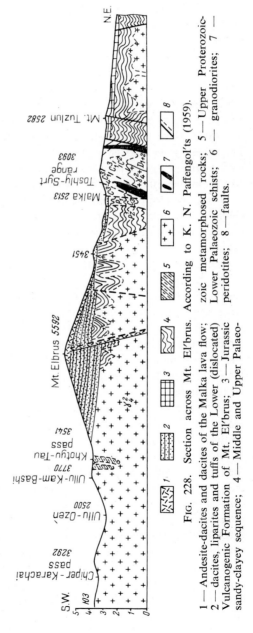

FIG. 228. Section across Mt. El'brus. According to K. N. Paffengol'ts (1959).

1 — Andesite-dacites and dacites of the Malka lava flow; 2 — dacites, liparites and tuffs of the Lower (dislocated) Vulcanogenic Formation of Mt. El'brus; 3 — Jurassic sandy-clayey sequence; 4 — Middle and Upper Palaeozoic metamorphosed rocks; 5 — Upper Proterozoic-Lower Palaeozoic schists; 6 — granodiorites; 7 — peridotites; 8 — faults.

Maikop strata and belong to the Middle or Upper Miocene. As examples one can select the Beshtau (Five Mountains) laccolith and that of Mt Zhelezmaya at the foot of which lie the watering spas of Pyatigorsk and Zheleznovodsk. The intrusions are shown on the map, and the relationships with the surrounding rocks on the sections, in Fig. 229. The sections show that magma in process of rising dragged a whole block of Upper Cretaceous deposits into a vertical position. According to G. D. Afanas'yev (1960) the age of laccolith is 25 myr. The valuable mineral waters of Pyatigorsk, Yessentuki and Zheleznovodsk, and the mineral-water 'Narzan' from Kislovodsk are associated with the young vulcanicity of the Mineralovodian region. The latter and many other mineral springs which are no longer exploited are situated near El'brus volcano.

Teplinian intrusions constitute a second group of distinctive bodies of Neogene and more probably Pliocene and Anthropogene age. They have acquired their name from Mt Tepli, situated in the upper reaches of the Cherek, a tributary of the Terek. The intrusions are found along the axial zone of the Main range, from the upper reaches of the Aragva in the east to the sources of the Byelaya in the west. Their outcrops are small, up to 4km^2 maximum, the bodies are either dykes or stocks, and are separated from each other. Their rocks are either holocrystalline granodiorites and diorites, or porphyritic dacites and andesites. The age of these intrusions is debatable, since they penetrate rocks of different ages, including Chegemian lavas. Those geologists who have studied them in detail consider them as of Pliocene or Anthropogene age. It is probable that these intrusions represent guiding channels of Anthropogene effusions. This explains their position at the axial part of the range.

The third group of younger intrusions is found on the southern slope, in western Georgia. The intrusions are sills of basalts and are closely associated with Anthropogene basalt lavas. The extrusive and the intrusive varieties together constitute a gabbro-diabase complex (traps and basalts) which is quite widespread. This complex is even more widespread in the Lesser Caucasus and has already been described.

In the Lesser Caucasus the Alpine magmatic cycle reaches an extraordinary development, since there were many eruptive episodes, the products of which cover a large area. Furthermore the volcanic formations are exceptionally well preserved and volcanic features are completely typical. The last manifestations of vulcanicity occurred in the Upper Anthropogene epoch and possibly in historical times. Powerful and diverse mineral springs are associated with the latest volcanic events. Some of the springs are internationally known, as is for instance Borzhomi, while Arzni and Dzhermuk in Armenia are becoming progressively better known.

In the Lesser Caucasus effusive rocks are particularly widespread and cover large areas. On the basis of age three large groups—Upper Cretaceous, Palaeogene and Anthropogene—are recognized.

Upper Cretaceous volcanic rocks can be subdivided into the Cenomanian and the Lower Senonian (Coniacian-Campanian). Cenomanian volcanic rocks are only found in northern Armenia and are represented by porphyrite flows, interbedded with tuffites and tuff breccias over 1000m thick. Lower Senonian volcanic rocks are found in isolated areas, throughout Armenia and the Nakhichevan A.S.S.R. The sequences normally consist of interbedded tuffs and more rarely lavas with sedimentary deposits, a few hundred metres in thickness. In places there is a thick volcanogenic formation (up to 1000m) which consists of tuff breccias, tuff conglomerates, tuffaceous sandstones, tuffs and porphyrite lavas. There are quite common horizons of sedimentary deposits such as sandstones and marls.

a

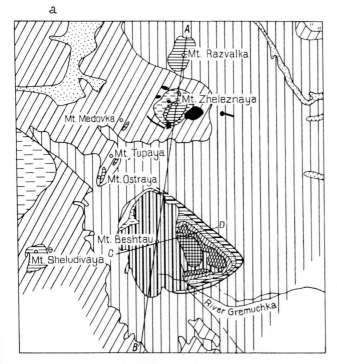

1. Quaternary.
2. Travertines.
3. A high terrace.
4. Biotite trachyliparites.
5. Pyroxene-biotite trachy-
   lipartites.
6. Pyroxene-amphibole
   trachyliparites.
7. Liparites.
8. The Maikopian Suite.
9. The Foraminiferal Suite.
10. Post-Senonian strata.
11. Upper Senonian.
12. Lower Senonian.
13. Turonian.
14. Albian.
15. Faults.

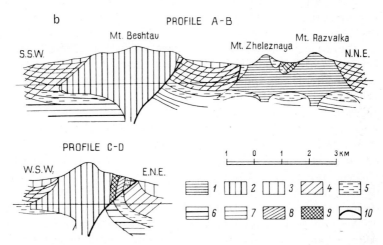

1 — biotite trachyliparites; 2 — pyroxene-amphibole trachyliparites: 3 —
the Maikopian Suite; 4 — the Foraminiferal Suite; 5 — post-Senonian
strata; 6 — Upper Senonian; 7 — Lower Senonian; 8 — Turonian; 9 —
Albian; 10 — faults.

FIG. 229. The geological map (a) and cross-section (b) of the Mineralovodian region. According to A. P. Gerasimov (1937).

The Cenomanian and the Lower Senonian volcanic sequences are very similar to each other and closely resemble Jurassic volcanics. The former sequence evidently brings a single cycle of vulcanicity to an end. Palaeogene volcanic rocks are of a different composition and resemble other Cainozoic lavas, thus beginning a new volcanic cycle, which lasted throughout Palaeogene times, at times becoming stronger, at others stopping altogether and all the time changing from district to district. The volcanic activity was conditioned by uplifts and depressions accompanying deep faulting. Two volcanic complexes, (1) Middle and Lower Eocene and (2) Oligocene and lower Miocene, are recognized.

The Eocene volcanic formation lies on the Palaeocene (Lower Eocene) flysch and is covered by marine Upper Eocene strata. It is, despite some breaks in continuity, found throughout the Lesser Caucasus. In the Lesser Caucasus anticlinorium it is detected from the Black Sea to the Caspian. In the west the Adzhar-Imeretinian and Trialetian ranges, the formation is widespread and reaches thicknesses of 1000–3000m. The Eocene volcanic formation is involved in complex folds. It consists of andesites interstratified with thick formations of tuff breccias and tuff sandstones. The coarse volcanic rocks are characteristic of it. The andesites are of pyroxenic type and vary from acid dacites to more basic varieties. Analogous andesitic tuffs of up to 1500m and more in thickness have been traced throughout northern Armenia. They are missing in Karabakh, but appear again in great thickness at Talysh (south-eastern corner of Azerbaidzhan), where there are 1000m of basal tuff breccias and tuffs, covered by Upper Eocene sandstones and shales with fish remains.

The Oligocene volcanic formation has a much more limited distribution and is only encountered in northern Armenia and in Talysh. Here clays with fish remains are overlain by the second volcanic formation (1200m) which consists of andesitic lavas, interbedded with tuff breccias, tuffs and tuff conglomerates. These rocks are overlain by tuff sandstones and shales with plant remains. A similar Oligocene volcanic formation of northern Armenia is found between lakes Sevan and Araks. The formation consists of andesites, tuff breccias, dacites and liparites of a total thickness of 1500–2000m.

The ancient volcano Aragats (Alagëz), which is 4095m high, is situated to the north-west of Erevan. This is the largest Transcaucasian volcano and its successive and numerous lavas and tuffs occupy a large area of 3000km² and have a complex structure (Fig. 230). The main mass of the volcano consists of alternating dacitic and andesitic lava horizons. At the foot of the volcano there are exposures of Anthropogenic flows covering Anthropogenic terraces. The distribution of the main types is shown on Fig. 231. The volcano was formerly considered to be of Oligocene age, but at present the dominant point of view is that it is younger and is probably Pliocene. It is quite probable that the present-day cone of Aragats is young, but that its base lies on dislocated Oligocene-Lower Miocene volcanic rocks, as has been shown on El'brus by K. N. Paffengol'ts (1959b).

Anthropogenic lavas which in places began outpouring in the Pliocene (the Akchagylian epoch) and whose outpourings ended in historic times in southern Transcaucasia are exceptionally widespread. In the west they form the large Akhalkalaian plateau which lies to the south of the Adzhar-Imeretinian and Trialetian ranges. In the central region, to the south of lake Sevan, Anthropogenic volcanic rocks give rise to a huge plateau, a part of which is known as the Agmangan plateau. Lastly in the extreme east, in Talysh there is the last basaltic plateau.

South of the Soviet frontier, in Turkey, the basalts cover even larger areas and can be referred to as the Asia Minor basalt province.

Anthropogenic lavas are of two types—centro-volcanic and fissure varieties. The former occur as typical volcanoes of composite type and are represented by trachytes and liparites. Such volcanoes are relatively rare and small. The second

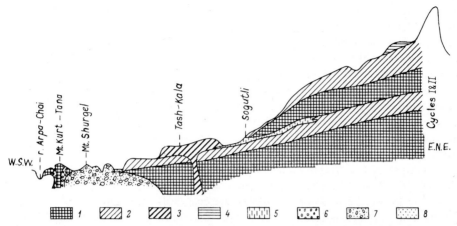

FIG. 230.   Section across the western part of Mt Aragats (Alagez). According to P. I. Lebedev (1937).

1 — basalts and andesite-basalts;   2 — dacites;   3 — Quaternary fissure dacites;   4 — alkaline dacites;   5 — tuffolavas;   6 — pumice;   7 — the Conglomerate-breccia Suite;   8 — tuffs.

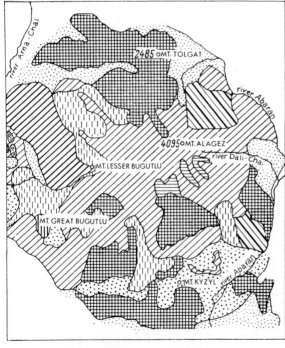

FIG. 231

The geological structure of Mt. Aragats (Alagez). According to P. I. Lebedev (1937).

1. Alluvium.

2. Andesite basalts.

3. Andesites.

4. Dacites.

5. Fissure dacites.

6. Alkaline dacites.

7. Liparates.

8. Arctic-type tuffolavas.

9. Black and red tuffs.

10. The Conglomeratic breccia Series

types of volcanic rocks are fissure basalts and andesite-basalts which were very mobile. Their flows occupy large areas despite small thickness of 20–40m, and less frequently of 60–100m. On the slopes of river valleys the basalts (Fig. 232) produce steep cliffs of sombre ash-black colour, which has become the colour of the national dress of Armenians, who wear black at any time of the year.

Paffengol'ts has studied Anthropogenic volcanic formations of Armenia in greatest detail He distinguishes five consecutive types denoted as *A*, *B*, *C*, *D*, *E*. This notation has acquired a wide range. Compositionally *A* represents basalts and andesite-basalts; *B* represents andesites; *C* again andesites; *D*—basalts approaching andesite-basalts; and *E* andesites. The volcanic rocks of Anthropogene age are especially well represented to the south of lake Sevan where they form the distinctive plateau of Agmangan, with isolated volcanoes reaching heights of 3400–3600m. On the map constructed by Paffengol'ts (Fig. 233), the linear distribution of volcanic centres, caused by a normal fault, their sizes and the outlines of lava flows are clearly shown.

Intrusions of the Alpine cycle are widespread and have a number of associated ore mineralizations. Two intrusive complexes, the Upper Eocene ophiolite belt and the Lower Miocene granodiorite complex, are especially important.

The ophiolite belt is found north of lake Sevan and can be traced to the east as far as Talysh where its last exposures are seen. The belt is shown on synoptic geological maps and consists of ultrabasic gabbro, peridotites and dunites which are often completely serpentized. Chromite ores are associated with dunites. All these rocks intrude Upper Cretaceous and Lower and Middle Eocene strata and are overlain by Oligocene sediments, thus pointing to the Upper Eocene age of the intrusions. The country rocks are strongly folded. The granodiorites of the Dash-kesan region and the large alkaline massif of Pambak (to the north-west of lake Sevan) can be regarded as final stage differentiates (Kotlyar, 1958).

Lower Miocene granodiorites have a wide distribution. Their massifs are relatively small and are sporadically found in a region which starts at Batumi and continues to the east of Kirovabad. The Merginskian Massif which margins the Araks to the east of Nakhichevan is the largest of these bodies. The massif has a variable composition and involves syenites, nepheline syenites and ultrabasic rocks, as well as the predominant granodiorites. Some of the Lower Miocene granodiorites are surrounded by Jurassic strata and used to be considered as Cimmeridian, but these bodies intrude not only Jurassic, but also Cretaceous and Palaeogene sediments.

Minor intrusions, represented by vein rocks, are notable for the absence of pegmatites and a widespread development of quartz-carbonate varieties, sometimes ore-bearing.

On ending the review of the magmatism in the Mediterranean Geosyncline one should once more underline its homogeneity, which is reflected in the preponderance of the young-Cimmeridian and Alpine cycles, and associated mineralization. The main debatable problem is the age of the youngest intrusive rocks. Paffengol'ts (1959a) and certain other geologists consider them as Upper Miocene. Paffengol'ts includes in this category the laccoliths of the Mineralovodian region, the Teplinian intrusions, the Tyrny-Auz intrusions and the granodiorites of the Lesser Caucasus. In a majority of cases this point of view is established and a widespread distribution of Lower Miocene intrusions does not provoke any doubts. Nevertheless, in some cases the relationships with sedimentary rocks remain incompletely determined. There are several indications that still younger intrusions exist in the region. The

FIG. 232. Anthropogenic basalts of Armenia. The basalt lava is 12m thick and lies on Miocene sands and clays. It is overlain by the present-day loams.

existence of young sub-volcanic intrusions representing feeding channels of Pliocene and Pleistocene lavas is incontrovertible. The possibility of the existence of young synorogenic intrusions cannot be excluded since young folding has been powerful.

USEFUL MINERAL DEPOSITS

Useful mineral deposits of the Carpathians, Crimea the Cis-Black Sea region, Ciscaucasia and especially the Caucasus have an all-Union importance. Some of these deposits such as the Baku oil, the manganese from Chiaturi and the Kerch iron ores are of international significance.

The Mediterranean Geosyncline is the health centre of the Soviet nation. Various mineral springs are of an especially outstanding significance. At the same time the value of agriculture must not be neglected. The grain of Kuban', the grapes and fruit of the Black Sea region and Crimea, the tea of Georgia, and the Transcaucasian citrus fruit are the most important products of the Soviet south.

Ores

The mineral wealth of the Mediterranean Geosyncline is less than that of Palaeozoic geosynclines but there are nevertheless several major deposits within it, including the Kerch iron deposit and the Chiaturi manganese deposit.

Ferrous Metals

Deposits in the region are well known and together with the Krivoi Rog iron ore, supply all the demands of the South. Some of the Kerch and Chiaturi ore is exported.

Iron. There are two main iron deposits—the Kerch Basin and the Dashkesan deposit.

The Kerch iron-ore basin occupies an area of 150km² and includes the deposits of the Taman Peninsula. Undoubtedly the ore is also present at the bottom of the Kerch straits. Altogether the basin is one of the largest deposits of oolitic iron ore in the world. The ore horizon is 25–30m thick and occurs in the middle of the sediments of the Cimmeridian Stage? The horizon is deformed into gentle brachy-anticlines which are largely eroded away, leaving behind the well-preserved brachy-synclines. The ore is oolitic and is interbedded with sideritic ores and thin layers of manganese ores. The iron content is 20–51 per cent, being on average 30–36 per cent. There is also 0·1–1·1 per cent of manganese, 0·4–1·5 per cent of phosphorus, 0·01–0·6 per cent of sulphur and a little vanadium. In places the mines can be worked along galleries, which is a useful feature.

Dashkesan is situated in Azerbaidzhan to the east of Kirovabad and represents a raw material source for the Transcaucasian metallurgical trust, situated in Rustavi to the west of Tiflis. The deposit consists of contact-metsomatic skarns formed as a result of the interaction, consequent upon intrusion of Eocene granodiorites into Upper Jurassic limestones. The zone of the ore-bearing skarns can be traced for 4km and their thickness is on average 40km. The iron content varies from 40 to 60 per cent. There are also associated hydrothermal ores of cobalt. The reserves of iron are considerable.

Manganese. Chiaturi is a major deposit of world renowned sedimentary oolitic ores, and is situated in north-western Georgia. The ore-bearing horizon is 2–5m thick, is traced over a large area and is flat-lying with a gentle dip to the east. It is found at the bottom of marine Oligocene sediments, which lie transgressively on Upper Cretaceous limestones. The oolitic ores consisting of oolitites of pyrolusite and psilomelane occurring amongst rubbly siliceous rocks are most important. The ores rarely get enriched, and produce concentrates with up to 52–58 per cent of manganese that are the best in the world, and also contain 0·15 per cent of phosphorus.

Chromium. Deposits of chromium are associated with dunits found in the Eocene massifs of the ophiolitic belt of the Lesser Caucasus. Two deposits are being worked.

Non-Ferrous Metals

Non-ferrous metals are of a considerable importance although there are few deposits, the largest of which are Sadon in the Northern Caucasus and Allaverdi, Shamlug and Zangezury (Kafan) in the Lesser Caucasus.

Copper. The Kafan vein deposit and the Allaverdi and Shamlug pyrite deposit are situated in the Armenian S.S.R. and represent one of the main cupriferous regions of the U.S.S.R.—the fourth in importance after the Uralian, the Kazakhian and the Uzhekian regions.

The Kafan (Zangezurian) deposit lies in the south-east of Armenia, near to the Araks. Its area is 25km², where there are several mines exploiting cupriferous veins, which lie amongst Lower Jurassic porphyrites. The ores are very rich and of hydrothermal origin (Mkrtchyan, 1958). Allaverdi and Shamlug lie to the south of Tblisi (Tiflis). Copper pyrites ores occur in a Jurassic tuff-porphyry formation, which are compositionally very similar to the ores of the Urals. The ores are smelted in the Allaverdi copper-smelting plant (Magakyan, 1947).

Polymetallic ores (lead, zinc, silver)

The Sadonian metasomatic vein deposit is situated in the Northern Caucasus, in the basin of the Ardon, to the south-west of Ordzhonikidze. The ore body consists of a complex vein crossing Palaeozoic granites and Liassic sediments. It is probably related to young keratophyres, which also intrude both types of rocks. The ore minerals are zinc-metal mineralization and in particular molybdenum and tungsten ores. The Tyrny Auz mineralization on the Northern Caucasus is of this type and is found to the south-west of Nal'chik. The deposit is of skarn type with both tungsten and molybdenum ores. At the contact between young granodiorites and Palaeozoic limestones grains of scheelite are found. Later quartz veins, with molybdenite and other sulphides, form a network in the skarn deposit. As a result the ores are complex and both metals are worked simultaneously.

In deposits of other type (copper-molybdenum) a space and genetic connection between moderately acid granitoids and molybdenum-copper ores is detected, while tungsten occurs only in small quantities. According to I. G. Magakyan (Magakyan and Mkrtchyan, 1957), who is one of the foremost specialists on the metallogeny of the Caucasus, regional faults and the associated joints constitute the main structural control of ore deposition and a whole series of deposits are associated with such features. Mineralization is found mainly amongst monzonites and syeno-diorites which form parts of the major Megrinian massif of a Lower Miocene age. A series of mineralized faults are accompanied by porphyry dykes. The main ore minerals are pyrite, chalcopyrite and molybdenite. The deposits are situated in south-eastern Armenia in the Kafan copper ore region and have been found and studied by Armenian geologists in the last few decades. Their reserves are considerable. Molybdenum reserves of the Caucasus and Transcaucasia are the most important in the U.S.S.R. (Amiraslanov, 1957).

Arsenic. Two deposits of realgar-orpiment type are known. They are low-temperature varieties normally connected with hot springs rich in arsenic causing the infilling of joints and fissures by various minerals, which are commonly carbonates. Amongst the ore minerals realgar and orpiment are most frequently encountered and there are admixtures of arsenopyrite, marcasite and sometimes cinnabar. The presence of the latter gives rise to the arsenic-mercury mineralization.

The Lukum deposit is situated on the southern slope of the Main range to the north of Kutausi. The ore body is represented by a complex branching vein lying in black Upper Jurassic shales (slates). Realgar, orpiment, antimonite and marcasite are found amongst quartz and calcite of the vein. Another mineral deposit is found in the Nakhichevan A.S.S.R. Its young age—Pliocene and Anthropogene—is interesting. A brachyanticline of folded Eocene marls and sandstones is faulted. There are hot arsenical springs, yielding arsenical gases, associated with the fault, while mineralization proceeds visibly.

Soviet geologists have done much in establishing the ore sources of the Mediterranean Geosyncline and doubtless will continue multiplying their efforts.

FUELS

The oil of the province is internationally known and sold, while gas is of a national significance. Coal is used by the local industry and population.

Oil

There are three main petroliferous provinces—eastern Azerbaidzhan, Groznyi and Maikop. These provinces were already exploited in pre-Revolutionary times, but during the several decades of Soviet regime their yield has increased several tenfold, some fields having been completely exhausted and new ones having come into operation. The Caucasus apart, oil has been found in Crimea where it has high prospects and in Western Ukraine the oil is extracted from the foothills of the Carpathians. The discovery of large Mesozoic oil and gas deposits in Ciscaucasia is very important indicating yet another promising province. The Krasnodar region is especially important.

Apsheron Peninsula. The pre-Revolutionary fields were situated near Baku. They were not numerous and relied only on the upper horizons of the productive formation down to the depth of 500–600m. Soviet geologists extended old fields and discovered new fields and deeper oil-bearing horizons. At present boreholes of 2000–3000m deep are not rare and the deepest (Zyrya) reaches 4812m.

The achievements of Soviet geologists and particularly Azerbaidzhan oil geologists are great. The latter learned their craft from such outstanding Russians as D. V. Golubyatnikov and M. V. Abramovich. The main achievements consist in finding new fields (anticlines) to the east and west from the oil-fields. In the west the Pirsagat deposit, to the south-west of Baku, is interesting. In the east a whole series of oil-bearing anticlines has also been found. Yet the most important achievements are in the discovery of submarine structures. The first major field was found in the Bay of Il'yitch. It is the continuation of the Bibi-Eibat structure found at the southern margin of Baku. A dam was built to drain the water and the consequent dry land was partly infilled and made a site for hundreds of new wells, which have yielded a great quantity of oil. An even more courageous project began with the boring and exploiting of the anticlinal structure found in the sea, to the east of the Apsheron Peninsula. The structure was partly exposed on Artem island. New boreholes were sunk in the sea from special metallic working platforms built above the sea waves. The platforms were connected to each other and to the shore of the island. Oil tankers are moored at such platforms and oil is pumped directly into their tanks. From an aeroplane the field makes an indescribable impression since numerous boreholes appear as minute patches lying in limitless sea. The exploitation of marine areas promises a great future.

The second field of expansion of geological exploration has advanced in the direction of deeper drilling. Spectacular results have already been obtained and the promise is good. The deepest parts of the productive Pliocene formation, which were formerly unsuspected, have been tapped at depths of over 4000m. Sarmatian oil has been found in laminated Middle Miocene sands of the Maikop Series. Indications of oil have been obtained in Upper and Lower Cretaceous strata, but the extraction from such deep horizons is a problem for the future. The bitumens of thick Upper Jurassic limestones should not be forgotten. Suggestions are made about the possibility of a connection of these with the Middle and Jurassic coal-bearing formations, but no further confirmation has yet been found. In general it can be said that although the Baku province is old it is still promising.

Other oil-fields of eastern Azerbaidzhan are adjacent to the Apsheron Peninsula. To the north-west of Baku (100–150km) there are the oil-fields of the 'Siazan'neft' establishment. Oil emerges in fountains from Maikopian sands. To the south-west of Baku (50–100km) there are also a series of oil-fields (Fig. 234). Some of these have been found in various sands of the productive formation and also in the lower part of the Apsheronian Stage. The discovery of gas and oil in Upper Cretaceous strata is important.

Groznyi. The province is situated in the north-eastern foothills of the Caucasus. The sections across the province are given in Figs. 223, 224. The tight anticlines found here stretch along the Main range and are complicated by thrusts. In relief each anticline appears as a low, narrow elongated ridge and often serves as an oil reservoir. Oil is found in sandy seams, lying amongst thick clays. There are altogether 18–22 such seams varying in thickness from a few to 50–60m. The total thickness of oil-bearing strata is 1000–1500m. Their lower part (1100m) belongs to the Chokrakian Stage and the upper part (400m) to the Karaganian Stage. As a whole the oil occurs in the Middle Miocene. The underlying Maikop Series is also promising. The yield of the Groznyi province is much less than that of the Baku province. B. A. Alferov, one of our major oil geologists, has contributed a lot towards the work of the Groznyi region. The 1958 discovery of oil and gas in Upper and Lower Cretaceous rocks of structures found to the west of Groznyi (Tsaturov, 1961) is exceptionally important.

The individuality of the oil-bearing horizon of recently found oil in a Lower Cretaceous sandstone is worthy of mention. The sandstone was drilled by a borehole at Ozek-Suat situated to the north-east of Groznyi and also in neighbouring regions.

Maikop oil region. This region is situated in the north-east foothills of the Caucasus in the basin of the river Byelaya, within the Krasnograd district. The yield is less than at Groznyi. The oil is found in Maikop sands folded into gentle anticlines or having a homoclinal dip. The sands sometimes form totally enclosed elongated bodies representing oil, sand-filled river courses. Such oil deposits are known as 'sleeve-like' [American 'shoestring']. Their existence has been proved by I. M. Gubkin. There are also important condensed gas deposits in Lower Cretaceous strata.

Ciscaucasia. The discovery of very large deposits of gas and oil on the platform structure of the Ciscaucasian Plain, which have often been indicated by Soviet geologists, is very important and promising. According to A. N. Mustafinov (1960) in the Krasnodar district, along the northern slope of the Azov-Kuban' part of the submontaine downwarp, gas fountains have originated from sands of Lower Cretaceous age. The Yeisko-Berezanian oil-gas region has especially bright prospects (Korotkov, 1959).

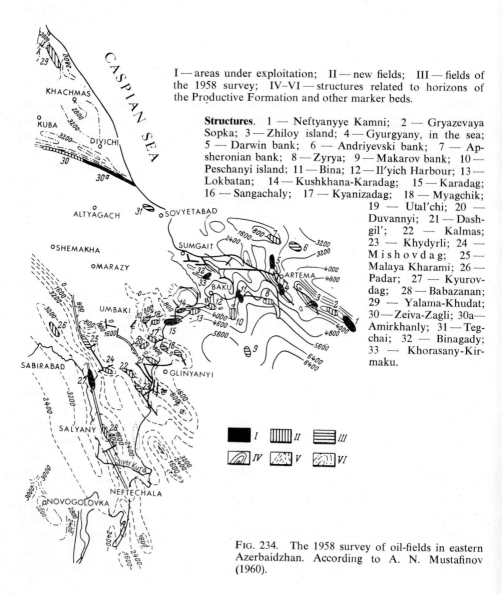

I — areas under exploitation; II — new fields; III — fields of the 1958 survey; IV–VI — structures related to horizons of the Productive Formation and other marker beds.

Structures. 1 — Neftyanyye Kamni; 2 — Gryazevaya Sopka; 3 — Zhiloy island; 4 — Gyurgyany, in the sea; 5 — Darwin bank; 6 — Andriyevski bank; 7 — Apsheronian bank; 8 — Zyrya; 9 — Makarov bank; 10 — Peschanyi island; 11 — Bina; 12 — Il'yich Harbour; 13 — Lokbatan; 14 — Kushkhana-Karadag; 15 — Karadag; 16 — Sangachaly; 17 — Kyanizadag; 18 — Myagchik; 19 — Utal'chi; 20 — Duvannyi; 21 — Dashgil'; 22 — Kalmas; 23 — Khydyrli; 24 — Mishovdag; 25 — Malaya Kharami; 26 — Padar; 27 — Kyurovdag; 28 — Babazanan; 29 — Yalama-Khudat; 30 — Zeiva-Zagli; 30a — Amirkhanly; 31 — Tegchai; 32 — Binagady; 33 — Khorasany-Kirmaku.

FIG. 234. The 1958 survey of oil-fields in eastern Azerbaidzhan. According to A. N. Mustafinov (1960).

In the Stavropol district, within the vale separating the lower reaches of the Kuma and the Terek, high-pressure stable gas fountains rise from rubbly Lower Cretaceous sandstones. The fountains obtained from sandy Middle Jurassic seams are especially interesting. Gas and oil deposits have also been found north of the Kuma, where they lie in sandstones at the very top of the Lower Cretaceous. The sandstones are folded into gentle folds over the buried Karpinskii Ridge. Further to the south in the Caucasian foothills, near Groznyi and in Dagestan, the oil has been found in fissured, Upper Cretaceous limestones and Lower Cretaceous sandstones.

Recently oil has been found in the Crimean Steppes to the north of Simferopol (Oktyabr'skoye).

Thus the problem of finding Mesozoic oil has been successfully solved by Soviet geologists.

Apart from the Mesozoic oil tens of new Palaeogene and Neogene deposits have been found in Ciscaucasia. Many of these deposits are situated in the old well-known Groznyi and Maikop provinces. A considerable number of other deposits is concentrated in three new oil provinces. The first is the Krasnodar Province and includes those deposits of Maikop age, which are found to the south, west and northwest of Krasnodar. The second province is in the Kuma region and includes the deposits of Palaeogene oil and, as already mentioned, of Mesozoic oil found near to the middle reaches of the Kuma. The third province is in Dagestan, where it stretches along the Caspian shore between the towns of Makharchkala and Derbent. The oil is found in Miocene, Maikop and Lower Cretaceous rocks.

The conversion of Ciscaucasia into an important oil and gas province is one of the important post-war achievements.

Transcaucasia. Three new provinces have been found in Transcaucasia—in the Lower reaches of the Rion, to the south-west of Tiflis and near Mirzaan, and to the south-east of Kirovabad near Naftalan. The oil is Neogenic.

Crimean Steppes. On the Tarkhankutian Peninsula oil has been found in Lower Cretaceous, Palaeogene and Miocene strata.

Kerch and Taman peninsulas. The two peninsulas represent a single system of brachyanticlinal folds, which are quite suitable for the accumulation of oil. The presence of large reserves of gas is confirmed by the numerous mud volcanoes. Drilling has yielded oil in sands of Maikopian and Lower Miocene ages, but there are no workable deposits. The main deposits of oil and gas must lie in the Mesozoic.

Eastern Carpathians. In the foothills of the Eastern Carpathians there are several oil-fields found to the south-west of Stanislav and near Borislav, to the south of Drogobych. Oil is found in faulted anticlines and occurs in Palaeogene sandstones (Figs. 238, 239).

Gas

Gas often accompanies oil and not infrequently, as in Baku, is jointly exploited. There are also independent gas deposits with small quantities of oil and then the latter is not exploited.

In the submontaine part of the Northern Caucasus, within the Stavropol plateau, lies the Stavropol gas region. At present the region has the largest reserves in the U.S.S.R. The seven years plan envisages gas pipes connecting Stavropol–Rostov–Moscow and Nevinnomyssk–Mineral'nyye Vody–Groznyi. The second largest gas field is the west-Ukrainian, with main deposits in the Lvov and Stanislav provinces.

This region has the highest yield in the U.S.S.R. The pipeline Dashava–Kiev is already built and will continue to Moscow and then Leningrad. Dashava and other deposits are situated near Lvov, in the Carpathian foothills. Gas is found in Upper Tortonian and Sarmatian sands.

The Dagestan gas region has been known and exploited for a long time. It is situated to the south of Makhachkala.

All regions with mud volcanoes are promising. They are especially numerous south of Baku, where their tops form small, ephemeral islands surrounded by sea and low cones on land. There are also many of them on the Kerch and Taman peninsulas. Large quantities of gas which ensure the continuous activity of such volcanoes, arise out of bituminous Upper Jurassic and Lower Cretaceous limestones, as can be judged on the basis of ejected fragments. The exploitation of this gas is an important and promising problem of fuel technology. Mud volcanoes are grand phenomena of nature, but it is better to use their gas for heating instead of being astonished by great fiery columns arising during their eruptions.

Recently to the south-west of Baku there have been found two gas fields—the Karaganian and the Kyanizadagian (Fig. 234). Their gas reserves are approximately equal to those of the Western Ukraine. There is thus a projected pipeline connecting Karadag–Kirovabad–Akstafa and Tiflis, with a branch from Akstafa to Erevan. The main line Karadag–Sumgait is already constructed thus altering the fuel balance of the Baku and Sumgait industrial regions.

Gas deposits of southern Central Asia are also considerable and so they are everywhere where there are Jurassic bituminous limestones, as for instance in western and southern Turkmenia, southern Uzbekistan and Tadzhikistan.

Very large gas deposits have recently been found below Stavropol plateau. Taken together they form 'Stavropol gas field', the largest in the world. Gas occurs in Khadumian sands (bottom of the Maikop Suite) which are 10–90m thick. Gas-bearing structures are of platform type, gentle (up to 1–2° dip), and large ($30 \times 75km^2$, $17 \times 35km^2$). The Stavropol structure to the north of Stavropol is being worked. Its gas has 98 per cent of methane and is pumped to several cities including Moscow and Leningrad. Recently the Northern Caucasus on the basis of prepared reserves of natural gas have become the most important region in the U.S.S.R. (Mustafinov, 1960). Gas deposits have also been found in the Crimea amongst Palaeocene strata on the Tarakhankutian Peninsula but Lower Cretaceous strata still enjoy the best prospects.

Thus the promise of the whole of the northern margin of the Mediterranean Geosyncline is good. Gas leaks have been obtained in boreholes sunk in the Melitopol region in the Manych region to the south-west of Astrakhan (Promyslovoye) but throughout this vast area so far little attention has been paid. Gas is the fuel of the near future.

Coal

As distinct from oil, coal deposits although numerous are small and no larger than average. Thus with the exception of the Tkvarchelian and Tkvibulian deposits, the cokes of which are of national importance, other coal-fields are only of local significance. Three groups of coal are distinguished: the Upper Palaeozoic, the Jurassic and the Palaeogene.

The Upper Palaeozoic Coal-bearing Formation is of the same age as that in the Donbas and is found along the upper reaches of the Kuban' and its tributaries. The formation is up to 1000m thick, badly faulted, and is found high in the mountains (at 2000m). It has substantially less coal than in the Donbas and there are few seams of Middle Carboniferous age and they are thin (up to 0·7–0·9m). The coal is of a high quality and is being worked. Deposits are situated at Kagar-Agur and Bogoslavskoye (Fig. 235).

Jurassic coal-bearing formations. These formations are most widespread and their position is shown in Fig. 235. On the northern slope the rocks are of a Lower Jurassic age and occur in the Dagestan, the Karachai and the Balkar coal-fields, which as yet are not economically important, and the Khumar coal-field in the upper reaches of the Kuban where seams of 0·4 to 1·0m are worked and produce a poor quality coke.

On the northern slope the coal-bearing formation is of Middle Jurassic age and forms a series of deposits shown on Fig. 235. The Tkvarcheli and Tkvibuli deposits are large and are being intensively worked. Coal seams in these deposits are on average 1–2m thick and sometimes reach 13 and even 20m. The structure is complex (Fig. 236). The quality of coals is high, producing good coke, and they are quite important in supplying the ferrous metal industry of Georgia. The total proven reserves are $0·5 \times 10^9$ tons and the yield in 1956 was $2·3 \times 10^6$ tons.

Palaeogene coal-bearing formations are found in the south-west of Georgia, in the Akhaltsykh region (Fig. 235) and in Western Ukraine. Akhaltsykhian coals are of Oligocene age. They are brown and occur in lensoid bodies, four of which are of a workable thickness. Brown coals of the western Ukraine are of Eocene age, thin, and found in a series of deposits.

NON-METALLIC DEPOSITS

Non-metallic useful mineral deposits are diverse and often represent considerable reserves. Magmatic deposits are usually of small dimensions, but are variable.

Asbestos. Two deposits of chrysotile asbestos are known, one in the valley of the Laba consists of asbestos-bearing veins of serpentinites cutting Jurassic peridotites and pyroxenites. The other deposits at Karachai is also associated with a Jurassic pyroxenite massif, where the mineral occurs as strands filling cracks and as crusts of serpentine blocks. The industrial importance of asbestos deposit is not great.

Talc. Talc is found in one commercial deposit in Southern Osetia.

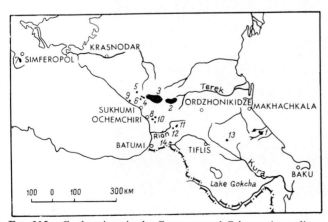

1. Dagestanian.
2. Balkarian.
3. Karachaiian.
4. Bogoslovskian.
5. Chernorechenskian.
6. Kafar-Agurian.
7. Beshuiian.
8. Tkuarchelian.
9. Bzybian.
10. Maganian.
11. Tkvibulian.
12. Gelatian.
13. Alandarisian.
14. Akhaltsikhian.

FIG. 235. Coal regions in the Caucasus and Crimea. According to A. A. Gapeyev (1949).

Alunite. The mineral has a composition of $KAl_3 (SO_4)_2 (OH)_6$ and is used for extracting aluminium. One deposit is known near Zaglik in Dashkesan. The deposit is associated with an Upper Jurassic volcanic formation.

Barytes is found in a number of deposits occurring along the southern slope of the Main range. The largest of these deposits are found to the north of Kutasi, where they occur as veins cutting a porphyrite-tuff sequence of a Middle Jurassic age. The absence of barytes veins from Upper Jurassic and Cretaceous rocks proves that they are associated with the Middle Jurassic episode of vulcanicity. The deposits are of national significance since they supply almost all industrial requirements.

Pyrites. The considerable deposit at Chigari-Dzor is situated south of Kirovabad and is associated with young granodiorites.

Lavas and tuffs constitute valuable building materials and are also used in other branches of industry. As building stone the lavas and tuffs are used in several regions, but are most intensively worked at Artik, on the western slope of Aragats (Alagez).

Andesitic lavas of Kazbek and the Borzhomi region and trachyliparites of Beshtau (Mineralovodian region) are used as acid-resisting stone.

Anthropogenic basalts of Erevan are being used as road metal.

Trass or compacted tuff, which if powdered can harden under water, is a

FIG. 236. Coal-bearing strata with coal seams at Tkvarcheli.

valuable cement material. Large deposits of these rocks are worked in Crimea on Mt Kara-Dag. Enormous deposits of volcanic ash are known in the Chegem region near Nal'chik—on the northern slope of the Caucasus.

Pumice, as porous volcanic glass, is used in abrasive, building and chemical industries. Many deposits are known, but the largest occur in Armenia near the town Ani, which is situated on the western slope of Aragats. These deposits (Anipemza) are worked intensively. Large reserves of pumice are also known from the Chegam region, near Nal'chik.

Fuller's earths locally known as gumbrite are worked in north-western Georgia, near the village of Gumbri.

Sedimentary deposits are no less abundant, variable and large.

Sodium and potash salts. The repetitive up and down movements in Palaeogene, Neogene and Anthropogene times produced considerable displacements of the shoreline causing the formation of gypsum, domestic salt and potash salts often in the same regions and sometimes in the same formations. They are especially widespread in the Carpathians—on the northern slope near Borislav and Kalush and on the southern slope near Solotvin. In the south of Transcaucasia large common salt deposits are known near Erevan. Present-day salt deposits are developed in many lakes or sea-shore lagoons along the Black and the Azov seas. Many such deposits are being worked.

Sulphur is exploited in Dagestan deposits where it was formed as a result of decomposition of Tithonian gypsum by bitumens, which rise out of Lusitanian bituminous limestones.

Iodine and Bromine are extracted out of iodine and bromine waters, which accompany oil deposits of Berekeya in Dagestan on the Caspian shore.

Diatomites of fresh-water origin occur in large deposits in Transcaucasia.

Cement raw materials. Major deposits of high quality cement raw materials are represented by flysch deposits, especially of the Novorossiisk region where they are being worked.

Building stone. Various shell limestones of ages varying from the Danian to the Anthropogene are valuable building stones. The Kerch and Odessa limestones are widely known. Naturally they are soft and easily cut, but harden on drying out.

Mineral waters. Mineral waters represent useful substances known not only in the Soviet Union but also abroad. Narzan, Yessentuki, Borshom and Arzni enjoy a well-deserved fame. Soviet geologists have discovered a series of other most valuable springs at Matsesta, Tskhaltubo, Sairme, Dzhirmuk, Dzau and many others. They help in curing many serious diseases. Most of the springs are associated with the young Anthropogenic vulcanicity.

In ending this review of useful mineral deposits, a review which is by no means exhaustive, one must comment on the great variety of deposits in the Mediterranean Geosyncline. Many of these deposits have been discovered by Soviet geologists, but many new ones are discovered every year.

REGIONAL REVIEWS

WESTERN UKRAINE

The geological structure of western Ukraine is very similar to that of the Caucasus and is continuous with it (Fig. 237). In the north there is the platform part of the

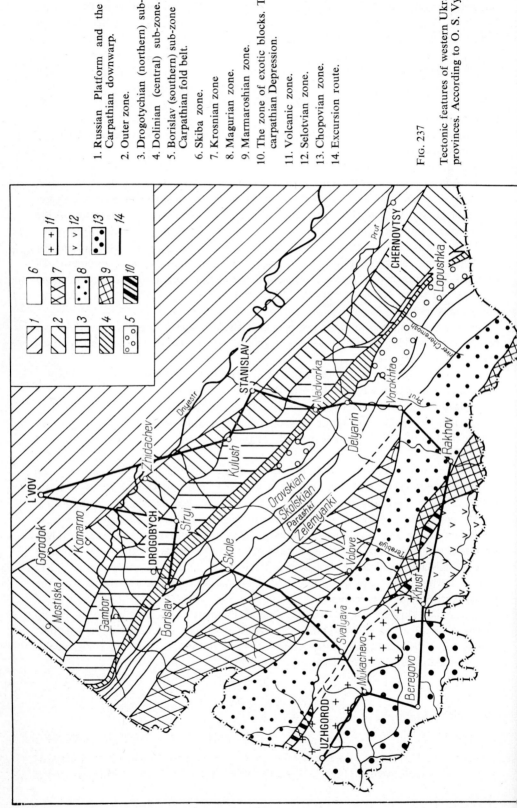

1. Russian Platform and the sub-Carpathian downwarp.
2. Outer zone.
3. Drogotychian (northern) sub-zone.
4. Dolinian (central) sub-zone.
5. Borislav (southern) sub-zone
   Carpathian fold belt.
6. Skiba zone.
7. Krosnian zone
8. Magurian zone.
9. Marmaroshian zone.
10. The zone of exotic blocks. Trans-carpathian Depression.
11. Volcanic zone.
12. Selotvian zone.
13. Chopovian zone.
14. Excursion route.

FIG. 237

Tectonic features of western Ukranian provinces. According to O. S. Vyalov.

Ukrainian Shield. To the south a relatively narrow sub-Carpathian downwarp passes into the folds of the Eastern Carpathian anticlinorium. In the extreme south there are the Transcarpathian plains. The four Carpathian zones completely correspond to the four Caucasian zones: the Ciscaucasian Plain, the Ciscaucasian downwarp, the Greater Caucasus anticlinorium and the Rion-Kura Depression.

In connection with the discovery of oil, gas, salt and other useful mineral deposits Soviet geologists developed their investigations, which resulted in the progress and accuracy of tectonic notions and have fundamentally changed the existing hypothesis of the geological structure of Western Ukraine. A whole series of new deposits, including gas at Dashava, the L'vov-Volyn' coal-field and a large deposit of native sulphur, have been found.

The platform represents the southern margin of the Ukrainian Shield which is very slightly inclined to the south-west. On the Precambrian lie Lower Cambrian and Upper Ordovician carbonate and sand-shale deposits covered by Silurian limestones, Lower Devonian red beds, sandstones, dolomites and Middle and Upper Devonian sands and shales. In places coal-bearing Lower Carboniferous rocks are present. Palaeozoic strata are exposed in the valley of the Dnyestr and to the north of it. In the north they are overstepped by Jurassic deposits, but normally Upper Cretaceous, Palaeogene and Neogene sediments rest on Palaeozoic strata. On the surface there are found Neogene and Anthropogene sediments.

Palaeozoic, Mesozoic and Cainozoic sediments are of a platform character, lie almost horizontally and are flexed in large, gentle brachyanticlines. In addition the strata are thin and weakly metamorphosed. Along the margin of the platform there are flexural monoclines associated with basement faults.

The sub-Carpathian downwarp, commonly known as cismontaine or the marginal downwarp, has a great economic significance, since there are oil and gas deposits within it. Its relief is that of a hilly plain. The downwarp is several tens of kilometres (15–20 to 60) wide and stretches along the Carpathians. In connection with the manifestations of the last stages of Alpine Orogeny the position of the downwarps changed, it shifted to the north and in the Sarmatian epoch became static (Fig. 237). Miocene sediments, which at the northern and southern part of the downwarp outcrop at the surface, are depressed in the centre to a depth of 2–3km. The top surface of Jurassic strata is depressed down to 4km and even 6km.

The outer (northern) and the inner (southern) zones of the Carpathians are clearly distinct. According to Vyalov (1949, 1956) one of the foremost specialists on the Carpathians, the outer zone of the downwarp consists of Upper Tortonian and Sarmatian deposits on the surface lying directly on Upper Cretaceous strata and folded into gentle brachyanticlines. Almost all of these folds are gas-bearing and the Dasheva is particularly noticeable since it holds large reserves of intensively exploited gas. Gas is present in Upper Tortonian, Sarmatian and Upper Cretaceous sands. In the Kalusha structure gas has been obtained since 1912. At the top of the Tortonian there is a gypsiferous suite with the largest deposit of sulphur in the U.S.S.R.

The inner zone has a full succession from the Sarmatian down to and including the Upper Cretaceous. The development of molasse and flysch formations several kilometres thick which belong to the Lower Miocene, Palaeogene and Upper Cretaceous, is particularly characteristic of the region. The region has also distinctively linear folds trending north-west, but often complicated by strike-thrusts.

At the southern margin of the submontaine downwarp in the so-called Borislav zone the structures are most deformed and complicated. The folds near Bitkov have been found on drilling and serve as an example (Fig. 238). The folds are overrun by a major thrust. Such thrusts show displacements of 10–12km. Within the inner zone of the sub-Carpathian downwarp there are major oil-fields, including that at Borislav and that at Dolina. Oil lies in sandstones of the Yamnenian (Palaeocene) Formation and in the Borislav sandstone of Upper Eocene age. Three new deposits of which one is considerable have recently been discovered (Palii, Petrov, 1961). In the inner zone of the sub-Carpathian downwarp there are deposits of ozokerite and potash and common salt found in a Miocene saliferous formation.

The Carpathian fold region is distinctive by a very complex tectonic structure affecting all the strata from the Neogene down to the Lower Palaeozoic and Precambrian (Riphean). In the Carpathian fold belt several elongated zones (Fig. 237) can be distinguished. The most northerly of these zones is adjacent to the submontaine downwarp (Borislav zone) and is known as the 'Skiba (imbricate) zone'. The most northerly 'scale' is pushed over the Borislav recumbent fold (Fig. 239). The most southerly zone of the Carpathian fold belt is known as the 'Blocky' on account of large numbers of blocks of Jurassic limestones, as already have been described (p. 640).

The central zone is low and corresponds to the main anticlinorium. Within this zone Jurassic, Trias and Palaeozoic rocks are exposed in the east, while in the Marmarosh (Rakhov) Massif there are even Precambrian (Riphean) rocks. The Carpathian fold region has some manifestations of metal ores.

The Transcarpathian Depression, or the Transcarpathian inner downwarp, as it is called by O. S. Vyalov, is a flat region. Nearer to the northern mountains it is slightly more elevated. A low transverse ridge divides the depression into two—the western, Alfoldian and the eastern Transylvanian or Upper Tissenian, since it is situated in the Upper reaches of the river Tissa.

Transcarpathia is composed of Neogene and Anthropogene strata, which have a great thickness of 5000–5500m. These sandy-conglomeratic rocks are of molasse type. In the Sarmatian there is a relatively thick coal-bearing formation with commercial coal seams. In places there is a saliferous Miocene formation with major deposits (Solotvin) of rock salt.

Neogenic lavas and tuffs are often very thick and characteristic. These deposits occupy large areas and are clearly seen on the synoptic 1:7 500 000 map. Eruptions occurred from deep faults margining and transecting the depression.

Neogene strata are gently inclined to the south, although in places there are open brachyanticlinal folds with large deposits of gas (methane) concentrated in Sarmatian sands.

SOUTHERN RUSSIAN DEPRESSION

To the south of the Russian Platform lies a distinctive feature which on the 1957 tectonic map of the U.S.S.R. is called 'the Great Donbas downwarp'. The downwarp begins in the west with the Przipetian (Polessian) Depression and after the Chernigov Saddle, which is about 1000m high, is followed by the Dnyepr and Donyets depressions which to the east pass into the Astrakhan lowlands. The latter in turn to the east joins the sub-Uralian downwarp (Fig. 4). All together these depressions form a gigantic trough, which is about 2500km long. Thus the term 'the Great Donbas downwarp' is inaccurate since the name Great Donbas refers to that area where a coal-bearing formation is well developed and this area is no more than one third of the downwarp as a whole. On the other hand the term 'Southern Russian' accurately expresses the position of the downwarp and does not refer to its coal. In describing the downwarp we will commence with the best investigated Donyets Basin.

The Donyets Basin

The Donyets Basin belongs to those regions which are tectonically disputable since they occupy intermediate positions between platforms and geosynclines. Such regions are Timan, the Volga-Emba salt region, the west-Siberian Lowlands. The

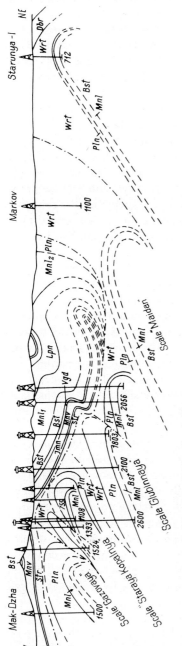

FIG. 238. Geological profile of the Bitkovian field. According to P. M. Novosiletskii (1958).

Suites: *Dbr* — Dobrotovian; *Urt* — Voro-tyshchian; *Pln*—Polyanitsian; *Mnl₂*—Upper Menilitovian; *Lpn* — Middle Menilitovian; *Mnl* — Lower Menilitovian; *Mnl* — Meni-litovian; *Vgd* — Vygodian; *Mnv* — Manyavian; *Jmn* — Yamnian; *Str* — Stryian.

FIG. 239. Section across the Borislav fold and the Skiba zone. According to A. A. Bogdanov.

Donyets Basin lies between the Russian Platform and the Ukrainian Massif and before the Middle Devonian epoch was itself a platform. Thus many geologists consider it as part of the Russian Platform. On the other hand Devonian, Carboniferous, Permian and Mesozoic strata of the basin have all the indications of geosynclinal sedimentation. The total thickness of these rocks is over 12km and beginning with the Upper Devonian and up to and including Upper Jurassic the beds are folded into a series of regionally trending linear structures, invaded by minor intrusions of the Hercynian and Cimmeridian magmatic cycles. These folds along the strike are associated with those of the Manych zone and the Mangyshlak. Even Upper Cretaceous and Palaeogene strata are folded into distinct flexures. Only Neogene and Anthropogene sediments remain unaffected. Thus the Donyets Basin is equivalent to the outer zone of the Mediterranean Geosyncline, and is its direct continuation. Consequently the basin is considered as a part of the outer geosynclinal zone.

The Donbas is a relatively wide, shallow depression situated between two flat, crystalline massifs. The shallow depth, the very gentle flanks the simple folding and the considerable thickness of infilling sediments are all diagnostic of the depression.

N. S. Shatskii has coined the term 'aulacogen' to cover the Donyets Basin and other similar tectonic structures. These structures were initially areas of graben-like depressions resembling geosynclines, but being of smaller dimensions. Subsequently the region became a fold belt with a low ridge-like relief. Thereafter the ridge was eroded away and the province converted into a platform adjoining and welded to an older platform, which in the case of the Donyets Basin was the Russian Platform.

History of Development

Up to and including the Eifelian epoch the Donyets Basin, like the whole of the Dnyepr-Donyets Depression, was a platform region. Together with the Voronezh, the Polessian and the Ukraine massifs it formed a single South Russian shield. The major part of the shield was elevated land, but in some regions, as in the western part of the Dnyepr-Donyets Depression, continental and near-shore marine Upper Proterozoic Valdaiian deposits did accumulate. These sediments are barren, thin sands and clays, which are neither folded nor metamorphosed. The sediments are typically platform and their discovery does not permit the claim, which is sometimes made, that the Donyets geosynclinal region had begun to form in the Upper Proterozoic.

During the first half of the Middle Palaeozoic era (Silurian, Lower Devonian and Eifelian) the Dnyepr and Donyets depressions were elevated land in process of erosion and non-deposition. This situation is demonstrated by the absence of Silurian, Lower Devonian and Eifelian deposits and is generally accepted. This view is, however, in need of revision. It is not understood how during such a prolonged period of erosion relatively thin and friable Lower Palaeozoic sediments remained preserved. Their discovery forces one to conclude that the epoch of erosion was not so prolonged and that during the Silurian and Lower Devonian sediments did accumulate but that these were then eroded away in Eifelian times. The existence of an elevated region, which in Eifelian times, at any rate, was being subject to erosion does not provoke any doubts.

In the succeeding Givetian epoch prolonged depression began ultimately resulting in the accumulation of no less than 10–12km of sediments in the Donyets geosynclinal region. The downward movement was in the main balanced and did not produce abrupt palaeogeographic changes over long intervals of time. During such intervals the Donyets Depression remained a near-shore alluvial plain or a shallow warm sea. On many occasions the balanced movement was interrupted by short-lived abrupt movements causing sudden palaeogeographic changes and the plain was converted into a sea bottom or *vice versa*. The sum total of balanced and fast movements conditioned the distinctive character of the Donbas Coal-bearing Formation and of other formations underlying or overlying it, as well as of its lateral equivalents.

In the Givetian, Frasnian and Fammenian epochs the Donyets Basin was a vast near-shore plain, where red sandy and clayey sediments were deposited. The slow downward movement and the arrival of terrigenous material brought in by temporary rain streams was the cause of the accumulation of 600m of red beds. The almost complete absence of plant remains indicates that the climate was hot and dry. In the Frasnian epoch the plain sometimes for a short period was submerged under the sea, producing conditions favourable for the accumulation of thin seams of muddy platy limestones with marine faunas.

At the very beginning of the Tournaisian epoch a fast downward movement produced sea on the site of the desert plain. The sea was shallow and epi-continental and soon became a zone of a fast accumulation of bedded brachiopod and coral limestones. The growth of sedimentary deposits should have led to the diminution of the depth of water and emergence of land. No such emergence, however, occurred and therefore the depth of the sea must have remained constant. Thus movements must have been slow and were balanced by the compensating sedimentation of limestones. This circumstance continued throughout the whole Tournaisian epoch and the first half of the Visean. During this interval of time some 600m of homogeneous dark and grey bedded limestones were found.

The Lower Carboniferous begins with the Tournaisian Stage which in Donbas is represented by 260m of limestones with beds of dolomite and clays towards the bottom. To the west of Donbas limestones pass laterally into sandstones and clays. The transition begins with the lower horizons and the total thickness of sediments decreases.

Ukrainian geologists divide the Visean Stage into three unequal suites. The Lower Suite (A) consists of limestones and is only 60–70m thick. The Middle Suite (B) underlies coal-bearing rocks, is terrigenous and practically coal-free. Its thickness is 500–600m. The Upper Suite (C) is up to 800m thick, is coal-bearing and contains 10–14 workable coal seams. The Coal-bearing Suite can, despite the diminution of its thickness, be traced well to the west throughout the Dnyepr-Donyets province and has coal seams everywhere.

The Namurian Stage lies transgressively with an angular unconformity on Upper Visean rocks. Its rocks are also sands and clays, but there are no workable coal seams and there are many marine bands. In the west Namurian strata rapidly wedge out, while in the centre of the basin they are 2500m thick.

Middle Carboniferous strata are divided into seven suites of a total thickness of 2000–6000m. The first two suites have no coal while the rest have the main workable seams (53) in Donbas. The rocks are sandstones and shales with beds of limestone.

Upper Carboniferous rocks (2000–4000m) are of the same lithological composition. Two lower suites are coal-bearing (up to 7 workable seams) while the top one consists of coal-free sandstones.

During the second half of the Visean epoch the important phenomenon of coal-accumulation began. Amongst the limestones with marine faunas sequences of grey sands and clays with plant remains and coal seams appeared. The number and thickness of these sequences soon increased and limestones fell back to a second position and occurred only as thin beds amongst continental coal-bearing sediments. At the same time the thickness of the latter rapidly increased becoming of the order of 3000–5000m in Lower Carboniferous times (Upper Visean and Namurian).

In the Middle Carboniferous the accumulation of coal-bearing deposits continued just as rapidly and the thickness of these sediments from west to east varies from 2000m to 6000m. It is interesting that the total thickness of limestone bands does not exceed 1 per cent and of coal seams, 1–2 per cent of these figures.

The lower part of Upper Carboniferous succession is of the same type and is 1500–3000m thick. The largest number of thick coal seams is found in the upper part of the Middle Carboniferous and lower part of the Upper Carboniferous.

When coals were accumulating the basin was a large near-shore alluvial plain, which was flat, swampy and traversed by a network of streams, which were tributaries of a large east–west flowing river. The major part of the plain was deltaic. In the north-west, north and east a sea surrounded the plain. During the numerous but short-lived episodes of downward movement the sea invaded the plain, laying down limestones with marine faunas.

Certain coal seams were found on the whole surface of the Donyets Basin indicating the great dimensions of the swamps in which plant remains accumulated. A great majority of the seams are less widespread suggesting smaller swamps.

Seams of shales with brackish-water lamellibranchs and gastropods originated in similarly large lagoons, while shales with fresh-water *Anthracosia* were lacustrine.

The unusual wealth of plant remains indicates a rich vegetation which could only have existed in a humid, tropical climate. In Devonian times no coals accumulated on a similar plain, thus suggesting that favourable conditions for the formation of coal were generated, by an abrupt change of climate. This change provoked the appearance of numerous rapidly growing plants. It is thought that 10–15m trees grew in one year.

At the end of the Upper Carboniferous the deposition of the Coal-bearing Formation stopped, the near-shore plain was overrun by sea and when the land emerged again the climate was dry, hot and of semi-desert type. The plants also changed and their number decreased abruptly. The sandstones and shales were deposited at this time and contain *Araucaria*. The so-called Araucarian Suite is 700–1000m thick and was formerly included in the Lower Permian.

In the Lower Permian epoch the climate became even drier and there was an onset of desert conditions. The near-shore plain continued sinking slowly and a suite of cypriferous sandstones with very rare plant remains was deposited. Its thickness is 500–1000m. The suite is succeeded by a gypsiferous-dolomitic suite of 130–800m in thickness. The latter suite has beds with a Sakmarian, marine fauna, but is in the main a terrestrial and lagoonal sequence. At the end of the Lower Permian a saliferous suite (up to 600m) was precipitated. It has seams of rock salt (up to 42m) which are worked at the well-known Artemian deposit.

In the Upper Permian epoch the sea retreated to the south of Crimea and a low, sun-drenched desert lay on the site of the Donyets Basin. Red clays, sandstones and marls which are normally 100–150m thick and in places 200–300m thick accumulated on the surface of the desert. Upper Permian red beds are succeeded without a break by Lower Triassic red beds which are up to 400–500m and sometimes much less thick. At the bottom of the sequence lie coarser sands and sandstones while clays with sandstone beds predominate at the top. After the Lower Trias a break in sedimentation implies that the low plain was elevated into a hilly upland area, which was subjected to erosion.

The formation of the region of erosion was caused by the onset of the Hercynian Orogeny. The movements began in the Upper Permian and ended in the Lower Trias, but were much weaker than the Hercynian Orogeny in the Urals, where molasse and flysch sediments were deposited at the margins of high mountains. In the Donyets Basin such formations are completely absent indicating that high ranges did not exist. In the Urals the Hercynian Orogeny is accompanied by massive granite intrusions, stretching for hundreds of kilometres. Such intrusions are absent from the Donyets Basin. In the Urals the sharp distinction in the degrees of metamorphism of the folded Palaeozoic strata and the horizontal Mesozoic sediments is characteristic, while in the Donyets Basin the moderately folded Permian beds are just as weakly metamorphosed as the overlying, moderately folded Jurassic sediments. Permian and Jurassic strata are sometimes involved in the same folds.

It should not be forgotten that the present-day structures affecting Palaeozoic rocks can equally be caused by orogenies later than the Hercynian. Even Jurassic, Upper Cretaceous and Palaeogene rocks are folded.

An interesting and as yet unsolved problem is: how could folds originate in deposits overlying a rigid basement of crystalline Precambrian rocks? There are three possible answers. In the first place at the depth of 12–15km Precambrian rocks may have acquired some plasticity and were also folded. The second answer is that Precambrian rocks are not folded but broken up into separate blocks, whose movements cause the folding of Palaeozoic strata, which is the most probable situation. Thirdly neither Precambrian rocks, nor the immediately overlying Palaeozoic horizons are folded. In such a case folds are only present in the Upper Palaeozoic and Mesozoic, and gradually die out downwards. Other answers are also possible but all of them can only be verified by deep drilling and geophysical investigations.

After an interval in sedimentation lasting throughout the Middle and a major part of the Upper Trias the Donyets Basin again became a vast plain. On the surface of the slowly sinking plain at first Upper Triassic red beds (100m) and then Jurassic sediments (300–350m) were deposited. The latter begin with a coal-bearing Lower Lias formation. In the north of the basin the break was more prolonged and Middle Jurassic strata rest unconformably on the Middle Trias.

Subsequently the plain became submerged under the sea and marine shales and sandstones of the upper part of Lias and the Middle Jurassic were laid down. At the beginning of the Callovian the plain became land (sandstones with plant remains) but it did not last long. From the Middle Callovian and to the end of Kimmeridgian marine sediments consisting mainly of limestones with Lusitanian coral reefs were again deposited.

The Tithonian and Lower Cretaceous epochs saw yet another break in sedimentation. Kimmeridigian strata are overlain unconformably by Cenomanian and Upper Albian sandstones. The break was caused by the last, relatively weak, phase of the Cimmeridian Orogeny.

The Trias includes the upper part of the continental suite of variegated sandstones and clays. In Donbas its thickness varies from tens to 200m, and considerably increases to the west. Stratigraphic boreholes have shown a thickness of

230m near Kupyansk, 313m near Chernigov, 382m in Reizerovo (south of Romny), and 412m in Smelyi (north of Romny). Trias sediments are divided into three suites —(1) the Upper Permian-Lower Trias Dronovskian Suite, consisting of sandstones and conglomerates; (2) the Lower Trias Serebryankian Suite of sands and clays; (3) the Upper Trias Protopivskian Suite of variegated clays. The Serebryankian Suite was formerly considered as Middle Triassic, but its Lower Triassic age has been proved.

Jurassic rocks are only developed in the north-west part of Donbas and in the Dnyepr Depression. The section begins with continental, variegated Lower and Middle Lias Novoraiskian Suite, which is lithologically almost indistinguishable from Protopivskian Red Beds. The Upper Lias (Toarcian Stage) is marine and consists of dark clays with sandstones reaching 15–80m. To the west marine deposits pass laterally into continental strata and further westwards wedge out altogether.

Marine Toarcian strata have a transitional passage into the Aalenian (30–50m), where the Upper Jurassic succession begins. Aalenian rocks are again black and grey clays and sandstones with foraminiferids and ammonites.

Bajocian strata rest unconformably on the Aalenian and begin with basal conglomerates with a marine fauna. The conglomerates are succeeded by grey clays with ammonites. The total thickness is 60–150m. To the north-west Aalenian and Bajocian rocks are replaced by continental sandstones and clays.

The Bathonian is represented by two formations: the Lower marine and muddy (25–140m) and the Upper continental and sandy (25–125m). The latter has also tuffogenous sandstones (up to 40–90m).

The Callovian (up to 60–70m) is represented by a formation of sandstones and clays, which are in places continental (Lower Callovian) and in other places marine.

Oxfordian and Kimmeridgian strata of Donbas are represented by a characteristic formation of limestones (up to 40m) which are sandy at the bottom and pure white and oolitic at the top. In places they are massive and coralline (Lusitanian). To the west these limestones thin rapidly and become replaced by clays and sandstones. Coralline limestones represent typical coastal reefs and it is understandable that in places they are succeeded by continental sandstones and variegated clays (up to 60m). The age of the continental formation is Volgian and lowermost Lower Cretaceous.

Owing to the former existence of hills Lower Cretaceous sediments are usually missing. In places there are sandstones and clays with spores and plant remains of a Lower Cretaceous age which cannot be specified more closely. Albian sandstones represent the beginning of the Upper Cretaceous sedimentary rhythm and have a thickness of 30–50m.

Upper Cretaceous strata encircle Donbas almost completely and spread out far to the west and east. The succession shows minor local variations, certain horizons being absent or some horizons changing their lithology. In the main these strata are very constant. The basal continental Upper Albian sandstones and clays (up to 20–40m) are overlain by grey, greenish glauconitic sandstones (40–60m) with a Cenomanian marine fauna. Higher up lies a thick formation (up to 600m) of chalky marls and chalk, which in places in some horizons are replaced by green-grey clayey gaizes. In such instances the thickness decreases to 150–200m. The fauna consists mainly of Foraminifera and permits the recognition of all stages up to and including the Maastrichtian. The Danian Stage so far has not been recognized.

The Palaeogene is just as widespread as the Upper Cretaceous and is closely associated with it. Their strata are folded into large flat folds together. The subdivisions of the Palaeogene are the same as on the Russian Platform. Facies changes are of some interest. In the south-west there are dominant sandstones and conglomerates of the Kanevian and Buchakian stages, which are often inseparable. These strata are thick (up to 60–90m in Chernigov). The Kievian and Kharkovian stages are lithologically normal and 40–60m thick. On going away from the Ukrainian Shield the thickness of the Kanevian and Buchakian stages rapidly decreases and along the northern margin of Donbas from Kupyansk onwards they are missing from the succession and Upper Cretaceous strata are overlain by basal sandstones of the Kievian Stage. In the eastern part of Donbas the depth of the sea increased and Volgian type successions with a thick Syzranian Stage (marine Palaeocene) appear.

At the end of the Oligocene the Donyets Basin again became land which is still existing. Neogene seas existed at the southern edge of this land.

Anthropogene deposits are available, all being continental fluvial and lacustrine sands and clays and aeolian, loessic loams.

The history of the Donyets Basin is interesting and distinctive. It differs sharply from the history of the Russian Platform and the history of the Urals, while showing a great similarity if not identity with the history of the outer zone of the Mediterranean Geosyncline, including for instance Ciscarpathia, Ciscaucasia, the Mangyshlak, the Ust'yurt and the Karakum.

Tectonic Structures

Structures affecting Palaeozoic strata are relatively simple and consist of normal anticlines and synclines, not infrequently complicated by thrusts and faults (Fig. 240). The small degree of deformation affecting Palaeozoic rocks is confirmed by the calculations made by P. I. Stepanov, who has shown that if all the folds are straightened out the width of the basin will increase by only 30–40 per cent.

The main anticlinorium and the four adjacent troughs are the principal structures of Donbas (Fig. 242). The major folds are complicated by folds of second and third orders. The southern troughs—Chistyakovskaya and Grushevskaya—are effectively one single so-called Southern syncline. The Bokovskian and Sulinskian troughs form the large 'Main' syncline, to the north of which lies the Northern anticline, which is broken up by a series of thrusts, which taken together form the Northern thrust zone. Within this zone folds are small, compressed, and dislocated by the thrusts (Fig. 241).

The relative simplicity of the tectonic structure of the southern major part of Donbas and the complexity of its narrower northern part are indicated on the tectonic map formulated by P. I. Stepanov and others (1937). The map indicates the two southern synclines, the Main anticline the northern synclines, the Northern anticline and the small, warped and dividing folds in the Northern thrust zone (Fig. 242).

Salt domes. To the north of the thrust zone along the northern margin of the Donyets Basin in several localities there are salt domes (Fig. 244). The cap breccia topping them was known for a long time as agglomeratic breccia, but its true identity was established by Soviet oil-geologists. The domes show indications of associated oil, while in the region of Romny they have yielded oil in commercial amounts. The breccia covering the salt plug has fragments of Frasnian limestones with *Spirifer* of the group *anossofi* Vern., of the same type as near Lake Baskunchak and the town of Romny.

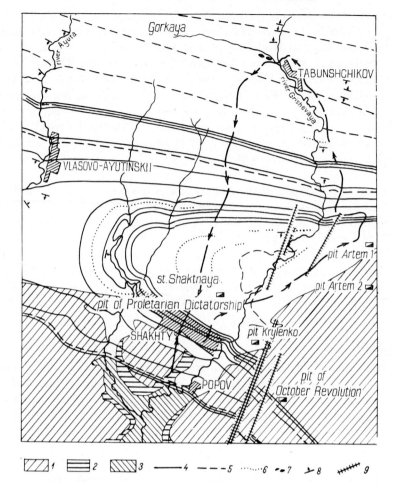

FIG. 240.   The structure of the Shakty region. According to P. I. Stepanov
and S. E. Verboloze (1937).

1 — Pontian   2 — Meotic;   3 — Kharkovian;   4 — coal;   5 — limestone;
6 — sandstone;   7 — igneous rocks;   8 — dip;   9 — faults.

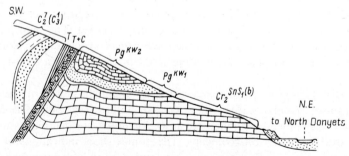

FIG. 241.   A section near the town of Lisichansk. Upper Campanian marls
are overlain by sands and marls of the Kiev Stage. The sequence is over-
ridden by a breccia of Triassic and Carboniferous blocks and by the
following Trias red beds and the Coal-bearing Formation. According to
V. S. Popov (1937).

Trias and Jurassic structures. Trias and Jurassic deposits developed in the north-western part of the basin are always folded regionally, but the folds are simple and gentler than those affecting Palaeozoic formations. Sometimes Mesozoic and Lower Permian strata are affected by the same folds. There are sub-Triassic, inter-Bajocian and sub-Kimmeridge unconformities.

Upper Cretaceous and Palaeogene structures. At first glance it appears that Upper Cretaceous and Palaeogene sediments lie horizontally, but on tracing boundaries and instrumental surveying it is easy to prove that these rocks are involved in very large flat brachyanticlines. Consequently the Neogene lies unconformably on the Upper Cretaceous and Palaeogene. In the Neogene and Athropogene only rare small faults can be found.

Greater Donbas

It has been demonstrated a long time ago that Palaeozoic outcrops did not embrace the whole of the area where coal-bearing rocks were developed. Drilling and geophysical explorations have shown that such strata spread to the west, east and especially north, but that they are concealed under a Mesozoic and Cainozoic cover. The actual area of distribution of the coal-bearing formation is known as the Greater Donbas, in the investigation of which much was done by P. I. Stepanov.

The boundary of Greater Donbas still demands further accurate work. This is especially so in the east where deep drilling has shown that in the Manych region and to the east of it despite the development of folded Upper Palaeozoic strata of Donyets type there are no commercial coal seams.

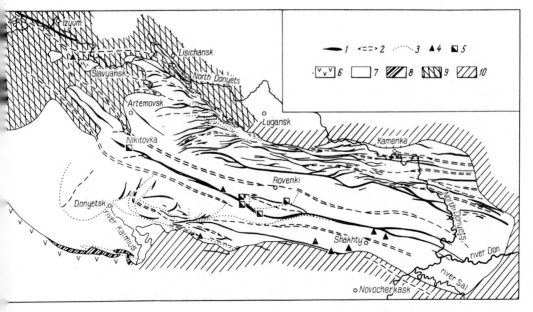

FIG. 242.   Main tectonic features of Donbas. According to P. I. Stepanov (1937).

1 — anticlines;   2 — synclines;   3 — faults;   4 — igneous rocks;   5 — ore;
6 — Precambrian;   7 — Carboniferous;   8 — Devonian;   9 — Trias and Jurassic;   10 — Upper Cretaceous, Palaeogene and Neogene.

The western continuation of Donbas has been more fully investigated and is known as the Cisdnyepr coal region, since it is situated between Dnyepropetrovsk and Kremenchug to the north of the Dnyepr. The Coal-bearing Formation of this region is hidden under a cover of Mesozoic and Cainozoic rocks which are 100m thick in the south and up to 600m in the north. Exploratory boreholes have shown interesting, important changes in the coal-bearing succession. The thickness of Lower Carboniferous beds decreases from 1700m in the east to 110m in the west, being on average about 1100m thick, thus several-fold thinner than in the Donbas proper. The thickness of Middle Carboniferous strata decreases to 600m, while the upper part of the Middle Carboniferous, the Upper Carboniferous and the Permian are missing. Thus Mesozoic rocks lie on various Carboniferous suites. The Lower Carboniferous succession here on the other hand has the largest number of coal seams. A large number of workable coal seams are the same as in the Donyets region.

Useful Mineral Deposits

The Donyets Basin is the most important coal region in the U.S.S.R. yielding much high-quality coking coal. Its importance is enhanced by its proximity to the Krivoi Rog and Kerch iron deposits and the Nikopol manganese deposit.

In addition and within the Donyets Basin there are deposits of mercury at Nikitovka and the large intensively worked deposit of rock salt at Artem. There are also other deposits of cement, fluxes and building materials. Thus Donbas is one of the most important mining districts of the U.S.S.R. Recently oil and gas have been found. Commercial oil has been found along the northern margin of the Dnyepr-Donyets Depression, in the cap breccia of the Romny dome, and along the southern margin of the depression—near Radchenko (west of Poltava) where oil occurs in Upper Visean and Lower Namurian sandstones.

According to the latest calculations of the Ukraine Scientific Geological Survey Institute (Baranov *et al.*, 1961), provisional estimates show that the Dnyepr-Donyets Depression is one of the most important petroliferous and gas regions of the U.S.S.R. All the oil and almost all of the gas is concentrated in Palaeozoic deposits and in particular in the Carboniferous (73·35 per cent oil and 66·5 per cent gas).

Deposits of gas are important and large. In the north of the depression the major Shebelinka deposit, situated to the south-west of Kharkov is being exploited. Here gas is found in the Lower Permian saliferous formation at a depth of 1300–1600m and in Upper Permian and Lower Trias red sandstones at depths of 75–850m. Nearby there are other gas deposits (Spivakovskaya and Volkovskaya). In the south of the depression commercial deposits of gas are found in the regions of Radchenkovo, Mikhailovka and Saigaidak, where it occurs in Lower Carboniferous and Visean sandstones.

THE DNYEPR DEPRESSION

General

The depression represents the continuation of the Donyets Basin to the west and differs from it by the absence of a coal bearing formation of Carboniferous age, which is replaced by marine and continental deposits of much lesser thickness. The second feature of the region is the discovery of oil and gas in salt domes and anticlinal structures.

Between the Donyets and Dneypr depressions there is a connecting zone, which has not been sufficiently investigated. The zone consists of a series of major faults in the basement. The faults have affected Palaeozoic deposits. Large faults can be seen in the Ukranian Shield and continuing northward across the depression from the connecting zone.

Between the Dnyepr and the Pripet Depression is the Chernigov Saddle which is also associated with deep faults in the crystalline basement. The faults were formed at the end of the Givetian Stage. The faults guided the basaltic magma, which produced 1160m of volcanic deposits. A deep borehole has demonstrated that along these faults a block of the crystalline basement has been elevated by 1000m above the bottom of the depression. The block thus connects the Russian Platform and the Ukrainian Massif.

As has been shown geophysically (Andreeyva and Chirvinskaya, 1958) the depth of the Dnyepr Depression gradually increases from 3–4km in the west to 6–7km in the east. In the Donyets Depression the depth increases still further, reaching 10–12km.

Stratigraphy

The cross-section across the Dnyepr Depression (Fig. 243) shows that the depth of the crystalline basement under the Visean of the flanks of the depression is 1426m at Reizerovo and 2218m at Smeloye. The structure of the central part is not completely known, but in Kolaidentsy there is a thickness of over 1000m of quartz-feldspar sandstones under the Tournaisian. The sandstones belong mainly to the Upper Devonian. Within cap breccias of Isachki and Romny there are flaggy Upper Frasnian limestones with *Spirifer anossofi* faunas. In Romny the limestones occur within a formation (160m) of clays with beds of sandstones. Higher up there are red beds and sandstones (100m) covered by Lower Carboniferous beds. The lowermost clays overlie the salt. Thus the presence of Upper and possibly Middle Devonian strata can be considered as established. On the southern flank the total thickness of the Devonian reaches 3500m. A characteristic feature of the Devonian is the presence

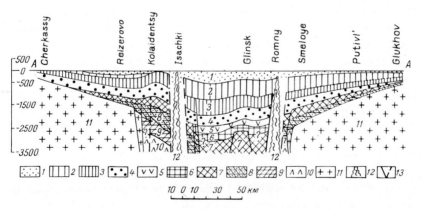

FIG. 243. Section across the Dnyepr Depression. According to I. Yu. Lapkin (1957).

1 — Cainozoic; 2 — Upper Cretaceous; 3 — Jurassic; 4 — Trias and Upper Permian; 5 — Lower Permian; 6 — Upper Carboniferous; 7 — Middle Carboniferous and Namurian; 8 — Visean (coal-bearing); 9 — Tournaisian; 10 — Devonian; 11 — Precambrian; 12 — salt domes; 13 — faults.

of salt domes and associated oil. Devonian rocks are underlain by platform Upper Proterozoic strata which to the south of Poltava (Mikhailovka) are up to 1500m in thickness. The quartz-sandstones of Ovruchian-type lie in the Chernigov borehole on weathered granites.

The Devonian is succeeded by sandy-clayey marine and continental Lower Carboniferous sediments, which are several hundred metres thick, but have no commercial coal seams. The Middle and Upper Carboniferous is also represented by terrigenous rocks, but in the uppermost part of the Upper Carboniferous certain limestones with a marine fauna have been encountered. Lower Permian strata are thin (200m) and consist of red beds with layers of gypsum. In the Shebelinka region there is a thick lower Permian saline formation (700m) with marine bands of Sakmarian age. The Upper Permian and the Trias consist of a thick cover, which overruns the whole of the depression. Lithologically the cover consists of continental red beds (sands and clays), which are subdivided into several suites. The average thickness of these beds is 300–350m, being 600–700m in the centre of the depression. In Shebelinka, to the south of Kharkov they contain an important deposit of gas.

The Jurassic is represented only by Middle and Upper divisions consisting of alternating continental and marine formations. The succession begins with basal Bajocian sandstones and clays, which are succeeded by marine Bathonian clays overlain by clays with siderite beds. The fauna of the latter beds indicates that they belong to the Callovian, Oxfordian and Lower Kimmeridgian, Upper Kimmeridgian and Volgian stages are either missing or represented by thin continental deposits. The total thickness of Jurassic strata is up to 300–400m (Table 13). A commercial deposit of gas has been found in Middle Jurassic rocks.

Lower Cretaceous strata are often missing (cf. Donbas), indicating a widespread emergence. Certain sandstones and carbonaceous clays (up to 120m), lacking faunas are thought to be Cretaceous.

The Upper Cretaceous succession begins with Upper Albian basal sandstones, succeeded by Cenomanian sandstones and clays of a total thickness of up to 60–75m. Higher up lies a sequence of white marl and chalk, which is typical of the Mesozoic. In the west (Chernigov) the section starts with marls (100m) and ends with chalk (44m). Further to the east, at Romny, Turonian and Santonian marls are 232m thick and Campanian-Maastrichtian chalk reaches 151m. Lastly in the Kharkov and Kupyansk region only chalk is developed and is 600–700m thick. Thereafter eastwards the thickness of Upper Cretaceous rocks decreases again.

Palaeogene and Upper Cretaceous strata are closely associated and occur ubiquitously. The thickness of the former is considerably less than that of the latter and does not exceed 150–200m. In the west there is a characteristic development of continental and near-shore sandstones of Kanevian and Buchakian ages. In Chernigov these strata are 92m thick, which is twice as much as the thickness of the Kiev Clays and the Kharkov Sands. In the Romny (Smeloye) region and elsewhere Montian (Palaeocene) deposits underlie Kanevian. In the east, near Kharkov and Kupyansk, the lower, sandstone formation is almost absent (1–2m) and the succession begins with the Kiev Clays (20m) and the Kharkov Sands (28m). In other regions of the Dnyepr Depression marine clays and marls of the Kievian Stage (15–30m), and sands and sandstones of the Kharkovian Stage (30–40m) are also widespread. The upper part of the continental variegated clays and sands of the Poltava Stage belong to the Miocene. The Poltava sequence and other Neogene continetal sediments are found in places. Their thickness is small, being of the order of 10–20m and is sometimes more. Anthropogene deposits are also thin and consist of lacustrine, fluviatile and terrestrial (loess) sediments.

Tectonics and Magmatism

The Dnyepr Depression began in the Givetian and Frasnian epochs, when deep and extensive faults conditioned the northern and in places the southern margin of the trough. Some faults also trended across it. The faults provided channels for extrusion of basaltic magma. In the Chernigov borehole the thickness of the volcanic formation is more than 1100m and basaltic lava flows and beds of tuff are found far to the east in the cap breccia of the Romny salt dome, and in the south of Donbas amongst Devonian deposits. The extrusion of basalts should have produced sub-volcanic intrusive rocks, which however remain undiscovered.

There are no synorogenic intrusions since folding has been weak. The suggestion that deep under Donbas there is a large Hercynian or younger synorogenic granite mass, which accounts for mineralized veins of the Nagol'nyi Ridge, is not adequately demonstrated and is unlikely since folds affecting Palaeozoic strata are gentle.

In the Dnyepr Depression (Fig. 243) folds are even simpler and gentler than in Donbas, but salt domes (Fig. 244) are interesting.

The useful mineral deposits of the Dnyepr Depression are mainly represented by large gas deposits (Shebelinka, Mikhailovka, Radchenkovo) and oil of the salt domes and other structures found in the Chernigov and Sumkian regions (Palii, Petrov, 1961). There are numerous, often large, deposits of various non-metallic deposits, such as constructional and building raw materials, ceramic and fire clays, gypsum, rock salt and many others.

THE PRIPET DEPRESSION

The Pripet (Polesskian) Depression is separated from the Dnyepr-Chernigov Depression by an elevation, but is in fact its continuation. This is proved by an almost complete identity of the two successions and tectonic structures. In connection with the discovery of oil and a thick salt horizon the Pripet Depression has been extensively drilled and studied. The relief of the crystalline basement is shown in Fig. 15. In the west the Pripet Depression is divided by the Mikashevich 'projection' into the

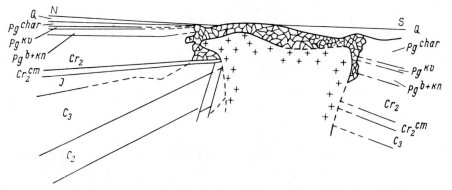

Fig. 244. Section through salt dome at Romny. The salt (ornamented in crosses) has its cap breccia exposed on the surface. The salt which is considered to be Devonian cuts across Carboniferous, Permian, Mesozoic and Kainozoic rocks. According to Yu. A. Kosygin (1952).

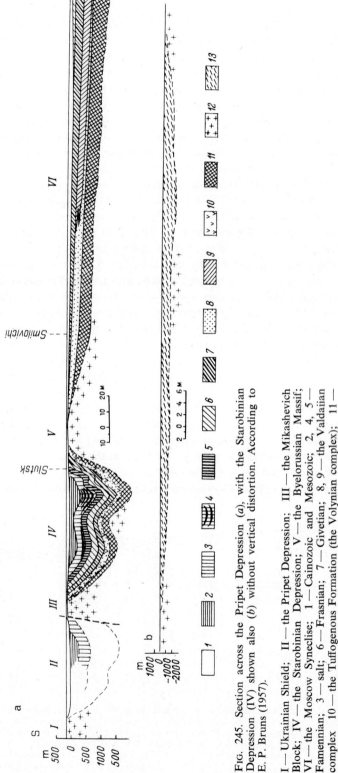

FIG. 245. Section across the Pripet Depression (*a*), with the Starobinian Depression (IV) shown also (*b*) without vertical distortion. According to E. P. Bruns (1957).

I — Ukrainian Shield; II — the Pripet Depression; III — the Mikashevich Block; IV — the Starobinian Depression; V — the Byelorussian Massif; VI — the Moscow Syneclise; 1 — Cainozoic and Mesozoic; 2, 4, 5 — Famennian; 3 — salt; 6 — Frasnian; 7 — Givetian; 8, 9 — the Valdaiian complex 10 — the Tuffogenous Formation (the Volynian complex); 11 — the Polessian complex (the Orshian Suite); 12 — crystalline basement; 13 — sedimentary deposits in the Starobinian Depression.

northern part (Starobinian) which closes against the Byelorussian-Lithuanian Massif, and the southern part (Naravlyankian) which beyond the Pinsk Elevation continues into the Kovel Depression. The greatest depth of the Pripet Depression is 3500–3800m, its greatest width at Mozyr is 150km, while its maximum length is 300–350km. On the flanks dips do not exceed 7°–10°, and are usually much less (Fig. 245). Fig. 245 shows the exaggerated vertical scale in the top sketch and a true section in the bottom sketch. The latter indicates that both the Pripet Depression and the equally deep Dnyepr Depression are hardly noticeable features in overall relief. The Donyets Depression is somewhat deeper, but is still of the same general type.

The Pripet succession begins with Upper Proterozoic complexes, such as the red Polesskian, the volcanic Volynian and the clastic Valdaiian, which are developed in the western, deeper parts of the trough. In the east these rocks are absent. The complexes have been described on p. 54. Their thickness is of the order of 400–450m.

Proterozoic deposits are overlain by Devonian strata (Table 51). The Eifelian Narovian Beds are represented by clays and dolomites (50m) and are overlain by the Givetian, red Luga Beds (up to 200m). The most important feature of the Pripet Depression is the extraordinary thickness of 3000m of Upper Devonian rocks, which include petroliferous sands and enormous deposits of salt reaching 800m in thickness (Table 51).

At the bottom there are variegated continental sandstones and clays which are 20–80m thick. Above these lies a marine formation of grey and greenish clays, marls and limestones reaching 150–400m. Its fauna indicates the presence of all horizons beginning with the Upper Shchigrovian and ending with Yevlanovian. Higher up is the saliferous formation (300m) belonging to the Upper Frasnian, Livenian Horizon. The intersaliferous carbonate-clay formation (500m) has a Lower Famennian marine fauna, which determines the age of both the lower and the upper saliferous formations. The Upper Saliferous Formation is up to 1500–1800m thick and corresponds to the Lebedyankian Horizon. The formation consists of rock salt with beds of clays and dolomites and is most complete in the western, deeper part of the depression. As one approaches to the margins of the depression the thickness of the salt decreases and beds of clays and dolomites become more abundant. The Upper Devonian succession is capped by a grey marl-clay formation (up to 300m) with an Upper Famennian flora.

The age of the two saliferous formations at one time provoked discussion. Some investigators considered them to belong to the Givetian Stage, while others attributed an Upper Devonian age. As sometimes happens, both points of view were correct. The two saliferous formations of the Pripet Depression are truly Upper Devonian—mainly Famennian. On the other hand the salt of the salt domes along the northern margin of the Donyets Depression (Slavyansk, Petrovsk) and the Dnyepr Depression (Romny, Isachki) is undoubtedly Givetian. The presence of two saliferous formations in the South Russian Depression corresponds to the existence of two such formations in the central provinces (p. 80).

Lower Carboniferous strata are only found in the eastern part of the depression and wedge out westwards. The Tournaisian Stage consists of grey clays and sands (200m) similar to those of the Upper Famennian but having a Tournaisian fauna and pollen. The Visean Stage begins by the analogues of the Moscow coal-field Coal-bearing Formation, clays and sands with brown coal seams with a flora. Higher up there are clays and sands with a marine fauna. These beds are overlain by 30–50m of limestones and clays. The total thickness of Visean sediments reaches 350m, but is usually no more than 60–100m. Namurian sediments have only been recognized at Chernigov. Middle Carboniferous strata are only found in the centre of the depression at Yel'sk where they consist of clays (45–80m) with beds of fossiliferous marine limestones. The marine Upper Carboniferous is unknown but certain continental red sandstones and clays (80m) which lie on Middle Carboniferous beds, may be Upper Carboniferous.

Permian and Lower Trias continental red beds are sometimes thick. The marine Lower Permian fauna is only found in the Chernigov borehole and is unknown

TABLE 51.

Devonian Succession of the Dnyepr-Pripet Depression

| Stage | Beds, horizons | Prip | |
		Starobin	Koreni
	Cover.	Tournaisian-Visean sandstones and clays.	Tournaisian-Visean sandstones and clays (332m).
Famennian	Dankovo-Lebedyankian.	Grey clays and marls (150m). Rock-salt (700m). Brown, cavernous dolomites with beds with marine faunas (100m).	Grey clays with beds of dolomite and salt (About 100m). Rock salt with beds of clays, sandstones and dolomites (1000m).
Famennian	Yeletsian and Zadonskian.	Grey limestones with beds of dolomite and gypsum, and a marine fauna (60m).	Unpenetrated.
Frasnian	Livenian, Yevlanovian and Voronezhian.	Dolomites with beds of sandstones.	
Frasnian	Semilukian.	Grey dolomites and gypsum. The strata are barren and 210m thick.	
Frasnian	Shchigrovian.	Variegated, dolomitic clays, with fish, *Estheria* and *Lingula* (20m).	
Givetian	Tartuian.	Variegated clays, silts and sandstones (190m).	
Eifelian	Narovian.	Grey dolomites and dolomitic marls (70m).	
Eifelian	Basement.	Lower Cambrian.	

further to the west. *Estheria*, ostracods and fish found in various horizons permit the identification of the Upper Permian (Tatarian Stage) and the Lower Trias (Vetlugian Stage). Clays and sands with beds of dolomites are provisionally included in the Lower Permian and reach a thickness of 150–200m. The thickness of Upper Permian red beds varies from 50–300m, while Lower Trias red beds are 100–500m thick. The total thickness is about 1000m. In the Pripet Depression Middle and Upper Trias or Lower Jurassic strata have not been found.

Middle and Upper Jurassic beds begin with a sandy-clayey formation (10–130m) found throughout the depression. The lower part of this sequence is continental and is considered as Middle Jurassic, then follow the sands and clays with a Lower

TABLE 51—continued

...olessian) Depression		Dnyepr Depression
Narovlya	Yel'sk	Chernigov
Tournaisian-Visean sandy-clayey formation.	Tournaisian clays and sandstones.	Visean sandstones and clays (169m).
Grey marly-clayey formation with a Famennian flora. In places a thick calcareous breccia (300m).	Grey clays with beds of dolomites and sandstones (250m). Upper Rock-salt with thin beds of clays and dolomites (1500m).	Crust of weathering (7m). Diabases, basalts, tuffs, tuff-breccias, tuffites (211m). Dark argillites with beds of sandstone and limestone. Marine lamellibranchs (*Buchiola cardiola*) are present (105m). Interbedded grey and dark argillites, tuffs and sandstones (165m).
	Intersaline clays with beds of marls and dolomites. A marine fauna is present (500m).	
Greenish, muddy limestones with a marine fauna (40m).	Lower Rock-salt with beds of clays and dolomites (300m).	
Grey breccia-like (pseudo-brecciated) limestones with a marine fauna (55m).	Argillites, sandstones, limestones and gypsum, tuffs and diabases (222m).	Basalts, diabases and porphyrites, interbedded with tuffs and tuff breccias (455m).
Above, grey and brown dolomites and marls; below, clays (85m).	Not penetrated.	Argillites, silts, sandstones, limestones, dolomitized limestones and gypsum-anhydrite. Beds of tuff, diabases and diabase porphyrites (222m).
Variegated sandstones and clays (80m).		
Crust of weathering on top of Archeozoic rocks.		Red quartzites cf. Ovruchian (6m). Granite gneisses (83m).

Callovian marine fauna. Higher up is a limestone-marl formation (20–100m) consisting of marls and clays with subordinate marine, fossiliferous limestones of Middle and Upper Callovian and Oxfordian ages. Younger Jurassic or Lower Cretaceous strata are unknown.

The Upper Cretaceous begins with quartz-glauconite sands of a suggested Cenomanian age, although there are indications that they may be Upper Albian. Higher up there is a formation of chalk-like marls (20–100m) considered as Turonian. To the east the marls thicken and chalk appears at the top (e.g. Chernigov borehole). In the borehole, Turonian, Coniacian and Santonian marls are covered by Santonian and Maastrichtian chalk and the total thickness of Upper Cretaceous sediments is 144m.

Palaeogene sediments cover the whole of the Dnyepr and Pripet depressions and continue without a break into the Polish-Lithuanian Depression as can be seen on synoptic maps. Neogene sediments remain at the surface as small remnants. The Palaeogene consists mainly of palaeontologically badly defined sands (about 100m) which have not been divided in detail. Towards the top there are clays with plant remains. The clays are in part Neogene.

The Anthropogene consists of continental deposits, like those in the Dnyepr Depression. The deposits are relatively thin reaching 50m.

The tectonics and magmatism are of the same type as in the Dnyepr Depression. Faults are much less frequent and fissure eruptions of basalts are only known to the south and north of the Mikashevich 'projection', along the margin of the depression. Useful mineral resources as in the Dnyepr Depression are typified by oil and Devonian salt, while non-metallic deposits are again of the same type.

THE ASTRAKHAN DEPRESSION

Donbas (Greater Donbas) is a coal region and its main diagnostic feature is the presence of commercial coal. The region ends where coal seams die out and consequently districts where there are no commercial coal seams cannot under any circumstances be considered as parts of Donbas. Thus the eastwardly continuation of the basin should not be called Eastern Donbas, since there is no coal. In the last

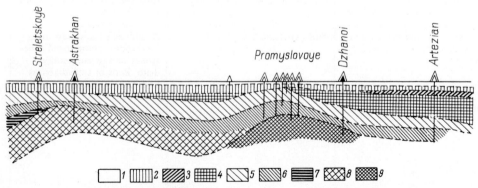

Fig. 246. Cross-section of the subterranean Karpinskii Ridge.
According to Ya. S. Eventov (1957).

1 — Quaternary; 2 — Pliocene; 3 — Miocene; 4 — Palaeogene; 5 — Cretaceous; 6 — Jurassic; 7 — Upper Permian and Trias; 8 — Lower Permian; 9 — Carboniferous.

few years boreholes have shown that folded Permian and Carboniferous sediments fill a narrow depression, which stretches towards Astrakhan. It is undoubted that a continuation of this depression can be traced further eastwards until it merges into the downwarp of the Uralian Geosyncline. The town Astrakhan lies in the central part of this depression. The most extensively investigated western part of the Astrakhan Depression lies to the north of the valley of Manych. At present the depression is entirely filled and covered by horizontally lying Palaeogene and Neogene sediments and is not reflected in any topographic features. Before Upper Cretaceous and Palaeogene times the depression was a low, hilly range, similar to the present-day Donyets range as it can be seen in borehole sections (Fig. 246). The hilly range is known as the Karpinskii Ridge, since A. P. Karpinskii first suggested it.

Like the Donyets range, the Karpinskii Ridge consists of folded Carboniferous and Lower Permian deposits. These deposits continue, as an unbroken belt, from Donbas through the town of Yelist and the village of Peschanoye to Astrakhan. At Astrakhan the southern margin of the depression lies somewhat to the south of the termination of the delta of the Volga. From Astrakhan onwards folded Palaeozoic strata are traced geophysically and in boreholes as far as the Urals.

The depression has an almost east–west trend, in the south adjacent to the Yaikian Massif, and joins the Urals to the north of the Sea of Aral. The eastern part of the Astrakhan Depression is shown schematically in Fig. 8. Its geological structure has not as yet been sufficiently studied. The depression represents a direct continuation of the Donyets Depression and is a typical aulacogen. In Upper Palaeozoic times the aulacogen became a low fold range. Its eastern part which merges into the Urals is known as the 'South Emba Uplift'.

11

THE PACIFIC OCEAN GEOSYNCLINE

GENERAL INTRODUCTION

The Pacific Ocean Geosyncline is the farthest and least studied in the U.S.S.R. A short while ago it was even less investigated and almost unknown, but owing to the discovery of major deposits of gold, tin, polymetallic ores, iron, coal and oil, mining has proceeded at a previously unknown speed. There is no doubt that the region will be soon as important as other geosynclines in the U.S.S.R.

BOUNDARIES. The Pacific Ocean Geosyncline is a vast region of powerful Cimmeridian and Alpine folding. The region margins the Pacific Ocean both from the west and the east and consequently an Eastern Pacific Geosyncline and a Western Pacific Geosyncline have been recognized.

A. A. Borisyak has suggested that the part of the fold region which diverges from the Pacific Ocean northwards to form the Verkhoyansk range should be called the Verkhoyansk Geosyncline. The suggestion has not been widely accepted, but nowadays deserves attention. The discovery of the submarine Lomonosov Ridge and its investigation have shown that it represents the continuation of the fold belt crossing the Arctic Ocean. Since such a fold belt cannot be referred to as a Pacific Ocean belt, as it does not lie in the Pacific Ocean, the term Verkhoyanskian as suggested by Borisyak for the southern part of it should be adduced to this zone. However, until such a suggestion is accepted the Verkhoyansk range is considered together with the Pacific Ocean Geosyncline. Thus the northern boundary of the geosyncline is the shore of the Arctic Ocean, or more accurately the edge of its shelf. The north-western margin of the geosyncline lies along the Siberian Platform and the southern margin is the boundary of the almost horizontal layer forming the abyssal bottom of the Pacific Ocean (Fig. 2).

SUB-MONTAINE (FRONT) DOWNWARPS. The clearly delineated sub-Verk-hoyansk downwarp is situated between the Siberian Platform and the Verkhoyansk range. The downwarp is relatively young and is filled by thick (2000–3000m) Creta-ceous coal-bearing deposits.

THE SOUTHERN DOWNWARP is situated between young volcanic ranges of the Aleutian islands, Kamchatka, the Kuril islands and Japan on one side and the Pacific Ocean block on the other. The downwarp is very young, is not entirely filled and consists of a series of deep oceanic depressions which trend continuously one after another. The Kuril-Kamchatka trough on the territory of the U.S.S.R. is over 10,000m deep.

CIMMERIDIAN AND ALPINE ZONES. The Cimmeridian zone is adjacent to the Siberian Platform, continues into the Lomonosov Ridge and includes all the ranges of the basins of the Yana, the Indigirka, the Kolyma and the Chukotian Peninsula. The zone shows strong manifestations of the Cimmeridian Orogeny including strongly folded and metamorphosed Jurassic strata, which are cut by young intrusive granites. The Alpine Orogeny was weak and only its oldest phases formed gentle folds affecting Cretaceous and Palaeogene beds. The later Alpine phases are absent and Neogene and Anthropogene sediments are almost horizontal.

The Alpine zone includes the rest of the geosyncline. Here both the Cimmeridian and the Alpine orogenies were powerful. All sediments, including Anthropogenic, are folded. Granites intrude not only Jurassic strata, but also Cretaceous, Palaeogene and even Neogene sediments. Several large ore deposits are related to the Alpine magmatism. The widespread Pliocene and Anthropogene volcanoes and flood basalts are characteristic and there is also Neogenic oil.

The boundary separating the Alpine and Cimmeridian zones is not sharp and is drawn arbitrarily along the Ussuri zone, the lower reaches of the Amur, through the Sea of Okhotsk towards the Gizhiginian harbour and the basin of the Anadyr.

Formerly the author used the terms 'outer' and 'inner' for the Cimmeridian and Alpine zones respectively, as is done in the case of the Mediterranean Geosyncline. The Cimmeridian zone can be considered as the outer since it forms the outer part of a young fold belt adjacent to an ancient platform. The Alpine zone which margins the Cimmeridian in the south can be quite naturally compared to the inner zone of the Mediterranean Geosyncline. Modern oceanographic work has, however, shown that such a comparison is incorrect. The Alpine zone of the Pacific Ocean Geosyncline is bounded to the south by a sub-montaine downwarp, followed by the Pacific Ocean Platform. Thus it is a marginal zone of a geosyncline and cannot be called the inner zone. If the Alpine zone of the Pacific Ocean Geosyncline cannot be called the inner, the Cimmeridian zone cannot be called the outer. In any case the outer zone of the Mediterranean Geosyncline differs considerably from the Cimmeridian zone of the Pacific Ocean Geosyncline since the former shows only weak manifestation of the Cimmeridian Orogeny such as gentle folding and absence of synorogenic intrusions and the associated mineralization.

Administrative Geography. The whole of the area is in the R.S.F.S.R. The greater northern part of the geosyncline is situated in the Yakutian A.S.S.R., the Magadan Province and the Khabarovsk region. The southern, more populated, part is situated in the Chita and Amur provinces, the southern part of the Khabarovsk region and the Primorian and Sakhalin provinces, including Sakhalin and the Kuril islands. Kamchatka belongs to the Khabarovsk region.

OROGRAPHY

Eastern Transbaikalia. This district is low but mountainous and near to the Amur hilly country. In the north it is bounded by the Yablonovoy range which has a height and relief similar to the Urals. Further on lies the wide populated valley of the Shilka with the towns of Chita, Nerchinsk and Sretensk along it. To the south of it lies the low Borshchevochnyi range (1500–2500m) and further to the south the Nerchinskian range (1500m), on the slopes of which are situated the Nerchinskian and Gazimurskian factories. Between the Nerchinskian range and the Argun lies a vast hilly plain cut in Upper Cretaceous and Palaeogene strata. The Shilka and Argun merge to form the Amur (Fig. 247).

Cisamuria. This vast geographic province situated between the scarp of the Stanovoy range and the valley of the Amur has the same relief as Eastern Transbaikalia of which it forms a continuation. Amongst the large hilly plains covered by a very rich luxuriant vegetation stand low forest-clad ranges (1500–2500m). Amongst the ranges flow full large rivers, such as the Zeya, the Bureya, and the Amgun which are all the tributaries of the Amur (Fig. 248). Amongst the ranges the Lesser Khingan, which consists of old, partly Precambrian metamorphic rocks and which is the continuation of the main massif with China, should be mentioned. The valley of the Bureya has a coal-field.

The Sea of Okhotsk. The Sea of Okhotsk is a typical marginal sea, open to the ocean and isolated by volcanic islands. Its northern part is shallow (down to 200–400m), but the southern part is deep (3300–3600m). The coasts are mountainous and in the west are represented by the escarpment of a young fold belt, known as

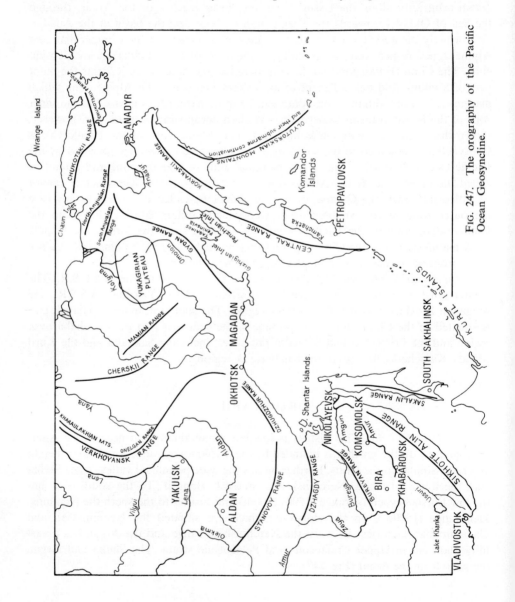

FIG. 247. The orography of the Pacific Ocean Geosyncline.

Dzhugdzhur (Fig. 249) and then by the scarps of the southern part of the Verk-hoyansk range.

Between the mouths of the rivers Uda and Amur the coast of the Sea of Okhotsk and the adjacent inland districts are often known as 'Western Cisokhotia'. The Shantar islands have a very similar structure to Western Cisokhotia (Fig. 250).

FIG. 248.   The lower reaches of the Amur. *Photograph by G. S. Ganeshin.*

FIG. 249.   The river Nemchun on the eastern slope of the Dzhugdzhur range.
Drawing by Yu. F. Chemekov.

FIG. 250.   The Shantar islands, consisting of Middle Devonian sandstones and shales. *Drawing by G. S. Ganeshin.*

In the northern part of the Sea of Okhotsk the Gizhigian and Penzhinian inlets, representing the lower drowned valleys of the Gizhiga and Penzhina, are of a large size. Between the inlets lies the Taigonos Peninsula. The new town of Magadan lies on the coast of the Sea of Okhotsk.

North-eastern Siberia. A large, mainly mountainous tract stretching from the Lena to the Bering straits for 3000km is known as North-eastern Siberia. Even as recently as 25–30 years ago it was almost uninvestigated and was shown on geological maps as a large blank patch. Owing to the work of Soviet geologists including S. S. Smirnov—an ore specialist—and Yu. A. Bilibin, a specialist on the geology of gold, large deposits of gold and tin have been found. As a result geological investigations were rapidly expanded and at present the basins of the Yana, the Indigirka, the Kolyma and the Omolon, the right tributary of the Kolyma have been studied. Each of these rivers with respect to its water capacity approaches the Volga or the Dnyepr. They flow in the main amongst mountain ranges and plateaux, and only near the Arctic Ocean through a vast tundra.

Amongst the mountain ranges the most important are: the Verkhoyansk range, up to 3000m high, the Cherskii range which crosses the watersheds between the Yana, the Indigirka and the Kolyma and also reaches over 3000m in height; and the Kolymian range which is up to 2500m high.

In the extreme north-east lies the Anadyr range and the Chukotian Peninsula highlands. The Verkhoyansk range terminates in the north-east by the Kharaulakhian Mts (Fig. 247).

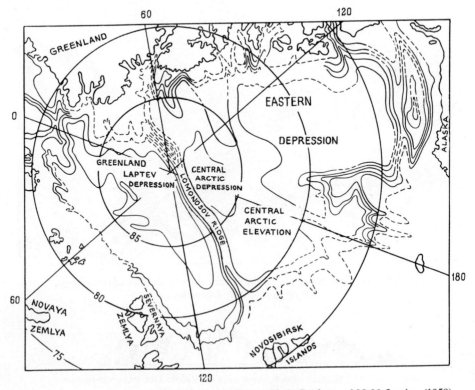

FIG. 251. Lomonosov oceanic ridge. According to N. A. Byelov and N. N. Lapina (1958).

The continuation of the Cherskian range is quite possibly represented by a recently discovered submarine ridge known as the Lomonosov Ridge (Fig. 251). The ridge arises out of the shelf to the north of Novosibirskian islands and rises 2–3km above the bottom of the sea. All the enumerated regions form parts of the Cimmeridian zone of the Pacific Ocean Geosyncline.

The Alpine zone includes the following regions.

The Sikhote Alin' Range. The range is low (1800–2000m), strongly dissected (Fig. 252) and completely overgrown by a dense forest, which continues from Vladivostok to the mouth of the Amur (Fig. 253). Its relief is considerably influenced by plateau basalts covering high, strongly dissected mountains and plateaux and by numerous valleys of rivers such as the Iman and the Anyui (Figs. 253, 254).

Sakhalin. The island of Sakhalin is long and narrow and stretches parallel to the dominant strike. In the south it is mountainous and in the north it is low and swampy. It has deposits of oil and coal.

The Kuril Islands. The Kuril islands are numerous narrow volcanic islands conecting the structures and volcanoes of Kamchatka to those of northern Japan.

Kamchatka. Kamchatka is a large mountainous peninsula of about 1200km in length along the tectonic strike. Along deep faults there are many volcanoes amongst which the Klyuchevskaya (4850m), the Koryakskaya (3464m) and the Kronotskaya (3528m) are the largest (Fig. 255).

The Koryakskian Range represents a continuation of Kamchatkan ranges (Fig. 256). The range is low (2115m) and is in places overgrown by forests. The rocks consist of folded mainly Palaeogene and Upper Cretaceous strata which are frequently coal-bearing. In places older Mesozoic and Palaeozoic deposits are exposed. The range is often referred to as the Koryakskian Highlands.

FIG. 252. The eastern slope of the Sikhote-Alin' range, at the head of the river Tetyukha. *Photograph by Yu. F. Chemekov.*

Fig. 253.   The western slope of the Sikhote-Alin' range, at the head of the river Iman (in the centre). The relief is highly dissected.

FIG. 254. The river Anyui on the western slope of the Sikhote-Alin' range.
In the distance is the highest mountain—Mt. Tardoka-Yani (2078m).
Photograph by G. S. Ganeshin.

FIG. 255. The sopka Klyuchevskaya.

The Anadyr Valley is a wide plain filled by Mesozoic and Cainozoic sediments, which over large areas are covered by Anthropogenic deposits.

The Komandor Islands. The islands are small and volcanic and have outcrops of Palaeogene and Neogene strata.

STRATIGRAPHY

In the Pacific Ocean Geosyncline all formations beginning from Archean are found, although Mesozoic and Cainozoic deposits predominate and are represented by marine continental (coal-bearing) and volcanic facies complexes. The investigation of the Pacific Ocean Geosyncline was guided by three regional conferences—at Khabarovsk in 1956, concerning the Far East; at Magadan in 1957, concerning the north-eastern part; and in Okha (Sakhalin) in 1959, concerning Sakhalin and Kamchatka. Papers, discussion data and unified schemes adopted at these conferences form the basis of the present chapter.

THE PRECAMBRIAN

DISTRIBUTION. Precambrian deposits are exposed in many regions. They cover large areas of ancient massifs, but are much less abundant amongst folded Mesozoic and Cainozoic strata. The main regions where Precambrian rocks are found are:

FIG. 256. The central part of the northern slope of the Koryakskii range. The helicopter of a geological expedition can be seen.

1. The Zeya-Bureyan Massif, which is large and includes the basins of the Zeya, the Selemdzha, the Bureya; in the north the basins of the Uda and the eastern part of the Stanovoy range and in the south the Lesser Khingan;
2. The Kolyma Massif and the Omolon Block;
3. The Okhotsk Massif;
4. The Chukotian (Chukotskian) Massif;
5. Eastern Transbaikalia (the Gazimur-Nerchinsk region);
6. The southern part of the Sikhote-Alin' range and the district of lake Khanka (Southern Primoria);
7. The Central Kamchatka range;
8. The Sakhalin ranges.

SUBDIVISIONS AND LITHOLOGY. The Archeozoic has been recognized in all the regions where the gneisses, schists and amphibolites are unconformably overlain by less metamorphosed Proterozoic strata, which are in turn unconformably overlain by Cambrian rocks·

In the northern part of the Zeya-Bureyan Massif (the Uda-Zeyan Province) Archeozoic rocks consist of gneisses, crystalline schists, marbles and amphibolites. In the north of Upper Primoria Lower and Upper Archeozoic rocks are recognized. Lower Archeozoic rocks consist of biotite and hornblende gneisses and migmatites, while Upper Archeozoic rocks include garnet gneisses and garnet-biotite gneisses with layers of amphibolite and marble. The total thickness is several kilometres.

Further to the south in the Khingan-Bureyan region Archeozoic rocks are divided into four suites (Fig. 257). The two lower suites consist of biotite paragneisses and various quartzites which are sometimes finely banded. The third (graphite-bearing) suite consists of paragneisses with almost ubiquitous flakes of graphite which sometimes concentrate into lenses. The fourth suite consists of augen gneisses and crystalline schists with bands of amphibolite and rare lenses of graphitic marbles. The total thickness of Archeozoic rocks exceeds 13km.

In Southern Primoria Archeozoic rocks are found in several regions, as for instance near lake Khanka and along the Ussuri. The rocks are again divided into four suites. The lowermost suite consists of gneisses, the next of schists and gneisses the third of marble and gneisses and the fourth of amphibolites. The total thickness of these deposits is about 9km.

In the Okhotsk Massif (to the north of Okhotsk) Archeozoic rocks consist of a charnockitic series, which is identical to the charnockitic series of the Siberian Platform. The series includes hypersthene gneisses, granitic gneisses and various schists and amphibolites. They are overlain by marine Cambrian sediments.

The Omolon Block shows biotitic, biotite-garnet hornblendic and hypersthenic gneisses and migmatites. The rocks are similar and analogous to those of the Okhotsk Massif, but are succeeded by Lower Proterozoic schists. On the Taigonos Peninsula similar rocks are found in identical relationships. Archeozoic rocks of the Taigones have been dated as 1700 myr, thus suggesting that they are Lower Proterozoic.

It is possible that gneisses of the Chukotian Peninsula, Wrangel island, the Kolyma Massif, Kamchatka and some other regions are also Archeozoic, despite the fact that they are normally considered as Proterozoic or undifferentiated Precambrian.

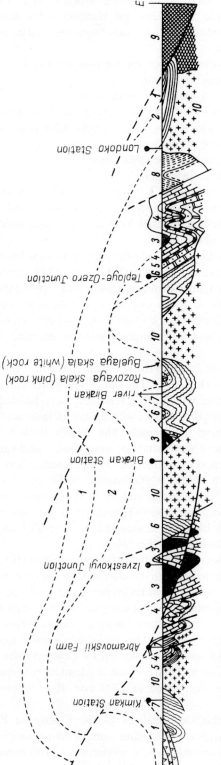

Fig. 257. Section across the Lesser Khingan range, showing the Kiman and the central anticlines. According to S. A. Muzylev (1938).

1 — Cretaceous porphyries; 2 — Jurassic; 3 — Palaeozoic siliceous shales; 4 — Londokovian limestones; 5 — the Ore-bearing Suite; 6 — Murandovian dolomites; 7 — Iginchian sandstones and shales; 8 — the graphite-bearing sequence; 9 — Archeozoic gneisses; 10 — Palaeozoic granites.

Proterozoic strata in several regions are divided into Lower and Upper. The Lower Proterozoic consists of metamorphic and crystalline schists, quartzites, marbles and volcanic rocks. The Upper Proterozoic is closely associated with the Lower Palaeozoic, forming homogeneous thick formations with typical massive limestones and dolomites. These formations were considered by some investigators as Upper Proterozoic and by others as Lower Cambrian and therefore they are shown as Upper Proterozoic or Lower Cambrian. In other regions these formations clearly embrace both the Proterozoic and the Cambrian. Not infrequently the division and dating of metamorphic Precambrian rocks has not been accomplished and they are shown as Precambrian.

In the Zeya-Bureyan Massif the Proterozoic is divided into lower and upper. Lower Proterozoic rocks are represented by a formation consisting of various schists, slates, meta-limestones and cleaved volcanic deposits. The total thickness in the Lesser Khingan region is no less than 7·5km. Migmatites are rare.

The Upper Proterozoic or Phyllitic Formation of the Khingan region is more than 6km thick and consists of three suites of mainly phyllites, phyllitic slates, cleaved sandstones and rarer green beds and jaspers. The upper or Murandovian Suite consists of dolomites, which are about 1000km thick (Fig. 257). Formerly the Ore-bearing Suite (200–400m) and the overlying Londokovian Suite of barren limestones and shales (1000m) were also considered as Proterozoic, but since the discovery of a single Lower Cambrian brachiopod in the former the two suites have been included in the Lower Cambrian. Whether the evidence of a brachiopod is adequate or not is difficult to assess since similar jaspilitic iron ores have been found amongst Proterozoic strata of other regions of the U.S.S.R. Yet the brachiopod is a brachiopod and the Ore-bearing Formation has to be considered as Lower Cambrian. Other Proterozoic successions of the Zeya-Bureyan Massif are similar to those of the Khingan-Bureyan region (Table 52).

A synoptic succession of the Precambrian of the Far East has been composed by V. V. Onikhimovskii (1956). Like other such successions it is theoretical and does not depict the full succession of any particular province. It is nevertheless an interesting testimony to the details which can be worked out for a complex of Precambrian formations.

Archean

The Okononian Series. Biotite-pyroxene (diopside) gneisses and marbles (2000–4500m). It is developed only in the Stanovoi range and Southern Primoria.

The Algominian Series. Biotitic gneisses and schists (2500–4000m). These are found in the Zeyan-Udaian, Upper Bureyan and Primorian regions.

The Afanas'yevkian Series. Biotitic gneisses and schists with subordinate hornblende gneisses (2000–4500m); found almost ubiquitously.

The Zolotogorskian Series. Hornblendic and biotitic gneisses with subordinate garnetiferous and graphitic gneisses (1000–1800m); found almost ubiquitously.

The Tastakhian Series. Ubiquitously typical (1100–2600m) and consists of graphitic marbles and various in places magnetite-bearing gneisses at the bottom, and at the top of gneisses with biotite, hornblende, graphite, garnet and in places magnetite.

The Salarinskian Series. Muscovitic, biotitic, augen and albite gneisses with subordinate schists and quartzites (up to 3000m). The series is only found in the Upper Bureyan and neighbouring regions.

TABLE 52. *Lower Cambrian and Precambrian Successions of the Far East*

Era	Upper Priamuria (Cisamuria) (I. A. Bogdanov, 1956)	The Zeya and the Selemdzha (D. A. Kirikov, 1956)	The Lesser Khingan and the Bureyan region (M. V. Chebotarev, 1956)	The Sikhote-Alin' range, southern part (N. A. Belyayevskii)
Lower Palaeozoic	Provisional Lower Palaeozoic—coarse sandstones and conglomerates (950m).	Lower Cambrian-quartzite sandstones and archeocyathid limestones (River Urmi).	Lower Cambrian. Londokian Suite—grey, siliceous shales (300m). Bituminous, sometimes banded limestones (400–500m).	Lower Cambrian. Limestones and shales (1500–2500m). Above, thick limestones and dolomites; below, repeated sequences of archeocyathid limestones and shales.
Upper Proterozoic	Phyllitic slates—Above, phyllitic slates and chlorite schists with a sequence of limestones; below, meta-quartzites (2000m).	Phyllites, cleaved sandstones, greenstones and limestones with algae.	Murandovian Suite of dolomites and shales, with magnesite (1000m). Iginchian Suite of phyllitic slates and sandstones (1000–2000m). Lower Carbonate (Diturian) Suite of flaggy limestones and phyllites (1000m).	Eocambrian-Phyllitized slates and sericitized marbles. Phyllites, graphitic schists and sericitized quartz sandstones (Up to 5000m).
Lower Proterozoic	Crystalline Schists Series. Chlorite-schist Suite—above, phyllitic schists with jaspers and dolomites; below, chloritic schists and lavas (3500m). Banded Slates Suite of green sericitized, towards the top, phyllitic slates.	Mica schists, marmarised limestones and cleaved lavas and quartzites.	Soyuznenskian Suite of mica schists and limestones. The suite is found in the south (2300m). In the north—above, mica schist and greenstones (3000m). Lower down, mica schists. At the bottom, aphibole schists (500m).	'Green' schists—chloritic and actinolitic; sericitized marbles. Below, mica schists and quartzites underlain by basal conglomerates (1500–2000m).
Archeozoic	Upper Archeozoic-garnet and biotite gneisses with bands of amphibolites and marbles. Lower Archeozoic-biotitic and hornblendic gneisses, in places migmatites.	Gneisses, granite gneisses and amphibolites.	Augen gneisses and crystalline schists (2500m). Graphitic gneisses (3700m). Biotite para-gneisses (3000m). Finely banded quartzites and albite gneisses (3900m). All above rocks, include migmatites.	Amphibolites and amphibole gneisses (1000m). Marbles, biotite gneisses and crystalline schists (2500 m). Biotite gneisses, interbanded with crystalline schists (3000m). Biotite gneisses and sillimanite schists (2500m).

Lower Proterozoic

The El'ginian Series. Amphibole and amphibole-pyroxene schists and amphibolites (400–800m). It has a restricted distribution and is commonly missing, although it can be found ubiquitously.

The Talmynskian Series. Muscovite and biotite schists with quartzites and conglomerates at the base. There are beds of magnetite-bearing rocks. Found in all regions (1600–3000m).

The Lower M'yenian Series. Strongly altered lavas and carbonates which have been converted into quartz-mica schist, muscovite schist, actinolite and talc schists, represent the series. The thickness is variable from a few hundred to 3700m. The schists occur in all regions.

The Soyuznenskian Suite. The suite shows a widespread occurrence of graphite. The lower part consists of micaceous and graphitic schists with amphibolites and schistose amphibolites. The thickness of this part is 1100–1800m. The middle part consists of graphitic marbles interbedded with sericitic and graphitic slates (200–900m). The upper part consists of garnet-mica schists and graphitic schists with bands of marble and quartzite. It is almost everywhere developed and is 600–700m thick.

The Zlatoustian Series. Quartz-muscovite and quartz-sericite schists with graphitic schists quartzites and marbles (300–2000m).

Upper Proterozoic

The Upper Bureyan Series begins with the Ortukian and Tuyunian suites of altered lavas and tuffs and subordinate sediments. The total thickness reaches 2000m, but the two suites are often missing out of the succession.

The Upper Diturskian, or the Lower Carbonate Suite of the Lesser Khingan, consists of flaggy, crystalline limestones, quartzites and phyllites (400–1000m).

The Malokhinganian Series is represented by two suites. At the bottom is the Iginchinian, which consists of dark phyllites, siliceous and clay slates, cleaved sandstones and beds of dolomites and limestones (1000m). The upper or Murandovian Suite consists of crystalline dolomites and limestones with lenses of magnesite and beds of slates (400–800m).

The Amnusskian Series is only found in the Selemdzhinian region, where it is 3900–6400m thick. At the bottom there are micaceous sandstones with bands of phyllite. In the middle part there are various shales and sandstones. At the top micaceous sandstones appear again. In the Lesser Khingan and other regions the series is absent.

The Diturskian Suite and the Malokhinganian and Amnusskian Series belong to the Sinian complex.

In Southern Primoria according to N. A. Belyayevskii the Proterozoic sequence also includes basal conglomerates and quartzites which pass upwards into micaceous and 'green' schists and marmorized limestones of a total thickness of 1·5–2km. The formation corresponds to Lower Proterozoic rocks of other regions. Still higher up there occur phyllitic slates, sericitized marbles and quartzites, graphitic slates and sandstones of a total thickness of 5km. N. A. Belyayevskii considers them as Eocambrian but they correspond to Upper Proterozoic strata of the Zeya-Bureyan Massif. Under such circumstances it is correct to denote all the rocks as the 'Sinian complex', which will include the Phyllitic Formation of the Zeya-Bureyan region.

The Sinian of the Far East represents a direct continuation of the Sinian of China, and in particular of its north-eastern part where it is typical. Thus Sinian strata of the Far East are type Sinian deposits in the U.S.S.R.

In Eastern Transbaikalia there is a belt of Precambrian rocks, situated in the south-eastern part of the area. According to G. I. Knyazev (Theses. . . ., Chita, 1961), at the bottom there are Lower Proterozoic strata (the Urovskian Suite) which are more than 1000m thick. The strata consist of crystalline schists and gneisses with bands of quartzite and marble. Higher up are Upper Proterozoic deposits which are divided into four suites of a total thickness of 4000–5000m and consisting of metamorphic schists, quartzites, conglomerates, porphyroids and amphibolites. After a break appear Sinian-Lower Cambrian dolomites and limestones (1700m). On these lie, with an unconformity, basal conglomerates and sandstones, succeeded by limestones and dolomites with beds of shales with a Lower Cambrian fauna.

In the Kolyma and Okhotsk massifs and Chikotian and Tagainos peninsulas, as well as in the basin of the Indigirka (the Selenyakhian Highlands), there are schists, slates, in places gneisses, quartzites, marmorized limestones and phyllites, all of which underlie Lower Palaeozoic strata and therefore can be included in the Proterozoic. The Lower part of these deposits is probably Archezoic.

In Central Prikolym'ya (the Kolyma Massif) Lower Proterozoic schists are overlain by the Sinian complex, which is subdivided into three suites and consists of quartzites and slates (1000m) at the bottom, algal limestones, dolomites and phyllitic slates (700–1000m) in the middle and dolomites, limestones and quartzites (1200–1400m) at the top. The Sinian complex is also found in the higher reaches of the Omolon, on the Taigonos and Chukotian peninsulas and near to Okhotsk.

In the south of the Central Kamchatkan range there are schists which are in places gneissose and are considered as Proterozoic and Cambrian. Some investigators suggested that they may be metamorphosed Upper Cretaceous shales and sandstones. Either dating is possible since the Alpine Orogeny was powerful and Alpine granites can produce a high metamorphic grade including the gneissification of Upper Cretaceous rocks. The problem can only be solved by finding the relationship of such rocks to the palaeontologically defined Lower Cambrian. Only those rocks which underlie Lower Cambrian sediments and have relevant spores and algae can be considered as Proterozoic. For example metamorphic schists of Eastern Transbaikalia used to be considered as Precambrian and even Archean on the basis of their high metamorphic grade. In the nineteen-thirties during the detailed geological survey they yielded organic remains. The geologist who found them could not believe his eyes since he correctly identified them as ammonites and these cannot be Archean. After a check by specialists the Liassic age of the ammonites was established. Thus a part of the metamorphic series was proved to be Lower Jurassic rather than Precambrian. Yet the other part of the series was found to be Precambrian.

It must also be noted that at the same time as some deposits, which used to be thought of as Precambrian, are much younger, certain other strata which were formerly considered as Palaeozoic or even Mesozoic are Precambrian. In this respect curious instances can be presented from the stratigraphy of Kamchatka. In the neighbourhood of Petropavlovsk there is a suite of metamorphic slates and siliceous slates, which on field relationships used to be included in the Upper Cretaceous. An investigation of its spores by B. V. Timofeyev (1956) has shown that it belongs to the lower part of the Sinian complex. A similar assemblage of spores has also been found by him in the aforementioned slates of the Central range proving their ancient age.

In other regions metamorphic slates of the Tukuring (upper Zeya) range, which were formerly considered as Devonian have been transferred into the Precambrian. The rocks of the Mynskian region on the river Selemdzha were formerly thought of as Triassic, the deposits of the Uchuro-Maian region as Lower Carboniferous, the metamorphic rocks of the Lesser Khingan and the basins of the Bureya and the Anyui as Devonian. All these strata as well as a part of 'Lower Cambrian' beds of the Kharaulakhian Mts have been relegated to the Precambrian. All the above examples indicate how carefully one should approach the dating of barren metamorphic formations. The study of the spores produced very important results, since such spores are well preserved in slates and phyllites and do not get recrystallized.

In Kamchatka V. P. Mokrousov (1959) includes the Kolpakovskian Series (4–5km) in the Archeozoic. The series consists of gneisses, crystalline schists, migmatites and amphibolites. It is followed by Proterozoic sediments—Kamchatka Series—of 4–5km in thickness, consisting of crystalline schists. The section ends with the 4–4·5km thick Sinian-Lower Palaeozoic Malkinskian Series of metamorphic slates and lavas, the former with Sinian spores. Higher up lies the Kikhchinian Series (over 3000m) consisting of sandstones and shales of Middle Palaeozoic-Mesozoic age. In the Eastern Sakhalin range (Decisions. . ., 1958) pre-Cretaceous deposits are divided into two series, the Valzinian and the Nabilskian. The former is provisionally considered to be of a Proterozoic-Lower Palaeozoic age and consists of metamorphosed shales and limestones with quartzite horizons. Its total thickness is 4500m. The Nabilskian Series is divided into the Gamonian and Ivashkinkian suites. The Gamonian Suite is provisionally thought of as Middle Palaeozoic and consists of phyllitic slates, cherts, crystalline limestones and tuffogeneous sandstones. No fauna has been found and the series is 850m thick. The Ivashkinkian Suite is Upper Palaeozoic, consisting of sandstones, argillites, siliceous rocks and lenses of limestones of 1550m in total thickness. The suite has yielded Upper Palaeozoic fusulinids. A. A. Kapitsa (1959) refers to the Valzinian Suite as the Susunaian and includes it in the Proterozoic-Devonian. He groups the Gamonian and Ivashkinkian suites together into the Gamonian Upper Palaeozoic Series.

THE PALAEOZOIC

Lower and Middle Palaeozoic deposits are rarely encountered and outcrop in small areas situated in the middle of most elevated and eroded structures of ancient massifs. In the Yablonovoi and Stanovoi ranges and their foothills there are three types of districts. In the first, the most northerly, marine Cambrian and coal-bearing Jurassic rocks lie horizontally being at the edge of the Siberian Platform. Further to the south Cambrian rocks are folded and metamorphosed, while Jurassic rocks lie horizontally being the eastern termination of the Palaeozoic Angara Geosyncline. Further to the south both Cambrian and Palaeozoic strata are folded, metamorphosed and intruded synorogenically, since they are in the Cimmeridian zone of the Pacific Ocean Geosyncline.

Lower Palaeozoic

Lower Palaeozoic rocks are comparatively badly exposed in the Cimmeridian zone, but occur in numerous districts. Cambrian sediments have long been known to occur in Eastern Transbaikalia. It has been shown by Chita geologists (Theses. . .,

Chita, 1961) that at the bottom of the Cambrian succession lie basal conglomerates and sandstones to a thickness of 200–300m, which are succeeded by thick (up to 2100m) limestones and dolomites and formations of shales. The strata are known as the Bystrinskian Series and their age is determined on the basis of archeocyathids and other faunas, as well as spores. The series is succeeded by the Altachian Suite (1700m) of slates and sandstones with beds of limestones and tuffaceous sandstones. The suite has yielded Middle Cambrian or Ordovician calcareous algae. Consequently it is considered as either Middle Cambrian or as Middle Cambrian to Ordovician inclusively. The first alternative is more likely since no Upper Cambrian or Ordovician fauna has been found in these rocks and the succeeding Nerchinskozavodian Suite with an Ordovician fauna is separated from the former by a major break. The absence of Upper Cambrian strata is also noticed in the regions adjacent to Transbaikalia. The Altachian Suite in places changes laterally into the shaly Ushmunian Suite (1600m).

The presence of Ordovician strata has been definitely proved although they were formerly considered to be absent. The Ordovician is represented by the Nerchinskozavodian Suite (1500m) of dolomites, limestones and shales. Ordovician bryozoans, algae and spores have been found and in the neighbouring provinces of China a rich fauna of trilobites, brachiopods and graptolites of Lower and Middle Ordovician age has been discovered. Ordovician strata are transitional into Silurian since at the top of the suite there is a Wenlock fauna. Higher up lies the unconformable shaly Blagodatskian Suite (700m) of a Lower Devonian age.

On strike to the east in Priamuria there have been found sandstones and shales with Ordovician trilobites, brachiopods and bryozoans, and still further eastwards—in the lower reaches of the Uda—Cambrian archeocyathid limestones. In the extreme north Lower Palaeozoic sediments are found in the Kharaulakhian Mts, thus forming the northern termination of the Verkhoyansk range. In a formation of dark grey bedded heavily dislocated limestones of over 1500m thickness, a rich trilobite-brachiopod fauna has been discovered. It indicates the development of all three divisions of the Cambrian and it is possible that Ordovician rocks are also present.

In the Novosibirsk islands, on Bennet island, the dark chloritic and carbonaceous shales have yielded a rich Middle Cambrian fauna. These rocks are overlain by dark graptolitic Lower Ordovician shales. It is interesting that in both the Kharaulakhian Mts and Novosibirsk islands the fauna is typically Western European rather than American.

In the north-eastern part of the Chukotian Peninsula and on Wrangel island Precambrian schists, quartzites and gneisses are overlain by a thick (3000–3500m) formation consisting of sandstones, quartzites and green schists at the bottom and quartz chlorite-sericite schists and ultimately carbonaceous slates towards the top. The formation is provisionally included in the Lower Palaeozoic. On Wrangel island it is overlain by Lower Permian strata and on the Chukotian Peninsula by Middle Devonian rocks. It is possible that the latter form uppermost horizons of the formation.

In the middle of the Yana-Kolyma Province Lower and Middle Palaeozoic deposits are widespread on the Yukagirian plateau occurring between the settlements of Verkhnekolymsk and Seimchan, as well as in the adjacent part of the Cherskian and Taskhayakhty ranges which are parts of the Kolyma Massif. The Lower Palaeozoic is represented by thick, strongly deformed and metamorphosed shales and sands with bands of limestones. The rarely encountered trilobite, brachiopod and grapto-

litic fauna indicates at the presence of Cambrian and all three divisions of Ordovician strata. The latter occupy the largest area. Graptolitic shales of Arenig age
should especially be pointed out. Marine Cambrian beds are found in the Okhotsk
Massif.

In the Alpine zone of the geosyncline Ordovician and Cambrian sediments are
sparse. In the valley of the Ussuri, to the north of Vladivostok, there are Lower
Cambrian deposits represented by a repetition of massive archeocyathid limestones
with suites of clay and siliceous shales reaching a total thickness of 1500–2500m.
In places under such limestones there are haematitic bedded ironstones analogous to
the Lower Khingan deposits. The ores are considered as Cambrian, but an Ordovician age cannot be excluded.

The Lower Cambrian limestone-shale formation is overlain by a coarse terrigenous conglomerate-breccia and sandstone formation. The absence of organic remains complicates the determination of its age. It is included in the Middle Cambrian but may be Middle Palaeozoic.

The Lesser Khingan succession is similar to that of Southern Primoria. Thick
Upper Proterozoic (Sinian) dolomites are here overlain by an ore-bearing formation
(250–400m) consisting of phyllitic siliceous and carbonaceous shales and ferri-
genous sandstones with groups of jaspilitic ironstones and manganiferous ores.
Higher up lie barren thick limestones and shales, which are also considered as
Lower Cambrian. The absence of archeocyathid limestones renders this suggestion
only provisional. Marine Upper Ordovician beds have been found in the Pendzhian
range.

Middle Palaeozoic

Middle Palaeozoic deposits are more widespread than the Lower Palaeozoic, but
less than Upper Palaeozoic and Mesozoic sediments. As a rule Middle Palaeozoic
rocks are intensely dislocated and strongly metamorphosed, and consequently full
sections and fossils are rarely found. Many thick clastic formations are either entirely barren or are palaeontologically badly characterized. Thus maps show only
major stratigraphical subdivisions such as: Palaeozoic, Middle Palaeozoic, Middle
and Lower Palaeozoic and Precambrian, Devonian and Silurian.

The presence of two zoogeographical provinces which have already been noticed
in the Lower Palaeozoic deserves attention. The boundary between the two provinces—the northern and southern—lies along the Yablonovoi and Stanovoi
ranges, then along the coastline of the Sea of Okhotsk and finally towards the
mouth of the Anadyr. To the north of this boundary Silurian, Devonian and Lower
Carboniferous faunas are identical to those of Western Europe, the Urals, Novaya
Zemlya and Taimyr. Here there are developed such facies as the Domanik limestones with *Stringocephalus burtini* and limestones with gigantoproductids. In
the south the faunas are equivalent to those of Kazakhstan, Kuzbas, China and
the southern part of North America. Here faunas of the type found in Hamilton,
Chemung, Kinderhook, Burlington and Heidelberg Limestone are widespread.

In the northern province the Caledonian type of Middle Carboniferous successions is predominant. Thus the Silurian succession is often complete, while the
lowermost Devonian sediments are transgressively overlain by those of the Givetian.
Higher up there are rocks of the widespread and complete Frasnian Stage and less
widespread but typical Famennian strata with *Cyrtospirifer archiaci* Vern. In places,
as for instance in the Kharaulakhian Mts, there are limestones with Tournaisian and

Visean faunas, but most frequently they are absent and Upper Palaeozoic strata rest directly on the Devonian. The absence of Lower Carboniferous rocks and sometimes of the whole of the Carboniferous from the succession is a diagnostic feature of the province. This feature must be explained by the length of the continental period which succeeded the Caledonian Orogeny.

The southern part of the Cherskii range can be taken as an example. The Silurian is represented by graptolitic shales at the bottom and various richly fossiliferous limestones at the top. The total thickness is 2400–2800m and the fauna indicates the presence of all stages. Silurian sediments are unconformably overlain by basal conglomerates and sandstones passing upwards into limestones and sandstones (650–850m) with a Givetian fauna. The Givetian is succeeded by overlapping Lower Frasnian muddy limestones and shales (450–500m) with *Spirifer novosibiricus* Toll, and are followed by dark bituminous Middle Frasnian shales and limestones (300–500m) which are at the top of Domanik type and have goniatites (*Gephyroceras*). The succession ends with barren greenish shales (200m).

In the north-east the Silurian is the most widespread of other Palaeozoic systems. Its sediments are widely found in the basins of the Indigirka and Kolyma, in Eastern Verkhoyania, on the Chukotian Peninsula and on the crest of the Sea of Okhotsk. The sediments consist of thick formations (2000–4000m) of limestones and shales. Terrigenous deposits are predominant only near the Okhotsk Massif and in the upper reaches of the Omolon. A rich fauna indicates at the presence of all stages. Graptolitic shales are quite common. In several districts as for instance on the Novosibirsk islands, the Kharaulakhian Mts, Central Ciskolymia, Devonian successions of Hercynian type have been found. Here the break between the Silurian and the Givetian is partly or completely bridged by marine limestones and shales with Lower Devonian and Eifelian faunas. Such relationships are also observed, and have been described, on the eastern slope of the Urals, in Novaya Zemlya and the south-west of Taimyr.

On the Omolon Block there is a thick (1200–1700m) formation of variegated volcanic and sedimentary rocks. The volcanic rocks are represented by dacites, spilites and tuffs, while bands of limestone shale and sandstone have yielded Givetian, Upper Devonian and Lower Carboniferous faunas and floras. A similar sedimentary-volcanic formation (the Dzhalkanian Suite) has been found at the south-western tip of the Verkhoyansk range (the Sette-Daban range). Its age ranges from the upper part of the Middle Devonian to the lower part of the Upper Devonian and its thickness is 1500–2000m. Lava flows alternate with limestones, sandstones and shales with marine faunas.

The Lower Carboniferous consists of limestones, sandstones and shales of a total thickness of the order of a few hundred metres. The junction with the Upper Devonian is transitional and Lower Carboniferous strata are widespread occurring almost everywhere where there are Upper Devonian rocks. A rich marine fauna consists mainly of brachiopods and corals.

Lower Carboniferous strata are unconformably overlain by the Lower Permian. The Chukotian Peninsula constitutes an easterly extreme of development of Middle Palaeozoic rocks of northern type. In the Chukotian Peninsula there are Silurian shales and limestones, Givetian limestones with *Stringochephlus* and Visean limestones with rugose corals. It is interesting that similar sediments are found along the Mackenzie river in Canada.

In the southern province Middle Palaeozoic sediments are most widely developed in Eastern Transbaikalia and the adjacent Priamuria. They are also quite wide-spread further to the east, along the upper reaches of the Zeya and its tributaries, outcropping as far as the shoreline of the Sea of Okhotsk. Silurian deposits consist of thick shales, limestones and sandstones which in places contain an as yet inade-quately collected brachiopod-coral fauna. The subdivision of the local Silurian is a problem for the future. So far only the Wenlockian has been established. The Devonian is well represented, its fauna which has been found in relatively large quantities has been well investigated, but no monographs so far have been published. The study conducted by E. A. Modzalevskaya has shown a close similarity of the fauna with the North American, and its extreme difference from the Western Euro-pean fauna. Lower Devonian rocks have been found in the region of the Gazimur-skia factory, where they consist of pale, massive limestones of Heidelberg type reaching a thickness of about 250m. Higher up lie dark and greenish-grey shales and siliceous sediments, becoming in places cherty and containing volcanic rocks. The total thickness of Lower Devonian strata is 700–1000m and in Priamuria it is of the same type (Fig. 258). The Middle and Upper Devonian is fully developed. In Priamuria the succession begins with pale, massive limestones overlain by sandstones and muddy limestones of the Hamilton Stage, which is of the same age as the Givetian and the lower part of the Frasnian (Fig. 259). The succession ends with shales and limestones of the Chemung Stage, which corresponds to the Famennian. The thickness of Devonian strata is no less than 4–5km. In the upper reaches of the Zeya the thickness of the Eifelian Stage alone is 1400–1600m. In the basin of the Uda Givetian sediments are 1500–2800m thick. The Devonian succession of Eastern Transbaikalia has been described in the papers of the Chita conference (Theses. . ., Chita 1961).

B. V. Timofeyev has found Upper Devonian spores in phyllitic slates of the Zeya-Selemdzha region where the age of these rocks was disputed, although they were very widespread.

The Lower Carboniferous of Eastern Transbaikalia is less commonly en-countered, although it is conformable on the Devonian, and consists of some 3000–4000m thick limestones and shales. The rich fauna of these rocks is peculiar and is of American type, indicating the presence of Kinderhook and Burlington stages, which corresponds to the Lower and Upper Tournaisian. The fauna has not yet been worked at a monograph level. The Keokuk Stage, equivalent to the lower part of the Visean Stage, has been recognized (Theses. . ., Chita, 1961). Lower Carboni-ferous rocks are well developed in the north of the Verkhoyansk range, in the Kharaulakhian Mts and on the shore of the Lena. The strata here consist of lime-stones (Fig. 260). Lower Carboniferous rocks are well developed in the north of the Verkhoyansk range, in the Kharaulakhian Mts and on the shore of the Lena. The strata here consist of limestones (Fig. 260). Lower Carboniferous rocks are also found in the Kolyma Massif, the upper part of the Omolon and the Chukotian Peninsula.

Further south beginning with the Lesser Khingan and up to the Okhotsk coast Devonian rocks of the same type as in Eastern Transbaikalia have been found to contain the same fauna.

In the Alpine zone the palaeontologically demonstrated Middle Palaeozoic is not widespread. Some of metamorphosed barren formations undoubtedly belong to the Middle Palaeozoic, since analogous deposits in Japan have yielded a Devonian fauna.

B. V. Timofeyev (Theses. . ., Khabarovsk, 1956) has found Upper Devonian spores in highly metamorphosed, graphitic quartz-sericitic and tuffogenous schists known as the Daldaginskian Suite (1000m) and occurring in Southern Sakhalin, where these rocks were formerly shown as undifferentiated Palaeozoic. The dis-

FIG. 258. Lower Devonian cleaved chloritic and sericitic siltstones on the river Amazar in Cisamuria. *Photograph by E. A. Mogzalevskaya.*

FIG. 259. 'The Brothers' — Middle Devonian sandstones on the Shantar islands. *Photograph by L. I. Krasnyi.*

covery once more underlines the importance of studying spores in Palaeozoic shales, which have no macrofauna. Middle Palaeozoic sediments have recently been found in the central part of the Koryakskii range where Lower Carboniferous reef limestones are found (Fig. 261). In the Penzhian range (north of Kamchatka) marine Silurian and Devonian sediments have been discovered.

FIG. 260. Cliffs of Lower Carboniferous limestones in the Kharaulakhian Mts. on the right of the Lena, some 15km from its mouth. *Photograph by A. A. Mezhvilk.*

FIG. 261. Middle Palaeozoic strata in the central part of the Koryakskii range. Byelaya Skala (the white rock) shows Visean reef limestones. In the background mountains, consisting of Upper Palaeozoic rocks.

Upper Palaeozoic

The Upper Palaeozoic is divided into two parts: (1) the Middle and Upper Carboniferous and (2) the Lower and Upper Permian.

The Middle and Upper Carboniferous

The Middle and Upper Carboniferous has a limited distribution and in a series of districts where the Permian is widespread Middle and Upper Carboniferous rocks are unknown. It has not as yet been established whether it is truly absent or that it has not been differentiated from thick, homogeneous poorly fossiliferous Lower Permian formations. Lithologically Middle and Upper Carboniferous rocks are diverse and consist of limestones, various shales, sandstones and conglomerates and locally volcanic rocks. The total thickness varies, but is normally no more than a few hundred metres. Lower Permian strata consist of homogeneous, kilometres-thick argillaceous formations of flysch type. The sudden lithological change was provoked by the occurrence of the Hercynian Orogeny. In Middle and Upper Carboniferous times orogenic movements did not occur; in Lower Permian times, on the other hand powerful and prolonged movements were accompanied by mountain building which caused the formation of flysch deposits. Middle and Upper Carboniferous rocks are most fully represented and palaeontologically defined in the Sikhote-Alin' range. Such rocks are also found in the Khabarovsk district in Verkhoyania and on Wrangel island (Table 53).

Throughout the north-east, except in Verkhoyania, Middle and Upper Carboniferous strata are rarely encountered and are represented by continental clastic deposits with plant remains (Zavodskii, Theses of papers..., Yakutsk, 1961).

In Verkhoyania the aforementioned suggestion was verified and lower horizons of the thick clastic Verkhoyansk complex have yielded Upper and Middle Carboniferous and even Namurian goniatites. These have been studied by Yu. N. Popov, who has contributed a lot towards our knowledge of the Permian and Trias in the north-east. His data and many others have been published in the 'Theses of papers' (Yakutsk, 1961). According to Popov Middle and Upper Carboniferous goniatites have also been found in the Tiksinskian Suite of the Orulgan range. In Verkhoyania the Imtandzhinskian Suite has the goniatites of the same type as in the Upper Carboniferous of the Urals, while in the Sette-Daban range there have been finds of Upper Carboniferous and Namurian goniatites. Middle Carboniferous goniatites and brachiopods have been found on Wrangel island. According to other investigators (Theses. . ., Yakutsk, 1961) the Verkhoyanian succession of Upper Palaeozoic strata often begins with basal conglomerates, followed by sandstones and shales with plant remains, reaching altogether 300–1000m. These rocks are known as the Atyrdakhian Suite, Uchaganian Suite and Bylykatian Suite, belonging to the Middle Carboniferous. Higher up there are thick, black sandstones and shales (900–2000m) with a marine Upper Carboniferous fauna. The latter formation is known as the Tiksinskian Suite (Table 53).

In the Sette-Daban (southern Verkhoyania) range according to B. S. Abramov (Theses. . ., Yakutsk, 1961), Tournaisian limestones, containing a marine fauna, are gradational into shales and siliceous rocks of Visean and Namurian stages. The thickness of these strata is 250–300m.

The Permian

The Permian is the most important and best developed Palaeozoic system.

DISTRIBUTION. Especially in the Cimmeridian zone Permian rocks are widely distributed. Enormous areas between the Khatanga and the Lena, the Verkhoyansk range, the southern part of the Cherskii range, the upper reaches of the Kolyma, the Yukagirian Highlands and the upper reaches of the Omolon are occupied by Permian rocks. It should be remembered that the term 'Permian' in these areas embraces kilometres-thick flysch-type strata. Permian faunas are relatively rarely found and consequently it is not impossible that the lower horizons of these strata may be Upper Carboniferous or even Lower Carboniferous while the Upper horizons may be Triassic or Jurassic.

The Permian succession of Wrangel island is very different in so far as thick limestones and dolomites with shales, which are sometimes phyllitic, predominate and reach a total thickness of 2500m. The relatively abundant fauna consists mainly of brachiopods and corals. Its study has shown that parallel with the dominantly Permian forms there are also Carboniferous and even Lower Carboniferous forms. The absence of fusulinids and a series of other critical Permian groups forces doubts as to its determination and suggests that the whole of the limestone formation may be Lower Carboniferous. Similar limestones on the Novosibirsk islands and in the Kharaulakhian Mts were formerly thought of as Upper Palaeozoic and are now included in the Lower Carboniferous. It should be pointed out that Lower Permian faunas of the Arctics include a series of forms which are similar to the Lower Carboniferous. The similarity is so close that a possibility remains of a converse solution; thus those faunas which are now considered to be Lower Carboniferous may be actually Lower Permian.

In Eastern Transbaikalia, Priamuria and the upper reaches of the Zeya and its tributaries a thick shale-sand formation (2000–3500m) is considered as Permian. In Eastern Transbaikalia, in the Borzin region, according to V. A. Bobrova and M. V. Kulikova (Theses. . ., Chita, 1961) there is a succession shown on Table 53. In the neighbouring regions other suites can be distinguished. The work of D. F. Maslennikov on the Permian of Transbaikalia is outstanding.

Lower and Upper Permian strata have been found in the Sikhote-Alin' range. Near to Vladivostok in addition to shales there are limestones, conglomerates and other rocks rendering the succession somewhat inhomogeneous. In central parts of Sakhalin and Kamchatka shaly formations shown as Upper Palaeozoic are lithologically homogeneous and fossils so far have not been found. It is probable that most of these rocks are Permian. In the Eastern Sakhalin range there is a thick (2600m) Gamonian Series of an Upper Palaeozoic age (Kapitsa, 1959). The series consists of radiolarites, sandstones and shales with limestones. At the top of the series fusulinids of probably Permian age have been found. Lower Permian marine deposits with foraminiferids and brachiopods have been found in the northern part of the Koryakskian range (Rusakov, Yegizarov, 1958). The deposits are represented by siliceous rocks, sandstones and lavas with lenses of limestones. The total thickness is 1200–1400m (Fig. 259).

SUBDIVISIONS. The small number of discoveries of faunas and floras and the peculiarity of the former have been the reason for the fact that so far the Permian system is only divided into Upper and Lower. More detailed subdivisions have been introduced only locally as in the south of the Sikhote-Alin' range since a rich and variable southern type fauna has been found.

According to the 'Theses of papers' (1957) given at the stratigraphic conference in 1957, in Magadan there are new and important data on subdividing the Permian of the Verkhoyansk type. Lower and upper divisions are further broken up into a series of local suites which are grouped into stages. The method of grouping however, follows two different approaches. The adherents of one approach attempt an equivalence with the stages of the Russian Platform and the Urals and try erecting Sakmarian, Artinskian, Kungurian, Kazanian and Tatarian stages. Despite the predominance of such a point of view, it cannot be considered correct, since in the north-east, Permian faunas differ so much from the European faunas that none of the European stage can be recognized. Following the second, more correct point of view, new local stages are proposed. According to A. S. Kashirtsev (1957), a member of the Yakutian branch of the Soviet Academy of Sciences, the Lower Permian should be divided into two stages: the Lower or Kharaulakhian and the Upper or Verkhoyanskian. The Verkhoyanskian Stage has a characteristic *Productus verkhoyanicus* fauna and corresponds approximately to the Artinskian and Kungurian stages. The Kharaulakhian Stage has an older, yet insufficiently studied fauna without *P. verkhoyanicus* and corresponds to the Sakmarian. The Upper Permian is also divided into two stages. The Lower Kolymian has a fauna with numerous large *Kolimia* and is correlated with the Kazanian Stage. The upper stage which is an analogue of the Tatarian so far remains unnamed. In the basin of the Gizhiga it has a rich marine fauna of southern type and consequently one can suggest the term Gizhigian.

Over the vast expanse of the Cimmeridian zone the Permian is divided into a lower division with a rich, mainly marine, and peculiar brachiopod fauna, and the upper division, which is predominantly continental, has plant remains and is in places coal-bearing. The strata of the upper division sometimes contain large flat

Fig. 262

Complexly folded Upper Permian sandstones and shales in the Kharaulakhian Mts.
Photograph by V. M. Lazurkin.

Inoceramus-like lamellibranchs which all belong to a single genus *Kolimia*, which may be either marine or lagoonal.

Further to the south in the Verkhoyansk range thicknesses increase to 3000–5000m (Fig. 262) but the general character of the succession remains unchanged. In places as a result of detailed surveying the Verkhoyanskian Permian has been divided into several suites.

Thus for instance in the north and south, in the upper reaches of the Allah-Una the succession has been established and is shown in Table 53.

The greatest expanse of Permian rocks occurs to the north of Khabarovsk, the upper reaches of the Bureya and the Kur-Urmiian region (the lower reaches of the Uda). A basic feature of these strata is the alternation of marine sediments, various volcanic rocks and continental deposits with plant remains.

In the Alpine zone the fullest Permian succession is developed in the Sikhote-Alin' range and especially in its southern part. The Lower Permian of the west consists of continental shales and sandstones with plant remains. To the east marine and volcanic intercalations appear. Here thick siliceous shales and beds of limestones with marine faunas are particularly characteristic. The study of the numerous foraminiferids over many years permitted M. A. Sosnina to show the presence of two zones: the lower, with *Pseudofusulina* and the upper with *Cancellina*.

Upper Permian strata are transitional with Lower Permian strata and are represented at the base by mainly marine deposits. Amongst such marine sediments there are massive reef limestones (200–250m) with an exceptionally variable fauna. Amongst the Foraminifera the critical form is *Neoschwagerina*. Sessile forms belonging to the genera *Richthofenia* and *Lyttonia* are typical brachiopods of this fauna. Formerly the fauna was considered as Lower Permian and so were the reef limestones. At present an Upper Permian age of the fauna is generally accepted.

Higher up lies a sandy-shaly formation (800–1000m) of flysch type and un-fossiliferous. The formation is succeeded by a volcanic-tuffogenous suite of variable thickness. The occasional beds of sandstones and limestones have a mainly brachiopod fauna with *Lyttonia*. The Permian succession ends with freshwater, terrestrial sandstones and shales (220–250m), which are in places coal-bearing and have plant remains (Table 53).

In Sakhalin the thick (over 7000m) Gamonian Series is included in the Upper Palaeozoic. The series consists of phyllitic shales, sandstones, radiolarites and siliceous shales. In the upper part of the series a marine Lower Permian fauna (Kapitsa, 1959) has been found. In the Koryakskii and Penzhian ranges there are found marine Lower and Upper Permian deposits of a total thickness of over 2000m and consisting of clastic and volcanic rocks (Mikhailov).

LITHOLOGY, FACIES AND PALAEOGEOGRAPHY. The Cimmeridian zone shows three basic features: a great thickness of Permian deposits, their clastic flysch-like lithology and their curious fauna. Furthermore, Permian deposits form only the lower part of a very thick sequence (up to 10–15km) of clastic rocks, which also range over the Trias, the Jurassic and the Lower Cretaceous and which are collectively known as the 'Verkhoyansk complex'. The thickness of the Permian part of this group of strata varies considerably. Where both the Upper and the Lower Permian are represented by marine sediments the total thickness is relatively small, reaching 1000m. Where, however, there are barren sandy-shaly continental deposits the total thickness approaches 3000–4000m and sometimes up to 5000m (Fig. 263). In such

TABLE 53.

Upper Palaeozoic Successions in the Far East

Division		The Sikhote–Alin' range (V. K. Yeliseyeva, 1957)				The Khabarovsk region (A. P. Glushkov, 1956)
		Central zone	Ol'ga-Tyetyukha zone	Southern Primoria	Region Grodekovo	
Upper Permian		Shetukhinskian Suite of basic lavas, tuffs and intercalations of shales (600–3000m). Lyudyanzinskian Suite of silts, argillites and sandstones (2300m). Kafenskian Suite of sandstones, shales and limestones with a rich marine fauna.	Sandstones with beds of shales, siliceous shales and limestones, with Foraminifera (450m).	Sitsian Suite of sandstones, carbonaceous shales, coals and conglomerates, with a flora (250–550m). Nakhodkian-Kaluzinian Suite of porphyrites, tuffs, tuffogenous rocks and with a rich marine fauna (800m). Lyudyanzinskian Suite of shales and sandstones, with a marine fauna (700m). Chandalazian Suite of repeated limestones, sandstones and shales, with an abundant marine fauna (700–800m).	Tufopeschanikovo-slantsevian Suite of tuffs, tuffites and carbonaceous shales. Volcanic-tuffogenous Suite (Effuzivno-tufogennaya), with lenses of limestones with a marine fauna (*Lyttonia*.) Sandy-shaley formation of sandstones silts and shales. Reef massifs with a rich marine fauna are present. Total thickness—2000m.	Siliceous shales, sandstones, limestones and porphyrites, with fusulinids. Seredukhinskian Suite of sandstones, shales and limestones with ammonites (1000m). Solonechnian Suite of sandstones and dark shales with ammonites (750m). Ungunian Suite of tuffogenous shales, sandstones and conglomerates, with a flora (800m).
Lower Permian		Khodiiskian Suite of abundant jaspers and siliceous shales. Marine Foraminifera are present (600–2500m). Vesnyanskian Suite of sandstones and carbonaceous shales with a flora (1000m). Sanzhagouian Suite of porphyrites, tuffs, limestones and sandstones, with a marine fauna (800–1500m).	Sibaigouian Suite of greywacke sandstones with beds of limestones and fusulinids (2200m). Zarodian Suite of porphyrites, tuffs, sandstones, limestones and cherty shales (1300m). Kavalerovskian Suite of sandstones, shales and siliceous shales. Below, limestones with *Pseudofusulina* (1500m).	Missing.	Continental, flora-bearing deposits—sandstones and shales.	Siliceous shales and sandstones (1000–1500m). Green Suite (Zelenotsvetnaya) of siliceous and tuffogenous shales, porphyrites and limestones with fusulinids (600–800m). Basal sandstones and conglomerates (300–400m).
Upper Carboniferous		Samarkinskian Suite of cherty shales and jaspers with beds of sandstones and limestones with *Triticites* (500–700m).	Sandstones and shales with beds of limestones with *Triticites*. (800–1000m).	Yuzagolian Suite of sandstones, shales and coals, with a flora (700–900m).	Suggestedly Upper Carboniferous barren shales and silts (600–1000m).	Limestones and calcareous shales with *Triticites*.
Middle Carboniferous		Merginskian Suite of sandstones, shales, limestones, porphyrites and tuffs. Fusulinids are present (1000–1200m).	Sandstones and shales with beds of siliceous rocks and limestones with fusulinids.	Not recognized.	?	?

TABLE 53—continued

Eastern Transbaikalia (V. A. Bobrov and M. V. Kulikov, 1961)	Eastern Taimyr (I. M. Rusakov, 1958)	The North-east (A. S. Kashirtsev, 1957)	Verkhoyansk range (L. A. Musalitin, 1958, 1961 and later)		Southern part of Allakh-Yun
			Northern part		
			Western slope	Eastern slope	
Borzian Suite of grey sandstones and silts; sequences of conglomerates and grey-wackes. A marine fauna is present (800m).	Tuffolava Suite (200m). Chernoyarian Suite—coal-bearing sandstones and clays, with a flora (400m).	Gizhigian Stage—sandstones and shales with beds of limestones. At the Gizhiga river a rich southern-type marine fauna is found.	Nerskian Suite of thick sandstones and shales with coal seams and a flora (1100m).	Nerskian Suite of sandstones and shales with coal seams and a flora. A fauna of lamellibranchs (Kolymia) is found (2000m).	Amgian Suite of continental shales with beds of limestones. A flora is present (1000–2000m). Allakhian Suite of carbonaceous shales and silts. The age is Upper-Lower Permian (700m). Dzhankangian Suite of shales and sandstones with a marine fauna (600m). Karskian Suite of tuffogenous sandstones and tuffs (800m). Satan'inskian Suite of sandstones and shales with marine Upper Carboniferous and Lower Permian faunas. Basal conglomerates and sandstones.
Belektuiian Suite of dark, greenish and grey grey-wackes and sandstones, with lenses of conglomerates and tuffs. The fauna s of marine brachiopods 900m).	Baikurian Suite—above, dark argillites and sandstones, with a marine fauna; below, basal coarse sandstones (500m).	Kolymian Stage—thick shales and sandstones with a fauna of Kolymia.	Endybalian Suite of flyschoid, repeated sandstones and shales. Beds with marine fauna, with Kolymia, occur (1000m).	Endybalian Suite of flyschoid, repeated shales and sandstones. Beds with marine fossils occur and have Kolymia (1600m).	
Kharanorian Suite of sandstones, grey-wackes and shales. Marine fossiliferous bands exist 600m).	Sokoliniyan, coal-bearing, Suite (500m). Byrrangian, flyschoid and barren, Suite (300m). Turuzian Suite of flyschoid sandstones and argillites (1000m).	Verkhoyanskian Stage—thick sandstones and shales with some beds having a rich marine fauna with Productus verkhoyanicus.	Egiian Suite of shales with beds of sandstone and a flora (100–800m), Kygyatasian Suite of rhythmically repeated shales and sandstones, with a fauna and a flora (700m).	Egiian Suite of shales with a marine fauna. Kygyatasian Suite of rhythmically repeated shales and sandstones, with a fauna and a flora (700m).	
Kundoiian Suite of siliceous and normal shales and sandstones. A fauna is present 900m).	Trautfetterian Suite—above, argillites and silts with a Sakmarian fauna; below, barren sandstones of a possibly Upper Carboniferous age (600m).	Kharaulakhian stage—sandstones and shales with marine, fossiliferous beds.	Imtandzhian Suite of shales and sandstones and faunas (1100m). Otoisuokhian Suite—sequences of shales and sandstones. Diabases are found (1700m).	Imtandzhian Suite. Above, shales and sandstones; below, sandstones and conglomerates. A marine fauna is present (2100m). At the bottom, Upper Carboniferous goniatites have been found.	
Missing.		Not differentiated.	Sopol'skian Suite of sandstones, silts and shales. The fauna is in places marine, and in places continental (lamellibranchs), and a flora is also found (2000m).	Tiksinian Suite of dark sandstones and shales with a marine fauna (1000–2000m).	Ovlachanskian Suite of dark shales with beds of limestones. Its marine fauna indicates at Middle and Upper Carboniferous ages (250–300m).
	Not recognized.		Bylykatian Suite of sandstones, carbonaceous shales and coal seams. A flora is present (400m).	Atyrdakhian Suite of sandstones and shales with a flora. Below, conglomerates of molasse type (300–1000m).	

cases it is possible that the lowermost horizons are Middle and Upper Carboniferous and the uppermost horizons are Triassic.

Lithologically Permian rocks are on the one hand homogeneous, being thick brown and dark grey sandy and clayey rocks, but on the other hand changing rapidly in both vertical and horizontal directions. The frequently observed repetition of shales and fine sandstones is typical and is probably caused by rhythmic sedimentation. The sandstones are dark-grey, grey, pale grey and bluish varying from fine-grained to medium grained and being quartzose, quartz-feldspathic, often muddy and calcareous. There are rare beds and lenses of coarse sandstones and conglomerates. The shales are of the same range of colours but often are darker and vary from compact fine-grained muddy varieties to sandy and calcareous types. There are rare seams of marls and limestones.

The thickness, rhythmic bedding, composition and poverty of faunas indicate the flysch character of Permian rocks in the Cimmeridian zone. Occasional suites are more similar to the molasse deposits and especially to the finer grain varieties. Flysch and molasse sediments get deposited at the edge of young rising fold ranges, on near shore plains and in adjacent marine basins. Such ranges existed at the margin of the Siberian Platform and the sediment was transported off them to the east and south into the area occupied by the present-day Verkhoyansk range, Priamuria and Eastern Transbaikalia. The absence of thick conglomeratic deposits indicates that the transporting rivers were small and slow and the ranges were not very high.

In the Alpine zone the thickness of Permian sediments is considerable and the sequence reaches 3000–4500m, but the facies were more homogeneous. The presence of large quantities of siliceous shales and cherts is characteristic, but there are also quite common limestones which are sometimes massive and have reefs. Amongst the rich fauna the sessile brachiopod *Richthofenia* is a critical Permian form. Numerous are fusulinids of Lower Permian and Upper Permian ages. Clastic facies—shales, sandstones and conglomerates—predominate and aspidic slates and siliceous shales are often very thick (2000m). In the Upper Permian there are common continental strata with plant remains and local coals. Volcanic rocks are quite frequently encountered. The facies in general indicate the predominance of the marine regime, but in Upper Permian times there were also swampy, near-shore plains, where coal accumulated. The plains were adjacent to a northerly land. Volcanic islands emerged out of the sea, but there were also terrestrial volcanoes.

TABLE 53—continued

Division	The Sikhote Alin range (V. K. Yeliseyeva, 1957)				The Khabarovsk region (A. P. Glushkov, 1956)
	Central zone	Ol′ga-Tyetyukha zone	Southern Primoria	Region Grodekovo	
Underlying rocks	Ariandinian Suite (Middle Carboniferous-upper Lower Carboniferous)—thick sandstones and shales with beds of limestones (4000m).	Lower Carboniferous. At the top, siliceous rocks; below, white limestones with Foraminifera and ultimately basal sandstones (950–1150m).	Lower Carboniferous —coal-bearing with sandstones, shales and coals (950–1150m).	Not recognized.	Not found.

THE MESOZOIC AND THE CAINOZOIC

One of the most important features of the Pacific Oceon Geosyncline, and of other Meso-Cainozoic geosynclines, was the vastness of Mesozoic deposition now reflected in complete successions of mainly marine deposits. Coal-bearing formations—the deposits of near-shore plains—and various volcanic complexes are widespread. In places there are thick conglomerates of molasse type, being typical deposits at the edges of high, snow-clad mountain ranges.

Trias

The Pacific Ocean Geosyncline is the only part of the U.S.S.R. where marine Trias, represented by all three divisions, has a wide distribution.

DISTRIBUTION. Triassic rocks are encountered in all provinces of the geosyncline. The Trias occupies especially large areas in the Yana-Kolyma basin where it occurs on the shore of the Khatanga Bay and the estuaries of the Anabara and the Olenek. To the east of the Verkhoyansk range Triassic rocks occur as a wide zone of 200–300km across and stretch for over 1500km from the Arctic Ocean almost as far as the Sea of Okhotsk. The belt then bends eastwards and north-eastwards and can be traced with interruptions to the east of the Kolyma Massif to the coast of the Eastern Siberian Sea, to the east of the Chaun inlet. A large area of Triassic deposits also occurs in Priamuria and near the coast of the Sea of Okhotsk. To the west there are some outcrops of Triassic rocks, as for instance in Eastern Transbaikalia, but it is not always differentiated from Permian and Jurassic sediments.

In the Sikhote-Alin' range the Trias is well developed in the south where its sediments succeed the Permian. There is no doubt that Triassic rocks exist in the north of the range, but here they have not been separated from Permian and Jurassic strata. In Sakhalin and Kamchatka no Triassic deposits are known, but it is possible that its beds form a part of barren metamorphic rocks provisionally included in the Upper Palaeozoic. This suggestion is confirmed by the recent demonstration (Mikhailov) of wide distribution of the marine Upper Trias in regions adjacent to northern Kamchatka, on the Taigonos Peninsula, the basin of the Penzhina and the middle reaches of the Anadyr. In these areas dark shales and sandstones predominate although tuffs are also frequent. The total thickness is considerable and the fauna is Norian with *Monotis ochotica*.

TABLE 53—continued

Eastern Transbaikalia V. A. Bolrov and M. V. Kulikov, 1961)	Eastern Taimyr (I. M. Rusakov, 1958)	The North-east (A. S. Kashirtsev, 1957)	Verkhoyansk range (L. A. Musalitin, 1958, 1961 and later)		
			Northern part		Southern part of Allakh-Yun
			Western slope	Eastern slope	
Lower Carbonferous.	Lower Carboniferous.	Lower Carboniferous.	Lower Carboniferous.	Lower Carboniferous.	Middle Palaeozoic.

SUBDIVISIONS. The occasional rich and variable marine fauna of ammonites and ceratites permits a detailed subdivision into main stages recognized in the Alps, Himalaya and other southern provinces forming a part of the Mediterranean Geosyncline. It is suggested that the Lower Trias should be divided into the Indian and Olenekian stages.

More frequently the Trias is represented by homogeneous thick almost barren clastic deposits in which not only stages but divisions are difficult to recognize. Consequently maps often employ such captions as 'Lower and Middle Trias', 'Lower Jurassic and Upper Triassic deposits', 'Trias', 'Norian and Carnian stages', 'Lower and Middle Jurassic and Rhaetic'. Subdivisions of the Trias in more investigated regions are shown in Table 54.

LITHOLOGY AND FACIES. Lower and Middle Triassic lithologies and facies depend to a large extent on the last phases of the Hercynian Orogeny while Upper Triassic lithologies and facies depend on initial phases of the Cimmeridian Orogeny. Consequently Lower and Middle Triassic strata form a single complex which is closely associated with the Permian, while Upper Triassic strata are no less closely bound to the Jurassic.

The lower complex is completely represented in the north-west—the lower reaches of the Olenek, the Anabara and the Khatanga—where its strata lie on the Permian, sometimes without but sometimes with a break. At the base of the Indian stage lies a formation of variegated tuffites (20–120m) with *Estheria*. The tuffites are succeeded by dark shales with lenses of limestones with a marine fauna (50–70m). The Olenek Stage (100–230m) consists of dark argillites with beds of sandstones and limestones with an abundant marine fauna (*Olenekites, Subcolumbites*). The Middle Trias begins with the Anisian Stage, which consists of pale and greenish sandstones and siltstones (200–400m) with intercalations of shales and limestones with a marine fauna. The Ladinian Stage is sometimes identified and is represented by sandstones and argillites. Normally this stage and the Upper Trias are absent (Sorokov, 1957).

FIG. 263. The Verkhoyansk complex in the central part of the Verkhoyansk range. In the front a reindeer caravan of the expedition. *Photograph by S. V. Obruchev*.

Triassic Succession in the Pacific Ocean Geosyncline

Stage	The Primorian region	Eastern Transbaikalia	The Yana Highlands (L. A. Musalitin, 1961)	The Upper Indigirka and the Upper Kolyma	The Chaun-Chukotian region
Rhaetic	Sandstones and shales with plant remains.	Not recognized.	Polymictic sandstones with beds of conglomerates and rarer shales. Plant remains are found.	Sandstones and shales with volcanic rocks. The marine fauna has brachiopods and lamellibranchs.	Sandstones and phyllitized shales with plants (*Phenicopsis, Neocalamites*).
Norian	Razdol'ninskian Suite of sandstones and shales with *Pseudomonotis, Monotis ochotica, Pecten* (150–200m). Upper Mongugaian Suite of grey sandstones with beds of shales, coal-seams and plant remains (400m).	Sandstones and shales with rare beds of limestones; the fauna is found in nests and consists mainly of lamellibranchs—*Monotis ochotica* and *Pseudomonotis*. Sandstones and shales with a lamellibranch fauna with *Pseudomonotis scutiformis*.	Sandstones with *Monotis ochotica, Pseudomonotis* and other lamellibranchs. Sandstones with beds of shales and a lamellibranch fauna with *Pseudomonotis scutiformis*.	Sandstones and rarer shales and tuffs with *Monotis ochotica* and other lamellibranchs (600–800m). Mainly shales with shell-bed intercalations of *Pseudomonotis scutiformis*. Shales with beds of sandstones; a fauna of *Pseudomonotis scutiformis, Halobia, Oxytoma, Pecten* is present.	Shales and sandstones, often phyllitized and with *Monotis ochotica* and *Pseudomonotis*. Shales and sandstones with *Pseudomonotis scutiformis* and *Monotis ochotica*. Dark-grey shales with *Halobia*.
Carnian	Suputinskian Suite of sandstones, silts and shales with a sequence of tuffogenous rocks. A rich lamellibranch fauna with *Pseudomonotis scutiformis, Oxytoma, Pecten* (300–500m). Tetyukhian massive limestones with *Megalodon* and corals. Lower Mongugaian Suite of sandstones with sequences of shales and coal seams (600m). Basal sandstones and conglomerates (5–60m).		Sandstones and shales with beds of conglomerates. *Pseudomonotis, Halobia, Oxytoma, Pecten* (250–300m). Shales and sandstones with beds of limestones and a fauna of lamellibranchs (*Halobia*) and ammonites (*Sirenites*.) Shales and silts with *Halobria* and *Daonella* and ammonites *Clionites* and *Arcestes* (300–500m).	Shales and sandstones with a rich fauna of lamellibranchs (*Hallobia*) and ammonites (*Sirenites*) (300–600m). Sandstones and shales with ammonites (*Nathorstites, Arcestes*), lamellibranchs and brachiopods (120–200m).	

Table 54 continued

Stage		The Primorian region	Eastern Transbaikalia	The Yana Highlands (L. A. Musalitin, 1961)	The Upper Indigirka and the Upper Kolyma	The Chaun-Chukotian region
Ladinian		Sandstones, alternating with silts and shales. A rich fauna with lamellibranchs (*Daonella*) and ammonites (*Hungarites*) is present (800m).	Not recognized.	Shales and sandstones with a poor marine fauna with *Daonella* (1100–1600m).	Sandstones and shales in places with plant remains; in places with a marine fauna with *Daonella* and ammonites.	—
Anisian		Banded and blotchy sandstones and silts with a rare fauna of ammonites (*Ptychites*) and *Sturia* (700m).		Shales and sandstones, in places marine, elsewhere continental, with a flora. The marine fauna has ammonites (1000–2000m).	Shales and sandstones; in the west, with plant remains, in the east with a marine, ammonite fauna.	—
Olenekian (Olenek)		Sandstones with a bed of limestone, possessing a fauna of ammonites (*Subcolumbites*) (100–150m).	Sandstones, dark shales with beds of limestones. A lamellibranch fauna with *Pseudomonotis* (*Claraia*) *clarai* and *Myophoria* is present.	Shales and sandstones with ammonites (*Sibirites*) (300–1500m).	Shales and sandstones with ammonites (*Olenekites, Keyserlingites*).	—
Indian		Sandstones and silts with lenses of ammonite-bearing limestones (100–200m). Basal sandstones and conglomerates (from 25 to 250m).		Sandstones and shales with a marine fauna (*Ophiroceras*) (150–400m).	Black shales with concretions and a fauna of ammonites (*Ophiceras, Hedenstroemia*).	—

In the zone of structures on the east and south-east margins of the Siberian Platform, the Lower and Middle Trias are to be sought in sandy-shaly formations of the Verkhoyansk complex. On the eastern slope of the Verkhoyansk range such sediments are widespread and occupy large areas, reaching a considerable thickness of 2000–3000m. Their distinguishing feature is a local presence of large numbers of intercalations with marine faunas indicating the presence of all Lower and Middle Triassic stages.

To the south, towards the Palaeozoic Massif situated to the north of Okhotsk, the succession is gradually reduced. First Lower Trias strata disappear, then Middle Trias strata die out and the Upper Trias rest directly on the Permian. To the west again almost all the marine Lower and Middle Trias is missing from the successions in the Uda and Zeya basins and Eastern Transbaikalia. Here almost barren, mainly continental deposits, which are not clearly differentiated from Permian sediments correspond to the Lower and Middle Trias since there are isolated discoveries of marine Lower Triassic faunas (Table 54).

The marine Lower and Middle Trias of the Verkhoyansk range continues along the southern slope of the Cherskii range and sweeps northward around the western margin of the Kolyma Palaeozoic Massif. Within the massif no Triassic sediments have been found but there are local Middle Triassic limestones, shales and tuffs with ammonites (*Nathorstites*). No Lower or Middle Triassic sediments are known and possibly they are missing further to the north and west. With the exception of the southern part of the Sikhote-Alin' range such sediments are unknown in the Alpine zone of the geosyncline. In the Sikhote-Alin' range Permian and older strata are overlain by the Lower Triassic complex, which begins with basal conglomerates and sandstones, succeeded by bedded greenish sandstones and dark shales, altogether reaching a thickness of 1000–1200m. The rich and diverse marine fauna indicates at the presence of the Lower Trias (*Proptychites, Meekoceras*) and of Middle Trias strata with ceratites and lamellibranchs (*Daonella*). The fauna is in general closer to the Indian (Mediterranean) than to the northern (Verkhoyanskian and Olenekian).

The Upper complex, which corresponds to the Upper Trias, is more widespread than the lower. It is everywhere associated with Jurassic deposits. Since the upper complex represents the commencement of a new sedimentary cycle, it is separated from the lower by a break associated with angular unconformities. Along the coast of the Arctic Ocean, from the lower reaches of the Kolyma to the Chukotian Peninsula, Upper Triassic and Jurassic strata form a single thick formation consisting of grey and brown bedded sandstones and shales with plant remains and basically of a continental origin. In places there are suites with marine faunas, as for instance to the east of the Chaun inlet. In these provinces the Upper Trias is separated from the Jurassic. Over the rest of the Pacific Ocean Geosyncline Upper Triassic strata are faunally and florally well defined and permit their separation from the Jurassic and their subdivision into stages. It is interesting and important that Upper Triassic strata almost everywhere consist of fine grained terrigeneous deposits such as shales and sandstones of dark grey, grey and greenish coloured and reaching a great thickness of 4000–5000m. Only in the extreme south-west—the Tetyukha region—are there Carnian limestones of considerable thickness.

If one considers the vast spread and thickness of Upper Triassic beds the total quantity of their clastic sediments is exceptionally large. The source of these sediments must have been young rising ranges. In front of such ranges but at some distance fine grained sediments of Upper Triassic horizons accumulated. The sediments are very similar to fine grained fractions of the Alpine Neogenic molasse.

The Upper Trias is especially fully developed in the basin of the Yana, near Verkhoyansk and to the south of it. Here the Carnian Stage is represented by black shales and sometimes sandy shales (flags) with beds of sandstones, the total thickness being 600m. In the upper horizons an abundant lamellibranch fauna with *Halobia* and *Pseudomonotis scutiformis* has been found. The conformably overlying Norian strata are represented by grey shales and sandy shales (flags) with intercalations of coarse sandstones and lenses of conglomerates. The sandstones become dominant toward the top. The total thickness is about 1000m. In the middle part of the succession a lamellibranch fauna with the ubiquitous but critical *Monotis ochotica* Keys. has been found. Recent investigators have established that the Upper Trias form which was formerly identified as *Pseudomonotis ochotica* in fact belongs to the genus *Monotis* and should be called *M. ochotica*. To the south in the upper reaches of the Indigirka and Kolyma the Upper Trias is represented by the same sandy and shaly formations, which, however, are 4000–5000m thick. In the upper part of the succession Rhaetic continental, sometimes coal-bearing, deposits of a few hundred metres in thickness have been recognized. In addition to sediments there are also quite abundant Triassic lavas, normally occurring in thin groups or suites.

In Eastern Transbaikalia in several localities Upper Triassic strata are found along the northern and western margins of the region. The strata (3000–4000m) consist of greenish, bluish, grey and dark grey shales and sandstones with basal conglomerates and sandstones at the bottom. The fauna is relatively rarely found and occurs in separate bands and formations. It consists mainly of lamellibranchs of a Norian age (*M. ochotica*), but is Carnian at the bottom of the succession. According to T. M. Okunyeva (Theses. . ., Chita, 1961) the succession (from top to bottom) is as follows:

Norian Stage

Bichektuian Suite. Siltstones and sandstones; of 500m thick. Varieties of *Monotis ochotica*.

Tuleyan Suite. Sandstones of up to 1000m. Varieties of *M. ochotica*.

Tyrgetuian Suite. Interbedded siltstones and sandstones with intercalations of shelly limestones, altogether reaching 700m. Abundant *Monotis jacutica*, varieties of *M. ochotica*.

Carnian Stage

Bodanovskian Suite. Sandstones with rare beds of siltstones and conglomerates; about 1000m thick. *Monotis scutiformis* and other lamellibranchs.

Baintsaganian Suite. Siltstones and sandstones; about 1500m. Fossils: *Sirenites. Halobia, Spiriferina.*

Upper Trias sediments are found on the shore of the Sea of Okhotsk, to the south of the Shantar islands and further to the south in the basin of the Bureya. The sediments are always associated with conformable Jurassic strata and consist of fine clastic shales, siliceous shales and greywackes of a large total thickness.

In the Sikhote-Alin' region Upper Trias rocks are found almost throughout the range. The sediments are more inhomogeneous than those of the more northerly provinces. In addition to the main, thick sandy-shaly rocks with *Monotis ochotica* in some beds, in the Tetyukha district there are thick reef limestones with hexacorals and *Halobia*.

The Mongugaiian Suite is distinctive. It consists of 1500–2000m of dark, brownish and greenish fine grained sandstones becoming coarse towards the top. The middle part of the suite has a group of marine deposits with *M. ochotica* indicating its Norian age. The flora supports this contention but also suggests that the upper horizons of the suite are Rhaetic and possibly even Liassic. The suite is coal-bearing and has thin coal seams at the bottom and anthracite seams at the top.

PALAEOGEOGRAPHY. Despite the wide distribution of Triassic deposits it is difficult to reconstruct the palaeogeography of Triassic times since marine faunas are often rare or are missing. The areas, where there are no marine faunas should be considered as predominantly sites of continental deposition on vast near-shore plains which were submerged under the sea only at times. Continental areas were adjacent to essentially marine regions with abundant marine faunas testifying as to their marine conditions but even these areas emerged now and again becoming parts of near-shore plains. Thus throughout Triassic times a continental regime was predominant in the struggle between land and sea.

The overwhelming predominance of clastic deposits reflects the existence of vast areas of denudation. Hercynian chains around the east and south-east of the Siberian Platform apart, isolated massifs of Middle and Lower Palaeozoic and Precambrian rocks formed islands in the west of the region. In Lower and Middle Triassic times such islands existed to the north of Okhotsk, on the site of the Kolyma and the Omolon massifs in the Chukotian Peninsula and possibly in other regions of the Alpine zone, such as Kamchatka and Sakhalin.

Upper Triassic times were marked by even less constant positions of shorelines and an even greater dominance of the continental regime. Only over short periods of time did the sea extend over the whole surface of the geosyncline, as happened during the *Monotis ochotica* epoch when only isolated islands remained unsubmerged. Upward and downward movements were associated by fissuring, leading to eruptions of magma and the consequent small volcanic centres. In the south near the shore coral reefs originated, as for instance in Tetyukha. The climate became more moist and gave rise to a luxuriant vegetation and formation of coals.

Jurassic

Jurassic, like Triassic times, saw a great accumulation of sediments over a large area. Three types of deposits: marine, continental (coal-bearing) and volcanic can be recognized. In the Cimmeridian zone Jurassic deposits form a part of the Verkhoyansk complex.

DISTRIBUTION. The most north-easterly outcrops of Jurassic rocks are found in the folds of the Khatanga inlet and the estuaries of the Anabara and the Olenek. Here the three divisions are developed, but in places Lower and Middle Jurassic strata are grouped into a single formation. Jurassic rocks thereafter are found on the western slope of the Verkhoyansk range, where in the north at any rate they remain as yet unsubdivided. To the south two formations can be recognized —the mainly marine Lower and Middle Jurassic and the continental, partly coal-bearing Upper Jurassic and Lower Cretaceous.

In the extreme north, on the Novosibirsk islands, the oceanic seashore and the lower reaches of the Yana and the Indigirka, Jurassic strata are widespread over large areas. Here Jurassic rocks are thick and of flysch type. Since they are almost barren there is no possibility of subdividing them. To the south in the Cherskii range Jurassic strata are predominant and can be divided into lower, middle and upper as well as a number of stages. In places Lower and Middle Jurassic sequences consist of an undifferentiated almost barren homogeneous sandy-shaly formation. Such Jurassic rocks are also found to the south-east, in the Kolyma range and the Omolon basin. The ring of volcanic rocks around these areas is of interest, since it reflects eruptions of miscellaneous lavas and tuffs of mainly an Upper Jurassic age.

Large areas of Jurassic rocks occur along the coast of the Arctic Ocean, stretching from the mouth of the Kolyma to the Chukotian Peninsula. The strata are barren, almost homogeneous clastic sediments the bottom part of which is Upper Triassic. Marine Upper Jurassic rocks are found in the north of the Alpine zone, in the basins of the Anadyr and the Penzhina.

Mainly Upper Jurassic lavas of intermediate and basic compositions are found on the shore of the Sea of Okhotsk—to the west and east of Magadan. Similar rocks also occur along the shore at the foot of the Dzhugdzhur range. Volcanic rocks are inter-bedded with sediments and stretch out into the Uda basin and the upper reaches of the Zeya and the Amur where they are of a considerable extent.

Large areas of ubiquitously occurring Jurassic rocks in the Far East allow the distinction of their three divisions, but on synoptic maps sometimes the Lower and Middle Jurassic are grouped together. In Eastern Transbaikalia Middle and Upper Jurassic rocks are well developed.

DIVISIONS AND LITHOLOGIES.   Whereas Triassic faunas have been studied on a monograph level both formerly and recently by L. D. Kiparisova, Jurassic faunas despite the outstanding work of V. I. Bodylevskii and N. S. Voronyets have not been so worked out. Thus the establishment of stratigraphic subdivisions and the correlation of distant succession is complicated. Further difficulties are introduced by the repetition and lateral passages of marine, near-shore continental (coal-bearing) and volcanic rocks. On the other hand the presence of extensive faunal horizons aids the correlation. In the Lower Jurassic there is a horizon with Middle Lias ammonites (*Amaltheus margaritatus* Mantf.); in the Middle Jurassic there is a horizon with Aalenian *Inoceramus retrorsus* Keys., while in the Upper Jurassic and the Valan-ginian there are aucellid marker horizons.

In the lower reaches of the Anabara and the Olenek there are Jurassic successions of platform type. The Lower Jurassic here begins with basal conglomerates, over-lain by grey and yellowish sands and clays of a total thickness of 150–700m and containing ammonites, gastropods, belemnites and wood fragments. *Amaltheus margaritatus* determines the age of these rocks. Middle Jurassic strata (200–550m) begin with clays and sands of the *Inoceramus* Horizon and contain *I. retrorsus*, of an Aalenian age. Higher up there are sandstones and clays varying from brownish to yellowish and possessing at the top a Bathonian fauna. The Upper Jurassic sequence is condensed (20–30m) owing to the presence of a 1·5–2m thick horizon of phoshorite and glauconite, which corresponds to Callovian, Oxfordian and Kim-meridgian stages. Valanginian dark-grey and grey clays with aucellids and ammonites succeed the condensed sequence and are followed by a Lower Cretaceous coal-bearing formation (500–750m).

In the north-east and particularly in the Cherskii range the Jurassic is represented by the rocks of the Verkhoyansk complex consisting of homogeneous sandstones and shales reaching a thickness of 4000–5000m. Normally the succession is almost unfossiliferous and only in a few horizons are there marine faunas indicating that

Lower, Middle and Upper Jurassic sediments are present and that Upper Jurassic strata are closely associated with Valanginian beds, which here reach a great thickness of 2000–3000m. There are quite common continental suites with plant remains and volcanic horizons. As is obvious from synoptic maps volcanic rocks reach their greatest development and thickness in the vicinity of volcanoes.

The succession on the Yana, the Indigirka and the upper reaches of the Kolyma (Tuchkov, 1957) serves as an example. Here Rhaetic sandstones and shales are overlain by Liassic strata (2000m) consisting of shales and sandstones with belemnite and ammonite-bearing beds. Middle Jurassic rocks (1400–2000m) consist of homogeneous shales and sandstones with fossiliferous bands with inoceramids and ammonites. A suite (400–500m) of sedimentary-volcanic rocks belongs to the Callovian and Oxfordian and is followed by shales and sandstones (1500–1600m) with intercalations of volcanic deposits and an aucellid fauna. The succession ends with a Lower Cretaceous coal-bearing formation (700m).

In Eastern Transbaikalia (Table 55) the Lower Jurassic sequence begins with basal conglomerates and sandstones (80m) overlain by dark shales and sandy shales (up to 1800m) which are sometimes metamorphosed and formerly taken as Precambrian. These rocks have a marine *Amaltheus margaritatus* fauna and Lower and Middle Lias plant remains. The Upper Lias starts with a sequence (400m) of conglomerates, which are overlain by sandstones and siltstones (up to 1250m) with ammonites. The Middle Jurassic commences with grey shales and beds of sandstones (470m) with an Aalenian marine fauna and plant remains. The overlying horizons have no marine fossils, but further to the east, in Priamuria, Aalenian-Bajocian, Bathonian-Callovian and Upper Jurassic marine faunas have been found. All Lower Cretaceous deposits are continental and the *Aucella* sea did not penetrate so far to the west. Jurassic strata are divided into several suites of local significance.

One of the most important results of investigations on the Jurassic has been the recent establishment of considerable facies variations (Theses . . . Chita, 1961). Along the strike, especially to the north marine deposits are succeeded by continental sediments. Amongst the latter the following marine facies complexes are noted: (1) coal-bearing facies, (2) volcanic-sedimentary facies, (3) foothill facies, (4) alluvial plains facies, (5) marine facies. The relationships of these five facies complexes are not entirely clear and are very complex. Nevertheless the possibility of mutual replacement and repetition at different times is quite important.

In the south the coal-bearing facies are normally Lower Cretaceous but can also be Upper Jurassic. Thus their flora has to be studied. The volcanic-sedimentary facies can also be of different ages. In Central Transbaikalia there are two volcanic sedimentary formations. The lower one is Middle Jurassic and consists of acid lavas and tuffs reaching 300–500m in thickness. The upper formation is of an Upper Jurassic age and consists of 600–700m of basic and intermediate lavas. In other regions a Lower Cretaceous volcanic-sedimentary complex has been found.

The foothill deposits consist of more or less massive conglomerates interbedded with shales and sandstones. These rocks are related to orogenic, frequently fold generating phases. Again their ages can undoubtedly vary. Alluvial deposits formed on near-shore plains or large river valleys are represented by brown and grey siltstones and sandstones with lenses of conglomerates. They are quite thick, almost unfossiliferous and have coal bearing formations.

Lastly there are the Turgaiian facies of typical lacustrine deposits. Since lakes undoubtedly existed in different epochs these facies can be not only Lower Cretaceous but also Jurassic. In the Far East the Jurassic succession is at its thickest reaching up to 8–10km and is most fully developed in the north, within the Cimmeridian zone where it has all the features of the Verkhoyansk complex. In the Alpine zone (the Sikhote-Alin' range) Jurassic rocks are less well developed but are rather distinctive.

In the north—the Bureyan coal basin—the Lower Jurassic sequence begins with basal sandstones and conglomerates (80–100m), which are overlain by dark fine-grained sandstones passing upwards into sandy shales, the total thickness being 500–650m. The fauna is marine and includes *Amaltheus margaritatus* and higher up *Hildoceras* and *Lioceras*. Middle Jurassic strata consist of thick (1200–1500m), dark argillites and fine-grained sandstones with the ammonite *Macrocephalites* at the bottom. The Middle Jurassic sequence ends with variegated arkosic sandstones (1200m) which in turn are succeeded by a thick formation (1800m) of dark-grey and grey shales with beds of sandstones. Throughout the formation there are dispersed plant remains and in its middle part there are intercalations with *Aucella* of an Upper Jurassic age.

In Cisokhotia the *Aucella* Formation is 2000–3000m thick and consists of shales and sandstones with isolated horizons with aucellid faunas of Upper Jurassic and Valanginian ages. There are also formations of conglomerates (Fig. 264). The

Fig. 264.   Upper Jurassic conglomerates in the river Gerbilak, the basin of the Uda. *Photograph by L. I. Krasnyi.*

Aucella Beds are overlain by a Lower Cretaceous coal-bearing formation (Table 55).

In the Sikhote-Alin′ range the Jurassic is represented by a thick (7000–8000m) homogeneous sequence of dark-grey and grey shales consisting of true shales, siliceous shales, argillites and fine-grained sandstones. In the north the fauna is rare and on synoptic maps the whole sequence is sometimes shown as undifferentiated Jurassic. In the south numerous fossiliferous horizons have permitted a subdivision into the Lower Jurassic, Middle Jurassic and Upper Jurassic. Both the *Inoceramus* and *Aucella* horizons are distinct.

In the south of Primoria—the Sandagan basin—a marine Lower Jurassic fauna with *A. margaritatus* has recently been found. In the Sikhote-Alin′ range there are local Liassic coals corresponding to the top part of the Mongugaiian coal-bearing formation.

In Sakhalin and Kamchatka no palaeontologically defined Jurassic has as yet been found, but it undoubtedly exists, which opinion is confirmed by the recent discovery of marine Lower and Upper Jurassic strata immediately to the north of Kamchatka (Mikhailov).

The principal successions of Far Eastern Jurassic strata are given in Table 55.

FACIES AND PALAEOGEOGRAPHY. In the Cimmeridian zone molasse formations (the Verkhoyansk complex) enjoy a vast distribution. Such formations consist of fragmentary terrigenous sediments produced as a result of the denudation of rising uplands and the deposition of the products of erosion on vast sub-montaine, near-shore, alluvial plains. More or less fine-grained sandstones, siltstones and shales are dominant lithologies. Sandstones and silts are often of a greenish-grey peppery variety of Artinskian type (the Western Urals). The coarser fractions are interbedded with siltstones and marls, thus producing the appearance of flysch deposits. The shales are usually dark grey or grey and can be sandy, argillaceous or calcareous. They contain large quantities of plant remains which are sometimes concentrated into coal laminae or even seams. Pure thick limestones are absent, but there are beds of muddy limestones and calcareous sandstones and shales with marine faunas (Fig. 265). In many regions there are formations of lavas and tuffogenous rocks which are normally thin (a few hundred metres) but have a vast distribution. The thick conglomerates (up to 2000–2500m) developed in Eastern Transbaikalia (Fig. 263) are peculiar. They wedge out rapidly and represent typical fan cones of mountain rivers emerging on to the sub-montaine plain. Pebbles of such conglomerates are of differing compositions, differing sizes and differing degrees of rounding. The conglomerates have beds of sandstones and shales. The identical conglomerates in the Alpine Molasse are known as the *Nagelfluh*.

Jurassic molasse deposits of the Pacific Ocean Geosyncline (a part of the Verkhoyansk complex) are identical with the Alpine Molasse, and are equally thick sequences of clastic, mainly continental sediments. Although they are not uncommonly freshwater deposits (freshwater molasse), they also quite frequently yield marine faunas (marine molasse). Such rocks are always deposited at the margins of rising ranges and massifs and this generalization applies to the Jurassic Molasse as well. A chain of mountain ranges margined the Siberian Platform. The Verkhoyansk range in particular belonged to this chain, but there were also such ranges within the geosyncline and their sites are indicated by the existing Palaeozoic massifs. The Chukotian Peninsula and adjacent districts remained all the time above sea level.

TABLE 55.

Jurassic and Lower Cretaceous Successions of the Far East
(According to K. M. Khudolei, 1956; M. S. Nagibin, 1956; L. I. Krasnov, 1956;
  Yu. A. Ivanov, 1956)

Division	Stage	Sikhote′ Alin′	Lower Priamuria (Cisamuria)	Western Okhotskian Seaboard (Cisokhotia)	
Lower Cretaceous	Albian. Barremian. Hauterivian.	Suchanian Series.	Dark shales with beds of sandstones and *Aucella* horizons (1100m).	Continental coal-bearing deposits.	A coal-bearing formation.
	Valanginian.	Dark sandstones, silts and shales with *Aucella* (1000–2000m).	Above, greenish shales; below, sandstones with *Aucella* (2500m).	Marine deposits (600–700m).	Continental sediments and lavas. A bed with *Aucella* is found (Up to 2000–3000m).
Upper Jurassic	Upper Volgian.	Molchanovskian Suite with *Aucella*.	—	Continental deposits (300–400m).	
	Lower Volgian.	Grey, greenish sandstones with *Aucella* (80–100m).		Sandstones with beds of tuff; *Aucella* occurs (3400m).	
	Kimmeridgian-Oxfordian.	Greenish sandstones with sequences of shales with ammonites (150–180m).		Fine–grained sandstones and tuffs with *Aucella* (800m).	
	Callovian.	Sandstones; towards the bottom, conglomerates with ammonites.		Siltstones and sandstones with a marine fauna (200m).	
Middle Jurassic	Bathonian. Bajocian.	Sandstones and shales with a marine fauna (700–1500m).	Dark sandstones and silts (1500m).	Sandstones and silts with a marine fauna (250–300m).	
	Aalenian.	Often rhythmically interbedded sandstones and shales with *Inoceramus* (Up to 2000–2500m).	Udylian Suite of black and siliceous shales with *Inoceramus* (1000–1500m).	Shales and sandstones with *Inoceramus* and *Aucella* (500–600m).	
Lower Jurassic	Upper Lias.	Sandstones and shales with a marine fauna (1000m).	Soleninskian Suite of coarse sandstones and conglomerates without a fauna (400–600m).	Micaceous sandstones and shales (800–900m).	
	Middle Lias.	Sandstones and shales—above, with a fauna; below, with coal seams (800–1200m).		Coarse sandstones; below, conglomerates (450–500m).	
	Lower Lias.	Siliceous and normal shales with a marine fauna; conglomerates towards the bottom (300–1000m).			

The Alpine zone shows a great variability of facies complexes. Molasse deposits are encountered within it, but are no longer dominant and are intercalated with

TABLE 55—continued

The Bureyian basin	The Leya region	The Upper Amur	The Ol'doi	Eastern Transbaikalia
Bureyan Coal-bearing Formation, consisting of five suites.	Molchanskian, coal-bearing, Suite of sandstones with beds of shales and coal seams (1000m).	Peremykinskian Suite—above, sandstones and argillites; below, conglomerates and sandstones (1500m).	Sandstones with beds of conglomerates and shales, without a fauna (300–350m).	Coal-bearing Suite (700m).
Sandstones and argillites with coal seams and plant remains (2000m).	Depskian, coal-bearing, Suite of sandstones and shales with beds of tuffs and coal seams (2300m).	Tolbuzinskian Suite (coal-bearing) of sandstones, argillites and coals (1350m).	Continental sandstones and shales with a flora (1000m).	Turgaian Suite—above, lacustrine, bituminous shales, marls, tuffs; below, sandstones and conglomerates (80–100m).
Chaganyian Suite occurring under the Coal-bearing Suite—shales and silts devoid of a marine fauna (600m).	Ayakian, coal-bearing, Suite of sandstones and, towards the top, carbonaceous shales (1000m).	Osekian Suite of sandstones and argillites with lavas and tuffs towards the bottom (1000m).	Interbedded grey sandstones and black shales with a marine Upper Jurassic fauna, and towards the top with *Aucella*.	An Upper Jurassic Suite of sediments and volcanic rocks—lavas and tuffs interbedded with sandstones and shales, with a flora.
Elgian Suite of interbedded sandstones, siltstones and argillites.	Sandstones and shales.	Interbedded sandstones, and argillites, with a marine fauna.		Continental deposits, sometimes coal-bearing. In places foothill conglomerates (About 1500m).
Above, barren; in the middle, with an Oxfordian-Kimmeridgian fauna; below, with a Bathonian-Callovian marine fauna (1800m).				
Epikanian Suite of arkosic sandstones and shales lacking a fauna (1200m).	—	—	Dark shales and sandstones with a marine fauna; above, Bathonian; below, Aalenian (200m).	Sandstones and shales with a Bathonian marine fauna (1000m). Upper Gazimurskian Suite of thick, foothill conglomerates and sandstones (Middle Jurassic) (1500–2000m).
Upper Umal'tian Suite of sandstones and silts with a fauna (1500m).	—	—		Onon-Borzian Suite of dark shales and sandstones—below, with an Upper Lias fauna; above, barren and of probably a Middle Jurassic age (1200–2000m).
Lower Umal'tian Suite of sandstones and silts; in places coal-bearing, elsewhere with marine faunas of Lower Middle and Upper Lias ages (500–650m).	—	—	Grey sandstones with beds of silts and a marine Lias fauna (250m).	Sivachinskian Suite of conglomerates and sandstones (Middle Lias) (400–800m).
				Ontagaiian Suite of shales and sandstones with basal conglomerates at the base. A marine Middle and Lower Lias fauna is present (1000–2000m).
At the base sandstones and conglomerates (80–100m).			Basal sandstones and conglomerates (50m).	

marine, volcanic and continental formations, in the Lower Jurassic, the latter being sometimes coal-bearing.

Configurations of land and sea were at all times complex and frequently changed a lot. Three main transgressions can be detected: the Middle Liassic, the Aalenian and the *Aucella* (Upper Jurassic-Valanginian). The latter was particularly extensive. Sandstones and shales with *Aucella* and other marine fossils of Upper Jurassic and Valanginian ages are ubiquitous. At this time all alluvial plains were submerged under a shallow sea, which approached the margins of mountain ranges. Palaeozoic massifs rose above the sea as islands of differing dimensions. In the north-east there was a wide belt of volcanic islands, which in the Lower Cretaceous epoch was especially widespread and can be clearly seen on synoptic maps. A belt of Upper Jurassic volcanic islands and land volcanoes surrounded the Kolyma Palaeozoic Massif.

The Lower Cretaceous

The Lower Cretaceous is closely associated with the Jurassic and its sediments represent the top part of a single sedimentary cycle. Thus Lower Cretaceous sediments are widespread, occur together with Jurassic strata, and in several regions participate in an unbroken Jurassic-Lower Cretaceous sequence. Lower Cretaceous lithologies, thicknesses and general characters of faunas are very similar to those of the Jurassic. A single difference lies in the circumstance of Lower Cretaceous strata representing the end of a sedimentary cycle, when marine basins were much reduced and alluvial plains most widespread. The surfaces of such plains were sites of numerous lakes and swamps where plant remains accumulated and produced the abundant coals of Lower Cretaceous deposits.

FIG. 265 *Aucella*-bearing Upper Jurassic sandstones and shales exposed in Ussuri Bay. Rhythmic layering and a general similarity to flysch deposits are characteristic. *Photograph by K. M. Khudolei.*

DISTRIBUTION AND SUBDIVISIONS.   The lower part of the Lower Cretaceous succession embraces most and sometimes all of the *Aucella* Beds, which in places are 2000–3000m thick. The beds are the deposits of mainly marine basins and consist of dark grey and greenish sandstones and shales with a marine fauna. Especially in the north-east there are also frequent lavas and tuffs. The *Aucella* Beds are overlain by continental, almost always coal-bearing, sediments. The coals of these sediments are important and are extracted in several regions. Lower Cretaceous successions are shown in Table 55.

In the Far East both the *Aucella* Beds and a coal-bearing formation are developed and the latter is known as the Nikanian Stage (Series). In the south of the Sikhote-Alin' range a large so-called Suchan coal basin is associated with this series. At the bottom of the series lies a basal formation of conglomerates and sandstones. It is in turn succeeded by fine-grained, bedded dark-grey, grey and brown sandstones and shales. Coal seams are concentrated in these coal-bearing suites separated by two coal-less suites. The total thickness of the coal-bearing formation is 1000–1500m (Fig. 266).

The Suchanian Coal-bearing Formation is conventionally divided into three suites: the Lower Suchanian at the bottom, the Older Suchanian in the middle and the Northern Suchanian at the top. Each suite includes a coal-bearing and an overlying coal-less sub-suite. The study of the flora has shown the Suchanian Formation to be distinctly Lower Cretaceous throughout. Faunas, however, indicate more complex relationships. Upper Cretaceous brackish-water lamellibranchs appear in the Older Suchanian Suite. They become much more abundant in the Northern Suchanian Suite and a series of marine Foraminifera, oysters and *Trigonia* make their appearance. Provisional determinations suggest the presence of several Cenomanian, Turonian and even Senonian species. Nevertheless since the total number of such individuals is small and they are badly preserved and also since there are Lower Cretaceous species as well the age of the Northern Suchanian Suite remains an open question.

In the Suchan basin the Suchanian Formation is overlain by the Korkinian Suite belonging also the Nikanian Stage (Series). The suite consists of chocolate-coloured shales and decorative sandstones with often a considerable admixture of tuffaceous material. The total thickness of the suite is 800–1000m. Its lower horizons have *Aucellina* of Lower Cretaceous type, which used to be taken as the index of its Lower Cretaceous age. At present many investigators consider the suite as Cenomanian since similar aucellids are also found in the Cenomanian.

It is interesting that in the north of the Sikhote-Alin' range the sediments corresponding to the Nikanian Stage are marine and consist of dark grey shales with intercalations of sandstones, which include bodies of porphyritic diabases passing laterally into tuffs and tuffaceous sandstones. The total thickness of the sequence is 1100m and the porphyrites alone reach sometimes 600m. The shales have a marine fauna of Aptian-Albian aucellids and Barremian ammonites.

The areally large Bureyan coal basin situated in the middle reaches of the Bureya and its tributaries has a similar Lower Cretaceous succession. The *Aucella* Beds are overlain by a coal-bearing formation, which is divided into five suites of a total thickness of about 2000m (Table 55).

In Eastern Transbaikalia the lower part of Cretaceous deposits is coal-bearing. In places its lowest horizons are replaced by the Turgaiian Horizon of lacustrine laminated clays and marls with freshwater fish, *Estheria*, fragments of insects, lamellibranchs and gastropods. The horizon is 80–100m thick and is overlain by the

Coal-bearing Suite. The Coal-bearing Suite has several commercial coal seams, which are sometimes thick and are worked in several localities. The suite begins with a sandy-gravelly horizon (200m) which is overlain by dark grey and brown sandstones, siltstones and shales with brown coal seams, plant remains and a freshwater fauna. The total thickness of the Coal-bearing Suite is 600–700m and more.

The Lower Cretaceous succession of the Argun Depression (Kutian coal-field) can be quoted as an example (Stasyukevich and Oleinikova; Theses . . . Chita, 1961).

1. A conglomerate-sandstone variegated formation (250–500m). It lies on Upper Jurassic lavas and has beds with freshwater faunas.

2. A volcanic-sedimentary formation with beds containing Turgaiian-type fauna (fish, *Estheria*, insects).

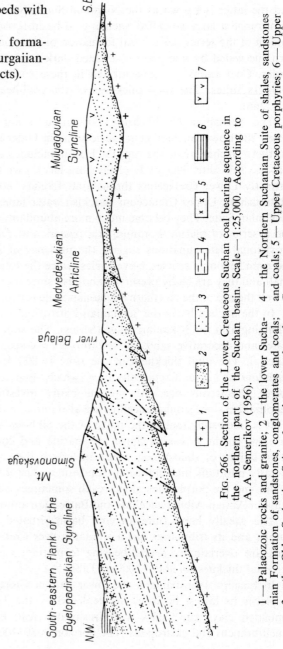

FIG. 266. Section of the Lower Cretaceous Suchan coal bearing sequence in the northern part of the Suchan basin. Scale — 1:25,000. According to A. A. Semerikov (1956).

1 — Palaeozoic rocks and granite; 2 — the lower Suchanian Formation of sandstones, conglomerates and coals; 3 — the Older Suchanian Suite of sandstones and coals; 4 — the Northern Suchanian Suite of shales, sandstones and coals; 5 — Upper Cretaceous porphyries; 6 — Upper Cretaceous; 7 — Palaeogene andesites.

3. A coal-bearing formation (750–800m) with freshwater lamellibranchs and a flora. There are interesting iron ores in the Lower Formation.

To the east lavas, reaching a considerable thickness, appear amongst Lower Cretaceous sediments. The lavas are intercalated amongst sandstones and shales, at times forming the whole of the succession as happens in the Dzhugdzhur range, the lower reaches of the Amur and the Amgun, the Okhotsk and Magadan districts, where volcanics are at times as much as 2000–3000m thick. Cretaceous lavas occupy a large area between the Penzhian, where they are partly of Lower Cretaceous and mainly of Upper Cretaceous age.

Lower Cretaceous strata are continuous along the whole of the western slope of the Verkhoyansk range and continue westwards towards the Khatanga estuary and the eastern tip of Taimyr. The strata are folded, moderately metamorphosed and reach a great thickness. The *Aucella* Beds (Valanginian) consist of shales, siltstones and sandstones and are of a variable thickness, ranging up to 400–600m. Higher up lies a coal-bearing formation with several important deposits of coal. In the most northernly Bulunskian coal region the total thickness of the formation is 2500–3000m and it has about 40 coal seams some of which are of a commercial thickness.

In the north-east the Lower Cretaceous coal belt deserves attention. It lies to the north of the Cherskii range and continues for a large distance between the Indigirka and the Kolyma, passing through the upper part of the Kolyma basin into the Omolon valley. The Coal-bearing Formation is of a normal composition, has workable coal seams and is several thousands of metres thick. In the basins of the Zyryanka and the Ozhogina (left tributaries of the Kolyma) according to I. K. Tuchkov (1957) the Palaeozoic is overlain by an Upper Jurassic lava-tuff suite (750m) with Callovian ammonites. Higher up lie *Aucella*-bearing shales and sandstones (250m) and higher still sandstones and shales (1700–1900m) with a Lower Cretaceous flora. Finally there is a thick (3000–5000m) coal-bearing formation.

Lastly in the Koryakskii range, in the north-west of Kamchatka, there are massive (5–9km) Lower Cretaceous sandy-shaly rocks which are predominantly marine. The fauna points to the Valanginian (sandstones and shales with *Aucella*) and the Aptian-Albian, again with aucellids (Fig. 267). The regional conference at Magadan, 1957, produced much useful material on Lower Cretaceous deposits of the north-east. In the 'Theses of papers' (1957) several successions are given. Valuable data can also be found in the Proceedings (Materialy) of stratigraphic conference in Okha (1961) and in the 'Theses of papers' of the stratigraphic conference in Yakutsk (1961).

FACIES AND PALAEOGRAPHY. In the Cimmeridian zone the *Aucella* Beds represent the last marine formation, while all the overlying deposits, including the modern ones, are continental, often coal-bearing or interbedded with volcanic lavas. In the Alpine zone the succession is very different. Lower Cretaceous and Upper Cretaceous rocks consist equally of alternating marine, coal-bearing and volcanic rocks, all of which are thick, folded and metamorphosed.

The Upper Cretaceous and Palaeogene

Upper Cretaceous and Palaeogene strata are closely associated, forming a single macrorhythm of sedimentation. Consequently their lithologies, metamorphism, magmatism, structures, useful mineral deposits and distributions are very similar.

Upper Cretaceous and Palaeogene strata of the Cimmeridian zone differ considerably from those in the Alpine zone of the geosyncline. In the Cimmeridian zone they are of a platform type, thin, weakly metamorphosed and almost entirely lacking marine sediments. Their folds are large and gentle and there are no synorogenic intrusions.

In the Alpine zone Upper Cretaceous and Palaeogene strata are geosynclinal, widely distributed and are of an enormous thickness (several kilometres). They are considerably metamorphosed, have complex structures and associated synorogenic intrusions; show a considerable development of marine sediments and have indications of oil. Their stratigraphy is discussed in the Theses of papers, Proceedings and transactions of conferences in Khabarovsk (1956), Magadan (1957), Okha (1959) and Yakutsk (1961).

DISTRIBUTION. Upper Cretaceous and Palaeogene strata are particularly widespread in the Alpine zone. In the Cimmeridian zone lavas and tuffs interbedded with sandstones and shales with plant remains are especially extensive. Between the Penzhian and the Chaun inlets and to the east of the latter the lower effusive rocks have a Lower Cretaceous flora and at the very bottom an *Aucella* fauna; the upper effusive rocks, however, have an Upper Cretaceous flora. The lower and the upper formations have only been distinguished in some localities, but in general they remain undifferentiated.

In the Alpine zone the major part of the Koryaksii range consists of thick formations (about 7000m) of sandstones and shales, which are folded, metamorphosed, faunally defined and divided into six suites. Palaeontological data indicate the presence of both lower and Upper Cretaceous deposits. Similar formations are developed in Kamchatka, while in Sakhalin and the Kuril islands only Upper Cretaceous strata are known.

FIG. 267. Cretaceous deposits of the Koryakskii range, belonging to Aptian-Albian Stage. The exposure is situated on northern slopes of the central part of the range. *Photograph by Yu. P. Degtyarenko (1958).*

SUBDIVISIONS AND LITHOLOGY. In the Cimmeridian zone of Eastern Transbaikalia only sparse continental deposits are present. Upper Cretaceous formations are so far unknown and to the Palaeogene are attributed friable sandstones and compacted clays with brown coal seams. The total thickness of the strata is 50–100m and they are folded into gentle folds developed in the depressions of the south-eastern part of Eastern Transbaikalia to the south of Borzi.

Classical exposures of continental Upper Cretaceous and Palaeogene strata are situated along the Amur, between the lower reaches of the Zeya and the Bureya. The Upper Cretaceous is represented by the Tsagayanskian Suite of yellowish, grey, white and greenish sands and sandstones which are often cross bedded and arkosic and not infrequently have intercalations of conglomerates. In places at the top there are formations of grey and brown clays with brown coal seams, all differentiated into the Kirdian Suite. Nearer to the Lesser Khingan there are outcrops of liparites and tuffs. The total thickness is about 100–150m.

The Tsagayanskian Suite is folded into large flexures with angles of dip being 4–5° and sometimes more. Its age—upper Senonian and Danian—is established on the basis of its flora, and has been confirmed by the discovery of Senonian dinosaur bones. The skeleton of one of the dinosaurs (*Mandschurosaurus amurensis*) is exhibited in the Central Geological Museum in Leningrad (Fig. 268). The thickness of the Tsagayanskian and the overlying Kivdian suites reaches 300–400km, but no

FIG. 268. *Mandschurosaurus amurensis* Riab. Photographed in the Central Geological Museum.

TABLE 56. *Upper Cretaceous, Palaeogene and Neogene Successions in the Far East* (According to V. N. Vereshchagin and G. I. Vlasov, 1956; N. M. Markin, 1959; I. I. Ratnovskii, 1959; A. A. Kapitsa, 1961)

System, Division	Sakhalin	North-western Kamchatka	Southern Primoria	The Central Sikhote-Alin' range	The Northern Sikhote-Alin' range	Middle Amur
Anthropogene	Terrigenous deposits.	Volcanic and terrigenous rocks.	Plateau basalts and terrigenous deposits.			Continental deposits.
Pliocene	Nutovian Suite of sands, clays and gravels with brown coals and marine bands (1500–3500m).	Kavranian Series of sandstones, tuffs, gaizes; at the bottom, conglomerates (1200–1800m).	Suifunian Suite of sands, gravels and tuffs (50–100m).	Intermediate and basic lavas and tuffs with diatomites and pollen.	—	Watershed sands (20–40m).
Miocene	Okobykaiian Suite of clays and silts, with a fauna (Up to 2500m). Sertunaiian Suite (800m). Upper Duian Suite—coal-bearing (1000–1400m). Khoindzhi Suite (Sergiyevskian) of lavas and sediments (Up to 1000m).	Vayampolkian Suite of andesites, basalts and tuff breccias (2000–2500m). Tuffogenous sandstones with a marine fauna (600m).	Ust'-Davydian, lignitic, Suite with dacites and andesites (600m).	A lignite formation Brusilovskian Suite of basalts, dacites and andesites (400–500m).	Rudaceous deposits.	—
Oligocene	Takaradaiian (Khandasian) Suite of siltstones, argillites and marls, with a rich marine fauna (Up to 1200m).	Kovachian Series of argillites, which are tuffogenous towards the top. A marine fauna is present (Up to 2000m).	Nadezhdinskian Suite of tuffs and shales (300m).	Liparites and tuffs.	Tuffs and clays.	Clays.

Eocene and Palaeocene	Senonian and Danian	Senonian and Turonian	Cenomanian and Turonian
Bazovskian Suite—coal-bearing (200–300m). Clays with a Raichikhian flora, overlying the Kivdian, coal-bearing, Formation.	Tsagayanian Suite of tuffs and lavas; below, coal-bearing (Kivdian Suite) (300–400m).	—	
Coal-bearing Formation of the river Nalyeo. Basalts and andesites. Palaeoliparites.	Ol'gian Suite of lavas and tuffs. (2000–3000m).	Udomian Suite of sandstones and tuffs (1700m).	El'gian Suite of shales and sandstones with a marine fauna (1500–2000m).
Kkhutsian coal-bearing Suite. Kuznetskian Suite of volcanic rocks with beds of sandstones (500m). Samarginian, volcanic Suite (Up to 1700m).	Takhobenian Suite of tuffs and tuffites. Olgian Suite of lavas. Selenchian Suite of black shales and sandstones (1000–1500m).	Udomian Suite of sandstones and silts with a marine fauna (1000m).	Largasinskian Suite of sandstones and silts, with a marine fauna (1000m).
Uglovian Suite—coal-bearing, with basalts (250m). Kuznetskian Suite of andesites and basalts and tuffs (500m). Nazimovskian Suite—coal-bearing (115m). Samarginskian Suite of andesites and dacites (400m).	Ol'gian Suite of quartz porphyries and tuffs, with a flora (Upper part of Senonian-Danian) (1000m and more).	Kangauzian Suite of tuffogenous silts and argillites (1000m).	
Tigilian Series of sandstones and argillites. Marine and coal-bearing sequences with basal sandstones and conglomerates (1500–3500m).	Tufo-slantsevian (tuff-sedimentary) Series of porphyrites and tuffogenous rocks. Cherty and normal shales with a marine fauna are also present (2500m). Palanskian Horizon of shell-beds (40–60m).	Lesnaian Series of thick grey and dark grey shales and siliceous shales with beds of sandstones; all unfossiliferous (Up to 2500m).	
Lower Duian (Nizhmeduian) Suite of sandstones, shales and conglomerates—above, marine (1000m); below, coal-bearing (600m). Conglomeratic (Kamenian) Suite of conglomerates, sandstones, shales and coals, with a flora (100–600m).	Orochenian Stage—porphyrites, sandstones and coal-bearing shales with marine fossils (>3000m).		Gilyakian Stage (Turonian) of sandstones and shales; to the north, continental and coal-bearing (Up to 5000m). Ainusian Stage (Cenomanian)—repeated shales and sandstones, with a marine fauna (900–1300m or 2900m). Pobedinskian Suite (Albian?) of shales, jaspers, sandstones and basal conglomerates (1500m).

more than 80–100m is exposed in the sections. The Kivdian Coal-bearing Formation (70–150m) contains coal seams of 3–8m in thickness, which are being worked in the Kivdian and Raichikha deposits. Formerly the Kivdian Suite was included in the Palaeogene (Palaeocene), but at present it is considered as Danian. The Palaeocene is represented by the Raichikha clays which have a flora younger than that of the Tsagayanskian and Kivdian suites.

The Eocene is represented by the Bazovskian Coal-bearing Suite found on the right side of the Amur to the east of Khabarovsk. The suite has a platform character, being thin (up to 200–300m) and weakly dislocated. It has seams of brown coal, which reach several metres in thickness and are being worked. The suite is similar to Palaeogene coal-bearing suites of the Sikhote-Alin′ range (Table 56), and is overlain by a formation of clays of provisional Oligocene age.

The platform type of the Upper Cretaceous-Palaeogene sequence persists in the north-east. In this region and especially in its southern part volcanic rocks show an exceptional development. The lavas are of differing compositions while the tuffs are interbedded with continental sands and clays with plant remains and reach a thickness of many hundreds of metres and sometimes as much as 2000–3000m. The area occupied by volcanic rocks is so large that it exceeds the area of the Siberian Traps. Starting from the Uda basin in the Dzhugdzhur range and the adjacent part of the Okhotsk coast Upper Cretaceous and Palaeogene volcanic form a continuous belt, trending north-east as far as the Gizhigian and the Penzhian inlets on the Sea of Okhotsk; then the belt diverges towards the Arctic Ocean and along it as far as the Chukotian Peninsula.

Palaeontologically uncertain fossils of the volcanic rocks—flora and freshwater fauna—do not permit their separation into suites, while at the same time, they point at ages, varying from the Lower Cretaceous to the Neogene. Volcanic rocks lie almost horizontally but in places form large open brachyfolds with dips no more than 10–15°.

In the north of the Cimmeridian zone the Upper Cretaceous and Palaeogene are represented by sparse continental deposits which are weakly folded, relatively thin (few hundred metres) and contain, sometimes in large quantities, plant remains, and even black and brown coals. There are quite frequent suites of volcanic rocks, but differences of opinion exist as to the age of the sediments. The flora was formerly thought to be Neogene, but later according to A. N. Krishtofovich it was placed in the Palaeogene. Yellow, grey and pink sands with chocolate-coloured clays and brown coals with a Palaeogene flora have been found in several localities in the Novosibirsk islands and on the continent, to the west of the lower reaches of the Indigirka.

Throughout the Alpine zone beginning with the Sikhote-Alin′ range, Sakhalin and the southern Kuril islands and ending with the valley of the Anadyr and the Koryakskii range Upper Cretaceous and Palaeogene strata have a wide distribution and are of a geosynclinal type. These strata are very thick (6–8km), intensely folded, metamorphosed and intruded by synorogenic massifs. The strata consist of marine, continental (coal-bearing) and volcanic suites. A rich and diagnostic Pacific Ocean marine fauna permits the distinction of all the analogues of Western European stages of the Upper Cretaceous and the Palaeogene. There are few common Pacific Ocean and Western European forms which renders the detailed correlation of the successions difficult and forces the usage of local Pacific Ocean subdivisions, which commonly do not correspond to the European.

V. N. Vereshchagin (Theses . . . Okha, 1961), who is a leading specialist on the Cretaceous of the Far East, divides the Upper Cretaceous succession into two series: the Gilyakian, corresponding to the Cenomanian and Turonian; and the Orochenian, corresponding to the Senonian and the Danian. The series are divided into several suites.

In Sakhalin Upper Cretaceous, Palaeogene and Neogene strata owing to their being coal-bearing and petroliferous have been fully studied in great detail. The Upper Cretaceous succession begins with basal conglomerates and sandstones, followed by shales, cherts and sandstones. All these deposits are grouped together (Kapitsa, 1961) as the Pobedinskian Suite (1500m). It is succeeded by the Ainuian Stage, a major part of which is Cenomanian. At the bottom there are shales and sandstones of terrestrial and marine origins, while at the top there is a suite with a marine *Trigonia* fauna. The total thickness according to some 900–1300m and according to others it is 3000m. On the suite lie the sediments of the Gilyakian Stage and consist of coal-bearing sandstones and shales, alternating with marine fossili-ferous strata of Upper Cenomanian (possibly) and Turonian ages. The thickness of the sediments is up to 5000m. The coal is black and there are 3–10 seams of it. The exploitation is complicated by intensively developed tectonic structures. The upper Orochenian Stage consists of repeated continental, coal-bearing deposits with coal seams and marine glauconitic sandstones and dark shales with a rich fauna of a Senonian age. It is possible that the uppermost barren sandstones and shales correspond to the Danian. The total thickness is up to 3000m.

In the Kuril islands Upper Cretaceous strata are divided into two suites: (1) the Matakatanian (Cenomanian and Turonian) consisting of lavas (500m) and (2) the Malokurilian (Coniac-Danian) consisting of argillites (200–300m) with beds of sandstones and a marine fauna.

The Palaeogene of Sakhalin has been studied in a more detailed fashion in connection with deposits of coal and especially oil. The lowermost Eocene and Palaeocene strata are in places divided into two suites: the lower Conglomeratic and the upper Nizhneduian. In the south the two are combined under the term 'Naibu-tain Suite'.

The Conglomeratic Suite varies in thickness, reaching 600m and consists of two horizons of conglomerates, separated by coal-bearing sandstones and shales with a Lower Eocene flora. Thus the lower conglomeratic horizon may be Palaeocene. It is possible that in places Palaeocene sediments are missing from the succession. The Nizhneduian Suite consists of continental dark sandstones and shales, which are locally coal-bearing. The thickness of the suite varies from 500–700m to 1200m.

Different suites in different parts of the island correspond to the Oligocene (Table 57). Compositionally these suites are similar to each other, only the Krasnopol'yevian Suite of Southern Sakhalin is peculiar since it consists of shales and sandstones with marine bands containing numerous Lower Oligocene oysters (*Ostrea*) and other faunas. The Gennoishian Suite and its analogues consist mainly of clays with numerous marine bands, reaching a total of 1200m. It is possible that its upper part is Lower Miocene.

A somewhat different succession in the south-eastern part of Northern Sakhalin is given by S. S. Razmyslova and K. P. Rakhmanov. The Conglomeratic Suite is included in the Lower Eocene and the Nizhneduian Suite (1000m) in the Middle Eocene-Middle Oligocene, while Upper Oligocene strata are separated into the Mutnyan Suite (1000m).

In the Sikhote-Alin' range Upper Cretaceous strata are mainly found in the northern part, and contain *Inoceramus, Trigonia* and other Cenomanian, Turonian and Senonian lamellibranchs and gastropods. The upper part of Upper Cretaceous deposits consists of volcanic rocks. The total thickness of Cretaceous rocks is about 4000m (Table 56).

In the northern part of the Sikhote-Alin' range there is the following sequence. At the base lie conglomerates and sandstones which pass upwards into sandstones and carbonaceous shales with a marine fauna at the bottom and plant remains at the top. The total thickness of these rocks is 1500–2000m and they form the Largasinskian (El'gian) Suite distinguished by E. B. Bel'tyenyev. The suite passes upwards into a thick (1700m) formation of siltstones, shales and sandstones (Udomian Suite) with plant remains and a marine fauna of Turonian age. On this suite lie conglomerates and sandstones tuff and tuff breccias with an abrupt, angular

TABLE 57. *Palaeogene and Neogene Suites of Sakhalin*
(According to I. I. Ratnovskii, 1959)

Age		Northern part of the island			Southern part of the island
		North-east	North-west	South-west	
Pliocene		Nutovian (500–2500m).	Rybnovskian (600m). Tamlevian (1300m).	Uandi (500m).	Orlovskian (300m). Maruyamian (3000m).
Miocene	Upper	Okobykaiian (1700m).	Nanivian (900m). Interval	Aleksandrovkian (800m). Sertunaiian (700m).	Kurasan (600m). Ausinskian (400m).
Miocene	Middle	Dagian (1900 m).	Upper Langerian (1400m). (Verkhnelangerian)	Upper Duian (1000 m). (Verkhneduian)	Uglegorian (200–1500m).
Miocene	Lower	Uininian (600m). Daekhurian (1200m).	Interval Lower Langerian (600 m). (Nizhnelangerian)	Agnevian (500m). Khoindzhian (800 m). Akhsnaiian (500m). Kuznetsovskian (700m).	Chekhovskian (600–900m). Nevelskian (800–1500m). Kholmian (Up to 1800m). Arakaian (Up to 1800m). Gastellovskian (Up to 800m).
Oligocene		—	—	Gennoishian (1000m).	Takaradaiian (1000m). Krasnopol'yevian (Up to 1200m).
Eocene and Palaeocene		—	—	Lower Duian (600m) (Nizhneduian). Conglomeratic (200 m).	Lower Duian (350–1200m). Conglomeratic (Up to 600m).

unconformity which pass upwards into a sequence of porphyrites. Beds of sandstones have a Senonian flora. The acid lavas at the top of the sequence are of a Danian age.

The formation of Senonian and Danian lavas (the Ol'gian Suite) is one of the most important stratigraphic horizons traceable throughout the Sikhote-Alin' range, and reaching a thickness of 2000–3000m. The lower part of the formation consists mainly of porphyrites and tuffs, while at the top quartz porphyries predominate. Quite often more detailed subdivisions of the suite can be recognized, and when such are termed 'suites' the whole formation should be referred to as the 'Ol'gian Series'. In the Suchan district the series lies on the Kangauzian Suite, but a little to north-west in the basin of Fudzin its lower part is replaced by black shales, siltstones and sandstones of a total thickness of 1000–1500m and containing a marine Lower Senonian fauna. V. N. Vereshchagin (Theses . . . Khabarovsk, 1956) has suggested the term 'Selenchaian Suite' for these rocks.

In the region of the Ol'ga-Tetyukha harbour between the porphyrites and quartz porphyrites lies the Partizanian Suite, consisting of shales and sandstones with plant remains, which are in places coal-bearing. Analogous coal-bearing suites have been found in other regions.

Thus the Senonian and the Danian stages mainly consist of a thick, widespread succession of lavas, with occasional suites, formations and sometimes individual beds of sands and clays occurring within plant remains and are coal bearing but more rarely have marine faunas.

Palaeogene rocks of the Sikhote-Alin' range are compositionally similar to Upper Cretaceous strata. They also consist of volcanic rocks with occasional coal-bearing formations of sandstones and shales, but marine deposits are absent since at that time the sea receded eastwards into Sakhalin.

In the regional conference in Khabarovsk (1956) a unified scheme composed by G. M. Vlasov was adopted for the Sikhote-Alin' range (Table 56). According to this scheme three volcanic suites are distinguished. The lower—Samarginian—reaches 1700m in thickness and consists of andesites and dacites at the bottom and liparites and felsites at the top. The following, Kuznetskian Suite (500m) consists of basalts and andesites with beds of sandstones with an Eocene flora. The upper or Nadezhinskian Suite (up to 340m) consists of dark, bedded argillites, tuffs and tuffites. In the north there is a single coal-bearing formation separating the Kuznetskian and the Nadezhinskian suites, while in the south there are two coal-bearing formations, the Nazimorskian (115m) and the Uglovian (Khasan) Suite (up to 250m).

In Kamchatka Upper Cretaceous and Palaeogene strata are widespread. Upper Cretaceous rocks occur on the western coast and in the south of the Central range. Palaeogene rocks are exposed on the eastern coast and western folds, as well as in the Central range. In connection with the search for oil Upper Cretaceous rocks have been most fully studied on the west coast. Here the Iruneiskian Suite consists of porphyrites and tuffs at the bottom and a thick tuff-shale siliceous sequence, with a marine Turonian-Senonian fauna at the top. The total thickness of the suite is more than 1000m. On the eastern coast the volcanic-siliceous formation is strongly folded and metamorphosed into green schist facies of chloritic and sericitic schists. Upper Cretaceous rocks of the Central range are even more strongly altered and it used to be said that even the crystalline schists and gneisses of a Precambrian age are really Upper Cretaceous and that they are strongly altered lavas and cherty and tuffaceous Upper Cretaceous shales. This conclusion was based on the discovery of

Inoceramus in strongly metamorphosed deposits, but it has not been confirmed since the strongly metamorphosed schists have yielded Sinian spores (p. 722).

Palaeogene rocks of the western coast lack a volcanic fraction and consist of kilometres thick terrigeneous sandy and shaly sediments, which are marine and continental and contain oil, and coal. At the bottom lies the Tigilian Suite (1000–2500m) consisting mainly of sandstones (Table 58). At its base lie thick conglomerates (100–300m) which are unconformable on older deposits. Higher up follow grey sandstones with dark clays and coal seams or marine bands with an Eocene fauna. The sandstones are in parts bituminous and oil-bearing. Still higher up lies the Kovachian Suite consisting mainly of clays and reaching a thickness of 1500–2000m. The suite consists of dark-grey and brown shales and muddy sandstones which are sometimes laminated, and have marly concretions. There are occasional indications of oil. Judging by rare occurrences of faunas the age is Oligocene.

In the last few years the study of Palaeogene and Neogene rocks of western Kamchatka has taken a sudden leap. The two aforementioned suites—the Tigilian and the Kovachian—have been subdivided into several sub-suites and even smaller units. Furthermore various facies acquired names and as a result the succession is of some complexity (Fig. 269) and has been formulated by B. F. D'yakov, the leading specialist on the Neogene and Palaeogene of Kamchatka. The analysis of his succession indicates that the Palaeogene and Neogene can be used as independent systems and consequently the Tigilian and Kovachian suites correspond to the divisions of a system and the more detailed subdivision to stages. As has already been mentioned their boundaries do not correspond to the boundaries of Western European stages. In any case the correlation is difficult and doubtful since the faunas are completely unlike. The succession (Fig. 269) formulated by D'yakov applies to the Tigil region in the centre of the western coast of Kamchatka. In more northerly areas of this coast it changes considerably (Table 58), owing to the appearance of volcanic rocks.

In the central part of Kamchatka the southern portion of the Central range was all the time a region of uplift and within it only a formation of conglomerates and tuffites with a Palaeocene fauna (Goryayev, 1959) has been found. In the central part of the Central range two volcanic suites are found: the lower is of a Palaeogene and Lower and Middle Miocene age and is 1500m thick, while the upper is of an Upper Miocene-Pliocene age (1300m).

The recognition of Alpine downwarps to the east of the Central range and to the east of the ranges formed by Mesozoic rocks (Gryaznov, 1959) is important and interesting. In the first downwarp, known as the Central Kamchatkan, the thickness of tuffogenous, volcanic and terrigenous deposits of Palaeogene and Lower and Middle Miocene ages reaches 7000m. In the second, Eastern Kamchatkan downwarp the thickness of Upper Oligocene and Miocene sediments reaches 8000–10,000m. The succession in the second downwarp is described on p. 771. The flysch character of some of the infilling formations indicates the uplift of the ranges adjacent to the downwarp.

The Kuril islands basically consist of volcanogenic and sedimentary Neogene and Anthropogene strata. Upper Cretaceous rocks are only found in the extreme south-east. On the Shikotan islands the Upper Cretaceous succession consists of andesites, agglomerates and tuffites at the bottom collectively forming the Shikotanian (Matakotanian) Suite of 400m in thickness, while towards the top lie the

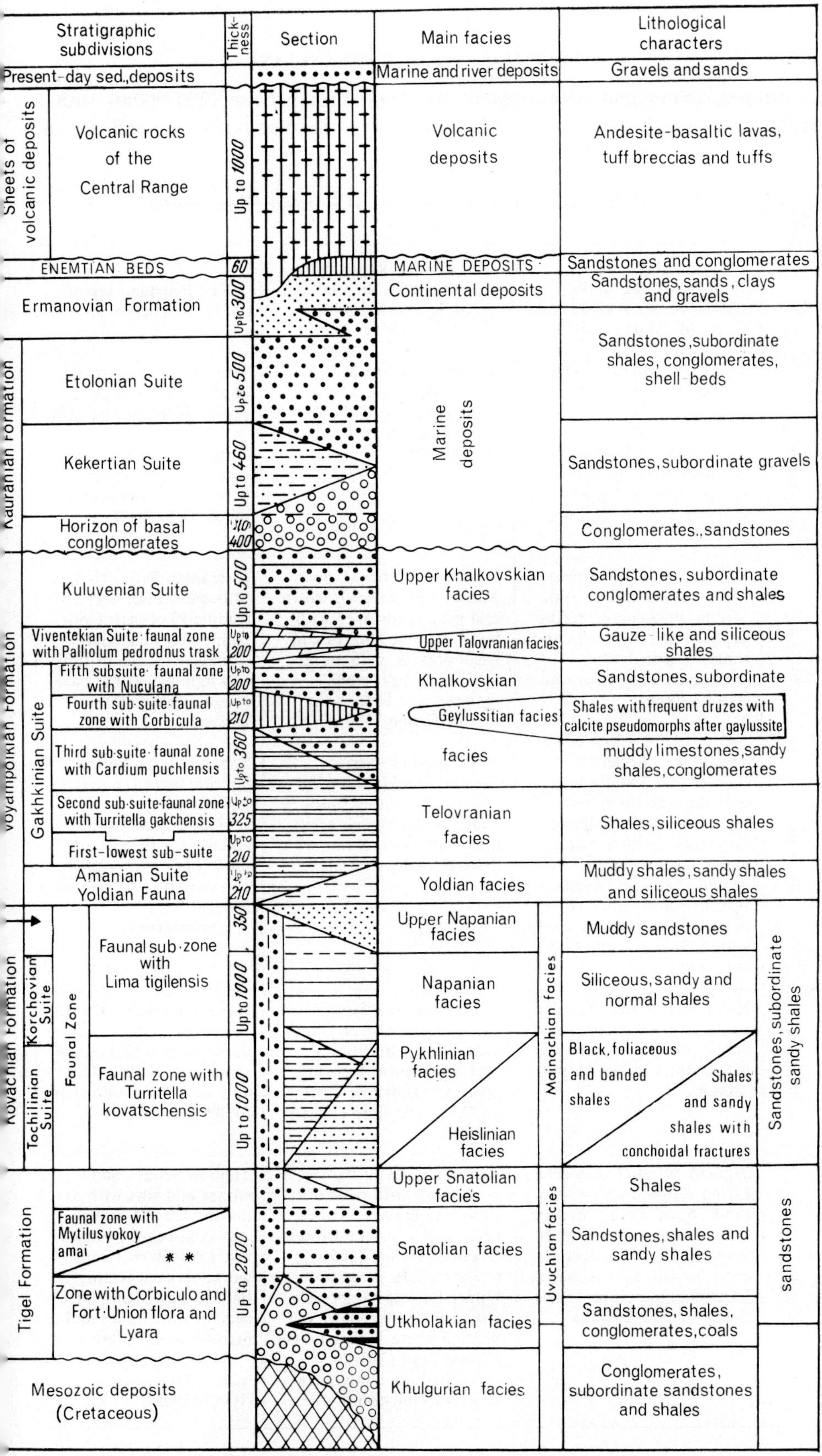

Stratigraphic subdivisions	Thick-ness	Section	Main facies	Lithological characters
Present-day sed., deposits			Marine and river deposits	Gravels and sands
Sheets of volcanic deposits — Volcanic rocks of the Central Range	Up to 1000		Volcanic deposits	Andesite-basaltic lavas, tuff breccias and tuffs
ENEMTIAN BEDS	60		MARINE DEPOSITS	Sandstones and conglomerates
Ermanovian Formation	Up to 300		Continental deposits	Sandstones, sands, clays and gravels
Kauranian formation — Etolonian Suite	Up to 500		Marine deposits	Sandstones, subordinate shales, conglomerates, shell-beds
Kekertian Suite	Up to 460			Sandstones, subordinate gravels
Horizon of basal conglomerates	up to 10 / 400			Conglomerates, sandstones
Voyampolian formation — Kuluvenian Suite	Up to 500		Upper Khalkovskian facies	Sandstones, subordinate conglomerates and shales
Viventekian Suite · faunal zone with Palliolum pedrodnus trask	Up to 200		Upper Talovranian facies	Gauze-like and siliceous shales
Gakhkinian Suite — Fifth subsuite · faunal zone with Nuculana	Up to 200		Khalkovskian	Sandstones, subordinate
Fourth sub-suite · faunal zone with Corbicula	Up to 210		Geylussitian facies	Shales with frequent druzes with calcite pseudomorphs after gaylussite
Third sub-suite · faunal zone with Cardium puchlensis	Up to 360		facies	muddy limestones, sandy shales, conglomerates
Second sub-suite · faunal zone with Turritella gakchensis	Up to 325		Telovranian facies	Shales, siliceous shales
First-lowest sub-suite	Up to 210			
Amanian Suite Yoldian Fauna	up to 210		Yoldian facies	Muddy shales, sandy shales and siliceous shales
Kovachian Formation — Korchovian Suite — Faunal sub-zone with Lima tigilensis	350		Upper Napanian facies	Muddy sandstones
Faunal Zone	Up to 1000		Napanian facies	Siliceous, sandy and normal shales
Tochilinian Suite — Faunal zone with Turritella kovatschensis	Up to 1000		Pykhlinian facies	Black, foliaceous and banded shales
			Heislinian facies	Shales and sandy shales with conchoidal fractures
Tigel Formation — Faunal zone with Mytilus yokoy amai	Up to 2000		Upper Snatolian facies	Shales
			Snatolian facies	Sandstones, shales and sandy shales
Zone with Corbiculo and Fort·Union flora and Lyara			Utkholakian facies	Sandstones, shales, conglomerates, coals
Mesozoic deposits (Cretaceous)			Khulgurian facies	Conglomerates, subordinate sandstones and shales

Additional facies labels in the diagram: Mainachian facies; Uvuchian facies; Sandstones, subordinate sandy shales; sandstones.

FIG. 269. Palaeogene and Neogene succession of the Tigil region, western coast of Kamchatka. According to B. F. D'yakov (1957).

sandstones, shales and limestones of the Malokurilian Suite (200–300m) with a Senonian fauna. No Palaeocene or Eocene deposits have so far been found. On the

TABLE 58.　*Cainozoic Succession of Western Kamchatka*

	The Tigil region (B. F. D'yakov, 1957; L. V. Krishtofovich and N. M. Markin, 1959)	The Palanskian region (M. F. Dvali, 1957; N. K. Arkhangel'skii, 1959)	The Penzhian region (A. D. Kochetkova, 1959)
Anthropogene	Volcanic formations (Up to 1000m).	Sedimentary fluviatile and marine deposits. The volcanic complex of the Central Kamchatkan Range (Up to 200m).	Fluviatile sands, gravels and sandy clays, deposits of marine terraces.
Anthropogene	Ermanovkian Formation of sandstones and clays; it is lignitiferous (Upper Pliocene-Lower Pleistocene (300m).		Ermanovkian Suite of continental sandstones, gaizes, tuffs and seams of lignite (300–1500m).
Pliocene	Kavranian Series (Upper Miocene and Lower and Middle Pliocene)—marine deposits of sandstones and shales (Up to 1000m). with basal conglomerates (Up to 400m).	Kavranian Series (Upper Miocene-Pliocene)—brown and grey sandstones with beds of clays, lignites and basal gravels, with an admixture of tuffaceous material and beds with marine faunas (300–350m).	Kavranian Series (Upper Miocene-Lower and Middle Pliocene). Conglomerates and sandstones with beds of tuffs and lignites. At the base, conglomerates (900–1600m).
Miocene	Vayampolkian Series (Lower and Middle Miocene and the lower part of Upper Miocene)—above, sandstones with a marine fauna; in the middle, repeated shales and sandstones; below, thick siliceous and normal shales with a marine fauna (Up to 2500m).	Vayampolkian Series (Lower and Middle Miocene)—alternating sandstones and greenish argillites, with concretions with marine fossils (600m).	Vayampolkian Series— liparitic lavas (andesites) (>700m). (Tuff breccias and andesites with beds of sandstones, diatomites and barren clays (Up to 2000m). Unconformity
Oligocene	Kovachian Series (Middle and Upper Oligocene)— above, muddy sandstones; below, thick shales (Up to 2300m).	Cape Kinkilskii Suite (Middle and Upper Oligocene)—lavas and tuff breccias; below, tuffs and greenish sandstones with an Oligocene flora (Up to 2500m).	Kovachian Series (Middle and Upper Oligocene)— dark shales with beds of sandstones with a flora and a marine fauna (Up to 2000m).
Palaeocene and Eocene	Tigilian Series (Palaeocene–Lower Oligocene)—above, sandy shales with a marine fauna; main part of sandstones with carbonaceous clays or with marine bands; basal conglomerates at the bottom (1500–2000m).	Tigilian Series of various sandstones with beds of clays and gravels. Above, beds with a marine Lower Oligocene fauna (500m.); in the middle, massive Upper Eocene sandstones (400m.); below, coal-bearing strata with a Lower Eocene flora (300m). At the base conglomerates and sandstones (70m).	Tigilian Series—sandstones and silts with a marine fauna (1400m). A coal-bearing formation (500–600m.) with workable coal seams. Below, dark shales with beds of sandstones and a marine fauna (1300m). At the base, sandstones with a Palaeocene flora (1000–1200m).

northernmost island a thick (2500m) Middle Paramushirian Suite of tuffs, tuff breccias, andesites and black clay and siliceous shales at the bottom, is thought to be of an Oligocene age. These rocks are overlain by the Iturupian Suite (1500–2000m) of a Middle Miocene age. The suite consists of rhythmically alternating tuffs and diatomites with tuff-braccia members. The Uper Miocene has not been recognized and the following Uttesian Suite (1000m) is of Lower and Middle Pliocene ages. The suite consists of andesites, andesite-basalts, tuff breccias and tuffaceous sandstones, which change rapidly and have a marine fauna.

The succession indicates that in the Kuril zone vulcanicity continued from the beginning of Upper Cretaceous times, through Palaeogene, Neogene and Anthropogene. Amongst clastic sediments marine fractions predominate, while rarer continental formations with floras also occur (Nikol'skii, 1959).

On the Komandor islands according to Yu. V. Zhegalov (1959) there is only a single Komandorskian Series of sedimentary-tuffogenous and volcanic deposits which is about 6000m thick. The series is divided into four suites. The lowest suite has an Upper Oligocene fauna, but the upper three are strongly dislocated and are attributed to the Miocene.

In the north-east of the Alpine zone in the region of the Penzhian inlet, the Koryakskii range, and the Anadyr Depression, Upper Cretaceous and Palaeogene strata are widespread, several kilometres thick, folded, metamorphosed and heavily intruded, this being typically geosynclinal. Their typical feature is the presence of well developed Upper Cretaceous and Palaeogene coal-bearing formations. Their fauna and flora have not been as yet sufficiently studied but indicate the presence of Senonian, Danian Palaeocene, Eocene and Oligocene strata. The Upper Cretaceous sequence involves frequent formations of siliceous and cherty-tuffogenous shales as well as sandstones and shales with black coal seams. Sandy and shaly sediments and lavas are dominant Palaeogene lithologies. The succession has been described in the Conference Proceedings (Materialy), held in Okha (1961).

FACIES AND PALAEOGEOGRAPHY. In the Cimmeridian zone only continental deposits are found. The absence of thick conglomerates and coarse sandstones suggests the absence of considerable highland areas and the flat nature of the land, often covered by lakes and swamps.

Towards the Alpine zone thick lava formations suggest large faults accompanying slow downward movements. Such movements did not produce any appreciable changes in the configuration of sea and land and were compensated by the accumulation of volcanic and terrestrial sedimentary deposits.

In the Alpine zone sea and land conditions often succeeded each other producing complex changes of the shoreline which, as has already been mentioned, did not move over the boundary between the Cimmeridian and Alpine zones of the geosyncline, either to the north or west (Fig. 270). Numerous vast coal-bearing formations were deposited over extensive near-shore plains and on low, swampy alluvial land. Bituminous and siliceous shales were deposited in stagnant basins constituting lagoons, separate embayments and straits. In marine sediments the negligible quantity of carbonates and the overwhelming preponderance of a fine-grained and medium-grained clastic fraction is typical. The many kilometres thick formations of clastic rocks were produced as a result of intensified transport of fragmentary material by a well-developed river system. Lastly the unusually widespread development of volcanic rocks depends on the large number of volcanoes of different types, but mostly terrestrial.

The Neogene and Anthropogene

Neogene and Anthropogene strata of the Cimmeridian zone are of a platform type, differing from Upper Cretaceous and Palaeogene rocks by the absence of folding and metamorphism and by a sharp reduction in the intensity of vulcanicity. The deposits are usually thin, not exceeding 100–200m, but in isolated cases as at the foot of rising mountains can reach 1000–1500m. Marine sediments are absent as a rule, but in the north large areas are overrun by the Boreal Transgression, which is still continuing.

In the Alpine zone Neogene and Athropogene strata are geosynclinal, differing from Upper Cretaceous and Palaeogene rocks only by less complex folds and a lower metamorphic grade. Otherwise both groups of rocks are affected by regional folding, by vast outpourings of lavas, by the presence of synorogenic intrusions of considerable size, by the repetition of marine, continental coal-bearing and volcanic rocks, by mineralization and by the generation of coal and oil.

DISTRIBUTION. Neogene strata are much less widespread than Upper Cretaceous or Palaeogene rocks.

In the north of the Far East, in the basins of the Zeya and its tributary the Selemdzha horizontal Pliocene sediments occur over vast areas. River valleys are covered by Upper Anthropogenic deposits, occupying especially vast areas near Khabarovsk at the confluence of the Amur and the Ussuri. Here Lower and Middle Anthropogenic deposits underlie Upper Anthropogenic sediments. Somewhat to the south, on the western slope of the Sikhote-Alin' range and in its southern part there are small areas where continental, in places coal-bearing, Pliocene deposits occur.

In the north-east very large areas of Anthropogenic strata occur along the shore of the Arctic Ocean, stretching from the lower reaches of the Khatanga to the Chaun inlet. Along the northern edge of the Cherskii range continental coal-bearing Pliocene sediments emerge from underneath a cover of Anthropogenic deposits. In certain depressions along the southern edge of the range such sediments have also

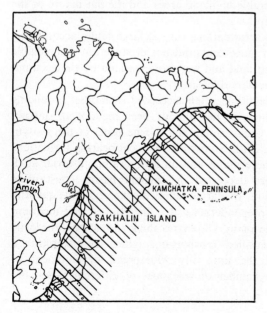

been preserved. In the valley of the Yana and its tributaries Anthropogenic glacial deposits are found over large areas.

It is interesting that at the shores of the Arctic Ocean there are almost no Anthropogenic marine deposits. This is explained by the fact that the northern, Boreal Transgression invades the continent and at present it is at

FIG. 270

Palaeography of Senonian times. According to V. N. Vereshchagin (1957).

its maximum. The Laptev and the Eastern-Siberian seas represent this transgression, which is obvious from submarine contours. The Novosibirsk islands and Wrangel island are upper parts of relatively low uplands, while the edge of the continent is outlined by 200 and 400m depth contours.

In the Alpine zone of Sakhalin, the Kuril islands, Kamchatka and the Koryak-skii range Miocene, Pliocene and Lower Anthropogene deposits are folded and ubiquitously exposed. A large area in Northern Sakhalin is occupied by Miocene and Pliocene strata and is a principal oil-bearing region.

SUBDIVISIONS AND LITHOLOGY. In the basin of the Zeya and its tributary the Selemdzha there are widespread Pliocene strata consisting of horizontally lying unconsolidated sands gravels and clays a few tens of metres and sometimes reaching 100m in thickness. Locally these rocks are known as 'Vodorazdel'nye Sands' and it is possible that a part of them is Anthropogenic. Analogous alluvial deposits of sug-gestedly Lower Anthropogene and Pliocene age are known from the region of Khabarovsk and along the lower reaches of the Amur. In the estuaries of the Zeya and the Bureya such sands lie on clays with a Palaeocene flora, the clays succeeding the coal-bearing sediments of the Zavitinskian coal region (the Kivda and the Raich-ikha). Miocene, Oligocene or Eocene strata have so far not been found. Miocene strata are even unknown from the valley of the Ussuri or the lower part of the Amur. In the north-east continental Neogene and Anthropogene formations are found in northern provinces, where they occur over large areas, but are relatively thin. As an example one can quote sands and clays, with seams and lenses of brown coals and lignites, developed as a sequence of a few tens of metres thick along the northern edge of the Cherskii range, side by side with a Lower Cretaceous coal-bearing formation. The Pliocene and Lower Anthropogene age of the sequence has now been proved.

Anthropogene deposits of the extreme north—the shore of the Arctic Ocean—are widespread and have two particular features: they have well developed Tundra deposits and very limited marine sediments. In addition there are also fluvial, lacustrine swampy and fluvioglacial deposits. Tundra deposits represent a relatively thick (up to 100–120m) formation of muds or silts, which are bedded and contain lensoid layers of ice, sometimes reaching 10–20m in thickness. Bodies of mammoth and woolly rhinoceros can be found in this ice. The bodies are so well preserved that soup was cooked from mammoth meat, but was not of a high quality. It is thought that these animals wandered over boggy ground, overgrown by moss and by sinking into hidden bogs froze to death. Later such bogs with their ice and mammoth were covered by clayey soil and overgrown by moss and forest and preserved the animals. The ice is probably the result of compaction of large quantities of snow accumulating over centuries on vast plains. Such a process of compaction occurs in places at the present time. Marshes and bogs, into which mammoths sank, originated on the surface of the previously formed ice sheet. The Quaternary buried ice is a specific feature of the Anthropogene of the extreme north. The first human beings in the area lived at the same time as the mammoth.

In mountain ranges and adjacent areas the Quaternary Glaciation was wide-spread and consequently glacial deposits occupy quite large areas to the north of the Cherskii range and on both slopes of the Verkhoyansk range, which is shown on synoptic maps. The deposits consist of morainic material, boulder clay and transi-tional fluvioglacial sands and clays with partly sorted mud and pebbles. The thick-ness of such deposits reaches 100–150m.

In Eastern Transbaikalia the Neogene is divided into three suites of a total thickness of the order of 100–150m. The lower (Undinskian) suite consists of red beds of alluvial, eluvial and deluvial origins and contains bones of horses and *Hipparion*.

The middle suite (Kholbonskian) is grey and consists of lacustrine—alluvial sediments. The upper suite (Ingurukian) is red reaching a thickness of 25m and consists of sands, gravels and pebble beds of deluvial and alluvial origins.

Lower and Middle Pleistocene rocks are not widespread and have not been much investigated. On the whole lacustrine and alluvial, sometimes relatively thick, sediments predominate. In places there are morainic sediments and basaltic lavas. Upper Pleistocene deposits on the other hand, are widespread and in several regions have a cold mammoth fauna. The alluvium of lower terraces, glacial deposits, loessic loams and lacustrine deposits belong to the Upper Pleistocene.

The Holocene is represented by the sediments of the present-day type such as peats and deposits of lower terraces of rivers and lakes.

In the Alpine zone the nature of Neogene and Anthropogene strata is very different since there are widespread marine and volcanic deposits of geosynclinal type.

In the Sikhote-Alin' range and along its western slope there are no marine deposits but there are widespread Pliocene and Anthropogene basalts and andesite basalts shown on synoptic maps. Along the sea shore there are local terraces with sediments yielding marine faunas. But such terraces are near to the shore and are low (6–10m). Sometimes sands with modern marine molluscs lie on basalts and sometimes such sands are in turn covered by basalts.

The Neogene is represented by continental, often lignite-bearing, deposits and associated lavas which in places reach a considerable thickness of up to 800m. To the Miocene belongs the lignitiferous Ust'-Davydovskian Suite of Southern Primoria. The suite consists of sandstones and siltstones with seams of lignite at the top and a Miocene flora. The suite is 650–700m thick and is folded into very gentle folds. Higher up lies the Pliocene Suifunian Suite, which consists of sands, gravels, diatomites and tuffs of a total thickness of 100m. The suite lies almost horizontally and in places passes laterally into lavas (liparites). Some palaeobotanists (A. N. Krishtofovich and M. I. Borsuk) consider its flora as of a Miocene age and similar to the flora of the Ust'-Davydovskian Suite, but its general relationships are more suggestive of a Pliocene age (Table 56).

Anthropogene strata are widespread, but only alluvial and volcanic complexes are thick. Alluvial deposits of Anthropogene age can be divided into three groups: the Lower and Middle Anthropogene high terrace (up to 150m) deposits, the Upper Anthropogene sediments and the modern sediments. Alluvial deposits are found in the basin of lake Khanka, along the valley of the Ussuri, at the confluence of the Ussuri and the Amur, and along the Amur. The alluvium consists of cross-bedded sands and gravels of river courses, horizontally-bedded sands, clays and peats of river valleys and fine muds and silts of lakes; as well as of other deposits. Volcanic formations consist of flood basalts, andesites and associated tuffs, reaching altogether 200–300m in thickness (p. 786).

On Sakhalin Neogene strata are petroliferous and thus have been studied in most detail. A comparison of successions in different regions is shown in Table 57. Like Palaeogene rocks, Neogene strata represent a repetition of three facies complexes: marine, coal-bearing (continental) and volcanic. The total thickness of Neogene deposits is up to 5000–6000m (Fig. 271) and they are folded and faulted. A rich and diagnostic fauna permits a detailed subdivision of the succession, which owing to the nature of the fauna cannot easily be correlated with the European. Coal-bearing formations contain numerous plant remains, which are again peculiar to the Far East, and brown coal and lignites. Coal, however, is in a subordinate position in comparison to oil.

Lower Miocene rocks consist mainly of submarine basalts and andesites and dominantly pyroclastic tuffs, tuffites and tuff-breccias. In the north of the island the thickness of these rocks is in places only 300m and elsewhere there are no Miocene strata at all. In the south Lower Miocene rocks thicken to 3900m and are divided here into four suites. The lower two suites, the Gastellovskian and the Arakaian (1000–2500m), consist of tuffs, tuff agglomerates, basalts and andesites with thin bands of argillites with marine faunas. The next, Kholmian Suite (800–1000m) is mainly clayey and has a marine fauna. The Nevelskian Suite (800–1500m) consists of a rhythmic, flysch-like sequence of thin tuffogenous sandstones, argillites and tuffs (Fig. 272). The Chekhovskian Suite which caps the succession again consists of tuffs, tuff conglomerates, basalts and andesites with beds of tuffaceous sandstones with marine faunas, belonging to the Lower Middle Miocene.

Middle Miocene strata are represented by four suites of different facies. The Verkhneduian Suite (south-west of the northern half of Sakhalin) consists of continental deposits richest in coals. In Southern Sakhalin the equivalent of this suite is known as the Uglegorian which is sometimes coal-bearing and sometimes marine. The Dagian Suite (north-east of the island) is most petroliferous and consists of marine sediments at the bottom, coal-bearing deposits in the middle and again marine sediments at the top. Its thickness reaches 1900m and altogether with the Upper Miocene Okobykaiian it forms the most important oil-bearing formation of Sakhalin. The Upper Langerian Suite in the north-west of the island is also basically marine.

The Upper Miocene is represented by a series of mainly marine suites. The Okobykaiian Suite (up to 1700m) consists of bluish-grey and dark grey clays with layers of sands which become dominant towards the top. The Sertunaiian and the Aleksandrovkian suites (1500m) are of the same lithology, but their rocks are paler and more sandy. The Kurasian Suite (600m) consists of grey and pale grey siliceous (diatomitic) and clayey rocks.

The Nutovian Suite belongs to the Pliocene, is thick (2500m) and sandy, the sands being often cross-bedded and having local beds of clay, fine grained conglomerates and brown coals. The rocks are basically continental but have marine bands. Other equivalent, contemporaneous suites such as the Maruyamian, the Tamlevian and the Rybnovskian are also mainly sandy and similar to the Nutovian Suite. The Orlovskian Suite of Southern Sakhalin consists of basalts and tuffs of a total thickness of 300m.

In the south-eastern part of Northern Sakhalin S. S. Razmyslova and K. F. Rakhmanov give a simplified succession. They place all Lower Miocene strata into a single Pilengian Suite (1500m), all Middle and Upper Miocene into a single Borian Suite (2500m) and divide the Pliocene into a lower (Uropaiian) suite (450m) and an upper (Khuzian) suite (800m).

The Pilengian Suite consists of flaggy, siliceous argillites which are places gaizes and are characteristically jointed. The joint planes are regarded as spaces for accumulation of oil. Lithologies of other suites are similar to those developed in other parts of Sakhalin.

FIG. 271

A thick Neogene sequence of sandy-shaly rocks is folded in Southern Sakhalin, giving rise to mountains.
Photograph by K. P. Yevseyev.

FIG. 272

Lower Miocene flysch-like rocks in Southern Sakhalin.
Photograph by K. P. Yevseyev.

In Kamchatka Neogene and Anthropogene strata are the same as in Sakhalin and especially so on the western shore of the peninsula. The eastern shore has widespread volcanic activity which is still continuous.

On the western shore the Oligocene Kovachian Series of shales is conformably succeeded by the Vayampolkian or Belesovatian Suite (Series) of 2500m in thickness, which consist of dark, whitish-weathering siliceous clays and muddy sandstones with beds of conglomerates and a marine Miocene fauna. In places these sediments pass into continental yellowish clays and sandstones with plant remains. The age of the suite is Lower and Middle Miocene. Higher up lies the unconformable Kavranian Suite of sandstones, conglomerates and clays which are succeeded by a coal-bearing sequence. In places the Ermanovkian Suite of continental sandstones, sands and clays with lenses of brown coals and lignite represents the whole of the Plio-Pleistocene sequence.

Anthropogenic deposits lie unconformably on Pliocene strata, are weakly folded, reach a thickness of over 100m, and consist of repeated marine and alluvial sediments, resembling the modern.

On the western shore the thickness of Neogene strata succeeded by the Lower Anthropogene reaches 8000–10 000m. The predominance of continental sands and gravels with rarer clay sequences is characteristic. The sediments are interstratified with lavas (mainly basalts) of up to 300m in thickness and with formations of tuffs and tuffogenous rocks. Marine faunas are relatively rare. The succession is divided into the Bogachevkian Oligocene-Lower Miocene Suite (6000–7000m) and the Tynshevkian Suite (3000m) of a Miocene-Lower Pliocene age (Gryaznov, 1959).

Anthropogenic strata of eastern and central Kamchatka consist almost exclusively of lavas, andesites, basalts and tuffs while beds of sedimentary deposits are rare. Along the shore there are terraces with marine fossiliferous sediments at heights of up to 400–1000m. Glacial deposits are interesting.

In the valley of the Anadyr and in the Koryakskii range the Neogene and Anthropogene succession is the same as in Kamchatka, but sections and faunas have not been sufficiently worked. It can be pointed out that in western regions there are repeated alluvial and marine deposits and subordinate volcanic formations, while in the east volcanic rocks predominate and are interbedded with continental sediments and marine strata are rare. Lavas and tuffs occur over large areas. In the Anadyr valley alluvial deposits are widespread and terraces with marine sediments are quite common. In the Koryakskii range widespread glacial and interglacial deposits are typical. The first glaciation occurred in the Middle Anthropogene epoch. The Upper Anthropogene begins with interglacial sediments, which are overlain by the moraines of the second glaciation. The present-day glaciation is on a small scale.

FACIES AND PALAEOGEOGRAPHY. The facies are basically the same as they were in Upper Cretaceous and Palaeogene times. Marine facies are shallow-water, terrigenous and rapidly alternating with the deposits of near-shore alluvial plains. Coal-bearing facies, associated with such plains, are inconstant. Volcanic rocks are generally terrestrial and are associated with central and fissure eruptions.

The palaeogeography of the Cimmeridian zone was governed by a continental regime. Thus at the end of Anthropogene times the continent moved down and the ocean transgressed from the north, giving rise to the present-day Laptev and Eastern-Siberian seas. At the same time the Bering straits, the Gulf of Anadyr and the Penshian and Gizhiga inlets as well as the shallow western part of the Sea of Okhotsk came into existence. All these stretches of sea have depths of less than 200m. In the Alpine zone sea and land interchanges continued. The shore line was subject to oscillations and volcanic activity intensified.

TECTONICS

The tectonics of the Pacific Ocean Geosyncline as of other Meso-Cainozoic geosynclines bear evidence of strong manifestations of the Cimmeridian and Alpine orogenies. Thus Cimmeridian and Alpine structures are intense, complex and very widespread.

STRUCTURES

Five structural storeys are distinguished in the area. The first includes Precambrian and Lower Palaeozoic strata. In some regions Lower Proterozoic and Archean gneisses and crystalline schists were found to be unconformably overlain by weaker metamorphosed Upper Proterozoic and Cambrian rocks. There is even an unconformity between Cambrian and Orovincian strata (the Salairian phase). The unconformities are considered to be surfaces of separation between several sub-storeys which as yet have not been sufficiently investigated.

The second storey is Middle Palaeozoic and is in places divided by the last phase of the Caledonian Orogeny, into two sub-storeys: the first consisting of Silurian and Lower Devonian strata and the second of Middle and Upper Devonian and Lower Carboniferous strata.

The third storey involves Upper Palaeozoic and Lower and Middle Trias rocks— a time span corresponding to the length of the Hercynian Orogeny. Middle and Upper Carboniferous sediments differ considerably from Permian and Triassic rocks, and can, therefore, be separated into a distinct sub-storey. Local unconformities between Upper Permian and Triassic strata suggest other sub-storeys.

The fourth storey consists of Upper Trias, Jurassic and Lower Cretaceous beds— a span of time corresponding to the duration of the Cimmeridian Orogeny. The rocks of this storey constitute a very thick (12–15km) homogeneous sand-shale sequence which as yet cannot be subdiveded into sub-storeys.

The fifth storey involves Upper Cretaceous to Recent rocks and represents the epoch of the Alpine Orogeny. In the Cimmeridian zone the storey is divided into two sub-storeys of which the lower sub-storey consisting of Upper Cretaceous and Palaeogene rocks shows distinct gentle large folds, while the rocks of the upper sub-storey—Neogene and Anthropogene strata—remain almost horizontal. In the Alpine zone the two sub-storeys also exist, but are much less distinctive and are only separated by minor unconformities. Such minor unconformities are also present within the two sub-storeys indicating that there were several phases of Alpine Orogeny. The phases and the resultant structural complexes have different manifestations from district to district. The last phase embraces Anthropogenic sediments, which are often involved in obvious, although simple and gentle folds.

The first storey structures are very complex and strongly deformed and are not shown in synoptic maps. The affected rocks are highly metamorphosed and are sometimes recrystallized. The folds shown in Fig. 257 have been produced in Precambrian rocks of the Lesser Khingan range and can serve as an example of first storey structures. The structures of the second storey are similar to those affecting Lower Palaeozoic and Upper Proterozoic strata, but the rocks are less metamorphosed. Again such structures are complex, broken up into separate blocks and on synoptic maps are shown rarely, although they can be seen on the Shantar islands.

The third or the Hercynian storey differs sharply from the preceding two. Its structures have a wide distribution and for instance in the Verkhoyansk range affect large areas. Along the margin of the range narrow regular folds, which have also been affected by the Cimmeridian Orogeny, can be clearly seen. Thus either the Hercynian Orogeny was weak or its structures have been later completely inherited by the fourth storey. Verkhoyanskian structures are clearly indicated on synoptic maps.

The fourth storey—the Cimmeridian—has a ubiquitous widespread development and can be clearly identified on synoptic maps of the Cherskii and Gydanskii ranges, and in the lower reaches of the Amgun. In the Cimmeridian zone the fourth storey structures are terminal, and occur as present-day folds. In the Alpine zone, however, they are overprinted by intense Alpine structures and are therefore distorted and complicated. The folds affecting the Lower Cretaceous coal-bearing formation of the Bureyan and Suchanian coal-fields are shown in Figs. 266 and 273.

The fifth storey differs from one zone to the other. In the Cimmeridian zone it is represented by large, gentle folds, affecting only weakly metamorphosed Upper Cretaceous and Palaeogene strata. Such folds are found in the Kivdinian coal-bearing series outcropping on the shore of the Amur to the east of Blagoveshchensk. Neogene and Anthropogene rocks are not folded and only in places are faulted or elevated and depressed on Mesozoic and Palaeozoic Blocks.

In the Alpine zone both sub-storeys are folded, often into complex isoclinal and fan folds and are broken by thrusts into separate scales. Such structures can clearly be seen on maps of Sakhalin and the western coast of Kamchatka and in cross-sections (Fig. 274). Folds apart, there are, especially in the Neogene, also common block structures, which can be clearly seen on maps of the eastern coast of Kamchatka.

Deep faults represent one of the most important features of the fifth storey. Such faults penetrate the Earth's crust for tens of kilometres and by tapping the magmatic belt become channels of vulcanicity accounting for the spread of volcanic rocks. On the surface the faults are covered over large areas by lavas and tuffs, which precludes the faults being shown on maps.

From the upper reaches of the Zeya to the Chukotian Peninsula there stretches the Eveneian volcanic belt of Upper Cretaceous and Palaeogene effusions. The belt is one of the largest volcanic zones on the Earth's crust. It is absolutely clear that contemporaneous faults, grouped as the Evenian belt of Upper Cretaceous-Palaeogene faults, were imperative for the origin of the volcanic zone. On synoptic geological maps these faults are only shown in places, as for instance to the north and east of Magadan.

A second—Primorian—belt of deep faults is of Pliocene-Anthropogene age and consists of two branches, the Sikhote-Alinian and the Kuril-Kamchatkan. The faults of these branches guided colossal eruptions of basaltic and andesitic lavas. The faults of the Sikhote-Alinian branch are at present inactive, but those of the Kuril-Kamchatkan branch are still open under a considerable number of active volcanoes. In the north the Kuril-Kamchatkan branch trends towards the south-east of the Koryakskii range and in the south it continues to Japan. The connection of the Komandor and Aleutian islands with Kamchatka is of interest. The Aleutian islands also represent a zone of deep faults with active volcanoes over it. In the Komandor islands (U.S.S.R.) there are no active volcanoes but there are abundant lava flows. It is possible that the eastern zone of Southern Kamchatkan volcanoes bends northeastwards and trends through the Komandor to the Aleutian islands (Fig. 275). The

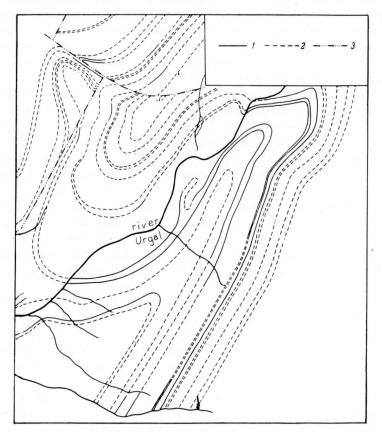

FIG. 273. Folds in the Lower Cretaceous coal-bearing sequence of the Middle Urgal deposit in the Bureyan basin. According to A. K. Matveyev.

1 — traced coal seam; 2 — conjectured (unsurveyed) coal seam; 3 — faults
The similarity to folds of Donbas (Fig. 240) attracts attention.

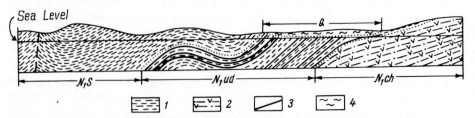

FIG. 274. Structures affecting Neogene strata of Sakhalin. According to E. M. Smekhov (1953).

1 — argillite; 2 — tuffogenous siltstone; 3 — coal; 4 — loams; Q — Quaternary; N_1s — the Sobolevskian Suite; N_1ud — the Upper Duiian (Verkhneduian Suite); N_1ch — the Khoindzho Suite.

western zone of Kamchatkan volcanoes continues northwards towards the Olyutorian faults and thereafter trends again south-eastwards around the deep depression of the Bering Sea (Fig. 275). Kamchatkan and Kuril structures have been described by P. I. Kropotkin (1961). The Kuril-Kamchatkan line is a part of a very large belt of deep faults which margin the northern part of the Pacific Ocean and cross it in the south and north as it is shown by the abundance of coral atolls and volcanic islands.

The metamorphism of the region is typical of Meso-Cainozoic geosynclines. In the Cimmeridian zone the Cimmeridian Orogeny produces in Mesozoic, say Jurassic, strata the same intensity of metamorphism as one associates with Middle Palaeaozoic strata of Palaeozoic geosynclines. The metamorphic grade of coal-bearing formations of the Bureyan coal-field and in Eastern Transbaikalia is the same as that of the Palaeozoic coal-bearing sequence of Kuzbas. The Palaeogene Danian coal-bearing formation in the lower reaches of the Bureya is as altered as the Jurassic coal-bearing formation of Kuzbas, and Karaganda. Neogene and Anthropogene strata

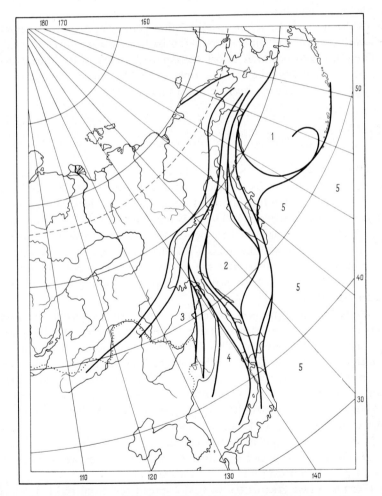

FIG. 275.   Zones of Alpine faults, fold belts and suggested massifs.

1 — Bering Massif;   2 — Okhotsk Massif;   3 — Khingan Massif;   4 — Japan Masif;   5 — Pacific Ocean Massif.

of the Bureyan basin are not folded or metamorphosed as is also the case with Neogene strata of Kuzbas.

In the Alpine zone the powerful Alpine Orogeny has caused an intensive metamorphism in Upper Cretaceous and Palaeogene rocks and a considerable or moderate metamorphism in Neogene and Anthropogene strata. Upper Cretaceous shales and sandstones of Sakhalin have been just as affected by metamorphism as Lower Carboniferous coal-bearing shales and sandstones of Karaganda. Miocene rocks of Sakhalin are often more altered than Jurassic rocks of Karaganda.

Such relationships of metamorphic grades can be considered normal, but in addition there are also deviations towards intensified metamorphism resulting in metamorphic and crystalline Mesozoic rocks. Such phenomena are identical to those observed in Palaeozoic geosynclines. On the eastern slope of the Central Urals Middle Palaeozoic strata are converted into green schists, their limestones have become marbles, their sandstones quartzites, and their shales and lavas are schists. A detailed investigation of green schists has shown that they pass laterally into normal Middle Palaeozoic deposits.

Such sudden localized rises of metamorphic grade can also be observed in the Pacific Ocean Geosyncline. In Eastern Transbaikalia Lower Jurassic strata are in places altered into schists which before the detailed survey in the thirties were considered to be Lower Palaeozoic and Precambrian and the Upper Palaeozoic and even Mesozoic age has only recently been demonstrated.

The History of Tectonic Movements

PRECAMBRIAN OROGENIES. The Pacific Ocean Geosyncline differs from the Mediterranean Geosyncline by a much larger exposure of its Precambrian rocks, of several complexes, differing amongst each other by their grade of metamorphism and structural complexity. Thus several ancient orogenies can be reliably identified. Archeozoic orogenies can be recognized from angular unconformities between gneisses and crystalline schists on one hand and the overlying Jaspilitic Series of a likely Lower Proterozoic age on the other. It is possible that such unconformities occur in the southern part of the Sikhote-Alin' range and in the Okhotsk and Omolon massifs.

The Karelidian Orogeny was widespread and can be recognized everywhere where two Proterozoic complexes differing by their intensities of folding and metamorphic grades, can be detected. This situation can be observed over a large area in the upper reaches of the Zeya, the Selemdzha and the Bureya where a complex of highly metamorphosed rocks is overlain by less metamorphosed rocks with a Cambrian fauna at the top. In the south of the Sikhote-Alin' range the Karelidian Orogeny conditions the possibility of recognizing the Sinian complex which is shown as Upper Proterozoic or Lower Cambrian.

The Baikalian Orogeny has not been recognized, but its occurrence is very likely. It should have produced an angular unconformity between diversely metamorphosed Upper Proterozoic and Lower Cambrian rocks.

THE CALEDONIAN OROGENY was widespread but from place to place manifests itself differently. In the north the last phase at the end of Silurian times results in missing Lower and Middle Devonian strata and the unconformable overstep of Givetian and even Frasnian rocks on the Silurian. In the south the complete absence of marine Ordovician strata is caused by the initial (Salairian) phase of the Caledonian Orogeny, producing widespread terrestrial conditions.

THE HERCYNIAN OROGENY was of a ubiquitous occurrence but was of a secondary importance in comparison with the Cimmeridian Orogeny. The effect of the Hercynian Orogeny in the Pacific Ocean Geosyncline was very different from that in Palaeozoic geosynclines. An almost complete absence of Upper and Middle Carboniferous rocks has not as yet been explained. It is very probable that the phenomenon was caused by the initial phases of the Hercynian Orogeny producing land on the site of the Lower Carboniferous sea. The last phases of the Palaeozoic geosynclines produced young mountain ranges, thus changing the character of sedimentation and palaeogeography. In the Cimmeridian zone of the Pacific Ocean Geosyncline the last Hercynian phases were responsible for the accumulation of the lower part of the Verkhoyansk complex, but most of the complex is associated with the Cimmeridian Orogeny. Evidently mountain building processes which conditioned the supply of the thick clastic sequence of the Verkhoyansk complex began in Lower Permian times and continued with brief intervals up to Lower Cretaceous times or namely throughout the whole period of accumulation of the Verkhoyansk complex.

In the Alpine zone at the Permo-Trias boundary there are angular unconformities, which are not associated with changes in palaeogeography. The general nature of Upper Permian and Lower Triassic sediments is the same; the sediments show similar distributions and similar structures. Such a close connection between the Upper Permian and the Lower and Middle Trias is also seen in the Cimmeridian zone. This circumstance forces one to include Lower and Middle Triassic rocks in the epoch of the Hercynian Orogeny. Upper Trias strata, on the other hand, are closely associated with Jurassic deposits and denote the beginning of the epoch of Cimmeridian Orogeny.

The Caledonian Orogeny, which drew to a close in Silurian times, is separated from the Hercynian Orogeny, which began in Middle Carboniferous times, by a long interval. The Cimmeridian Orogeny, on the other hand, follows the Hercynian and is in turn followed by the Alpine without a break.

THE CIMMERIDIAN OROGENY was, as already mentioned, the main orogeny in the Cimmeridian zone. In the Alpine zone it shares this role with the Alpine Orogeny.

The Cimmeridian Orogeny began in Upper Triassic and ended in Lower Cretaceous times. All deposits corresponding to this interval of time have the same composition and are affected by the same structures. Although unconformities and breaks are observed in this interval of time they are of local nature and are not accompanied by major differences in successions and palaeogeography. Consequently the identification of separate orogenic phases is complicated and they remain unnamed. The only exception is the Kolymian phase of folding and intrusion of synorogenic granites in Lower Cretaceous, post-Valanginian times. The main deposits of gold, tin and rare metals are associated with these granites.

The Kolymian phase is sometimes referred to as the Valanginian, but this is unlikely to be true. In Valanginian times almost everywhere one observes a transgressive *Aucella* sea. The post-Aucellan coal-bearing formation is, however, typical of a period of upward movements, mountain building and intensive erosion and accumulation of clastic material. Consequently it is more correct to associate mountain building, folding and intrusions not with the Valanginian but with the subsequent epochs, probably Hauterevian and Barremian, to which the Shilka Conglomerates are also attributed.

Apart from the Kolymian phase an intensification of mountain building and vulcanicity occurred in Upper Jurassic times. In Eastern Transbaikalia a formation of conglomerates (up to 2000–2500m) represents typical alluvial cones at the margin of young rising mountains. A vast development of volcanic rocks indicates the intensity of deep faulting. Synorogenic intrusions are also widespread. All these features separate Middle from Upper Jurassic strata and as a result they are denoted separately on maps. Consequently those phases which occurred in Upper Triassic and Lower and Middle Jurassic times are known as Older Cimmeridian and those which occurred in Upper Jurassic and Lower Cretaceous times as Younger Cimmeridian.

THE ALPINE OROGENY succeeds the Cimmeridian without a break, and consequently sometimes the two are combined and designated as the Alpine Orogeny. In practice such a usage is incorrect since it obscures the actual extent of younger orogenic phases. On such a basis one should consider that the Alpine Orogeny was powerful in both zones whereas in fact this conclusion only applies to the Cimmeridian Orogeny. The Alpine Orogeny was only powerful in the Alpine zone. In the Cimmeridian zone during the Upper Cretaceous-Palaeogene epoch it was weak and during the Neogene-Anthropogene epoch altogether inactive.

The start of the Alpine Orogeny also demands a clarification. Normally it is considered that it began in Cenomanian times. Several successions, however, indicate that the new sedimentary macrorhythm and the new transgression as well as the causative tectonic movements began not in Cenomanian but at the end of Albian times. Consequently the Alpine Orogeny began during the Upper Albian. This feature is particularly clear in the Mediterranean Geosyncline, but it is also true for the Alpine zone of the Pacific Ocean Geosyncline. For instance in the Sikhote-Alin' and Suchan coal-field the first beds with marine faunas occurring at the top of the coal-bearing formation are considered as Upper Cretaceous, but it is probable that the lowermost of them are Albian.

Alpine phases are numerous and do not always correspond with the boundaries of Western European subdivisions, but rather correspond to the boundaries of local suites. In the Sikhote-Alin' range, according to N. A. Belyayevskii, an orogenic phase precedes the Upper Albian-Cenomanian transgression. This phase is analogous to, but does not correspond completely with, the Austrian phase. In the south the Sikhote-Alin' phase is weak and sometimes unnoticeable but in the north it is powerful and is accompanied by synorogenic intrusions. The phase is provisionally called Albian. The subsequent phase, analogous to the Laramidian but again not corresponding accurately, occurred at the end of the Senonian. This phase N. A. Belyayevskii considers as the main formative phase. Subsequent phases were weaker and sometimes of only local significance. They occurred at (1) the end of the Palaeocene; (2) in the Middle Eocene; (3) the end of the Eocene; (4) in the Middle Miocene and (5) the beginning of the Anthropogene. All these phases were dated on the basis of angular unconformities and breaks of a limited extent. The absence of marine faunas renders their dating provisional.

In Sakhalin, according to E. M. Smekhov, who worked there a long time, the marine macrorhythm began in Upper Albian times and continued in the Cenomanian and Turonian epoch. A powerful orogenic phase, equivalent to the Laramide, occurred at the end of the Upper Cretaceous, but it is not manifest everywhere. The phase provoked a widespread extension of continental conditions. A subsequent macrorhythm embraces Palaeogene and Miocene strata. It was terminated by the succeeding Upper Miocene orogenic phase which is reflected in unconformities between the Okobykaiian and the Nutovian suites, causing widespread geographic changes. Then follows the end-Pliocene phase reflected in the unconformity between the Nutovian and the Supra-Nutovian suites. The phase was responsible for the land connection between Sakhalin and Sikhote-Alin' structures. At the end of Anthropogene times isolated massifs moved along faults and got submerged under the sea, thus giving rise to the present-day shoreline. A downward movement of a block on the suite of the Tartary straits caused the conversion of Sakhalin into an island. At the same time other depressions caused the formation of the western and northern, shallow parts of the Sea of Okhotsk, where drowned river valleys form submarine canyons. Contemporaneous large scale downward movements gave rise to the Bering Sea, and Bering straits, thus separating Asia from America, which before that formed a single continent.

Large faults of deep penetration became channels for numerous volcanoes. These faults are still moving and with them are associated all present-day volcanoes of Kamchatka and the Kuril islands. Tectonic phases in Kamchatka basically correspond with such phases in Sakhalin.

The present-day condition of the Pacific Ocean Geosyncline is quite interesting. Its Cimmeridian zone represents a young Neogene platform. As has already been mentioned its southern boundary continues along the Ussuri and the lower reaches of the Amur, crossing the Sea of Okhotsk towards the Penzhian inlet and then proceeding along the Penzhina and the Anadyr. Within the zone regional folding ceased at the end of Oligocene times. Only relatively shallow faults, unassociated with vulcanicity, appears later. Blocks bounded by such faults sometimes rose and sometimes were depressed. Rising blocks suffered erosion giving rise to considerable sequences of conglomerates overlain by clays and sands. It is possible that block movements were partly responsible in the formation of the present-day relief of such ranges as the Verkhoyansk, the Cherskii, the Taskhayakhty and the Gydan. It is not entirely clear what has caused large scale downward movements along the north-eastern margin of Asia resulting in the present-day Boreal Transgression and the formation of the Laptev and Eastern-Siberian seas, and also what caused the transgression resulting in the western and northern shallow-water parts of the Sea of Okhotsk. It is possible that block movements were the cause, but the possibility of a really extensive small-amplitude upwarps and downwarps cannot be excluded.

The Alpine zone serves as a typical example of a present-day geosyncline. Powerful and diverse tectonic movements reinforced the destruction of rising tectonic structures. The products of erosion accumulated on near-shore alluvial plains and in shallow-water coastal seas, giving rise to multi-kilometre clastic formations. Lavas emerged out of deep faults and powerful eruptions produced volcanic cones of several kilometres in height. The erosion of such cones produced tuffs, tuffites, and tuffogenous rocks. In places carbonate deposits, future limestones and dolomites, were formed in the shallow sea. On near-shore plains enormous swamps became sites of accumulation of plant remains to be converted into brown coal and lignite.

The existence of regions of contemporaneous accumulation of continental coal-bearing marine shallow-water terrestrial and more rarely submarine volcanic rocks is characteristic. Identical formations were deposited throughout Mesozoic and Cainozoic times alternating in space and time. Altogether they have an enormous thickness of about 30–40km. These stratified rocks are at present being compressed into folds and turn in separate scales and blocks. Faults control frequent and powerful earthquakes.

An important feature of the present-day epoch is the existence of oceanic troughs, which are downwarps of the bottom of the sea, reaching depths of up to 8–9km. These features render the present-day epoch unlike the former ones. The troughs (deeps) are situated between the platform—the bottom of the Pacific Ocean—and the mountainous Kamchatka, Kuril islands and Japan. In their position and morphology the deeps are equivalent to former submontaine downwarps but they receive deep-water sediments unknown amongst Cainozoic and Mesozoic strata. There is no doubt that the latter feature should not be explained by the absence of such sediments amongst shallow-water sedimentary rocks. A detailed comparison with the present-day deep-water sediments would enable the recognition of their Cainozoic and Mesozoic analogues.

MAGMATISM

All magmatic cycles are found in the Pacific Ocean Geosyncline, but various lavas and intrusions belonging to the Cimmeridian and Alpine magmatic cycles are especially typical.

PRECAMBRIAN MAGMATIC CYCLES. There are quite common, sometimes extensively exposed intrusions, which in ancient massifs are of great significance. Thus, for instance, in the Lesser Khingan, the Bureyan Massif and the Zeyan Massif Precambrian and Caledonian granites together form huge batholiths. The length of the Khingan-Bureyan Massif is more than 500km while its width is 200–250km. The age of the batholith forming granites varies from one to another, but can only be provisionally suggested. The oldest, undoubtedly Archean granites are gneissose, and cataclased alaskites are also considered Archeozoic. Two mica-tourmaline-microcline granites intruding all the previous ones are considered as Caledonian. Caledonian granites occupy the largest areas (Fig. 257).

Precambrian granites of the upper reaches of the Zeya continue eastwards into the Dzhugdzhur range and thereafter into the Okhotsk Massif. These granites are exposed in the basin of the Omolon. In addition to the dominantly acid intrusions small massifs of basic rocks are also encountered. More or less metamorphosed lavas are not distinguished on synoptic maps.

THE CALEDONIAN CYCLE. The limited distribution of Lower Palaeozoic and Silurian strata conditions the limited occurrence of Caledonian lavas. In comparison with sediments volcanic rocks are subordinate and are not shown on synoptic maps.

Caledonian intrusions are not very widespread and are usually small. Only in the Khingan-Bureyan Massif do they occur over large areas, but their age here is provisional and a Proterozoic age cannot be excluded. Nepheline syenites of the upper reaches of the Omolon are interesting. Their pebbles are found in Upper Devonian strata.

THE HERCYNIAN CYCLE. Rocks of this cycle are ubiquitously found and in places occupy large areas, as for instance do granites occurring in the upper reaches of the Zeya and in Southern Primoria. There are also Devonian lavas in the Omolon basin.

The volcanic phase begins with Upper Devonian porphyrites and tuffs, found in the basins of the Omolon and the Uda. Lower Carboniferous volcanic rocks are very rare. Equally rare are Upper Carboniferous volcanic rocks, although such are known from the Ol'ginskii region of the Sikhote-Alin'. A vast extent and a geosynclinal character of Permian strata lead one to expect extensive volcanicity, but the sediments of the outer zone are represented by flysch and molasse type deposits of the Verkhoyansk complex. It is interesting that despite the enormous thickness of several kilometres and their vast extent there are no volcanic rocks amongst them. In the inner zone Permian strata are of a normal type and volcanic rocks which are at first basic and later acid are quite commonly found.

Hercynian granites, granodiorites and diorites are encountered in Eastern Transbaikalia, in the basin of the Uda, in the Dzhugdzhur range, in the Lesser Khingan, in the region of lake Khanka, the south of Sikhote-Alin' and in the Kolyma Massif. It must be pointed out that in the circumstances of country rocks being palaeontologically badly defined, the separation of Late Hercynian from Cimmeridian granites presents great difficulties and there is a possibility that those granites which intrude Lower Permian strata and lie within them are Hercynian and not, as is normally accepted, Cimmeridian. In such a way one can explain an otherwise unintelligible phenomenon, that in the Far East and Eastern Transbaikalia Hercynian intrusions are widespread, while in the North-East in the same stratigraphic and tectonic conditions they are almost entirely absent.

Nepheline syenites, which lie within Middle Palaeozoic strata and intrude the Lower Carboniferous of the Chukotian Peninsula, are distinctive. Their age is not definitely known. They are quite often considered as Older Cimmeridian but a younger Hercynian age (Upper Permian-Lower Trias) is more likely. Between the Chaun inlet and the Chukotian Peninsula there are widespread, numerous but small intrusions of dark greenish olivine-pyroxene gabbros occurring in dykes and sills emplaced in Upper Palaeozoic and Triassic strata. These intrusions are very similar to the Siberian Traps.

THE CIMMERIDIAN CYCLE. The cycle is the main dominant manifestation of magmatism in the Cimmeridian zone and its rocks are also widespread in the Alpine zone.

Vulcanicity occurred ubiquitously and throughout the whole period from the Upper Trias up to and includng the Lower Cretaceous volcanic rocks were most abundant. It was an epoch of maximal orogenic activity.

Lower Mesozoic volcanic rocks normally occur in suites of variable thickness, which lie within thick sandy-shaly formations. Such suites can be exemplified by Lower and Middle Jurassic porphyrites of the Sikhote-Alin' range and by tuffs, porphyrites, porphyries and keratophyres of a Norian age occurring in the upper reaches of the Kolyma. On synoptic maps they are not shown. Only in the central part of the Kolyma Massif, where Lower and Middle Jurassic sediments are thin, do volcanic rocks become dominant; they are shown on synoptic maps.

Middle Mesozoic, Upper Jurassic and Lower Cretaceous volcanic rocks occur in vast areas, and especially in the Cimmeridian zone. In Eastern Transbaikalia there are Upper Jurassic volcanic rocks of different types, including various porphyries, porphyrites, tuffs and breccia interbedded with sedimentary rocks. In places more basic spilites can be recognized. Further to the east in Priamuria, spilites occur amongst Lower Cretaceous deposits, while on the southern slope of the Dzhugdzhur

Massif Upper Jurassic and Lower Cretaceous volcanic and sedimentary rocks form a single undifferentiated formation of a considerable thickness.

To the north-east of the Dzhugdzhur range Lower Cretaceous volcanic rocks are 2000–3000m thick and often merge with Upper Cretaceous volcanic rocks, forming a single Cretaceous complex. Jurassic volcanic rocks are separate from the Cretaceous and occur in two zones—the Magadan zone of mainly basic rocks and the Indigiro-Kolyma Upper Jurassic zone with rocks of a mixed composition. It is interesting that Upper Jurassic lavas surround the Kolyma Massif from all sides, indicating that the massif as a large host began rising in the Upper Jurassic epoch and was bounded by deep faults through which magma reached the surface. Amongst the rocks surrounding the massif there are both basic and acid varieties and amongst the latter quartz porphyries, felsites and tuffs are present. Volcanic rocks have beds and intercalations of sedimentary rocks with faunas and floras and consequently their age is determined.

Cimmeridian intrusions are of great economic importance since large deposits of gold, tin, tungsten, molybdenum and polymetallic ores are associated with them.

Cimmeridian intrusions occur ubiquitously and in several provinces form enormous massifs of hundreds of kilometres in length, as for instance in the Cherskii and Taskhayakhty ranges. Normally the intrusions are small, being 10–30km across and even smaller (Fig. 276).

The age of Cimmeridian granites is not always definitely established. There are cases when Cimmeridian and Hercynian intrusions are indistinguishable and their ages remain provisional. Nevertheless, in a majority of cases there are undoubted active contacts against Jurassic and Valanginian strata and transgressive contacts against overlying Upper Cretaceous or Palaeogene sediments. The discovery of pebbles and grains of Cimmeridian granites in Upper Cretaceous conglomerates and sandstones is also significant.

The relationships between ancient pre-Mesozoic granites, young 'Kolymian' granites and Jurassic and Lower Cretaceous dacites are clear in the Bureyan coalfield (Fig. 277).

In Eastern Transbaikalia and the Sikhote-Alin' range two generations of granites have been found: the Older Cimmeridian (Jurassic) and the Younger Cimmeridian (post-Valanginian). In the North-East only Younger Cimmeridian intrusions of a Hauterivian-Aptian age have been found and are normally know as 'Kolymian' since they accompany the Kolymian orogenic phase (see p. 777).

Absolute age determinations by N. I. Polevaya indicate that amongst Older Cimmeridian granites both younger ages of 130–145 myr (Upper Jurassic) and older ages of 180–190 myr (Upper Triassic) can be recognized. Both types are found in Transbaikalia. Younger Cimmeridian intrusions are 115–125 myr.

Older Cimmeridian intrusions are represented by granites and granodiorites. Kolymian, Younger Cimmeridian intrusions are more variable. In Eastern Transbaikalia granites predominate, which are sometimes very acid, but there are also diorites, porphyries, granite-porphyries and in places syenites peripherally transitional into monzonites. The intrusions of Cisokhotia are of the same type. In the Sikhote-Alin' range according to N. A. Belyayevskii the order of intrusions is as follows: first gabbros and alkaine syenites, then acid rocks up to potassic granodiorites. In the north-east granodiorites and granites predominate but around the Kolyma Massif small bodies of intermediate and basic rocks are encountered. From the Chaun inlet to the Chukotian Peninsula a zone of minor intrusions of mainly olivine-pyroxene gabbro occurs essentially within Triassic and Jurassic strata and consequently may be assumed to be Cimmeridian.

In the Omolon Massif alkaline syenites, teschenites and essexites have been found intruded into Jurassic sediments. Alkaline granites are encountered to the north of Okhotsk. Minor intrusions intersecting Cimmeridian granites have been found in several regions, but are relatively rare.

THE ALPINE CYCLE. The cycle is governed by Alpine orogenic movements and consequently differs in its manifestations in the outer from the inner zone.

The outer zone shows only weak indications of initial phases of the Alpine Orogeny and no sign of later phases, consequently synorogenic intrusions are absent from it. Volcanic activity is very diverse in time and space. In the Upper Cretaceous and Palaeogene epochs vulcanicity was very widespread in southern provinces adjacent to the inner zone, while in northern provinces vulcanicity was almost absent. In the Neogene and Palaeogene epochs the whole outer zone consolidated into a platform. Deep faults were rare and so were the lavas.

The region of Upper Cretaceous and locally Palaeogene eruptions was exceptionally large. It is a continuous belt 3000km long and up to 200–250km wide and can be traced from the northern and southern slopes of the Dzhugdzhur range along the

FIG. 276

Cimmeridian granites in rapids of the river Kolymą. The boat belongs to the expedition. *Photograph by S. V. Obruchev.*

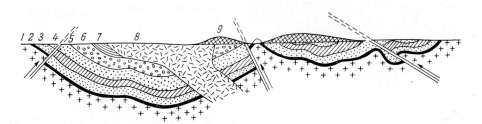

FIG. 277. The relationships between older and younger granites in the Bureyan coal region (the Buro-Turukian syncline). According to S. A. Muzylev (1938).

1 — ancient; pre-Mesozoic granites: 2–7 — Jurassic suites; 8 — Younger Cimmeridian, Lower Cretaceous granites; 9 — Upper Cretaceous dacite.

shore of the Sea of Okhotsk (Fig. 278) up to the Gizhigian inlet and then in a north-easterly direction as far as the shore of the Chukotian Sea and along the Chukotian Peninsula to the Bering Sea. The thickness of volcanic rocks interbedded with continental sediments containing plant remains reaches 2000–3000m, but is often considerably less (hundreds of metres). The geologists of the Far East have suggested the term 'Evenian volcanic arc' for these volcanic deposits. Since the arc changes its curvature it should be called the Evenian volcanic belt.

The age and composition of volcanic rocks within the belt change from district to district. The age varies from the Cenomanian (Upper Albian) to the Eocene (Oligocene) but most volcanic eruptions fall within the interval Cenomanian-Palaeocene. Volcanic rocks can be shown to have a succession from basic andesites to acid dacites and liparites. Lavas, especially when they are acid, are accompanied by large quantities of tuffs and tuff-breccias. There are quite common subvolcanic intrusive rocks filling in necks and dykes.

Neogenic and Athropogenic volcanic rocks are rare and occupy such small areas that they are not shown on synoptic maps. Their largest outcrops occur in the Novosibirsk islands and along the oceanic coast from the Kharaulakhian Mts to the mouth of the Indigirka. In the Svyatoi Nos region basalts form a series of table-like hills rising to 400m. Some hills still have volcanic form, with existing craters. According to V. A. Obruchev the region is an independent volcanic district with basic and alkaline lavas of Pliocene and Quaternary ages. Near to this region in the valley of the Moma, a left tributary of the Indigirka, there is the extinct Balagantakh (tent-rock) volcano with which is associated a minor lava flow. The volcano is shown on

FIG. 278.   The 'Morzh' (Sea-lion) sea stack in Northern Sakhalin, consisting of Alpine lavas. *Photograph by K. F. Rakhmanov.*

synoptic maps. Another extinct volcano—Anyuiskii—is found far to the east in the valley of the Great Anyui, a left tributary of the lower Kolyma. Here a lava flow is 60km long and together with the volcano is shown on synoptic maps. A fairly extensive tract of young basic lavas is also found in this region.

The inner zone has an unusually fully representative Alpine magmatic cycle with widely distributed varieties of rocks. Here, in Kamchatka and the Kuril islands there are the only active volcanoes of the U.S.S.R.

The most powerful phases of the Alpine Orogeny occurred at the end of Upper Cretaceous and the end of Miocene times. During these periods widespread synorogenic intrusions were formed. Consequently the Alpine cycle can be divided into three sub-cycles.

The first sub-cycle began with Upper Cretaceous volcanic rocks and ended with the so-called Laramidian intrusions. The second sub-cycle began with Palaeogene and Miocene volcanic rocks and ended with Upper Miocene intrusions. The third sub-cycle is not as yet complete and is represented by Pliocene and Anthropogenic basic lavas, while no synorogenic intrusions or acid volcanic rocks are known. In other regions, as for instance in Kamchatka, basic lavas are dominant but acid varieties also occur, although in very subordinate quantities.

The largest areas of Upper Cretaceous lavas are found in Kamchatka and the eastern part of the Koryakskii range. Volcanic sequences are represented by basalts and porphyrites forming suites and members lying amongst thick sedimentary intensely folded rocks. In Sakhalin such members are relatively thin and are encountered quite frequently but not on a scale to be shown on synoptic maps. Eruptions are known to have occurred in Upper Albian and Cenomanian times, but the most intense volcanic activity is assigned to the Senonian. In Senonian strata of the western coast there are quite common porphyries, porphyrites and diabase tuffs and andesitic lava flows. In the Sikhote-Alin' range Upper Cretaceous basic and less frequently acid volcanic rocks are ubiquitous, but their area of distribution or thickness is much less than those of Palaeogene volcanic deposits. At the base lies a Senonian porphyritic formation (200–400m) which is mainly pyroclastic and is unconformable on Cenomanian strata. The formation is overlain by acid lavas and tuffs of a Danian age, which may partly belong to the Palaeocene. Quartz porphyries of the Ol'ginskii region serve as an example. The total thickness varies, reaching up to 1200–1800m.

The largest area of Palaeogene and Miocene volcanic rocks is in the Sikhote-Alin' range, where they are also at their thickest. These rocks differ compositionally from Upper Cretaceous porphyrites and lie unconformably on them. The break associated with the unconformity was minor and frequently Maastrichtian and Lower Palaeocene quartz porphyries are succeeded by end-Palaeocene and Eocene volcanic rocks. In Southern Primoria according to Melyayevskii there were five volcanic rhythms, each beginning with basic andesites and ending with acid liparites and dacites. The first rhythm, represented by the Ol'gian Suite, begins with Senonian porphyrites and ends with quartz porphyries, tuffs and tuff breccias of Danian age. The second rhythm corresponds to the Palaeocene, Samarginian Suite. It begins with andesites and andesite-dacites and ends with dacites. The third rhythm consists of Eocene (Kuznetskian Suite) andesites and Oligocene (Nadezhdinskian Suite) liparites and liparitic tuffs. The fourth rhythm consists mainly of basalts at the bottom and liparitic tuffs at the top, the sequence being of an Upper Miocene age. The fifth rhythm is incomplete and consists of basic Pliocene and Anthropogene andesites and basalts. In the case of the last and possibly preceding rhythm, a connection with deep faults, traceable along the coast of the Sea of Japan, can be noted. Volcanic rocks are interbedded with continental, often coal-bearing deposits. The total thickness is considerable, reaching 2–3km and in some localities even more.

In Sakhalin lavas form suites of medium and small thicknesses and different compositions which lie amongst marine and continental Palaeogene and Miocene sediments. The sequence is intensely folded and consequently volcanic rocks have not been shown on synoptic maps.

In Kamchatka and the Koryakskii range both Palaeogene and Miocene volcanic rocks are developed. Palaeogene andesites and tuffs form suites of up to 500m in thickness, lying amongst sediments. Miocene volcanic formations which are shown on maps, contain mainly andesites, dacites, liparites and tuffs with beds and sequences of sedimentary rocks amongst volcanic deposits.

Pliocene and Anthropogene effusions are widespread throughout the inner zone of the geosyncline but vary according to the distance from the Neogene platform. Nearer to it, as in the Sikhote-Alin′ range and on Sakhalin, there are Upper Miocene-Lower Pliocene volcanic rocks at the bottom and Pliocene-Anthropogene volcanic rocks at the top. The former consist of andesites and dacites, lying almost horizontally while the latter in the north form hills reminiscent of volcanoes. Upper Pliocene-Anthropogene effusions are represented by the so-called 'plateau basalts' consisting of sombre basaltic sheets forming scarps of several tens of metres (Fig. 279) and vast flat or slightly hilly plateaux clearly identifiable in relief. A characteristic feature of basalts is the absence of volcanoes or pyroclastic formations. Eruptions ended at the beginning or in the middle of Anthropogene times. At present there are no active or extinct volcanoes.

In Kamchatka and on the Kuril islands Pliocene and Anthropogene effusions are not distinct and vulcanicity is still continuous. In Kamchatka is the highest volcano in the U.S.S.R.—Klyuchevskaya Sopka (4850m, Fig. 255)—and other large volcanoes such as Koryakskaya Sopka (3464m), Kronotskaya Sopka (3528m) and many others of smaller dimensions. Their distribution is suggestive of association with large deep faults, which is also observed on the Kuril islands which represent a volcanic chain with peaks rising up to 1200–1800m above the sea level. Amongst Anthropogenic and present-day volcanic rocks there are dominant pyroxene and hornblende andesites and corresponding tuffs, while basalts are subordinate.

Alpine intrusions of the inner and outer zones are very different. In the outer zone as has already been mentioned there are no synorgenic intrusions.

In places, as for instance in western Cisokhotia, small bodies of granites and granodiorites are associated with faults and have active contracts against Jurassic and Lower Cretaceous sediments and Upper Cretaceous lavas. A possibility cannot be excluded that these intrusions are infillings of sub-volcanic channels.

In the inner zone synorgenic intrusions are frequently found and are represented by small massifs. They have been investigated most completely in the Sikhote-Alin′ range, since here at the end of Cretaceous times (Senonian-Danian interval) they reached considerable dimensions. Granitoids, gabbros and diorites predominate. Absolute age determinations have shown them to be no older than 90 myr. Of the same group are intrusions at the Tetyukha, the central zone, Southern Primoria, Iman and Ol′ginskii region. A majority of synorogenic intrusions is associated with the Laramide orogenic phase. Granitoids of 50–60 myr in age are associated with Eocene movements, while the youngest Oligocene granodiorite porphyries are 27–35 myr old. The existence of synorogenic intrusions has been indicated by N. A. Belyayevskii. Neogenic synorogenic intrusions are unknown. There are indications of small sub-volcanic intrusive bodies to which possibly belong the small granodioritic Neogene intrusions in the Koryakskii range and the Kuril islands.

Outcrops of Older Alpine granodiorites are known in Kamchatka and on the Koryakskii range. Small ultrabasic massifs of dunites, serpentinites and harzburgites form a belt which begins to the east of the Penzhian inlet and ends in the north-east of Kamchatka. These bodies are sometimes considered as Younger Cimmeridian but an Upper Cretaceous age is more likely.

Fig. 279

A scarp of the basalt plateau on the eastern slope of the Sikhote-Alin' range.
Photograph by Yu. F. Chemekov.

USEFUL MINERAL DEPOSITS

The Pacific Ocean Geosyncline possesses the largest deposits in the U.S.S.R. of precious and rare metals and has also considerable reserves of black and brown coal and oil. Soviet geologists have multiplied the previously known reserves by finding new deposits of sometimes world significance. Such is the gold of the north-east. Other deposits of considerable magnitudes are: tin in the north-east and the Sikhote-Alin' range, gold in Darasun and Balei and in Eastern Transbaikalia; coals of the Bureyan basin.

ORES

Precious Metals

Alluvial gold was already panned in the distant past. Soviet geologists not only helped in finding new reserves in old deposits, but also discovered a series of new often large mineralizations. Such is for instance the large gold region in the basins of the Kolyma, the Indigirka and the Yana. The region can justifiably be called the Soviet Alaska. The entire Pacific Ocean Geosyncline is characterized by an extraordinary development of acid granodiorites of different ages beginning with the

Precambrian and ending with the Palaeogene. This feature is responsible for ubiquitously distributed gold deposits.

The largest and most widespread deposits are associated with Cimmeridian granites and in particular with those belonging to the 'Kolymian phase'. Precambrian granites in this respect fall in the second place and are followed by Alpine and Cimmeridian intrusions. The largest gold-bearing province, situated in the upper reaches of the Kolyma, the Indigirka and the Yana, is associated with 'Kolymian' granites. As a result several new townships and an administrative centre of Magadan—on the Okhotsk coast—have been built. Yu. A. Bilibin was first responsible for the exploration of this region. Several Eastern Transbaikalian deposits, such as those at Darasun and Balei, are associated with 'Kolymian' granites.

Numerous syngenetic and alluvial deposits have been found in the upper reaches of the Amur, in the basin of the Zeya and its tributaries, in the upper reaches of the Bureya, in the Lesser Khingan range and lower reaches of the Amgun' (river Kerbi). In these regions gold is associated not only with Cimmeridian, but also with Precambrian and Hercynian granites. The same can be said about deposits to the north of Okhotsk.

Deposits of the Sikhote-Alin' range are not only Cimmeridian, but also younger Alpine, Upper Cretaceous, Palaeogenic and possibly even Neogenic (Byelaya Gora near Nikolayevsk). Gold of the Anadyr basin is Alpine. A description of gold deposits and suggested problems for future exploration have been given by P. Yu. Antropov (1959).

Non-Ferrous Metals

The second feature of the Pacific Ocean Geosyncline is abundance of tin. Large deposits of world significance have been found by Soviet geologists in the north-east part of the geosyncline. In their discovery the work of S. S. Smirnov, a leading ore-geologist, and G. L. Padalka, a specialist on Far Eastern ores, was particularly outstanding. The geology of tin deposits in the north-east has been described in a monograph by V. K. Chaikovskii (1960).

Considerable deposits of tin have been found in Eastern Transbaikalia (Khapcheranga, Sherlova Gora) and in the Bureya-Lesser Khingan region, and very recently in Southern Primoria. Then there are the deposits of the upper reaches of the Yana, the Indigirka and the Kolyma, the deposits of the south Kolymian range, the Chaun group, and lastly those of eastern Chukotka.

Tin deposits are so widespread that one cannot mention any single belt of mineralization. The whole vast area is characterized by the presence of Cimmeridian granites, with which tin is even more closely associated than gold.

Rare Metals

Tungsten is an associate of tin and is found in the same regions, albeit more rarely. The most important groups of tungsten deposits are: the Eastern Transbaikalian and the Eastern Chukotian, where mineralization is related to Cimmeridian granites. The Sherlova Gora deposit in Eastern Transbaikalia consists of a series of veins and nests emplaced in a massif of greisenized biotite granite of a Cimmeridian age. The massif feeds several tungsten-bearing alluvial deposits.

MOLYBDENUM. The most important and best-known molybdenum deposits are found in Eastern Transbaikalia.

ANTIMONY. Antimony occurs in the Lenin deposit situated in the central part of the Far East.

MERCURY. Second-rate deposits of mercury are known in the south of the Sikhote-Alin' range and in Eastern Transbaikalia.

ARSENIC. A rather large Zapokrovskoye deposit of arsenic in Eastern Transbaikalia is a vein-like body emplaced in a skarn limestone.

Deposits of antimony, mercury and arsenic are associated with Cimmeridian granites which are in general very metalliferous. There are also mineralizations associated with Hercynian and Alpine granites.

Polymetallic Ores

Lead, zinc and silver are widespread in all the regions where younger intrusions are developed. Their deposits are normally small, but groups such as the Tetyukhian and the Nerchinskian have considerable reserves.

The Tetyukhian is situated in the south of the eastern slope of the Sikhote-Alin' range. Mineralization is hypothermal and is one of the largest in the U.S.S.R. The ore consists of sulphides (galena, zinc blends, chalcopyrite) and lies at the contact of Triassic limestones and Alpine (Palaeogene) porphyries.

Nerchinskian deposits are typical silver-lead-zinc hydrothermal ores. They are situated in the eastern part of Eastern Transbaikalia and are very numerous. Mineralization is related to Cimmeridian granites and has been studied by S. S. Smirnov.

Ferrous Metals

Iron. Until quite recently iron reserves of the Pacific Ocean Geosyncline were relatively small. At present they have been rapidly expanded and supply all demands of Far Eastern metal industry. Deposits are quite numerous, sometimes large and are of different origins. The most important are deposits of jaspilites found in ancient metamorphic rocks. They have been studied most intensively in the Lesser Khingan range. Here the Kimkan deposit of hydrothermally altered primary sedimentary ores serves as an example. The age of the ore-bearing formation, as has already been mentioned, is debatable. Formerly it was thought to be Proterozoic in conformity with other deposits of this type. Recently a Cambrian fauna has been found in the ore-bearing formation, but the fauna is so poor that it cannot be used unequivocally. Another region (Gar') where a major deposit of iron ores associated with ancient metamorphic rocks has been found is the basin of the river Gar', a right tributary of the Selemdzha.

With Lower Devonian dolomites is associated the Pobednoye deposit which is situated on the right shore of the Kolyma at a distance of some 50km from the town Verkhne-Kolymsk, and within the eastern part of the Palaeozoic Kolyma Massif. Here iron ores form a horizon of 3–6m and sometimes up to 20m thick and consist of massive rather homogeneous haematite with 50–70 per cent iron content. The deposit is sedimentary and is not associated with igneous rocks.

The small contact deposits of the Ol'ga harbour, in the south of the Sikhote-Alin' range, are associated with Cimmeridian or possibly Alpine granodiorites. A major part of iron-ore deposits of the southern part of Eastern Transbaikalia and in particular the Berezovskoye deposit are of the same type and age.

The youngest deposits are small accumulations of Neogenic and Lower

Anthropogenic brown bog iron ores found on the right shore of the lower reaches of the Amur.

The aforementioned types of iron ores do not exhaust the great variety of such deposits.

MANGANESE. Thick bedded manganese ores are associated with lower horizons of the Lesser Khingan iron-ore formation.

It should be pointed out that metal reserves of the Pacific Ocean Geosyncline grow annually and will continue doing so.

FUELS

Fuel reserves are abundant and diverse, but have not been sufficiently investigated. The main fuels are black and brown coals and considerable deposits of oils. Deposits of gas or oil shales have not been studied.

Black and Brown Coals

The oldest coal-bearing formations are encountered amongst Upper Carboniferous, Lower Permian, Upper Permian and Lower Trias sediments. Coal seams are not numerous, but are thin and therefore remain almost unexploited.

The second rhythm of accumulation of coal begins in the Upper Triassic Mangugnaiian coal-bearing formation of the Sikhote-Alin' range. It is continuous in Jurassic strata of more northerly regions, such as Eastern Transbaikalia, the upper reaches of the Zeya and the Uda basin. The rhythm terminates with Lower Cretaceous coal-bearing formations of minor industrial significance. Coals of the second rhythm occur in the Nikanian Coal-bearing Formation of Southern Primoria, the Bureyan coal-field and the northern margin of the Moma Massif, which lies to the north of the Cherskii range, between the Indigirka and the Kolyma.

The third rhythm which includes the Upper Cretaceous and the Palaeogene is dominant in the south and in the Alpine zone including Sakhalin, Kamchatka and the Koryakskii range, but its coals are also found in the south of the Cimmeridian zone.

The fourth rhythm (Neogene and Anthropogene) is only represented by small deposits of lignite and peat, occurring mainly in the Alpine zone.

All industrial (commercial) deposits of coal in Eastern Transbaikalia belong to the second rhythm. Until recently they were compared with the Cheremkhovskian coal-field and were therefore considered as Middle Jurassic. At present it is thought that Eastern Transbaikalian coals are equivalent to those of the Bureyan coal-field. The Bureyan coal-bearing sequence used also to be considered as Jurassic, but after the demonstration of its post-*Aucella* position it has been transferred into the Lower Cretaceous. Consequently all coal-bearing sequences of Eastern Transbaikalia, which used to be considered as Jurassic, are now included in the Lower Cretaceous. However, there still is a possibility that a part of them is Jurassic. Coal-bearing formations have been preserved in graben within Palaeozoic and Precambrian rocks and therefore occupy only small areas. Their thickness varies from the usual several hundred metres to 1000m and more. The lower part of the succession often consists of coarse clastic sediments, indicating that the surrounding Palaeozoic massifs were rising at the same time as deposition took place. Thus conditions of accumulation of Eastern Transbaikalian coal-bearing sequences were the same as conditions of accumulation of Triassic coal measures of the Chelyabinsk coal-field and of coal-bearing sediments of the Vale of Fergana.

The Bureyan coal basin is situated in the upper reaches of the Bureya. Here at the base occurs a thick (5000m) formation of sandstones and shales with a marine fauna of Jurassic and possibly Upper Triassic ages. The formation is overlain by the coal-bearing sequence (2000m) which belongs to the Nikanian Stage. The sediments are folded into relatively gentle folds and are metamorphosed. Coals are of coking variety but are highly ashy. There are 25 workable seams of 0·75m to 12m in thickness. Geological reserves are 25×10^9 tons and proven reserves are $5·4 \times 10^9$ tons.

The Suchanian coal basin is situated in the south-east of Primoria, to the east of Vladivostok. It has coals varying from brown to anthracitic varieties. The basin is an important coal-field of the Far East.

At the bottom lie Permian partly coal-bearing sediments with coal seams reaching 0·5m in thickness. They are unconformably overlain by coal-bearing rocks of mainly Lower Cretaceous (Nikanian) age. In their upper parts the rocks have marine bands, with a fauna of possibly Upper Cretaceous or Aptian-Albian ages. The coal-bearing sequence is succeeded by Upper Cretaceous deposits with marine faunas. The total thickness of coal-bearing beds is 1000–1500m, and they represent deposits of a vast near-shore plain with surface swamps and lakes. The plain was enclosed by southerly and westerly uplands, but was open to the sea in the north and east. The coal-bearing formation is folded into large gentle folds and the rocks are moderately metamorphosed. There are 13–16 workable seams of 0·5–5m in thickness. To the north, near to the sea the number of coal seams increases. Coals are generally of a high quality. Proven resources are $0·2 \times 10^9$ tons and the yield in 1956 was $1·4 \times 10^6$ tons.

The Upper Suifun coal region is situated in the southern part of the Sikhote-Alin' range near to the town of Ussuriisk and in the basin of the river Suifun. The coal-bearing sequence is of the same age as in the Suchanian coal basin or may be slightly younger (Aptian-Albian) and is 300m thick. It is folded into one large gentle trough and has 2–5 workable seams of 0·7–7m in thickness. The coal is black and its yield in 1956 was $0·7 \times 10^6$ tons.

The Moma-Zyryankian coal basin is situated on the northern slope of the Momian range and in the basins of the Moma and the Zyryanka. The coal-field has only been recently discovered by Soviet geologists. Its coal-bearing sequence has 18 workable seams of 0·5–7·0m in thickness. The age of the sequence is Lower Cretaceous and its outcrops by virtue of size are shown on synoptic maps. Within the basin five coal regions are recognized. The fullest section of the coal-bearing formation occurs in the Zyryanka-Silyanka region where it is 5000m thick. The two lower suites are of Valanginian-Hauterivian age, while two upper suites— the Conglomeratic and the Upper Coal-bearing—are of a Barremian-Aptian age. The Coal-bearing Formation lies in tectonic depressions.

In the southern part of the Cherskii range in the Balygychan-Sugoi region a Lower Cretaceous coal-bearing sequence is again preserved in a tectonic depression. With this depression is associated the Galimoskoye deposit.

In the south, the region of the Tauian inlet there is yet another development of the Lower Cretaceous coal-bearing formation with the Khasyn coal-field in it.

In the north-east several deposits associated with isolated depressions filled with coal-bearing strata are attributed to the third, upper Cretaceous-Palaeogene rhythm.

In the upper reaches of the Kolyma there is the Arkagala deposit, where the coal-bearing formations are Cenomanian age, and 600m thick. To the north of Magadan, on the river Kheta, there is the Pervomaiskoye deposit, where the coal-bearing sequence is Danian.

The Upper Miocene or Pliocene Kykhtui deposit to the north of Okhotsk is assigned to the fourth rhythm.

In the Far East the Kivdin and Raichikha deposits of the Zavitinskii coal region situated in the lower reaches of the Bureya are considered to belong to the third rhythm. Here coal-bearing strata are Danian, 300–400m thick, folded into large gentle folds and remain unmetamorphosed. Individual sections of aparently horizontal rocks are no more than 100m thick. Coals are brown, of a good quality and are used as a high class fuel. In each pit only a single seam is being worked. The one at Kivda is 2m and at Raichikha 3–8m thick. Proven reserves are 0.5×10^9 tons and the yield in 1956 has been the highest in the Far East amounting to 12.1×10^6 tons.

In the Sikhote-Alin' range Palaeogene brown coals have been found in several localities, but the Uglovskian coal-field, which is situated in the extreme south of the range near Vladivostok, is the most important coal region. The total thickness of Palaeogenic continental deposits of the coal-field reaches 1300m. The coal-bearing Lower Suite is 230m thick and is succeeded by two coal-less suites. The succession ends with the Upper, Lignitic, Suite (330m) which may be in fact of a Neogene age. The Lower Suite has 2–8 workable brown coal seams of 2–10m in thickness. The suite is folded into large gentle folds and is moderately metamorphosed. Proven reserves are 0.9×10^9 tons and the yield in 1956 was 3.2×10^6 tons.

The enumerated coal-fields and basins are numerous and have considerable extents, nevertheless they do not exhaust the reserves of Cretaceous and Palaeogene coals of the Pacific Ocean Geosyncline. In 1955–56, for instance, a large coal-field was inaugurated in the upper reaches of the Selemdzha (on river Shevli) and two major deposits were opened along the northern slope of the Sikhote-Alin' range (Bazovskoye and Iman) as well as several minor ones.

Coal-bearing strata of Sakhalin are of a different type and can be called geosynclinal. Here several coal-bearing formations lie amongst marine and volcanic Upper Cretaceous and Palaeogene rocks. Each coal-bearing sequence is several hundred metres thick, and is intensely folded, strongly metamorphosed and invaded by intrusions. The occurrence of coals in Upper Cretaceous strata, where several workable seams have been found, is characteristic. Some of the coals are of coking type and are worked for local needs. Upper Cretaceous coals so far have not been sufficiently investigated. Palaeogene coals are better known and are being exploited in Northern and Southern Sakhalin, where they are more abundant. The Lower Duiian (500–600m) and the Upper Duiian (1000–1200m) suites of Northern Sakhalin are coal-bearing as are their analogues in Southern Sakhalin where they are 1000m and 1500m thick, respectively. The number of workable seams in both suites is no less than 11 in Northern Sakhalin and 24 in Southern Sakhalin, reaching a total thickness of 50m. Sakhalin coals are very variable and include brown, black and even anthracitic varieties. Some of them are of coking type. Palaeogenic coal-bearing strata are just as strongly folded and metamorphosed as the Upper Cretaceous. Geological reserves of Sakhalin coals are 19.4×10^9 tons and proven reserves are 4.3×10^9 tons. The yield in 1956 was 3.6×10^6 tons.

Upper Cretaceous and Palaeogene strata of the western and eastern coasts of Kamchatka also contain coals which have long been worked in the Korer harbour.

These strata are no less coal-bearing in the Koryakskii range, in the northern part of which one township is known as Ugol'nyi (coaly). Finally Palaeogene coals are known in the basin of the Anadyr. All these deposits so far remain insufficiently investigated but even our present knowledge permits an emphasis on coal capacity of the Pacific Ocean Geosyncline.

Oil

So far commercial oil has only been found in Northern Sakhalin, but indications of it and non-commercial seepages have been obtained from Southern Sakhalin and on the west coast of Kamchatka. These facts augur good prospects for investigating the entire Alpine zone of the geosyncline where it often has not been much studied (Fig. 280).

Prospects of the Cimmeridian zone are not so good since here Upper Cretaceous and Cainozoic strata are completely continental and contain no sequences with high bitumen content, such as oil shales. One such sequence is known to occur amongst Lower Cretaceous lacustrine deposits and is known as the Turgaian Suite. It consists of finely laminated marly shales with fish remains and often an elevated bitumen content. The suite is found in Eastern Transbaikalia and to it are evidently related small deposits of oil in the north-eastern part of China. The widespread and thick Verkhoyansk complex, which includes Permian, Triassic and Jurassic strata, is unsuitable for the accumulation of bitumens, since clastic sediments were deposited very rapidly.

Northern Sakhalin oil has good prospects of future expansion of reserves if new horizons and fields are found. Upper Cretaceous deposits so far have not yielded any oil but widespread indications of gas suggest its presence. Lines of mud volcanoes situated over faults in Upper Cretaceous rocks are of interest.

Oil is unknown in Palaeocene and Eocene strata, but the first oil-bearing suite (the Khandsaiian) is of an Oligocene-Lower Miocene age. The suite is thick (1200m) and consists of shales with beds of sandstones, which despite their poor porosity in places contain oil and gas.

FIG. 280.   Northern Sakhalin. *Photograph by K. P. Rakhmanov.*

The Middle Miocene Dagian Suite is closely associated with the Upper Miocene Okobykaiian Suite forming a single oil-bearing formation, which includes the most important productive horizons of Sakhalin, exploited in Okhinskii, Ekhabinskii and other oil-fields. Oil is found in faulted anticlines (Fig. 281).

The Pliocene, Nutovian Suite has only second-rate, commercially unimportant deposits.

Oil-shales

Oil-shales have not been much investigated and are not being worked. The fresh-water lacustrine Turgaian oil-shales occurring amongst Lower Cretaceous strata of Eastern Transbaikalia are best known and represent beds (tens of metres thick) of calcareous and marly paper shales with numerous remains of fish and insects and in places having a high bitumen content. The shales are considered as the parent rock of 'Jurassic' freshwater oil found in small quantities in continental Lower Cretaceous deposits, which formerly used to be considered as Jurassic

NON-METALLIC DEPOSITS

Non-metallic ores are diverse and quantitatively considerable although they have not been sufficiently studied or utilized. The absence of rock salt is a negative feature of the region.

Graphite. A major metosomatic deposit of flaky graphite is known from graphitic gneisses and crystalline schists of the Lesser Khingan.

Magnesite. A sedimentary deposit of magnesite is associated with Proterozoic carbonate strata of the Lesser Khingan range.

Fluorite. In Eastern Transbaikalia there are 20 known deposits of fluorite, several of which are being worked. These deposits occur in veins of hydrothermal origin and accompany younger Cimmeridian granites. Large deposits are known from Southern Primoria.

Sulphur. Deposits of native sulphur are found around extinct Kamchatkan volcanoes. The element is a secondary product of volcanic activity.

Building Materials. Building materials are widespread and variable and ensure all industrial demands. The new materials include limestones, flaggy sandstones, quartzites, slate, cement raw materials, clay, and gravel. In the future various igneous rocks such as basalts, granites, diabases and tuffs may acquire importance.

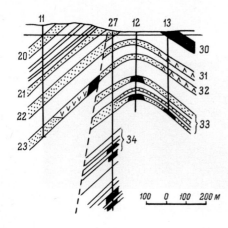

FIG. 281

A section across an oil-bearing structure in Northern Sakhalin. According to N. P. Budnikov. Oil in black sands and sandstones are dotted.

Rock Salt. The problem of rock salt remains unsolved. The ever-growing population and the fishing industry require increasingly large quantities of salt which has to be transported either from the Irkutsk province or Odessa. Search for salt has been systematically carried out for the last 25 years, but so far not a single deposit has been discovered. The main reasons have been the conditions of accumulation of coal-bearing formations, which do not permit the accumulation of salt. Coal-bearing formations require moist climate and a uniformally distributed annual precipitation. Salt-bearing formations originate in dry climates when precipitation is light and there are long periods of drought. Thus where salt precipitates, coals cannot form and *vice versa.*

REGIONS

The North-East

The term North-East refers to that part of the U.S.S.R. which is bounded by the 58° parallel in the south, by the valleys of the Lena, the Aldan and the Maya in the west and by the state frontier in the east. Administratively, the minor north-eastern part of this region is a portion of the Yakutian A.S.S.R. and the major, south-eastern part belongs to the Khabarovsk region.

Owing to the difficulty of access and the distance involved the geology of the North-East has not been uniformly and sufficiently investigated. The middle and upper reaches of the Yana, the Indigirka the Kolyma and their tributaries have been studied to a greater or lesser extent. Northern and eastern regions, along the coast of the Arctic and Pacific Oceans have not been much studied and were sometimes approached in isolated traverses.

The economy of the North-East is based on gold and rare metal deposits and on tin. The exploitation of these deposits changed this wild, formerly almost uninhabited province. New towns and settlements have grown and are connected by highways. The centre of the region is the new town of Magadan, situated on the shore of the Sea of Okhotsk.

Orography

The west, the south and the east of the North-East are mountainous regions, crossed by a series of highly dissected mountain ranges and plateaux. In the north, the lower reaches of the Yana, the Indigirka and the Kolyma there are vast lowlands, the tundra, with thousands of lakes and marshes. Within the tundra there are minor ridges and plateaux of which the largest is the Alazeiian plateau with heights of 800–900m. A smaller, but nevertheless considerable, low plain is situated in the lower reaches of the Anadyr.

The longest mountain chain is the Verkhoyansk range (Fig. 247), which separates the basins of the Lena, the Aldan and the Maya from those of the Yana and the upper reaches of the Indigirka. The range is about 1500km long and is low in the north (1200–1500m) but rises to the south (up to 2500–3000m). The northern part of the range is known as the Kharaulakhian Mts (Fig. 282).

To the south of the Verkhoyansk range, along the coast of the Sea of Okhotsk, lies a narrow but relatively high (up to 2000–2300m) range known as the Dzhugdzhur.

To the east and north of the Verkhoyansk range is the second largest Cherskii range, called so after Cherskii, a Polish geologist who was exiled by the Tsarist regime and has done much for the elucidation of the structure of the Cherskii range. The range begins between the Yana and the Indigirka as a low Taskhayakhty range, then across the Indigirka continues towards the upper reaches of the Kolyma. The range is about 1000km long, up to 2000–3000m high and its highest point is Mt Pobéda which reaches 3147m.

The third major range is the Kolymskii (Kolymian) which bounds the Kolyma basin from south and east. The range begins to the north of Magadan and continues to the basin of the Greater Anui for a distance of over 800km. Its height reaches 1500–2200m.

To the north of the central part of the Cherskii range lies the minor, but high (over 2000m) Momian range, which is separated from the former by the valley of the Moma.

Between the middle reaches of the Kolyma and the Kolymian range lies a wide, but low Yukagirian plateau. In the east it is bounded by the great river Omolon (800km long) which is a right tributary of the Kolyma. To the north of the Kolymian range (from the Chaun inlet) and to the Chukotskii point lies the meandering Chukotian or Anadyr range with heights of 1500–2000m. The range lies to the north of the basin of the Anadyr, which is the largest river in the extreme east.

Along the Bering Sea, from Kamchatka to the mouth of the Anadyr, lies a low Koryakskii range (uplands) which is the least investigated in the Soviet Union. To the east of its central part the Olyutorian Mts stretch far to the south under the surface of the sea. The length of the submarine range is about 400km and its height above the oceanic floor is over 3000m. The range lies almost parallel to Kamchatka.

Between the Koryakskii and Kolymian ranges lie valleys of small rivers such as

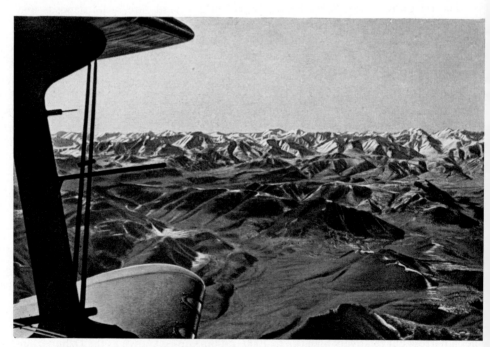

Fig. 282. Northern part of the Verkhoyansk range, where the Verkhoyansk complex involving Permian, Triassic and Jurassic strata is developed. *Aerial photograph by V. M. Lazurkin.*

the Gizhiga and the Penzhina. These rivers have deep inlets separated by the Taigonos Peninsula.

Geological Structure

The geological structure shows clearly defined zoning. The zones are determined by the strike of major structures and are obvious on geological maps. The zones are divided into two groups, based on dates of structural events, and are known as the Cimmeridian and the Alpine.

Cimmeridian zones occupy western and northern parts of the North-East and include the Kolyma and the Okhotsk Palaeozoic massifs, which influence the trend of the zones. According to the tectonic scheme of the North-East compiled by L. A. Snyatkov (1959) Cimmeridian zones can be collectively called the Kolyma-Chukotian belt (Fig. 283). Snyatkov's scheme constitutes in a simplified form the basis of the following discussion.

The Kolyma-Chukotian belt is a part of the Cimmeridian zone of the Pacific Ocean Geosyncline. The western part of the belt is divided into two zones—the Verkhoyanskian and the Cherskian—so called after the main mountain ranges within the zones. The Verkhoyanskian zone from the mouth of the Lena bends east-wards, while in the south it converges and joins the Dzhugdzhur range and there-after continues into the upper reaches of the Zeya and Eastern Transbaikalia. Throughout its extent it margins the Siberian Platform and the adjacent, southern Palaeozoic structures. The zone originated at the end of the Hercynian Orogeny, but its formation continued at the beginning of the Cimmeridian orogenic period. Consequently within the zone there are quite common late Hercynian and early Cimmeridian granites and relatively rare late Cimmeridian (Lower Cretaceous) granites. Permian, Lower Triassic and Middle Triassic rocks occur over vast areas.

The Cherskian zone divides into two. One part of it margins the massif repre-sented by the Novosibirsk islands and continues into the Lomonosov submarine ridge. The other part branches to the east, margins the Kolyma Massif, and trends towards the Greater Anyui basin, where it is recognized as the Anyuian north-west trending zone. In the south the Cherskian zone lies between the Okhotsk and the Kolyma massifs and margins the south of both the Kolyma Massif and the Omolon 'projection'. In the east the Cherskian zone is abruptly terminated by the structures of the Alpine belt. The Cherskian zone is younger than the Verkhoyanskian and shows mainly Upper Triassic, Jurassic and Lower Cretaceous deposits. Within it late Cimmeridian (Kolymian) intrusions are especially widespread.

In the North-East there is the third main zone—the Chaunian (Fig. 283) which like the Anyuian zone has a north-westerly trend and is closely associated with it. It is probably more correct to consider the Anyuian zone as part of the Chaunian, rather than Cherskian. The Chaunian zone has the same type of rocks as the Cher-skian with predominantly Upper Triassic, Jurassic and Cretaceous strata intruded by Kolymian granites. A belt of small olivine-pyroxene gabbros of Cimmeridian age is a distinctive feature of the zone.

Palaeozoic massifs represent a definite feature of the Cimmeridian zones. Such massifs are found amongst folded Mesozoic structures, influencing their trends and at the same time also being involved with them. Palaeozoic and Precambrian strata of such massifs are strongly folded and metamorphosed. Even Mesozoic strata within such massifs are folded and metamorphosed. This phenomenon led Snyatkov to

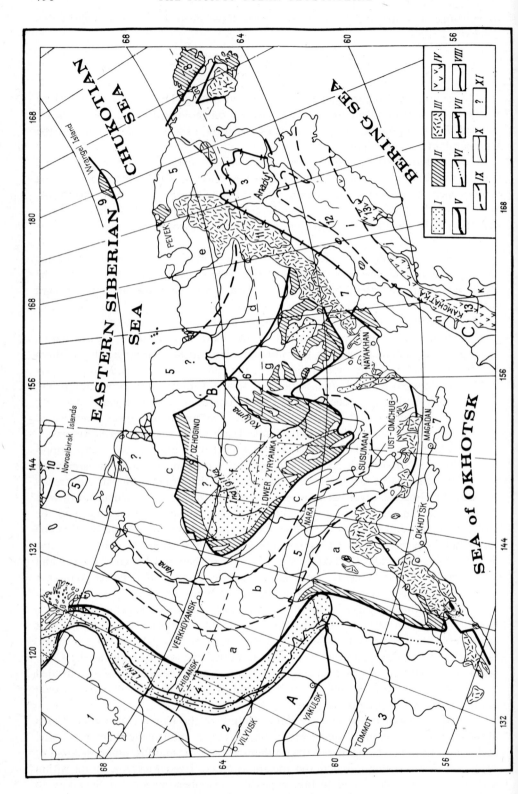

FIG. 283. The tectonics of the North-East. According to A. S. Vas'kovskii and L. A. Snyatkov (1957).

A — the Siberian Shield: 1 — the Olenek Massif; 2 — the Vilyui synclise; 3 — the Aldan Massif; 4 — the Lena-Khatanga Depression.

B — the North-East region of Mesozoic deformation: 5 — the Kolyma-Chukotian belt: a — the Verkhoyansk anticlinal zone, b — the Kularo-Sugoian neutral zone, c — the Debin-Poloussnian synclinal zone, d — the Anyuian synclinal zone, e — the Chaunian neutral zone; 6 — the Kolymian deformed Massif; f — the central Kolymian Block, g — the Omolon Block; 7 — the Okhotskian belt; 8 — the Uzleno-C'yuordian deformed Massif; 9 — the Wrangel island deformed Massif; 10 — the Novosibirsk islands deformed Massif; 11 — the Evenian volcanic arc.

C — the Nipponian region of Kainozoic deformation: 12 — the Kamchatka belt: h — the Anadyr-Tigil synclinal zone; i — the Koryakskian anticlinal zone, j — the Olyutorskian synclinal zone; 13 — the Itelmian volcanic arc. I — Cretaceous downwarps; II — outcrops of Middle and Upper Palaeozoic and Precambrian rocks in deformed regions; III — Upper Cretaceous and Palaeogene volcanic rocks; IV — Anthropogenic volcanic rocks; V-X — boundaries to various tectonic regions; XI — ground hidden under Anthropogenic deposits.

name them 'the deformed Palaeozoic massifs'. Five massifs have been recognized, namely: the Kolymian, the Chukotian, the Novosibirskian, the Okhotskian and the Taigonos (Fig. 283).

The Kolyma Massif and the Omolon Block, continuous with it, are the largest, being 1200km long and 700km wide. Palaeozoic and Precambrian strata occur within it ubiquitously. Middle and Lower Palaeozoic deposits, represented by all systems, predominate. The strata are thick, intensely dislocated, metamorphosed and intruded by Palaeozoic granites. Volcanic rocks, especially of Upper Devonian age, are widespread. Thus Lower and Middle Palaeozoic sediments have a geosynclinal character indicating that the massif was formed as a result of the Caledonian and the Hercynian orogenies. Precambrian rocks are not very widespread and neither are Upper Palaeozoic (Permian) strata.

The widespread Mesozoic strata of the Kolyma Massif consist mainly of Upper Jurassic and coal-bearing Lower Cretaceous rocks, while in the Omolon Massif Lower and Middle Jurassic and Triassic deposits are uncommon and consist exclusively of continental and volcanic rocks, which are moderately folded. All Mesozoic rocks beginning with Lower Cretaceous strata are as folded and metamorphosed as in fold belts adjacent to the massif. Nevertheless Mesozoic facies within the Palaeozoic massif are different from those outside it. This feature and also the clear deflection of fold zones round the massif indicate that such massifs were basically formed at the end of the Palaeozoic Era, during the Hercynian Orogeny. The Cimmeridian Orogeny nevertheless has had a considerable influence on the massifs, fragmenting them into separate blocks and displacing such blocks together with their cover of Triassic and Jurassic strata, and deforming such strata into folds. Such displacements are responsible for the 'deformed massifs'

The other massifs are much smaller and are commonly referred to as blocks. Southern massifs—the Okhotsk, the Taigonos and the Chukotian—show a widespread development of Precambrian formations which occupy somewhat larger areas than Middle and Lower Palaeozoic strata. Northern massifs represented by Wrangel island and the Novosibirsk islands consist entirely of Palaeozoic strata and Precambrian rocks are unknown. Mesozoic and Cainozoic deposits are of the same type as in the Kolyma Massif.

The structure of the Kolymian and other massifs of the North-East is very different from the structure of the Ukrainian Shield, which is also surrounded by fold zones. The Ukraine Shield has approximately the same size as the Kolyma Massif, but Palaeozoic and younger rocks lying within it are not folded and are almost unmetamorphosed, and the shield cannot be called a deformed massif. The similarity with Palaeozoic and Precambrian massifs of the Caucasus such as the axial massif of the Greater Caucasus, the Dzirul Massif and other ancient massifs of the Lesser Caucasus is much closer, but an identity cannot be claimed. The axial massif of the Greater Caucasus is surrounded by the oldest Mesozoic deposits and does not deflect the trend of structures affecting them. Conversely the youngest Mesozoic structures are adjacent to the Kolyma Massif and there are Lower Cretaceous and even Neogene rocks within it. A closer similarity exists between the Dzirul, the Nakhichevan (Armenian) and other massifs of the Lesser Caucasus and southern massifs of the North-East—the Okhotskian, the Taigonos and the Chukotian. Both groups of massifs are formed by Palaeozoic and Precambrian rocks and are surrounded by Mesozoic and Cainozoic strata. Nevertheless a complete identity still cannot be claimed, and it is best to regard the ancient massifs of the North-East as peculiar structures, still unidentified in other regions.

The establishment of tectonic zones has helped the study of metallogeny and the metallogenic map of the North-East to a large extent is based on tectonic data, as has been pointed out by V. T. Matveyenko (1959) at a metallogenic conference held in 1958 in Alma-Ata. Five intrusive metal-bearing complexes can be recognized in Cimmeridian zones.

The first, Lower Triassic complex, is represented by diabases which are not widespread and which are accompanied by an only weak ilmenite-magnetite and copper mineralizations.

The second complex involves a pre-batholithic series of stocks and dykes of diorite porphyrites and quartz-albite porphyries. The intrusive bodies form a belt 1100km long, 200km wide, known as the Yana-Indigirka-Kolyma gold belt. It has a great importance since main syngenetic and alluvial gold deposits are associated with this complex.

The third or batholithic complex is represented by major batholiths of leucocratic Lower Cretaceous granites which occur around the Kolyma Massif. Quartz veins gneisses, skarns and pegmatites associated with the batholiths have a rare metal, tin-tungsten-molybdenum mineralization of some considerable importance.

The fourth or synorogenic complex is represented by small granite bodies, developed mainly in the Chukotian zone and the associated mineralization is negligible.

The fifth complex is anorogenic, is not accompanied by folding and consists of intrusive bodies which fill fissures in Jurassic and Lower Cretaceous strata. The age of intrusive rocks is Upper Cretaceous and with them is associated a fairly widespread rare-metal mineralization of the Yana-Kolymian zone.

The distinction of these complexes and the investigation of the associated mineralization have enabled the recognition of a major mining region. Fruitful ideas and constructions advanced by S. S. Smirnov and Yu. A. Bilibin have been useful in this respect.

Alpine zones represent fold belts produced by the Alpine Orogeny. The orogeny has a long duration. Thus beginning with Lower Cretaceous times onwards a series of Alpine phases, separated by periods of relative calm, affected the region giving rise to various, non-contemporaneous fold belts.

In the North-East from the coast of the Sea of Okhotsk and to the mouth of the Anadyr all zones converge and join each other. To the south they diverge. Two zones, the Okhotskian and the Anadyrian, deviate to the north and margin the deep part of the Sea of Okhotsk from the west. Further south these zones continue into the Sikhote-Alin' range and Sakhalin (Fig. 275). Two other zones, the Koryakskian and the Olyutorian, continue into Kamchatka and the Kuril islands margining the deep part of the Sea of Okhotsk from the east (Fig. 284).

The Sea of Okhotsk depression with depths of 500 to more than 3000m lies amongst fold zones and is surrounded by them like the Kolyma Massif. Thus it is quite possible that the depression is a Palaeozoic massif. It is, however, also possible that the depression consists of divergent widened and depressed fold zones and is a peculiar structure (a trough) which occurs in contradistinction to the 'knot' situated to the north.

An important feature of Alpine zones is their different trend from Cimmeridian zones of the north. For instance the Chukotian and the Anyuiian zones which have a north-westerly trend are overprinted and deformed by the Okhotskian and Anadyrian zones with north-easterly trends. Such an overprinting and deformation of structures is also observed along the coast of the Sea of Okhotsk, to the west of Magadan. In the south of the Far East trends become gradually convergent and become almost parallel at the Sikhote-Alin' latitude. It is interesting that further to the south the Sikhote-Alin' and Sakhalin trends again diverge and the Sakhalin zone which corresponds to the Anadyr zone is again adjacent to the Kamchatkan zone represented by the Kuril islands. Between the Sikhote-Alin' and the Sikhote zones lies the Japan Sea Depression which, like the Sea of Okhotsk, possibly represents an ancient Palaeozoic massif.

Let us now consider briefly the individual Alpine zones. The Okhotskian zone is

FIG. 284.   Cape Olyutorskii on the Bering Sea. *Photograph by Yu. P. Degtyarenko.*

adjacent to Cimmeridian zones, at times intersecting and overprinting them, and at times trending in a parallel fashion. In the extreme north the Okhotskian zone lies parallel to the Chaunian zone and then overprints the Anyuiian and the Cherskian zones, lying almost at right angles to them. Approximately near Okhotsk the Okhotskian zone bends southwards, crosses the sea of Okhotsk, and continues in the Sikhote-Alin' range. Throughout its great length of over 3000km the zone has several distinctive features. The first is the unusual development of Upper Cretaceous and Palaeogene volcanicity. Andesitic and liparitic lavas and tuffs occupy large areas reaching 3km in thickness. The second feature is the interstratification of continental coal-bearing (rarely marine) and volcanic deposits. The third feature is an almost complete absence of marine Neogene strata. Deposits of gold, mercury, tin, lead and zinc are associated with volcanic rocks. An intrusive (Okhotskian) complex of rather large grandiorite bodies is characteristic of the zone. The complex is pre-Upper Cretaceous and is accompanied by gold, tungsten, molybdenum and tin.

The Koryakskian zone includes the whole of the Koryakskii range and western and central parts of Kamchatka. This is the oldest uplift and is in many respects similar to the Anadyrian zone. Within the Koryakskian zone there are the oldest, Upper Proterozoic (Sinian) schists, which to the north of the Koryakskii range of Kamchatka have yielded spores. There are also indications of the presence of Silurian and Devonian strata. Lower Carboniferous strata (1200–1400m) are found in the Koryakskii range and consist of lavas, sandstones and reef limestones, with Visean and Namurian faunas (Fig. 261), uncomformably overlain by sandy-shaly carbonaceous rocks (1200m) with an Upper Palaeozoic flora. The later sequence is succeeded by *Aucella*-bearing Upper Jurassic Lower Cretaceous shales (650–750m). Upper Cretaceous, Palaeogene and Neogene sediments are of a great thickness of up to 25km. The sediments are clastic and are invaded by ultrabasic intrusions, with deposits of platinum, chromite and asbestos, and by Miocene granodiorites accompanied by an antimony-mercury mineralization.

The Olyutorskian zone includes: all north-eastern foothills of the Koryakskii range, the region of the Olyutorskii gulf, the eastern part of Kamchatka and the

FIG. 285. An Anthropogenic sequence (Anukian) of tuffolavas in the Koryakskii range. Pale rocks at the bottom are tuffs, while darker top cover consists of lavas. *Photograph by Yu. P. Degtyarenko.*

Kuril islands. To the east of the southern part of the zone lies one of the largest oceanic trenches. A distinctive feature of the zone is an unusual development of young, Pliocene, Anthropogene and Recent vulcanicity, including the largest active volcanoes in the U.S.S.R. Areas occupied by lavas and tuffs are enormous (Fig. 285). This is associated with an unusual development of deep faults which provide channels for rising magma. The whole zone shows high seismicity. Amongst older strata, sometimes appearing from under young volcanic rocks, marine continental (coal-bearing and volcanic facies of a total thickness of 4–6km) can be recognized.

The brief outline of the Olyutorskian zone ends the description of the geological structure of the Pacific Ocean Geosyncline, which is the most distant, least studied and a very interesting geological region of the U.S.S.R.

The description of the Pacific Ocean Geosyncline also ends the description of the geology of the Soviet Union.

BIBLIOGRAPHY

RUSSIAN ABBREVIATIONS

A.A.—Academy of Sciences (Akademiya Nauk).
B.M.O.I.P.—Bull. Moscow Society of Naturalists.
V.I.M.S.—All-Union Institute of Raw Minerals, Moscow.
V.N.I.G.N.I.—All-Union Scientific-Investigatory Petroleum Institute, Moscow.
V.N.I.G.R.I.—All-Union Scientific-Investigatory Institute of Surveying, Leningrad.
V.S.E.G.E.I.—All-Union Institute of Geology, Leningrad.
G.G.R.V.—Main Geological Office (later reorganized into Committee of Geological Affairs), Moscow–Leningrad.
Geol. kom.—Geological Committee (now V.S.E.G.E.I.).
G.I.N.—Geological Institute of the Academy of Sciences, U.S.S.R.
D.A.N.—Doklady of Academy of Sciences, U.S.S.R.
I.A.N.—Izvestiya of Academy of Sciences, U.S.S.R.
I.G.E.M.—Institute of the Geology of Ores, Petrography, Mineralogy and Geochemistry, Moscow.
I.G.N.—Institute of Geological Sciences, Academy of Sciences, S.S.S.R.
N.I.I.G.A.—Scientific-Investigatory Institute of the Geology of the Arctics, Leningrad.
Sov. geol.—Soviet Geology (a journal).
Ts.N.I.G.R.I.—Central Scientific-Investigatory Surveying Institute (now V.S.E.G.E.I.).

AALOE, A. O., MARK, E. YU., *et al.*, 1958. *A Review of the Stratigraphy of Palaeozoic and Quaternary Deposits of the Estonian S.S.R.* Tallin. 46 pp.
AFANAS'YEV, G. D., 1955. The problem of the age of magmatic rocks of the Northern Caucasus. *Izv. Akad. Nauk, ser. geol.,* No. 4, 57–79.
——, 1957. On Cainozoic magmatism in the Caucasus and certain conclusions of absolute age determinations of Caucasian rocks by K-Ar method. *Izv. Akad. Nauk, ser. geol.,* No. 6, 30–54.
AFANAS'YEV, G. D., *et al.*, 1960. Results of geochronological investigations of magmatic rocks in the Caucasus, pp. 161–94 in the book: *The Determination of Absolute Age.* Moscow.
AIZENSHTADT, G. E.-A., 1956. A classification of Southern Emba salt domes. *Trudy V.N.I.G.R.I.,* 2, **95,** 214–21.
ALEINIKOV, A. A., 1960. *Main Problems of Investigation of Quaternary (Anthropogenic) Deposits of the North-West of the U.S.S.R.* Leningrad. 66 pp.
ALEKSEICHIK, S. K., *et al.*, 1959. The geological structure and gas and oil in the northern part of Sakhalin. *Trudy V.N.I.G.R.I.,* **135,** 233 pp.
ALEKSIN, A. G. & TSATUROV, A. I., 1957. Main results of reference drilling in the Terek-Kuma Plain. *Trudy V.N.I.G.R.I.,* **111,** 232–53.
ALESHKOV, A. N., 1935. In the north of the Arctic Urals, pp. 150–76, in the book: *Trudy of the Ice Expedition,* 4th issue. Leningrad.
ALESKEROVA, Z. T., Li, P. F., Osyko, T. I., Rostovtsev, N. N., and Tolstikhina, M. A., 1957. The stratigraphy of Mesozoic and Tertiary deposits of the Western Siberian Lowlands. *Sov. Geol.,* **55,** 145–72.

ALFEROV, B. A., *et al.*, 1961. Uvatskian borehole. *Trudy V.N.I.G.R.I.*, **178**, 91 pp.

ALIKHOVA, T. N., 1960. *The Stratigraphy of Ordovician Deposits of the Russian Platform.* Moscow. 76 pp.

ALIYEV, M. M., 1939. Inoceramids of Cretaceous deposits of the north-eastern part of the Lesser Caucasus. *Trudy geol. Inst. Azerbaidzhan Fil.*, **12** (63).

ALIYEV, M. M., ALI-ZADE, K. A., *et al.*, 1960. The palaeography of eastern Transcaucasia during the Cretaceous period and in Cainozoic times, pp. 138–46 in the book: *Regional Palaeogeography. Doklady of Soviet Geologists, XXI Intern. Geol. Cong.* Moscow.

ALI-ZADE, A. A., 1945. *The Maikop Suite of Azerbaidzhan and its Oil.* Moscow.

——, 1954. *The Akchagylian Stage of Azerbaidzhan.* Baku. 344 pp.

AMIRASLANOV, A. A., 1957. The development of the centres of non-ferrous metals from 1956 to 1960. *Izv. Akad. Nauk, ser. geol.*, No. 4, 59–65.

ANDREYEV, YU. F. & BYELORUSOVA, ZH . M., 1961. The geological structure of the Tazovskian Peninsula. *Trudy V.N.I.G.R.I.*, **186**, 176–202.

ANDREYEVA, F. I. & CHIRVINSKAYA, M. V., 1959. Hypsometry of the basement to the Dnyepr-Donyets Depression. *Geologiya Nefti*, **6**, 55–61.

ANDRONOV, S. M., 1961. Devonian deposits of the eastern slope of the Southern Urals. *Dokl. Akad. Nauk*, **141**, 4, 925–8.

ANDRUSOV, N. I., 1893. The geotectonics of the Kerch Peninsula. *Mater. Geol. Ross.*, **16**, 63–335.

ANKINOVICH, S. G., 1960. Kazakhstan—a new centre of Vanadium ores, pp. 342–6 in the book: *Basic Ideas of N. G. Kassin*, Alma-Ata.

ANTROPOV, P. YA., 1959. *Prospects of Acquisition of Natural Resources of the U.S.S.R.* Moscow. 159 pp.

——, 1960. To improve further the geoexploration. *Soviet Geol.*, No. 1, 3–9.

ARAKELYAN, R. A., 1952. *Palaeozoic deposits of Armenia and the Nakhichevan A.S.S.R.* (pp. 5–12). Baku.

——, 1957. The stratigraphy of the ancient metamorphic complex of Armenia. *Izv. Akad. Nauk armyan. SSR*, **10**, issue 5–6, 3–16.

ARKHANGEL'SKII, A. D., 1910. Upper Cretaceous deposits of the eastern part of European Russia. *Mater. Geol. Ross.*, **25**, 631 pp.

——, 1932. *The Geological Structure of the U.S.S.R.* Moscow. 424 pp.

——, 1941. *The Geological Structure and the Geological History of the U.S.S.R.* Moscow. 376 pp.

ARKHANGEL'SKII, N. K., 1959. Tertiary deposits of the Palanskian region, Kamchatka, pp. 101–2 in the book: *Theses of Papers of the Stratigraphical Conference at Okha.* Okha.

ASATKIN, B. P., 1932. Ecardines from the Lower Silurian of the Siberian Platforms. *Izv. vses. geol. razved. Ob"ed.*, **51**, 32, 483–95.

ASLANYAN, A. T., 1958. *The Regional Geology of Armenia.* Erevan. 429 pp.

AZIZBEKOV, SH. A., 1947. *The Geology and Petrography of the North-Eastern Part of the Lesser Caucasus. Izd. Akad. Nauk Azerbaijan S.S.R.*, Baku, 300 pp.

BABAYEV, A. G., 1957. The facies and the geological history of Western Uzbekistan. *Bull. Mosc. Soc. Nat. geol. Div.*, **32**, 3, 3–31.

BAKIROV, A. A., 1948. *The Geological Structure and Prospects of Gas and Oil in Palaeozoic Deposits of the Central Russian Syneclise.* Moscow. 282 pp.

——, 1951. The main features of the geotectonic evolution of the interior of the Russian Platform, pp. 3–46 in the book: *Contributions to the Geology of Central Regions of the Russian Platform.* Moscow.

BARANOV, I. G., VITENKO, V. A., ZAV'YALOV, V. M., & MUROMTSEV, A. S., 1961. Prognosed reserves of oil and gas in the Dnyepr-Donyets Depression. *Geol. Nefti*, **8**, 17–19.

BARKHATOV, G. V., VASIL'YEV, V. G. *et al.*, 1958. *Oil and Gas in the Eastern Siberian Platform.* Moscow. 131 pp.

BARTSEVA, M. N. & PERFIL'YEV, YU. S., 1957. Data on the stratigraphy of the Ordovician and Silurian of the north-western Altais. *Trudy vses. aérogeol. Tresta*, **3**, 5–12.

BEL'SKAYA, T. N., 1960. The late Devonian sea of the Kuznyetsk Depression, the history of its development, life and sediments. *Trudy paleont. Inst.*, **82**, 184.

BELYAYEVSKII, N. A. 1956. A description of the geology of Primorie. *Mater. V.S.E.G.E.I.*, **15**, 5–12.

BENSH, F. R., 1958. A stratigraphic scheme of Permian deposits of Central Asia, pp. 86–91 in the book: *Theses of Papers of Stratigraphic Conference on Central Asia.* Moscow.

BENSH, F. R., SERGUN'KOVA, O. I. & SOLOV'YEVA, M. K., 1958. A scheme of stratigraphic sub-division of the Central Asiatic Carboniferous, pp. 55–66 in book: *These of Papers of Stratigraphic Conference on Central Asia.* Moscow.

BEZRUKOV, P. L., 1936. Danian stage of the Eastern European Platform. *Izv. Akad. Nauk S.S.S.R.*, **5**, 657–88.

BILIBIN, YU. A., 1938. *Principles of the Geology of Unconsolidated Sediments.* Moscow. 504 pp.

BOBROV, A. K., 1960. The geological structure of the Yakutian A.S.S.R. and its prospects for oil and gas. *Trudy V.N.I.G.R.I.*, **63**, 72–139.

BODYLEVSKII, V. I., 1944. Marine Jurassic strata of the Urals, pp. 266–78 in the book: *Geology of the U.S.S.R.*, vol. XII, part 1. Moscow.

BODYLEVSKII, V. I. & SHULGINA, N. I. 1958. Jurassic and Cretaceous fauna of the Lower Yenisei. *Trudy nauchno-issled. Inst. Geol. Arkt.*, **93**, 196.

BOGACHEV, V. V., 1961. *Data of the History of Fresh-Water Fauna of Eurasia.* Kiev. 341 pp.

BOGDANOV, A. A., 1949. Main features of the tectonics of the Eastern Carpathians. *Soviet Geol.*, **40**, 11–12.

——, 1959. Main features of the Palaeozoic structure of Central Kazakhstan. *Bull. Mosc. Soc. Nat. geol. Div.*, **34**, 1, 3–38.

BOGOMAZOV, V. M., 1961. The stratigraphy and conditions of formation of pre-coal and coal-bearing Carboniferous and Permian deposits of the Minusinskian basin, pp. 99–117 in the book: *Problems of the Geology of Coal-bearing Deposits.* Moscow and Leningrad.

BORISYAK, M. A., 1957. *The Stratigraphy and Brachiopods of Silurian Deposits of the Range Chingiz Region.* Thesis. Leningrad. 15 pp.

BOROVIKOV, L. I., 1955. The Lower Palaeozoic of the Dzhezkazgan-Ulutau region of the western part of Central Kazakhstan. *Trudy V.S.E.G.E.I.*, new ser., **6**, 252.

BORUKAYEV, R. A., 1955. *The Pre-Palaeozoic and the Lower Palaeozoic of the North-East of Central Kazakhstan.* Alma-Ata. 408 pp.

BRAZHNIKOVA, N. E., NOVIK, E. L., SHUL'GA P. L. *et al.*, 1956. The fauna and flora of Car-boniferous deposits in the Galitsia-Volyn Depression. *Trudy Inst. Geol. Akad. Nauk Ukraine S.S.R.*, **10**, 410.

BREDIKHIN, I. S., 1961. The stratigraphy of Mesozoic deposits of the Southern Yakutian coal basin, pp. 133–4 in the book: *Theses of Papers of Conference on the Development of Stratigraphic Schemes in the Yakutian A.S.S.R., Yakutsk, 1961.* Leningrad.

BRUNS, E. P., 1956. The history of the development of the Pripet downwarp in Palaeozoic times (In: *Data on the Geology of the European Territory of the U.S.S.R., Materialy V.S.E.G.E.I.*, **14**, 185–208.

——, 1957. The stratigraphy of ancient Pre-Ordovician deposits in western regions of the Russian Platform. *Sov. Geol.*, **59**, 3–24.

BUDANOV, N. V., 1957. The role of recent tectonics and the resultant fissures in the hydro-geology of the Urals. *Soviet Geol.*, **58**, 25–39.

BULEISHVILI, D. A., 1957. The geological structure of the intramontane depression of Georgia and prospects of oil in it. *Trudy V.N.I.G.R.I.*, **111**, 293–351.

BYELOUSOV, V. V., 1938. *The Great Caucasus (Trudy Ts.N.I.G.R.I.)*: 1938. **108**, 99; 1940. **121**, 175; 1939. **126**, 91.

——, 1944. The Facies and Thicknesses of Sedimentary Formations in the European Part of the U.S.S.R., *Trudy Inst. geol. Sci. Akad. Nauk S.S.S.R.*, **76**, 116.

——, 1948. *General Geotectonics.* Moscow, 599 pp.

——, 1962. *General Problems of Geotectonics.* Moscow, 608 pp.

BYELOUSOV, A. F. and Sennikov, V. M., 1960. The Cambrian of the north-eastern Altais. *Trudy Middle Asia Sci. Inst. geol. ores*, **13**, 123–35.

CHAIKOVSKII, V. K., 1960. *The Geology of Tin Deposits in the North-East U.S.S.R.* Moscow. 335 pp.

CHERNYSHEV, F. N., 1902. Upper Carboniferous brachiopods of the Urals and Timan. *Trudy geol. Com.*, **16**, 2, 749.

——, 1889. General Geological Map, Sheet 139. *Trudy Geol. Com.*, **3**, issue 4.

CHERNYSHEVA, N. E., 1955. The stratigraphy of Cambrian deposits of the south-eastern margin of the Siberian Platform. *Mater. V.S.E.G.E.I.*, **7**, 29–40.

——, 1957. A contribution to the problem of dividing Cambrian deposits of the Siberian Platform. *Soviet Geol.*, **55**, 78–92.

CHIRKOVA-ZALESSKAYA, E. F. *The Differentiation of Terrigenous Devonian Strata in the Ural-Volga Region on the Basis of Fossil Plants.* Moscow. 139 pp.

CHIRVINSKAYA, M. V., 1958. The tectonic structure of the Dnyepr-Donyets Depression and the Pripet downwarp, in the book: *Tectonics of Petroliferous Regions*, vol. 2.

CHOCHIA, N. G., 1955. The geological structure of the Kolva-Vishera district. *Trudy V.N.I.G.R.I.*, **91**, 392–404.

CHOCHIA, N. G. & ADRIONOVA, K. I., 1952. The Devonian of the Kolva-Vishera region. *Trudy V.N.I.G.R.I.*, **61**, 122–99.

CHOCHIA, N. G., KRASNOV, V. I. & IPATOVA, Z. N., 1956. Minusinskian troughs, in the book: *Descriptions of the Geology of the U.S.S.R.* (Vol. 1). *Trudy V.N.I.G.R.I.*, **96**, 215–34.

CHOCHIA, N. G., BELYAKOVA, E. E. & BOROVSKAYA, I. C., 1958. *The Geological Structure of the Minusinskian Intramontane Depressions and their Prospects for Oil and Gas.* Leningrad. 296 pp.

CHOCHIA, N. G., et al., 1961. The Muzhinskian Urals and their geological structure. *Trudy V.N.I.G.R.I.*, **186**.

CHURAKOV, A. N., 1941. The Proterozoic of the north-western part of the Eastern Sayans. *Trudy Geol. Inst., Akad. Nauk*, 59, 92–4.

DAVITASHVILI, L. SH., 1933. *A Review of Molluscs in Tertiary and post-Tertiary Deposits of the Crimean-Caucasian Oil Province*, Moscow, 168 pp.

——, 1956. On the evolution of Black Sea faunas during Pliocene. *Akad. Nauk Georgian S.S.R.*, **17** (3), 227–34.

DAVYDOVA, T. N., 1961. On the positioning of the main break in the Cambrian and Ordovician succession of the northern part of Soviet Baltic countries. *Izv. Akad. Nauk S.S.S.R. ser. geol.*, No. 12, 58–70.

DEMOKIDOV, K. K., 1958. On stratigraphic subdivision of Cambrian deposits in the north of the Siberian Platform. *Trudy nauchno-issled. Inst. Geol. Arkt.*, **67**, 3–12.

DEMOKIDOV, K. K. & LAZARENKO, N. P., 1961. Stratigraphic subdivisions of Cambrian deposits in north-western Yakutia, a table, pp. 31–3 in the book: *Theses of Papers*, Yakutsk.

DEMOKIDOV, K. K., SAVITSKII, V. E., et al., 1959. The stratigraphy of Sinian and Cambrian deposits of the north-east of the Siberian Platform. *Trudy nauchno-issled. Inst. Geol. Arkt.*, **101**, 211.

DENISOVICH, V. V., DIKENSHTEIN, G. KH., et al., 1961. Main results of geological explorations for oil and gas in Central Asiatic republics. *Geol. Nefti*, **10**, 11–17.

DIKENSHTEIN, G. KH., 1953. Achievements of investigations of Palaeozoic deposits in western regions of the Ukrainian S.S.R. in the past ten years (1939–49). *Trudy L'vov. geol. Soc.*, **2**, 3–17.

——, 1957. The problem of finding oil and gas in lower Palaeozoic deposits of the north-western part of the Russian Platform. *Geol. Nefti*, **9**, 7–13.

DIKENSHTEIN, G. KH., BEZNOSOV, N. G., et al., 1958. *The Geology and Oil and Gas of Steppe and Submontane Crimea.* Moscow. 147 pp.

DIKENSHTEIN, G. KH., et al., 1959. *The Gazlinskian Gas-Oil Deposit.* Moscow. 46 pp.

DNEPROV, V. S., 1962. The geological structure of the South Emba elevation. *Trudy V.N.I.G.R.I.*, **194**, 124.

DOMRACHEV, S. M., 1952. The Devonian of the Kara Tau range and of adjacent regions in the Southern Urals. *Trudy V.N.I.G.R.I.*, **61**, 5–121.

DOVZHIKOV, A. E., et al., 1958. Main types of Silurian deposits in the Tyan-Shan, pp. 33–5 in the book: *Theses of Papers of Stratigraphic Conference on Central Asia.* Moscow.

DROBYSHEV, D. V., 1951. Upper Cretaceous and the Palaeogene carbonate deposits in the Northern Caucasus. *Trudy V.N.I.G.R.I.*, **42**, 225.

DROBYSHEV, D. V. & KAZARINOV, V. P., 1958. Geology and oil in the Western Siberian Lowlands. *Trudy V.N.I.G.R.I.*, **114**, 273.

DRONOV, V. I., 1958. The scheme of subdivision of Triassic and Jurassic deposits of the Pamirs, pp. 105–10 in the book: *Theses of Papers of Stratigraphic Conference on Central Asia*. Moscow.

DRUMYA, A. V. & IVANCHUK, P. K., 1962. On the geological structure of Zmeinyi Island (Black Sea). *Bull. Moscow Nat. Hist. Soc.*, **37**, 1, 125–30.

DUBINSKII, A. YA., 1956. *Eastern Donbas*. Thesis. Leningrad. 22 pp.

DVALI, M. F., 1955, *The Geological Structure and Oil of Eastern Kamchatka*. Leningrad. 268 pp.

——, 1957. The geological structure of the Palanskian region (western Kamchatka). *Trudy V.N.I.G.R.I.*, **102**, 95–179.

D'YAKOV, B. F., 1955. The geological structure and oil of western Kamchatka. *Trudy V.N.I.G.R.I.*, **14**, 255.

——, 1957. Facies analysis of Tertiary deposits and main features of the Palaeogeography of the Tigil region of western Kamchatka. *Trudy V.N.I.G.R.I.*, **102**, 180–241.

D'YAKOV, B. F., IMASHEV, N. U., KRUCHININ, K. V., et al., 1961. The Southern Mangyshlaks—a new major oil province. *Geol. Nefti*, **12**, 4–12.

D'YAKOV, B. F. & TIMOFEYEV, B. V., 1956. On the age of metamorphic rocks of Kamchatka. *Trudy V.N.I.G.R.I.*, **95**, 165–70.

DZEVANOVSKII, YU. K., 1956. The stratigraphy of Jurassic continental deposits of the Aldan Shield, pp. 23–4 in *Theses of papers of the Conference on the Stratigraphy of Siberia*. Leningrad.

——, 1958. Archean metamorphic complex of the Aldan Shield, pp. 37–42 in the book: *Trudy of the Conference on the Stratigraphy of Siberia*. Leningrad.

——, 1961. Archean formations of the Aldan Shield, pp. 6–9 in the book: *Theses of Papers of the Conference on the Development of Stratigraphic Schemes of the Yakutian A.S.S.R.* Yakutsk.

DZHANELIDZE, A. I., 1953. The territory of Georgia in the system of the Alpine orogen. *Trudy Inst. geol. Akad. Nauk Georgian S.S.R.*, **7**, 121–29.

EBERZIN, A. G., 1957. *The Unified Stratigraphic Scheme of Neogenic Successions in Southern Regions of the European Part of the U.S.S.R.* Moscow.

EINOR, O. L., 1960. *Principles of the Geology of the U.S.S.R.* Kiev. 337 pp.

EVENTOV, YA. S., 1957. Palaeozoic deposits of the western part of the Ciscaspian Depression. *Sov. Geol.*, **57**, 130–53.

FAVORSKAYA, M. A., 1950. Stages of development of young vulcanism in Southern Primoria. *Izv. Akad. Nauk, ser. geol.* **3**, 133–47.

——, 1956. Upper Cretaceous and Cainozoic magmatism of the eastern slope of the Sikhote Alin range. *Trudy I.G.E.M.*, **7**, 308.

FEDOROV, P. V., 1952. *The Quaternary History of the Caspian Sea*. Moscow. 298 pp.

FLEROVA, O. V. & GUROVA, A. D., 1958. Upper Cretaceous deposits in central regions of the Russian Platform, pp. 185–226 in the book: *Mesozoic and Cainozoic Deposits*. Moscow.

FLORENSOV, N. A., 1956. Certain structural features of coal bearing formations of Cisbaikalia. *Trudy Lab. geol. coal*, **6**, 558–67.

——, 1960. Mesozoic and Cainozoic depressions in Cisbaikalia. *Trudy East. Siber. Fil.*, **19**, 258.

FLOROV, B. M., 1961. The stratigraphy of Sinian deposits in the basin of the river Sarma (Central Cisbaikalia). *Trudy V.N.I.G.R.I.*, **186**, 101–8.

FOMICHEV, V. D. & ALEKSEYEVA, L. E., 1961. A geological description of Salair. *Trudy V.S.E.G.E.I.*, new ser., **63**, 217.

FORSH, N. N., 1955. Permian deposits—the Ufa Suite and the Kazanian Stage. *Trudy V.N.I.G.R.I.*, **92**, 156.

FRUKHT, D. L., 1959. Triassic deposits in central regions of the Russian Platform, pp. 5–29 in the book: *Mesozoic and Tertiary deposits in Central Regions of the Russian Platform*. Moscow.

FURSENKO, A. V., 1953. On Upper Devonian deposits of the Pripet Polessia. *Dokl. Akad. Nauk*, **90**, 2, 239–42.

——, 1957. On the stratigraphy of Devonian deposits of the Pripet Depression. *Trudy leningr. Obshch. Estest.*, **69**, issue 2, 5–23.

GABRIL'YAN, A. M., 1957. *The Lithology, the Palaeogeography and Problems of Finding Oil in the Upper Cretaceous and Palaeogene of the Fergana Depression.* Izd. Akad. Nauk Uzbekian S.S.R., Tashkent, 397 pp.

GAMKRELIDZE, P. D., 1951. General ideas on the geotectonic structure of Georgia. *Trudy Inst. geol., Akad. Nauk Georgian S.S.R.*, pp. 405–18.

GAR'KOVETS, V. G., DIKENSHTEIN, G. KH., *et al.*, 1961. A contribution to the problem of search for oil in western Tadzhikistan. *Geol. Nefti*, **8**, 7–12.

GEDROITS, N. A., 1951. The Taimyr Lowlands, *Trudy nauchno-issled. Inst. Geol. Arkt.*, **13**, 166.

GEISLER, A. N., 1956. New data on the stratigraphy and tectonics of lower Palaeozoic strata in the North-Western part of the Russian Platform, pp. 174–85 in the book: *Data on the Geology of the European Territory of the U.S.S.R.* Leningrad.

GEKKER, R. F., OSIPOVA, A. I. & BEL'SKAYA, T. N., 1960. The Palaeogene Bay of Fergana, pp. 147–63 in the book: *Regional Palaeogeography.* Moscow.

GERASIMOV, A. P., 1939. The stratigraphy of Precambrian formations of the Caucasus, pp. 183–89 in the book: *The Stratigraphy of the U.S.S.R.*, vol. 1—*The Precambrian of the U.S.S.R.* Moscow and Leningrad.

GERASIMOVA, P. A., MIGACHEVA, E. E., NAIDIN, D. P. & STRELIN, B. P., 1962. *Jurassic and Cretaceous Deposits of the Russian Platform.* Moscow. 195 pp.

GLUSHKOV, A. P., 1956. A stratigraphical scheme of Upper Palaeozoic deposits of Primorie, pp. 27–9 in the book: *Theses of Papers of the Stratigraphic Conference at Khabarovsk.* Khabarovsk.

GODIN, YU. N., LUPPOV, N. P., SYTIN, YU. I. & CHIKHACHEV, P. K., 1958. Main features of tectonic structure in the territory of the Turkmenian S.S.R. *Soviet Geol.*, **1**, 3–24.

GOLUBTSOV, V. K. & MAKHNACH, A. S., 1961. *Palaeozoic and Early Mesozoic Facies in Byelorussia.* Minsk. 177 pp.

GOLUBYATNIKOV, V. D., 1956. On the tectonics of Central Ciscaucasia. *Mater. V.S.E.G.I.*, **8**, 285–93.

GORBACHEV, I. F., 1961. Rybinskian reference borehole (the Krasnoyar region). *Trudy V.N.I.G.R.I.*, **175**, 119.

GORODETSKAYA, N. S., 1948. The structure of the coal-bearing formation of the Kizelian basin, *Bull. Moscow Nat. Hist. Soc.*, **14**, 3, 52–64.

GORSKII, I. I., 1943. Geotectonic conditions of formation of coal deposits in the Urals. *Izv. Akad. Nauk, ser. geol.*, Nos. 4–5, 12–40.

——, 1948. Older Cimmeridian tectonic movements in the Urals. *Izv. Akad Nauk, ser. geol.*, No. 4, 67–82.

GORSKI, I. I. & LEONENOK, N. I., 1960. Lower Mesozoic accumulation of coal in Kazakhstan, pp. 401–11 in the book: *Basic Ideas of N. G. Kassin.* Alma-Ata.

GORYANOV, V. B., MIKLUKHO-MAKLAI, A. D., PORSHNYAKOV, G. S. & YAGOVKIN, A. I., 1961. Palaeozoic stratigraphy in the South Fergana antimony-mercury belt. *Sci. Zap., Middle Asiatic Sci. Inst. geol. mineral*, **2**, 7–28.

GORYAYEV, M. I., 1959. The stratigraphy of Tertiary deposits of the central part of Kamchatka, pp. 127–8 in: *Theses of Papers of the Stratigraphic Conference at Okha.* Okha.

GRAMBERG, I. S., 1958. The stratigraphy and lithology of Permian deposits of the north-east margin to the Siberian Platform. *Trudy nauchno-issled. Inst. Geol. Arkt.*, **84**, 215 pp.

GRAMM, M. I., 1958. A stratigraphic scheme of Neogenic continental deposits of the Central Asia, pp. 168–72 in the book: *Theses of Papers of Stratigraphic Conference on Central Asia.* Moscow.

GRIGOR'YEV, V. N., 1956. On the character of the Lower Cambrian flysch of the north-eastern margin of the Yenisei ridge. *Bull. Moscow Nat. Hist. Soc.*, **31**, 4, 55–64.

GRIGOR'YEV, V. N. & SEMIKHATOV, M. A., 1958. A contribution to the problem of age and origin of the so-called tillites of the northern part of the Yenisei ridge. *Izv. Akad. Nauk, ser. geol.*, No. 11, 44–58.

GROMOV, V. I., 1948. The palaeontological and archaeological basis of the stratigraphy of continental Quaternary deposits on the territory of the U.S.S.R. *Trudy geol. Inst. Akad. Nauk S.S.S.R.*, **64**, 520.

——, 1960. On the scheme of division of the Quaternary System on the territory of the U.S.S.R. and abroad. *Trudy geol. Inst. Akad. Nauk S.S.S.R.*, **26**, 3–10.

GROMOV, YU. YA. & PUTINTSEV, V. K., 1961. Main features of the Geology of Precambrian in the south of the Far East and adjacent territories. *Dokl. Akad. Nauk*, **138**, 6, 1409–12.

GROSSGEIM, V. A., 1948. Dibrarian rocks of the north-eastern Caucasus. *Izv. Akad. Nauk*, No. 2, 105–20.

——, 1961. The history of terrigenous minerals in Mesozoic and Cainozoic strata of the Northern Caucasus and Ciscaucasia. *Trudy V.N.I.G.R.I.*, **180**, 376.

GRYAZNOV, L. P., 1959. The stratigraphy of Tertiary deposits of eastern Kamchatka, pp. 119–22 in: *Theses of Papers of the Stratigraphic Conference at Okha*. Okha.

GRYAZNOV, V. I. & SELIN, YU. I., 1959. Main features of the geology of the Bolshetokmakian deposit. *Geol. Rudn. Mest.*, **1**, 35–55.

GUSEV, A. I., 1958. The stratigraphy of coal-bearing deposits of the Lena coal basin, pp. 85–88 in the book: *Trudy of the Conference on the Stratigraphy of Siberia, Mesozoic and Cainozoic*. Leningrad.

IL'YIN, V. D., 1959. The stratigraphy of Upper Cretaceous deposits of western Uzbekistan and adjacent regions of Turkmenia. *Trudy V.N.I.G.N.I.*, **23**, 181–222.

IL'YINA, T. G., 1962. On similarities and differences of Upper Permian and Lower Triassic coral faunas of Dzhul'fa. Thesis. *Bull. Moscow Nat. Hist. Soc.*, **37**, 1, 155–6.

IVANCHUK, P. K., 1957. The Geological structure of south-western and southern Cis-Black Sea regions. *Ocherki geol. S.S.S.R.*, **3**, 162–208.

IVANOV, A. A., 1950. Potassic salts in the Angara-Lena salt fields. *Zapiski, Min. Soc.*, **79**, 4, 297.

——, 1956. Certain new data on salt of the south-eastern margin of the Siberian Platform, pp. 268–85 in the book: *Material (Data) on the Geology and Useful Mineral Deposits*, vol. 1.

IVANOV, A. A. & LEVITSKII, YU. F., 1960. *The Geology of Halogenous Deposits in the U.S.S.R.* Moscow. 424 pp.

IVANOV, A. N. & MYAGKOVA, E. I., 1950a. The stratigraphy of Lower and Middle Palaeozoic rocks of the western slope of the Central Urals. *Trudy Mining geol. Inst. Ural. filial*, **17**, 3–20.

——, 1950b. The determination of the Ordovician fauna, of the western slope of the Central Urals. *Trudy Mining geol. Inst. Ural, filial*, **18**, 32.

IVANOV, B. A., 1949. Coal-bearing and other Mesozoic continental deposits of Transbaikalia. *Trudy East. Siber. geol. Dept.*, **32**, 1–193.

IVANOV, N. V. & LYUFANOV, L. E., 1961. A contribution to the stratigraphy of Jurassic deposits of the southern part of the Lena basin, pp. 324–40 in the book: *Problems of the Geology of Coal bearing Deposits*. Moscow and Leningrad.

IVANOV, YU. A., 1956. The stratigraphy of Jurassic and Lower Cretaceous deposits of Lower Priamurie (Cisamuria), pp. 47–9 in: *Theses of Papers of the Stratigraphic Conference at Khabarovsk*. Khabarovsk.

IVANOVA, A. M., 1958. *Upper Cambrian and Ordovician Deposits of the Pai-Khoi Range, in the Northern Part of the Arctic Urals.* Thesis. Leningrad. 15 pp.

KACHARAVA, I. V., 1944. Rachinsko-Lechkhumian Basin in Palaeogene times. *Trudy Inst. geol. Akad. Nauk Georgian S.S.R.*, **2**(7). 144 pp.

KALEDA, G. A., 1960. Devonian deposits of southern Fergana, pp. 183–240 in the book: *Problems of the Geology of southern Tian-Shians*, vol. 2, L'vov.

KALINKO, M. K., 1959. The history of geological development and prospects for finding oil and gas in the Khatanga Depression. *Trudy nauchno-issled, Inst. geol. Arkt.*, **104**, 360.

KALMYKOVA, M. A., 1958. The succession of microfaunal zones of Darvaz, pp. 94–7 in the book: *Theses of Papers of Stratigraphic Conference on Central Asia*. Moscow.

KALYUYZHNYI, V. A., 1959. Ancient metamorphic formations and metallogenic features of Timan. *Izv. Akad. Nauk, ser. geol.*, No. 6, 62–83.

KALYUZHNYI, V. A. & IVANOVA, K. P., 1959. Productive middle and Upper Devonian deposits in Southern Timan, pp. 32–60 in the book: *The Geology and Oil in the Timan-Pechora Region*. Leningrad.

KAMENSKII, S. N., TOLSTIKHINA, M. M. & TOLSTIKHIN, N. I., 1959. *The Hydrogeology of the U.S.S.R.* Moscow. 366 pp.

KAPITSA, A. A., 1959. Pre Upper Cretaceous deposits of Sakhalin Island, pp. 11–16 in the book: *Theses of Papers of the Stratigraphic Conference at Okha.* Okha.

KAPLUN, L. I. & RUKAVISHNIKOVA, T. B., 1958. The Silurian-Devonian boundary in the north-eastern Cisbalkhashia. *Izv. Akad. Nauk, ser. geol.*, No. 11, 59–70.

KARASEV, I. P., 1959. Litho-stratigraphical and geochemical characters of rocks in the southern part of the Siberian Platform, pp. 8–186 in the book: *The Geology and Oil in Eastern Siberia.* Moscow.

KAREVA, E. A., 1958. The stratigraphic scheme of the southern part of the Chelyabinsk brown coal basin. *Trudy V.N.I.G.R.I.*, **126**, 225–68.

——, 1959. The Upper Palaeozoic and Lower Mesozoic of the eastern slope of the Urals and the western part of the Western Siberian Lowlands. *Trudy V.N.I.G.R.I.*, **140**, 40–61.

KARLOV, N. N., 1957. A contribution to the history of investigations of volcanic ashes of the European part of the U.S.S.R. *Bull. Moscow Nat. Hist. Soc.*, **32**, 2, 25–47.

KARPINSKII, A. P., 1874. Geological investigations in the Orenburg district. *Zap. Min. Soc.*, issue 9, 101.

——, 1887. A description of physico-geographical conditions of European Russia in past geological periods. *Zap. Akad. Nauk*, **55**, Part 8, 1–36.

——, 1891. On ammonites of the Artinskian stage and on certain similar to their Carboniferous forms. *Zap. Min. Soc.*, issue 27, 15–208.

——, 1894. The general character of oscillations of the earth's crust within European Russia. *Izv. Akad. Nauk ser. geol.*, **1**, issue 1, 1–19. (This and his 1887 article were frequently re-issued in 1919 under the title: *Descriptions of the geological past of European Russia.*)

KASHIRTSEV, A. S., 1957. The biostratigraphy of Upper Palaeozoic deposits of Verkhoyania, pp. 39–41 in the book: *Theses of Papers of the Stratigraphic Conference at Magadan.* Magadan.

KAZARINOV, V. P., 1958. *Mesozoic and Kainozoic Deposits of Western Siberia.* Gostoptekhizdat. 324 pp.

KELLER, B. M., 1949. The Palaeozoic flysch formation in the Zilairian Synclinorium in the Southern Urals and other similar deposits. *Trudy Inst. geol. Sciences, Akad. Nauk S.S.S.R.*, **104**, 164.

——, 1952. Reef deposits of marginal downwarps of the Russian Platform. *Trudy Inst. geol. Nauk, Akad. Nauk S.S.S.R.*, **109**, 61.

KHAIN, V. E., 1950. *The Geotectonic Development of the South-eastern Caucasus.* Gostoptekhizdat. Baku. 224 pp.

——, 1954, *Geotectonic Principles of the Search for Oil.* Baku. 692 pp.

KHAIN, V. E, & AKHMEDOV. G. A., 1957. The geological structure of the Azerbaidzhan S.S.R. on the basis of reference drilling data (in: *Descriptions of the Geology of the U.S.S.R.*, vol. 3). *Trudy V.N.I.G.R.I.*, **111**, 254–79.

KHAIN, V. E. & LEONT'YEV, L. N., 1950. Main stages of the geotectonic development of the Caucasus. *Bull. Moscow Nat. Hist. Soc.*, **55**, 3, 30–64.

KHAITSER, L. L., 1962. New dada on the stratigraphy of the Permian and the Trias of river Ad'zva (northern part of the Chernyshev ridge). *Bull. Moscow Nat. Hist. Soc.*, **37**, 1, 57–71.

KHALETSKAYA, O. N., 1958. A stratigraphic scheme of Silurian deposits of Central Asia, pp. 29–33 in the book: *Theses of Papers of the Stratigraphic Conference on Central Asia.* Moscow.

KHERASKOV, N. P., 1952. The Southern Urals—the Precambrian. *Trudy Lab. geol. Precambrian*, **1**, 9–16.

KHIZHNYAKOV, A. V., 1957. Devonian deposits of the south-western margin of the Russian Platform. *Geol. Nefti*, **9**, 21–31.

KHODALEVICH, A. N., 1939. *Upper Silurian Brachiopods of the Western Slope of the Urals.* Trudy Geological Department of the Urals. Sverdlovsk. 134 pp.

——, 1949. A contribution to the stratigraphy of the Silurian and Devonian carbonate formations of the western slope of the Central Urals. *Soviet Geol.*, **39**, 99–107.

——, 1951. Lower Devonian and Eifelian Brachiopods of Ivdel and Serovsk regions in the Sverdlovsk province. *Trudy Sverdlovsk Mining Inst.*, **18**, 168.

KHOREV, I. A., 1955. The features of pre-Riphean folding. *Izv. Akad. Nauk*, No. 2, 72–81.

KHUDOLEI, K. M., 1956. A stratigraphic scheme for the Jurassic of the Far East, pp. 39–42 in the book: *Theses of Papers of the Stratigraphic Conference at Khabarovsk*, Khabarovsk.

KIPARISOVA, L. D., 1947. Triassic deposits in the U.S.S.R., pp. 5–51 in the book: *The Atlas of Main Forms*, Vol. VII. Leningrad.

——, 1956. A correlation of Triassic deposits of the Far East, pp. 32–5 in the book: *Theses of Papers of the Stratigraphic Conference at Khabarovsk*. Khabarovsk.

KIRICHENKO, G. I., 1958. Relationships between Cambrian and Sinian Deposits of the Western margin of the Siberian Platform, pp. 93–8 in the book: *Trudy of the Conference on the Stratigraphy of Siberia*. Leningrad.

KIRICHENKO, G. I. & TURGANOVA, E. V., 1955. A contribution to the problem of age and composition of 'watershed gravels' in the south of the Siberian Platform. *Mater. V.S.E.G.E.I.*, **7**, 148–59.

KIRSANOV, N. V., 1957. On the character and composition of the terrigenous part of lower Akchagylian deposits of the Vyatka-Kama district. *Izv. Kzath. Filial, ser. geol.*, issue 6, 141–49.

KLITOCHENKO, I. F., MUROMTSEV, A. S., et al., 1957. Prospects for gas and oil in the eastern part of the Dnyepr-Donyets Depression. *Geol. Nefti*, **9**, 1–7.

KLYAROVSKII, V. M., DMITRIYEV, A. N., et al., 1961. The absolute age of Cretaceous and Tertiary deposits of the Western Siberian iron-ore field, as indicated by glauconite, pp. 216–27 in the book: *Trudy Ninth Session of Commission for the Determination of Absolute Age*. Moscow.

KNYAZEVA, L. N., 1958. Results of investigations on bauxite deposits of the Northern Urals. *Mater. geol., useful min. dep. Urals*, **6**, 81–90.

KOCHETKOVA, A. D., 1959. The stratigraphy of Tertiary deposits of the eastern coast of the Penzhian inlet, pp. 99–100 in the book: *Theses of Papers of the Stratigraphic Conference at Okha*. Okha.

KOMAR, V. A., 1957. The stratigraphy of Devonian deposits of the Ore Altais. *Trudy vses. aérogeol. Tresta*, **3**, 15–45.

KONDRAT'YEVA, Z. A., 1960. Results of reference drilling in the Irkutsk amphitheatre and Western Transbaikalia. *Trudy V.N.I.G.R.I.*, **163**, 5–71.

KORNEYEVA, V. G., 1959. The geological structure and oil of south-western Ciscarpathia and the adjacent part of the Soviet Carpathians. *Trudy V.N.I.G.R.I.*, **141**, 199.

KOROBKOV, I. A., 1955. Middle Eocene molluscs of the Northern Caucasus and conditions of their habitat. *Zap. Leningrad Univ., ser. geol.*, **189**, 158–230.

KOROLEV, V. G., 1957. *Ancient Rocks of the Terskian Alataus and the Ranges Adjacent from the South*. Thesis. Frunze. 24 pp.

KOROLEVA, A. S., NIKOLAYEV, V. A., SHUMILOVA, E. V., 1952. A contribution to the stratigraphy and lithology of Miocene deposits on the river Tym. *Trudy Inst. Mining and Geol. —West. Siber. filial*, **12**, 119–36.

KOROTKOV, S. T., 1959. New data on the geological structure of western Ciscaucasia and prospects of finding new major deposits of oil. *Geol. Nefti*, **11**, 6–12.

KOROVIN, M. K., 1952. On the ancient massif of Tobolia in Western Siberia. *Trudy Inst. Mining and Geol.—West. Siber. filial*, **12**, 3–8.

KORZHINSKII, D. S., 1936. The petrology of the Archean complex of the Aldan Shield. *Trudy Ts.N.I.G.R.I.*, **86**, 75.

KOSTENKO, N. N., 1960. A short description of the Anthropogene in Kazakhstan, pp. 187–216 in the book: *Basic Ideas of N. G. Kassin*. Alma-Ata.

KOSYGIN, YU. A., 1952. *Principles of Tectonics of Petroliferous Regions*. Moscow and Leningrad. 509 pp.

KOTLYAR, V. N., 1958. *Pambak. The Geology, Intrusions and the Metallogeny of the Pambak Range*. Erevan. 228 pp.

KOVALEVSKII, S. A., 1935. *Continental Strata of Adzhinaur*. Baku. 180 pp.

KOZLOV, I. G., et al., 1961. The Khanty-Mansiian borehole. *Trudy V.N.I.G.R.I.*, **176**, 76.

KRASHENINNIKOVA, O. V., 1956. *Ancient Suites of the Western Slope of the Ukrainian Crystalline Shield*. Kiev. 195 pp.

KRASNOV, I. I. & MASAITIS, V. L., 1955. The tectonics of the Olenek-Vilyui watershed in connection with the structure of marginal zones of the Tunguska syneclise. *Mater. V.S.E.G.E.I.*, **7**, 217–33.

KRASNYI, L. I. & NIKIFOROVA, I. K., 1956. The stratigraphy and lithology of Jurassic and Lower Cretaceous deposits of western Cisokhotia, pp. 43–5 in the bok: *Theses of Papers of the Stratigraphic Conference at Khabarovsk*. Khabarovsk.

KRATS, K. O., 1958. The Pre-Cambrian—Karelia and the Kola Peninsula, pp. 60–8 in the book: *The Geological Structure of the U.S.S.R.* Leningrad.

KREMS, A. YA., 1958. Main features of the geological structure of the Timan-Pechora province and prospects of exploration for gas and oil. *Geol. Nefti*, **10**, 1–8.

——, 1961. Main results of geological exploration for oil and gas in 1960 in the Timan-Pechora region. *Geol. Nefti*, **7**, 11–16.

KRISHTOFOVICH, A. N., 1932. *A Geological Review of Far-Eastern Countries*. Leningrad. 332 pp.

KRISHTOFOVICH, L. V., 1959. Correlation of Tertiary deposits of the north-eastern part of the Pacific Ocean girdle, pp. 153–64 in the book: *Theses of Papers of the Stratigraphic Conference at Okha*. Okha.

KRISHTOFOVICH, L. V. & IL'YINA, A. P., 1960. The biostratigraphy of Tertiary deposits of western Kamchatka. *Bull. Moscow Nat. Hist. Soc.*, **31**, 1, 98–109.

KRIVTSOV, A. I., 1955. New data on the Lower Carboniferous stratigraphy of the eastern slope of the Baltic Shield, pp. 40–6 in the book: *Information Manual V.S.E.G.E.I.*, **2**.

KROPOTKIN, P. I., 1961. Main neotectonic features of Kamchatka, the Koryakskii range and the Kuril islands, pp. 278–89 in the book: *The Neotectonics of the U.S.S.R.* Moscow.

KUDINOVA, E. A., 1939. The geological structure of the Puchezhsko-Chkalovsk Povolzhia district *Bull. Moscow Nat. Hist. Soc.*, **17**, issue 4–5, 93–122.

KUMPAN, A. S., 1960. Main features of the stratigraphy and palaeogeography of Upper Palaeozoic rocks of Eastern Kazakhstan, in the book: *Basic Ideas of N. G. Kassin*. Alma-Ata.

KUSHNAREVA, T. I., 1959. The stratigraphy, lithology and oil of Devonian deposits, pp. 81–93 in the book: *The Geology and Oil of the Timan-Pechora Region*. Leningrad.

KUZICHKINA, YU. M., REIMAN, E. A. & SIKSTEL', T. A., 1958. The stratigraphic scheme of Jurassic deposits of Central Asia, pp. 112–21 in the book: *Theses of Papers of Stratigraphic Conference on Central Asia*. Moscow.

KUZNETSOV, I. G., 1951. The tectonics, vulcanism and stages in the formation of the structure of the Central Caucasus. *Trudy Inst. geol. Sci. Akad, Nauk S.S.S.R.*, **131**.

KURZNETSOV, V. A., 1952. Basic stages of the geotectonic development of the south of the Altai-Sayan mountain region. *Trudy Mining geol. Inst. West. Siber. filial*, **12**, 9–41.

——, 1954. Geotectonic regionalization in the Altai-Sayan fold province, pp. 202–28 in the book: *Problems of the Geology of Asia.*, Vol. 1. Moscow.

KUZNETSOV, YU. A., 1960. The geological structure and the origin of the relief in south-eastern Fergana, pp. 3–183 in the book: *Problems of the Geology of the Southern Tian-Shians*, vol. 2, L'vov.

LAVROV, V. V., 1956. Tertiary coal accumulation in Kazakhstan and the south of Western Siberia. *Trudy Lab. geol. coal*, **6**, 489–98.

——, 1957. *Marine Palaeogene rocks of Transuralian Plains and their Continental Equivalents*. Alma-Ata. 117 pp.

LAZURKIN, V. M., 1957. The geological structure of the lower Lena. *Trudy nauchno-issled. Inst. geol. Arkt.*, **81**, 461–83.

LEBEDEV, A. P., 1951. Certain problems of the geology of Siberian traps in the light of new data. *Izv. Akad. Nauk, ser. geol.*, No. 4, 48–56.

LEONOV, N. N., 1961. *The Tectonics and Seismicity of the Pamir-Alai Zone*. Moscow. 163 pp.

LESGAFT, A. V., 1958. The stratigraphy of Pre-Cambrian Deposits of the Yenisei ridge, pp. 112–16 in the book: *Trudy of the Conference on the Stratigraphy of Siberia*. Leningrad.

LEVEN, E. YA., 1958. Permian deposits of the south-eastern Pamirs, pp. 95–100 in the book: *Theses of Papers of the Stratigraphic Conference on Central Asia*. Moscow.

LIBROVICH, L. S., 1947. Goniatite, Carboniferous faunas of the U.S.S.R. and their significance in stratigraphy. *Bull Moscow Nat. Hist. Soc.*, **22**, 5, 51–68.

LIEPIN'SH, P. P., 1954. Devonian brachiopods of Baltic countries. *Izv. Azad. Nauk Latvian S.S.R.* No. 12/89, 87–112.

——, 1955. On lower beds of the Devonian in the western part of the Eastern-European platform, *Dokl. Akad. Nauk S.S.S.R.*, **103**, 2, 113–17.

LIKHAREV, B. K., 1946. Data on the brachiopod fauna of the Upper and Middle Carboniferous and Lower Permian of Fergana. *Mater. V.S.E.G.E.I.*, **7**, 91–110.

LOGVINENKO, N. V., 1952. On facies of coal bearing strata of the Donyets Basin. *Trudy Kharkov gorn. Inst.*, **1**, 31–56.

LOSEV, A. P., 1955. Coal deposits of the Tuva autonomous region. *Soviet Geol.*, **46**, 44–65.

LUPPOV, N. P., 1959. The stratigraphy of Upper Cretaceous deposits of south-western branches of the Gissar range, *Trudy V.N.I.G.N.I.*, **23**, 167–80.

LUR'YE, M. A. & OBRUCHEV, S. V., 1950. The Pre-Cambrian of the Eastern Sayans and the Khamar-Dabans (stratigraphy and magmatism. *Izv. Akad. Nauk, ser. geol.*, No. 6, 77–91.

——. 1952. Cambrian stratigraphy of the Eastern Sayans and the basin of the Dzhida. *Izv. Akad. Nauk, ser. geol.*, No. 1, 89–108.

LUR'YE, M. L., 1954. Cainozoic basalts of Eastern Sayans, pp. 343–56 in the book: *Problems of the Geology of Asia.* Vol. 1. Moscow.

L'VOV, K. A., 1939. *Precambrian and Lower Palaeozoic deposits of the Urals. Explanatory notes to 1:500,000 geological map of the Urals.* Leningrad. 228 pp.

——, 1958. The Proterozoic and the Lower Palaeozoic in the Urals. *Bull. All Union Sci., geol. Inst.*, No. 1, 58–71.

LYASHENKO, A. I., 1956. The Biostratigraphy of Middle Devonian and Frasnian deposits of central regions of the Russian Platform, pp. 135–71 in: *Trudy of the Conference on Problems of Finding Oil in the Ural-Volga Regions.* Moscow.

LYUTKEVICH, E. M., 1959. Oil and gas of the Cisverkhoyansk downwarp and of the Vilyui Syneclise. *Trudy V.N.I.G.R.I.*, **130**, 234–69.

MAGAK'YAN, I. G., 1947. *Alaverdi Type of Mineralisation and its Ores.* Erevan. 102 pp.

MAGAK'YAN, I. G. & MKRTCHYAN, S. S., 1957. The interdependence of structure, magmatism and metallogeny on the example of the Lesser Caucasus. *Izv. Akad. Nauk Armenian S.S.R.*, **10**, 4, 67–76.

MAKHNACH, A. S., 1958. *Lower Palaeozoic Deposits of Byelorussia.* Minsk. 226 pp.

MAKHNACH, A. S., STEFANENKO, A. YA., TSAPENKO, M. M. & KOSLOV, M. F., 1957. *A Short Description of the Geology of Byelorussia.* Minsk. 214 pp.

MALYAVKINA, V. S. & KAREVA, E. A., 1956. A contribution to the problem of the stratigraphy of the Chelyabinsk brown coal basin. *Dokl. Akad. Nauk*, **110**, 5, 828–30.

MALYUTIN, V. L., 1956. Brown coal deposits of the Southern Urals. *Trudy Lab. geol. coal*, **6**, 454–64.

MARKIN, N. M., 1957. Geological investigations and Tertiary deposits of the Penzhian inlet. *Trudy V.N.I.G.R.I.*, **102**, 5–94.

MARKOV, F. G., 1954. The stratigraphy of Palaeozoic deposits of the Taimyr Peninsula. *Trudy nauchno-issled. Inst. geol. Arkt.*, **69**, 435.

MARKOV, F. G., RAVICH, M. G. & VAKAR, V. A., 1957. The geological structure of Taimyr. *Trudy nauchno-issled. Inst. geol. Arkt.*, **81**, 313–87.

MARKOVSKII, N. I., 1956. Coal capacity of Lower Carboniferous rocks in the Central Volga and Transvolga districts. *Trudy Lab. coal geol.*, **6**, 366–79.

MARTINSON, G. G., 1956. The significance of various groups of fresh-water fauna in biostratigraphy of continental coal-bearing deposits in the book: *Theses of Papers of the Conference on the Stratigraphy of Siberia.* Leningrad.

MARTYNOVA, M. V., 1961. *The Famennian Stratigraphy and Brachiopods of Central Kazakhstan.* Moscow. 151 pp.

MATVEYENKO, V. T., 1959. The 1:2,500,000 metallogenic map of the North-East, U.S.S.R., pp. 124–7 in the book: *Metallogenic and Prognostic Maps.* Alma-Ata.

MATVEYEV, A. K., 1957. Descriptions of coal basins, pp. 192–446 in the book: *Mining—an Encyclopaedic Reference Book*, Vol. 2—*The Geological Description of Coal Basins.* Moscow.

MEFFERT, B. F., 1941. The Miocene of Western Transcaucasia, pp. 283–97 in the book: *The Geology of the U.S.S.R.*, Vol. 10, pt. 1. Moscow.

MENNER, V. V., 1961. The stratigraphy of Devonian deposits in the north-western part of the Siberian Platform. *Dokl. Akad. Nauk*, **141**, 6, 1441–4.

MESROPYAN, A. I., 1957. The geological structure of Armenia and its prospects for oil. *Trudy V.N.I.G.R.I.*, **111**, 280–92.

MIKANOVA, N. E., 1958. The stratigraphy of Palaeogene deposits of Central Asia, pp. 149–57 in the book: *Theses of Papers of the Stratigraphic Conference on Central Asia.* Moscow.

MIKLUKHO-MAKLAI, A. D., 1955. The zoogeographical regionalisation of marine Permian strata in the U.S.S.R. and the correlation of Permian successions. *Uchen. Zap Leningr. gos. Univ. ser. geol.*, **6**, 3–20.

——, 1958. The stratigraphy of marine Permian deposits of Central Asia, pp. 93–4 in the book: *Theses of Papers of the Conference on Central Asia.* Moscow.

——, 1960. The stratigraphy of Carboniferous deposits of Central Asia. *Vest. Leningr. gos. Univ.*, **6**, 20–30.

MIKLUKHO-MAKLAI, K. V., 1952. New data on the Palaeozoic stratigraphy of the region of the north-eastern depression of the Main Caucasian Range. *Dokl. Akad. Nauk S.S.S.R.*, **83**, 2, 277–9.

——, 1956. Upper Permian deposits of the north-western Caucasus. *Mater. V.S.E.G.E.I.*, **14**, 60–78.

MIKRAMALOVA, S. KH., 1958. *The Stratigraphy of Palaeogene Deposits of Fergana and Tashkent Regions.* Moscow. 128 pp.

MIKUNOV, M. F., 1957. New data on Upper Palaeozoic deposits of the Ore Altais. *Trudy vses. aérogeol. Tresta*, **3**, 61–9.

MIKUTSKII, S. P., 1961. The stratigraphy of pre-Upper Palaeozoic deposits of the Cis-Yenisei part of the Siberian Platform. *Trudy sib. nauchno-issled. Inst. Geol. Geofiz. miner. Syr.*, **13**, 90–109.

MIRCHINK, M. F. *et al.* 1961. The Matych–Karatau ridge. *Dokl. Akad. Nauk*, **141**, 4, 138–941.

MIRONOV, S. I. (editor), 1956. *Materialy (data) on the Geology and Oil of Georgia.* Moscow. 162 pp.

MKRTCHYAN, S. S., 1958. *The Zangezurian Ore Field.* Erevan. 287 pp.

MOKRINSKII, V. V., 1956. Features in the formation of structural forms of coal bearing sediments of southern Yakutia. *Trudy Lab. geol. coal*, **6**, 568–79.

——, 1958. The southern Yakutian coal field and its prospects. *Trudy Lab. geol. coal*, **7**, 124–43.

MOKROUSOV, V. P., 1959. The stratigraphy of pre-Cretaceous deposits of Kamchatka, pp. 71–5 in the book: *Theses of Papers of the Stratigraphic Conference at Okha.* Okha.

MOOR, G. G., 1940. Charnockitic Series of the Anabaran Precambrian. *Izv. Akad. Nauk, ser. geol.*, **6**, 3–19.

MORDVILKO, T. A., 1956. A unified scheme of the stratigraphy of Lower Cretaceous deposits of the Northern Caucasus and Ciscaucasia, pp. 112–35 in the book: *Trudy of Conference on the Stratigraphy of the Mesozoic of the Russian Platform.* Leningrad.

MOSKVIN, A. I., 1960. The experience of applying a unified stratigraphic scheme to Quaternary Deposits of Western Siberia. *Trudy geol. Inst. Akad. Nauk*, **26**, 11–36.

MOSKVITIN, A. I., 1954. A stratigraphic scheme of the Quaternary period in the U.S.S.R.. *Izv. Akad. Nauk*, No. 3, 20–50.

MURATOV, M. V., 1949. The tectonics and the history of the development of Alpine, geosynclinal region of the south of the European part of the U.S.S.R. and adjacent countries, in the book: *The Tectonics of the U.S.S.R.*, Vol. II, Moscow. 510 pp.

——, 1955a. The history of tectonic development of the deep Black Sea depression and its posible origin. *Bull. Moscow Nat. Hist. Soc.*, **30**, 5, 27–50.

——, 1955b. The tectonic structure and the history of planar regions separating the Russian Platform from the mountains of Crimea and Caucasus. *Soviet Geol.*, **48**, 36–66.

——, 1960. *A Brief Description of the Geological Structure of the Crimean Peninsula.* Moscow. 207 pp.

MURATOV, M. V. *et al.*, 1960. The stratigraphy, facies and formation of Jurassic deposits of Crimea. *Bull. Moscow Nat Hist. Soc.*, **35**, 1, 87–97.

MURATOV M. V. & ARKHIPOV, I. V., 1961. On the tectonic position of the Pamirs in the system of fold mountains in the south-western and Central Asia. *Bull. Moscow Nat. Hist. Soc.*, **36**, 4, 97–121.

MUROMTSEV, V. S., 1956. The Kuznetsk Depression (In: *Descriptions of the Geology of the U.S.S.R.*, Vol. 1). *Trudy V.N.I.G.R.I.*, **96**, 186–214.

MUSTAFINOV, A. N., 1958. Main results of geoexploration on the territory of R.S.F.S.R. in 1957 and problems for 1958. *Geol. Nefti*, **5**, 1–7.

——, 1960. New data on oil and gas content of the *R.S.F.S.R.*, on the basis of geological explorations. *Geol. Nefti*, **5**, 1–5.

MUZYLEV, S. A., 1938. A geological section across the Lesser Khingan, pp. 123–35 in the book: *Sbornik in Honour of Akademician Obruchev*. Moscow and Leningrad.

NAGIBINA, M. S., 1956. A stratigraphic scheme of the Jurassic and Cretaceous of Upper Cisamuria, pp. 45–7 in the book: *Theses of Papers of the Stratigraphic Conference at Khabarovsk*. Khabarovsk.

NALIVKIN, V. D., 1949. The stratigraphy and tectonics of the Ufa plateau and the Yurezano-Sylvenian depression. *Trudy V.N.I.G.R.I.*, **46**, 204.

——, 1950. The facies and the geological history of the Ufa Plateau in the Yurezano-Sylvenian Depression. *Trudy V.N.I.G.R.I.*, **47**, 127.

——, 1958. New data on geology and gas and oil. *Trudy V.N.I.G.R.I.*, **126**, 309–24.

NALIVKIN, V. D., ROZANOV, L. N. *et al.*, 1956. *The Volga-Ural Oil Province: Its Tectonics*. Leningrad. 312 pp.

NEDZVETSKII, A. P., 1957. Achievements of the science of geology in Tadzhikistan over the past forty years. *Trudy Inst. geol. Akad. Nauk Tadzhikian S.S.R.*, **2**, 3–15.

NEKHOROSHEV, V. P., 1956. Altaiian zones of deformation, the peculiarities and practical significance. *Information sbornik V.S.E.G.E.I.*, **3**, 50–61.

——, 1958. *The Geology of the Altais*. Moscow. 261 pp.

——, 1960. Ordovician and Silurian bryozoans of the Siberian Platform. *Trudy V.S.E.G.E.I.*, **41**, 163–5.

NEUSTRUYEVA, I. YU. 1961. On the age of continental, variegated deposits of the Kan-Taseyevian Depression, pp. 250–5 in the book: *Problems of the Geology of Coal bearing Deposits*. Moscow and Leningrad.

NIKIFIROVA, K. V., 1960. The Cainozoic of the Golodnaya (Hungry) Steppe of Central Kazakhstan. *Trudy geol. Inst. Akad. Nauk S.S.S.R.*, **45**, 255.

NIKIFIROVA, O. I., 1937. Upper Silurian brachiopods of Central Asia. *Palaeontol. Monographs S.S.S.R.*, **35**, 1, 92 pp.

——, 1949. *The Atlas of Main Forms of Fossil Faunas in the U.S.S.R.*, vol. 2, *The Silurian System*. Leningrad and Moscow. 375 pp.

——, 1954. *The Stratigraphy and the Brachiopods of Silurian deposits in Podolia*. Moscow. 218 pp.

——, 1955. New data on the stratigraphy and palaeogeography of the Ordovician and the Silurian of the Siberian Platform. *Mater. V.S.E.G.E.I.*, **7**, 50–106.

NIKIFOROVA, O. I. & ANDREYEVA, O. N., 1961. The Ordovician and Silurian stratigraphy of the Siberian Platform and its palaeontological basis (brachiopods). *Trudy V.S.E.G.E.I.*, **56**, 411.

NIKIFIROVA, K. V., GREBOVA & KONSTANFIROVA, N. A., 1960. The stratigraphy of continental Cainozoic deposits of Central Kazakhstan and their correlation (In: *The Stratigraphy of Quaternary Deposits*), *Trudy geol. Inst. Akad. Nauk S.S.S.R.*, **26**, 204–47.

NIKITIN, I. F., 1956. *Cambrian and Lower Ordovician Brachiopods in the North-East of Central Kazakhstan*. Alma-Ata. 141 pp.

NIKOLAYEV, V. A., 1952. Central Asia—the Precambrian. *Trudy Lab. geol. Precambrian*, **1**, 38–44.

NIKOL'SKII, A. P., 1959. New data on the Precambrian of Krivoi Rog. *Trudy Lab. Precambrian*, **2**, 72–97.

NIKOL'SKII, V. M., 1959. The stratigraphy of the Kuril Islands, pp. 137–40 in the book: *Theses of Papers of the Stratigraphic Conference at Okha*. Okha.

NOVIK, E. O., 1954. The stratigraphy of Devonian deposits of the Dnyepr-Donyets Depression. *Izv. Akad. Nauk*, No. 2, 44–54.

OBRUCHEV, V. A., 1934. *Ore Deposits—Descriptive Part*. Moscow. 596 pp.

——, 1935–8. *The Geology of Siberia*, Vols. 1–3. Izd. Akad. Nauk S.S.S.R. Moscow and Leningrad.

——, 1941. *Data on Devonian Fish of the U.S.S.R.*, pp. 20–44. Moscow.

OBUT, A. M., 1958. Zonal division of the Ordovician and Silurian in the U.S.S.R. on the basis of graptolites, pp. 35–42 in the book: *Theses of Papers of the Stratigraphic Conference on Central Asia.* Moscow.

ODINTSOV, M. M., 1961. A new Jurassic basin in the south-west of the Siberian Platform. *Dokl. Akad. Nauk*, **138**, 5, 1170–1.

OLLI, A. I. & ROMANOV, V. A., 1959. *Explanatory Notes to the Tectonic 1:650,000 Map of Bashkiria.* Ufa. 37 pp.

——, 1960. The pre-Ordovician history of the tectonic development of the Southern Urals. *Voprosy (Problems) geol. east. margin Russian Platform and the Southern Urals.* **7**, 3–33.

ONIKHIMOVSKII, V. V., 1956. The stratigraphy of the Precambrian and Cambrian of the Khingano-Bureiian and Kur-Urmiian regions, pp. 14–17 in the book: *Theses of Papers of the Stratigraphic Conference at Khabarovsk.* Khabarovsk.

ORVIKU, K. K., 1958. The Anthropogenic system, pp. 35–41 in the book: *A Review of the Stratigraphy of Palaeozoic and Quaternary Deposits in the Estonian S.S.R.* Tallin.

OVANESOV, G. P., 1960. Main results of geological explorations for oil and gas in the Bashkirian A.S.S.R. *Geol. Nefti*, **7**, 1–5.

PAFFENGOL'TS, K. N., 1931. The stratigraphy of Quaternary lavas in Eastern Armenia. *Zap. Min. Soc.*, **60**, 2, 237–58.

——, 1956. New data on the age of effusions in the Central Caucasus (the El'brus, the Chegen-Nal'chik, the Kazbek), the laccoliths of Pyatigoria and the granites of the Main range. *Mater. V.S.E.G.E.I.*, **14**, 5–24.

——, 1959a. *A Geological Description of the Caucasus.* Erevan. 506 pp.

——, 1959b. El'brus—a Geological Description. *Izv. Akad. Nauk*, No. 2, 3–23.

PALII, A. M. & PETROV, L. A., 1961. Results of geoexploration for oil and gas in the Ukrainian S.S.R. over the period between the 21st and 22nd Congresses of the Communist Party of the Soviet Union, *Geol. Nefti*, **10**, 17–21.

PAVLOVSKII, E. V., 1939. The Precambrian of Cisbaikalia, in the book: *The Stratigraphy of the U.S.S.R.*, Vol. I, *Precambrian.* Moscow and Leningrad.

PERGAMENT, M. A., 1958. *The Stratigraphy of Upper Cretaceous Deposits in North-western Kamchatka (The Penzhina region).* Thesis. Moscow. 24 pp.

PETRUSHEVSKII, B. A., 1951. Mesozoic Cainozoic structure of the Western Siberian Lowlands. *Bull. Moscow Nat. Hist. Soc.*, **26**, 4, 3–40.

——, 1955. *The Ural-Siberian Epihercynian Platform and the Tian Shians.* Moscow. 552 pp.

PLESHAKOV, I. B., 1969. Tertiary deposits of Sakhalin and Kamchatka, pp. 151–2 in the book: *Theses of Papers of the Stratigraphic Conference at Okha.* Okha.

POLEVAYA, N. I., MURINA, G. A., SPIRINTSSON, V. D. & KAZAKOV, G. A., 1960. The determination of absolute age of sedimentary and volcanogenic formations, pp. 32–54 in the book: *The Determination of Absolute Age of Pre-Quaternary Formations. Intern. Geol. Cong.* Moscow.

POLKANOV, A. A., 1937. A brief review of the pre-Quaternary geology of the Kola Peninsula, pp. 12–23 in the book: *Guide of the Northern Excursion, XVII International Geological Congress.* Leningrad.

——, 1947. The Geology of Karelia and the Kola Peninsula, pp. 45–53 in the book: *Soviet Geology in the Last Thirty Years.* Moscow.

——, 1956. The Hoglandian-Jotnian geology of the Baltic Shield. *Trudy Lab. Precambrian*, **6**, 122.

POLKANOV, A. A. & GERLING, E. K., 1960. The geochronology of the Precambrian of the Baltic Shield, 57–82. Book: *The Determination of Absolute Age, Intern. Geol. Congress.* Moscow.

POLOVINKINA, YU. I., 1953. The stratigraphy, magmatism and tectonics of the Precambrian in the Ukrainian S.S.R. *Trudy Lab. Precambrian*, **2**, 44–68.

POPOV, G. I., 1955. The history of Manych straits. *Bull. Moscow Nat. Hist. Soc.*, **30**, 2, 31–49.

POPOV, V. I., 1938. *The History of Depression and Uplift of the Western Tian-Shians.* Tashkent. 415 pp.

POPOV, V. I. et al., 1958. A complex bio- and rhythmo-stratigraphic division of Cretaceous and Tertiary deposits of the Kyzylkum and Bukhara Depression, pp. 142–8 in the book: *Theses of Papers of the Stratigraphic Conference on Central Asia.* Moscow.

Popov, V. V., 1960. The Anthropogene stratigraphy of the Tian Shians. *Trudy Geol. Inst. Akad. Nauk S.S.S.R.*, **26**, 116–26.

Poyarkova, Z. N., 1961. The Chulyma reference borehole. *Trudy V.N.I.G.R.I.*, **183**, 138.

Pozner, V. M., 1955. The stratigraphy of terrigenous, Lower Carboniferous strata in the Kama-Kinelian depression. *Dokl. Akad. Nauk*, **104**, issue 6, 892–4.

Predtechenskii, A. A., 1960. An ancient uplift of Southern Siberia. *Trudy Siber. Sci. Inst. geofiz. Min. Resources*, **13**, 65–77.

Rabkin, M. I., 1956. The Precambrian of the Anabara Shield, pp. 19–21 in the book: *Theses of Papers of the Conference on the Stratigraphy of Siberia—Stratigraphy of the Pre-Cambrian*. Leningrad.

Radchenko, G. P., 1940. On the distinction of phytostratigraphic zones in the Palaeozoic sequence of the Kuznyetsk Basin. *Vest. West. Siber. geol. Dept.*, **3–4**, 30–8.

Radugin, K. V., 1952. Mountain Shoria, the Kuznyetskian Alataus and the Eastern Sayans. *Trudy Lab., geol. Precambrian*, **1**, 58–83.

Rasskazova, E. S., 1958. A contribution to the stratigraphy of Upper Palaeozoic deposits of the Tunguska Basin. *Bull. Moscow. Nat. Hist. Soc.*, **28**, 5, 91–109.

Ratnovskii, I. I., 1959. The Palaeozoic stratigraphy of Sakhalin, pp. 21–4 in the book: *Thesis of Papers of the Stratigraphic Conference at Okha*. Okha.

Rauzer-Chernousova, D. M., 1937. On fusulinids and the stratigraphy of Upper Carboniferous strata and the Artinskian stage of the western slope of the Urals. *Bull. Moscow. Nat. Hist. Soc.*, **15**, 5, 478–80.

——, 1940. The Upper Carboniferous and Artinskian stratigraphy of the western slope of the Urals and data on the fusulinid fauna. *Trudy Inst. geol. Sci. Akad. Nauk S.S.S.R.* **7**, 37–104.

——, 1949. The stratigraphy of Upper Carboniferous and Artinskian deposits of Bashkirian Cisuralia. *Trudy Inst. geol. Sci. Akad. Nauk S.S.S.R.*, **105**, 3–21.

Ravich, M. G., 1954. The Pre-Cambrian of Taimyr. *Trudy nauchno-issled. Inst. geol. Arkt.*, **76**, 311.

Ravskii, I., 1960. A contribution to the stratigraphy of Quaternary (Anthropogenic) deposits of the south and east of the Siberian Platform. *Trudy geol. Inst. Akad. Nauk*, **26**, 37–95.

Raznitsyn, V. A., 1959. The stratigraphy and gas and oil content of Lower Carboniferous rocks in Southern Timan, pp. 134–53 in the book: *The Geology and Oil of the Timan-Pechora Region*. Leningrad.

Reingard, A. L., 1932. A few words on the chronological connection between Caucasian glaciations and Caspian transgressions. *Zap. Min. Soc.*, **61**, 1, 151–8.

——, 1937. A contribution to the problem of the glacial period in the Caucasus. *Trudy Assoc. invest. Quatern. period Europe*, **1**, 9–30.

Reinin, I. V., 1961. On marine Quaternary deposits of the north-western part of the Western Siberian Lowlands. *Trudy V.N.I.G.R.I.*, **186**, 212–23.

Rengarten, V. P. (editor), 1941. *Geology of the U.S.S.R.*, Vol. 10, *Transcaucasia*, part 1. *Geological Description*. 614 pp.

——, 1956, A stratigraphic scheme of Upper Cretaceous deposits of the Northern Caucasus, pp. 74–83 in the book: *Trudy of the Conference on the Stratigraphy of the Mesozoic of the Russian Platform*. Leningrad.

Repina, L. N. & Khomentovskii, V. V., 1961. On Lower Cambrian subdivisions. *Izv. Akad. Nauk, ser. geol.*, No. 8, 83–7.

Robinson, V. N., 1932. A geological review of the Trias and the Palaeozoic in the basins of the Laba and the Byelaya in the Northern Caucasus. *Trudy V.G.R.O.*, **226**, 55–9.

——, 1956. The Trias of the Caucasus, pp. 201–6 in the book: *Trudy of the Conference on the Stratigraphy of the Mesozoic of the Russian Platform*. Leningrad.

Rostovtsev, N. N., 1956. The Western Siberian Lowlands. Descriptions of the Geology of the U.S.S.R. (on reference drilling data), Vol. 1. *Trudy V.N.I.G.R.I.*, **96**, 107–52.

Rostovtsev, N. N. *et al.*, 1957. See Aleskerova, Z. T. *et al.*, 1957.

Rubinshtein, M. M., 1961. On time of formation of the crystalline substratum of the Caucasus. *Bull. Comm. on Determination Absol. Age*, **4**, 59–63.

Rusakov, I. M. & Yegizarov, B. Kh., 1958. The stratigraphy of Precambrian and Palaeozoic deposits in the eastern part of the Koryakskii range. *Trudy N.I.I.G.A.*, **85**, 3–19.

RZHONSNITSKAYA, M. A., 1948. Devonian deposits of Transcaucasia, *Dokl. Akad. Nauk*, **59**, 8, 1477–80.

——, 1955. On stratigraphic significance of sporiferids in Devonian deposits of the U.S.S.R. *Mater. vses. nauchno-issled. geol. Inst., new ser.*, **9**, 207–10.

SAKS, V. I., 1953. The Quaternary period in the Soviet Arctic. *Trudy nauchno-issled. Inst. geol. Arkt.*, **77**, 626.

SAKS, V. N., 1960. The geological history of the Arctic Ocean during the Mesozoic era, pp. 108–24 in: *Regional Palaeogeography, Reports of XXI International geol. Cong.*

——, 1961. The palaeogeography of the Arctic during Jurassic and Cretaceous periods, pp. 20–48 in: *Doklady (Papers) in Commemoration of V. A. Obruchev.*

SAKS, V. N. & RONKINA, Z. Z., 1957. Jurassic and Cretaceous deposits of the Ust'-Yenisei Depression. *Trudy nauchno-issled. Inst. geol. Arkt.*, **90**, 232.

SAKS, V. N., GRAMBERG, I. S., RONKINA, Z. Z. & APOLONOVA, E. K., 1959. Mesozoic deposits of the Khatanga Depression. *Trudy nauchno-issled. Inst. Geol. Arkt.*, **99**, 226.

SALOP, L. I., 1954. The Lower Palaeozoic of the Central Vitim mountain country. *Trudy V.S.E.G.E.I.*, **1**, 72.

——, 1958a. The Precambrian stratigraphy of the Baikal mountainous region, pp. 170–206 in the book: *Trudy of the Conference on the Stratigraphy of Siberia, Precambrian.* Moscow and Leningrad.

——, 1958b. The geological structure and useful mineral deposits of the Baikalian mountain region, pp. 55–70 in the book: *The Geological Structure and Useful Mineral Deposits of Eastern Siberia.* Moscow.

SAZONOV, N. T., 1957. *Jurassic Deposits of Central Regions of the Russian Platform.* Leningrad. 155 pp.

——, 1958. The geological history of the Jurassic period in central regions of the Russian Platform. *Bull. Moscow Nat. Hist. Soc.*, **33**, 1, 43–6.

SAZONOVA, I. G., 1958. Lower Cretaceous deposits in central regions of the Russian Platform, pp. 31–184 in the book: *Mesozoic and Tertiary Deposits of Central Regions.* Moscow.

SELIN, YU. I., 1962. Oligocene deposits of the Bol'shetomakian manganese deposit. *Bull. Moscow Nat. Hist. Soc.*, **37**, 1, 72–84.

SEMENENKO, N. P., 1951. The structure of the Ukrainian crystalline massif and the history of its formation. *Izv. Akad. Nauk, ser. geol.*, No. 1, 54–9.

——, 1953. The Precambrian of the Ukrainian S.S.R. *Trudy Lab. Precambrian*, **2**, 24–43.

SEMENENKO, N. P., BURKSER, E. S. & IVANTISHIN, M. N., 1960. Age groups of mineralisation of Ukrainian rocks on an absolute age basis, pp. 112–31 in the book: *The Determination of Absolute Age.* Moscow.

SENCHENKO, G. S., 1960. Relationships between the reef and the conglomeratic facies in Cisuralia. *Voprosy geol., Southern Urals*, **7**, 73–83. (*Vop. Geol. vost. Okr. russk. Platf.*)

SERGIYEVSKII, V. M., 1948. Middle Palaeozoic vulcanism and the history of formation of tectonic structures of the eastern slope of the Urals. *Mater V.S.E.G.E.I.*, **8**, 3–21.

SERGUN'KOVA, O. I., 1950. Lower Carboniferous brachiopods of the eastern part of the Tian-Shian geosyncline. *Trudy Inst. geol. Uzbekian Akad. Naut*, **5**, 1, 48–89.

——, 1958. The stratigraphy of Devonian deposits of the western part of the Northern Tian-Shian, pp. 42–50 in the book: *Theses of Papers of the Stratigraphic Conference on Central Asia.* Moscow.

SHANTSER, E. V., 1955. The present-day position of the doctrine of the Quaternary glaciation. *Trudy Comm. invest. Quatern. Period*, **12**, 5–21.

SHATSKII, N. S., 1927. Notes on the tectonics of Tertiary foothills of the north-eastern Caucasus. *Bull. Moscow Nat. Hist. Soc.*, **5**, 321–69.

——, 1952. On ancient deposits of the sedimentary cover to the Russian Platform and on its structures in Older Palaeozoic times. *Izv. Akad. Nauk, ser. geol.*, No. 1, 17–32.

——, 1937. On the tectonics of the Eastern European Platform. *Bull. Moscow Nat. Hist. Soc.*, **15**, 1, 4–26.

——, 1945. Tectonic descriptions of the Volga-Ural region and the adjacent part of the western slope of the Urals. *Bull. Moscow Nat. Hist. Soc.*, **2**, 6, 129.

——, 1946. Main features of the structure and development of the Eastern European Platform. *Izv. Akad. Nauk, ser. geol.*, No. 1, 5–62.

——, 1947. On structural connections of the platform with folded, geosynclinal regions. *Izv. Akad. Nauk*, No. 5, 37–56.

SHCHERIK, E. H., 1958. Main types of structures of the north-western slope of the Greater Caucasus and western Ciscaucasia and the conditions of their formation in the Tertiary period. *Trudy V.N.I.G.N.I.*, **12**, 274–301.

SHCHUKINA, E. N., 1960. Lavas of the distribution of Quaternary deposits and their stratigraphy on the territory of the Altais. *Trudy geol. Inst. Akad. Nauk S.S.S.R.*, **26**, 127–164.

SHIKHALIBEILI, E. SH., 1956. *The Geological Structure and the Development of the Azerbaidzhan Part of the Southern Slope of the Greater Caucasus*. Baku. 223 pp.

SHIROKOV, A. V., 1956. Lower Carboniferous strata of the north-western continuation of Donbas and their coals. *Trudy Lab. geol. coal*, **6**, 319–26.

SHLYGIN, E. D., 1959. *A Short Course on the Geology of the U.S.S.R.* Moscow. 271 pp.

SHMIDT, N. G., 1957. Geophysical methods applied to geological map constructions of the crystalline basement of the Kursk Magnetic Anomaly (K.M.A.). *Sov. Geol.*, **58**, 138–149.

SHORYGINA, L. D., 1960. The stratigraphy of Cainozoic deposits of Western Tuva. *Trudy geol. Inst. Akad. Nauk S.S.S.R.*, **26**, 165–203.

SHTREIS, N. A. & KOLOTUKHINA, 1948. The geological structure of the Ortau and Kos-Muran Mountains (the Karaganda region). *Trudy Inst. geol. Sci. Akad. Nauk S.S.S.R.*, **101**, 69–104.

SHUL'GA, P. L., 1952. Palaeozoic stratigraphy of the south-western margin of the Russian Platform (Volyn and Podolia). *Geol. J., Akad. Nauk Ukrainian S.S.R.*, **12**, 4, 22–40.

SHUMILOVA, E. V., 1952. Contributions to the Lithology and Mineralogy of Neogenic deposits of the central part of the Western Siberian Lowlands. *Trudy Mining geol. Inst. West. Siber. filial*, **12**, 109–18.

SIDOROV, D. P. & SLASTENOV, YU. L., 1961. A contribution to the stratigraphy of Mesozoic coal-bearing deposits of the Ust'-Vilyni gas region. *Trudy V.N.I.G.R.I.*, **186**, 32–43.

SIGOV, A. P., 1956. Tertiary deposits of the eastern slope of the Urals, pp. 25–8, in the book: *Theses of Papers of the Stratigraphic Conference in Sverdlovsk*.

SIMAKOV, S. N. *et al.*, 1957. The geological structure and oil of Fergana. *Trudy V.N.I.G.R.I.*, **110**, 605.

SIMORIN, A. M., 1956. *The Stratigraphy and Brachiopods of the Karaganda Basin*. Alma-Ata. 300 pp.

SLAVIN, V. I., 1956. A stratigraphic scheme of the Mesozoic in western regions of Ukraine, pp. 87–100 in the book: *Trudy of the Conference on the Stratigraphy of the Mesozoic of the Russian Platform*. Leningrad.

SLEPAKOVA, T. L., 1961. Sub-salt structures of the Ciscaspian Depression on the basis of geophysical data. *Trudy V.N.I.G.R.I.*, **186**, 253–72.

SLODKEVICH, V. S., 1938. Tertiary lamellibranchs of the Far East, of the book: *The Palaeontology of the U.S.S.R.*, Vol. 10, pt. 19. Moscow and Leningrad.

SMEKHOV, E. M., 1953. *The Geological Structure of Sakhalin Island and its Oil and Gas*. Moscow. 320 pp.

SNYATKOV, L. A., 1957. The geological structure and stages of the development of the North-East, S.S.S.R., in the book: *Theses of Papers of the Stratigraphic Conference at Magadan*. Magadan.

SOBOLEV, N. D., 1952. *Ultrabasites of the Greater Caucasus*. Gosgeolizdat, Moscow. 240 pp.

SOBOLEVSKAYA, V. N., 1951. The Palaeogeography and structures of the Russian Platform in Upper Cretaceous times, pp. 67–123 in the book: *Papers in memory of A. D. Arkhangelskii*. Moscow.

SOKOLOV, B. S., 1952. The age of the oldest sedimentary cover of the Russian Platform. *Izv. Akad. Nauk, ser geol.*, No. 5.

——, 1953. A stratigraphic scheme of pre-Devonian deposits of the north-west of the Russian Platform, pp. 16–38 in the book: *The Devonian of the Russian Platform*. Leningrad and Moscow.

SOKOLOV, B. S. & DZEVANOVSKII, YU. K., 1957. On stratigraphic position and age of sedimentary strata of the late Precambrian. *Soviet Geol.*, **55**, 31–50.

SOLNTSEV, O. A., 1959. Metamorphosed shales, pp. 5–18 in the book: *The Geology and Oil in the Timan-Pechora Region*. Leningrad.

SOLNTSEV, O. A. & KUSHNAREVA, T. I., 1957. The Timan-Pechora region. (In: *Papers on the Geology of the U.S.S.R.*, Vol. 2). *Trudy V.N.I.G.R.I.*, **101**, 5–48.

SOLOVKIN, A. N., 1939. *Intrusions and Intrusive Cycles of the Azerbaidzhan S.S.R.* Izd. Az. F.A.N. Baku. 138 pp.

SOLOV'YEV, S. P., 1952. *The Distribution of Magmatic Rocks in the U.S.S.R. and Certain Problems of Petrology.* Gosgeolizdat, Moscow. 216 pp.

SOROKOV, D. S., 1957. The stratigraphy of the Novosibirsk Archipelago, pp. 18–19 in the book: *Thesis of Papers of the Stratigraphic Conference at Magadan.* Magadan.

——, 1957. The stratigraphy and facies of Mesozoic deposits of the Lena-Olensk region. *Trudy geol. Inst. Arkt.*, **85** (*Geology of the Arctic*), **9**, 20–36.

SPIZHARSKII, T. N., 1958. The geological regionalisation of the Siberian Platform and main generalisations about the distribution of useful mineral deposits on its territory, pp. 22–41 in the book: *The Geological Structure and Useful Mineral Deposits of Eastern Siberia.* Moscow.

STARIK, I. E., 1961. *Nuclear Geochronology.* Moscow. 630 pp.

STEFANENKO, A. YA. & MAKHNACH, A. S., 1952a. Lower Palaeozoic deposits of Byelorussia. *Izv. Akad. Nauk beloruss. S.S.R.*, No. 1, 75–85.

——, 1952b. Devonian deposits of Byelorussia. *Izv. Akad. Nauk beloruss. S.S.R.*, No. 4, 135–150.

STEPANOV, P. I. *et al.*, 1937. The geological structure of the Donyets coal-basin, pp. 5–43 in the book: *Guide to the Southern Excursion of the 17th International Geological Congress.* Leningrad and Moscow.

STEPANOV, D. L., 1941. The Upper Palaeozoic of the Bashkirian A.S.S.R. *Trudy V.N.I.G.R.I.*, **20**, 98.

——, 1948. Upper Carboniferous brachiopods of Bashkiria. *Trudy V.N.I.G.R.I.*, **22**, 63.

——, 1951. *The Upper Palaeozoic of the Western Slope of the Urals.* Leningrad. 246 pp.

STILLE, H., 1924. *Principles of Comparative Tectonics.*

STRAKHOV, N. M., 1948. *Principles of Historical Geology* (part I—250 pp., part II—391 pp.). Moscow.

STRUGOV, A. S., 1956. Prospects for coal and the nature of coals in the western part of the Vilyuian Depression. *Trudy Lab. geol. coal*, **6**, 580–90.

STUKALO, A. P. & STOVNOVOI V. N., 1955. On the geological structure and coal of the western continuation of Donbas. *Soviet geol.*, **46**, 66–81.

SUDOVIKOV, N. G., 1937. A geological description of the Kandalaksha region, pp. 24–39 in the book: *Guide to the Northern Excursion, XVII International Geol. Cong.* Leningrad and Moscow.

——, 1939. A review of the Precambrian stratigraphy, tectonics and magmatism in the Karelian A.S.S.R., pp. 57–80 in the book: *Stratigraphy of the U.S.S.R.*, vol. I. Izd. Akad. Nauk S.S.S.R.

SULIMOV, I. M., 1961. The Zhigalov borehole (Eastern Siberia). *Trudy V.N.I.G.R.I.*, **173**, 92.

SULTANAYEV, A. A. and Goldberg, I. S., 1960. The Geology and Oil of Palaeozoic deposits in the zone of the join between the eastern slope of the Southern Urals and the Turgai Bay. *Sci. Research V.N.I.G.R.I., for 1956*, **18**, 61–9.

SULTANOV, K. M., 1956. The Apsheronian Stage of Azerbaidzhan. *Trudy Inst. geol. Akad. Nauk Azerbaidzhan S.S.R.*

SUVOROV, A. I., 1961. Uspenskian zone in Central Kazakhstan and its certain analogues. *Izv. Akad. Nauk, ser. geol.*, No. 8, 67–82.

TEBEN'KOV, V. P., GANTMAN, D. S., and EINOR, O. L., 1939. The geological structure and the coal of the Lower Tunguska river. *Trudy Inst. Arkt.*, **126**, 11–189.

TEODOROVICH, G. I. *et al.*, 1959. *Data on the Geology and Oil and Gas in the Region of Minysinskian Depressions.* Moscow. 176 pp.

TIKHII, V. N., 1957. Devonian deposits: the Volga-Ural region. *Trudy V.N.I.G.R.I.*, **106**, 241.

TIKHOMIROV, YU. P., 1961. The present-day situation and problems of exploration for oil and gas in Yakutia. *Geol. Nefti.* **11**, 9–13.

TIMERGAZIN, K. R., 1958. Pre-Devonian deposits of western Bashkira. *Vop. (Problems) geol. east, margin Russian Platform*, **1**, 5–26.

——, 1959. *Pre-Devonian Formations of Western Bashkiria and Other Prospects for Oil.* Ufa. 312 pp.

Timofeyev, B. V., 1955. Discoveries of spores in Cambrian and Precambrian deposits of Eastern Siberia. *Dokl. Akad. Nauk*, **105**, 547–50.

——, 1958. Discoveries of spores in sedimentary metamorphic formations of Central Asia, p. 16 in the book: *Theses of Papers of the Stratigraphic Conference on Central Asia.* Moscow.

——, 1959. The ancient flora of Baltic Countries. *Trudy V.N.I.G.R.I.*, **129**, 320.

——, 1960. On the age of sedimentary-metamorphic rocks of Eastern Transbaikalia. *Trudy V.N.I.G.R.I.*, **163**, 486–92.

Tkvalchrelidze, G. A., 1957. Metallogenic epochs in the Caucasus. *Soviet Geol.*, **59**, 152–69.

Trofimuk, A. A., 1950. *Oil in the Palaeozoic of Bashkiria.* Moscow. 246 pp.

Tsagareli, A. A., 1954. The Upper Cretaceous of Georgia. *Inst. geol. Akad. Nauk Georgian S.S.R., Monograph 5*, 463 pp.

Tsaturov, A. I., 1961. Biennial results and the future direction of geoexploration for oil and gas in the Checheno-Ingushetia and Kabardino-Balkaria. *Geol. Nefti*, **9**, 1–9.

Tsibovskii, N. I., 1956. The Palaeozoic of Central Ciscaucasia. *Sbor. V.S.E.G.E.I.*, **14**, 52–59.

Tuayev, I. P., 1941. A description of the geology and oil bearing properties of the Western Siberian Lowlands. *Trudy V.N.I.G.R.I.*, **4**, 95.

Tuchkov, I. K., 1957. A new stratigraphic scheme for the Upper Trias and Jurassic of the North-East U.S.S.R. *Izv. Akad. Nauk, ser. geol.*, No. 5, 56–63.

Tuzhikova, V. I. & Arkhangel'skii, I. I., 1961. New ideas on Triassic bauxites of the eastern slope of the Urals. *Dokl. Akad. Nauk*, **140**, 4, 916–18.

Ul'yanov, A. V., 1954. *The Geological History of Western Georgia in Tertiary Times.* Izd. Akad. Nauk. S.S.S.R. Moscow. 108 pp.

Usov, M. A., 1936. *Phases and Cycles of Tectogenesis in the Western Siberian Region.* Tomsk. 209 pp.

——, 1949. The vulcanism and the metamorphism of the sediments of the Kuznetsk Basin: the Kuznetsk Basin, pp. 224–51 in the book: *The Geology of the U.S.S.R.*, Vol. 16. Moscow.

Vakar, V. A., 1957. The distribution of useful mineral deposits in the Yenisei-Lena region, pp. 58–63 in the book: *Jubilee Session of Scientific Institute of the Geology of the Arctic.* Leningrad.

Vakar, V. A. *et al.*, 1953. The geological structure and useful mineral deposits of the Lake Taimyr region. *Trudy nauchno-issled. Inst. geol. Arkt.*, **63**.

Vakhromeyev, V. A., 1958. The Stratigraphy and fossil flora of Jurassic and Cretaceous deposits of the Vilyui depression, in the book: *Regional Stratigraphy*, Vol. 3. 136 pp.

Vakhromeyev, V. A., Peive, A. V. and Kheraskov, N. P., 1936. *The Mesozoic of Tadzhikistan. Trudy of Tadzhik-Pamir Expedition of Akad, Nauk S.S.R.* Moscow and Leningrad. 195 pp.

Vala, A. I., Dalinkevichyus *et al.*, 1959. *A Brief Description of the Geology of the Lithuanian S.S.R.* Vilnyuo. 79 pp.

Vangengeim, E. A., 1961. The palaeontological basis of the stratigraphy of Anthropogenic deposits of the north of Eastern Siberia. *Trudy geol. Inst. Akad. Nauk S.S.S.R.*, **48**, 183.

Vardanyants, L. A., 1948. *The post-Pliocene History of the Region of the Caucasus, the Black Sea and the Caspian Sea.* Erevan. 184 pp.

Varentsov, M. I., 1950. *The Geological Structure of the Western Part of the Kurinskian Depression.* Moscow. 257 pp.

——, 1956a. The Zaisan Depression (pp. 178–85 in the book: *Description of the Geology of the U.S.S.R.*, Vol. 1), *Trudy V.N.I.G.R.I.*, **96**.

——, 1956b. The Iliian Depression (In: *Descriptions of the Geology of the U.S.S.R.*, Vol. 1). *Trudy V.N.I.G.R.I.*, **96**, 169–77.

Vasil'kovskii, N. P., 1952. *The Stratigraphy and Vulcanism in the Upper Palaeozoic of South-Western Branches of the Northern Tyan-Shans.* Tashkent. 302 pp.

Vassoyevich, N. B., 1948. *Flysch and the Methods of Investigating It.* Leningrad. 216 pp.

——, 1951. *Conditions of Formation of Flysch.* Leningrad. 239 pp.

VEBER, V. N., 1934. The Geological Map of Central Asia, Sheet VIII-6 (Isfara). *Trudy V.G.R.O.*, **194**, 279.

VERESHCHAGIN, V. N., 1957. Main problems of the Cretaceous stratigraphy of the Far East. *Soviet Geol.*, **55**, 124–44.

——, 1959. The correlation of Cretaceous deposits of eastern margins of the Soviet Union and foreign countries, pp. 143–9 in the book: *Theses of Papers of the Stratigraphic Conference at Okha*. Okha.

VINOGRADOV, A. P., *et. al.*, 1960a. The absolute geochronology of the Ukrainian Precambrian, pp. 83–112 in the book: *The Determination of Absolute Age*. Moscow.

VINOGRADOV, A. P., *et al.*, 1960b. The age of the Crystalline Basement of the Russian Platform, pp. 132–48 in the book: *The Determination of Absolute Age*. Moscow.

VINOKUROVA, E. G., 1958. A stratigraphic scheme of Cretaceous deposits of Central Asia, pp. 122–8 in the book: *Theses of Papers of Stratigraphic Conference on Central Asia*. Moscow.

VISSARIONOVA, A. Ya., 1959. The stratigraphy and the facies of middle and lower Carboniferous deposits of Bashkiria and their oil content. *Trudy Ufa Neft. Scientific Institute*, **5**, 221.

VLASOV, G. M., 1956. A scheme of the stratigraphy of Tertiary deposits of the Far East, pp. 70–2 in the book: *Theses of Papers of the Stratigraphic Conference at Khabarovsk*. Khabarovsk.

VYALOV, O. S., 1949. The structure of the Carpathians and the Transcarpathian region of the Ukrainian S.S.R., pp. 211–310 in the book: *Trudy and the Conference on oil in Ukraine*. Kiev.

——, 1956. A short review of the history and conditions of formation of sediments in western regions of the Ukrainian S.S.R., pp. 233–46 in the book: *Trudy of the Conference on the Tectonics of the South of the Ukrainian S.S.R.*, Vol. 2.

——, 1961. *The Palaeogene Flysch of the Northern Slope of the Carpathians*. Kiev. 136 pp.

VYALOV, O. S., PISHVAKOVA, L. S., PETRUSHKEVICH, M. I. and GRISHKEWICH, G. M., 1961. A stratigraphical scheme of the Transcarpathian Miocene. *Dokov. Akad. Nauk Ukrainian R.S.R.*, **10**, 1338–41.

YAKHIMOVICH, V. L., 1958a. *The Cainozoic of the Bashkirian Cisuralia*. Vol. 2. Ufa. 175 pp.

——, 1958b. *The Cainozoic of the Bashkirian Cisuralia*. Vol. 1. Ufa. 134 pp.

YAKOVLEV, S. A., 1955. *The History of the Black Sea in connection with Climatic Fluctuations of the Quaternary Period*. Leningrad.

——, 1956. Principles of the geology of Quaternary deposits of the Russian Plain. *Trudy V.S.E.G.E.I.*, **17**, 313.

——, 1958. The Quaternary scheme, the stratigraphical scheme, the Russian Plain, pp. 537–48 in the book: *The Geological Structure of the U.S.S.R.* Leningrad.

YANSHIN, A. L., 1953. *The Geology of Northern Cisaralia*. Moscow. 736 pp.

YASKOVICH, B. V., 1958. New data on the stratigraphy of Cambrian Deposits in southern Fergana, pp. 21–3 in the book: *Theses of Papers of the Stratigraphic Conference on Central Asia*. Moscow.

YAVORSKII, V. I., 1957. Conditions of formation of coal-bearing deposits of the Kuznyetsk Basin and their tectonics. *Trudy V.S.E.G.E.I.*, **19**, 74.

YEGIZAROV, B. KH., 1957. A geological description of the Severnaya Zemlya archipelago. *Trudy nauchno-issled. Inst. geol. Arkt.*, **81**, 388–423.

YELISEYEVA, V. K. & SOSNOVA, M. I., 1956. The stratigraphy of Upper Palaeozoic deposits of Primoria, pp. 29–32 in the book: *Theses of Papers of the Stratigraphic Conference at Khabarovsk*. Khabarovsk.

YEMELYANTSEV, T. M., 1954. The geological structure and the prospects for oil on the eastern shore of the Anabara inlet. *Trudy nauchno-issled., Inst. geol. Arkt.*, **78**, 76–100.

ZABALUYEV, V. V., 1959. The geological structure and the history of the development of the Cisverkhoyansk marginal downwarp. *Trudy V.N.I.G.R.I.*, **130**, 158–82.

ZARIDZE, G. M., 1955. On the Origin of Caucasian Granitoids and their Ores, pp. 392–9 in the book: *Magmatism and Associated Useful Mineral Deposits*. Moscow.

ZAVARITSKII, A. N., 1939. *Magmatic and Metamorphic Rocks of the Urals. Explanatory Notes to 1:500,000 Geological Map of the Urals*. Leningrad.

ZAVIDONOVA, A. G., 1951. Lithological-stratigraphic Characteristics of Devonian and Lower Palaeozoic deposits of the Kaluga region, pp. 121–38 in the book: *Contributions to the Geology of Central Regions of the Russian Platform*. Moscow.

ZHEGALOV, YU. V., 1959. The stratigraphy of Quaternary deposits of the Komandor Islands, pp. 141–2 in the book: *Theses of Papers of the Stratigraphic Conference at Okha*. Okha.

ZHEIBA, S. I., 1960. *The stratigraphy and Fauna of Famennian Deposits of the Lithuanian S.S.R.* Thesis, Vilnyuo. 19 pp.

ZHIZHCHENKO, B. P., 1953. Data for the development of a unified scheme of division of Cainozoic deposits in the south of the European part of the U.S.S.R., and the Northern Caucasus, pp. 183–224 in the book: *Problems of the Geology and Geochemistry of Oil and Gas*. Moscow and Leningrad.

——, 1958. *Principles of Stratigraphy and the Unified Scheme of Division of Cainozoic Deposits of the Northern Caucasus and Adjacent Regions*. Moscow. 312 pp.

ZHUKOVSKII, L. G., *et al.*, 1957. New deposits of gas and oil in the Bukhara-Khiva Depression. *Geol. Nefti*, **11**, 63–72.

ZHURAVLEVA, Z. A., KOMAR, V. A. & CHUMAKOV, N. M., 1961. The structure and the age of the deposits assigned to the Tolbinian Suite (south-eastern Yakutia). *Dokl. Akad. Nauk*, **140**, 3, 658–9.

ZIMKIN, A. V., 1957. A contribution to the stratigraphy of Permian deposits of the North-East of the U.S.S.R. *Mater. geol. and useful minerals North-East S.S.S.R.*, **11**, 31–64.

ZARICHEVA, A. I., 1956. A contribution to the stratigraphy of Palaeozoic deposits of the northern part of the Russian Platform, pp. 153–68 in the book: *Data on the Geology of the European territory of the U.S.S.R.* Leningrad.

GENERAL BIBLIOGRAPHY

Atlas of Main Forms of Fossil Fauna of the U.S.S.R., Vol. VII. *The Trias System.* 1947. Leningrad and Moscow. 251 pp.

Basic Ideas of N. G. Kassin on the Geology of Kazakhstan. 1960. Alma-Ata. 421 pp.

Bauxites, their Mineralogy and Genesis. 1958. Moscow. 488 pp.

Decisions of the All Union Conference on the Development of a Unified Scheme of Stratigraphy of Devonian and Pre-Devonian Deposits of the Russian Platform and the Western Slope of the Urals. 1951. Leningrad. 19 pp.

—— *of the All Union Conference on the Development of a Unified Stratigraphy of Carboniferous Deposits of the Russian Platform and the Western Slope of the Urals.* 1951. Leningrad. 12 pp.

—— *of a Conference on the Development of Unified Stratigraphic Schemes for Central Asia.* 1959. Tashkent. 130 pp.

—— *of a Conference on the Palaeozoic of the Volga-Ural Region.* 1960. Moscow. 63 pp.

—— *of the Conference on Stratigraphic Schemes for the Western Siberian Lowlands.* 1960. Leningrad. 27 pp.

—— *of the Regional Conference on the Development of Unified Stratigraphic Schemes for the Far East.* 1958. Khabarovsk. 51 pp.

—— *of the Regional Conference on the Development of Unified Stratigraphic Schemes for Sakhalin, Kamchatka, the Kuril and Komandor Is'ands, held at Okha.* 1959. Leningrad. 21 pp.

—— *of the Regional Conference on Unified Stratigraphic Schemes of Siberia.* 1959. Leningrad. 35 pp.

Descriptions of the Geology of the U.S.S.R. (on borehole data), Vol. 1, 1956. *Trudy V.N.I.G.R.I.* **96**, 234 pp.

—— *of the geology of the north of Western Siberian Lowlands.* 1960. *Trudy V.N.I.G.R.I.*, **158**, 267 pp.

—— *of Regional Geology of the U.S.S.R.* Moscow State University—A series being published by the Geological Faculty, with A. A. Bogdanov and M. V. Muratov as editors.

Determination of absolute age of pre-Quaternary formations. 1960. *Papers of Soviet Geologists, 21st Intern. Geol. Congr.* Moscow.

Explanatory Notes to the 1:500,000 Geological Map of the Urals, 1939. Leningrad.

—— Notes to the 1:5,000,000 Tectonic map of the U.S.S.R. and Adjacent Countries. 1957. Moscow. 78 pp.

Field Atlas of Ordovician and Silurian Faunas of the Siberian Platform. 1955. Edited by O. I. Nikifirova. Moscow and Leningrad.

Gas Resources of the U.S.S.R. 1959. Moscow. 350 pp.

The Geology of Azerbaidzhan. I. Petrography. 1952. Baku. 828 pp.

—— *of Azerbaidzan. II. Useful Non-Metallic Mineral Deposits.* 1957. Baku. 560 pp.

—— of Kamchatka, 1957. *Trudy V.N.I.G.R.I.*, **102**, 242 pp. (Papers by N. M. Markin, M. F. Dvali and B. F. D'yakov).

—— of the Soviet Arctic, 1959. *Trudy nauchno-issled. Inst. geol. Arkt.*, **81**, 520 pp.

—— *of the U.S.S.R., the Urals, part I—the Geological Description.* Moscow and Leningrad. 688 pp.

—— and oil in the Western Siberian Lowlands. 1958. *Trudy V.N.I.G.R.I.*, **114**, 273 pp.

—— and oil in the Timan-Pechora region, 1959. *Trudy V.N.I.G.R.I.*, **133**, 419 pp.

—— and oil in the west of the Western Siberian Lowlands. 1959. *Trudy V.N.I.G.R.I.*, **140**, 387 pp.

—— and Oil and Gas in Eastern Siberia—*Papers of Vost. sibneft-geologia.* 1959. Moscow. 487 pp.

—— *and Oil and Gas in Eastern Siberia—Papers of Vostsibneft-geologia.* 1959. Moscow. 487 pp.

—— *and Useful Minerals of the Central Part of the Northern Caucasus.* 1956. Moscow. 286 pp.

(*The*) *Geological Structure and Gas and Oil of the Dnyepr-Donyets Depression.* 1954. Kiev. 829 pp.

—— *Structure of Central and Southern Kazakhstan.* 1961. *Mater. V.S.E.G.E.I.*, **41**, 500 pp.

—— *Structure and Useful Mineral Deposits—a Collection of Papers.* 1958. Moscow. 119 pp.

—— *Structure of the U.S.S.R.*, 1958. 1—Stratigraphy, 588 pp.; 2—Magmatism, 331 pp.; 3—Tectonics, 384 pp. Moscow.

Guides to Caucasian Excursions: XVII intern. Geol. Congr., 1937. Moscow and Leningrad.

(*The*) History of Lower Cretaceous accumulation of coal on the territory of the Kenderlykian trough. 1961. *Trudy Lab. geol. coal*, **12**, 187–252.

(*The*) *Ice Age on the Territory of the European Part of the U.S.S.R. and Siberia.* 1959. Izd. M.G.U. (Moscow State University). 569 pp.

Iron Ore Deposits of the Altai-Sayan Mountain Region, Vol. 1. 1958. Moscow. 331 pp.

Deposits of the Altai-Sayan Mountain Region, Vol. 2. 1959. Moscow, 609 pp.

Material (*Data*) *of the Conference on the Development of Unified Stratigraphic Schemes in Sakhalin, Kamchatka, the Kuril and Komandor Islands* (*25 May–2nd June 1959 at Okha*). 1961. Moscow. 297 pp.

—— of Scientific Session on Metallogenic and Prognostic Maps. 1958. Alma-Ata 319 pp.

—— for the Middle Asiatic and Kazakhstan conference on the investigation of the Quaternary Period (Theses of Papers). 1960. *Uch. Zap.* (*Scientific Notices*) *Middle Asia, Sci. inst. geol. minerals*, **4**, 155 pp.

—— on the Geology, Hydrogeology, Geophysics and Useful Mineral Resources of Western Siberia. 1959. Leningrad. 178 pp.

—— on the Geology and Gas and Oil of Central Asia. 1959. Moscow. 222 pp.

—— to the Central Asiatic and Kazakhstan Conference on the Investigation of the Quaternary Period. 1960. *Sci. Zap. Central Asiatic Sci. Inst. geol. and raw minerals*, **4**, 155 pp.

(*The*) Mesozoic and the Cainozoic of southern Baltic countries and in Byelorussia. 1960. *Inst. geol. and geog. Scientific Notices, Acad. Nauk Lithuanian S.S.R.*, **12**, 246 pp.

Mesozoic and Tertiary Deposits of Central Regions of the Russian Platform. 1959. Moscow. 292 pp.

Metallogenic and Prognostic Maps (*Charts*). Alma-Ata. 395 pp.

Metallogeny and Prognostic Maps (*Charts*). 1959. Alma-Ata.

(*The*) *Neotectonics of the U.S.S.R.* 1961. Riga. 336 pp.

Oil and Gas in the North of Siberia. 1958. Leningrad. 217 pp.

(*The*) *Problem of Industrial Oil and Gas Content of Western Regions of the Ukr. S.S.R.* 1955. Kiev.

Prospects of oil and gas in the western Siberian Lowlands. 1961. *Geol. Nefti,* **11,** 1–8.

Regional Palaeogeography. 1960. *Papers by Soviet Geologists, XXI Intern. Geol. Congr.* Moscow and Leningrad. 191 pp.

Resolutions of Conference on Unification of Stratigraphic Schemes of Pre-Palaeozoic and Palaeozoic formations of Eastern Kazakhstan. 1958. Alma-Ata. 39 pp.

Review of the Stratigraphy of Palaeozoic and Quaternary Deposits of the Estonian S.S.R. 1958. Tallin. 46 pp.

Stratigraphic Classification and Terminology. 1956. Mezhved. Stratigr. Committee. Leningrad. 28 pp.

Stratigraphic Dictionary of the U.S.S.R. 1956. Moscow. 1283 pp.

(The) Stratigraphy and Correlations of Middle and Upper Devonian Strata of the Volya-Ural Oil Province. 1959. Ufa. 110 pp.

—— *of Mesozoic and Cainozoic rocks of the Western Siberian Lowlands.* 1957. Moscow. 354 pp.

—— of Quaternary *(Anthropogenic)* deposits of the Asiatic part of the U.S.S.R. and their correlation with European deposits. 1960. *Trudy geol. Inst. Akad. Nauk S.S.S.R.,* **26,** 281 pp.

Tectonics of Petroliferous Regions, Vol. 1. 1958. Moscow. 516 pp.

—— *of Petroliferous Regions,* Vol. 2. 1958. Moscow. 613 pp.

Theses of Papers and Notices of the Conference on the Development of Unified Stratigraphic Schemes for the North-East, U.S.S.R. (10–20 May 1957) at Magadan. 1957. Magadan. 137 pp.

—— *of a Scientific Session in Honour of the 50th Anniversay of the Death of Academician F. B. Shmidt (The Geology of Estonia).* 1958. Tallin. 54 pp.

—— *of the Conference on the Development of a Unified Stratigraphic Scale of Tertiary deposits of the Crimean-Caucasian Province.* 1955. Baku. 115 pp.

—— *of the Conference on the Development of Stratigraphic Schemes for Transbaikalia.* 1961. Chita. 108 pp.

—— *of the Conference on the Development of Stratigraphic Schemes for Yakutian A.S.S.R., at Yakutsk in 1961.* 1961. Leningrad. 209 pp.

—— *of the Conference on the Development of Unified Stratigraphic Schemes for Sakhalin, Kamchatka, the Kuril and Komandor Islands, in 1959 at Okha (Sakhalin), 1959.* Okha. 169 pp.

—— *of the Conference on the Development of Unified Stratigraphic Schemes for Central Asia.* 1958. Tashkent. 193 pp.

—— *of the Conference on the Development of Unified Stratigraphic Schemes for the Far East (10–20 May 1956 at Khabarovsk).* 1956. Khabarovsk. 94 pp.

—— *of the Conference on the Development of Unified Stratigraphic Schemes in Siberia.* 1956. Session on the Pre-Cambrian 32 pp. Session on the Cambrian, Ordovician and Silurian —40 pp. Session on the Middle and Upper Palaeozoic—64 pp. Session on Mesozoic and Tertiary deposits—46 pp. Session on Quaternary deposits—24 pp. Leningrad.

—— *of the Conference on the Mesozoic and Cainozoic of Southern Baltic Countries and Byelorussia.* 1959. Vilnyus. 43 pp.

—— *of the Conference on Unification of Stratigraphic Schemes of the Urals and Correlation of Ancient Strata of the Urals with those of the Russian Platform (Sverdlovsk, 13–18th February 1956).* 1956. Leningrad. 575 pp.

—— *of the Conference on Unification of Stratigraphic Schemes of pre-Palaeozoic and Palaeozoic Rocks of Eastern Kazakhstan.* 1957. Alma-Ata. 155 pp.

—— *of the Regional Conference on the Investigation of the Quaternary Period.* 1957. Moscow.

Trudy (Transactions) of the Scientific-Geological Conference on Oil, Osokerite and Gas of the Ukrainian S.S.R. 1949. Izd. Akad. Nauk Ukrainian S.S.R. Kiev. 400 pp.

—— *of an All Union Conference on the Development of a Unified Stratigraphic Scheme of Mesozoic Deposits of the Russian Platform.* 1956. Leningrad. 383 pp.

—— the 1956 Regional Conference on the Development of Unified Stratigraphic schemes in Siberia. *Doklady on the Mesozoic and Cainozoic.* 1957. Leningrad. 575 pp.; *Doklady on the Stratigraphy of Precambrian Deposits.* 1958. Leningrad. 253 pp.

Voprosy (Problems) of Geology and Geochemistry of Oil and Gas. 1949. Moscow. 384 pp.

—— *of the Geology of Asia,* Vol. I. 1954. Moscow. 797 pp.

—— of regional geology and the method of geological investigations *(Papers on the Geology of the Russian Platform).* 1957. *Trudy Moscow Neft. Inst.* **19,** 1–173.

INDEX OF AUTHORS

STRATIGRAPHICAL INDEX

In this index the pagination refers to either page or table facing it

INDEX OF STRUCTURAL TERMS
AND DEPOSITS